AUTOMOTIVE SOFTWARE

Other SAE books in this series:

Multiplexing and Networking, Volume 2
by Ronald K. Jurgen
(Order No. PT-128)

Electronic Engine Control Technologies, 2nd Edition
by Ronald K. Jurgen
(Order No. PT-110)

Sensors and Transducers, Second Edition
by Ronald K. Jurgen
(Order No. PT-105)

Electric and Hybrid-Electric Vehicles
by Ronald K. Jurgen
(Order No. PT-85)

On- and Off-Board Diagnostics
by Ronald K. Jurgen
(Order No. PT-81)

Electronic Transmission Controls
Ronald K. Jurgen
(Order No. PT-79)

Multiplexing and Networking
Ronald K. Jurgen
(Order No. PT-78)

Electronic Steering and Suspension Systems
Ronald K. Jurgen
(Order No. PT-77)

Navigation and Intelligent Transportation Systems
Ronald K. Jurgen
(Order No. PT-72)

For information on these or other related books, contact SAE by phone at (724)776-4970, fax (724)772-7090, e-mail: publications@sae.org, or the SAE website at www.sae.org.

AUTOMOTIVE SOFTWARE

PT-127

Edited by

Ronald K. Jurgen

Published by
SAE International
400 Commonwealth Drive
Warrendale, PA 15096-0001
U.S.A.
Phone (724)776-4841
Fax (724)776-0790

For permission and licensing requests contact:

SAE Permissions
400 Commonwealth Drive
Warrendale, PA 15096-0001-USA
Email: permissions@sae.org
Fax: 724-776-3036
Tel: 724-772-4028

For multiple print copies contact:

SAE Customer Service
Tel: 877-606-7323 (inside USA and Canada)
Tel: 724-776-4970 (outside USA)
Fax: 724-772-0790
Email: CustomerService@sae.org

ISBN 0-7680-1714-9
Library of Congress Catalog Number: 2005937533
SAE/PT-127

Printed in USA

INTRODUCTION

Complexity Mandates Rapid Software Development

The need for rapid software development necessitated by ever increasing automotive electronics complexity is a recurring theme throughout this compilation of recent SAE papers on various aspects of software development, methodology, and application.

The 73 papers herein are grouped into 8 different categories: Overviews, Software in Embedded Control Systems, Virtual Prototypes and Computer Simulation Software, Safety Critical Applications, Software for Modeling, Software for Testing, Software Source Codes, and Miscellaneous Software Applications. Note that placement of papers in these categories has been somewhat arbitrary. It is hoped that the placement decision, however, has been logical and helpful, although it is by no means mandatory.

The following selected quotations from these papers reflect current software engineering practices, problems, and solutions:

- “Automotive powertrain and safety systems under design today are highly complex, incorporating more than one CPU core, running with more than 100 MHz and consisting of several million transistors. Software complexity increases similarly, making new methodologies and tools mandatory to manage the overall system.” – (2005-01-1342)
- “Companies in industries from automotive to consumer products have infused analysis software into their design cycle the same way writers use spell check to prepare documents.” – (2005-01-1563)
- It is estimated that software that monitors the health of an ECU now takes up about 60% of the total ECU software code to monitor, diagnose, and announce the problems an ECU may have during its normal or abnormal operational modes.” – (2005-01-0327)
- “With the increasing number of ECUs in modern automotive applications (70 in high-end cars), designers are facing the challenge of managing complex design tasks, such as the allocation of software tasks over a network of ECUs.” – (2004-01-0757)
- “A systematic approach for the selection of test scenarios and a notation for their appropriate description must therefore form the core elements of a safety testing approach to automotive control software. In order to allow testing to begin at an early stage, test design should build upon development artifacts available early on, such as the specification of an executable model.” – (2005-01-0750)

- “In recent years, automobiles have come to have more and more electronic control components. Further complexity and multi-functionality of such modern in-vehicle components have required a great amount of control software development work in a short period. System engineers have had to respond to the demand accordingly, developing high-quality software with high efficiency.” – (2004-01-0707)
- “Current automotive software development does not pay enough attention to non-functional issues . . . It is only near the end of the software development process that the engineer downloads code to the target microcontroller. Timing violations at this stage often result in expensive redesigns and schedule overruns.” – (2004-01-0279)
- Unfortunately, display hardware technology has outpaced software technology, and there are currently no established, predictable methods for developing graphics software for vehicles.” – (2004-01-0270)
- One element of safety-critical design is to help verify that the software and microcontroller are operating correctly. The task of incorporating failsafe capability within an embedded microcontroller design may be achieved via hardware or software techniques.” – (2005-01-0779)
- In many areas of engine control and systems monitoring, it is now impracticable to use anything other than electronic solutions, and the development and testing of software for Electronic Control Modules/Engine Management System is a substantial and key part of engine development programmes.” – (2005-01-0056).

* * * * * * * * * * * * * * *

This book and the entire Automotive Electronics Series are dedicated to my friend Larry Givens, a former editor of SAE’s monthly publication, *Automotive Engineering International.*

Ronald K. Jurgen, Editor

TABLE OF CONTENTS

Introduction

Overviews

Software in Embedded Control Systems

Virtual Prototypes and Computer Simulation Software

Safety Critical Applications

Software for Modeling

Software for Testing

Software Source Codes

Miscellaneous Software Applications

Future Software Trends

OVERVIEWS

2005-01-1563

Upfront Analysis in the Product Development Process

Suchit Jain
Analysis Products, SolidWoks

Tin Bui, Jason Frick, Alain Khella, John Mason, Richard Naddaf and Stewart Prince
California State University, Northridge

ABSTRACT

Companies in industries from automotive to consumer products have infused analysis software into their design cycle the same way writers use spell check to prepare documents. Analysis tools, closely integrated with 3D CAD applications, help catch errors earlier in the design cycle and optimize designs for better performance and more efficient material use. The direct involvement of design engineers in analyzing their own designs allows for quick turnaround times and ensures that modifications indicated by analysis results are promptly implemented in the design process. Used properly, it yields trustworthy results that are already driving efficiencies and cost savings. This paper will describe the analysis process and the benefits of sophisticated analysis functionality that is available to engineers of any skill level. It will showcase a case study of the Formula SAE (FSAE) team at California State University, Northridge (CSUN) and how they utilized the computational capabilities of integrated 3D CAD and design analysis software to design, analyze and simulate a formula race car.

INTRODUCTION

Most analysis software on the market today employs the finite element analysis (FEA) method. The FEA process consists of subdividing all systems into individual components or "elements" whose behavior is easily understood and then reconstructing the original system from these components. This is a natural way performing analyses in engineering and even in other analytical fields, such as economics. The approach of using discrete components to solve full systems has been used in structural mechanics since early 1940s, and the term FEA was first used in 1950s. Aerospace engineers adopted FEA in the 1960s to analyze aircraft designs.

FEA moved from manual calculations in the '50s to FORTRAN applications running on mainframes and Cray supercomputers during the 1960s and '70s. Analysts ran design data through the applications, which yielded numeric results the analysts would interpret for the designers. By the '80s, analysis software came out of the back room to run on high-powered workstations that displayed 3D images rather than numeric data. Analysis was, however, still mainly the domain of specialized analysts who did nothing else except analyze designs.

Though it was an improvement over the mainframe days, it was still an awkward system. Analysts would suggest modifications to the design, but designers would not always know how to apply those modifications.

The picture improved dramatically in the mid '90s. As desktop CAD became a design industry staple, software vendors responded by integrating analysis with CAD software so users could model and test designs in the same environment. That eliminated re-creating a design in an analysis application, which was time consuming and often caused errors.

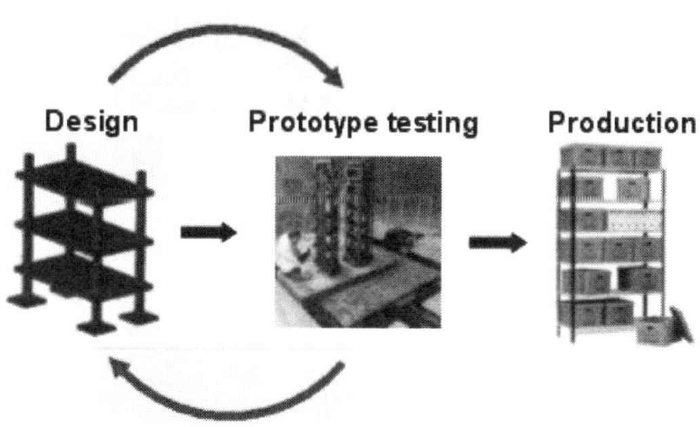

Figure 1 - Design-prototype-test iterations.

NOMENCLATURE

FEA - Finite Element Analysis, CAD - Computer Aided Design, CAE – Computer Aided Engineering, CFD – Computational fluid Dynamics, CSUN – California State University, Northridge,

CHANGING ROLE OF DESIGNERS AND ENGINEERS

For many years, design analysis was the exclusive domain of highly specialized analysts who would analyze designs after an engineer had finished his or her work. This approach wasted a lot of time with back-and-forth interaction between designers and analysts, especially when solid models had to be recreated in a separate analysis package.

But in recent years, the benefits of using design analysis as part of conceptual design have become obvious. When properly trained to use today's modern, integrated analysis systems, design engineers are better positioned than analysts to leverage analysis results to modify solid models as design iterations progress. Design engineers have far greater product expertise and are closer to the product development process than analysts.

The direct involvement of design engineers in analyzing their own designs speeds turnaround times and ensures that design modifications indicated by analysis results are promptly implemented into the design in progress.

A growing number of companies are asking their designers and engineers to do more than just design work. These designers and engineers are being tasked to come up with better designs not only for form and fit, but also to meet functional specifications. Analysis software has started to show up on desks of industrial designers and project managers.

ADVANCES IN FEA TECHNOLOGY MAKING THESE CHANGES POSSIBLE

Meshing

FEA software breaks a solid model down into geometric "elements," which are mathematically represented on the computer as a 3D mesh overlaying and permeating the solid model, to solve differential equations that govern physical phenomena as they apply to simulated geometries. Each element is represented by a bunch of "nodes" connected via element "edges."

In earlier days of FEA meshing was limited to manually creating nodes and element connectivity by specifying x, y, and z coordinates of the nodes. This painstakingly laborious method limited users to analyzing very simple geometries such as plates or rods. FEA technology took a giant leap when automatic meshers were developed which could mesh complex geometries, allowing analysts to conduct analysis on real life models.

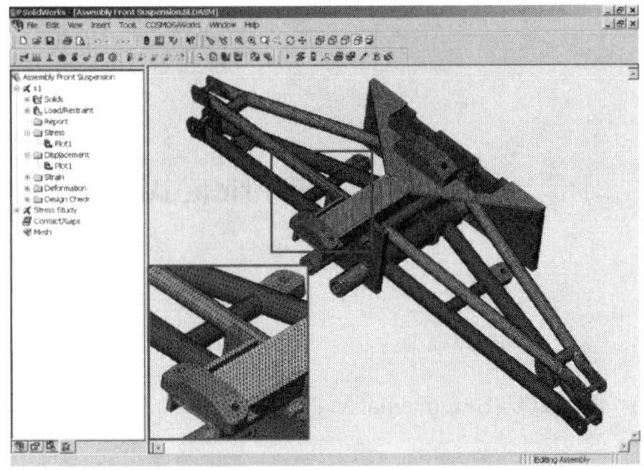

Figure 3 - FEA mesh for front suspension assembly of a snowmobile in COSMOSWorks and SolidWorks.

However, there were problems with these automatic meshers in meshing just any geometry. Very small geometric features such as small holes and fillets caused mesh failures. The geometry's source, such as the CAD systems it was created in, representation in NURBS (Non uniform rational B-spline) or analytical format, also lead to mesh failures. A bunch of utility software program sprung up in early '90s which would take the raw geometry and "heel" it for meshing purposes. These heeling software programs would try to remove geometric inaccuracies such as gaps between surfaces and slivers and make the model air tight for meshing.

Modern FEA software has reduced the complex art of meshing to simple push of a button. All the heeling is incorporated inside the mesher and is done in the background. Meshing problems have been reduced drastically allowing many more designers to start conducting analysis on their models.

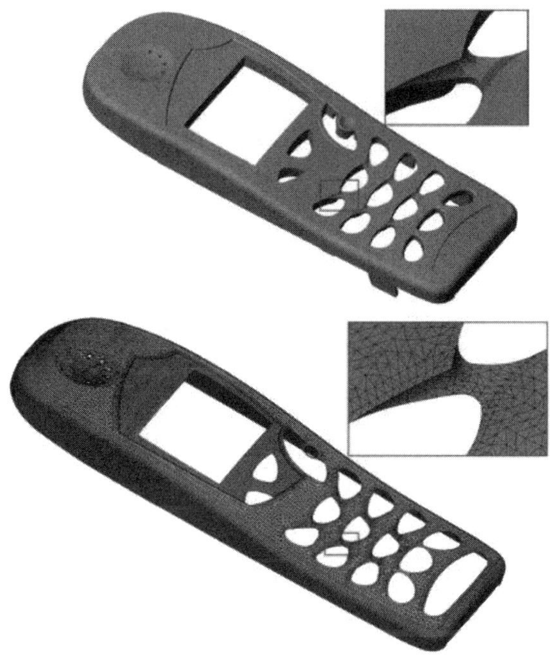

Figure 4 – FEA mesh of cell phone cover. Small geometry and slivers are automatically heeled by the mesher.

Future trend in FEA software is to build load and restraint templates which can be customized by company experts. These templates will provide a framework for less experienced users to conduct analysis and reduce errors in defining loads and restraints.

Assembly Analysis

Analyzing assemblies is more complex than single parts as the analysis needs to take into account the interaction between the different components. Since each part can deform, the stresses developed in the whole assembly depend on how the parts are connected to each other. For example, in an assembly different parts can be welded, bolted, joined by a pin. They can also come in contact which each other upon loading. Modern FEA software provides easy ways of modeling these interactions between parts easily and intuitively. Contact analysis which used to be the domain of specialized analysts is now available standard in most FEA software.

Modern FEA software allow user to directly input or simulate connections such as pins, springs, bearings, bolts in one step rather than using combination of several inputs.

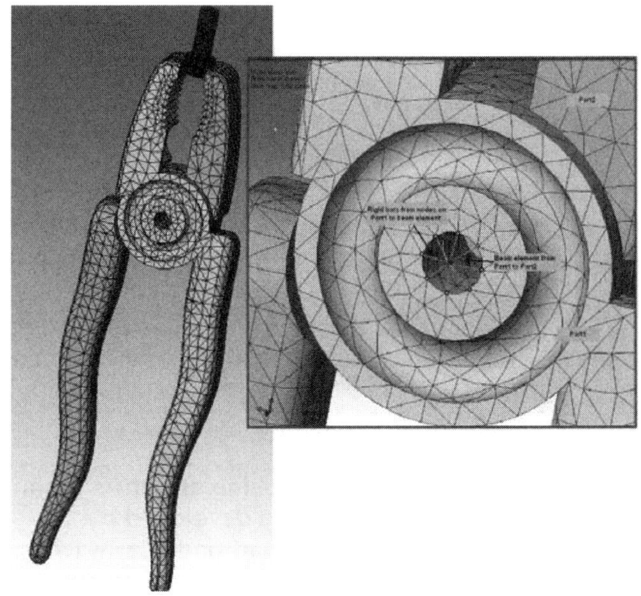

Figure 7 – Simulating a pin connection in COSMOSWorks is as simple as selecting the two cylindrical faces of the two arms of the pliers. Inset shows the complex procedure by defining beams in traditional FEA software.

Solution Speeds

The computing power of the mainframe computers of the 1980s is now available on the desktop at a fraction of the original price. The developments of fast equation solvers have enabled users to take advantage of the availability of affordable computing power and reap the benefits of FEA. Today fast iterative solvers are standard in any professional FEA software and can solve several million degrees of freedom simulation in hours if not minutes on a laptop. Quick solution times have allowed engineers to study performance of very complex parts and large assemblies with finer geometric details in matter of minutes. Specialized techniques and abstract concepts developed by FEA vendors to speed up solution times, such symmetry concepts of slicing models into halves for analysis or removing fillets or small geometric details, are no longer required.

FEA solvers are now also optimized to take advantage of dual processors and distributed computing systems.

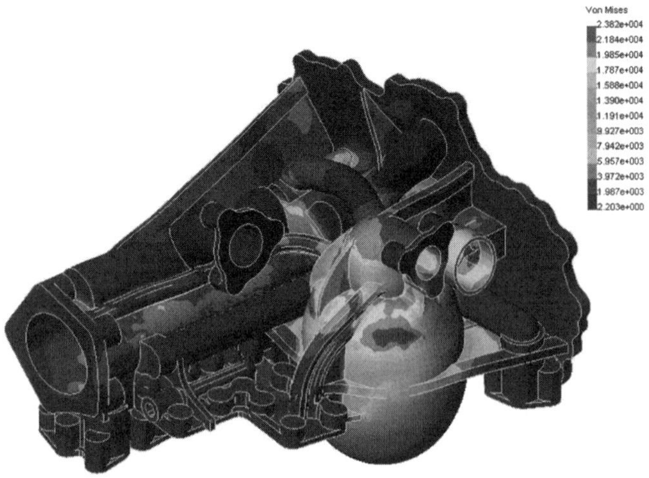

Figure 8 – Stresses in a housing structure for an aircraft engine lubricating pump developed by Nichols Aircraft Division. This part had nearly 1.7 million degrees of freedom and was meshed and solved in couple of hours.

<u>**From General Purpose to Specialized Tasks**</u>

Other emerging trends include the customization of analysis technology to address specific tasks and the adoption of FEA by non-traditional industries. Until recently, FEA technology has been packaged as a general-purpose tool that can simulate the behavior of just about any type of product design. While some analysis applications provide specific physics functionality, such as mechanical, thermal, and computational fluid dynamics (CFD), specialization within each of these areas is increasing and will continue at an accelerated rate. Just as consumers are demanding customization in everything from automobiles to computers, engineers require analysis capabilities that address a narrow range of problems that are particular to their specific industry.

By focusing analysis technology on specific tasks, such as "drop test" analysis for hand-held products, cooling analysis for electronics systems design, and multiphysics interaction analysis (mechanical, thermal, and electromagnetic) for micro-electromechanical systems (MEMS) design, analysis vendors are pushing advances in ease of use even further. When addressing specific types of problems, analysis developers can leverage interface wizards to automate analysis setup and effectively reduce what might require several steps with a general-purpose FEA package to a single step.

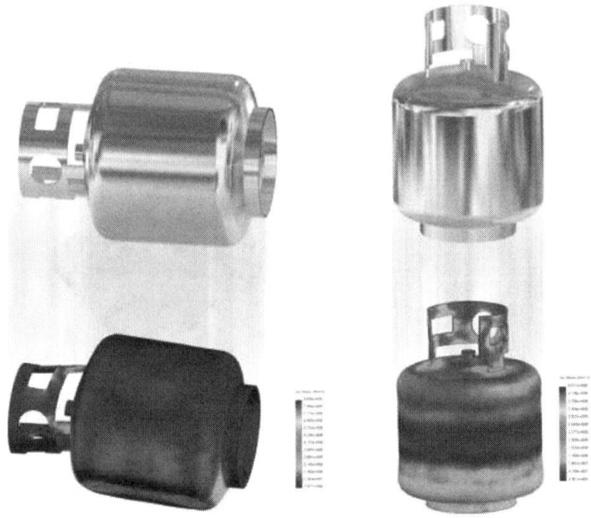

Figure 9 – Drop test simulation inside COSMOSWorks of a propane tank when dropped in different orientations.

RACECAR DESIGN FOR FORMULA SAE COMPETETION

One of the graduation requirements for the mechanical engineering degree at California State University, Northridge (CSUN) is to undergo a senior design project. This is where senior students take all the theoretical learning from past courses and apply it to a practical problem over the course of two semesters or about 9 months. One such problem that a group of 15 students encountered was the design and fabrication of formula style racecar for the annual Formula SAE (FSAE) competition held in Detroit, Michigan.

Normally in the automotive industry, the typical time frame to design and build a working prototype is to 3 – 5 years. Unfortunately, the CSUN FSAE team has only 9 months to design, build, and race a formula racecar. The only way that the CSUN FSAE team and other teams across the world are able to complete this assignment is through the use of advanced computational aided engineering (CAE) programs. The 2003 – 2004 CSUN FSAE team utilized the computational capabilities of SolidWorks®, COSMOSFloWorks™, COSMOSWorks®, and COSMOSMotion™ to design, analyze, simulate, and build a formula racecar.

<u>**What is FSAE?**</u>

The FSAE competition is an annual competition spread out over five days in May. The competition is hosted by the Society of Automotive Engineers (SAE) and sponsored by DaimlerChrysler, Ford Motor Company, and General Motors. Solidworks is also a main sponsor of the event.

The competition challenges student's creativity, imagination, and knowledge to build a working prototype that meets the rules and regulations of the competition. The engine displacement is limited to 610cc and many restrictions are placed on chassis design. The competition consists of static and dynamic events.

The competition begins with the static events, which consist of the cost analysis, sales presentation, and engineering design events. The second half of the competition begins the dynamic events, which include the acceleration run, skidpad, autocross, and endurance coupled with fuel economy. The total amount of possible points is 1000.

Modeling of a Formula Race Car

The CSUN FSAE team mirrors industry in the way the design process is commonly used. The team will begin the semester with a preliminary design event (PDR) followed by an internal design event (IDR). The critical design event (CDR) is held at the end of the fall semester where the team presents all final designs and analysis for approval. The spring semester begins the manufacturing phase with design for manufacturing (DFM) to weight the benefits of manufacture vs. buy.

The CSUN FSAE team structure also mirrors industry with a CEO filled by the faculty advisor and a project manager reporting to the CEO and overseeing seven different departments making up the racecar. Each department will model their respective parts and assemble all required components including hardware. Then the project manager will combine the entire sub – assemblies to a top level assembly to produce what will be the model for the racecar as shown in figure 10.

Figure 10 – Top level assembly of racecar with 7 main sub – assemblies and over 1000 parts!

The chassis department implemented a new design from traditional designs. Most of teams at the competition use a steel space frame consisting of tubes welded together. The new design is a hybrid space frame/monocoque that consisted of a steel space frame in the front and mid – sections of the racecar with an aluminum box in the rear that mounts to the engine. The use of SolidWorks aided with the modeling of all of the components that make up the hybrid chassis design and integrating components from other departments that connect to the hybrid chassis. Figure 11 below shows the hybrid chassis design.

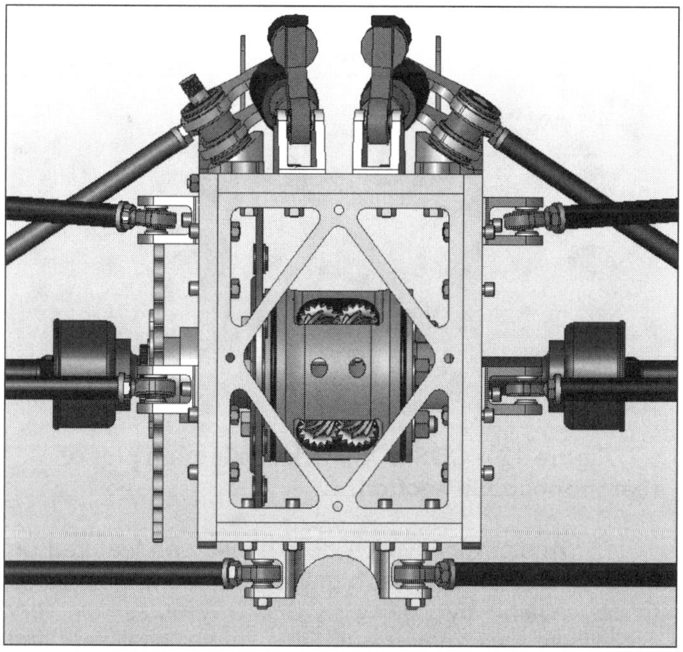

Figure 11a (top) – Isometric view of rear end of racecar including drivetrain, suspension, and chassis departments. Figure 11b (bottom) – Rear view of vehicle detailing the aluminum moncoque box, aluminum differential inside of box, and the suspension components.

Analysis of a Formula Race Car

With students designing intricate parts, the task of analyzing these parts by hand becomes complicated and cumbersome. This is where the COSMOS® suite of analysis products is effectively utilized to reduce design time and perform thousands of calculations in fractions of a second. With the short learning curve of COSMOS®, students were able to learn how to use the software in about one week and begin their analysis.

The formula racecar was designed to last the whole race and nothing more to make it as light as possible. In this case, the most critical event is the endurance event which lasts about thirty minutes. The formula racecar was designed to survive this event and not for infinite life.

After the aluminum box was modeled and fit into the assembly, the question of can the aluminum box withstand the forces succumbed during the endurance event was addressed. The aluminum box consists of 5 plates bolted together. COSMOSWorks® was used to design the aluminum box as well as other parts designed by students that were subjected to high stresses. Figure 12 below shows a deflection plot, which allowed the calculation of the torsional rigidity of the monocoque.

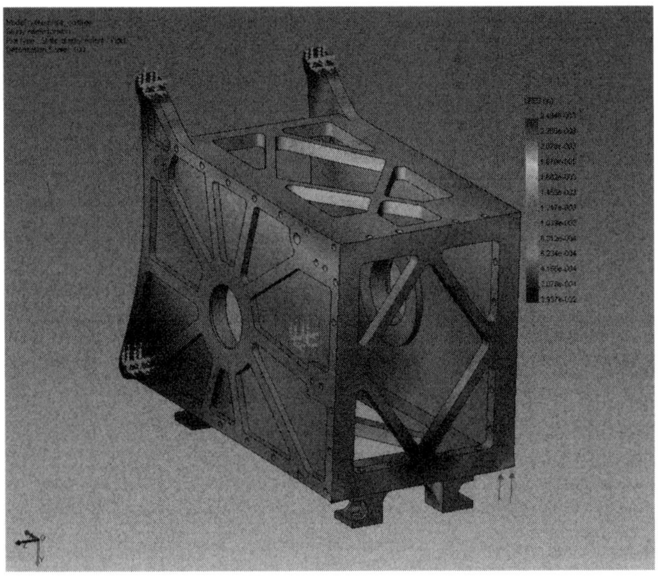

Figure 12 – COSMOSWorks deflection plot of rear monocoque section.

Another innovative design that was featured on the formula car was an aluminum differential housing. Once again by using COSMOSWorks, a very complicated geometry was able to be analyzed and optimized within the very short time frame of the FSAE competition.

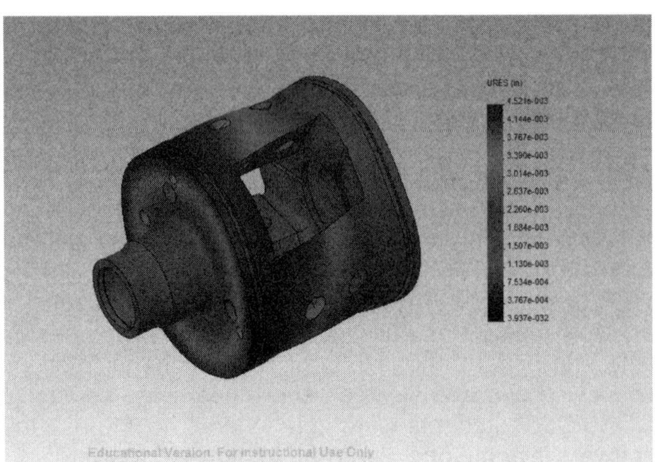

Fig 13 – COSMOSWorks deflection plot of the differential as the car undergoes maximum acceleration.

In addition to FEA stress analysis, the CFD capabilities of COSMOSFloWorks were also utilized. Several components on the engine carry airflow, and optimization of these designs helps improve maximum horsepower.

One rule specific to the FSAE competition is the use of an engine restrictor. As the airflow moves through this restriction, the airflow accelerates greatly, making compressibility and boundary layer effects very significant. While some analysis could be performed by hand as the geometry of the design becomes increasingly complex, CFD is the only feasible option. Fig 14 show the pressure and velocity plots for the converging-diverging restrictor designed using COSMOSFloWorks.

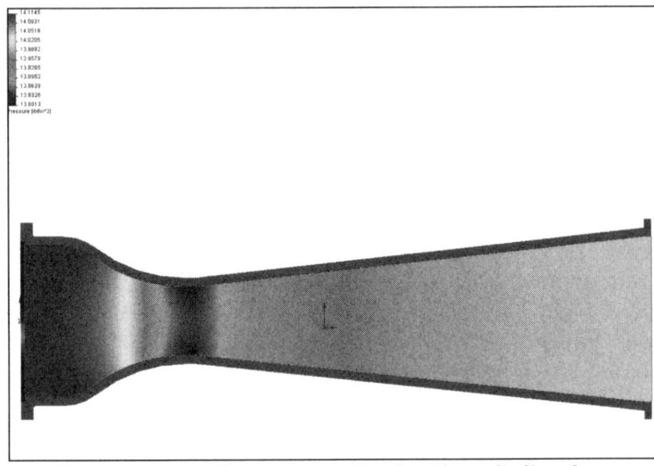

Figure 14a – Pressure plot for the air flowing though the engine restrictor using COSMOSFloWorks.

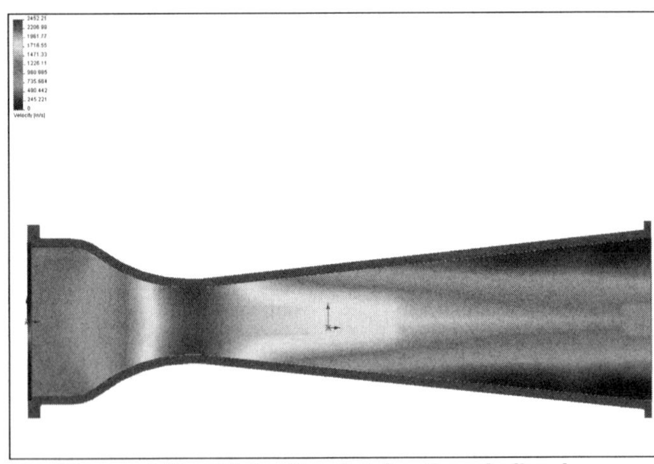

Figure 14b – Velocity plot for the air flowing through the engine restrictor using COSMOSFloWorks.

In addition to both FEA and CFD analysis, a kinematics package was also required in order to reduce the amount of time required to design linkages and mechanisms. Using COSMOSMotion, every mechanism from the suspension to the shifter was able to be analyzed and optimized in an amazingly short time. Accuracy also improved as the motion was shown in full 3-D, as opposed to a 2-D approximation used to simplify hand calculations. Below in figure 15 is an image of the shifter linkage, showing the restraints and joints as defined through COSMOSMotion.

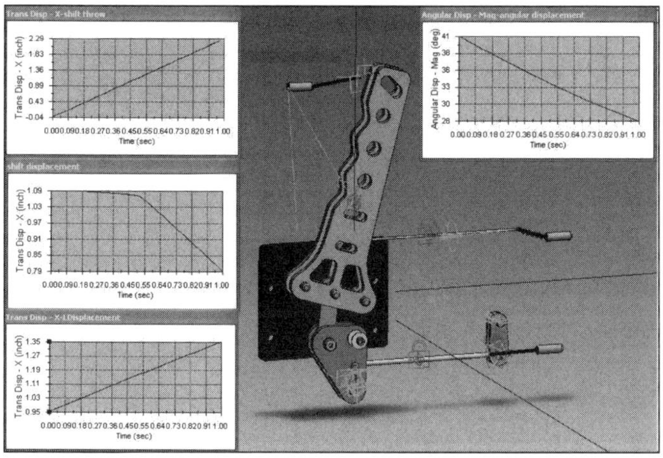

Figure 15 – COSMOSMotion analysis of the shifter linkage.

Competition Results

At the beginning of the spring semester, all of the designs and analysis was complete. The spring semester marked the beginning of the critical manufacturing and testing phase. The formula racecar was fabricated within three months with about one month of testing time. The CSUN FSAE team arrived at the annual competition in Detroit with the necessary skills and preparation for success. At the conclusion of the competition, the CSUN FSAE team scored its highest ever points in history with 683.599 points. This earned the team a 14th place finish out of 140 schools worldwide, which was CSUN's third best finish in history. The team also earned 3rd place in the Ricardo Powertrain award for the excellent designs in the engine and drivetrain.

The CSUN FSAE team arrived in Detroit with confidence in their engineering designs as evidenced by the team's success. The use of CAE software provided novice students with the ability to design and analyze parts within three months efficiently and quickly. The skills learned by the students working with the CAE software will be carried for years to come in their professional careers and has become an invaluable part of their collegiate career.

CONCLUSION

As analysis software is becoming easier to use, more people in product development who are not experts at analysis are using it. This is especially true for product designers who create the initial part and assembly models. By using analysis during the initial design stage, designers not only minimize the probability of failure but also can leverage analysis results to design better products, faster, and at lower cost. Effective product design requires much more than simply creating a geometric shape for a particular function. Designers

Figure 16 – The 2003 – 2004 CSUN FSAE team at the conclusion of the competition in Detroit in May 2004.

have to balance a range of design variables and options, from the properties of available materials to size, weight, and loading constraints. Quick and easy design analysis results, even if they only show the approximate deflection or stresses instead of the exact results demanded by analysts, can help designers make prudent decisions at the beginning of the design cycle that minimize problems, delays, and costs later on.

This paper has talked about some of the advancements in technology which has moved up FEA in the initial design stages and is enabling increasing number of product designers and engineers to verify their design. However, there are products and designs which require more specialized and in-depth analysis and need someone specialized in those fields.

REFERENCES

1. Suchit Jain, 2003, "Making the business case for analysis software", 17th Reliability, Stress Analysis and Failure Prevention Conference, Chicago, IL, vice president of analysis products, SolidWorks corporation
2. Suchit Jain, 2004, "The changing role of FEA in product development process", 2004 ASME International Mechanical Engineering Congress, Anaheim, CA, vice president of analysis products, SolidWorks corporation
3. Mason, John. Formula SAE Fall Semester Engine Design Report. CSUN, 2003
4. Bauer, Horst. Bosch Automotive Handbook. Robert Bosch GmbH, 2000
5. SolidWorks Corporation, http://www.solidworks.com/
6. SRAC, http://www.cosmosm.com/
7. CSUN FSAE, http://www.ecs.csun.edu/sae

2005-01-0327

A Software Component Architecture for Improving Vehicle Software Quality and Integration

Brendan Jackman
Centre for Automotive Research, Waterford Institute of Technology

Shepherd Sanyanga
TRW Automotive

ABSTRACT

It is estimated that the software which monitors the health of an ECU now takes up about 60% of the total ECU software code to monitor, diagnose and announce the problems an ECU may have during its normal or abnormal operational modes. It is the abnormal operation of the system which is the constant problem for vehicle OEMs because this side of the system operation is not easily defined or simulated. The integration of Failure Mode and Effects Analysis (FMEA) to normal design is now becoming central to tackling these issues head-on, such that FMEA is now used as part of the integration process. Having between 10 and 20 different ECUs on a vehicle network still leaves the integration of software from many different suppliers a difficult task. The main issues are incompatible interfaces, misunderstandings of vehicle OEM internal software requirements and a general lack of time to carry out rigorous and methodological integration testing at the Data and Physical Layers before proper vehicle production commences.

The vehicle OEMs have attempted to alleviate these problems by specifying common ECU infrastructures, providing standard outsourced software modules to their suppliers and by taking part in standardisation efforts such as OSEK/VDX and AUTOSAR. However due to location, perception and language differences, this has not reaped the benefits that were being sought and has created a new set of problems.

This paper describes an object-oriented component-based approach to vehicle software development to try to better solve the above issues. The work of the Object Management Group (OMG) is examined, and it is shown how the Object Request Broker (ORB) concept can be applied to existing real-time embedded automotive software systems to ease integration and simulation. The Object Request Broker concept has been successful in enabling reusable software components in the commercial software world. An ORB architecture supports integration and reuse of legacy software by separating component interfaces from their definition.

INTRODUCTION

Figure 1 depicts the typical information flow between a vehicle manufacturer and Tier 1 supplier during the development of an ECU subsystem. Due to the fact that early-on during the development of a vehicle the OEM is still unclear on what the final vehicle architecture will be in terms of requirements, the requirements will continuously evolve as the program progresses. This tends to be problematic for suppliers, since they need a static view of the vehicle architecture in order to give realistic system quotations and time scales for their ECU development. To manage this situation, most suppliers take a snapshot of requirements and quote based on that version of requirements. However this still does not mitigate the suppliers' costs in the development phase, so budget over-runs are common.

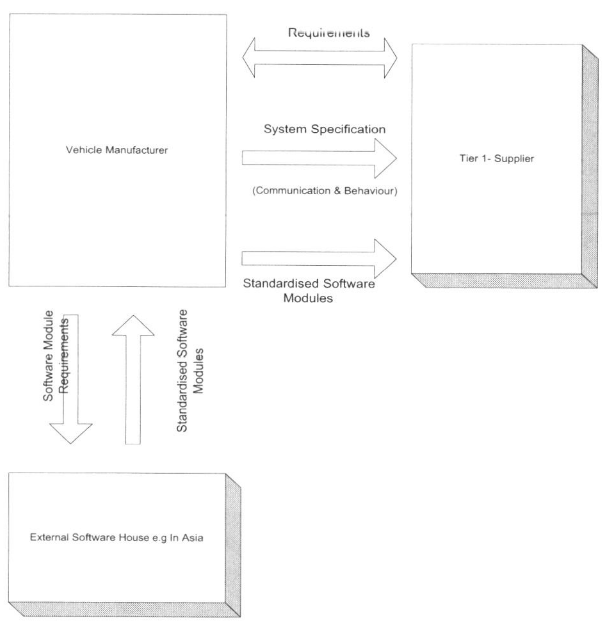

Figure 1. Information Flow and Interaction between OEMs and Supplier.

These budget overruns happen because most suppliers follow a methodological approach to their ECU system design process. This is usually called the 'V' cycle mode of ECU development, and is shown in Figure 2. It is after the architecture partition phase that suppliers identify where the standardized OEM software modules are going to be co-located in their subsystem design. It is at the module testing stage that inconsistencies between operational requirements and the OEM supplied modules are first seen. This also includes problems related to interface links to the supplier's proprietary code. To resolve these conflicts and to implement these amended requirement changes, suppliers end up expending extra resources to ensure that all parts of the design are updated with these final agreed code changes. In most systems with the OEM the cost implication of these changes can be potentially under-estimated in the agreed piece cost of the ECU subsystem.

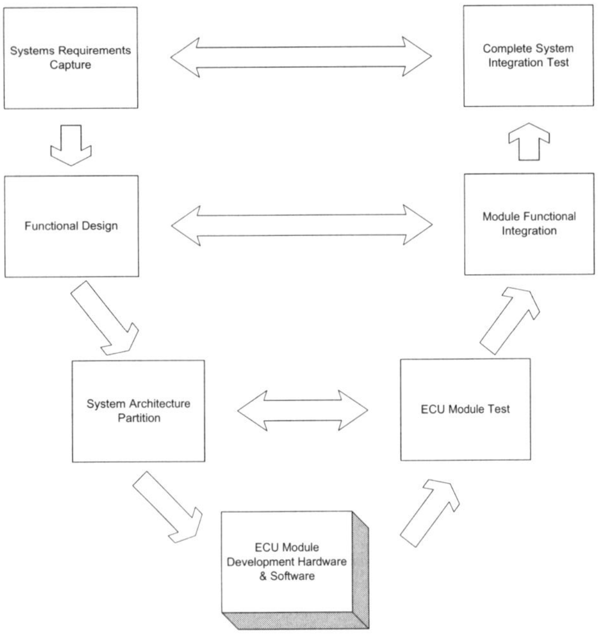

Figure 2. "V Cycle" ECU Subsystem Development Process.

SYSTEM INTEGRATION

The vehicle OEM usually builds what is termed a bench car. This is the electrical & electronic architecture representation of the final vehicle with all the electrical components that will exist on the real vehicle. This bench car integrates all ECU subsystems on every network communication on the target vehicle platform. This is the first time that the vehicle manufacturer is able to see how each ECU on the vehicle will interact with its neighbour and hence serves as an a early warning system to program teams of vehicle system problems. This bench car is very important, especially for suppliers, since it is used to develop the ECU subsystem software for the different ECUs before the pre-production prototype vehicles become available, as well as provide a platform for suppliers to really test their software in a real vehicle environment.

Due to the complexity of the vehicle architecture network and its interaction with its operating environment, it is here where the majority of vehicle system-wide problems are detected and solutions are sought. These problems may take the following form: Interface compatibility issues due to distributed functions; timing and data latency problems caused by incorrectly coded OEM supplied standardized software modules; harness connectivity issues; data corruption issues due to current and voltage distortion slew rates; ECU software configuration issues due to ECUs with the wrong level of software functionality on a network; vehicle ignition start-up and shut-down issues due to incorrect implementation of network management procedure requirements and ignition after-run requirements.

AUTOMOTIVE SOFTWARE INITIATIVES

The rapid increase in software functionality and additional ECUs being added to vehicles brings additional complexity to the system integration process. System integration can be eased by managing the complexity of the distributed vehicle control system. A common approach to the management of complexity has traditionally been standardization, the use of interchangeable parts that provide compatible interfaces and services. OEMs have usually standardized aspects such as the choice of microprocessors, ROM/RAM size, development tools and programming languages to simplify ECU application development and facilitate software reuse between ECUs and vehicle variants. ECU component standardization has also meant big cost reductions for suppliers.

Various industry-wide attempts at standardization have been initiated over the past decade with varying degrees of success. The widespread use of Matlab/Simulink as an application modeling and development tool has been a large factor in the proliferation of vehicle software functions. Matlab/Simulink allows engineers to model new applications graphically and therefore more productively than using conventional programming languages. Matlab/Simulink is well supported by Hardware in the Loop (HIL) tools and automatic code generators, allowing production ECU code to be generated directly from the application models. Furthermore, Simulink models can be easily shared and reused between applications.

The use of high- and low-speed CAN as a standard vehicle network has allowed for the easy exchange of data between ECUs that were previously interconnected with OEM proprietary serial point-to-point protocols such as ISO9141 (equivalent to K-line), UBP (UART-Based Protocol). Higher levels of vehicle control are now possible by exchanging data between several existing subsystems. For example, Vehicle Stability Control (VSC) systems depend on the interactions between Traction Control, ABS and Engine Management systems for successful operation.

In 1993 some major German automotive companies formed the OSEK group [1] with the purpose of defining a common software standard for ECU operating systems and associated network interfaces. The idea was to specify an open system standard that software vendors can comply with. OSEK essentially provides a set of microprocessor-independent services to ECU application software. Services include a real-time task scheduling system, interrupt handling, Alarm/Event handling and an inter-task communications mechanism that allows tasks on the same or different ECUs to exchange data. OSEK has been widely adopted in Europe and has significant support from independent software vendors. OSEK is also addressing the need for fault-tolerant and time-triggered services in subsequent versions of the standard.

The biggest weakness in the OSEK concept has been the lack of hardware device driver standards. A lot of ECU application code is required to interact with microprocessor peripherals, and this is still the biggest differentiating factor among microprocessors. So although OSEK goes a long way towards increasing the portability and reusability of ECU application software, the onus is still very much on the software designer to decouple the device handlers from the application code. The success with which this is done determines the level of application portability and reuse.

The ASAM group [2] has been successful in defining a set of standards for implementing measurement, calibration and diagnostic systems. The CAN Calibration Protocol (CCP) and its successor XCP is widely used for calibration and flash programming of vehicle ECUs. Standards such as these help the system integration process by providing a single data exchange format for calibrating and diagnosing all ECUs in a distributed vehicle control system.

The International Standards Organization (ISO) has provided a set of standards for retrieving diagnostic data from ECUs across K-Line, LIN and CAN networks. The most widely used standards are KWP2000 (ISO 14230, 1-4) and Diagnostics on CAN (ISO 15765, 1-4).

The latest standardization initiative is AUTOSAR [3], founded in 2003, which is made up of some of the largest OEMs and a whole host of Tier-1 suppliers. The goal of AUTOSAR is to establish an open standard for automotive Electrical/Electronic architecture. AUTOSAR intends to provide a complete run-time environment for ECU application software that will provide complete hardware independence. This run-time environment will be compatible with and make use of established automotive standards such as OSEK and CAN. One of the most significant aspects of the AUTOSAR proposal is the development of software interface standards for all functional aspects of a vehicle. For example, there will be a set of predefined software interfaces for powertrain control functions. This will ensure a high level of compatibility between different suppliers' systems and simplify the interchange and integration of these software systems. The first validation tests for AUTOSAR are expected in 2006.

These standardization efforts have brought the advantages of simplification and cost reduction to many OEMs and suppliers. However, the integration of co-operating ECUs on a distributed vehicle network is still a major issue in terms of network message interpretation, bus loading, and proper handling of system failure modes. The AUTOSAR standardization effort holds much promise, but widespread agreement and tool support for the standard, if successful, cannot be realistically expected until the end of the decade. In the meantime OEMs and suppliers cannot stand still in the face of increasing vehicle software complexity driven by customer demand and hence must look to alternative approaches to contain the integration problems.

AUTOMOTIVE SOFTWARE ARCHITECTURE

One of the main causes of integration problems has been the piecemeal development of system ECUs in isolation from one another. The vehicle control systems are usually partitioned into a set of networked ECUs early on in the vehicle design cycle before the complete software system requirements are known and properly understood. This early decision on hardware configurations has a detrimental constraining effect on the resulting software architecture. In effect, the choice of hardware predetermines the software architecture, since it influences the functional partitioning across the available ECUs and therefore dictates the network communications requirements. The emphasis is currently on hardware module design rather than software function design. While the hardware design costs have to be restrained because of the large volumes involved, there is a growing realization among OEMs that software costs are more significant than hardware costs. What must be considered is not just the software development cost alone, but the total software cost across the life cycle of the vehicle. This should include test and calibration costs, reprogramming/flashing costs, warranty costs and the potential cost of poor customer satisfaction caused by software failures. Surveys increasingly show that the majority of vehicle breakdowns are caused by software or electronic faults, not mechanical failures. The goal therefore must be to achieve a high level of robustness in the vehicle software/electronic systems.

What many OEMs lack is a software architecture for their vehicles. While they may have an Electrical/Electronic (E/E) Architecture, this is not quite the same thing. In the first instance, E/E architectures are not completely portable between different vehicle variants because of differences in body types and vehicle features. If software functions are closely coupled to individual ECU hardware, and a specific ECU cannot be fitted to a vehicle variant, then all associated software functions are lost to that vehicle variant. Usually some of the software functionality is required, which the supplier then tries to integrate onto a different ECU with resulting cost and integration problems. If the software architecture is completely independent of the hardware architecture, then none of these problems would arise. The OEM would just choose the subset of software functions to be delivered on the vehicle variant and partition the functions across whichever vehicle ECUs are available. For this scheme to work there must be a software infrastructure available to decouple the software functions from the ECU hardware and vehicle network. The remainder of this paper outlines ideas for such a software infrastructure based on existing software standards.

OBJECT-ORIENTED DEVELOPMENT

Most automotive software is developed using structured or procedural approaches (usually written in C) which view the software as a collection of separate code functions that share some common data. These code functions tend to be tightly coupled to the shared data and to each other so that any requirements change has a significant impact on the structure of the software.

Object-Oriented development is the process of developing software systems by considering the system to be made up of a set of co-operating *objects*. An object contains a set of data values called attributes that describe the object, together with a set of methods or services that can be invoked on the object.

Objects invoke the methods of other objects to carry out the functionality of the system. The overall system functionality is thus shared among the various objects in much the same way that work is shared among the employees of a company. Objects are a natural, intuitive way to view software systems, particularly systems that interact with real world objects. Software objects can be developed to represent each real world object being manipulated. For example, an engine control ECU could have a separate object for each fuel injector being controlled. Each fuel injector object would have attributes detailing the injector firing angles and duration, plus methods to fire and switch off the injector. The methods are in effect the responsibilities of the object. It should be noted that an object represents a specific instance, so that a fuel injector object represents a specific fuel injector. If an engine ECU is controlling four fuel injectors, then it would have four Fuel Injector objects, one for each physical injector. Object-Oriented methodologies use the concept of a class to generalize about objects, so the fuel injector class would represent the common traits (attributes and methods) of all fuel injector objects. Another way of looking at it is that classes represent the static structure of the system, whereas objects provide a dynamic, run-time view.

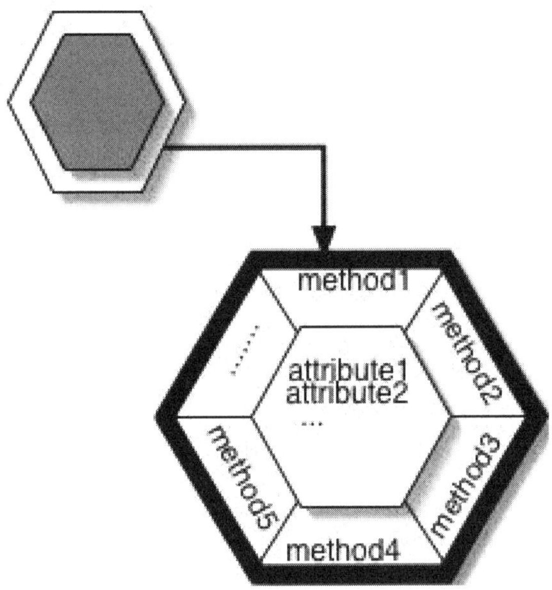

Figure 3. Object Structure and Method Invocation

The Unified Modeling Language (UML) is widely used for object-oriented modeling of systems. The use of object-oriented modeling techniques for automotive real-time systems has been investigated by Volvo [4] and DaimlerChrysler [5]. A detailed treatment of UML and object-oriented software development can be found in the many books available on the subject, such as the one by Craig Larman [6]. Figure 4 shows an example of how the fuel injector class would be modeled in UML.

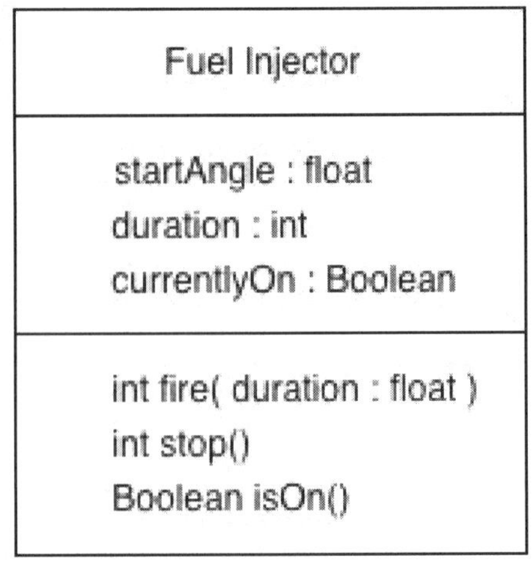

Figure 4. UML Class Representation.

ENCAPSULATION

Objects support the idea of encapsulation or information hiding. This means that an object's attributes are not directly accessible by other objects. An object's attributes may only be read from or written to by the methods of that object. This provides a complete separation of interface from implementation, allowing the data types of attributes and the implementation algorithms of methods to be modified without affecting client objects that invoke an object's methods. As long as the method interface remains unchanged the client object remains unaffected. This powerful mechanism allows software objects to be easily upgraded without impacting the rest of the system. The objects are very much self-contained and loosely coupled with one another.

INHERITANCE

When defining systems in terms of objects it is quickly realized that many objects are very similar to other objects. The concept of inheritance allows one class to be specified in terms of another class. The new class is known as a subtype of the existing base class and only additional attributes and methods over and above those in the base class need to be specified for the new class. This is a powerful technique for describing variants in ECUs and vehicles.

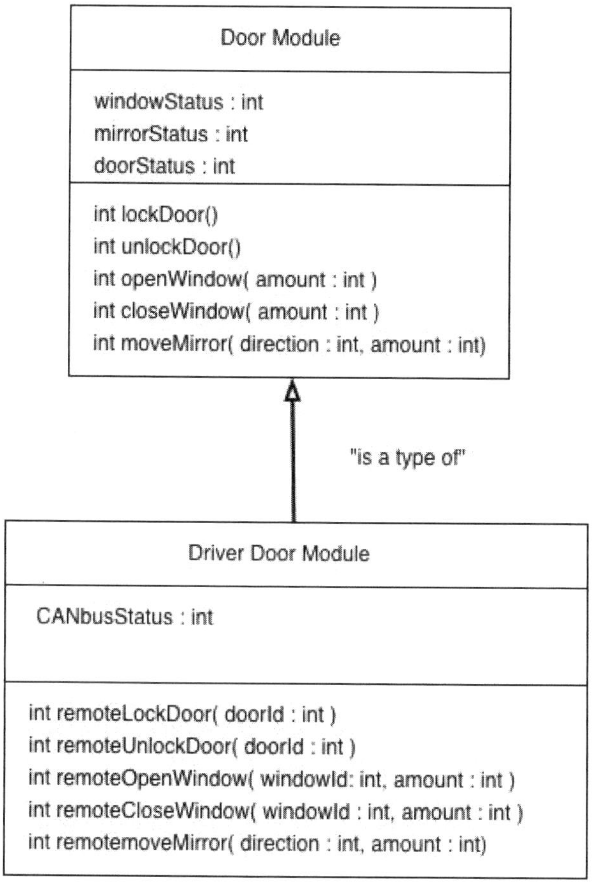

Figure 5. Inheritance between UML Classes.

In Figure 5 UML notation is used to show that a driver's door module is a specialization of a standard door module. Inheritance is an easy way to reuse software objects. Only the additional functionality needs to be added to the subclass. In the Driver Door Module example, when developing the Driver Door Module class just the additional functionality over and above the Door Module class needs to be implemented.

AGGREGATION

Aggregation is the process of using existing objects in the implementation of new objects. It is an alternative to inheritance for reusing objects. Aggregation is similar in concept to an assembly-subassembly type structure. Aggregation has the advantage over inheritance as a reuse mechanism because the reused objects are completely encapsulated within the new object and are not exposed to any client objects. With inheritance the methods of any inherited class may also be called by client objects, even if that is not the intention. Figure 6 illustrates the use of aggregation in UML, where the Driver Door Module class contains an instance of a Door Module class. Notice how the Driver Door Module class now needs to have a more comprehensive interface, since it no longer inherits the interface of the Driver Door Module class.

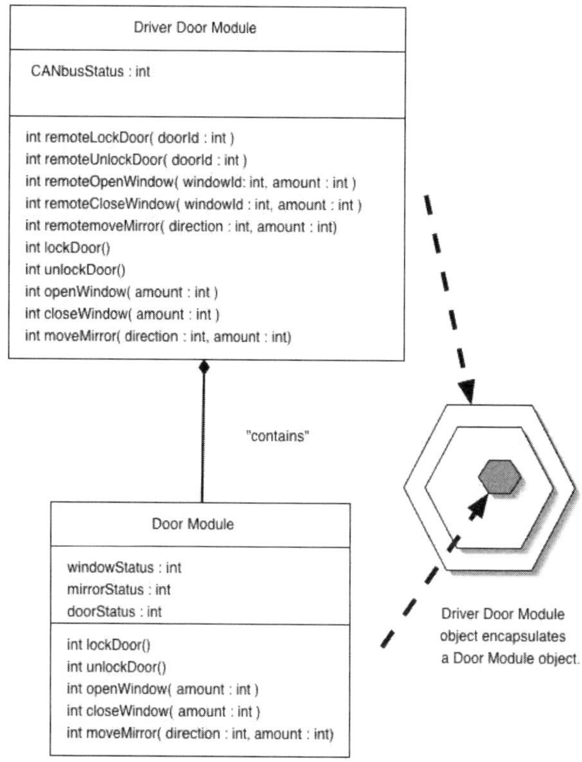

Figure 6. Aggregation between UML Classes.

POLYMORPHISM

Polymorphism is a feature of object-oriented systems that allows the same method name to be used in more than one class of object. For example, the method *stop* could be used with various classes such as Motor, Engine, Fuel Injector and so on. Polymorphism allows the name *stop* to be used with all of these classes. The correct piece of code is executed based on the class of object in question.

Polymorphism has a number of advantages for the software developer. First of all, meaningful names can be given to methods regardless of whether the name has already been used. Secondly, the developer can generalize methods in the base class of an inheritance hierarchy without knowing in advance what subtypes may be defined. This makes the resulting objects more reusable.

SOFTWARE COMPONENTS

Component-based software extends the object-oriented concept to software applications as a whole. It is at a higher level of granularity, such that a single software user function exhibits the same characteristics as a software object. That is, it has certain well-defined interfaces that can be invoked, and it is possible that multiple instances of the software component exist. It should be emphasized that even though a software component has an object-oriented behavior it does not have to be implemented using an object-oriented programming language. The internal implementation could be in a language such as C or even Simulink code such as S-functions.

The Object Management Group (OMG) [7] is an industry consortium whose goal is to define a set of interfaces for interoperable software. It has been in existence since 1989 and has specified the Common Object Request Broker Architecture (CORBA) that defines a high-level facility for distributed computing. CORBA uses an object-oriented approach to hide the differences in programming languages, operating systems and object location in a distributed application. It provides an open distributed computing environment that facilitates component integration and reuse.

The central component of CORBA is the Object Request Broker (ORB) which works as a software bus that transparently relays object requests across the various implementation technologies. Application objects interact only with the ORB in a client-server fashion. A client object locates the required server object, invokes operations on it and is notified of the occurrence of any errors through a standard exception handling mechanism. The ORB is responsible for routing the request from the client object to the server object and returning any results. CORBA can handle both synchronous and asynchronous requests. A component may act as both a client and server simultaneously. The ORB implementations include code stubs and skeletons known as Object Adaptors that map the object interfaces to specific implementation languages. Figure 7 shows the structure of a CORBA-based software application.

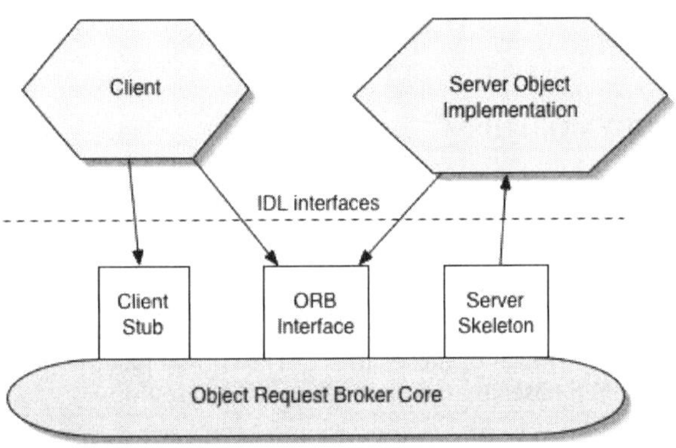

Figure 7. CORBA Distributed Application Structure.

The OMG has defined a range of object interfaces for commonly used services such as Object Location and Naming, Event Notification, Persistence, Concurrency, Transactions etc. These are known as CORBAservices and are provided by ORB vendors. CORBAservices are similar to many of the services provided by operating systems. There is also a set of object standards called CORBAfacilities that define functionality that would be common across many types of application. Examples are Printing, Task Management and User Interface. There are also standards known as CORBAdomains that define objects that would be useful in specific vertical markets. Some of the domains currently addressed include Air Traffic Control, Telecommunications, Data Acquisition and Computer Aided Design.

The key to integrating application objects is the specification of standard interfaces using the Interface Definition Language (IDL). The IDL is the mechanism that separates interface from implementation, and it provides for communication between objects that is independent of programming language, hardware platform, networking protocols and physical location.

INTERFACE DEFINITION LANGUAGE (IDL)

The IDL is a notation for defining application programming interfaces. It is independent of programming languages and defines the boundary between client code and object implementations of services. The IDL is a fundamental part of OMG standards and provides platform-independent definitions of software interfaces.

The OMG has defined standard language bindings for the IDL that enable an automatic translation of IDL specifications to programming languages such as C, Java, C++, Ada etc. The IDL is a pure specification language and does not dictate the implementation of the software object. The object may be implemented as a library function on the same computer as the client object, or may even be implemented on a remote networked computer. As long as the client and server objects adhere to the IDL specifications, communication will be successful. The following is an example of IDL for the door module objects.

```
module VehicleDoorSystems {

   interface Door_Module {

       // attribute declaration
       attribute unsigned long windowStatus;
       attribute unsigned long mirrorStatus;
       attribute unsigned long doorStatus;

       // possible exceptions that may be
       // raised when executing methods

       enum Faults { DoorFaultDetected,
              mirrorFaultDetected };

       // method declaration
       unsigned long lockDoor();
          raises ( DoorFaultDetected );
       unsigned long unlockDoor()
          raises ( DoorFaultDetected );
       unsigned long openWindow (
          in unsigned long amount );
       unsigned long closeWindow (
          in unsigned long amount );
       enum Direction {UP, DOWN, LEFT, RIGHT};
       unsigned long moveMirror (
          in Direction directionToGo,
          in unsigned long amount )
          raises ( MirrorFaultDetected );
   }

   // define Driver Door Module to be a subtype
   // of the basic Door Module

   interface Driver_Door_Module : Door_Module {

       // additional attributes to basic door
       // module
       attribute unsigned long CANbusStatus;

       // additional methods
       enum DoorType { PASSENGER, DRIVER, ALL };
       enum WindowType ( PASSENGER, DRIVER, ALL,
              BACKLEFT, BACKRIGHT };
       unsigned long remoteLockDoor(
              in DoorType door )
          raises ( DoorFaultDetected );
       unsigned long remoteUnlockDoor(
              in DoorType door )
          raises ( DoorFaultDetected );
       unsigned long remoteOpenWindow (
          in WindowType windowed,
          in unsigned long amount );
       unsigned long remoteCloseWindow (
          in WindowType windowed,
          in unsigned long amount );
   }
}
```

The example illustrates the use of enumerated types in IDL to define allowable values for items such as door types and windows types. Using enumerated types in the IDL definitions provides greater clarity on the expected usage of parameters. IDL also has an exception handling mechanism that allows methods to raise exceptions when an operational error occurs. The client application is notified of server or ORB exceptions synchronously through the calling interface.

The use of IDL provides a hardware-independent interface between the software functions and thereby simplifies the task of integrating software components. The quality of the IDL designs is very important for ensuring that the components are interoperable with one another and are widely reusable. For example, although the IDL will define the number and types of method parameters, the client and server objects must both agree on the semantics of how the parameters are used. CORBA and IDL do give the immediate advantage of hardware independence for the vehicle software architecture. Even ECU device drivers could be given an IDL object interface to make them part of the software architecture, free to be implemented on the appropriate ECU. The principles of high cohesion and loose coupling between IDL-defined objects will ensure greater reusability. Design patterns for use with object-oriented technology are well documented [6].

AUTOMOTIVE OBJECT REQUEST BROKER

The CORBA standards grew out of a need to provide seamless interfaces between the heterogeneous technologies used to implement enterprise-wide applications. The use of the internet and client-server computing paradigms as a means of implementing distributed applications motivated the design of the platform-independent CORBA architecture. CORBA has been hugely successful in the enterprise computing domain, and it has delivered on the promises of reusability, portability and simplified integration.

The CORBA specifications were originally designed with network servers and a PC environment in mind, but this does not mean that the concepts are not applicable to real-time distributed computing applications. Indeed, there is a subcommittee of the OMG tasked with the application of CORBA to real-time applications. The result is two additional specifications; the Minimum CORBA specification and the Real-Time CORBA specification. The Minimum CORBA specification describes the basic features needed to implement CORBA on resource-limited systems such as embedded systems. The Real-Time CORBA specification describes an implementation of CORBA on systems where end-to-end request timing constraints must be met and where real-time task scheduling policies are used. Although these specifications are a step in the right direction for showing how CORBA can be used in automotive applications, currently available real-time ORBs such as Borland's VisiBroker-RT [8] seem to be limited to the real-time Linux, Windows CE and VxWorks operating systems. As such they are not directly usable with OSEK systems that are typically used on vehicle ECUs. They might however be useful for vehicle infotainment and telematics applications where more computing power is usually available.

The key feature that underpins CORBA is the IDL specification. The authors suggest that most of the benefits of simplified integration and portability can be achieved by using IDL to specify software interfaces. Vehicle software components can be defined at a subsystem level using IDL early on in the design process. The structure and interactions among components can be verified functionally at an early stage using IDL and object-oriented specification methodologies without regard to the communication mechanisms. The software components could be functionally simulated on a PC using stub components. Once the OEM is satisfied, the subsystems can then be further decomposed into sets of software components (also specified in IDL) to be allocated to suppliers. The result is that the interactions among the components have already been functionally verified and failure mode responses explored before the supplier implements the software component. As long as the IDL interfaces remain unchanged, the process of integrating the finished components should be much easier. Using IDL as part of the requirements specification process removes much of the ambiguity of natural language specification, since the IDL is a compilable notation. The IDL can be supplemented with UML sequence diagrams that indicate the dynamic use of the software components. Software modules delivered by suppliers can be verified against the IDL specification using software test harnesses.

Once the overall software requirements have been defined and expressed using IDL, the next step is to map the software components to the hardware architecture. Some components may have to be mapped to a specific ECU because of sensor or other hardware requirements. Techniques such as cluster analysis [9] can be used to map remaining software modules to available ECUs. In this way the allocation of software to ECUs is neither predetermined nor constrained.

The issue of mapping IDL calls to specific implementation languages and tools needs to be addressed. The main problem to be overcome is the mapping of the dynamic run-time ORB operation to the static run-time operation of automotive ECUs. The Object Request Broker normally includes facilities to dynamically locate server objects, to invoke additional server objects to deal with increased loading and to dynamically prioritize client requests. Sufficient computing resources do not exist for this highly dynamic mode of operation in an automotive environment. On the other hand, the software components to be used in an automotive system are static and are defined during the software design phase, so the client-server object interactions are also known in advance. This prior knowledge allows the Object Request Broker implementation to be hard-coded to optimize the use of resources. This can be accomplished using a software configuration process prior to the final software build, in the same way that OSEK systems are statically configured at build time using the OSEK Implementation Language (OIL) file. This process is summarized in Figure 8.

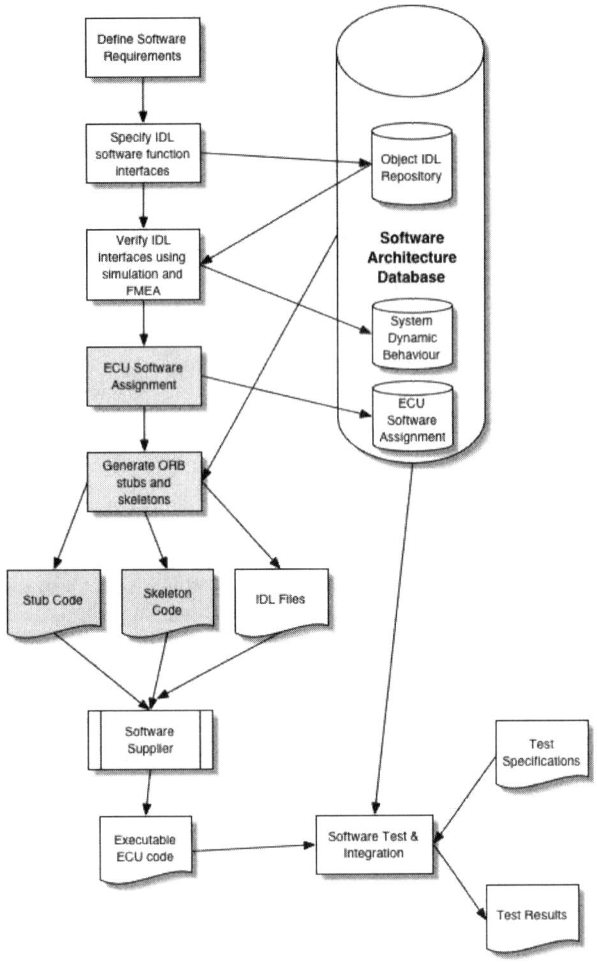

Figure 8. Development Process using CORBA/IDL.

The process of configuring the Object Request Broker consists of the following activities

1. ECU software assignment. The assignment of individual software functions to ECUs must be carried out first so that the system generation process knows whether each IDL call is to a server object on the same ECU or to a different ECU on the network. The actual client and server source code does not need to be changed as a result of reassigning software functions to different ECUs because the following process steps will take care of the interfacing details. Thus a significant simplification of ECU software integration can be realized using IDL interfaces.

2. Client Stub generation. CORBA IDL calls contained within client application code must be replaced with appropriate stub code that converts the IDL call to a proper call of the server object code. This works like a normal pre-compilation process by replacing IDL references in application code with appropriate stub routine calls. In a full-featured ORB the standard stub would call the ORB library, which would in turn locate and call the server object. In a custom static automotive implementation the configuration tool would generate code based on the target server object identification, the server object location and the desired communication mechanism. For example, if the server object is on the same ECU as the client, then a simple library call might be sufficient. If the server is located on another ECU then the stub would have to map the IDL call to a network message using OSEK COM or direct CAN messages, transmit the request to an ORB request handler on the remote ECU, and wait for a response before returning control to the client application. The mapping of CORBA requests to CAN messages has already been addressed in existing research [10].

3. Server Skeleton generation. The server side of the ORB implementation consists of code that unpacks the IDL parameters from remote CAN-based requests or else maps IDL parameters from local client requests to the correct calling interface for the server code. The skeleton code is responsible for returning any results to the client object according to the IDL specification. Various approaches may be taken to implement server code. Where only one instance of an object exists, the implementation can exist as a library routine in ROM. Where more than one instance of an object exists, the methods of the object could be implemented as reentrant routines in ROM with object-specific attribute values and other data stored in RAM. Techniques for mapping UML object-oriented designs to procedural C code have been described by James Rumbaugh [11]. The server object methods may be implemented in any suitable language. The skeleton code does the mapping from the object-oriented IDL interface to the ECU-specific implementation.

4. Target system build. The generated client stubs and server skeletons are linked with the application software functions for each ECU in the vehicle. The vendors of CORBA technology normally provide implementations of CORBAservices and CORBAfacilities with their ORB implementations. While most of these services are of little use in an automotive environment, one that is quite useful is the Event Notification service. This provides a publish-and-subscribe type interface between event sources and event consumers. It provides both push and pull type interfaces to event sources and event consumers. This service could be developed as a custom implementation based on the CAN bus. The result of doing this would be a hardware-independent distributed event notification service that would further decouple software functions in a predominately event-driven environment. A description of a CAN-based CORBA Event Notification service is described by Finocchiaro et al [12].

The OMG has specified a standard mapping of IDL to C and provides basic implementations of stubs and skeletons (Object Adaptors) that can be used as the basis for custom implementations.

CONCLUSION

This paper has examined the problems of software integration of ECUs on a vehicle network. Most of the problems are due to the tight coupling between the software functions and ECU hardware. The integration problems are compounded by the fact that the integration between ECUs currently takes place at the CAN network layer by means of message transfer. Integration testing occurs at a very low level, i.e. at the Physical Layer, which is in turn constrained by what information can be deduced, mapped and translated to the higher level network layer faults. This is further hampered by the fact that the Physical Layer has its own inherent problems due to its electrical characteristics in terms timing, transmission errors, latency problems, so it is difficult to debug and understand the problems at the higher layers due to the masking effect of these lower level issues. Using the techniques suggested in this paper, the integration process can be moved up to a higher application level where ECU software interactions take the form of client-server requests and responses. At this level the software design is easier to understand and debug. In addition, the intended operation of the system can be simulated and the failure mode operation analyzed and understood prior to programming. Using object-oriented modeling techniques for designing the software architecture will help achieve more flexible and reusable vehicle software functions.

It is hoped that the suggestions described in this paper for implementing CORBA technology in an automotive environment might be considered for future AUTOSAR and OSEK standardization, thereby bringing the associated benefits to a wider group of automotive software developers.

REFERENCES

1. http://www.osek-vdx.com. OSEK Specifications.
2. http://www.asam.de. ASAM Specifications.
3. http://www.autosar.org. AUTOSAR group details.
4. Axelsson, J., "Holistic Object-Oriented Modelling of Distributed Automotive Real-Time Control Applications". IEEE 1999. Proceedings of 2nd IEEE Symposium on Object-Oriented Real-Time Distributed Computing.
5. http://www.dc-cc.com. Automotive UML. DaimlerChrysler Competence Centre.
6. Larman, C. (2002). Applying UML and Patterns. Prentice Hall PTR. ISBN 0-13-092569-1.
7. http://www.omg.org. OMG, CORBA, IDL information and specifications.
8. http://www.borland.com/visibroker. VisiBroker-RT ORB product information.
9. Rushton, G., Zakarian, A. and Grigoryan, T. "Development of Modular Electrical, Electronic, and Software System Architectures for Multiple Vehicle Platforms". SAE paper 2003-01-0139.
10. Lankes, S., Jabs, A. and Bemmerl, T. "Integration of a CAN-based Connection-oriented Communication Model into Real-Time CORBA". IEEE Proceedings of the International Parallel and Distributed Processing Symposium 2003.
11. Rumbaugh, J. et al. Object-Oriented Modeling and Design. Prentice Hall. ISBN 0-13-630054-5.
12. Finocchiara, R., Lankes, S. and Jabs, A. "Design of a Real-Time CORBA Event Service customised for the CAN Bus". IEEE Proceedings of the 18th International Parallel and Distributed Processing Symposium 2004.

CONTACT

Brendan Jackman B.Sc. M.Tech.

Brendan is the founder and Director of the Centre for Automotive Research at Waterford Institute of Technology, where he supervises postgraduate students working on automotive software development, diagnostics and vehicle networking research.. Brendan also lectures in Automotive Software Development to undergraduates on the B.Sc. in Applied Computing Degree at Waterford Institute of Technology. Brendan has extensive experience in the implementation of real-time control systems, having worked previously with Digital Equipment Corporation, Ireland and Logica BV in The Netherlands.

Email: bjackman@wit.ie

Website: http://www.wit.ie/car

Shepherd Sanyanga (BEng BSc MSc PhD CEng EurIng MIEE)

Shepherd is a Technical Specialist in the design of Subsystem Communication Protocols & Diagnostics and the verification of such systems at TRW Automotive in the United Kingdom, working with the major OEMs in Europe and North America. He has had extensive experience in the system design of real-time embedded control systems for safety critical and comfort applications in both the Automotive and Aerospace industry. He has worked in organizations such the United Nations, Lucas Aerospace Systems, Sagem Automotive and Ford Motor Company. He is also an external examiner at an automotive research establishment in Ireland which runs MSc research courses for the automotive industry.

Email: shepherd.sanyanga@trw.com

2005-01-0312

Why Switch to an OSEK RTOS and How to Address the Associated Challenges

Thierry Rolina
ETAS Inc.

Nigel Tracey
LiveDevices (ETAS Group)

ABSTRACT

Most automotive software systems are still built today using a cyclical scheduler that runs tasks at fixed time intervals. This approach provides excellent insight into and control of the real-time behavior of a system, as tasks repeatedly run one at a time and in the same order.

Given that so many operating systems are using an approach that provides the necessary real-time control and transparency, why change to an OSEK operating system? In other words, what value does an OSEK operating system deliver compared to one that uses the traditional approach to task scheduling?

In this paper, we answer precisely that question. An OSEK OS can offer significant efficiency advantages that ultimately save cost in development and production. However, an organization faces certain challenges when it decides to replace its traditional OS with an OSEK OS. We identify and discuss these challenges.

INTRODUCTION

"We build a lot of software today, and we just keep on doing it. Over, and over, and over again," Stephen Mellor rightly complained in a recent article in Embedded Computing Design.(1) In his remark, Mellor points out that application software is often redone rather than reused, and that this is due to the fact that application software code is bound to all the layers the software relies on. Mellor suggested that a new way of thinking about application software is necessary in order to put an end to the waste of time and money in development. The solution according to Mellor is to view and value application software as an asset.

Ten years ago, the OSEK initiative reflected similar thinking in terms of non-application software. Those who created OSEK saw a need to specify uniform services, interfaces, and protocols for the operating system, communication, and network management of distributed embedded real-time systems. While OSEK standardization efforts have been successful and are ongoing, the automotive industry has not yet implemented many of the OSEK standards. In this paper we will show why using the OSEK approach to developing real-time embedded software offers great advantages over the traditional approach.

THE AUTOMOTIVE SUPPLY CHAIN

The automotive industry builds vehicles in large quantities (up to tens of millions of units in some cases). Small problems may therefore have a great impact, since they have the potential of manifesting themselves in a great number of vehicles. If we assume a population of 1 million vehicles, with each vehicle being used 1 hour a day, 300 days a year, the potential impact of a small problem becomes obvious. A quick calculation leads to 300 million hours of use per year, which poses a heavy functional reliability constraint on any system. To ascertain good functional reliability for their products, OEMs usually assign the task of function specification design to their own engineers. The task of implementing these functions, however, is usually outsourced to suppliers.

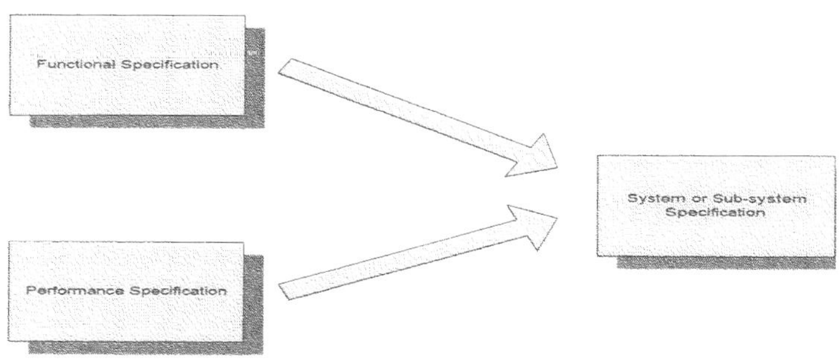

Similarly, OEMs define the performance requirements that embedded systems intended for their vehicles must fulfill. Then the OEMs' suppliers must provide evidence that their embedded systems satisfy their throughput requirements, or that they will respond to inputs within the allocated response time.

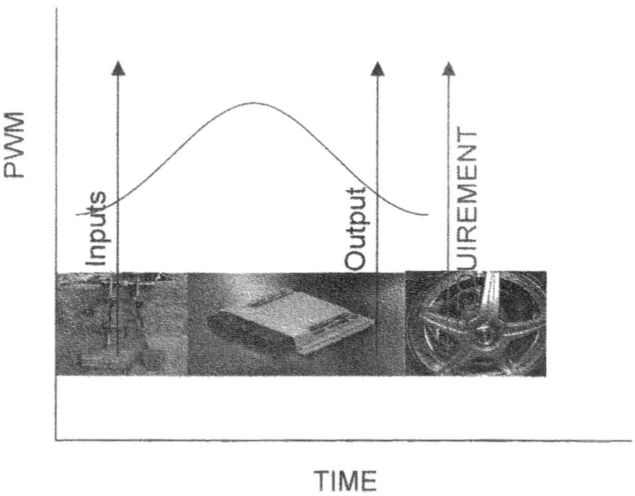

Figure 1.

To put it in general terms, automotive OEMs specify the systems and suppliers implement them. Integration is done later at the OEM or the supplier. In the area of powertrain controls in North America, both the hardware platform (microcontroller) and infrastructure software (scheduler API, HWIL) are specified by the OEM. In other areas, such as chassis or body controls, a different distribution of tasks gives suppliers greater flexibility. The reason for this distribution of development tasks in the automotive industry comes from the necessity to reduce costs.

A WELL PROVEN APPROACH?

Embedded real-time automotive software runs on microcontrollers, i.e., silicon. Function developers, however, typically model the functions that will later run on a microcontroller in a PC environment. Controls engineers implement these functions, and software engineers fit them into the confined environment of a microcontroller, where instruction size, memory size, CPU load, etc. are at a premium.

Most embedded real-time systems use a simple microcontroller that repeatedly runs one main loop: Whenever external inputs request some activity, the program code branches off to service each request and then returns to the main loop. This structure is commonly called "sequential organization". In real life, we most often see systems dependent on multiple interrupts. The advantage here is to eliminate the tedious polling of interrupts, which would result in faster response to an external event. The following illustrates "sequential computing":

```
void main(void)
{
  do_init();
  while (1) {
    t1();
    t2();
    t3();
    delay_until_cycle_start();
  }
}
```

Functions are computed sequentially, providing a clean and simple software implementation. If we now add a periodic interrupt, the system is almost fully loaded (see figure 2.). Since 3ms defines the "timing granularity", every function must be processed every 3ms. The load can then be calculated from the requirements table:

Task	Sample rate	Processing time	Requirement	Effective load
T1	3ms	0.5ms	16.6%	16.6%
T2	6ms	0.75ms	12.5%	25%
T3	14ms	1.25ms	8.9%	41.7%
isr	10ms	0.5ms	5%	16.6%
Total CPU load			48%	99.9%

In its simplest form of cyclic implementation, the system is fully loaded after adding the interrupt service routine, and can no longer accommodate new functionality. Adding new functionality will require hardware modifications.

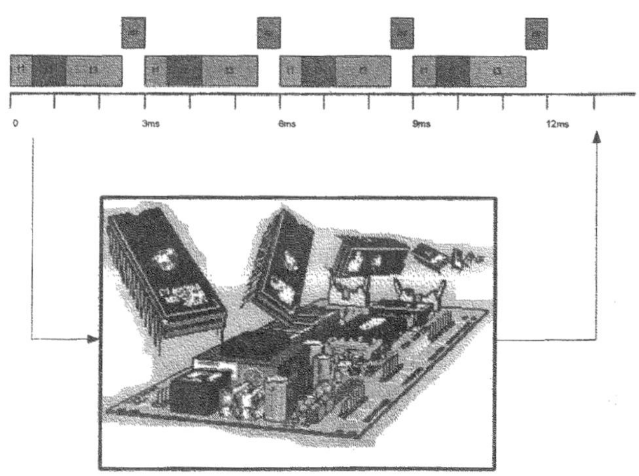

Figure 3.

There are more sophisticated implementations of the cyclic scheduler using the concept of major and minor cycles. In our example, a major/minor cycle approach would free some CPU because the less demanding tasks (T2 and T3) would not be required to run at the 3ms rate. As we are freeing the CPU, adding new functionality becomes possible, but it will have to be partitioned so that it fits into the available cycles. We must also point out that a tight connection between software and hardware exists in such an implementation, making any retargeting to a different processor difficult, and reuse of application code impossible. It is also very difficult to deal with requirement changes from the customer which often lead to patching the scheduler as well. On the positive side, the cyclic implementation is free, and the source code is available. Also, in such an implementation, design will always meet performance requirements.

TOWARD A MORE FLEXIBLE APPROACH: OSEK SCHEDULING

The answer to such problems is to separate the application software from the infrastructure. It will be especially important, because distribution of functionality has made its way into vehicles, causing the number of devices to grow considerably over the years. Distribution of functionality is achieved though data networks such as CAN, LIN, Flexray..., which introduce another level of complexity.

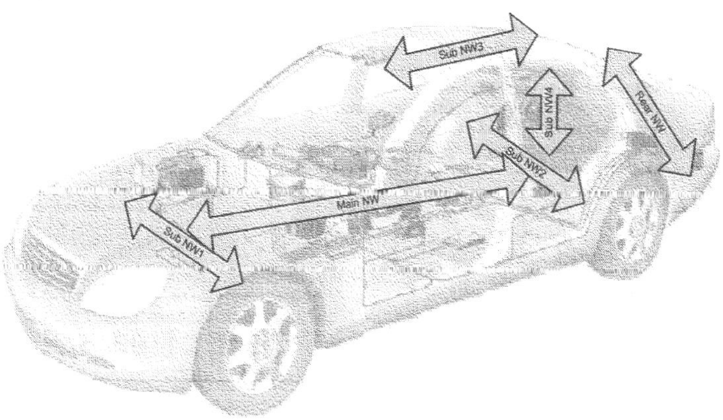

Figure 4.

Each module connected to a network has to include specific network drivers. For simple ECUs, an architecture where direct hardware and network access has been granted to the application software may still be appropriate (figure 5). Such an approach certainly falls short with more complex ECUs, where the software load needs to be carefully designed and controlled.

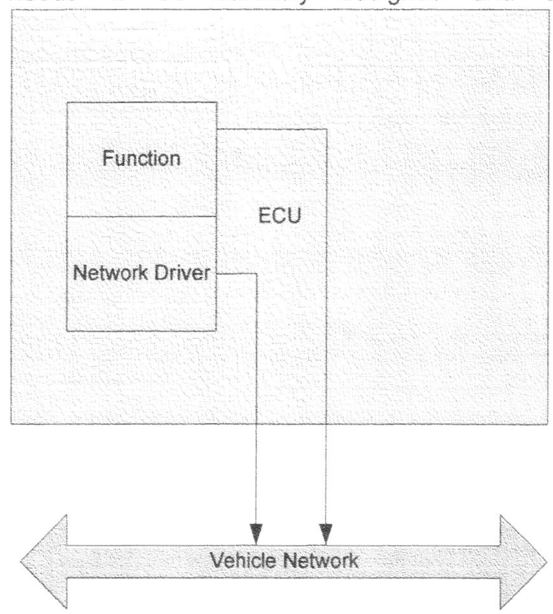

Figure 5.

Scheduling is a vital component of infrastructure software. Careful scheduling enables the ECU to meet its performance requirements. It is the heart of the system. An often cited argument against using an RTOS is that writing and debugging the application will increase in complexity and most developers want to stay away from error-prone complexity. Another argument against using an RTOS is that it comes at a price, while other operating systems are free.

However, as we will reinforce in the return on investment analysis, the advantages of using an RTOS clearly outweigh the drawbacks. An RTOS comes with a set of validated API routines that are ready to use, giving everybody the opportunity to use the same language, regardless of what microcontroller they use. An RTOS also provides the embedded software engineers with a higher level of abstraction, allowing them to focus on their task, i.e., the application. The OSEK-VDX initiative specifies all this.

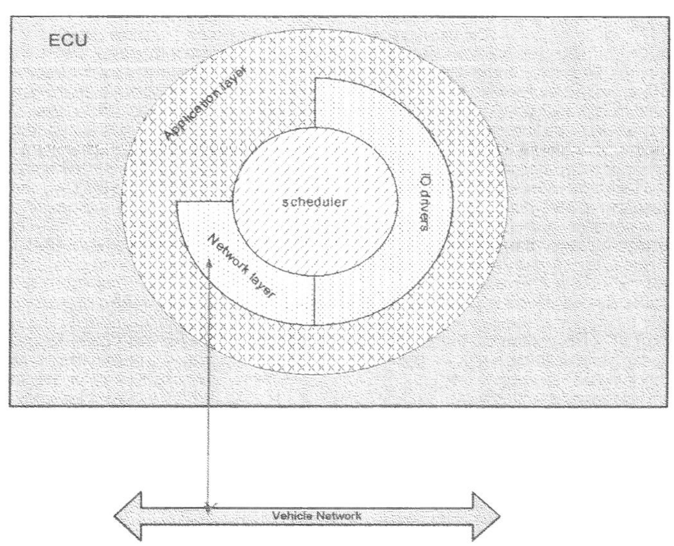

Figure 6.

Figure 6 illustrates the more complex ECU mentioned previously. The application layer now accesses hardware and network resources though abstract layers, allowing designers to build more abstract and reusable application components.

OSEK Scheduling

The OSEK scheduling mechanism is based on events, which will guarantee the most efficient processing of interrupts in the application. Instead of polling, the ECU responds to a set of events, where time is simply one particular type of event. The OSEK RTOS specifies a scheduler that uses a fixed priority scheduling policy. Under this policy, each task is assigned a fixed priority. The scheduler will always run the highest priority task that is ready. The higher priority task will then preempt any lower priority task running at that time. When the higher priority task is finished, the lower priority task is resumed at the point of preemption. Interrupts are processed in the same manner. If we now go back to the 3 tasks system described in 2 (T1 running every 3ms, T2 every 6ms, T3 every 14ms, and isr every10ms), an OSEK scheduler will process the various tasks at their required rate, thereby freeing up CPU time. This CPU time is available for additional tasks that may have to be implemented in response to modifications made to the function requirements.

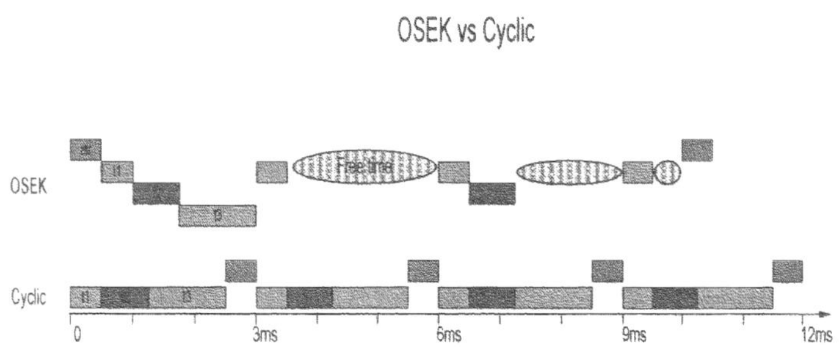

Figure 7.

In addition, the OSEK scheduling is scalable. Notions such as multiple task activations are supported, giving the ability to retrigger functions if need be. Tasks can also wait for events. This can be particularly useful in applications where a higher degree of user interaction is required. For lower-end applications, the basic set of attributes provides the necessary capabilities to run with full preemption, even in an 8-bit environment. Alarms are the mechanism by which OSEK scheduling handles periodic behavior. Again, looking at time is just looking at a different type of event. Sporadic behavior is handled via interrupts which could be done inside or outside the RTOS. Support of critical sections can be handled by using resources, and safe access to resources is handled via a priority ceiling protocol (best response/no deadlocks).

THE CHALLENGES OF USING AN OSEK RTOS

Using an OSEK scheduler offers a lot of benefits, the biggest ones probably being the increased traceability from the specification to the implementation, and the ability to design software at a level that is less target-dependent. On the other hand, when selecting a scheduler, consideration must be given to the following facts. Using preemption introduces memory overheads, which could drive the price of the ECU up. Also, the RTOS is now part of the ECU performance equation.

Memory Overheads

Preemption requires RAM space, because each time a high priority task preempts a lower priority task, the context of the low priority task must be saved on the stack so it can be restored later. In some OSEK RTOS implementations, the RAM demand will grow proportionally to the number of tasks. This could be a problem, especially if production cost is an issue, since RAM is costly. This is particularly true for RTOSs that have not been engineered so that all the tasks can share a single stack. This will be an important point in the production cost section of our return on investment analysis.

Systems' Performance Determination

The other area where software engineers must be careful is the overall performance of the system, as the system's performance is dependent on the RTOS performance. Figure 8 shows an example of preemption where the scheduler has to run in addition to the performance hit caused by the high priority task switch in and task switch out. Special care must be taken here, and one must make sure that the scheduler's performance data have been published, that it is deterministic, and that the implementation is efficient enough to allow the user to take advantage of the RTOS flexibility. We have actually seen instances where the RTOS implementation was so inefficient that users could not use any of the system services. Overall, the RTOS implementation should provide an improved system response compared to a cyclic implementation.

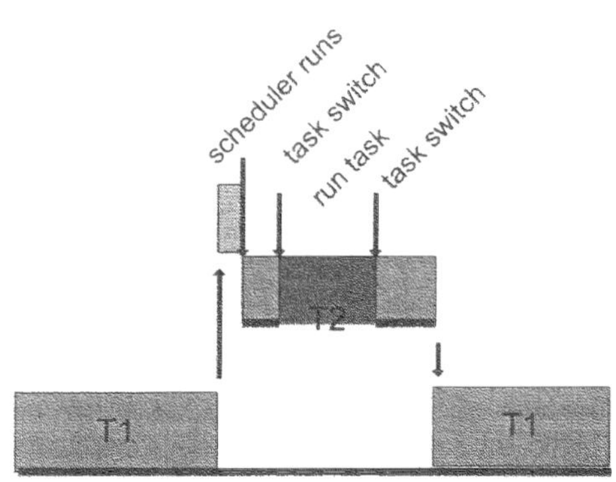

Figure 8.

A quick comparison of how response time is determined in both cyclic and priority-based scheduling reveals which approach leads to the best system response time. In cyclic scheduling, the response time (R) of each task is simply given by:

R = C + Polling_delay, where C is the computation time.

The system will meet its performance requirements if all the tasks meet their performance requirements. It is typical to poll more frequently in such designs, as this is the only possible way to improve the system's response time. Frequent polling will increase the interrupt rate, and will decrease the system's overall throughput.

With a priority-based scheduler, the response time is a function of the RTOS, plus any interference (preemption) and blocking (if resources are used) caused to the task:

R = C+ f(RTOS) + I + B

Research shows that OSEK scheduling fulfils the assumptions of the deadline monotonic theory (2). This mathematical approach to preemptive scheduling gives us the ability to compute the response time (R_i) for each task:

$$R_i = S_i + B_i + C_{sw} + \sum_{\forall k \in hp(i)} \lceil R_i / T_k \rceil (C_{sw} + C_k)$$

Where:

- Ci is the worst case execution time of a given task Ci
- Tk is the period of task k
- Csw is the cost of switching to and back from a preempting task
- Bi is the blocking time of task i
- Si is the scheduler overhead for task i
- Hp is the set of tasks of higher priority than task i

Assuming that we have access to the performance data of the scheduler, i.e. Csw, we are now able to determine the response time of each task in the system, and hence the overall performance of the system.

THE COST OF BUILDING EMBEDDED REAL-TIME SOFTWARE

There are two cost components when building ECU software: development cost - which covers design, implementation, and testing - and production cost. In the automotive industry, development cost constitutes a significant amount of the overall cost. We will examine this issue in the return on investment analysis at the end of this paper. Minimizing development cost will imply:

- Minimizing the impact of requirement changes at all stages (this can be done by reducing the interdependency of infrastructure software and application software).
- Minimizing development time.
- Minimizing the cost of infrastructure development, since the latter is not part of the core business.
- Improving the product integration phase.

The size and quality of the software will have an impact on the overall production cost when

- the application or the infrastructure software demands a large amount of RAM and when
- system performance response becomes unpredictable due to added functionality.

In both cases, the solution involves over engineering, either by adding unnecessary RAM space, or by running at a higher clock speed (which could affect other parts of the system). We will consider these factors in the context of our return on investment analysis at the end of this paper.

Using a Cyclic Scheduler

As we highlighted previously, a cyclic scheduler may be appropriate for small applications. It will quickly fall short with more complex systems as the functionality has to be split to fit into the scheduler's time slices. With this approach, the application software will be less reusable. Reusability is also reduced by the more hardware-dependent nature of the application software. On the process side, it will be more difficult to use modern approaches to software design, especially autocoding. And, as we outlined earlier, a cyclic approach makes inefficient use of the CPU. To conclude, opting for a cyclic scheduler comes at a price after all, as this seemingly smart decision will drive up the cost of developing and producing embedded systems

Using an OSEK Scheduler

An OSEK-based implementation offers organizations a better approach to software design. Distributed development is possible because the architecture is typically in place at the beginning of the project. A number of teams can then work on the various subsystems. Since all the teams share the same insight into the scheduler, requirement changes can be easily assessed. In addition, this approach promotes reusability of the software, as its design is less target-dependent. Using an OSEK scheduler makes the development cost of the ECU software manageable.

On the production side, implementation of an OSEK OS can increase the cost (larger RAM and increased execution time). To evaluate an OSEK-based approach, it is imperative to answer the following questions:

- How has the OSEK RTOS been designed to handle preemption with respect to RAM usage?

- What is the inherent performance of the scheduler (especially in terms of context switches)?

The following table sums up the pros and cons of each implementation.

	Cyclic	OSEK-RTOS
Infrastructure effort	some	none
Maintenance	Difficult	easy
Reuse	Limited	high
Verification/Validation	In-house testing	Certified
CPU requirements	high	Should be low
Memory requirements	Initially low, increases with maintenance	Should change little with the number of tasks

Comparison of scheduler possibilities

A RETURN ON INVESTMENT ANALYSIS (ROI)

Let us take a look at a concrete example. The following sample project is based on past experience with a customer project. We will look at the development and production of an ECU (including application software):

- Development takes 5 man years at $80k/year = $400k.
- Production will be 1 Million ECUs/year for 5 years.
- Microcontroller cost is $2/ECU.

The total baseline cost will be $10.4M. We will assume no differences in application development between the cyclic and OSEK-based approaches. We will assume that we have 5 requirement changes throughout the project.

Cyclic implementation

Let us first look at the cyclic implementation. Restructuring software involves identifying how functions can be split up to meet the performance requirements of the system. If we assume 2 man weeks/change, this will require a total of 10 man weeks, or $20,000. Tuning the software to remove timing problems takes 4 man weeks/changes, so it will require 20 man weeks, or $40,000. Finally, an upgrade to a more powerful controller will be necessary. If the cost is $0.5/controller, total CPU upgrade cost is 5x1Mx0.5= $2.5 Million.

There is no infrastructure cost involved (purchasing tools, training...).

We estimate that the total cost for the changes will be $2.56 Million.

The total cost for this project will be $12.96 Million.

OSEK-RTOS

Let us next look at an OSEK-based implementation. Restructuring software will take 1 day/change, which leads to a total cost of $2,000. It will not be necessary to test software for timing problems if the proper design tool suite is used, and it will not be necessary to tune the software to avoid timing problems. Additional tasks do not increase RAM requirements.

Tools will be necessary to support this approach, so there will an infrastructure cost. For this project, it will be below $100,000.

The total cost for the changes will be $2000, if we consider the infrastructure cost part of the initial development.

The total cost for this project will be $10.5 Million.

When doing such an ROI analysis, it is imperative to look at both the development and the production cost of embedded software. In our OSEK-based approach, we were able to save a tremendous amount of money and time to release the production software, as each modification of functionality can be assessed, and implemented without disturbing the overall ECU performance.

CONCLUSION

This paper described the advantages of using an OSEK-based approach to scheduling real-time embedded software. This approach is time and cost effective. It promotes reusability, and provides control on the overall system's performance.

REFERENCES

1. "Software as assets", fall 2004 Embedded computing design http://www.embeddedcomputing.com/articles/mellor/
2. Deadline Monotonic Analysis http://www.embedded.com/2000/0006/0006feat1.htm
3. OSEK-VDX http://www.osek-vdx.org

CONTACT

Thierry Rolina is Business Development Manager for the Software Components product line of ETAS in North America. He can be reached at Thierry.rolina@etas.us

Dr. Nigel Tracey is Director of Product Management at LiveDevices (ETAS group). He can be reached at Nigel.tracey@livedevices.com

2004-01-0757

Constraint-Driven Simulation-Based Automatic Task Allocation on ECU Networks

Paolo Giusto
Automotive Team - Cadence Design Systems, Inc.

Gary Rushton
E/E Vehicle Systems – Visteon Corporation

ABSTRACT

With the increasing number of ECUs in modern automotive applications (70 in high end cars), designers are facing the challenge of managing complex design tasks, such as the allocation of software tasks over a network of ECUs. The allocation may be dictated by different attributes (performance, cost, size, etc.). The task of validating a given allocation can be achieved either via static analysis (e.g., for cost, size) and/or dynamic analysis (e.g. via performance simulation – for timing constraints). This paper brings together two key concepts: algorithmic and optimization techniques to be used during static analysis and virtual integration platforms for simulation-based exploration. The two concepts together provide the pillars for a constraint-driven / simulation-based approach, tailored to optimize the entire ECU network according to a cost function defined by the user.

INTRODUCTION

In order to satisfy different drivers' needs such as performance, comfort, safety, and to meet stringent time to market and cost requirements, the automotive industry is increasingly using *configurable platforms* (mechanical, electrical/electronic) to implement safety-critical sophisticated applications such as adaptive cruise controls (Figure 1). These applications are implemented on complex distributed architectures where data messages are exchanged between Electronic Control Units (ECUs) via high-speed, fault-tolerant, serial buses (e.g., Flexray, TTP). The increasing functional complexity and chassis requirements (sensors/actuators allocation) have led to an increase in the number of ECUs (between 6 for a low-end and 35 for a high-end model, see Figure 2) and to the usage of multiple buses in terms of different

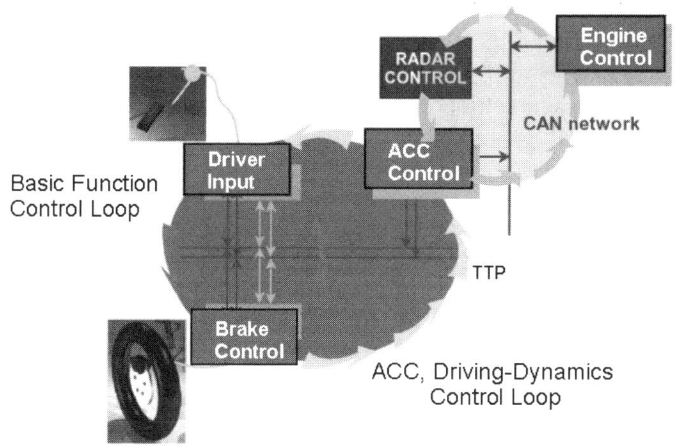

Figure 1: Adaptive Cruise Control

bandwidth requirements (10Mbs for a dependable fault tolerant communication protocol).

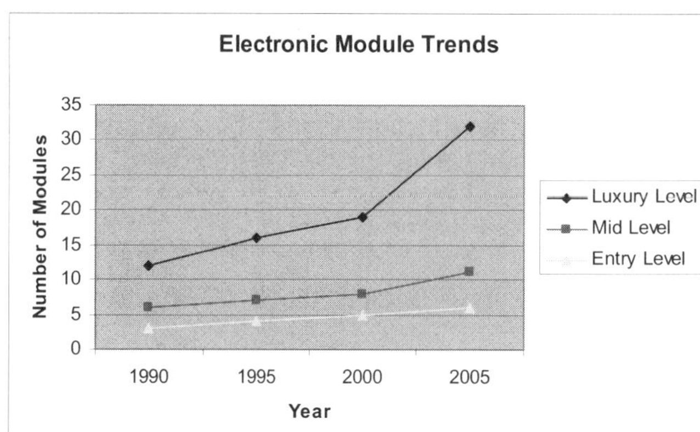

Figure 2: Electronic Module Trends

In this paper, we have identified two design issues. First, the OEM system architect often decides the distributed target electronic architecture (ECUs, buses, sensors, actuators, etc) *very early* in the design cycle, when not much information is available (HW is often not available), while the *validation of it is performed quite late,* only after sub-systems' HW/SW implementations become available. Thus, integration issues are discovered when changes are very expensive. Moreover, resilience to faults can only be checked via expensive and not easily repeatable tests on physical prototypes. Secondly, the communication between the car manufacturers (OEM) - playing the role of system integrator once the sub-systems have been delivered - and the providers of the sub-systems (e.g. engine control), is often a source of misunderstandings regarding the intended functional and non-functional requirements the sub-system has to provide in terms of timing, cost, and safety.

Besides, OEM and Tier1 system architects are facing the challenge of managing complex design tasks, such as the allocation of software tasks over a network of ECUs (SW binding). The trend toward the standardization of SW platform APIs consistent with OSEK specifications for communication among ECUs (e.g. COM and FTCom) and with application software processing and scheduling (e.g. OSEK Time) makes it possible to distribute the functionality (application SW) across different ECUs.

When looking at the vehicle as a system, the distributed nature of these applications also provides potential for optimizations that can be achieved via an efficient use of HW resources (e.g. a smaller number of ECUs). This optimization, carried out by system architects, is applicable to both OEM and Tier1 suppliers. While the Tier1 goal is to provide an optimized platform that can be re-used for different products by different OEMs, the OEM is more concerned with creating a platform that can be shared across vehicles.

In this paper we illustrate how two key concepts, algorithmic and optimization techniques for static analysis [1][2][3], and simulation-based virtual integration platforms [4][5][6] coupled together, provide the pillars for a constraint-driven and simulation-based approach, tailored to optimize the entire ECU network according to a cost function defined by the user. In the rest of the paper, we describe the concepts and the technologies aimed at realizing such an approach into an innovative tool flow in more detail.

THE VIRTUAL INTEGRATION PLATFORM

In [6], the concept of a virtual integration platform was introduced. In a nutshell, the Cadence Automotive System Design Platform (Figure 3) supports a distributed model-based system design process based upon several orthogonal concepts:

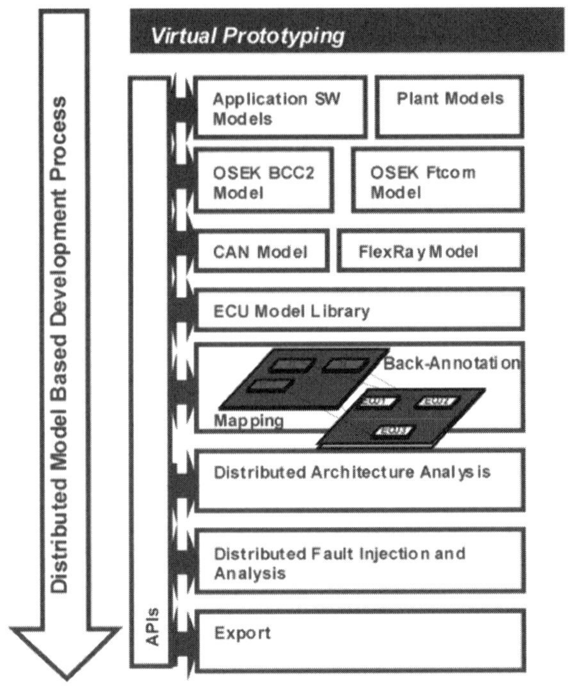

Figure 3: Cadence Automotive Platform

- A design shift from physical integration to virtual integration of models (Figure 4)

The extension of a single ECU model-based system design paradigm, in which the transformation from a design representation to its implementation is automatic.

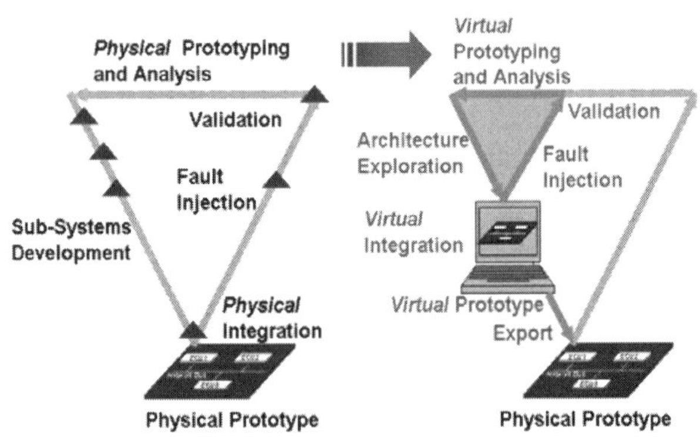

Figure 4: Design Shift

- The separation between executable models of control algorithms and the high level/abstract performance/functional models of the architectural resources including the network communication stack
- Binding control algorithm tasks to HW resources
- Binding functional abstract communications (signals and shared variables) to network protocol stack models

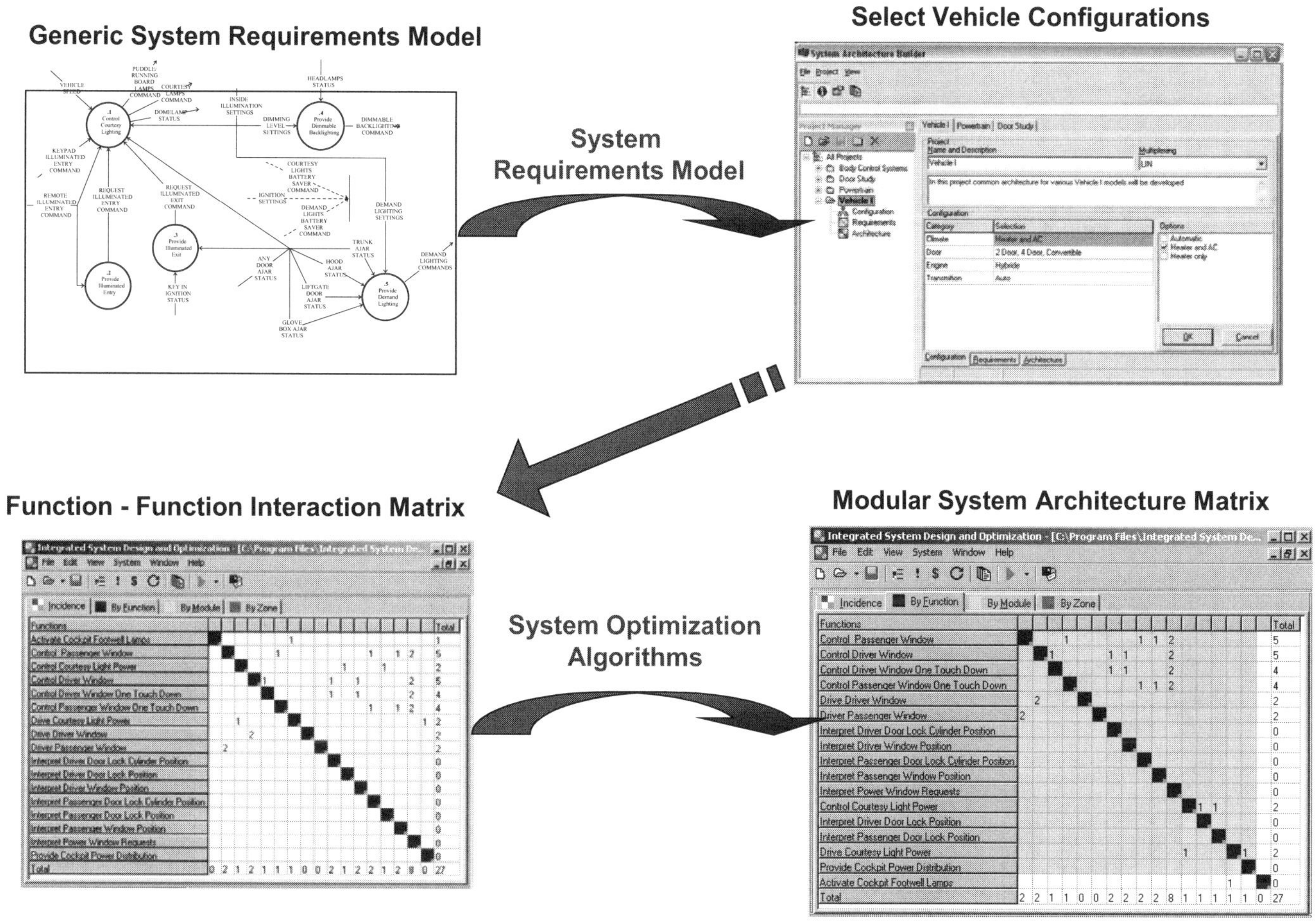

Figure 5: ISDO Tool

- Automatically annotating the control algorithm tasks and communications with timing performance formulae for dynamic computation (at simulation time) of the message transmission latencies over the network bus as well as the SW scheduling execution times due to shared resources (buses, CPUs)

- Simulating the bound virtual platform (e.g. representing a possible candidate distribution of functionality over the network) for verification purposes under regular conditions and/or with fault injection (e.g. data corruption, abnormal task delay, etc)

Notice that the virtual integration platform relies on providing library elements for both functionalities and architectural resources. The platform provides links with the most popular tools for algorithmic development and simulation, thus enabling the seamless import of such models and their composition in the overall distributed complex control function. The platform also provides models of popular communication protocols such as CAN. One important aspect, the XML-based programming capabilities provided by the platform, constitutes the core of the linkage between the Cadence Automotive Platform itself and the Visteon Integrated System design and Optimization (ISDO) static analysis tool (described in the next section). This aspect is detailed later in the paper.

ISDO TOOL FLOW

In [3], new approaches and software algorithms were presented that allow vehicle Electrical, Electronic and Software (EES) system design engineers to develop modular architectures / modules that can be shared across vehicle platforms (for OEMs) and across OEMs (for suppliers). The methodology uses matrix clustering and graph-based techniques. The ISDO software tool (Figure 5) allows system design experts to build a low cost EES architecture that can be shared across multiple vehicle platforms. The ISDO software builds an optimal system architecture for a given set of system features and feature take rates by grouping (integrating) system functions into physical modules and determining the best possible tradeoffs between system overhead costs, give away costs, and wiring costs. Also in [3], a new approach is presented that allows system developers to identify common modules in EES architectures that can be shared across multiple vehicle platforms. In a nutshell, the ISDO tool flow can be described as follows:

- Import the system requirements model into a specified database using predefined templates to create various vehicle configurations
- Automatic creation of the *function-function interaction matrix* once the requirements for each vehicle configuration are identified
- User pre-defined, or carryover, modules: the user can select functions from the incidence matrix and group them into predefined modules. This step makes sure that in the final vehicle architecture the functions of carryover, or predefined modules, will appear in a single cluster/module.
- Applying optimization algorithms to the function-function interaction matrix to identify optimal functional grouping in the vehicle architecture.

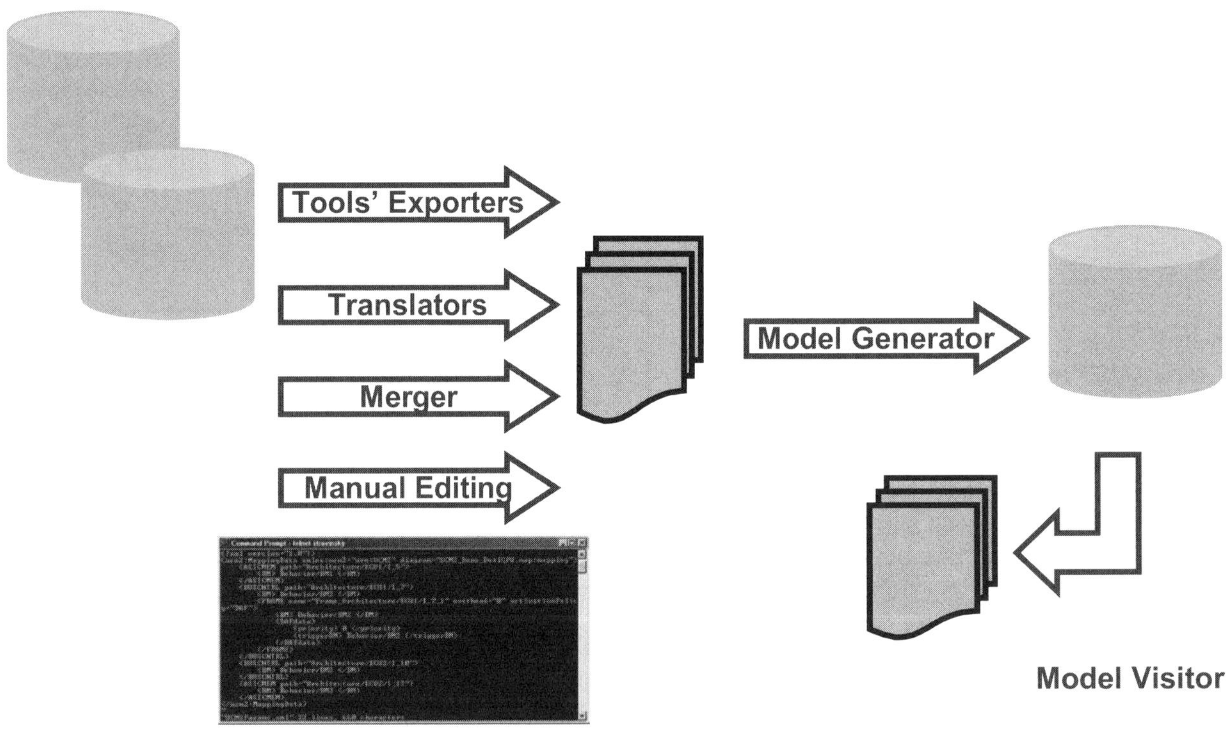

Figure 6: XML-Based Design Representation

INTEGRATED TOOL FLOW

The Cadence Automotive Platform and the Visteon ISDO tool are complementary in that, while the former provides a powerful dynamic analysis environment, the latter provides powerful static allocation mechanisms. Therefore, good results are produced when the tools are coupled together in the following sequence:

- The system design expert starts from a description of the overall complex control algorithm by performing the functional decomposition and function interface definition (signals and shared variables)
- Next, the ISDO tool is used to allocate functions to modules
- Next, the allocation is exported by the ISDO tool and imported by the Cadence Automotive Platform environment and simulated
- If the simulation results are satisfactory, the designer can (optionally) refine the design, provided that he/she does not modify the function interfaces, by replacing the coarse grain functional models used during the simulation in the Cadence Automotive Platform with more detailed models imported from other tools (e.g. Simulink) and re-importing them with the Cadence Platform

XML-BASED DESIGN REPRESENTATION

In the Cadence Automotive Platform, a design is represented in an isomorphic fashion. First, an XML-based representation is used to describe the functionality (excluding the primitive block models), the architecture, the software execution architecture (SEA), meaning the set of tasks with their activation policies and priorities, the software and communication binding information, the fault scenarios' descriptions, and the simulation set-up. The XML representation is manipulated via a Cadence-defined scripting language, to, for instance, define/update/change a binding of tasks to a set of ECUs. Once all the needed manipulations have been performed, the XML representation is then

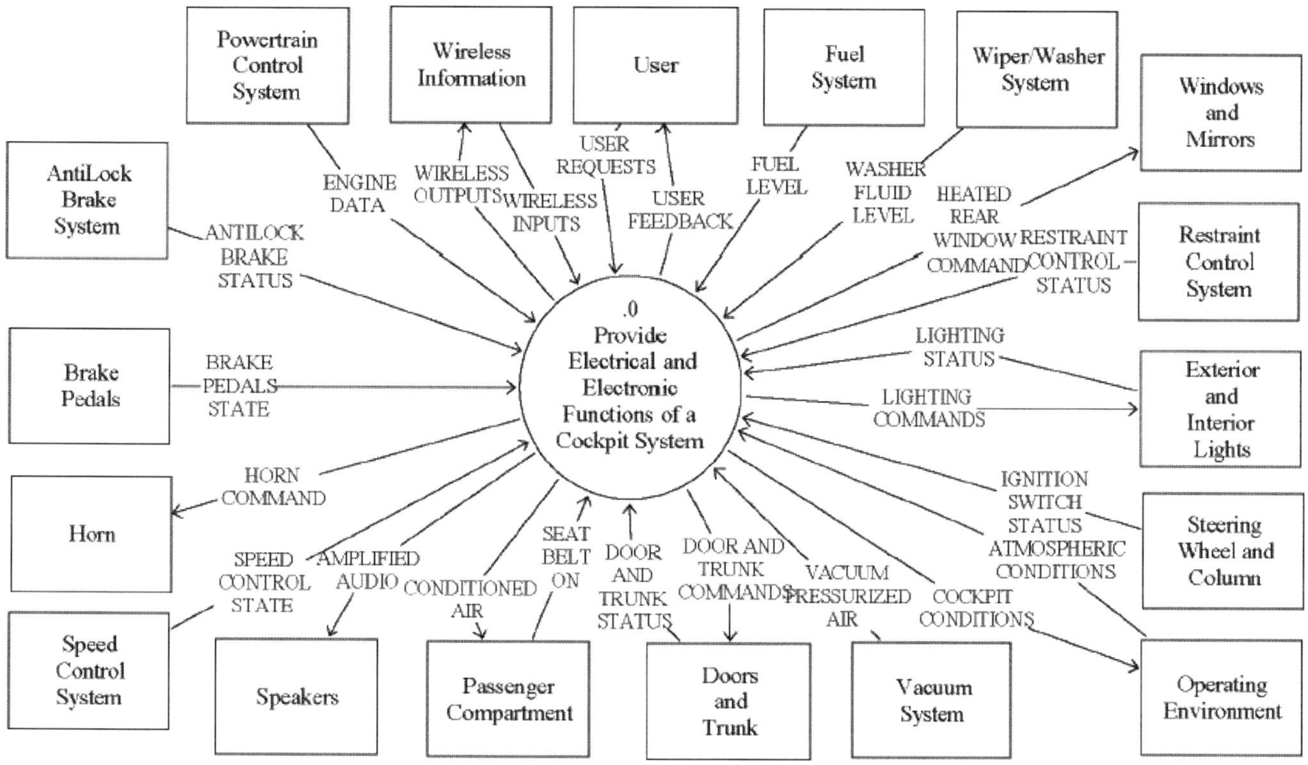

Figure 6: Top Level Hierarchical Context Diagram

compiled into a simulatable representation (Figure 6) [6]. Notice that the XML-based representation includes information about the functions' interfaces with their activation mechanisms (via signals or periodic timers) and shared variables. A very important piece of information is represented by the Software Execution Architecture (SEA). In fact, in order to be scheduled and simulated, the functions have to be assigned to tasks, which in turn, are assigned to ECU scheduler models during the binding step (OsekTime for the Cadence Automotive Platform). Once the user has programmed with scripting language the type of communication protocol model that has to be used to simulate the network traffic, the Cadence Automotive Platform engine automatically determines the network bus communication matrix. In fact, at this point of the design flow, it is known which messages are sent by an ECU to which other ECUs, because the software tasks that send messages to the bus controllers and receive them from them have been assigned to the ECUs themselves within the Visteon ISDO tool. The engine provides the user with an automatic configuration of the network. For each bus cluster, a communication cycle with arbitration is provided, while each frame is activated via a triggering mechanism. Notice that the user can modify the configuration, as part of the XML manipulations, via the scripting language.

YOURDON DE MARCO REPRESENTATION

In the ISDO tool, a design is translated from hierarchical context diagrams (Yourdon-DeMarco methodology) (Figures 7-8) to an XML-based format. The hierarchy is used to manage complexity. Notice that the context diagrams are not executable, since the wired connections between functions do not have any execution semantics (e.g. hardware interrupts). They are meant to represent the logical flow of data (information, material, or energy) between functions. Moreover, there is no definition of SEA and no executable models for the functions themselves. Since the purpose of the Visteon ISDO tool is to perform static analysis and allocation of functions to modules (hereafter either the term *modules* or ECUs are used with the same meaning) the tool does not need executable semantics for its design representation, since no simulation is performed. It is therefore important to implement a linkage between the ISDO design representation and the Cadence Platform representation to simulate a given allocation. This is explained in the next section.

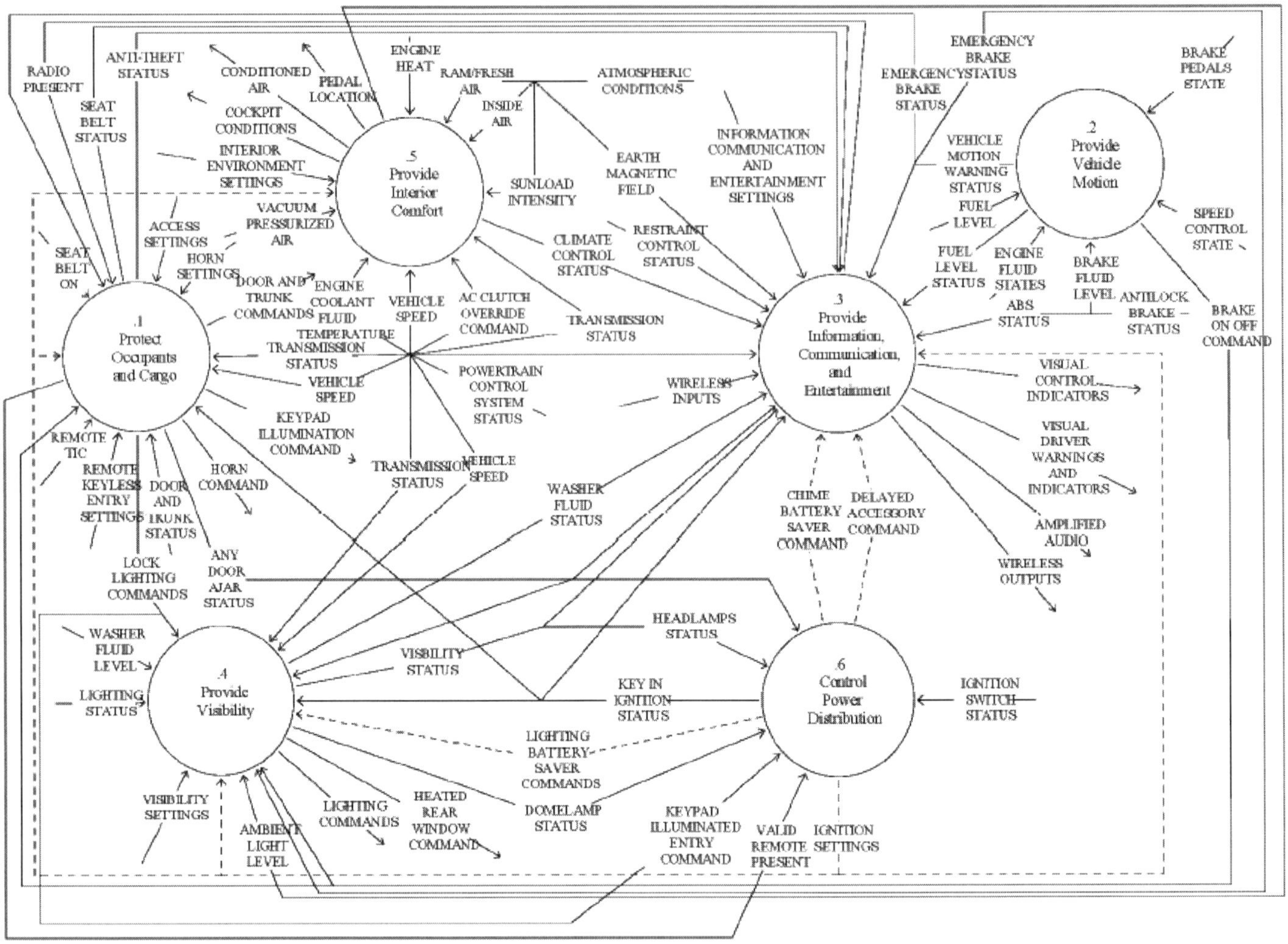

Figure 8: Next Level Hierarchical Context Diagram

COMMUNICATION MATRIX IMPORT

An efficient way of importing the Visteon ISDO design representation to simulate a given functional allocation is via the communication matrix import. The communication matrix is a representation that does not take into account the original design hierarchy – the communications are represented with respect to the ECUs instantiated in the network cluster. Therefore, if one is able to annotate the communication matrix by itself with information about the activation policy of the messages being broadcast, then this representation can be used to automatically create a simulatable design in the Cadence Automotive Platform. Notice that the usage of the communication matrix is equivalent to flattening the original design hierarchy – which was irrelevant for simulation purposes anyway, since it was only used for managing design complexity. Therefore, we envision the following flow (Figure 9 at the end of the paper) between the two environments:

- First, the ISDO tool determines the allocation of functions to modules
- Second, the allocation is transformed by the ISDO tool into an equivalent XML-based communication matrix
- Third, the designer annotates (for instance with an XML editor of choice or via Java API based command) the messages in the communication matrix with activation policy information (e.g. periodic vs. triggered): this is needed in order to simulate the network traffic.
- Fourth, the Cadence Automotive Platform reads in the communication matrix and compiles into a set of XML based representations (again see Figure 9)
    - An architecture XML that represents a single-cluster network of ECUs connected via a network bus (ECU=Module)
    - A functional XML that represents a trivial functional network, in which a set of *transactor* behaviors are used to generate network traffic, one transactor per ECU

- A SEA XML that represents the trivial assignment of one transactor to a task
- A binding XML that represents the mapping of each single transactor task to one ECU
- A simulation set-up XML initially empty, later on updated to include any probing of given metrics (e.g. bus load)

- Fifth, the five XML representations are compiled into a simulatable representation and the simulation with data collection and analysis can take place.

Notice that in this flow design exploration is provided with respect to analyzing a given allocation of functionality vs. specific attributes of interest, such as the network traffic. It is important to notice that since the original hierarchy is not preserved, should the user want to explore a different task organization, this step must be performed in the Visteon ISDO tool. The steps described above should then be repeated. However, it is also important to note that, since there is no function executable specification in ISDO, once a specific allocation has been validated via simulation, it makes sense to refine the coarse grain transactor models by importing finer models from a tool such as Mathworks/Simulink via dSPACE Target Link code generation. At this point, the designer would be able to explore different software execution architectures and further refine the target implementation model. Note also that the flow is independent of the communication protocol model being used for the simulation – the user can replace a general yet highly programmable model such as the Universal Communication Model (UCM) [4] with a more refined model for CAN (included in the Cadence Automotive Platform) and utilize error injection capabilities provided by the Cadence Automotive Platform.

CONCLUSION

In this paper, we have presented a novel flow that couples together a static analysis tool for functional allocation and a simulation environment to verify a given allocation. The flow is aimed at reducing the design cycle time by using highly integrated and programmable simulation models and algorithms for static allocation. The final result is a validated functional allocation that can be handed over to sub-system software developers. We are envisioning extending the automatic generation of the simulatable design to multi-cluster networks (e.g. LIN plus CAN) by incrementally importing communication matrixes and composition via gateways to realize a complete virtual car analysis environment.

ACKNOWLEDGMENTS

The authors would like to thank Pascal Bornat and Jean-Yves Brunel from Cadence Design Systems, Velizy, France, Alberto Ferrari from Parades, Rome, and Luciano Lavagno from Politecnico di Torino, Turin, Italy for their contributions to the paper.

REFERENCES

1. Zakarian, A. and Rushton G. J. (2001), “Development of Modular Electrical Systems”, IEEE/ASME Transaction on Mechatronic,
2. Rushton, G., Zakarian, A., and Grigoryan, T. (2002), “Algorithms and Software for Development of Modular Vehicle Architectures”, SAE World Congress 2002-01-0140.
3. Rushton, G., Zakarian, A., and Grigoryan, T. (2003), “Development of Modular Electrical, Electronic, and Software System Architectures for Multiple Vehicle Platforms “, SAE 2003-01-0139
4. Thilo Demmeler, Paolo Giusto (2001), “A Universal Communication Model for an Automotive System Integration Platform”, Design Automation and Test Europe Congress 2001
5. Thilo Demmeler, Barry O’Rourke, Paolo Giusto (2002) “Enabling Rapid Design Exploration through Virtual Integration and Simulation of Fault Tolerant Automotive Application”, SAE World Congress 2002, SAE 02AE-76.
6. Paolo Giusto, Jean-Yves Brunel, Alberto Ferrari, Eliane Fourgeau, Luciano Lavagno, Alberto Sangiovanni Vincentelli (2003), “Virtual Integration Platforms for Automotive Safety Critical Distributed Applications”, Conference International Su les Systemes Temps Reel (RTS) 2003

CONTACTS

Paolo Giusto has a Bachelor Degree in Computer Science from the University Of Turin, Italy and a Masters Degree in Information Technology from CEFRIEL, Milan, Italy. He has over 12 years of industrial experience and is currently working as the Automotive Team marketing director with Cadence Design Systems. Previously, with Magneti Marelli, he visited the EECS Department at University of California at Berkeley as an Industrial Fellow, working on hw-sw co-design methodologies for embedded systems. Previously, with Cadence, he worked as technical leader on system level design methodologies and tool sets with particular focus on safety-critical automotive distributed applications. (giusto@cadence.com).

Gary Rushton has over 18 years of commercial and military electrical/electronic systems engineering experience. He has an MS in Automotive Systems Engineering from the University of Michigan. He is currently working as an electrical/electronic systems engineering technical fellow with Visteon Corporation. As an engineer with Visteon Corporation, he has worked on audio software, subsystem product development/design, diagnostics, vehicle system architectures, and cockpit system design. Previously, with General Dynamics, he worked on avionics systems for the F-16 and vetronics systems for the Abrams M1A2 tank. (grushton@visteon.com).

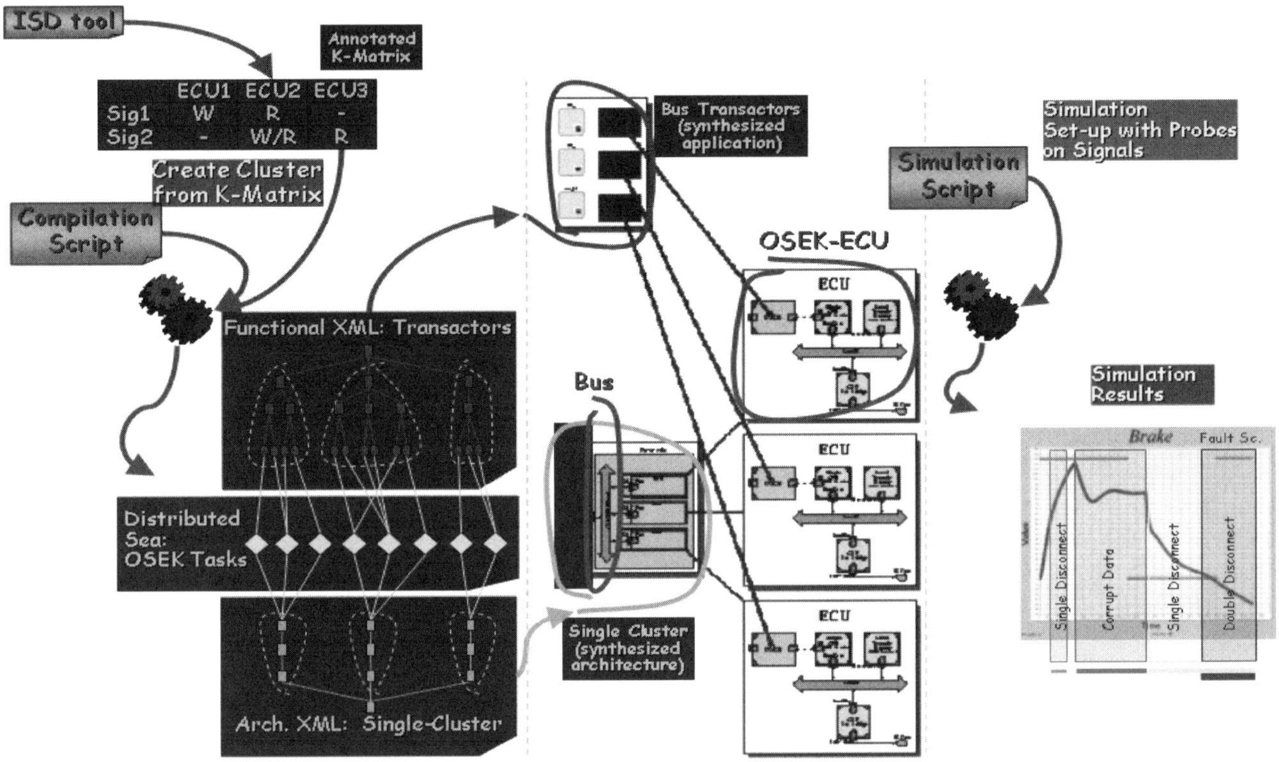

Figure 9: The Novel Flow

2004-01-0719

Solving the Technology Strategy Riddle - Using TRIZ to Guide the Evolution of Automotive Software and Electronics

Alex Shoshiev
Yazaki North America

Victor Fey
The TRIZ Group, LLC

ABSTRACT

Significant resources are spent by the OEMs and suppliers to determine the most promising directions for the development of next-generation vehicle electronics and software. Existing technology planning methods, while useful, still leave a large margin of error.

The authors describe a new structured approach to technology and product planning based on TRIZ methodology. This approach largely removes guesswork from pinning down the next most likely evolutions of automotive software and electronics and provides objective justification for investment decisions.

INTRODUCTION

In a nonstop race to conquer the market, automotive companies seek answers to key questions like:

1. What new vehicle features may please our customers?
2. How should the existing features be improved to boost their customer appeal?
3. How can we make sure that our R&D efforts will provide a robust competitive advantage?

Various methods are used to answer these questions: market research, competitive benchmarking, customer clinics, QFD, just to name a few. These methods work reasonably well for incremental improvements, but they are rarely effective when used to predict radically new innovations.

Delay in the development and introduction of a new market-grabber may cost a company dearly. Considering that an automotive product development cycle runs 2-4 years, an OEM that first introduced a successful feature could enjoy healthy non-competitive profits for at least that long. Speed to market has become increasingly key as a strategic advantage.

Making a wrong decision may therefore be costly. With an always-limited R&D budget, money spent on a "wrong" feature is taken away from the "right" feature that is not developed at just the right time.

Here are some features that have been recently floating around in the automotive community:

42V - after enthusiastic introduction and considerable expenditures by many OEMs and suppliers, the development has stagnated, allegedly due to a slow economy. However, recently it has been questioned if 42V is ever going to be realized as a platform, or it will be passed over towards higher-voltage systems?

Software tools vendors have been promoting code generation tools for several years. Proposed benefits are acceleration of the software development, creation of verifiable executable models, and reduction of the overall software effort. The tools are expensive, as is related training, and many are opposed to the idea. Is this a right direction for the OEMs and suppliers?

Answers to these and similar product planning questions have been highly subjective – they are, essentially, individual opinions. Various existing product planning tools process these opinions in different ways, but do not change their subjective nature.

There is a tangible need in a systematic approach to evaluate future technological opportunities. Such an approach exists, and it is called TRIZ

TRIZ TECHNOLOGY AND PRODUCT PLANNING

Although not yet widely used in the US, TRIZ is not a newcomer among product development methods. It has been developed for over 50 years in Russia by Genrikh Altshuller and his followers [1-3], verified and perfected in military and space programs there. Over the last few years, TRIZ has been rapidly gaining popularity in the West. It has helped several leading companies in the

US, Japan, and Europe to develop many new product and technology innovations [4].

TRIZ is a Russian acronym for Theory of Inventive Problem Solving. The main postulate of TRIZ is that *evolution of technological systems is governed by objective laws*. Just like laws of nature, laws of technology evolution operate regardless of our knowledge about them. These laws describe "life tracks" of a system's evolution. Knowing the current system's design, the most likely future designs can be predicted using the laws.

By analyzing tens of thousands of patented and commercialized product and process innovations, and selecting and examining the most effective designs, Altshuller formulated several *laws of technological system evolution* (or *laws of evolution* for short). These laws of evolution make up a theoretical base of the comprehensive TRIZ methodology that contains numerous powerful tools for resolving conflicts between system elements, tools for developing conceptual designs, methods for problem formulation, etc.

TRIZ is both a theory of technology evolution and a methodology for effective development of new technological systems. It contains two major components: tools to identify and develop next-generation technologies and products, and methods for developing conceptual system designs. The structure of TRIZ is shown in Fig. 1.1

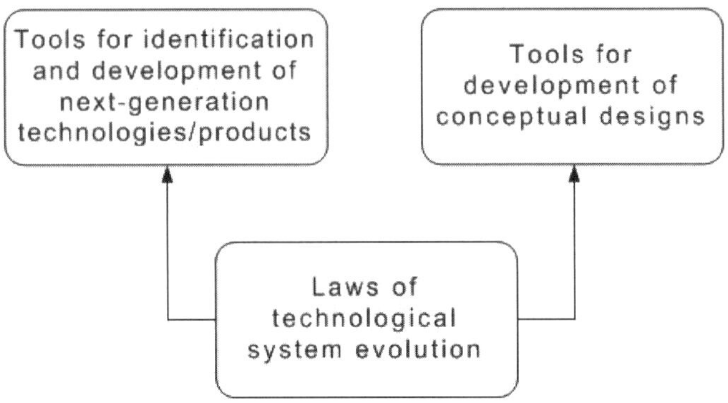

Fig. 1.1: The structure of TRIZ.

Due to its limited scope, this paper describes the subset of tools of TRIZ that can be used to predict the evolution of automotive software and electronics.

LAW OF INCREASING DEGREE OF IDEALITY

This law states that **evolution of technological systems proceeds in the direction of increasing their degree of ideality**.

The degree of ideality is a ratio of the system's functionality to the sum of expenditures associated with building and making the system function. In essence, it is a benefit-to-cost ratio.

In the qualitative formula in Fig. 1.2, functionality refers to the number of functions performed by the system and to the level of performance of these functions. "Cost" means all expenditures (monetary and others (e.g., parts count, weight, etc)) associated with the system's creation and maintenance. Problems are measurable parameters, such as level of noise, intensity of wear, as well as intangible ones (complexity of operation, discomfort, etc).

$$\textbf{Degree of Ideality} = \frac{\textbf{Functionality}}{\sum \textbf{Costs} + \sum \textbf{Problems}}$$

Fig. 1.2. Degree of ideality.

This law shows the prevailing dynamics of the benefit-to-cost ratio: as systems evolve, we pay less for the improved old and added new functions. Today's cars are more powerful and much more comfortable than cars made a few decades ago, yet they cost (in dollars adjusted to inflation) less than their predecessors.

This law also implies that for a new system to survive a market selection process (in the long run), its degree of ideality has to be higher than that of the incumbent system. The law of ideality in TRIZ plays a role similar to that of the compass in navigation. As the compass provides a traveler with a sense of direction, the law of ideality shows a principal vector, along which the best solutions can be found. Most other laws of evolution can be compared to meridians that lead to these best solutions and converge at the "ideality pole."

From the formula in Fig 1.2, it is clear that the system's

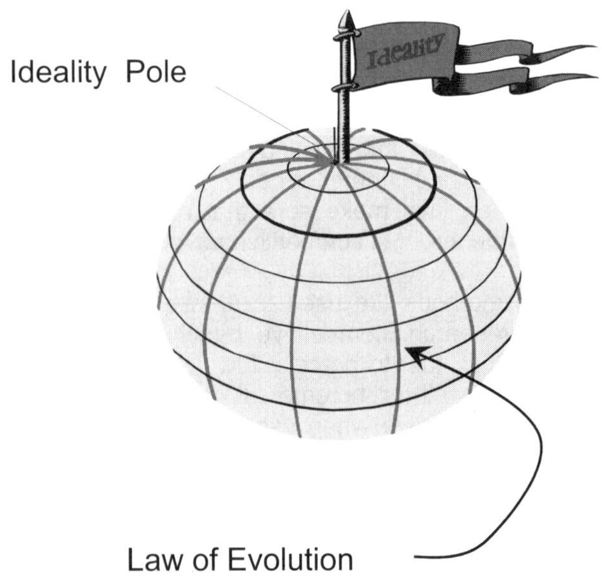

Fig 1.3 Laws of evolution and ideality.

degree of ideality can be changed by the following three approaches:

• Increasing system functionality (i.e., the number of its functions and/or level of their performance), while keeping the cost of the system unchanged.

• Reducing the cost, while maintaining its functionality.

• Integration of the first two approaches.

LAW OF NON-UNIFORM EVOLUTION OF SUBSYSTEMS

This law states that **systems components evolve at different paces; the more complicated the system is, the more non-uniform the evolution of its components.**

The disparity of rates of development of different components produces a situation in which improving one component (function) makes some other components (functions) inadequate. Such a situation is called in TRIZ a **system conflict.**

A problem associated with a system conflict can be solved either by finding a compromise between opposing demands, or by satisfying them. Compromises or trade-offs are perceived to be so unavoidable that finding the most optimal trade-offs is a typical part of many system engineering activities! Trade-offs do not eliminate system conflicts, but rather alleviate them. Conflicting requirements keep "stretching" the system. Then the time comes when further improvement of the system's performance becomes impossible without eliminating the system conflict.

From a TRIZ standpoint, to make a breakthrough means to resolve a system conflict by satisfying all of the opposing demands. TRIZ offers several methods for compromise-free system conflict resolutions.

Example 1.

The earliest cars had a start-stop handle and no windshield – they were moving slowly. Increased car speed created a system conflict between the speed and the need to stop quickly, as well as between the speed and comfort of the occupants (wind in the face). Resolution of those conflicts resulted in introduction of the brake system and the windshield.

Example 2.

When embedded software was limited in functionality and size, all software testing was done by hand. Increase in size and complexity of the software resulted in the exponential increase in the magnitude of software testing. That system conflict was resolved by the development of a system that performs auto- generation of test vectors and auto-test execution of embedded software.

To use the law of non-uniform evolution of subsystems, we recommend the following steps:

1. Identify the major components of your system and the primary functions they perform.
2. Select a component of interest and mentally improve its primary function significantly.
3. Identify problems (resulting from this improvement) in other components. Formulate system conflicts.
4. If necessary, repeat the previous steps for other major components.

Example 3.

Consider the vehicle power distribution system. Delivery of power is the main function of this vehicle subsystem. Suppose that we want to double the budget for vehicle power (for features like electrical power steering, air conditioner, or the like) – a task that is challenging today's vehicle designers. Doubling the power budget would mean doubling the output capacity of the alternator. This will increase the size of the alternator, causing it to fight for space inside the tightly packed engine compartment (system conflict #1). If the alternator technology doesn't change, power losses in the alternator would also increase, possibly requiring a move from the air to liquid cooling (system conflict #2). If we then try not to increase the power distribution losses and retain the conventional 12V system voltage, the wire and contact resistance would need to be reduced fourfold ($P=I^2*R$). That, in its turn, would cause an unacceptable increase in the wire's size and mass and complicate wire harness routing (system conflict #4). If we decide to move to a higher system voltage (e.g., to the proposed 42V system), we face a different set of conflicts: 42V light bulbs do not withstand vehicle vibration (system conflict #5); disconnecting "live" connectors causes arcs that destroy the terminals (system conflict #6). Issues of jump start and external infrastructure would also need to be addressed (system conflict #7).

All of these system conflicts are known today, but they could have been foreseen and investigated much earlier had the law of non-uniform evolution of subsystems been used.

LAW OF INCREASING FLEXIBILITY

This law states that technological systems evolve toward increasing adaptation to changing environmental conditions and varying performance regimes. The law reveals itself through several "lines of increasing flexibility". We will describe one of these lines called "the line of transition to continuously variable systems."

A new system is usually developed to perform a particular set of functions in a specific environment. It is a "rigid," one-state system. As the environment changes, the system adapts to new or additional usage, and becomes multi-state. Ultimately it becomes continuously variable, i.e., a system with – theoretically - an infinite number of states.

Example 4

At first automotive modules with embedded microprocessors had programs stored in a "maskable" PROM. Software engineers were sending the code containing the program to the microprocessor manufacturer, where the "mask" was created and the code was "burned-in" into the microcontroller chip during microprocessor production. Once burned in, the code could not been changed. The process was very rigid - in case of a software defect, the whole microprocessor had to be scrapped, and a new mask had to be developed. The process was also slow – it would take some 12 weeks between code submission and microprocessor release.

To accelerate development and reduce product development expenses, microprocessor vendors introduced OTPs – one-time programmable micros. That allowed microprocessors to be programmed at the supplier's manufacturing plant, thus slashing 12 weeks from the development cycle time. Next, introduction of flash memory allowed multiple reprogramming of the microprocessor memory, significantly reducing expenses caused by software defects. Now instead of removing and replacing the module, the dealer could simply re-program it using service tools at the dealership.

Example 5

One of the earlier software development processes was the model called "waterfall." In this process, the software development progressed through well-defined phases: Requirements, Design, Coding, Test, and Maintenance. (Fig. 1.4) The process was rigid – the software development could only be in one of the phases at any time. For example, if the requirements were not fully gathered, organized and reviewed, the Design phase could not begin. If a design mistake was found during the Coding phase, the project would be returned to the Design phase and all coding stopped until the design was fixed and reviewed.

Such a rigid process was impractical for many software applications and a "modified waterfall" or "sashimi model" was soon introduced. (Fig. 1.5) In this model, software development was allowed to move into the next

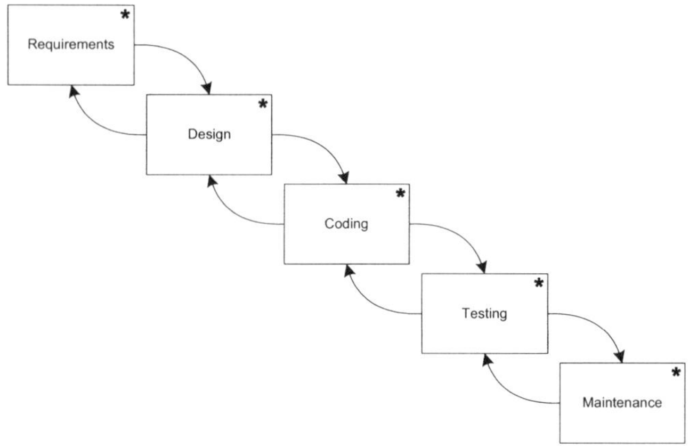

Fig 1.4 Waterfall model

phase after most of the work in the current phase has been completed. Thus, the software development could span two, or sometimes three phases, depending on the application.

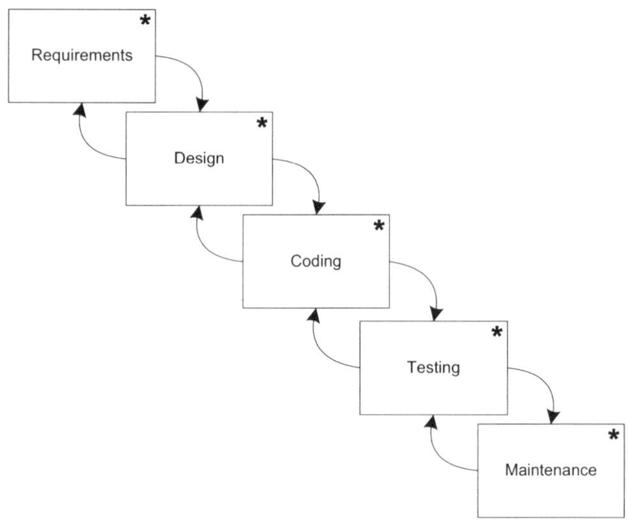

Fig. 1.5 Modified Waterfall model

Further evolution led to the introduction of a multitude of life-cycles, each adapted toward specific software development circumstances.

LAW OF TRANSITION TO HIGHER-LEVEL SYSTEMS

This law describes the evolution of systems from simple to more complex (higher-level) ones by merging various objects. The law states that ***technological systems evolve from mono-systems to bi- or poly-systems***.

A mono-system performs one primary function; two or more mono-systems (similar or different) can merge to create bi- and poly-systems. In other words, as systems evolve, they absorb functions that were previously performed by other systems. This makes bi- and poly-systems have the higher degree of ideality than that of the constituent mono-systems.

Example 6.

Automotive electronics modules used to employ three mono-systems: a voltage regulator providing 5V power to the microcontroller; a watchdog - to reset the microcontroller in the event it gets "stuck" or enters an "endless" loop; and a CAN or J1850 transceiver (Physical Layer protocol interface) to assure proper communications. Originally these functions were implemented on separate chips. Then the voltage regulator was combined with the watchdog, and lately all three functions have been integrated into one chip. This integration allowed power consumption of the circuitry to be reduced in sleep mode through on-chip coordination between the components (Fig. 1.6)

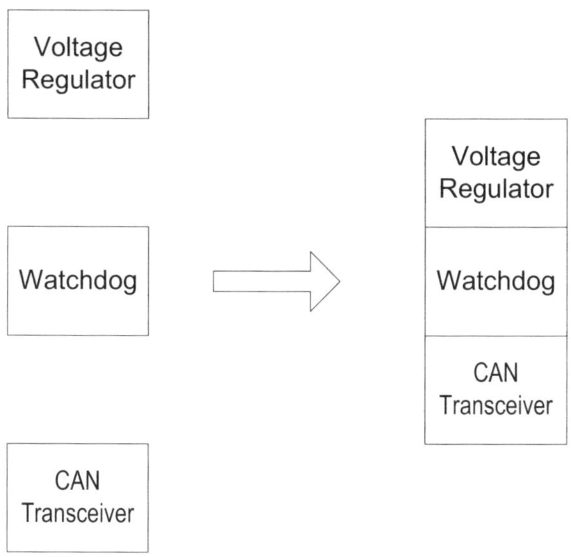

Fig 1.6 Service components integration

Example 7.

Similar transitions are happening in the power semiconductor field, too. Originally, FETs (Field Effect Transistors), used to switch power, required a multitude of peripheral logic to protect them from over-current, reverse voltage, etc. Addition of those devices was complicating the board layout and was also costly. The latest FETs (sometimes called Pro-FETs or smart-FETs) integrate this logic on the same silicon with the FETs. This integration allows for instantaneous on-chip thermal protection, current measuring capabilities, and other functions, while saving real estate on the PCB board and the cost of multiple packages. (Fig. 1.7)

Example 8.

Microprocessors in general went down a similar path of development. First microprocessors, used for industrial control purposes, required an additional "peripheral" circuitry on the board. Most of this circuitry soon moved onto the microprocessor silicon, thus giving birth to the microcontroller.

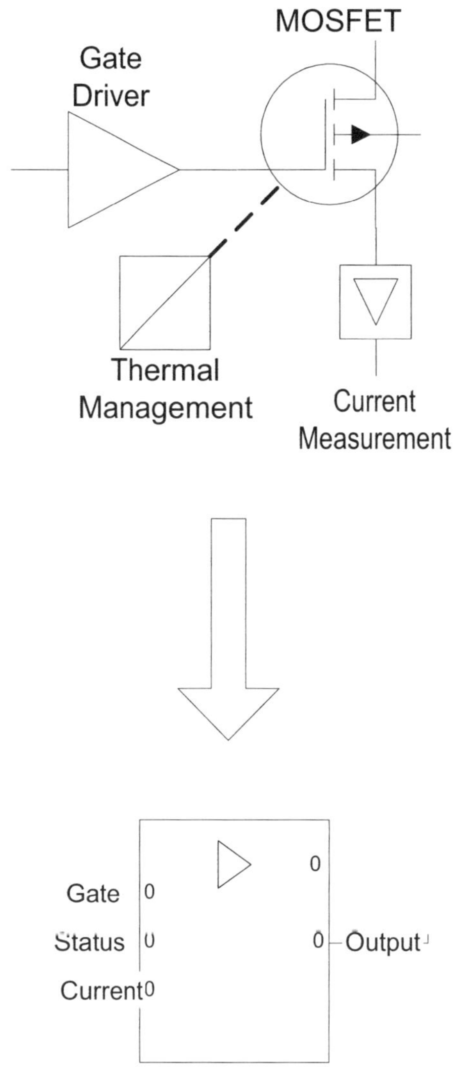

Fig. 1.7 FET Integration

Example 9.

The same law of evolution is largely "responsible" for shaping modern software development tools. The first programs were written directly in the machine language. Those programs could be executed directly, but to create a program the software developer had to remember multitudes of codes. That practice was acceptable as long as programs were short, limited in functionality, and were run on a few types of computers. The situation became cumbersome and error-prone when both program size and complexity increased. Assembler language came to the rescue. It allowed the software developer to write programs not in binary digits but in mnemonic expressions more meaningful to humans. To be executed on the computer, the programs were "assembled" – translated (by a special program) line-by-line from the mnemonic language into the machine language. The computer not only did the translation but also checked the program's syntax.

Next came compilers and high-level languages. Unlike Assembler, one statement in a high-level language could represent many instructions in the machine language. The software developer was now writing shorter programs in more English-language-like fashion. Preparing programs for a computer also became faster. Compilers then translated the source code into instructions in Assembler, and then an Assembler program converted the instructions into the machine language. In addition to syntax and data type checking provided by a compiler, programs became machine-independent: the same high-level language program could potentially run on many computers – adaptation of the program to the specific computer was done by Compiler and Assembler programs.

Recently we witnessed the emergence of the first code-generation tools. Provided that software design is done in a certain predefined fashion, these tools can generate the source code by themselves. This will eliminate the whole class of coding-related errors and allow the software developer to concentrate on the software design. At the moment these tools are not yet perfect – they only work adequately for limited applications, but that was also true of the first compilers.

LAW OF COMPLETENESS

Automation (i.e., elimination of human involvement) is one of the prevailing trends of technological evolution. The law of completeness states that **to be fully autonomous, any technological system should have four principal parts: a "working means," "transmission," "engine," and "control means."** (Fig. 1.8)

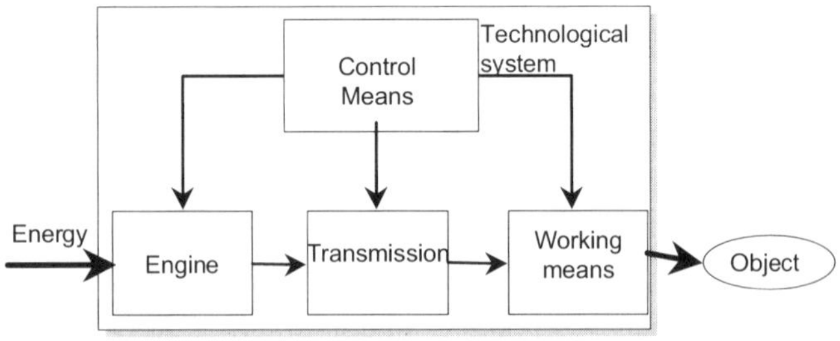

Fig. 1.8 Autonomous system

A "working means" is a component that directly performs the primary function of the system. A "transmission" transforms the energy produced by the "engine" into the energy that controls the "working means." A "control means" allows the user to vary the performance parameters of the other principal parts.

In early stages of a system's evolution, some of these principal components are performed by "human elements". Gradually these "human elements" are eliminated, being replaced by technological components performing the same or enhanced functions.

Example 10.
Consider the evolution of the side-view mirror adjustment. The primary function of a side-view mirror is to expand the driver's field of view. This, naturally, makes the mirror a "working means." For many years the driver had to change the field of view by reaching out through the window and move the mirror in horizontal and vertical directions. (Fig. 1.9) Thus, the driver ("human element") performed functions of the other principal parts of the mirror adjusting system – "engine," "transmission," and "control means."

Fig 1.9 Manual adjustment mirror

Next came "remote" mirror adjustment – a cable became a "transmission", while the "engine" and "control means" remained unchanged. (Fig. 1.10)

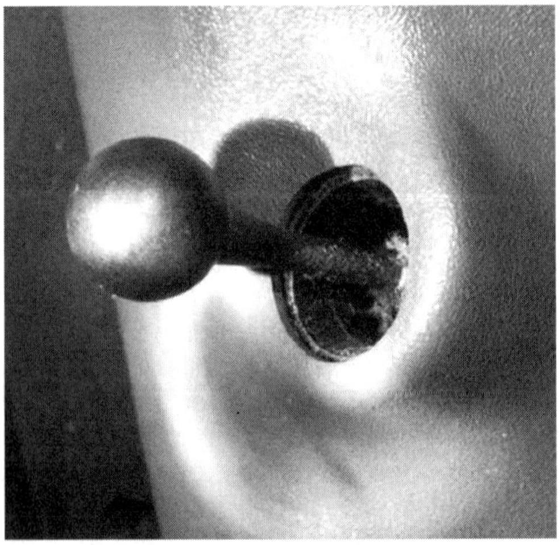

Fig. 1.10 Remote adjustment mirror

Subsequent introduction of an electric motor ("engine") left only the function of control still performed by the "human element."(Fig. 1.11)

Fig. 1.11 Power adjustment mirror

In some of today's cars, the mirror is adjusted automatically in certain situations (e.g., when shifting into reverse). As the functionality of the side-view mirror continues to advance, the mirror adjustment system will become fully autonomous.

TRIZ TECHNOLOGY FORECASTING

PROCESS SUMMARY

Just as X-ray machines fundamentally changed the field of medicine, the use of the laws of evolution can dramatically change product-planning processes, including those in the automotive industry. It is now possible to see the inner workings of a technology's evolution and predict where it will go next.

A technology planning process using TRIZ involves the following steps:

1. Product Positioning: Analysis of the past and current states of the product's evolution (determining where the product's technology is in relation to its limits).
2. Setting Directions – Identification of high-potential directions for the evolution of technology.
3. Concept Development – TRIZ concept development methods are applied to develop the most promising concepts that will move the technology along the identified directions.
4. Concept Selection –Various technical and business criteria, as well as risk mitigation tools, are used to select the best concepts.

Since the primary goal of this paper is to describe the tools of TRIZ used for technology planning, next we will describe Steps 1 and 2 of this process. Steps 3 and 4 are described in [5].

PRODUCT POSITIONING

No technology develops at an even pace. Introduction of a technology is usually slow - there are many unknowns, the development is expensive, efforts frequently fail, and supporters are few. More often than not, during the initial stages, costs (development, operational, etc.) far outweigh benefits. Then, at some point, the technology gets acceptance and undergoes a phase of rapid growth – critical issues have been resolved, investors sponsor further development, and the technology enters the markets. After a while, the rate of growth slows down – the manufacturing processes have been perfected, and available market niches have been filled. The technology has matured. Finally, the technology stagnates and frequently disappears, being overtaken by a new, more beneficial one.

The plot shows the dynamic "performance-to-cost ratio" (or "benefit-to-cost ratio") is shaped as a "lazy" S, and is commonly called the S-curve. There are distinctive points on the curve marking milestones in the technology's life cycle.

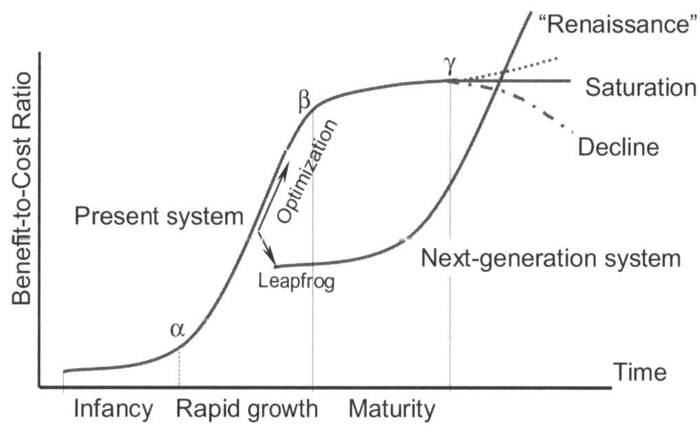

Fig 2.1 S-Curve plot

The non-uniform rate of technology development has been discussed in business literature for some time. Richard Foster uses the term "limit" as related to the barrier, beyond which a particular technology cannot be improved. The closer the technology is to the limit, the harder it is to improve it. "If you are at the limit, no matter how hard you try you cannot make progress. As you approach limits, the cost of making progress accelerates dramatically. Therefore, knowing the limit is crucial for a company if it is to anticipate change or at least stop pouring money into something that can't be improved. The problem for most companies is that they never know their limits". [6]

Most companies cannot be blamed for not knowing the limits of their technologies:

"Mistakes can be made, even by scientists, about limits – particularly the limits of a competitor's technologies. Even if the limits are clearly defined, the breakthrough idea about how to reach them may be missing." [6]

To help position a technology on its S-curve, TRIZ uses a correlation between the technology's life cycle and inventive activity that was discovered by Altshuller. He suggested that all inventions (both patented and otherwise) be classified into 5 novelty levels:

- Level 1 – Slight modifications of the existing system
- Level 2 – Solutions derived from similar systems.
- Level 3 – Breakthroughs within a single engineering discipline.
- Level 4 – Radical developments stemming from multi-discipline approaches.
- Level 5 – Pioneering inventions resulting from new scientific discoveries (radio, laser, etc.)

Altshuller showed that for any given technology, both invention activity (number of inventions) and levels of invention closely correlate with the technology's S-curve:

- The plot "number of inventions vs. time" (Fig. 2.2) has two inflection points. Point A is associated with efforts to promote the technology from a concept to production; point C reflects the drop of interest after unsuccessful attempts to improve the stagnating technology.
- After pioneering inventions gave birth to a new breakthrough, the levels of invention drop and then rise again (point A in Fig. 2.2). This dynamic is due to the resolution of significant problems that were keeping the concept from becoming a viable product. Once the technology goes into production (the idea turns into a product), the main development efforts shift to resolving relatively smaller issues (increasing efficiency, minimizing costs, etc.), and the level of invention drops.

By conducting inventive activity analysis, described above, on the company's technology/product as well as on those of its competitors, the company can reliably determine their positions on the respective S-curves. Then the management can decide whether incremental improvement of the product will be adequate or a leap to another technology is required.

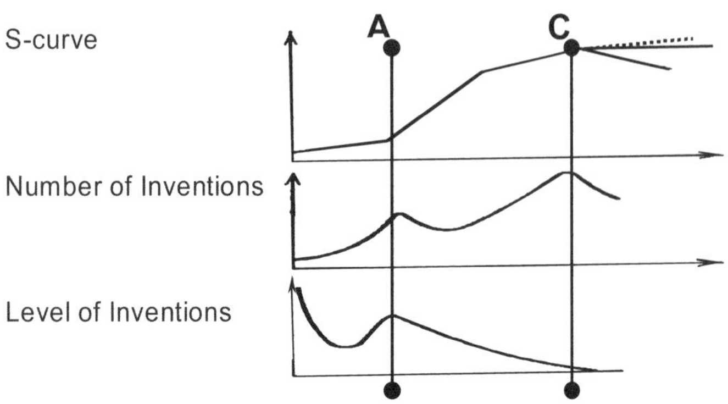

Fig 2.2 S-Curve and Inventions correlation

SETTING DIRECTIONS

Having reliably positioned the current product on its S-curve, the company can now proceed to the next step – determining the directions for improvement. Depending on whether the product needs to be improved incrementally or radically, certain changes need to be envisioned: *How do we move the product up its S-curve?* or *What new product technology may replace the current one?*

For an OEM it is usually useful to start with the concept of domination of higher-level systems: Evolutionary trends of a system (e.g., an automobile) determine the directions of evolution of its components (i.e., vehicle subsystems). Determining what new features and what performance the next automobile will have will determine which subsystems will be involved in the delivery of these features and what level of performance each subsystem should achieve.

The main activity at this step is applying the laws of technological system evolution to the features (functions) and/or subsystems of the vehicle. First, the current state of the selected feature is identified (with respect to any law of evolution). Then conceptual changes "prescribed" by this law are visualized.

Consider, for example, the "cruise control" feature and the law of increasing flexibility. From a standpoint of this law, conventional cruise control is a "rigid" system: it's either "On" or "Off." The law instructs us that the next-generation system has to be more adaptive: the vehicle should adjust its speed according to the distance to the vehicle ahead. This requires a capability to determine distances. Recent advances in radar technology may provide just that. In fact, some OEMs have already started introduction of an adaptive cruise control.

After the feature/product has been forecasted using one law of evolution, the next step is to apply another law to the same feature/product, and then another law, and so on. The more laws of evolution that are used, the better the "opportunity space" will be explored.

CONCLUSION

Like a radar in the fog, the TRIZ method helps managers of technology to dramatically reduce the uncertainty of the decision process when navigating in the ever shifting technology landscape. This allows the company to objectively justify its technology plan and lower the risks of technology investments.

Introduction of TRIZ in automotive product planning processes would help the industry to remove guesswork from the "fuzzy front end" of product and technology forecasting. Judicious use of the laws of evolution would significantly improve the effectiveness of new feature development.

TRIZ does not provide information on when the predicted change of technology will take place. Time of the development depends on many non-technological factors: economic situation, influence of special interests, etc. Those forces can delay evolution, but they cannot change its principal direction.

REFERENCES

1. Altshuller, G.S., Shapiro, R.B. (1956). On the psychology of engineering creativity, in Problems of Psychology, Vol. 6, pp. 37-49 (in Russian).

2. Altshuller, G.S. (1988). *Creativity as an Exact Science*, New York: Gordon and Breach.

3. Altshuller, G.S. (1999). *The Innovation Algorithm*, Worcester, MA: The Altshuller Institute for TRIZ Studies.

4. Raskin, A. (2003). A Higher Plane of Problem-Solving, Business 2.0, June, pp. 54-56.

5. Clausing, D., Fey, V. (2004). *Effective Innovation: The Development of Winning Technologies,* NY: ASME Press.

6. Foster, R., (1986) *Innovation: the attacker's advantage*, NY: Summit Books

CONTACT

Victor Fey has over 25 years of experience in the fields of TRIZ research and application. He was a student and close collaborator of Genrikh Altshuller, has had numerous successful projects in the area of new product and technology development with leading corporations and has conducted courses for industry and academia. He is an Adjunct Professor at Wayne State University and is a partner in The TRIZ Group, LLC. He can be reached at 248-538-0136 and fey@trizgroup.com.

2004-01-0295

A Backbone in Automotive Software Development Based on XML and ASAM/MSR

Bernhard Weichel and Martin Herrmann
Robert Bosch GmbH

ABSTRACT

The development of future automotive electronic systems requires new concepts in the software architecture, development methodology and information exchange.

At Bosch an XML and MSR based technology is applied to achieve a consistent information handling throughout the entire software development process. This approach enables the tool independent exchange of information and documentation between the involved development partners.

This paper presents the software architecture, the specification of software components in XML, the process steps, an example and an exchange scenario with an external development partner.

INTRODUCTION

The amount of software in vehicle control units is growing for years. Software covers more and more system functionality. This growing complexity of functional requirements as well as non functional requirements of the automobile industry such as integration capabilities, reusability and portability require new concepts in software architecture but also in development methodology, in particular regarding data exchange and documentation.

More and more specific components have to be integrated into e.g. engine software, some from companies with particular know-how, some from OEM to apply a maker specific "branding" to the vehicle. To support this trend and the accompanying requirements of black box integration and the protection of intellectual property, it must be possible to exchange tailored information at different steps during the development process.

The structure of this paper is as follows. First the software architecture of the engine control unit is illustrated. Subsequently the usage of XML [1] and MSR [2] as an exchange format are outlined. The paper finally gives a particular example.

EDC/ME(D)17 SOFTWARE ARCHITECTURE

The Bosch software architecture of electronic control units (ECU) for gasoline and diesel engines EDC/ME(D)17 [3] supports the development process in several ways:

- consequent separation of the application software and the base software
- clear modularity of the application software, according to the vehicle domain model (CARTRONIC [4])
- definition of interfaces for the system's components

The development process of ECU software is characterized by different roles and groups within appropriate phases of the development process. These different roles have their own view on the system and therefore require an appropriate representation of the software according to the process step in question.

The EDC/ME(D)17 software architecture supports this requirement by functional view, dynamic view and static view of the software system (see figure 1).

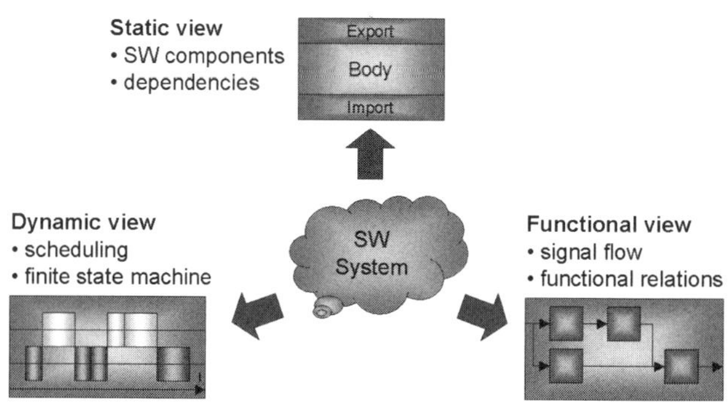

Figure 1: The different views on the software system

- The functional view describes the overall relationships in the system.
- An engine control unit is an embedded system with high real time requirements. The aspects concerning dynamic behavior are covered by the dynamic view of the EDC/ME(D)17 software architecture.
- The static view describes the software structure at development time via a layer model and a component model (see figure 2).

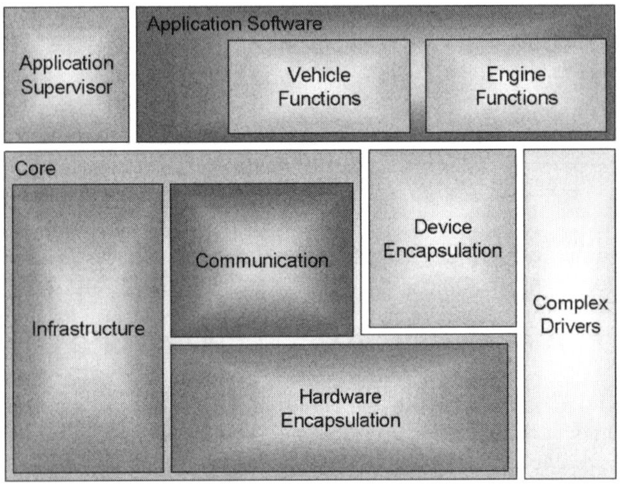

Figure 2: static view/component model of the EDC/ME(D)17 software architecture

USING XML AND MSR AS EXCHANGE FORMAT

MSR, a consortium of car manufacturers and suppliers, supports common developments between car manufacturers and their electronic system suppliers by enabling process synchronization and proper information exchange based on XML.

For the practical implementation of an architecture driven software development (and co-operation) it is necessary to be able to handle the resulting work products consistently and with the possibility of a flexible workload distribution. Results of the MSR project can be successfully applied here.

Using an MSR based information storage it is easy to join and split information as needed for any process step. The sum of all MSR information about a system is called the MSR-Backbone.

Figure 3 and 4 use the V-Model [5] to show some example steps where data exchange takes place. Figure 3 illustrates exchange based on specific incompatible formats as they were used in the past. Figure 4 displays the exchange used now based on MSR information and storage in the MSR-Backbone.

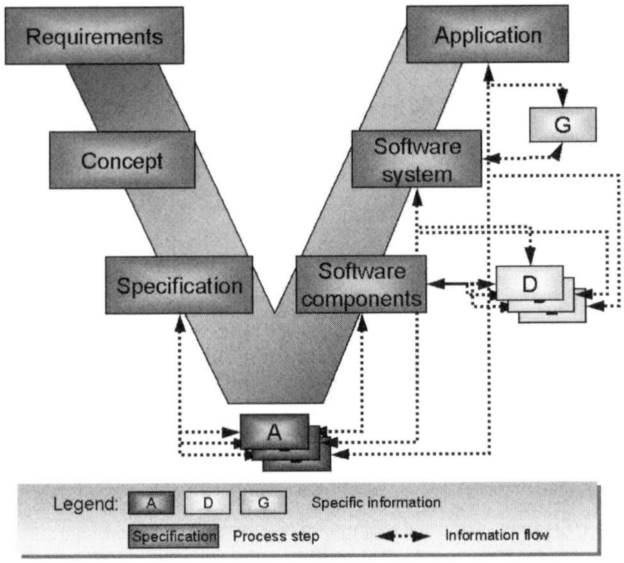

Figure 3: In the past - information exchange in the development cycle based on specific incompatible formats

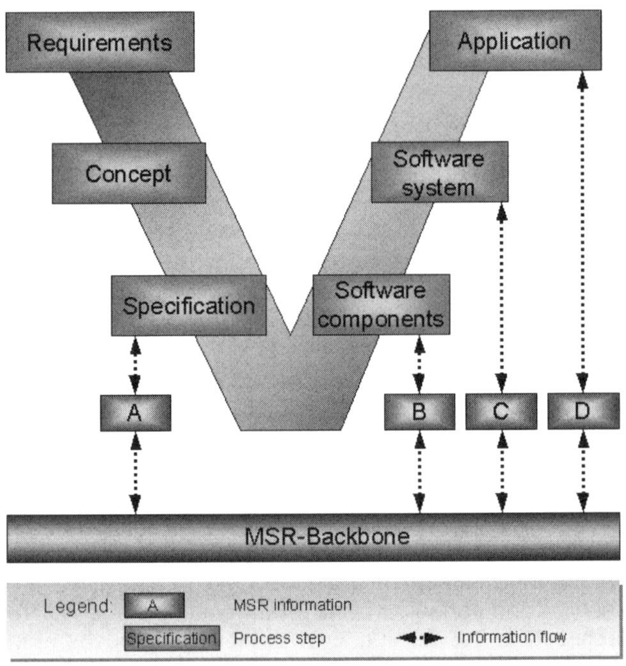

Figure 4: Now - information exchange in the development cycle using the MSR-Backbone

The exchange of information based on MSR may be done in any point of the development process.

There have been several specific exchange formats in the past, but with MSR there is a consistent format for definition, exchange and administration of all relevant information.

USED XML DOCUMENT TYPE DEFINITIONS

The information management of MSR is based on XML. MSR developed several XML document type definitions (DTD), based on a common set of definitions and practices. Thus they all follow a uniform design concept allowing to create a comprehensive and consistent data management. This is the base for well defined data exchange between the development partners.

For this process, several well matching DTDs of the MSR family are in use. As an example the MSR Software DTD (MSRSW.DTD) is described in more detail below. It permits - for instance - the representation of the substantial aspects of the architectural views:

- Components of a software system (static view) with points of variation and interfaces
- Functionality of a component (functional view)
- Details of the interfaces, such as data structures, services, physical characteristics on different abstraction levels, from the C-Typedef up to an abstract class definition (static view)
- Dynamic aspects, such as tasks, processes, operating conditions (dynamic view)

Even if there are specific description techniques for all these aspects, the integrated approach of MSRSW offers substantial advantages regarding reduction of redundancy, consistency and handling as files.

Data exchange itself is executed by delivery of exchange files and their administrative data. For this the container catalog (MSRCC) is used. MSRCC can carry all meta data necessary for the loose coupling of the involved configuration management systems. Thus MSRCC supports coordinated and well defined data exchange between the involved parties.

SPECIFICATION OF SOFTWARE COMPONENTS USING MSRSW

Software components are specified using XML with the MSRSW.DTD.

The specification is focused on the interfaces of the component. The interfaces define:

- offerings of the software component to the system.
- consumptions of the component from the system.

Following this approach only the compiled object and the accompanying XML files are required in order to build a component based software system.

The information about the component offerings to the system is specified in it's **export interface**. The information about the required consumptions is specified in the **import interface** (see figure 7).

The offerings and consumptions may be elements such as variables, messages, services, calibration parameters, but also meta data such as computation methods[1] or base definitions. The imported elements may itself be exported thus providing a propagation of interfaces. Therefore it is necessary to explicitly denote the *owning* component, which determines the properties of such elements. The elements are specified in a very detailed way in the data dictionary (see chapter EXAMPLE).

This information model enables to implement a thorough consistency check between the software components[2]. These checks allow to detect much more problems (e.g. inconsistent physical value ranges of a variable between a consumer and a provider) than tools like compiler and linker, due to the fact, that they operate on an architectural level.

USE CASES

By the use of the consistent data structure provided by the MSR DTDs it is now possible to arrange the information flow with more flexibility within the development process by defining and combining appropriate use cases. An use case thereby covers the information handled within one particular process step.

In the case of a change of the process the process steps and the responsibilities change. However the logical data structure remains stable. The information can thereby be captured once, be improved in the course of the process and kept permanently consistent. Even with new development cycles the results of earlier cycles will be reused and updated accordingly. The information content thereby remains independent of the distribution on files, responsibilities and process steps.

Figure 5 shows the possible Use Cases (with the internal naming and the regarded contents).

[1] A computation method describes how to convert a physical value to the internal representation and vice versa.

[2] These checks may even take place, before the implementation is finished, when using the interface contract files.

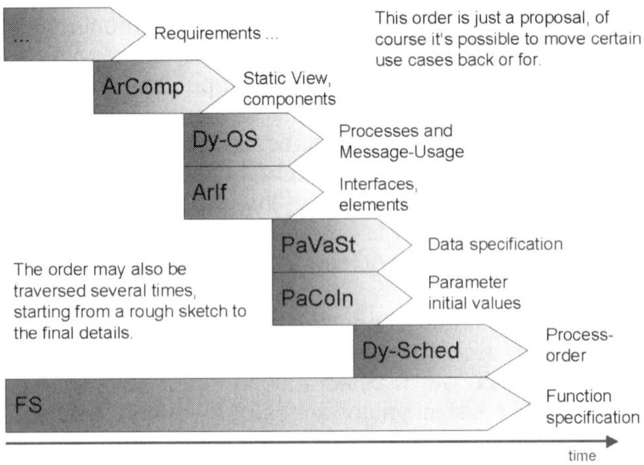

Figure 5: The Use Case process

Figure 6 shows a scenario with an external development partner. Bosch delivers a file with the Use Case contents of:

- the static view **ArComp**[3] and
- the planned interface of the appointed component **ArIf**[4]

to the development partner (OEM).

With these information the development partner is able to develop his own software components. These components are then delivered to Bosch.

The delivery consists of two files:

- the implemented interface specification **ArIF**, the function specification **FS**[5], the data specification **PaVaSt**[6], all in one MSRSW file
- the object file

The implemented interface may now be checked against the planned interface specification (interface contract).

When all checks are successful the components can be integrated into the entire software system.

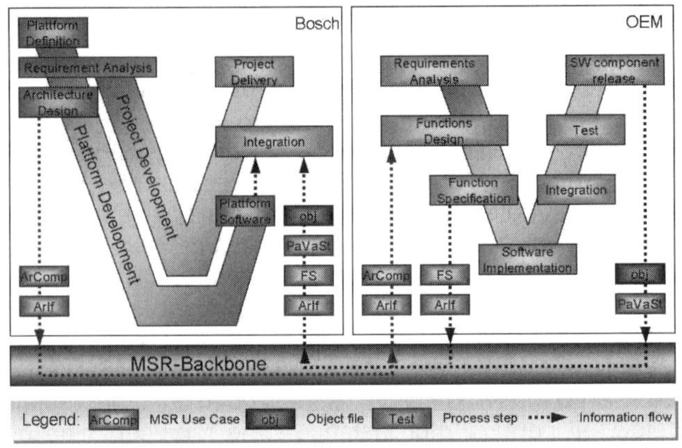

Figure 6: Exchange scenario with an external development partner

EXAMPLE

This chapter shows a brief example of how to specify a software component in MSRSW.

The example component is named K1. K1 is exporting a variable (message) named Var_b and a service (function) named Srv_Y (see figure 7).

Exported elements are listed in the export interface of the component and may be used by other components of the system.

To fulfill its task the component K1 needs some resources from the system: a variable named Var_a and a service Srv_X, which must be exported by some other component of the system. These elements are part of the import interface of K1 (see figure 7).

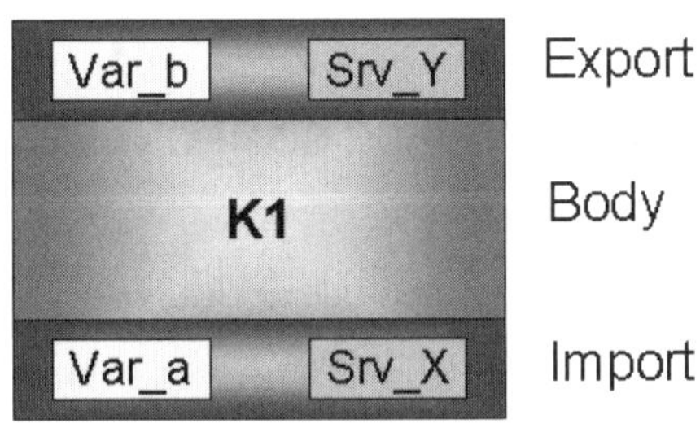

Figure 7: Example software component, exporting a variable Var_b and a service Srv_Y, importing a variable Var_a and a service Srv_X

The example component K1 may also be displayed in a UML representation (see figure 8).

---
[3] Architectural Components
[4] Architectural Interfaces
[5] Functional Specification
[6] Parameters, Variables, Structures

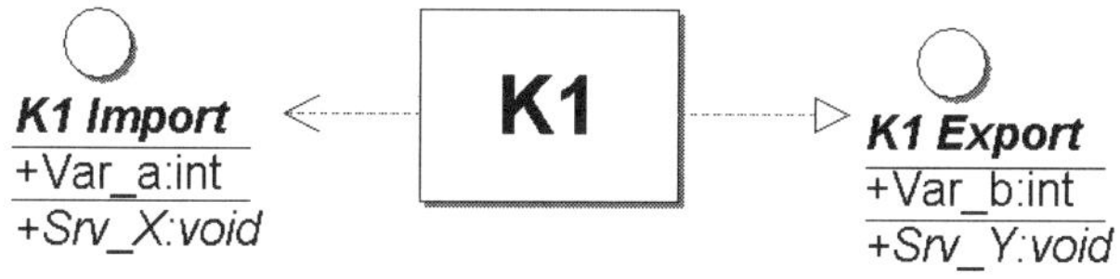

Figure 8: The UML representation of the above example

The available information about K1 is mapped on the MSRSW-XML tree as shown in figure 9. Of course it is possible to specify much more information about every element (e.g. implementation details, documentation, process information, value ranges, scope of exported elements, etc.) which is not shown here.

Figure 9: The MSRSW-XML tree of the example component

This is an illustration of the MSRSW specification for the example software component K1:

```
<MSRSW>
 <CATEGORY>PaVaSt</CATEGORY>
 <SW-SYSTEMS>
  <SW-SYSTEM>
   <LONG-NAME>EDC/ME(D)17</LONG-NAME>
   <SHORT-NAME>MEDC17</SHORT-NAME>
   <INTRODUCTION>
    <P>This is the name of the system for which the
pavast file is applied. As it is intended and sold
as a common system, we name it MEDC17.</P>
   </INTRODUCTION>
   <SW-DATA-DICTIONARY-SPEC>
    <SW-VARIABLES>
     <SW-VARIABLE>
      <LONG-NAME>Variable A</LONG-NAME>
      <SHORT-NAME>Var_a</SHORT-NAME>
      <CATEGORY>VALUE</CATEGORY>
      <SW-DATA-DEF-PROPS>
      <SW-ADDR-METHOD-REF>near</SW-ADDR-METHOD-REF>
       <SW-BASE-TYPE-REF>uint8</SW-BASE-TYPE-REF>
      <SW-CALIBRATION-ACCESS>
       CALIBRATION</SW-CALIBRATION-ACCESS>
      <SW-CODE-SYNTAX-REF>
       variables</SW-CODE-SYNTAX-REF>
      <SW-COMPU-METHOD-REF>q40</SW-COMPU-METHOD-REF>
      <SW-IMPL-POLICY>STANDARD</SW-IMPL-POLICY>
     </SW-DATA-DEF-PROPS>
    </SW-VARIABLE>
    <SW-VARIABLE>
      <LONG-NAME>Variable B</LONG-NAME>
      <SHORT-NAME>Var_b</SHORT-NAME>
      <CATEGORY>VALUE</CATEGORY>
      <SW-DATA-DEF-PROPS>
      <SW-ADDR-METHOD-REF>near
      </SW-ADDR-METHOD-REF>
      <SW-BASE-TYPE-REF>sint16</SW-BASE-TYPE-REF>
      <SW-CALIBRATION-ACCESS>
       READ-ONLY</SW-CALIBRATION-ACCESS>
      <SW-CODE-SYNTAX-REF>
       Msg_s16</SW-CODE-SYNTAX-REF>
      <SW-COMPU-METHOD-REF>
       speedComp1</SW-COMPU-METHOD-REF>
      <SW-IMPL-POLICY>MESSAGE</SW-IMPL-POLICY>
     </SW-DATA-DEF-PROPS>
    </SW-VARIABLE>
  </SW-VARIABLES>
    <SW-SERVICES>
     <SW-SERVICE>
      <LONG-NAME>Service X</LONG-NAME>
      <SHORT-NAME>Srv_X</SHORT-NAME>
      <CATEGORY>PRIMITIVE</CATEGORY>
     </SW-SERVICE>
     <SW-SERVICE>
      <LONG-NAME>Service Y</LONG-NAME>
      <SHORT-NAME>Srv_Y</SHORT-NAME>
      <CATEGORY>PRIMITIVE</CATEGORY>
     </SW-SERVICE>
    </SW-SERVICES>
   </SW-DATA-DICTIONARY-SPEC>
   <SW-COMPONENT-SPEC>
    <SW-COMPONENTS>
     <SW-FEATURE>
      <LONG-NAME>Component No 1</LONG-NAME>
      <SHORT-NAME>K1</SHORT-NAME>
      <SW-FEATURE-OWNED-ELEMENTS>
       <SW-FEATURE-ELEMENTS>
        <SW-SERVICE-REFS>
         <SW-SERVICE-REF>Srv_Y</SW-SERVICE-REF>
        </SW-SERVICE-REFS>
        <SW-VARIABLE-REFS>
         <SW-VARIABLE-REF>Var_b</SW-VARIABLE-REF>
        </SW-VARIABLE-REFS>
       </SW-FEATURE-ELEMENTS>
      </SW-FEATURE-OWNED-ELEMENTS>
      <SW-FEATURE-INTERFACES>
      <SW-FEATURE-INTERFACE>
        <SHORT-NAME>K1_IF1</SHORT-NAME>
        <SW-INTERFACE-EXPORTS>
         <SW-INTERFACE-EXPORT>
          <SW-INTERFACE-EXPORT-SCOPE>
           <SW-INTERFACE-EXPORT-LEVEL>
            PARENT</SW-INTERFACE-EXPORT-LEVEL>
          </SW-INTERFACE-EXPORT-SCOPE>
          <SW-FEATURE-ELEMENTS>
           <SW-SERVICE-REFS>
            <SW-SERVICE-REF >Srv_Y</SW-SERVICE-REF>
           </SW-SERVICE-REFS>
           <SW-VARIABLE-REFS>
            <SW-VARIABLE-REF>Var_b</SW-VARIABLE-REF>
           </SW-VARIABLE-REFS>
          </SW-FEATURE-ELEMENTS>
         </SW-INTERFACE-EXPORT>
        </SW-INTERFACE-EXPORTS>
        <SW-INTERFACE-IMPORTS>
         <SW-INTERFACE-IMPORT>
          <SW-FEATURE-ELEMENTS>
           <SW-SERVICE-REFS>
            <SW-SERVICE-REF>Srv_X</SW-SERVICE-REF>
            </SW-SERVICE-REFS>
            <SW-VARIABLE-REFS>
            <SW-VARIABLE-REF>Var_a</SW-VARIABLE-REF>
           </SW-VARIABLE-REFS>
          </SW-FEATURE-ELEMENTS>
         </SW-INTERFACE-IMPORT>
        </SW-INTERFACE-IMPORTS>
       </SW-FEATURE-INTERFACE>
      </SW-FEATURE-INTERFACES>
     </SW-FEATURE>
    </SW-COMPONENTS>
   </SW-COMPONENT-SPEC>
  </SW-SYSTEM>
 </SW-SYSTEMS>
</MSRSW>
```

TOOL SUPPORT

The decision to use XML as a base makes it possible to apply available XML tools such as XML editors or XML converters. Tool customizations such as style files, reference managers, converters can be used again and again cause of the MSR application profile. For frequently performed processes it is of course possible to use specifically optimized programs. The produced results remain manageable in any case.

The tools for EDC/ME(D)17 are implemented on three levels:

- Adjustment of existing internally developed tools.
- Use and adjustment of standard (Commercial off the shelf) tools.
- Development of new domain-specific tools.

BENEFIT

The presented approach results in substantial improvements in the information flow of the development process:

- Quality increase by automated consistency checks during the entire process and a single point of information acquisition
- Better understanding for the development partners by common and well defined MSR data structures.
- Comprehensive, integrated data processing and documentation production. Optimization of the processes by flexible allocation of the information on processing units and tool independence of the data formats.

OUTLOOK

The presented approach takes the requirements of the involved development partners into account, with respect to the software architecture and the exchange methods. In order to fulfill the requirements on multilateral co-operation models, further premises must be created, in particular the coordination and standardization of structures and interfaces over manufacturers and suppliers.

The presented MSR based exchange formats are to be established internationally over standardization committees and be made available in standard development tools. For this purpose the MSR project is now integrated into the ASAM e.V. [6].

REFERENCES

[1] XML - Extensible Markup Language. www.w3.org/XML

[2] MSR Project - Manufacturer Supplier Relationship. www.msr-wg.de

[3] C. HAMMEL, H. JESSEN, B. BOSS, A. TRAUB, C. TISCHER and H. HÖNNINGER - A common Software Architecture for Diesel and Gasoline Engine Control Systems of the New generation EDC/ME(D)17. In SAE World Congress. 2003.

[4] T.BERTRAM, R.BITZER, R.MAYER and A.VOLKART. CARTRONIC – An Open Architecture for Networking the Control Systems of an Automobile. In SAE World Congress. 1998.

[5] V-MODEL'97. Development Standard for IT-Systems of the Federal Republic of Germany – Lifecycle Process Model. www.v-model-iabg.de

[6] ASAM e.V. – Association for Standardisation of Automation- and Measuring Systems. www.asam.net

CONTACT

Bernhard Weichel
Robert Bosch GmbH, GS-EC/EMT4
P.O. Box 30 02 40
70442 Stuttgart
Germany

Phone: +49 711 811 8322
E-Mail: Bernhard.Weichel@de.bosch.com

Martin Herrmann
Robert Bosch GmbH, FV/SLI
P.O. Box 10 60 50
70049 Stuttgart
Germany

Phone: +49 711 811 7309
Fax: +49 711 811 7602
E-Mail: Martin.Herrmann@de.bosch.com

DEFINITIONS, ACRONYMS, ABBREVIATIONS

DTD: Document Type Definition

ECU: Electronic Control Unit

EDC/ME(D)17: Control unit for gasoline and diesel engines

MEDC17: Control unit for gasoline and diesel engines

MSR: Manufacturer Supplier Relationship

MSRCC: MSR Container Catalog

MSRSW: MSR-Software (abbrev. for the MSRSW.DTD)

OEM: Original Equipment Manufacturer

Use Case: One step in the MSR based development process

XML: Extensible Markup Language

2003-01-0139

Development of Modular Electrical, Electronic, and Software System Architectures for Multiple Vehicle Platforms

Gary Rushton
Visteon Corporation

Armen Zakarian and Tigran Grigoryan
University of Michigan – Dearborn

ABSTRACT

Rising costs continue to be a problem within the automotive industry. One way to address these rising costs is through modularity. Modular systems provide the ability to achieve product variety through the combination and standardization of components. Modular design approaches used in development of vehicle electrical, electronic, and software (EES) systems allow sharing of architectures/modules between different product lines (vehicles). This modular design approach may provide economies of scale, reduced development time, reduced order lead-time, easier product diagnostics, maintenance and repair. Other benefits of this design approach include development of a variety of EES systems through component swapping and component sharing. In this paper, new optimization algorithms and software tools are presented that allow vehicle EES system design engineers to develop modular architectures/modules that can be shared across vehicle platforms (for OEMs) and across OEMs (for suppliers). Approaches presented in this paper use matrix clustering and graph based techniques. The application of the approach is illustrated with an example from the automotive industry on the development of a modular EES system that can be shared across multiple vehicle platforms.

INTRODUCTION

Modularity arises from the way a product is physically divided into components and refers to the use of interchangeable components to create product variants. Two major benefits associated with modular products are the standardization of components and the ability to achieve product variety through the combination of components. Ulrich and Tung (1991) describe five different ways that modularity is used in current industry to exploit component standardization and to achieve product variety, i.e., component swapping, component sharing, fabricate-to-fit, bus, and sectional modularity. Ulrich and Tung (1991) also defined product modularity and explored the benefits and costs associated with modular products. O'Grady (1999) provides an in-depth description of modularity and showed how companies can use modularity to reduce product development time, costs, and capital investments. Pimmler and Eppinger (1994) used product decomposition to address the integration problems in development of modular products. Kusiak and Huang (1996) presented a methodology for development of modular products while considering product cost and performance. The product modularity problem was represented with a graph, while the module components of a product set were determined by a heuristic approach. Huang and Kusiak (1998) also presented a matrix representation of the modularity problem. A decomposition approach was used to determine modules for different products. Zakarian and Rushton (2001) presented a methodology that combines the system modeling, integration analysis, and optimization techniques for development of modular electrical, electronic, and software (EES) systems. A new clustering technique was developed to identify clusters in the incidence matrix, group the functions, and create EES system modules. The ability to produce a variety of systems through the combination of modular components is a meaningful benefit of modularity. Other potential benefits of modularity include economies of scale, reduced order lead-time, easier product diagnostics, maintenance and repair, and decoupling of tasks. In this paper, new approaches and software algorithms are presented that allow vehicle EES system design engineers to develop modular architectures/ modules that can be shared across vehicle platforms (for OEMs) and across OEMs (for suppliers). The methodology presented in this paper uses matrix clustering and graph based techniques. The **I**ntegrated **S**ystem **D**evelopment (**ISD**) software tool presented in this paper allows system design experts to build a low cost EES architecture that can be shared across multiple vehicle platforms. The **ISD** software builds an optimal system architecture for a given set of system features and feature take rates by grouping (integrating) system functions into physical modules and determining the best possible tradeoffs between system overhead, give away, and wiring costs (Rushton, G., Zakarian, A., and Grigoryan, T. 2002). Also in this paper a new approach is presented that allows system developers to identify common modules in EES architectures that can be shared across multiple vehicle platforms.

DEVELOPMENT OF COMMON SYSTEM ARCHITECTURES

In this section we present the approach and **ISD** software that allows vehicle system design engineers to determine a low cost EES architecture in a multi vehicle development environment using modular design concepts. The overall design approach is shown in Figure 1 (Zakarian and Rushton 2001). System modeling is used to develop the functional requirements model of a vehicle system. The system requirements model defines the interfaces (interactions) between the functional elements (primitives) to support the functions of various vehicle configurations. Within this system model, a system design engineer may identify the requirements of various vehicle configurations (see Figure 2). For example, the system requirements model identifies requirements for three different configurations: 1) four-door body style, automatic transmission, and diesel engine; 2) two-door body style, automatic transmission and hybrid engine and 3) convertible body style, automatic transmission and gas engine (see Figure 2). It is clear that some of the functions of the EES architectures that support the above three configurations are mutually exclusive. For example, an EES architecture that supports a two-door body style configuration may not have some of the functions that are required for the EES architectures that supports a four door or convertible body style configuration. Once the requirements models of various vehicle configurations are built and the interfaces among the functional elements that support all these different configurations are identified, a function-function interaction (incidence) matrix of the interfaces is developed (see Figure 3). Each row (column) in the interaction matrix corresponds to a function (primitive process), and each none zero entry in the matrix represents a data flow (interaction) between the processes (functions). Clustering techniques and optimization algorithms are used to identify a low cost EES architecture that can support all three vehicle configurations. The objective of the clustering algorithm is to determine an optimal (low cost) common system architecture for the given features and feature take rates by grouping (integrating) system functions into physical modules and determining the best possible tradeoffs between system overhead, give away, and wiring costs. It should be noted that the approach also optimizes system wiring costs by considering various multiplex network options.

The **ISD** software presented in this paper allows system design experts to import the system requirements model into a specified database and using predefined templates create various vehicle configurations (see Figure 2). Once the requirements for each vehicle configuration are identified the software tool automatically builds a function – function interaction matrix (see Figure 3). A function - function interaction matrix $[a_{ij}]$ is shown in Figure 3 and includes integer, as well as "blank" entries, where an integer entry indicates the information, material or energy link (signal flow) between functions i and j, and the direction of the link (flow) is from j to i. The number also indicates the number of signals or flows sent from function j to function i. One may see that the matrix in Figure 3 is not binary but weighted.

Before the clustering algorithm is applied to the function – function interaction matrix, the software tool allows the user to build predefined, or carryover, modules. The software tool allows the user to select functions from the incidence matrix and group them into predefined modules. The functions grouped into a predefined module will not be moved into separate modules. In other words, this step makes sure that in the final vehicle architecture the functions of carryover, or predefined modules, will appear in a single cluster/module. Once carryover modules are identified, the user can cluster the matrix to identify optimal functional grouping in the vehicle architecture (see Figure 4).

The EES architecture matrix is shown in Figure 4. Note, the matrix in Figure 4 is a small portion of the example EES system architecture matrix that includes 190 functions and 253 interactions. From Figure 4 one may see that the total cost of the common EES architecture is $512. 93, vehicle architecture is partitioned into 12 zones and there are 49 modules in this architecture. One may also see the cost break down between architecture overhead, giveaway and wiring/multiplexing costs. It was explained earlier that various vehicle configurations may have mutually exclusive functions. Functions that support a four-door body style configuration may not, or should not, appear in convertible body style configuration. The EES architecture in Figure 4 supports all three body style configurations. Therefore, when the EES architecture in Figure 4 is implemented for two door body style configuration, functions that are mutually exclusive in the four door and convertible body style configurations will not be included and will contribute to the giveaway costs.

Next, we present literature review on clustering techniques and briefly explain the basic approach of the optimization algorithm used in the **ISD** software.

CLUSTER ANALYSIS TECHNIQUES – Clustering techniques are used to group objects into homogenous clusters based on object features. Clustering is also a generic name for a variety of mathematical methods that are used to find natural groupings in a given object/data sets. Cluster analysis have been widely used for solving various engineering problems, e.g., design of modular systems (see Zakarian and Rushton 2001, Pimmler and Eppinger 1994, Kusiak and Huang 1996, 1994), group technology (see Kusiak and Chow 1998, King 1980, Ng 1991, pattern recognition (Ni and Jain 1998, Tambouratzis 2002), and so on. The technique has also been widely used in the natural sciences (see Sneath and Sokal 1973) and increasingly in social and management sciences (see Birnbaum 1977, Mahajan and Jain 1978).

Traditional clustering algorithms can be classified into two main categories: hierarchical and partitioned methods. Partitioned methods create a unique partition of objects or data sets while hierarchical methods produce several nested partitions of objects.

Hierarchical clustering techniques construct tree structures reflecting the underlying patterns in a given data set. The trees obtained by this clustering method are called dendograms and are typically binary. A dendogram consists of several layers of nodes representing different clusters and are typically used to highlight the similarity between clusters. From the dendogram data analysis different partitions can be built by cutting the tree horizontally using similarity values. The two most commonly used hierarchical clustering approaches are single link and complete link methods (Jain and Dubes 1988, Sneath and Sokal 1973).

Partitioned methods create a single partition of points. Approaches in partitioned clustering include error square clustering (Cheng and Tong 1991), clustering based graph theory (Tarjan 1972), and density estimation clustering. Error square clustering methods minimize the square error for a fixed number of clusters. Graph based clustering methods examine graph structure, for example using spanning tree method, to identify and remove inconsistent edges of the graph and determine strongly connected components.

Several algorithms and mathematical models have been developed for clustering binary matrices. Most of the mathematical models, i.e., *p*–median, generalized *p*–median (Kusiak 2000), require calculations of similarity distances between columns (rows) of the interaction matrix. Other algorithms, i.e., similarity coefficient method (McAuley 1972), rank order clustering algorithms (King 1980), and cluster identification algorithms (Kusiak and Chow 1987) use matrix row and column swapping techniques to obtain clusters in the binary matrix.

Incidence matrices of vehicle system models are large in size (more than 1,000 rows/columns) and may have binary and non-binary structures. Most of the clustering techniques developed in the literature are primarily designed for clustering binary matrices. Furthermore, algorithms presented in the literature are designed for decomposing binary matrices into mutually separable clusters. System level interaction matrices are large and practically impossible to decompose into mutually separable clusters. In other words, for large matrices, like the ones developed in this research, bottleneck elements (functions) will always exist. An element (entry x_{ij} = integer number) is considered a bottleneck, when it does not allow the decomposition of the function – function interaction matrix into mutually separable clusters.

The **ISD** software presented in this paper uses a new clustering algorithm and cost optimization techniques for development of low cost, modular vehicle architectures that can be shared across various vehicle platforms. The algorithm can be used for clustering both binary and non-binary (weighted) matrices. The clustering algorithm identifies clusters in the $n \times n$ incidence matrix. For the function - function incidence matrix the algorithm first creates n clusters. Once initial clusters are obtained the algorithm continuously improves the initial solution by moving the bottleneck elements to a cluster, if such an assignment improves the quality of the solution, i.e., minimizes average cost of the EES architecture that supports multiple vehicle platforms (configurations).

EES ARCHITECTURE DEVELOPMENT AROUND COMMON MODULES – In this section we present an approach that allows system design engineers to identify common modules between EES architectures that support various vehicle platforms (configurations). These common modules are then used in developing optimal low cost EES architectures for each vehicle platform (configuration). Using design process similar to one described in Figure 1 one may develop a system requirements model for various vehicle configurations and construct function – function interaction matrix for each configuration. Once interaction matrices for various vehicle configurations are constructed, clustering and graph based techniques may be used to identify common modules (chunks) in the incidence matrices. For example, Figure 5 represents incidence matrices of two different configurations with 14 and 18 functions, respectively. Using the clustering and graph based optimization techniques one may find two clusters that are identical (common) in both incidence matrices (i.e., EES architectures) (see Figure 6). One may see in Figure 6 that clusters (modules) 2 and 3 are in both matrices. Once common modules are determined one may define these common modules as predefined or carryover modules in the ISD software and cluster each matrix to obtain a low cost EES architecture using the techniques presented earlier. For example, after making modules 2 and 3 predefined one can further cluster the remaining functions shown in the matrices in Figure 6 to obtain EES architectures (see Figure 7). One may see that in both clustered matrices in Figure 7, structure of modules 2 and 3 are identical.

CONCLUSION

Modular design approaches used in development of vehicle electrical, electronic, and software (EES) systems allow sharing of architectures/modules between different product lines (vehicles). This modular design approach may provide economies of scale, reduced development time, reduced order lead-time, easier product diagnostics, maintenance and repair. Other benefits of this design approach include development of a variety of EES systems through component swapping and component sharing. In this paper, new approaches and software tools were presented that allow vehicle EES system design engineers to develop modular architectures/modules that can be shared across vehicle platforms (for OEMs) and across OEMs (for suppliers). Approaches presented in this paper used matrix clustering and graph based techniques. The application of the approach was illustrated with an example from the automotive industry

ACKNOWLEDGMENTS

This research was supported with a grant from Visteon Corporation.

REFERENCES

1. Birnbaum, P. H., (1977), "Assessment of Alternative Measurements Forms in Academic Interdisciplinary Research Projects", Management Science, Vol. 24, pp. 272-284.
2. Cheng, H. D. and Tong, C. (1991), "Clustering Analyzer", IEEE Transactions on Circuits and Systems, Vol. 38, No. 1, pp. 124 - 128.
3. Jain, A. K. and Dubes, R. C., "Algorithms for Clustering Data", Englewood Cliffs, New Jersey, Prentice Hall, 1988.
4. King, J. R. (1980), "Machine-Component Group Formation in Production Flow Analysis: An Approach Using a Rank Order Clustering Algorithm", International Journal of Production Research, vol. 18, no. 2, pp. 213-232.
5. Kusiak, A. (2000), "Computational Intelligent in Design and Manufacturing", Wiley, New York, NY.
6. Kusiak, A. and Chow, W. S. (1987), "Efficient Solving of the Group Technology Problem", Journal of Manufacturing Systems, vol. 6, no. 2, pp. 117-124.
7. Kusiak, A. and Chow, W. (1988), "Decomposition of Manufacturing Systems," IEEE Journal of Robotics and Automation, Vol. 4, No. 5, pp. 457-471.
8. Kusiak, A. and Huang, C. C. (1996), "Development of Modular Products", IEEE Transactions on Components, Packaging, and Manufacturing Technology – Part A, vol. 19, no. 4, pp. 523-538.
9. Kusiak, A. and Huang, C. C. (1998), "Modularity in Design of Products and Systems", IEEE Transactions on Systems, Man, and Cybernetics-Part A: Systems and Humans, vol. 28, no. 1, pp. 66-77.
10. Mahajan, V. and Jain, A. K. (1978), "An Approach to Normative Segmentation", Journal of Market Research, Vol. 15, 338-345.
11. McAuley, J. (1972), "Machine Grouping for Efficient Production", The Production Engineer, February, pp. 53-57.
12. Ng, S. M. (1991), "Bond Energy, Rectilinear Distance and Worse-case Bound for the Group Technology Problem," J. of Operations Research, Vol. 42, No. 7, pp. 571-578.
13. Ni, L. M. and Jain, A. K. (1985), "A VLSI Systolic Architecture for Pattern Clustering". IEEE Transactions on Pattern Analysis and Machine Intelligence, Vol. 7, pp. 80-89.
14. O'Grady, P. (1999), "The Age of Modularity: Using the New World of Modular Products to Revolutionize Your Corporation", Wiley, New York, NY.
15. Pimmler, T. U. and Eppinger, S. D. (1994), "Integration Analysis of Product Decomposition", Design Theory and Methodology – DTM, DE-vol. 68, ASME.
16. Sneat, P. H. and Sokal, R. R., "Numerical Taxonomy. San Francisco", CA. W. H. Freeman, 1973.
17. Tambouratzis, G. (2002), "Improving the Clustering Performance of the Scanning n-Tuple Method by Using Self Supervised Algorithms to Introduce Subclasses," IEEE Transactions on Pattern Analysis and Machine Intelligence, Vol. 24, No. 6, pp. 722-733.
18. Tarjan, R. (1972), "Depth-First Search and Linear Graph Algorithms", SIAM Journal of Computing, vol. 1, no. 2, pp. 146-160.
19. Ulrich, K. and Tung, K. (1991), "Fundamentals of Product Modularity", DE-vol. 39, Issues in Design Manufacture/Integration, ASME.
20. Zakarian, A. and Rushton G. J. (2001), "Development of Modular Electrical Systems", IEEE/ASME Transaction on Mechatronics, (to appear).
21. Rushton, G., Zakarian, A., and Grigoryan, T. (2002), "Algorithms and Software for Development of Modular Vehicle Architectures", SAE World Congress 2002-01-0140.

CONTACT

Gary Rushton has over 17 years of commercial and military electrical/electronic systems engineering experience. He has an MS in Automotive Systems Engineering from the University of Michigan. He is currently working as an electrical/electronic systems engineer specialist with Visteon Corporation. As an engineer with Visteon Corporation, he has worked on audio software, subsystem product development/design, diagnostics, vehicle system architectures, and cockpit system design. Previously, with General Dynamics, he worked on avionics systems for the F-16 and vetronics systems for the Abrams M1A2 tank. (grushton@visteon.com).

Armen Zakarian received his B.S. degree in mechanical engineering from Yerevan Polytechnic University, Yerevan, Armenia, his M.S. degree in industrial and systems engineering from the University of Southern California, Los Angeles, California and his Ph.D. degree in industrial engineering from the University of Iowa, Iowa City, Iowa, in 1997. He is an Assistant Professor of Industrial and Manufacturing Systems Engineering at the University of Michigan - Dearborn. His research interests are in development of integrated products and system, process modeling and analysis and manufacturing system. He has published papers in journals sponsored by ASME, IEEE and IIE societies. (zakarian@umich.edu).

Tigran Grigoryan received his B.S. degree in applied mathematics and computer science from Yerevan State University and his M.S. degree in applied mathematics from Yerevan State Engineering University. He is currently M.S. degree candidate in Industrial and Manufacturing Systems Engineering Department, at University of Michigan - Dearborn. He is interested in operations research and computational intelligence. (tigr@umich.edu)

APPENDIX

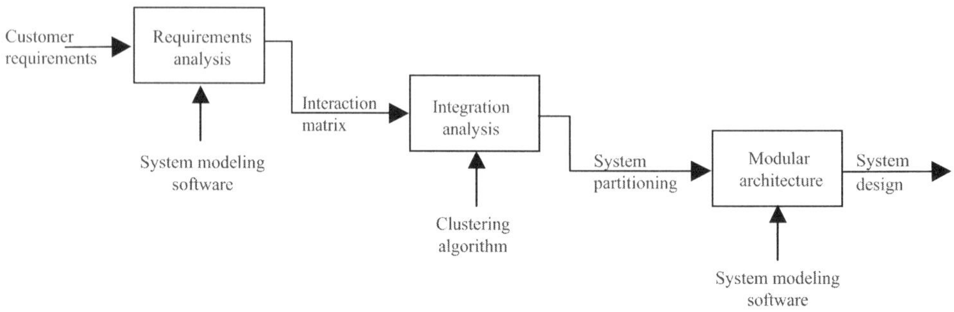

Figure 1: System Design Process

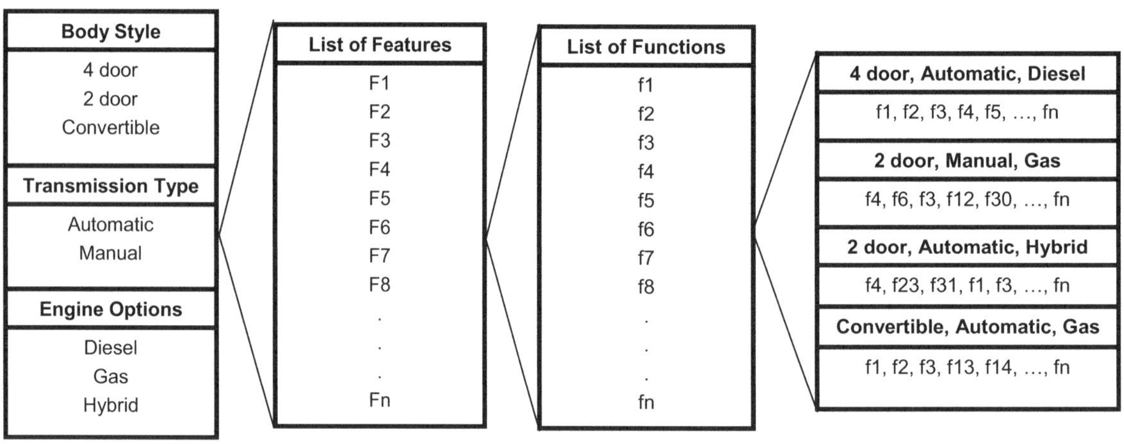

Figure 2: Developing Requirements Model for Various Vehicle Configurations

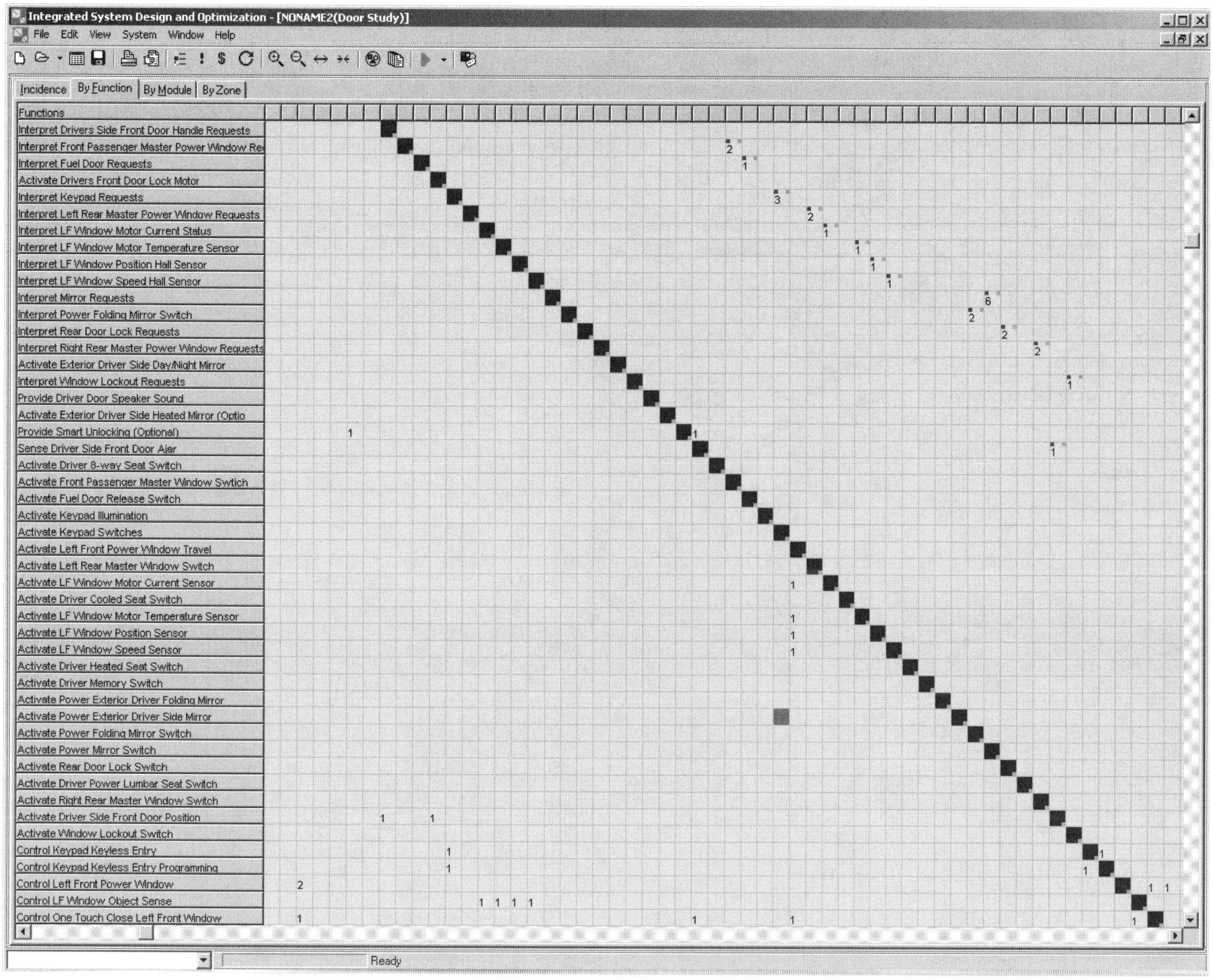

Figure 3. Function – Function Interaction Matrix

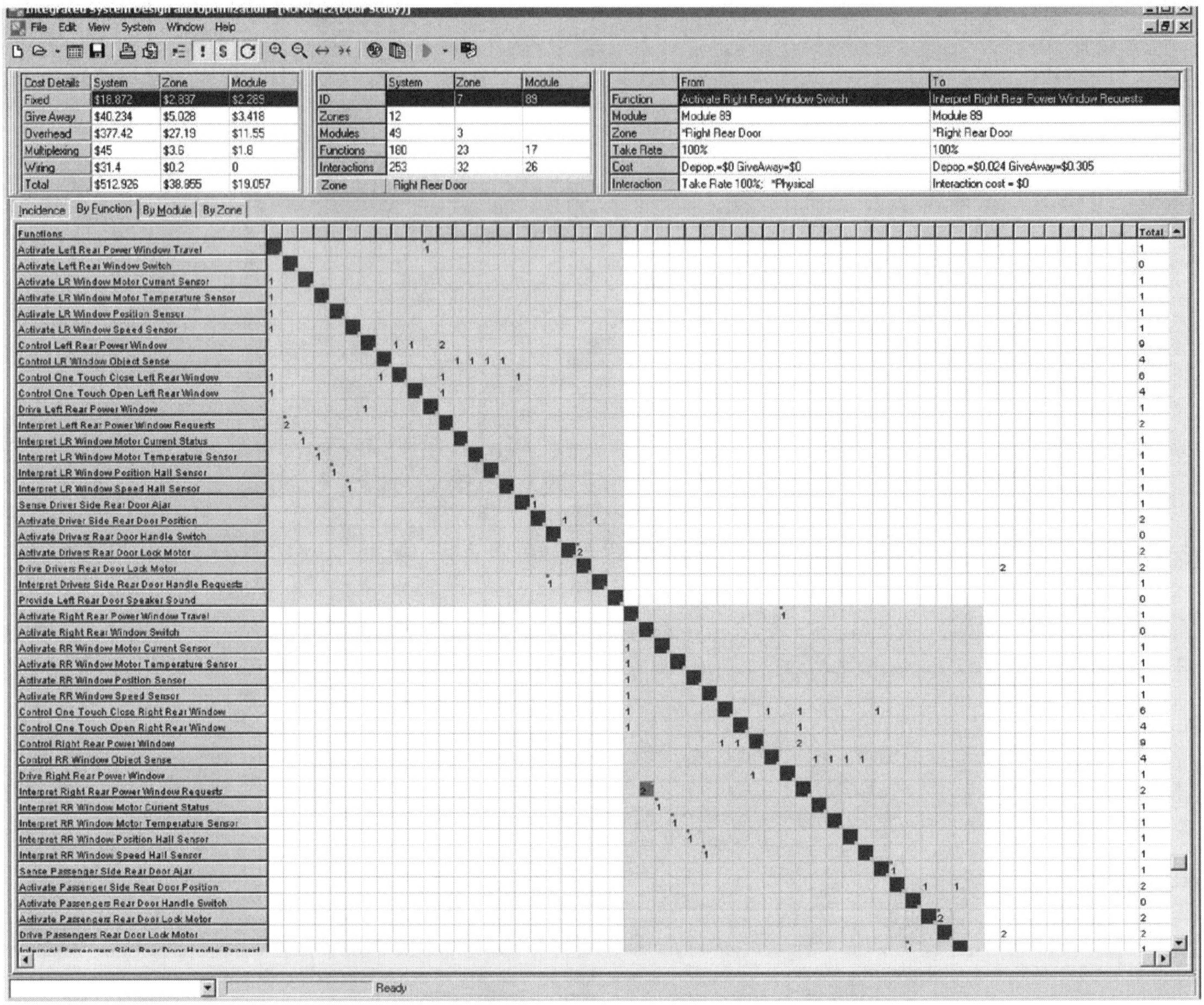

Figure 4. Common EES Architecture Matrix

	1	2	3	4	5	6	7	8	9	10	11	12	13	14
1	*		2					2						
2		*												
3			*			1								
4				*		1								
5					*		1				2			
6						*		3						
7							*							
8	2		2					*		1				
9							2		*					
10				3						*		1		2
11							2		1		*			
12				2								*		
13		2											*	2
14											1			*

	1	2	3	4	5	6	7	8	9	10	11	12	13	14	15	16	17	18
1	*		2					2										
2		*																
3			*			1												
4				*		1												
5					*						2					1		2
6						*		3										
7					1		*											
8	2		2					*		1								
9		2							*				3					
10				3						*		1		2				
11					2		2				*				1			
12				2								*						
13							1						*					
14											1			*			2	
15															*			
16	2															*		
17														3			*	
18				1	1													*

Figure 5. Function-Function Incidence Matrices of Two Different Systems

	2	5	7	9	1	3	6	8	4	10	12	11	13	14
2	*												1	
5		*	1									2		
7			*											
9		3	2	*										
1					*	2		2						
3						*	1							
6							*	3						
8					2	2		*		1				
4							1		*					
10									3	*	1			2
12									2		*			
11			2	1								*		
13	2												*	2
14												1		*

Module 2

Module 3

	2	5	7	9	1	3	6	8	4	10	12	11	13	14	15	16	17	18
2	*																	
5		*										2				1		2
7		1	*															
9	2			*									3					
1					*	2		2										
3						*	1											
6							*	3										
8					2	2		*		1								
4							1		*									
10									3	*	1			2				
12									2		*							
11		2	2									*			1			
13			1										*					
14												1		*			2	
15															*			
16					2											*		
17														3			*	
18		1							1									*

Module 2

Module 3

Figure 6. EES Architecture Matrices with Two Common Modules

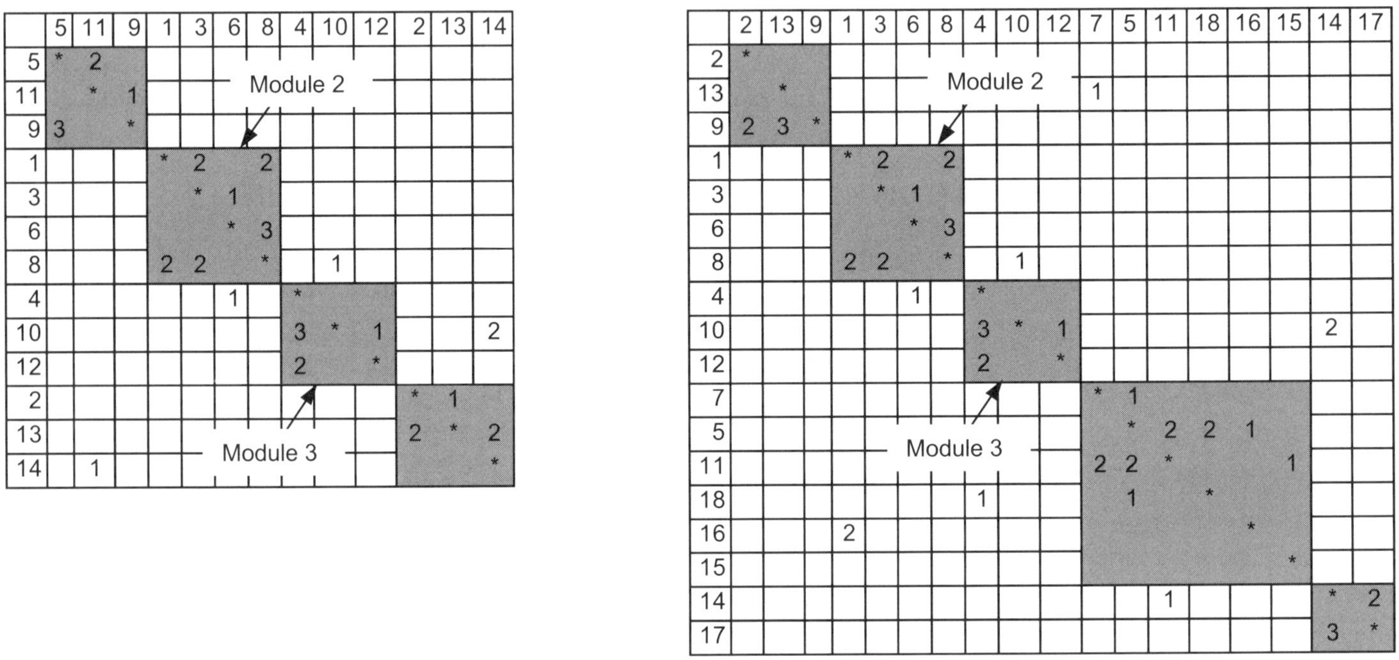

Figure 7. Clustered EES Architecture Matrices with Two Common Modules

2003-01-0131

A New Calibration System for ECU Development

Andre Rolfsmeier, Jobst Richert and Robert Leinfellner
dSPACE GmbH

ABSTRACT

Automotive manufacturers and suppliers of electronic control units (ECUs) will be challenged more and more in the future to reduce costs and the time needed for ECU development. At the same time, increasing requirements concerning exhaust gas emissions, drivability, onboard diagnostics and fuel consumption have led to the growing complexity of modern engines and the associated management systems.

As a result, the number and complexity of control parameters and look-up tables in the ECU software is increasing dramatically. Thus, in powertrain applications especially, calibration development has become a time-consuming and cost-intensive stage in the overall ECU development process.

This paper describes the current situation in calibration development and shows how the new dSPACE Calibration System will face this situation. It provides an overview of the main benefits of the tool, which has been designed in close cooperation with calibration engineers. A special focus will be on how the tool contributes to saving calibration costs and development time.

Further emphasis will be placed on integration into the dSPACE tool chain, providing a complete environment for the development of ECU software. Interaction with tools for control design, function prototyping, target implementation and ECU testing will be detailed.

In addition, the paper describes how international standards like ASAM/ASAP and NEXUS facilitate the calibration process and how these standards are supported by dSPACE.

INTRODUCTION

Automotive control has become a mainspring in automotive innovation. Vehicle manufactures and ECU suppliers have widely identified controller strategy development as a means of differentiating themselves from the competition. In order to keep pace with the increasing requirements from both legislators and customers, a streamlined development process is necessary.

The steadily growing development effort for electronic control software and the increasing time-to-market pressure demand highly efficient tool support during the development phases. Apart from powerful tools, integration into a complete development environment or tool chain is becoming more and more important.

Today, dSPACE already provides development tools for rapid control prototyping, automatic production code generation and hardware-in-the loop simulation. The company is continuing to extend its product range to cover all major phases of the ECU software development process. The next major step has been taken with the new dSPACE Calibration System.

SITUATION IN CALIBRATION DEVELOPMENT

A look at the last two decades of automotive engine control reveals the evolution of engine controller complexity and – closely related to this – calibration development. In the 1980s most systems were primarily controlled mechanically and hydraulically with some kind of electronic assistance. Simple strategies for major systems like fuel, ignition and exhaust gas recirculation found their way into the vehicle. By the early 1990s, several hundred parameters had to be tuned in the ECU software. By and by the strategies became large and complex, with an increasing amount of interaction between the systems. With the advent of torque-based control in 1997 [1], more than 1000 parameters had to be calibrated. Today, the amount of calibration data in modern engine control units already exceeds 5000 variables, and several hundred look-up tables have to be filled during calibration development.

In the past, the calibration process typically started at the test bed, where approximately 30 to 50 percent of all parameters were set, largely in the form of look-up tables, while the engine was run in a steady state. About 50 to 70 percent of the calibration data was tuned in the vehicle. Typically, this stage began with optimizing the drivability, followed by adjustment of emissions and fuel consumption, calibration of functions for on-board diagnostics and finally fine-tuning and validation during fleet tests. This calibration work was often done by trial and error.

The evolution of engine control was closely related to the progress in processor technology. In the last few years there has been a move from 16- to 32-bit processors and from CISC to RISC architectures [2]. The increasing complexity of automotive control required more and more flash memory and faster clock frequencies. Modern microcontrollers with up to 1 MB of on-chip flash and clock frequencies up to 56 MHz are commonplace today.

The predominant technique used in the past to perform calibration with powertrain control units is external memory emulation. This method is also called *parallel calibration* and characterized by a parallel link between the microcontroller bus and an external device emulating an ECU flash or ROM memory. But with new, high-performance microcontrollers with on-chip flash, this task is becoming more and more challenging. Typically, these new microcontrollers must run in a special mode in order to give the external emulation device full access to the microprocessor bus. In this mode some input and output pins may not be available. To regain this I/O, it is usually necessary to design additional circuitry, a so-called port replacement unit, on the development ECU.

As a result, development ECUs for calibration are often different in terms of hardware compared to the final production ECUs. Some extra calibration effort may thus be necessary if parameters need to be adapted to the final production hardware. Differences in the ground plane, for example, may require filter parameters to be recalibrated.

Moreover, providing full access to the microcontroller bus may not allow software code to be executed at full speed. In addition, the emulation of on-chip memory is only possible if the processor features some kind of overlay mechanism, that is, hardware-controlled redirection of data accesses from the on-chip flash to the external emulation RAM. But this is also accompanied by decreased performance depending on the access time of the emulation RAM. Consequently, in modern ECUs the code often resides in the on-chip flash, while the calibration data is placed in a flash device external to the microprocessor.

Parallel calibration has thus become hardware and cost intensive and difficult to design. However, it provides the greatest capabilities in terms of full memory emulation and data acquisition rates.

In transmission and chassis control especially, serial interfaces have also been used for calibration development. As for CAN based applications, the CAN Calibration Protocol (CCP) from ASAM e.V. has been widely accepted as a standard protocol [3]. With CCP, an extra service in the ECU code is necessary to handle the data transmission. This extra service provides an additional load to the processor, which may not be acceptable for some powertrain developers. Due to the limited RAM on today's microcontrollers and the maximal baud rate of 1 Mbit/s with CAN, CCP was typically used in less demanding calibration scenarios, even if special development ECUs with additional calibration RAM were available. However, it provides a cost-effective way to do calibration development, especially for exercises with ECUs in a temperature critical environment, for example, in a gearbox.

As a result of this, silicon vendors have been asked by OEMs and ECU suppliers to provide high-speed, on-chip interfaces that do not influence microcontroller operation. The solution was to enhance the processor on-chip debug ports with special capabilities supporting rapid control prototyping (bypassing) and calibration development. Different silicon vendors came up with different solutions like the READI interface (Motorola), AUD (Hitachi), JTAG/OCDS (Infineon) or the SDI interface (Mitsubishi). Parallel to this, a consortium of competing companies created a new standard debug interface for embedded control applications called NEXUS, also known as IEEE-ISTO 5001™ –1999 [4].

WHAT WILL THE FUTURE BRING?

Systems are becoming more complex and networked. Features like electronic stability, variable valve timing, camless engines, or X-by-wire technologies will become a common thing. Electronic control units and controller strategies will provide more flexibility so that they can easily be adapted to different engine and vehicle variants.

As a result, the number and complexity of control parameters and look-up tables is expected to increase further. Due to the growing competition and time-to market pressure, OEMs and ECU suppliers are forced to review conventional calibration and design approaches against the background of a reduction of development costs and time.

The need for reduced calibration effort will motivate new methods of controller design, like model-based control or artificial neural networks. In contrast to map-based control, where the actuator settings come directly from look-up tables, model-based control and artificial neural networks do not require the time-consuming calibration task of confirming each data point in the table.

More calibration work has to be done off the vehicle at the test bench and in earlier stages of the development, for example, when no vehicle or engine prototypes are available. In order to handle the increasing complexity of look-up tables and engine maps, improved optimization and evaluation strategies will be necessary [5]. Intelligent test bench systems that allow dynamic engine behavior to be measured and calibrated automatically and real driving cycles to be simulated will further facilitate the calibration work. Moreover, the ability to tune ECU parameters by means of engine or vehicle simulation can noticeably reduce the amount of time and money needed for development. Finally, the calibration process in the vehicle also provides enough potential for improvement. Routine tasks may be automated so that the calibration engineer can concentrate on the calibration process itself rather than on routine work. Trial and error has to be replaced by systematic optimization methods. However, using the vehicle only for the purpose of validation will be the ultimate goal.

Not only the calibration process itself is developing further, processor technologies also continue to progress. In the future there will be a need for high-speed, high-performance, but cost-effective processors with higher levels of system integration. Solutions with clock frequency over one hundred MHz, megabytes of on-chip flash, and several hundred kilobytes of integrated RAM are already on the horizon. The trend towards miniaturization, i.e., smaller and engine-mounted ECUs, is also promoting hybrid technologies using ceramics as a medium for ECU substrates. Reduced pin counts of microcontrollers, deep instruction pipelines and on-chip caches will further limit the visibility via the external bus. These trends make it even more difficult, if not impossible, to attach external devices for memory emulation.

While the need for memory emulation may never go away fully, calibration development via cost-effective, less invasive interfaces will become more relevant. New processor generations will further increase the capabilities for serial calibration by means of high-speed interfaces and integrated emulation concepts.

For OEMs and ECU suppliers to reduce the cost of tool instrumentation, standardized protocols and interfaces are necessary. Moreover, proprietary system-on-silicon (SOC) solutions call for a uniform, reliable tool interface independent of the individual SOC design. The XCP activities by the ASAM e.V. are an important step towards a widely accepted protocol. This protocol is independent of the physical layer and thus can be adapted to a variety of interfaces like CAN, Ethernet, and USB. NEXUS is also expected to play an important role in the future as a standardized interface to microcontrollers for debugging, rapid control prototyping and calibration.

The new dSPACE Calibration System has been especially designed to face the future situation in calibration development.

THE dSPACE CALIBRATION SYSTEM

A lot of discussions with calibration engineers and tool managers have shown that the basic requirements of OEMs and system suppliers for a new calibration and measurement system were very similar. It was no surprise that the items at the top of the list mostly pointed in the same direction. A cost-effective and reliable solution for calibration and measurement, without compromising basic functionality, was the most important point for the majority of customers. Easy and intuitive operation was rated equally important in order to increase user acceptance and to reduce the overall training and teaching effort. Another major aspect was integration into the development process at the customer and the functionality of the calibration system itself in order to meet future scenarios.

The requirements were discussed and detailed in close cooperation with calibration engineers. Figure 1 gives an overview of the resulting system.

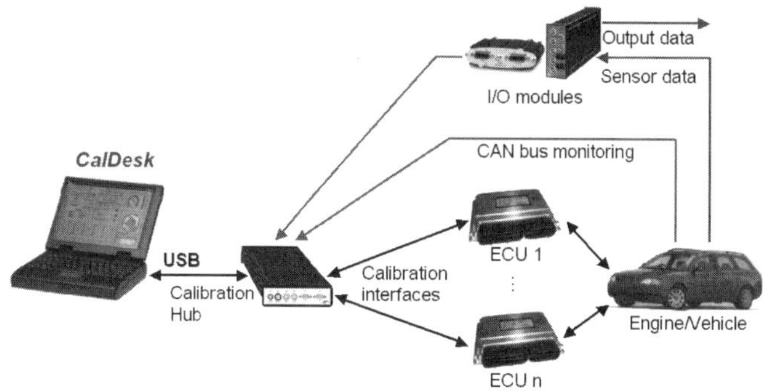

Figure 1: Overview of the dSPACE Calibration System

Multiple devices can be connected to the host PC for calibration and measurement including almost any combination of memory emulators, XCP devices, I/O modules or CAN interfaces. Each individual device can also be linked directly to the host PC without any additional box in between. This modular and scalable architecture guarantees that the system can be tailored to an individual calibration scenario without imposing extra costs on the customer.

THE EXPERIMENT SOFTWARE

Typically, the experiment software and its graphical user interface are of paramount importance to calibration engineers. The dSPACE experiment software for calibration and measurement is called *CalDesk*. In several discussions it has been pointed out that easy and intuitive operation is the key to success. The tool must, by its design, prevent operating errors and guide users through typical calibration tasks. The basic functionality is expected to be accessible within a minimum number of steps.

CalDesk is completely based on the latest Microsoft Windows standards, i.e., all toolbars, menus, shortcuts and even keystrokes are purposely designed to have the Windows look and feel in every possible way. This makes the tool very intuitive to operate since today almost everybody is familiar with this standard. For in-vehicle scenarios, CalDesk is optimized for keyboard only operation. Acoustic signals for trigger and threshold conditions will further support calibration work during mobile applications. Keyboard shortcuts can be customized as required. This can be done either manually, as with all Windows tools, or via a shortcut configuration file. This mechanism means the tool can be quickly configured according to the individual preferences of the calibration engineer. Figure 2 displays the main window of CalDesk.

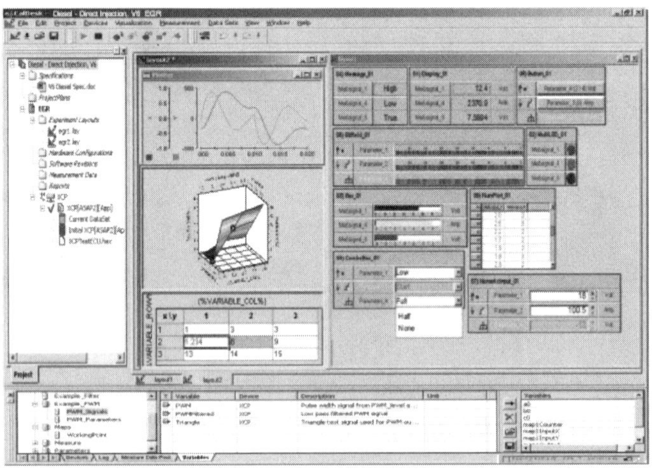

Figure 2: Main window of CalDesk

In order to further facilitate operation, wizard and template mechanisms are used to guide the user and to predefine basic settings. An example is given in Figure 3.

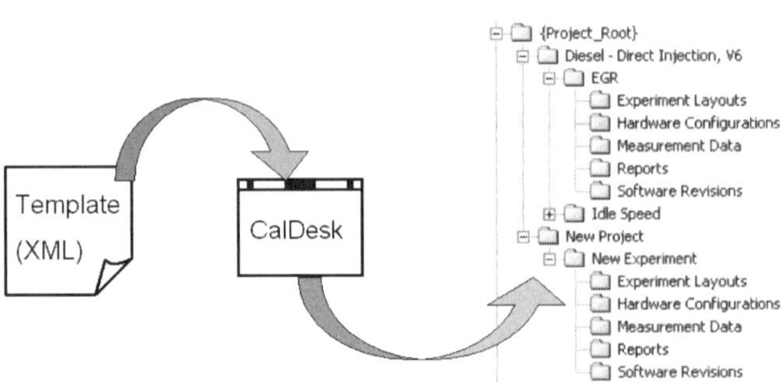

Figure 3: Template mechanism in CalDesk

The folder structure for a complete project can be automatically generated by means of an XML template file. This serves to automatically configure the default structure and can of course be adapted to individual needs. File dialogs are largely avoided and users can go through the configuration process quickly, in a minimum number of steps. Users are guided through the complete configuration process. In the example in Figure 3, they are required only to enter the labels for <New Project> and <New Experiment>; the tree structure and the other folders are set up automatically.

Via the template mechanism it is also possible to define default folders, for example, for measurement data, calibration data sets and reports. All files generated during a calibration exercise will then automatically be assigned to the appropriate folder without bothering the user with unnecessary file dialogs.

Basic Functionality

The core system handles all the tasks necessary during a typical calibration exercise including project, experiment and hardware configuration, simultaneous calibration, measurement and data recording, data analysis and data set management.

All measurement data is time-correlated, no matter whether the data comes from the memory emulator, the CAN bus or measurement modules. The number of data acquisition rates, event or time-triggered, and the number of measurement channels are not limited by the tool itself. The only constraint is the data throughput of the host interface. In order to prevent data losses and malfunctions, CalDesk features an automatic bandwidth check. Even during offline configuration the calibration engineer will receive a warning if the number of measurement channels configured exceeds a critical level.

As for data recording, many customers asked for a mechanism to record not only measurement variables, but also information about the active data set and parameter changes during the measurement interval. In addition to this, it will be possible in CalDesk to create new variables, so-called virtual variables, for measurement and post-data analysis. These variables are based on basic mathematical operations and can be configured flexibly via a formula editor. In order for the calibration engineer to analyze captured data directly in the vehicle, a synchronous cursor mechanism has been implemented. This feature allows scrolling through the captured data step by step while the cursor positions are synchronized and the data values updated in all the instruments. In addition, the plotter utility will allow reference data to be loaded for easy comparison with the current trace capture and to be moved vertically and horizontally in order to overlay different captures precisely.

CalDesk will support working with several calibration data sets. Typically there is one write-protected data set and one or more working data sets. No matter whether you are calibrating via a memory emulator or XCP on CAN, the data set handling is always the same. The write-protected data set is typically used for reference and to provide a quick fallback solution. Calibration changes can be done online and offline. This allows parameters to be calibrated in the office and activated later in the vehicle. A multilevel undo/redo functionality makes it possible to revoke parameter changes step by step. 2-D and 3-D look-up tables and maps are supported by convenient instruments featuring visual editing and operating point display. It is also planned to trace the operating point in order to display the current area of operation to the user and to zoom in fast.

When thinking of multiple-ECU scenarios, it will be more and more important in the future to calibrate several ECUs simultaneously. CalDesk will allow the grouping of any number of ECUs or calibration interfaces in order to perform precisely synchronous calibration. Apart from grouping devices, CalDesk also provides a mode to activate multiple parameter changes on one or more ECUs at a time. Related parameters of a control algorithm often need to be changed at the same time to ensure proper behavior.

Integration into Development Process

Open interfaces and the support of standards are crucial for the integration of a calibration tool into the development process at the customer. CalDesk fully supports the ASAM MCD (formerly ASAP) standards concerning measurement and calibration. The main goal of these standards is to provide uniform interfaces for data exchange and compatibility between different tools.

ASAM MCD 1 (ASAP1) defines a driver interface for the communication between the ECU and the calibration and measurement system. The CAN Calibration Protocol CCP has been widely accepted as a standard here. CalDesk supports *XCP on CAN,* which can be seen as a further development or the successor of CCP. ASAM MCD 2MC (ASAP2) provides a standardized ECU data description format. It is widely established in Europe and is also starting to gain acceptance in the USA and Japan. Finally there is the ASAM MCD 3MC (ASAP3) interface, which allows the calibration tool to be remote controlled by an automation system, for example, for test bed calibration.

Mainly driven by US companies, a uniform, tool-vendor-independent calibration data file (CDF) format has been defined within ASAM. This file format is based on XML and serves to interchange calibration data between different systems. A secondary function is to allow calibration engineers to open the file in a text or XML editor and to easily change values. CalDesk will support the CDF format with the integrated data set manager. With this utility it is possible to compare and merge any number of data sets and to generate XML-based reports. The XML approach allows the report information to be transformed into different output formats like HTML, PDF and RTF via so-called stylesheets. This mechanism enables the output format and style to be customized flexibly.

The interface to an automation or optimization tool will be another crucial point in the future since more and more calibration work will be shifted from the vehicle to test benches or earlier stages in the development process. Apart from ASAP3, CalDesk provides a COM interface that has been designed in coordination with the ASAM e.V. It is planned to introduce a new ASAM MCD 3 standard based on an object-oriented approach. This approach covers different forms of implementation, for example, (D)COM, Java-RMI, and ASAM-GDI. Via the COM interface it will be possible to get access to the main functionality in CalDesk and to automate calibration scenarios from external tools like MATLAB® or Microsoft Excel. The major benefit compared to the old standard will be a distinct improvement in measurement and data recording capabilities.

There will also be an integrated script language available in order to automate routine calibration tasks in CalDesk itself.

The dSPACE Data Dictionary

Despite the ASAM MCD standards, there are still a lot of proprietary file formats at OEMs and system suppliers. Thus, in several calibration projects the tool vendors are challenged to maintain compatibility with the proprietary environment. With the new dSPACE Data Dictionary concept illustrated in Figure 4, CalDesk will provide open and documented API interfaces, which make it easy to adapt to company-specific environments and to make use of company-wide data pools. Proprietary file formats, no matter whether binary or ASCII, can be easily integrated. Various import formats, including XML, ASAP2 and the CAN database format DBC will be supported by default.

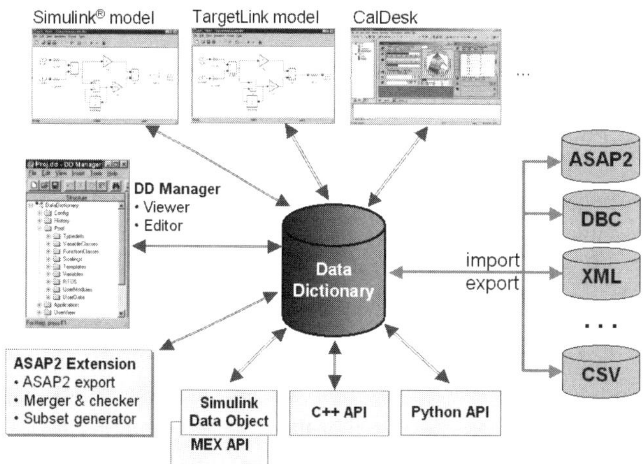

Figure 4: dSPACE Data Dictionary

The dSPACE Data Dictionary is the central data container within CalDesk and also within the dSPACE tool chain. It holds all relevant information on calibration and measurement items. It contains data description information of the ECU, the CAN bus and external I/O modules. Even conversion formulas for virtual variables can be stored in the data pool. Typical instrument settings like display colors, units, ranges, data formats and data representations can be saved with each individual variable. Based on this information, the calibration and measurement instruments in CalDesk are configured automatically. The number of steps needed to set up a complete experiment environment properly is reduced dramatically by this auto-configuration mechanism.

As for the tool chain, the data dictionary is meant to define and manage all variables during the complete development process. Its main purpose is to keep the data consistent throughout the different stages and to relate the respective representations. For example, the stored information on cross-references allows calibration data from CalDesk to be fed back to the underlying model.

The Data Dictionary Manager makes it possible to view the different information and to edit and create variables. An extra utility will also allow the export of ASAP2 files, generation of subsets and the merging and checking of multiple files. In addition, it will be possible to update address and data type information via linker map files or debug-information files.

Integration into the dSPACE Tool Chain

Although the dSPACE Calibration System has been especially tailored for test bed and in-vehicle calibration, another focus has been on integration into the dSPACE tool chain. Customers highly value a complete and seamless tool chain as an important means to reduce development costs and time. The V-cycle has become a widely accepted model for a streamlined and efficient software design process in the automotive industry [6]. The dSPACE tools have been developed according to this process.

MATLAB®/Simulink®/Stateflow® has become a quasi-standard for controller design and modeling in this area. As with all dSPACE tools, CalDesk provides a high-level interface to this environment. Calibration engineers will be able to use CalDesk as an experiment environment for offline simulations and, via the data dictionary, to reference subsystems in the underlying Simulink model. Moreover, CalDesk scripts for automated parameter tuning can be tested offline by means of Simulink models. This way calibration engineers can make sure that the automated calibration task works properly before applying the script to the real engine or vehicle.

Not only calibration interfaces and I/O modules are supported by CalDesk but also dSPACE rapid prototyping platforms. This will allow customers to perform function development (bypassing) in parallel with calibration and measurement. Direct access to any ECU internal variables provides a high degree of flexibility and convenience and thus facilitates controller design. Additionally, parameters in the ECU software which are related to the function being bypassed can be tuned as early as the controller design phase. By this means, the calibration effort can be reduced in the later development stages.

Close-knit interaction between the code generator and the calibration tool is essential for a streamlined development process. For this reason, the dSPACE production code generator, TargetLink, automatically implements service calls for the dSPACE memory emulator or for XCP in the generated code. Parameters of the calibration interface can be configured in TargetLink and output as an ASAP2/A2ML file. The data dictionary concept also allows feedback of calibration data to the TargetLink model and the exchange of specific information like memory structures of look-up tables or maps, which cannot be described by the standard ASAM MCD 2MC (ASAP2) format.

Finally the calibration tool may also aid the customer during the ECU test phase in connection with hardware-in-the-loop (HIL) simulation. Often a calibration tool is beneficial in HIL scenarios in order to double-check variables on the simulator and verify the overall set-up. Moreover, HIL simulation will be increasingly used for pre-calibrating ECU parameters. A special version of CalDesk, a kind of kernel version without a graphical user interface, will be available for these applications. The version will provide full access to ECU variables directly from within the dSPACE instrumentation software for HIL. Thus, engineers will not have to deal with two different user interfaces, so they can concentrate on their primary task.

Figure 5 describes the integration of CalDesk into the dSPACE development environment.

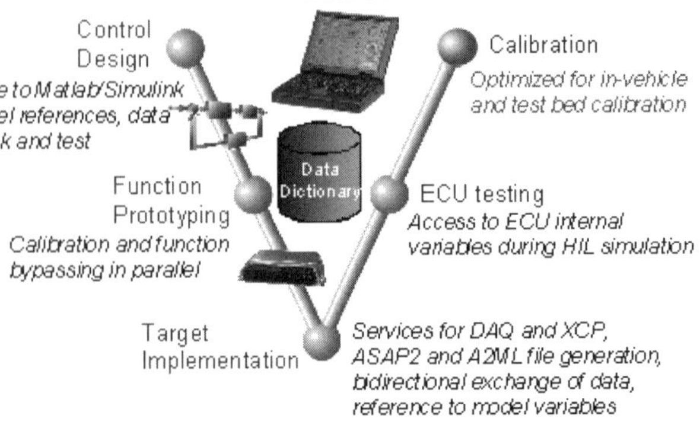

Figure 5: Integration of CalDesk into the dSPACE tool chain

CALIBRATION INTERFACES AND HARDWARE

The guideline for the hardware development was to design cost-effective solutions for calibration and measurement without compromising functionality. The resulting concept is reflected by two development goals:

1. The hardware must be modular and scalable in order for users to tailor the system exactly to their needs.
2. There must be a high degree of reusability and flexibility in order to guarantee a high return-on-invest.

It goes without saying that reliability is of paramount importance. All hardware components are designed for harsh automotive conditions. For example, all devices are designed for the temperature range –40 to 85°C.

Host Interface

Easy handling, high data throughput and security, and availability on any laptop have been rated by customers as the most important criteria for the host interface. In order to prevent clutter and confusion in the harness, another requirement has been postulated: Regardless of the calibration and measurement scenario, there will be only one connection to the laptop. USB is the ideal interface for this. It is possible to connect up to 127 devices and there is a variety of gateways on the market providing connections to other serial interfaces like RS232, CAN or Ethernet.

The USB2.0 standard has been chosen for the dSPACE Calibration System. It provides a data throughput of up to 480 Mbit/s and is completely compatible with the USB1.1 standard. Particular emphasis has been put on adapting the USB interface to the harsh automotive environment. For this, a dSPACE proprietary USB protocol has been implemented in order to ensure a high level of data security by means of checksum, error correction, acknowledge and auto-repeat mechanisms. Moreover, the complete USB harness will rest on flexible, silicone-based cables with special connectors and an integrated opto-isolation. The only commercial USB device used will be the USB connector to the host PC. Benchmarks revealed that with USB1.1 a net data rate of more than 4.9 Mbit/s can be guaranteed, which is equal to one hundred and fifty 32-bit measurement variables at 1 kHz. With USB2.0 the net data rate even exceeds 22 Mbit/s.

In a multiple-ECU scenario with 10- and more-cylinder engines or redundant ECUs, another USB feature is important: The USB round interrupt mechanism allows synchronization of multiple devices like memory emulators. Via the round interrupt, for example, it is possible to simultaneously calibrate several ECUs at exactly the same time and to synchronize the clocks on the memory emulators for time-stamping.

dSPACE Calibration Interfaces

Today there are various ECUs on the market providing different interfaces for calibration depending on the capabilities of the microcontroller and the calibration scenario itself. The dSPACE Calibration System primarily focuses on powertrain scenarios with engine and/or transmission control units. Relevant interfaces for these scenarios are given in Figure 6.

Although the prevalence of memory emulation will decrease in the future, it is today still the dominant calibration technique - at least in the USA and Europe. The only exception may be Japan, where NBD/AUD interfaces for calibration and measurement are very common. dSPACE provides a generic memory emulator (GME) supporting different microcontrollers and ECUs. No matter whether the ECU provides an 8-, 16- or 32-bit architecture, a multiplexed or non-multiplexed bus, the GME can be easily adapted to ECU-specific needs via firmware. This approach guarantees the customer a high return-on-invest, since the GME can be reused in different scenarios. Adaptation to the ECU is done via a customer-specific target adapter, which serves only for mapping the connector pin-out on the ECU to the pin-out on the GME. This concept ensures a short adaptation time to new ECU versions.

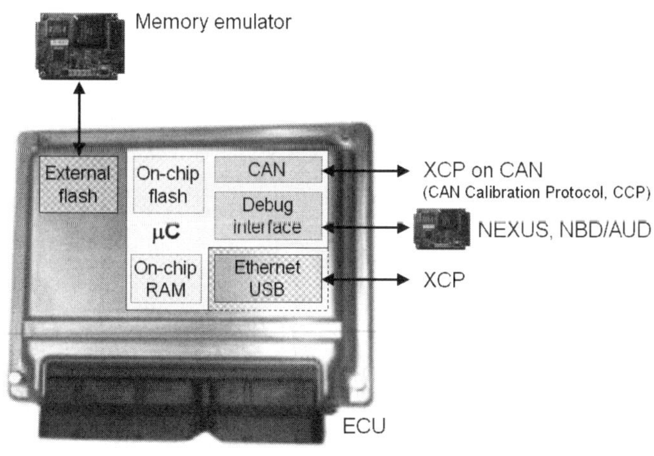

Figure 6: Calibration interfaces in powertrain scenarios

Using an ECU with a socketed Motorola MPC555 microcontroller as an example, the basic principle is reflected in Figure 7. The small dimensions will allow the GME to be integrated into nearly any ECU housing. Since it is a completely self-contained device, it can be directly connected to the host PC. For this reason, the GME interface cable provides an integrated USB opto-isolation in order to prevent ground loops when multiple USB devices are linked together.

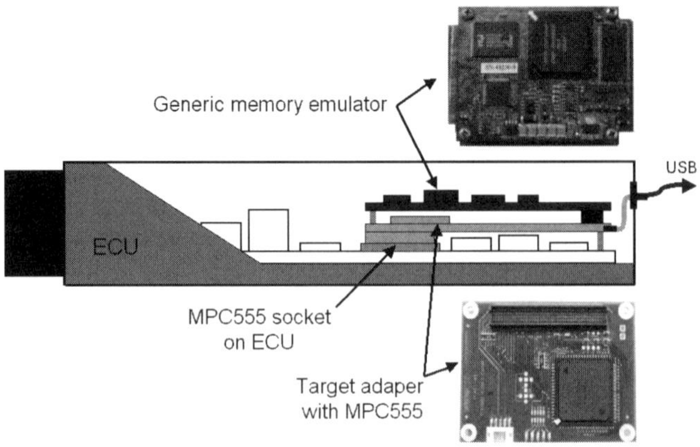

Figure 7: Generic memory emulator with target adapter

The design of the GME guarantees that no measurement data is lost if the USB connector is temporarily unplugged, or due to corrupted data transmission. Measurement data can be temporarily stored in an SRAM of several megabytes. As soon as the USB connection is re-established, all measurement data is sent to the host PC in order to provide contiguous data capture. The SRAM may also be used to download several calibration data sets in order to test and directly compare different variants. Additionally, the on-board flash serves to store calibration and measurement data permanently in the case of a power down. Figure 8 displays the basic architecture of the GME.

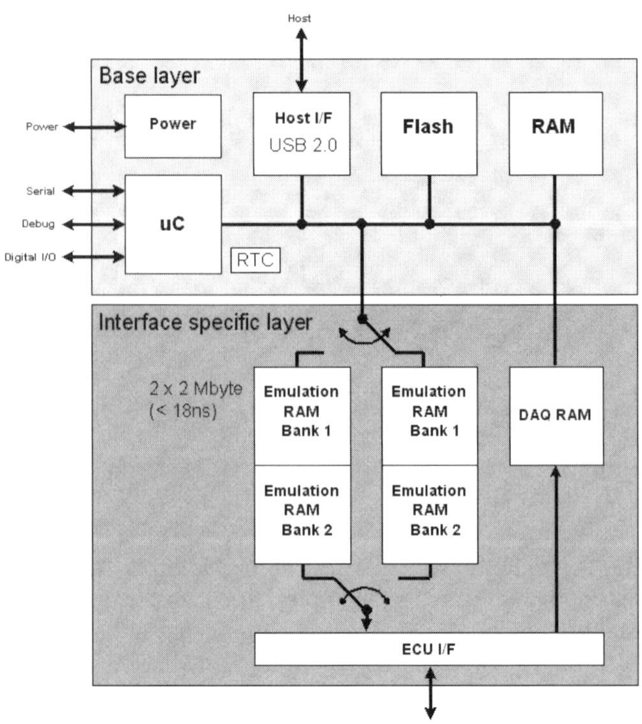

Figure 8: Basic architecture of GME

With future processor architectures there is less external or off-chip visibility of the processor bus. Thus, calibration and measurement techniques via debug ports will be increasingly interesting for OEMs and ECU suppliers. While NBD/AUD has been established on the Japanese market for some years now, the NEXUS standard recently started to penetrate the automotive industry. Motorola has launched in the form of its 32-bit MPC56x RISC family and the integrated READI interface for the first time a NEXUS-compliant solution, and there are more to come. The next generation MPC5500 family will be compatible as well, and other silicon companies have also committed to supporting NEXUS in the future.

dSPACE is responding to this situation with a NEXUS-compliant calibration interface. In order to address the Japanese market as well, the same device can be adapted to NBD/AUD by changing the on-board firmware. The GME and the dSPACE NEXUS interface are based on the same hardware concept. Only the interface-specific layer (see Figure 8) needs to be changed. Due to this generic hardware approach the NEXUS interface can also be directly attached to the host PC via USB.

High-speed interfaces like USB, Ethernet and time-triggered bus systems are already appearing on the horizon as future physical layers for serial calibration. The CAN bus meanwhile is widely established in the automotive industry and microcontrollers for powertrain applications typically feature one or more on-chip CAN interfaces. The CAN Calibration Protocol CCP has been standardized by ASAM in order to provide a standard calibration and measurement protocol via CAN.

Based on experience with CCP, the specification has been further developed under the name XCP. The aim of the XCP activities is to provide a universal measurement and calibration protocol which is suitable for different physical layers like CAN, Ethernet and USB. Compared to CCP, XCP provides major benefits. The specification itself has been improved in order to guarantee compatibility between different implementations. Furthermore, much effort has been spent on a better service configuration, flash programming capabilities as well as on improved service efficiency and data throughput. New features, like an optional ECU description upload, time-stamped and synchronized data acquisition and an optional block transfer mode for transferring data without protocol overhead have also been tackled. Last but not least, XCP has been tailored not only for calibration and measurement, but also for rapid control prototyping, i.e., bypassing. Optional commands for writing data synchronously to the ECU allows XCP to be used for closed-loop real-time applications.

dSPACE is an active member of the XCP work group in ASAM. In a pilot implementation based on CAN, dSPACE already verified the bypass capabilities of XCP. The new calibration tool will support *XCP on CAN* right from the beginning. For this reason, dSPACE provides its own XCP service to be implemented onto the ECU. This service can be flexibly configured according to the individual needs of the customer. Further implementations for *XCP on USB* and *XCP on Ethernet* are already planned for the future.

In order to connect multiple memory emulators and to provide an interface to CAN or other serial busses, a special hub module, called *Calibration Hub,* has been developed. The number of interfaces can be easily multiplied by cascading the modules. The *Calibration Hub* provides a cost-effective solution to link multiple devices in a calibration scenario. By default, the hub features two opto-isolated, high-speed CAN interfaces and two USB 2.0 output ports. One of these ports can be reconfigured via an additional plug-in-module in order to provide an interface, for example, to the Serial Measurement Bus (SMB), Ethernet, or the K-line. This concept is illustrated in Figure 9.

The CAN channels of the *Calibration Hub* can also be used for low-speed CAN via a small converter module. This module has to be attached to the hub externally. Additionally, there is a single USB-to-CAN gateway available in a separate housing. A well thought-out cable concept makes it possible to connect the different devices quickly and intuitively. Moreover, it protects the user from accidental polarity inversion.

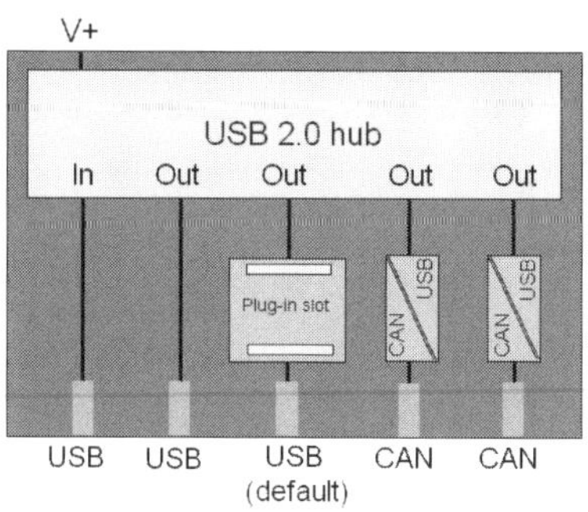

Figure 9: Schematic of Calibration Hub

The open concept of the Calibration Hub allows a variety of measurement modules to be interfaced. Customers are therefore not forced to purchase new measurement equipment when they opt for the dSPACE calibration solution. However, in order to offer a complete package including I/O modules to customers, robust and proven data acquisition and data output modules from IMC [7] and IPETRONIK [8] have been optimally integrated into the dSPACE Calibration System.

CONCLUSION

The almost exponentially increasing complexity in engine control and the steadily growing time-to-market pressure force vehicle manufactures and ECU suppliers to rethink the conventional calibration process. A complete tool chain with open interfaces and standards can dramatically save time and money for both controller strategy and calibration development. This is especially true with a consistent data and workflow between the different development phases.

Automated calibration and evaluation strategies at the test bench or by means of engine and vehicle simulation will be commonplace in the future. The dSPACE Calibration System with its open architecture has been prepared for this. The progress in processor and ECU development also requires new calibration interfaces and cost-effective hardware solutions. Apart from the conventional memory emulator, standards like XCP and NEXUS will gain momentum. dSPACE is already responding to this situation by cost-effective concepts and a generic calibration approach. The intuitive and easy operation reduces the overall training and teaching effort. In initial pilot projects with lead customers the usability and robustness of both the software and hardware has already been proven.

dSPACE is an independent tool supplier and provides a seamless tool chain based on MATLAB/Simulink/ Stateflow. The data dictionary concept guarantees a consistent data flow and an optimal integration of the calibration tool into the overall development process. The result is a streamlined calibration process and a reduction in development costs and time.

REFERENCES

1. J. Gerhardt, N. Benninger, W. Heß, Drehmoment-orientierte Funktionsstruktur der elektronischen Motorsteuerung als neue Basis für Triebsysteme, 6. Aachener Kolloquium Fahrzeug- und Motorentechnik 1997
2. G. Miller, K. Hall, W. Willis, W. Pless, The Evolution Of Powertrain Microcontrollers And Its Impact On Development Processes And Tools, White Paper, The Nexus 5001™ Forum
3. ASAM e.V., www.asam.net
4. The NEXUS 5001™ Forum, www.nexus5001.org/
5. W. Nietschke, Applikationen beschleunigen mit Rapid Calibration, Automotive Engineering Partners, 1/2002, page 16-17
6. H. Hanselmann, Development Speed-up for Electronic Control Systems, Convergence International Congress on Transportation Electronics, Dearborn, October 19-21, 1998
7. imc Messsystem GmbH, Berlin, Germany, www.imc-berlin.de/

8. IPETRONIK GmbH & Co.KG, Baden-Baden, Germany, www.ipetronik.de/

CONTACT

For more information, please contact:

Dipl.-Ing. Andre Rolfsmeier
Product Manager Calibration
dSPACE GmbH
Technologiepark25
D-33100 Paderborn
Germany

Phone: +49 5251 1638-648
Fax: +49 5251 16198-648
Email: arolfsmeier@dspace.de

2002-01-0878

Extensible and Upgradeable Vehicle Electrical, Electronic, and Software Architectures

Peter Abowd and Gary Rushton
Visteon Corp.

ABSTRACT

The rapid growth of electronic feature content within the vehicle continues to challenge the automotive industry. Customers want cutting edge consumer electronics features in a vehicle before the features are obsolete. However, automotive manufacturers continue to struggle with introducing new features into vehicles before they become obsolete to the customer. The ability for automotive manufacturers to seamlessly upgrade existing products with new and improved products continues to plague the automotive industry. Vehicles traditionally take 4 plus years to design and manufacture. Automotive manufacturers need to plan consumer electronics features early, but not actually integrate those into the vehicle until late in the design cycle, possibly on the production line. This would help facilitate providing the most recent features. Also, automotive manufacturers need the ability to upgrade existing vehicle features and add the latest and greatest consumer electronics features after the vehicle has left the show room. The challenge for automotive manufacturers and Tier 1 suppliers is to design and develop the vehicle Electrical, Electronic, and Software (EES) architecture in order to provide a more effective means of introducing new features expediently and efficiently, then support this strategy with appropriate business and financial decisions.

The automotive manufacturers should uncouple the dependencies between a vehicle's powertrain, chassis and sheet metal from the electrical, electronic and software system. These dependencies have historically interfered with an automotive OEM's ability to provide current electronic and software based features into a vehicle which is 2 to 4 years into design but not yet in production. This typically results in a *"new"* vehicle which is unable to provide currently available consumer electronics features. Standardized interfaces, better development processes, methods, and tools which allow engineers to go from features to product much faster will enable automotive manufacturers and Tier 1 suppliers to meet the customer wants. *Open Systems* appear to offer many of these advantages, but how well does this approach suit automotive development and what does it mean to the business strategy.

INTRODUCTION

Traditional vehicle EES systems lack the ability to easily upgrade software or electronics functionality. Additionally these same systems are often incapable of significant aftermarket feature additions. To establish terms and understanding for this paper, the first of these issues we will refer to as an **Upgradeability** issue, the second an **Extensibility** issue. To clarify, an upgrade is simply a new version of the same feature; extensibility provides new features that did not exist when the product was originally manufactured. Both of these situations have benefits to automotive embedded systems but may have distinct implementation differences. As demonstrated with the explosive growth of vehicle electronics and software features, cause for more elegant design solutions to enable upgrades and extensions are required. In addition, there are advantages of separating the control systems from the human machine interface of embedded automotive EES systems [2]. We will build upon this fundamental separation of concern to discuss the relevant design problems presented by the desire to upgrade and extend automobile EES features.

THE PROBLEM

Simply put, customers want new cars with new features, not new cars with perceived *"old"* electronics. The consumer electronics product turnover has altered and raised consumer expectations. Automobile manufacturers need to respond with exciting new flexible and personal features in vehicles without reductions in quality or reliability. In many cases this problem is far more difficult in the automotive and aircraft industries than the consumer electronics field. In combination with the tremendous consumer demand for up-to-date vehicle electronics, general design solutions for traditional vehicle features are involving more and more electronics and software. Figure 1 depicts a simple analysis of production vehicles to demonstrate the growth of EES features.

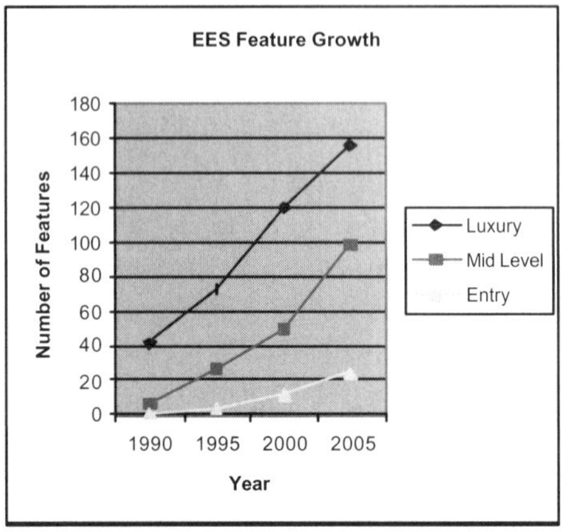

Figure 1

At the root of the problem, sits the relatively short design cycles in consumer electronics versus the long design cycles in automotive development [2]. Unless carefully planned and executed, automotive electronic designs and features are often several years old when a vehicle begins production. In some cases this is purposeful and necessary. It would be unwise to create parts of a vehicle EES system with the same quality and reliability of a cellular phone, PDA or desktop computer. Simply reflect upon the last "dropped" cellular phone call, dropped PDA or desktop "Blue Screen" you experienced. Then reflect upon the last time your car didn't start, stop, turn or accelerate. There is a significant difference that necessitates considerably more design robustness for automotive applications than consumer electronics.

In total, the automotive engineers must create EES feature designs quickly. Provide variously composable systems to the vehicle manufacturing plant. Allow for simple upgradeability and configuration at the dealership and finally provide for personalization by the consumer. Each of these groups, automotive design engineers, vehicle manufacturing, dealership sales and service, and finally the consumer all benefit differently from extensible and upgradeable systems.

CHALLENGES FOR THE SOLUTION

Recently many automotive engineers have focused more seriously upon the topic of Open Systems. This discussion is often problematic due to the ambiguity of the term *open*. For reference, included below are two useful definitions of Open Systems and the Open Systems Approach from the Open System Joint Task Force created by the US Department of Defense. In addition, the Software Engineering Institute's definition of Open System follows.

OPEN SYSTEM [1]

A system that implements sufficient open standards for interfaces, services, and supporting formats to enable properly engineered components to be utilized across a wide range of systems with minimal changes, to interoperate with other components on local and remote systems, and to interact with users in a style that facilitates portability. An open system is characterized by the following:

- *Well defined, widely used, preferably non-proprietary interfaces/protocols;*
- *Use of standards which are developed/adopted by recognized standards bodies or the commercial market place;*
- *Definition of all aspects of system interfaces to facilitate new or additional systems capabilities for a wide range of applications;*
- *Explicit provision for expansion or upgrading through the incorporation of additional or higher performance elements with minimal impact on the system.*

OPEN SYSTEMS APPROACH [1]

The open systems approach is an integrated business and technical strategy to (1) choose commercially supported specifications and standards for selected system interfaces (external, internal, functional, and physical), products, practices, and tools, and (2) build systems based on modular hardware and software design. In order to achieve an integrated technical and business strategy, an integrated product team (IPT) process is needed that involves all interested parties, e.g. engineering, logistics, finance, contracting, industry, etc. Selection of commercial specifications and standards shall be based on:

- *Those adopted by industry consensus based standards bodies or de facto standards (those successful in the market place);*
- *Market research that evaluates the short and long term availability of products;*
- *A disciplined systems engineering process that examines tradeoffs of performance;*
- *Supportability and upgrade potential within defined cost constraint; and*
- *Allowance for continued access to technological innovation supported by many customers and a broad industrial base.*

WHAT IS AN OPEN SYSTEM? [3]

An open system is a collection of interacting software, hardware, and human components

- *designed to satisfy stated needs*

- *with interface specifications of its components that are*
    - *fully defined*
    - *available to the public*
    - *maintained according to group consensus*
- *in which the implementations of the components conform to the interface specification*

All of these definitions have interesting implications to the application of "Open System" technology to the automotive industry. The automobile, has traditionally been regarded, and implemented, as a "closed" embedded system, defined for clarity as follows:

EMBEDDED SYSTEM [4]

A combination of computer hardware and software, and perhaps additional mechanical or other parts, designed to perform a dedicated function. In some cases, embedded systems are part of a larger system or product, as is the case of an anti-lock braking system in a car.

AUTOMOTIVE OPEN SYSTEMS

What is the motivation for the automotive industry to develop an "Open System"? The next section will address what may motivate the EES Architect and the automotive manufacturer towards an "Open System" design.

THE AUTOMOTIVE EES ARCHITECT

The software engineer has created the "open system" concept. The automotive design engineer is simply attempting to utilize technology that has already achieved success in other domains. The advantage of open systems development to the automotive design engineer can be summarized largely as reduced average development time per EES **feature** [2]. The engineer achieves reduction through the use of third party developed components. Since open systems are built upon public specification, a source of previously developed features could be available. Even in the case of immediately unavailable components, the nature of the open specification allows the development of some system functionality to be outsourced, providing for increased parallel development. Therefore, larger systems with more features may be built in a shorter number of days and possibly with less effort. Unfortunately, engineers like to create standards and debate their merits as a daily practice. The decision to create and implement standards that contribute to an open system remains as a more difficult business decision for the automotive OEMs to consider. As Edison learned well, the adoption of a standard is not always based upon the technical merits, but rather a party of investors' and stakeholders' commitment to its profitability.

THE AUTOMOTIVE MANUFACTURER

The automotive manufacturer would like to offer more features to the customer without significantly increasing manufacturing complexity; desirably with a reduction in complexity. Furthermore, the automotive manufacturer would like to reduce the amount of time between consumer electronics features and the appearance of similar features in the automobile. Additionally, the automobile, unlike the desktop workstation, must survive months of no-maintenance and full operation in harsh, noisy, and vibrating environments. How can open systems assist in achieving these goals?

OPEN STANDARDS

Few vehicle EES features rely solely upon software or hardware. This means, that in order to build various configurations of the same vehicle, software and hardware elements must change, as well as their physical and logical interconnections. To an open system this interchangeability should not be a problem, assuming the interfaces have been developed to an accepted standard. However, what is the *standard* for automotive electrical, electronic and software interfaces. In the case of open software systems there are many solutions that fit the definitions previously supplied (OSEK, JAVA, LINUX, OSGi ...etc), upon deeper investigation an ample number of electrical and electronic standard interfaces may be cited (TITUS). However the choice of "open" technology standards demands a decision is made regarding the scope of *openness*. If the scope is too narrow, such as a single vehicle, will an "open" design payoff? Would it be possible to distinguish between vehicle brands if all vehicles utilized the same open system? Would all the cars end up looking and feeling the same; similarly in the manner that all Windows™ applications look and feel alike?

Fundamentally, an automotive OEM has to make a decision regarding which systems can benefit from being "open" and which would not. Certainly chassis, body and Powertrain subsystems would benefit significantly from being upgradeable in the field, but is extensibility necessary or even safe? New protocols such as Time Triggered Protocol (TTP) [6] may make these systems more reliable and provide a more dependable upgrade path, but are these the types of systems consumers will want extended? Entertainment and communications systems certainly would benefit from extensibility in today's quick changing connected lifestyle, but could automotive engineers design and manufacture a system whose extension costs are significantly lower than complete replacement? Historically, the best examples of "open" systems are desktop workstations whose upgradeability and extensibility decline rapidly after 2 years in lieu of complete replacement. Properly engineered, a vehicle can have a decade of reliable service. During that time the consumer could benefit from owning an upgradeable, extensible vehicle. However, the automobile manufacturer would be obligated to maintain the "open" standard for that vehicle throughout that lifetime.

To make the open system effective, the automotive OEM would need a broad implementation plan. This would require the supply base to support the open standard as well. Ultimately, if the standards are to become profitable for suppliers and the automotive OEM, they must be adopted by multiple OEMs. In this fashion the suppliers are prevented from supporting multiple *standards* that often drive costs up rather than down. However, with a true "open" standard, new organizations should appear, particularly in the aftermarket. Historically this is where automotive OEM's have successfully controlled free enterprise by keeping many vehicle parts of proprietary design and locking out the competition. As a means of controlling brand quality and reliability, this strategy pays off well; as an automotive OEM can better control these variables for vehicle parts. Unfortunately this strategy typically doesn't lead to commodity market pricing and encourage widespread purchasing of significant extensible features by end consumers.

The most appropriate plan seems to be consistent with several slow moving industry activities. Critical performance and safety real-time systems in the vehicle could utilize an open standard shared by OEMs and suppliers for use in designing and implementing these types of systems. Due to the critical nature of these systems to the performance of the vehicle and its compliance to legal standards, these systems would benefit from upgradeability to lower the cost of potential recalls. However, extensibility may not be as value added as the quality and reliability of systems decreases with the increase in complexity. In contrast, interior systems controlling information, entertainment, comfort and convenience may have a stronger case for upgradeability AND extensibility. Particularly in the case of multimedia and telematics systems, many OEMs have placed their bets on AMI-C to develop an "open" standard that will effectively accomplish this task. Furthermore, the automobile, similar to an aircraft, could experience various interior refreshes through its lifetime adding to customer satisfaction through the addition of new features [2].

CONCLUSION

A properly implemented open system is profitable. The profitability stems from the ease and speed of keeping the system "fresh". Offering the consumers new and exciting ways to add value to their automobile, will keep them returning to the dealership for new features. Furthermore, providing the ability for consumers to add custom extensions to their own vehicles provides greater personalization opportunities. The details of the technical solution prove irrelevant if the net result is financially unacceptable. However, when a product proves to be profitable, increased merit falls upon the technical solution, sometimes in spite of its *technical* merits. Ultimately, the open systems approach requires a significant business commitment from automotive OEMs and suppliers to develop and exploit the technical standards.

REFERENCES

1. Open Systems Joint Task Force 1998 Terms and Definitions. http://www.acq.osd.mil/osjtf
2. Abowd, P., Rushton, G. and Merchant, V. (2000), "Optimizing Distributed Systems for Automotive E/E Architectures", SAE Paper 2000-01-C083.
3. Software Engineering Institute Open Systems Approach. http://www.sei.cmu.edu/opensystems
4. Barr, M., "Programming Embedded Systems", O'Reilly ©1999-2001 Netrino L.L.C.
5. Freund, U., Burst, A., "TITUS - a Graphical Design Methodology for Embedded Automotive Software", ETAS GmbH Stuttgart
6. http://www.tttech.com

CONTACT

Mr. Abowd has an undergraduate degree in electrical engineering (BSEE'88) from the University of Notre Dame and a Master of Software Engineering (MSE '94) from Carnegie Mellon University. Mr. Abowd has been employed by Visteon Corporation for 13 years. As a manager, he is currently involved in developing Visteon's Advanced Architectures and Safety Products. In this capacity he is leading projects for developing electronic/software distributed architectures and tele-immersive rapid prototyping environments. (pabowd@visteon.com)

Gary Rushton has over 17 years of commercial and military electrical/electronic systems engineering experience. He has an MS in Automotive Systems Engineering from the University of Michigan. He is currently working as an electrical/electronic systems engineer specialist with Visteon Corporation. Previously, with General Dynamics, he worked on avionics systems for the F-16 and Vetronics systems for the Abrams M1A2 tank.(grushton@visteon.com)

2002-01-0754

A Rapid Prototyping Methodology for the Decision Making Algorithms in Automotive Electronic Systems

Michele Ornato, Rosanna Bray, Massimo Carignano and Valter Quenda
Centro Ricerche Fiat

Francesco Mariniello
Elasis

ABSTRACT

The importance of the numerical simulation and testing techniques for the software specifications development of on-board automotive systems design is the paper main issue. In order to promote flexible and rapid procedures improving software specifications, new methodologies are necessary. The proposed procedure is based on the design, simulation, validation, software compiling and rapid prototyping algorithms concerning management strategies of automotive electronic systems.

The new feature of this methodology is provided by the comparison between two prototyping environment outputs: rapid prototyping tool outputs represented by strategies running in DSpace® using powerful microprocessors and CPU outputs characterized by limited calculation resources of an almost real one.

INTRODUCTION

The vehicle is a complex system consisting of many electrical, electronic and mechanical subsystems, mostly independently designed by different suppliers. The current issues show how the management strategies specifications drafting up, the debug and the final implementation on a real microprocessor are fundamental contributions to build up a unique methodology related to different automotive subsystems design.

Automotive systems design needs robust methodologies for decision making algorithms. These techniques are described in terms of validated specifications that can be simulated and tested in iterative processes. The robustness is represented by an efficient procedure which outlines the design process through automatic procedures. During the algorithms drawing up and implementation, re-engineering may occur in order to achieve improved management performances.

APPROACH METHODOLOGY

The approach methodology starts from the definition of system functional requirements and consists of three main steps integrated in a design process able to provide the specified algorithms. Their test is performed on the field directly through rapid prototyping techniques.

The starting point consists of the specifications definition in terms of boolean truth tables and mathematical relationships that represent the desired requirements.

The first step of the methodology deals with the implementation of strategy specifications in terms of finite states machines (FSM) into a high level simulation and rapid prototyping environment such as Matlab®/Simulink®/StateFlow®. The testing and debug algorithms applied to FSM constitute a formal approach for the reengineering of system management strategies. The FSM has to be considered an "event driven system" represented by a sequence recognizer evolving through inputs variations and transitions among states. With the use of simulated FSM inputs sequences, it is possible to observe the step by step evolving specification model behavior.

The second step is involved with the validation phase and concerns the definition of a method for the debugging of strategies specifications. The correctness of strategies is verified through iterative algorithms that represent an efficient and exhaustive way to release software specifications respecting the functional requirements. During this step it has brought into evidence how the specifications debug phase has to be considered a fundamental design step. Exhaustive methods are necessary in order to promote flexible and rapid procedures improving software specifications. The correspondence between state diagram outputs and the previously debugged truth table has been checked in the final step: during the simulation the "step by step" checkout occurs and the possible mismatching has to be analyzed and corrected to provide a robust re-engineering of the state machine in terms of a new strategy specifications definition.

The debug process consists of the following steps. In the initialization Matlab® file it is necessary to specify:

- states and inputs number;
- logical conditions determining the transitions among the states defined;
- states label description;
- reduced truth table in terms of old state, inputs sequences and new state.

The procedure consists of the following phases running after the initialization one.

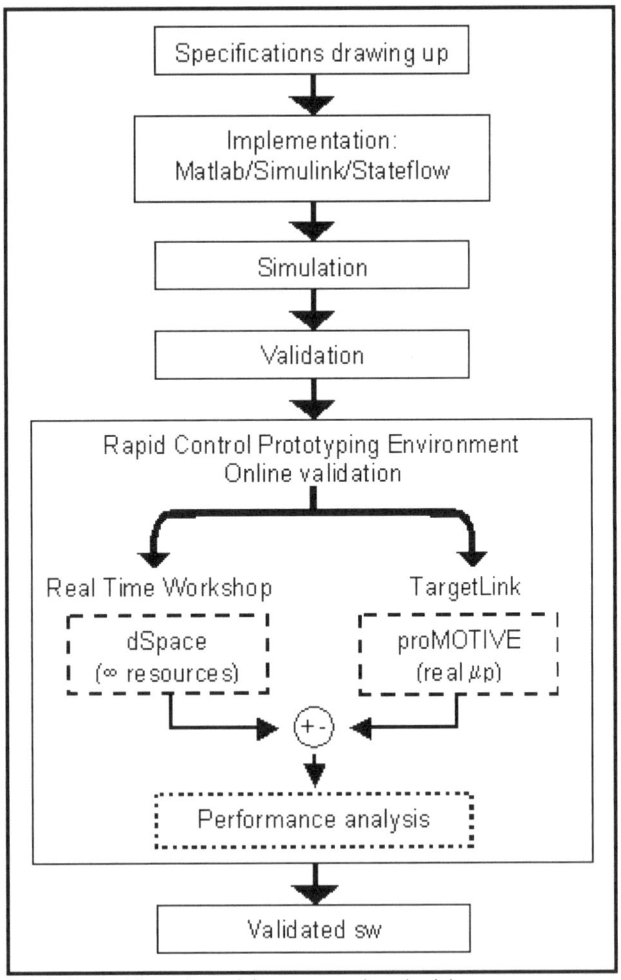

Fig. 1. Logical flow chart of the presented methodology.

The first phase is involved in the modification of the initialization file in respect of the management strategy considered in terms of the transition conditions matrix definition and the ordered states matrix.

The second phase runs the execution of the following operations: being each reduced truth table rows scanned and corrected, the possible errors are searched and debugged. At the end of this phase the algorithm provides the list of the corrections that have been executed.

The third phase consists of the new Stateflow® statechart creation and the final one is the verification between corrected truth table and automaton outputs [2] respecting the exhaustive criterion.

Having performed the FSM correction and the Simulink® model creation, the exhaustive analysis [2] could be considered finished and the specification of the strategy could be considered validated.

The last step includes online testing in a rapid prototyping layout, with the comparison between two different hardware platforms. The former is a DSpace® rapid prototyping hardware, with a floating point CPU, highly powerful calculation resources and several low level real-time monitoring tools. The latter consists of a rapid prototyping ECU developed in Centro Ricerche Fiat, running TargetLink® automatically generated fixed-point software.

SPECIFICATIONS DEFINITION AND VALIDATION

The main requirement of the validation methodology phase, previously described, is an exhaustive test.

It consists of generating and inputting to the FSM all the possible input combinations in relation to each state, verifying that all the possible paths among the states are performed [2].

Two different types of algorithms can be applied:

- random input algorithm
- fixed input algorithm

The main feature of the first algorithm is that it causes all possible "walks" among the states in a "free way" giving to FSM random inputs sequences. The uniform random probability distribution and the no equi-probability of active states could cause no simulation convergence and no respect of exhaustive criterion. The algorithm guarantees the convergence and it respects the exhaustive criterion creating and inputting to FSM all possible input combinations related to each state, verifying that all paths between the states are performed. In order to provide FSM overall paths exhaustive coverage, the iterative procedure continues till the complete domain of input sequence combinations is explored.

In the fixed input algorithm, the inputs sequence is prefixed and the consequent transitions are based on optimal walk, i.e. no path repetition occurs in correspondence of the same inputs configuration. The most important feature of this solution is concerned with optimal input sequences given to each FSM state: boolean increasing input combinations values should not repeat more than once. Sequences already inputted are stored for each state. When the state is reactivated, the procedure gives as input sequence the next stored value and so on till occurs a transition condition providing the exit from the current state.

Both algorithms have to provide the same results but while the fixed input algorithm has proved to be efficient in case of complex FSM (inputs/states high order), the random one is characterised by the "one shot" execution in terms of overall "walks" among the FSM states.

An example of the truth table referred to automatic transmission management and the description of application of methodology is in Appendix.

The described procedures guarantee the verification between corrected truth table and automaton outputs and represent a robust method to debug and validate the specifications of a strategy.

METHODOLOGY IMPLEMENTATION

After the FSM has been validated and exhaustively tested through a formally corrected procedure, the implementation phase can take place. Such a step includes online testing in a rapid prototyping layout, with the comparison between two different hardware platforms.

The former is DSpace® MicroAutoBox®, a high-performance processor on which the FSM can be calculated in real time. It is connected to the outer world with numerous I/O units interfaces already included, and designed for typical automotive applications. Among the I/O units the CAN bus has been chosen for this application. The box even contains signal conditioning for automotive signal levels.

The latter consists of a rapid prototyping ECU developed in Centro Ricerche Fiat, running TargetLink® automatically generated fixed-point software.

DSPACE® ENVIRONMENT IMPLEMENTATION

MicroAutoBox is a real-time hardware concept to perform rapid function prototyping. Thanks to its small size MicroAutoBox has been designed as an in-vehicle device that operates with or without user intervention. It can be installed at the same locations as the production ECU. But in addition MicroAutoBox provides all the benefits of a DSpace® real-time system, including the link with modeling and design complete software.

It consists of a rapid prototyping hardware, with a floating point CPU characterized by highly powerful calculation resources and several low level real-time monitoring tools. So with the Dspace® MicroAutoBox® becomes possible to use a powerful processor and an ample memory that permits long-term data acquisitions. An integrated data recorder serves for storing application program and vehicle mission acquisitions in a non-volatile memory. A PC or notebook can be connected temporarily for program download and data analysis.

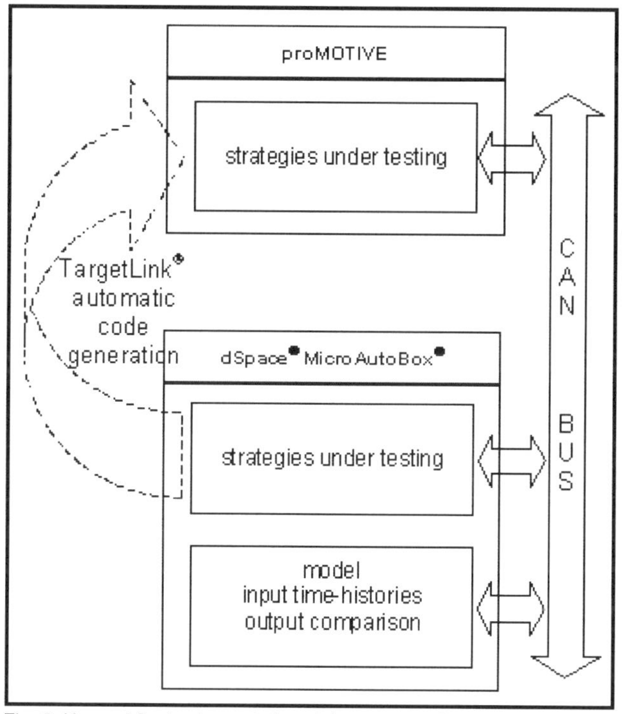

Fig. 2. Hw and Sw implementation of the presented methodology.

PROMOTIVE® ENVIRONMENT IMPLEMENTATION

PROMOTIVE® is a system for ECU programming using a simplified high-level language. The PROMOTIVE® program is written on PC, using a specific programming environment. In the same environment the user can compile his program, generating a binary file containing a meta-code. This code is then transferred to the ECU via a serial link. On the ECU a specific task performs the interpretation of the meta-code, executing the user program. This task implements a kind of virtual machine having as instruction set the meta-code generated by the PC compiler.

The PROMOTIVE® language is very simplified, and looks like a sort of BASIC. This allows the user to write simple programs, accessing hardware resources of the ECU in a high-level way, without having to worry about the details of the low-level embedded programming. The devices to which PROMOTIVE® gives an easy access are digital, analog and frequency I/Os, the CAN buses and the serial port of the ECU.

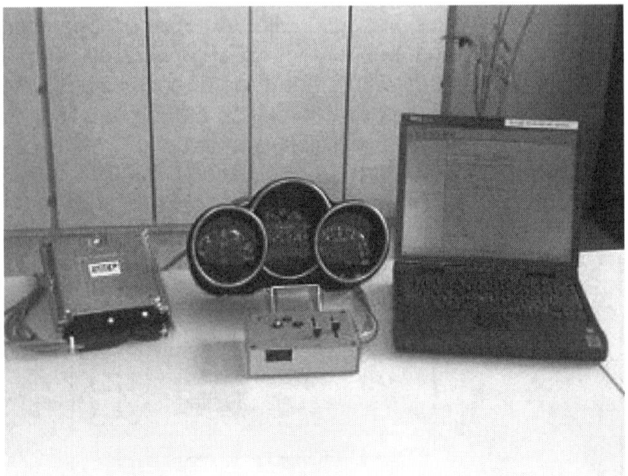

Fig 3. PROMOTIVE® hardware

The advantage of this kind of approach is that a skilled embedded programmer is not required anymore to write simple programs. In addition, being the PROMOTIVE® virtual machine extensively debugged and quite stable, the possibility of errors when accessing the low-level devices is greatly reduced; this reduces the development and test time.

On the other side, the extreme simplification of the language leads to less expressive power, which means less capabilities available to the programmer. In other words, it is not possible to write complex programs, such as programs implementing controls behavior or finite state machines.

The solution to this limitations has been found in the possibility of writing C libraries which can be accessed from PROMOTIVE® programs. These are pre-compiled libraries, whose functions can be called from PROMOTIVE® programs. Libraries are written in C and then compiled with a standard embedded compiler, using specific compile settings.

The advantage of this solution is that, having the function representing the behaviour of a particular system, it is possible to make it run on a real ECU in short time, simply compiling the function as a PROMOTIVE® library and calling it from a program.

In our case study we generated, using TargetLink from DSpace®, the C code representing the behaviour of a finite state machine. This code has been compiled as a PROMOTIVE® library and then a short program has been written to call the function providing all the inputs and reading the outputs. As a result, we obtained an ECU running the finite state machine code in a short time, ready to test without having to worry about the hardware and low-level details.

EXPERIMENT LAYOUT

The experiment is mainly managed by the DSpace® MicroAutoBox®, that plays two roles: the simulation master and the rapid prototyping hardware platform (see Fig. 2).

As a simulation master, it generates the CAN signals inputted to the strategies being tested, collects the CAN signals outputted by the strategies themselves and performs the output comparison. As a rapid prototyping hardware platform, receives the input signals via CAN line, runs the strategies and outputs results via the same CAN line.

The output provided by DSpace® and PROMOTIVE® are compared in DSpace® using ControlDesk graphical environment.

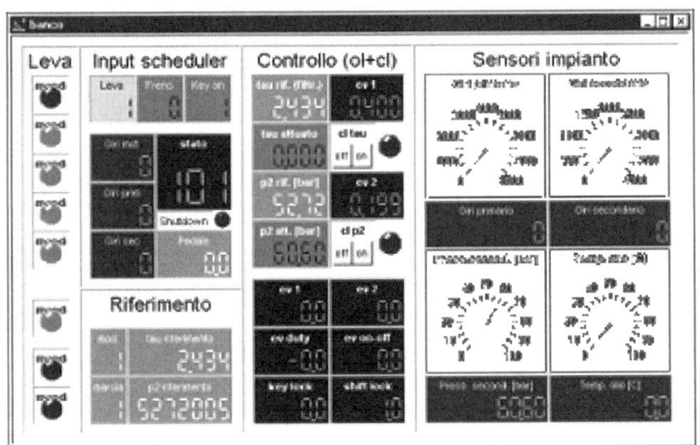

Fig 4. Output comparison in a DSpace® environment

A CASE STUDY: AUTOMATIC TRASMISSION MANAGEMENT

A finite state machine managing the task scheduler of an automatic transmission has been chosen as a case study for the methodology here presented. The related decision making algorithm is implemented by the automaton displayed in Fig 5. During the simulation all transitions are executed by several different input conditions that have been reported in Appendix.

The output of this algorithm is represented by the state coding value, according to the following table:

Init	:	State coding = 1
KeyOn	:	State coding = 10
Parking	:	State coding = 21
Reverse	:	State coding = 22
Neutral	:	State coding = 23
Auto	:	State coding = 40
Manual	:	State coding = 30
PowerLatch	:	State coding = 50
PowerOff	:	State coding = 60.

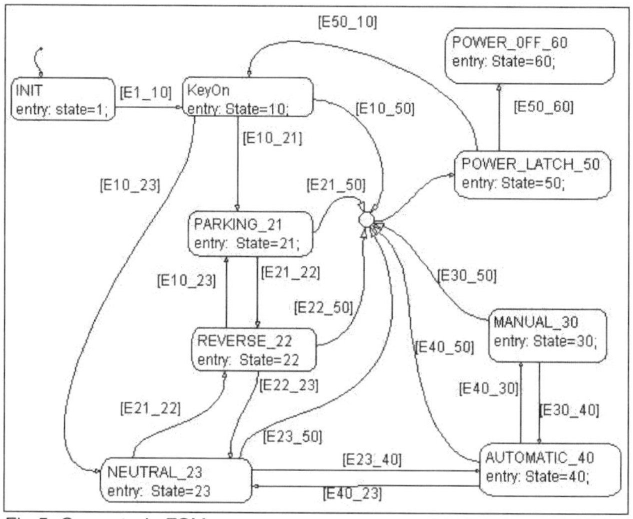

Fig 5. Case study FSM

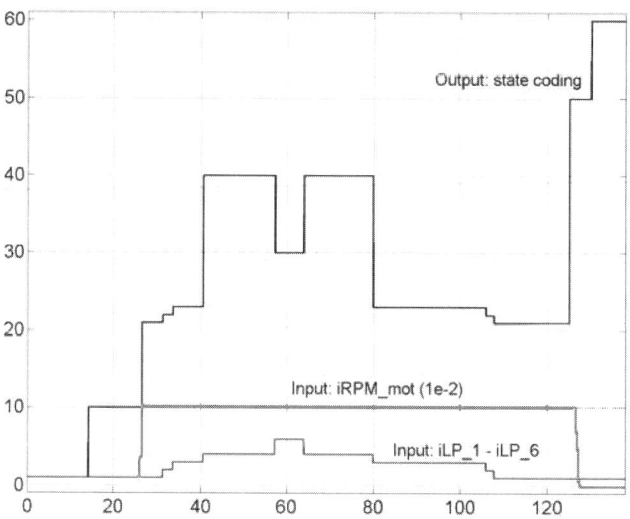

Fig 6. Sample output plot .

The transitions shown in Fig 5 are listed below:

$$E_{1\_10} = (S_1, S_{10}) : S_1 \rightarrow S_{10} \text{; Init} \rightarrow \text{KeyOn}$$
$$E_{10\_21} = (S_{10}, S_{21}) : S_{10} \rightarrow S_{21} \text{; KeyOn} \rightarrow \text{Parking}$$
$$E_{10\_23} = (S_{10}, S_{23}) : S_{10} \rightarrow S_{23} \text{; KeyOn} \rightarrow \text{Neutral}$$
$$E_{21\_22} = (S_{21}, S_{22}) : S_{21} \rightarrow S_{22} \text{; Parking} \rightarrow \text{Reverse}$$
$$E_{22\_21} = (S_{22}, S_{21}) : S_{22} \rightarrow S_{21} \text{; Reverse} \rightarrow \text{Parking}$$
$$E_{22\_23} = (S_{22}, S_{23}) : S_{22} \rightarrow S_{23} \text{; Reverse} \rightarrow \text{Neutral}$$
$$E_{23\_22} = (S_{23}, S_{22}) : S_{23} \rightarrow S_{22} \text{; Neutral} \rightarrow \text{Reverse}$$
$$E_{40\_30} = (S_{40}, S_{30}) : S_{40} \rightarrow S_{30} \text{; Automatic} \rightarrow \text{Manual}$$
$$E_{30\_40} = (S_{30}, S_{40}) : S_{30} \rightarrow S_{40} \text{; Manual} \rightarrow \text{Automatic}$$
$$E_{23\_40} = (S_{23}, S_{40}) : S_{23} \rightarrow S_{40} \text{;Neutral} \rightarrow \text{Automatic}$$
$$E_{40\_23} = (S_{40}, S_{23}) : S_{40} \rightarrow S_{23} \text{;Automatic} \rightarrow \text{Neutral}$$
$$E_{10\_50} = (S_{10}, S_{50}) : S_{10} \rightarrow S_{50} \text{;KeyOn} \rightarrow \text{PowerLatch}$$
$$E_{21\_50} = (S_{21}, S_{50}) : S_{21} \rightarrow S_{50} \text{; Parking} \rightarrow \text{PowerLatch}$$
$$E_{22\_50} = (S_{22}, S_{50}) : S_{22} \rightarrow S_{50} \text{; Reverse} \rightarrow \text{PowerLatch}$$
$$E_{23\_50} = (S_{23}, S_{50}) : S_{23} \rightarrow S_{50} \text{; Neutral} \rightarrow \text{PowerLatch}$$
$$E_{30\_50} = (S_{30}, S_{50}) : S_{30} \rightarrow S_{50} \text{; Manual} \rightarrow \text{PowerLatch}$$
$$E_{40\_50} = (S_{40}, S_{50}) : S_{40} \rightarrow S_{50} \text{; Automatic} \rightarrow \text{PowerLatch}$$
$$E_{50\_10} = (S_{50}, S_{10}) : S_{50} \rightarrow S_{10} \text{; PowerLatch} \rightarrow \text{KeyOn}$$
$$E_{50\_60} = (S_{50}, S_{60}) : S_{50} \rightarrow S_{60} \text{; PowerLatch} \rightarrow \text{PowerOff}$$

The machine implementation result is displayed in Fig 6, where the output is the FSM state coding as previously described, while the only reported inputs are engine speed and a combination of gear shift lever positions.

It is remarkable that such an example FSM has a quite simple structure, so the compiled code execution is not expected to give timing problems or suffer limitations due to the finite resources of a real microprocessor. Actually, neither it needs huge calculations nor it is affected by floating point/fixed point limitations. So no discrepancies can be appreciated in the comparison of the outputted state value between the execution of this FSM on PROMOTIVE® and DSpace®.

The performance analysis is managed by comparing the actual synchronism between the output CAN signals generated by the two ECUs. The examination of the different delay times on transition events has shown that the DSpace® hardware takes 2 μs to run a calculation cycle, while PROMOTIVE® needs 2.44 ms. These values are deeply different and show the calculation power provided by the two ECUs, but what's important is that both values are quite lower than the maximum expected value, set at 100 ms for this FSM.

Summarizing, the methodology provides the comparison between two turnaround time values: $\Delta t_{DSpace®}$ and $\Delta t_{PROMOTIVE®}$. The users sets the reference turnaround time Δt_{ref} according to the system specification requirements, and the acceptance test is represented by the following inequality

$$\Delta t_{DSpace} \leq \Delta t_{proMOTIVE} \leq \Delta t_{ref}$$

In this expression, it is worth spotting that the first part is always verified and is reported only for completeness sake.

CONCLUSIONS

It has been shown how the numerical simulation, testing techniques and rapid prototyping for the decision making algorithms bring a large contribution in the development of strategies that has to be implemented in real vehicle systems with limited calculation resources.

This method paves the way for more extensive studies involved with the development of tools for the embedded programming using rapid prototyping techniques applied to automotive electronic subsystems.

1. It has been underlined the importance of the numerical simulation for the specification development of on-board automotive systems design to perform the designer targets fixed during the drawing up functional specifications phase.

2. In order to promote flexible and rapid procedures improving software specifications a new design methodology has been developed that allows the strategy designer to implement and test algorithms derived from the specifications.

3. The proposed procedure is based on the drawing up, design, simulation, validation, software compiling and rapid prototyping of algorithms concerning management strategies of automotive electronic systems: considering specifications in terms of FSM, the efficient methodology here proposed provides time reduction in the validation and software testing.

4. A validation method using numerical simulation environment like Stateflow® and FSM formalism has been described: the correction procedure proposed provides an iterative debugging algorithm.

5. The proposed method checks, corrects and re-design the strategies expressing the management strategies in terms of simulating models to be implemented in the vehicle electronic control units.

6. The new feature of this methodology is provided by the comparison between two rapid prototyping environments outputs, in such a way that the debugging tools of a more powerful DSpace® CPU works together with the limited resources of an almost real processor.

7. The procedure employed and the approach taken in the development of the algorithm has resulted in a simple and reliable methodology enabling a complete iterative method to re-engineer algorithms for the vehicle management.

ACKNOWLEDGMENTS

The authors wish to gratefully acknowledge the contribution provided by their colleagues: R. Aimasso, D. Albero, A. Borrione, P. Borodani, C. D'Ambrosio, A. Gallione , D. Idone, M. Lupo, V. Murdocco, G.L. Morra, R. Vay.

REFERENCES

1. D. Harel. A Visual Formalism for Complex Systems. *Science of Computer Programming,* 1987.
2. R. Bray, M. Carignano, C. Giordano and F. Pastorelli. A New Approach for the Validation of Engine Management Strategies Specifications through Simulation Algorithms. *SAE Paper 2001-01-1223*, 2001.
3. C. Giordano, R. Bray, M. Carignano and F. Pastorelli. The Reengineering of specifications expressed in terms of Finite State Machines through simulations and testing techniques. *Proceeding of the 1st* IMEchE Automobile Division Southern Centre Conference on *Total Vehicle Technology*, Brighton UK, 2001.
4. P. Tabuada, and G. Pappas. Hybrid Abstraction that preserve Timed Languages. *Proceedings 4th International Workshop Hybrid Systems,* Lecture Notes in Computer Science , Springer-Verlag, Rome, Italy (2001).
5. R. Alur, R. Grosu, I. Lee, and O. Sokolsky. Compositional Refinement for Hierarchical Hybrid Systems. *Proceedings 4th International Workshop Hybrid Systems,* Lecture Notes in Computer Science, Springer-Verlag, Rome, Italy (2001).
6. M.A.Lynch. Microprogrammed State Machine Design. *CRC press*, 1993.
7. Erkkinen, T. A Software Engineering Framework for Electronic Engine Controllers. *SAE Paper, 2000-01-0297,* 2000.
8. MathWorks, Stateflow® User's Guide Version 2, 1999.

CONTACT

The contact people for any type of questions are:

Michele Ornato
E-mail: m.ornato@crf.it

Rosanna Bray
E-mail: r.bray@crf.it
Massimo Carignano
E-mail: m.carignano@crf.it
Valter Quenda
E-mail: v.quenda@crf.it
Francesco Mariniello
E-mail: francesco.mariniello@elasis.fiat.it

APPENDIX

The state machine formalism includes the following sets, defined later:

Plant FSM variables set $\underline{V} = \{V_0, V_1, .., V_K\}$

Primary inputs thresholds set $\underline{T} = \{T_0, T_1, ..., T_W\}$

FSM primary inputs $\underline{I} = \{I_0, I_1, ..., I_N\}$

Primary inputs sequence set $\underline{PI} = \{PI_{11}, PI_{12}, ...PI_{NN}\}$

FSM states $\underline{S} = \{S_0, S_1, ..., S_N\}$

If a particular primary inputs sequence exists

$$PI_{ij} = PI_{ij}(I_1, I_2, ...I_N)$$

derived from particulars input values that generates a logical condition causing the transition between the states S_j and S_i : $S_j \rightarrow S_i \Rightarrow Pi_{ij} \rightarrow L(E_{ij})$, $L(E_{ij})$ represents the transitions set

$$\forall(i,j) \in \underline{S} \rightarrow \underline{L}(E_{ij}) \; ; \; \underline{L}(E_{ij}) = \{E_{11}, E_{12}, ..., E_{NN}\}$$

The plant FSM variables set is

$$\underline{V} = \left\{ \sum_{j=1}^{6} iLP\_(j), iBrake, iRpm\_Mot, iRpm\_Sec, iKey, \right\}$$

In the following table can be focused the the FSM input list. Note that they are all digital values, except for the speed variables, which are analogue values.

iLP_1	:	Gear Shift Lever Position 1
iLP_2	:	Gear Shift Lever Position 2
iLP_3	:	Gear Shift Lever Position 3
iLP_4	:	Gear Shift Lever Position 4
iLP_5	:	Gear Shift Lever Position 5
iLP_6	:	Gear Shift Lever Position 6
iBrake	:	Brake Pedal
iRpm_Mot	:	Engine Speed
iRpm_Sec	:	Secondary Shaft Speed
iKey	:	Key Signal
iS_old_1	:	State Old 1
iS_old_21	:	State Old 21
iS_old_22	:	State Old 22
iS_old_23	:	State Old 23
iS_old_30	:	State Old 30
iS_old_40	:	State Old 40
iS_old_50	:	State Old 50
iS_old_60	:	State Old 60

The threshold set is as follows:

$$\underline{T} = \{T_0, T_1, T_2, T_3, T_4\} = \left\{ \begin{matrix} CThreshold1, CThreshold2, PThreshold, \\ RThreshold, PLTime \end{matrix} \right\}$$

The logical conditions are then translated in the binary values of the primary inputs represented by:

$$\underline{I} = \left\{ \begin{matrix} I_0, I_1, I_2, I_3, I_4, I_5, I_6, I_7, I_8, I_9 I_{10}, \\ I_{11}, I_{12}, I_{13}, I_{14} I_{15}, I_{16}, I_{17}, I_{18} \end{matrix} \right\} =$$

$$\left\{ \begin{matrix} iLP\_1, iLP\_2, iLP\_3, iLP\_4, iLP\_5, iLP\_6, \\ iBrake, iRpm\_Mot > CThresold1, iRpm\_Sec > CThresold2, \\ iKey, iS\_old\_1, iS\_old\_21, iS\_old\_22, iS\_old\_23, \\ iS\_old\_30, iS\_old\_40, iS\_old\_50, iS\_old\_60 \end{matrix} \right\}$$

If these logical conditions are true, they could be translated in a binary value equal to one, if false equal to zero, otherwise they will be represented by the don't care logical value. In this way it becomes possible to draft the reduced truth table to be debugged. The binary sequences of these inputs represented by the truth table rows determines the primary inputs sequences set:

$$\underline{PI} = \left\{ \begin{matrix} PI_{1\_10}, PI_{10\_21}, PI_{10\_23}, PI_{21\_22}, PI_{22\_21}, \\ PI_{22\_23}, PI_{23\_22}, PI_{40\_30}, PI_{30\_40}, PI_{23\_40}, \\ PI_{40\_23}, PI_{50\_10}, PI_{50\_60}, PI_{10\_50}, PI_{21\_50}, \\ PI_{22\_50}, PI_{23\_50}, PI_{30\_50}, PI_{40\_50} \end{matrix} \right\}$$

and the states transition sets are expressed by:

$$L(E_{ij}) = \left\{ \begin{matrix} E_{1\_10}, E_{10\_21}, E_{10\_23}, E_{21\_22}, E_{22\_21}, \\ E_{22\_23}, E_{23\_22}, E_{40\_30}, E_{30\_40}, E_{23\_40}, \\ E_{40\_23}, E_{50\_10}, E_{50\_60}, E_{10\_50}, E_{21\_50}, \\ E_{22\_50}, E_{23\_50}, E_{30\_50}, E_{40\_50} \end{matrix} \right\}$$

According to the listed steps, the state machine truth table is listed in the following. The first column is the decimal value of starting state (**old state**). The next 18 columns are state machine inputs. The last column is the decimal value of arrival state (**new state**). Every row describes input combination enabling a transition between two states. The symbol **x** is the "don't care" value.

```
[1 x x x x x x x 0 x 1 x x x x x x x x 10;
10 1 x x x x x x 1 x 1 x x x x x x x x 21;
10 x x 1 x x x x 1 x 1 x x x x x x x x 23;
21 x 1 x x x x 1 1 0 1 x 1 x x x x x x 22;
22 1 x x x x x x 1 0 1 x x 1 x x x x x 21;
22 x x 1 x x x x 1 x 1 x x 1 x x x x x 23;
23 x 1 x x x x 1 1 0 1 x x x 1 x x x x 22;
40 x x x x x 1 x 1 x 1 x x x x x 1 x x 30;
30 x x x 1 x x x 1 x 1 x x x x 1 x x x 40;
23 x x x 1 x x x 1 x 1 x x x 1 x x x x 40;
40 x x 1 x x x x 1 x 1 x x x x x 1 x x 23;
50 x x x x x x x x x 1 x x x x x x 1 x 10;
50 x x x x x x x x x 0 x x x x x x 1 x 60;
10 x x x x x x x x x 0 x x x x x x x x 50;
21 x x x x x x x x x 0 x x x x x x x x 50;
22 x x x x x x x x x 0 x x x x x x x x 50;
23 x x x x x x x x x 0 x x x x x x x x 50;
30 x x x x x x x x x 0 x x x x x x x x 50;
40 x x x x x x x x x 0 x x x x x x x x 50]
```

SOFTWARE IN EMBEDDED CONTROL SYSTEMS

2005-01-1430

Entire Embedded Control System Simulation Using a Mixed-Signal Mixed-Technology Simulator

Ken G. Ruan
Synopsys Inc.

ABSTRACT

An embedded control system is commonly used in the automotive industry to achieve complex and accurate control functionality. An embedded control system consists of three portions including a control object, i.e. the peripheral under control, a micro-controller and control software that is executed on the micro-controller. This paper presents an approach that meets well the challenge in entire embedded control system simulation. Two examples are presented to illustrate how an embedded control system can be simulated as an entity to explore the interaction among the three elements, including the customer code, the micro-controller and the control object of the system. The entire embedded control systems are implemented in Saber, a mixed-signal, mixed-technology simulator.

INTRODUCTION

An embedded control system is commonly used in the automotive industry to achieve complex and accurate control functionality. An embedded control system consists of three portions including a control object, i.e. the peripheral under control, a micro-controller and control software that is executed on the micro-controller.

In the past years, instruction set description languages [1-2] have been developed for micro-controller modeling. Advances in instruction set simulation technology have made it possible to run control software on a virtual micro-controller. However, the focus of instruction set simulation (ISS) is mainly on the CPU (central processing unit) of a micro-controller.

The connection between a micro-controller and peripheral is accomplished using I/O ports. ISS does not cover I/O ports of micro-controller. There is still a significant coupling gap between control software and the control object. I/O port modeling becomes an important issue in embedded system simulation.

A micro-controller normally consists of a set of I/O ports, including analog-to-digital converter, compare/capture, synchronous/asynchronous serial ports, parallel data I/O ports, etc. [3-4]. Although the basic functionality is similar, I/O ports vary from micro-controller to micro-controller. Differences are often found in the special function registers and some aspects of operations.

In order to address the modeling issues in I/O ports of micro-controller, new model architecture was proposed [5]. This model architecture makes it more efficient for I/O port modeling and the control software developed by the designer for a particular control system can be integrated with I/O ports automatically.

In this paper, two examples are described to illustrate the approach for entire embedded control system simulation using the Saber simulator capable of simulating mixed-signal, mixed-technology systems.

The next two sections describe the proposed model architecture for I/O ports and the simulation procedure. Two embedded control systems are represented as examples. One is a controller for a DC brushless motor that is driven by 3-phase sinusoidal signals. The second is also a DC brushless motor controller, however, the motor is driven by 3-phase, 6-step inverter signals.

The summary is presented in the CONCLUSIONS section. Portions of customer code and configuration file for the first example are provided in appendices for information.

THE MODEL ARCHITECTURE FOR I/O PORTS

The model architecture proposed in [5] is briefly described here for explanatory purposes.

An I/O port model consists of four portions including a MAST port template, a port partner, customer software and an interface module. Figure 1 illustrates the model architecture of an I/O port, where the blocks are numbered and are described below.

Block 1 is the template portion of an I/O port model described in MAST hardware description language. Connection pins are provided (depicted by the arrow on the left side of block 1) for connections to other templates in schematic design.

Block 2 is the partner portion of the I/O port that is described in C++, called port class. The connection between template and port class is accomplished using the mixed-mode interface [7].

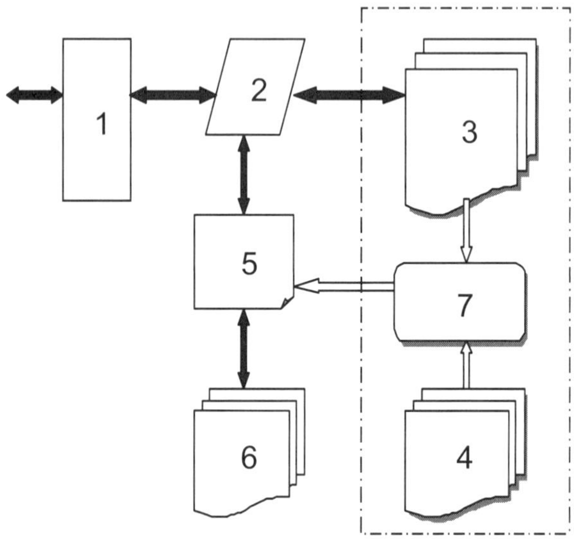

Notes on blocks:

1. The MAST template of a port;
2. The partner portion of a port model, a class in c++;
3. Adapters for the port;
4. Design configurations. One per design.
5. Interface module for links to customer software;
6. Customer software designed for driving the port;
7. The code generation module;

Figure 1. Model architecture of an I/O port

Block 3 represents the adapter of an I/O port. Any micro-controller specific functionality is implemented in the adapter. Adapter varies with micro-controller.

Block 4 is the configuration file that is specific to a particular design. This will be further discussed in the examples.

Block 5 is the interface module between the port class and customer functions represented by block 6. Interface module is automatically generated (block 7) based on the configuration specifications (block 4). The interface module and the customer software are compiled to create a shared object library. Functions in this library are used to support the port class and are dynamically loaded during simulation. During simulation, only blocks 1, 2, 3, 5 and 6 are active.

SIMULATION PROCEDURE

The following is a typical procedure for simulation of an embedded control system using the proposed approach.

1. Select the I/O ports needed by the system and place them on the schematic diagram. Multiple instances of the same I/O port may be used;
2. Design the schematic diagram to represent the hardware portion (excluding the micro-controller) of the system. Make necessary connections between the I/O port symbols and the rest of the blocks to complete the hardware design;
3. Develop customer code for those I/O ports and the control functions;
4. Create a configuration file for this design to specify the relations between I/O port instances and customer functions;
5. Run simulation on the entire control system.

A set of commands is provided for execution of simulation [8].

In case the system does not behave as expected, the designer may modify the schematic diagram and/or the customer code to fix the problem.

The debugger provided by the host computer may be used for debugging customer code while the simulation on the control system is running. Any code change may impact the behavior of the system and can be easily observed by inspecting the simulation signal waveforms.

Two embedded control system simulation examples are discussed below.

A CONTROLLER FOR BRUSHLESS DC MOTOR DRIVEN BY SINUSOIDAL SIGNALS

In this example, three portions, the control object, the micro-controller and the customer software, of this DC motor speed control system are individually described in following sub-sections.

CONTROL OBJECT

The control object, i.e. the peripheral, is represented in schematic diagram. The schematic diagram of the sinusoidal brushless dc motor controller is shown in Figure 2.

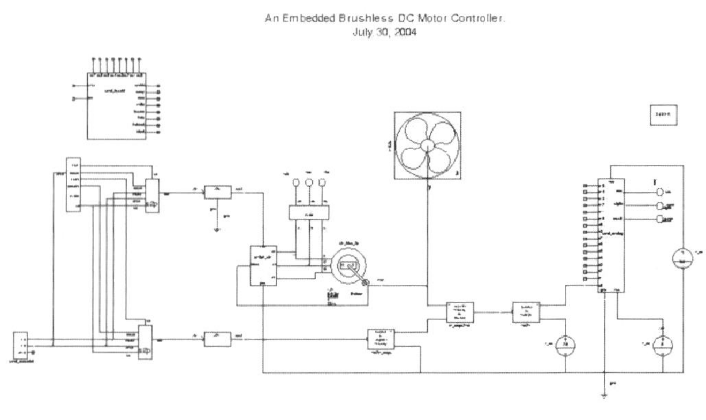

Figure 2. A sinusoidal DC motor controller

A synchronous/asynchronous serial I/O port **smci_ascserial** of a micro-controller is used in the design. The outputs of the serial port feed two instances of **s2p** template. The **s2p** template converts the serial digital inputs into parallel digital signal and then converts it again into an analog state. Outputs of the upper s2p instance are used as power supply voltage of **drv3ph_sin** template, a three-phase driver. Outputs of the lower s2p instance are used as the command speed to control the rotational speed of the DC motor **sin_bldc_3p** that is driven by the **drv3ph_sin** driver. The outputs of **drv3ph_sin** are 3-phase sinusoidal signals va, vb and vc. The DC motor is connected to an electric fan as its load.

The outputs of the serial port of the micro-controller are in byte, i.e. 8 bits. It needs two bytes to form a 16-bite binary. The **s2p_ctrl** template, a data controller, is used to control the outputs of smci_ascserial. In each cycle, **s2p_ctrl** loads two bytes smci_ascserial outputs to the upper **s2p** instance for power supply voltage and then load two bytes outputs to the lower **s2p** instance for command speed.

The rotational speed of the DC motor is compared with the command speed. The error signal is fed to an A/D converter port of the micro-controller.

MICRO-CONTROLLER

The micro-controller is implicitly represented in the selected adapter. There is no actual micro-controller model used in simulation. However, the special function registers and all supporting functions are micro-controller specific. The adapter used in this example is developed according to the specifications of C166 micro-controller [4].

Some of the special function registers used by the synchronous/asynchronous serial port and the A/D converter port are listed in APPENDIX A.

CUSTOMER CODE AND SIMULATION RESULTS

The operations of the serial port and A/D converter port are controlled by customer software.

To the A/D converter, the input signal is the rotational speed error of the DC motor with respect to the command speed. It is used to control the power supply voltage of the inverter in the **drv3ph_sin** template. The rmctrl function is designed for this purpose. See case 0 branch of rmctrl function, APPENDIX B. This function is an interruption service function associated with the interruption signal eoc of the A/D converter. It is invoked when an A/D conversion is finished and eoc becomes logic high. A configuration file is provided for specifications on the usage of the rmctrl function and is shown in APPENDIX D.

Each time when rmctrl function is invoked, the error signal is available in the result register result.adres of the A/D converter port. Taking into account the conversion resolution, the actual error rmerr in rotational speed can be obtained. The incremental in power supply voltage dvpm is calculated based on the speed error. The power supply voltage vpm of the inverter is then obtained by accumulating dvpm via a second order compensation filter implemented with the pwmpid function. The value of vpm is transmitted to the inverter via the smci_ascserial template. Since vpm is an integer in 16-bit and the serial port transmits only one byte a time, two transmission cycles are needed for transmitting one instance vpm value. See synTrans1 and synTrans3Con functions in APPENDIX C.

The command speed is programmed in the drvSignal array. The speed at any given time rmRef is calculated by the refOmega function. The calculation is also accomplished at the same time when vpm is evaluated. Since rmRef is in 16-bit resolution, it also needs two transmission cycles for the serial port to transmit its value. See also APPENDIX C.

By adequately adjusting the filter coefficients including the gain factor **vGain**, damping factor **damp**, **nPeriod** and **nOmega**, the control system becomes stable and the overshot in transient behavior of the rotational speed is within an acceptable range.

The outputs va, vb and vc of the inverter are shown in Figure 3, where the amplitude and frequency of the sinusoidal waveforms vary due to the change in power supply of the inverter.

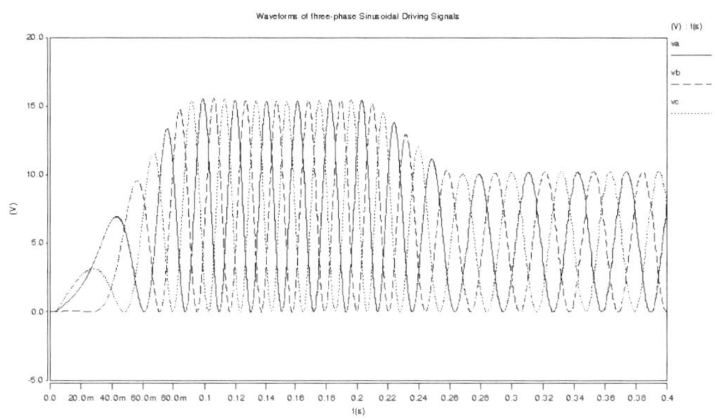

Figure 3. Three-phase outputs of Inverter

The phase voltages of the DC motor are shown in Figure 4, where the amplitude and frequency vary, accordingly.

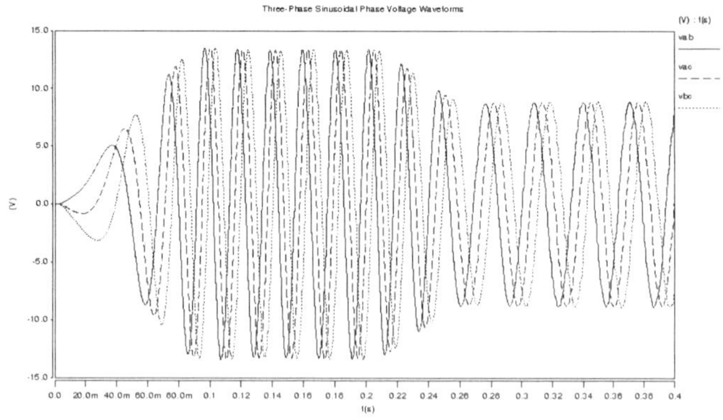

Figure 4. Phase voltage waveforms

The command speed wrm_ref and the rotational speed of the DC motor under control are shown in Figure 5.

It can be observed that the rotational speed approaches the command speed at 150 (rad/sec) after a transition time. When the command speed drops to a new value 100 (rad/sec), the rotational speed also drops and eventually approaches the new command speed.

The error in rotational speed in the steady state is less than 0.5% of the full range value.

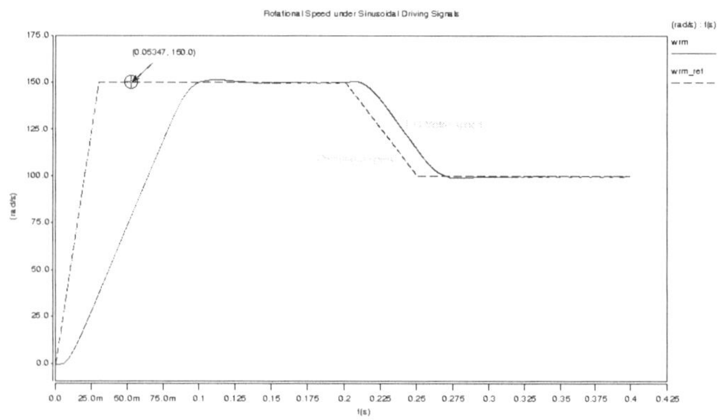

Figure 5. Command speed and motor speed

A CONTROLLER FOR BRUSHLESS DC MOTOR DRIVEN BY THREE-PHASE SIX-STEP SIGNALS

This example is similar to the previous one. The difference is that a six-step inverter is in place of the sinusoidal inverter. The schematic diagram is shown in Figure 6.

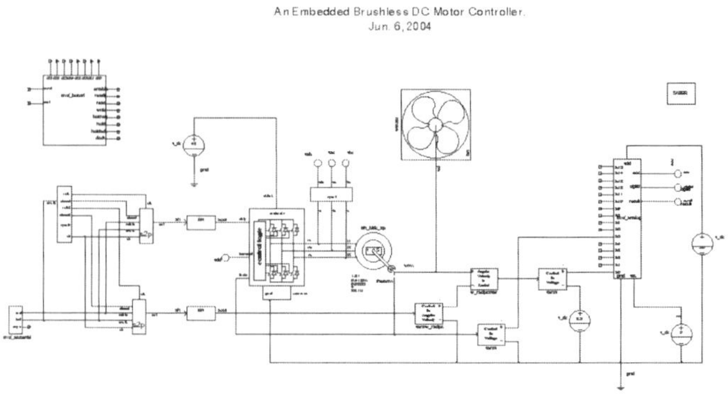

Figure 6. A six-step DC motor controller

Another change in this controller is that the initial angular position of the DC motor is taken as the injection input of the A/D converter port [4].

The outputs of the three-phase 6-step signal inverter are controlled by the duty cycle input that is provided by the serial port outputs. The angular position of the DC motor is fed back to the inverter for controlling the phase of 6-step signals.

The serial port of the micro-controller provides both command speed signal and the duty cycle signal as described in the previous example. However, the outputs of the serial port are now the duty cycle and command speed, instead of power supply voltage and command speed. The error speed is now used in calculation of the duty cycle, instead of power supply. Therefore, the customer functions rmctrl and syntrans1 are changed, accordingly.

A second order compensation filter is also used in evaluation.

The inverter outputs and the phase voltage of the DC motor are shown in Figure 7 and Figure 8, where a section of phase voltage waveforms are also included to illustrate waveform details.

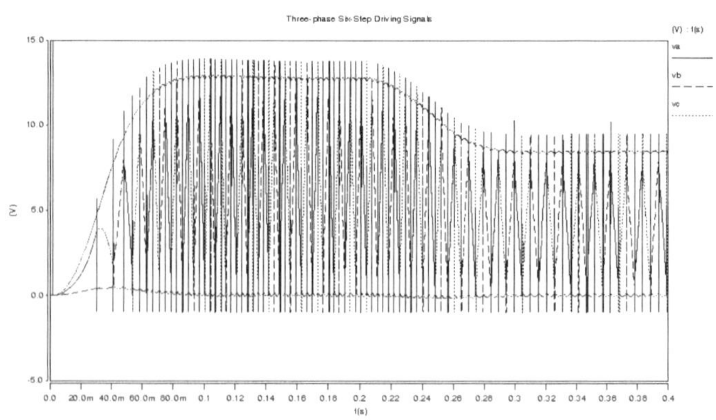

Figure 7. Six-step inverter outputs

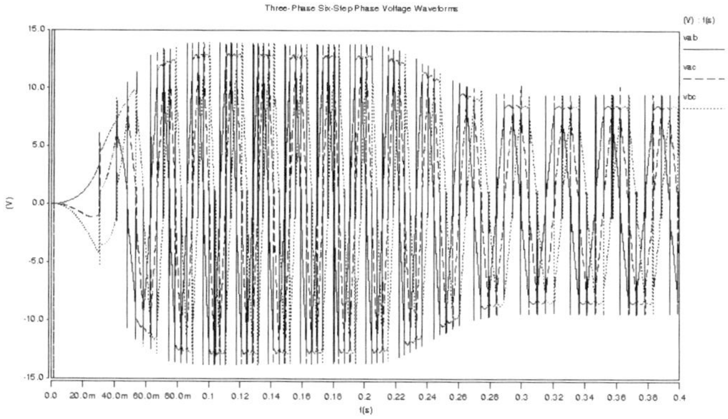

(a) Entire phase voltage waveforms

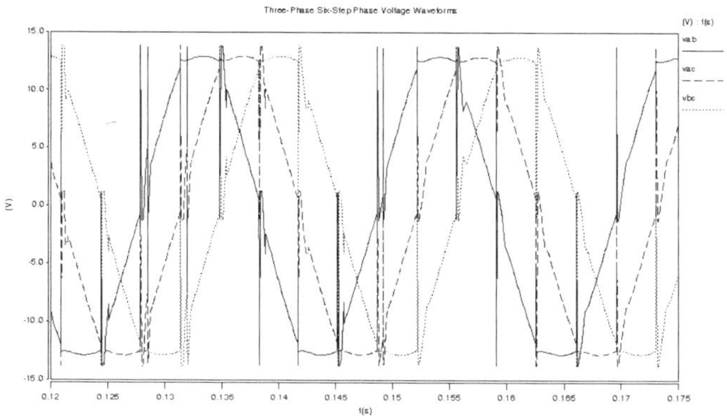

(b) A section of phase voltage waveforms

Figure 8. Phase voltage waveforms

The command speed and rotational speed of the DC motor are shown in Figure 9.

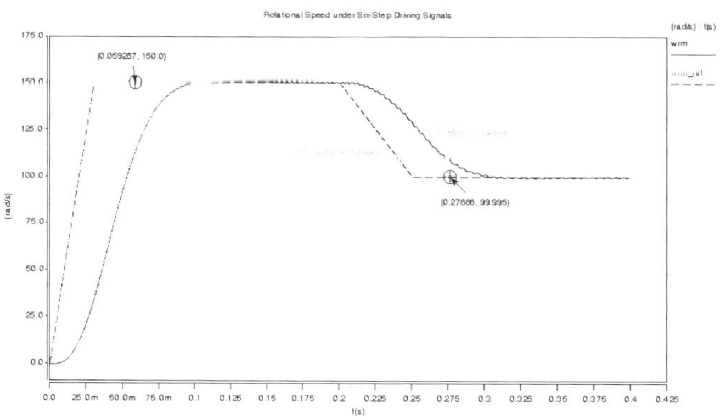

Figure 9. Command speed and motor speed

It can be observed that the rotational speed still approaches the command speed after the transition time.

CONCLUSION

A new simulation approach for entire embedded control system is presented. In this approach, the hardware portion of the system is represented in a schematic diagram, where the selected port instances are included for connection. The customer code (control algorithm) is seamlessly and automatically integrated with the port models for simulation. No instruction set simulator is required.

Two DC brushless motor controller examples are presented. Simulation results show that the rotational speed of the DC motor under control approaches the command speed as expected after a certain transition period of time.

In these examples, an adapter for c166 micro-controller is used. Additional adapters for different micro-controllers may be developed based on demands. Currently developing a micro-controller adapter is a time consuming task. However, additional software tools may make the adapter development more efficient and feasible to users. This may be considered as future work.

REFERENCES

1. R. Cmelik, D. Keppel: "Shade: A Fast Instruction Set Simulator for Execution Profiling," Proc. SIGMETRICS, ACM, Nashville, TN, 1994, pp 128-137.
2. G. Hadjiyiannis, S Hanono, S. Davadas: "ISDL: An Instruction Set Description Language for Retargetability," Technical report, MIT, 1996.
3. Peter Spasov: "Microcontroller Technology, The 68HC11", Prentice-Hall, Inc. 1999.
4. "XC161 Derivatives Peripheral Units", Draft User's Manual, V1.1, 2002-02, Infineon.
5. Ken G. Ruan: "A New Model Architecture for Customer Software Integration", IEEE ISCAS'2004, May 23-26, 2004, Vancouver BC, Canada.
6. "MAST Language Reference Manual", Synopsys, Inc. 1994.
7. "Mixed-Mode Interface", Synopsys, Inc. 1989.
8. Ken G. Ruan: "Functional Specifications on Embedded System Simulation", Synopsys, Inc. 2003.

APPENDIX A. SPECIAL FUNCTION REGISTERS

For each port of a micro-controller, there is a set of predefined special function registers. The operation of a port is controlled by the contents of these special function registers [4]. Definitions of some of the special function registers (SFR) for the a/d converter port and the synchronous/asynchronous serial port are presented here for information. These definitions are referenced by the customer code for the examples.

<u>SFRs for a/d converter port</u>

Listed below are SFR declarations for the a/d converter port. These are bit-addressable.

```
struct c166_adcon1_t {
  c166Reg icst:1,   // conversion and sample timing:
                    // 0: standard
```

```
                    // 1: improved
    sample:1,   // sample:
                // 0: not in sample phase
                // 1: in sample phase
    cal:1,      // calibration:
                // 0: not in calibration phase
                // 1: in calibration phase
    res:1,      // conversion resolution:
                // 0: 10-bit;
                // 1: 8-bit
    adctc:6,    // conversion time control
    adstc:6;    // sample time control
};

struct c166_adcdat_t {
  c166Reg chnr:4,       // channel number
    adres:12;           // a/d conversion results
};

struct c166_adctr0_t {
  c166Reg md:1,     // Mode
                    // 0: compatibility mode
                    // 1: enhanced mode
    sample:1,       // sample:
                    // 0: not in sample phase
                    // 1: in sample phase
    adcts:2,    // channel injection trigger input
selection
                // 00     no trigger input by hardware
                // 01     trigger input capcom1/2
selected
                // 10     trigger input cc6 selected
                // 11     reserved
    adcrq:1,    // channel injection request flag
    adcin:1,    // channel injection enable
    adwr:1,     // wait for read control
    adbsy:1,    // converter busy
    adst:1,     // start bit
    adm:2,      // Mode selection
    caloff:1,   // calibration disable:
                // 0: executed
                // 1: disabled (off)
    adch:4;     // a/d channel selection
};

struct c166_adctr2_t {
  c166Reg msb:2,  // reserved
    res:2,      // conversion resolution:
                // 00:    10-bit;
                // 01:    8-bit,
                // 10, 11: reserved.
    adctc:6,    // conversion time control
    adstc:6;    // sample time control
};
```

SFRs for serial port

Listed below are SFR declarations for the asynchronous, synchronous serial port. Except for c166_asctbuf and c166_ascvbuf, the rest are bit-addressable.

```
struct c166_asccon_t {  // ASC control register
  c166Reg r:1, // Baud rate Generator Run Control Bit
               // 0 - Baud rate Generator disabled
               // 1 - Baud rate Generator enabled
    lb:1,      // Loop back mode Enabled
               // 0 - loop back mode disabled, standard
transmit/receive mode
               // 1 - loop back mode enabled
    brs:1,  // Baud rate Selection
            // 0 - Baud rate timer prescaler divided by 2
            // 1 - Baud rate timer prescaler divided by 3
    odd:1,  // Parity Selection
            // 0 - Even parity selected
            // 1 - Odd parity selected
    fde:1,  // Fraction divider enabled
            // 0 - Fraction divider disabled
            // 1 - Fraction divider enabled
    oe:1,   // Overrun Error Flag
    fe:1,   // Framing error flag
    pe:1,   // Parity error flag
    oen:1,  // Overrun check enable
            // 0 - Ignore overrun errors
            // 1 - check overrun errors
    fen:1,  // Framing check enable
            // 0 - Ignore framing errors
            // 1 - check framing errors
    pen:1,  // Parity check enable
            // 0 - Ignore parity
            // 1 - check parity
    ren:1,  // Receiver enable bit
            // 0 - receiver disabled
            // 0 - One stop bit
            // 1 - two stop bit
    m:3;    // Mode control // 1 - receiver enabled
    stp:1,  // number of stop bits selection
        // 000 - 8 bit data for synchronous operation
        // 001 - 8 bit data for asynchronous operation
        // 010 - 8 bit data IrDA mode for synchronous
operation
        // 011 - 7 bit data and parity for asynchronous
operation
        // 100 - 9 bit data for asynchronous operation
        // 101 - 8 bit data and wake up bit for
asynchronous operation
        // 110 - Reserved, Don't use.
        // 111 - 8 bit data and parity for asynchronous
operation
};

struct c166_ascbg_t {
        // ASC Baud rate Timer/ Reload Register
  c166Reg res:3,    // Reserved for future use
    br_value:13;    // Baud rate Timer/ Reload value
    };

struct c166_ascfdv_t {  // Fractional Divider Register
  c166Reg res:8,    // not used
        fd_value:8; // Fraction Divider Register Value
    };

typedef c166Reg c166_asctbuf;
typedef c166Reg c166_ascrbuf;
```

APPENDIX B. FUNCTION FOR POWER SYPPLY

```
int rmctrl(double t, c166_adcon con, c166_adcon1 con1)
{
    /*
     Obtain the analog input quantities from saber.
     siglIdx is the index of an analog signal on Saber
side.
     Since the input signal is the error signal, it may be
either positive or negative. The binary code should be
nob (normal offset binary). An offset is included in the
input.
   */
  double dvpm = 0;
  double t_old = 0.0;
  double tmpGain, ratio, ratioX = 0.5;
  int siglIdx = result.chnr;

  switch (siglIdx) {
    case 0:{
        /*
         Calculate the command wrm of the motor shaft
         */
        wrm = refOmega(drvSignal,t);
        rmRef = wrm*vScale/(2.0*wrmMax);
        /*
        A/D converter input in this channel is the signal
corresponding to the rotational error. It can be either
positive, or negative. Therefore, nob code is used.
         */
        rm = result.adres;
        rmerr = (double)(rm-bias)* rmScale/fullScale;

        /*
         rmerr represents the error in rm. If error > 0, the
rotational speed is higher than the command signal, the
duty cycle should be reduced. Otherwise, it should be
increased. Therefore, a minus sign is included in the
evaluation of dvpm.
         */
        tmpGain = vGain;
        ratio = rmerr/rmScale;
        ratio = abs(ratio);
        if (ratio > ratioX) {
            tmpGain = vGain*(1.0+gFact*(ratio-
ratioX)*(ratio-ratioX));
        }
        dvpm = -rmerr*tmpGain;

        /*
         * vpm represents the power supply to the
inverter. It final value transmitted to the inverter is
calculated in the case 0 and case 1 branches in
synTrans1.
         */
        vpm = pwmpid(kNum,numer,kDen,
                                  denom,dvpm);
        break;
    }
    case 5:{
        break;
    }
  }
  return 0;
}

double pwmpid(int n, double num[], int m, double den[],
double verr)
{
  double retVal=0.0;
  int i;
  for(i=n; i>0; i--) {
    x[i] = x[i-1];
    retVal += num[i]*x[i];
  }
  /*
   * Since the input is an error signal, the full signal
should be
   * constructed by adding the previous value.
   */
  x[0] = y[0] + verr;
    retVal += num[0]*x[0];

  for(i=m; i>0; i--) {
    y[i] = y[i-1];
    retVal -= den[i]*y[i];
  }

  y[0] = retVal/den[0];
  return y[0];
}
```

APPENDIX C. FUNCTION FOR SERIAL PORT

```
int synTrans1(double t)
{
    /*
     * Use 16-bit data for vpm.
     * If transIdx = 0, transmit low-byte of duty cycle to
asctbuf;
     * If transIdx = 1, transmit high-byte of duty cycle to
asctbuf;
     * The peripheral will combine these two bytes to
form a word for
     * power supply of the inverter template.
     *
     * If transIdx = 2, transmit low-byte of reference
speed to asctbuf;
     * If transIdx = 3, transmit high-byte of reference
speed to asctbuf;
     * The peripheral will combine these two bytes to
form a word for
     * command speed of the motor.
     */
    switch (transIdx) {
        case 0: {
            tmpVpm = (vpm + vbias);
            if (tmpVpm > vScale) tmpVpm = vScale;
            asctbuf = tmpVpm & 255;
            break;
        } case 1: {
            asctbuf = tmpVpm/256;
            break;
        } case 2: {
```

```
            tmpWrm = rmRef + vbias;
            if (rmRef + vbias > vScale) {
                tmpWrm = vScale;
            }
            asctbuf = tmpWrm & 255;
            break;
        } case 3: {
            asctbuf = tmpWrm/256;
            break;
        }
    }
    return 0;
}

int synTrans3Con(double t)
{
    asccon.r = 1;
    asccon.ren = 0;

    /*
     * transIdx swings between 0 and 3. Different data
contents are selected based on transIdx. See synTrans1
function.
     */
    transIdx += 1;
    if (transIdx > 3) {
        transIdx = 0;
    }

    return 0;
}
```

APPENDIX D. THE CONFIGURATION FILE

The configuration file of the first example is provided here for information. One design a configuration file.

```
Design Name : ctrl5_c166 ;
Part Name : c166 ;

# If master is partner, this simulation will be executed
# using Smc command.
Master : partner ;

# Specify the customer code in simulation
Program file : rmctrl5.c rmctrl.h synTrans1.h ;

# Virtual mode is the choice of simulation mode.
# Use host for running debugger on customer code.
Virtual Mode : host ;

Clock Config : noClock;

# By design, one smci_bisctrl model for each
# micro-controller must be included in the Saber netlist.
ebc mode: c166_ebcMode0  ebcmod0 ;
ebc mode: c166_ebcMode1  ebcmod1 ;

# In this example, only one serial port and one a/d
# converter port are used. Therefore, the specifications
# on these two ports are provided here.
# Both a/d converter and ascserial port are internal ports
# of the micro-controller.
# Port functions such as rmctrl, synTrans1 and
# symTransCon  are interruption service
# functions and are activated when the specified
# interruption signal is active.
Channel Config : internal ;
SMCI Interface :  smci_analog ;

# Arguments are required by adapter functions and
# customer functions. Error message will be issued if
# any required arguments are missing.
Arguments : c166_adcon adcon ;
Arguments : c166_adcon1 adcon1 ;
Arguments : c166_adctr0 adctr0 ;
Arguments : c166_adctr2 adctr2 ;
Arguments : c166_adctr2in adctr2in ;
Arguments : c166_adcdat2 inject ;
Arguments : c166_adcdat result ;
Arguments : c166_adceic adc_eic;
Arguments : c166_adccic adc_cic;
Arguments : c166_adceoc eoc ;

# Register is the variable in customer code that
# corresponds to the I/O signal of a port.
Registers : int rm;
Function  : sim : rminit();
Function  : irc : rmctrl(smcTime, adcon, adcon1) : eoc ;
Function  : irc : adccic(&adcon, &adcon1) : overrun ;
Config Ends : ;

# This section is for the async/sync serial port. Multiple
# instances of a port are supported. However, in this
# example, only one instance is used.
Channel Config : internal ;
SMCI Interface :  smci_ascserial ;
Arguments : c166_asccon asccon ;
Arguments : c166_asctbuf asctbuf ;
Arguments : c166_ascrbuf ascrbuf ;
Arguments : c166_ascbg ascbg ;
Arguments : c166_ascfdv ascfdv;
Function  : irc : synTrans1(smcTime) : asctrans ;
Function  : irc : synTrans3Con(smcTime) : asctransbuf ;
Config Ends : ;

Part Ends : ;

# A dctr command is defined for running simulation.
dctrcmd : dctr (sigl va vb vc vab vac vbc eoc vdd vpm
result wrm theta wrm_ref rmerr asclk sdata zclk zclk2
dload dload2, pf tr, ts 10n, te 0.4) ;
```

2005-01-0785

Effective Application of Software Safety Techniques for Automotive Embedded Control Systems

Barbara J. Czerny, Joseph G. D'Ambrosio, Brian T. Murray and Padma Sundaram
Delphi Corporation

ABSTRACT

Execution of a software safety program is an accepted best practice to help verify that potential software hazards are identified and their associated risks are mitigated. Successful execution of a software safety program involves selecting and applying effective analysis methods and tasks that are appropriate for the specific needs of the development project and that satisfy software safety program requirements. This paper describes the effective application of a set of software safety methods and tasks that satisfy software safety program requirements for many applications. A key element of this approach is a tightly coupled fault tree analysis and failure modes and effects analysis. The approach has been successfully applied to several automotive embedded control systems with positive results.

INTRODUCTION

The last decade has seen rapid growth of automotive safety-critical systems controlled by embedded software. Embedded processors are used to achieve enhancements in vehicle comfort, feel, fuel efficiency, and safety. In these new embedded systems, software is increasingly controlling essential vehicle functions such as steering and braking independently of the driver. Although many of these systems help provide significant improvements in vehicle safety, unexpected interactions among the software, the hardware, and the environment may lead to potentially hazardous situations. As part of an overall system safety program, system safety analysis techniques can be applied to help verify that potential system hazards are identified and mitigated.

During the execution of a system safety program, developers of embedded control systems recognize the need to protect against potential software failures. Unlike mechanical or electrical/electronic hardware, software does not wear out over time, and it can be argued that software does not fail. However, software is stored and executed by electronic hardware, and the intended system functionality that is specified by the software may not be provided by an embedded system if potential electronic hardware failures occur or if the software is incorrect.

In this paper, we define a *software failure* as any deviation from the intended behavior of the software of a system. There are three main categories of potential causes of software failure modes: hardware failures, software logic errors, and support software (e.g. compiler) errors.

Typical sources of potential hardware failures, which can be either internal or external to the controller the software executes on, include:

- Memory failures in either the code space or variable space,
- CPU failures (ALU, registers), and
- Peripheral failures (I/O ports, A/D, CAN, SPI, watchdog, interrupt manager, timers).

For example, memory cell failures can cause conditions where the software inadvertently jumps to the end of a routine or into the middle of another routine. Interrupt failure modes, such as return of incorrect priority or failure to return (thereby blocking lower priority interrupts), can also be caused by memory corruption.

Software logic errors may arise due to incomplete or inconsistent requirements, errors in software design, or errors in code implementation. Software logic errors can lead to failure conditions such as infinite loops, incorrect calculations, abrupt returns, taking a longer time to complete routine execution, etc. In addition, software stored in an embedded system may not be correct if the tools necessary to configure, compile and download the software do not function as expected.

Similar to the effective best-practice approach applied to help prevent potential system hazards due to hardware failures, embedded system developers can apply system safety engineering methods to protect against software failures. However, the unique potential failure modes and the overall complexity of software warrant that additional software-specific analysis methods and tasks be included in the overall system safety program. To address this need, the system safety program should include a software safety component as well. A software safety program involves the execution of a number of software-related tasks intended to help identify and mitigate potential software failures.

Although requirements for an automotive software safety program can be derived from existing software safety guidelines and published sources [1,2,3,4], efficient methods and tasks for satisfying these requirements, that are appropriate for the automotive domain, are still needed. In this paper, we present a set of methods and tasks that we have effectively applied to several automotive embedded control systems to satisfy automotive software safety program requirements. First, we describe a generic software life cycle and its relation to a software safety process proposed by Delphi [5]. Next, we provide details on specific analysis methods and tasks that we applied for each of the major steps in the life cycle. Finally, we present our conclusions.

SOFTWARE SAFETY LIFE CYCLE OVERVIEW

Table 1 shows the typical software development life cycle phases and corresponding software safety tasks performed during each phase. The tasks shown satisfy the requirements of a proposed Delphi software safety program procedure [5]. Note that the Conceptual Design phase is actually part of the system development process, but is included here for completeness. In general, there may be more than one set of methods that can be applied to satisfy the required tasks. The specific set of methods selected depends on the target product's stage of development and on any unique aspects of the product.

Table 1: Relation Between Software Development Phases and Software Safety Tasks.

Software Development Phase	Typical Software Safety Tasks
Conceptual Design	Preliminary Hazard Analysis and SW Safety Planning
SW Requirements Analysis	SW Safety Requirements Analysis
SW Architecture Design	SW Safety Architecture Design Analysis
SW Detailed Design and Coding	SW Safety Detailed Analysis and SW Safety Code Analysis
SW Verification and Validation	SW Safety Testing, SW Safety Test Analysis, SW Safety Case

Figure 1 shows the six primary software life-cycle phases, where each phase has associated detailed software safety inputs, outputs, and tasks. These inputs, outputs, and tasks satisfy Delphi's proposed software safety program requirements for advanced automotive safety-critical systems and are consistent with part 3 of the IEC 61508 [2] standard that addresses software safety. The methods and tasks shown represent a tailored subset of those suggested by the Delphi software safety program requirements and by IEC 61508. Given that individual projects have unique aspects to them, the selected set of methods and tasks described in this paper may not be appropriate for all projects. In the following sections, we provide details of the specific software safety methods that we applied during the different software development phases of several of our automotive embedded control systems.

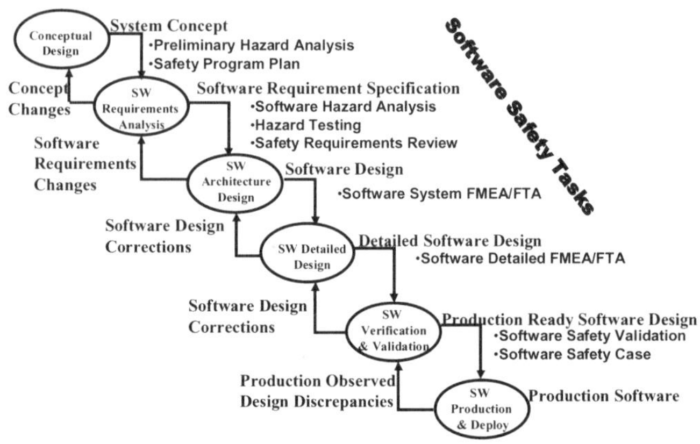

Figure 1: Software Life Cycle with Associated Software Safety Tasks.

CONCEPT DESIGN PHASE

During this phase of system and software development, project leaders must determine if a system safety program is required for the product concept. This decision is typically made based on past product knowledge or based on the results of a preliminary hazard analysis (PHA). Regardless of how the decision is made, a preliminary hazard analysis and system safety program plan are typically completed if a system safety program is required. If the preliminary hazard analysis identifies any potential hazards that may arise due to potential software failures, then a software safety program plan is developed as well.

PRELIMINARY HAZARD ANALYSIS

The goal of PHA is to identify potential high-level system hazards and to determine the criticality of potential mishaps that may arise. PHA is performed in the early stages of system development so that safety requirements for controlling the identified hazards can be determined and incorporated into the design early on. The PHA tends to quickly focus the design team's attention on the true potential safety issues of a product concept. The basic steps for performing a PHA are:

1. Perform brainstorming or review existing potential hazard lists to identify potential hazards associated with the system,
2. Provide a description of the potential hazards and potential mishap scenarios associated with them,
3. Identify potential causes of the potential hazards,
4. Determine the risk of the potential hazards and mishap scenarios, and

5. Determine if system hazard-avoidance requirements need to be added to the system specification to eliminate or mitigate the potential risks.

If time is a factor for a potential hazard or potential mishap occurrence, then the timing constraints that the potential hazard places on the design may be investigated as well.

Consider the hypothetical control system shown in Figure 2. A sensor provides the needed input signal to the system ECU. The system ECU then computes the actuator command satisfying the system function.

Figure 2: Example Control System.

One potential hazard of such a system is an unintended system function. Unintended system function can result in undesirable system behavior that could potentially be hazardous. Some of the causes of unintended system function would include potential ECU failures or sensor failures.

Table 2: Example Control System PHA.

Pot. Hazard	Pot. Hazard Risk	Pot. Causes	Safety Strategy	Revised Pot. Hazard Risk
Unintended System function	High	Sensor Fault; ECU Fault; Motor driver fault; Actuator fault;	High integrity sensor signals High Integrity ECU Operation; High Integrity Mechanical Actuator	Low

Table 2 shows a portion of the PHA for this example. For the system without a safety strategy implemented, the potential risk is high, because an unintended event may occur. However, once appropriate safety features are incorporated as specified by the safety strategy, the revised potential risk is low. In this example, high integrity sensor, ECU, and actuator design strategies will be implemented to help ensure potential failures are detected and handled appropriately. For example, one method to provide a high integrity-sensor signal value is to use two sensors and compare the output of the sensors for consistency. A sensor fault is detected if the values from the sensors do not agree within some tolerance, and when this occurs, the system transitions to a fail-safe state (e.g., controlled shutdown of the system).

To achieve high integrity ECU operation, the design team must consider the potential for unintended system function due to software failures. As previously described, software failures may occur if hardware faults exist.

SOFTWARE SAFETY PROGRAM PLAN

A software safety program plan is the plan for carrying out the software safety program for a project. This plan typically includes the software safety activities deemed necessary for the project and the resources and timing associated with the activities. In effect, this plan defines the software safety life cycle for the project. The plan typically evolves during the software safety life cycle to reflect newly identified needs.

SOFTWARE REQUIREMENTS ANALYSIS PHASE

In this phase of software development, the goals of the software safety program include identifying software safety requirements to eliminate, mitigate, or control potential hazards related to potential software failures. Software safety requirements may also stem from government regulations, applicable national/international standards, customer requirements, or internal corporate requirements. A matrix identifying software safety requirements may be initiated to track the requirements throughout the development process.

Methods used to satisfy the software safety goals include:

1. Software Hazard Analysis,
2. Hazard Testing, and
3. Software Safety Requirements Review.

Software hazard analysis identifies possible software states that may lead to the potential hazards identified during the PHA. Using the link established between software states and potential hazards, software *hazard-avoidance requirements* are developed and included in the software safety requirements specification. To help quantify these hazard-avoidance requirements, hazard testing identifies specific *fault response times* that must be provided by the software functionality to help ensure that potential hazards are avoided. In general, all of these activities are tightly coupled, with interim results from one activity feeding into the others. Finally, software safety requirements review helps ensure that safety requirements are complete and consistent. The following sections provide more detailed descriptions of the software safety analysis methods that may be applied to satisfy the goals of the software safety program during this software development phase.

SOFTWARE HAZARD ANALYSIS

Software hazard analysis consists of identifying the potential software failures that may lead to potential system hazards. For each potential system hazard, possible software states leading to the potential hazard are identified. Based on the link established between the potential hazards of the system and the potential software causes, any identified system hazard-avoidance requirements are translated into corresponding software hazard-avoidance requirements.

The most common technique applied to accomplish this task is fault tree analysis, which is a top-down (deductive) analysis method that identifies potential causes for some top-level undesired event. The immediate causes for the top-level event are identified, and the process is repeated, such that the causes are considered events, and their associated causes are identified. The analysis continues until a base set of causes is identified. For system-level software hazard analysis, these base causes are software states. It is important to note that at this point a software architecture or detailed design does not exist, so the software states identified in the FTA are anticipated. As described later, the analysis must be updated to reflect the actual software architecture and detailed design.

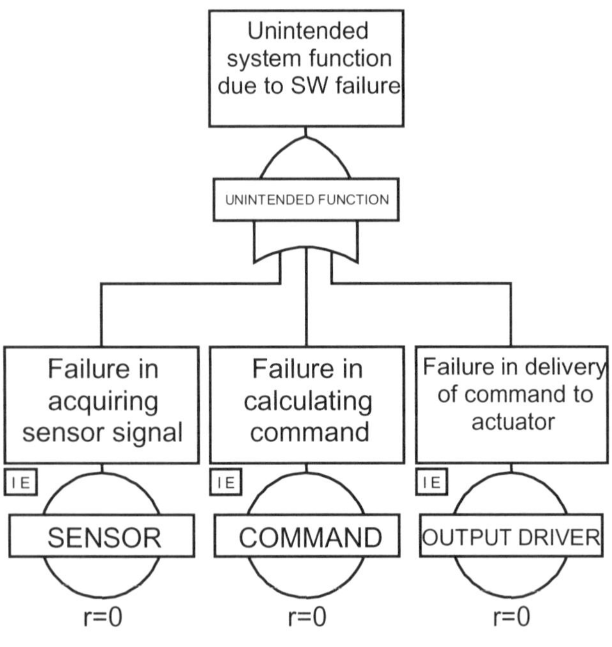

Figure 3: System-Level Fault Tree for the Example Control System.

For the example control system, the unintended system function fault tree in Figure 3 shows the identified potential software failures. For each of these potential software failures, high-level software safety requirements are specified (Table 3).

HAZARD TESTING

During hazard testing, requirements to test the actual behavior of the system under the potential hazard conditions are developed. The results of this testing provide the fault response times and signal deviation levels required by the system to avoid a potential hazard before it occurs. The fault response times drive the design of the software diagnostics. This timing is very critical in designing the software tasks schedule and may also give some insight into whether the chosen controller has enough processing power and throughput to handle the various tasks within the given time. Although these tests may initially be performed using simulation models or bench setups, the identified fault response times should be confirmed by testing the actual system in a vehicle if possible.

Table 3: Example Software Safety Requirements.

Req. No.	Software Safety Requirement
SW-SAFETY-1	Software sensor diagnostics shall detect deviations of actual vs. measured sensor signal.
SW-SAFETY-2	Software shall detect deviations of computed actuator command.
SW-SAFETY-3	Software shall detect actuator control errors resulting in a deviation of delivered vs. computed command

For the example control system, hazard testing using simulation and vehicle testing might lead to the hypothetical requirement that the undesired behavior produced in the vehicle due to failures shall not exceed a specific amount within a specific amount of time. Although the software does not yet exist, it may be possible to use this quantitative vehicle level requirement to quantify the software safety requirements based on the vehicle and system simulation model. In this paper, we assume that the vehicle requirement corresponds to the following ECU requirement: *the ECU output command delivered to the actuator shall not deviate from the desired value by X amount for more than Y ms.*

SOFTWARE SAFETY REQUIREMENTS REVIEW

Software safety requirements review examines the software safety requirements identified by software hazard analysis, to help ensure they are complete and consistent. Early identification of missing, incorrect or inconsistent software safety requirements allows the requirements to be modified with little or no impact to program schedule or cost. Late identification of software safety requirements deficiencies can result in expensive, schedule impacting changes to the overall design. The software safety requirements analysis process also evaluates the software functional requirements for their impact on safety. The end product of this task is a set of software safety requirements for the software design. These requirements will be based on the earlier developed system level safety requirements and the results of the hazard analyses and hazard tests. The requirements also may include general software safety coding guidelines and industry, government, or international coding standards that must be followed by the software development team.

Table 4: Revised Software Safety Requirements.

Req. No.	Requirement
SW-SAFETY-1	Software sensor diagnostics shall detect deviations of actual vs. measured sensor signals of TBD amount within TBD ms.
SW-SAFETY-2	Software command diagnostics shall detect deviations of computed Actuator command of X amount within Y ms.
SW-SAFETY-3	Software communication/driver diagnostics shall detect actuator communication errors within Y ms
SW-SAFETY-4	Software failure management routine shall initiate controlled shutdown of the system immediately after a diagnostic detects a failure.
SW-SAFETY-5	All software shall conform to the MISRA C coding guidelines

For the example control system, the safety analysis results & requirements shown in Table 2, Figure 3, and Table 3 are reviewed for consistency and completeness. Table 4 above shows the updated requirements. The existing SW-SAFETY-2 requirement is revised based on the ECU integrity requirements obtained from hazard testing, with the specific limit values being directly assigned. The SW-SAFETY-1 requirement is revised to reflect that a TBD level and TBD detection time will be specified once the relationship between the sensor signal and command output is better defined. Since a communication error could result in a bad command being delivered to the actuator, SW-SAFETY-3 requirement is revised to reflect the detection time determined by hazard testing. At this point, there are no requirements on what should happen after a fault is detected. To address this, another requirement, SW-SAFETY-4 is added to specify system behavior once a fault is detected. It is also common to identify existing external or internal corporate standards that will be followed. Finally, a fifth requirement, SW-SAFETY-5, is added indicating that the software shall adhere to the MISRA coding guidelines [3] to help ensure best practice coding techniques are followed.

SOFTWARE ARCHITECTURE DESIGN PHASE

In this phase of software development, the goals of the software safety program include identifying the safety-critical software components and functions, and applying appropriate high-level analysis methods to these components and functions to help ensure potential hazards are avoided or mitigated. The software development team specifies the software components and functions that are needed to create a functional system that satisfy identified software requirements (including software safety requirements). From the existing software hazard analysis, an *integrity* or *criticality level* can be assessed for each software component or function. The criticality level depends on the potential hazards that could arise from a malfunction of the software component or function. The higher the criticality, the greater the level of analysis required. There are various schemes for quantifying criticality or integrity, with the simplest being to label software components or functions as either safety critical (if they can lead to a potential hazard) or non-safety critical (if they cannot lead to a potential hazard).

To satisfy these goals, the existing fault tree analysis is extended to identify the specific software components or functions that produce software states that may lead to potential hazards, and a system-level software Failure Modes and Effects Analysis (FMEA) is performed to provide broad coverage of potential failures. Software component or function criticality is assigned based on the highest risk potential hazard that is linked to potential software causes in the developed fault trees.

```
INIT:
        PowerUpTest();
MAIN_LOOP:
        DetermineSystemMode();
        AcquireSensorInput();
        DiagnoseSensorInput();
        ComputeOutput();
        Check&SendOutput();
BACKGROUND_LOOP:
        ECUDiagnostics();
SHUTDOWN:
        Shutdown();
```

Figure 4: Example Control System Software Architecture.

To help understand the analysis methods presented in this section, a software architecture for the example control system is shown in Figure 4. This software architecture must accommodate identified safety requirements (Table 4), and in some cases specific software modules need to be included (e.g., DiagnoseSensorInput()). The architecture includes an initialization task, which is run at power up, a main loop and a low priority background loop, both of which are run during normal execution (after the initialization task is complete), and a shutdown task that is executed based on the results of the DetermineSystemMode() function.

FAULT TREE ANALYSIS

At this stage of development, the existing fault tree is revised such that specific software modules are included in the fault tree. This typically involves replacing the existing software portion of the fault tree, which to this point has been developed based on knowledge of the necessary software function but not on the software structure, with a new software sub-tree based on a structured analysis of the software architecture. The newly developed software sub-tree is compared to the old sub-tree to be sure no knowledge is lost.

For the example control system, the software portion of the fault tree shown in Figure 3, is replaced with a tree developed by identifying the immediate causes of the quantified top software event. The tree is created by stepping through the software architecture shown in Figure 4 to identify relevant software failures of the software components. Event descriptions in the tree are quantified based on the requirements in Table 4. A portion of the revised tree is given in Figure 5

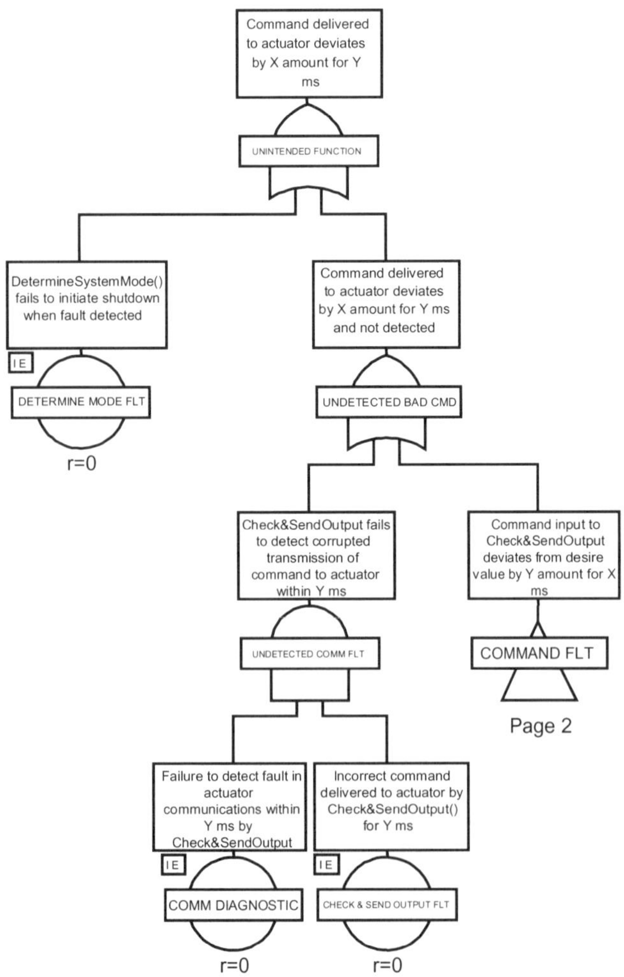

Figure 5: Revised Software System-Level Fault Tree.

Two immediate causes of delivery of a bad command to the actuator are identified:

1. Command delivered to actuator deviates by X amount for Y ms and is not detected, and

2. DetermineSystemMode() fails to initiate shutdown when fault detected.

The branch of the fault tree for the first cause includes potential failures of each of the software modules needed to produce and deliver the command to the actuator, and potential failures in the associated diagnostics intended to detect deviations in the desired command. Including the diagnostics in the tree results in the introduction of AND gates in this branch. The branch of the fault tree for the second cause includes the failure of the software module that initiates system shutdown if a fault is detected. This branch does not contain any AND gates because there are no identified diagnostics as of yet to detect this type of failure.

Potential failures of all of the software modules in the MAIN LOOP in Figure 4 appear in the fault tree, so all of these modules can be considered safety critical software modules. However, the only single point failure that may cause the top software event is failure of the DetermineSystemMode() module, so this module is assigned a higher criticality level than the others. During the detailed design phase, this software module will be analyzed in more detail due to its higher criticality.

SYSTEM LEVEL SOFTWARE FMEA

Software FMEA aids in identifying structural weaknesses in the software design and also helps reveal weak or missing requirements and latent software non-conformances. A software FMEA can be performed at different design phases to match the system design process. The goal of the software FMEA performed during the software safety architecture analysis is to examine the structure and basic protection design of the system. The PHA and the hazard testing results are key inputs to the system-level software FMEA. The FMEA techniques described in this paper are consistent with the recommendations of SAE ARP 5580 [6]. In contrast to SAE J-1739 [7], SAE ARP 5580 provides specific guidance for software FMEAs.

Analysis of the software components and functions assumes that a high-level design description of the software architecture is available. The analyst performing the software FMEA needs to have a complete understanding of the software design, the underlying hardware structure, interfaces between the software and hardware elements, the software language to be used and specifics of the software tools being used. If possible, the system development program should use compilers that are certified to a standard for the language to be used. Thus, early involvement in the software design FMEA will allow needed compiler and language restrictions to be imposed on the design process at a cost-effective time [8].

System-level software FMEA uses inductive reasoning to examine the effect on the system of a software component or function failing to perform its intended behavior in a particular mode. Generic failure modes (guide words) are applied to the top-level software components and the impacts are analyzed. In an approach consistent with SAE ARP 5580 there are four failure modes for all components and two additional failure modes for interrupt service routines (ISRs). The four common failure modes are:

- Failure to execute,
- Executes incompletely,
- Executes with incorrect timing which includes incorrect activation and execution time (including endless loop), and
- Erroneous execution.

The two additional software failure modes for ISRs are:

- Failure to return, thus blocking lower priority interrupts from executing, and
- Returns incorrect priority.

The failure to return failure mode for an ISR also includes the condition where an ISR fails to complete, and thus goes into an endless loop.

System-level software FMEA is performed by assessing the effects of the relevant failure modes for each functional subroutine. The effects of the failure modes on the software outputs are analyzed to identify any potentially hazardous outcomes. If potentially hazardous software failure events are identified then either a software safety requirement was not provided or a safety requirement was not adequately translated into the software design. In these cases, a software safety requirement is added and the software design is modified to accommodate this change. In order to assess the changes made to the software, the system-level software FMEA is updated when changes are made.

For each component or function, failure mode guidewords are applied and the local and system-level impacts are analyzed, including assigning a severity. This is documented in a tabular form. Recommendations to improve the safety of the software design are documented and passed on to the software design team. Table 7 shows a portion of the system-level software FMEA documented in a tabular form for the example control system. Safety-related software requirements identified by the FMEA are added to the software safety requirements for the design. To maintain consistency between the FMEA and FTA, specific software failure modes and new diagnostics identified by the FMEA can be included in the system fault tree.

SOFTWARE DETAILED DESIGN AND CODING PHASE

In this phase of software development, the goals of the software safety program include analyzing the detailed software design, and analyzing the implemented software to help ensure software safety requirements are satisfied. Subsystem interfaces may be analyzed to identify potential hazards related to subsystems. The analysis may check for potentially unsafe states that may be caused by I/O timing, out-of-sequence events, adverse environments, etc. Two methods that may be used to achieve the software safety program goals are detailed software FTA and detailed software FMEA.

These activities can be performed in a coordinated manner. The software hazard analysis that was performed at a high-level using FTA during the requirements and architecture phases, can be further extended to decompose the identified potential hazards into software variables and states. A detailed FMEA can be applied to all or higher risk software modules by tracing potential failures in input variables and processing logic through the software to determine the effect of the failure. These effects are then compared against those that cause each of the potential hazards to occur to determine if the individual potential failure can lead to a potential hazard. Potential failures, which can lead to one of the potential hazards, are identified along with appropriate software design corrective actions.

Typically the level of analysis performed depends on the criticality level (or potential risk) of individual software modules, and the product design's overall stage of development (e.g., prototype vs. production). In the following sections, the process for performing a complete FTA and FMEA at the detailed design and code level is described. In cases where less analysis is required, a subset of the methods described in this paper can be applied.

```
DetermineSystemMode(Boolean Flag1, Boolean Flag2)
        {
        Enumerated SystemState = (NORMAL,
                        FAILED);

        SystemState = LookUpState(Flag1, Flag2);
        If (SystemState == FAILED) Then
                CallShutDownTask();
        }
```

Figure 6: Example Code for Determining System Mode.

To help understand the analysis methods presented in this section, a hypothetical coded procedure for the example control system is provided in Figure 6. This software code for the DetermineSystemMode() procedure looks up a system state based on diagnostic flags that have been set by other routines. If a critical failure has occurred, the system transitions to a safe state (in our example system, the controller shuts down).

DETAILED SOFTWARE FAULT TREE ANALYSIS

From the software architecture phase, the existing fault tree links top-level software components and functions to the potential hazards. With the software detailed design and code now available, the fault tree can be extended to identify lower-level software components that directly assign the output of the top-level components already in the fault tree. These lower-level software components can be tagged as safety critical and any additional software hazard avoidance requirements that are needed can be specified. As the results from the detailed software FMEA technique become available, the FTA and FMEA results can be compared for consistency and completeness.

DETAILED SOFTWARE FMEA TECHNIQUE

Detailed software FMEA is a systematic examination of the real time software of a product to determine the effects of the potential failures in the individual variables implemented in the software. The detailed FMEA allows a more thorough assessment of the degree to which the system design remains vulnerable to potential individual failures, such as single point memory failures. In addition, a detailed FMEA may be used to assess the protection provided by the diagnostic approach to potential dormant software non-conformances. This detailed analysis is time consuming, and is typically only applied to high criticality software components. For distributed systems with redundant controllers, the need for detailed software FMEA is reduced, because by design, potential software failures due to hardware faults typically do not lead to potential hazards.

Table 5: Example Variable Mapping.

Variable	Routines				
	Acquire-Sensor-Input	Diagnose Sensor-Input	Compute-Output	Check & Send Output	Deter. System Mode
Variable-1	Output	Input			
Varaible-2	Local				
Variable-3			Output	Input	
...	...	...	...	...	...
Variable-n				Output	Input

To support the FMEA, a variable mapping is developed to map all the input, output, local, and global variables of the software to their corresponding software routines. Thus each variable, which is either an input or an output, has a mapping. Each input variable is either a hardware input or is an output of another routine. Table 5 shows a variable mapping for a portion of the example control system.

Once the mapping is in place, failure modes are developed both for the variables used and for the software processing logic. The variable failure modes for input variables and the failure effects for output variables are based on the variable type.

Variable Failure Modes

Three basic variable types are recognized: Analog, Enumerated, and Boolean. An analog type variable is any variable that measures quantity in a continuous manner. Enumerated variables are those, which can have a limited number of discrete values, each with a unique meaning. All variables with only two possible values are treated as Boolean variables. Variables are stored in memory locations, and if the memory locations, buses, and data registers do not contain data integrity protection (e.g. parity), any variable may be corrupted during operation. Thus, the potential failure modes for each variable type, shown in Table 6 below, must be considered as possible input failure modes to every routine that uses the variable. The following list contains an example of potential variable failure modes for a portion of a software routine "DetermineSystemMode()" shown in Figure 6:

- Flag1 set to TRUE when it should be FALSE,
- Flag1 set to FALSE when it should be TRUE,
- Flag2 set to TRUE when it should be FALSE,
- Flag2 set to FALSE when it should be TRUE,
- SystemState set to FAILED when it should be NORMAL, and
- SystemState set to NORMAL when it should be FAILED.

Table 6: Failure Modes for Different Variable Type.

Variable Type	Failure Modes
Analog	High
	Low
Boolean	True when False
	False when True
Enumerated Example Values: A, B, C	A when it should be B
	A when it should be C
	B when it should be C
	B when it should be A
	C when it should be A
	C when it should be B

Software Processing Logic Failure Modes

In addition to potential variable failure modes, potential software processing logic failure modes may be considered. This type of analysis involves examining the operators (e.g., addition, subtraction, comparison) in the code to determine possible negative effects that must be addressed.

Integrating Results

Once an FMEA has been performed on each of the software modules, the output variables are used to provide a mapping between the modules. The failure effect on an output of one module is traced to the corresponding input variable failure modes at the succeeding module. This variable failure mode/effect tracing is repeated until the top level routines are reached. To help support this activity, software threads that link software modules and variables from data acquisition to final output may be created. Once the set of effects of a failure have been traced to the top-level routines, the mapping of the failure to the hazards is determined.

The detailed software FMEA is analogous to the component level hardware FMEA process except that variables are substituted for signals and signal paths of the electronic hardware [8]. A portion of the detailed software FMEA is given in Table 8 for the example control system.

Finally, when the detailed FMEA is completed, a mapping will exist from the top-level potential hazards to the top-level critical variables. The top-level critical variables are those variables that are necessary and sufficient to enable a potentially hazardous software state. Figure 7 provides an example of a set of top-level critical variables identified by a detailed hazard analysis.

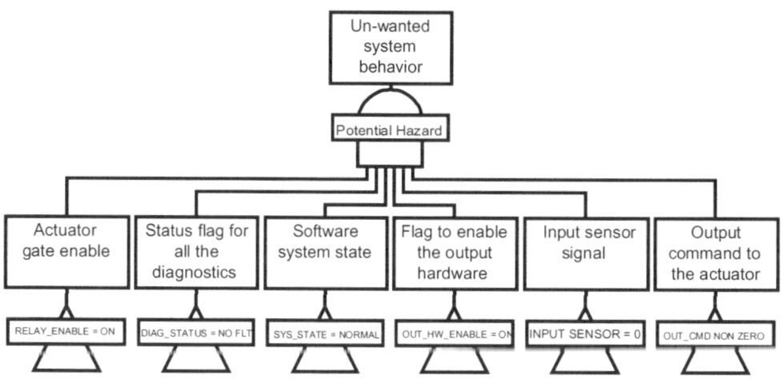

Figure 7: Example Fault Tree.

If the detailed FMEA identifies potential failure modes that trace to the identified hazards, then missing or incorrectly implemented software safety requirements are identified and corrected. Similar to the system-level FMEA, the software design deficiencies must be identified and the requirements documentation updated. The safety test plan document is updated with additional software safety testing requirements during the detailed design and coding phase.

DEFENSIVE PROGRAMMING

In addition to FTA and FMEA methods applied during this phase of software development, adopted or developed coding guidelines (e.g., [3]) often recommend that developers implement defensive programming techniques. Critical functions may be separated from non-critical functions in the code to reduce the likelihood that non-critical potential faults can lead to potential hazards. Using a logic “1” and “0” to denote states or decision results for safety-critical functions is not recommended due to bit-flip concerns. Software engineers should consider implementing reasonableness checks and sanity checks for critical signals.

SOFTWARE VERIFICATION AND VALIDATION PHASE

In this phase of software development, the goal of the software safety program is to execute safety test plans to help ensure that the software satisfies all software safety requirements. This typically involves performing unit testing and integration testing in any of the following environments: simulation, bench, and in-vehicle. The developed safety test plans demonstrate that fault detection and fault handling capabilities (e.g., see Table 4) are functioning as expected. In addition, software stress testing may be applied to help ensure the software is robust to changing inputs. Finally, compliance with any applicable government and international standards or relevant guidelines is assessed in this phase.

Although FTA and FMEA are primarily performed before the verification phase of product development, the detailed examination of requirements, design, and code they afford can be a significant help in verifying that the software satisfies specified requirements. FTA and FMEA results should be compared to those of actual testing during the verification phase to help ensure any assumptions or conclusions made during these analyses were correct.

SUMMARY AND DISCUSSION

In this paper, we have presented software safety methods and techniques that we have successfully applied to several advanced automotive systems. These methods and techniques satisfy the task requirements of a proposed Delphi software safety program procedure. A key component of this methodology is an integrated FTA/FMEA approach for investigating potential software causes of system hazards. The chief difference between the FMEA approach and the FTA approach is a matter of depth. Wherein the FMEA looks at all failures and their effects, the FTA is applied only to those effects that are potentially safety related and that are of the highest criticality [4]. The broad coverage provided by an inductive FMEA is combined with a deductive FTA to focus the analysis. Experience has shown that the FTA/FMEA approach has been effective in identifying and mitigating potential hazards. Initiating software safety activities at the beginning of the product development life cycle facilitates the implementation of identified corrective actions such that the impact on program timing and cost is minimized.

Since FTA and FMEA are static analysis techniques, they have certain limitations. Although they may focus attention on identified safety-critical modules, they assume that the software provides desired behavior in the absence of potential failures. Thus, design or code reviews should be performed on safety-critical modules. Software FTA and FMEA do not verify the correctness or stability of control algorithms. For these evaluations, appropriate modeling and simulation tools need to be used to verify stability and correctness of control algorithms.

Table 7: Example System-Level Software FMEA.

Software Element	Failure Mode	Local Effect	System Effect	Potential Severity	Recommendation	Projected Severity
ACQUIRE SENSOR INPUT	Fails to execute	No sensor signals are read	System will continue to use the last read sensor signal value and output calculated based on that value. Since the last read signals are within range, DIAGNOSE INPUT function will not detect the fault. If system is in Normal Operation mode, this could potentially be hazardous if desired output is different from the system calculated output. If system is in startup mode, then default values will be used. Potentially there could be no output in that case.	10	A software execution monitor that checks the execution of this software element needs to be employed	8
ACQUIRE SENSOR INPUT	Erroneous Execution	Some or all sensor signals incorrect	DIAGNOSE INPUT function will catch any out--of-range signal values. However if the sensor signals values are within range, system will continue to use the erroneous sensor signal value and hence output will be incorrect. Potentially incorrect output command send to the output hardware leading to unwanted behavior of the system.	10	This could be caused due to either erroneous behavior of the A/D peripheral, sensor failure or any memory byte corruption. Need to have checks that monitor the ADC peripheral and the related controller memory cells.	8

Table 8: Example Detailed Software FMEA.

Variables	Failure Modes	Variable Type	Software Modules Affected	Local Effect	System Effect	Potential Severity	Recommendation	Projected Severity
Flag1	TRUE when should be FALSE	Input Global	Determine SystemMode	May cause SystemState to be FAILED when it should be NORMAL, resulting in unwanted call to initiate shutdown.	System will shutdown thus causing loss of function	8		8
	FALSE when should be TRUE	Input Global	Determine SystemMode	May cause SystemState to be NORMAL when it should be FAILED, resulting in no call to shutdown when there should be one	System may provide incorrect output	10	Replace Boolean flags with enumerated data type such that '00' is FALSE and '11' is TRUE; Diverse programming of diagnostic routines; Comprehensive fault injection testing to verify diagnostics	8

REFERENCES

1. Leveson, N.G., Safeware: System Safety And Computers, ISBN 0-201-11972-2, 1995.
2. IEC 61508-3, Functional Safety Of Electrical/Electronic Programmable Electronic Safety Related Systems – Part 3 Software Requirements First Edition, 1998-12.
3. MISRA Guidelines For the Use Of The C Language In Vehicle Based Software, April 1998.
4. FAA System Safety Handbook, Dec. 2000.
5. Czerny, B.J., et al., *An Adaptable Software Safety Process for Automotive Safety-Critical Systems*, SAE World Congress 2004.
6. SAE Aerospace Recommended Practice ARP-5580, Recommended Failure Modes and Effects Analysis (FMEA) for Non-Automobile Applications, SAE International, July 2001.
7. SAE J1739, Potential Failure Modes and Effects Analysis Reference Manual, SAE International, June 2000.
8. Goddard, P.L., “Software FMEA techniques,” *Proceedings of the Annual R&M Symposium 2000.*

CONTACT

Padma Sundaram
Delphi Corporation
Innovation Center
12501 E. Grand River
Brighton, Michigan 48116-8326
Phone: 810-494-2453
Email: padma.sundaram@delphi.com

2005-01-0750

Evolutionary Safety Testing of Embedded Control Software by Automatically Generating Compact Test Data Sequences

Hartmut Pohlheim and Mirko Conrad
DaimlerChrysler AG

Arne Griep
IAV GmbH

ABSTRACT

Whereas the verification of non-safety-related, embedded software typically focuses on demonstrating that the implementation fulfills its functional requirements, this is not sufficient for safety-relevant systems. In this case, the control software must also meet application-specific safety requirements.

Safety requirements typically arise from the application of hazard and/or safety analysis techniques, e.g. FMEA, FTA or SHARD. During the downstream development process it must be shown that these requirements cannot be violated. This can be achieved utilizing different techniques. One way of providing evidence that violations of the safety properties identified cannot occur is to thoroughly test each of the safety requirements.

This paper introduces Evolutionary Safety Testing (EST), a fully automated procedure for the safety testing of embedded control software. EST employs extended evolutionary algorithms in an optimization process which aggressively tries to find test data sequences that cause the test object to violate a given safety requirement.

A compact description formalism for input sequences for safety testing is presented, which is compatible with description techniques used during other test process stages. This compact description allows 1) an efficient application of evolutionary algorithms (and other optimization techniques) and 2) the description of long test sequences necessary for the adequate stimulation of real-world systems. The objective function is designed in such a way that optimal values represent test data sequences which violate a given safety requirement. By means of repeated input sequence generation, software execution and the subsequent evaluation of the objective function each safety requirement is extensively tested.

The use of EST for the safety testing of automotive control software is demonstrated using safety requirements of an adaptive cruise control (ACC) system.

The EST approach can easily be integrated into an overall software test strategy which combines different test design techniques with specific test objectives.

INTRODUCTION

Many of the innovations in the automotive field are results of the use of software-intensive systems in vehicles. The proportion of functions which are activated for regulatory or controlling purposes when the vehicle is in motion has risen steeply in the last few years. Especially in the case of safety-relevant functions, the quality of the control software is of the utmost importance.

Software development is not yet at the stage at which the quality of embedded control software can be ensured using constructive procedures alone. Their use must be supplemented with analytical procedures which detect errors in the software. Since there are currently no universally recognized measurement procedures for software quality, measuring is often replaced with dynamic testing [HV98]. Functional (black-box) test design techniques are mainly used for testing embedded control software, i.e. testing that focuses on demonstrating that the implementation fulfills its functional requirements. The test cases or test scenarios are predominantly created manually. In some cases, the functional test design techniques are also complemented with structural (white-box) test methods in order to demonstrate the structural integrity of the software.

For safety-related systems this approach is not sufficient. In this case, the control software must also meet application-specific safety requirements. Such safety requirements typically arise from the application of hazard and/or safety analysis techniques, e.g. Failure Mode and Effect Analysis (FMEA) [IEC60812], Fault Tree Analysis (FTA)

[IEC 61025] or Software Hazard Analysis and Resolution in Design (SHARD) [FMN+94]. During the downstream development process it must be shown that these safety requirements cannot be violated. This can be achieved utilizing different techniques. One way of providing evidence that violations of the safety properties identified cannot occur is to thoroughly test each of the safety requirements. As is the case for any other test problem, the search for suitable test cases/test scenarios and their appropriate description is of decisive importance.

Being universally applicable strategies for improvement and optimization, Evolutionary Algorithms (EA) have become widespread in a broad range of searching problems. Their application is based on a seemingly simple and easily comprehensible principle, namely 'Darwin's evolution paradigm' or, in other words, the principle of the 'survival of the fittest'. Like in nature, solutions are improved step-by-step, and optima are identified or approximated by applying variation and selection to a population of alternative solutions [VDI/VDE 3550].

Prerequisites for the application of EA are the definition of the search space, in which solutions are searched for and of an objective function, with which the fitness of the solution proposals found can be evaluated.

The interpretation of the search for suitable test scenarios, which violate a given safety requirement, as an optimization problem and the subsequent use of EA to solve this problem leads to the concept of Evolutionary Safety Testing (EST) [Weg01]. Thereby, the search space is defined by a compact description formalism for possible input sequences.

In principle, the EST approach can be applied to different embedded control software development stages. In order to be able to incorporate the results of evolutionary safety testing early in the development process, it is, however, advisable to already apply the procedure to the executable model of the future software.

Here, EST should not be the only test design technique which is employed but rather it should be used together with others as part of an overall test strategy for automotive control software developed in a model-based way.

The remainder of the paper is structured as follows: Section 1 explains why sequence testing is necessary for cyclic software components which are typical for embedded automotive controls. Section 2 shows how evolutionary algorithms can be used to automatically generate real-world input sequences for the safety testing of embedded automotive control systems. An example is used to illustrate the entire process. In Section 3 we show how the evolutionary safety testing is incorporated into a model-based test strategy. Related work is presented in Section 4.

1 CYCLIC SOFTWARE COMPONENTS AND SEQUENCE TESTING

1.1 CYCLIC SOFTWARE COMPONENTS

As embedded controls mostly interact with the outside world continuously via sensors and actuators, they must be able to process time-variable sensor signals and produce time-variable outputs.

When analyzing the software of those systems, a large class of implementations can be identified which use a implementation scheme whereby the control algorithm is triggered at regular time intervals by its environment to compute output values and an internal state from given inputs. This class is referred to as cyclic software components (cf. [GHD98], [MK99]). In the world of automotive controls it is very common for a software function to be based on an initialization and step function. Whereas the initialization function is only called once at the beginning, the step function is executed periodically e.g. every 10 ms. An adaptive cruise control system (ACC) which checks the velocity and distance to a preceding vehicle at regular intervals to find out if a safe distance is maintained, is an example of this system class.

Example ACC

If the driver so desires, the ACC system controls the speed of the vehicle whilst maintaining a safe distance from the preceding vehicle. The activated system monitors the section of road in front of the vehicle. If there is no preceding vehicle ('target') on this section, the system regulates the longitudinal vehicle speed in the same way as a conventional cruise control system. As soon as a preceding vehicle is recognized, the distance control function is activated which makes sure that the vehicle follows the vehicle in front at the same speed and at a safe distance.

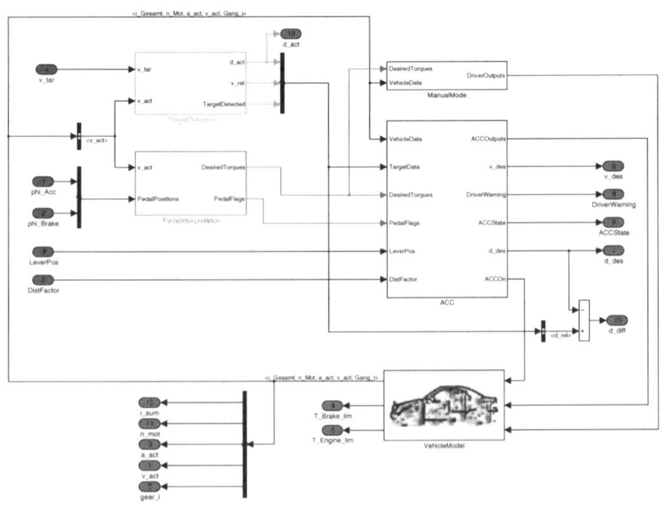

Figure 1. Simulink/Stateflow model of the ACC system

In the model-based development process [Rau03], [KCF+04], the ACC is first described using an executable Simulink/Stateflow [SLSF] model. The structure of the overall system is presented in Figure 1. Besides the actual ACC system, there are two preprocessing systems for the recognition of preceding vehicles (TargetDetermination) and the evaluation of the current pedal values (PedalInterpretation). A system for parts of the plant model (VehicleModel, ManualMode) also exists. These make a closed loop test of the entire system, including the vehicle and the vehicle in front, possible. A more detailed description is contained in [CH98] and [Con04].

If a cyclic software component such as the ACC is to be simulated with realistic input data during the test and if a system reaction is to be induced, the test inputs must be given as time-variable signal waveforms.

1.2 SEQUENCE TESTING

In principle, the testing of cyclic software components can be performed in different ways.

Single input-state-pairs

The approach traditionally used is to generate pairs of internal states and input situations as shown in Figure 2. This means that a test case represents a certain point in time within a test sequence. For this, internal states have to be set directly. This approach works well for simple software components but raises several problems.

The direct setting of internal states is not possible in all cases and requires changes to be made to the test object. When generating the internal states, the test environment has to make sure that the states produced are valid according to the specification. Forcing the system into a generated state is, in most cases, not useful for the tester. This is because the tester has to use the test data generated to analyze software bugs. He first needs to ascertain how to produce the initial state which caused the problem. In other words, he has to manually reconstruct the initial part of the relevant sequence. This information is not provided by the test environment.

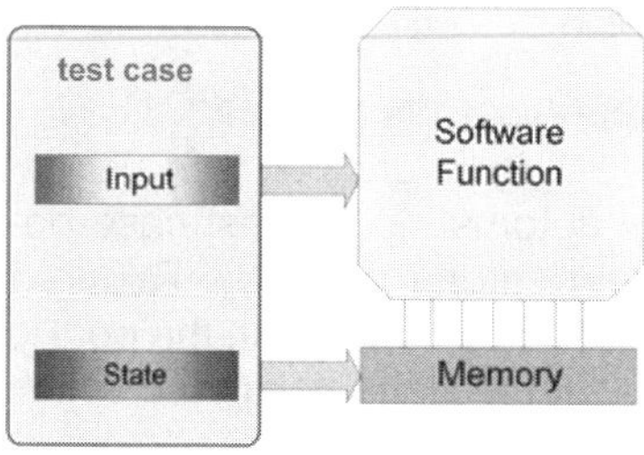

Figure 2. Automatic testing by generating (state, input) pairs

Input sequences

For complex systems it is necessary to provide real input sequences.

(a) Extensional input sequences: Within this group, one approach is to create lists of inputs for sequential calls of the function under test. This means that a test sequence is formed using a list element for each point in time in the test sequence. In other words, the test sequence is represented by enumerating / listing all of its elements (Figure 3).

As a rule, this approach leads to long and low-level test descriptions, but is often sufficient for software components based on pure state models.

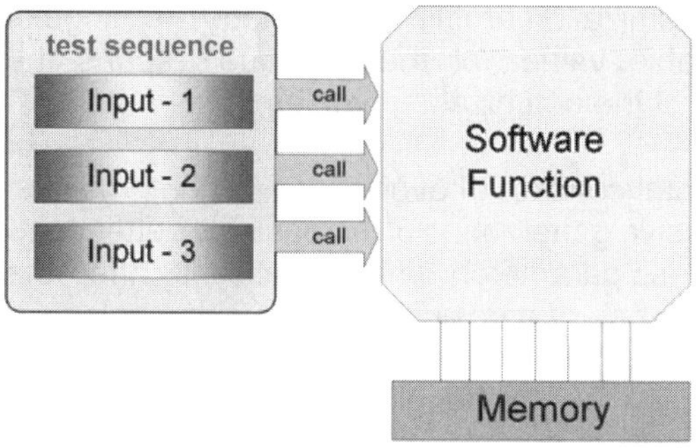

Figure 3. Automatic testing by generating list of inputs

(b) Intensional input sequences: For control systems requiring long 'real world' input sequences, an abstraction from the extensional sequence description is indispensable. Test sequences are described by their main characteristics. One way of doing this is to describe only selected points in time in a test sequence (time tags) and to specify transition functions for those values in between. This leads to an intensional or encoded sequence description (Figure 4).

Intensional descriptions can be very compact and comprehensive and are also often easier to understand for humans. A test sequence can be described using substantially less parameters than is the case for extensional input sequences.

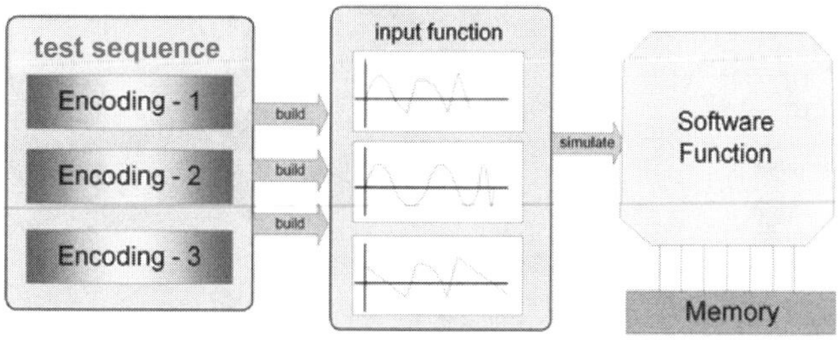

Figure 4. Automatic testing by generating input sequences from encoded input functions

The two variants for describing tests with input sequences are, from the tester's perspective, the better solution, since they guarantee that the software component is tested in the same way as it will later be used.

Test approaches which use test sequences (as opposed to single input-state-pairs) in order to stimulate the actual test object will be referred to as sequence testing (approaches) from here on. Within sequence testing, test inputs and system reactions are not considered to be static values but rather time-variable signal waveforms.

2 EVOLUTIONARY SAFETY TESTING

Evolutionary algorithms (EA) have been used to search for data for a wide range of applications. EA is an iterative search procedure using variation and selection to copy the behavior of biological evolution. They work in parallel on a number of potential solutions, the population of individuals. In every individual, permissible solution values for the variables of the optimization problem are coded. The range of possible variable values of the individuals spans the search space of the optimization problem.

The fundamental concept of evolutionary algorithms is to evolve successive generations of increasingly better combinations of those parameters which significantly affect the overall performance of a design. Starting with a selection of good individuals, the evolutionary algorithm achieves the optimum solution by exchanging information between these increasingly fit samples (recombination) and introducing a probability of independent random change (mutation). The adaptation of the evolutionary algorithm is achieved by the selection and reinsertion procedures used because these are based on the fitness of the individuals. The selection procedures control which individuals are selected for reproduction depending on the individuals' fitness values. The reinsertion strategy determines how many and which individuals are taken from the parent and the offspring population to form the next generation. The fitness value is a numerical value that expresses the performance of an individual with regard to the other individuals in the population so that different designs may be compared. The notion of fitness is fundamental to the application of evolutionary algorithms.

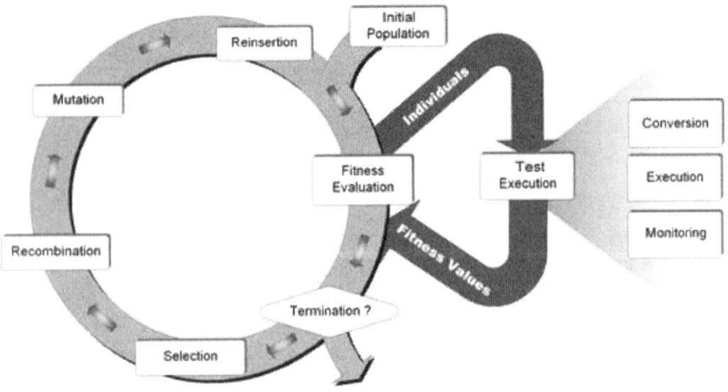

Figure 5. The structure of Evolutionary Testing

Figure 5 gives an overview of a typical procedure for evolutionary algorithms. First, an initial population of guesses as to the solution of a problem is initialized, usually at random. Each individual in the population is evaluated by calculating its objective value. Afterwards, the fitness of each individual is calculated by ranking the objective values of all the individuals of the population. The remainder of the algorithm is iterated until the optimum is achieved, or another stopping condition is fulfilled. Pairs of individuals are selected from the population according to the pre-defined selection strategy, and are combined in some way to produce a new guess in an analogous way to biological reproduction. Combination algorithms are many and varied. Additionally, mutation is applied. The new individuals are evaluated for their objective value (and fitness), and survivors into the next generation are chosen from the parents and offspring, often according to fitness though it is important to maintain diversity in the population to prevent premature convergence to a sub-optimal solution.

For the optimization we employed extended evolutionary algorithms. This includes the use of multiple subpopulations, each using a different search strategy. We also used competition for limited resources between these subpopulations.

For a detailed discussion of evolutionary algorithms and the extensions used see [Poh99]. The algorithms employed are implemented in the widely used MATLAB based 'Genetic and Evolutionary Algorithm Toolbox for use with Matlab – GEATbx' [Poh05]. It consists of a large set of operators e.g. real and integer parameter, migration and competition strategies.

When using EA for a search problem, it is necessary to define the search space and the objective (fitness) function.

2.1 EVOLUTIONARY ALGORITHMS FOR SEQUENCE TESTING

Evolutionary Testing uses EA to test software automatically. The different software test objectives formulate requirements for a test case / test sequence to be generated. Until now, the generation of such a set of test data usually had to be carried out manually. Automatic software testing generates a test data set automatically, aiming to fulfill the requirements in order to increase efficiency and resulting in an enormous cost reduction.

The general idea is to divide the test into individual test objectives and to use EA to search for test data that fulfills each of the test objectives.

In the past, evolutionary testing has proven itself to be of value for different testing tasks:

- Evolutionary temporal behavior testing [Weg01]
- Evolutionary structural testing: Evolutionary Structural Testing has the goal of automating the test case design for white-box testing criteria [BSS02]. Taking a test object, namely the software under test, the goal is to find a test case set (selection of inputs) which achieves full structural coverage

Depending on the task and test objective, both single input-state-pairs and test sequences can be created by the evolutionary test as stimuli for the test object.

Since complex dynamic systems, like the ACC, must be evaluated over a long time period (longer than the highest internal dead time or time constant), such systems must be tested by using input sequences. Furthermore, they must not be stimulated for only a few simulation steps, but rather the input signals must be up to hundreds or thousands of time steps long. Long input sequences are therefore necessary in order to realistically simulate these systems. As a consequence, the output sequences to be monitored and evaluated have the same duration.

Example ACC

In order to check the correct reaction of the ACC system with regard to speed changes of the vehicle in front, the length of realistic test sequences has to be in the magnitude of some 10s. Assuming a sampling rate of 10ms for the cyclic ACC software component, this results in input (and subsequently) output sequences which are some 1000 time steps long.

Several disadvantages result from the length of the sequences necessary: using extensional sequence descriptions, the number of variables to be optimized and the correlation between the variables is very high. For this reason, one of the most important aims is the development of a very compact, intensional description for the input sequences. Such a description has to contain as few elements as possible but, at the same time, offer a sufficient amount of variety to stimulate the system under test as much as necessary.

Moreover, possibilities for automatically evaluating the system reactions must be developed, which allow differentiation between the quality of the individual input sequences.

The generation of test sequences for dynamic real-world systems poses a number of challenges:

- The input sequences must be long enough to stimulate the system realistically.
- The input sequences must possess the right qualities in order to stimulate the system adequately. This concerns aspects such as data type of the signal, speed and rate of signal changes.
- The output sequences must be evaluated regarding a number of different conditions / properties. These conditions/properties are often contradictory and several of them can be active simultaneously.

In order to develop a test environment for the functional test of dynamic systems which can be applied in practice, the following steps must be completed:

- definition of the test objective, i.e. the safety property to be violated,
- description of the search space, i.e. the description of the input sequences,
- description of the objective (fitness) function in order to evaluate output sequences regarding the test objective,
- assessment of counter examples generated

2.2 DEFINITION OF THE TEST OBJECTIVE

In order to utilize evolutionary sequence testing for safety testing, the test objective is to violate a safety requirement resulting from a hazard and/or safety analysis.

Example ACC

A major requirement for the ACC system is to maintain a 'safe' distance to the vehicle in front. More precisely the desired distance d_{des} between one's own and the target vehicle is defined according to the German 'Tacho-halbe-Regel' by:

$$d_{des}\,[m] = v_{act}\,[m/s] * 3.6/2 * DistFactor$$

For normal road conditions the distance factor is 1, for wet or icy roads >1.

A derived safety requirement could be that the actual distance d_{act} is not allowed to go below the d_{des} by more than 10m. In other words:

$$d_{act}\,[m] > d_{des}\,[m] - 10$$

2.3 DESCRIPTION OF THE SEARCH SPACE

In order to define the search space, we have to deal with a number of descriptions. At the beginning, there is the compact description of the search space given by the safety tester (compact search space description). Finally, we need the individual (sampled) signals, which are used as inputs for the simulation (extensional test sequences). In between, there are the boundary descriptions of the variables to be optimized (optimization variable boundaries) and the instantiation of the individual sequences (intensional test sequences). These different sequence description levels are illustrated in Figure 6.

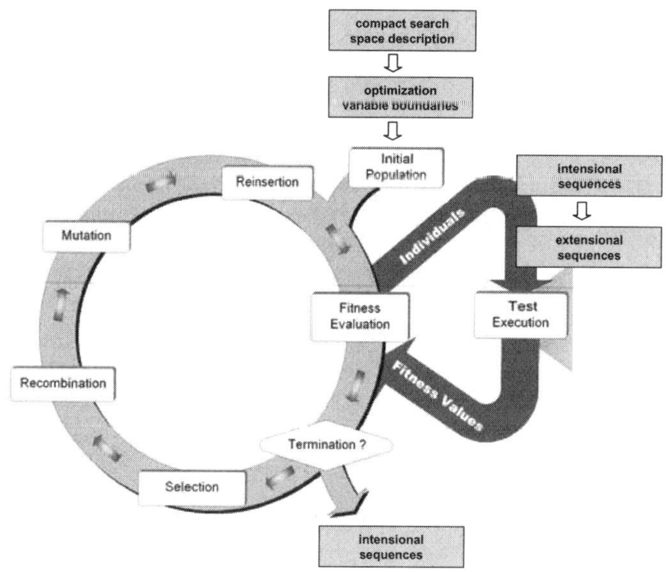

Figure 6. Different levels of input sequence description

For an intensional sequence description, each signal of a long simulation sequence is subdivided into consecutive sections. Each section is parameterized by the variables: section length, signal amplitude (at the beginning of the section) and interpolation function. Typical interpolation functions are, for instance, step, ramp (linear), impulse, sine and spline.

Only the admissible ranges for these parameters are specified for the optimization. In this way, the boundaries are defined, within which the optimization generates solutions which are subsequently evaluated by the objective function with regard to their fitness.

The signal amplitude of each section can be defined absolutely or relatively. The length of the section is always defined relatively (to ensure the monotony of the signal). The interpolation functions are given as an enumeration of possible types. These boundaries must be defined for each input of the system. For nearly every real-world system these boundaries can be derived from the specification of the system under test.

Example ACC

An example of an input sequence description in the MATLAB .m scripting language is provided in Figure 7. Figure 8 provides an example of an extensional sequence generated by the optimization based on the settings given in Figure 7.

```
% Input settings

% input names and order: 'phi_Acc', 'phi_Brake', 'Lever-
Pos', 'v_tar', 'DistFactor'

% number of sections / base points for each input signal
BasePoints = [10 10 10 10 10];

% relative section length
BasisBounds = [ 1  1  1  1  1; ...
               10 10 10 10 10];

% min. / max. amplitude for each input signal
AmplitudeBounds = [0 0 0 20 1; ...
                   0 0 2 45 1];

% possible interpolation functions for each input signal
TransitionPool = {{}; {}; {'impulse'}; {'spline'}; {}};
```

Figure 7. Textual description of input sequences

The ACC system under test has 5 inputs (see Figure 1). We use 10 sections for each sequence. The amplitude for the brake and accelerator pedal can change between 0 and 100, the control lever can only have the values 0, 1, 2 or 3.

In real-world applications the variety of an input signal is often constrained with regard to possible base signal types. An example of this is the control lever in Figure 7. This input contains only impulses as the base signal type.

To generate a further bounded description, it is possible to define identical lower and upper bounds for some of the other parameters. In this case, the optimization has an empty search space for this variable – this input parameter is a constant. An example is the distance factor in Figure 8. This input signal is always set to a constant value of 1. Thus, it is not part of the optimization (but is used as a constant value for the generation of the internal simulation sequence).

The different descriptions ensure that the requirements for an adequate simulation of the system and a very compact and comprehensible description by the tester are fulfilled. The compact description is used for the optimization, ensuring a small number of variables.

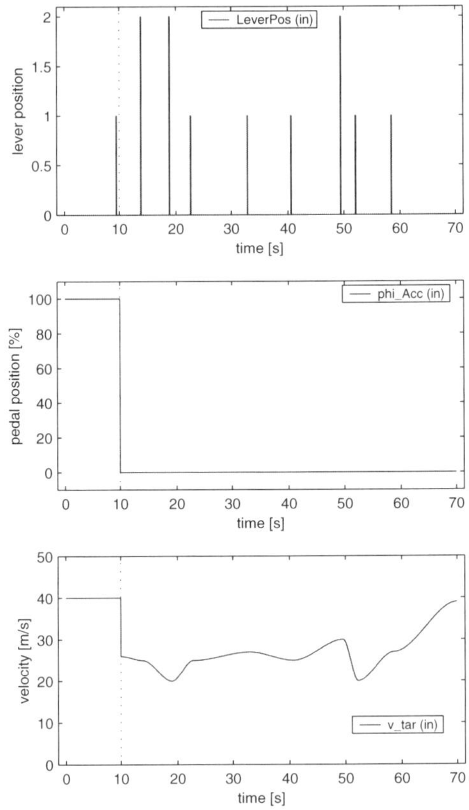

Figure 8. Simulation sequence generated by the optimization based on the textual description in Figure 7; top: throttle pedal, middle: control lever, bottom: velocity of target car

Example ACC

When comparing the size of both descriptions for the example dynamic system used (5 inputs – 3 of the inputs are constant, 10 signal sections, 2 variable parameters for each section (section length and amplitude), 60s simulation time, sampling rate 0.01s) the differences are enormous:

$$
\begin{aligned}
NumParameter_{Extensional} &= 5 \cdot 60 \cdot (1/0.01) = 30000 \\
NumParameter_{Compact} &= (5-3) \cdot 10 \cdot 2 = 40 \\
CompressionRatio &= \frac{30000}{40} = 750
\end{aligned}
\tag{1}
$$

Only this compact description opens up the opportunity to optimize and test real-world dynamic systems within a realistic time frame.

2.4 DESCRIPTION OF THE OBJECTIVE FUNCTION

The test environment must perform an evaluation of the output sequences generated by the simulation of the dynamic system. These output sequences must be evaluated regarding the optimization objectives. During the test, we always search for violations of the defined requirements. Possible aims of the test are to check for violations of:

- signal amplitude boundaries,
- maximal overshoot or maximal settlement time.

Each of these checks must be evaluated over the whole or a part of the signal lengths and an objective value has to be generated.

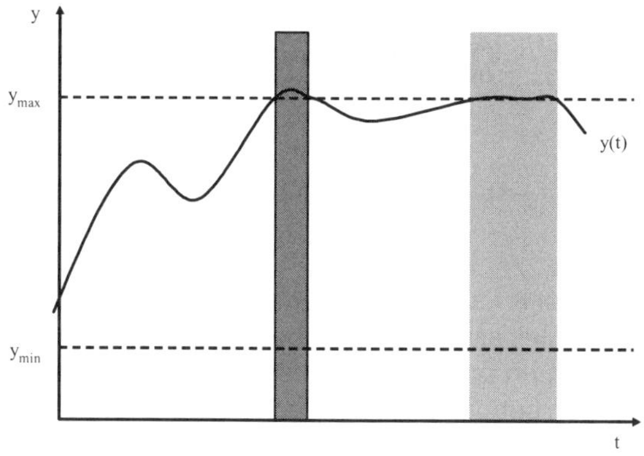

Figure 9. Violation of maximum amplitude

An example of the violation of signal amplitude boundaries is given in Figure 9. A minimal and maximal amplitude value is defined for the output signal y. In this example the output signal violates the upper boundary. The first violation is a serious violation as the signal transgresses the bound by more than a critical value y_c (parameter for this requirement). In this case, a special value indicating a severe violation is returned as the objective value (value -1), see equation (2). The second violation is less severe, as the extension over the bound is not critical. At this point an objective value indicating the closeness of the maximal value to the defined boundary is calculated. This two-level concept allows a differentiation in quality between multiple output signals violating the bounds defined. The direct calculation of the objective value is provided in equation (2).

$$signal_{max} = \max(y(t)) \qquad ObjVal_{max} = \begin{cases} -1 & signal_{max} \geq y_{max} + y_c \\ \left(\dfrac{signal_{max}}{y_{max}}\right)^6 & signal_{max} < y_{max} + y_c \end{cases} \tag{2}$$

<u>Example ACC</u>

Equation (3) shows the objective function for the test objective defined in Section 2.2.

A similar assessment is used for calculating the objective value of the overshoot of an output signal after a step in the respective reference signal. First, the maximal overshoot value is calculated. Next, the relative height of the overshoot is assessed. A severe overshoot outside the specification returns a special value (again –1). This special value is used to terminate the current optimization. The test was successful, as we were able to find a violation of the specification and thus reach the aim of the optimization. In all other cases, an objective value equivalent to the value of the overshoot is calculated (similar to equation (2)).

$$d\_act_{min} = \min(d\_act(t)) \qquad ObjVal = \begin{cases} -1 & d\_act_{min} \leq d\_des - 10 \\ \left(\dfrac{-d\_act_{min}}{-10}\right)^6 & d\_act_{min} > d\_des - 10 \end{cases} \tag{3}$$

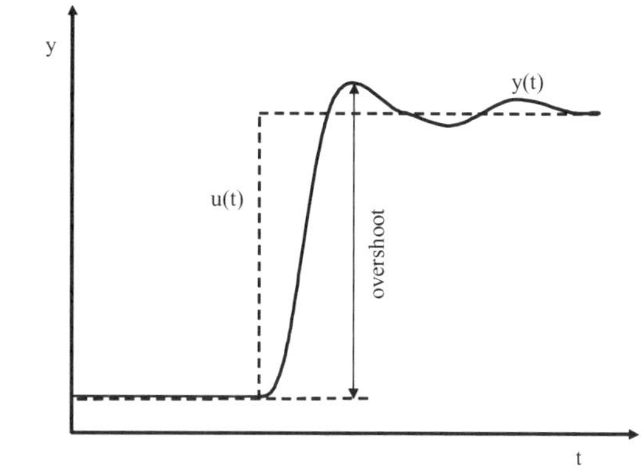

Figure 10. Assessment of signal overshoot

Each of the requirements tested produces one objective value. For nearly every realistic system test we receive multiple objective values. In order to assess the quality of all objectives tested we employ multi-objective ranking as supported by the GEATbx [Poh05]. This includes Pareto-ranking, goal attainment, fitness sharing and an archive of previously found solutions.

<u>Example ACC</u>

With an active ACC, the car can be accelerated only by pushing the control lever upwards (the respective input value is 1). The car is decelerated by pushing the control lever downwards (input value: 2). Besides the amplitude of the input control lever, the relative length of the signal sections could be changed between 1 and10. The results of a successful optimization are shown in Figures 11 and 12.

The optimization process is visualized in Figure 11. The left graph presents the progress of the best objective value over all the generations. The optimization continually finds better values. In the 83rd generation a serious violation is detected and an objective value of –1 returned. The middle graph presents the variables of the best individual in a color quilt. Each row represents the value of one variable over the optimization. The graph on the right visualizes the objective values of all the individuals during the optimization (generations 1 to 82).

The graphs in Figure 12 provide a much better insight into the quality of the results, visualizing the problem-specific results of the best individual of each respective generation. The input of the control lever is shown in the two top most graphs. The resulting velocity of the car is presented below. The graphs are taken from the 3rd (left) and 83rd generation (right). At the beginning the actual distance d_{act} (bottom graph left) is far above the critical boundary. During the optimization, the velocity is increased (the control lever is pushed up more often and at an earlier stage as

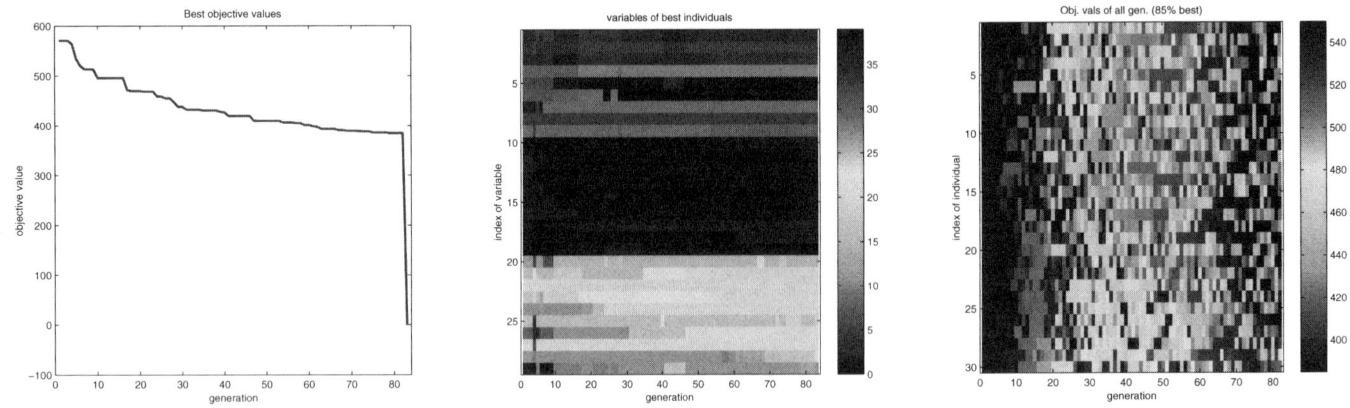

Figure 11. Visualization of the optimization process, left: best objective value; middle: variables of the best individual; right: objective value of all individuals

well as being pushed down less frequently). Additionally, the velocity of the target car is low (nearly the whole time at the defined lower boundary) during the whole scenario (3rd graph from the top). At the end an input situation is found, in which the actual distance d_{act} is smaller than the bound specified (bottom graph right). By looking at the respective input signals the developer can check and change the implementation of the system.

During optimization we employed the following evolutionary parameters: 20 individuals in 1 population, discrete recombination and real valued mutation with medium sized mutation steps, linear ranking with a selection pressure of 1.7 as well as a generation gap of 0.9. Other tests with a

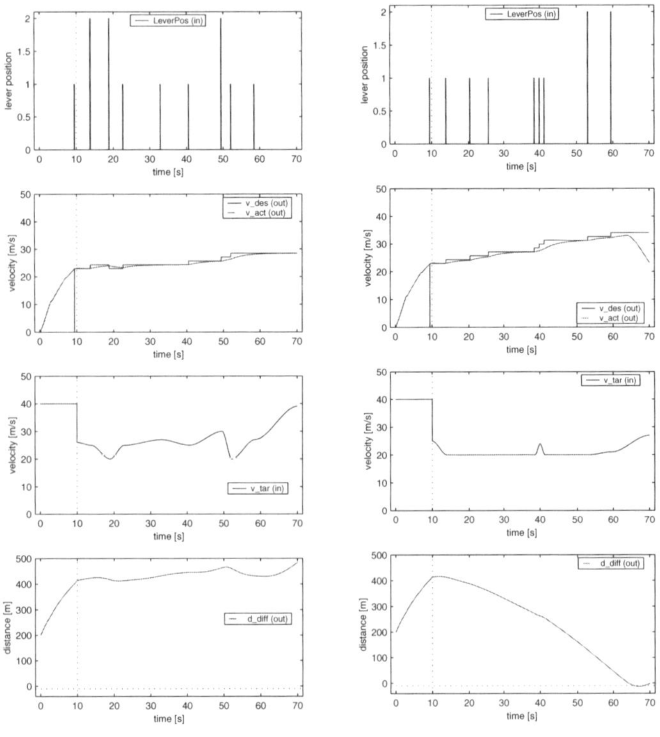

Figure 12. Problem-specific visualization of the best individual during the optimization, left: 3rd generation, right: 83rd generation; from top to bottom: input of the control lever, vehicle velocity (desired and actual), velocity of target, distance

higher number of relevant input signals and thus more optimization variables employ 4-10 subpopulations with 20-50 individuals each. In this case, we use migration and competition between subpopulations. Each subpopulation uses a different strategy by employing different parameters (most of the time differently sized mutation steps).

2.5 ASSESSMENT OF COUNTER EXAMPLES

In the context of an automatic safety test it is of central importance that the test sequences which were generated automatically (counter examples) undergo an intensive analysis (assessment) carried out by human experts. Therefore, it is useful to provide a graphical notation for the generated input sequences which caters for the human tester [Con01]. Moreover, for the seamless integration of the test sequences resulting from EST with test sequences generated by other test design techniques, it is desirable to have input sequences depicted in the same way as other functional test scenarios.

For this purpose, one possibility would be to apply the extended classification-tree notation, which is used in the context of the Classification-tree Method for Embedded Systems (CTM_{EMB}) [Con04], [Con04a].

The CTM_{EMB} notation allows a comprehensive graphical description of time-dependent test sequences by means of abstract signal waveforms that are defined stepwise for each input.

The classifications of the tree represent the input variables of the functional model. Its classes are obtained by dividing the range of each variable into a set of non-overlapping values or intervals. Based on the input variable partitions, the test scenarios describe the course of these inputs over time in a comprehensive, abstract manner.

Time dependent behavior, i.e. changing the values of a model input over time in successive test steps can be modeled by marking different classes of the classification corresponding to that input. Continuous changes are defined by transitions (solid connecting lines) between marks in subsequent test.

Example ACC

Figure 13 shows an example of the visualization of the counter example given in Figure 12 by means of the CTM_{EMB} notation. The manual assessment shows that the input sequence which leads to the violation of the given safety requirement could occur in reality too. This means the algorithms or parameters of the ACC function must be modified and the evolutionary safety test has to be repeated.

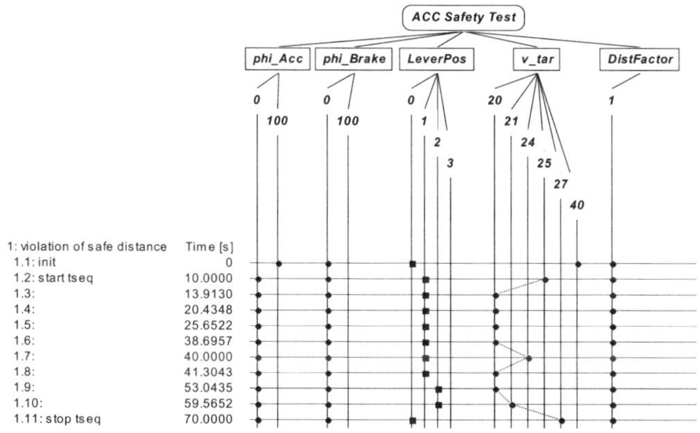

Figure 13. Visualization of a counter example generated by the EST

3 MODEL-BASED TEST STRATEGY

A singular testing technique does not generally lead to sufficient test coverage. Therefore, in practice, the aim is for complimentary test design techniques to be combined in the most adequate way to form a test strategy. The aim of an effective test strategy is to guarantee a high probability of error detection by combining appropriate testing techniques. An effective model-based test strategy must adequately take into account the specifics of model-based development and especially the existence of an executable model.

A model-based test strategy which integrates evolutionary safety testing is outlined in the following. The systematic functional model test based on the functional specification, the interfaces, and the executable model forms the focal point of such a model-based test strategy. In addition, an adequate structural test criterion is defined on model level, with the help of which the coverage of the tests thus determined can be evaluated and the definition of complementary tests can be controlled [Con04a]. If sufficient test coverage has thus been achieved on model level, the model should be mature enough to start safety testing.

Based on the safety requirements, the EST approach should be applied to every given safety requirement in order to aggressively try to generate counter examples.

If the EST process leads to one or more counter examples, those examples have to be assessed carefully. If the counter examples are impossible in practice, the search space description has to be adapted in order to avoid those scenarios in future. If scenarios described by the counter examples could happen in practice, the model or its parameter has to be corrected. After that, the model test has to be repeated: all existing model tests have to be repeated. By doing this, the tester has to ensure that the structural coverage is still sufficient and that all the tests lead to valid system reactions. Finally, the EST process has to be repeated. If no new counter examples are generated, the code generation can be started. All model test sequences can be reused for testing the control software generated from the model and the control unit within the framework of back-to-back tests. In this way, the functional equivalence between the executable model and derived forms of representation can be verified ([BCS03], [CSW04]).

The proposed model-based test strategy can be tool supported by integrating the different tools for creating test sequences into the model-based testing environment MTest ([LBE+04], [MTest]).

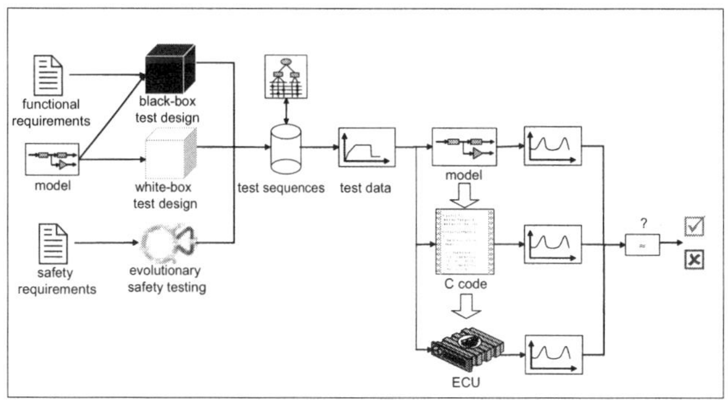

Figure 14. Model-based test strategy

4 RELATED WORK

Within the framework of model-based development, the derivation of safety requirements on model level can be supported by special model-based hazard/safety analysis techniques ([GC04], [PMS01]).

The instrumentation of the executable models with watchdogs/assertions (see e.g.[Rau99], [Reactis], [SL-VV]) or model checking of safety properties (cf. [EmbVal]) are alternative or supplementary techniques for checking the fulfillment of safety requirements.

Test design techniques which can be applied to black-box model testing are described in [HPP+03], [CFS04], [Con04], [Con04a], [HSM+04], and [LBE+04]. [HSM+04]White-box test design techniques for models are discussed in [Ran03], [Lin04], and [HSP05].

An approach for safeguarding the transformation from the tested model into code, i.e. the code generation, is described in [SC03].

CONCLUSION

In this paper we have presented the use of evolutionary sequence testing for the safety testing of embedded automotive control systems.

We have presented a small selection of the results. The results show the new test method to be promising. It was possible to find test sequences, without the need for user interaction, for problems which could previously not be solved automatically.

During the experiments a number of issues were identified which could further improve the efficiency of the test method presented. It is necessary to include as much of the existing problem-specific knowledge in the optimization process as possible.

A systematic approach for the selection of test scenarios and a notation for their appropriate description must therefore form the core elements of a safety testing approach for automotive control software. In order to allow testing to begin at an early stage, test design should build upon development artifacts available early on, such as the specification or an executable model.

ACKNOWLEDGMENTS

The work described was partially performed as part of the IMMOS project funded by the German Federal Ministry of Education and Research (project ref. 01ISC31D), http://www.immos-project.de

REFERENCES

[BCS03] Baresel, A., Conrad, M., Sadeghipour, S.: The Interplay between Model Coverage and Code Coverage. Proc. 11. Europ. Int. Conf. on Software Testing, Analysis and Review (EuroSTAR 03), 2003.

[BPS03] Baresel, A., Pohlheim, H., Sadeghipour, S.: Structural and Functional Sequence Testing of Dynamic and State-Based Software with Evolutionary Algorithms. Proc. Genetic and Evolutionary Computation Conf. (GECCO 03), pp. 2428 – 2441, 2003.

[BSS02] Baresel, A., Sthamer, H., Schmidt, M.: Fitness Function Design to improve Evolutionary Structural Testing. Proc. Genetic and Evolutionary Computation Conf. (GECCO 02), pp. 1329-1336, 2002.

[Bei83] Beizer, B.: Software Testing Techniques. New York: Van Nostrand Reinhold, 1983.

[CFS04] Conrad, M., Fey, I., Sadeghipour, S.: Systematic Model-Based Testing of Embedded Control Software: The MB3T Approach. Proc. ICSE 2004 Workshop W14S on Software Engineering for Automotive Systems (SEAS 04), pp.17-25, 2004.

[CH98] Conrad, M., Hötzer, D.: Selective Integration of Formal Methods in the Development of Electronic Control Units. Proc. 2. IEEE Int. Conf. on Formal Engineering Methods (ICFEM 98), IEEE Computer Society, pp. 144-155, 1998.

[Con01] Conrad, M.: Beschreibung von Testszenarien für Steuergerätesoftware - Vergleichskriterien und deren Anwendung. VDI Berichte, Vol. 1646, VDI Verlag, pp. 381-398, 2001.

[Con04] Conrad, M.: A Systematic Approach to Testing Automotive Control Software. Proc. Convergence 2004, SAE paper 2004-21-0039, 2004.

[Con04a] Conrad M.: Modell-basierter Test eingebetteter Software im Automobil – Auswahl und Beschreibung von Testszenarien. PhD thesis, Deutscher Universitäts-Verlag, 2004.

[CSW04] Conrad, M., Sadeghipour, S., Wiesbrock, H.-W.: Automatic Evaluation of ECU Software Tests. Proc. SAE World Congress 2005, SAE paper 2005-01-1659, 2005.

[EmbVal] EmbeddedValidator (product information). OSC – Embedded Systems AG, www.osc-es.de/products/en/embeddedvalidator.php

[FMN+94] Fenelon, P., McDermid, J.A., Nicholson, M., Pumfrey, D.J.: Towards Integrated Safety Analysis and Design. ACM Computing Reviews, Aug. 1994, pp. 21-32, 1994.

[GHD98] Grieskamp, W., Heisel, W., and Doerr, H.: Specifying embedded systems with statecharts and Z - An agenda for cyclic software components. Proc. Formal Aspects of Software Engineering (FASE 98), LNCS 1382, Springer-Verlag, 1998.

[GC04] Grönberg, R., Conrad. M.: Werkzeugunterstützung für Sicherheitsanalysen von Hardware- und Software-Systemen. Technical Report FT3/A-2000-007, DaimlerChrysler AG, Forschung und Technologie 3, Berlin, Germany, 2000.

[HH+01] Harman, M., Hu, L., Munro, M., Zhang, X.: Side-Effect Removal Transformation. Proc. IEEE Int. Workshop on Program Comprehension (IWPC), Toronto, Canada, 2001.

[HPP+03] Hahn, G., Philipps, J., Pretschner, A., Stauner, T.: Prototype-Based Tests for Hybrid Reactive Systems. Proc. 14. IEEE Int. Workshop on Rapid System Prototyping, San Diego, US, 2003.

[HSM+04] Horstmann, M., Schnieder, E., Mäder, P., Nienaber, S., Schulz, H.-M.: A framework for interlacing Test and/with Design. Proc. ICSE 2004 Workshop W14S on Software Engineering for Automotive Systems (SEAS 04), 2004.

[HSP05] Hermes, T., Schultze, A., Predelli, O.: Software Quality is not a Coincidence - A Model-Based Test Case Generator. Proc. SAE World Congress 2005, SAE paper 2005-01-1664, 2005.

[HV98] Hohler, B., Villinger, U.: Normen und Richtlinien zur Qualitätssicherung von Steuerungssoftware. Informatik-Spektrum, 21, pp. 63-72, 1998.

[IEC60812] IEC 60812:2003: Analysis techniques for system reliability - Procedure for failure mode and effects analysis (FMEA)

[IEC 61025] IEC 61025:2004: Fault tree analysis (FTA), 2004.

[JSE96] Jones, B.-F., Sthamer, H., Eyres, D.: Automatic structural testing using genetic algorithms.

Software Engineering Journal, vol. 11, no. 5, pp. 299–306, 1996.
[KCF+04] Klein, T., Conrad, M., Fey, I., Grochtmann, M.: Modellbasierte Entwicklung eingebetteter Fahrzeugsoftware bei DaimlerChrysler. Lecture Notes in Informatics (LNI), Vol. P-45, Köllen Verlag, pp. 31-41, 2004.
[Kor90] Korel, B.: Automated Test Data Generation. IEEE Transactions on Software Engineering, Vol. 16 no. 8, pp.870-879, 1990.
[LBE+04] Lamberg, K., Beine, M., Eschmann, M., Otterbach, R., Conrad, M., Fey, I.: Model-based Testing of Embedded Automotive Software using MTest. Proc. SAE World Congress 2004, SAE paper 2004-01-1593, 2004.
[Lin04] Linder, P.: Modellbasiertes Testen von eingebetteter Software - Ein Ansatz auf der Grundlage von Signalflussplänen. Proc. Automotive Safety & Security 2004, Stuttgart, Germany, 2004.
[MK99] Mackenthun, R., Kelling, C.: An Agenda for the Safety Analysis of Cyclic Software Components. In: ESPRESS: Final Reports, Berlin, Germany, June 1999.
[MTest] MTest (product information), dSPACE GmbH, http://www.dspaceinc.com
[PMS01] Papadopoulos, Y., McDermid J., Mavrides, A., Scheidler S., Maruhn, M.: Model-Based Semiautomatic Safety Analysis Of Programmable Systems In Automotive Applications. Proc. Int. Conf. on Advanced Driver Assistance Systems (ADAS 01), 2001.
[Poh99] Pohlheim, H.: Evolutionäre Algorithmen - Verfahren, Operatoren, Hinweise aus der Praxis. Berlin, Heidelberg: Springer-Verlag, 1999. http://www.pohlheim.com/eavoh/index.html
[Poh05] Pohlheim, H.: GEATbx - Genetic and Evolutionary Algorithm Toolbox for Matlab. http://www.geatbx.com/, 1994-2005.
[Ran03] Ranville, S.: MCDC Unit Test Vectors From Matlab Models – Automatically. Proc. Embedded Systems Conference, 2003.
[Rau99] Rau, A.: Verwendung von Zusicherungen in einem modellbasierten Entwicklungsprozess. It+ti 3/2002, pp. 137-144, 2002.
[Rau03] Rau, A.: Model-based Development of Embedded Automotive Control Systems. PhD thesis, http://dissertation.de, 2003.
[Reactis] Reactis Tester / Validator (product information). Reactive Systems Inc., http://www.reactive-systems.com
[SC03] Stürmer, I. and Conrad, M.: Test Suite Design for Code Generation Tools. 18. IEEE Int. Conf. on Automated Software Engineering (ASE 03), 2003.
[SLSF] Simulink/Stateflow (product information). The MathWorks Inc., http://www.mathworks.com/products.
[SL-VV] Simulink Verification and Validation Toolbox (product information). The MathWorks Inc., www.mathworks.com/products/simverification
[Sth96] Sthamer, H.: The Automatic Generation of Software Test Data Using Genetic Algorithms. PhD Thesis, University of Glamorgan, Pontyprid, UK, 1996.
[TCC+98] Tracey, N., Clark, J., Mander, K., McDermid, J.: An Automated Framework for Structural Test-Data Generation. Proc. 13. IEEE Conf. on Automated Software Engineering, 1998.
[VDI/VDE 3550] Computational Intelligence, Evolutionary algorithms - Terms and definitions. VDI/VDE guideline VDI/VDE 3550, Part 3, Verein Deutscher Ingenieure, 2003.
[Weg01] Wegener, J.: Evolutionärer Test des Zeitverhaltens von Realzeit-Systemen. PhD thesis, Shaker-Verlag, 2001.
[WSB01] Wegener, J., Sthamer, H., Baresel, A.: Evolutionary Test Environment for Automatic Structural Testing. Special Issue of Information and Software Technology, vol. 43, pp. 851–854, 2001.
[WSJ+97] Wegener, J., Sthamer, H., Jones, B., Eyres, D.: Testing Real-time Systems using Genetic Algorithms. Software Quality Journal, vol. 6, no. 2, pp. 127–135, 1997.

CONTACT

Hartmut Pohlheim received a Diploma in Systems Engineering in 1993 from the University of Technology Ilmenau (Germany). In 1998 he earned his PhD at the University of Technology Ilmenau with a dissertation titled “Development and Engineering Application of Evolutionary Algorithms”. Since 1995 he has been a research scientist at the Software Technology Lab of DaimlerChrysler Research & Technology in Berlin.

In 1995 Mirko Conrad earned a Diploma degree in Computer Science from the Technical University Berlin (Germany). In 2004 he received his PhD from the TU Berlin for his work on model-based testing of embedded automotive software. Since 1995 he has been a project manager and research scientist at the Software Technology Lab of DaimlerChrysler Research & Technology. He is member of the Special Interest Group for Testing, Analysis and Verification of Software in the German Computer Society (GI TAV) and the MathWorks Automotive Advisory Board (MAAB).

Arne Griep received a Diploma in Systems Engineering in 2002 from the University of Technology Ilmenau (Germany). Since 2003 he has been a software developer at IAV GmbH (Germany).

2004-01-1768

Supporting Model-Based Development with Unambiguous Specifications, Formal Verification and Correct-By-Construction Embedded Software

Wolfram Hohmann
Esterel Technologies

ABSTRACT

In this paper we will explore how 15 years after being introduced into avionics systems, "by-wire" technologies have entered the automotive world. The use of software within safety-relevant application areas like restraint systems, braking, steering and vehicle dynamics support and control systems, is requiring changes in the processes and methodologies used for embedded software development.

INTRODUCTION

Critical embedded systems in automotive applications must operate correctly under all circumstances. Erroneous behavior may not only cause increased costs and delay time to market, but also endanger the safety of passengers and put the business at risk through legal liability.

It is a well-known fact that software quality cannot be assured by system testing alone. The entire design process must also be defined by a "correct-by-construction" methodology and automation. Experiences from the avionics industry have shown that the use of a formal, deterministic and automated solution both reduces cost and improves quality.

This paper will describe a seamless development flow that combines design methodologies for dataflow and finite state machines with formal verification and correct-by-construction code generation.

The tool suite SCADE Drive™ from Esterel Technologies will be examined to illustrate a correct-by-construction methodology with safe automated implementation.

Conventional software development workflows typically follow this pattern: textual specification, manual translation of the specification into a design, review and verification of the translation, manual translation of the design into source code, manual generation of test cases and test pattern, and finally testing the product. The "conventional" method has no future for the development of critical embedded software in the automotive industry. Not only must the previously-mentioned safety objectives be achieved, but also the industry needs methodology and tools that improve communication between members of the project team, between departments within the automotive organization and between the automotive organization and its suppliers.

A model-based workflow, when automated with a correct-by-construction methodology, solves many of the aforementioned problems. "Correct by construction" means that the code produced is guaranteed to be a 100% match with the software specification. The remainder of this paper will explore this method, from creating the software specification to implementing it as embedded code.

SCADE Drive, Esterel Technologies' model-based programming tool suite uses two semantics that complement each other and result in a complete application-level system behavior description.

THE DATAFLOW SOLUTION

DATAFLOWS

The description of data flows connects internal and external data with logical, mathematical and combinatorial operations. Operations can be further combined into user-defined operators.

The graphical symbols used are well-known to electrical engineers from the areas of control engineering and digital signal processing.

Although the description is created using a graphical methodology, the basis for the models is the formal language Lustre, developed at the University of

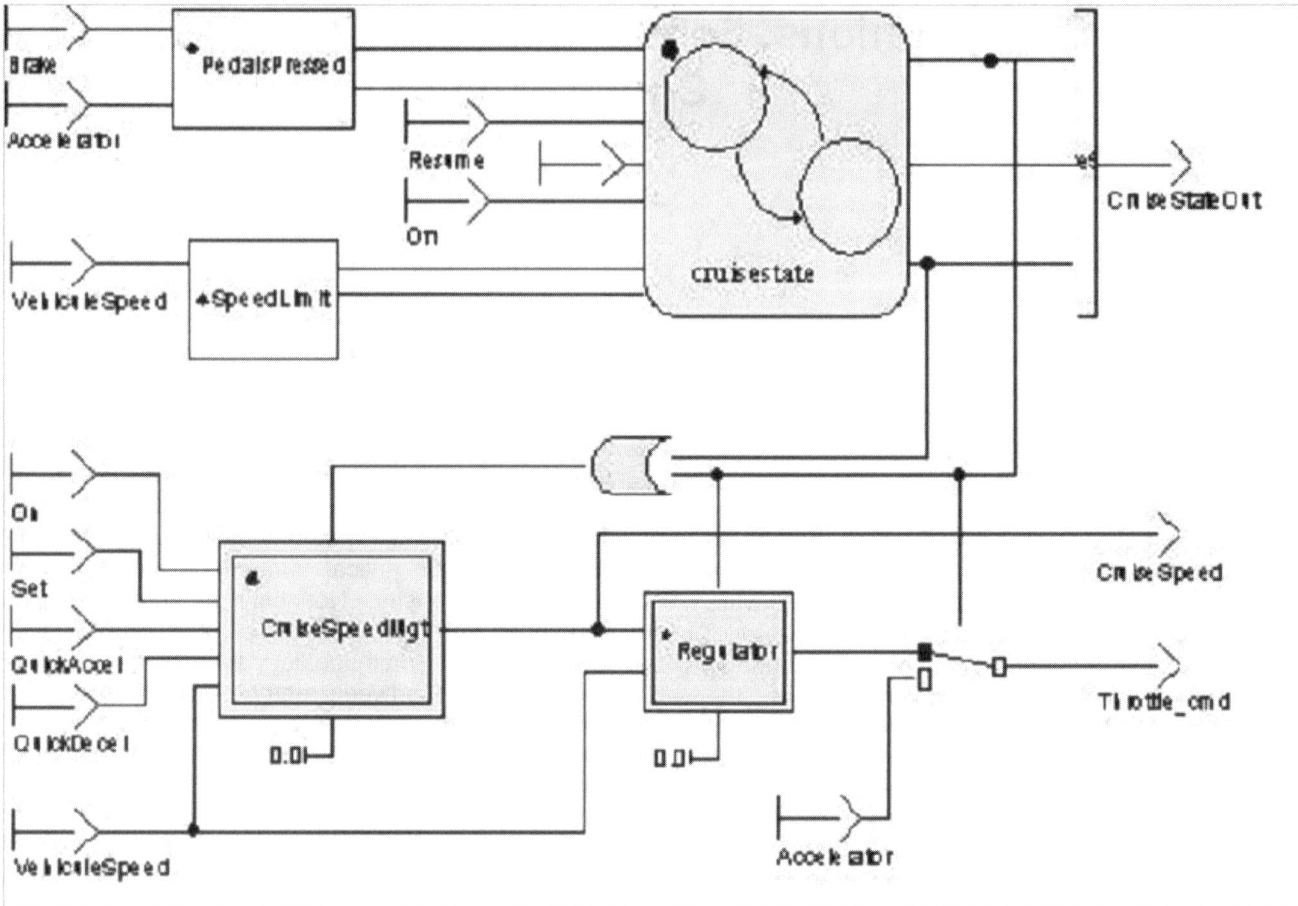

Figure 1 – Data Flows Connect Internal and External Data

Grenoble by Paul Caspi et. al. Data flow diagrams can also contain finite state machines.

STATE MACHINES

The description of finite state machines is accomplished using the formalism of "synch charts". Synch charts have been transformed by Esterel Technologies into a robust modeling method called "Safe State Machines". The finite state machines designed with this formalism can include hierarchies and true parallelisms.

Compared to other methods and tools, designers programming with Safe State Machines use a precise, unambiguous formalism, including the clear assignment of priorities in branched transitions rather than an automatic assignment of priorities based upon the graphical ordering of design elements. It is this level of formalism that assures that the eventually embedded software matches its specification 100%. Tools and methods that use informal processes cannot guarantee this.

Safe State Machines are based upon the formal, synchronous language Esterel originally defined at INRIA and Ecole de Mines by Gerard Berry, now Chief Technical Officer of Esterel Technologies.

DECLARATIVE LANGUAGES, DETERMINISM AND SYNCHRONICITY

The main technological difference between SCADE Drive and competing methodologies is not the look and feel of the tool or even its sophisticated use of symbols. The core of the differentiation is the underlying synchronous languages, Lustre and Esterel, that form the basis of the models. Only the use of these languages enables predictability of simulation, formal verification and correct-by-construction code generation. We will now highlight some of the important specific

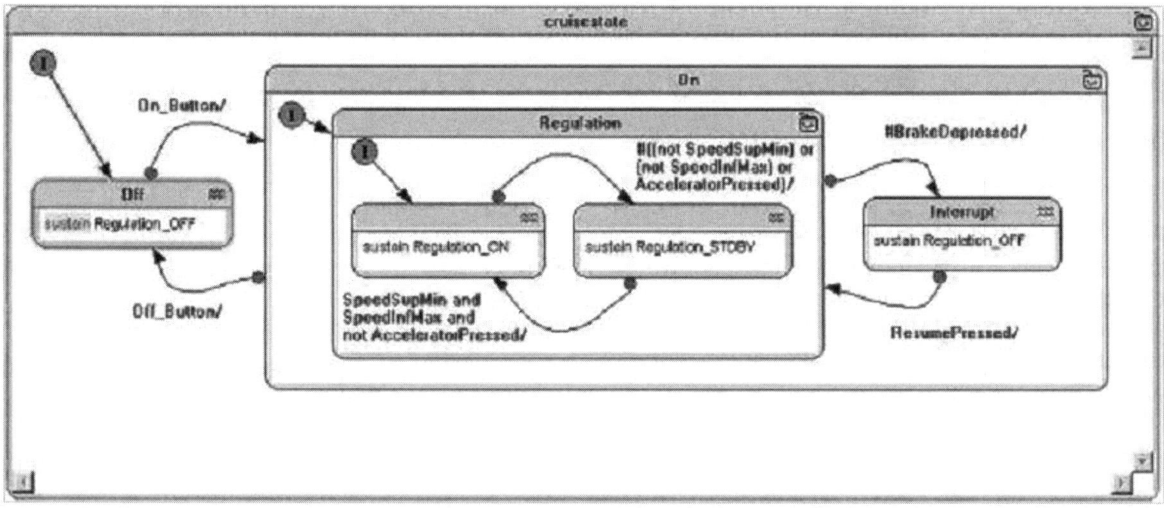

Figure 2 – State Machines

differences between these declarative languages and compare them to imperative languages.

Declarative Languages

Declarative languages differ from imperative languages (like C, C++ or Java) in that statements in declarative languages don't describe the sequence of the program flows such as *first add a to b, then if x <z add a to c.*

Instead, declarative languages describe a set of output values of a system O (t) and the set of the system state S(t) by means of the set of input values I(t) and the set of the previous state. S(t-T). For example:

O(t) = F1 (I(t), S(t-T)) and

S(t) = F2 (I(t), S(t-T))

Synchronicity

Synchronicity is the basis of the computational model in the Esterel and Lustre languages. It is built upon the theory that all computations are executed at a certain logical moment and that **all** computations (calculating the outputs of all blocks by means of their inputs, determining the new state of **all** finite state machines) are executed in the **same** logical moment. Internal events that are generated from "firing" transitions are not inserted into event queues; they will be calculated in the same logical moment when they are generated. The same model is used for data flows that are generated by executing mathematical or logical operations. All operations are virtually executed at the same time, which implies that all feedback loops have to be explicitly calculated based upon the dataflow value that was valid at an earlier moment. This is one of the reasons why time-related operators like y= **Pre**(x) which assigns the value of x to y at a previous moment in time are so important in Lustre. In fact, all types of circular assignments in Safe State Machines will be detected by the tool in a similar way and the programmer notified of these errors.

It should be mentioned that circular event assignments as well as feedbacks lead to potential deadlock situations in asynchronous programming environments, especially if they use "event queues". As a consequence, customers using these tools have defined design guidelines that restrict many unsafe practices and forbid things these tools cannot do well, including parallel state machines and hierarchy. Because the SCADE methodology and tool handle both parallelism and hierarchy elegantly, no such restrictions are necessary.

Determinism

The **Determinism** of behavior is a direct result of synchronicity. Synchronicity guarantees that in a given system state the same sequence of input data always results in the same set of output data and new system state, respectively. Therefore it is possible to clearly predict the system behavior of the embedded software from the simulation results.

SIMULATION AND FORMAL VERIFICATION

SYSTEM AND SOFTWARE TESTS

An important objective of model-based methods for embedded systems is the earliest possible verification of the correctness. Simulation and formal verification are proven techniques that have a long and successful history in hardware development, saving time and effort for verification and preventing cost-intensive rework. Software development, including the development of critical embedded software, has lagged far behind.

Historically, the following issues have bedeviled software development:

1. Testing always starts late, compounded by the fact that massive rework is typically necessary after the first tests.

2. In the traditional flow, hardware development and software development remain sequential, which means that most of the software development work starts with the first hardware-software integration. This typically puts software development on the critical path of the entire system development.

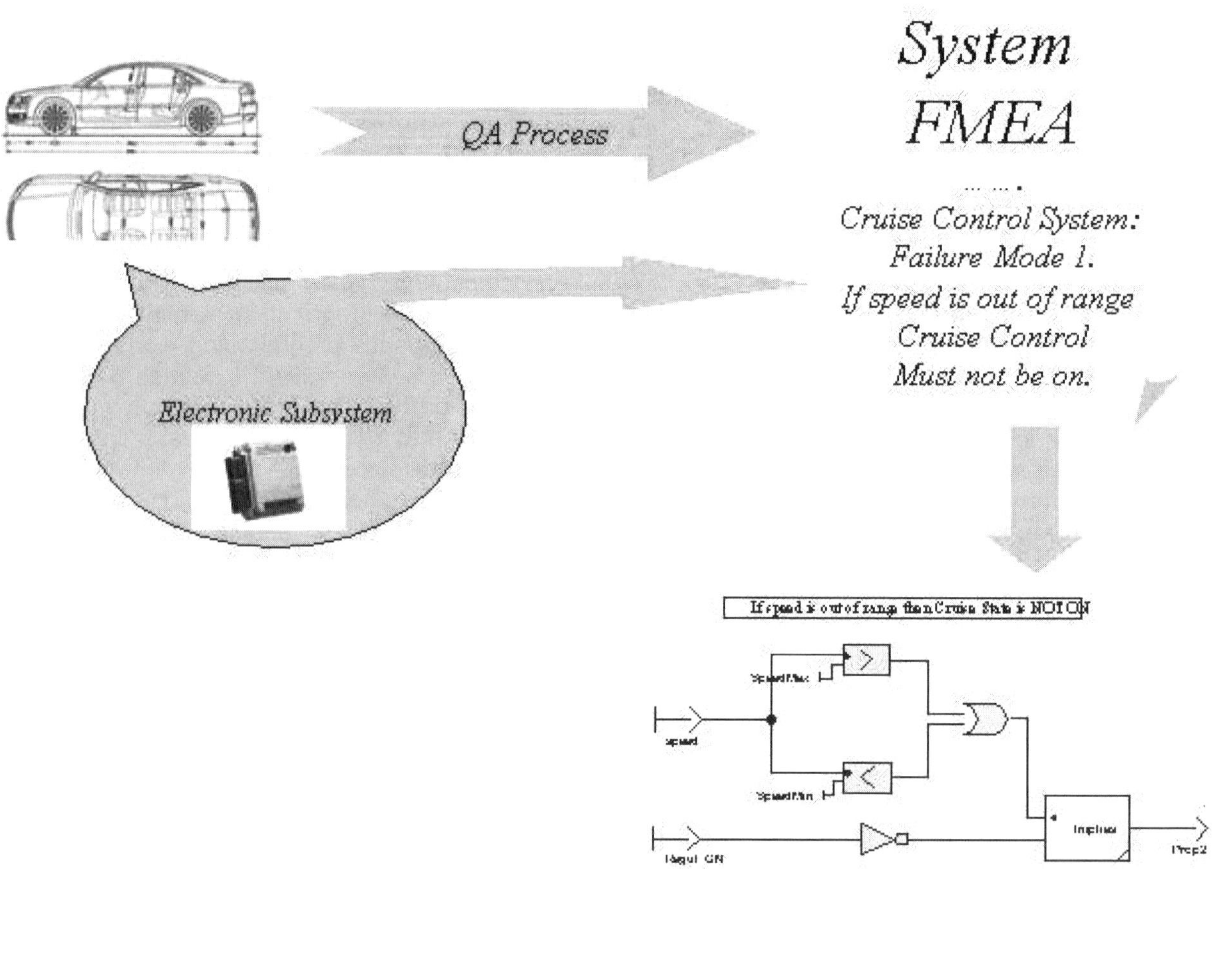

Figure 3 – FMEA System

3. The correct analysis of errors is very difficult due to the fact that during testing the software is embedded into the entire target system.

4. Test coverage measurement is very complicated, and absolute coverage may be nearly impossible to predict. History has shown that even months and months of traditional testing cannot prevent the failure of software components just a few hours after delivery.

SIMULATION

One of the most effective solutions to the need for early design verification is simulation. One basic aspect of this issue is especially important:

The simulated behavior needs to be a clear and unique prediction of the final system behavior. As discussed previously in this paper, this requirement can be achieved best through the use of deterministic models.

RAPID PROTOTYPING

Rapid Prototyping, defined here as simulating system behavior using a prototype version, is a good development strategy IF this same software can then be reused for the embedded code. Otherwise prototyping is a waste of resources and risks the introduction of new errors during the reimplementation of the code. The correct-by-construction methods described in this paper achieve the objectives of rapid prototyping but create embeddable code from the start.

FORMAL VERIFICATION

Traditional test methods define test cases in a positive and forward manner. This means they are based upon the system specification and are intended to prove that if certain paths of stimulation (forward) are followed, the function behaves correctly per the specification (positive). Test coverage is then estimated using various methods that have been created through theoretical and practical studies.

In addition to traditional testing, a quality assurance process based upon Failure Mode Effect Analysis (FMEA) is a de-facto standard in the automotive world and uses a “negative” and “backward” approach. First the application is analyzed from the outside and fatal situations that must be avoided under all circumstances are defined (negative). Next, tests are specified to determine that these situations never occur (backward).

Formal Verification offers an excellent method to fulfill the negative, backward approach of FMEA, creating also the real possibility of guaranteeing total test coverage. Instead of designing test cases by hand, executing them and then trying to estimate the test coverage afterwards, the mathematical correctness of the model itself is checked with formal verification. The fatal situation defined in an FMEA is then proven mathematically to determine if the situation to be avoided will ever occur.

An example: A cruise control system shall only regulate the speed if the speed itself is inside the bounds of minimum and maximum speed. This specific requirement derived from an FMEA can be modeled using an automated formal design and verification method such as the one provided by the SCADE Drive tool.

SCADE automates formal design and verification with Design Verifier*. Design Verifier allows the designer to detect specification errors and validate the design by proving that the required properties hold in all the possible cases and that:

The potential fatal situation can never happen. The test coverage of this result is 100 %, or under certain circumstances the fatal situation can happen. Design Verifier will give precise information about the erroneous scenario.

Formal verification is an excellent enhancement to a standard testing plan and can greatly improve the quality of testing. In fact, for safety-critical applications, formal verification provides such a large improvement in testing quality, it is very probable that we will see the continuation of today’s trend where formal verification is defined to be a state-of-the-art process for automotive embedded software quality assurance.

AUTOMATIC CODE GENERATION

REQUIREMENTS FOR AUTOMATIC CODE GENERATION

The entire workflow can only be seen as being seamless if it is possible to generate portable C code from the designed, simulated and verified model. To reach this objective, the code generator has to fulfill certain requirements:

- The identicalness between the model and the generated code has to be proven.
- The generated code has to obey the highest quality standards.
- The generated code must not have too large an overhead with respect to code size.

In order to fulfill the first requirement, we decided to certify its code generator according to the avionics guideline DO-178B at the highest level of that guideline, Level A. The result is a “qualifiable” code generator. The customer using this code generator must produce a “qualification kit” for the specialists working on the project. The qualification kit is provided by Esterel Technologies. The code generator is qualified on the project and the resulting code can then be certified. Although this extremely strict guideline is not required in the automotive industry the automotive industry is moving toward these same types of guidelines and certification requirements for safety-critical embedded software.

We have been closely watching the automotive software development standard IEC-61508, and fully expect it to become the ubiquitous standard for safety-relevant software development in the automotive industry. In 2003, we began a certification project of its code generator with the German certification authority, TÜV. The goal is to achieve certification of the code generator as being certified for IEC-61508 SIL 4 (the highest level). We have already analyzed its generated C code relative to another emerging standard for development, MISRA, and found it to be in compliance. This analysis has been documented and is available from Esterel Technologies upon request.

One of the concerns sometimes voiced about automatic code generation is that the generated code has too much overhead. While this is true of older, primitive code generators, the latest benchmarks and comparisons of the SCADE code generator have shown that although the output of the code generator is dependent upon the design quality of the model (a model created using “brute force and ignorance” will always result in a suboptimal code), models designed with emphasis on design quality when generated with SCADE produce code that is absolutely comparable in size and performance to hand-coded software.

CONCLUSION

The proliferation of embedded software in safety and business-critical automotive applications, coupled with persistent cost and time to market pressures, has mandated new methods and tools for the development of embedded software.

Lessons can be learned from the Avionics industry, which began going through a similar transition more than 10 years ago and has pioneered the development of methods and tools that automatically produce embeddable code for even the most sensitive of safety-critical applications.

This paper has shown that model-based workflows, formal verification and correct-by-construction automatic code generation maintains or increases the quality and productivity of embedded software development and can enhance existing processes like FMEA.

REFERENCES

1. G. Berry and G. Gonthier. The Esterel synchronous programming language: Design, semantics, implementation. *Science of Computer Programming*, 19(2):87–152, 1992.
2. G. Berry. The foundations of Esterel. In G. Plotkin, C. Stirling und M. Tofte, Editors, *Proof, Language and Interaction: Essays in Honour of Robin Milner*. MIT Press, 1998.
3. Ch. Andre. Synccharts: A visual representation of reactive behaviors. Rapport de recherche tr95-52, Université de Nice-Sophia Antipolis, 1995.
4. N. Halbwachs, P. Caspi, P. Raymond und D. Pilaud. The synchronous dataflow programming language LUSTRE. *Proceedings of the IEEE*, 79(9):1305–1320, 1991.

* Design Verifier is powered by Prover Plug-In. Prover Plug-In is a trademark of Prover Technology AB in Sweden, the United States and other countries.

CONTACT

Wolfram Hohmann, wolfram.hohmann@esterel-technologies.com

2004-01-0720

Managing the Challenges of Automotive Embedded Software Development Using Model-Based Methods for Design and Specification

Mark Yeaton
The MathWorks, Inc.

ABSTRACT

This paper will discuss the issues associated with the creation of embedded software for automotive electronic control systems and show how these issues can be addressed using model-based methods to design, test and implement these systems. Model-based methods are already in use for many automotive applications, and there are potentially many more areas where they could be used, especially as the number and complexity of automotive embedded control systems increase. This paper will cite several examples of the successful use of model-based design.

AUTOMOTIVE EMBEDDED SOFTWARE

Embedded software will continue to be an important factor affecting the cost and development time of automotive electronic systems. In a recent article in *Automotive Engineering International* magazine, the president of a large automotive electronic components supplier states that the total value of software will more than triple, to 13% of a vehicle's cost, between 2000 and 2010. Because of consumer demand and heavy government regulation centered about environmental, safety, and performance issues, it is likely that a fair amount of this software will fall into the class of "control systems," with the remainder classified into other categories such as convenience, entertainment, and similar functions. The expansion of software applications will also be fueled by faster processor speeds and cheaper, more abundant memory. Design teams will need to increase in size and number as the amount of software continues to grow. Information exchange within and between these teams will need to occur rapidly and flawlessly.

Clearly, software development methods will need to keep pace so that vehicle, vehicle subsystem, and component delivery schedules and design budgets are not at risk of being compromised by software faults and failures. Well known research data (*Software Engineering Economics*, Dr. B. Boehm, Prentice-Hall 1981) has shown that the relative cost of fixing software defects increases one hundred times from initial requirements analysis to in-service use. In his book *Software Verification and Validation for Practitioners and Managers* (Artech House 2001), Steven Rakitin cites other supporting data that justifies the economic need to improve design and development techniques so that any defects that do exist are found and fixed as early as possible in the design effort.

EMBEDDED SOFTWARE DESIGN AND DEVELOPMENT CHALLENGES

The increased use of embedded systems in all types of products has challenged the ability of embedded systems software designers to successfully manage their design, development, and implementation. An article in the December 2002 issue of *Software Development Times* magazine cites a market study by Venture Development Corp., a technology market research and consulting firm that specializes in industrial and commercial electronics, computing, communications, software, and power systems markets. Their research data shows that over half (51.6%) of all embedded software design projects are late. Of the 48.4% of the projects that did not make it into the late category, almost one fifth were cancelled and almost one quarter were ahead of schedule, leaving only a small fraction that were judged to be on schedule.

The survey categorized the reasons for the projects being late into five responses: changes in specifications, complexity of the application, inadequate specifications, too few developers, and too few testing personnel. While this data did not specifically address the automotive industry or control system software, it is reasonable to assume that embedded control system software development in the automotive industry cannot be any less of a challenge than the norm in the embedded industry as a whole. This paper will examine how a "model-based design" approach can address the issues cited in the survey in the context of embedded control system software design and implementation.

THE NEED FOR MODEL-BASED DESIGN

Many manufacturers and component suppliers have already reached the point where their embedded control system projects cannot be undertaken using existing practices. These traditional methods typically require that designers write and implement software on prototype systems in order to understand the design. The need to begin the design before product prototypes and target hardware platforms are available has led to the adoption of model-based design methods. Model-based methods let designers construct detailed mathematical models of the product and software behavior. Using simulation, they evaluate their control system software designs on a desktop workstation, rather than in the lab or on a test track.

Historically, model-based design methods have been used in the aerospace industry for years, where the ability to construct working prototypes and test articles is severely limited by cost, safety, and security concerns. Early implementations of model-based design methods required engineers to create models using FORTRAN code and develop solution techniques for the equations that defined the models. While it is still possible that code could be written to implement a model of the physical product and the software itself could provide a model of the functionality of the final application, this is not the normal practice today.

A VISUAL DESIGN ENVIRONMENT AIDS IN UNDERSTANDING COMPLEX SYSTEMS

Today, most model-based design efforts are implemented in commercially available design packages that have the ability to abstract the product model and control system design information from a textual level to a visual level. Using shapes, color, size, and position, visual icons are used to describe the models of both the product itself and the embedded control system software. In addition, hierarchical design partitioning is a natural extension of the visual paradigm, and allows even the largest systems to be represented in a compact and meaningful way. As an added benefit, these hierarchical modeling elements can be easily reused in subsequent design efforts. The ability to visualize these complex software designs and the mathematical models of product behavior allows designers to easily understand and evaluate the design's correctness and performance.

Avoiding the need to write software early in the design process does not give designers free reign to create models and implement designs that are of no use to anyone but themselves. Successful implementations of model-based methods using visual design environments are carefully planned so that all team members adhere to standards and agree to certain practices for creating the models. This leads not only to efficiency within the design team itself, but also facilitates information exchange between vendors and suppliers.

THE EVALUATION OF SYSTEM SPECIFICATIONS BY MODEL SIMULATION

Model-based methods help to address the problem of inadequate specifications, since models can be simulated to determine if the design will meet requirements. Each hierarchical model element can stand on its own and design testing and evaluation can start very early in the model building process. The model becomes more complex and fully defined through the addition of more hierarchical elements or the substitution of "placeholder" or "stub" models with fully defined models. At every step in the construction of the model, simulation continues to give the designer confidence that the design is progressing as expected. This "build a little, test a little" philosophy reduces the likelihood of "surprises" at or near the end of the design effort.

THE MODEL IS THE SPECIFICATION

As visual methods for model-based design have matured, their ability to capture more than just mathematical and behavioral information about the system has improved. Very specific information that defines the actual software implementation, such as data typing, data structures, timing, and software structure can be defined and, where appropriate, evaluated by simulation. Most design tools provide the ability to capture textual information, which is used by members of the design team as a roadmap for design development and for formal documentation purposes. With these abilities inherent in the modeling environment, the model can become the specification for the design, allowing management of changes to be accomplished in the visual environment, rather than in a textual document.

Using the model information as a specification, the automatic creation of documentation and software becomes possible. With flexible and capable code generation tools, the automatically generated code can serve multlple purposes, such as prototyping, calibration, and even actual production implementation, all using the same model.

The ability to automatically generate code from the fully tested specification defined by the model frees software developers from the need to perform the manual task of creating code from a specification. They are able to use their time more effectively for more challenging systems integration tasks and to manually develop the code that is needed for hardware device drivers, communications, and operating system interfaces.

SOFTWARE TESTING CONSIDERATIONS

Since the complete system behavioral model, consisting of the physical product and the embedded software, is fully exercised by simulation very early in the design process, many errors are found and fixed long before final integration and testing takes place. Testing personnel will find fewer errors and will likely work more efficiently.

EXAMPLES OF MODEL-BASED DESIGN IN PROTOTYPE AND PRODUCTION APPLICATIONS

Automotive manufacturers and suppliers have largely adopted model-based design methods, but to different degrees. Each design situation is unique, with varying levels of legacy and opportunity for change that drive the degree to which a model-based design approach can be adopted.

Implementing a model-based design approach does not require complete reengineering of existing software development practices. Prudent business practices generally favor risk mitigation when adopting new technologies, and organizations can accomplish the insertion of model-based design into their existing environment in a staged manner. This can be as simple as "shadowing" a design effort or carving out a small project where model-based design can prove itself in a well-bounded and low-risk way. Then, as confidence is gained and the relationship of model-based design methods to existing practices is understood, full projects can be undertaken.

A key to the success of model-based design methods is the flexibility to integrate easily with existing design and implementation environments. Many well-known manufacturers have made considerable progress using model-based design techniques to the point where real-time prototyping and tuning of control system designs does not require the designer to manually develop any application-specific source code. Some manufacturers are currently using the model information to create software applications that are used directly in production vehicles. Others have built complete hardware-in-the-loop facilities where production embedded engine controllers can be tested without the need for real engines. The real-time software for these simulated engines comes directly and automatically from the models.

A few actual examples from the automotive industry are provided below, and many more are available. The range of automotive application areas where model-based design is used spans everything from the most basic to the most complex. In addition to the examples shown, model-based design methods have proved themselves in applications such as climate control, chassis control, transmission control, motorcycle engine control, and diesel engine control for heavy construction equipment. Even the design and production of equipment for automotive system testing, such as road simulators, has been aided with the use of model-based design.

With new technologies such as fuel cell power plants, engineers are taking advantage of a model-based approach to designing the embedded control and fuel production systems that make them practical for automotive use. Certainly, there will be no shortage of new technologies in the future of the automotive industry. Electric valve actuation, 42V electrical systems, and intelligent vehicles will require their developers to use model-based design approaches or face competitors who do.

EATON CORPORATION

Hybrid electric powertrains for trucks have the potential to reduce total cost and improve performance while also reducing emissions. They combine a traditional internal combustion engine, electric traction motors, a transmission, and energy storage devices that both propel the vehicle and extract and store energy during braking. The operation of these components is coordinated by a central powertrain control unit. Designing and building a prototype hybrid vehicle is a significant challenge. The internal combustion engine, electric motor, transmission, and energy storage devices can be connected in several different configurations, depending on how the vehicle will be used, and these components need to be sized for optimal performance over various routes.

A model-based design approach allowed Eaton to design the system before prototype hardware was available, then gradually integrate the real components and embedded control system using a real-time hardware-in-the-loop approach, where component models provided the specification for real-time software that could accurately simulate the behavior of the components that were not yet integrated.

JAGUAR

To meet demands for increasingly complex new vehicles while continuing to reduce costs, where possible, Jaguar develops and tests new functionality using existing production vehicles instead of building expensive prototypes.

Using general-purpose electronic control units (ECUs) and a model-based approach, Jaguar can support a variety of application areas including transmission, driver entertainment, and body system. The model-based design approach allows initial development to take place off-line on the designer's desktop computer, then easily transition to real systems, allowing more complete and accurate specifications to be sent to suppliers for them to develop the actual system.

MOTOROLA

The Motorola Automotive Group recently developed and optimized software for a battery management controller using a model-based design approach. Developing the controller was difficult because the interfaces among various system components were fluid and subject to change. A model-based design approach proved to be well suited to managing the changing requirements and providing fast design turnaround times. Code was automatically generated from the system model for the fixed-point HC12 microcontroller, using scaling factors that were automatically selected during model simulation.

RICARDO CONSULTING ENGINEERS

Gasoline direct-injection engines differ from other fuel-injected systems in that the fuel is injected directly into the combustion chamber, where it can form a stratified (non-uniform) mixture with the air. This system has the potential to give the engine improved performance and economy while still meeting emission standards. However, these advantages can only be realized with control strategies developed to optimize the appropriate fuel/air mixture.

Ricardo was able to model and simulate the six operational engine modes required, then design a control system for each one, plus manage the transition between each. The final design reduced emissions by approximately 50% and was completed with considerable cost savings.

TOYOTA

Engineers at Toyota needed an alternative to traditional design methods that were neither cost-effective nor efficient and that were hampered by costly or incomplete hardware prototypes and a design process that required re-engineering and re-programming at several steps along the way. A model-based design approach was selected as a way to bridge the gaps in their traditional automotive electronics development and create executable specifications to consolidate the work of spec writers, control designers, and programmers.

Using the model as an executable specification, real-time software for both hardware-in-the-loop testing and electronic control unit prototyping is automatically generated. When the system design was debugged and calibrated, software for the production control unit is automatically generated from the same model used to produce the code for prototyping and calibration. Toyota has shortened design cycles, reduced the amount of hardware prototypes required, and developed innovative new products as a result of adopting a model-based design approach.

SUMMARY

Designers of embedded system software in all industries face challenges as their designs progress from requirements to implementation. A model-based approach to embedded control system design is a proven technique for addressing challenges faced by automotive embedded systems engineers. This paper has shown how each challenge is addressed by the model-based design approach and has described actual usage of model-based design methods in commercial environments.

Born out of necessity in the aerospace industry, the concept of model-based design has fully evolved from a primitive, do-it-yourself endeavor to a well-accepted, commercial, off-the-shelf practice that can be used to the extent needed by a given project or organization. Today's modeling environments for embedded control systems are flexible, adaptable and can contain all the information necessary to define the complete embedded control system design and implementation. This gives organizations the immediate benefits of modeling and simulation, as well as the ability to fully exploit the model information for real-time software and documentation when they are ready.

Significant savings in time and cost over traditional design methods are possible. Model-based methods allow designers to begin design work earlier and confirm requirements and performance without the need to create hardware prototypes and write software. The ability to express the model in a visual environment lets designers easily share and reuse design information, avoiding the introduction of errors due to misinterpretation. The net result is that fewer errors are found during final testing and implementation of the software design and better products are produced.

2004-01-0709

A Development Method for Object-Oriented Automotive Control Software Embedded with Automatically Generated Program from Controller Models

Kentaro Yoshimura and Taizo Miyazaki
Hitachi Research Laboratory, Hitachi, Ltd.

Takanori Yokoyama, Toru Irie and Shinya Fujimoto
Automotive Systems, Hitachi, Ltd.

ABSTRACT

This paper describes a development method for object-oriented automotive control software embedded with automatically generated programs from controller models. We have designed control software with object-oriented application frameworks. An application framework consists of objects. Each object is composed of the attribute of a control system and a function to calculate the status. The function is a C function, which is automatically generated from the block diagram of a controller model designed with a computer aided engineering tool. An object is implemented with a wrapper and the generated C function. The wrapper defines the interface and attribute of the object. The wrapper is automatically generated referring to the generated C function.

We have applied this method to develop some parts of automotive engine control software and have found that 95% of the source code could be generated automatically from controller models.

1. INTRODUCTION

The functions of an automotive electronic control unit (ECU) are increasing to meet environmental regulations. Since ECUs are also networked together, the amount of embedded software being developed is increasing, so there is an increasing demand for improving development efficiency by reusing software.

However, many kinds of conventional embedded control software systems do not have software architectures that encourage the reuse of software components, and the many operations required to reuse these kinds of software has become a major problem. Therefore, object-oriented software development, which excels in the reuse of software components, has been attracting attention [1].

In the field of control design, model-based design is becoming more important [2]. For example, an engine controller model is designed with a computer aided design (CAD) tool, and the controller model is checked by simulation on the CAD tool. C code can also be generated from the controller model. Recently, the quality and efficiency of the code is becoming the level to apply a production [3], but a development method has not been established that generates embedded code automatically for production. The establishment of such a method is greatly needed.

The objective of our research is to propose a development method that integrates object-oriented software design and automatic code generation. This paper describes a way to develop object-oriented automotive control software embedded with an automatic generated program from controller models. The main feature of this method is that a wrapper wraps an automatically generated function, which is handled as an object, and the wrapper is automatically generated, too. Therefore, the automatically generated function can be embedded efficiently.

In this paper, we first propose an architecture for automotive control systems. Next, we explain a development method with some examples. We also evaluate the effect of the proposed method. Finally, we discuss the effectiveness of the development method.

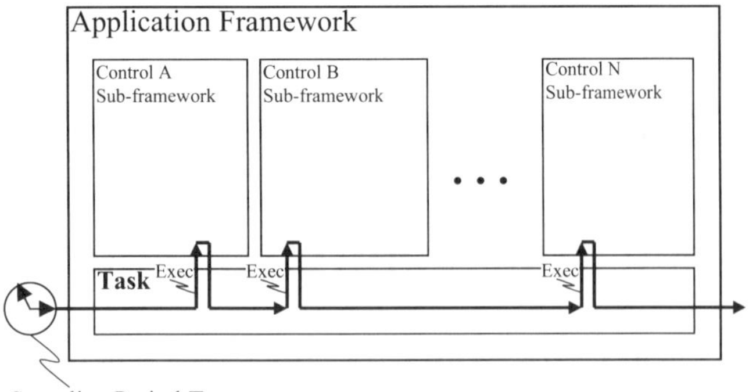

Fig.1 Application framework

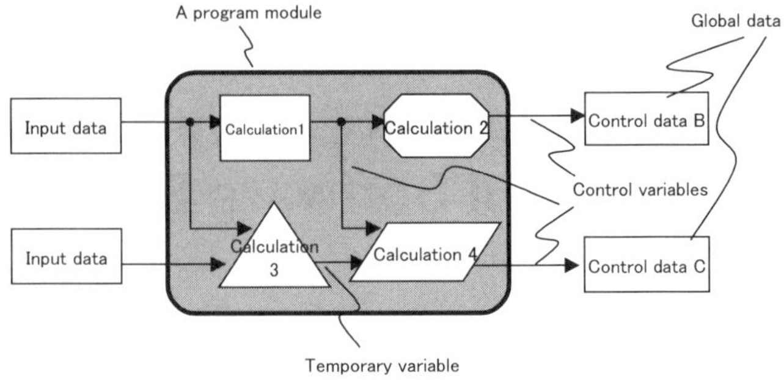

Fig.3 Conventional Software Component

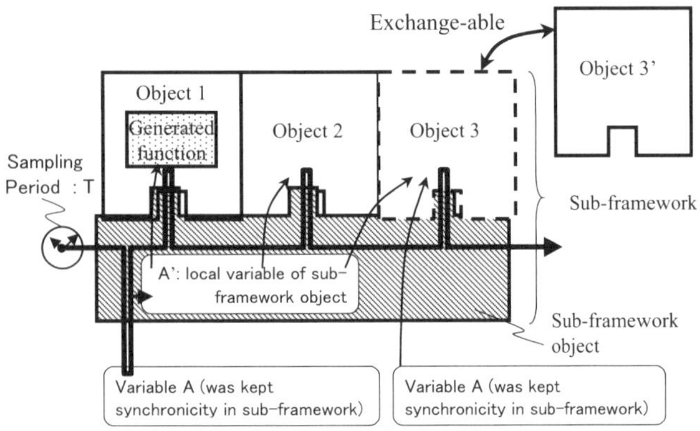

Fig.2 Sub-framework

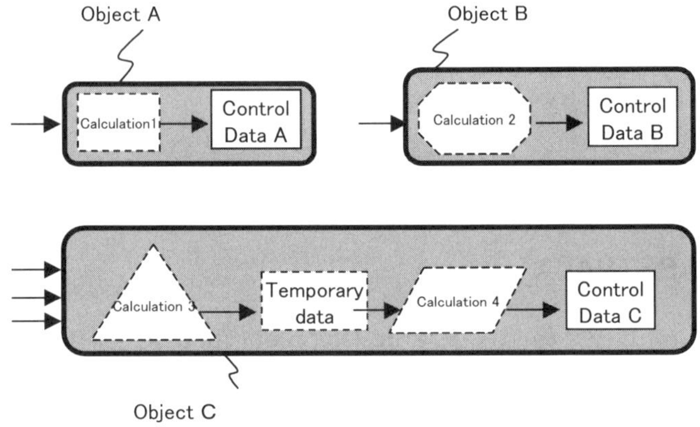

Fig.4 Proposed Software Component

2. SOFTWARE ARCHITECTURE

2.1 APPLICATION FRAMEWORK

We used the concept of an "application framework" for constructing the control software architecture. An application framework is a standard pattern of objects to implement the functions of the application [4]. We have proposed a model of time-triggered object-oriented software for embedded control systems [5]. The framework to be presented in this paper is based on this model.

A control system is a combination of software components that defines a control function. As Fig.1 shows, a framework consists of several "sub-frameworks". Each sub-framework defines a control function. Fig.2 shows a basic model of a sub-framework. The sub-framework consists of one or more objects and a sub-framework object. An object calculates and stores a data value. A sub-framework object performs a control function by activating the value objects in the order designed by the control system designer. We can modify the control logic easily by exchanging value objects.

An embedded control system is generally a multi-task system that consists of several control functions. Each function has sampling periods and priorities. The sub-framework object keeps the synchronicity of the data value, in case several value objects in the same sub-framework refer to the same value of an object. So the sub-framework can operate the control function under the preemption control. We can limit the number of data values to be synchronized to reduce the overhead of synchronizations.

GRANULARITY OF VALUE OBJECTS

Application software should consist of reusable components. Therefore, the independence of the objects is important. We should make the objects reusable in terms of size and consider the meaning of the object in the control logic.

We adopt a data-centered method to make reusable components. In other words, we choose the controller system's status variables as the granularity of objects, e.g., input/output values, system's observed status variable, target values of the system, etc. These data are rarely deleted or added when the control logic is changed. Therefore, we can build a stable structure of objects and we can modify the control logic by exchanging the value objects.

The application software is usually decomposed into components by functions, to make the implementation easy. This makes the initial components coarse. This is the reason why several control functions and values are combined in a component, as shown in Fig.3. To be decomposed into functions is the reason why control software is complicated and not very reusable.

We adopt the data-centered method to take an application to reusable components. As shown in Fig.4, an important control value is chosen and an object is

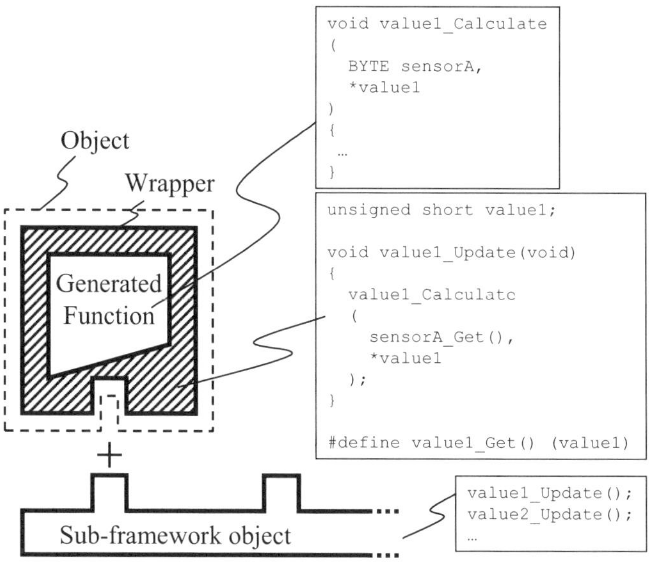

Fig.5 Wrapping

defined as the object calculates and stores the control value. "Important control values" are necessities of the control system for all implementation methods of the control logic. Although the control logic is modified, exchanging the object is easy. Moreover, the size of the object is small and the effect range of modification is clear. So unit tests of objects are easy.

2.2 IMPLEMENTATION OF OBJECTS

Objects are written in the C language, which is not an object-oriented language, for efficiency of implementation size [6]. A method for calculating the attributes of objects is automatically generated as a C function by commercial code generation software [7], from block diagram models of the CAD tool. The calculation methods are generated as functions with the syntax "*value_name*_Calculate()". Input values are assigned as arguments of the C function. The output value's address is assigned the argument pointer of the C function.

To implement automatically generated functions as objects, we use wrappers. Each wrapper corresponds to a function, as shown in Fig.5. A wrapper declares a public attribute (= data) of the corresponding object and defines the data update method and data access method.

Declaring an attribute

Each wrapper declares a public attribute (data) of an object to store the result of a calculation. The attribute is a variable that is named as the value name (i.e. value1 as shown in Fig.5).

Data update method

The data update method executes a calculation for the attribute of an object. The data update method is a function named "*value name*_Update()", as shown in

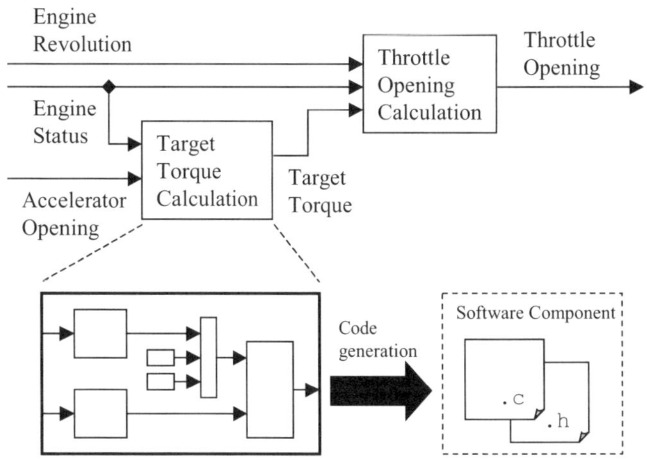

Fig.6 An example of the block diagrams

Fig.5. This method gets input argument(s) from automatically generated functions by calling the data access method of the object that calculates the variables as arguments. This method also assigns the variables as input arguments of the function "*value name*_Update()" . This method assigns the pointer of the(an) attribute as the output argument of the function.

Data access method

The data access method is called when another object refers to the result of the calculation of the object. This method is implemented as the macro "*value_name*_Get()" for efficiency of size and execution.

We have developed the software that generates a wrapper automatically. The software generates a wrapper by analyzing the C source code of an automatically generated function. We call the software "Wrapper Maker".

One of the features of our method is that the interface of the data update method doesn't have any arguments. When the control logic is modified, a set of data is often added or deleted, whose data each object refers to during calculations. Our implementation method can hide the modification of the data set inside the wrapper's data update method, which was generated automatically. Therefore, we don't need to modify a sub-framework object that calls the data update methods of value objects. We can reuse the sub-framework object without any modifications.

3. DEVELOPMENT METHOD

3.1 DEVELOPMENT FLOW

We propose a development method for application frameworks for embedded control systems that is continuous using CAE/CAD. We will explain the proposed method by giving an example of an application for an automotive engine control system.

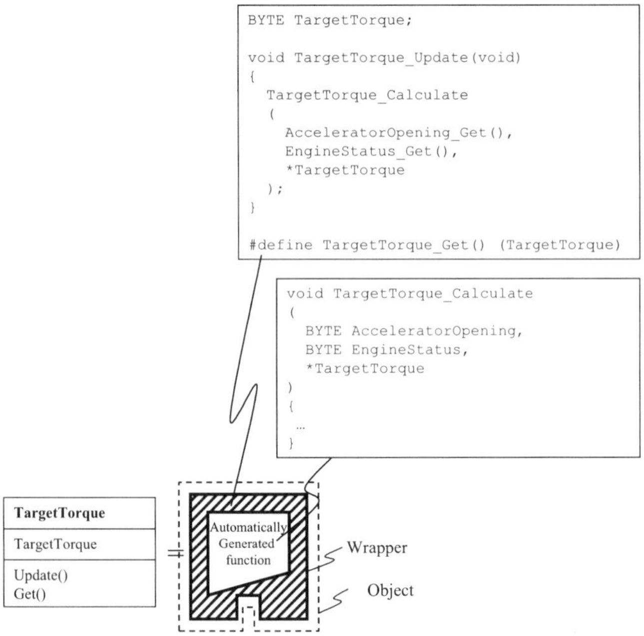

Fig.7 An example of the objects

3.2 CONTROLLER MODEL DESIGN

Fig.6 shows an example of a block diagram model of a controller. In Fig.6, the block shows a calculation, and arrows show the direction of data flow. This example shows calculations for the target engine torque and the target throttle opening. Details of calculations are described inside each block. In Fig.6, the values of engine revolution, engine status and throttle opening are input values, which are calculated by other control functions. Control is realized by executing the control logic periodically, which is described by the controller model.

Controller models are designed with CAE/CAD tools for control [8]. The controller models can be simulated in CAE/CAD tools, so we can check the validity of the control logic with a simulation. After the validation of the control logic, we design the quantization of fixed-point numbers.

3.3 AUTOMATIC GENERATION OF OBJECTS

After the controller model has been validated, we generate objects automatically with the code generation tool and Wrapper Maker. The granularity of value objects is the status variable of the controller systems, e.g., Input/Output values, the system's observed status variable, the target values of the system, etc. In Fig.7, "Target Torque" and "Throttle Opening" are value objects. Fig.7 shows a Target Torque object as an example of an automatically generated object.

First, a calculation function is automatically generated from a block of the controller model with the code generation tool. The calculation function is an ANSI-C function named "TargetTorque_Calculate( )" for implementation. This function has the input values "Accelerator Opening" and "Engine Status". This function

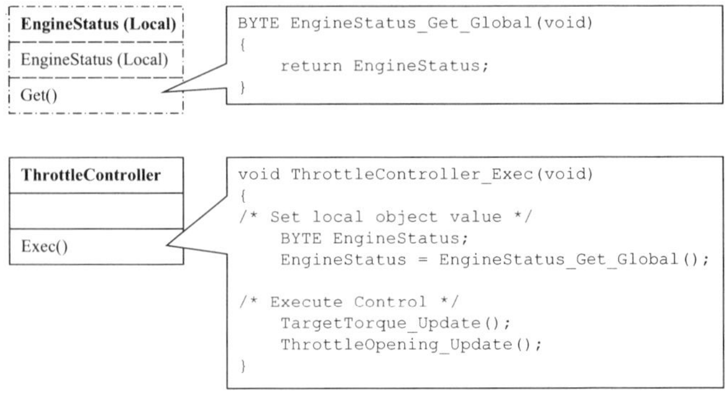

Fig.8 An example of the sub-framework object

outputs a calculated value to "TargetTorque" that is assigned as a pointer. As for the other block of the controller model, we can also generate the calculation functions automatically.

Next, we generate objects from automatically generated functions with the wrapper. A wrapper corresponds to an automatically generated function. The Wrapper Maker generates the wrappers by analyzing the input values of the function "TargetTorque_Calculate()". The object name is the value name "TargetTorque". The wrapper declares a variable "TargetTorque" as an attribute of the object. The wrapper defines the "Update" method by executing the calculation and updating the attribute. The update method gets the values of the input parameters "AcceleratorOpening" and "EngineStatus", using a data access method "Get()" of the objects that calculate the input data. Then the update method assigns the input data to the automatically generated function "TargetTorque_Calculate()". The wrapper defines the data access method "TargetTorque_Get()" that returns the attribute value. This method is a macro to manage the values' synchronicity. This management will be explained after.

3.4 DESIGN OF A SUB-FRAMEWORK OBJECT

Fig.8 shows an example of a sub-framework object. In Fig.8, the sub-framework object consists of a sub-framework object (ThrottleController) and a local object (EngineStatus(Local)). The sub-framework object controls the order of execution of value objects. The local object manages the synchronicity of values.

The sub-framework object calls the data update method of the value object. We enter the data update method in the "*sub-framework*_Exec()" method of the sub-framework object. The sub-framework object has to activate the value objects in a procedure that doesn't contradict the data flow of the block diagram of the controller model as shown in Fig.6. The sub-framework object calls the data update methods from the head to the tail of the block diagram.

A local object is generated for data that is managed for synchronicity of values. In Fig.6, although "EngineStatus" is updated in other sub-frameworks, 2 objects in this sub-framework refer to "EngineStatus". If

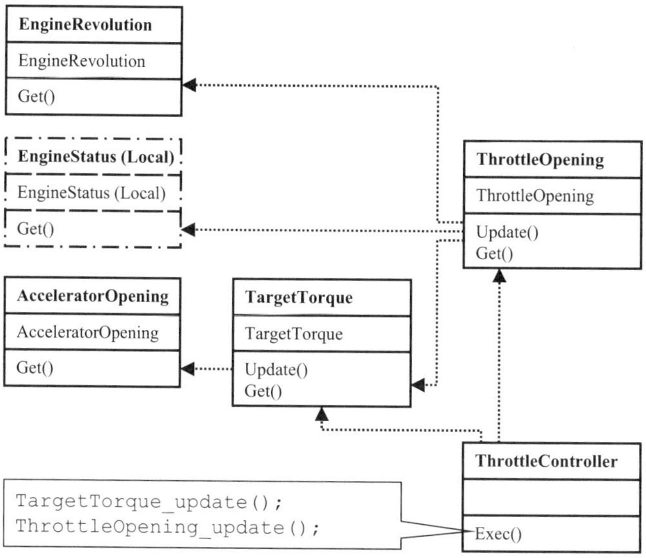

Fig.9 An example pattern of the sub-framework

the priority of the sub-framework that updates "EngineStatus" is higher than the priority of the sub-framework as shown in Fig.6, "EngineStatus" could be updated, interrupting the execution of the sub-framework. So a local object of "EngineStatus(Local)" is generated to manage the synchronicity of values. The local object declares a local variable "EngineStatus(Local)" and copies the attribute of "EngineStatus" to the local variable. The access method "Get()" is a macro, so the scope of the access method "EngineStatus_Get()" is moved to the local variable.

Fig.9 shows the pattern of the sub-framework of this example. The sub-framework object "ThrottleController" calls the update methods "TargetTorque_Update()" and "ThrottleController()" in turn. "TargetTorque_Update()" calls the access methods "EngineStatus_Get()"(Local) and "AcceleratorOpening_Get()", and then calculates the value of "TargetTorque". "ThrottleOpening_Update()" calls the access methods "EngineRevolution_Get()", "EngineStatus_Get()"(Local) and "TargetTorque_Get()", and then calculates the value of "ThrottleOpening".

3.5 SYSTEM ARCHITECTURE

Fig.10 shows an example of the proposed embedded control systems. The software consists of an application framework and platform software.

The application framework has periodic tasks (ex. 10[ms], 4[ms]) and event triggered tasks (ex. Engine synchronous task). Tasks call the "Exec()" method of the sub-frameworks.

The platform software consists of a real-time OS and an Input/Output driver. We have adapted an OSEK-OS [9] for the real-time OS. To enhance the portability of the application software, the interface of the I/O driver is independent of the hardware.

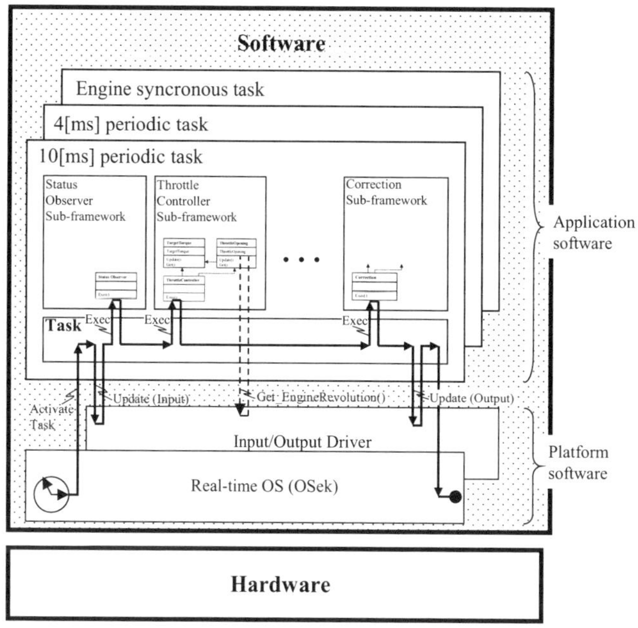

Fig.10 An example of the proposed control system architecture

Fig.10 outlines a behavior of the proposed systems. At first, the real-time OS activates a periodic task (ex.10 [ms]) when a timer interrupt occurs. At the beginning of the task, the task requests the input driver to measure the external input data. Next, the task calls "Exec()" methods of the sub-framework controller object in this order: "Status Observer", "Throttle Controller", ... ,"Compensation". Each sub-framework object calls the "Update()" methods of value objects. Each value object refers to input data that are attributes of the other value objects and external input data. The value object calculates and updates its attribute. At the end of the task, the task requests the output driver to control an actuator.

3.6 REUSE OF SOFTWARE

Basically, a control software that has been created in the past is the basis for a new control software, and only the addition / correction part of the control software is changed. Because the changes are inside of objects, a sub framework object can be used as is. That part of an object, which does not have to change the control logic, can be reused as is. Only the objects to be changed will be replaced.

On the other hand, to add new controls, changing a sub-framework object may be needed. That is, the sub-framework object is modified to define the order to call a new object. The generation of a local object corresponding to the new control may be added. A sub-framework object, which does not have change of control, is reusable as it is.

The new control software is completed when the objects of the new and changed parts are made to unite with the control software used as a base.

As shown above, efficiency can be increased by generating an object with a wrapper to implement the automatically generated program in an object-oriented embedded control system. Furthermore, by constructing a library of objects, software reusability is expected to improve.

4. APPLICATION AND EVALUATION

4.1 CASE STUDY

We applied the proposed system to some control factors of an automotive engine, and we evaluated the rate of automatic code generation, and reusability. The target controls are an estimation of the quantity of intake air, wall flow compensation control, and torque-based control.

4.2 EVALUATION

We generated the implementation code from a controller model, and we evaluated the rate of automatic code generation. We only need to code manually the sub-framework object. In this case, we can generate more than 95 % of the target control software.

Furthermore, to evaluate reusability, we have changed the control logic of the torque-based control, whose development was completed once. Since we generated objects in the unit of the variable, which is meaningful in control with little additional change, parts with more changes can be concealed inside objects with this proposed method. Therefore, most objects and sub-framework objects are reusable. As a result of changing the torque-based control logic, we modified only a few value objects, and changing the composition of the objects was unnecessary. Therefore, we didn't have to change the value object and sub-framework object. We could reuse them.

As shown above, it is possible to improve the productivity of an embedded control system with the proposed method.

5. DISCUSSION

During the development process of a conventional embedded control system, the application software programmer understands the control specifications that the control designer drew up, and the programmer creates the software based on the control specifications. That is, the software which represents the intention of a control designer, can be created when that control designer and an application software programmer have a common understanding. To do this, a control designer has to write the control specification document correctly in natural language, and a software programmer has to understand the document correctly.

Recently, the control development which used the Computer Aided Design tools, is coming into use in the control field [2]. CAD tools express control logic with a block diagram or UML with sufficient readability. CAD tools make it possible to simulate and check the contents of the control logic before software is coded.

In response to such a background, a technology which generates an implementing code automatically from a controller model is also being developed [3]. By using automatic code generation, human error can be eliminated from the disagreement in the specification between a control designer and an application software programmer. Also, a mistake in the coding can be eliminated. In the latest tool, efficiency of code size is also reaching a level useful for production. Not only improvement in software productivity but also agreement between the control specification and program can be reached by generating a program automatically from controller models. However, it is not easy to replace the whole application by an automatic generated program.

Moreover, since the scale of embedded control software in the field is increasing, improvement in productivity and quality has been a demand in recent years.

The reuse of software is seen as the leading solution [1]. Development of the object-oriented software structure gives reusability, but since advanced software techniques are required, it has not yet come into widespread practical use.

The proposed method integrates object-oriented programming and automatic code generation, based on an application framework technology, developing the new technology of automatic wrapper generation, and making construction of the embedded control system feasible. We have thereby made it possible to easily implement the automatically generated program from the control model, and we expect that we can improve software productivity.

6. CONCLUSION

We have proposed a development method that integrates object-oriented programming and automatic program generation. Objects are generated automatically from the controller models with wrappers. Wrappers can also be automatically generated.

Next, we will study the automatic generation of a sub-framework object from a control model block diagram. Moreover, application to a real product is due to be evaluated in terms of quality and cost.

REFERENCES

1. Hermsen, W. and Neumann, K. J. Application of the Object-Oriented Modeling Concept OMOS for Signal Conditioning of Vehicle Control Units. SAE technical paper series, No. 2000-01-0717, 2000.
2. Freund, U., von der Beeck, M., Braun, P. and Rappl, M. Architecture Centric modeling of Automotive

Control Software. SAE technical paper series, No. 2003-01-0856, 2003, Detroit.
3. Thomsen, T. Integration of International Standards for Production Code Generation. SAE technical paper series, No. 2003-01-0855, 2003, Detroit.
4. Johnson, R. E. Frameworks = (Components+ Patterns). Communications of the ACM, Vol. 40, No. 10, pp. 39-42, 1997.
5. Takanori Yokoyama, Hidemitsu Naya, Fumio Narisawa, Satoru Kuragaki, Wataru Nagaura, Takaaki Imai and Shoji Suzuki. A Development Method of Time-Triggered Object-Oriented Software for Embedded Control Systems. Systems and Computers in Japan, Vol. 34, No.2, pp.43-54, 2003.
6. Narisawa, F., Naya, H. and Yokoyama, T. A code generator with application-oriented size optimization for object-oriented embedded control software. Object-Oriented Technology: ECOOP'98 Workshop Reader, LNCS-1543, Springer, pp.507-510, 1998.
7. dSPACE GmbH. Production Code Generation Software. Solutions for Control, pp.100-113, 2002.
8. MathWorks. The MathWorks - Simulink, www.mathworks.com/products/simulink/.
9. OSEK/VDX: Open systems and the corresponding interfaces for automotive electronics, http://www.osek-vdx.org/osekvdx_OS.html.

CONTACT

Kentaro Yoshimura - Vehicle Control Unit, Hitachi Research Laboratory, Hitachi, Ltd.

(MD#104) 2520 Takaba, Hitachinaka-shi, Ibaraki-ken, 312-8503 Japan

E-mail: yosimura@gm.hrl.hitachi.co.jp

2004-01-0707

Development of an Engineering Training System in Hybrid Control System Design Using Unified Modeling Language (UML)

Hisahiro Miura, Masahiro Ohba, Masashi Tsuboya and Atsuko Higashi
DENSO E&TS TRAINING CENTER CORPORATION

Masayuki Shoji
Nippon System Gijutsu Corporation

ABSTRACT

In recent years, automobiles have come to have more and more electronic control components. Further complexity and multi-functionality of such modern in-vehicle components have required a great amount of control software development work in a short period. System engineers have to respond to the demand accordingly, developing high quality software with high efficiency. The Unified Modeling Language (UML), one of the object-oriented technologies, has a great potential for providing a better solution.

This paper presents the effectiveness of the object-oriented method in an embedded system development and examples of analysis and design processes in the system. We have developed the hybrid control system paradigm using UML. Hybrid vehicles include a motor generator that functions as an electric motor as well as a generator, and it can provide auxiliary driving power even when the engine is stopped. This system indicates that object-oriented technology is useful in every phase of system developments: analysis, design and programming. Each object is a reusable software component due to its independence in the system.

This system model shows that UML can be advantageously used in automotive control system developments. The result enables us to develop an engineering training system in "Control system design using object-oriented development technology" to achieve high quality system developments.

INTRODUCTION

The number of automotive electronic control units is increasing to cope with environmental issues and to enhance comfort in the vehicle. Each unit has a lot of targets to control, and controlled objects are increasing in complexity. The amount of control software development is also rapidly increasing to implement many functions, and thus process methodologies with high efficiency are required to maintain a high quality of software.

Object-oriented technology is expected as one of the methodologies to solve these problems due to reusability of the existing software and the high independence of software components. The Unified Modeling Language (UML) can visualize and specify a target system throughout the development process using the unified modeling notation, and is especially expected as a solution to eliminate repetition of development procedures or to create a system without errors. Reuse of software components such as abstracted classes in UML allows the software to be developed more efficiently.

The methodology can greatly help us to develop high quality software, but it is impossible to complete the process only depending on CAD tools. After all, engineers who understand the characteristics of the methodology, and are able to analyze user requirements and promote system design can improve system development.

The UML methodology has already been established and introduced in books or training courses. The only problem is that these materials do not always eliminate system-specific difficulties in practical systems. Usually UML education only covers case examples of business models. When UML methodology is applied to embedded systems, knowledge of such education is insufficient to analyze a system or clarify associations, and the development may be conclusively terminated.

We conclude that we need to use a development example tightly focusing on a target system to educate engineers who can develop a practical embedded system. This paper presents the effectiveness of object-oriented technology, and the processes of analysis and design appropriate for embedded system developments using UML.

DEVELOPMENT PROCESS USING UML

The software development using UML consists of the following steps:

-Modeling by a use case diagram

A use case diagram visualizes requirements in a system containing actors and use cases, and specifies operations of the system.

-Identification of objects

This step identifies things that have functions to implement the system.

-Identification of classes and definition of association between classes

A class diagram involves structural features of the system. There are two structures: a logical structure, and a physical structure that depends on the hardware.

-Definition of attributes and behaviors of each class

This step defines information necessary to achieve roles which are associations between classes, and ways to process the information.

-Implementation of classes

The program of the process is implemented.

-Allocation to system tasks

The control system is implemented by combining functions to send messages and give events to the classes.

These steps can show the system using a diagram corresponding to the level of abstraction. A way of analysis in each step should be changed for a more suitable process according to the characteristics of a target system.

APPLICATION TO EMBEDDED SYSTEM

An embedded system does not standardize an interface to transfer information coming into and going out of the system. Additionally, sensors or actuators are frequently replaced to improve the system.

It is important for us to establish a way to associate artifacts with different levels of abstraction produced in each development process in order to educate engineers who can deal with the characteristics of the system and develop it using UML.

USE CASE WITH "ADAPTATION LAYER"

A use case diagram shows services that a system provides for its actors. A straight line is drawn from the actors to the use cases to represent an association. The lines indicate that the actors and the system exchange information. GUI fulfills the same role in enterprise systems. GUI is a standard library and is available as a steady software component. On the other hand, in embedded systems, an actor which is a driver does not communicate with control systems directly. The driver gives operational information about the steering or the throttle through sensors, and actuators drive the vehicle in accordance with operations on the control systems. Because those automotive components constantly improve their performance, their control software also needs to be changed.

When a use case describes a system, there should be an application layer to achieve the driver's requirements logically, and a hardware layer to inform the application layer of the driver's operations or the system requirements in order to describe concrete use cases.

This paper defines the scope of communication which is controlled as an "adaptation layer". The adaptation layer is a subsystem to control system input or output. The definition of this layer can separate logical processes of the system in the application layer, and input or output in the hardware layer.

In the development of the application layer, we can focus on the functions for essential controls in the system. As a result, the developed application can be generalized as one of the software components including control algorithms, even if hardware devices are changed.

IDENTIFICATION OF OBJECTS

Objects are identified based on a use case in text using robustness analysis which can show information flows. The robustness diagram consists of three elements as follows:

-*A boundary element* is an interface.

-*A control element* processes and evaluates information.

-*An entity element* maintains information.

These elements are organized into groups and identified as objects by the following guidelines:

-Control elements can be objects.

-Control elements are organized with the entity elements associated with them.

-In case there are no associated entity elements, control elements are organized with boundary elements.

-When boundary elements are objects like a timer in a system, the elements are organized with entity elements.

In embedded systems, because a boundary element which depends on an external device is frequently changed, we make it so that such an outside modification does not influence the system by specifying the boundary element as an interface class of UML.

DEFINITION OF BEHAVIORS OF OBJECTS

A sequence diagram to describe messages between objects shows behaviors of objects identified in a use case. The objects receive messages from the actors or other objects, behave accordingly and change their states. The change of the state is shown in a statechart diagram. The diagram must contain two elements: 1) all states of an object and 2) the action and the state transition corresponding to a message. Because one use case describes a certain scene, the sequence diagram based on the use case naturally shows the behaviors there. A behavior of a certain object can be defined by organizing the same one in sequence diagrams based on different use cases. A message that an object receives in a sequence diagram is the event in the statechart diagram, and, similarly, a process or a sent message becomes an action. According to the rules, a statechart diagram can be constructed using a sequence diagram and show a behavior of an object.

DEVELOPMENT OF EDUCATIONAL MATERIAL BASED ON A HYBRID VEHICLE SYSTEM (HVS)

The above development procedure using UML verifies that an automotive embedded system can be readily developed, and enables us to develop educational material for a software development process using HVS.

TARGET OF CONTROL

Educational material for an embedded system is required so that course participants can examine outcomes of practices and actually carry out operations on a real automotive embedded system. To meet these requirements, we selected a hybrid control system as a target, which involves a function to stop the engine while idling, an auxiliary driving power while the engine is stopped, and a function of "motor assist " when starting a vehicle, and implemented the controls to stop idling and to regenerate energy during deceleration in the system.

ANALYSIS OF SYSTEM REQUIREMENTS

We specified the use case diagram based on the instruction manual of the hybrid system. The diagram shows the requirements of the actor, a driver (refer to Figure 1).

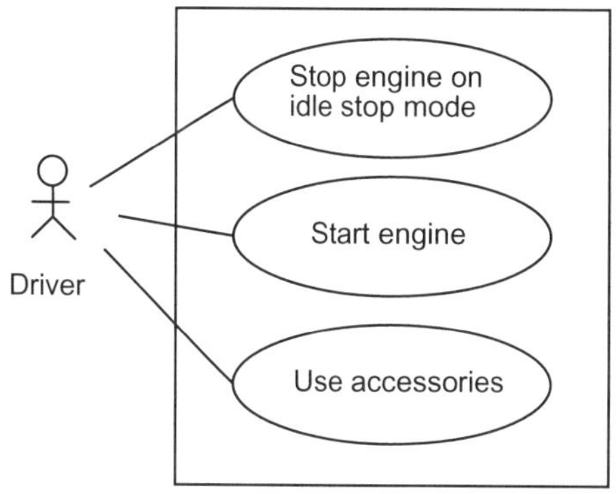

Figure 1: Use cases of idling stop system

To clarify associations between the actor and the use cases, we introduced an adaptation layer. It defines a sensor subsystem with behaviors and states corresponding to the driver's operations, and an actuator subsystem which definitely sends the application requirements to the vehicle (refer to Figure 2).

The application layer is an energy saving application. The adaptation layer contains four subsystems: a driver's operation, an electric motor, a mode indicator, and a battery monitor.

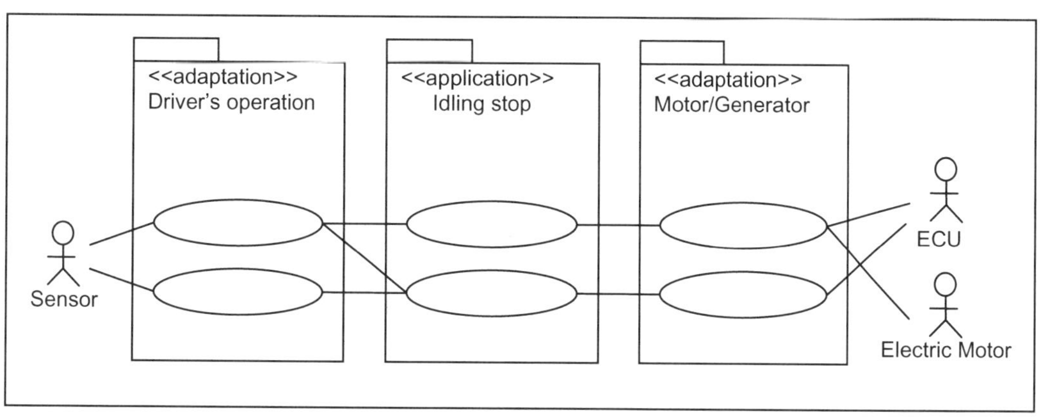

Figure 2: Example of adaptation layer

The use cases of each subsystem in the adaptation layer describe the devices and the application layer as their actors (refer to Figure 3).

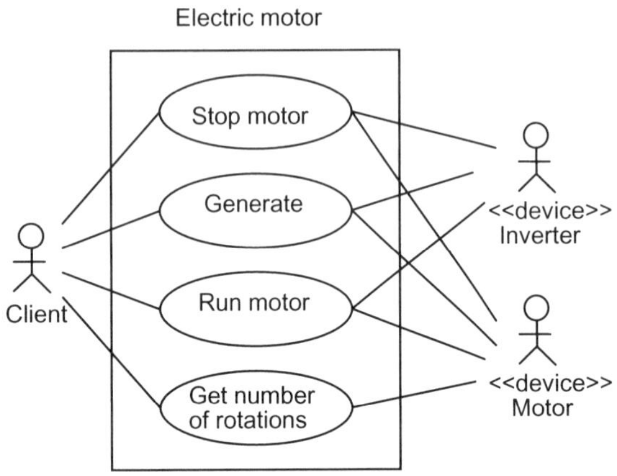

Figure 3: Use cases of adaptation layer

USE CASE

The use cases describe the various behaviors of the application. When describing targets of operations or determination of conditions, use cases in the adaptation layer allow the more concrete description of driver's operations to share the roles of the application without misunderstandings (refer to Table 1).

IDENTIFICATION OF OBJECTS

A robustness analysis diagram is created based on the flows and scenarios of the use cases in text.

The driver's operation, "brake", contains three elements: a boundary element to show braking equipment, a control element to get the braking state, and an entity element to keep the information of the state. Elements for objects in the adaptation layer are identified in the same way. The application to determine the "idle stop" state is identified as a control element (refer to Figure 4).

Table 1:Use Case Description

USE CASE DESCRIPTION

Use case:	Energy-saving function stops the engine.
Purpose:	The engine is stopped to improve fuel efficiency in "idle stop".

Precondition:
1. The shift lever is in Drive position.
2. The engine has finished warming up.
3. The engine has started.
4. The electromagnetic clutch has been connected.

Main flow of events:
1. The driver brakes.
2. The system detects the "braking" state.
3. The system gets the speed of the vehicle and evaluates stopping.
4. The system determines that the vehicle has stopped based on this evaluation.
5. The system counts the specified time of braking using a timer and determines the availability of the "idle stop".
6. The system stops the engine.
7. The system disconnects the electromagnetic clutch.
8. The system determines that the engine has stopped and turns on the "idle stop" lamp.

Exceptional flow:
5a. When the driver changes the shift lever position from D to P or N, the system does not count the specified time and skips to Step 6.

Postcondition:
1. Energy saving function stops the engine.
2. The electromagnetic clutch is disconnected.

At the second step, elements are organized into groups to identify objects in accordance with the guidelines for grouping.

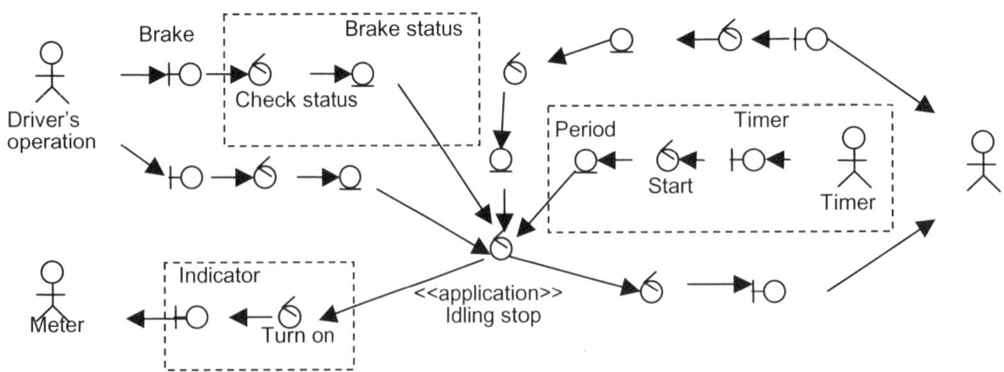

Figure 4: Robustness analysis and identifying objects

-The entity element and control element are organized as a "brake state" object.

-The boundary element, the display of the "idle stop" state, and the control element, the turning on/off controller, are organized as an "idle stop" indicator object.

-The timer boundary, timer controller and timing entity in the system are organized as one object.

Nine objects are identified in the robustness analysis diagram.

ANALYSIS OF INTERACTIONS AND DEFINITION OF BEHAVIORS

A sequence diagram can analyze the interaction between identified objects. Figure 5 shows the messages between them based on the use case. Because the "idle stop" evaluation object has many messages and becomes complicated, behaviors of the object can be clarified using the statechart diagram shown in Figure 6. Merging the diagram with a statechart diagram based on another use case completes the diagram for this object.

The sequence diagram can represent the interaction shown on the class diagram. In this case example, every object strictly corresponds to its class (refer to Figure 7).

IMPLEMENTATION

This system program is implemented on a microcontroller where RTOS compliant with μITORON is installed and classes are used on the OS as tasks. All programs are written by C++.

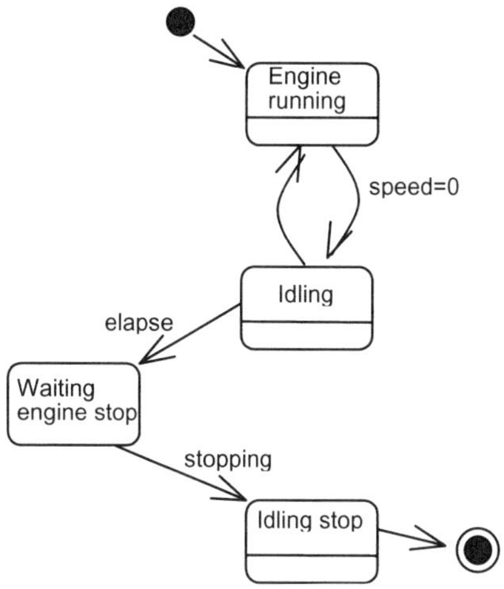

Figure 6: Part of statechart diagram for idle stop evaluation object

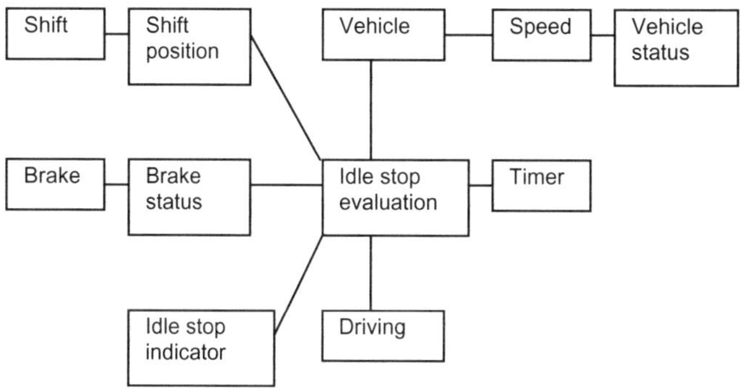

Figure 7: Relation between classes

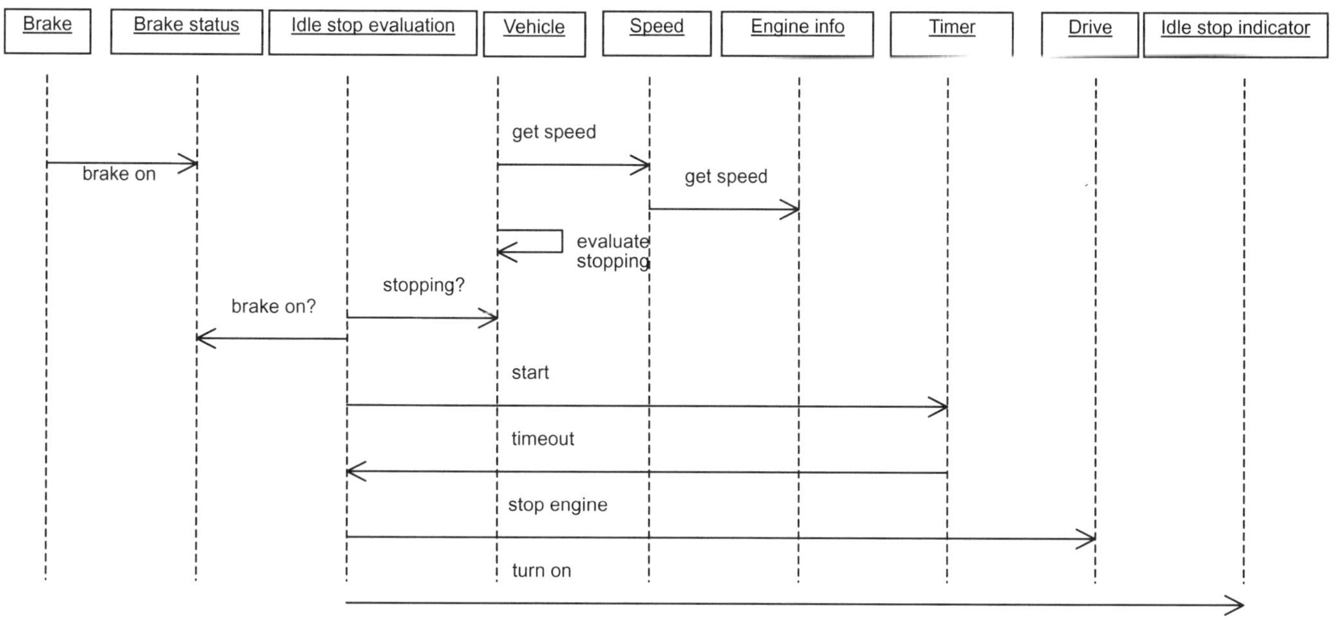

Figure 5: Sequence diagram in one scene

After defining the detailed roles of all classes on the class diagram, a UML translation tool converts the model to the executable C++ program.

A class that is generated when the system starts is assigned to the task on RTOS as an active class. A dynamically generated class is scheduled to be executable in the context of the task where the class is generated.

RESULT

We have analyzed, designed and implemented a control system in HVS according to the UML development process where the "adaptation layer" is introduced to analyze embedded systems that frequently change input or output devices. As a result, the adaptation layer can promote reusability of a software component.

This paper describes how object-oriented technology and UML methodology can implement embedded systems. We can emphasize the effectiveness and efficiency of UML to course participants in our educational material.

CONCLUSION

This paper presents the necessary process to develop an automotive embedded system using UML. The educational course using this HVS example educates engineers who can then promote innovative system developments in the future.

This result is expected to facilitate new embedded software developments with high quality and high efficiency.

ACKNOWLEDGMENTS

We thank our colleagues for informative advice on the hybrid vehicle system and willingly providing test equipment. This work is supported by their cooperation.

REFERENCES

1. James Rumbaugh, Ivar Jacobson, Grady Booch, *The Unified Modeling Language Reference Manual*, Addison Wesley, Massachusetts, U.S.A, 1998
2. Martin Fowler, *UML Distilled*, Addison Wesley, Massachusetts, U.S.A, 1997
3. Doug Rosenberg, Kendall Scott, *Applying Use Case Driven Object Modeling with UML: An Annotated E-Commerce Example*, Addison Wesley, Massachusetts, U.S.A, 2001
4. Hiroyuki Watanabe, Masahiko Watanabe, Kazuto Horimatsu, Kazunori Tomotake, *Embedded UML*, Shoei-sya, Japan, 2002
5. Grady Booch, James Rumbaugh, Ivar Jacobson, *The Unified Modeling Language User Guide*, Addison Wesley, Massachusetts, U.S.A, 1999

CONTACT

The author's address:

Hisahiro Miura
DENSO E&TS TRAINING CENTER CORPORATION
Engineering Education Institute
1-11, Nakou, Yokone-cho, Obu-shi, Aichi-ken,
474-0011 Japan
HISAHIRO_MIURA@denso.co.jp

2004-01-0360

Building Blocks Approach for the Design of Automotive Real-Time Embedded Software

Thierry Rolina
Business Development Group, Strategic Marketing & Communications, ETAS

ABSTRACT

Software content is undoubtedly increasing in vehicles with more and more functionality being implemented as real-time embedded features. This paper reviews the traditional approach to implementing such features, discusses the challenges that approach poses and describes a better method to overcome them. We will explore the components of a "building blocks" approach, which will constitute the basis of a process for future real-time automotive embedded software development.

INTRODUCTION

Lowering manufacturing cost for automotive components has been one of the driving forces in the development of automotive systems, but improving vehicle safety is today emerging as another such force. In the US, e.g., recent government legislation mandates that automotive manufacturers implement advanced airbag safety systems as well as tire pressure monitoring systems for future vehicle models. Given such pressures, the automotive industry is struggling to bring vehicles with an ever increasing array of new functionality to market ever more quickly. Unfortunately, the implementation of these functions--in terms of the software running on electronic control units--is causing increased maintenance cost for vehicles under warranty, as a recent article in "USA Today" underscored. Citing the results of a J. D. Power and Associates survey, the writer of the article commented that "fancy new cars [with a large amount of embedded control units] spend more time in the shop". The survey found that a great many customers were bringing their vehicles under warranty to the dealership for repairs. What appeared to the vehicle owners as a repair issue (may be a check engine light was lit up) really was not. These problems most often go away, after dealer report "no fault found". Many other examples of " no fault found" reports by mechanics underscore the fact that minor flaws in the ECU software were causing major cost in terms of warranty and maintenance. Developing reliable software where such "minor flaws" are guaranteed not to occur should therefore be an essential goal for vehicle manufacturers who want to reign in runaway warranty cost.

This paper revisits the requirements for real-time automotive embedded software, discusses the traditional development process and explores a way to improve it. The paper shows in conclusion that reliable software depends on a well-designed core that also enables a safe and modular implementation of new functionality and components.

Embedded Software Requirements

What are the essential requirements for embedded software systems?
They must

- ensure sufficient reliability
- fit within limited resources
- have very low production cost
- support evolving requirements and handle increasing function complexity
- support quicker time to market
- cope with limited engineering resources

Some of these requirements are already being fulfilled today, as companies have done a very good job over the past ten years of improving their development processes. Many of them have removed one of the proverbial brick walls, i.e. the communication-disconnect between engineers involved in the requirements phase and those involved in the design phase.

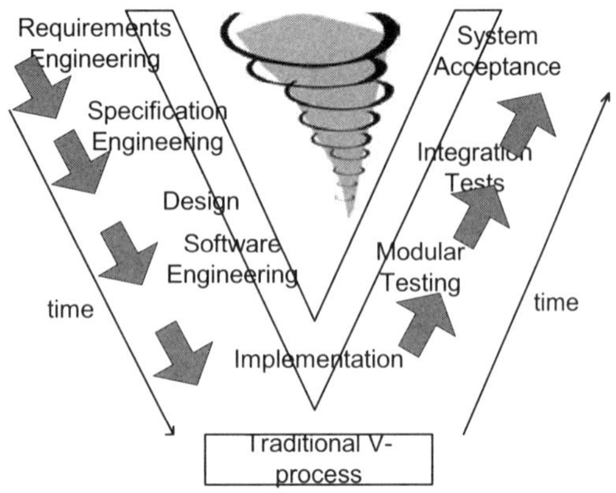

Figure 1.

Anyone involved in the development of automotive embedded software is familiar with the V-process model (figure 1.), or one of its derivatives. It is widely used to represent a model-based approach to the development of embedded software. While the approach is often maintained through the specification phase, "brick-walls" still exist between subsequent phases moving down the left side of the V-cycle. If we then look at the development process as a whole, inconsistencies are introduced due to the lack of flexibility of tools and complicated implementation of changes developers are facing in various development phases. The model-based approach is often abandoned once the system is in the design phase. In other words, models are simply translated into code, which often introduces additional inconsistencies. Closer inspection of the development process outside of functional aspects of systems--where modeling is in fact widely used--will also reveal areas where it is hardly ever used. One of these is the area of implementing and guaranteeing system timing behavior. For the most part, the control unit software designed by an ECU-supplier will fulfill its functional requirements, and the supplier will have tested the software in order to be able to guarantee this. However, it is important for a supplier of a generic ECU to guarantee the proper timing behavior of the system, i.e., that the ECU will meet its deadlines under all circumstances.

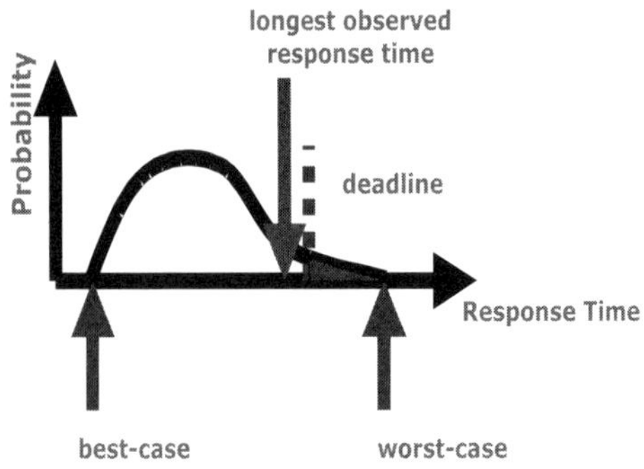

Figure 2.

Figure 2 shows the probability distribution of response time of an embedded system. We can typically observe the best case response, and test the embedded unit for some time, and observe a "longest observed response time". The typical process of testing the system will involve determining weather or not the observed response time is smaller than the deadline assigned to the ECU, i.e. the unit shall respond within a given time. If it is, the system will have qualified to "meet its deadline". End of test. The ECU is then ready for delivery. However, this approach does not explicitly test for the worst case scenario in terms of timing behavior, which could then occur after the ECU has been deployed in the field.

A simple calculation will reveal the impact of task deadline problems on warranty cost. If we assume that a given control unit is produced for three years at a production rate of 500.000 units/year, then the total volume is 1.5 million. If we further assume that the unit is used one hour per day on average, the unit will be used 547 million hours once deployed. The likelihood of the worst case scenario to occur is very high indeed, which will have a decided impact on cost, causing a warranty problem, or perhaps a recall.

Towards an integrated approach

In order to guarantee that our embedded controller will meet its timing requirements, let us look at its implementation. In the most general terms, the embedded software program consists of tasks that represent the implementation of functions. Scheduling tasks appropriately is the only way to ensure that they will execute within their deadlines. If all the system deadlines are met, the system is said to be schedulable. In a perfect world, we should be able to separate the notion of scheduling from the other layers in the system: IO drivers, network communication, and application layer…Figure 3 shows a conceptual diagram where in fact, the scheduler is the "heart" of the system, the other layers accessing it by the means of an API (application programming interface)

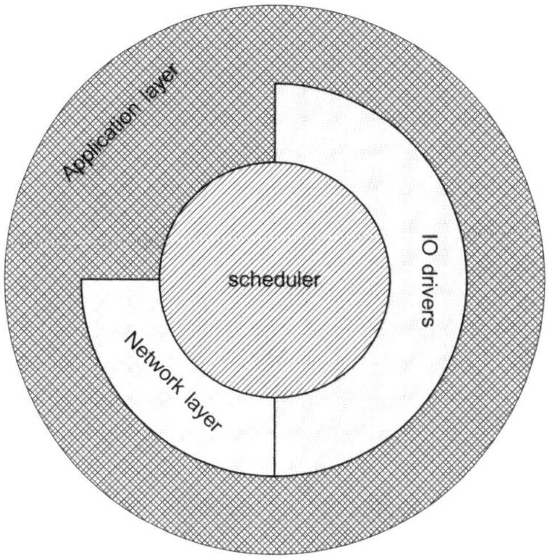

Figure 3.

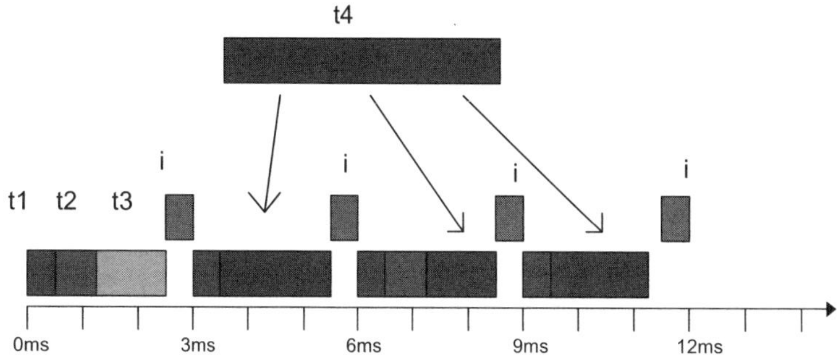

Figure 5.

Traditional approach to scheduling

In cyclic scheduling--a widely accepted approach --the available CPU time is divided into time slices. The time slices define the granularity at which we want to operate. Figure 4 shows the implementation of a cyclic system with four tasks t1, t2, t3, and t4 and one interrupt I.

Task/ISR	Period	Processing time
t1	3ms	0.5ms
t2	6ms	0.75ms
t3	14ms	1.25ms
t4	14ms	5ms
I	10ms	0.5ms

Figure 4

A look at the processing time for t4 reveals that this function will not fit neatly into the available time slices within the cycle. The way this is typically remedied in cyclic scheduling is to break the function down into code fragments that are distributed over several time slices. While this is a proper solution, it will complicate code maintenance because the small code fragments of functions will have to be coordinated. In addition to this issue, interrupt handlers need execution time as well, which will make less processing time available for the tasks themselves.

Figure 5 illustrates that the system is indeed using almost all the processing time available for each cycle. Cyclic scheduling does have very clear advantages, however. Scheduling is very easy to do, and deadlines can be verified at design time, statically. We can thus safely say that all the tasks in our example will meet their implicit deadlines (their periods). However, as we have seen, the approach also has clear drawbacks. CPU processing time is not used efficiently; and the necessity to fragment functions complicates maintenance. Furthermore, fragmenting functions prohibits the use of design tools that provide automated C-Code generation capabilities.

Nonetheless, the cyclic approach to scheduling lets us establish the worst-case response time for our tasks in the design phase: we simply add the polling delay and proper computation time and thus establish the worst-case response time for each task. Still, this approach leaves a lot of room for improvement.

Fixed priority scheduling

To resolve the issue of function code fragmentation and inefficient use of processing time, let us now explore another popular approach to scheduling. In fixed priority scheduling, each task and interrupt is assigned a priority. This attribute reflects the degree of urgency of each task and interrupt. The scheduler now executes the ready task/interrupt in the system that has the highest priority.

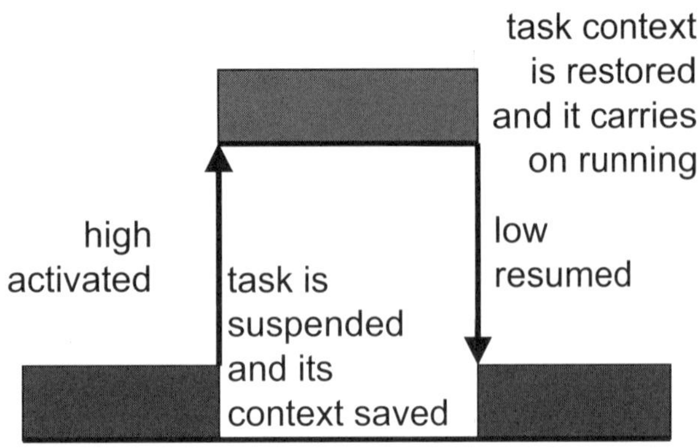

Figure 6.

Figure 6 illustrates the effect of preemption: upon activation of the higher priority task, the lower priority task is preempted. On task termination, the preempted task resumes. This approach improves the global system response time; the scheduler is handling the task partitioning dynamically; and the implementation truly matches the specification, all of which takes care of some of our issues.

Our next goal is to use a necessary and sufficient set of criteria to prove that the system is schedulable. In order to calculate the worst-case response times of tasks and interrupts, we will use a mathematical method called Deadline Monotonic Analysis (DMA), which we summarize in the following section.

Let:

- Ri be the response time of task i
- Tk be the period of task k
- Di be the deadline of task i
- Ci be the computation time of task i
- hp be the ensemble of tasks potentially preempting Ti
- $\lceil \ \rceil$, the round-up function

$$R_j^{n+1} = C_j + \sum_{\forall k \in hp(i)} \left\lceil \frac{R_j^n}{T_k} \right\rceil C_k$$

Equation 1.

We can solve equation 1 by recurrence. So, after calculating Ri for each task, if Ri $\leq$ Di, the system will be guaranteed to always meet its deadlines.
Figure 7 shows a graphical representation of the worst case execution time for the system. However, so far we have made the assumption that the scheduler needs zero time to switch tasks, which is of course unrealistic. Figure 8 shows the impact of the scheduler on task preemption. We can remove this assumption by extending equation 1 and write equation 2. In equation 2, we added the corresponding terms to the right hand side.

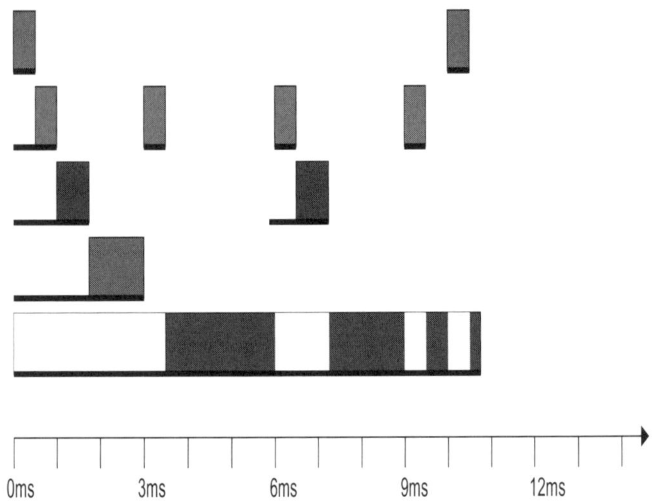

Figure 7.

Likewise, if we want to allow the use of semaphores, we must account for potential blocking time. Again, the blocking term Bi will also be added to the right hand side of equation 1. Overall, we can then state that:

$$R_j^{n+1} = S_j + B_j + C_{sw} + C_j + \sum_{\forall k \in hp(i)} \left\lceil \frac{R_j^n}{T_k} \right\rceil C_k$$

Equation 2.

Where:

- Sj is the scheduler overhead for task j
- Bj is the blocking time of task j
- Csw is the cost of switching to and from a preempting task

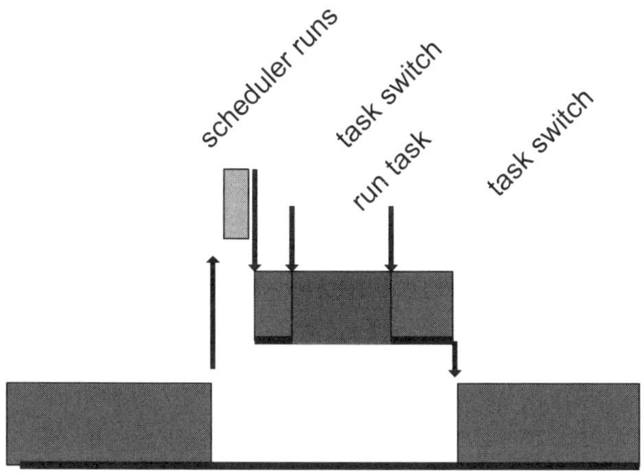

Figure 8.

Lessons learned from using a fixed priority scheduler

Using a preemptive scheduler offers a wide range of benefits, the biggest one being probably the implementation matching the specification. On the other hand, it also poses some challenges: The indirect cost of a fixed priority scheduler often surpasses its direct cost (purchase price of the scheduler itself). Here is why:

- Preemptive scheduling requires increased RAM space. Therefore, the scheduler has to be engineered in such a way that not only preemption can be achieved with minimal RAM usage, but also guarantee that RAM demand will not grow proportionally to the number of tasks.
- The scheduler introduces overheads. Not only do these overheads need to be bounded, but the scheduler has to be engineered to minimize them.
- Timing behavior is complex, and we need to make sure that the scheduler is deterministic and has been engineered so that timing analysis can be done.

Designing for schedulability

In order to reap the most benefits from the timing analysis environment, it is crucial to capture the worst case execution time for each task identified in the design. If we are doing a redesign, this number is a roughly known entity, based on past experience. For a new design, this number will have to be "budgeted" by the design team in addition to the functional specification. It will be important to have the architecture definition team keep in tight contact with the design team, so that the impact of going over the estimated execution time will be evaluated early. This approach cleanly delineates the responsibilities of the OEM and the supplier, as the OEM not only delivers a functional specification but a timing contract as well. This timing contract guarantees that the unit to be delivered will meet its deadlines under all circumstances.
Implementing a timing contract brings a number of benefits realized as deliverables, such as a use cases based timing contract, functional and timing models, test vectors. These benefits are appreciated throughout the development lifecycle of the vehicle, and they improve the supply-chain management as well. In fact these benefits are cross-functional, and will be appreciated by both sides of the OEM-Supplier relationship. Ultimately, they will help improve the ECU specification, increase the system reliability and flexibility, and avoid a situation where a simple fault causes a costly recall.
Figure 9 shows a desired implementation of the building blocks approach, the middle part of the V showing some of the deliverables that can be produced and can then be carried along to the next design phase.

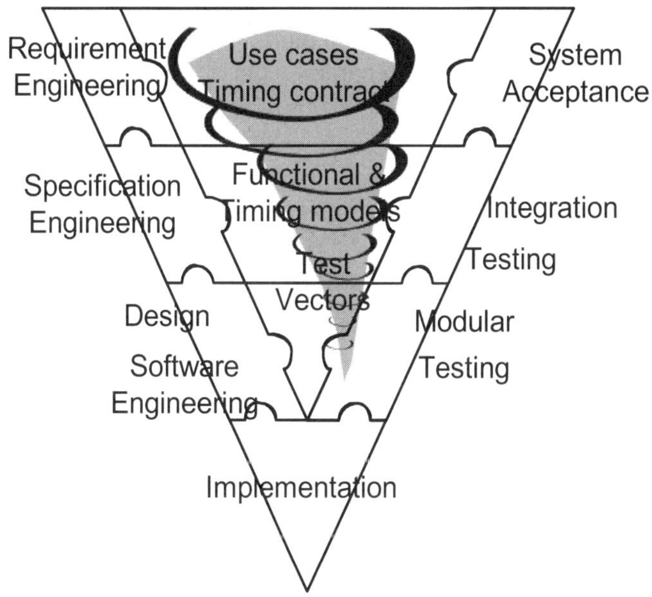

Figure 9.

Getting more functions into the microprocessor

The fixed priority preemtive scheduling approach that we took ultimately allows us to approach the design iteratively and incrementally. Iterative and incremental approaches help reduce risks by assessing them earlier in the design process. Typically, moving to a fixed priority scheduling topology helps in dealing with adding functionality: adding a new feature set may consist in adding one or more tasks to the application. As we have seen, granted that we can perform the timing analysis at the macro level (system), we will be able to assess whether or not the system will meet its deadlines. From the risk stand point, it is also interesting to look at the micro level (task), and how sensitive the overall system response is to each of the task computation time. In other words, how much can we add within each task, and still meet the deadlines. For this, we can use the recurrence relation defined in equation 2, and vary C_j to its maximum value so that all the deadlines in the system are still met. We can then get a precise idea of where adding new computations to existing tasks presents the least risks.

What we are looking for at each stage of the development process

Let us review the main phases of the embedded development process:

Requirements engineering

This phase is usually accomplished using a high level language that can be understood by most people. This is especially important, as this phase constitutes the early phase of the project. Text, Use Cases, and Message sequence charts are common practice here.

Specification engineering

The specification is an elaboration of the requirements. It constitutes the contract that will bind the customer and the supplier. While the functional specification consists of a "what", a "when", and a "how", the timing specification shall state the worst-case response characteristics of the unit under design.

Design

Design comprises an outline of the main features of the system. Decisions relevant to their implementation will be made at this phase for the hardware, software, and control levels.

Software Engineering

In this phase, application software will be generated. In addition to meeting its functional requirements, the application software must comply to its timing contract.

Implementation

The application software and the implementation of the control strategies will now run on the target hardware.

Modular testing

Each embedded control unit will be functionally tested individually against its requirements.

Integration testing

Multiple units will be tested in context in the vehicle, in a bench environment.

System acceptance

The system will be calibrated and delivered to the customer at this stage.

CONCLUSION

The building blocks approach addresses the requirements inherent to embedded software development. In this paper, we have drilled down to the way embedded software is scheduled, and concluded that cyclic scheduling will cause to systems to be brittle, and not reusable. Using a fixed priority preemptive scheduling approach can solve these problems, however, special care needs to be taken when selecting a new scheduler, because of the added overhead in response time and RAM space. It is also important to stay as open as possible. The OSEK consortium defines an automotive standard for scheduling real-time applications. The OSEK standard is currently undergoing ISO certification. OSEK is also fully analyzable using deadline monotonic analysis, which is necessary and sufficient to guarantee that a system meets its deadlines.

Finally, the analysis can be leveraged at the specification stage with the notion of introducing a "timing contract". This contract is designed to ensure that embedded devices not only fulfill their functional requirements, but that will also meet timing requirements under all circumstances.

CONTACT

Thierry Rolina Strategic Marketing & Communications, ETAS, 3021 Miller Road, Ann Arbor, MI 48104
Thierry.rolina@etas.us

ADDITIONAL SOURCES

Verification of Systems Specifications, a case study in the automotive industry, Thierry Rolina, CE98 conference, Tokyo 1998.

Class 306, Dr. Ken Tindell, True real-time embedded systems engineering, ESC 2000.

Embedded Systems Programming, Dr Ken Tindell, Deadline Monotonic Analysis, June 2000

OSEK-OS specification, www.osek-vdx.org

2004-01-0279

Integrated Modeling and Analysis of Automotive Embedded Control Systems with Real-Time Scheduling

Zonghua Gu, Shige Wang, Jeong Chan Kim and Kang G. Shin
University of Michigan at Ann Arbor

ABSTRACT

As the complexity of automotive embedded software grows, it is desirable to bring as much automation into the development process as possible, as evidenced by efficient code generators from modeling tools such as Simulink. However, the current development process does not pay enough attention to non-functional issues such as real-time requirements. Modeling tools such as Simulink do not allow representation of the target platform, so it is not possible to analyze real-time behavior in the early design stage. The typical approach is to download the executable to the target micro-controller and measure to see if there are any missed deadlines. We describe a tool that can be used to model system-level software and hardware architectures and analyze system timing behavior at early design stages, and assess the impact of the target platform on control system performance.

INTRODUCTION

Embedded control systems such as automotive engine control have stringent real-time requirements. They present numerous challenges to software designers, since they have to ensure not only *functional* correctness (the correct result is produced), but also *non-functional* correctness (the result is produced at the correct time). Many such control systems are safety-critical and require a high level of confidence in their correctness. It is a recent trend to replace traditional mechanical sub-systems with electronically controlled systems with no mechanical backup, as exemplified by the *x-by-wire* project, where *x* stands for brake, steer, etc. In such systems, a late result is a wrong result, and may result in severe harm to humans and/or property.

Current automotive software development does not pay enough attention to non-functional issues. A typical design process follows the V-cycle, as shown in Figure 1. Modeling tools such as Simulink [9] are used to design control algorithms, and then automatic code generators such as Real-Time Workshop (RTW) are used to generate C code, which can be compiled and downloaded to the target micro-controller. Much attention has been paid to improving the efficiency of code generators, and as a result the generated code sometimes outperforms hand-written code in terms of compactness and efficiency. However, there is no tool support for assessing system-level non-functional properties at an early design stage. It is only near the end of the software development process that the engineer downloads code to the target micro-controller. Timing violations detected at this stage often result in expensive redesigns and schedule overruns.

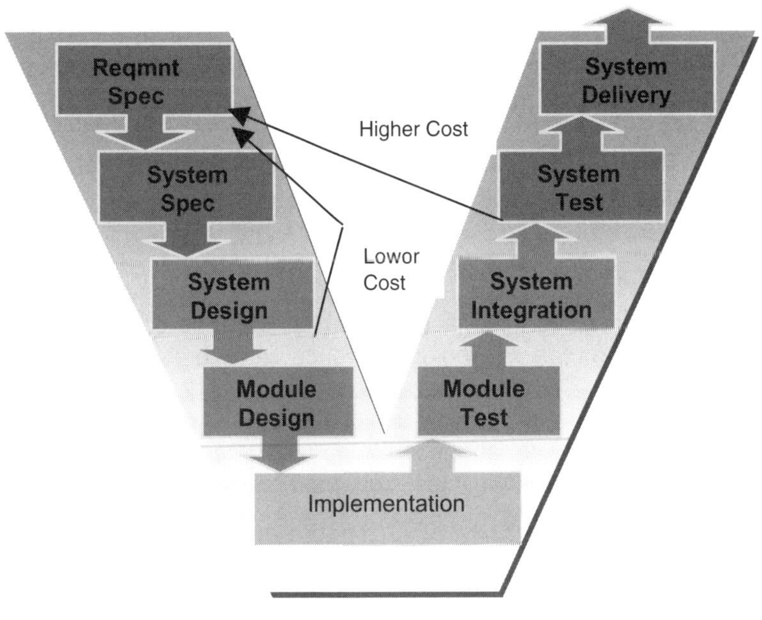

Figure 1. Problems detected at the system test stage are very costly to be fixed, while problems detected at early system design stages are less costly. Therefore it is desirable to model and analyze the system during early design stages.

The DARPA MoBIES (Model-Based Integration of Embedded Software) program, started in 2000, has been exploring model-based approaches to embedded software composition and analysis, especially emphasizing non-functional issues such as timing, synchronization, dependability and real-time constraints.

We have developed a software tool called AIRES (Automatic Integration of Reusable Embedded Software) that is targeted towards both the automotive control and Avionics Mission Computing domains. Its main purpose is to bring research results developed in the real-time computing community into industry practice, and help the engineers perform early analysis on system timing behavior.

Simulink is mainly targeted towards control engineers instead of software engineers. Therefore it lacks certain concepts related to software architecture such as tasks, operating system, target execution platforms, etc. There have been some attempts at bridging the gap by defining architectural description languages such as the new SAE standard AADL (Avionics ADL) based on MetaH from Honeywell. However the target domain seems to be avionics embedded systems. AIRES can be viewed as an ADL toolset specialized for the domain of automotive embedded control systems.

REAL-TIME SCHEDULING THEORY: SOME DEFINITIONS

Here we provide some basic definitions of real-time scheduling theory.

Real-Time Task: A process or thread in a real-time operating system that is scheduled for execution on the processor. It may be *periodic*, triggered by a periodic timer, or *aperiodic*, triggered by an external interrupt.

WCET: Worst-Case Execution Time. Worst-case delay from when a task is triggered to when it completes execution *without* interference from other tasks.

WCRT: Worst-Case Response Time. Worst-case delay from when a task is triggered to when it completes execution *considering* interference from higher-priority tasks.

Delay Jitter: Variations in the current delay value during multiple execution cycles, from best-case delay to worst-case delay.

Schedulable: A task is schedulable when its WCRT is less than its deadline. For a periodic task, its deadline is typically, but not always, the same as its period. The system is schedulable if all tasks in the system can be scheduled.

Given a set of periodic tasks with periods and WCETs defined, real-time scheduling theory can be used to calculate task response times, hence determine system schedulability. For aperiodic tasks, a minimum inter-arrival time of task triggers must be provided in order to treat the aperiodic task as a periodic one.

In this paper we assume that task WCETs are known, and use scheduling theory to determine WCRTs. The typical way to determine WCET is to measure the target with representative system inputs. There are also tools that use static analysis techniques to determine the WCET. While we acknowledge that WCET determination is an important issue, our focus is on system-level timing analysis given WCETs for software components.

THE AIRES TOOL

AIRES is built on top of a meta-modeling tool called Generic Modeling Environment (GME) [5] from Vanderbilt University. Based on the Model-Integrated Computing (MIC) approach, GME is a configurable tool set for creating domain-specific modeling and program synthesis environments through a *meta-model* that specifies the modeling paradigm of the application domain. The meta-model captures all the syntactic, semantic and presentation information regarding the application domain, and defines the family of models that can be created using the resulting modeling environment. It contains descriptions of the entities, attributes, and relationships that are available in the modeling environment, and the constraints that define which modeling constructs are legal. A meta-model is also often called a *modeling paradigm*.

As an example, the meta-model in Figure 1 captures a language used to model the composition of neighborhoods. It shows that each Neighborhood may contain zero or more Buildings, where each Building may either be a House or a Store. Each House may contain Residents, and each Store may contain Patrons. The Neighborhood may also contain Walkways that connect the Buildings to each other.

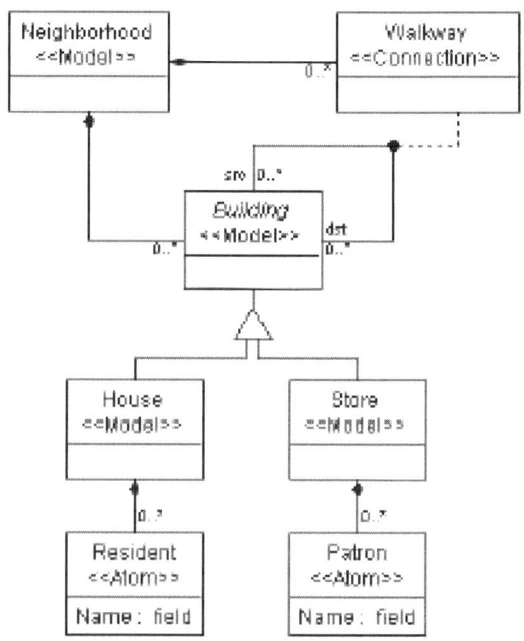

Figure 2. UML description of a domain-specific modeling language, used to model neighborhoods.

Given this meta-model, GME can be used to synthesize a domain-specific visual modeling environment for neighborhoods, which can be used to create concrete instances of neighborhood models. Of course we can always use a generic drawing tool to construct models of neighborhoods, but using a domain-specific modeling tool offers several advantages:

- The user is constrained by the modeling environment not to create illegal models that violate the constraints specified by the meta-model.
- Instead of just pretty pictures, GME provides powerful APIs to write *semantic translators*, for example, a code generator from graphic models to C code. *Semantic translation* is defined as transforming a model conforming to its meta-model *A* to another model conforming to a different meta-model *B*. It is generally a more difficult problem than *syntactic translation,* such as file format conversion from postscript to PDF, where document syntax is changed while the semantics remains the same.

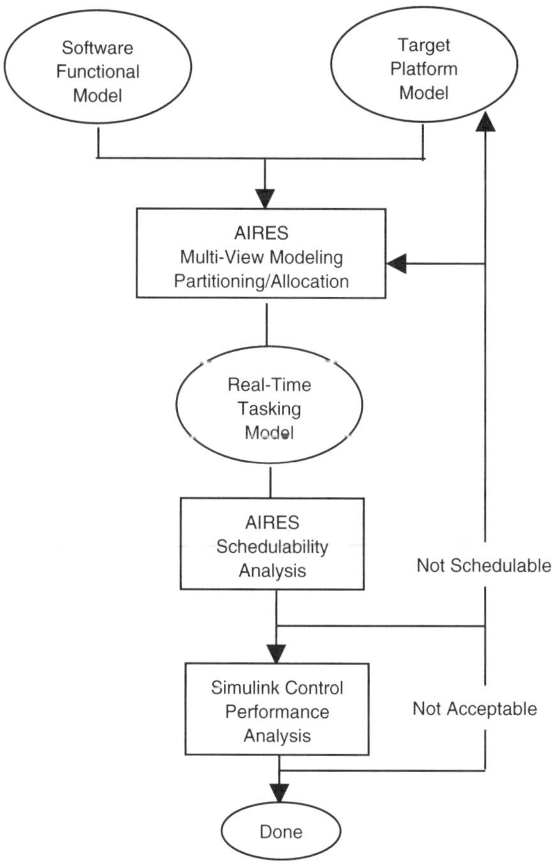

Figure 3. Workflow for the AIRES tool.

The meta-model for AIRES specifies modeling constructs for various aspects of an embedded system, including several sub-models:

- *Software functional model*: similar to traditional block diagrams such as Simulink.
- *Target platform model:* models for the CPUs and networks the software runs on.
- *Real-time tasking model:* the real-time tasks running on the target platform, as a result of considering timing and triggering in the software functional model.

Figure 3 shows the workflow for the AIRES tool. Simulink blocks are imported into AIRES to form the *software functional model*. The Simulink importer is based on mapping between constructs defined in the Simulink/Stateflow meta-model and those in the AIRES meta-model. The system designer manually constructs the *target platform model*, consisting of processors and networks such as a CAN bus. He then assigns functional blocks to processors, if the target platform has multiple processors. Taking into account timing and triggering information present in the software functional model, AIRES can be used to construct the *real-time tasking model* and perform schedulability analysis to determine the response times of the tasks. If the task set is not schedulable, i.e., if some task has a response time longer than its deadline (typically the same as its period), then the designer has to go back to AIRES and take one or more of the following measures:

- Adjust task priority attributes.
- Re-allocate tasks to processors.
- Enhance the target platform, for example, add a new processor or switch to a more powerful/expensive processor, etc.

The process is iterated until the system becomes schedulable. However, schedulability may not be the only system design criterion. Sometimes the designer may want to leave some slack in the system for purposes of future feature enhancements and upgrades, or just to leave a safety margin. AIRES is also able to give the designer some measurement of system slack. For CPU scheduling, AIRES uses the Rate Monotonic Analysis theory to calculate worst-case response times of tasks. Based on block execution rates and data message size information, the message sets that are transmitted on the CAN bus can be determined, and analysis techniques in [6] can be used to analyze the schedulability of the CAN bus.

When the system is deemed schedulable, the designer then inserts special *delay* blocks into the original Simulink model and analyzes the resulting control performance. This step is elaborated in the next section.

In order to perform automated transformation from a software functional model to a real-time tasking model, we traverse the system dependency graph from a timer trigger, marking all blocks reachable from the timer as executing at the rate of the timer. All blocks that are assigned the same rate are contained in a real-time task of that rate. For example, the 5Hz thread contains all blocks executing at 5Hz. It is possible for dependency trees from multiple timers to intersect one another. The blocks at the intersection are *multi-rate* blocks. A *control loop* may also be multi-rate. For example, suppose we have a linear chain of blocks A → B → C, where block A performs sensor sampling, block B performs algorithm calculation, and block C performs actuator control. It is not uncommon for blocks A and C to run at a higher rate than block B. In this case, automated rate assignment cannot be performed, and the designer has to manually annotate the model in GME in order to construct the tasking model.

CONTROL PERFORMANCE ANALYSIS

Control design typically starts with designing a control law in the continuous time domain and then discretizing the controller to reflect implementation on a digital microprocessor. Selection of task execution rates or periods is a heuristic process, with the objective of maximizing control performance while maintaining a reasonable processor load. Once control design is completed, the set of control laws and task execution rates are handed over to the software engineer, who then must implement the control laws on a target microprocessor while respecting the given timing constraints, i.e., all tasks must finish execution within their periods.

In the control analysis stage, the designer typically assumes an ideal execution platform while designing control laws. Even though task execution rates are taken into account by discretization of continuous time control laws, task execution time is assumed to be 0, that is, the control task reads sensor inputs, performs computation, and produces actuator outputs all at the same instant, the start of a task period. The rationale for this assumption is that task execution rate has a much greater impact on control performance than delay and jitter. Even though this observation is true in most cases, it is desirable to explicitly model delay and jitter at the modeling stage in order to have a more accurate assessment of such implementation effects on control performance.

Figures 4, 5, 6 and 7 show the effects of controller period (execution rate) and execution time on control performance of a PID controller. The reference signal *u* periodically alternates between values −1 and 1, and the control objective is to make the actual plant output *y* follow *u* as closely as possible. Control performance refers to how well *y* tracks *u*, and can be quantitatively defined with the usual metrics such as overshoot, settling time and steady state error. We use visual assessment instead of explicitly calculating values for these metrics. As we can see from the figures, control performance deteriorates with an increase of period or execution time. As expected, the controller period seems to have a greater impact on control performance than execution time. As the period increases from 6ms to 8ms and then to 14ms, while keeping execution time at a constant 1ms, plant output *y* deviates more and more from the reference signal *u* and eventually diverges when the controller period is set to 14ms, i.e., the system becomes unstable. The effects of the controller task execution time cannot be ignored either. When the period is set to 8ms, an increase in execution time from 1ms to 7ms also has a marked impact on control performance.

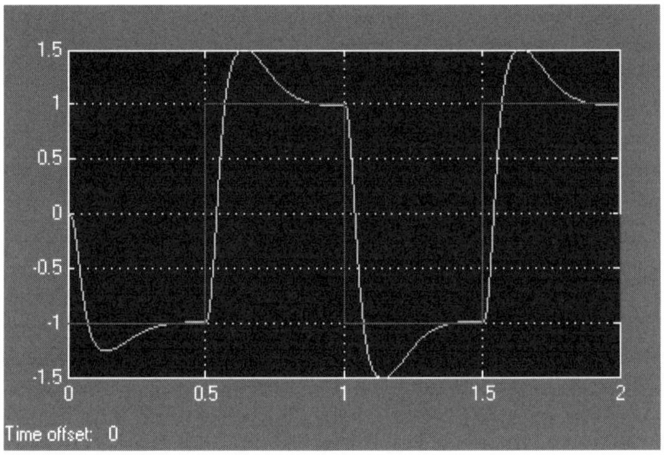

Figure 4. Period = 6ms, Execution time = 1ms.

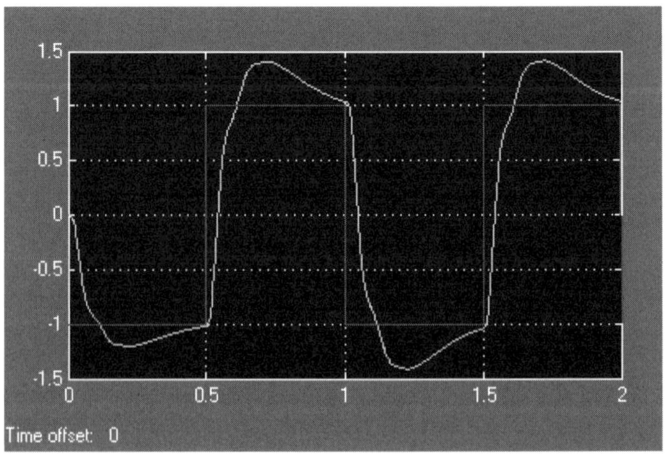

Figure 5. Period = 8ms, Execution Time = 1ms

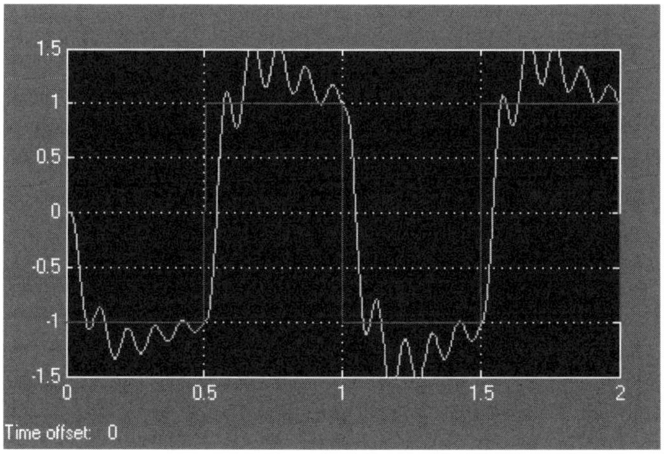

Figure 6. Period = 8ms, Execution Time = 7ms

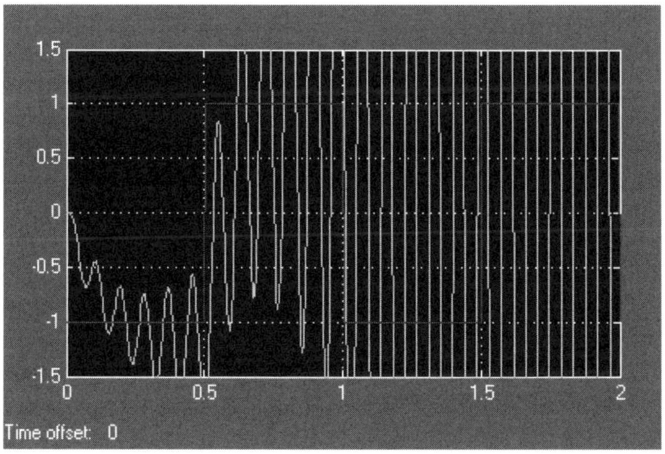

Figure 7. Period = 14ms, Execution time = 1ms

In this small example, the controller task is the only task that is executing on the CPU. In a multi-tasking system, the task may suffer preemption delays from higher-priority tasks executing on the same CPU, and we have to use worst-case response time (WCRT) instead of worst-case execution time (WCET) in the experiments performed above.

One approach [11] to modeling scheduling effects within Simulink is to design special Simulink blocks that simulate the scheduling behavior, and can be inserted into the original functional Simulink models for co-simulation of control and scheduling. This approach does not take into account the worst-case situation, since it is not guaranteed that the system experiences worst-case delay during the finite length of simulation time. We use a different approach of inserting a *delay block* into the original Simulink model, which is set to be the worst-case delay calculated from real-time scheduling theory. One drawback of this approach is that delay jitter effects cannot be studied. Currently a delay block is manually inserted, but it is possible to automate this process by generating a Matlab script from AIRES. The delay cannot exceed the task period; if it does, the task set is not schedulable.

APPLICATION CASE STUDIES

CASE STUDY I

We first consider the Electronic Throttle Control (ETC) example provided to us by researchers at The University of California, Berkeley, available for download at [8]. The ETC system is a drive-by-wire system in which the direct linkages between the accelerator and the throttle or the steering wheel and the steering gear are replaced with pairs of sensors and actuators. The system consists of three tasks, as shown in Figure 8. In the Simulink model, the tasks *Monitor*, *Manager* and *Servo Control* are subsystems triggered by the scheduler, which is a Stateflow block. The scheduler is triggered at a period of 1ms, and, based on a counter, triggers the *Monitor* task every 30 periods, the *Manager* task every 10 periods, and the *Servo Control* task every 3 periods. Therefore, the *Monitor*, *Manager* and *Servo Control* tasks have periods of 30ms, 10ms, and 3 ms, respectively. The tasks run with priority-based preemptive scheduling on a single CPU running an OSEK-compliant RTOS, and standard rate monotonic scheduling techniques can be used to determine task WCRT given WCET, which can be obtained via measurement on the target processor. Figure 9 shows the task execution timeline as the analysis results with AIRES. Results from control performance analysis indicate that task execution delays do not have any significant impact, since the most important task, Servo Control, has the shortest period and highest priority, so it is not affected by the other 2 tasks. Its own execution time does not matter much as long as it finishes within its period, since its period is already so small.

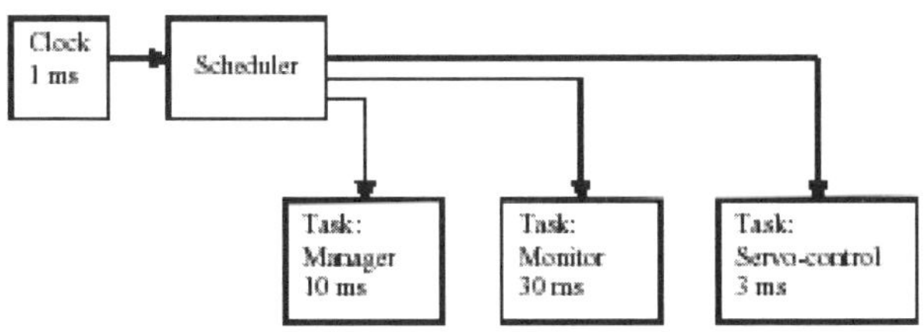

Figure 8. Real-time tasking model in ETC example.

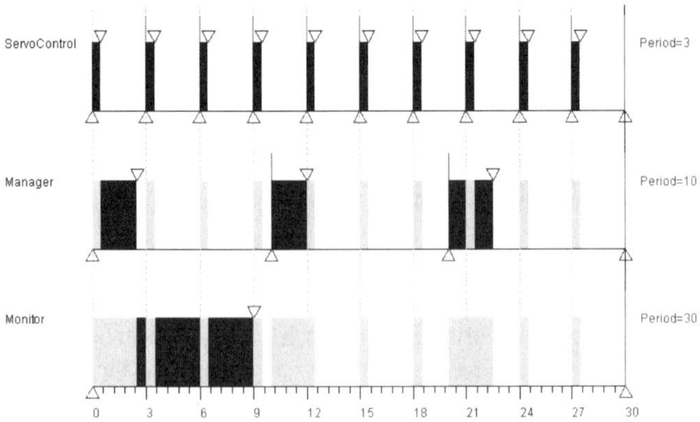

Figure 9. Analysis results of AIRES showing the task execution timelines for Monitor, Manager and Servo Control tasks. Blue indicates task execution; yellow indicates preemption; white indicates task idling.

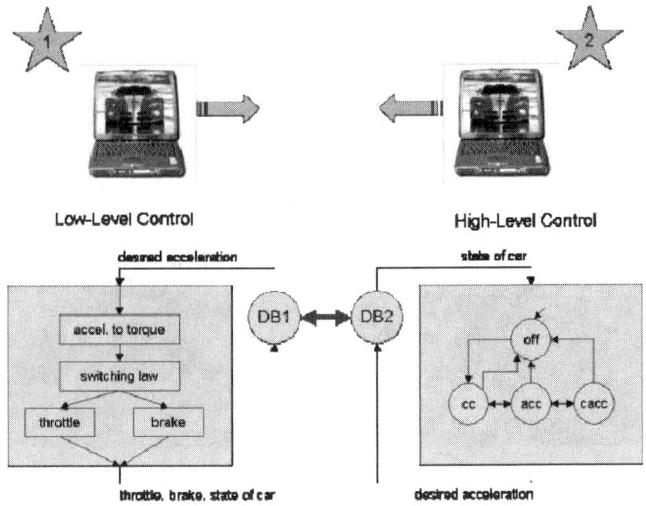

Figure 10. System architecture for V2V control.

CASE STUDY II

We consider another application example provided to us by US Berkeley [8]: vehicle-to-vehicle cooperative adaptive cruise control, abbreviated as *V2V control* in the following discussions. The modeling tool used here is Teja [7] instead of Simulink. (Since AIRES is currently only integrated with Simulink, we have to manually enter the models into AIRES, and we cannot analyze control performance. Note this is only a limitation in tool implementation, not an inherent limitation of our approach) The goal is to maintain a platoon of vehicles traveling on the highway in tight formation and high speed, while avoiding rear-end collisions, taking into account various situations such as road curves, cut-ins by other vehicles, dense traffic, rapid acceleration and deceleration, etc. The platoon consists of a leader vehicle and a number of follower vehicles communicating through a wireless link.

The software environment is split onto two CPUs, P1 and P2, running QNX RTOS, as shown in Figure 10. The two computers communicate via an RS-232 serial port. P1 deals with low-level control. It receives state information from the longitudinal sensors and acceleration commands from P2, and sends outputs to throttle and brake actuators. In addition, status and position information is sent via an RS-485 serial port to the Human Machine Interface Computer, not shown in the figure. P2 deals with high-level control. It receives state inputs from P1, and additional inputs from a Doppler radar, a vision system and a GPS. It switches between high-level modes such as *off*, *cruise control*, *adaptive cruise control,* and *collision-avoidance adaptive cruise control.* The communication paradigm is publish/subscribe via two shared databases residing in main memory. The messages that are exchanged between CPUs are the database variables node1_to_node2 (26 bytes) on P1 and node2_to_node1 (12 bytes) on P2. Every time the node1_to_node2 variable is modified on P1 (on average every 21 ms), updates are sent to P2, which does some calculation and sends a response back to P1,

Analysis results in Tables 1 and 2 provide worst-case response time, resource consumption (utilization, computational resource only), and number of context switches experienced during the worst-case execution scenario for each task. As can be seen, all tasks have WCRT less than their periods, therefore, the task sets on both processors are deemed schedulable. AIRES further provide the overall resource utilizations used for the application and system. Given that the timer resolution for both processors is set to 1ms, the computation resource used by application tasks, timer, and scheduler on P1 are 23.15%, 0.56%, and 0.0015%, respectively. The resource consumption for application tasks, timer, and scheduler on P2 are 0.43%, 0.56%, and 0, respectively. This indicates that the workloads on these two processors are not balanced. The scheduler overhead on P2 is not really 0, but is rounded to 0 since it is extremely small (as the number of context switches and number of tasks running on P2 is small). The overhead introduced by the timer depends only on the timer resolution (how frequently is the timer signal processed), so it is the same on both processors. It is possible to experiment with different allocations of application tasks to the target platform and assess the CPU load and system slack. We do not describe this due to space limitations.

Note that some of the task periods are not integers. This is because these tasks are not really periodic, so we take the average inter-arrival time of task triggers as an approximation for the task period.

Task	Period (ms)	Execution Time (ms)	Response Time (ms)	Utilization	# Context Switches	Priority	Processor
V2V Controller-P1.atmio16	21	0.139	2.578	0.007	18	10	Processor 1
V2V Controller-P1.atmioe	2	0.114	0.231	0.057	2	27	Processor 1
V2V Controller-P1.button	30	0.024	2.672	0.001	20	8	Processor 1
V2V Controller-P1.canbrake	8	0.001	0.337	0	5	23	Processor 1
V2V Controller-P1.canfix	20	0.001	0.338	0	6	22	Processor 1
V2V Controller-P1.cani	10000	0	2.799	0	23	3	Processor 1
V2V Controller-P1.canread	7.7	0.104	0.336	0.014	4	24	Processor 1
V2V Controller-P1.cansteer	4	0.001	0.232	0	3	26	Processor 1
V2V Controller-P1.db_slv	1.5	0.037	0.043	0.025	0	29	Processor 1
V2V Controller-P1.DL	1953.9	0.065	2.799	0	22	5	Processor 1
V2V Controller-P1.hmi	200	0.062	2.734	0	21	6	Processor 1
V2V Controller-P1.moblong	21	0.76	2.439	0.036	17	11	Processor 1
V2V Controller-P1.NL	75160	1	3.842	0	25	2	Processor 1
V2V Controller-P1.node1rd	21	0.262	1.448	0.012	13	12	Processor 1
V2V Controller-P1.node1rw	21	0.083	1.186	0.004	12	13	Processor 1
V2V Controller-P1.path101	21	0.047	1.103	0.002	11	14	Processor 1
V2V Controller-P1.pctio10	21	0.107	1.056	0.005	10	15	Processor 1
V2V Controller-P1.radioDriver	29.5	0.07	2.648	0.002	19	9	Processor 1
V2V Controller-P1.regulation	21	0.159	0.943	0.008	9	16	Processor 1
V2V Controller-P1.supervisor	21	0.101	0.784	0.005	8	17	Processor 1
V2V Controller-P1.TL	150108	0	3.842	0	26	1	Processor 1
V2V Controller-P1.veh_iols	21	0.345	0.683	0.016	7	18	Processor 1
V2V Controller-P1.veh_lat	2	0.074	0.117	0.037	1	28	Processor 1

Table 1. Analysis results for Processor 1 for V2V control.

Task	Period (ms)	Execution Time (ms)	Response Time (ms)	Utilization	# Context Switches	Priority	Processor
V2V Controller-P2.db_slv	6.4	0.005	0.011	0.001	0	25	Processor 2
V2V Controller-P2.evt300	65.356	0.055	0.215	0.001	1	10	Processor 2
V2V Controller-P2.hi	21	0.007	0.223	0	1	10	Processor 2
V2V Controller-P2.node2rd	21	0.042	0.223	0.002	1	10	Processor 2
V2V Controller-P2.node2wr	21	0.006	0.223	0	1	10	Processor 2
V2V Controller-P2.veh_iomb	5000	0.047	0.223	0	1	10	Processor 2

Table 2. Analysis results for Processor 2 for V2V control.

RELATED WORK

There are a number of tools based on real-time scheduling theory, such as TimeWiz [1] and RapidRMA [2], for analyzing schedulability of a real-time task set. However there is no seamless integration with control design tools such as Simulink, and the designer has to manually manage mapping from Simulink blocks to real-time tasks. This is probably the reason why these tools have not been widely adopted in the automotive industry.

In the control theory community, there is some literature dealing with analysis of delay effects on control performance and design of delay or jitter-compensating controllers to take into account these effects [3]. However these techniques are rarely adopted in industry since they require expert knowledge in control theory. We believe a simulation-based approach as proposed here is more likely to receive wide acceptance.

There are a number of tools for CAN bus analysis and simulation [4]. CANoe is CAN simulation software that can be integrated with Simulink models. Our approach is complementary to theirs, since we rely on analytical analysis in order to assure worst-case performance, while their tool gives the designer more intuition as to the system behavior via simulation. It is also conceivable to use the integration of CANoe and Simulink to study the impact of CAN bus message delays on control performance.

Automotive Modeling Language (AML) [10] is similar to AIRES in that it allows modeling of software and system architectures. Its focus is on system modeling with UML at an abstract level and mapping into ASCET-SD at a more concrete level. There are no concepts of real-time or control performance analysis.

CONCLUSION

In this paper, we have described a tool called AIRES that is designed to model embedded software and the target execution platform, and perform real-time analysis to give the system designer early feedback on system timing behavior. We believe AIRES fills a key gap in the current system development process in assessing and fulfilling non-functional requirements, and serves a complementary role to automatic code generators from modeling tools such as Simulink. Future work includes integration of more analysis techniques or third-party tools into AIRES, such as safety and reliability analysis.

REFERENCES

1. TimeSys website: http://www.timesys.com
2. Tripacific website: http://www.tripac.com
3. P. Marti, J.M. Fuertes, G. Fohler, K. Ramamritham, "Jitter Compensation for Real-Time Control Systems", Proceedings of Real-Time Systems Symposium, 2001
4. Vector CanTech website: http://www.vector-cantech.com/
5. ISIS website: http://www.isis.vanderbilt.edu/
6. K. Tindell, A. Burns, and A. Wellings, "Analysis of Hard Real-Time Communications", Real-Time Systems, 9(2), 1995
7. Teja website: http://www.teja.com
8. Berkeley Automotive OEP website: http://vehicle.me.berkeley.edu/mobies/
9. Mathworks website: http://www.mathworks.com
10. U. Freund, M. von der Beeck, P Braun and M. Rappl, "Architecture Centric Modeling of Automotive Control Software", SAE Congress paper 2003-01-0856.
11. D. Henriksson, A. Cervin, K.E. Arzen, "TrueTime: Simulation of Control Loops under Shared Computer Resources", Proceedings of the 15th IFAC World Congress on Automatic Control, 2002

CONTACT

Zonghua Gu
Real-Time Computing Laboratory
Electrical Engineering and Computer Science Dept.
University of Michigan
Ann Arbor, MI 48109, USA
Email: zgu@umich.edu

2004-01-0270

A Practical, C Programming Architecture for Developing Graphics for In-Vehicle Displays

Michael T. Juran
Altia, Inc

ABSTRACT

This paper presents a practical, C programming architecture for developing graphics and graphical user interface (GUI) software for in-vehicle displays such as those found in automotive telematics, shipboard engine control panels and airplane glass cockpits. This paper will compare C graphics to several other methodologies for writing in-vehicle display software. It will also discuss techniques for mixing C graphics with application code written in object oriented languages such as C++ and Java.

INTRODUCTION

Improvements in display technology have produced an explosion of uses for graphics displays in vehicles. Figure 1 illustrates a few potential uses in the automobile cockpit – from the center stack and instrument cluster to rear seat entertainment systems and heads up displays.

Unfortunately, display hardware technology has outpaced software technology and there are currently no established, predictable methods for developing graphics software for vehicles. Consequently, it takes an exorbitant amount of time to design, test and deploy graphics software. Worse yet, a change in the underlying hardware or real-time operating system (RTOS) could require a complete rewrite of the graphics code.

Other industries, such as consumer electronics, information technology, and telecom offer solutions for building GUIs quickly, but these are rarely appropriate for the tight cost, size, performance, reliability and safety constraints found in land, sea and air vehicles.

Thus, there exists a need to alleviate today's graphics development bottleneck for vehicles. This paper explores the issues surrounding embedded systems graphics development and offers some potential solutions.

Figure 1: *Example of in-vehicle graphics displays.*

THE HISTORY OF THE EMBEDDED GUI

In the early days of microprocessor control, embedded systems were faceless. The computer under the hood did its thing and the user never knew about it. If a user needed status or information, he looked at the mechanical gauges. If a user needed to provide input, he would turn knobs, press buttons and step on the gas pedal. Everyone was happy and this arrangement worked for decades.

As solid state LED display technology became available and cost effective, someone got the bright idea of hooking up a simple 7 segment numeric display to a 4-bit microprocessor. Figure 2 is a sample of such a display. With an I/O port and a few well written assembly language commands, the digital odometer was conceived. It didn't take much memory or computer power, yet it was far more reliable than its mechanical counterpart.

Later, embedded processors got a little more powerful and programmers had a few more bytes to spend on graphics. The first pixilated, multipurpose displays became available and programmers built reusable graphics libraries using the C programming language.

In the meantime, desktop computing power has gone through the roof. At the turn of the millennium, 32 bit processors became the norm. Memory hit the gigabyte mark. High-speed graphics boards became standard. A single desktop OS dominated. Graphics programming in this environment was "no holds barred." In the context of graphics development, desktop computers became virtually limitless.

Figure 2: *A first generation embedded display.*

Embedded systems, however, have not experienced this exponential growth. To be sure, they are endowed with more memory and power than their early predecessors, but the market tendency is to keep these systems as small, lightweight and inexpensive as possible. In addition, processor longevity, reliability and low power consumption are far more important than cutting edge speed and memory size[1]. Consequently, the speed and capability gap between the desktop PC and embedded systems has widened significantly. As a result, the tools and techniques needed to build embedded systems must be quite different.

JAVA and C++ are currently the standard programming languages for desktop, workstation and server applications. These computers have ample power and resources to support the large object oriented graphics widget sets and Java virtual machines. On the embedded system, however, C programming still rules the roost.

Unfortunately, the expectation for GUIs in the embedded world is driven by what people see on their desktops, where there exists more memory and computer power. This disparity has created a problem because, on the low end, displays may have gotten cheap enough to provide lots of pixels, but there still isn't the memory and processor power for massive graphics code. How do we create sophisticated GUIs within the embedded systems limitations? To answer this, we need to first explore the way people build embedded GUIs today.

CURRENT METHODS AND TOOLS

Today there still exist a wide range of embedded systems. The methods used to program such systems range widely as well – from very simple, bare-bones techniques to using toolkits and libraries. Murphy offers an excellent book on this subject.[2] This paper simply introduce a few basic techniques.

BARE-BONES EMBEDDED GUI PROGRAMMING

When programmers have an extremely small amount of computer memory with which to work, the approach to graphics is often attacked from the bottom up. It usually starts with a frame buffer that stores graphics as individual pixels. When the programmer needs to change something on the display, such as an icon or piece of text, he determines which individual pixels needs to be turned off and on, and writes the appropriate data to the frame buffer. Specialized hardware and driver software then refreshes the display at regular intervals.

This method, of course, can be accomplished using any programming language from assembly to Java but is most typically done in C because efficient access to the hardware is crucial.

THE GRAPHICS LIBRARY (GL)

After most programmers have built at least one "bare-bones" system, they discover that it is useful to create routines that do typical tasks like draw lines, circles and text. Instead of writing directly to the frame buffer and setting individual pixels, they simply call these routines with the appropriate parameters and let the routine fill in the frame buffer. These collections of reusable graphics routines are called Graphics Libraries, GLs or drawing libraries.

Many programmers build their own GLs. There are also a myriad of commercially available GLs. GLs tend to be specific to the target operating system ("OS") because the way each OS draws to the screen or fills its frame buffer is different. For example, the GL for MS Windows is Win32. For Unix it is Xlib. OpenGL is a GL ported to multiple operating systems. And there are many more. Like bare-bones programming, GLs can be accessed via C or object oriented languages such as C++ and Java. For efficiency's sake, GLs themselves are almost always written in C.

WIDGETS AND TOOLKITS

GLs usually focus on basic primitive drawing routines for objects such as circles, polygons, lines or text. Of course, many of today's GUIs require more sophisticated objects (widgets) such as menus, buttons and windows. It is possible to build these objects from the basic primitives and many programmers still successfully achieve this on their own. However, like GLs, there are numerous commercially available widget sets and toolkits that are built on top of GLs. For example, Motif is built on Xlib. MS Foundation classes are built on Win32. Widgets and toolkits can be object oriented (Swing, AWT, SWT, Qt, and Interviews) or they can simply be C based (Motif).

JAVA

This section will explore the confusion surrounding the relationship between Java and graphics development.

A significant distinction between Java and other languages is that it is interpreted and thus requires a target specific interpreter – Java Virtual Machine ("JVM") – for each target computer or embedded system. It is often thought that Java magically draws graphics without the need for target specific GLs, widget toolkits, frame buffers or drivers. Somehow the JVM simply handles this and there is no need to port the graphics to different targets. This is not the case. Java is a programming language like C++, C# and C. There are widget sets that are specific to Java. SWT, Swing, and AWT are examples. These Java widgets plug into GLs just like any other.

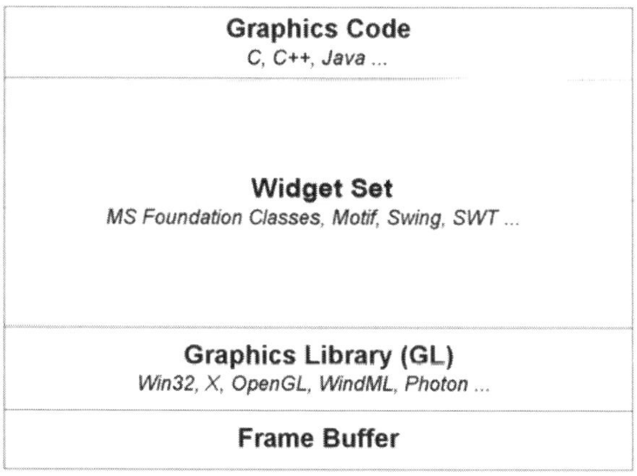

Figure 3: *Graphics layers from the bottom up.*

On the desktop, this transformation may seem magical because most desktops can afford the addition of a JVM and multiple GLs. Therefore it is transparent to the programmer. In the embedded world, the extra resources needed to run Java rarely reside naturally on the system. An explicit decision must be made to include it, which leads to significant concern over JVM and OS compatibility. Memory size and performance are also a major concern when including a JVM in an embedded system.

Figure 3 illustrates a bottom-up view of the tools and libraries embedded programmers use today.

COMPARISON OF CURRENT METHODS

With a basic understanding of some of the options, the embedded programmer is faced with the question: "What is the graphics strategy for my next project?" As is so often the case, the answer is: "It depends." It depends on (1) available memory, (2) the processor power, and (3) the company's future product strategy. Williamson explores this question in "Choosing the Right Graphics Development Approach"[3]. This section will explore the strengths and weaknesses of each approach.

Bare-bones programming strengths are (1) small code size, (2) high performance, and (3) minimum requirements for additional target software and hardware layers. Disadvantages include the facts that it can be (1) very time consuming to code, (2) difficult to change, and (3) often impossible to build sophisticated GUIs.

GLs have the advantage of separating the programmer from the hardware and providing a large percentage of the graphics and drawing functionality needed. GLs improve programmer productivity while providing a degree of portability. The disadvantages of GLs are that they increase the memory requirements and are often not full-featured enough for very sophisticated GUIs such as in-car navigation systems or complex control screens.

Widgets and toolkits have the advantage of providing very sophisticated graphics objects with less programming effort. Their disadvantages are that they add significant code size – on the order of megabytes – and require much more computing power. In addition, many embedded systems require custom, unique widgets that are specific only to that product – such as an artificial horizon, speedometer or animated graphic of battlefield status. Widgets only provide standard desktop GUI objects such as menus or buttons. Programming custom widgets can be time-consuming. Finally, widgets typically are delivered in binary form, which means they cannot be changed and bugs cannot be fixed.

Java's strengths include ease of use, the advantages of object-oriented programming, and portability, even on the fly. Java's disadvantages are its large code size, slow performance and requirement of a virtual machine.

Figure 4 lists the types of libraries and toolkits and approximate sizes that might be used depending on the programming technique chosen. As shown on Figure 4, there is a much higher price to pay the more a programmer takes advantage of higher productivity solutions.

Component	Memory
Java Virtual Machine	**1 MB**
Java Widget Set (AWT, Swing, SWT, etc.)	**2 MB**
C++ Widget Set	**1.5 MB**
C Widget Set (Motif)	**1 MB**
Graphics Library (GL)	**50 KB**

Figure 4: *Approximate memory requirements for components you might use to develop you graphics application.*

A PRACTICAL C GRAPHICS ARCHITECTURE

Knowing the strengths and weaknesses, it is safe to say that no single approach fits all. Therefore, this paper focuses on a mixture of approaches that could satisfy the widest variety of applications.

In a nutshell, this approach suggests developing all of the graphics using the C programming language and a generic, portable, GL interface. The proposed architecture also allows for the possibility of developing application, logic, and control code in languages other than C.

This architecture is chosen because it is the lowest risk and highest performance. That is, it is guaranteed to run on any target. It does not require a lot of memory, a 32-bit processor, or a virtual machine. Equally important, it has the best chance of being ported if the target needs to change because of last minute system specification changes. The C architecture allows for some breathing room. It doesn't eat up every last bit of processor power or memory, leaving room for application code or last minute changes and bug fixes.

Finally, the proposed approach also does not require a high level toolkit or widget set – although it does not preclude it and those could be added later if necessary.

THE ARCHITECTURE

To develop this architecture, it is first useful to address the weaknesses of the C programming approach. This offers an opportunity to create an architecture that minimizes these weaknesses. The primary disadvantage of using C is its relatively long development time. That is, the time needed to write the initial code and to make changes.

Some ideas that can be implemented to minimize C coding inefficiency include the following:

1. Critical to minimizing the amount of C code needed is the choice of a GL. The GL should have ample functionality, but be capable of scaling down if not every function is needed. (Note: Make sure you have the rights to modify the source for this GL. This will allow you to make changes and tradeoffs necessary for code reduction.)

2. Provide the ability for other applications to draw on the display. Many applications, such as mapping software, already have drawing capability and routines. There is no need to reinvent the wheel. Design graphics libraries so these off-the-shelf applications can draw directly into the frame buffer. This way, items like map drawing routines won't have to be built from scratch.

3. Build the graphics code to allow import of graphics generated from other tools. These tools will greatly improve graphics development efficiency. (Image editing tools that produce bitmaps are a good example.)

4. When possible, use graphics code generators – as long as they output portable C code and you have access and rights to the generated C source code.

5. Create your own graphics routines, libraries and modules and use them from project to project. This will improve C programming productivity and reliability.

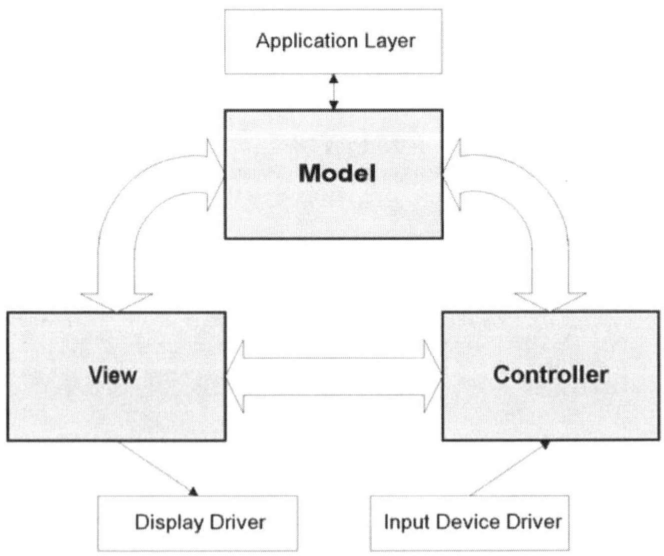

Figure 5: *Classic Model-View-Controller architecture.*

6. Build functions and graphics structures that make sense for the current application. Don't try to make them too general or plan too far ahead for future projects. If functions get too general at the early stages, code size will grow too large. As you move onto the next project, you can continue to generalize your graphics libraries and interfaces to suit your needs.

7. When the schedule gets tight, simplify the GUI and eliminate snazzy graphics. This may even have the positive side effect of creating a more usable GUI.

In general, improving C coding efficiency requires that you reuse existing code, add layers as needed, and code other parts of the application in structures and languages appropriate to the application domain. To achieve this, the architecture must distinctly separate the graphics code from the logic, control, algorithmic, and application code. This idea is, of course, not new. The Smalltalk folks have been using this separation principle since the 80's. Figure 5 illustrates the classic model-view-controller architecture proposed by Smalltalk.

Figure 6 is a detailed view of the recommended C based architecture. Critical to this architecture's success is a sufficient event API that adequately handles the communication between the application code and the graphics. This API is the key to programming flexibility and the ability to add off-the-shelf program modules as needed. It is also important to separate the physical display driver and input driver from the rest of the code. This enables the portability needed as hardware and OS requirements change.

Separating the graphics from the behavior makes intuitive sense. Graphics and behavior are very distinct problems that require different technologies and solutions to solve them. For example, there are software development tools that are very good at designing application behavior, but not good at creating graphics code. Likewise, there are good graphics development tools that don't do a good job describing logic and algorithms. If graphics are intertwined with behavior, it

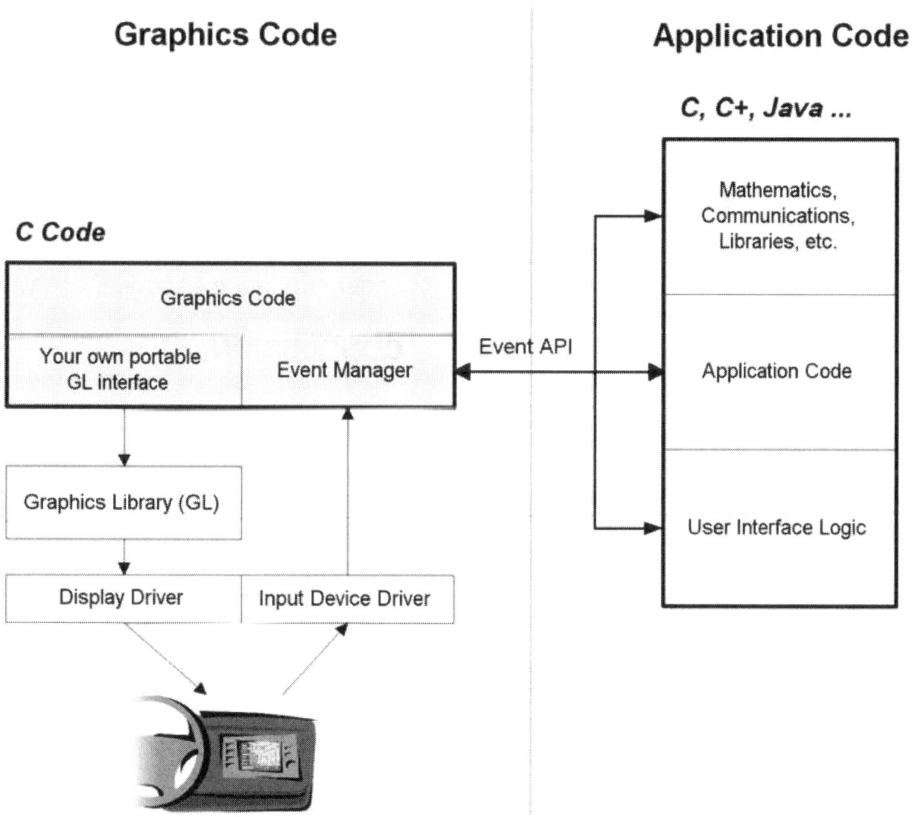

Figure 6: *Architecture that separates your graphics code from application code. This allows you to take advantage of C's small size and portability while still using other programming methodologies and languages that may be more appropriate for your application.*

becomes impossible to choose best in class tools to do the job.

Finally, building graphics applications is very different than behavior. Graphics and GUIs have a very clean transition from the what-you-see-is-what-you-get ("WYSIWYG") representation devised by the graphics designers, artists and human factors engineers. Application code, on the other hand, starts from a more abstract form (flow charts, UML diagrams, and statecharts) and must be converted to explicit lines of code.

The tools and languages used to accomplish these distinct tasks can, and should, be very different. For example, C is a perfectly fine language to capture pixels from an artist's drawing and render them on an embedded system display screen. It's a very straightforward translation. However, building complex communications, engine control, body control and navigation system user interface logic requires languages with more structure and methodology – such as that found in UML based languages like Java and C++. The architecture proposed in this paper will allow the programmer to save processing power and memory for the complex application code. The architecture judiciously uses the austerity of C for the basic graphics grunt work.

MIXING C GRAPHICS WITH C++ AND JAVA

In many cases, memory and processing power will be sufficient only to allow use of C code throughout the entire application. However, there are times when a programmer may be able to introduce some C++ or Java code into the system. There might be libraries or subsystems that will be much more efficiently programmed in C++ or Java. For that reason there may be a need to make C API calls from C++ or Java. Making C calls from C++ is pretty straightforward and there is no need to discuss that here. Making C calls from Java is not too difficult either, but it requires calls through the Java native interface ("JNI"). Pont provides some useful tips on how to achieve this. [4]

FUTURE DIRECTION

To be sure, memory and processing power within embedded systems will, on average, increase. However, there will always be low-end systems because of the ever-present drive to produce smaller and cheaper devices. For this reason, the industry will never be relieved of the need to handcraft software and squeeze it into tight spaces.

However, some relief may be available from tools targeted at this problem. Code generators are an area to consider for this purpose.

CODE GENERATORS

Code generators automatically turn high-level descriptions and views of a design into lines of programming code. Code generation is a relatively new technology that promises to significantly raise the productivity of programmers. Of course, these tools are not without risks and overhead – just as early compilers were met with skepticism from a generation of grizzled assembly programmers. However, modern tool vendors claim that code generators produce error-free code that exactly matches the specifications and is more compact than handwritten code[5]. While the jury is still out, early results show promise.

For purposes of this paper, a distinction must be drawn between graphics code generation and application code generation. Application code generators typically take some abstract view of a system – flowcharts, UML diagrams, statecharts, etc. – and turn it into lines of compilable code. Graphics code generators are a bit more straightforward, taking the exact 2D graphical representations of GUIs from a host development system and turning them into code that renders the exact same graphical representation on the target display. This "translation" is less prone to errors or misinterpretation of the designer's intent.

Because of the direct one-to-one translation, graphics code generators may be more viable in the short term – provided they can generate efficient C code that comfortably fits into the target memory space. You may or may not be able to use a code generator for your system today, but if the system has been architected as described in this paper, you'll be ready to drop in the generated graphics code if and when you're ready.

CONCLUSION

Given the stringent, unique requirements of in-vehicle software, this paper concludes that a C programming based architecture is the most practical and efficient method of building embedded displays. New methodologies and languages are emerging, but they are still not appropriate for most embedded systems. The C based architecture is efficient, reliable and portable across multiple projects and it can be used on both low and high-end hardware. This portability provides significant reuse within an organization, raising reliability and reducing development time.

REFERENCES

1. Turley, Jim. "Motoring with Microprocessors," Embedded Systems Programming, September 2003, p. 43.
2. Murphy, Niall. *Front Panel, Designing Software for the Embedded User Interface,* Emeryville, CA: Publishers Group West. ISBN:0-87930-528-2

3. Williamson, Jason and Henry, Hannah. "Choosing the Right Graphics Development Approach," *Embedded Developers Journal*, June 2001, p.19.
4. Pont, Mick. "Calling C Library Routines From Java," Dr. Dobb's Journal, July 2003, p. 28.
5. Webb, Warren. "Virtual Programmers Build Embedded Code," *EDN,* February 21, 2002, p. 32.

CONTACT

Mr. Juran began developing embedded systems hardware and software in 1985. He holds BS and MS degrees in Electrical Engineering from Carnegie Mellon University. After graduating, he joined Bell Labs to develop microprocessors, single board computers and microprocessor emulators. In 1989, Mr. Juran joined Hewlett Packard as an embedded software engineer. In 1992, he co-founded Altia, Inc.

Email: mikej@altia.com Web: www.altia.com

2002-01-0873

Robust Embedded Software Begins With High-Quality Requirements

Ronald P. Brombach, James M. Weinfurther, Allen E. Fenderson and Daniel M. King
Ford Motor Co.

ABSTRACT

In an effort to improve the quality of software and take advantage of *Lessons Learned*, Ford Motor Company has created a generic list of software requirements to help prevent software design errors, mistakes and faults from being delivered to our customer in our vehicles. Ford's intent of publishing these requirements is to provide a basis for an SAE Recommended Practice. Ford's goal is to encourage the software community to participate in the development of a recommended practice that can benefit all software developers. These particular requirements were developed for Automotive Body Features.

INTRODUCTION

Expanding functional complexity (more functions in the same box) has led to an increase in software defects and design problems. Management at Ford Motor Company requested us to assess the quality of software and its associated risk to vehicle programs. We started off the assessment with reviews after suppliers have answered a series of checklist questions. During the reviews we found many designs were fraught with potential problems. Each of the suppliers had their own methods for developing software. Some of the processes and coding standards were good others were not. This made it difficult to assess the quality of the supplier's software product.

Over the past few years, Ford had collected a set of *Lessons Learned* in Body Feature Modules. *Lessons Learned* are great teaching tools if they are read and used by the software developers Generally, *Lessons Learned* are an eclectic mix of disjointed ideas; each *Lesson Learned* can be totally unrelated to others. Also, each supplier will have knowledge of only a subset of those *Lessons Learned*. Ford might have the *Lessons Learned* cataloged, but our suppliers do not know that the *Lessons Learned* exist. It became obvious that Ford needed to convert the *Lessons Learned* catalog into requirements in order to avoid repeating the same mistakes.

So, Ford's catalog of *Lessons Learned*, or "good" software practices, were compiled, organized and converted to requirements. Each supplier is now provided with this set of requirements, and must assess their compliance to these requirements. A review process was also created to help the supplier understand the robustness of their software design and assess compliance to the software requirements.
The forerunner to this document is a SAE paper titled "Lessons Learned the Hard Way" [#2001-01-0019]. The requirement examples used in "Lessons Learned the Hard Way" paper, are a subset of the requirements detailed in this document.

REQUIREMENT CATEGORIES

There are a total of 56 requirements that have been partitioned into nine categories:

1. Determinism and Interrupts
2. Microprocessor Set-Up
3. Microprocessor Selection
4. Design and Coding Practice
5. EEPROM Management
6. Hardware Interface
7. Diagnostic Trouble Codes and Software Monitoring
8. Input Debounce
9. Sleep/Wake (low power mode)

DOCUMENT FORMAT

This paper is laid out into tables of requirements. Each table represents one software requirement and contains the following four elements:

1. **Requirement – (R)** Describes the requirement.
2. **Exceptions – (E)** Lists any acceptable deviations to the requirement.
3. **Justification – (J)** Supports the reasons why the requirement is needed.
4. **Review Questions – (Q)** These are typical questions that will be asked at the design review.

1.0 TEMPORAL DETERMINISM AND INTERRUPTS

Temporal determinism is a measure of the software's ability to perform a function within a known, short (milliseconds) time frame regardless of processor load. The software program is considered non-deterministic if it can ever exceed this required time frame (loop-time).

If the software is non-deterministic errors may be introduced in two ways: a) The software design does not account for all of the possible worst-case scenarios resulting in time frame overrun, and b) Inconsistent sample periods can create control system errors resulting in unexpected behaviors.

Requirements for temporal determinism are: a) Counting software instructions, b) CPU chronometrics.

Interrupts also impact temporal determinism and can be a problem if not handled very carefully. Interrupt related requirements include: a) What actions to take for external interrupt, b) Software (internal) interrupts c) Nesting interrupts, and d) Overlapping interrupts.

1.1 DETERMINISM and INTERRUPTS – Software Timing	
R	Software shall never rely on counting instructions for timing purposes.
E	Instruction counting can be used for low-level hardware interfacing. The reason for the exception must be clearly documented and code must be thoroughly commented. List all exceptions. Each exception must be justified in the design review. Include timing calculation, microprocessor type, and clock speed.
J	The practice of timing by instruction counting makes the subsystem sensitive to hardware and software changes.
Q	Describe all instances where you have used counting instructions for timing purposes.

1.2 DETERMINISM and INTERRUPTS – CPU Load	
R	Worst-case CPU load shall not exceed 80% sixteen months before production and must be less than 90% at product launch.
E	None
J	Guarantees determinism, accommodates feature creep, and improves margin management.
Q	1. **Prove** that the hardware low-pass filter will protect the microprocessor bandwidth from excessive interrupts. Proof might include; worst-case voltage analysis, worst-case corner frequency, worst-case cut-off frequency… 2. Is interrupt overload behavior defined? 3. What is the expected subsystem behavior if interrupt rate is exceeded? (I.e., FORD-9141 is 960 μseconds.) 4. What is the worst-case CPU load due to all interrupts running at their worst possible frequency? 5. How many milliseconds of the basic scheduler looptime can be consumed by these worst-case interrupts? 6. **Prove** the code is deterministic and that the software does not exceed the scheduler loop time. 7. For all external interrupts prove the design compensates for expected sources of input signal degradation. Sources of degradation include: time/usage, operational temperature range, mechanical vibration environment, EMC, harness routing, connectors, daughter cards, in-module jumpers, water, dust, salt, etc. 8. How is the CPU idle time consumed? 9. If an input arrives unexpectedly, is a response specified? 10. Present your Worst-Case CPU Load Analysis. 11. What is the worst-case interrupt timing analysis? 12. Have you determined whether main loop timing has exceeded 80% or 90% of available CPU bandwidth?

1.3 DETERMINISM and INTERRUPTS – CPU Idle Time	
R	CPU idle time must be consumed by the scheduler routine only.
E	Sleep mode may have it's own idle loop time.
J	Enables determinism measurements, promotes good software design and common software architecture.
Q	See questions in Requirement 1.2

1.4 DETERMINISM and INTERRUPTS - External Inputs	
R	No external (to the module) inputs shall cause an interrupt in the microprocessor.
E	If external interrupts are needed, then: • Low-pass hardware filtering must be used. • Clipping frequency must be above the anticipated worst-case frequency by at least 5%. • Input frequencies above the clipping frequency must not cause any interrupts. • Interrupts occurring at the clipping frequency (as defined above) will not exceed worst-case CPU load as defined in Requirement 1.3. External interrupts may be used to wake the module from low power state. However, the input is configured as a digital input once the module is out of low power mode. (i.e., Door Ajar switch is used to wake the module up, then it is configured to a digital input and the signal is debounced.)
J	Interrupts can cause the software to be non-deterministic and unstable. They can also consume 100% of the processing bandwidth.

1.4 DETERMINISM and INTERRUPTS - External Inputs	
Q	1. If interrupts are disabled or masked, can events be lost? If yes, explain what events are lost. 2. See questions in Requirement 1.2

1.5 DETERMINISM and INTERRUPTS – Internal Module Interrupts	
R	Interrupts occurring at the worst-case frequency will not exceed worst-case CPU load as defined in Requirement 1.3. Internal module interrupts include; CAN controller, SCP controller, UARTs, and PCB mounted Hall-effect sensors.
E	None
J	Interrupts can cause the software to be non-deterministic and unstable.
Q	See questions in Requirement 1.2

1.6 DETERMINISM and INTERRUPTS - Nesting	
R	Supplier must ensure that nested interrupts cannot occur.
E	None
J	Reduced complexity and improved design robustness. Also, the system is more deterministic. Nested interrupts are not warranted at this time for body modules.
Q	How do you prevent nested interrupts?

1.7 DETERMINISM and INTERRUPTS – Overlapping Interrupts	
R	Overlapping interrupts are a condition when the CPU is servicing Interrupt A and another Interrupt A becomes pending. The design must take into account that overlapping interrupts can occur. There are two methods of managing this condition. Preferred Method: • If an overlapping interrupt is pending, increment a counter • Take appropriate fault management for this application (Maybe disable interrupts for a period of time.) • Clear the overlapping interrupt Alternate Method: • Take appropriate fault management for this application (Maybe disable interrupts for a period of time.) • Clear overlapping interrupt
E	None
J	If interrupts are occurring faster than the processor is able to service them, the system is unstable. Not clearing the pending interrupts may cause infinite loops. Increases robustness and fault tolerance.
Q	See questions in Requirement 1.2

2.0 MICROPROCESSOR SET UP

Microprocessors have many control registers that must be initialized upon reset. It has been Ford's experience that many software developers do not realize that these values may change over time due to minor flaws in silicon. This forces the scrupulous software developer to periodically refresh the control registers. These requirements address microprocessor control registers and memory usages: a) Control register refresh, b) Clock pre-scalar, c) Unused memory, and d) Keep alive memory.

2.1 MICROPROCESSOR SET UP – Control Registers	
R	Microprocessor control registers shall be periodically refreshed.
E	A specific control register does not have to be refreshed if it could de-stabilize the microprocessor or cause undesired side effects.
J	Improves robustness by protecting control registers from corruption via inadvertent modification or degradation, etc. This is especially important if the module is always powered.
Q	Show the microprocessor-programming model and explain how each register is configured and how often the register is refreshed.

2.2 MICROPROCESSOR SET-UP - Clock Pre-Scalar	
R	Microprocessor clock pre-scalar shall not be dynamically changed, e.g., when entering low-power mode, clock pre-scalar value shall not be altered.
E	The clock pre-scalar may be changed *immediately* before entering low-power (Sleep) mode and *immediately* after exiting low-power mode.
J	Simplifies testing. Prevents design errors. Lessons learned – previous use has caused a misinterpretation of input signal durations and incorrect functional operation.
Q	Is the Micro clock prescalar changed at any time other than immediately after reset?

2.3 MICROPROCESSOR SET UP – RAM/ROM/FLASH/EEPROM Management – Unused Memory	
R	Execution of unused memory location(s) shall result in a microprocessor hardware reset. This memory space must be filled with the smallest size instructions that will cause the microprocessor to reset. Do not use an interrupt to meet this requirement (the vector table may be damaged). One solution is to fill unused memory with NOPs except the last instruction, which is a branch to itself. This assumes the watchdog timer is running. See requirements 4.1 and 4.2.
E	None

2.3 MICROPROCESSOR SET UP – RAM/ROM/FLASH/EEPROM Management – Unused Memory	
J	Increases robustness. Since the program counter is outside of the program address range, a hardware reset is the safest recovery method. This strategy requires the microprocessor to burn up time waiting for the watchdog reset to occur.
Q	1. Is RAM cleared on reset? 2. What is done with unused RAM/ROM/FLASH/ EEPROM?

2.4 MICROPROCESSOR SET UP – Keep Alive RAM Management	
R	If Keep Alive RAM is used, a data integrity verification method must be used. E.g., checksum. If verification fails, force a watchdog reset. After the reset, Keep Alive RAM must be initialized.
E	None
J	Keep Alive RAM data values can become corrupted leading to faulty feature operation. This is especially true with modules that are never reset i.e., modules with low power mode that monitors inputs.
Q	1. What is the result of corrupted data in KAM? 2. How is KAM integrity verified?

2.5 MICROPROCESSOR SET UP – Control Registers	
R	Upon reset, all control registers must be set to some value regardless if it is used or not.
E	None
J	Never trust the microprocessor defaults. The microprocessor potentially resets from any number of different operating states/voltages. The microprocessor is only tested/verified for a limited number of operating states/voltages.
Q	List all control registers, their default values and what value they are set to after reset.

3.0 MICROPROCESSOR SELECTION

Selecting the microprocessor is very important to meeting all the functional, operational, and software requirements. Problems arise when the microprocessor is undersized and printed circuit board changes are needed to upgrade the microprocessor. These requirements apply when choosing the microprocessor: a) Custom microprocessors, b) Production volumes, and c) Memory sizing.

3.1 MICROPROCESSOR SELECTION – Upgrade Path – Custom Microprocessor	
R	Custom microprocessors shall not be used.
E	None
J	Custom microprocessor upgrade path is limited compared to standard microprocessors. Microprocessor manufacturers provide greater support (manufacturing and engineering) to standard devices.

3.1 MICROPROCESSOR SELECTION – Upgrade Path – Custom Microprocessor	
Q	Is this microprocessor an off-the-shelf product?

3.2 MICROPROCESSOR SELECTION – Production Volumes	
R	Microprocessor must be commercially available in production volumes through the microprocessor manufacturer on a two-week notice. If FLASH is **not** used, an emergency plan must be developed to address late software changes
E	None
J	Vehicle campaign response time constraints. Improved risk management.
Q	Is there a version of the Micro that supports software changes available in production volumes on a two-week notice? (E.g. OTP or FLASH)

3.3 MICROPROCESSOR SELECTION – Upgrade Path – Memory Sizing	
R	The initial microprocessor selection must allow for RAM and ROM/Flash size increases without changing the printed circuit board footprint.
E	None
J	Can help avoid late board layout changes. Accommodates feature creep and improves margin management. Improved risk management.
Q	What is the next RAM and ROM/FLASH microprocessor size that has the same footprint?

4.0 DESIGN AND CODING PRACTICE

The practices defined in this section are related to the design and coding of software. MISRA (Motor Industry Software Reliability Association) has listed 127 rules in the Guide For The Use Of The C Language In Vehicle Based Software [2] that Ford also requires suppliers to follow and are not listed as part of these requirements. This section discusses: a) Watchdog timer, b) Software loops, c) Checking critical data values, d) Usage of include files, e) Data comparison, f) Statement Side-effects, and g) Usage of Assembly and Object Oriented Language.

4.1 DESIGN AND CODING PRACTICE - Watchdog timer	
R	Watchdog timer must be used.
E	None
J	Allows recovery from microprocessor lock-ups due to software and/or hardware errors.
Q	What is the watchdog timeout period?

4.2 DESIGN AND CODING PRACTICE - Watchdog timer – Timeout Duration	
R	Watchdog timer period shall be longer than the worst-case (longest) task scheduler period by at least 20%, including any tolerance stack ups and worst-case interrupt performance.

4.2 DESIGN AND CODING PRACTICE - Watchdog timer – Timeout Duration	
E	None
J	Increased design robustness. This ensures that the watchdog will not time-out during normal operation.
Q	Explain the criteria used to determine the watchdog timer period?

4.3 DESIGN AND CODING PRACTICE - Watchdog timer - Servicing	
R	Watchdog timer shall be serviced within the task scheduler in one software location only, and not within an interrupt service routine.
E	Low-power mode might be another location where the watchdog timer is serviced.
J	Ensures the watchdog isn't serviced within an unintended infinite loop.
Q	In what location is the watchdog timer serviced?

4.4 DESIGN AND CODING PRACTICE - Software loops	
R	All loops waiting for an event must have an alternate escape mechanism other than the watchdog timer reset.
E	Scheduler loop and intentionally forcing a watchdog reset.
J	Allows improved exception handling, design practice, and promotes control independence.
Q	What is the escape mechanism for any potential endless loop?

4.5 DESIGN AND CODING PRACTICE - Run Time Boundary Checking	
R	Check critical inputs and data for valid boundary values based on software FMEA.
E	None
J	Improves design robustness
Q	1. List all critical data variables that require boundary checking. 2. Explain how boundary checking is accomplished. 3. What is result of an out-of-bounds condition?

4.6 DESIGN AND CODING PRACTICE – Dynamic Memory Allocation	
R	Dynamic memory allocation shall not be used. A high-level language may allocate local variables on the stack.
E	Compiler generated dynamic memory allocation is allowed.
J	Improves robustness by eliminating: memory fragmentation, memory leaks, out-of-memory condition, stack/heap collision, and heap management.
Q	Is dynamic memory allocation used? (Except stack operations.)

4.7 DESIGN AND CODING PRACTICE - Date format for human readability only.	
R	If date codes are used to track module software versions, the format of the date shall be DAY-MONTH-YEAR in DD-MMM-YYYY format, e.g. 12-NOV-2000.
E	None
J	Prevents confusion between European and American date formats which swap the day and month locations.
Q	What date format is used for tracking software versions?

4.8 DESIGN AND CODING PRACTICE – Include Files	
R	The supplier shall not include files that contain executable statements. Only header files may be included.
E	None
J	Including executable statements results in decreased testability / readability, and increased complexity. The programmer should investigate creating an "interface abstraction" and using included header files instead.
Q	Explain the use of any executable statements in include files.

4.9 DESIGN AND CODING PRACTICE - Comparison of Values	
R	For variables with a contiguous range of values, only check (VAR == value) if (value+1) and (value-1) are valid values. Otherwise, use ≠ , ≥, ≤, or use a CASE statement with a DEFAULT clause. E.g., if Duration is defined to only have the values 1, 2, 3 and 4, then you can use *if (Duration == 3)* but not *if (Duration == 4)*. In the latter case, use *if (Duration >= 4)*. However, pick a convention and be consistent.
E	None
J	Robust software designed to handle data integrity problems.
Q	Show me the results of the Static analyzer, such as QAC.

4.10 DESIGN AND CODING PRACTICE - Conditional Statements	
R	No conditional statement shall contain side effect, e.g. assignments, bit shifts, etc. inside IF, CASE or loop controls.
E	Initial assignment on iterative loop controls
J	Reduces the potential for unintended side effects. Increases software readability and maintainability.
Q	1. Describe the criteria and inspection methods used to prevent side effects in the subroutines/function? 2. Did you do assignments in control statements such as, if, while, until, switch, etc? 3. Did you do bit-wise ANDing or ORing in control statements?

4.10 DESIGN AND CODING PRACTICE - Conditional Statements	
	4. Did you do bit-shifts in control statements? 5. Do you avoid the ++ and -- operators in control statements?

4.11 DESIGN AND CODING PRACTICE – Language - Assembly	
R	Use of Assembly language shall be limited to hardware access, or isolated in a subroutine or macro. E.g. disable interrupts, enable interrupts, NOP, parts of device drivers, initialization etc.
E	If any assembly language is used, it must be clearly documented, code must be thoroughly commented. The assembly language listing must be presented at the design review.
J	Improves software reliability, readability, and maintainability.
Q	1. List of locations in the code that use assembly language. 2. Explain any use of assembly language.

4.12 DESIGN AND CODING PRACTICE – Language – Object-Oriented	
R	Object-Oriented languages shall not be used.
E	None
J	Body control computers do not need the capabilities of the language.
Q	What programming language are you using?

5.0 EEPROM MANAGEMENT

The good news about EEPROM is that memory values are preserved between resets – the bad news is that memory values are preserved between resets. A microprocessor reset may temporarily "fix" the software problems (corrupted RAM, obscure infinite loops) in modules that do not have EEPROM.

EEPROM is special though and poorly managed EEPROM can wreak havoc even between resets. For example, care must exercised to ensure that all EEPROM writing will finish even if a reset occurs. The following requirements address problems with EEPROM management: a) Data integrity, b) Waiting on updates, c) Life expectancy d) EEPROM initialization, e) Guaranteed write cycle, and f) Memory access.

5.1 EEPROM MANAGEMENT - Data Integrity - Detection	
R	EEPROM management strategy must provide a mechanism to verify integrity of data stored in the EEPROM, e.g., parity bit or checksum.
E	None
J	Helps ensure that the content of the EEPROM does not change due to corruption and degradation.
Q	How is EEPROM integrity verified?

5.2 EEPROM MANAGEMENT - Data Integrity - Defaults	
R	Default values from MROM/FLASH or a duplicate EEPROM shall be used, if EEPROM integrity check or range check fails.
E	The following condition shall always take precedence: If the EEPROM integrity check fails in the presence of Regulatory and/or Severity Mitigating Software (SMS) configuration flags, the configuration flags shall be checked for error. If they are faulty and their proper value cannot be determined, then they must be set to the default value (based upon hazard analysis) only.
J	Data integrity checking or range checking are of no value without data recovery.
Q	What is the result of corrupted data in EEPROM?

5.3 EEPROM MANAGEMENT – Data Integrity – Update	
R	Revised EEPROM data shall be stored as soon as possible (preferably within a single scheduler loop time) from the time it's acquired.
E	Multiple data bytes may take longer than one loop time
J	Prevent loss of updated diagnostic information, memory seat and RKE rolling code data by forcing EEPROM to be updated as soon as possible instead of delaying until an operator induced event (ignition OFF).
Q	How long does it take to update EEPROM once a new value is ready for storage?

5.4 EEPROM MANAGEMENT - Data Integrity – Range Check	
R	Each time data is accessed in EEPROM (read or write), a range check shall be performed
E	None
J	Design robustness. Lessons learned - A zero value in EEPROM on an MROM part resulted in a timer value of zero that locked up the microprocessor (microprocessor entered low-power mode immediately). This was not identified when using OTPs for design verification.
Q	How and when are range checks performed on EEPROM data?

5.5 EEPROM MANAGEMENT - Data Integrity - Recovery	
R	Regulatory or Severity Mitigating Software (SMS) configuration flags which are stored in EEPROM must be written using redundant bits that will allow recovery from one-bit failures at a minimum, e.g., AA, 55, F0, 0F. Hazard analysis and system safety analysis will identify these critical flags and the level of redundancy needed.
E	None

5.5 EEPROM MANAGEMENT - Data Integrity - Recovery	
J	SMS regulatory and safety feature configurations must be immune to EEPROM corruption.
Q	1. List all of the Regulatory and Severity Mitigating Software configuration flags. 2. Explain the recovery process of a one-bit error in the Regulatory and Severity Mitigating Software configuration flags.

5.6 EEPROM MANAGEMENT - Waiting On Update	
R	Software shall not perform a busy-wait operation (tight loop burning time) while waiting for the EEPROM read/write completion.
E	If a few NOPs are used, the reason for the exception must be clearly documented and code must be thoroughly commented. This must be included in the list of software exceptions that are presented at a design review. Include timing calculation, microprocessor type, and clock speed.
J	Maximize CPU resources. Avoid infinite loops.
Q	Explain how EEPROM read and write contention is resolved.

5.7 EEPROM MANAGEMENT - Life expectancy	
R	Number of write cycles to any EEPROM location over the life cycle of the product shall not exceed a maximum allowable number of write cycles specified by the EEPROM manufacturer. The operating temperature may have a negative effect on the life expectancy of the EEPROM (by a factor of 10).
E	None
J	Prevents the degradation of feature performance over time. Reduces potential for data corruption.
Q	1. What is the maximum allowable number of write cycles for a memory location as defined by the EEPROM manufacturer? 2. Provide the data used to determine the expected number of write cycles for each parameter over the life of the product.

5.8 EEPROM MANAGEMENT - External EEPROM Initialization	
R	If external EEPROM is used, the module supplier must initialize it to defined values before the module leaves the module manufacturing facility.
E	None
J	Design robustness.
Q	At what point is the EEPROM initialized?

5.9 EEPROM MANAGEMENT - Guaranteed Write Cycles	
R	System design shall ensure successful completion of any EEPROM write cycle even in the event of a module power failure.
E	None
J	Avoids data corruption.

5.9 EEPROM MANAGEMENT - Guaranteed Write Cycles	
Q	Explain how software and hardware ensures that the EEPROM data write cycle is not interrupted in the event of a power failure.

5.10 EEPROM MANAGEMENT - Guaranteed Access	
R	If there is a potential for concurrent EEPROM read/write cycles, then provision for EEPROM contention resolution shall be made.
E	None
J	Avoids data corruption for reads and writes from EEPROM.
Q	Explain how EEPROM read and write contention is resolved.

6.0 HARDWARE INTERFACE

To be able to design software that interfaces to the hardware correctly, the software engineer must completely understand the hardware design and how to control it. The following requirements address two functional areas: a) Analog to digital conversions, and b) Software controlled power supplies.

6.1 HARDWARE INTERFACE – A/D Sampling Relative to battery voltage – Software Ratiometric Conversions	
R	The raw battery voltage value (software unfiltered) must be used on software ratiometric conversions for other analog channels that are referenced to battery voltage.
E	None
J	Lessons Learned – Use of the filtered battery voltage will cause large errors in the ratiometric conversion. Additionally, these errors are difficult to detect via testing.
Q	Are analog channel ratiometric conversions done in hardware or software?

6.2 HARDWARE INTERFACE - External Power	
R	Any external device shall be accessed only while its power supply is stable.
E	None
J	Lessons learned – Reading an analog channel after its power was removed resulted in incorrect feature behavior.
Q	1. Does the design compensate for expected sources of input signal degradation? Sources of degradation include: time/usage, operational temperature range, mechanical vibration environment, EMC, harness routing, connectors, daughter cards, in-module jumpers, water, dust, salt, etc. 2. How much time is given in software to stabilize the power supplies before data is read or outputted?

6.3 HARDWARE INTERFACE – A/D Sampling Relative to battery voltage – Filter Matching	
R	Input filters between the measured battery voltage and the A/D channel being sampled must be matched such that there is a maximum of a 0.5% error (one-bit error with an eight-bit A/D) due to maximum battery voltage slew rate.
E	None
J	Lessons Learned – Measuring a battery voltage with a time constant of 45 milliseconds against a channel with a time constant of 4 milliseconds cause large errors when the battery voltage changes. Battery voltage is, in general, a very noisy environment. The one-bit A/D error is included to accommodate the mismatch in the hardware design.
Q	1. Explain general input operation i.e., de-bounce strategy, analog data processing, PWM processing. 2. Explain the voltage mapping technique for battery voltage. How is a logic level "1' and logic level "0" determined? How does high/low battery voltage affect mapping? A/D conversion or comparator circuit. 3. Is rate of change used or specified for the analog signal? 4. Are analog channel ratiometric conversions done in hardware or software? 5. Is a minimum and maximum arrival rate specified for each input, including communications?

6.4 HARDWARE INTERFACE – A/D Sampling Relative to battery voltage – Time Delay	
R	If the sampling of battery voltage and the A/D channel are separated in time, the time separation must be limited to at most a 0.5% error (one-bit error with an eight-bit A/D).
E	None
J	Delays in time have the same effect as mismatched input filters.
Q	1. Explain general input operation i.e., de-bounce strategy, analog data processing, PWM processing. 2. Explain the voltage mapping technique for battery voltage. How is logic level "1' and logic level "0" determined? How does high/low battery voltage affect mapping? A/D conversion or comparator circuit. 3. Are analog channel ratiometric conversions done in hardware or software?

7.0 DIAGNOSTIC TROUBLE CODES AND SOFTWARE MONITORING

Monitoring the health of the software and reporting false failures are described below: a) Ignition switch position, b) Monitoring critical software parameters, c) Exiting diagnostics operation, and d) Battery voltage monitoring.

7.1 DIAGNOSTIC TROUBLE CODES and SOFTWARE MONITORING – Ignition Switch Position	
R	Cycling the ignition switch between positions within four seconds shall not cause false DTCs to be stored. E.g., Moving the ignition switch from RUN to START such that it takes 4 seconds to complete the transition.
E	None
J	The customer, under normal operation, must not be able to cause a DTC.
Q	Can a customer cause a Ford Diagnostic Trouble Code to be set through normal (or even abnormal) use?

7.2 DIAGNOSTIC TROUBLE CODES and SOFTWARE MONITORING - Critical Software Parameter Monitoring	
R	Software shall support monitoring and reporting of critical software parameters. Diagnostic PID reads shall be used to transfer the information on a diagnostic communication link. The monitored data shall be stored in non-volatile memory and the **Diagnostic Command F800hex** shall reset all monitored values to zero. Each monitor counter value must be clamped at its maximum value: 255 for one-byte counters and 65,535 for 2-byte counters. **PID REQUEST C950hex** Byte#1 Power ON reset: This counter shall be incremented with each Power ON reset. Power ON reset is defined as a battery re-connect, watchdog timer reset, or any unexpected event that causes software to execute the power-on (battery re-connect) initialization routine. Byte#2 Illegal op code: If the microprocessor traps illegal op codes, this counter shall be incremented each time the microprocessor attempts to execute an illegal instruction. Byte#3 Watchdog timer reset: If the software has the capability to distinguish between Power ON reset and Watchdog Timer reset, this counter shall be incremented each time a Watchdog timer reset occurs. Byte#4 Reserved. **PID REQUEST C951hex** Byte#1 Stack overflow: Every time the software encounters a stack overflow, this counter shall be incremented. Byte#2 Loop time overflow or scheduler idle time overflow: This counter shall be incremented each time the software exceeds its maximum loop time. Byte#3 Minimum Idle-Time: This represents the shortest remaining idle time identified before entering the idle-loop – scaled to fit into an eight-bit byte. The scalar value must be defined by the supplier and documented. The supplier must

7.2 DIAGNOSTIC TROUBLE CODES and SOFTWARE MONITORING - Critical Software Parameter Monitoring	
	attempt to define the scalar to provide an appropriate timer resolution that accounts for both expected idle-time and potential lack of floating point operations. Immediately before the microprocessor enters the idle-loop, it must determine the actual duration of remaining loop-time and compare it against the current value for this byte; if the value is less, then Byte #3 acquires this new value. Byte#3 will not be updated during Diagnostic Mode (unless it is explicitly cleared by a diagnostic request). Zero is used only when loop-time has been exceeded (Byte#2 is incremented in this case). 255 represents any value larger than the value represented by 254. Byte#3 will be set to 255 by the supplier when it is shipped to the OEM. This value will be used to verify calculated idle-time. Byte#4 Reserved. **PID REQUEST C952hex** Byte#1 & #2 wake-up counter due to RKE: Every time the module is in sleep (low power) mode and receives an RKE signal, this two-byte counter shall be incremented. The counter shall be incremented regardless of whether the RKE signal is valid, causing the module to wake-up, or invalid, allowing the module to remain in sleep mode. The intent is to aid in identification of situations in which RF noise causes excessive processing of RKE events. The noise may be unintentionally generated internally or external to the vehicle. The noise may be unavoidable but the RKE receiver may have inadequate rejection. A significantly large count would be expected due to transmissions from both associated and unassociated in-range key fobs could indicate a concern. Byte #1 is the most significant byte of the counter and Byte #2 is the least significant byte. Byte#3 & Byte#4 Wake-up counter due to discrete input: Every time the module wakes up due to an input change, this two-byte counter shall be incremented. Reset and Watchdog events are excluded. Byte #3 is the most significant byte of the counter and Byte #4 is the least significant byte
E	Not required if a module doesn't have any diagnostics or communication link
J	Provides a method for collecting software operating metrics. Metrics provide the following benefits: 1. Provides a level of "state of health" 2. Helps isolate sources of problems or abnormal behavior. Be it hardware, software or hardware/software interface problems.

7.2 DIAGNOSTIC TROUBLE CODES and SOFTWARE MONITORING - Critical Software Parameter Monitoring	
Q	How many times has the module experienced the following during testing? 1. Power on reset 2. Illegal op code 3. Watchdog timer reset 4. Divide by zero 5. Stack overflow 6. Loop time overflow 7. Wake-up events due to Remote Key Entry 8. Wake-up events due to discrete inputs

7.3 DIAGNOSTIC TROUBLE CODES and SOFTWARE MONITORING – Exiting Diagnostics	
R	When exiting Diagnostics State the ECU shall perform a hardware reset by allowing the watchdog to time-out.
E	The diagnostic command to enter Sleep mode does not require a reset when the wake event occurs, however, it is recommended if that's part of "normal" behavior.
J	While performing diagnostic operations the integrity of the inputs and outputs versus the internal states of the features can be compromised. Going through resets initializes the I/O and internal software states.
Q	How do you ensure the integrity of the feature states when exiting diagnostics?

7.4 DIAGNOSTIC TROUBLE CODES and SOFTWARE MONITORING –DTC Logging and VBatt	
R	The micro controller shall only log DTCs when the battery voltage (measured at the module pin) is within the *DTC_Logging_Voltage_Range*. The *DTC_Logging_Voltage_Range* is defined as 10.0 - 15.0 volts inclusive. Logging is suspended immediately when battery voltage is outside the *DTC_Logging_Voltage_Range.* Upon reset, assume that the battery voltage is outside the *DTC_Logging_Voltage_Range*. In order to start/resume logging DTCs, Vbatt must be continuously inside the *DTC_Logging_Voltage_Range* for four seconds.
E	These DTCs may be logged, if appropriate, when VBatt exceeds the DTC_Logging_Voltage_Range: • Low Battery Voltage • High Battery Voltage • Defective ECU
J	When VBatt exceeds the DTC_Logging_Voltage _Range, there is a much higher probability of detecting DTCs incorrectly and these false DTCs should not be logged.
Q	1. Based on your worst-case analysis, in what voltage range do you log DTCs? 2. Describe any time-based hysteresis associated with the DTC Logging Voltage Range.

8.0 INPUT DEBOUNCE

Input signals must be stable before the software acts on it. The following requirements describe methods for stabilizing the input signal: a) Input signal sample period, b) Single digital inputs, c) Multiple digital inputs, d) Analog inputs, and e) Pulse train inputs.

8.1 INPUT DEBOUNCE – Maximum Sampling Period	
R	Sampling period must be no more than 11 milliseconds. This applies to the following inputs: Single Digital, Multiple Digital, Continuous Analog, and Discrete Analog.
E	Does not apply to Pulse Trains, Bus/Protocol Inputs, or where any time constants greater than 100 milliseconds are used.
J	Based on a Key-Life Test for mechanical switches, the worst case settling time is 15 milliseconds. A three times margin results in a 45 milliseconds debounce duration. 45 milliseconds divided by four sample periods yields 11.25 milliseconds per sample. Most low-frequency electrical noise is associated with body vibration (10Hz – 30Hz). This noise can cause higher frequency resonance in other components (the steering column for one) up to 50Hz. Also, the highest frequency periodic electrical noise (very uncommon) can generate frequencies up to 700Hz. An 11 msec sample period avoids aliasing the 50Hz and lower body vibration noise.
Q	1. Explain general input operation i.e., de-bounce strategy, analog data processing, PWM processing. 2. For all inputs (except interrupts), ***prove*** the design compensates for expected sources of input signal degradation. Sources of degradation include: time/usage, operational temperature range, mechanical vibration environment, EMC, harness routing, connectors, daughter cards, in-module jumpers, water, dust, salt, etc.

8.2 INPUT DEBOUNCE - Single Digital Input	
R	Debounce duration shall be 34 – 56 milliseconds with a minimum of five consecutive samples. All samples must be identical in order to assert a new debounced value.
E	If this input is tied to an interrupt, see requirement 1.4 and 1.7 Some switches require a longer debounce duration but may not exceed 120 milliseconds for human perception reasons. Inputs that are latched in software (e.g. Heated Backlight, One Touch Down) are not included in this exception. Based upon the type of digital input device and location (i.e. solid state mounted on same board as CPU versus a daughter board) or subsystem response time requirements (e.g. fast response to pinch strips), shorter durations and/or fewer samples may be used. However, a list of sampling frequencies and the number of samples must be presented at the design review.
J	Based on a Key-Life Test for mechanical switches, the worst case settling time is 15 milliseconds. A three times margin results in a 45 milliseconds debounce duration. The calculation is based on four sample *periods* resulting in improved noise immunity. However, five samples are required to guarantee four sample periods.
Q	See questions in Requirement 8.1

8.3 INPUT DEBOUNCE - Multiple Digital Input	
R	A multiple digital input is a set of single digital inputs that are combined to form a single result with more than two values (e.g. Ignition switch, Wiper switch). A two-stage filter must be used to debounce these inputs. The first stage will debounce each single input individually. The debounce duration for the first stage shall be 22 – 44 milliseconds with a minimum of four consecutive samples. The individually debounced values are combined to form the composite value. The second stage must debounce the composite value for three times the number of samples as the first stage at the same sampling frequency. For example, using an 8-millisecond sampling period, the first stage requires 4 samples and the second stage requires 12 samples. Total debounce duration will be (4 + 12 –1) * 8 = 120 milliseconds.
E	If this input is tied to an interrupt, see Requirement 1.4 and 1.7. Based upon the type of digital input device and location (i.e. solid state mounted on same board as CPU versus a daughter board) or subsystem response time requirements (e.g. fast response to pinch strips), shorter durations and/or fewer samples may be used. However, a list of sampling frequencies and the number of samples for each filter stage must be provided. Also, the minimum 3 to 1 ratio between the first and second stage must be maintained
J	This filters out low-frequency electrical noise and intermediate switch positions that are dependent upon speed of operator actuation. The first stage filters out most electrical noise. The second stage filters out intermediate switch positions and higher frequency noise. The 3 to 1 ratio on debounce duration allows the shortest possible pulse that will be passed by the first stage to be filtered out by the second stage. Another advantage is that this technique is very robust and will return information

8.3 INPUT DEBOUNCE - Multiple Digital Input	
	even when there is a lot of noise on several input lines.
Q	1. Explain how the design compensates for expected sources of input signal degradation. Sources of degradation include: time/usage, operational temperature range, mechanical vibration environment, EMC, harness routing, connectors, daughter cards, in-module jumpers, water, dust, salt, etc. 2. Explain general input operation i.e., de-bounce strategy, analog data processing, PWM processing. 3. See questions in Requirement 8.1

8.4 INPUT DEBOUNCE – Discrete Analog Input – Maximum Number of Values	
R	A discrete analog input is a signal that is quantized into a finite number of expected values. The maximum number of values including short circuits, open circuits, and dead bands will be limited to six.
E	If more than six discrete values are used, the supplier must use Monte Carlo simulation to prove that overlapping ranges will occur in no more than three subsystems per million. A subsystem includes the module, wiring, switches, connectors, etc.
J	With the current accuracy of the electronics (10% accuracy for battery voltage), higher numbers of values results in overlapping ranges when doing worst-case analysis. An 8-bit A/D is not the limiting factor at this time.
Q	1. See questions in Requirement 8.1 2. For any discrete analog inputs, list the possible state for each channel (including shorts and open circuits). Provide a worst-case analysis showing state boundaries and nominal values for minimum, nominal, and maximum operational voltages. List each boundary/nominal value in ohms, volts at the device (not the micro pins), and A/D counts.

8.5 INPUT DEBOUNCE – Discrete Analog Input – Range Distribution	
R	Every discrete value must occupy a minimum of 16% of the entire A/D range without overlap.
E	Error values or dead bands may use a range of less than 16%.
J	Lessons learned. Wider, evenly distributed ranges minimize false input changes. The 16% A/D range is based on the 1999 analog front wiper switch campaign.
Q	See questions in Requirement 8.1.

8.6 INPUT DEBOUNCE – Discrete Analog Input – Stabilization	
R	A signal is considered to be stable if it varies less than 5% of the total A/D range for at least 100 ms (minimum of 10 samples). The signal must be stabilized before a new debounced value is asserted. Averaging must not be used to stabilize the signal.
E	When there are only two ranges (digital input read on an analog channel), the debounce time may be reduced to 5 samples.
J	Averaging can allow a noise signal to cause false activation. Stabilization filters out intermediate switch positions and noise. 100 ms is specified because all known inputs of this type are human activated.
Q	See questions in Requirement 8.1

8.7 INPUT DEBOUNCE – Discrete Analog Input – Dead Bands	
R	Although dead bands are discouraged, if they must be used the dead band strategy must be reviewed and approved by the OEM.
E	None
J	In most cases, dead bands add unnecessary complexity. In addition, when the input stabilizes to a dead band that is not adjacent to the current range, the system response may be feature dependent.
Q	See questions in Requirement 8.1

8.8 INPUT DEBOUNCE – Pulse Train Input – Over-Sampling	
R	Over-sampling can be used to detect a pulse only if at least 5 samples can be obtained during the pulse occurrence. Over-sampling can also be used to detect a difference between two pulses only if 5 samples can be obtained during the shortest difference in the pulse durations. This applies under all module operating modes (sleep versus awake).
E	None
J	Lessons learned. The Nyquist sampling period for detecting the existence of a pulse is pulse-width/2. This sample rate is susceptible to noise, therefore, 5 samples are specified for noise rejection.
Q	See questions in Requirement 8.1

9.0 SLEEP/AWAKE (LOW POWER MODE)

Body module computers must be able to operate in a low power mode or when the key is in the OFF position. These requirements address wake and sleep functions.

9.1 SLEEP/AWAKE – No Reset	
R	When transitioning from Sleep mode to Awake mode the ECU shall not force a hardware reset. Sleep mode is not the same as unpowered.
E	None

9.1 SLEEP/AWAKE – No Reset	
J	The features are specified to operate independently of Sleep mode and using a reset to exit Sleep mode adversely affects behavior of many features, i.e. Autolamps.
Q	Explain the use of reset on the Sleep/Awake transition.

9.2 SLEEP/AWAKE – Refreshing Registers	
R	Microprocessor control registers shall be refreshed when transitioning from Sleep to Awake mode.
E	A specific control register does not have to be refreshed if it could de-stabilize the microprocessor or cause undesired side effects.
J	Improves robustness by protecting control registers from corruption via inadvertent degradation, etc. This is especially important if the module has been asleep for an extended duration.
Q	When are registers refreshed on the Sleep/Awake transition?

9.3 SLEEP/AWAKE – Unintended Wakeups	
R	On a detected transition of a wakeup input, the modules shall temporarily wakeup and debounce the input. If the newly debounced value is the same as the pre-sleep value, the module will return to sleep within one minute.
E	None
J	Minimize the key-off load caused by noisy inputs waking the module.
Q	1. Explain the conditions/actions for each transition to/from each of these major modes (Please use Finite State Machine Diagram/Table to document this). 2. How is a wake-up signal debounced? 3. Is the algorithm different from the normal debounce? This may be different if a pulse width is a critical wake-up input. 4. What's the worst-case time it takes to debounce the wake-up inputs? 5. When the module is in Sleep mode and a wake-up input wakes the module but the debounce indicates that there was no change of state on the input, how long does it take to re-enter Sleep mode? 6. Include a detailed list of operation/functions performed during this event. E.g., Set a wake-up timer, Write to EEPROM, etc. Include any exception handling.

9.4 SLEEP/AWAKE – Wakeup Input Scanning	
R	While "asleep," the module will scan all polled wakeup inputs using a sampling period no longer than 50 milliseconds.
E	None
J	Minimize increases in function response time upon module wakeup. Minimizes potential for missed momentary switch activation.
Q	1. How are wake-up inputs monitored when in Sleep mode? 2. What is the sampling period used during sleep for polling wake-up inputs?

CONCLUSION

These software requirements were developed to mitigate software defects in automotive body module computers. This paper provides a starting point for an SAE Recommended Practice by examining mistakes that have caused issues in previous projects at Ford Motor Company. This list of requirements is based on Ford's collection of lessons learned. From each lesson, a requirement was written to prevent the same type of mistake from occurring in future projects. As requirements were written, review questions were added to the review process to help ensure compliance to the requirement.

Ford has already implemented the listed requirements and design review process. These and other software requirements now apply to all body module computers in the 2004 model year programs. As we discover new lessons learned, we will continue to update or add new requirements.

New and unique mistakes are understandable, but repeating the old ones can and must be avoided. It is our contention that to avoid repeating Lessons Learned you must convert them into requirements. Our hope is that these requirements and others are adopted by SAE as a recommended practice.

ABOUT THE AUTHORS

Ronald Brombach has 20 years of experience in the computer field. He has worked in the steel industry, industrial automation design, and for the past 15 years, at Ford Motor Company. Ron has been an active member of SAE since 1992. Ron has earned a B.S. in electrical engineering from Oakland University in Rochester, Michigan and a M.S. in Computer Control Systems from Wayne State University in Detroit, Mi.

James Weinfurther has worked in the computer industry for over 25 years with over 17 years experience in real-time applications. Jim was worked at Ford Motor Company for the past 11 years and has recently been active in the SAE. He earned a B.E. in Computer Engineering from The University of Michigan in Ann Arbor, MI and also has an M.S. in Computer Control Systems from Wayne State University in Detroit, Mi.

Allen Fenderson has 24 years of experience in the computer field at Ford Motor Company. He has real-time and analytical software experience. Allen holds a B.S.

and an M.S. in Physics from State University of N.Y. at Buffalo.

Daniel King has 7 years of experience in the automotive software field at Ford Motor Company. Dan has earned a B.S. on Electrical Engineering and an M.S. in Electrical Engineering from Michigan Technological University in Houghton, Mi.

REFERENCES

1.J. Weinfuther, A. Fenderson, D. King, R. Brombach, *"Lessons Learned the Hard Way", SAE 01PC-345,* March 2000.

2. Motor Industry Software Reliability Association, "*Guide For The Use Of The C Language In Vehicle Based Software",* The Motor Industry Research Association, April 1998.

VIRTUAL PROTOTYPES AND COMPUTER SIMULATION SOFTWARE

2005-01-2458

Optimization of Accessory Drive System of the V6 Engine Using Computer Simulation and Dynamic Measurements

Jaspal S. Sandhu, Antoni Szatkowski, Brad A. Rose and Fong Lau
DaimlerChrysler Corp.

ABSTRACT

At the initial accessory drive system design stage, a model was created using commercial CAE software to predict the dynamic response of the pulleys, tensioner motion and pulley slip. In a typical 6 cylinder automotive accessory drive systems, the first system torsional mode is near the engine idle speed. The combination of these two events could generate numerous undesirable noise and vibration effects in the system. Data acquisition on a firing engine with a powertrain dynamometer confirmed the computer model's results. Correlations are then developed and established based on results between the firing engine to the CAE model to increase confidence in the generated model. Further system optimization through design modifications are used to tune the system to minimize the overall system dynamics.

INTRODUCTION

Single serpentine belts with tensioners are used frequently in passenger and commercial vehicles to drive the engine accessories. The single belt drive and tensioner is an enhancement over earlier multiple belt designs because they provide improved packaging, durability, serviceability, and reduce manufacturing complexity.

In order to achieve the best performance and cost efficient system, it is required that the dynamic characteristics of the system be understood early in the design process. Due to the inertias of the accessories (alternator, A/C compressor, water pump and power steering pump) and belt modulus, most automotive systems have the first rotational mode between 20 and 50 Hz. This frequency range aligns to the firing frequency at idle for a typical 6 cylinder engine. This combined with the highest crankshaft rigid body torsional input leads to very high rotational vibrations of the pulleys and belt tension fluctuations.

The dynamic performance of accessory drive system can be enhanced through strategic positioning of the accessories in the system. But this often conflicts with other constraints such as packaging, servicing and safety requirements. In addition, the positioning of the accessories is determined in the early stages of engine compartment packaging and becomes difficult to relocate them at the later stages of product development. Therefore, a simulation tool is needed to predict the dynamic behavior of serpentine belt system to help develop a stable low vibrating system. Design Kit for Accessory Drives (DKAD) and SIMDRIVE software are used for this purpose. These tools allow a user to quickly predict accessory drive dynamic performance and assess alternative designs.

Dynamic behavior of an accessory drive can be categorized into two major groups: rotational vibration of the accessory drive system and lateral belt span vibration.

Rotational response involves the accessory pulley rotation with the belt span acting as axial springs. Rotational modes are primarily excited by the crankshaft excitation but A/C compressor and power steering pump torque fluctuations can also excite these system modes. Excessive vibration leads to tension fluctuations in the belt spans which can cause slip, and in some situations, can excite lateral vibrations parametrically.

Lateral belt span vibrations act similar as to a guitar string which has several modes. Lateral belt modes can be excited by pulley run-out, bumps on the belt or axial tension fluctuations.

The Objective of this paper is to demonstrate the process of evaluating the accessory drive behavior with a CAE model, measuring the dynamic behavior, and then optimize the system.

METHODOLOGY

Rotational analysis involves a model based on the belt behaving as an axial spring and rotation of pulley inertias under applied torque or prescribed motion. The belt modulus, pulley and tensioner properties are defined. The equilibrium state is then established based on steady engine speed and static torque loads of accessories. The equations of motion are linearized about this equilibrium state to obtain the associated eigenvalue problem. Forced response analysis is conducted to predict the rotational response due to crankshaft induced harmonic response. Once individual pulley response is known, the tension fluctuations in different belt spans can be obtained. In the presence of non-linearity due to dry friction tensioner, or one way clutch (OWC) pulley, the equations of motion are solved by numerical integration. With belt tension across each pulley known, the likelihood of belt slip can be calculated.

Belt lateral vibrations are based on equations governing an axially moving string. The lateral natural frequencies decrease with engine speed. For a given pulley run-out for each pulley, the amplitude of lateral vibration is calculated. Tension induced lateral vibration response is computed from tension fluctuations stored during rotational forced response analysis.

ASSUMPTIONS

Rotational and lateral response analysis includes following assumptions:

1. Belt stretches axially in a quasi-static manner and has no bending stiffness.
2. Uniform belt elastic and mass properties.
3. Pulleys except the tensioner have fixed axis of rotations.
4. Belt/pulley contact points are fixed and do not change.
5. Belt coefficient of friction is assumed constant.
6. Belt slip on the pulley is negligible.
7. Modal damping is assumed for the belt while dry friction or viscous dissipation is assumed for the tensioner arm.

DISCUSSION

Schematic diagram for the accessory drive is shown in Figure 1. Crankshaft pulley is used to power AC compressor, power steering pump, and alternator. Idler pulley between the power steering and the alternator is used to increase the adjacent pulley's wrap and to maximize the tight side belt length between the crankshaft and alternator. Coulomb friction based tensioner is used in the slack belt side to maintain the initial tension of 310 N.

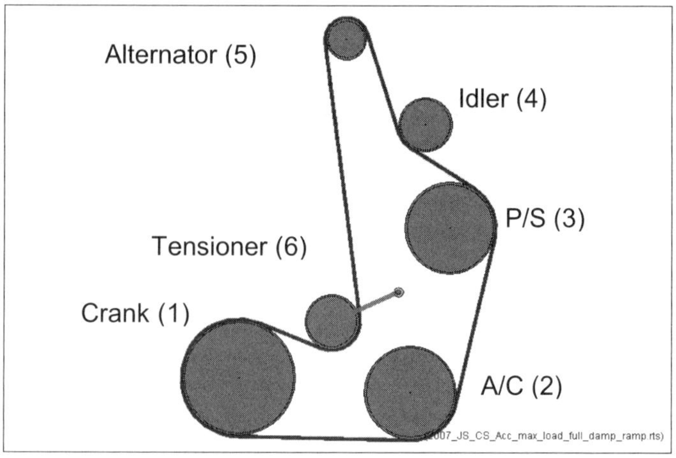

Figure 1: Accessory Drive Layout

EQUILIBRIUM ANALYSIS

The belt span tensions in all six spans, under idle operating condition, are shown in Figure 2. Since there is no steady torque applied at the idler and tensioner pulley locations, the tension across these pulleys is constant. The tension in the span next to the tensioner is nearly 300 N. This shows the benefit of the tensioner in maintaining minimum tension in the serpentine belt drive system.

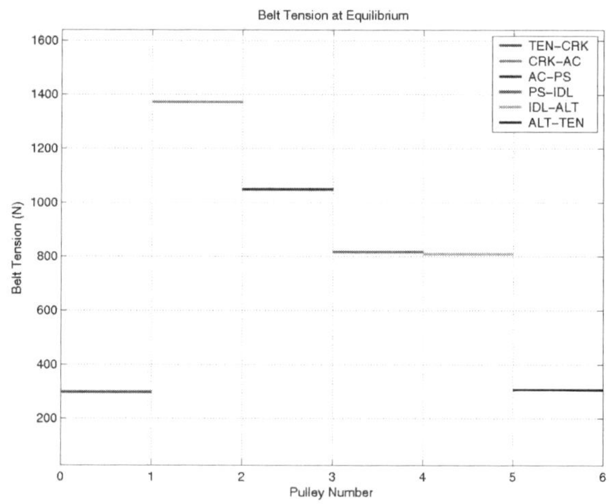

Figure 2: Equilibrium belt span tension

NATURAL FREQUENCY ANALYSIS

The equations of rotational motion are linearized about the equilibrium state to form an eigenvalue problem. It is desired to estimate equivalent viscous damping for the tensioner based on energy loss per cycle. Viscous damping of 35 N.mm.sec/deg is used under idle condition. The rotational modes under operating conditions from DKAD are shown in Figure 3. First mode occurs at 32 Hz, and is dominated by the alternator pulley response. All the pulleys including the tensioner arm are in phase with each other. This mode will resonate with the third order of the engine operating at 640 RPM. Power steering to alternator span is expected to experience large tension fluctuations during this mode.

The second mode is at 172 Hz and this mode has AC and power steering pulleys in phase while the alternator is out of phase. This mode is typically not a concern because engine inputs at this frequency are relatively low. Its mode shape also indicates that the tensioner arm has very small angular motion during this mode and therefore tensioner damping will not be very effective. Other significant mode is at 252 Hz and it's dominated by the tensioner arm motion and again this mode is not a concern for the same reasons as the 172 Hz mode.

The natural frequencies of the accessory drive are used to define engine system setup during calibration. Extended engine dwelling near accessory modes should be avoided to reduce noise and belt/tensioner wear.

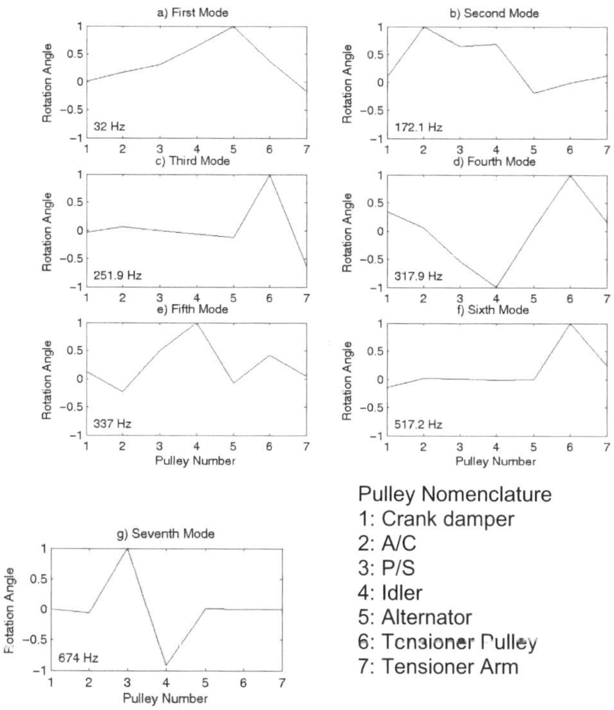

Figure 3: Natural frequencies and Mode shapes

ROTATIONAL RESPONSE ANALYSIS

Once the natural frequencies and mode shapes are known, rotational response due to harmonic excitation can be calculated. Equivalent viscous damping is assumed for the tensioner arm. Figure 4 from SIMDRIVE shows the forced response of all the pulleys, and the tensioner arm due to crankshaft excitation. Excitation amplitude of 40 rpm is held constant through the operating speed of 400 to 4000 engine rpm. At 600 rpm the crank pulley has amplitude of nearly 1° and all other accessory pulleys have higher amplitude. Alternator pulley amplitude is about 10°. This mode is of concern since it can be excited during idle engine conditions. At 3100 rpm all pulleys exhibit smaller amplitudes. The amplitude of power-steering, AC compressor rises to near 1°. This corresponds to the second mode from Figure 3.

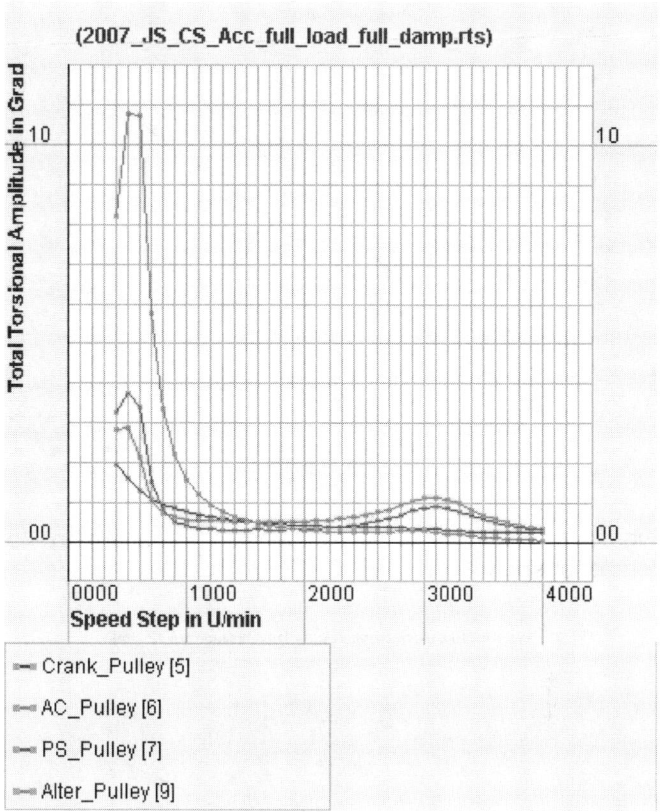

Figure 4: Forced response analysis, 400 – 4000 rpm, and ±40 rpm

BELT LATERAL RESPONSE

The belt lateral response can be excited by pulley run-out or parametrically by tension fluctuation. Once equilibrium tension is known the dynamic characteristics of each belt span is established based on belt speed and mass per unit length. The first four lateral modes for alternator tensioner spans are shown in Figure 5. Note that the natural frequency (fundamental) decreases with engine speed. The first lateral natural frequency for alternator-tensioner span is 75 Hz at 1000 rpm but drops to 54 Hz at 6400 rpm. Critical speed where the lateral natural frequency falls to zero is above maximum operating speed of engine.

The lateral modes shown will be excited when the natural frequency coincides with the alternator or tensioner pulley run-out. The first order lines for alternator and tensioner pulley are also shown in Figure 5. Estimate of maximum belt deflection can be determined by forced response analysis.

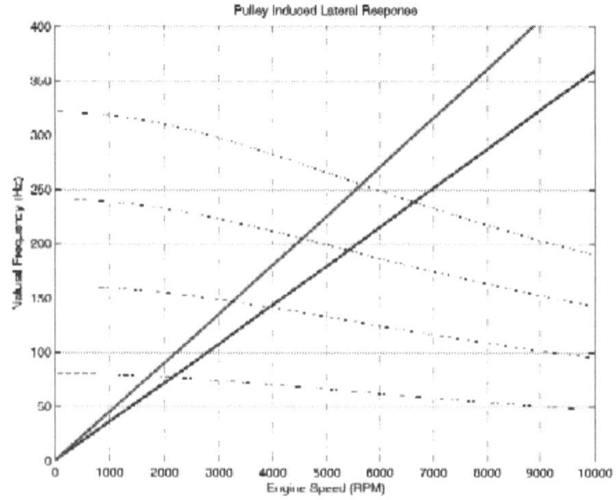

Figure 5: Lateral natural frequencies for alternator to tensioner span

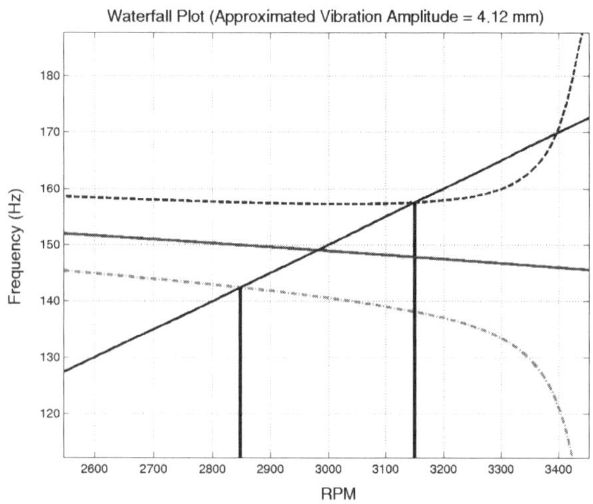

Figure 6: Parametric vibration prediction. Model predicts vibration of 4.12 mm between 2850 rpm and 3150 rpm

As mentioned before, the tension variation can excite the belt lateral response when the tension variation frequency is twice the belt span natural frequency. This type of lateral vibration is called a parametric vibration. Figure 6 from DKAD predicts a parametric vibration will occur between 2850 rpm and 3150 rpm and have a response of 4.12 mm. Resonance band widens as the engine speed approaches the second rotational mode. In this case, if the third order engine firing frequency intersects with the second belt harmonics near the second rotational mode, one would expect amplification of belt span vibrations.

MODEL VERIFICAITON

In order to correlate the CAE model parameters to the engine, detailed measurements were made on the accessory drive system. The most important plots are presented below. The accessory drive first rotational mode was verified to be 34 Hz. The measured and simulated 3rd order vibration of the tensioner arm is shown in the Figure 7. Maximum tensioner response is 6.8 mm at 680 rpm, and the maximum simulation response is 5.4 mm is at 620 rpm.

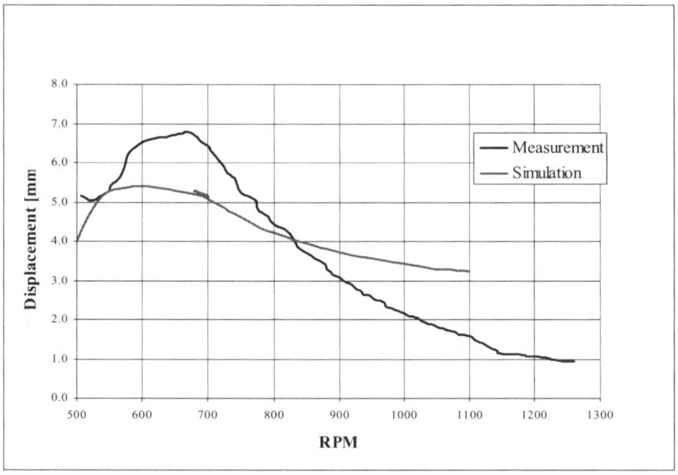

Figure 7: Accessory belt tensioner 3rd order vibration.

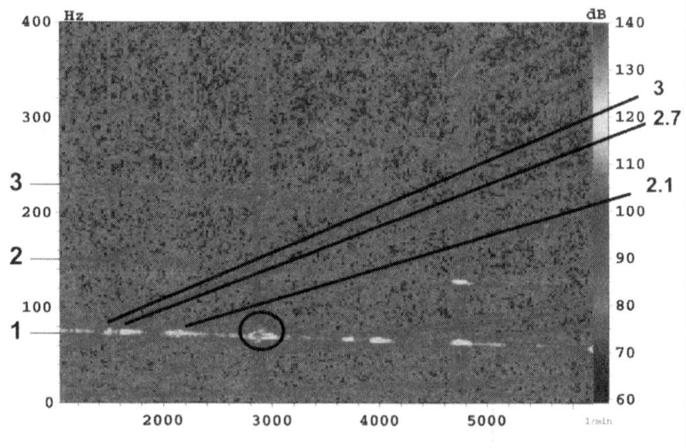

Figure 8; Alternator to tensioner belt span lateral vibration.

The belt vibrations were measured with non contacting laser displacement probes during an engine speed sweep. The output from the laser was processed in frequency domain with the displacement range set low. An example of the belt span vibration using this technique is shown in the Figure 8. This is the alternator to tensioner span and it shows correlation with the simulated data for the same span, shown in the Figure 5 & 6.

The results showed in Figure 8 are a map of the lateral vibrations for this specific span. The span fundamental natural frequencies and its harmonics are easily identified, along with the forcing functions, as orders of the engine speed. When the forcing functions intersect the span natural frequencies, the belt vibrates.

Forcing functions originate from the harmonics of the engine crankshaft vibrations, power steering and A/C compressor torque fluctuations, and pulley run outs. In figure 8, the alternator and tensioner pulley orders are 2.7 and 2.1 respectively.

A parametric excitation occurs at about 2850 rpm which generates a broad band response by the belt. This parametric vibration is caused by the 3rd order belt tension fluctuations at 143 Hz parametrically exciting the belt natural frequency at 71.5 Hz. This correlates well with figure 6.

ACCESSORY DRIVE SYSTEM OPTIMIZATION

Developed CAE model of the accessory drive system shown in the Figure 1 was used to optimize the design. The first torsional mode was causing the excessive belt slip at the power steering (PS) and alternator pulleys. This could lead to noise and excessive belt wear.

A new location of the idler pulley was proposed and analyzed as shown in figure 9. The new location of the idler pulley solved the belt slip issue and was proposed for production. In addition, the model was used to evaluate lowering belt modulus. The model showed that lowering the modulus would add robustness by lowering the dynamic belt tensions at idle.

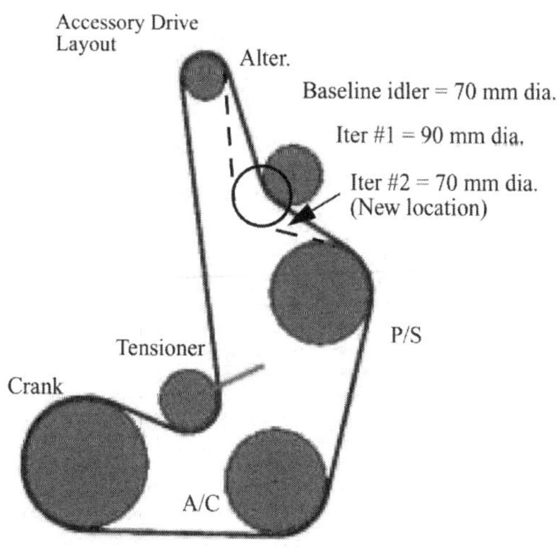

Figure 9: Optimized location of the idler pulley.

CONCLUSION

1. The CAE model of the accessory drive system was developed using DKAD and SIMDRIVE software. The model parameters were optimized and correlation was established between measured and simulated parameters.
2. First rotational mode is dominated by the alternator pulley response and occurs due to 3rd order crankshaft vibration at 34 Hz (680 rpm).
3. Second rotational mode at 172 Hz is dominated by the AC pulley and can cause increase tension variation in crank-AC, and power steering (PS)-alternator span. This mode will be excited by power steering or A/C compressor torsional vibrations above idle engine speeds but it is unlikely to cause a problem.
4. Tensioner damping is not effective for attenuating the second rotational mode because of the small tensioner motion in this mode.
5. Critical speed (belt tension falls to zero) for all spans is well above the maximum operating conditions of engine.
6. Parametric lateral vibrations for alternator-tensioner span can occur over a narrow engine operating range of 2800 to 2900 rpm. The amplitude of 4 mm will not cause a problem.
7. Originally proposed first accessory drive system layout leads to the belt slip over the power steering and the alternator pulleys during some loading conditions.
8. A new position for the idler pulley was identified using the computer model that reduces belt slip possibility over power steering and alternator pulleys in all loading conditions.
9. Simulating results from lower belt modulus showed lower dynamic tension and more slip robustness.

ACKNOWLEDGMENTS

The authors acknowledge support provided by William F. Resh of Powertrain CAE\Simulation, DaimlerChrysler. They also acknowledge Jeff Orzechowski of Powertrain NVH, DaimlerChrysler for support on this project.

REFERENCES

1. Ma, Z.-D., and Perkins, N., User Mannual for DKAD, 2002.
2. Mote, C. D., Jr., "A Study of Band Saw Vibrations", Journal of the Franklin Institute, Vol. 279, No. 6, 1965, pp. 430-444.
3. Mockensturm, E. M., Perkins, N. C., and Ulsoy, A. G., "Stability and Limit Cycles of Parametrically Excited, Axially Moving Strings", Journal of Vibration and Acoustics, Vol. 118, July 1996, pp. 346-351.
4. Barker, C. R., Oliver, L. R., and Breig, W. F., 1991, "Dynamic Analysis of Belt Drive Tension Forces during Rapid Engine Acceleration", SAE Peper No. 910687.
5. Hwang, S. -J., Perkins, N. C., Ulsoy, A. G., and Meckstroth, R. J., "Rotational Response and Slip Prediction of Serpentine Belt Drive Systems", Journal of Vibration and Acoustics, Vol. 116, No. 1, pp. 71-78.

6. Leamy, M. J., and Perkins, N. C., "Nonlinear Periodic Response of Engine Accessory Drives with Dry Friction Tensioners", Journal of Vibration and Acoustics, Vol. 120, October 1998, pp. 909-916.
7. Sandhu, J. S., Wehrly, M. K., Resh, W. F., and Rose, B. A., "An Investigation of Rotational Response of Accessory Drive Serpentine Belt System", 2001 SAE Noise & Vibration Conference, 01NVC-227.

2005-01-2392

A Tool for the Simulation and Optimization of the Damping Material Treatment of a Car Body

M. Danti and D. Vigè
FIAT Research Center

G. V. Nierop
FIAT Auto

ABSTRACT

The cost and weight reduction requirements in automotive applications are very important targets in the design of a new car. For this reason all the components of the vehicle should be optimized and the design of the damping material layout needs thoroughly analyzing, in order to have a good NVH performance with the minimum weight and cost. A tool for optimizing the damping material layout has been implemented and tested here - the need to explore the entire design space with a large number of variables suggested the use of a multi-objective genetic algorithm. These algorithms require a large number of calculations, and so the solution of the complete NVH model would be too expensive in terms of computational time. For this reason a new software tool has been developed, based on simulation of the damping material treatments by means of auxiliary mass and stiffness matrices added to the baseline modal base. Using this procedure, the required simulation time for each damping material layout configuration is reduced to a few minutes, by exploiting the capability of the genetic algorithm to efficiently explore the design space. As a result, some configurations with a significant weight reduction or a much better acoustic performance have been found.
Once the most effective damping areas have been identified, a second developed optimization tool is able to further refine the shape of the previously found damping patches, in order to improve the performance/cost ratio. This second procedure uses the superelement approach for the car body surrounding the panel in question, and the acoustic response is calculated by a simplified approach based on a unilateral fluid-structure coupling.

INTRODUCTION

One of the major complaints about cars regards passenger comfort, and its impact on the stress of the driver. To improve this comfort factor, several tools and products such as absorption and damping materials are currently available on the market but, unfortunately, they are very heavy and expensive. Moreover, the acquired noise and vibration benefits must be paid for in terms of additional weight, which results in higher fuel consumption and consequently, pollution. The challenge from the last decade is to balance any increase in the vibro-acoustic performance with the need to reduce the weight of the car. This paper outlines the implementation of these ideas in the field of noise reduction, by reducing the acoustic transfer functions of a trimmed car.

1 – ACOUSTIC TRANSFER FUNCTION CALCULATION

A common way to assess the acoustic performance of a vehicle is to evaluate experimentally the acoustic transfer function P_i/F_j (where P is the acoustic pressure and F is the input force) from the *jth* point of excitation (which can be a suspension or engine attachment point) to the *ith* receiver or microphone, usually located at coordinates corresponding to the passengers' ears. The same results can be obtained by means of numerical models and finite element analysis, which are able to speed up the problem-solving phase for the acoustic performance.
In order to estimate the vibro–acoustic behavior of a structure, it is necessary to model both the solid structure and the fluid cavity.
The vehicle trimmed body FE model (Fig. 1) consists of the complete vehicle minus the powertrain, exhaust

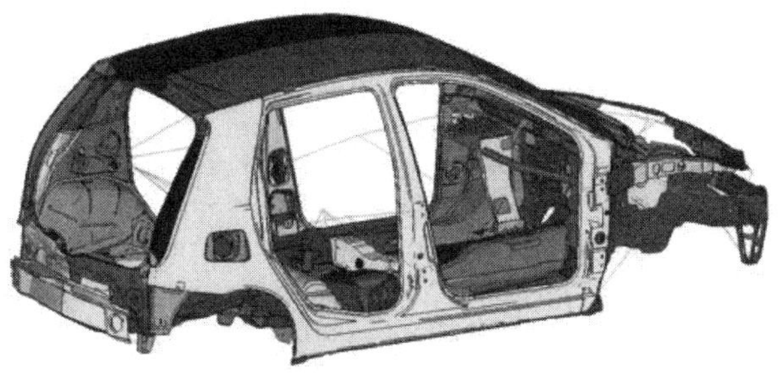

Fig. 1
FE trimmed model of a passenger car

system and suspensions; all the parts isolated from the vehicle body are excluded. The vehicle body-in-white (BIW), consisting of all the structural metal parts of the vehicle, is modeled with shell elements. The average shell element size is selected to best simulate the mode shapes of the entire vehicle up to 300 Hz. As a result, the size of the BIW model is usually larger than 3000000 degrees of freedom. For the analysis of the vibro-acoustic behavior, the BIW model has to be completed with all the non-structural components (carpets, seats, battery etc.) and damping coefficients. If some of the damping values are unavailable, a modal damping approach is used.

The internal cavity, modeled with solid elements, is obtained using the cavity generator of the commercial software package SFE/AKUSMOD (Fig. 2). The model

Fig. 2
Internal cavity FE model

includes the seats and the trunk compartment, where each of these are independent models with non-coincident boundary nodes. Their connections between each other are obtained by means of a large set of equations governing the transmissions of the pressure wave across the cavity (the equations are related to the pressure coordinate – see Eq.1). Suitable physical characteristics of the air are defined for the fluid elements.

The last step in the modeling phase for a cavity consists of selecting the pressure sample points (nodes) inside the passenger compartment, at useful locations for monitoring and optimizing the noise of the car. Usually the occupants' ears locations are taken into account to qualify and measure the noise inside a car: the most important are the driver's ears as the driver is the customer of the car maker and the end-user of the vehicle.

The driver's ear position is chosen by taking into account the seat position (the distances from the dashboard and the floor) and the variety of driving positions for drivers of different heights (using a statistical distribution). Usually the mean position of the driver's ear is chosen.

Once all the FE models have been created, the analysis can be prepared and performed. At this point the model consists of different sub-models, as described previously, that require connecting to one another. The calculation of the P/F curves can be done by either MSC/Nastran and SFE/AKUSMOD DMAP, or by a software procedure developed by Fiat Research Centre (CRF) called CRF/VEIPROD. The latter is much more efficient and allows the user to deepen the analysis of the structure.

Both approaches exploit the FSI (fluid-structure interaction) technique, consisting of a numerical algorithm that couples the respective dynamics of the structure and cavity. The pressure *P*, which is the variable for the cavity model, is coupled with the structural displacements *u*, by means of a set of equations governing the transmission of a unit area vibration to the near fluid pressure (Eq.1). Those coefficients are stored in the A coupling matrix, which is calculated by means of SFE/AKUSMOD.

$$\begin{bmatrix} -M_s \cdot \omega^2 + K_S & -A^T \\ -\omega^2 \cdot A & -M_F \cdot \omega^2 + K_F \end{bmatrix} \cdot \begin{Bmatrix} u \\ P \end{Bmatrix} = \begin{Bmatrix} F \\ 0 \end{Bmatrix} \quad \text{(Eq.1)}$$

M and K are the respective mass and stiffness matrices (S=structure, F=fluid), while F is the force applied to the structure (at rotational frequency ω). Due to the huge dimensions of the systems involved in the acoustic calculation, the modal approach has been chosen as the standard calculation and this has proved to be valid. MSC/Nastran and SFE DMAP also permit direct calculations to be ran, but this issue is beyond the scope of this paper.

The phases of the modal approach are now outlined.

Every subsystem (structure and fluid model) is solved separately by modal analysis, in order to extract the real eigenvalues and eigenvectors (φ_S and φ_F matrices).

Once the two modal bases are available, the A matrix is converted into this new modal space and the number of coupling equations decreases significantly. The final two steps of the procedure are the solution of this reduced set of equations in modal coordinates and its back-transformation into the physical coordinates.

$$\begin{bmatrix} -m_s \cdot \omega^2 + k_S & -\Phi^T \\ -\omega^2 \cdot \Phi & -m_F \cdot \omega^2 + k_F \end{bmatrix} \cdot \begin{Bmatrix} \xi_S \\ \xi_F \end{Bmatrix} = \begin{Bmatrix} f_S \\ 0 \end{Bmatrix} \quad \text{(Eq.2)}$$

where

$\{u\} = [\varphi_S] \cdot \{\xi_S\}$, where ξ_S = structural modal coordinates

$\{p\} = [\varphi_F] \cdot \{\xi_F\}$, where ξ_F = fluid modal coordinates

$\{f_S\} = [\varphi_S]^T \cdot \{F\}$ = modal force

$[\Phi] = [\varphi_F]^T \cdot [A] \cdot [\varphi_S]$

$[m_s] = [\varphi_S]^T \cdot [M] \cdot [\varphi_S]$ = structural modal mass

$[k_S] = [\varphi_S]^T \cdot [K] \cdot [\varphi_S]$ = structural modal stiffness

$[m_F] = [\varphi_F]^T \cdot [M_F] \cdot [\varphi_F]$ = fluid modal mass

$[k_F] = [\varphi_F]^T \cdot [K_F] \cdot [\varphi_F]$ = fluid modal stiffness

This approach is the counterpart of the NASTRAN modal analysis (SOL 111) with SFE/AKUSMOD DMAP, and has the same disadvantage of modal truncation (truncation of the high frequency modes). Therefore a

"residual vector" approach has been employed to reduce modal truncation error in dynamic analysis. This vibro-acoustic analysis is usually performed in the frequency domain, but can also be effective in the time domain.
The outputs of this FEA phase are the vibro-acoustic sensitivities of the vehicle body (P/F and A/F, where A is the acceleration) due to unit force inputs at the trimmed body attachment points (Fig. 3).
In order to more realistically simulate road or engine excitations, a set of force inputs with their relative

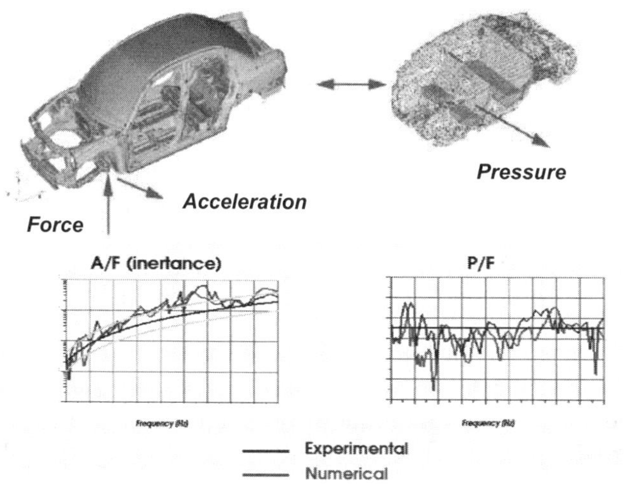

Fig. 3: Fluid structure interaction

phases and magnitudes are used. In fact the vibro-acoustic transfer functions can be regarded as weighting functions of the exterior operational forces and, for this reason, are deeply controlled and analysed.
A relevant advantage of the P/F (or A/F) calculation is that by applying the ***Noise path analysis*** method to these numerical curves, it is possible to assess the critical transfer paths to the total interior noise level. In fact, the method determines the total interior sound pressure level P_{tot} at rotational frequency ω by summing the contributions at the individual attachment points. This is given by summing the products of the acoustical sensitivities $(P/F)_i$ and the operational forces F_i at each of the N attachment points :

$$P_{tot}(\omega) = \sum_i^N P_i(\omega) = \sum_i^N F_i(\omega) \cdot \left(\frac{P}{F}(\omega) \right)_i \quad \text{(Eq.3)}$$

2 – IMPROVEMENTS OF THE ACOUSTIC TRANSFER FUNCTIONS

The challenge facing engineers and technicians is to improve both the operational forces and the acoustic transfer functions, subject to several constraints ranging from weight to packaging. Noise level reduction in cars can be accomplished in different ways. For an acoustic problem in the high frequency range, absorption material placed at the right panels can decrease noise emission and increase noise absorption effectively. If the problem consists of large operational forces, a new softer arrangement of bushings can solve the critical summation terms in Eq. 3. The 20 – 200Hz frequency range is one of the most critical however, because the absorption material is not effective in this range. The operational forces due to the road profile and the second-order forces of the standard four-cylinder gasoline engine are also relevant in this range. Reducing the P/F functions becomes vital in designing a quieter car, and this can be achieved by the proper design of stiffeners, or by the attachment of damping patches to the structure with structural adhesives.
The subsequent methodology outlined in this paper refers to the latter case and assumes that the structure has been already optimized or stiffened, as this normally takes place in the early design phase.

DAMPING MATERIAL

The damping patches consist of a viscoelastic layer exhibiting damping properties. These properties can be modeled by means of a complex Young modulus or by an equivalent modal damping coefficient ζ, which is simply twice the loss factor η of the damping properties. In modal theory, the modal damping coefficient represents the coefficient of the exponential decay ($2\pi\zeta$) exhibited by a single degree of freedom system when released from some perturbed condition (i.e. non-zero initial conditions). There are two major types of damping material currently available on the market:

- Free layer damping treatment (FLDT)
- Constrained layer damping treatment (CLDT)

A complete description of these treatments is beyond the scope of this paper (see [1]). Although both damping types are efficient, the first type is the more widely used within the automotive industry. Therefore, the procedure for optimizing the layout of the damping patches has been developed for a free layer damping treatment.
One feature to highlight at this stage is that both the Young Modulus and the loss factor of the FLDT are highly frequency-dependent, and this cannot be adequately modeled by MSC/NASTRAN.

EQUIVALENT PROPERTIES

The need for a better simulation of the damping patch dynamic behavior, coupled with the need to decrease the calculation time, led to the use of the RKU theory (see [1] and [2]).This theory is able to represent the behavior of the complete system metal sheet plus FLDT by a simple equivalent single-layer system. The RKU theory states that for flexural waves, the effect of any kind of damping material bonded to a shell is equivalent to the metal structure with a modified damping loss factor η_{EQ}. This is shown in Eq 4, where the damping coefficient of the metal sub-layer can be viewed as negligible.

$$\eta_{equiv} = \eta_{damping} \cdot \frac{e_2 h_2 \cdot (3 + 6h_2 + 4h_2^2 + 2e_2 h_2^3 + e_2^2 h_2^4)}{(1 + e_2 h_2) \cdot (1 + 4e_2 h_2 + 6e_2 h_2^2 + 4e_2 h_2^3 + e_2^2 h_2^4)} \quad \text{(Eq.4)}$$

where

$$e_2 = \frac{E_{damping}}{E_{structure}}$$

$$h_2 = \frac{H_{thickness\_damping}}{H_{thickness\_structure}}$$

Eq.4: Equivalent properties

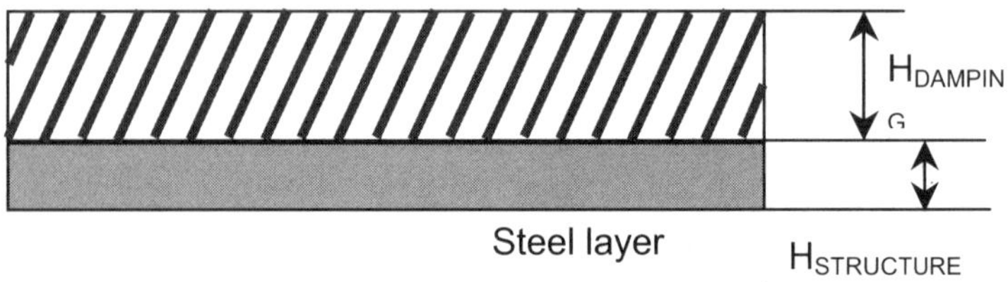

Fig 4: sketch of the system

These equivalent properties depend upon the ratio between the metal layer and damping layer thicknesses, the ratio between the two Young Moduli and, naturally, on the damping properties of the FLDT added to the panel.

The advantage of applying this theory to a full trimmed car body is that, by using a reduced set of parameters, it is possible to estimate the damping properties from a set of parameters which are frequency dependent (stiffness and inertial properties will be explained shortly). The variation of these parameters with respect to the thickness ratio is shown in Fig. 5 for a steel sheet.

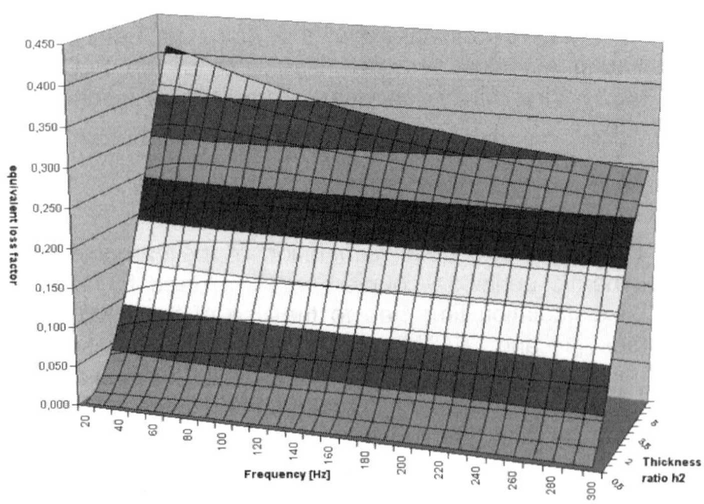

Fig. 5: frequency variation of the equivalent damping coefficient

3 – METHODOLOGY TO SIMULATE DAMPING PATCHES

Several methodologies are available for analyzing the impact of the damping patch properties on the dynamics of the complete trimmed vehicle and cavity. These consist of:

- Full FE approach (both direct and modal solution)
- FE approach by means of the superelement technique (both direct and modal solution)
- Approximated modal technique

The first method involves meshing every single part of the model (including the damping) and running a complete simulation of the structure with a certain arrangement of damping patch thicknesses. Consequently, it is the best way to assess the impact of the structural stiffness and mass variation due to the presence of the FLDT. The full FE technique has many disadvantages though - very long calculation time, no possibility of taking into account loss factor variability with respect to frequency, and it has a time-intensive meshing phase that requires great care from the user to avoid errors. The second approach uses the superelement of a part of the structure to overcome the problem of the calculation time, but it excludes that part from the potential optimization areas. Moreover, the SE calculation time is usually high, even if the SE modal or direct response function calculation is so significantly reduced that it can be joined to an optimization procedure. In fact, any optimization algorithm generally requires a large number of iterations (hence calculations), so the calculation time is crucial for the success of the methodology (see [11]).

The last technique is based only on the modal approach, however it exhibits many advantages, though it approximates the physical behavior of the FLDT.

This technique consists of calculating the real structural eigenvalues through a standard NASTRAN calculation (SOL 103). The results can be used to display the eigenmodes of the structure, or to synthesize the dynamic frequency response function of the studied vehicle.

A similar calculation is simultaneously ran for the cavity (with the same benefits of the structure). Then the modal coupling matrix is calculated with the CRF-developed software tool CRF/VEIPROD and the acoustic transfer functions of the two coupled systems are evaluated and analyzed, including the modal and panel participation factors. This is the standard procedure for the evaluation of the P/F curves.

The additional effort in implementing the new procedure arises from extracting the K and M matrices at a location of the metal structure where it is possible or convenient to put an FLDT layer. Once this data is available in a suitable NASTRAN DMIG format, the modal representations of the original K and M matrices are evaluated and stored. This is the most time consuming phase of the entire process.

As a final step, the equivalent properties of the metal sheet with viscoleastic layer are evaluated with respect to the chosen thickness and mechanical properties of the FLDT layer and metal sheet of the proper area. This calculation is quite straightforward to implement in a MATLAB or C routine, where the parameters can be quickly calculated when information is provided on the FLDT parameters (thickness, E,ν,ρ,η and frequency variation law for the damping patch) and base metal sheet parameters (E, ν,ρ and thickness).

From this point of view, any layout of the FLDT (thicknesses and positions) is simply one extra matrix that needs taking into account during the calculation of the P/F function.

This approach has several advantages:

- The calculation is extremely fast in comparison to the full FE approach (a few minutes versus several hours) and is also faster than the superelement FE technique
- Every area of the car can be treated as a variable. Moreover, it is easy to add new patches to the structure during the optimization process, without needing to run any additional calculation (this is not possible with the SE – FEM approach)
- With the frequency dependent law, it is possible to simulate the dynamic behavior of the component in a more realistic way (within the limit of the approximation), and to use a mixture of several damping materials with different frequency-dependent properties (whereas NASTRAN does not allow this).

The major disadvantage of this procedure is that it exploits a fixed modal base and consequently, when the thickness of the damping layers increases, the effects of the added mass, the damping layer stiffness and the shift of the neutral plane are not negligible. Therefore the original modal base of the reference structure cannot be employed without introducing a certain degree of error. As a rule of thumb, the original modal base can be exploited up to a thickness ratio equal to 3.5.

For these reasons this approach is a powerful one, and can easily explore all of the structural zones to find the most effective locations for the damping patches. It can save money by avoiding an experimental trial and error procedure, or save time by eliminating the need for a tedious complete FE approach of every proposed damping layout. The research to improve this methodology is still ongoing, with future studies focusing on the addition of some correction matrices to increase the constraint of the approach (the suggested limiting thickness ratio mentioned above), and on the implementation of the same procedure for the constrained layer damping material type.

4 - TEST CASE

This methodology has been exploited in the design process for the damping treatment in a fastback vehicle. The design requirement was for the car to be the best in

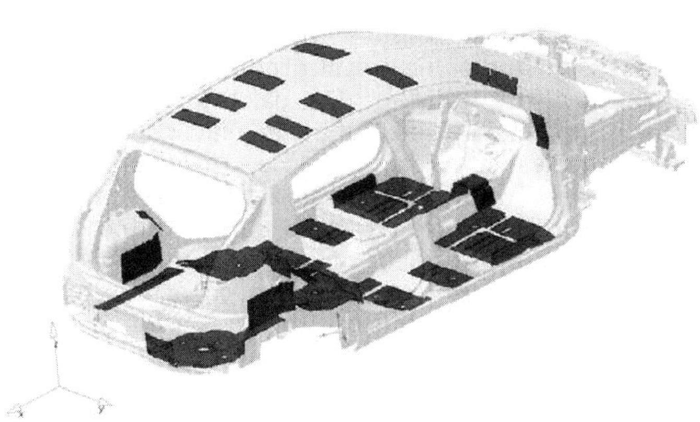

Fig. 6: layout of the reference configuration of the damping patches on the fastback vehicle

its class with respect to noise performance, and particular attention was paid to the low frequency range. The original configuration of the FLDT was designed to cover all the parts of the body where the accessibility was guaranteed, and where previous empirical experience showed that damping patches were effective. In particular, many patches were added to the front and rear floor, as depicted in Fig. 6. All the patches were 2 mm thick and their total weight was approximately 6.5 kg.

DEFINING THE PROBLEM

The first phase was to correctly define the problem in terms of targets and possible constraints. The targets used were the weight of the FLDT configuration, and an overall acoustic performance index (API) defined by the following equation.

$$API = \alpha \cdot (\beta - \sum_{j=1}^{Number\_P/F} (P_J - P_{TARGET})) \quad \text{(Eq. 5)}$$

Eq. 5: Acoustic index, defined as the summation of off-target pressure values

With the specific α and β values used here, an API value of 30 meant that all of the responses were within target over the entire frequency range of interest. The choice of two targets required the use of a multi-objective algorithm, one that was able to explore the entire design space in a robust and reliable way. This kind of algorithm can be found in many commercial tools (from MODE/Frontier to MATLAB) but a robust CRF-developed genetic algorithm was chosen, due to its high suitability for this particular optimization (more details can be found in [12] and [13]). This MATLAB-based software maximizes the objective function so that the higher the API value, the better the configuration. No constraints were defined, so that the algorithm was completely free to explore the entire variable space.

SETTING THE OPTIMIZATION

The second step of the optimization procedure was the definition of the variables. The patch thickness is the most important parameter for the free layer damping, so the variables consisted of the respective patch thicknesses. Two separate optimizations were ran - the first one with the same layout as the normal production (NP) structure, in order to improve the thickness distribution among the baseline configuration (26 variables, Fig. 6). The second optimization used larger area patches, to test if there were any additional zones where damping could be effective (34 variables were chosen for this latter case). The considered patches for the second optimization are illustrated in Fig. 7.
The range of the variables was set as 0 - 7.5 mm for all the areas involved in the optimization.

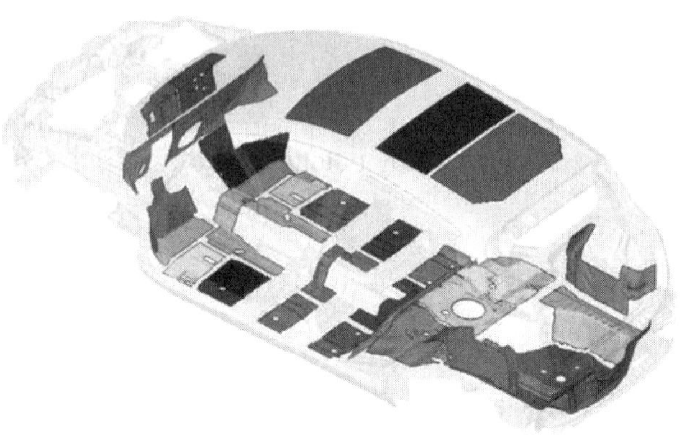

Fig.7: 34 variables considered for the second optimization run

RESULTS AND ANALYSIS

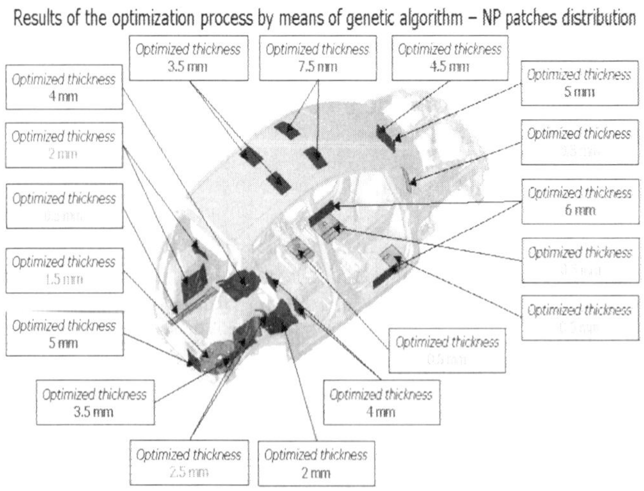

Fig. 8: distribution of the optimized configuration - red patches are the most effective ones

Each optimization calculation ran for approximately one week on an AIX IBM server. The results of the first optimization (with the NP layout) showed that the damping patches on the front and middle floor were ineffective, and suggested that thicker layers on the roof and rear part of the floor should be used instead.
This result can be explained by studying the floor in detail, as the front part of it contains a large number of stiffeners and beams relating to crash performance. As a result, the layouts with patches on the front and middle floor are not efficient when compared to other layouts, as they damp the vibrations to a much lesser extent, but with the same increase in weight. The optimization allowed a better solution to be found (Fig. 8).
In particular, the optimization procedure was able to find a Pareto frontier (see [12] and Fig. 9) of optimal configurations, from which it was possible to find two extreme optimal designs. One of these had the same weight as the NP reference layout and a better API index, whilst the other had the same overall API index, but with much less weight added to the car.

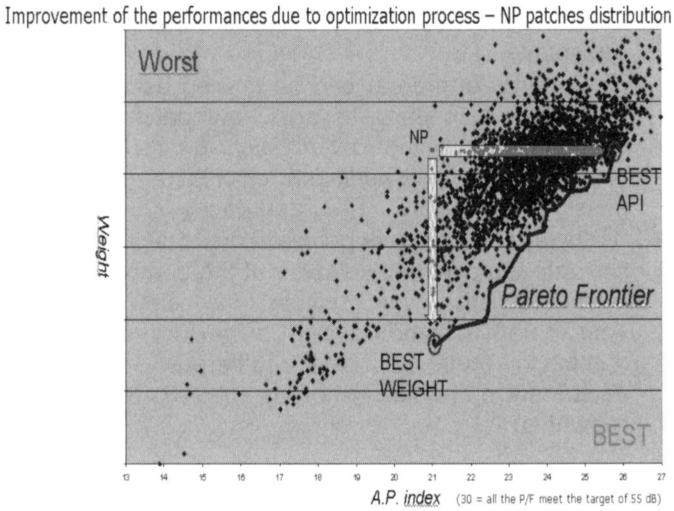

Fig 9: Design space found with first optimization

The figures are reported in the following table:

configuration	Weight [kg]	API	Improvement Weight	Improvement API
NP layout	6,5	21,00	Ref	Ref
ISO weight BEST API	6,49	25,80	Similar to Ref	+ 22,8 %
ISO API BEST Weight	2,61	21,03	- 59,8 %	Similar to ref

Table 1: results of the first optimization

The effect of the BEST API configuration on the P/F transfer functions shows the benefits of this optimization approach. When considering all of the excitation points from the engine and suspension attachments, 80% of the curves were significantly improved, 15% of them remained at the same level and only 5% were worse, in relation to the reference configuration. An example of an improved curve is shown in Fig. 10, where the target has been satisfied for the majority of the considered frequency range, without increasing the number of stiffeners or the metal sheet thickness.

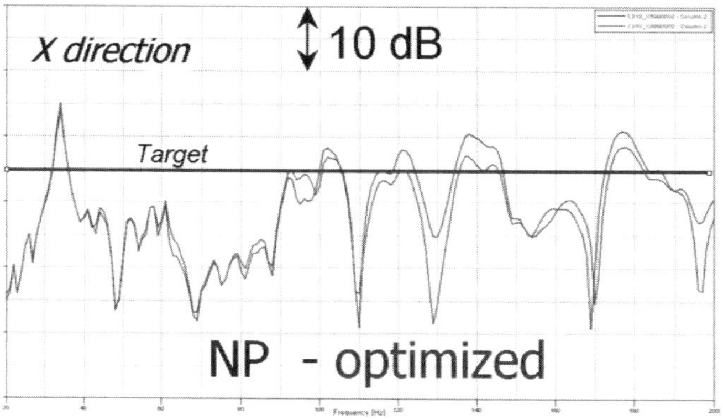

Fig. 10: Example of improved P/F function (red)

The second optimization confirmed the suggestions made by the first one. It also suggested that with 34 larger damping patches, it is possible to find a significantly improved solution with respect to the reference design.

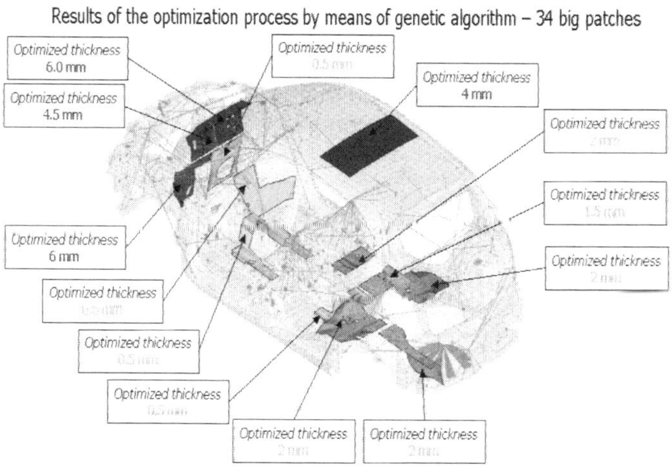

Fig. 11: Results of the second optimized configuration - red patches are the most effective ones

The results of the first optimization were been tested on a real car, which exhibited the same improvements in noise performance. At the end of the damping patches process, a configuration that inherits all the information of the optimization best configuration has been approved.

5 – LOCAL OPTIMIZATION OF THE DAMPING PATCHES

The previous methodology indicates the regions where the application of damping materials is most effective. The performance/cost ratio (or performance/weight) can be yet further improved by a local optimization of the damping patch materials, with a refinement of their local shapes. In order to do this, a methodology and a dedicated optimization tool have been developed.

To efficiently apply the optimization scheme, this methodology needs to simulate each configuration with a very short calculation time, and without losing reliability in the obtained results.

For these reasons a superelement approach was chosen, applied to only one patch at a time. The steel component at the location of the damping patch was retained, and all the residual body structure was placed in the superelement, excluding the load nodes. The damping patch here was simulated by a detailed model, using solid elements with their own Young modulus, Poisson coefficient, density and loss factor.

For the acoustic simulation (P/F transfer functions) a simplified approach was chosen, taking into account only the noise emitted by the considered panel, and using an unilateral fluid-structure coupling. In this way the simulation could be split into two parts, the first of which was the dynamic simulation of the panel with the damping material connected to the superelement. The second part was the noise emission obtained by the accelerations of the previous panel nodes and corresponding surface coupling areas. In this way, if the superelement is built retaining a sufficient number of modes then the only effective approximation is the unilateral approach, and this doesn't take into account the effect of the acoustic pressure on the structural vibrations. For a car body structure however, the difference is not as significant.

Each different geometrical configuration of the patch was obtained by activating or deactivating specific solid damping material elements.

Thanks to this approach, the acoustic simulation of the P/F for each damping patch geometrical configuration requires only a few minutes calculation, allowing an efficient application of the genetic algorithm-based optimization procedure described previously.

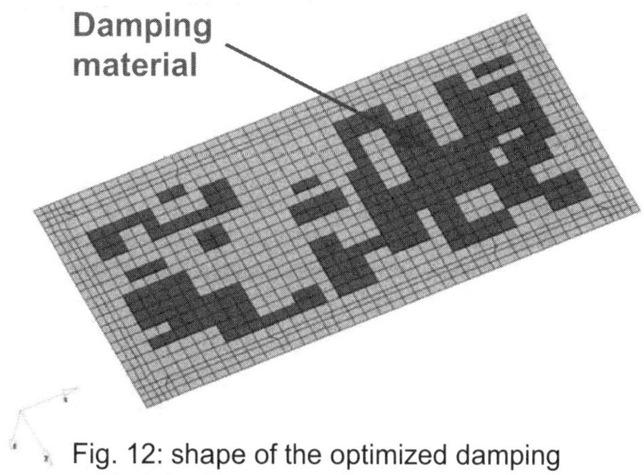

Fig. 12: shape of the optimized damping material patch

To illustrate the results that can be obtained by this procedure, the damping material applied to a rectangular steel panel has been optimized, giving the final geometrical configuration shown in Fig. 12. The corresponding results are showed in Fig. 13, with a comparison between the bare configuration (no damping material applied to the panel), the completely covered panel and the optimized configuration. The optimized solution give better results than the completely damped case, but uses only half as much damping material. The orange curve represents the result of a second optimization obtained by modifying certain optimization parameters, and doesn't significantly differ with respect

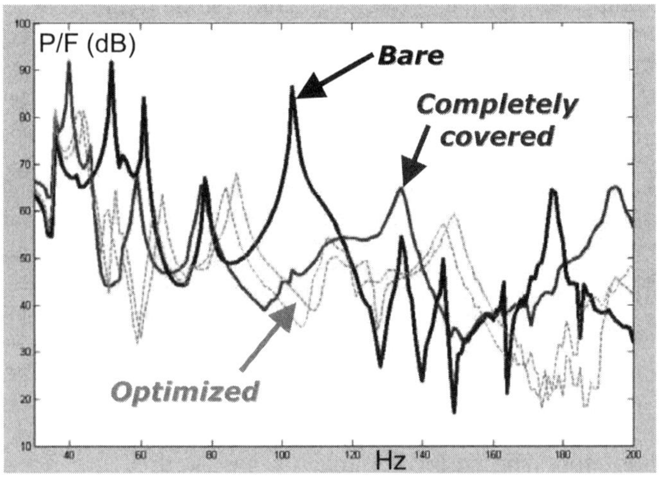

Fig. 13: P/F comparison with bare, completely covered and optimized configuration

to the first optimization.

CONCLUSION

An approximated methodology to simulate the behavior of damping patches was proposed and tested here. When implemented with a multi-objective genetic algorithm, the procedure successfully obtained the optimum layout of damping patches on a trimmed car, where the target was to increase the acoustic performance whilst minimizing the added weight. The procedure was able to find an optimal solution after one week of calculation time and, although approximated, can reveal the most important panel regions for effectively decreasing the acoustic transfer functions.

ACKNOWLEDGMENTS

The authors gratefully thank Dr Meneguzzo for his patience and reliability, and Dr Ottonello for his unrivalled experience in the field of structural dynamics.

REFERENCES

[1] Nashif A. D., Jones D. I. G., Henderson J. P.: "Vibration damping", Wiley Interscience, 1985

[2] Ruzicka J. E.: "Structural damping", ASME, 1959

[3] Burfeindt H., Zimmer H.: "Calculating Sound Pressure In Car Interiors", 16th MSC Eur. Users Conf., 1989

[4] Cavaliere M., De Rosa S., Lecce L., Marulo F.: "Acoustic-Structural Interaction with MSC/NASTRAN: A Review", The 1989 MSC World Users Conf., 1989.

[5] Nova M., Berria C., Tamburro A., Pisino E.: "Noise and Vibration Reduction for Small/Medium Car-Market Segment: An Innovative Approach for Engineering Design and Manufacturing", IMechE, Birmingham, 1997

[6] Ottonello G., Pidello A., Preve A., Zimmer H.: "Comparison of Analytical and Numerical Formulations of Fluid-Structure Interaction with FEM Different Coupling Approaches", 18th MSC Eur. Users Conf., 1991

[7] Pisino E., Preve A.: "Multibody and FEM Approaches as an Integrated Tool in order to Predict Acoustical and Vibrational Comfort", ATA 3rd International Conference on Vehicle Comfort and Ergonomics, Bologna, 1995

[8] Zimmer H., Hovelmann A.: "Practical Applications of Acoustic Calculations with SFE/AKUSMOD and MSC/NASTRAN", 20th MSC European Users Conf., 1993

[9] Danti M., Nierop G., Vigé D.: "FE simulation of structure-borne road noise in a passenger car", NAFEMS World Congress 2001, Lake Como, 2001

[10] Campanile P., Baret C.E.: "NVH perceived quality of passenger cars", Tag der Fahrzeug-akustik, Aachen, 2001

[11] Ferrali L., Caprioli D.: "Gold for lightweight damping treatment", Autotechnology., 05/2003

[12] Fonseca C. M., Fleming P. J.: "Genetic Algorithms for multiobjective optimisation: formulation, discussion and generalization", Autotechnology., 05/2003

[13] Van Veldhuizen D. A., Lamont G. B.: "Multiobjective Evolutionary algorithms: analysing the state of the art", Evolutionary computation Vol 8 number 2

[14] Goldberg D. "Genetic Algorithms in search, optimisation and machine learning", Addison Wesley Publishing Company, 1989

DEFINITIONS, ACRONYMS, ABBREVIATIONS

NVH: Noise, Vibration and Harshness

CRF: Centro Ricerche FIAT

FE: Finite Element

FEM: Finite Element Model

FEA: Finite Element Analysis

SE: Superelement

BIW: Body In White

DMAP: Direct Matrix Abstraction Program (Nastran internal language)

FSI: Fluid Structure Interaction

FLDT: Free Layer Damping Treatment

CLDT: Constrained Layer Damping Treatment

API: Acoustic Performance Index

NP: Reference (Normal Production) Configuration

RKU: Ross, Kervin, Ungar Theory

2005-01-1657

How to Do Hardware-in-the-Loop Simulation Right

Susanne Köhl and Dirk Jegminat
dSPACE GmbH

ABSTRACT

Not only is the number of electronic control units (ECUs) in modern vehicles constantly increasing, the software of the ECUs is also becoming more complex. Both make testing a central task within the development of automotive electronics.

Testing ECUs in real vehicles is time-consuming and costly, and comes very late in the automotive development process. It is therefore increasingly being replaced by laboratory tests using hardware-in-the-loop (HIL) simulation. While new software functions are still being developed or optimized, other functions are already undergoing certain tests, mostly on module level but also on system and integration level. To achieve the highest quality, testing must be done as early as possible within the development process.

This paper describes the various test phases during the development of automotive electronics (from single function testing to network testing of all the ECUs of a vehicle). The requirements for the test system and corresponding concepts are described. The paper also focuses on test methods and technology, and on the options for anchoring HIL simulation into the development process.

HARDWARE-IN-THE-LOOP SIMULATION: AN ESTABLISHED PART OF THE CONTROL DEVELOPMENT PROCESS

Time to market is speeding up, especially in automotive electronics. 90% of automotive innovations are currently connected with new electronics. Test drives can scarcely cope with the volume of systematic testing needed, especially just before start of production. The growing number of recall campaigns is a clear indication of this. It is little wonder that testing and error finding have become key tasks in the development process. [5]

ECU testing typically is done using hardware-in-the-loop simulation. The ECU (prototype) is connected to a real-time simulation system simulating the plant (engine, vehicle dynamics, transmission, etc.) or even the whole vehicle.

One means of reducing development times is to schedule early availability of the test system, which can be achieved by integrating HIL into the development process and involving the HIL system supplier as soon as ECU specification is available. This allows the simulator to be up and running shortly after receipt of A-, B-, and C-sample ECUs.

Automated tests increase test coverage and shorten testing times by running complete test suites and overnight tests. HIL systems testing 24 hours, 7 days per week are not fiction but reality.

Another measure taken by the OEMs is to transfer testing responsibility to the suppliers. Nowadays suppliers are more and more forced to perform early HIL tests. This not only includes function tests during function design but also complete integration and acceptance tests. The need for suppliers and OEMs to exchange tests, test results, models, etc., is important in this context.

DIFFERENT USERS - DIFFERENT NEEDS

As HIL has become a standard method for testing ECUs and control strategies during the whole development cycle (i.e., not only after availability of the final ECUs), different needs of different users have to be addressed by the various test systems. Figure 1 shows the various HIL applications and the resulting test contents of the different phases.

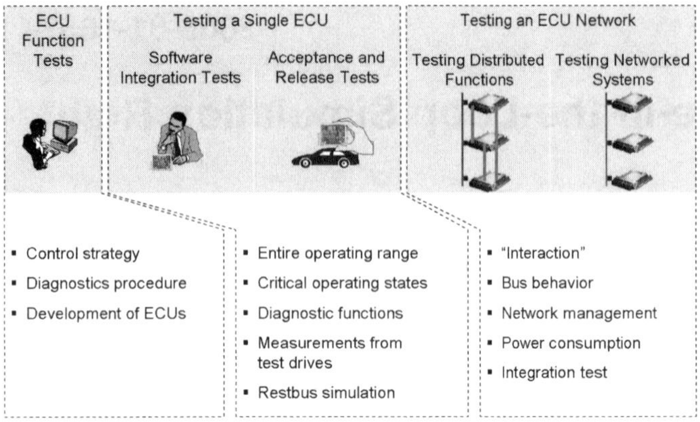

Figure 1.: Usage of HIL during the development process.

FUNCTION DEVELOPER AT ECU SUPPLIER (OR AT OEM)

Typically, only prototype ECUs are available during function development. Microscopic tests on a function are essential at this stage. The control strategy itself needs to be validated. Flexible, interactive operation needs to be possible. The simulator hardware also needs to be flexible for easy adaptation to changes in the ECUs or its peripherals. Low automation is typically required. During this development phase, test scripts are often set up in parallel to ECU/function development, or even after ECU/function development has finished.

Even though the diagnostics procedure must also be tested, it might be necessary to deactivate some diagnostics of the ECU during function testing. The diagnostics depend on signal values that have to be calibrated. Often the calibration of the ECU functions is done prior to calibrating the diagnostics, which necessitates deactivation..

The typical objective of this phase is function acceptance testing. Ideally, this is automated by running the test scripts for all modules.

Reusing control functions for different OEMs requires flexible HIL systems that can be adapted to the different ECU variants. Administration of the HIL software components, such as partial models and test scripts, is also needed.

To avoid redundancy, tests successfully performed during function development should not have to be repeated during integration testing. While at this stage, functions are verified by HIL tests, it is important to test the proper interaction of all functions during integration testing. Close cooperation between supplier and OEM is therefore desirable, to exchange test protocols on the one hand and models (which are typically available at the OEM) on the other.

In the case of function development at the OEMs (which typically still requires function integration into the final ECU provided by a supplier), the test process is quite similar to the process described above. Exchanging models and tests is far easier, however, as the exchanged data remains under the same roof.

ECU PROJECT MANAGER AT THE OEM (OR SUPPLIER) – RELEASE AND ACCEPTANCE TEST

Once all the functions have been integrated together with the lower software levels (operating system, I/O drivers), macroscopic testing of the complete ECU and/or its functions needs to be performed. This includes tests on overlapping administration layers (handling of diagnostic memory).

Either the manufacturer or supplier performs an ECU release test. Automated tests are indispensable at this stage. The HIL should only be used interactively to find the cause in the event of an unexpected error.

Manufacturers definitely need to repeat tests on ECUs that are provided by different suppliers (second source).

Flexible systems that can be adapted to various ECU types are required at this stage. However, the administration of the simulator's software components (partial models, test scripts, etc.) is even more important. Experiment software layouts represent the functionality of the test system.

The objective is to release the complete ECU as error-free including diagnostics.

VEHICLE ELECTRONICS SYSTEM MANAGER AT THE OEM

As already mentioned, tests that were already finished on component level should not be repeated when networked systems are tested, for efficiency reasons. In an examination of release tests for the complete vehicle electronics, the focus lies explicitly on testing distributed functions and testing bus communication. Network management is also one function under test in this context.

Another important issue comes into play at this stage, if it has not already done so: variant handling. Country-specific variants for a worldwide market presence, as well as different equipment variants and frequent revisions in model cycles, make it necessary to handle different configurations.

As a result, combinatorial tests for the various ECU/vehicle variants are required. This again requires a flexible system based on hardware and software that support different variants in plant models, I/O channels, and bus communication.

Automated tests are indispensable here. Tests designed for the system can easily be replayed for all ECU/vehicle variants. The higher the degree of automation, the higher the test coverage. Only a few of the tests established on function level should be reused here.

Another important aspect of automated tests is that tests, which verified performance during the development phase, can also be used to investigate warranty issues after series production has started.

The complete system must be error-free including diagnostics.

WHAT NEEDS TO BE CONSIDERED WHEN CONFIGURING/SELECTING A SIMULATOR?

Instead of being connected to an actual vehicle, the electronic control unit(s) to be tested are connected to a hardware-in-the-loop simulation system. Software and hardware models simulate the behavior of the vehicle and related sensors and actuators. The models were typically developed with a suitable modeling tool, such as MATLAB®/Simulink®. C code is generated automatically and downloaded to real-time processors for execution. I/O boards, together with signal conditioning for level adaptation to the automotive voltages required by the ECU, provide the interface to the ECU pins. Figure 2 shows a typical hardware-in-the-loop system architecture [1]. The most important components of an HIL system are described below.

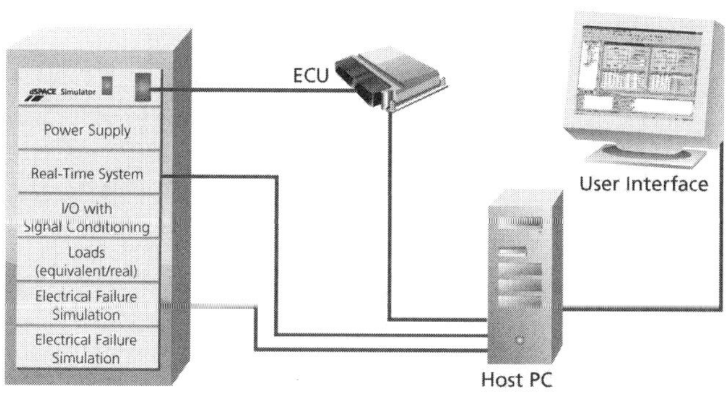

Figure 2.: Typical hardware-in-the-loop system architecture.

PROCESSING POWER

HIL simulation with complex, precisely detailed simulation models requires enormous real-time computing power. Common automotive HIL models typically need sampling times of 1 ms or less to meet real-time requirements. In Formula One applications, engine and vehicle dynamics simulations are typically performed with sampling times of 0.5 or 0.25 ms.

The complexity of HIL models has rapidly increased in the past 5 years due to a widening range of applications:

In the beginning, simple models were sufficient to keep the ECU running in normal operation modes, i.e., without switching into failure modes. Today's ECUs are far more sensitive. One example: While previously it was sufficient to run a complex powertrain model with a simplified model of the exhaust system, today's engine ECUs require detailed data from the exhaust system to control the engine properly. Hence a comprehensive model of the exhaust system is necessary for testing the most modern engine ECUs.

On-board diagnostics are becoming more and more complex, which again results in more complex simulation and tests with HIL simulators.

Increasingly, customers are doing precalibration with their HIL systems. This again requires very precise models.

There is also "the chicken and the egg problem": Due to increasing computing power, customers have reused complex models (available from offline computing) for HIL and (obviously) now want to stay with this degree of complexity.

HIL models typically are configured in just one task (no matter how large and complex they are). While there might be a solution for splitting a simple mean-value model, there is no sense in splitting a complex engine model that calculates each cycle separately. In this case the only option would be to split the engine from the powertrain. This always requires a good working knowledge of simulation dynamics.

There is a clear trend in PC technology towards multi-CPU cores that make use of hyper-threading in symmetric multiprocessing (SMP). This allows performance to be increased by running multiple concurrent threads. The higher the degree of multi-threading, the more performance an application can wring out of the hardware. This is not helpful for typical HIL models as described above. Even if the model allows partitioning into two or more tasks, true parallelization can only be achieved by a comprehensive software environment - taking care of priority handling with regard to accessing the shared memory interfaces and the shared I/O bus.

Multiprocessing and Scalability

Besides the need for increased processing power, there are other aspects that necessitate scalability:

In the past, HIL simulation was often set up within one application, typically testing engine controllers, vehicle dynamic controllers, and body electronics separately. But nowadays, there are more and more control functions being distributed on several ECUs. For example, ESP affects the engine, transmission, and brakes.

Hence, existing HIL simulators need to be combined. For example, an engine ECU needs to be coupled with an HIL simulator for ESP to test the interplay of the two

ECUs. It must still be possible to use both systems as stand-alone-systems. The system to be extended needs to be designed for scalability, however.

Multiprocessing and Spatial Distance

While performance is one issue, the second is spatial distance.

Systems designed for testing networked ECUs also need to be capable of performing component tests. To avoid downtime for the rest of the simulator while component tests are run, true separation of the logical units (if necessary even with spatial distance) needs to be considered. To find errors on just one of the networked ECUs, it still makes sense to run the simulators separately, i.e., parallel to one another.

Moreover, installation of the simulators in different locations is also often desired to minimize the length of cables to real components and ECUs. Multiprocessor systems need to be designed for spatial distance to cope with this requirement.

True Multiprocessing for the Greatest Flexibility in Performance and Scalability

The above-described applications require flexible multiprocessor systems

- Where comprehensive software is responsible for complex jobs such as task and I/O synchronization, and data transfer
- Where spatial distance between the CPUs is possible while keeping high-speed interprocessor communication
- Which can be used either stand-alone or in a master-slave environment.

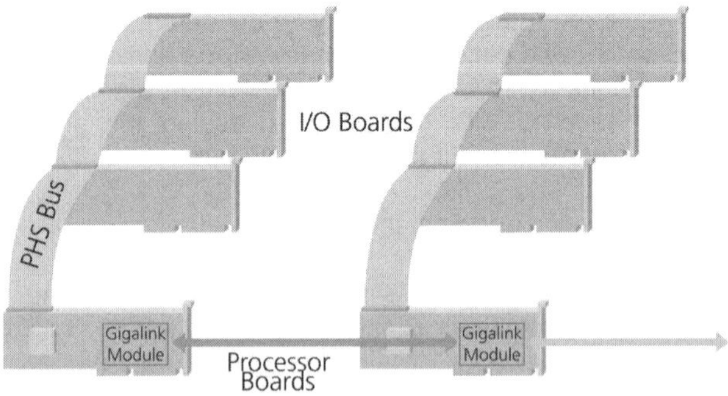

Figure 3.: Multiprocessor Systems

With one and the same MP-concept/technology, dSPACE offers scalability in terms of increasing performance and/or spatial distribution (Fig. 3). Customers do not have to worry about losing earlier investments, as systems remain expandable at later stages. It is even possible to couple a DS1005 running at 480 MHz with a DS1005 running at 1 GHz.

dSPACE multiprocessor systems achieve a net transfer rate of >600 megabit/s (after deducting the protocol overhead) with the help of the fiber-optic 1.25-gigabit/s technology. This way, multiple processor boards can be connected in one system over distances exceeding 100 meters. Using a high-performance processor results in high computing power with efficient multiprocessing capability for maximum utilization.

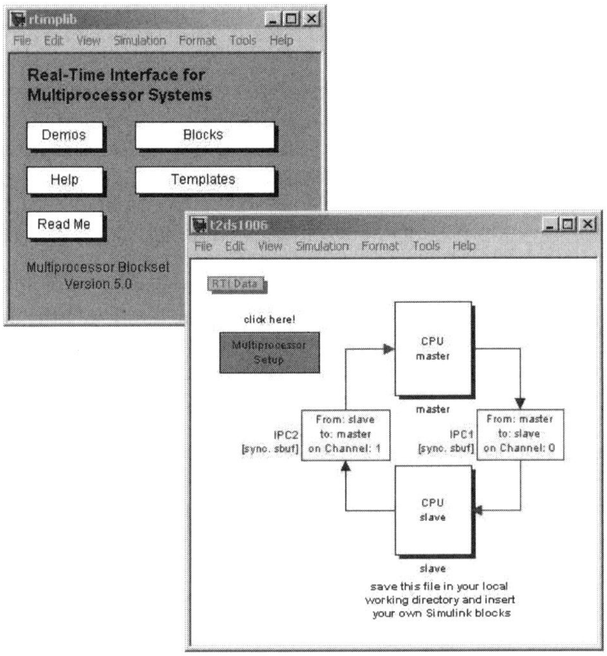

Figure 4.: RTI-MP

With RTI-MP (Fig. 4), the multiprocessor structure is defined within Simulink®, which allows developers to design system dynamics within Simulink®, and set up the structure of the multiprocessing network, including the communication channels between the processors. Automatic code generation takes care of the communication code for the processor network as well as task handling and synchronization.

ELECTRICAL FAILURE SIMULATION

In HIL simulation, the ECU functions are stimulated and the ECU outputs values are monitored to check the behavior. A real-time model closes the loop. Besides testing normal operation, it is especially interesting to check the ECU's performance during exceptional situations such as faulty operation of components such as sensors or actuators, and errors in bus communication. Simulation of these errors can be performed within the model and/or with the help of additional hardware that inserts errors. Nearly every HIL simulator is equipped with relay boards for electrical failure simulation. Failure insertion hardware is typically required to test diagnostic functionality and the reaction of an ECU or even the entire network to electrical faults.

To be able to introduce electrical failures, ECU output pins are wired to the load and to the HIL input channel via relays on a failure insertion unit (FIU). It is then possible to modify the electrical potential on actuator pins. In normal operation, the drivers of the ECU's power stage themselves check the potential on the actuator pin. An error will be detected if the ECU activates a low-side switch but the FIU connects the pin to battery voltage.

Failure insertion units can simulate the following failure conditions: open circuit, short to ground, short to battery voltage, and short between different ECU pins. If the ECU diagnostics have to check the current through the load, it might be necessary for the load (equivalent or real) to remain connected throughout failure simulation. More detailed tests can be performed by inserting a resistor to the reference in series or between ECU pins. By changing the actual value of the resistor, it is possible to check the threshold of the diagnostics integrated in the driver of the power stage.

Electrical failure simulation can also be necessary on sensor and bus protocol pins, for example, to verify/analyze the ECU's functionality in the case of a broken CAN high wire. For sensors which are connected to the ECU via a differential input, disconnecting one of the input lines can stimulate floating ground effects.

To simulate realistic switch behavior or a loose contact, high-frequency pulse patterns on ECU inputs can be simulated with electrical failure simulation hardware based on CMOS switches.

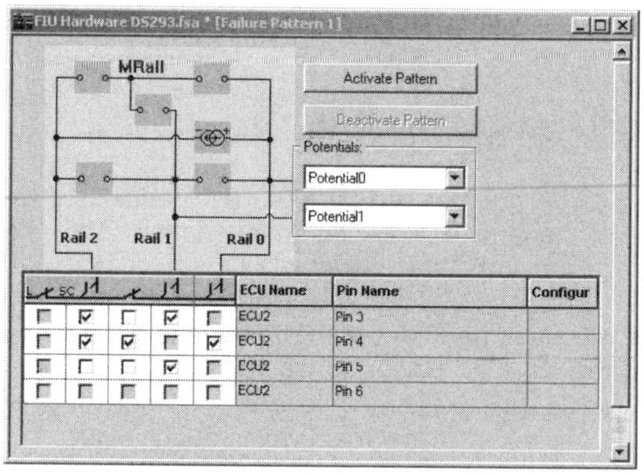

Figure 5.: Failure simulation control of dSPACE HIL Simulators

As a rule, the ECU's behavior when failures occur can be tested independently of the state in the simulation model. Fig. 5 shows a screenshot of a failure pattern to be simulated in the wiring of an ECU. However, there are cases where it is necessary to analyze the ECU's reaction to failures in conjunction with other real-time signals coming from the simulation model, such as faults on the wheel speed sensor during ESP intervention. In such a case, an electrical failure has to be inserted in relation to a real-time variable. The control of the electric fault simulation has to be done on the real-time system. It might be necessary to capture the failure entry in the ECU (possibly even on a time base) to check that the failure entry was made on time and that the control reacted appropriately.

SIMULATION AND TESTING BUS COMMUNICATION

The ECUs in modern cars communicate via different bus systems, such as LIN, CAN, and FlexRay. Normally, the information provided on the bus is necessary for each ECU to operate properly. Hence, there is often the need to simulate bus nodes and/or check the behavior in the event of erroneous bus communication by means of HIL simulation.

Monitoring the communication between the ECUs (CAN, LIN, etc.) is an essential precondition for performing network tests. Behavior in normal operation mode is important, and behavior in the event of failures even more so – for example, missing bus nodes, erroneous message content, and electrical faults (short circuits) on the bus line. Here are some possible issues:

- How does the ECU or the distributed function behave when an expected CAN message is absent?
- How does the ECU react when certain CAN messages contain implausible signals?

These are therefore the requirements for the test system:

- It must be possible to suppress either one or more targeted CAN messages of any ECU.
- It must be possible to manipulate either one or more targeted messages of any ECU in the network.

Restbus simulation

When ECUs are tested separately, or if not all the ECUs are available for network testing, rest-bus simulation comes into play. Here the simulator emulates the missing bus nodes. To make this possible, the communication (including signal scaling) must be specified, for example, in MATLAB®/Simulink®, on the basis of a CAN or LIN database.

Sometimes it is sufficient to generate messages with synthetic signals (independently of the simulation environment). This suffices when the ECU has no plausibility check between the signals in the message and further input or model signals. Nevertheless, it is necessary to satisfy typical check mechanisms such as the message counter, checksum, and toggle or parity bits, and to fill the signals with proper values.

Flexible manipulation options for switching off a whole message or replacing a single signal for a defined number of messages are required. Synthetic rest-bus simulation allows verification of the on-board diagnostics with regard to alive and structure checks. For ECUs/functions with a plausibility check, the relevant signals have to get their current values from the real-time model.

Gateway

Signal manipulation via a failure gateway has proven its usefulness in the investigation of network failures. The bus lines of a device on the bus are switched to a failure bus on the simulator, and the messages are manipulated (if required) and then transferred back to the original CAN bus. Changes to individual bus signals (such as checksums), entire messages (missing, wrong timing), and even the complete absence of a bus node can be performed and their effects on the rest of the network can be investigated.

Fig. 6 shows two CAN controllers for the CAN bus available in the simulator. Each ECU can be connected separately to either of the controllers. Flexible bus termination needs to be carried out for each sub-bus to allow switching during run time. Via software, the simulator functions as a (fault) gateway between the two controllers. All the messages received on one controller are immediately sent to the other controller. This ensures that each ECU receives the CAN messages of the other ECUs. The delays that occur are so slight that the ECUs are not affected by them. Software manipulation blocks can now be used to generate additional messages or signals [2].

It is often necessary to replace some signals, e.g., from an integrated sensor of one ECU, with a value coming from the real-time model to properly simulate the environment before the second real bus node receives the message.

The interaction between the two sub-busses also includes messages with variable structures and confidential contents. No manipulation is necessary for these messages. They must only be sent immediately to the other controller.

The combination of rest-bus simulation (described above) and gateway functionality qualifies an HIL system for nearly all use cases in conjunction with bus communication. The diagnostics in the different ECUs can be checked individually. In addition, missing functionality in one ECU can be added via the real-time system. Hence early testing of single ECUs is possible, even if other bus participants are missing.

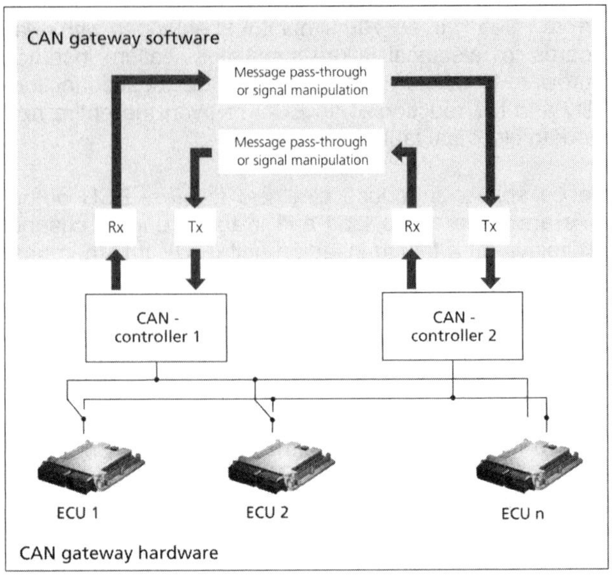

Figure 6.: CAN Gateway [2].

DIAGNOSTIC INTERFACE

Testing diagnostic functions of course requires the ability to access and read the diagnostic memory of the ECUs.

There are various ways of accessing the diagnostic memory of ECUs from within an HIL environment. One is to make use of calibration or diagnostic tools with adequate interfaces that can be remote-controlled by the HIL and hence integrated into the automated test procedure.

The host PC remote-controls the calibration tool. The ASAM-MCD 3MC interface [1, 4] is widely used for coupling the calibration or diagnostic tool to the HIL simulator.

TESTING NETWORK MANAGEMENT

Having a large number of ECUs leads to additional requirements regarding network management and the power consumption of the individual ECUs.

When a vehicle is parked, the ECUs have to enter sleep mode to reduce their power consumption to a minimum (typically <300 μA). Separate wake-up channels or special CAN transceivers allow reactivation at any time.

Network management can only be properly tested if all the ECUs are networked. The HIL simulators that are used need to behave neutrally on the CAN network. This can be achieved by using CAN transceivers identical to the transceivers in the ECUs to be tested. Besides testing bus activities, the power consumption of every single ECU needs to be measured. This serves as an indicator of proper switching between operating modes [3].

FLEXIBLE AND MODULAR HARDWARE AND SOFTWARE CONCEPT

As already discussed, to successfully run dynamic models of engine, transmission, vehicle dynamics, or chassis, powerful processor boards are required. Typically, execution times of less than 1 ms are necessary for real-time simulation.

dSPACE's DS1006 Processor Board has an AMD Opteron® processor with a clock rate of 2.2 GHz, allowing it to compute a mean-value engine model, a brake hydraulics model, and a vehicle dynamics model, including the entire I/O for the engine and ESP ECUs, in less than 350 μs. For even more complex models, or to connect several simulators, the boards can be networked to form distributed multiprocessor systems.

I/O Hardware for Highly Dynamic Signal Processing

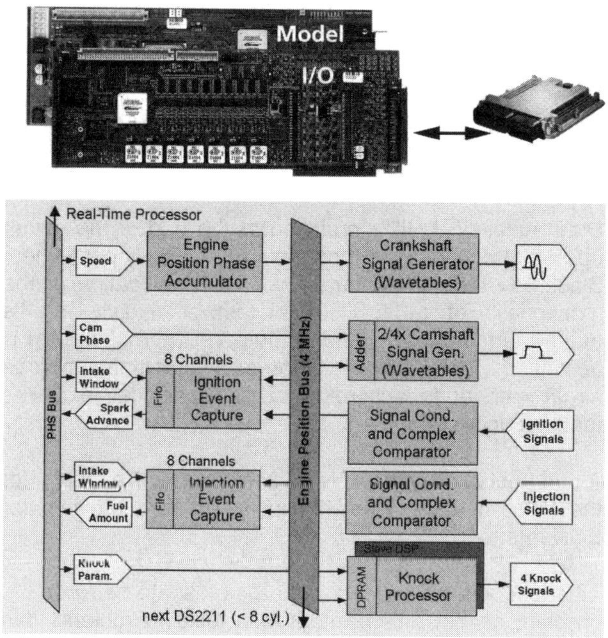

Figure 7.: DS1006 Processor Board for simulating dynamic models and the DS2211 HIL I/O Board form the basis for various HIL testing systems. [3]

Engine simulation involves generating crankshaft, camshaft and knock signals synchronously to the engine angle, while injection and ignition signals are measured synchronously to the crankshaft angle. Special hardware is generally used for this task, for example, the DS2211 HIL I/O Board (Fig. 7), which is in widespread use in the automotive industry. The board is cascadable and provides the entire I/O for an 8-cylinder engine including signal conditioning, for example. Moreover, there are two operating voltages, allowing up to 42 volts (nominal) to be used (utility vehicles and 2-voltage electrical systems). Simple I/O interfaces and complex angle-synchronous functions can be specified in MATLAB®/Simulink® together with the dynamic model, and configured in test operation. This combination of processor board and HIL I/O board now forms the basis for hundreds of HIL test systems throughout the world. [3]

HARDWARE-IN-THE-LOOP-SIMULATORS

Hardware-in-the-loop simulators are built from hardware and software components:

Hardware Components

- Processor boards
- I/O fulfilling specific HIL requirements (algorithm and waveform-based signal generation, angle-based measurement of injection and ignition pulses, etc.)
- Simulation of automotive busses such as CAN, LIN, MOST, and FlexRay, including rest-bus simulation
- Signal conditioning for level adaptations to automotive voltages (12 V, 24 V, 36 V, 42 V)
- Electrical failure simulation
- Load simulation (dummy loads, electrically equivalent loads, real loads, I-to-U conversion for current controlled valves, etc.)

Software Components

- Implementation software (for implementation and real-time execution of the simulation model and the corresponding I/O connections)
- Software to establish and monitor bus communication
- Real-time models
- Experiment management software
- Test software to (graphically) program and administrate automated tests

Optionally:

- 3-dimensional real-time animation
- Integration (and synchronization) of additional tools such as tools for diagnostics or calibration

dSPACE Simulator Concepts – Different Systems for Different Tasks

The dSPACE software components are standardized and can be integrated in any dSPACE simulator. The tight integration of dSPACE software and the modeling tool MATLAB®/Simulink® from The MathWorks provides a powerful development environment.

dSPACE Simulator's graphical user interface provides a convenient and flexible environment. Simulated driving cycles, data acquisition, instrumentation, monitoring, test automation and all other tasks are executed graphically within dSPACE Simulator. [5]

The hardware requirements, however, vary immensely depending on the HIL application. For example, function tests typically are executed with simulators that have a fixed (super)set of I/O, and adaptations to the ECU are most often made in the cable harness. In contrast, acceptance tests call for flexible and combinable simulator setups.

dSPACE Simulator Mid-Size

dSPACE Simulator Mid-Size is a standardized off-the-shelf HIL simulator. Its hardware is based on a DS100x processor board and the DS2211 HIL I/O Board. Other I/O boards can be added if required.

In standard configuration, dSPACE Simulator Mid-Size contains a failure insertion unit that allows electrical failures to be simulated on all ECU output pins connected to the DS2211. A hardware extension allows electrical failures to be simulated on ECU inputs as well. With this "sensor FIU" hardware, it is even possible to simulate loose contacts. Real or equivalent loads can also be connected to the ECU outputs.

The transparent system allows new users a quick start.

dSPACE Simulator Full-Size

dSPACE Simulator Full-Size is a modular simulator concept that is assembled from off-the-shelf components according to the specific needs of a project. It features enormous extension capabilities.

dSPACE standard processor and I/O hardware is adapted to project-specific needs. Signal conditioning for all dSPACE I/O boards is available, based on a modular signal-conditioning concept. Failure insertion units can be installed for ECU inputs and outputs. Combined with a modular load concept, this allows a customized simulator to be set up. Free grouping of I/O (e.g., to connect different ECU types), easy integration of bus gateways, good integration of drawers to store ECUs or real loads, are important advantages of this concept. Expandability during the project is also provided.

dSPACE Simulator Full-Size offers enhanced failure insertion units that allow the load (equivalent or real) to remain connected throughout failure simulation, for example.

Simulator networks

Independently of the chosen simulator concept, several units of dSPACE Simulator can be connected to set up a networked simulator environment. The flexible multiprocessing feature of the processor hardware especially supports this. Both monolithic and modular setups are possible with dSPACE Simulator Mid- and Full-Size.

dSPACE Simulator Full-Size allows network management to be tested by power switch modules. The power switch modules have a high-precision meter for measuring an ECU's entire supply current. Five different measurement ranges allow precise current measurement during different operating modes. [3]

TEST METHOD AND TECHNIQUE

An appropriate test strategy is the key to getting maximum benefit from an HIL simulator. While the first tests during function development are typically performed initially on a manual basis, the function developer soon goes over to automated tests. Typically, very detailed tests are created at this stage. A thorough knowledge of the implemented structure of the software is required. These so-called white-box tests are based on a thorough knowledge of the internals of an ECU. They use not only input and output variables, but also internal variables (model variables and internal ECU variables). At this stage, measuring internal ECU variables is indispensable, as described in DIAGNOSTIC INTERFACE.

White-box tests typically are applied during function development. They have proven successful in error finding, for example, when problems occur during release and acceptance tests.

During classical HIL simulation at the end of the development process, black-box tests are typically performed. Black-box tests concentrate on the specification of the functionality of the ECU under test, so usually only its outer interface (inputs and outputs, no internal values) is accessed. The test description and implementation can already be done according to the specification of the function under test.

A test pool of all types of tests allows recursive testing for the different ECU versions, including the final test for ECU release.

This test pool also allows white-box tests to be rerun if a problem occurs during integration tests. Problems can be narrowed down to their source with the help of existing tests.

For each of the above-mentioned areas, the HIL test specifications are developed at the same time as the performance specifications.

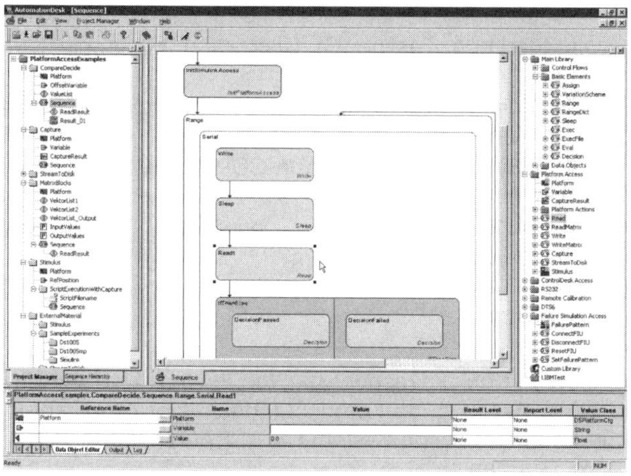

Figure 8.: Automatic testing by means of dedicated software support

A test management system that handles all the different tests and types of tests is necessary (Fig. 8). It must provide structured applicability criteria handling. This allows the different users of the different development steps to select tests that make sense in the current scenario. This is especially important for large test projects, where it is not possible for the test operator to check the applicability of each test manually [1].

EXPERIENCE AND COMPETENCE

To achieve the highest efficiency, it is now essential to define the HIL test process as an indispensable part of a vehicle's development process. Close cooperation between OEM, ECU supplier, and HIL system supplier also increases the benefit.

INTEGRATION INTO THE DEVELOPMENT PROCESS:

Integration into the Project Plan

The increasing number of ECUs in modern cars combined with short development cycles results in a tight project plan at the OEM and its supplier. The critical point for HIL simulation, especially for network tests, is the tight time frame between the availability of all the necessary components and the start-up of the HIL simulator.

A modular and flexible hardware concept can meet this challenge when the ECU-independent and the ECU-specific parts are separated. The ECU-independent components, e.g., the real-time system with the processor and I/O-boards and the signal conditioning, can be set up very early on the basis of very rough information on the ECU(s).

The ECU-specific components, e.g., the load boards, the ECU mount, and the wiring harness, are designed and constructed gradually after specification and development of the ECU is completed. These components can be added at short notice.

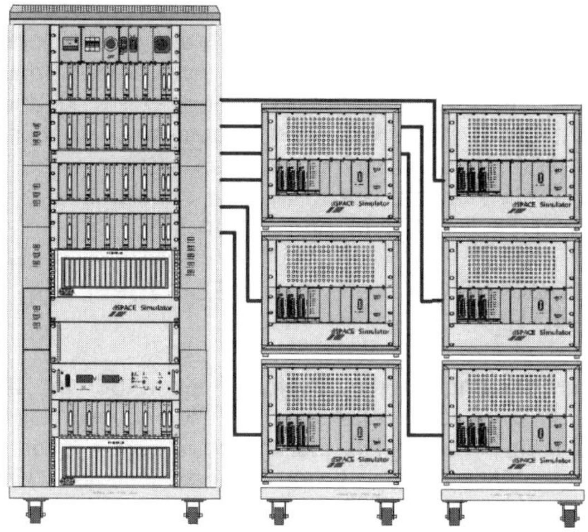

Figure 9.:Plattform HIL

A similar strategy is setting up simulators to test various ECUs with basically the same functionality, either by following the second source principle or by testing engine control units for diesel engines with the same simulator setup as ECUs for gasoline engines. Fig. 9 shows a simulator concept well prepared for handling different ECU variants by replacing ECU-specific units. The reusability of the platform hosting the ECU-independent parts is very high.

Requirements Management

A second area of process integration is the OEM's requirements management. Tools like DOORS and Test-Director are commonly used to collect and administrate ECU requirements and specifications. These tools also handle test specifications and test results in a smooth process. Not only the different documents have to be managed, but also the relation between the functional requirements and the test result.

PROJECT CONTEXT

The development process for automotive control units has become complex, and so has the test process, with numerous interactions between different departments and companies. The result is a huge amount of documents and data, for example, describing different interfaces. The larger a project, the more important is its structure.

A common project context for all development and test aspects used in the test environments for A-, B-, and C-sample ECUs will contribute to a more efficient process. The project can be structured according to the different functions, typically represented in trees form. The OEM and the supplier can use the sources for function development, module testing, and integration testing.

In a project, all the required data, such as specifications of functions and protocols, variant configuration, and ECU software (hex files, a2l files), are collected and handled together with experiment layouts for calibration, the real-time model, the parameter files, experiment layouts for the interactive use of the HIL simulator, a test library, and resulting test reports.

CLOSE COOPERATION BETWEEN ECU SUPPLIER, OEM, AND HIL SYSTEM SUPPLIER:

Automotive manufacturers typically allow 36 to 42 months for developing a new vehicle. Close consultation with the supplier of the HIL test system is highly recommended to define function scope at an early stage. This allows parallel work on ECU design and HIL simulator setup. Ideally, the HIL simulator should become available at the same time as the first prototype ECU. On principle, setting up, operating, and modifying the HIL simulator should be integrated into the development process to ensure maximum output. Set-up and maintenance of the modeling part is basically independent of the availability of the ECU(s) and should therefore be handled independently as well.

The creation of automatic tests can already be started during the planning stage for the test system(s) and should be systematically worked into the project schedule. Optimum efficiency can be achieved if the ECU supplier has already tested the individual ECU by means of an HIL system that is also present as a “sub test system” in the OEM’s network simulator.

CONCLUSION

The increasing complexity of vehicle electronics implies a high demand for innovation, time saving, and quality during the development and testing of electronic control units.

Testing the overall ECUs as well as single functionality is increasingly becoming a key task during all development phases. Hardware-in-the-loop has meanwhile become well established, both after availability of the units under test and during function development.

Good interplay between flexible hardware and software is indispensable to supporting this demanding task.

REFERENCES

1. Lamberg, K.; Richert, J.; Rasche, R.: A New Environment for Integrated Development and Management of ECU Tests, SAE2003 , Detroit, USA
2. Lemp, D.: Köhl, S.; Plöger, M., ECU Network Testing by Hardware-in-the-Loop Simulation, ATZ/MTZ extra "Automotive Electronics" 10/2003
3. Wältermann, P.; Schütte, H.; Diekstall, K.: Hardware-in-the-Loop Testing Of Distributed Electronic Systems, ATZ 5/2004
4. Association for Standardisation of Automation- and Measuring Systems, ASAM: http://www.asam.net/docs/MCD-18-3MC-SP-R-020101-E.pdf
5. dSPACE Simulator product information, http://www.dspaceinc.com

CONTACT

Susanne Köhl is responsible for the product strategy, product planning, and product launches of Hardware-in-the-Loop Simulation Systems at dSPACE GmbH, Paderborn, Germany.

E-mail: skoehl@dspace.de
Web: http://www.dspace.de

2005-01-1342

Virtual Prototypes as Part of the Design Flow of Highly Complex ECUs

Joachim Krech
ARM

Albrecht Mayer and Gerlinde Raab
Infineon Technologies AG

ABSTRACT

Automotive powertrain and safety systems under design today are highly complex, incorporating more than one CPU core, running with more than 100 MHz and consisting of several 10 million transistors. Software complexity increases similarly making new methodologies and tools mandatory to manage the overall system. The use of accurate virtual prototypes improves the quality of systems with respect to system architecture design and software development. This approach is demonstrated with the example of the PCP/GPTA subsystem for Infineon's AUDO-NG powertrain controllers.

INTRODUCTION

According to a study [1] 77% of all electronic failures of cars are caused by software. Detecting these bugs early prevents reputation loss ("car parks, driver walks"), reduces cost and improves time-to-market.

For complex systems like powertrain control, it is of particular importance to understand and analyze the behavior in all possible scenarios thus increasing the software quality. Cycle accurate virtual prototypes of the main components are a perfect environment for stimulation of any use case and for a detailed analysis of the resulting system behavior.

When changing to the 90nm process the costs for the chip architecture development started to exceed the costs for layout for the first time. A virtual prototype allows to do architecture exploration during the concept phase to efficiently select the most suitable IP blocks and reliably dimension system resources based on quantitative measurements.

Software development based on virtual prototypes enables an early development start, improved visibility of internal resources and ease of debugging as well as the ability of generating arbitrary system stimulation. The latter allows to analyze corner cases and to increase the software test coverage.

There are a number of known modeling approaches mostly differing in modeling abstraction levels, the language as well as the tooling. Common to all approaches is the requirement for models that are supposed to be fast, accurate and require only little effort to be developed, verified and maintained. Obviously these are contradicting requirements, however methodology and tools may help to optimize all three requirements at same time. The actual accuracy requirement heavily depends on the use case of the virtual prototype and therefore needs to be carefully selected for the sake of the other requirements.

C/C++- BASED VIRTUAL PROTOTYPES (VP)

C/C++ models of hardware IP have been mostly used for functional and cycle based simulation in the past, mostly known as Instruction Set Simulator (ISS). The introduction of SystemC being a C++ class library has extended the scope to include fully time accurate models offering an event driven simulation paradigm as well as the concept of constructing a system model from component models.

At the functional end the use of Matlab/Simulink and UML is continuing into even more abstract models which rather model applications and algorithms. At the end of full timing accuracy the hardware description languages like VHDL and Verilog are used. Whereas hardware description languages as well as SystemC have been specifically designed to describe and model hardware, C/C++ is a general purpose language offering no specific support for modeling.

The main disadvantage of HDL is the limited simulation performance. This is due to the fundamental computation model that is based on events and processes which enforces the usage of a complex simulation engine. This also applies to SystemC in case the event driven and the process language constructs are being used. C/C++ is a procedural language that enforces the serialization of parallel processes taking place in the hardware when described in the model thus almost eliminating the scheduling overhead at runtime. The cycle based simulation approach offers the most efficient way to synchronize multiple components. Furthermore C/C++ is well known and accepted by the hardware and software designer providing a common language.

VIRTUAL PROTOTYPE APPLICATIONS

Virtual Prototypes are used for:

- Architecture exploration
- Executable specification
- Software development and test
- System analysis and test

ARCHITECTURE EXPLORATION

At the beginning of the design of a complex system the partitioning between software and hardware is not yet determined. Also the specific selection of IP blocks and the system interaction as well as their system performance are mostly unknown and are usually based upon some very high level estimates. In order to explore the system design space based on some quantitative and reliable measurements, a virtual prototype of the IP blocks need be developed. The simulation of the most critical software kernel routines then allows generating meaningful system profiles. Based on these profiles the system can be optimized to reach compliance with the specification without wasting of hardware resources, area as well as power.

In order to be able to design a significant amount of architectural alternatives it is important to start from rather abstract models with only little granularity. The IP block's behavior is modeled only in a purely functional manner where latencies are only approximated by a fixed average value. Instead of modeling bus protocols to the last cycle, the interaction of IP blocks is simplified to read and write accesses accounting fixed latencies only.

The effort of creating such abstract models has to be extremely low, using a standard programming language, strong modeling and interface guidance and automated component framework generation avoiding repeating modeling tasks.

Only after a certain number of design choices have been made, the accuracy of the prototypes needs to be refined in order to provide more accurate system performance measures and providing a higher degree of corner case coverage and overall completeness of the modeled IP's functionality.

Once the refinement of the components of the virtual prototype has been completed and the architecture has been finalized the virtual prototype provides the means of functional correctness as well as of system performance.

EXECUTABLE SPECIFICATION

In the development process of complex systems (e.g. chip design) there is usually a customer and application oriented concept engineering group which specifies all parts of the system. These specifications are implemented by the design team, which creates the microarchitecture and focuses on cost and performance optimizations. The problem is, that native language, used for these specifications, always is ambiguous, leaving room for interpretation. The misinterpretation is sometimes only detected, once the new chip is plugged into the board one or two years later.

In case the concept engineering team provides models as executable specifications of components or even for whole systems to the customer as well as the design team, a significant improvement of the link between requirement and implementation is reached:

1. The customer can check, whether the requirements have been fully understood and considered by the concept engineering. An additional motivation is the fact that the development of (low level) software can be started immediately.
2. The process of developing an executable specification by definition includes a check for consistency and completeness whereas the use of a native language requires additional measures. Even if the model is solely used within the concept team, it will improve the native language specification and avoid iterations with the design team.
3. From the design team's point of view, the model is an unambiguous starting point for the IP implementation. In some cases it can be even used as a golden reference for the verification. An alternative approach is to rerun the functional test cases of the model for the implementation.

SOFTWARE DEVELOPMENT AND TEST

Time to Market is very much impacted by the speed in which software is developed and integrated into the hardware. Without the use of virtual prototypes, the software development and test depends on the availability of evaluation boards or other kinds of physical prototypes which usually become available late in the design cycle. A virtual prototype that has been used already as an executable specification is available much earlier. Even if certain modifications become necessary during hardware development, once the design is frozen, a stable and reliable prototype is available. Due to the fact that software models are able to expose internal resources without any limitations the visibility into the modules can be superior in comparison to hardware. The flexibility of software models also allows adding further debug capabilities, provided that the tool environment offers respective support.

Enabling software development on the basis of a virtual prototype in parallel to hardware design can help in achieving higher quality software in the same time frame.

Virtual prototypes furthermore add the ability to feed specific system level test patterns to a subsystem thus allowing system stimulation and creation of arbitrary scenarios easily.

Due to the compromise between model performance, completeness (also of the environment model) and accuracy, software testing on a model will always be in addition to conventional ways of software engineering based on hardware prototypes.

SYSTEM ANALYSIS AND TEST

The analysis of system level interdependencies is the key for the ability to break up a complex system into smaller, less complex pieces (divide and conquer). Such smaller pieces of the system can then be tested individually using a realistic stimuli and response behavior of the missing part of the system. Ideally complex systems are designed in a hierarchical manner using well defined interfaces between subsystems.

The advantages of dealing with subsystems are:

- Focus on less components and hardware effects
- Higher performance allows to use cycle accurate models
- Stimuli and system response can be used to create arbitrary stimulation and response

PCP/GPTA SUBSYSTEM AS VP EXAMPLE

In order to demonstrate the development and testing of ECU software the PCP subsystem of Infineon's AUDO-NG family is used as an example. The PCP subsystem consists of multiple closely interacting components. The PCP2 is a programmable, interrupt driven processor that handles service requests from all system peripherals independently of the main Infineon TriCore CPU. The General Purpose Timer (GPTA) is a powerful and complex timer peripheral executing the most time-critical tasks. Its sophisticated network of timer cells is programmed via memory mapped registers which are accessed over the System Peripheral Bus (SPB). The SPB is a multi master bus using priority based arbitration producing variable access latencies. Whenever the GPTA has completed one of its parallel running tasks requiring reprogramming, it initiates the execution of service routines by the PCP2. For this purpose it sends service requests to the PCP's arbitrating interrupt system. The interrupt latencies heavily depend on the priorities and duration of higher priority pending services.

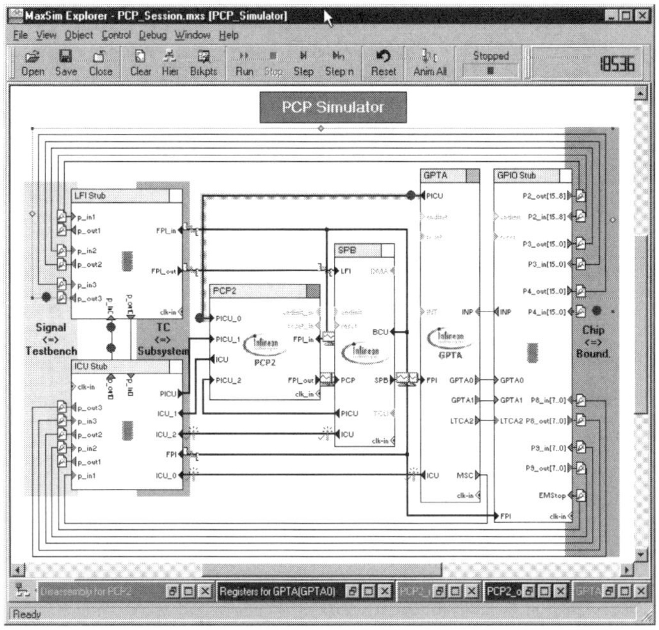

Figure 1 PCP/GPTA Subsystem

The external input and output events observed or generated by the GPTA are multiplexed by the General Purpose Input Output (GPIO) component which connects on-chip and off-chip signals. The interaction between the PCP and TriCore subsystems takes place via the interfaces of the TriCore Interrupt Control Unit (ICU) and the Local Memory Bus to Flexible Peripheral Interconnect Bus Bridge (LFI).

From the software point of view the programming of the PCP2 represents a challenge because of the functional complexity and numerous firm real-time conditions. Due to the multifaceted interdependencies of all involved modules (on-chip and off-chip) testing and debugging on the real hardware can be time intensive, inefficient and error prone, especially when it comes to timing issues. In the sequel details of the software development methodology based on the cycle-accurate virtual prototype of the PCP ECU subsystems are given.

In order to manage the functional complexity it is desirable to first reduce the interdependencies to a minimum, e.g. by simulating the GPTA on its own running as a single task using scripts for the register programming, stimuli generation and checking against expected results. This assumes the availability of a scripting language allowing architectural resource access, stimuli generation as well as testing capabilities. The virtual prototyping tools should also support comfortable monitoring and debugging of the script *(Figure 2)*.

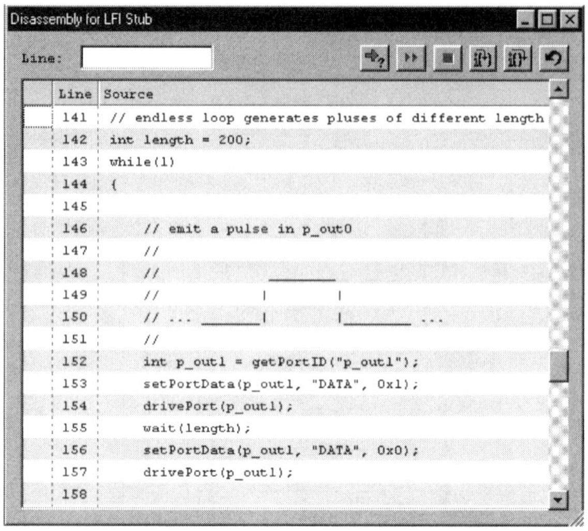

Figure 2 Script Debug View

In order to understand and analyze the behavior of a highly complex component like the GPTA, full visibility of architectural details is the key. The programmer needs to inspect component register subfields, symbolic values and visualize the internal state like clock busses, event output state of timer cells and other hidden states that further complement the understanding of the internal functionality. In the case of large and complex register structures, like in the GPTA, it is helpful to have access to online documentation with hints about the functionality of the bits and the address mapping *(Figure 4)*.

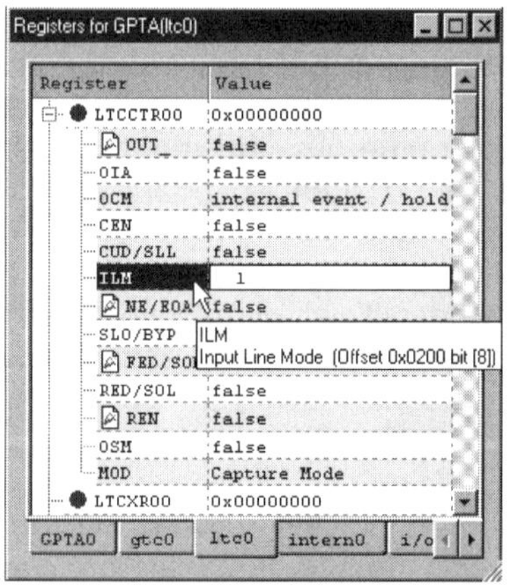

Figure 3 Register View with Online Documentation

In addition, the logic analyzer tool helps to analyze signals and values over time. Advanced execution control using conditional breakpoints on register, fffmemory or signals enhances the ability to efficiently track and locate specific events. The above functionality becomes even more important when dealing with functional interdependencies between parallel software tasks. The programmer needs to deal with the proper use of shared resources concurrently accessed by the tasks. E.g. it must be avoided that shared resources like the clock divider of the clock bus gets accidentally reconfigured or that different tasks route their output to the same pin. Single cycle execution is required to enable detailed inspection of the system resources on any clock edge.

As the next step, we propose to take the relevant component inter-dependencies into account and therefore gradually increase the scope of the system. Timing related interdependencies mostly deal with the problem that shared resources like the SPB as well as the PCP2 introduce variable latencies depending on the overall system conditions. Furthermore, it has to be ensured that various scenarios of off-chip closed loops are properly handled by the software. The analysis of all such interdependencies requires system level visibility and controllability as well as flexibility to efficiently develop meaningful scenarios. It is of significant benefit to the programmer if the tool offers protocol aware visualization of the component communication and provides a history of transactions using bus monitors *(Figure 4)*.

Figure 4 Bus Monitor

Enhanced execution control of the virtual prototype allows the programmer to set breakpoints on communication events between components, thus being able to track cycle by cycle the GPTA interrupt request resulting in a PCP channel program response reconfiguring the GPTA.

If suspicious behavior is detected, it is helpful to analyze the trace information of the current and past input and output signals e.g. produced by the GPTA or GPIO, side-by-side with traces of registers, bus accesses and interrupt requests *(Figure 5)*. In this way the interdependencies can be easily tracked down within a single view window allowing precise measurement of the latencies between events.

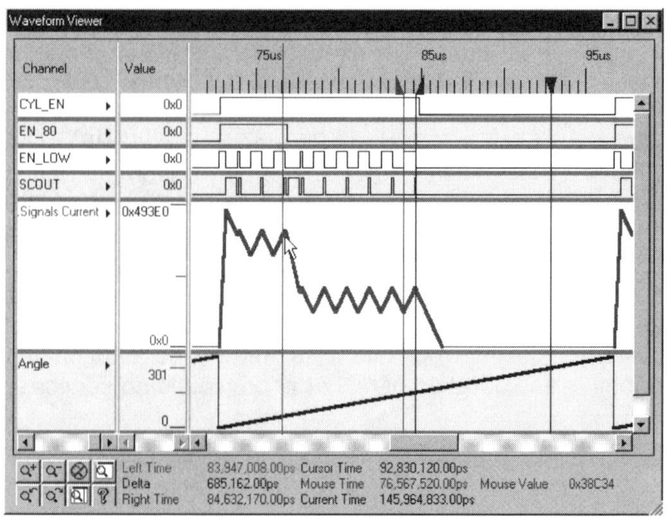

Figure 5 Trace Analyzer

Once a satisfactory basic functionality of the system is reached, the next step is to evaluate the performance of the system as a whole, so further dedicated optimizations can be conducted. In the case of the PCP, the latency information of the SPB bus is an indicator of bus performance and potential bottlenecks which could seriously limit the performance of the system at critical times. In order to generate quantitative bus traffic data and identify worst case conditions, the information about bus transactions needs to be collected and visualized *(Figure 6)*.

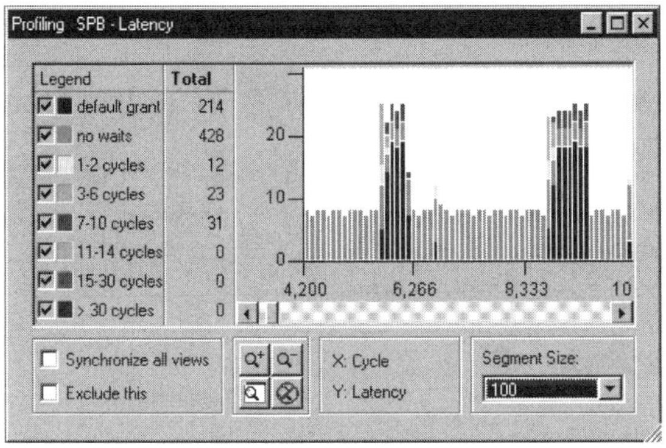

Figure 6 Bus Latency Profiling

The solutions for bus load problems can range from simple changes in bus priority schedule up to the need to restructure the system or make use of a more powerful bus system.

CREATING A NEW MODEL

Before launching any modeling activities the most basic decision has to be made on the simulation environment. This initial decision will determine the long term success of the modeling strategy. Setting up any detail of modeling methodology as well as the simulator environment from scratch requires the highest possible efforts. Mostly this high effort can be brought down to lack in maintenance, documentation and support for semi-professional in-house solutions. If any recommendation towards simulators can be given, then it is to rely on buy-in of methodology (including simulator support) and IP blocks to the most possible extend. Public, license free solutions may seem promising considering the aspect of initial investment, however may suffer from not offering a wide spectrum of features like GUI, visualization widgets and debug capabilities.

The higher the complexity of the component to be modelled, the more important it becomes to interpret the specification for the hardware "first time right". Therefore an ideal team setup has to include software as well as hardware design experts.

If the accuracy of models regarding the aspects functionality and/or timing is of importance no way leads around verification of models. And as in hardware verification, it is the chosen methodology that will determine the quality of verification results. A verification strategy that runs completely with disregard to the related hardware module cannot deliver sufficient quality. The past has proven, that only equivalence checks at model boundary against the reference hardware module is not enough. In an ideal world, model and hardware verification should share a single verification environment to allow any judgment of accuracy.

GENERAL MODELING CONSIDERATIONS

Modeling methodologies and languages (e.g. C++) impose nearly no limit on what to describe and how to describe it. This section contains some basic considerations how to find the sweet spot between accuracy, completeness, speed and effort.

SPEED VERSUS ACCURACY

For certain types of models, there is a clear trade-off between accuracy and simulation speed. For instance a very straightforward CPU model can be based on a big case statement in ANSI-C, which computes the operands based on the instruction opcode. This model will be very fast and execute the behavior of the CPU functionally correct, however it will not reflect any timing effects due to pipelining, instruction interdependencies, memory latencies, etc. Adding such details in a fully accurate fashion may require to model combinatorial logic, representing every gate by an if-statement, which needs to be reevaluated in each simulation cycle. Obviously such a model will be much slower.

For some CPU cores there is the possibility to have a very fast and efficient model by just "mapping" the instructions of the simulated CPU to appropriate instructions of the simulation host computer. For other types of hardware (e.g. random logic) this type of abstraction is impossible.

The overall simulation speed (frequency f_s) of a system can be approximately calculated from the simulation frequency f_i of the components:

$$f_s = \frac{1}{\sum \frac{1}{f_i}}$$

Obviously a system can never be faster than the slowest component. Therefore only the performance optimization of the slowest components results in significant overall performance improvements.

SYSTEM COMPLEXITY

According to Moore's law the size (number of transistors) of SoCs grows by a factor of 2 every 18 months. An optimistic estimation is, that a system with twice the size needs also twice as much functional evaluation and twice as many tests in terms of clock cycles (doubled system simulation depth in cycles). However a system with twice the size has only half the simulation speed on the same computer. As a conclusion, the system simulation performance requirements grow at least by a factor of 4 every 18 months or 160% per year. The problem is, that the speed of workstations or PCs only increase by roughly 50% every year.

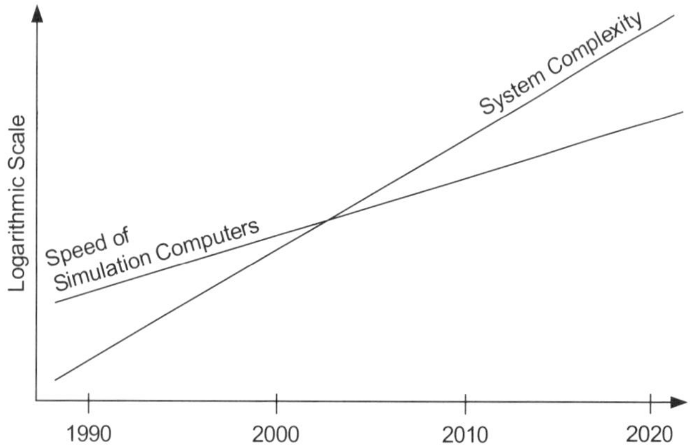

Figure 7 : Simulation Speed Requirements grow faster than the Performance of Simulation Computers

This fundamental trend needs to be taken into account especially for complex systems, when planning a long term simulation strategy. The only way out of this dilemma is to either raise the abstraction level or restrict the scope of the simulation to reasonably sized subsystems, or to do both at once.

CREATION AND VERIFICATION EFFORT

The obvious approach to reduce the modeling and verification effort is to use components from a modeling library. If this is not possible, it needs to be decided for which abstraction and accuracy level the model is to be created. If the IP exists for example as RTL design an RTL to C converter tool could be used, ideally being correct and verified by construction. However such a generated model tends to be nearly as slow as running the RTL description on a HDL-simulator. Another option is the development of a model in a more abstract, less structural and more efficient way and reuse the functional test bench of the RTL design for verification. This greatly reduces the initial verification and thus overall effort, requires however usually a very accurate model.

PUSHING THE LIMITS

Since a model can be considered just as a piece of software, all software performance optimization methods and tricks can be applied. For instance the architecture and the model of computation can be completely different (e.g. functional event based) from the RTL design. Going further the simulator scheduling, structural architecture and functional behavior can be mingled together in the model code or parts can be even written in assembly language. The drawback of these approaches is, that they make the model much harder to create, verify and to maintain and the risk of functional bugs is growing strongly.

On the other hand a model is the more valuable the sooner it becomes available. If the RTL design or even silicon already exists, the benefit of a model is limited. Since the overall speed of a simulation is determined by the slowest component(s), the optimization of one part has only a limited effect. In essence it has to be carefully decided, where to spend effort on speed optimizations.

SUMMARY OF THE MODELING CONSIDERATIONS

The art of a modeling project is to identify in the beginning the sweet spot between accuracy, completeness and speed and to reach this point with the available resources on time.

CONCLUSION

The use of fast and accurate models for architecture exploration, executable specification, software development and system analysis strongly improve the system design cycle in terms of quality, risk reduction and time to market. For maximum benefit the models of the components and the system have to be available as part of the specification phase. This can only be achieved by using a professional, efficient and flexible simulation tool including a broad model library as well as through reuse of existing component models to the greatest extend. The reward is the fact, that the software will be already available, once the first silicon comes back from the fab and having it up and running in no time.

REFERENCES

1. G. Jacobi, “Software ist im Auto ein Knackpunkt”, VDI Nachrichten, 28th February 2003.
2. Snapshots: RealView ESL Tools, MaxSim Explorer by ARM

CONTACT

Joachim Krech
Development Systems (ESL)
Tel: +49 (0) 2407 908620
Email: Joachim.Krech@arm.com

Albrecht Mayer
Principal Advanced Emulation Concepts
Tel: +49 (0) 89 234 83267
Email: albrecht.mayer@infineon.com

Gerlinde Raab
Staff Engineer
Tel: +49 (0) 89 234 87166
Email: gerlinde.raab@infineon.com

DEFINITIONS, ACRONYMS, ABBREVIATIONS

HDL: Hardware description language

ISS: Instruction set simulator

RTL: Register Transfer Language

VP: Virtual Prototype

ANSI: American National Standards Institute

2004-01-1344

Nonlinear FE Centric Approach for Vehicle Structural Integrity Study

Cong Wang and Narendra Kota
General Motors Corp.

ABSTRACT

This report summarizes the methodology used in automotive industry for virtual evaluation of vehicle structural integrity under abusive load cases. In particular, the development of a nonlinear finite element (FE) centric approach is covered that is based on the functions implemented in ABAQUS (by ABAQUS Inc.). An overview is also given for comparative study of the ABAQUS capability with the existing ADAMS (MSC Software) based methods.

INTRODUCTION

It is necessary in vehicle development process to assess the performance of vehicle structures over a range of abusive events to ensure structural strength capacity, in addition to surviving long duration fatigue tests. Examples of such abusive events include subjecting the vehicle to extreme potholes, panic braking on chatter bumps, and curb impact. These events typically are highly transient, short duration, and induce buckling rather than fatigue cracks. As such, they often define the section sizing of key components of the suspension and their interfaces with the vehicle. Several initiatives have been taken to develop an integrated CAE approach to evaluating structural integrity for these types of load cases, particularly during the virtual validation stage. This report covers the development of a nonlinear FE centric approach that is based on implicit time integration scheme. This approach has been implemented in the FE software code ABAQUS/Standard.

TECHNICAL BACKGROUND

In order to capture the structural response in severe events with reasonable accuracy, the CAE simulation needs to be performed at vehicle system level including tire and road interactions. The tools and methodology used should include the following fundamentals:

- Strong capability to handle transient dynamics.
- Ability to model the vehicle suspension system including large displacement/rotation in joints and structural members.
- Capability to deal with nonlinear response in bushings, struts, and tires.
- Capability to accurately evaluate stress/strain distribution in structural members including the plasticity effects.

In Figure 1, four approaches currently in use for CAE structural integrity analyses at vehicle system level are presented, including tire and road interaction.

Column (A) represents the most widely used approach in the current practice. It separates the structural integrity evaluation to: (1) loads-prediction done by rigid multi-body (RMB) dynamics code, and (2) a subsequent finite element stress analysis under the predicted loads.

Typically, in RMB dynamics simulation, the structural components are modeled by rigid bodies connected with joints or bushings. Only hard interface points and lumped mass/inertia parameters are needed from the structural components. The interaction between tire patch and road surface is modeled by data driven empirical models. The problem size is usually small due to the limited number of degrees-of-freedom (DOF). The main advantage of the RMB simulation lies in the fast turn-around time with no stringent requirement for detailed component geometry and properties, and thus making it ideal for design synthesis and loads management. The structural integrity is then checked at component level by applying the predicted loads. However, the drawback of this approach is the uncertainty in predicting stress/strain distribution. To illustrate the point, equation (1) is a generic form of Newton's equation of motion at the component level in terms of FE nodal DOF:

$$\{F_e\} - [M]\{a\} + \{F_b\} = \{F_i\} \qquad (1)$$

where $\{F_e\}$ is the external loads at the connecting nodes; $\{a\}$ is the acceleration field; and $\{F_b\}$ is the constraint force due to the imposed boundary constraints. The sum of the terms in the left hand side needs to be balanced by the internal force $\{F_i\}$ on the right hand side. The internal forces are implicit functions of nodal displacements (or displacement increments) $\{u\}$, that takes into consideration the material responses (evaluated from strain distribution based on interpolation of $\{u\}$) and geometric nonlinear effects.

It should be noted that the predicted $\{F_e\}$ from RMB code might be compromised by not considering material responses. In general, $\{F_e\}$ is not self-balanced. To prevent erroneous motion, it is a common practice to add artificial boundary constraints and/or impose assumptions on acceleration field by inertia relief or rigid body approximations. The net effect of all these compromises could lead to significant error in stress/strain distribution in the structural parts, particularly if the design is on the marginal side.

In recent years, multi-body dynamics codes have added the capability to accommodate structural compliance using component mode synthesis (CMS) approach. The stress/strain is also directly recovered by superposition of the modal responses without having to perform subsequent FE analysis. However, the flexible multi-body implementation is strictly limited to linear material responses. It is also known that CMS is less efficient in the impact-type of problems due to the wide frequency content.

The other three approaches presented in Figure 1 are finite element transient analyses based on different types of solution techniques. They are called FE (Finite Element) centric approaches to emphasize the fact that a FE solver is used to directly obtain the stress/strain history in the parts as well as the transient dynamics of the vehicle system. The solution is performed in a unified solver environment irrespective of whether the parts are represented in the form of rigid body, component modes, finite elements, or subsystem matrices. Due to the need for detailed part information and the large problem size, the application of FE centric simulation is often limited to the later validation phase instead of design synthesis in the vehicle development process.

Column (B) is a transient dynamic analysis approach based on explicit nonlinear finite element codes such as LSTC/DYNA. An example is the Virtual Proving Ground (VPG) kit offered by ETA. The explicit FEA based approach is attractive for not having to deal with convergence difficulties and being more scalable in parallel computations. In the ETA/VPG implementation, the tires are modeled by FE shell/solid elements in pure Lagrange configuration and the road surface by rigid facets. Their interaction is handled by taking advantage of the strong contact capability of slave/master pair in LSTC/DYNA. This approach has the potential to overcome some weaknesses of the empirical tire models, particularly in the lateral response.

However, the general applicability of explicit dynamic code is limited by the stringent stress analysis requirements in dealing with the severe load cases:

- The vehicle system needs to be settled into a static equilibrium state (static trim height) before the transient analysis.
- The mesh refinement required at the critical local regions can reduce the size of stable time increment to the order of 10^{-7} second or less, making the problem computationally intensive.
- Due to lack of rigorous equilibrium iterations in explicit dynamics algorithm, error accumulation, even at a slow creeping rate, could grow significantly after a large number (on the order of 10^7) of time steps.
- In the case of modal based super-elements (Component Modal Synthesis) for materially linear elastic parts in finite rotations, the mass inertia matrix usually contains many off-diagonal terms. Further approximation is necessary to satisfy the diagonal mass inertia requirement in explicit dynamics scheme.
- Fine meshes are needed to mimic the tire construction and capture the interaction at tire/road contact patch. This causes the size of tire model to be at the same order of or even larger than the vehicle structural model.
- The effort of tuning the parameters of tire model to match real test data is also non-trivial.

The approach in Column (C) is the focus of this report. It is a nonlinear FE transient analysis based on implicit time-integration algorithm. Its application at vehicle or subsystem level is only recently enabled by a few technology enhancements. Compared with the explicit algorithm, it has the following advantages:

- Allows flexibility in modeling the structural members as an arbitrary combination of rigid body, finite elements, component modes, and subsystem matrices, depending on the need.

- Capable to include initial static step(s) to determine the static equilibrium state (trim height) before the dynamic simulation.

- Allows local fine FE meshes for stress/strain evaluation.

- Ensures accuracy of the solution through rigorous equilibrium iterations.

- Takes advantage of the empirical tire model.

The disadvantage of the implicit based approach is that it is computationally intensive, and requires skill and expertise in dealing with potential convergence problems. The application of this approach will be demonstrated in the subsequent sections.

Column (D) is a dynamics analysis practice based on linear FE analysis with modal representation of tire/road interface. This approach is mostly for small amplitude vibration and ride analysis with limited application in durability analysis but inappropriate for the abusive load cases.

In summary, the authors' believe that, in addition to the multi-body approach for design synthesis, there is room for FE centric simulation for virtual validation of structural integrity in the impact type of road events. The implicit nonlinear FE approach has the most potential for such tasks.

ENHANCEMENT IN ABAQUS

The ABAQUS/Standard software by Abaqus, Inc. is chosen for the effort in this work. This code features relatively robust convergence performance and a rigorous implementation of nonlinear kinematics in its elements, constraints, and constitutive formulations. However, for specific application in the severe load cases, two additional ABAQUS/Standard features prove to be critical: the solver's strong capability to deal with Lagrange multipliers and the extensive interface for taking in user elements.

The main barrier to a straight use of existing implicit codes for suspension systems is the lack of general capability in dealing with constraints from joints. For historical reasons, implicit codes typically use direct elimination method to deal with such constraints in an effort to limit the problem size. This leads to the following undesirable consequences:

- Only a limited type of simple joints can be handled in this way.

- Cannot allow rigid loops and cannot resolve duplicate constraints that are unavoidable in modeling suspension systems.

- It is inconvenient to describe joint motion and recover constraint forces.

To overcome the barrier, Abaqus Inc. added a class of new elements called 'connector elements' in ABAQUS version 6.1. These elements are intended for a wide range of simple and compound joint constraints enforced by Lagrange multipliers formulation. The relative joint DOFs and motions are described in the local joint coordinates that will co-rotate with the motion of the system. The Lagrange multipliers are added as additional solution variables in exchange for general applicability, thus augmenting the FE problem with additional constraint equations. The implementation is similar to the ones in multi-body dynamic code such as ADAMS (by MSC Software). Details regarding the connector elements can be found the ABAQUS User's Manual.

In the meantime, a parallel effort by Abaqus Inc., not covered in this report, added the capability that allows finite rotation of substructures to accommodate CMS and superelement representation of structure components or assemblies.

TIRE MODEL

A GM proprietary durability tire model, named GMTire, is ported to ABAQUS for tire/road interaction in this study. Subroutines are coded to replace the ADAMS routine and functional calls in the original type model and to interface it as a user-defined element in ABAQUS. Since different rotational measures are used in ABAQUS and the original GM Tire model (Euler Parameters vs. a combination of orientation matrices and Euler Angles, respectively), the key is to construct a one-to-one mapping of rotational measures between ABAQUS and the original tire model.

For the implicit code, the Jacobian matrices of the tire model are required in iterations of the nonlinear solver. A central difference scheme is used to construct the Jacobian matrices by a perturbation of the displacement and velocity terms of the six DOFs (three translational plus toe-rolling-steering angles) at the wheel center.

THE EXAMPLE MODEL

A vehicle model based on a concept passenger car program that is non-production intended is used to demonstrate the technical capability of the FE centric approach. The body of the car is kept as rigid. The main focus is on the structural responses in the suspension system.

Figures 2 & 3 display the front and rear suspension system. For the front suspension, the vehicle model included MacPherson struts, steering knuckles, drop links, stabilizer bar, lower control arms, rack-and-pinion, cradle with built-in powertrain, radiator, and body mount brackets. The rear suspension is composed of trailing arms, twist-beam, a tie-bar, spring seats, coil springs, and shock absorbers.

All the hard points, joints and bushings are translated from a corresponding ADAMS RMB model. The suspension components are replaced by finite element meshes except for the inner and outer tubes of MacPherson struts, the knuckles, and the rack-and-pinion assembly that remain as rigid members. A tire type of P195/75 R14 X5.5 RIM is used with the example model.

LOADCASES

STATIC DESIGN FACTOR (SDF) STUDY

To validate the suspension kinematics of the ABAQUS model and evaluate the new connector elements in ABAQUS, an SDF study is performed to benchmark against the results from the original ADAMS rigid body model. Also included are the results from an ADAMS flexible body models. The same FE mesh is used to provide component modes for the latter. Also included is a special ABAQUS case in which all the FE meshed parts are turned to rigids to create an equivalent RMB model in ABAQUS. The details of this benchmark study can be found in a separate report.

In Figures 4 and 5, wheel force, toe-in, and camber responses in SDF ride analysis were plotted for the front and rear suspension respectively. Except for a small difference in the initial calculation of toe-in angle, the results from the front suspension are basically the same for all the four methods due to very stiff structural members. At the rear end, including the flexibility effect leads to significant differences in responses. The ABAQUS FE model even reveals minor development of plastic deformation.

INITIAL TRIM HEIGHT

The vehicle suspension system typically includes pre-compressed coil springs to be counter-balanced by the gravity of the vehicle structures and payload. The initial configuration from the CAD drawing is typically way out of balance. It is very important to have the vehicle starting at the static equilibrium configuration (or trim height) before the transient dynamic analysis.

Although ABAQUS/Standard can allow as many static steps as needed, it is not uncommon that due to the high out-of-balance forces, the initial state could be out of the convergence region in the Newton-Raphson algorithm of the solver. This will cause convergence difficulties in achieving initial static equilibrium. In addition, the vehicle model has at least 10 rigid body degrees of freedom for motion (three translational plus roll-pitch-yaw DOFs at vehicle C.G. and four rolling DOFs at the at the wheel centers). This means the stiffness matrix of the system is not fully ranked, posing a numerical challenge to implicit code in static steps.

In order to overcome these difficulties, the following techniques are used to guide the solution to the equivalent positions:

- Add a time-dependent duplicate spring parallel to each of the pre-compressed coil springs and ramped the stiffness down to zero at the end of the step.
- Add torque springs between the wheel and knuckle at each of the four spindle centers and ramp the stiffness down to zero at the end of the step.
- Apply constraints at the vehicle body C.G. to suppress the fore-aft, lateral, and yaw rigid body motion.
- Make sure the gravity load is applied gradually through out the whole static step.

All these measures are meant to smooth out the solution process and improve the numerical conditions for the static analysis without compromising the accuracy of the solution.

In a transient dynamics step, following the static step(s), the ‘model change’ capability can be used to remove the duplicate springs and the remaining constraints at the body CG.

EXTREME-POTHOLE RESPONSES

An extreme-pothole simulation is performed with no applied steering, driving or braking inputs. The vehicle moves at a prescribed initial speed of 11173.6 mm/second (or 25 mph) to go over the specified pothole laid on the left track. A transient dynamic simulation time of 1 second is long enough to cover the event.

In Figure 6, the Z-component of front shock tower forces were plotted. The time history, except for the difference in peak value between ADAMS and ABAQUS runs, are very similar from all the four methods. This is consistent with the observation in the SDF study. Other forces transmitted at the key attachment points of suspension also exhibit the similar trend.

Figure 7 and 8 show the plasticity region (at a threshold of 0.2%) experienced in the suspension structure members from the ABAQUS FE simulation of the event.

OTHER EVENTS

The events of bumper road and cross ditches were also tried out with the vehicle model. These simulations are completed without any difficulty.

PERFORMANCE

In an implicit nonlinear FE analysis, the turnaround time is not only dependent on the problem size and simulation time, but also on severity of the events that affect the size of time increment and the number of iterations for each increment. The comparison of computer wall-clock time is reported in tables 1. The timing of ABAQUS jobs was originally generated in January 2002 on HP J5000 workstation (single thread) using v6.2-1.

Table 1. Comparison of Turnaround Time

Road Events	Traverse speed	Simul. tiime	Turnaround time (in hours:minutes)			
	(MPH)	(sec)	ADAMS RMB	ADAMS FLEX	ABAQUS RMB	ABAQUS FE
SDF Ride	NA	Static	0:02	0:12	0:11	2:51
SDF Roll	NA	Static	0:02	0:09	0:16	2:31
Flat Road	25	2	0.01	0.1	0:20	6:47
Pothole	25	1	0:03	0:29	0:27	13:30
Bumper Road	25	1	0.02	0:13	0:19	13:52
Cross Ditches	5	10	0:21	2:30	0:44	31:49

It is important to note that the ADAMS runs and the ABAQUS RMB run only cover the loads prediction. To get the stress/strain histories in all the critical structure parts, a significant mount of time is needed to process the loads information, set up and perform subsequent FE analysis.

The authors would also like to point out the rapid pace of improvement in the performance of the FE centric approach due to continued computer hardware and software upgrade. The same set of ABAQUS FE models was rerun 15 months later (in April, 2003) with ABAQUS v6.3-2 on HP J5600 workstation (single thread). The comparison of turnaround time is shown in Table 2.

Table 2. Improvement of ABAQUS FE Simulation Performance (from January 2002 to April 2003)

Road Events	Traverse speed	Simul. tiime	Turnaround time (in hours:minutes)	
	(MPH)	(sec)	ABAQUS v6.2-1 on HP J5000 Workstation (January, 2002)	ABAQUS v6.3-2 on HP J5600 Workstation (April, 2003)
SDF Ride	NA	Static	2:51	2:12
SDF Roll	NA	Static	2:30	2:01
Flat Road	25	2	6:47	3:31
Pothole	25	1	13:30	7:31
Bumper Road	25	1	13:52	6:44
Cross Ditches	5	10	31:49	18:17

CONCLUSION AND FURTHER WORK

From the comparison of numerical results and turnaround performance, it can be concluded that:

- The ABAQUS FE approach can be effective for short duration severe events that are stress-sensitive.
- From the perspective of computer resource requirement, rapid progress in computer hardware and software will position this FE centric approach within the reach of routine simulation work in the near future
- The ABAQUS FE approach is currently not cost-effective for loads prediction unless the material responses affect the load path.

With the introduction of new capability in ABAQUS/Standard allows CMS and super-element representation of structure components or assemblies that stay in linear material response range but allow finite rotations. A combination of FE and such CMS representation would be ideal to fully account for the structural compliance while achieving a reasonable turnaround performance.

ACKNOWLEDGEMENT

The authors would like to fully acknowledge the contribution to this effort from the following colleagues:

- Lee Krispin, C.G Liang, Wayne Nack, Jim Cao, and Brandon Bai for participation and input during the early preparation;

- Maolin Tsai, Fan Li, Robert Geisler, and Todd Vest for input, cooperation, and continued support.

The authors would appreciate the excellent cooperation from Abaqus, Inc. in this effort. Part of this work was delivered to GM by Abaqus Inc. under the service contract #11ZZZDY00 .

REFERENCES

1. Shih-Chin Wu, Edward J, Haug, and Sung-Soo Kim, 'A Variational Approach to Dynamics of Flexible Multibody Systems', Mech. Struct. & Mach., 17(1), 3-32, 1989.

2. *ABAQUS Users' Manual*, Version 6.2, by Hibbitt, Karlsson & Sorensen, Inc., 2000.

3. Narendra Kota and Cong Wang, 'Comparative Study of Delta-A Car SDF Analysis Using ADAMS/Flex and ABAQUS', GM VSASPC Evaluation Report (draft), 2000

CONTACT

Communication regarding to this paper can be sent to

Cong Wang, Global Performance Integration, General Motors Corp, 6400 E 12 Miles Road, Mail Code 480-305-200, Warren Michigan 48090-9000

Or by E-mail to: cong.wang@gm.com

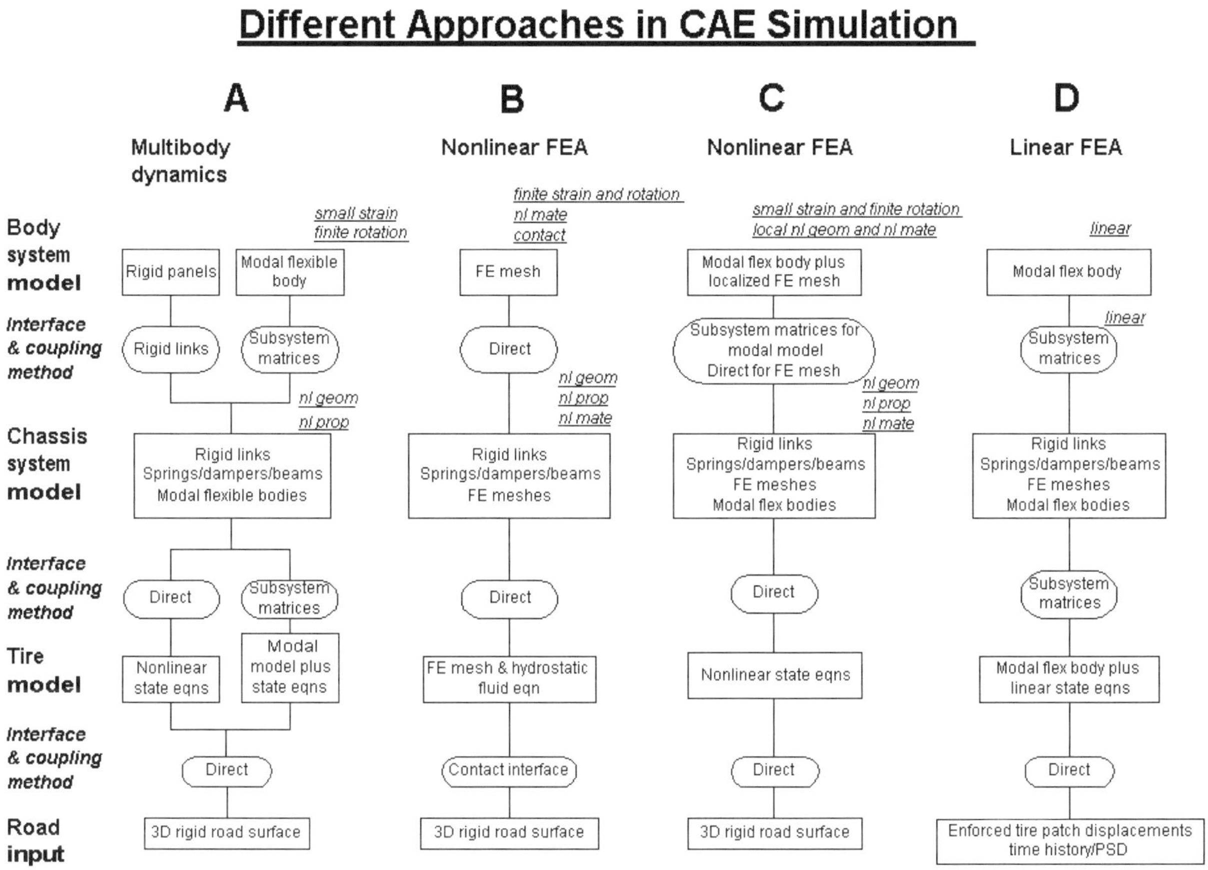

Figure 1, *Four different CAE simulation approaches currently used in practice inside GM structural durability analyis at vehicle system level.*

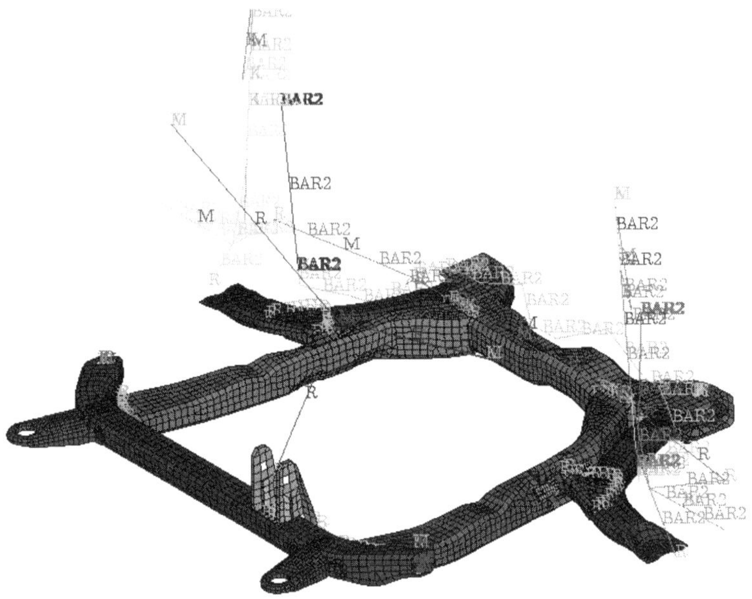

Figure 2. The FE mesh of the front suspension of the Delta-A car

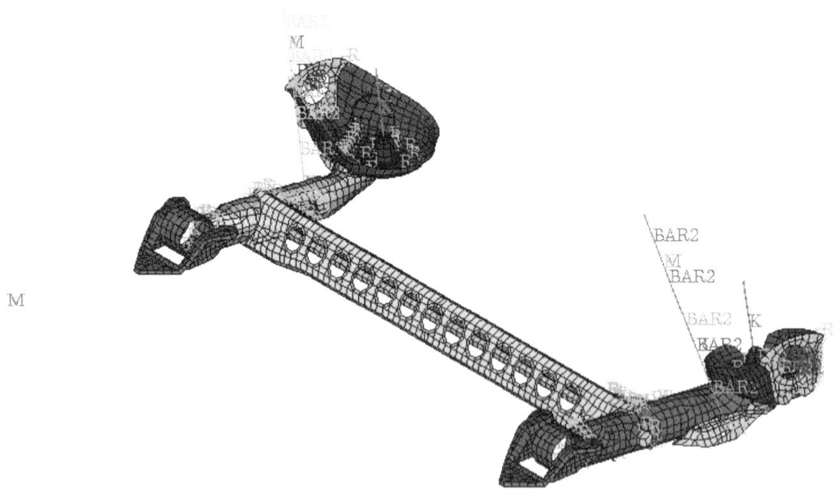

Figure 3. The FE mesh of the rear suspension of the Delta car

Figure 4. Front Ride SDF Analysis Results

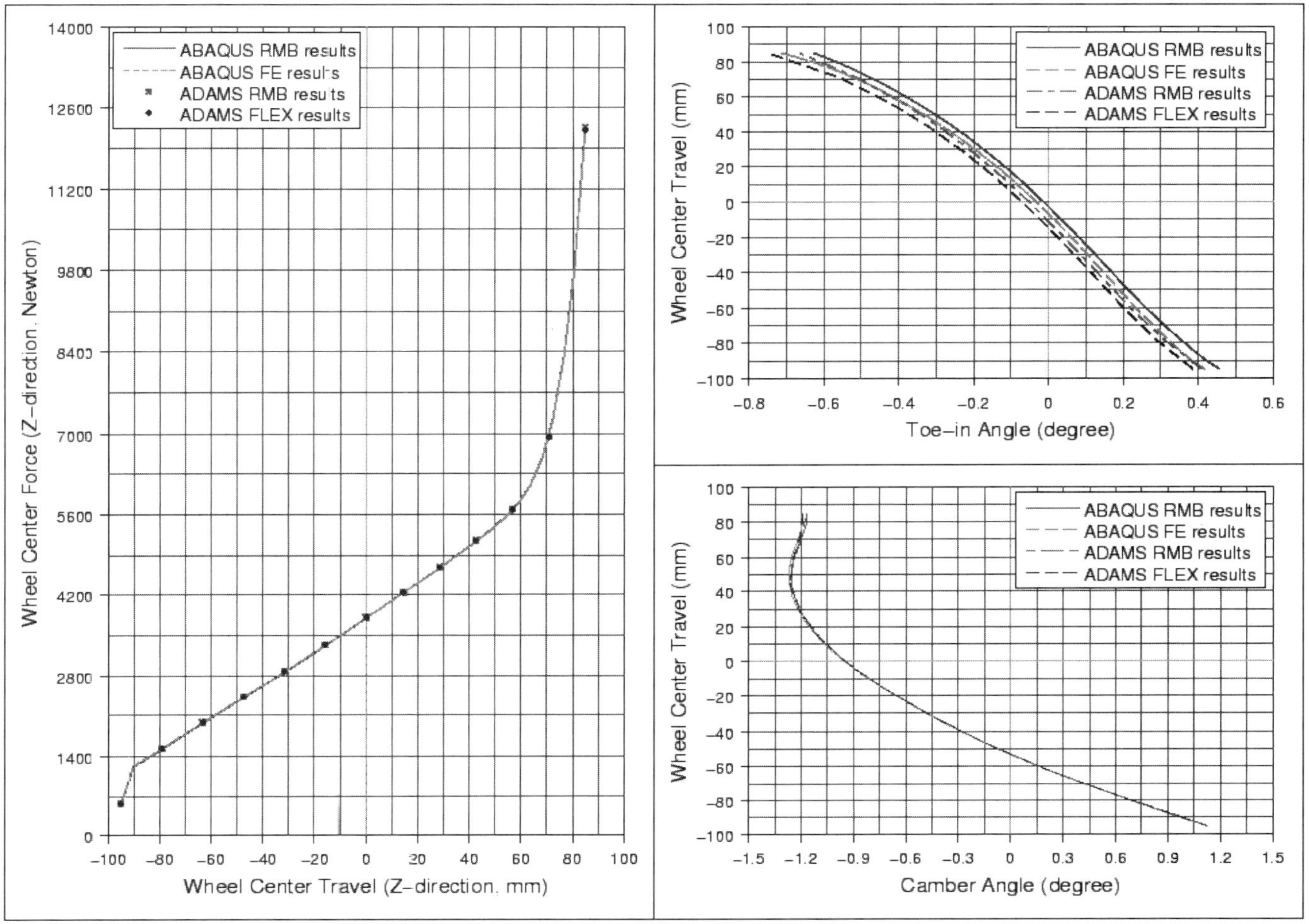

Figure 5. Rear Ride SDF Analysis Results

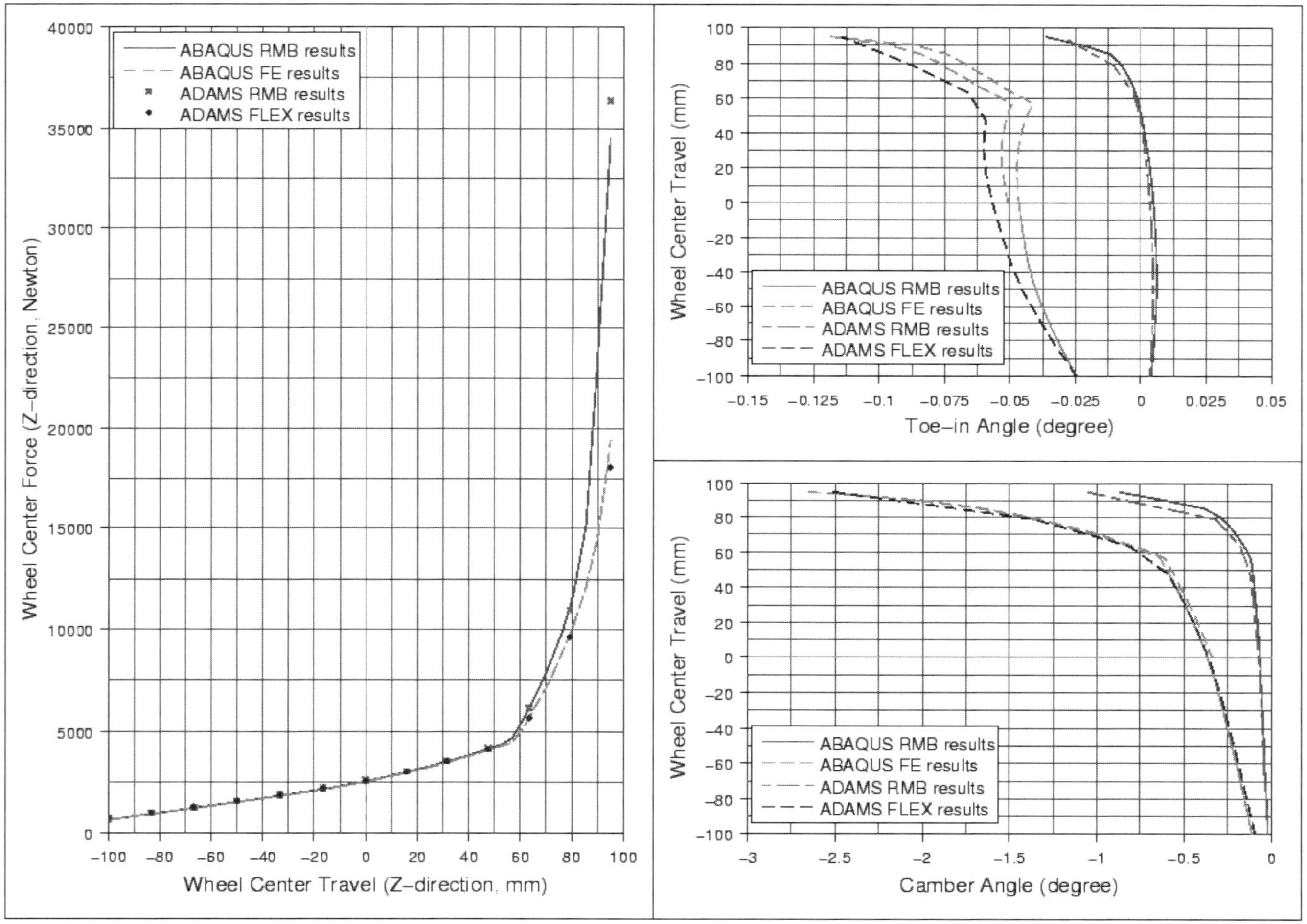

Figure 7. Front shocktower force histories (Z-component)

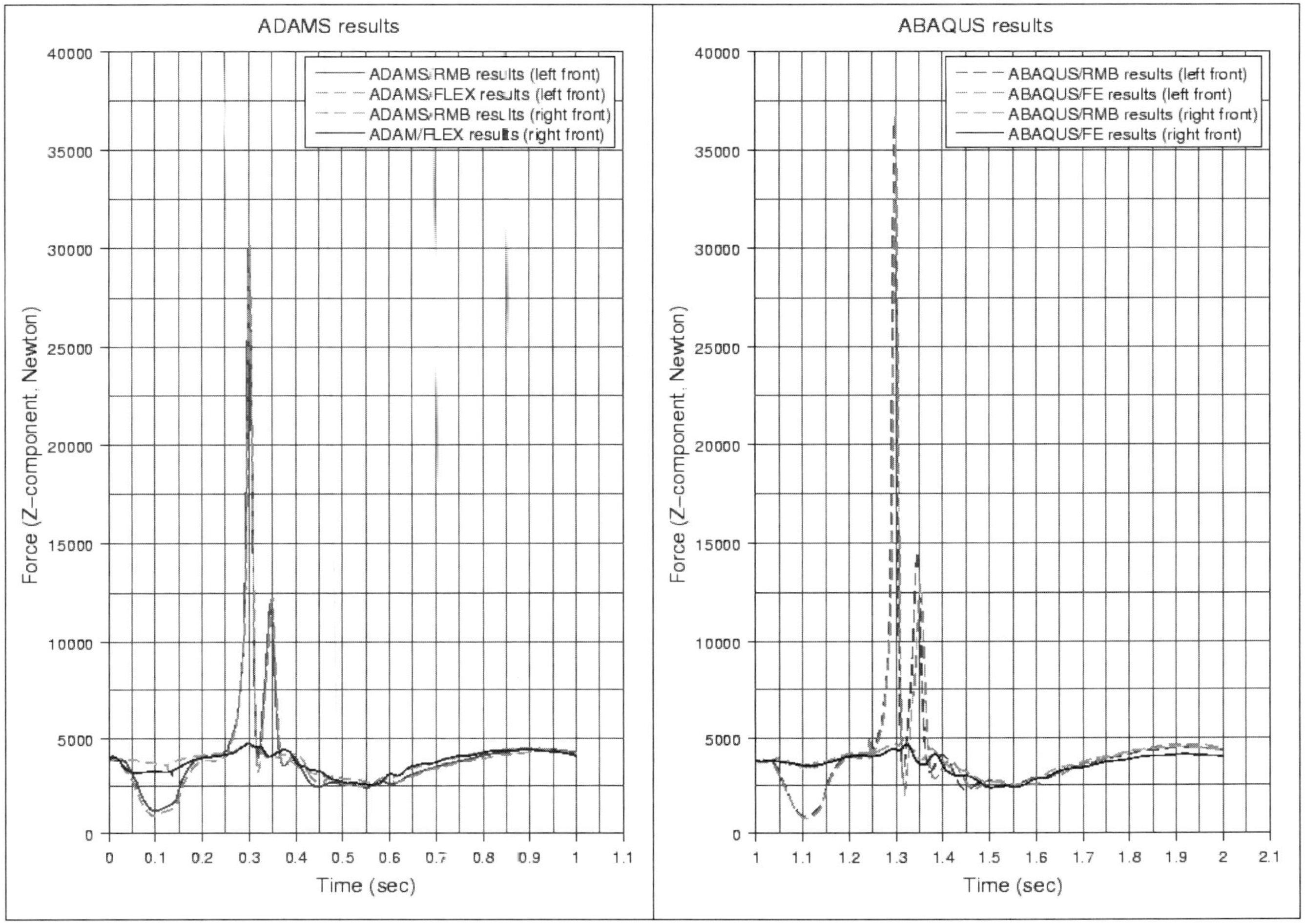

Figure7. The plastic region of the front suspension incurred in the Max-Pothole event

Figure 8. The plastic region in the rear suspension incurred in the Max-Pothole event

2004-01-0188

Virtual Aided Development Process According To FMVSS201u

Christoph Knotz, Bernd Mlekusch
CONCEPT TECHNOLOGIE GmbH

ABSTRACT

Many safety regulations in the automotive engineering use impactor testing (e.g. FMVSS201 in the US; Pedestrian Protection, ECE-R21, proposal for EEVC WG13 in Europe) in the certification process. Through the increasing demand for very short development times virtual engineering has become an inevitable tool.

We show a complete virtual development process for the Free-Motion-Headform (FMH) regulation (FMVSS201u), where we use a combination of self-developed and standard software. The process starts with the definition of the target-points, the possible and allowed positioning of the FMH, the detection of worst case angles, the automated generation of section cuts, the Finite-Elements (FE) analysis and the web based documentation of the results. Our self-developed tools play an important role in the FMH-positioning/worst case detection area as well as in the result analysis and documentation.

Since all impactor testing follows in principle the same outlines, we are now working on carrying over this process systematically to other regulations.

INTRODUCTION

Impactor testing has become very popular in recent years. Heads (free motion or guided), legs and upper legs (Pedestrian Protection) are fired against different areas on the interior and exterior of the car structure. The general advantages of this kind of testing is that these tests are

- very reproducible
- cheap in comparison to full-size tests and
- large areas and zones of the structure can be covered.

On the other hand you have to carry out tests on many different points and locations, leading to a large amount of development activities.

In the early stages of a modern development cycle all parts and structures exist only on CAD. To guarantee 201-FMH functional design possessing sufficient deformation elements we followed the work flow of our most experienced testing engineers. We recognised that we had to develop missing software-tools to overcome otherwise time-consuming work. Following the guidelines described below it is possible to perform the virtual computer aided testing in a highly automated way and give a maximum reassurance for the functionality of the design.

VIRTUAL DEVELOPMNET PROCESS ACCORDING TO FMVSS201U

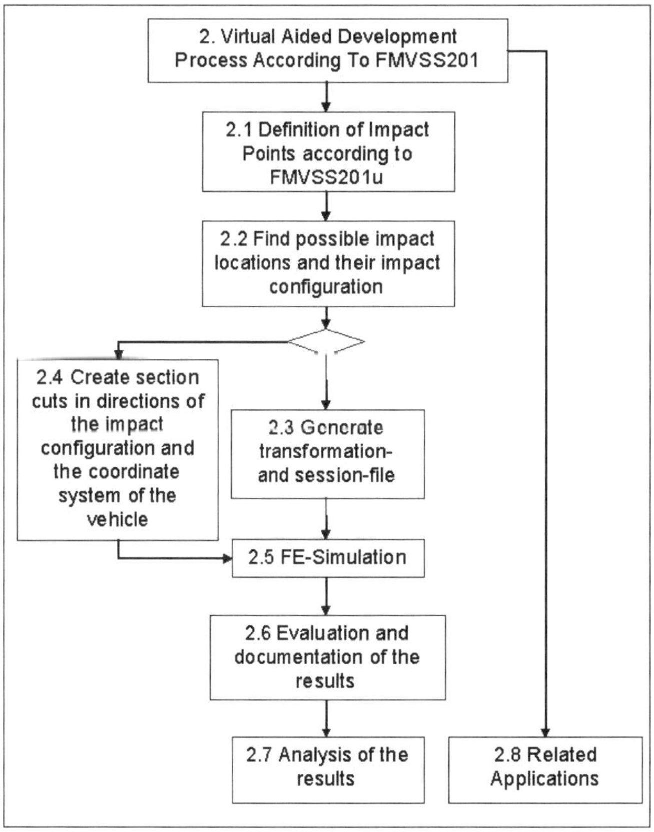

Figure 1: General Process Diagram.

Figure 1 shows the general process diagram. In the course of the virtually supported development process according to FMVSS201u, the application of various tools for the support of individual partial stages is implemented:

Definition of Impact Points according to FMVSS201

The impact points (AP1, AP2...) are generated automatically on CAD. This is done by a programmed application in Catia V5. Alternatively, neutral formats like 'stl' can also be used. The coordinates of the various impact points serve together with the related components as input for the next following stages:

Find possible impact locations and their impact configuration

When carrying out simulations according to the FMVSS201u, we often come across the problem that there are many possibilities for the positioning of the head in relation to the target. The combination of the forehead impact zone with the horizontal and vertical test angle areas theoretically opens infinite possibilities with regard to positions and angles of the head to the target. On the real vehicle, the technician tries to find the worst case and relevant combinations based on his experience. However, with the possibilities numerical simulation offers, we can check a wide variety of variants in order to obtain a large statistical sample and search for the worst case(s) to assure a robust design. With the FPT (FMH Positioning Tool) software, which is described in the flow chart of Figure 2, a large quantity of possible (and by the regulation allowed) solutions regarding the position and angle of the head to the target is calculated completely automatically. In the second step one seeks then to find the worst case(s). E.g. the FPT-software will calculate 200 (a number given by the user) possible and regulation conform solutions, whereby then 5 are selected to be investigated with FEM-analysis.

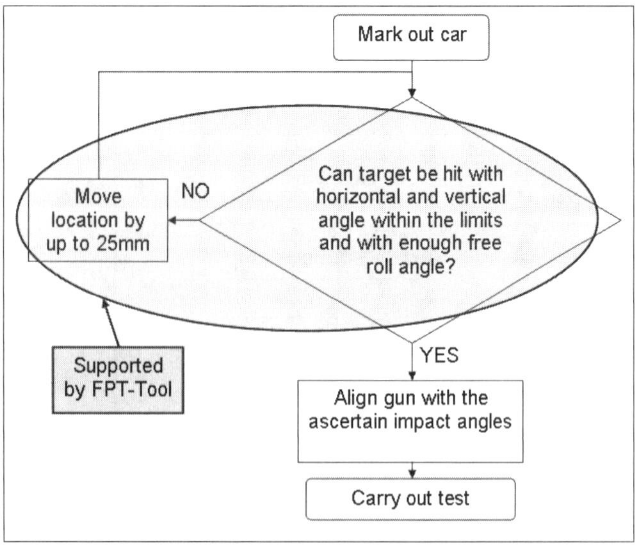

Figure 2: FMH Positioning Tool (FPT) Flowchart.

To do this, the FPT-software offers the following functions:

- Calculation of a predetermined number of positions and angle combinations for a certain target in compliance with the 201-regulations (e.g. considering the correct free-rotation angle).
- Independent relocation of targets when not reached according to the regulations.
- Illustration of input and output in a 3D viewer for visual monitoring (Figure 3).
- Different sorting algorithms (like critical distance, minimum free rotation angle, point location on target zone etc.) to find the worst case(s).

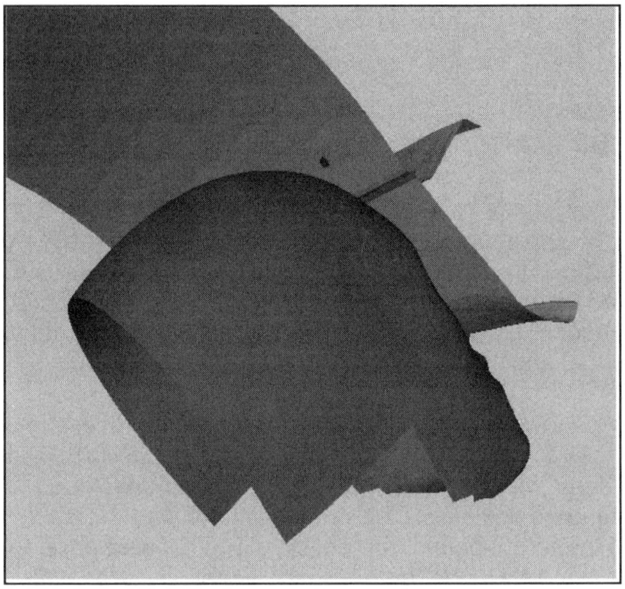

Figure 3: FPT 3D viewer for visual monitoring

A very useful tool to find troublesome configurations is the so-called "critical distance" algorithm. In addition to the relevant trim parts, hard ("critical") components (e.g. metal sheets, wiring harnesses, weatherstrips, etc.) are defined in the software. Then the impact direction of the displacement of the head to these critical components is measured automatically across the entire head, and the shortest configuration for each direction is stored. As a result a coloured graphic is calculated representing the distance as colour.

As mentioned above only a number of three to five configurations are handed over to the FEM-analysis.

Generation of transformation- and session-files

Once the configurations have been chosen that should be virtually tested in detail in FE-simulations to calculate the HIC(d)-values, the FE-transformation and session files are automatically generated from the output of the FPT (Figure 4). Thus it is possible to avoid any potential typing errors.

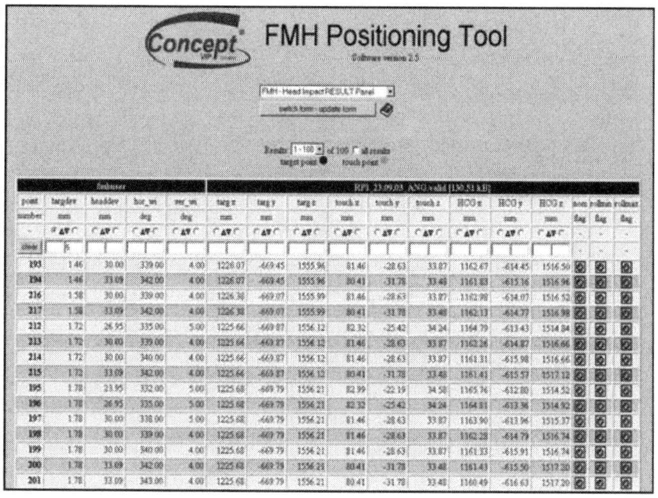

Figure 4: FPT output of possible target configurations

Create section cuts in directions of the impact configuration and the coordinate system of the vehicle

Furthermore, for the previously chosen configurations, automatic section cuts for evaluation of the package situation at the target-points and respective directions can be generated. This task is once again done automatically by a Catia V5 application program.

FE-Simulation

Standard FE-Simulations are carried out next. As Software-packages LS-Dyna and PAM-Crash are used at CONCEPT depending on the customers demand.
At CONCEPT critical polymeric materials are mechanically characterised routinely, whereby the measurement technique is based on dynamic bending-tests [1]. The model of the FMH-head was developed in house and passed several validation tests.

Evaluation and documentation of the results

For the purpose of analysis, videos and diagrams are generated fully automatically. The cutting direction is directly taken from the FPT-software.
The simulation-diagrams are presented in a way which is equal to that of the real testing lab. For the momentum-diagram (acceleration over displacement) that means, that the displacement is calculated from the x-acceleration by double integration as in the testing lab. Therefore virtual and real testing results are optimally comparable.
An in-house developed database is used to store the results (both from testing and simulation). Different search functions are implemented and therefore the development of critical points/areas can be followed easily over time.
All data can be accessed via the web. This allows an optimal following of the project for our clients and also helps in communication over different development locations.

Analysis of the results

All results, which are not clearly positive (HIC(d)<600) or negative (HIC(d)>1300), are analysed in detail. Two different algorithms are employed:

- $HIC\text{-}^{opt}$: This tool uses ASCII-Data to calculate the HIC and the related time window. The acceleration pulse can then be modified very easily (directly via drag and drop in the diagram) to see the influence of possible modifications and improvements. The pulse is recalculated in a way, that the overall energy (velocity difference between incoming and outgoing impactor speed) as well as the given displacement stays the same. Optionally, these values can be altered by the user. Therefore the influence of mechanical modifications, like stiffer deformation elements, on the complex value HIC can be judged in advance.
- V^{return}: We also calculate the return velocity of the impactor routinely. Based on this value we determine if the impact occurred is elastic (impact coefficient = 1) or plastic (impact coefficient = 0). A more elastic impact shows room for further improvements while a high HIC together with a plastic impact implies a need for more deformation space.

Related Applications

Since many other regulations are also based on impactor testing, we are now working to implement similar schemas to address them (Pedestrian Protection: EEVC WG17, ACEA, EuroNCAP and Japan NCAP;). Adaptations are necessary in the target location and definition process, although general algorithms stay the same. This is also true for the analysing tools and the database system.

CONCLUSION

A reliable, fast and robust methodology is shown to deal with impactor-testing on the virtual testing side. A combination of self-developed and standard software is employed. The whole schema is demonstrated at the FMVSS201u, but can easily be adapted to other regulations.

REFERENCES

1. Martin Fritz, "Determination of material data for automotive crash analysis using dynamic bending tests", Diploma-Thesis, University of Leoben, 2003.

2. U.S. Department of Transportation, National Highway Traffic Safety Administration, "Laboratory Test Procedure for FMVSS 201U", April 1998

3. European Enhanced Vehicle-Safety Committee "Development Of A European Side Impact Interior Headform Test Procedure" May 2003

4. European Enhanced Vehicle-Safety Committee "EEVC Working Group 17 Report, Improved Test Methods To Evaluate Pedestrian Protection Afforded By Passenger Cars, September 2002

2003-01-3667

A Semi-Analytical Method to Generate Load Cases For CAE Durability Using Virtual Vehicle Prototypes

Joselito Menezes da Cruz
Ivan Lima do Espírito Santo
Adilson Aparecido de Oliveira
Ford Motor Company

ABSTRACT

The growing competition of the automotive market makes more and more necessary the reduction of development time and consequently, the increase of the capacity to quickly respond to the launching of the competitors. One of the most costly phases on the vehicle development process is the field durability test, both in function of the number of prototypes employed and the time needed to its execution.

More and more diffused, the fatigue life prediction methods have played an important part in the durability analysis via CAE. Nevertheless, in order they can be reliable and really being able to reduce the development time and cost, they need to be provided with load cases that can accurately represent the field durability tests.

This paper presents an efficient method to generate such a load cases, called Semi-Analytical Road Loads. It consists of a process to cascade loads acquired from the hub of a vehicle instrumented with set of wheel force transducers (WFT) to the relevant points of the suspension/body interface with the use of models generated into the ADAMS, a multi-body dynamics simulation software.

INTRODUCTION

There are a number of reasons to use CAE methods in the modern vehicle development process. Foremost the need to cut cost increases the usage of virtual testing versus costly hardware testing. The general goal is to achieve the level of "zero prototype" in the development process.

Another important argument is the fact that certain characteristics of the system cannot always be measured without influencing the system (intrusive measurement). The characteristics, which can be measured with relative low effort, may be used to correlate and to verify the validity of the calculation models. In return these models deliver a large number of characteristics who's measurement would be too time consuming and costly in the frame of a regular development project.

Since hardware prototypes are not available early in the design process, especially in the "pre-prototype phase", the usage of virtual models is required and represents the most important scenario. During this period, the results from calculations and the engineering experience of the designers are the only basis for design decision, which influence the entire design process until production. The more innovative the design is, the less engineering expertise is required; hence the importance of the calculation increases dramatically in the concept phase.

Depending on the need to consider the interaction between models, a modular approach is required. Component level analysis as well as subsystem and full system analysis must be possible.

BACKGROUND

Load cascading for calculating component durability is becoming a common practice among worldwide car manufacturers. Currently there are three major methods for obtaining vehicle and component loads.
They are:

- Direct Measurement;
- Hybrid or Semi Analytical Road Load Data
- Fully Analytical Road Load Data

The first and second methods are largely used and the third one could be considered as the state-of-the-art in the subject. The Direct Measurement is a very common method and requires dedicated prototypes, which need to be extensively instrumented. Such a method demands a hard and frequent work of inspection and typically it takes several moths of development.

Semi-Analytical road load (Semi-ARL) is a hybrid method, which is a mix of measured data and simulation of analytical models to determine loads. It involves a limited set of data in order to support analytical determination of the remaining required loads. Its main characteristics are listed bellow:

- Measuring of some loads and calculating of the rest;
- No dedicated prototype necessary (Requires a re-usable prototype);
- Requires lesser degree of instrumentation (extensive intrusive instrumentation not necessary);
- Provides better quality;
- Requires typically two months.

Within the scheme of hybrid loads, depending on the specific problem, two approaches are possible:

a) Component free-body method
    - Measure some component loads;
    - Compute the other loads based on free-body equilibrium equations.

b) Sub-system loads method
    - Measure wheel spindle loads;
    - Develop ADAMS models of suspension system;
    - Compute all chassis loads.

Compared to the Direct Measurement, the hybrid method presents a number of improvements. The reduction in instrumentation can be quite dramatic. The previous process requires a variety of load measuring devices, while hybrid sub-system loads methods uses only four off-the-shelf load cells called Wheel Force Transducer (WFT), as shown on figure 1, and some individual sensors that will be used as correlation channels. The WFT is a sophisticated transducer able to generate in real-time a history of the 6-load components: The Longitudinal (Fx), Lateral (Fy) and Vertical (Fz) forces, plus the Turnover Moment (Mx), Torque (My) and Steering Moment (Mz). All of them are described at the spindle reference.

The acquisition system which the WFTs are connected to, makes automatically the analog to digital conversion and storages the data.

Figure 1: Wheel Force Transducer

Once the data is obtained, a number of mathematical treatments should be done on them in order to eliminate spikes, drifts and other effects, that do not represent a real input from the test tracks, but are inherent from the acquisition systems. The final generated file is in RPCIII format, a binary ADAMS compatible format that has all the acquired channels in it.

Another requirement for the use of Semi ARL is a numeric model of half suspension that is going to receive as input the data collected. There is no need for tire models, since the input come straightly from the spindle.

Fully analytical road load is a step ahead. Similar to the semi-ARL approach but needs full vehicle models. Additional requirements are the tire durability model, which represents the contact between tire and track as well as a digitized representation of the events that compose the durability track.

The method used in this paper is the Semi-Analytical Road Load.

OBJECTIVE

The objective on this work was to cascade forces and moments coming from the wheels of an SUV vehicle to the complete interface suspension/body/chassis using a rigid body model with a multibody dynamic software (ADAMS/pre) in order to make feasible through the generated load case, the vehicle and the component strength and fatigue evaluation via CAE analysis.

METHODOLOGY

THE NUMERIC MODEL - To accomplish the task, which the Semi-ARL approach is assigned to, it was necessary to build a half suspension ground attached model. Such a kind of model when combined with the input wheel load and submitted to the dynamic load case event existing in the ADAMS/ pre software yields the time history of loads in the hardpoints of interest, that normally are the interface between suspension, body and chassis. More over the model does not contemplates neither the powertrain nor the chassis flexibility.

The vehicle used was an SUV car with a McPherson front suspension (figure 2) and an independent multi-link rear suspension (figure 3).

The reliability of the derived loads is highly dependent on the quality of the ADAMS model, which in turn depends on accurate correlated models. Thus, it was necessary to test and extract the force-deflection characteristics of the following components: springs, jounce bumpers, rebound bumpers, suspension bushings, shock absorbers, etc.

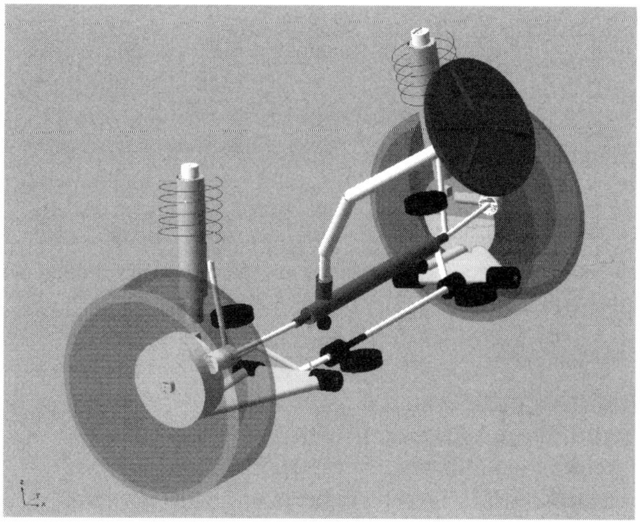

Figure 2: Front Suspension scheme - *McPherson.*

Typically 0.01% to 0.1% of the bushing stiffness was used for bushing damping specifications.

In order the model could be up to date and representative, some other important parameters like component mass, geometric location of hardpoints (measured through a CMM), passanger and luggage mass and location as well as the prototype load configuration, attitude and clearances had to be measured.

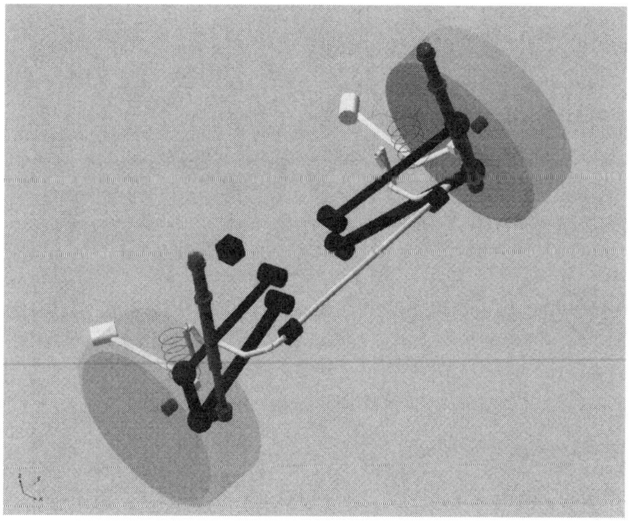

Figure 3: Rear Suspension scheme – *Multi-Link.*

A special analysis was carried out to study the influence of the local stiffness on the places where the suspension is attached to the body and sub-frame. With the help of a FEA model, using Nastran, it was possible to obtain the stiffness matrix from where it was taken its X, Y and Z components. The analysis involved the comparison of each coordinate component of the local stiffness with the bushing component in the same direction. However, the bushings are written at the local coordinate system (its own coordinate) while the body is written at the global system, that most of times are different. To solve that problem it was needed to make use of a transformation matrix called Euler Matrix showed on the Figure 4. With the Euler matrix one can perform a simultaneous rotation around the X, Y and Z directions with the angles α, β and γ respectively becoming possible to match one coordinate system to the other.

$$\mathrm{EulerRotation}[\alpha,\beta,\gamma]=\begin{pmatrix}\cos(\alpha)\cos(\beta)\cos(\gamma)-\sin(\alpha)\sin(\gamma) & \cos(\beta)\cos(\gamma)\sin(\alpha)+\cos(\alpha)\sin(\gamma) & -\cos(\gamma)\sin(\beta)\\ -\cos(\gamma)\sin(\alpha)-\cos(\alpha)\cos(\beta)\sin(\gamma) & \cos(\alpha)\cos(\gamma)-\cos(\beta)\sin(\alpha)\sin(\gamma) & \sin(\beta)\sin(\gamma)\\ \cos(\alpha)\sin(\beta) & \sin(\alpha)\sin(\beta) & \cos(\beta)\end{pmatrix}$$

Figure 4. A Representation of the Euler Matrix

The criterion of influence adopted was to regard the local stiffness if its value, on each direction, is below than 10 times the bushing stiffness at the same direction, after the Euler transformation. Once considered, the local stiffness was added to the bushing stiffness.

THE TEST TRACKS - The traditional durability analysis used at Ford Motor Company requires the prototype to follow a specific route composed of several special events to perform an advanced life test. One of them is showed in the figure 5.

Figure 5. Special Event Track – Belgian Blocks

The objective is to foresee and correct any stress or fatigue problem during the phase of development. Such a route is composed by a number of cycles and each cycle, in turn, contemplates several track events like cobblestone, belgian blocks and so on. The numeric durability analysis methods that will be fed with the output of the Semi-ARL employs the same approach, i.e., it's necessary to have the loads from each track event and moreover, the number of cycles that the entire route performs.

MODEL VALIDATION

A crucial point for the entire Semi-ARL process is the model validation. It's necessary to adjust the model so that it presents the same responses of the prototype when submitted to the same excitation. In this way, the strategy adopted was to instrument the prototype with suitable sensors in specific points that knowingly would suffer excitation during the maneuver chosen for correlation. Likewise, the model was virtually instrumented with the same type of sensors used in the vehicle. They also needed to be positioned at the same location they were put in the car. This is not always possible. The objective was to investigate the correlation between both systems.

The figures 6 and 7 show the efforts on the Tie Rod measured in the prototype for the curve-to-right event compared with the forces observed in the multibody model. The difference noticed at the beginning of the event is more due to the positioning of the strain gage. The ADAMS/Pre works with rigid body and doesn't take in account the element flexibility, so, the derived loads are calculated on the joints while the experimental ones are the real loads on the place where the sensor is, ie, the ADAMS loads are always written at one joint, which frequently has a little different position from the sensor position. Such a restriction causes the difference between the transient behaviour of the prototype and the model.

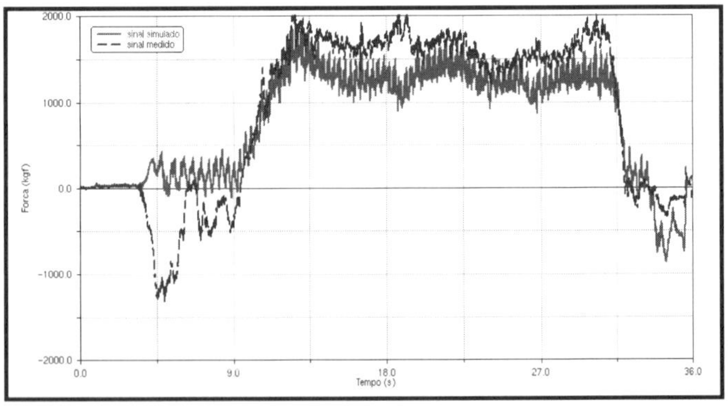

Figure 6. Loads on the left Tie Rod at the curve-to-right event

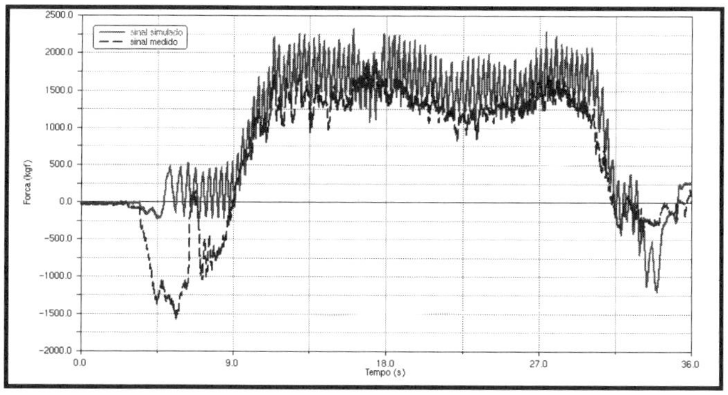

Figure 7. Loads on the Right Tie Rod at the curve-to-right event

The figures 8 and 9 illustrate the displacement of the right rear shock-absorber in the curve-to-right event. The results obtained from the simulation show a displacement with amplitude greater than those one measured in the prototype, although the peaks occur in very similar amplitudes. That happens because the multibody model does not represent the chassis flexibility and it is ground attached. That channel is very important because it is directly related to the suspension bumpers in which the biggest force peaks appears.

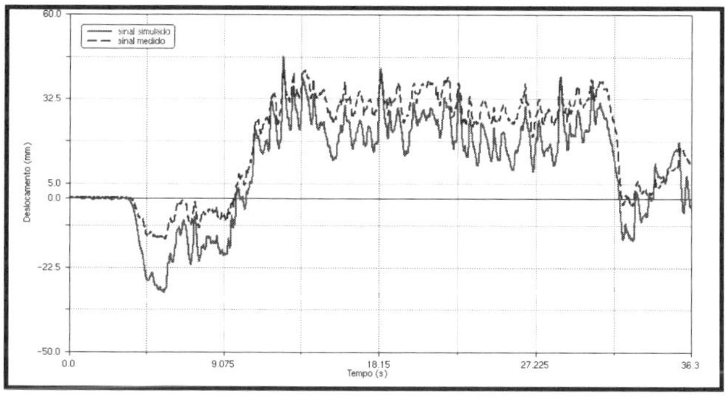

Figure 8. Front left shock displacement at curve-to-right event.

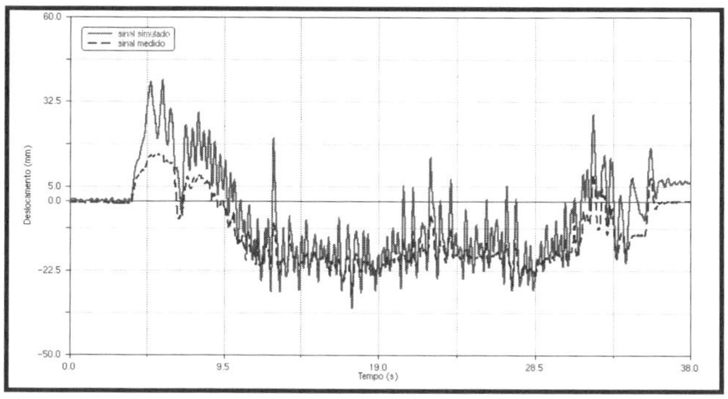

Figure 9. Front right shock displacement at curve-to-right event.

The figure 10 shows the effort on the upper control arm of the rear suspension at the WOT event. A satisfactory correlation can be verified for this channel in the referred event.

For a good correlation it's important to have in mind two main aspects: the choice of both, the suitable channels and the suitable events. Since it is about a half vehicle multibody model solidary to the ground, the torques measured by WFT derived from the powertrain reaction should be eliminated from the input signal (driving torque) as well as channels directly relateded to movements of sprung mass also cannot be used in the correlation.

Regarding the events, it's important to choose those ones simple and able to excitate each one of the correlation channels in a manner, as much as possible, decoppled. The reason for that is because it's much easier to correlate and adjust a model for a simple event and moreover, the vehicle response, until a certain piont, is predictable.

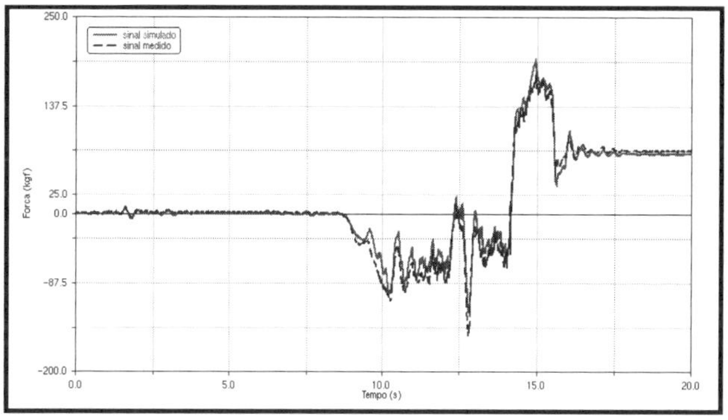

Figure 10. Loads on the rear UCL at the WOT event

It is also possible to work, at the same time, with more then one model, each one fitted to some channels and not necessarilly to all of them. However, in this work this has not been done.

The Semi-Analytic method of load cascading via half vehicle model ground attached involves basically 4 distinct stages: (1) measurement of the prototype, (2) model update, (3) model correlation and (4) cascading. All the stages are very important for the reliability of the results. In the measurement stage of the prototype, besides of the correlation channels, one should measure the vehicle attitude and axis weight at both, the Curb condition, and at the load condition in which the acquisition was done, as well as the jounce and rebound bumpers clearances and displacement of the spring. All components of the suspension should be measured so that the model can be up-to-date in the next stage (shock-absorbers update, bumper, etc). Once properly measured the data are used to update the model. The phase of Model correlation is, undoubtedly, the most important. In that

2003-01-1606

Tools for Integration of Analysis and Testing

Shawn You and Christoph Leser
MTS Systems Corporation

Eric Young
Thermo King Corporation

ABSTRACT

The automotive vehicle design process has relied for many years on both analytical studies and physical testing. Testing remains to be required due to the inherent complexities of structures and systems and the simplifications made in analytical studies. Simulation test methods, i.e. tests that load components with forces derived from actual operating conditions, have become the accepted standard. Advanced simulation tools like iterative deconvolution methods have been developed to address this need. Analytical techniques, such as multi body simulation have advanced to the degree that it is practical to investigate the dynamic behavior of components and even full vehicles under the influence of operational loads. However, the approach of testing and analysis are quite unique and no seamless bridge between the two exists.

This paper demonstrates an integrated approach to combine testing and analysis together in the form of virtual testing. Multi body simulation software [1] was used for multi body simulation of both the component under investigation as well as the test equipment used for physical testing. Road load simulation software [2] was used to reproduce field observed data on both the physical and virtual test rigs. There are two main advantages to this approach. The integrated application of physical and virtual tools allows the user to conduct virtual tests prior to having physical prototypes available. This accelerates the design process and can reduce cost. Secondly, by using a common framework for all physical and virtual investigations, the results remain comparable and trouble shooting of both domains is facilitated. This paper presents the results of a study of a vehicle mounted refrigeration unit. Observed failures in both physical and virtual tests corresponded very closely with respect to location and time to failure.

INTRODUCTION

Good vehicle design requires extensive analysis and adequate testing. In the past, the analysis and testing were separate entities. The approaches of test engineers and analysts to study vehicle behavior are quite different. In the lab, accelerated testing methods have been developed and proven to provide accurate durability and performance information. These methods usually involve special test equipment and simulation software [3], [4], [5]. On the other hand, analysts usually only simulate the vehicle on the road rather than in the lab. As a result, physical and virtual results do not agree due to:

- Load paths are not the same
- Boundary Conditions are not the same
- Test procedures differ
- Test setup and transducer locations differ
- File formats are different
- Result processing and display differ
- Inaccuracy in modeling in non-linear systems.

The desire, therefore, is to integrate the testing and analysis by conducting virtual tests. Virtual testing in this context is defined as the simulation of physical tests by using a variety of analysis tools. The following are advantages of the virtual test method:

- Virtual tests can be conducted at a very early design stage.
- The design can be evaluated before an expensive prototype is built.
- A test can be simulated before the test equipment is available.
- The simulation can provide information regarding the type of test equipment needed.
- Load and boundary conditions can be investigated.
- Virtual tests can be conducted faster, easier, and at lower cost than physical tests.
- Additional information can be obtained from a virtual model that is not readily available from a physical test. For example, "requests" can be made for displacement, velocity, acceleration and load at any point in the structure.
- A number of design alternatives can be rapidly evaluated in "what if" studies. This allows optimized design with respect to cost, weight, durability, etc.

PROBLEM DESCRIPTION

A refrigeration unit (often configured to provide heating and cooling) is used for different types of truck trailers. The refrigeration unit of Figure 1 is mounted in front of a trailer. During the 3 million kilometer projected lifetime of a refrigeration unit, the trailer will travel on different types of road surfaces that will induce specific loads.

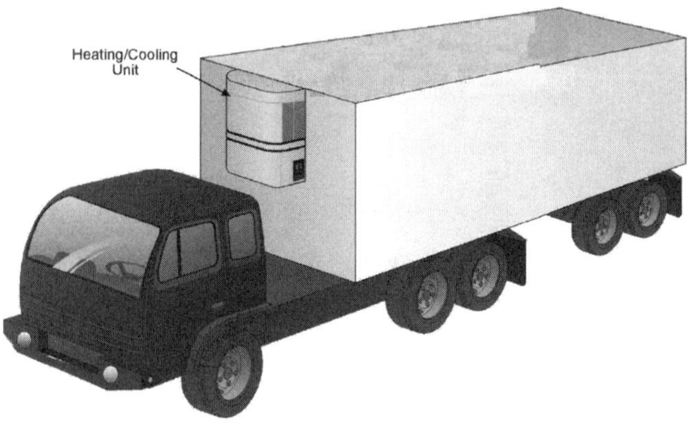

Figure 1. Truck, trailer and heating/cooling unit

Through the virtual test method, the frame design was evaluated with respect to fatigue life. The virtual test result was then validated by a physical test.

MODAL ANALYSIS

To understand the dynamic behavior of the refrigeration unit, modal analysis was conducted (Figure 2). Natural frequencies and corresponding mode shapes were found. The lowest natural frequency of the frame-bending mode was about 80 Hz. The frame design is considered sufficiently rigid. The result helped to understand the dynamic behavior of the refrigeration unit and provided guidance to select the fatigue analysis method. In an ideal situation, analytical results are correlated with modal test results to assure FEA accuracy, this step was not performed due to time constraints.

SIMULATION TESTING [6]

With the advent of minicomputers and lower cost array processor technology in the mid-seventies, a technique was developed [7] that used an iterative technique to converge toward accurate recreation of measured service responses in the laboratory. This technique was based on the deconvolution of response errors with a linear frequency domain estimate of the system and the frequency response function (FRF). Due to the advantages of this compensation technique for accurate reproduction of both amplitude and phase in the response of non-linear, coupled, multiple-input systems several commercial versions of the algorithm were developed.

Initially the application of the iterative deconvolution method was the control of laboratory full vehicle automotive test systems, either tire or wheel spindle coupled. The control challenge is to simulate or reproduce service-loading conditions on these test systems (the desired specimen responses, accelerations, loads and displacements) by applying loads remote from the point of measurement. Service loads that can be measured and recorded on a vehicle component in service or on the proving ground are measured on the body or on the suspension components but the actual loading into the vehicle is through the tire contact patch with the road. At a minimum a non-linear tire spring is introduced into the control scheme. Additionally, the responses measured are frequently due to load inputs from more than one tire patch. By using an iterative deconvolution approach, both the non-linear spring effects and the cross-coupling can be compensated and accurate vehicle component responses can be simulated in the laboratory. Subsequently, the technique has found wide application in automotive sub-system and component testing where testing of the specimen involves recreation of specimen loading due to multiple inputs applied through a non-linear system. This technique can also be applied for the execution of virtual tests.

It is important to note that both the amplitude and phase of both the multi-axial input and output in the laboratory test have to be preserved to allow the multi-axial service loading effects on the specimen to be accurately reproduced. This limits the options available to the laboratory test engineer to "accelerate" the test beyond what would take place in service. The most common method is to examine the desired response loads and strains and use a time history editor to remove those sections of the service or proving ground load histories that provide low damage to the specimen. The software tools to perform this task, either manually or through some fatigue sensitive editing routine, are usually provided as part of the iterative deconvolution package. Depending on the mix of severity of the service or proving ground road surfaces test, "accelerations" in the order of five to ten times real time are possible. Insofar as the endurance life testing performed on such units in Europe, acceleration rates approach 275:1. This per the fatigue analysis provided by the test facility when comparing damage between "typical" road surfaces and the specified track regime.

The development of a typical iterative deconvolution compensation test takes place in six steps.

1. Record Service or Proving Ground Data - The specimen is instrumented with low frequency sensing accelerometers to measure its response to service loading, where the service loading is represented by running the vehicle at a proving ground. The specimen responses are simultaneously recorded as a time history on a tape recorder or equivalent

recording device. "Control" transducers are applied at the specimen mount interface for the purpose of recording the desired response motions. "Correlation" transducers are applied to obtain additional information to judge the quality of the simulation test and provide redundancy in case any of the control transducers fail [8].

2. Digitize and Analyze Data - The recorded time history is transferred to the computer-based analysis system. This may involve digitizing an analog time history recording. Using the analysis tools provided in the iterative deconvolution software package, the data is checked for accuracy and possibly reduced in length using the editing tools described earlier. The output of this step is a set of multi-channel digitized time history records or files containing the "desired response" of the laboratory test system.
3. Measure the System Frequency Response Function (FRF) - The test specimen is connected to a laboratory or virtual test system. A set of drive signals is developed to excite the specimen in the test system and the specimen response is measured through the same instrumentation used to measure the service load data. Typically the drive signal developed is a set of uncorrelated shaped random multi-channel time histories although a single channel may be subsequently excited. In this application, multi-channel excitation was used. Using FFT based spectral analysis methods a linear estimate of the frequency domain response function of the complete test system plus specimen is computed. These linear Multiple-Input Multiple-Output (MIMO) models of the system contain the amplitude and phase of the input-output characteristic of the system between all inputs and outputs over the frequency control band of interest. This system model is then mathematically inverted to become an output-input model in preparation for the next step in the process. In some cases a system model where the number of outputs exceeds the number of inputs may be useful. In these cases the system is over-determined and the general approach is to compute a "pseudo-inverse" where residual errors are minimized in a "least squares" sense. This takes the response of more locations into account and averages their respective influence.
4. Apply Drive Estimate to the System - Each of the desired response time histories is convolved with the inverted system model, which is deconvolved with the forward system model to provide an estimate of the system drive signal required to produce the response. Note that this estimate is based on a linear model and therefore may be substantially in error. From safety considerations (in case of a lab test, not required for virtual tests) this drive estimate is scaled, typically by half, and applied to the test system. The resulting response of the specimen is measured.
5. Calculate Error and Iterate - The desired specimen response is subtracted from the response achieved on the test system due to the drive signal and a calculated response error signal. The response error signal is convolved with the inverse system model and a linear estimate of the drive error signal results. A scaled proportion of this drive error is added to the previous drive signal to produce a better estimate of the drive signal required to produce the desired response, scaling being employed to reduce the possibility of overshoot and instability in the iterative process. The modified drive signal is then used to run the test system and the new response is recorded. A new response error is calculated as described above and the process repeated ("iterated") until the response error is reduced to an acceptable value, that is, when the achieved response on the test system has converged onto the desired service load response. This process is repeated for all separate service load recordings made. The final sets of drive files for each response signal are combined into a durability test schedule.
6. Execute Durability Schedules - The final step is to run the durability test schedule into the test system and monitor the performance of the test specimen over the laboratory simulated service life. For a virtual test fatigue analysis can be performed to estimate the fatigue life for a given durability schedule of loads.

Attributes of iterative deconvolution control:

- Can over-program, i.e. during the iteration process loads higher than field measured loads may be applied
- Requires use of a computer (typically a PC) and an analog-to-digital, digital-to-analog conversion device to drive the test system and measure the responses.
- Pre-training required through measurements of the system transfer function using multi-channel orthogonal white noise or input by input excitation.
- Works with non-linear systems that exhibit cross coupling between inputs and outputs.
- Matches achieved system amplitude and phase to the desired responses.
- Number of iterations required for a given accuracy of reproduction can be reduced through modification of system frequency response function at each iteration step.
- Applicable to both physical and virtual tests.

DATA ACQUISITION AND ANALYSIS

In this application, the refrigeration unit was mounted to the trailer frame through seven mounting locations. Four tri-axial accelerometers were positioned at four corners of the refrigeration unit. Two vertical, three longitudinal, and one lateral acceleration channel were used as control channels for a total of 6 degrees of freedom. Additional acceleration channels were acquired as correlation channels. Data was collected directly in digital format using a sample and hold circuit prior to the analog-to-digital (A/D) converter. The data acquisition sample rate was 512 points per second. Because the highest frequency to be reproduced was less than 50 Hz, the sampling rate was considered high enough to provide good peak resolution. The truck trailer was towed over five selected proving ground surfaces to measure the response-time histories that were representative for a variety of operating conditions. Low-pass analog filters, designed to roll off before the Nyquist frequency, were applied to the data prior to digitizing to prevent aliasing on each recorded channel. The data collected from the proving ground was analyzed to determine the most significant and required inputs for the specimen under consideration. Non damaging portions were removed and remaining sections of the files were then smoothed on either side of the deletion region to remove any discontinuity prior to joining them to a new history.

The edited time history was then band pass filtered. The pass band for the acceleration signals is from 1 Hz to 50 Hz. The reason for this selection is that the testing system is an inertially reacted system for which it is difficult to achieve low frequency control without large actuator displacements. The upper frequency is chosen based on the experience that little road-induced damage is induced for frequencies above 50 Hz.

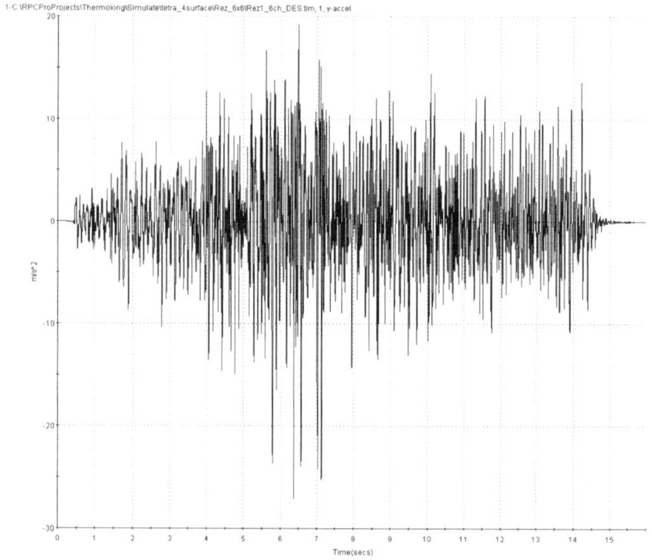

Figure 3. Representative Edited and Filtered Acceleration Time History

PHYSICAL TEST RIG

One commonly used test system for components that experience loads due to their inertia, such as the refrigeration unit, is a multi-axial simulation table, shown in Figure 4. This type of system allows the excitation of each of the six degrees of freedom (namely translation in x, y, z and rotation around these axes). Typically, the simulation range is up to 50 Hz.

Figure 4. Multi-axial simulation table

VIRTUAL TEST RIG MODEL

Functional Virtual Prototypes are based on three-dimensional component solid models and modal representations of component finite element models to accurately predict the operating performance of the product. Depending on the level of detail desired individual components will be modeled with a varying degree of complexity and resulting accuracy [9].

The multi body dynamics model of the test system originated from the CAD model. Care was taken to replicate key features such as geometry, mass properties, and global stiffness properties. All of the appropriate communicators were set up so that the test rig model would couple directly to a multi body dynamics model of the specimen.

For this initial study, the test stand model was kept as simple as possible. All of the test stand components were assumed to be rigid bodies. While this is a simplification of the existing physical test rig it was believed that all relevant static and dynamic properties required for a comparison of responses from analytical and physical prototypes were retained. In most test cases, this can be assumed, in that the Eigen-frequencies of the fixture are typically above 50 Hz. Model development always involves trade-offs between model accuracy and computational efficiency. The development of a validated model is elusive at best. A model is only valid over a specific range of applications.

A multi body simulation model of the test rig, fixture and the refrigeration unit was built (Figure 5). The fixture and part of the refrigeration unit was modeled by beam elements to consider the flexibility of the system. The refrigeration unit was connected to the fixture through seven fixed joints at the actual frame mounting locations. Requests were set up to output the acceleration information at the accelerometer mounting locations. The power train assembly was modeled as a rigid body. The three engine mounts were modeled by bushing elements. The three snubbers, which restrict engine travel, were modeled by force elements.

In this study the virtual test process is predicated on data gathered from an existing production unit with similar structural and mass characteristics.

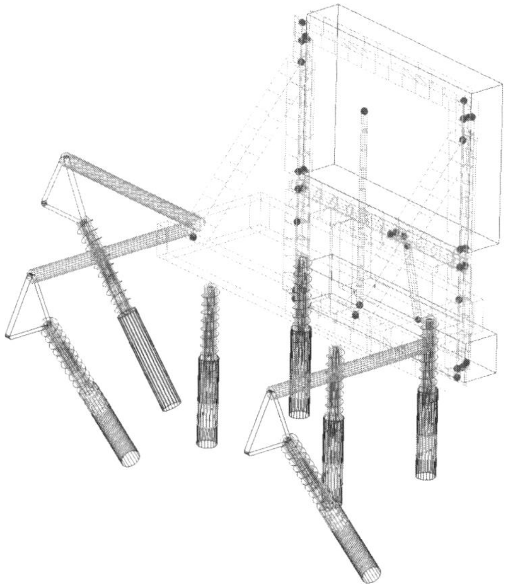

Figure 5. Multi Body Simulation Model of the Test Rig and Specimen

VIRTUAL TEST SERVER

For a complicated multi body simulation model, it usually takes a long time to obtain a simulation result. In this case, it took about 60 minutes for a 1 GHz, 1 GB RAM PC computer to obtain the solution for a sixteen second event. Therefore, the iteration process with a virtual system could be quite lengthy especially when the system is quite nonlinear and many iterations are needed to minimize the error.

Software has been developed that allows for seamless communication between the road load test rig simulation software and the multi-body simulation software. This allows for the exchange of files and multiple iterations to be conducted automatically. This facilitates batch type execution of the iteration process without requiring user interaction and thereby saving time.

DURABILITY TEST

PHYSICAL TEST

The drive files are nested together in a sequence/block that reproduces a lap around the proving ground. The block is repeated until a target goal has been met. In this test, a block sequence was defined that contained five different road surfaces. The physical durability test then repeated this block sequence. At 10,070 repeats a failure was observed at a snubber mounting location (Figure 6). In addition a bracket and bolt failure were observed. It is assumed that these failures were independent of each other.

Figure 6. Frame damage at the snubber mounting location

VIRTUAL TEST

Unlike the physical system, the virtual system is perfectly repeatable and therefore a single repetition and average building was sufficient to estimate the frequency response function (Figure7).

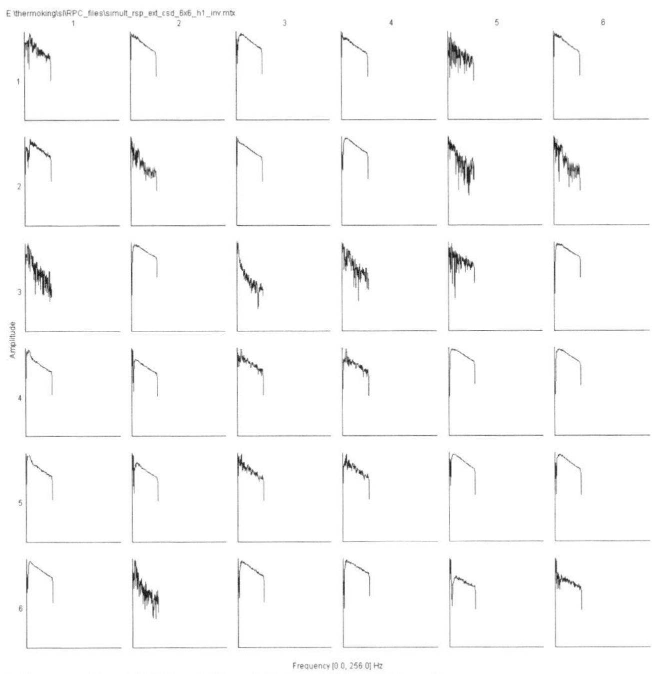

Figure 7. FRF of the Virtual Test System

Like for the physical system, iterative improvements to the drive signals were performed to reproduce the desired response signals. In this application, two iterations were conducted for each road surface. After two iterations, the

RMS errors for the six acceleration channels were all below 15% for all events.

Figure 8 shows the desired vs. achieved time history of one such event. This demonstrates that a virtual test system in conjunction with iterative road load simulation can reproduce the road measured acceleration signals accurately.

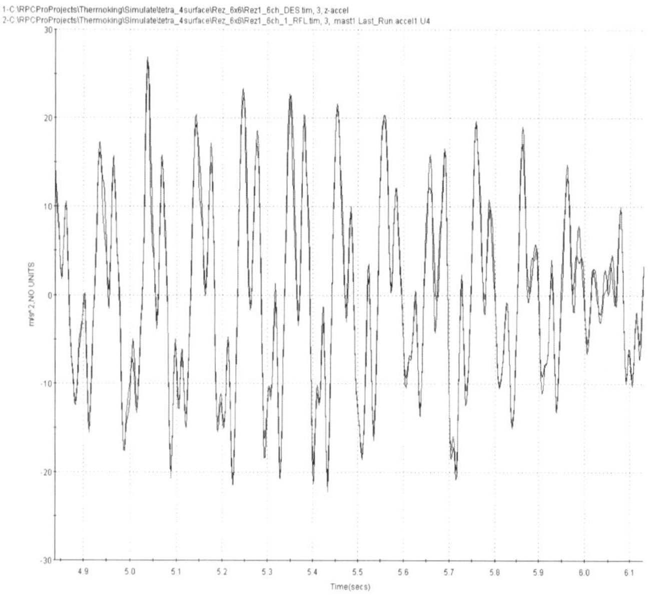

Figure 8. Desired and Achieved Time History after Iterations

LOAD OUTPUT

After the drive files that could reproduce the measured acceleration signals were developed, they were played out through the virtual model. The loads transferred through each frame mounting location, each engine mount location, and each snubber location were recorded in the for the subsequent fatigue analysis. At each frame mounting location and each engine mount location, six load channels of data were exported (three forces and three moments). At each snubber mounting location, there is one force output channel. Therefore, a total of sixty-three channels of loads were exported from the virtual prototype model for each road event. The load time history files can also be used as input for a dynamic stress analysis.

STRESS ANALYSIS

To calculate the stress distribution, an Finite Element Analysis (FEA) model was built. A total of sixty-three load cases were considered. These consisted of seven mounting locations, three engine mount locations, and three engine snubber locations with a total of sixty-three load components. For each case a unit load was applied. Inertial relief calculations were conducted for each load case to obtain the stress distribution. In the inertial relief calculation, the FEA software finds an acceleration field to balance the applied load and then finds a static solution to obtain the stress distribution due to the applied load and the balancing acceleration field. An inertial relief calculation was chosen because it does not require any boundary conditions to be applied.

After solving the sixty-three load cases, the result was exported for fatigue analysis.

FATIGUE ANALYSIS

The sixty-three load cases of stress information from the FEA model and the sixty-three load time histories from the virtual prototype model were imported into the fatigue analysis software [10]. Based upon the schedule of the accelerated test, a static strain life calculation was conducted to estimate the life of the structure. The static fatigue analysis was chosen because the modal analysis shows that the lowest frame natural frequency is above the frequency range interested. Smith-Watson-Topper mean stress correction and Neuber elastic-plastic correction was used.

After fatigue calculation, the result file was output in FEA model, where the life of the refrigeration unit was plotted. In this way, the life estimation of the refrigeration frame can be observed at all locations.

A physical durability test takes at least several weeks even for the accelerated test. However, the virtual durability test is only a fatigue analysis, which may only take several hours. The time saving fatigue analysis is a big advantage of the virtual test.

VIRTUAL VS. PHYSICAL TEST RESULT COMPARISON

Virtual testing predicts that the snubber mounting locations would fail first which matches the lab test observation. Figure 9 shows the virtual life prediction plot that corresponds to the failure location in Figure 6.

From the fatigue life plot, one can see the fatigue life of the virtual test = $10^{4.243}$ = 17,500 repeats. In the lab test, failures were observed at 10,070 repeats. The ratio between virtual and physical fatigue life is 1.7, which can be considered a close match.

The bracket and the bolt failures that occurred in the lab test were not predicted, as these parts were not modeled due to their complexity. These failures may have been caused by the snubber failure. This reinforces the continuing need for physical testing as not every aspect can be efficiently modeled with current technologies.

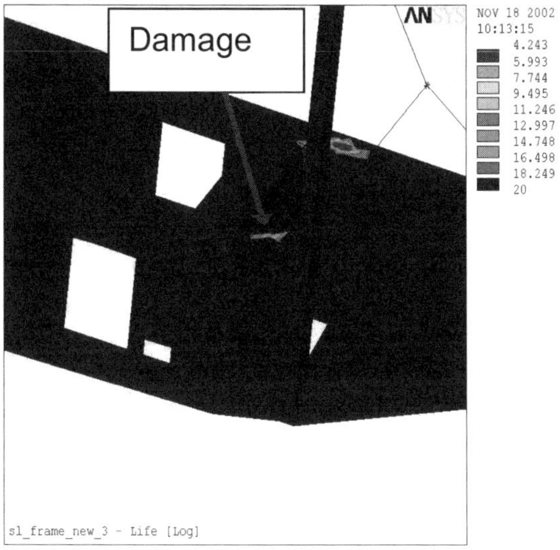

Figure 9. Fatigue life prediction at the snubber mounting location, same location as failure in Figure 5

SOUND QUALITY ANALYSIS OUTLOOK

A range of additional predictive analysis can be performed within the virtual prototype environment once the level of correlation to a physical prototype is shown. One such study that the authors are contemplating is the prediction, and comparison to measured results, of sound power emitted from the structure. The pre-requisite of predicting structural response accurately with respect to phase and amplitude has been met as can be seen from the close fatigue life prediction and waveform agreement as shown in Figure 8. The validated finite element model can provide input information for an acoustic analysis model. Sound power emission of the evaporator unit can then be predicted by the acoustic analysis. From the actual sound power a host of subsequent analysis can be performed with respect to measured and perceived sound characteristics. After the acoustic model is validated, what if type of analysis can be conducted to evaluate options to insulate the system.

CONCLUSION

By conducting a virtual and physical test of a truck trailer refrigeration unit, we have demonstrated an integrated approach to evaluate a vehicle component design. This approach involves both physical testing and virtual testing. Simulation of road load data using iterative road load simulation software was used for both the physical and virtual tests. The virtual test can help to evaluate the design at the very early design stage.

Due to the inherent complexities (especially when flexible bodies are considered) and the fact that not all dynamic effects and details of the design will be modeled accurately, uncertainties arise. Therefore, it is highly recommended to validate the models before signing off on a design for production. The optimal method for validating an analytical prediction is to perform a physical and virtual test that correspond to one another and reflect realistic operating conditions. Virtual testing as presented in this paper is intended as a design development tool and not as a substitution for final acceptance testing of a design.

The virtual test can provide information to design the test setup and conduct better tests. The virtual and physical integrated approach can accelerate the vehicle design process significantly.

ACKNOWLEDGEMENTS

The authors are grateful for the assistance given by Phil Berling, Eric Little, Bob Pope and Jake Rawn at MTS Systems Corporation and Jerry Brownfield, David Dykes, Robert Lattin and Steve Gleason at Thermo King Corporation.

REFERENCES

1. MSC.Software Corporation, MSC.ADAMS User Guide, 2002.
2. MTS Systems Corporation, RPC Pro User Manual, 2002.
3. Grote, P., Grenier, G. "Taking the Test Track to the Lab", *Automotive Engineering*, June 1987, Volume 95, Number 6.
4. Grote, P. Fash, J. W., "Integration of Structural Dynamics Testing and Durability Analysis," *Sound and Vibration*, April 1987.
5. Fash, J. W., Goode, J. G., Brown, R. G., "Advanced Simulation Testing Capabilities", SAE 921066, 1992.
6. Soderling, S., Sharp, M., Leser, C., "On Servocontroller Compensation methods," SAE1999-01-3000, 1999.
7. B.W. Cryer, P.E. Nawrocki and R.A. Lund: "A Road Simulation System for Heavy Duty Vehicles", SAE Paper 760361, Automotive Engineering Congress and Exposition, February 1976.
8. Englerth, M. Dutton, D. Grenier, G., Leese, G., "The Use of Fatigue Sensitive Critical Location in Correlation of Vehicle Simulation and In-Service Environments," SAE Technical Paper, 880807, 1988.
9. Dittmann, K. J., Albright, F. J., Leser, C., " Validation of Virtual Prototypes via a Virtual Test Laboratory," 1st European MSC.ADAMS User Conference, 13-14 November, 2002, London, England.
10. nCode International, FE Fatigue User Manual, 2002.

2003-01-1217

Simulation Based Reliability Assessment of Repairable Systems

Animesh Dey, Robert Tryon and Loren Nasser
VEXTEC Corporation

ABSTRACT

This paper presents a simulation software allowing the automotive designer to predict repairable system reliability during vehicle concept development. The software uses Monte Carlo simulation to virtually test many systems. Reliability is estimated using a Non-Homogeneous Poisson Process (NHPP). The software program computes rate of occurrence of failure (ROCOF), incidents per 100 (I/100), reliability and the trip reliability. These parameters are calculated either considering all the test samples as a single ensemble or allowing the user to group the systems into smaller test sample sizes.

INTRODUCTION

Historically, reliability prediction has been based on either data from vehicles of the past or data acquired through prototype testing, which is a slow and expensive process. For example, a reliability test of just one automotive system could take up to 6 months and cost upwards of $200,000. Management has to make a cost/benefit decision as to how much product testing is required prior to launch of a new vehicle model. Most automotive companies do not address vehicle or system reliability as a structured component during the development phase. The lack of such an approach typically leads to a time lag of about 12 to 15 months after vehicle production, before sufficient customer warranty data is accumulated and remedial design changes are undertaken.

This paper describes a reliability assessment technique that can be integrated in the development process of large repairable systems. Recognizing that vehicle warranty incidents are neither independent nor identically distributed, a Non-Homogeneous Poisson Process (NHPP) was used to predict reliability. A MS-Windows computer program called SimSARRS-System was developed to simulate the warranty incidents and predict the system reliability.

RELIABILITY ASSESMENT OF REPAIRABLE SYSTEMS

For repairable systems, failures are not necessarily independent or identically distributed. For example, if the system under test is a charging and cranking system, consisting of a starter-motor, battery, alternator, ignition switch and wiring, failure of an alternator could result in a battery failure. Failure of the ignition switch could result in a burned starter-motor. Also since the system under test consists of electrical, mechanical and chemical components, the failure times (or failure mileages) would not fit one single distribution. Therefore, a stochastic process is necessary to predict system reliability. For the present study, a Non-Homogeneous Poisson Process (NHPP) was considered for reliability estimation. A complete discussion on NHPP is provided in Lu and Rudy (1994).

In a NHPP model, the quantity of interest is the intensity function also defined as the Rate of Occurrence of Failure (ROCOF). The intensity function (or ROCOF) in a NHPP is not constant and varies with time (or mileage). Mathematically, ROCOF is computed as:

$$\text{ROCOF} = \lambda\beta t^{(\beta-1)}$$

where, λ and β are parameters that are estimated from actual or simulated tests and t is the failure time (or failure mileage).

When β > 1, ROCOF is monotonically increasing and successive mileages between failures will decrease stochastically. This represents typical wear-out of a system in a vehicle. When β < 1, ROCOF is monotonically decreasing and the successive mileages between failures will increase stochastically. This represents a debugging situation in a system.

The parameters λ and β of ROCOF can be computed as follows (Crow, 1974).

$$\lambda = \frac{\sum_{i=1}^{k} N_i}{\sum_{i=1}^{k} (T_i^\beta - S_i^\beta)}$$

$$\beta = \frac{\sum_{i=1}^{k} N_i}{\lambda \sum_{i=1}^{k} \left(T_i^\beta \ln T_i - S_i^\beta \ln S_i \right) - \sum_{i=1}^{k} \sum_{j=1}^{N_i} \ln x_{ij}}$$

where,

k = number of systems being tested

N_i = number of incidents experienced by the i-th system

T_i = ending test mileage of the i-th vehicle

S_i = starting test mileage of the i-th vehicle

x_{ij} = mileage of the i-th vehicle at the j-th occurrence of an incident.

Having computed λ and β, other parameters of interest such as I/100, reliability and trip reliability can also be estimated (Lu and Rudy, 1994).

SIMULATION PROCESS

To implement the above mentioned reliability estimation theory a Monte Carlo simulation technique was developed. The simulation process requires the following input.

(a) The number of systems under test (number of simulations)

(b) The mileage until which each system is to be tested (target mileage)

(c) The number of components in each system

(d) Statistical description of the incident history of each component in the system.

(e) Group number of each component (explained in the next section).

Based on the above input, the incident miles for each component are generated using the inverse-CDF method (Dey and Tryon, 1998) and tabulated. This simulation process is repeated until the incident mile of any component exceeds or equals the target mile. In the event a component experiences a failure before the target mile, it is assumed that the component is replaced with a new component and the simulation proceeds. At the completion of the simulation process, λ and β are computed using the formulae mentioned above. Having determined λ and β, reliability, trip reliability and I/100 can be estimated.

SIMULATION SOFTWARE

A MS windows based program called Sim SARRS-System was developed to perform the simulation for system reliability analysis. The program is divided into three major units – (a) Preprocessor, (b) Simulation engine, which is described in the previous section, and (c) Post processor

PREPROCESSOR

The Sim SARRS-System preprocessing unit collects the data required for executing the simulation process. Figure 1 shows the preprocessing unit input screen. User input requirements include the total number of simulations, the target mileage, a name for the system under test and the number of components in the system. Additionally, the user has to input a complete statistical

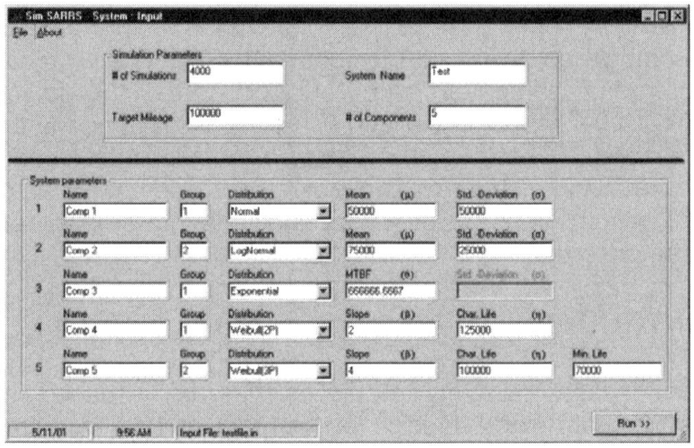

Figure 1: Program input screen

description for each component in the system. The software provides a choice of four statistical distributions – Normal (or Gaussian), LogNormal, Exponential and Weibull. The required input statistical parameters depend on the selection of the distribution-type.

Besides the staististical failure distribution of each component, the user can also enter a group number for selected components. The group number box for each component is left blank by default. No input for each of these boxes indicates the failure of each component is independent of all other components in the system. However, two or more components sharing the same group number, indicated by filling in the group number box, means that if any component in the group experiences a failure event and is therefore replaced, all of the other components in that group are also replaced regardless of whether they failed. Such situations are encountered in vehicle design quite commonly. For example, typically, when the sun-roof window in a vehicle experiences problems, the dealer replaces the whole assembly instead of just replacing the faulty part.

The preprocessor also allows the user to save complete input description for a system, which can be used to execute the program later.

POSTPROCESSOR

The postprocessing unit displays the results of the simulation graphically and in a tabular form. The software provides three output choices – (a) System, (b) Test Sample Size, and (c) Pareto.

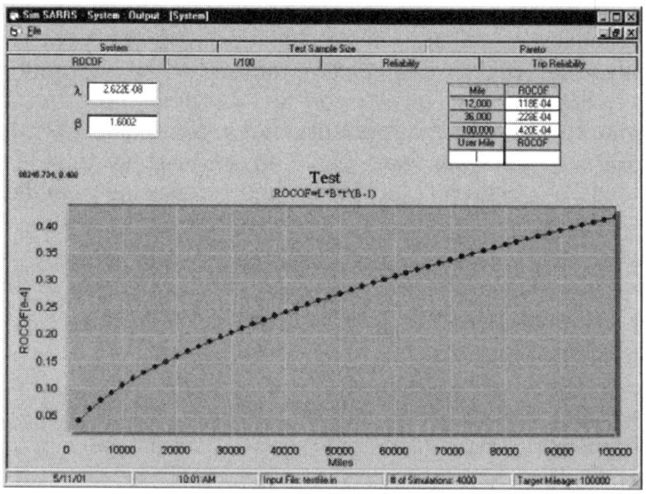

Figure 2: Program system output

System

A typical system output is shown in Figure 2. System analysis output presents values of λ and β for the system as well as plots of ROCOF, I/100, reliability and trip-reliability, from zero until the target mileage. For each plot, the values of the corresponding parameter are also tabulated for 12000, 36000 and at target mileage. An option has also been included for entering a different mileage to determine the exact value of the parameter being plotted. Conversely the user can input values for ROCOF, I/100, reliability or trip-reliability and determine the mileage at which those values are obtained.

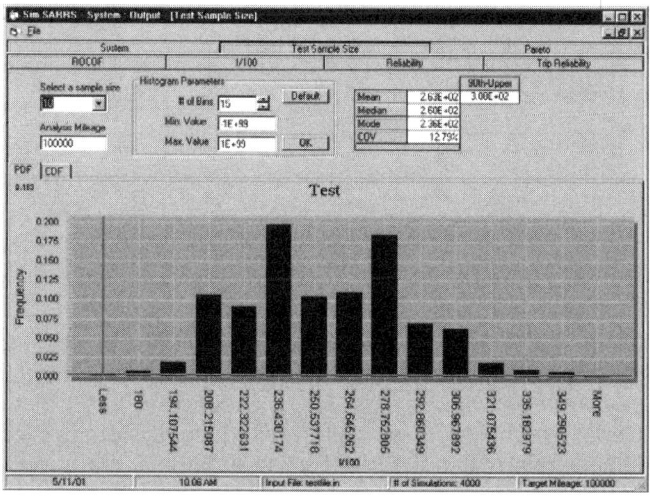

Figure 3: Histogram plot of I/100

Test Sample Size

In most practical automotive test situations, it would be unusual that thousands of systems are tested at one time. This would be an expensive as well as time-consuming proposition. In most cases a sample of 5-6 systems are tested to estimate the functions parameters such as λ, β, I/100 etc. To simulate such a scenario the "test sample size" post processing option was integrated into the program.

This option allows the user to group the total number of systems into smaller test sample sizes. For example, 4000 systems under simulated test can be grouped into 400 groups of 10 systems each. The function parameters and the associated reliability measures are found for each group to produce 400 sets of output. The output is then plotted either as a histogram chart (Figure 3) or as a cumulative distribution function (CDF) plot (Figure 4). The variation of these parameters about their respective mean values are also determined and tabulated as shown in Figure 4.

The test sample size option is useful in determining the expected differences between actual performance of the vehicle fleet compared with the performance inferred from the testing of a limited number of systems. For example, the histogram chart in Figure 3 indicates that for the system under consideration, if a test of 10 systems were performed to 100,000 miles, it would be reasonable for the I/100 to range from 180 to 350.

The test sample size option is also useful in test plan preparation. By changing the sample size, the engineer can determine the expected benefit in the confidence of the test results expected versus the cost of adding another system to the test program.

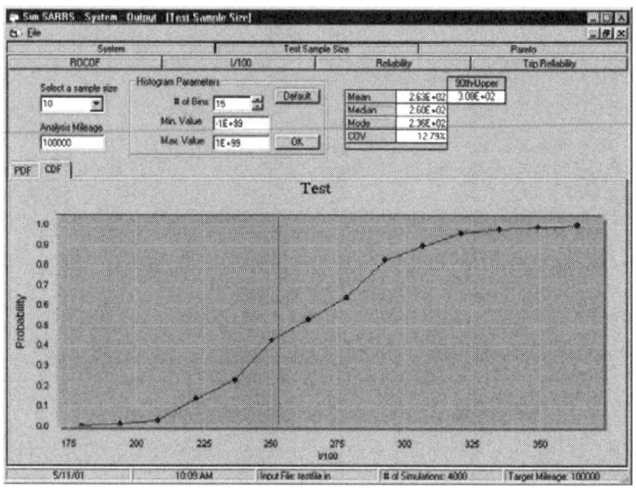

Figure 4: CDF plot of I/100

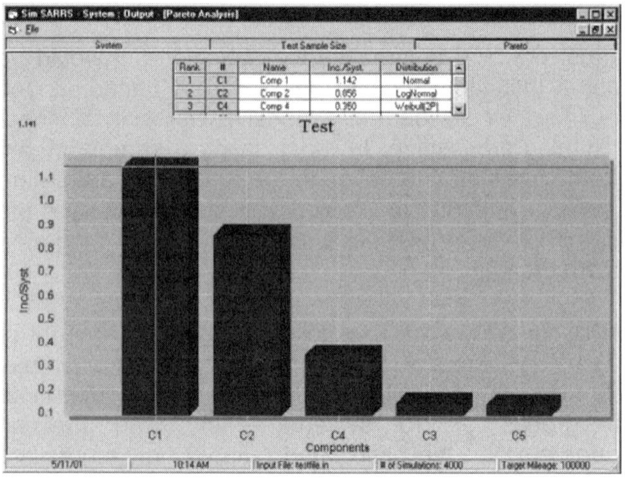

Figure 5: Pareto analysis output

Pareto Analysis

Pareto analysis determines which components in the system experience most warranty incidents. This option plots as well as tabulates the component incidences in descending order (Figure 5). The values tabulated and plotted are normalized incidences, i.e., incidents divided by the number of simulations. If the total number of components exceeds 15, the chart only plots the top 15 components with the highest number of incidents.

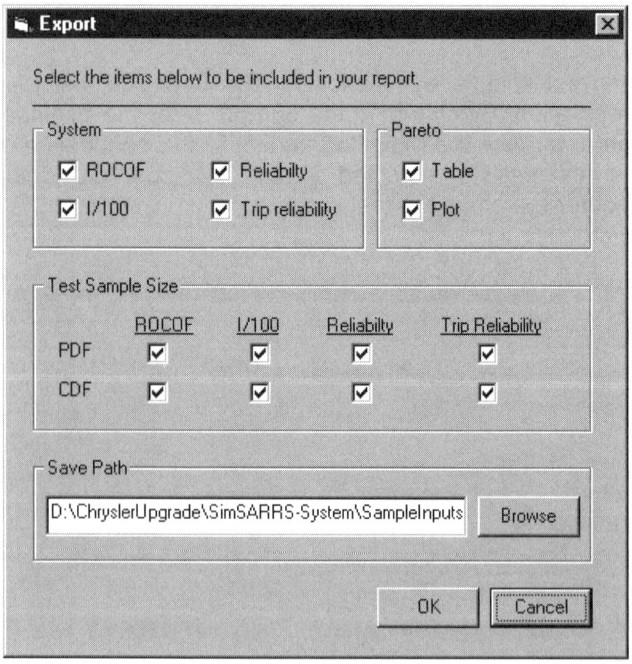

Figure 6: Report export dialog box

Report Writer

Along with displaying the results of the analysis with ample plots and tables the program also provides the user with the convenient feature of exporting and saving all the relevant details of a particular analysis into a MS Word document. The document provides a complete snapshot of the analysis input and the results and can then be shared and emailed to various people within a department or across multiple departments.

Figure 6 shows how the user has the flexibility of exporting specific plots and tables into a report. By default, all plots and tables are exported into the Word document.

SOFTWARE PREDICTION RESULTS

Sim SARRS – System has proven to be an effective reliability prediction tool. DaimlerChrysler is in the process of validating the software by comparing predictions with actual vehicle warranty data. As an example of prediction effectiveness, an engine from the 1997 model year was input as the test system. The engine system included the following thirteen components.

- Oil pan
- Cylinder head
- Valve tappet
- Crankcase vent
- Crankshaft
- Timing chain
- Engine assy/short
- Intake manifold
- Oil filter
- Valve
- Exhaust manifold
- Engine support
- Piston

The actual incidents per 100 vehicles experienced through historic warranty records compared closely with software predictions as shown in Table 1.

Table 1: Predicted results versus actual data

Incidents per 100 vehicles	Sim SARRS Predicted	Historic Warranty Records
12,000 miles	7.51	7.86
36,000 miles	24.2	27.39
100,000 miles	71.9	N/A

CONCLUSION

A simulation software allowing the automotive designer to predict system warranty incidents during product development was presented. The software uses Monte Carlo simulation to virtually test many systems. Since this software can be readily integrated in the product development cycle, it helps reduce design time and design cost without having to wait for accumulation of sufficient warranty data.

The software can also be used to design optimum test plans by considering the trade-off between adding another system in the test sample versus the confidence (or the variation) in the simulated test results. The software is easy to use and allows the user the option of displaying the results both graphically as well as in a tabular format.

REFERENCES

[1] Crow, L. H., 1974, "Reliability Analysis for Complex Repairable Systems", *Reliability and Biometry, Statistical Analysis of Lifelength (Ed. F. Proschan and R. J. Serfling)*, Society for Industrial and Applied Mathematics, pp. 379-410.

[2] Dey, A. and Tryon, R. G. 1998. "QuickFORM User Manual", VEXTEC Corp., Brentwood, TN.

[3] Lu, M. and Rudy, R. J. 1994. "Vehicle or System Reliability Assessment Procedure". In *Proceedings of the IASTED International Conference on Reliability Engineering and its Applications*, (Honolulu, Hawaii), 37-42.

CONTACT

Dr. Animesh Dey is Vice-President of VEXTEC responsible for development of software and probabilistic-based prediction methods for automotive, aerospace and other industrial clients Dr. Dey received his Ph.D. from Vanderbilt University. His email address is adey@vextec.com

2003-01-0124

ACE Driving Simulator and Its Applications to Evaluate Driver Interfaces

Vivek Bhise
University of Michigan- Dearborn

Edzko Smid
Oakland University

James Dowd
Collins & Aikman

ABSTRACT

A fixed base driving simulator called the VVDS (Virtual Vehicle Driving Simulator), its operating procedure and software system have been developed by a team of automotive suppliers (called ACE -- Advanced Cockpit Enabler) for quick evaluations of early working prototypes of driver interfaces. The system is designed to provide quick feedback to the product designers in early concept generation and validation phases of new automotive HMI architecture strategies and interfaces of various in-vehicle devices. The simulator consists of a reconfigurable cab with quick-change attachments to mount various controls and displays in package positions. A number of drivers are asked to drive the simulator and perform a number of tasks when prompted by pre-recorded voice commands. The entire data collection and data analysis procedure is developed such that new experiments can be configured, implemented and analyzed quickly and with the least amount of a human analyst's involvement. The system generates reports showing graphs of driver behavior, performance measures (e.g. driver inputs and outputs as functions of time, number of glances, total visual time, lane position standard deviation, velocity standard deviation, etc.) and subjective impressions of drivers (e.g. ratings on work load, control operating feel, surface tactile feel of control surfaces, etc.) for different tasks associated in operating/using various in-vehicle controls and displays.

The paper presents the description of the simulator system, illustrations on its evaluation procedure and results of two experiments conducted to evaluate different radio designs by involving different radio and non-radio tasks (such as increase/decrease radio volume, adjust base and treble, select/seek/tune a given radio station, change CD track, eject a CD and insert a new CD, answer a cell phone, dial a phone number, etc.).

PURPOSE

To provide a quick method to measure usability and customer feedback on alternate driver interface concepts.

INTRODUCTION

With the rapid advances in information technologies, a number of safety, communication and comfort-convenience features are being developed by various inventors for incorporation in future vehicles. Incorporation and operation of these new features involves designing interfaces, i.e. controls and displays, in the already crowded cockpits. Minimizing complexity of these interfaces is critical to assure that the driver is not overloaded. The complexity of many of the driver interfaces has increased to the point that it is simply not possible for a design team to develop the interfaces through the team interactions and "gut feel". A tool is needed to evaluate various driver interface proposals. The VVDS was developed for this purpose. It allows testing of early working prototypes of different driver interfaces in various simulated driving situations and provides automated data collection for quick feedback to the designers.

Figure 1. Aluminum Framed Cab

The ACE project, created and managed under the leadership of Collins & Aikman, include the following three basic teams: a mechanical team, an electrical team and an HMI (Human Machine Interface) team. Each team involves cross-functional co-located engineers and designers from a group of seven automotive interior suppliers, an auto manufacturer, and two universities. The teams develop different proposals for developing improved and cost efficient cockpit systems and their related driver interfaces. Working prototypes of promising designs are made and refined with the data obtained from testing in the VVDS. More detailed description of the ACE HMI process is available in *(1)*.

Use of a simulator for quick evaluations has a number of advantages. Some key advantages of using the simulator are:

- Precise repetition of driving test situation so that behavior of many drivers can be observed and recorded.
- Safety- The elimination of possibility of accidents and injuries to the test subjects.
- Quick feedback to the designers and engineers
- Generally less costly than testing on an actual highway

DESCRIPTION OF THE DRIVING SIMULATOR

The simulator consists of: a) an aluminum framed vehicle buck with front seats, a fully adjustable steering column with steering wheel and stalk controls, and an aluminum channel framework to mount instrument panel, door and console mounted controls, b) a large 3.5 m wide and 3 m high screen mounted at 3 m distance from the driver's eye, c) an overhead mounted video projector for projection of the driving scene, d) driving scene generator SGI computer with the driving simulation software, e) three PCs to issue voice commands to the driver and collect data on lane position, velocity, hands off the wheel time, and f) a synchronized video camera to record driver's eye glances. (See Figures 1, 2 and 3).

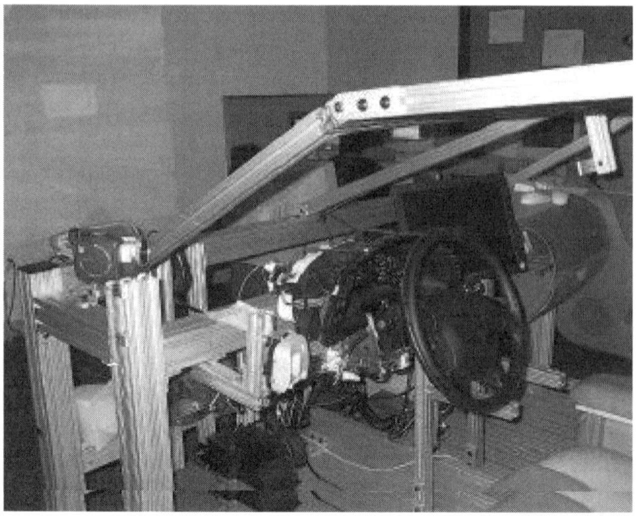

Figure 1. Aluminum Framed Cab

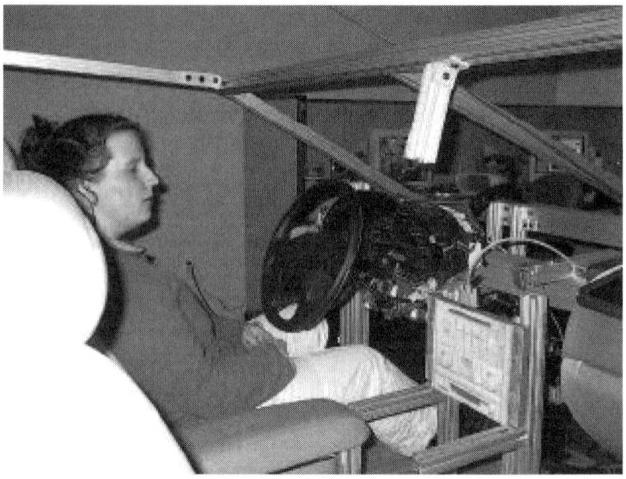

Figure 2. Driver in the Cab with a Prototyped Radio and Climate Control Unit

Figure 3A. Driver's Visual Screen (Rural undivided highway)

Figure 3B. Driver's Visual Screen (Urban Streets)

Figure 3B. Driver's Visual Screen (Urban Streets)

The buck is made to resemble the interior of an actual vehicle. All interior components can be easily relocated for optimization. This setup enables us to optimize a cockpit at a much lesser expense compared to testing on an actual vehicle.

The driving simulator uses an SGI server and a number of Windows based PC's to run applications and collect data. The SGI server simulates the interactive driving environment based on the Virtual Vehicle Simulating Software (VVSS) originally developed in Oakland University for suspension testing *(2, 3, 4).* The MATLAB and Simulink software provides data collection and processing capabilities.

Hardware Setup

The VVSS environment consists of a Silicon Graphics high performance computer for rendering the virtual scenery (see Figure 4). This host station provides network services for 3 PC workstations on the Local Area Network (LAN) to interact on-line with the simulation execution.

The human-in-the-loop capability is provided by PC-1, which is configured to capture the driver inputs for steering, throttle and brake. These driver inputs are sent to the simulation execution through the LAN using Matlab/Simulink design interface. PC-1 also monitors the lane deviation from the road median, and generates engine sound corresponding to the speed of the virtual vehicle.

The vehicle location is monitored by PC-2. When the driver reaches certain pre-programmed locations in the scenery, PC-2 will issue verbal task commands through the audio system.

Driver information is provided through a conventional automotive cluster, which is controlled by PC-3. This workstation also maintains a truth signal for hands-on-the-wheel and eyes-on-the-road. PC-3 is equipped with a multi-channel analog and digital input/output card.
A fourth workstation is used to record digital video data on disk of the driver subject. This information is later used to confirm hands-on-the-wheel and eyes-on-the-road information.

Unique about the VVSS is that it utilizes custom automotive commercial hardware that is provided by Collins & Aikman. The VVSS can be easily adapted to any other hardware for driver inputs.

Software

All the data from PC's 1-3 is logged and stored in one Matlab binary file. Matlab is used to conduct primary analysis on the data by using a script file that generates an HTML document that includes graphic plots and charts. Together with the observation sheets and the rating scales, this information then provides the basis for the higher-level driver performance evaluation and HMI quality assessment.
The configuration of the VVSS is displayed in Figure 4. The mathematical dynamics model of the vehicle contains 24 degrees of freedom and has over 90 states.
A special feature of the simulation environment is the multi-access modularity through the LAN.

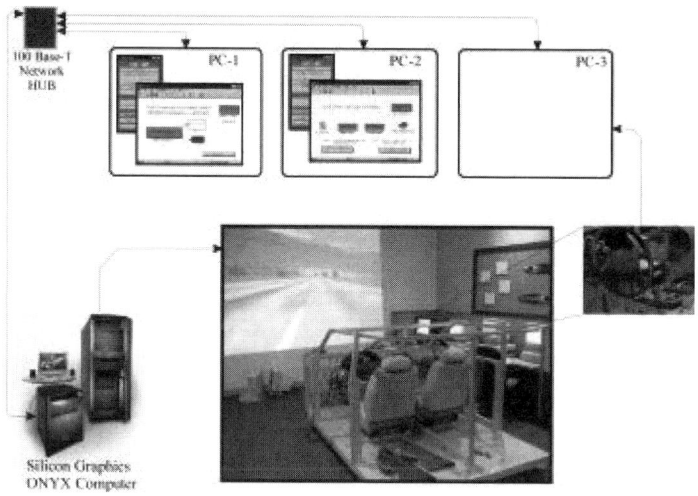

Figure 4. Hardware Configuration

Access to the state values of the mathematical model is implemented through a directional vector table. This allows for external host stations to replace the states of the system by virtually replacing part of the mathematical model. For example, the suspension height of the left front wheel is represented as a state in the vehicle dynamics model. An external model for suspension can be activated, that will replace the model formulation of the left front suspension height and insert a run-time value, derived from the other states and inputs. Elsewhere in the vehicle model where left front suspension height is used to derive other states, this new value from the external model will be used.
Likewise, also the driver input model, vehicle behavior and additional scenery characteristics are accessible for monitoring and altering the driving simulator models. This feature of the VVSS is exploited for the HMI evaluation process described in this paper.

Test Scenarios

In order to develop test scenarios, there is a dedicated scenario manager. The scenario manager stores the route, positions and tasks, and the association for each of the experiment laps in a single file. Each test consists of 6 trips (or laps)—first 2 practice trips followed with the next 4 trips for evaluations runs.

Each trip consists of a course route through the virtual scenery of approximately 20 minutes. The trajectory contains an average number of turns, stop signs, traffic lights, intersections and so on. Associated with the course are a series of assigned locations in the trajectory, which are used to trigger tasks. These locations are stored in the scenario manager. In addition to the course, there are an independent number of task definitions defined in the scenario manager. The task definitions consist of a name and a sound file, which contains the task description. The scenario manager contains an association matrix that defines which task is to be executed at which location for which experiment.

Test Procedure

A standardized test procedure is incorporated to minimize time required to set-up the simulator to conduct evaluation of driver interfaces. The standardized test procedure involves:

1. Pre-Test Procedure
   - Each subject is given basic information on the test.
   - The subject signs a consent form.
   - The subject is asked to fill-out a demographic information form.
   - The subject is given a reaction time test to determine his/her information processing rate in a choice decision situation, vision tests, and his/her relevant anthropometric characteristics (e.g. stature, weight) are measured.

2. Familiarization:
   - The experimenter shows all controls (primary and secondary controls and displays (including feedback LEDs, sounds, etc.), HVAC, Radio, etc.) to the subject.
   - The subject is asked to adjust seat, steering wheel and pedals to his/her preferred settings.
   - The experimenter checks to assure that hands-off-the-wheel measurements work properly.
   - The experimenter makes sure that the "engine sound" is turned on.
   - The experimenter adjusts lighting to avoid any discomforting/disturbing glare from internal or external light sources.
   - Each subject is then asked to drive the simulator for at least 5 minutes before any data collection commences (subjects unfamiliar with the simulator are given longer time until they feel comfortable with the simulator driving).

3. Driving Tests and Data Collection
   - The experimenter provides more specific driving instructions about performing selected tasks involving usage of certain interior controls/displays.

Sample Driver Instructions:

This study will investigate driver performance in operating the driving simulator while performing secondary tasks. You will be asked to drive the simulator through several routes. During these routes we will give you verbal instructions for you to operate different controls in the vehicle.

Keep in mind that your primary task is to keep the vehicle in control and maintain the proper speed and lane position. Your performance on the primary task will be measured. Drive in the left lane (closest to the yellow line) at all times, even when making turns. The recommended speed is 35 to 40 mph. You should attempt to maintain this speed as closely as possible at all times during the drive. You should also maintain your lane position so that the vehicle does not cross the lane boundaries. Just as you drive a car, you should try to minimize the vehicles' lateral -- i.e. left/right movement, and maintain a stable position.

It is important that you always keep your hands in the location that the experimenter tells you on the steering wheel, except when you will need to reach and operate controls to perform the instructed tasks. There are wires on the steering wheel. They are not energized and you cannot receive a shock through them.

The test consists of six laps. The first two laps are practice laps; the other four are experimental laps. Each lap takes 15-20 minutes and the whole test is about 2 hours. You may take breaks between the laps and if at any time you have questions or wish to stop participating in this experiment for any reason, please feel free to notify the experimenter.

   - Pre-recorded instructions are played to enable the subject to practice performing each task. This is conducted in a random order and in a static condition to assure that they are familiar with the driver interface.

   - Each subject is then asked to drive six laps on a pre-selected driving course. During each lap the subjects is asked to perform 8 different tasks. When the subject arrives at each pre-selected location, verbal instructions from pre-recorded audio files are automatically played through a computer's sound system.

   - The experimenter observes the subject and fills out a pre-developed form to record subject's behavior, control operation errors and ratings for each task.

A video camera records the facial behavior of the driver during the complete test. This video data are used to measure eye-movement and eyes-on-the-road information (see Figure 5).

During each lap, the 3 PC's create three data files, each containing the time data and associated log of the driving behavior and vehicle outputs. The data files contain the local time of the PC during the simulation, the synchronized reference time of the simulation environment and the application data associated with each PC. The X-Y position of the vehicle is collected from the driving simulator. The throttle, brake and steering angle are captured from the human driver interface, and the task number is generated from the scenario manager. The lane deviation is computed by comparing the current X-Y position of the vehicle with a pre-recorded center-lane trajectory.

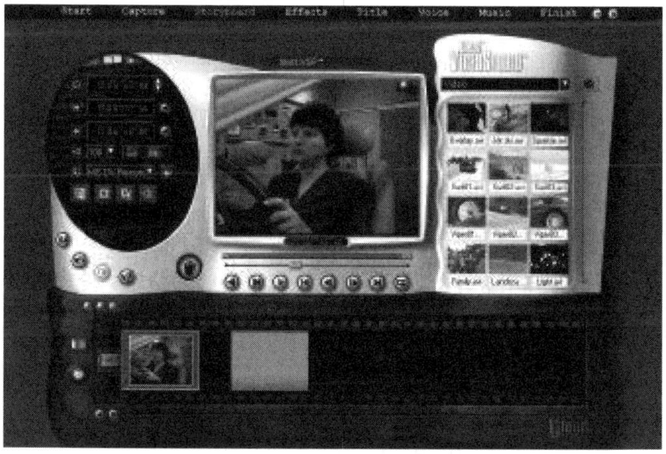

Figure 5. Illustrations of Video Data Collected to Obtain Driver Eye Glance Data

This provides a running difference between the actual and the desired lane position. The hands-on-the-wheel signal is generated from an electrical touch-contact on the steering wheel, and the eyes-on-the-road signal is generated by a push button from an external observer. The external observer is dedicated to push a button whenever the driver is not directly looking at the road in the scenery. The local time is automatically stored from Simulink along with the logged application data. The reference time from the VVSS is also considered as application data. Immediately after the experiment, the 3 data files are collected and consolidated in a special Matlab array format, and stored as the complete experimental data from the test.
In the next section, this data will be subjected to a report generation tool, which extracts the relevant driver performance information from the data in a useful format for further analysis.

Automatic Report Generation

Report generation is conducted by post-processing the collected data files in Matlab. A script is designed that will interpret and analyze the data files and generate 196 graphs and an associated HTML document. This document contains a table of contents, all the graphs and the support documentation for the data. It also contains the scenario definition, task definition and test subject information for the current test
The report generator provides graphs for lane deviation and vehicle velocity, both in actual form and standard deviation and averaged standard deviation.
The final stage of the report generation consists of the additional appendix with documentation on the current test scenario configuration and test subject data.
During this process of the report generation, an additional table is being generated with statistical values for lane deviation and vehicle velocity. Based on pre-defined threshold values for this data, a total count is accumulated per each task. This count provides an indication for the amount of driver distraction for the particular task. The total count of the test provides an indication of the quality of the prototype interface in terms of minimizing the driver distraction during task execution. More details on the report generation are provided in (5).

STUDY 1: DRIVING SIMULATOR STUDY

This study was conducted to get baseline data on a production radio and to compare the data with a new prototyped radio. Twelve drivers (8 males and 4 females, ages 16-48) participated in this study. Figure 6 shows a picture of the production radio (called Radio 1). Figure 7 shows a prototyped radio that is somewhat similar to the actual prototyped radio (called Radio 2) used for the actual evaluation. To maintain confidentiality the picture of the Radio 2 is not presented here.

Figure 6. Production Radio (Radio 1) Used for Evaluation

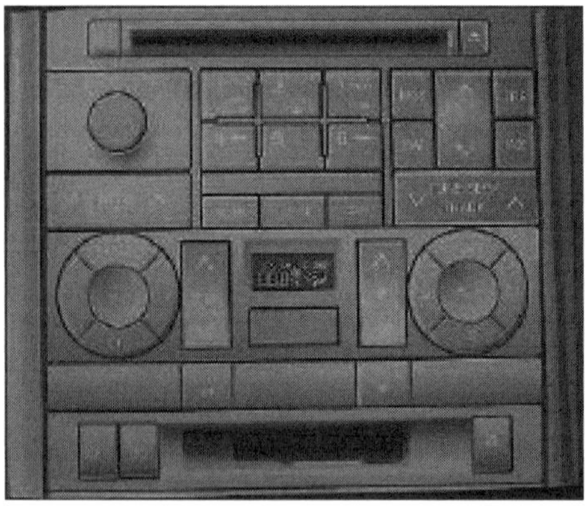

Figure 7. An Illustration of a Prototyped Interface of a Radio and Climate Control Unit

Procedure:

After familiarization with the vehicle controls, and a 20-minute simulator and test route familiarization, each driver drove six complete laps on the test course. A lap consisted of an eight-mile rural 2-lane road, which took about 15 minutes to complete. Sixteen locations were pre-selected on the test route. In any given lap, the test drivers were asked to perform eight of the sixteen randomly selected in-vehicle tasks. The instructions for each of the sixteen tasks (described below) were recorded in voice files, which were played when a subject arrived at each pre-selected location. The tasks were so arranged that during each successive two laps (i.e. laps 1 and 2, laps 3 and 4, and laps 5 and 6), all the sixteen tasks were presented. Thus, by the time the drivers performed the tasks in laps 5 and 6, they had already experienced in performing all the tasks (i.e. performing each task at least two times in the first 4 laps). All the 12 subjects performed exactly the same tasks, except half the subjects were tested with a production radio (referred as Radio 1) and other half used a prototype radio (referred as Radio 2) with similar functionality. Both the Radio 1 and Radio 2 had a CD player (located higher up), a rotary volume control and six preset stations. The following instructions were provided through pre-recorded audio files at the initiation of each task:

Task 1: Press FM (push button) and select preset 6 (push button).
Task 2: Press CD (push button), eject CD in the radio, and insert the "Billboard Top Hits" CD.
Task 3: Press FM, listen to the first three presets, and then select the music of your choice.
Task 4: Adjust the bass and treble to your liking.
Task 5: What is the answer to the following math problem: Five plus nine, minus six, times twenty-three?
Task 6: Press CD, seek track 4.
Task 7: Press FM, tune to 95.5.
Task 8: Turn volume up.
Task 9: Find the cell phone, and dial your home phone number backwards.
Task 10: Press FM, tune to 107.5.
Task 11: Turn the volume down.
Task 12: Press FM, tune to 93.1.
Task 13: Press FM, seek to 105.1.
Task 14: Press CD, seek to track 2.
Task 15: Press CD, eject the CD in the radio, and insert the "Paula Abdul" CD.
Task 16: Ring....Ring.....Ring (Answer the cell phone when it rings).

The cell phone used for Tasks 9 and 16 was kept on front passenger's seat. The phone did not have a flipping door to access the keys. These two cell phone tasks along with Task 5 (which required no visual involvement) were used as reference or control tasks.

Data Collection and Analyses:

The outputs of the simulator runs were recorded in computer files. The files consisted of time and distance based values of lane position, velocity, pedal positions and steering wheel angle. In addition, the driver's face was video taped using a digital video camera, which was synchronized with simulator data. The eye glance data were reduced manually from the digital video files. The experimenter also kept track of any control operational errors (e.g. long looks at the controls, pushed wrong button, etc.) and erratic vehicle maneuvers (e.g. lane deviations, slowed down, etc.). The data were processed to obtain values of the following performance measures for each task performed by each subject in each lap.

Performance Measures:

Number of glances: Total number of glances made away from the forward road scene to perform a given task of operating an in-vehicle device.

Total eye time: Total time spent in glances made away from the forward road scene to perform a given task of operating an in-vehicle device. (Sum of eye glances in seconds)

Total visual time: Total time elapsed between the beginning of the first glance and the end of the last glance made in performing a task (measured in seconds).

Lane position standard deviation: standard deviation of lane position (measured in meters) obtained from 90 lane position data samples (which included: 5 samples/sec x 6-second time interval x 3 6-second intervals) over an 18 seconds interval measured from the location at which the verbal instruction was given to perform a task.

Velocity standard deviation: standard deviation of velocity (measured in m/sec) obtained from 90 velocity data samples (which included: 5 samples/sec x 6-second time interval x 3 6-second intervals) over 18 seconds interval measured from the location at which the verbal instruction was given to perform a task.

Driving Performance Score: Number of erratic events in the 18-second interval (defined above). Erratic events were defined by: 1) when the lane position standard deviation in any of the first three 6-second intervals (following the verbal instructions) exceeded 0.6 m, or
2) when the velocity standard deviation in any of the first three 6-second intervals (following the verbal instructions) exceeded 1.8 m/sec. The values 0.6 m and 1.8 m/sec were determined by the authors to represent high likelihood of traffic hazards due to lane intrusions and sudden slow downs in traffic streams respectively.

Results:

Figures 8 and 9 provide summaries of eye glance based data obtained from the video recording of the drivers' faces during the 16 tasks. The data show that simple tasks such as using preset stations and volume up/down are performed in less than 2 eye glances and total visual time of about 2 secs. Whereas, complex visual tasks such as tuning to a desired station or adjusting sound base and treble took about 4 to 9 eye glances and about 9 to 18 secs of total visual time. The tasks of ejecting a CD, retrieving a new CD from a CD case and inserting it was very demanding taking about 4-5 glances and about 10-15 secs of total visual time.

Figures 10, 11, 12 and 13 show the effect of the radios, i.e. the differences due to the radios 1 and 2 on the four measures. The data plotted in these figures are from the last two laps (laps 5 and 6) where the subjects had the most familiarity with the tasks.

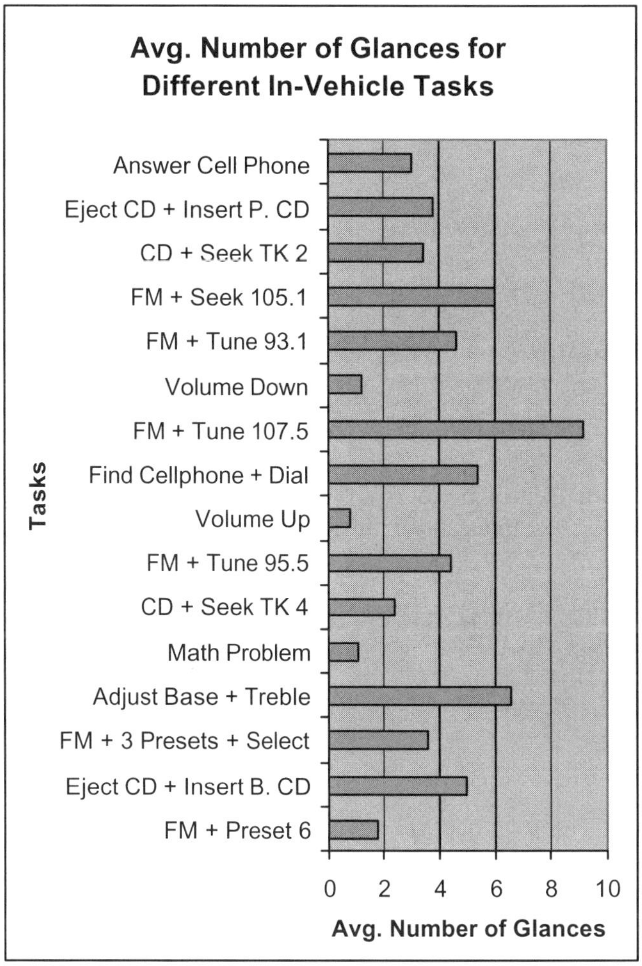

Figure 8. Average Number of Glances Made in Performing the 16 Tasks (from data combined over the evaluation of both radios).

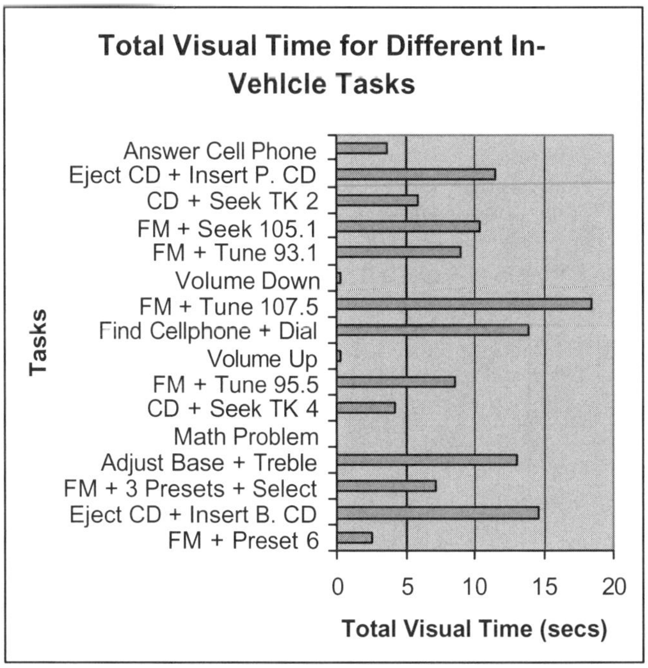

Figure 9. Average Total Visual Time for the 16 Tasks (from data combined over the evaluation of both radios).

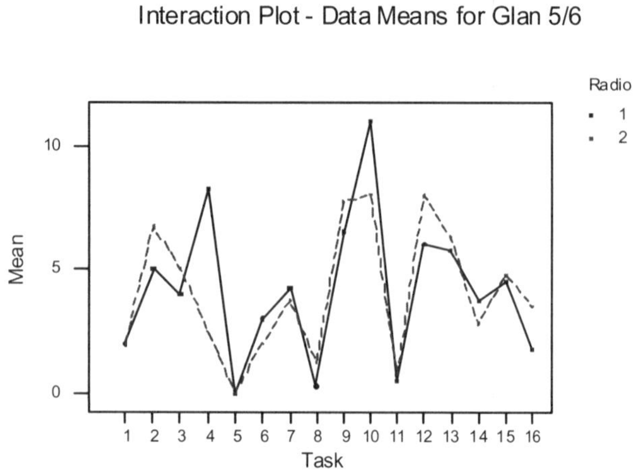

Figure 10. Number of Eye Glances Made during the Evaluations of Radio 1 and Radio 2.

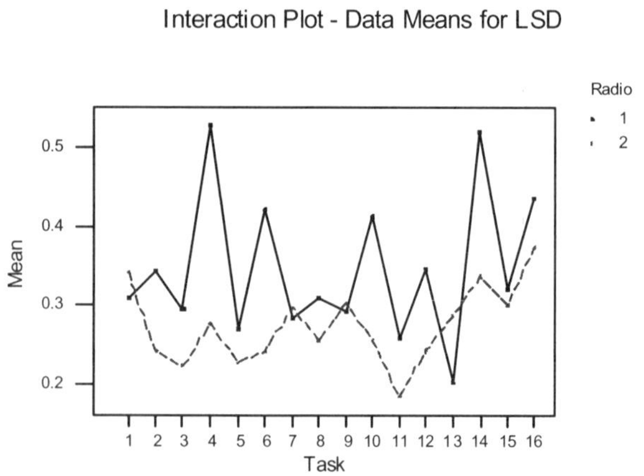

Figure 11. Lane Position Standard Deviation (measured in meters) during the Evaluations of Radio 1 and Radio 2.

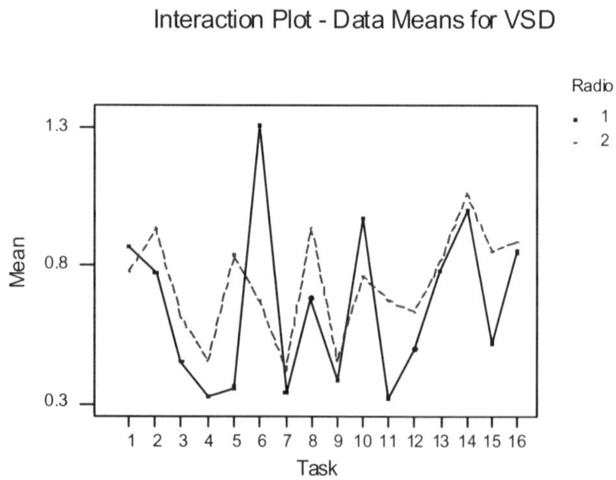

Figure 12. Velocity Standard Deviation (measured in meters) during the Evaluations of Radio 1 and Radio 2.

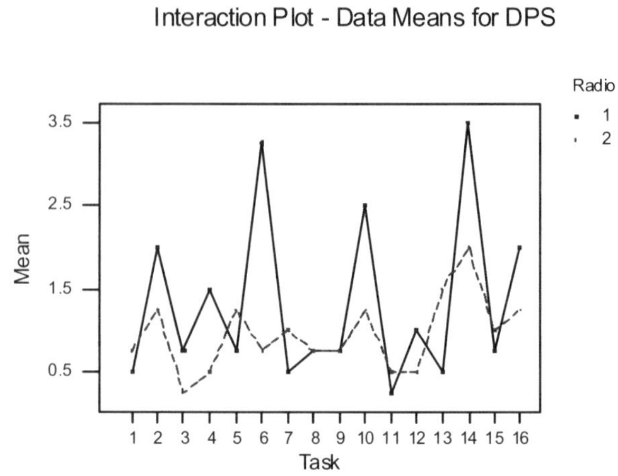

Figure 13. Driving Performance Measures for the Evaluations of Radio 1 and Radio 2.

The data presented in the above figures showed that tasks using Radio 2 were easier to perform than with Radio 1. The figures also show that Tasks 5, 9, and 16 which were not related to the radio operation, in general, did not show significant differences in values of the corresponding measures.

STUDY 2: EVALUATION OF TWO RADIOS

In this study, the driving simulator was used to evaluate an additional radio called Radio 3 using the same 16 tasks (described earlier) and 5 additional subjects. The photograph of the radio is provided in Figure 14.

Figure 14. Photograph of Radio 3 Used in Study 2

Since the eye glance data was not available for Radio 3, only the standard deviation of lane position and velocity data were processed. Figures 15 and 16 present comparison of the data obtained from testing with Radio 3 with Radio 1. The ANOVA tests on the measures showed that the differences between radios were significant at p=.0005.

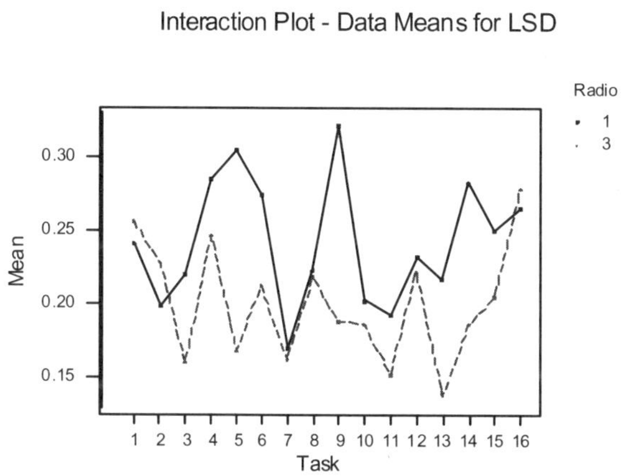

Figure 15. Lane Position Standard Deviation (measured in meters) during the Evaluations of Radio 1 and Radio 3.

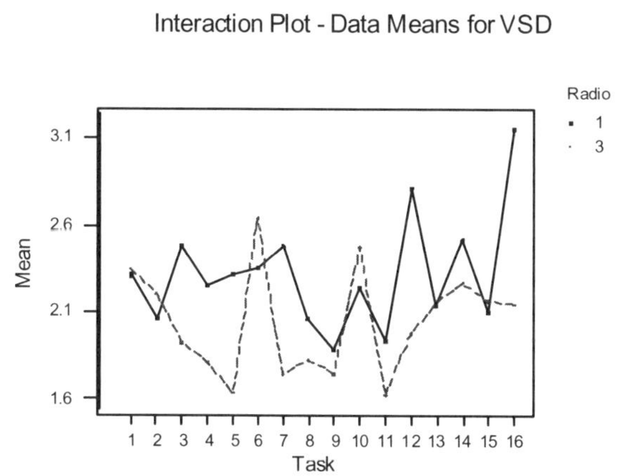

Figure 16. Velocity Standard Deviation (measured in meters) during the Evaluations of Radio 1 and Radio 3.

CONCLUSIONS

The simulator and the HMI evaluation process provided a very powerful method to evaluate tasks associated with new in-vehicle devices and to compare the measures obtained from the tests with similar data obtained from the evaluation of other in-vehicle devices. The evaluation results of Radio 1, Radio 2 and Radio 3 presented in this paper illustrate that the simulator is an excellent HMI evaluation tool to evaluate differences in different interface designs. Recently, two on-road studies in actual vehicles were conducted using the same procedure. In general the data obtained in the simulator showed similar characteristics to the driver behavior observed on the public roads (*6*). In future, the simulator will be used to evaluate various alternate designs of different centerstack mounted devices, reconfigurable displays and multi-function controls at different locations (e.g. on steering wheel, instrument panel, stalks, etc.).

AKNOWLEDGEMENTS

The authors wish to acknowledge and thank the following members of the ACE team for their support and technical assistance during this project: Collins & Aikman (formerly Textron Automotive Company), Sanyo FMS Audio Sdn. Bhd., Nippon Seiki International, Ltd., Douglas Autotech Corporation, Valeo Switch and Detection Systems, KSR International Company, Alcoa Fujikura, Ltd., University of Michigan-Dearborn and Oakland University. We also would like to thank Mr. Scott Davis for his support and persistence throughout the projects and the students for their efforts and enthusiasm.

REFERENCES

1. Bhise, V., Dowd, J., Davis, S. and Smid, E. A Comprehensive HMI Evaluation Process for Automotive Cockpit Design. SAE Paper number 2003-01-0126. To be presented at the SAE Annual Congress, Detroit, Michigan, March 2003.

2. G. E. Smid, "Virtual Vehicle Systems Simulation. A modular approach in real-time". Ph.D. thesis, Department of Electrical and Systems Engineering, Oakland University, 1999.

3. G. E. Smid, Ka C. Cheok, K. Kobayashi. Simulation of Vehicle Dynamics using Matrix-Vector Oriented Calculation in Matlab. Proceedings of CAINE '96, pp 115-120. ISCA, Orlando FL, December 1996.

4. G. E. Smid and Ka C. Cheok. Multi-Computer Real-Time Simulation of Vehicle Control Systems. Proc. of the 7th International Conference on Intelligent Systems (ICIS '98), Paris July 12-15 1998.

5 Smid, E., Dowd, J. and Bhise, V. Design and Implementation of a Driving Simulator Facility for the Optimization of Human-Machine Interface. To be presented at the Transportation Research Board for a presentation at the Annual Meeting in January 2003.

6. Bhise, V., Dowd, J. and Smid, E., Driver Behavior While Operating In-vehicle Devices. To be presented at the Annual Meeting of the Transportation Research Board, Washington, D. C., January 2003.

2003-01-0647

Development and Correlation of Internal Heat Test Simulation Using CFD

Corey T. Halgren and Frances K. Hilburger
Guide Corporation

ABSTRACT

Two primary focuses of the automotive industry have been cost reduction and lead time reduction. At the same time, automobiles have grown in complexity. Tier One suppliers must be able to provide less expensive, higher quality products faster. In this light, many suppliers have developed virtual simulation techniques in order to expedite the development process and insure the products can meet customer and legal specifications.

An essential predictive tool being employed is computational fluid dynamics (CFD). CFD is used to predict lamp temperatures and flow behavior of the air inside the lamp. Many OEM's have specific tests that require the lighting product to be cycled in a chamber. Lighting suppliers are required to show their products can meet these test requirements. CFD has been used to simulate heat tests in order to ensure the lighting product can pass the physical test portion. The simulation was developed using commercially available software and proprietary CFD analysis methods. The model is built as a fully coupled natural convection/radiation model using a transient analysis. The simulation provides the ability to predict component temperatures prior to building prototypes. This allows for changes in the lamp components and expedites the development process.

The intent of this discussion is to show the contents of the simulation, the technical aspects of the simulation, and the results including correlation data.

INTRODUCTION

In the past decade, automotive forward and signal lighting has changed from being primarily a safety and operational function to also being a major focus for the overall styling of the vehicle. As Tier One manufacturers of automotive lighting products, we are constantly being driven toward designs that, while exciting stylistically, can be quite difficult to execute and still achieve the thermal performance required by customer specifications and governmental regulations. In order to meet these requirements and still work toward the goal of 'Better, Faster, Cheaper' that is so prevalent in industry today, companies have turned to the use of state-of-the-art computational fluid dynamics (CFD) software to meet these requirements and to delight their customers.

Many OE manufacturers, as well as various government bodies, have defined specific tests to assess the thermal performance of automotive lighting products. Some of these tests require a cyclic application of power for certain lamp functions (e.g. a duty cycle of 5 minutes on and 5 minutes off for an hour) with the lamp enclosed in an environmental chamber.

The simulation we will discuss here utilizes the ADINA-F™ computational fluid dynamics software package to solve a model that combines conduction, natural convection, and wavelength-dependent radiation [1]. The model will include multiple sources of energy that will be cycled at various rates. The results of the simulation will be compared against experimental test data from actual parts, with a discussion as to their suitability for 'production' use to follow.

COMPUTATIONAL METHOD

This analysis utilizes a fully coupled radiation and natural convection model as its basis, and uses a transient solver in order to capture the cyclic nature of the power application to the model. The heat sources (in this particular case, the bulb filaments) emit radiation that is absorbed by the bulb glass envelopes, transmitted through the glass to the inner surfaces of the lamp, and reflected (a very small portion) back into the bulb. The heating of the filaments also induces convection of the bulb fill gases. In addition to the energy that is transmitted through the bulb glass, the bulb glass also re-radiates absorbed energy into the lamp. The heating of the bulb glass induces natural convection of the lamp air. Some lamp components are highly reflective surfaces, causing a portion of the incident energy to be reflected specularly within the lamp. The lens transmits a portion of its incident energy to the environment. Energy is also absorbed by the lamp components. This energy is then conducted to the exterior of the lamp where it is radiated and convected to the environment.

The basic model described above is based upon the analysis methods described by an earlier paper [2]. As such, a brief summary of the major computational components of the model should suffice. For greater detail, please refer to this earlier paper.

NATURAL CONVECTION

The bulb fill gases and the lamp air have had a viscous, incompressible model applied to them. Viscosity and thermal conductivity have been defined as temperature-dependent, and density is maintained as a constant. In order to obtain the induced convection, the Boussinesq approximation is utilized to capture variations in density as temperature varies throughout each fluid continuum:

$$\rho-\rho_\infty = \rho \cdot \beta(\theta-\theta_\infty) \quad (1)$$

External lamp convection is handled with the basic equation:

$$Q_{conv} = h(\theta-\theta_{amb}) \quad (2)$$

RADIATION

As noted above, this model includes emission, absorption, reflection, and transmission of radiation within the lamp. The surfaces within the lamp perform some combination of each of these functions. Emission of energy from surfaces within the lamp is caluclated as a net heat flux into (or out of) the surfaces. Net heat flux is caluclated using the equation:

$$q = \varepsilon(G-\sigma\theta^4) \quad (3)$$

The calculation of external lamp radiation is handled by using the same equation.

OVERALL ENERGY

The primary change that has been made to this analysis method is the inclusion of transient effects, which allows us to accurately model the effect of turning the bulbs on and off. The primary equation that is affected is the incompressible flow Navier-Stokes energy equation [3]:

$$\rho C_v(\partial\theta/\partial t) + \rho C_v v \bullet \nabla\theta + \nabla \bullet q = 2\mu D^2 + q^B \quad (4)$$

The first term ($\rho C_v(\partial\theta/\partial t)$) is the heat capacitance term, and is nominally the only transient term in this equation. Under the basic steady-state analysis method, this term would drop out. Now that we have included the transient solver, this term will remain in the calculation. The primary effect we are adding, however, is to allow the use of a transient source term, $q^B(t)$. This is accomplished by defining a step function for the source term, since the on/off switching is in effect instantaneous. The convection/conduction and viscous dissipation terms remain the same.

SPECIFIC EXAMPLE AND COMPARISON WITH EXPERIMENTAL DATA

The lamp selected is a combination forward signal lamp. There are two bulbs in the lamp, with a total of three functions available. The lower bulb is a Phillips 898 tungsten-halogen fog lamp bulb, while the upper is a Wagner 3757KA amber signal bulb. An aluminum heat shield is included which acts as a divider between the upper and lower cavities. The lower cavity also has a separate metalized reflector made of a high-heat plastic material. The housing and lens are plastic, and inner surfaces of the housing are coated with a highly reflective material that required the use of the specular radiation model included in this analysis. Each function is cycled in a different manner in order to see the difference in results between the three functions.

Figure 1 - Lamp Geometry w/Mesh Applied (no lens)

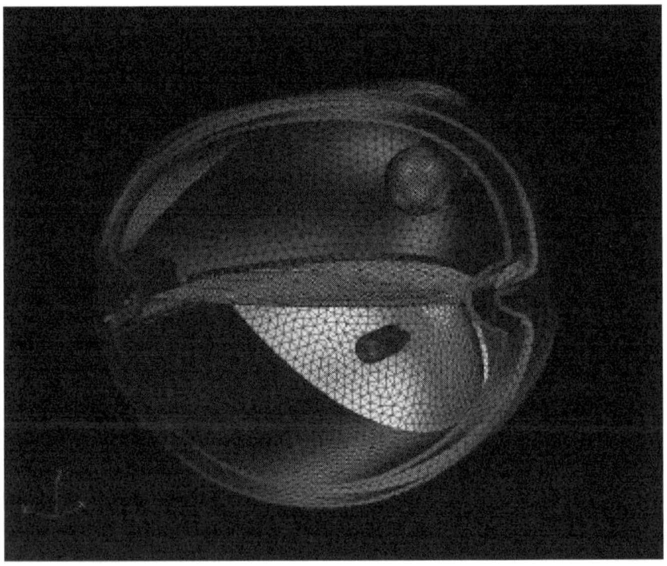

The basic finite element model is an unstructured tetrahedral mesh consisting of approximately 634,000 elements used to define the lamp air, lamp components, and bulb components. A total of about 13,000 specular nodes are defined in the model.

Figure 2 - Lamp Air Mesh, Section View

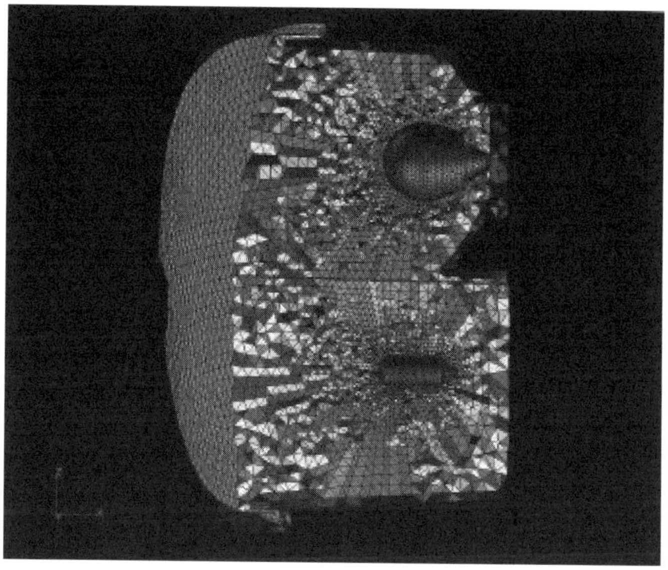

To support this analysis, an experimental run was also performed. The test lamp was placed in an environment chamber at 35°C. Thermocouples were applied to the back of the lower reflector (#2) and to the top of the heat shield (#1), and thermal camera images of the lens and upper rear area of the housing were used to acquire temperature data. Various images and tables are used to compare and communicate the results.

Figure 3 – Thermocouple Placement

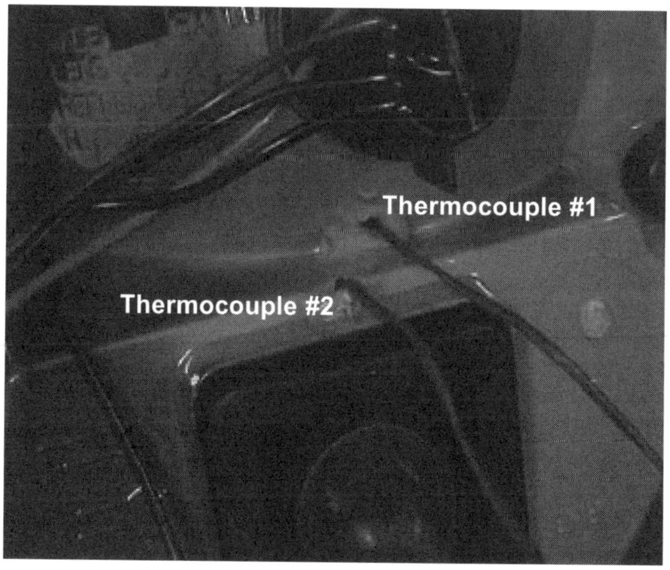

The combination of energy sources produces a very complex model with the following resultant temperatures.

Figure 4 – Thermal Camera Image, 30'

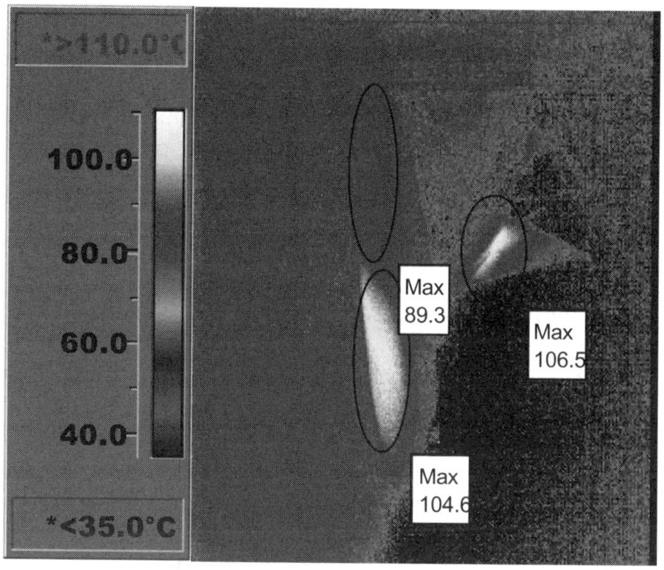

Figure 5 – Lens Temperature Comparison

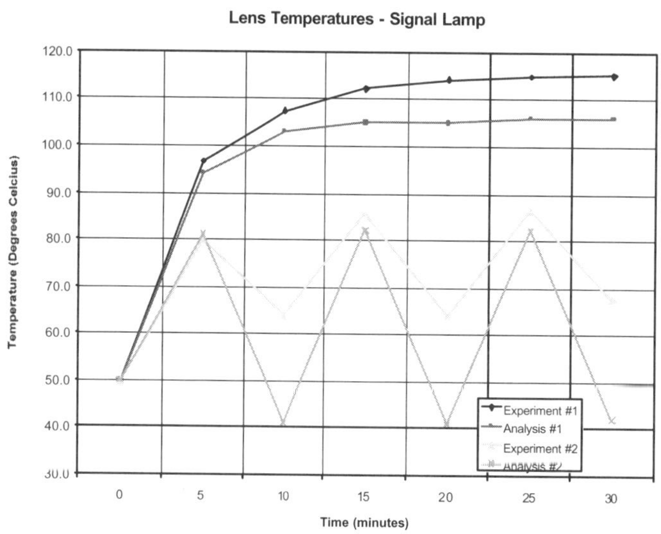

As can be seen from the chart, the results of lens temperatures compare favorably between test and experiment. In both cases, the simulation underpredicts the experimental result by between 5 and 8 percent.

DISCUSSION

The above data shows that this analysis model demonstrates good correlation with the temperature results obtained through experiment. This allows us to make decisions on the use of materials comprising the lamp as follows.

The lens temperatures are such that the use of a PMMA material would likely be problematic. This would direct us to look at the use of a polycarbonate material for the lens. Housing temperatures in the lamp are such that using polycarbonate for the housing will not work as well. A higher-heat material needs to be used here, as well as for the reflector in the lower cavity. Using higher-heat materials overall will allow this aggressive design to become a legitimate production lamp.

The internal air flow pattern can also be examined to identify potential stagnation zones and to provide insight as to the potential location of venting apertures.

CONCLUSION

Using proprietary modeling techniques, a simulation has been developed which allows the cycling of multiple energy sources to produce temperature results that correlate well with experimental data. This method can be applied to any appropriate test specification that requires power cycling. The simulation has been successfully utilized on other lamps to identify potential thermal issues that have later been realized through testing of early prototype parts. This has allowed us to communicate such information to product development teams, allowing them to take action to prevent thermal problems from continuing into production-level parts. As this analysis is the evolution of an existing analysis methodology, it stands to reason that further improvements will be made in the future.

ACKNOWLEDGMENTS

We wish to thank Mr. Ron Hilliard of the Guide Corporation Engineering Test Laboratory for his assistance in performing the experimental data acquisition and testing of lamps for this paper.

REFERENCES

1. ADINA-F v7.4, v7.5, ADINA R&D Inc., Watertown, MA
2. Moore, William I., et al, "Temperature Predictions for Automotive Headlamps Using a Coupled Specular Radiation and Natural Convection Model", 1999 SAE World Congress, Detroit, MI
3. ADINA Theory & Modeling Guide, Volume III: ADINA-F, Report ARD 00-9, Section 2.6, ADINA R&D Inc., Watertown, MA, August 2000

ADDITIONAL SOURCES

1. Incropera, Frank P., et al, "Fundamentals of Heat and Mass Transfer, 5th Ed.", ©2002 John Wiley and Sons, Inc.
2. White, Frank M., "Fluid Mechanics, 2nd Ed.", ©1986 McGraw-Hill, Inc.

DEFINITIONS, ACRONYMS, ABBREVIATIONS

CFD	Computational Fluid Dynamics
OEM	Original Equipment Manufacturer
ρ	Density
ρ_∞	Free Stream Density
β	Thermal Expansion Coefficient
Q_{conv}	Convective Energy
h	Heat Transfer Coefficient
ε	Emissivity
G	Incident Radiation
σ	Stefan-Boltzmann Constant
C_v	Specific Heat at Constant Volume
θ	Temperature
θ_∞	Free Stream Temperature
t	Time
v	Velocity
q	Energy
μ	Viscosity
D	Shear Rate
q^B	Energy Source Term
°C	Celcius Degree
PMMA	Polymethylmethacrylate (acrylic)

APPENDIX A

The following chart of temperature over time is for the thermocouples used in the lab testing. As noted above, thermocouple #1 is located on the upper surface of the aluminum shield, while thermocouple #2 is on the upper back side of the lower reflector.

Figure A-1: Thermocouple Data, Model #1

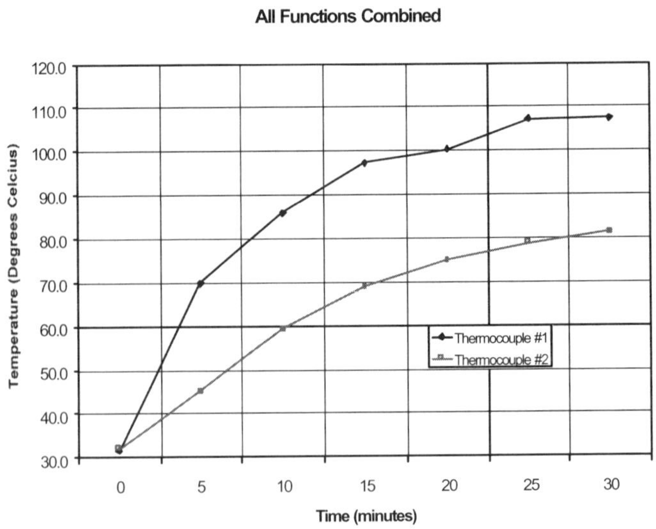

2002-01-3388

Virtual Reality Technology for the Automotive Engineering Area

Antonio Valerio Netto
University of São Paulo

Arnaldo Marin Penachio
Anésio Tarcisio Anitelle
T-Systems of Brazil Ltd.

ABSTRACT

The appearance of new equipment and software for implementation of Virtual Reality environment are providing a lot of opportunities of applications in several industrial areas and services. One of the area more beneficiaries with the progress to wide steps of this new technology are the automotive engineering. This article aims to present a brief description of what is Virtual Reality, differentiating of other technologies as animation, CAD or multimedia and where this technology this being used today in the area of the automotive engineering. The text also exposes the justifications of using this new technology in that area, which the forms of using its and that it forms its can improve the quality, to reduce costs and to reduce the time of vehicle development. The text finishes with final considerations and adaptations regarding the theme, always adapting the information to the automobile industrial market.

KEYWORDS

Virtual Reality, Visual Simulation, Virtual Prototype, Training, Marketing/Sale and Product Development.

INTRODUCTION

The emergence of new equipment and software for Virtual Reality (VR) environment implementation is providing many application opportunities in different industrial and service areas. One of the most favored with the fast evolution of this new technology is automotive engineering.

One of the main objectives of this new technology is to minimize problems that would only be detected after the physical prototype construction. Today, with virtual environment development software and appropriate interaction devices, it is possible to model machinery, land vehicles, boats, and planes aiming at simulating the effective behavior of the equipment. This saves money and development cycles, besides allowing training sessions and validation with the virtual prototype, in addition to enabling the presentation and sale of the final product, even before its effective existence in the real world.

But, what is Virtual Reality? The word is credited to Jaron Lanier, founder of the VPL Research Inc, who coined it in the early 80's to distinguish the traditional computer-aided simulations involving multiple users in a shared environment [1]. Researches such as Myron Krueger's in the middle of the 70's already used the term artificial reality, and William Gibson employed the term cyberspace[1] in 1984 in his fiction novel Neuromancer [10][19]. Cybernetic space (cyberspace) was the term used to designate a graphic data representation abstracted from the databases of all the human system's computers.

The term VR is rather comprehensive, and academics, software developers, and researchers tend to define it based on their own experiences, generating different definitions in the literature. It is possible to say, in an oversimplified way, that VR is the most advanced form of interface between user and computers available to date [11]. It is an interface that simulates a real environment allowing participants to interact with it [16], enabling people to view, manipulate, and interact with extremely complex representations [2]. Is its a paradigm according to which a computer is employed to interact with something that is not real, but that may be considered real while it is being used [12].

Practically, VR allows the user to navigate and watch a three-dimensional world in real time, and with six Degrees Of Freedom (6DOF). This demands the software ability to define and the hardware capacity to recognize six kinds of movement: forward/backward, up/down, right/left, up/down pitch, left/right yaw, and left/right spin. In essence, VR is a "mirror" of physical reality in which the individual exists in three dimensions, has the sensation of real time, and the capacity to interact with the world around him/her. VR equipment simulates these conditions, up to the point where users can "touch" objects in the virtual world, and have them answer or change in accordance with his/her actions [27].

[1] A metaphor related with the non-physical space where the user can execute actions (interaction with the environment) with this non-real environment.

VR interfaces involve a highly interactive 3D control of computing processes. The user enters the virtual space of applications and views, handles and explores the application data in real time using his/her senses, in particular the body's natural three-dimensional movements. The great advantage is that the user's intuitive knowledge of the physical world may be transferred to the virtual world. To support this interaction type, the user employs non-conventional devices such as viewing and controlling helmets, and data gloves. The use of these devices offers the user the impression that the application is running in a real 3D environment, allowing the environment exploration and natural handling of objects with the hand [14].

VR is often mistaken for animation, CAD (Computer Aided Design), or multimedia. VR differs in relation to these technologies because it is [18]:

- Oriented to the user, the observer of the virtual scene;
- More immersive, as it offers a strong sensation of presence inside the virtual world;
- More interactive, as the user may change and influence the object behavior;
- More intuitive, as there is little or no difficulty to manipulate the computing interfaces between user and machine.

Besides, VR presumes rendering (updating of object texture and geometry) in real time, that is, images are updated as soon as they suffer any type of change, and includes a functional description of the objects, extending CAD's topological and geometrical description.

The development of a VR system demands studies and resources linked with sensorial perception, hardware, software, user interface, human factors, and applications [3]. It is also necessary some mastering of non-conventional I/O devices (Input/Output), high performance computers, parallel and distributed systems, 3D geometrical modeling, real time simulation, navigation, collision detection, assessment, interface project, and social impact.

JUSTIFICATION OF VR IN THE INDUSTRY

It is possible to verify that along the time there was change in market demand. For example, in the automobile industry, in the past it existed a long product life cycle, small model variety and a new car every 10 years. In the present, it exists a reduced product life cycle, more product variety and a new car every 4-6 years. In the future, it will be necessary short product life cycle, explosion of variety and increased individualization. Therefore, the companies need to create a new form of supplying this need. It is obvious, it changes in market demands, it changes the way products are developed.

In a world of technological next "big things", Virtual Reality is expected to be the next "big thing". For engineers, this means that within the next decade designing on a flat computer screen will no longer be the norm. Expect to at least touch and manipulate a virtual part as you design and probably even walk inside and around a projection of the design in progress [29].

The rise of microchip power and fall in computer prices have ushered in what some analysts call the next stage in engineering technology-greater and better visualization capabilities. And this next stage takes many forms. The most accessible visualization technology at present is the virtual prototyping software now in use at many engineering firms and often used in conjunction with CAD systems and analysis software. Virtual prototyping software allows engineers to test their designs on a computer, rather than building an actual prototype and testing it in real life.

The virtual prototypes can be made to simulate operation that is, the computer simulation can be powered to predict the way the part would act in real world conditions. If the virtual part doesn't pass a test or fails an analysis, engineers have a good idea how to tweak the design in the CAD system to ensure that a second prototype will perform up to par. Usually, a physical prototype enters near the end of the process, to confirm or correct the CAD model.

In the Figure 1, it is possible to see the influence of modifications during product development process. The more lately a modification is accomplished in the project, more expensive is to accomplish it. One modification in planning phase costs U$100.000, in production phase will cost U$3.675 Million.

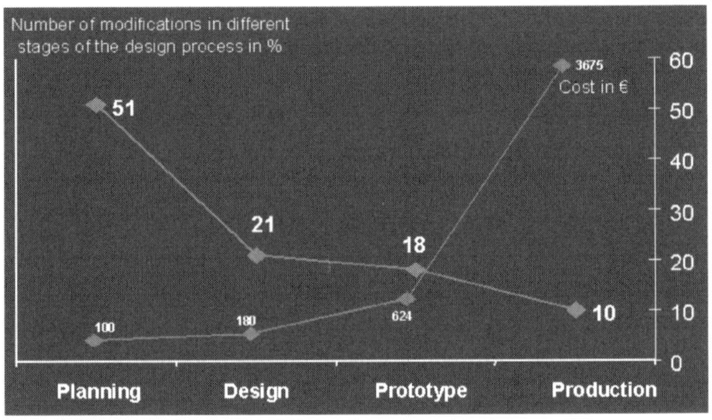

Figure 1 – Influence of modifications during product development process.

VR is the intuitively correct man machine interface for all stages of the product design process. This technology is a fast method to recognize design errors instantly. With VR is possible to have an error reduction, time-gains and cost-reduction.

Using a VR system is possible to enable the customers to cut development costs and time; to maintain financial and organizational control of the entire development process and digitally to evaluate product before expensive hardware is created or ordered.

In the Figure 2, it is possible to observe how the old product development process is executed and the difference between old and new approach. In the new product development process is a completely digital product for "Job 1", with a high degree of quality, which can be produced and assembled according to the planed process.

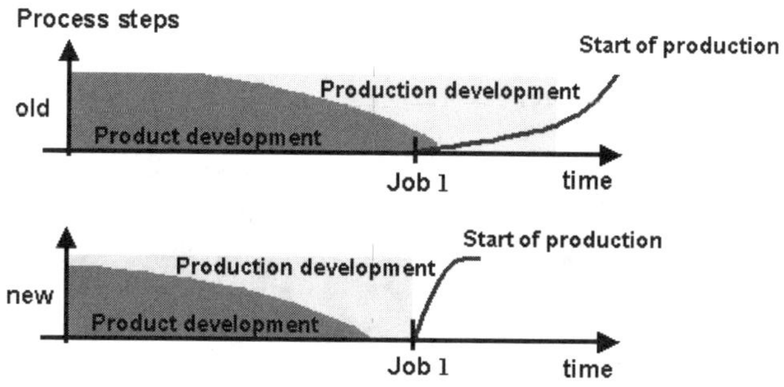

Figure 2 – Difference between old and new approach of product development process.

Industrial VR fields of application are diverse. It is possible to use in CAD-design, sales, training, production-planning, styling, development, ergonomics studies, mass data visualization, etc. Always with the objective of reduction of "time to market" and, improved quality and time of decisions.

However, it is important to point out that VR is not a stand-alone solution. It is necessary the integration in the process chain.

USE OF THE TECHNOLOGY IN THE INDUSTRIAL AREA

Several enterprises have been using VR in many fields such as project automation, marketing and sales, planning and maintenance, training, simulation, and data conception and viewing. However, new applications appear every time in most different knowledge areas and in a very diversified manner due to the demand and people's creative capacity. In many cases, VR is revolutionizing the way people interact with complex systems, offering better performance, and saving costs.

Several articles [13] [15] [4] [25] mention the advantages and facilities of RV usage in industry, mainly in the manufacturing sector. For instance, VR may be used [8]:

- To design machines that may have their structural and functional properties assessed and tested;
- To develop functional and reliable ergonomics, without the need of building a model in real scale;
- To conceive products with an esthetical design in accordance with each client's preference;
- To ensure that manufactures equipment is within the established norms issued by government offices;
- To facilitate remote operation and equipment control (Tele-manufacturing and Tele-robotics);
- To develop and assess processes that ensure manufacturing feasibility, without effectively producing the product in commercial scale;
- To develop production plans and routes, and simulate if they are correct;
- To educate employees in advanced manufacturing techniques focused mainly on working safety.

Virtual environments may be applied for prototyping, thus helping the product development cycle. From information on the project geometry and topology, simulation results obtained by modeling tools combined with kinematics calculation, material, tolerance and other available data on the product, it is possible to generate realistic prototypes in the computer, reducing costs with real prototypes and time to make them available for testing [22]. A virtual prototype also allows interactions with the product even in the initial development stages.

For Leston [18], virtual prototyping is one of the most important areas of project automation using VR. Some articles [24] [16] [5] [6] [4] expose the justification of the virtual prototyping usage mainly in the automotive area [20] [21].

The main advantages of virtual prototyping for industrial processes are:

- *Time reduction*: the time parameter today is one of the most important factors for industry. Time-to-market is the marketing key that distinguishes competitors.
- *Cost reduction*: virtual prototypes may reduce the need of making a large number of physical prototypes, also allowing development time reduction and human work employed in the project. There is also a reduction in the number of tools and materials employed in the making of physical prototypes. The results of virtual prototyping are obtained more quickly, thus allowing project feedback before the definition of production costs.
- *Quality improvement*: the application of different alternatives to a project may be performed more quickly, enabling an improvement in the validation of appropriate solutions that comply with the parameters specified by the client at a lower cost.

As previously observed, a virtual environment enables quick product development, adds high quality to it, and enables system-based decisions. One of the areas in which this technology can be applied is related with the possibility of viewing a huge and complex volume of graphical data.

In many cases, projects involving virtual environments have their starting point in models created with the Digital Mock-Up (DMU). DMU is an important step in geometrical modeling that will be used later on for the generation of virtual interactive prototypes. However, it is important to point out that DMU does not necessarily use virtual environments. However, immersion, interaction and involvement may facilitate and extend the possibilities of tests, analysis, and later on the change of the digital prototype. Generally, a 3D model is imported from a CAD 3D software and displayed using special viewing equipment and interaction (design review) (Figure 3).

Figure 3 – Display and interaction with a vehicle prototype [23].

The interactive design review can also be used as product presentation tool (Figure 4) and as a new way to leverage sales. Besides, design decisions on the color and style variation or concept supported through VR have their processes speeded up with a quality higher than the traditional method (Figure 5).

Figure 4 – Interactive Design Review used for product presentation [23].

A virtual environment can also be applied to ergonomic studies (Figure 6). The planned vehicle may be displayed and assessed as if it were real. Besides, with the execution of simple commands it is possible to make changes to enable comparisons between ergonomic projects. There is also the possibility of studying the glare and reflex effects on console instruments that are relevant to truck ergonomics. A solution process in this area demands dynamic calculation of physical reflexes. These calculations are presently corrected only through static processes or programs (Rayshade, Radiance or PovRay). Solution bases have been prepared for the generation of lighting, glare, and reflex effects in virtual environments, thus allowing an interactive digital solution to solve problems of this nature.

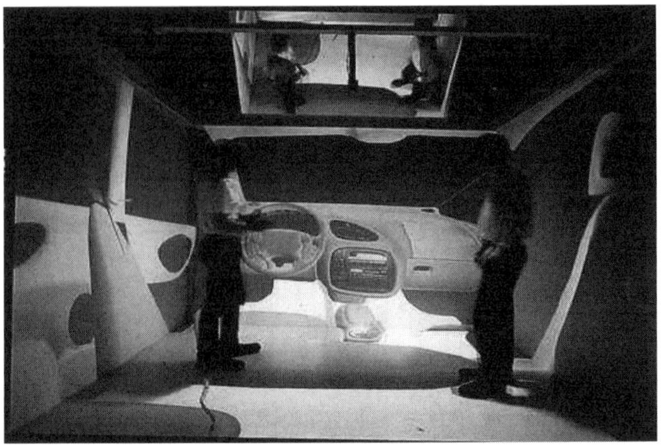

Figure 5 – Interactive Design Review for decision-making on style and color [23].

Figure 6 – Virtual environment used for ergonomic studies.

There are also virtual environments used for ergonomic studies and interior design, and also for the installation of components inside vehicles. This way, much of the design work detailing may be executed before the vehicle is created. Users can assess the visibility of the design and instrument positioning while seating exactly where the seats were planned for the vehicle. Once the engineers are satisfied with the virtual model, the project skips to another stage where a partially built vehicle is assembled in front of a curved screen (200°) with 7m x 4.4m cylindrical projection. In this projection, besides the generated stereo sounds, there are images displayed on screen, thus facilitating the user's immersion [9].

Virtual environments are also employed to assess the behavior of life-sized crash (impact simulation) tests. The vehicle inside the virtual environment collides against a wall that may be built of different material types. Results may be analyzed in 3D, and offer a better view of the deformation. Some parts (of the vehicle deformed by the crash) that could impair a more accurate deformation analysis might be gradually erased. This technique, currently reveals more details than a high-speed camera, and a great variety of tests may be conducted as if in the real world, and without the cost of destroying the vehicle at each test.

There are interactive and immersive environments already for product/parts assembly and disassembly processes (Packaging Studies). This allows the assembly and maintenance assessment through engineering, ergonomics, and efficiency criteria in a shorter period, with the opportunity of testing different alternatives until the ideal model is achieved. Besides, virtual environments may be used for the study of fluid dynamics, for instance, to view airflow conditions inside an automobile, and for external aerodynamics studies (Figure 7).

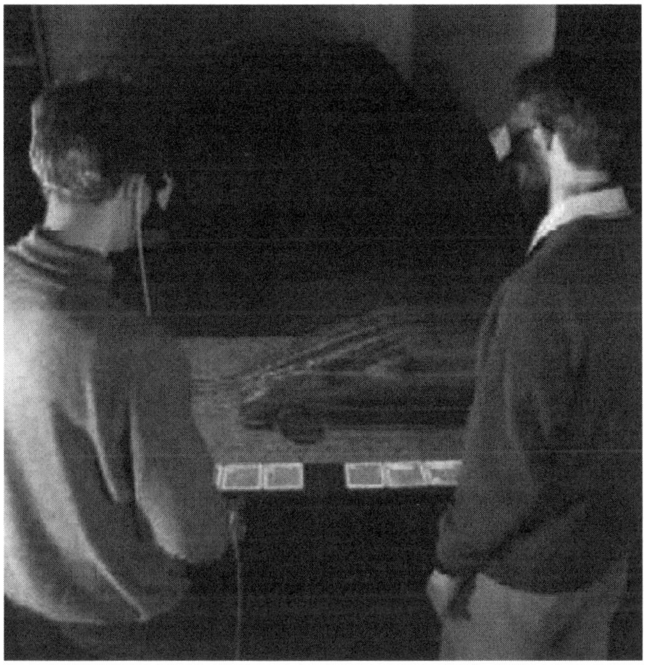

Fig. 7 – Aerodynamics studies using virtual environments [23].

FINAL CONSIDERATIONS

Over the years, VR has raised the interest of the industrial and technical service sectors. The purpose of this technology is to help the development of new products and even improve the existing products in the marketplace.

Designers and engineers can interact, handle, and validate their processes and/or products with the easiness offered by this technology. They are capable of performing an assessment of the new projects much more quickly than in the old methodology. They can also operate the equipment and to assess the assembly and obstructions without having to build a physical prototype. These assessments may be performed for different purposes, such as design improvement, styling, ergonomics, or functionality. That represents a lower cost, as there is no expenses with parts, material, or assembly hours of physical prototypes. With a virtual system, it is also possible to reduce the analysis time of the new project conception, and incorporate it to the production process on a short time basis.

However, the industry in general still considers with reserve the use of this new technology due to its initial cost, as it is necessary to acquire or even lease proper equipment and software for modeling and virtual environment integration and development. Besides, of course, training and hiring of expert labor to work in this field.

It is clear that for different tasks, companies used different types of models and organized the use of VR in different ways. For professional tasks, the model could be quite abstract, and the ability to interact with the model was more important than the quality of the images. For tasks that involve interaction with people who are non-professionals, such as design review with clients, the quality of images and the ability to move through the model in real time were more important. These different uses were seen as distinct, and on one project, a company created two separate VR models from the CAD data. One was used for design review (communication with non-professionals) and the other was used for co-ordination of detail design (professional use). For example, the first model, for presentation of design, showing surface finishes and details; and the second, for improving co-ordination of design required for clash detection and engineering.

The lack of literature on the theme, and few examples published for practical purposes, associated with the inexistence of deep research reporting the performance results against traditional system in terms of time and cost savings in the product development process, discourages industrial investment in this area. There are some notable similarities and some significant differences between the findings of the study and the academic literature. Example as the users companies in the construction sector are applying VR to a range of business tasks, which can be clearly split into professional tasks and those for wider interactions with non-professionals [28].

But this picture has changed over the last years regarding investments made and emergence of successful cases. The advance in researches in the VR area provides increasingly powerful hardware and software tools, and with more sophisticated immersion and interaction feelings, and that is producing a major interest in several segments of the industry, and an increasing number of users and applications. Besides, information reporting the success of RV use began to appear in product development processes. This is the case of the German DaimlerChrysler that has the most sophisticated set of graphical computers in the automotive industry, located in Sindelfingen, namely the Virtual Reality Center (VRC). The use of these new tools allows a series of conveniences leading to a cost reduction higher than 20% regarding a modeling project for new vehicles due to the reduction of physical prototypes and model building, besides reducing its development time [7] [9].

Obviously, the acceptance of these new processes and methodologies using VR by engineers and technicians is not immediate, and there is a natural and understandable resistance to change. This way, an enormous effort in information, publicity, and usage justification for this new technology is necessary.

In spite of the fact that the use of virtual system in larger scale depends on the evolution of equipment and software technology to allow increasing approaches of the situations to be simulated with what we really are able to simulate, the credibility of this technique is crucial to the decision of a change procedure.

Virtual reality allows a new working style where decisions are made based on a reality that does not physically exist, but that can be seen and interacted with. The most frequent question asked by people involved in the technical-industrial sector is: how to believe in data supplied by a virtual environment. But it is important that enterprises be aware of this new technology, as its incorporation does not occur with the same speed at which a product or process loses competitiveness.

REFERENCES

[1] ARAÚJO, R. B. (1996). *Specification and analysis of a distributed system of virtual reality*, São Paulo, June, 144 Pp., Thesis (Doctorate), Department of Engineering of Computation and Digital Systems, Polytechnic school of the University of São Paulo.

[2] AUKSTAKALNIS, S. & BLATNER, D. (1992). *Silicon mirage: the art and science of virtual reality*, Berkeley, CA.

[3] BISHOP, G. et al. (1992). Research directions in VR environments, *Computer Graphics - ACM*, 26(3):153-177, Aug.

[4] BRUNETTI, G. et al. (2000). Virtual reality techniques supporting the product and process development, *5° International Seminar of High Technology*, UNIMEP, Santa Barbara d´oeste, pp. 83-98, October.

[5] DUPONT, P. (1996). Virtual reality today, *Computer Bulletin*, pp. 14-15, June.

[6] DVORAK, P. (1997). Engineering puts virtual reality to work, *Machine Design*, pp. 69-73, February.

[7] EDITORIAL (2001) Virtual Reality, Magazine Auto Style - special edition, Reviewed of the dealerships of Mercedes-Benz, y. 1, no. 1, pp. 27-29.

[8] EXHIBITORS (1997). *Virtual reality in manufacturing research and education*. http://www_ivri.me.uic.edu/symp96/preface.html (August).

[9] GAVINE, A. (2000) Cave men, *Testing Technology International*, November, pp. 16-17

[10] GIBSON, W. (1984). *Neuromancer*. New York, ACE Books.

[11] HANCOCK, D. (1995). Viewpoint: virtual reality in search of middle ground, *IEEE Spectrum*, 32(1):68, Jan.

[12] HAND, C. (1994). Other faces of virtual reality, *First International Conference MHVR'94 - Lecture Notes in Computer Science n.1077*, pp. 107-116, Ed. Springer, Moscow, Russia, September.

[13] INTELLIGENT MANUFACTURING (1995). *Virtual Reality is for real*, vol. 1, n. 12. http://lionhrtpub.com/IM/IM-12-95/IM-12-vr.html (December).

[14] KIRNER, C. (1996). *Cycle of lectures of virtual reality*, Activity of the Project AVVIC- CNPq (Protem - CC - phase III) - DC/UFSCar, São Carlos, pp. 1-10, Out.

[15] KREITLER, M. et al. (1995). Virtual environments for *design* and analysis of production facilities, *IFIP WG 5.7 Working Conference on Managing Concurrence Manufacturing to Improve Industrial Performance*, Washington - USA. http://weber.u.washington.edu/~jheim/VirtualManufacturing/vrPaperIFIPS.html (September).

[16] LATTA, J. N. & OBERG, D. J. (1994). A conceptual virtual reality model, *IEEE Computer Graphics & Applications*, pp. 23-29, Jan.

[17] TEMPLEMAN, M. (1996). VR is the business for land rover, *Computer Bulletin*, pp. 16-18, June.

[18] LESTON, J. (1996). Virtual reality: the it perspective, *Computer Bulletin*, pp. 12-13, June.

[19] MACHOVER, C. & TICE, S. E. (1994). Virtual reality, *IEEE Computer Graphics and Application*, pp. 15-16, January.

[20] MAHONEY, D. P. (1995). Driving VR, *Computer Graphics World*, pp.22-33, May.

[21] RESSLER, S. (1997). *Virtual reality for manufacturing - case studies*, National Institute of Standards and Technology. http://www.nist.gov/itl/div894/ovrt/projects/mfg/mfgVRcases.html (September).

[22] RIX, J. et al. (1995). *Virtual prototyping - virtual environments and the product design process*, IFIP Chapman & Hall, 348 Pp..

[23] TAN (2002). TAN Projktionstechnologie. http://www.tan.de (Janeiro).

[24] TERESKO, J. (1995). Customers transform virtual prototyping, *IW Electronics & Technology*, pp. 35-37, May.

[25] VILELA, J. B. et al. (2000). Virtual development of product, *II Brazilian Congress of Administration of Development of Product*, São Carlos, SP, pp. 187-190, August.

[27] VON SCHWEBER, L. & VON SCHWEBER, E. (1995). Cover story: virtual reality, *PC Magazine Brazil*, pp. 50-73, v. 5, n. 6, June.

[28] WHYTE, J. (2001). *Business drivers for the use of virtual reality in the construction sector*, AVR II and CONVR 2001, Conference at Chalmers, Gothenburg, Sweden, October.

[29] THILMANY, J. (2001). Electronic spelunkers, *Mechanical Engineering Magazine*, June. http://www.plmsoluctionseds.com/publications/articles/ me_evis_0701/ (january/2002)

2002-01-0563

Enabling Rapid Design Exploration through Virtual Integration and Simulation of Fault Tolerant Automotive Application

Thilo Demmeler
BMW AG - BMW Technology Office

Barry O'Rourke
Methodology Services - Cadence Design Systems, Ltd.

Paolo Giusto
SFV - R & D - Cadence Design Systems, Inc.

ABSTRACT

Modern automotive applications such as X-by-Wire are implemented over distributed architectures where electronic control units (ECU's) communicate via broadcast buses. In this paper, we present a framework for quick exploration of design alternatives in terms of HW/SW architectures for distributed applications. The exploration is carried out on a virtual integration platform that allows the distribution of embedded software onto ECU's. The framework shortens design turn-around time by supporting semi-automatic communication protocol model configuration (e.g. frame packaging, redundancy level, etc.), and then by allowing the designer to run fast yet accurate simulations of a virtual prototype of the distributed architecture that includes models of the application software and the bus communication protocols. As a result, design errors can be found earlier in the design process, before the system integration in the car, therefore resulting in savings in both production and development costs. Simulation results show that the method is scalable in terms of simulation performance degradation.

INTRODUCTION

The entire automotive industry is trying to move tests from cars to labs, where real conditions can be emulated/simulated at a much lower cost[1]. Having a virtual environment rather than prototyping HW for designing and testing can significantly reduce development and production costs, since faster time-to-market can be achieved - designers are able to simulate the distributed application on their host workstations rather than in a test track. Hence, redundancy and fail-safe system tests can be repeated after every change in the design. Besides, this method provides more flexibility because derivative designs (variants) of the same application can be easily supported - there is no need to wait for the next hardware prototype on which to load and run the application SW.

Car manufacturer goals are time to market and reduction of development and component costs. This can be achieved by finding a near-to-optimal solution for automotive electronic systems and distributed networked applications. The solution is found by applying cutting-edge electronic system design methodologies and a tool sets, which deploy the move towards a virtual integration platform and enable the evaluation of distributed systems in an early development stage [19][20][21][22][23]. Investigating and improving the partitioning design of these systems in early stage can

[1] The cost for setting up an experiment on a car is about $120-$500 per hour. The time needed to set it up is about 1 hour. The number of tests that can performed every day is ~2 [26]

optimize functional networks[2] and target architectures located in the car chassis.

Today's car electronics systems can be classified in the following categories:

- Cabin/*Infotainment*[3]/Telematics. Main features are wide-band, soft real time constraints, non-critical (e.g. power windows, conditioning)

- Powertrain/Chassis. Main features are hard real-time constraints, safety critical, fault tolerant, low band (e.g. engine, brakes, steering), with sub-systems being isolated from one another mostly for historical reasons

Our proposal focuses onto the latter category of applications. The first one is left for further studies.

Today's car electronics systems are implemented over distributed architectures that include (among others):

- Several ECU's communicating via

- One or more (for fault tolerant systems) communication protocols (e.g. CAN [1], TTP[3][4], LIN [24], and FlexRay) over networked broadcast buses.

In turn, each ECU includes (among others):

- Application and diagnostic SW

- Base SW (RTOS, Communication layers, etc)

- One or more micro-controllers with local memories

- Bus controller(s) with one or multiple channels to support redundancy for fault tolerant systems.

- Dual ported RAM's for communications between bus controllers and micro-controllers

Since *in-car* networking enables extensive exploitation of add-on-functions and therefore increases the customer value of embedded systems, automotive applications have been becoming more and more distributed in nature since the late 80ties. New design challenges such as *X-By-Wire* for steering and braking have introduced a new design dimension – *fault tolerance* – that provides additional complexities yet potentials for optimizations such as, for example, reduction of the number of needed ECU's. In fact, a better utilization of each ECU may potentially reduce the number of ECU's itself[4]. In a nutshell, the problem, as described in [8][17][19][23], consists of distributing a pool of functions over the target architecture with a goal of satisfying the requirements in terms of cost, safety, and real-time. It is extremely important for the designer to set up experiments quickly in order to reduce the development process (and hence its cost), achieve the time-to-market[5], and ultimately come up with a cheaper implementation. Because of the distributed nature of these applications, the communication protocol must also be considered. Such a type of investigation is only possible by addressing the integration step at the virtual level, and not on the car itself as the current design practices dictate.

Therefore our main proposal can be summarized as following:

- Usage of a virtual platform, for system testing and prototyping (HW/SW architecture) via simulation, based upon the Virtual Component Co-design tool (VCC) by Cadence Design Systems, Inc.

- Usage of virtual models of the application SW and the target HW/SW architecture (bus controllers, CPUs, RTOS schedulers, communication protocols) to create a virtual prototype the *entire* distributed application. The application SW models are *imported* from other tools [8], or can be authored within VCC. The architectural models are developed within VCC (the communication protocol model is the subject of further chapters) using a standard C++ API.

- A major shift from a "per-ECU" tool-supported design style where each ECU is considered separately – the design exploration is limited to one ECU at the time and the integration step is done later in the design process on the car – to a virtual platform based design style where the integration is done at the *virtual* level

The V-model is a quite popular method in automotive for describing the development process of Control Software. The idea is that it has to be possible, at every level of abstraction to simulate the model, in order to check its correctness, and then generate the code for an appropriate target. Therefore, the V-model links the information "horizontally", by simulation and code generation, and "vertically", between the different layers of abstraction. With our proposal (Figure 1[6]), we are essentially adding a new branch to the traditional V-model.

[2] A functional network includes the overall system functionality with the definition of the sub-systems and their interfaces independent from the target architecture

[3] Horrible neologism for the electronic subsystems devoted to information processing, communication with outside world and entertainment [22]

[4] Notice that the re-distribution is not always possible since in some applications the SW is tied to a specific ECU

[5] This is of importance to higher-end cars while reduction of costs might be the predominant factor for lower end segments

[6] Source BMW

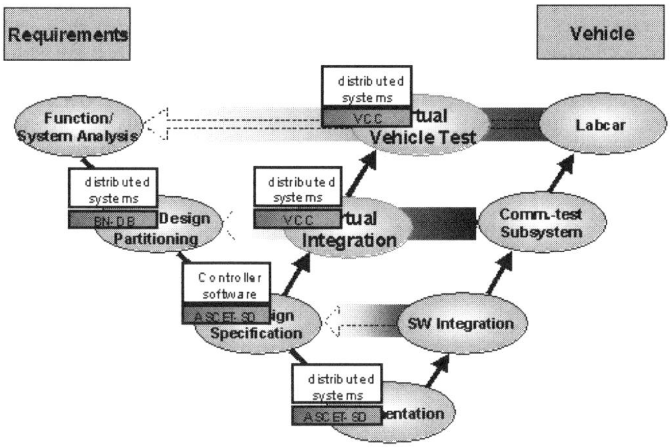

Figure 1: The Enhanced V-Model

Once the assessment is made in VCC in terms of near-to-optimal HW/SW architecture, this information can be exported/used by the downstream tools. For example, once the designer has been able to decide in VCC the SW distribution on each ECU, then, a downstream tool for code generation can use the information such as number of tasks needed, scheduling policies, etc, to generate the SW (RTOS configuration plus app SW) , for either physical prototyping or real implementation on a real chip. At the same time, the downstream tools used to for communication protocol analysis can be configured based upon the configuration data determined at the virtual level (type of protocol, frame packaging, communication cycle, redundancy management policies, etc.). Thus, a step that would require costly human intervention can be completely automated. Note that a by-product of this methodology is that the designer can both model off-the-shelf existing protocols and also rapidly prototyping novel ones.

In this paper, we focus on describing one essential key element of the methodology, the *universal* communication (protocol) model (UCM), the highly configurable framework used for abstract modeling of most common features of communication protocols. Our claim is that the model provides a good trade-off between accuracy in the results and simulation performance therefore enabling the move from test tracks to virtual simulation and testing environments.

The paper is organized as follows. The next section describes our concept of *virtual integration platform* based design. After that, we describe the UCM. Finally we draw some conclusions and show ideas for future developments.

VIRTUAL INTEGRATION PLATFORM BASED DESIGN

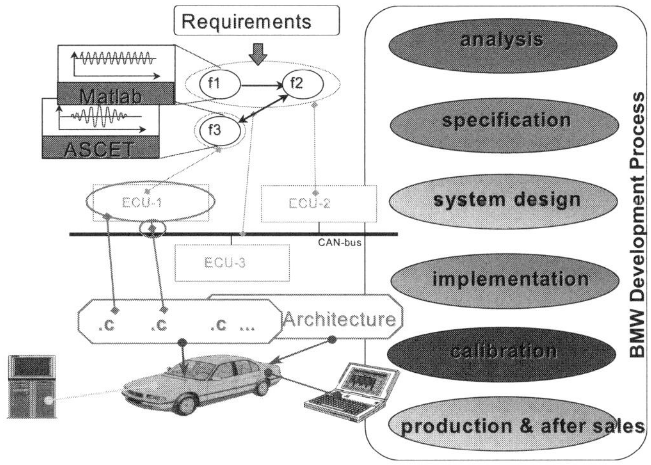

Figure 2: "Per-ECU" style Methodology

Figure 2 shows the ideal design flow for a car manufacturer. The development process starts with the analysis phase, where a functional network is developed, and continues with the specification phase, where algorithms for each of the functional sub-components are *defined.* The responsibility for the functional components can reside either in the system suppliers that deliver control algorithms and the hardware or in the car manufacturer - that is often the case when competitive differentiation is crucial. Nevertheless, the car manufacturer is responsible for the entire system and for proper integration of all the different subcomponents. In the system design phase the distribution of the functionality onto the target architectural network is determined. In the next phase, a composition of functional components is implemented onto the target hardware and finally the system is calibrated in the car.

Based on the common methods and tools available today this flow poses several issues:

- *Lack of continuity*: i.e. a big gap exists between the requirement analysis and the definition of the functional network - there is no formal way of proving that the functional network implements the requirements. Several efforts are under way in this direction [26][27]. This is not the focus of the paper.
- *Lack of integration tools* : the methods and tools deployed today usually support only a "per-ECU" design style. The design exploration is concerned to sub-systems only.
- *Long design turnaround time*: the exploration and validation of the overall distributed system including the communication protocols[7] and the scheduling is

performed very late in the design cycle when the HW components and the system partitioning is already defined.

Because of the above issues, the development and production costs are obviously affected. "Notice that vehicle manufacturers traditionally focus on production cost rather than on development cost – the sensors and the actuators along with the bare ECU represent almost the whole amount of costs for electronics spent. Through software does not have a direct amount of production cost it is not for free!"[25].

CURRENT "PER-ECU" DESIGN PRACTICES - In the specification and the implementation phases, the ASCET-SD tool is commonly used in the automotive domain for algorithm development and code generation for single processor units. ASCET-SD [12] is a typical *ideal world* representative tool set that assumes no execution delays during the simulation. An ASCET-SD/VCC automated import flow that preserves the functional specification details such as hierarchy, s and scheduling information is one of the essential components of the virtual integration platform. An imported ASCET-SD model is represented in VCC as a hierarchy with the project at the top level that comprises the functionality of one entire ECU. The modules at the next lower level state the functional components, which are the smallest mapping unit that can be distributed over the system network. The processes, included in the modules, are the smallest schedulable unit and constitute the leaf blocks in the hierarchy. Finally, tasks are the aggregation of processes, which have the same scheduling policy. To enable the VCC performance estimation, the source code of the processes, which share a considerable amount of data within the module, are imported as VCC white boxes. Furthermore, the scheduling information of the functional ASCET-SD model in terms of ordering, timing, priorities and properties can be preserved, as well as the data exchange and the interfaces of the processes, the modules and the entire project. The communications between the software components are *naturally* modeled as shared memories, called behavioral memories (BM) in VCC.The non-consuming data access is realized through BM-read or BM-write function calls, which are invoked in the process source code that is generated in the import/export step. ASCET differentiates between global variables and messages, which are preemption-protected global variables. The protection mechanism of the messages is re-modeled in VCC within separate additional processes, which are automatically generated by VCC at the import step [8]. The aggregation of processes in the modules differs from the aggregation of processes in the tasks, which results in a dependency between the system mapping step and the message protection mechanism. This dependency is dissolved in the proposed design workflow by the VCC tool that re-generates the message protection processes after (and therefore in dependency of) each new system mapping or after tuning the scheduling. After importing *incrementally* either single modules or complete projects, the I/O interface of the top-level blocks is determined through unbound behavioral memory references. Alternatively behavioral models can be imported manually as "C" white or black boxes of plain C or from the Matlab tool set.

THE NOVEL METHODOLOGY - The proposed virtual integration platform is built within the Virtual Component Co-Design (VCC) tool set as shown in Figure 3. The basic concept [7][8][17][15][23] is to have a behavioral model of the system with ideal world assumptions in terms of zero software execution time and communication delays, which is separated and independent from an architectural model that represents an implementation variant. By mapping the functionality onto the architecture, a specific system partitioning is chosen. The system models are transformed into performance models that include close-to-real execution and communication delays. The system models (functionality, architecture, and mapping) are *automatically* transformed by VCC into performance models that include close-to-real software execution time and communication delays (virtual prototype). After performance simulation results in satisfactory results, the information about the HW/SW architecture can be exported back to ASCET-SD. This can then be used to automatically generate the software for each ECU (including the communication layers) and hence a complete physical prototype of the whole system.

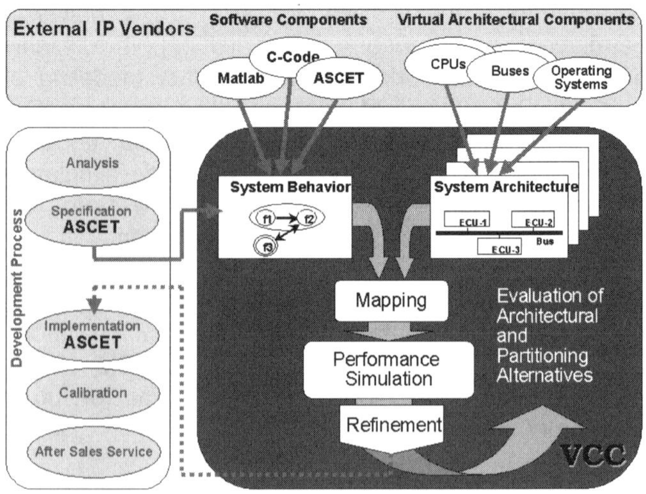

Figure 3: Novel Methodology

THE ECU MODEL IN VCC - An automotive system network consists of several ECU's that are connected to at least one bus. For fault tolerance reasons, redundancy may be introduced by using multiple bus channels. As shown in Figure 4, the software that is running on the host is usually separated in hardware independent application software and communications layer that is hardware (and application) dependent [14]. Also, a real time operating system (RTOS) that provides services to the SW depends on the underlying architecture.

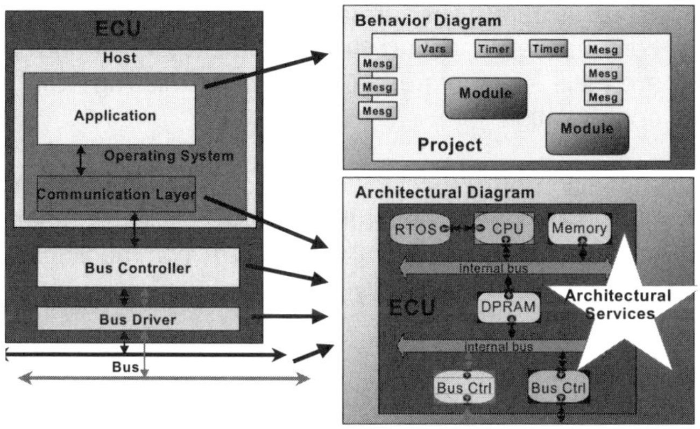

Figure 4: ECU model in VCC

In order to achieve behavior/architecture independence and therefore the re-usability of functional and architectural virtual components, only the application software is appropriate to be imported and modeled as a behavior in VCC. All other layers are modeled as architecture models in VCC. In the proposed design methodology the interface between host application and architecture is independent from a specific bus implementation in the first stages of the design. Later in the refinement process, hardware specific features like certain error modes may be addressed that influence the host interface. . The messages that *produce* the bus traffic are automatically determined from the functional mapping of the system. A VCC mapping diagram represents a specific HW/SW partitioning plus SW scheduling and protocol configuration thus represents each ECU.

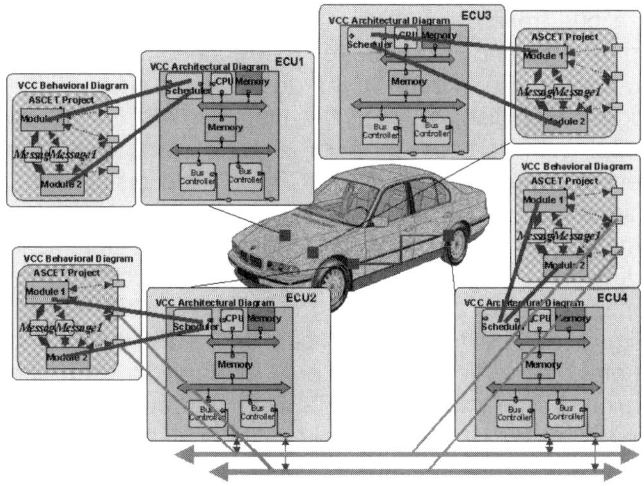

Figure 5: Virtual Car – First Evaluation Step

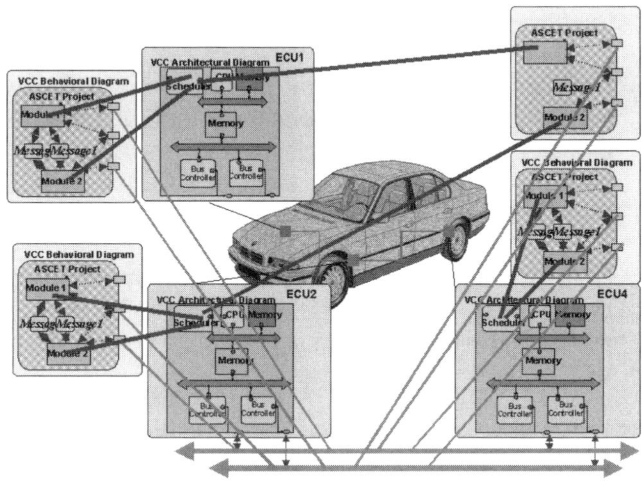

Figure 6: Virtual Car – Next Evaluation Step

In Figures 5 and 6 the red mapping links represent the SW scheduling assignment of imported ASCET-SD processes to task whereas the green links represent the mapping of ideal-zero-time communications to protocol models. These aspects are explained in more detail later in the paper.

The automotive systems that we are considering consist of several ECU's and one or more bus protocols. In the virtual prototype of the system, the ECU models (represented by mapping diagrams) are connected to several models of the communication protocols. Each protocol model is defined via the UCM framework that providesabstract modeling for the most common features of the protocol of interest (see next section for more details). By modeling the *entire* distributed application on the host workstation, the designer can, for example, reduce the number of ECU's and verify that the requirements for safety and real time are satisfied. VCC provides visualization tools such as *gantt* charts displaying scheduling information (such as task execution time) and protocol information (such as delay of a bus transaction, delay of a message tranmission). Figures 5 and 6 illustrate the concept. In a first evaluation step (Figure 5), the designer can simulate the original SW distribution of the imported ASCET-SD by mapping the processes in exactly the same way they were grouped in tasks in ASCET-SD (no distribution). This enables the benchmarking of the performance simulation results, which includes the automatic estimation of task run time and modelling of communication delays. In the further design steps (Figure 6), the functionality can be re-distributed and different kind of architecture alternatives can be explored. The former ASCET-SD project structure (one project per ECU) might dissolve as modules that belong to the same projects can be mapped to different ECU's. This can lead to savings, in for instance, the number of needed ECU's (Figure 6).

DESIGN WORKFLOW – In a nutshell, the design workflow consists of the following steps carried out on a host workstation:

1.) Definition of a behavioral diagram in VCC by importing functional components in form of software projects and modules (this creates the functional network) or by authoring them in VCC *black box* C++/*white box* C [11]

2.) Definition of a test-bench (environment) by importing models (e.g. from Matworks/Simulink[8]) or by authoring them in black box C++/white box C within VCC [11]

3.) Generation of an ideal communication between the functional components in the behavioral model and of the functional network with the test-benches, which does not consider delay or error handling

4.) Creation of an architectural diagram in VCC by usage of virtual models of ECU's and communication protocols – for derivative designs it is of course possible to re-use existing architectural diagrams

5.) Mapping of the software modules onto the CPU of a cluster by either retaining the mapping of all modules of the project to the according ECU (vis-à-vis) (Figure 5) or not (Figure 6)

6.) Generation of the CPU scheduling. This step can be done either manually or automatically by the import step if the original scheduling information is preserved

7.) First performance simulation of the network. No communication performance is considered yet. The software execution time is estimated. [15] Only the scheduling performances are modeled.

8.) Design iteration by re-distribution of the functionality and tuning of the scheduling of single CPUs.

9.) Initialization of the UCM performance model and automated generation of an initial communication matrix that carries the dependency of the functional system mapping (this detail is explained later in the paper)

10.) More accurate performance simulation. The bus communication delays are estimated. Bus latencies are still inaccurate due to the missing UCM configuration

11.) Definition of a specific bus protocol implementation by UCM parameterization - the UCM is user-configured to model the desired communication protocol more accurately in terms of:

- Communication cycle layout (event-driven, time-triggered, mixed-mode)
- Data Frame Definition (message packaging in frames)
- Bus controllers' configuration:
    - Activation policy: event-driven, time-triggered, mixed-mode
    - Timing, e.g. time period of the bus transactions (for time-triggered)

12.) Accurate performance simulation including the bus latencies

13.) Design iteration by re-distribution of the functionality and tuning of the scheduling of single CPUs and/or protocol configuration until a satisfactory (in terms of time performance, cost, etc) HW/SW architecture is determined

14.) Hardware dependent features that are not already covered, like specific error or start up mechanisms, can be modeled and refined.

Phases 1, 4, 5, 6 and 7 are described in [8] in detail.

MODELING COMMUNICATION PROTOCOLS WITH THE UCM

REQUIREMENTS – Abstract modeling of (automotive) communication protocols is dictated by the following requirements:

[8] An import mechanism will be available soon as part of VCC

Broadcast – Each ECU is the master of the bus at a certain point in time. When the master ECU sends a data frame[9], the frame is to be received by the other ECU's in the cluster. The latency of the frame with the respect to:

- the sending and the receiving application SW
- the sending and receiving ECUs

has to be modeled.

Time Triggered with no Arbitration – Each ECU is the master at a specific point in time determined statically – each ECU is assigned a specific slot within the communication cycle. This aspect is modeled in order to simulate time-triggered bus protocols that are increasingly deployed for safety critical systems due to their deterministic behavior. .

Event-Driven with Arbitration – In order to be able to model CAN-like type of protocol configurations, it has to be possible to model arbitration where several ECU's race for taking the owner ship of the bus.

Mixed-type of Communication Protocol - Modeling communication cycles where dynamic parts with arbitration and static parts without it are defined is mandatory to model high end protocols like FlexRay.

Redundancy – Redundancy has to be supported such as hardware redundancy via multiple channels and bus controllers.A-synchronicity – The SW running on the host is totally independent from the communication protocol running on the bus. In practice, this means that when the application SW is sending a data frame, the transmission over the bus does not necessarily happen synchronously with that thread – it may happen later in time, for example, if the bus controller has not taken ownership of the bus. Hence, in the real application, the bus transactions and the functionality running on the CPU, are parallel activities.

Synchronicity – The SW running on the host can be synchronized via interrupt to the communication that is going on the bus. To take full performance advantage this is usually done for systems that deploy a time triggered protocol.

Local Inter-CPU communication – In modern automotive architectures, a ECU may include more than one CPU especially if some redundancy is deployed. Therefore, communications between SW modules running on different CPU's take place, for example, via a dual-ported RAM.

[9] By frame we mean a generic aggregation of data, regardless the communication protocol (e.g. it can be an aggregation of messages, a.k.a. telegrams)

Complex Bus Topologies – This includes redundancy, star constellations with star couplers and gate ways between different busses and bus protocols.

THE UNIVERSAL COMMUNICATION MODEL (UCM) FRAMEWORK - This section provides details of the implementation and the use model.

High Level Overview of the Implementation - Once the UCM settings are matching a specific communication bus protocol, we expect accurate performance results to be obtained that allow qualitative assessments, at least for the system running properly under no fault conditions. Although VCC provides such a capability, we do not model the delays of HW/SW interrupts (i.e. pulses coming from the crank shaft) since they are *negligible* with respect to the overall system performance, *dominated* by the *latencies* of frames sent over the network bus. The latency of the host/bus controller interface is not considered yet in the sense that we have not modeled the performance related to access data from the application SW to the bus controller. The reason being, this delay is negligible (usually in the order of nano-sec's) with respect to the global performance of the entire system (SW scheduling and communication protocol – usually in the order of micro-sec's). This more refined performance model of the distributed application is left to future developments. In our first implementation of the virtual integration platform, we first have focused on the communications that involve VCC behavioral memories (BM's) that are utilized to model message exchange among different ECU's. We concentrate on modeling the latencies related to *frames* (specifically in the UCM a frame is a set of messages packaged together each of them modeled as BM's [17]). The mapping of the memory accesses (reads and writes) onto *communication patterns* (that model different communication delays providing the performance model of the communication) is performed automatically by VCC [17]. A *dynamic* performance model is assigned next. In fact, depending on the VCC mapping of the imported ASCET-SD modules onto architectural resources, the same functional data exchange in the behavior may take different communication delays depending whether the modules are mapped to different ECU's or to the same one. In the latter case, the communication would be purely SW to SW on the same ECU (either on the same CPU or through different CPU's on the same ECU via, for example, dual-ported RAM). In the former case, a bus is involved and therefore the modeled delay must be different. As a result of mapping the functional components onto the architectural components, the mapping of a communication arcs for the BM references to a *communication pattern* is automatically inferred by VCC [17]. Notice that we have implemented a wizard, that depending upon the user mapping determines the mapping of the communication to the correct performance models. Therefore, this provides the automatic generation of an initial communication matrix that carries the dependency of the functional system mapping.

Only BM's that are connected to modules that are mapped onto different ECU's, participate to the bus traffic. This states an essential information for the VCC user, which is revealed by VCC after (and dependent on) the distribution of the functional components. We differentiate between two different types of behavioral memories:

- ***Register type (RT)*** behavioral memories that represent messages or global variables that are not sent between ECU's [8]
- ***Bus type*** **(BT)** behavioral memories that are sent over the bus [8].

The UCM covers communication delays that are bus protocol specific, such as packaging the messages into frames, the frame transmission policy and the queuing mechanism of data frames.

Architectural Services and Delay Equations – An architectural component in VCC contains architectural services that are virtual C++ functions, which model both, the performance and also some of the functionality of the component. As shown in Figure 7 the *delay equation* that dynamically models the performance of the communications is implemented by a stack of the needed architectural services modeling the single bus components of the network cluster. VCC determines the path from the architectural topology net-list and links the necessary architectural service components into the UCM. Changing the topology does not require remodeling the UCM services. In fact, the components are assembled together depending on whether RT BM's or BT BM's are involved hence modeling the correct performance.

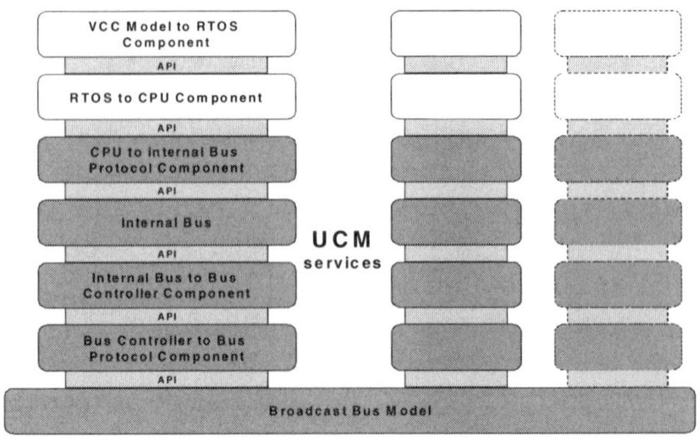

Figure 7. Delay Equation with architectural services

Time-Triggered, Arbitrated, Mixed-type – In the UCM, we allow the user to define the communication cycle that assigns static parts for time-driven data frames and dynamic parts for event-driven frames, as shown in figure 8. This leads the designer to explore the performance of different communications.

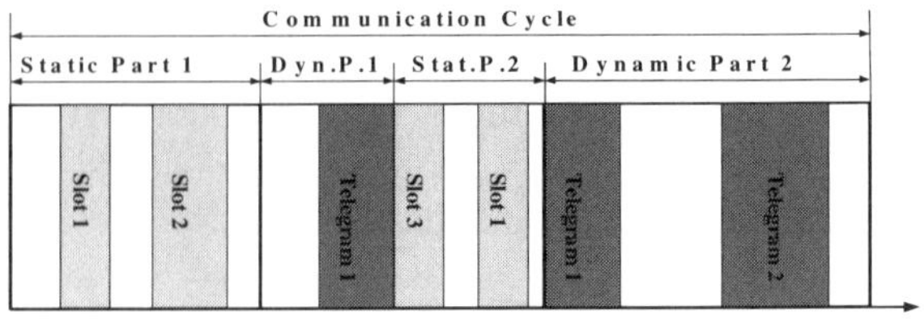

Figure 8: Communication Cycle of the UCM

In the static parts of the communication cycle, the time-triggered frames are transmitted in slots according to a statically defined scheme. In the dynamic parts event-driven frames, called telegrams, are transmitted according to an arbitration algorithm. The communication cycle, represented by state machine in Figure 9, is controlled by a global synchronous time.

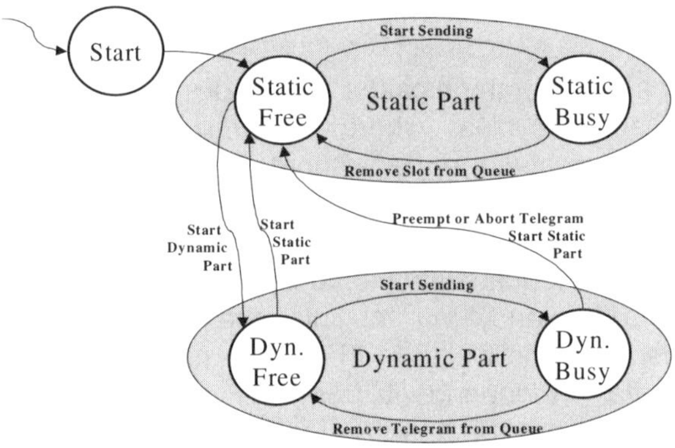

Figure 9: State Machine of the Bus Model

The bus state is "Busy" while a frame is sent. If the transaction is over, the data frame is removed from the sending queue and the bus mode changes back to "Free". Cycle changes are usually performed in the state "Free", except telegrams that can be interrupted to ensure a proper start of time driven frames. The interrupted telegram remains in a queue and is sent later in the next dynamic part.

A-synchronicity, Time-Triggered, Event-Driven, Mixed-Mode - Notice that communication cycle and bus controller activation policies are kept separate, hence promoting a plug-and-play methodology. Moreover, the operation of the UCM is driven by two sets of entities that are independent of each other (Figure 10). First, by the behavioral models of the application SW, which runs on the various ECU's of the network. These interact with the UCM whenever they write to a BM. This causes either a register transfer, if the recipient of the message is mapped to the same ECU, and takes negligible time or a bus transfer, which will begin to call the service stack which implements the communication.

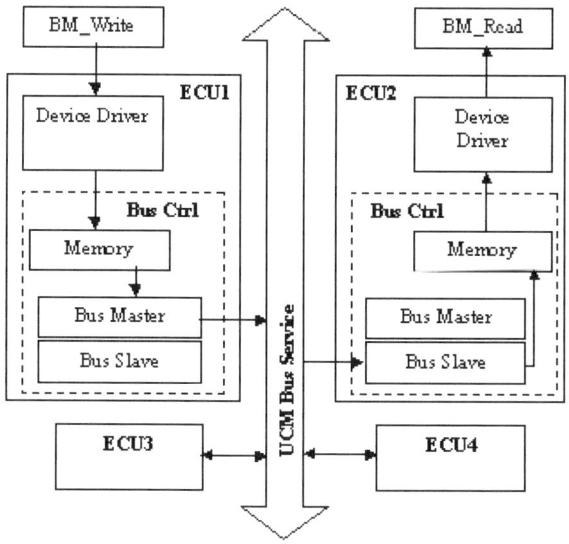

Figure 10: UCM stack

A BM *"write"* operation, however, only causes the services required to store the required data into the ECU's local memory. Notice that, as illustrated in Figure 11, the ECU's local memory is used in conjunction with the "ECU Local Memory BT" in order to support redundancy management. This aspect is explained in more detail in the sub-chapter. The initiation of the actual bus communication is triggered by the Bus Arbiter Architectural service resident on the bus model itself asynchronously from the application behavior. The bus controller attempts to send messages from its local memory depending on which mode of operation the bus arbiter is currently in. The three modes of operation are:

- Static Time Triggered Frames (STTF). These always get sent when the global bus arbitration dictates that it is time for the message's slot. Their queue size should never be greater than one, since their transmission period should be exactly equal to their production and consumption period, both instantaneously and on average.
- Dynamic Periodic Frames (DPF). These are sent at regular intervals. Time comes from the global state but each ECU has locally programmed triggers for its messages to write into the frame. The telegram is queued until it wins control of the bus – only thenthe bus controller can actually send it to the recipient.
- Dynamic A-periodic Frames (DAF). Any time the application SW is running and writes onto a behavioral memory, this causes an event trigger to be set into the bus controller model. This functionality is facilitated by the inclusion of an interrupt register on the bus controller that the Device Driver service writes to as the message data is transferred into local memory. The bus controller detects this register write and immediately puts the message in the queue for transmission. The transmission will take place the next time the bus is free according to the arbitration policy..

Broadcast and Redundancy – The broadcast and redundancy concepts are supported by the following architectural model (Figure 11).

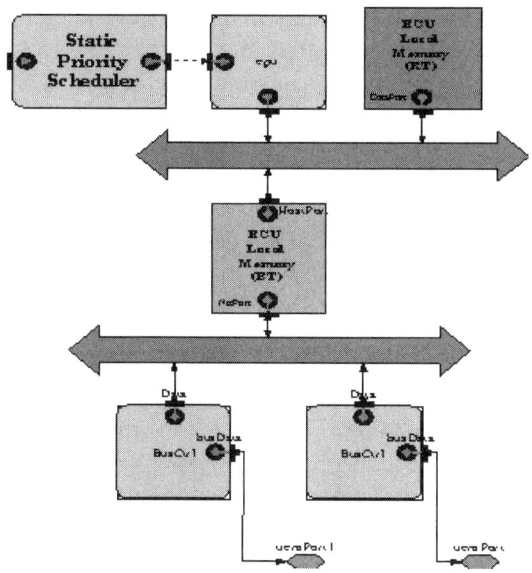

Figure 11: ELM BT AND ELM RT

In the architecture, the designer uses two ECU's local memories, called ""ECU Local Memory BT" (ELM BT) and ""ECU Local Memory RT" (ELM RT). Each RT BM gets copied on the ELM RT and each BT BM is copied on the ELM BT. The essential operation of the ELM BT block is to apply the redundancy management function to all writes to BT BM's from the other ECU's via UCM so that the memory service on the block contains the 'correct' values for each of the BT BM's on the receiving ECU. Notice that, the example below has only two bus controllers. However, in our general modeling technique, the designer can place as many bus controllers as needed in the diagram. Generally, the ECU has the following internal transactions:

- SW to SW communications (within the same CPU) via RT BM's read/write data from/to the ELM RT only – essentially, behaviors write and read data to the BM's. However, ELM RT that holds the copies of the BM's provides the delay of the communication (next to zero). This is necessary in order to support broadcast at the functional network level (one SW sender and more than one SW reader on the same CPU).
- Inter-ECU Broadcast communications over the bus:
  - Sender side: each time the application SW wants to send data via broadcast bus, the data is copied onto the ELM BT only. The ELM BT block then copies this data into the memories on the bus controllers (of the same sending ECU) so that event driven frames can be triggered or frames can be sent in the appropriate slot. The

UCM service will then make a copy of the data on each bus controller memory of the receiving ECU's as explained below.

- Receiver side: each time a new frame is received by an ECU, the data is copied into the memories on the bus controllers. Then a redundancy management function (customizable by the user) copies the 'correct" data onto the ELM BT block. Subsequently, each application SW reads (according to the scheduling) the data from the ELM BT. Notice that, according to the topology, data is broadcast to all the ECU's attached via the UCM to the sending ECU. However, the actual data read by application SW is determined by the fact that the SW itself is mapped on to the receiving ECU.

Figure 12 illustrates the proposed default redundancy management scheme. Each time line refers to a different UCM broadcast bus (upper part of the figure). There are two writes and three reads on a BM. "UCM 2" suffers from periodic interference that corrupts some frames sent over it. Making the assumption that corruption of a frame means that the BM data on the receiving bus controller is not updated until some sort of error recover takes place, the first BM read will discover that the data from "UCM 1" is more recent than that from "UCM 2". Therefore, the "UCM 1" copy of the BM data is returned to the reading behavior.
On the second read "UCM 2" has determined that it has valid data for the BM and has updated its memory service copy accordingly. Now, the "UCM 2" copy has the most recent timestamp so its BM is returned to the read behavior. Note that this value should be the same as that on "UCM 1" (if the error recovery scheme is operating correctly). By the third read, there has been a first time success BM update on both buses so both copies of the BM have the same timestamp and the same data so it does not matter which one is returned and an arbitrary choice is made ("UCM 1" in our example).
This modeling technique seems to exhibit the best simulation performance since the frequency of the bus transactions is lower or equal of the frequency at which the behavior that reads the memories. Therefore, this check shouldn't affect too heavily the simulation performance. In fact, the redundancy policy is implemented at the bus interface rather than onto the RTOS. It is left to further customizations the following:

1. Frame validity: in the bus controllers a check that the frames are valid can be introduced
2. Redundancy policy: being defined as a C++ virtual function the designer can replace the current default implementation to provide a more sophisticated redundancy function.

Complex Bus Topologies – The user can place on the diagram as many bus controllers within the same ECU as desired. Each bus controllers can use a different UCM bus therefore complex bus topologies can be represented.

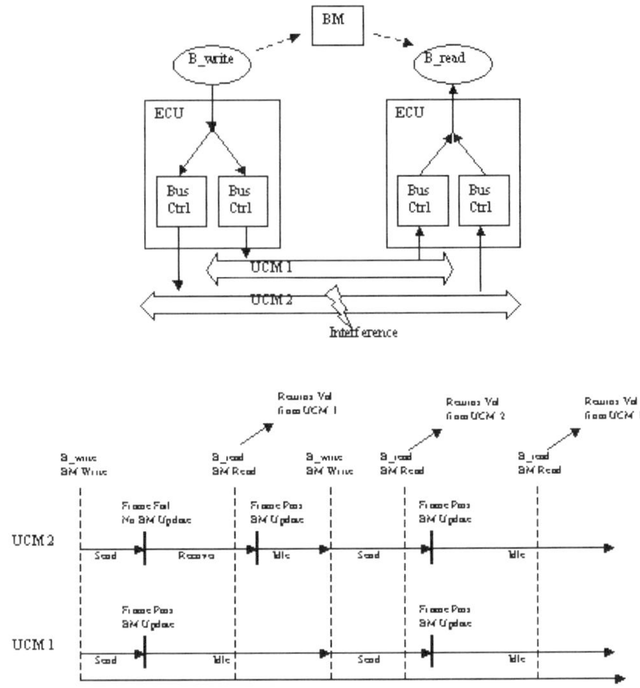

Figure 12: Redundancy Management

Local Inter-CPU communication - For the time being, since the performance of the overall system are dominated by the message latencies over the inter-ECU buses, we have left the inter-CPU communication performance modeling to the next release of the virtual automotive architectural platform. Several configurations are being investigated and a solution will be provided soon.

UCM Use Model - The user must specify the number of different frames (*Frame Count*) that the ECU may send on the bus. The *Frame List* is an array of frame data-types that define all the properties required for the UCM. For example, an ECU could have two frames to send (since an ECU might have more than one bus controller, the discussion applies to each bus controller). For example, the first (FrameList [0]) is a Slot frame which is sent once every communication cycle [17] at an offset of 1 millisecond and consists of two BM's with a packaging overhead of 8 bits. The second frame (FrameList [1]) is a Telegram of a single BM, which is triggered every time the BM is written to. The frame has an overhead of 24 bits (since telegrams have to carry more data then slots) and is of priority 2.
The user can specify the bus arbiter parameters as well. For example, a particular system transmits at 1 Mbps and the communication cycle repeats every half-second. There is a single static part that starts at the beginning of the cycle and lasts for 100 milliseconds therefore leaving the remaining 400 milliseconds of the cycle for

dynamic transactions. The minimum time allowed between frames transmitted across the bus is 0.2 milliseconds and the *Arbitration Factor* value defines a linear relationship between the number of nodes racing for control of the bus and how long the arbitration phase of a telegram communication lasts for.

Experimental Results – The implementation of UCM has just ended the first phase. The second phase with redundancy management is under heavy implementation and testing. Thus, we only have some preliminary results related to the first phase. We ran a simulation of a Drive by Wire imported design in VCC on an NT workstation with 512Mb of RAM and a Pentium 3 processor. The design has 4 ECU's, the UCM configured in the pure event-driven mode, and 100 mapped behaviors. To simulate 1 minute of real time, the elapsed time was about 5 hours at the rate of 2K frames/sec (88 bytes each). Notice that 1 driving maneuver is between 30 seconds and 5 Minutes. More results will be available in the final version of the paper.

CONCLUSIONS

The UCM framework constitutes the key component of a virtual automotive architectural platform where behavioral models of the application SW can be imported from different tools such as ASCET-SD and Matworks/Simulink [13] in order to define the functional network which can be distributed over an architectural network. The supported levels of abstraction enable a seamless flow in the design process, from a broad variety of partitioning possibilities to refinement stages that allow qualitative performance assessments. The UCM framework provides a high grade of automation, although the designer has still to adapt the bus protocol properties and refine the communication matrix.

The communication matrix of real automotive systems is a result of an extended development process, where many developers and external partners are involved. It requires a lot of experience and knowledge about the functional requirements and the system behavior. The import of the required bus model properties as, for instance, the communication matrix, which is available for specific designs from external databases, would add considerable value to the flow. We are considering this aspect as part of future developments. Also, a high grade of automation will be required to support the re-use and maintenance of communication configurations that are already refined in VCC, thus enabling an iterative design process and its implementation.

The UCM framework, as part of our virtual integration platform based methodology, because of its high grade of re-configurability enables rapid design exploration of car electronic systems. In fact, the framework shortens turn-around time by supporting semi automatic protocol configuration, and then by allowing the designer to run fast yet accurate simulations of the distributed architecture including the application software and bus protocol. Therefore, errors can be found earlier in the design stage resulting in savings in both production and development costs. .

Once a performance model for the distributed functional components and their communication is generated, hardware relevant protocol characteristics such as specific failure mechanisms can be introduced as additional architectural features to the UCM. As future work, specific bus protocol features can be implemented in VCC by either refining the architectural service models, where for example failure states could easily be implemented, or by explicitly importing specific bus protocol models e.g. from silicon suppliers. This step will become more relevant when a potential system partitioning is found and finally a safety analysis should be performed to confirm the failure concept of the application.

ACKNOWLEDGMENTS

The authors would like to thank Luciano Lavagno from the Cadence Design Systems, Inc., Berkeley Labs, Steve Wisniewski and Claudio Zizzo from Cadence Design Systems, Methodology Services, Livingston, UK for providing useful suggestions for the methodology.

REFERENCES

[1] Robert Bosch, CAN Specification Version 2.0, Technical Report ISO 11898, Robert Bosch GmbH, 1991.

[2] ByteFlight homepage, http://www.byteflight.com/

[3] TTP Forum, TTP/C specification V0.5. http://www.ttpforum.org/, 1998.

[4] H. Kopetz, R. Hexel, A. Kruger, D. Millinger, R. Nossal, A. Steininger, C. Temple, T. Fuhrer, R. Pallierer, and M. Krug, A prototype implementation of a ttp/c controller. Proceedings of SAE Congress and Exhibition, Feb. 1997.

[5] E. Dilger , L.Å. Johansson, H. Kopetz, M. Krug, P. Lidén, G. McCall, P. Mortara, B. Müller. Towards An Architecture For Safety Related Fault Tolerant Systems In Vehicles. ERSEL - European Conference on Safety and Reliability

[6] E. Dilger, T. Führer, B. Müller, S. Poledna, T. Thurner. X-By-Wire: Design of Distributed Fault Tolerant and Safety Critical Applications in Modern Vehicles. VDI - Verein Deutscher Ingenieure

[7] P. Schiele. Transition Methodology from Specifications to a Network of ECUs Exemplarily with

Ascet-SD and VCC. SAE Technical Paper Series Nr. 2000-01-0720, 2000.

[8] Translating Models of Computation for Design Exploration of Real-Time Distributed Automotive Applications – Submitted – 2001

[9] S. Edwards, L. Lavagno, E. Lee, A. Sangiovanni-Vincentelli. Design of Embedded Systems: Formal Methods, Validation and Synthesis. Proceedings of the IEEE, vol. 85 (n.3) - March 1997, p366-290.

[10] Vector Informatik. Calibration of Electronic Control Units via CAN. http://www.vector-informatik.de/english/products/index.html, Canape, 2000.

[11] Cadence Inc. Virtual Component Codesign Product Documentation. Cadence Inc., 1998.

[12] ETAS GmbH, White paper ASCET-SD-, ETAS GmbH, 1998.

[13] Matlab. Homepage Technical report, <UPDATE>.

[14] OSEK/VDX Organisation. OSEK/VDX Operating System Specification 2.1. http//www.osex-vdx.org.

[15] L. Lavagno, A. Sangiovanni-Vincentelli and E. Sentovich. Models of Computation for Embedded System Design. 1998 NATO ASI, Proceedings on System Synthesis, Il Ciocco, 1998.

[16] E. Lee, A. Sangiovanni-Vincentelli. Comparing Models of Computation. Proceedings of ICCAD, 1996.

[17] T. Demmeler, P. Giusto. A Universal Communication Model for an Automotive System Integration Platform. DATE, 2001.

[18] I. Gutkin, P. Giusto, J. Ehret. Modelling the CAN bus within the VCC environment. Proceedings of the International Conference on Parallel and Distributed, Processing Techniques and Applications, 1999

[19] Chris Edwards, Carmakers plan to spread software around vehicles, http://www.electronicstimes.com, March 2001

[20] Charles J. Murray, Auto industry braces for media revolution, http://www.electronicstimes.com, March 2001

[21] Stefan Poledna, Markus Novak, TTP scheme fuels safer drive-by-wire, http://www.electronicstimes.com, March 2001

[22] Alberto Sangiovanni-Vincentelli, Automotive Electronics: Trends and Challenges, Convergence 2000, Detroit, October 2000

[23] Paolo Giusto, The VCC experience in automotive domain, Stuttgart International Symposium, February 2001

[24] Clay Merritt, An Introduction to the LIN Protocol, Digital DNA – Motorola

[25] Ulrich Freund, Alexander Burst, ETAS Gmbh, Graphical Programming of ECU Software – An Interface Based Approach

[26] Roland Jutter, General Manager ETAS GMbh, Current Trends in the Design of Automotive Electronic Systems, Automotive day at DATE2001, March 2001

[27] Automotive UML homepage, http://www.automotive-uml.com/

[28] Grant Martin, Luciano Lavagno, Jean Louis-Guerin, “Embedded UML: a merger of real time UML and co-design”, CODES 2001, Denmark, April 2001

SAFETY CRITICAL APPLICATIONS

2005-01-0784

Software Certification for a Time-Triggered Operating System

Peter S. Groessinger
TTTech Computertechnik AG

ABSTRACT

This paper presents the software certification activities carried out on TTP-OS to make this hard real-time, fault-tolerant operating system available for safety-critical applications in the automotive and aerospace industries requiring certification. The steps and measures, while specifically tailored to make an RTOS certifiable, were defined in accordance with the RTCA/DO-178B [1] guideline.

The major single goal of these activities is to achieve traceability of requirements. Requirements are traced from the Software Requirements Document all the way down through the software lifecycle to the test-cases ensure consistency and accuracy of a mature software development approach. The steps and milestones along the lifecycle are described, offering an insight into the software certification efforts required.

INTRODUCTION

This document reflects certification activities carried out on TTP-OS, a fault-tolerant real-time operating system which is certifiable according to Software Considerations in Airborne Systems and Equipment Certification, RTCA/DO-178B – Level A. The term "certifiable" means that the operating system is not certified as a stand-alone component; rather, it is certified as a system together with the hardware platform it runs on. Projects using a certifiable operating system can benefit from the certification effort already invested by incorporating the certification documents into their own certification.

An introduction to TTP-OS is given in the first section. The main focus is drawn on the software lifecycle as carried out at TTTech Computertechnik AG. The software lifecycle covers requirements engineering, software design and software verification.

MAIN CHARACTERISTICS

TTP-OS is a real-time operating system for time-triggered embedded computers running hard real-time software with high safety and reliability requirements. It was developed according to DO-178B [1] Level A in order to create a dependable operating system certifiable for the aerospace and automotive industries. Level A is the highest level defined in the DO-178B guideline. It defines the requirements for certification of software that might produce catastrophic consequences upon a failure.

TTP-OS is an OSEKtime-compliant [4] operating system well suited for safety-critical real-time applications. Together with a fault-tolerant communication layer (FT-COM) the operating system provides an ideal runtime environment for safety-critical distributed real-time software. Even though FT-COM is an essential part for a fault tolerant distributed real-time system developed according to OSEKtime, FT-COM and its services are not covered by this presentation, which focuses only on the software certification for TTP-OS.

TASK ACTIVATION

The main purpose of the operating system is to deterministically activate time-triggered tasks according to a static schedule, and to run non-time-triggered software whenever no time-triggered task is active. As specified by OSEKtime [4], TTP-OS uses a preemptive scheduling strategy. A task is regarded as the smallest unit of computation that can be activated. There is no mutual synchronization of tasks via blocking mechanisms; timing and resource relations and constraints between time-triggered tasks are specified and resolved at design time.

ERROR DETECTION

For safety critical operation, fast and reliable error detection is required. TTP-OS performs on-line checks to monitor the safe state of operation. For its own configuration data and execution tables, TTP-OS performs signature checking to ensure that it is operating on valid data. For each time-triggered task, a statically defined deadline is monitored by TTP-OS to detect if any task violates the schedule properties specified at design time. In this way, time-triggered software running on an ECU will behave fully deterministic in the fault-free case, and otherwise an exception is raised.

EFFICIENCY

TTP-OS is a very compact operating system supporting the deterministic execution of hard real-time and fault-tolerant applications. Short task switching times and off-line chaining of tasks allow keeping the execution overhead low and efficiency high. The services provided are restricted to task activation and error monitoring; drivers (e.g. CAN) and scheduling/dispatching of asynchronous tasks are added individually and require independent certification, if applicable.

SOFTWARE LIFECYCLE

PROJECT PLANNING

The Software Project Plan (SPP) is based on the Software Development Manual (SDM) and the Software Development Standards (SDS). The SDM specifies guidelines for the production of airborne systems software, with the SDS providing software development standards, e.g., coding standards.

Development Guidelines

The SDM comprises the following non-functional process requirements:

- Project management
- Organizational requirements
- Software verification plan
- Software configuration plan
- Software quality plan
- Customer liaison

Development Standards

The SDS comprises the following software development standards:

- Software document standards

  All software lifecycle documents are created using the LaTeX [3] typesetting environment together with a set of user-specific style files to ensure consistency. This offers a programmable environment that allows for flexible automatic generation of documents of all sorts.

- Software requirement standards

  Functional product-specific requirements are written in structured natural language, which is an approach defining forms and templates to express the requirements specification.

- Software design standards

  Methods and rules are defined to be used to develop software architecture and detailed design definitions.

- Software coding standards

  Coding standards for software development in any programming language used are defined.

- Software test case and procedure standards

  Standards for both test cases and test procedures are defined.

Development Plan

The Software Project Plan defines the project-specific rules and methods to be applied to the development of a certifiable software product. Furthermore, the Software Project Plan defines the project responsibilities and includes all the planning data. The planning data implies the delivery dates as well.

The planning phase results in requirements engineering.

The following diagram shows software lifecycle processes (in oval boxes) and the resulting artifacts. The diagram indicates that Configuration Management (CM) and Software Quality Activities (SQA) accompany all processes.

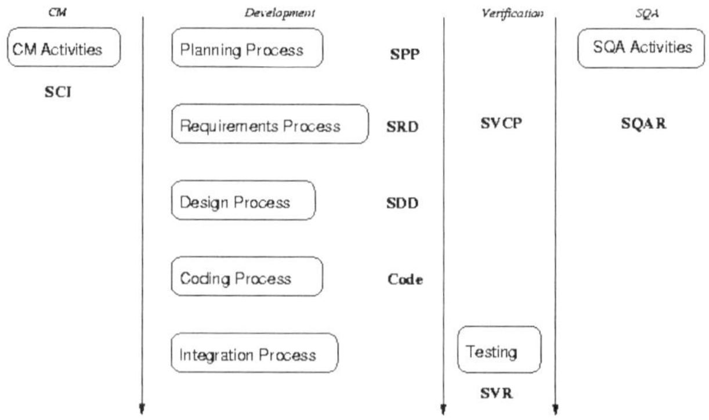

Figure1: Software Lifecycle Processes And Artifacts

REQUIREMENTS ENGINEERING

This subsection covers the definition and specification of functional high-level requirements. Functional low-level requirements are described in the SOFTWARE DESIGN section.

Requirements Specification

The functional high-level requirements specification is based on the respective requirements definition (system requirements), which is a contract between the customer and the software developers.

Non-Functional Product Requirements

Non-functional requirements describe constraints on the services or functionality provided by the system. For example, they list constraints on memory use and execution times of OS services.

Functional Product Requirements

These requirements state the services and/or functionality the system shall provide. For example, they define the services for OS startup, task activation, error handling, OS shutdown and OS restart.

Each functional product requirement is labeled with a unique identifier, a tag, which is used to create a table listing all the functional product requirements defined.

Reviews

The functional high-level requirements are reviewed to prevent ambiguity using The Bender Ambiguity Review Process [2] guidelines and DO-178B related checklists. This review is aimed at improving the quality of requirements to make them deterministic, complete, unambiguous and correct. Changing high-level requirements later in the software lifecycle typically incurs high additional efforts; therefore, a stable high-level requirements document is the necessary prerequisite for the next phase, the software design.

SOFTWARE DESIGN

Once the architecture of the system to be developed is specified by decomposing the system into subsystems and further into modules, functional low-level requirements are to be defined. These requirements are either directly traceable to or derived from the functional high-level requirements.

The DO-178B guidelines [1] define the term "derived" as follows:

"Derived requirements are requirements that are not directly traceable to higher level requirements."

Derived low-level requirements are newly introduced in this stage of the software lifecycle. The architecture and the low-level requirements are documented in the Software Design Document (SDD).

Software Design Document Inputs

The software design inputs are the Software Development Plan, the Software Development Standards and the Software Requirements Document.

Software Design Document Overview

The SDD comprises the following main parts:

- Architecture

  This part documents the system that is decomposed into subsystems, for example, the OS Startup subsystem.

- Data structures and data types

  This defines the configuration data structures of the operating system, for example the Time Source Configuration and data types.

- Detailed Design

  The OS subsystems are further decomposed into modules, like 'Start Schedule Table'. An activity diagram, its corresponding low-level requirements, and a table that identifies the module callers specify each module.

- Application Specific Requirements

  This part identifies the requirements that need to be satisfied by the application for compliance with the operating system.

- DO-178B Compliance Table

  This table summarizes DO-178B compliance of the Software Design Document.

- Traceability Matrices

  The traceability matrices reflect the high-level to low-level requirements traceability, the low-level to high-level requirements traceability, and provide an index of derived low-level requirements.

Traceability

The traceability between low-level and high-level requirements is provided to make derived requirements visible, which allows verifying the complete implementation of the high-level requirements.

Review

The review of the design document focuses on DO-178B compliance and configuration management to check if the document is reproducible, but also on review of all compliance matrices listed in the Software Design Document.

At this point, the requirement tags are available as lists/matrices, and a comprehensive check method is essential to find out if all requirements are traceable. Simple issues such as a typo in a requirement tag can break traceability. Here, automation by LaTeX programs ("scripts") is extensively used: Scripts are used to parse the requirements specified in the SRD and SDD and support the review of the compliance matrices listed in these documents by automatically providing traceability tables and lists of unmatched requirements for correction. Later in the software lifecycle the source code and the test cases are also checked for traceability.

SOFTWARE CODING

The source code is implemented according to the software architecture, which consists of the subsystems and the modules into which these subsystems have been decomposed, and the low-level requirements.

Software Coding Inputs

The inputs for this phase are the software architecture, the low-level requirements for the operating system, the Software Development Manual, and the Software Design Standards.

Coding Guidelines

The Software Development Standards define the coding guidelines to be used, specifying not only the guidelines that do not depend on a certain programming language, but also the guidelines that actually do (for example, primitive data types).

Traceability Data

Traceability data precedes each function code. Traceability information is implemented in the form comments in the source code to allow automatic traceability checking of the low-level requirements specified in the software design. Script-supported traceability checking supports, but does not replace, manual traceability checks.

Code Review

The source code is reviewed for compliance with software standards, low-level requirements and architecture, testability, verifiability, accuracy, and consistency (for example, division by zero).

SOFTWARE INTEGRATION

The executable object code is loaded into the target for software-hardware integration.

Qualified Compiler

A qualified compiler is used to get trusted output in the form of executable object code. This trusted executable object code is an input to the verification phase.

Integration Review

Integration review is a procedure used to verify the make process, the optimization options, compiler warnings that may occur, the completeness of the software components, and whether the memory-mapped addressing is consistent with the hardware and software design (for example, memory areas to be used by the startup of the operating system).

The outputs resulting from the software design and software integration design might be used as an input to the low-level design or even the high-level requirements engineering phases, if necessary.

THE SOFTWARE VERIFICATION PROCESS

The software verification process is a combination of

- Reviews and analyses
- Tests

The goal of this process is to evidence the requirements to have been correctly implemented. Verification is a combination of the points mentioned above, because testing alone cannot prove the absence of errors. Reviews and analyses are performed manually.

The purpose of this process is to detect errors that might have been introduced during the software development processes – any of this data is used as an input to the respective phase in the development process.

Verification Process Inputs

The inputs to the software verification process are system requirements, high-level software requirements, software architecture, traceability data, source code and executable object code.

Software Testing

Software testing is based on the software requirements specified. Module tests are based on low-level requirements, whereas integration tests are based on high-level requirements.

Software testing is conducted by defining

- Test Procedures

  Test procedures provide a step-by-step guide on how to execute test cases. Test procedures are part of the Software Verification and Procedures Document (SVCP), which provides a detailed description of the test environment.

- Test Cases

  Each test case is defined by a set of input conditions, the results expected to achieve code coverage, information on low-level requirements traceability, and pass/fail criteria. There are two kinds of test cases: robustness tests and normal test cases. Robustness tests show how the system responds to abnormal input conditions.

Test Environment

An automatic certification test environment is used to carry out test cases. The test environment running on a host PC executes test cases on the target platform and records whether a test case has passed or failed.

Structural Code Coverage

Code and decision coverage is achieved by executing test cases. Full Modified Condition/Decision Coverage (MC/DC) is required for Level-A software, as defined in DO-178B [1]:

"*Every condition in a decision in the program has taken all possible outcomes at least once, every decision in the program has taken all possible outcomes at least once, and each condition in a decision has been shown to independently affect that decision's outcome. A condition is shown to independently affect a decision's outcome by varying just that condition while holding fixed all other possible conditions.*"

Requirements Reviews And Analyses

It might be infeasible to record all the results obtained from requirements testing; for example, a reboot after an exception where "reboot" is defined as the consequence is a correct result, but difficult to record. In these – rather rare – cases, the behavior is analyzed.

Traceability

A traceability matrix providing the traceability between requirements and test cases concludes the SVCP. This is vital information, giving evidence that the requirements have been thoroughly tested.

Software Verification Results

The Software Verification Results (SVR) documents all results obtained from the automatic execution of test cases and analyses.

THE REVIEW PROCESS

All the software lifecycle artifacts and source codes are subject to review.

Entry Criteria

Before an artifact – e.g. the Software Requirements Document – enters a formal review, it needs to be internally reviewed according to the process defined in the Software Development Manual and Software Design Standards.

Before a peer review starts, the roles of those participating in this review are defined as follows:

- Project Manager
- Peer Reviewer/ Reviewer

  The peer reviewer checks the entire artifact or all code under review. This is to ensure consistency. Reviewers check predefined sections of artifacts or code modules.

- Author

  The author changes the artifacts or source code and responds to review comments accordingly.

- Recorder

  The recorder summarizes the entries made by the reviewers and the peer reviewer and calls a peer review meeting, if required.

Formal Peer Reviews

Formal reviews are conducted using in-house news groups. Entries are made maintaining the following order:

- Review entry: it defines the roles and criteria according to which a review is conducted.
- Announcement of the artifact and its version.
- Review/Peer review entries are categorized as
  - Stoppers, which means that a complete redesign is required
  - Major findings
  - Minor findings
- Responses to deficiencies found by each peer reviewer and responses to reviewer entries, which is followed by an announcement of a new version of the respective artifact/code
- Reviewer's replies to what the author's replied to in order to confirm a finding to be fixed or obsolete
- A summary of the review process by the recorder and a peer review announcement, if required. The outcome of such a peer review meeting is documented in the in-house newsgroup mentioned above

If there are open issues remaining after the peer review is concluded, these issues require additional development effort and are accordingly put into a problem tracking system and categorized.

CONCLUSION

DO-178B provides guidance as employed on the design, specification, development, testing, and deployment of TTP-OS. The DO-178B guidance has been used to create a software certification process for the development of TTP-OS. The development of TTP-OS according to this process has yielded a certifiable real-time operating system.

DO-178B covers software life cycles, software planning processes, software development processes, software verification processes, software configuration management processes, software quality assurance processes, and other aspects of creating quality software for a safety-critical environment.

Future projects using this certifiable operating system will benefit from the certification effort already invested.

REFERENCES

[1] RTCA. *DO-178B - Software Considerations in Airborne Systems and Equipment Certification*. 1992.

[2] Bender, Richard. *The Bender Ambiguity Review Process*. Bender RBT Inc. 2003

[3] Lamport, Leslie. *LaTeX: A Document Preparation System*. Addison-Wesley Professional. 2nd ed. 1994

[4] OSEK group. OSEK/VDX Time-Triggered Operating System, Version 1.0. 2001

CONTACT

Peter S Groessinger
TTTech Computertechnik AG
Schoenbrunnerstrasse 7
1040 Vienna, Austria
Tel : +43-1-5853434-0
Fax : +43-1-5853434-90
Email : peter.groessinger@tttech.com

DEFINITIONS, ACRONYMS, ABBREVIATIONS

SDM: Software Development Manual. This document provides a guideline for the production of airborne-systems software.

SPP: Software Project Plan.

SCI: Software Configuration Index. This document contains lifecycle data collected during the whole software lifecycle.

SQAR: Software Quality Assurance Records. The software quality assurance activities include process evaluation activities and the software conformity review.

SRD: Software Requirements Document. This document contains the high-level functional product-specific requirements.

SDD: Software Design Document. This document contains the low-level functional product-specific requirements.

SVCP: Software Verification Cases And Procedures. This document contains the test cases and procedures for both high-level and low-level requirements and analysis of requirements.

SVR: Software Verification Results. This document contains the results of test runs and analysis.

2005-01-0779

Survey of Software Failsafe Techniques for Safety-Critical Automotive Applications

Eldon G. Leaphart, Barbara J. Czerny, Joseph G. D'Ambrosio, Christopher L. Denlinger and Deron Littlejohn
Delphi Corporation

ABSTRACT

A requirement of many modern safety-critical automotive applications is to provide failsafe operation. Several analysis methods are available to help confirm that automotive safety-critical systems are designed properly and operate as intended to prevent potential hazards from occurring in the event of system failures. One element of safety-critical system design is to help verify that the software and microcontroller are operating correctly. The task of incorporating failsafe capability within an embedded microcontroller design may be achieved via hardware or software techniques. This paper surveys software failsafe techniques that are available for application within a microcontroller design suitable for use with safety-critical automotive systems. Safety analysis techniques are discussed in terms of how to identify adequate failsafe coverage. Software failsafe techniques are surveyed relative to their targeted failure detection, architecture dependencies, and implementation tradeoffs. Lastly, certain failsafe strategies for a Delphi Brake Controls application are presented as examples.

INTRODUCTION

Delphi has been involved with development and production of numerous vehicle systems that may be classified safety critical with respect to their operation on the vehicle. Technological advances associated with these systems may require corresponding advances in techniques to help verify safe operation of these systems. One such technological advancement is the inclusion of electronics to aid in the control and safety aspects of vehicles. Such systems as Throttle-by-wire, Controlled Braking, Controlled Steering, and Supplemental Inflatable Restraint systems are commonly recognized as being integral to the safety aspects of the vehicle. These systems have advanced tremendously in their capabilities and application across a wide number of vehicles. As these types of systems continue to evolve, Delphi has been involved with helping to determine the proper methods and techniques for evaluating these systems and understanding the safety and reliability aspects at all levels of the design – be it within the whole system, a sub-system, or at a component level.

In the overall consideration of available techniques, the product teams need to understand the trade-offs between utilizing these techniques within their system hardware designs, and more and more commonly, within their software designs. With today's systems, a particular concern may be addressed by any of these design methods or by a combination of design methods. The software techniques and analysis methods described here do not represent an exhaustive list when compared to all techniques available within the broader embedded controls community, but they do represent sound methods that design teams may choose to utilize for their products.

ANALYSIS METHODS FOR IDENTIFYING NEEDED FAILSAFE TECHNIQUES

Software failsafe techniques are primarily developed to detect potential Electronic Control Unit (ECU) or peripheral hardware failures, thus enabling the system to initiate a transition to a safe state if any such potential failures occur. These techniques are important for safety-critical systems, because system developers must help verify that potential failures will not lead to any potential system hazards. There are many possible techniques to apply in helping to identify potential failures and needed failsafe techniques, but of these, fault tree analysis (FTA) and failure modes and effects analysis (FMEA) are the most commonly applied. In this section, we review these methods, as well as two others that we have found useful: preliminary hazard analysis (PHA) and fault coverage matrix.

A PHA is a high-level hazard analysis performed during the early stages of development to help identify the potential high-level hazards of the system, and to identify the potential risks of the high-level hazards. During PHA, the potential hazards are identified and described, potential worst-case mishap scenarios are determined, potential causes are identified, and the risk associated with the potential hazards and mishap scenarios is determined.

For potential high-risk items, the design team identifies ways to eliminate or mitigate the potential hazards. The mitigating actions become safety requirements for the system and may be implemented in hardware, in software, or in both. The safety requirements identified by the PHA are typically high-level, and as a result, don't necessarily identify individual failsafe techniques. Instead, these high-level requirements often provide direction on identifying an overall ECU integrity strategy. The strategy may include specific ECU hardware features to support high integrity operation and an initial list of software failsafe techniques appropriate for the targeted ECU integrity strategy. This initial list of software failsafe techniques would be primarily based on past development experience with similar ECU integrity strategies.

FTA is a deductive analysis method used to identify the specific causes of potential hazards, rather than to identify potential hazards. The top event in a fault tree is a previously identified potential system hazard, such as unwanted apply of the brakes. The goal of a FTA is to work downward from this top event to determine the potential credible ways in which the undesired top-level event could occur, given the system operating characteristics and environment. The fault tree is a graphical model of the parallel and sequential combinations of faults that could result in the occurrence of the top-level hazard. FTA uses Boolean logic (AND and OR gates) to depict these combinations of individual faults that can lead to the top-level potential hazard.

Each of the specific potential failures or classes of potential failures identified by a FTA is reviewed, and if necessary, appropriate hardware and software mitigation techniques are identified to reduce the likelihood that the top-level potential hazard will occur. One possible output of this activity is a list of software failsafe techniques needed to mitigate the identified potential hazards. While developing the fault tree, the initial list of software failsafe techniques, identified by the PHA, can be included in the analysis. Development of the fault tree can also identify additional software failsafe techniques that may be necessary as well as eliminating unnecessary techniques that add no value. If the initial list is based upon previously developed failsafe techniques, then the revised list will mostly likely be made up of well-understood techniques that require little development effort.

FMEA is an inductive analysis method used to:

- Identify and evaluate potential failure modes of a product design and document their system effects
- Determine actions or controls which eliminate or reduce the risk of the potential failure
- Document the process.

FMEAs are widely used in the automotive industry where they have served as a general-purpose tool for enhancing reliability, trouble-shooting product and process issues, and analyzing potential hazards.

Each of the potential failures or classes of failures identified by the FMEA is reviewed, and similar to FTA, appropriate hardware and software mitigation techniques are identified. Thus, a possible output of FMEA is a list of software failsafe techniques needed to mitigate those potential failure modes that may lead to potential system hazards.

Since FTA focuses on only those potential failures related to known potential hazards and FMEA considers all potential failures independently, it is probable that the FMEA will generate a larger set of potential failures to consider. However, the FTA may also contain potential failures or combinations of failures that are not identified by the FMEA process.

Another tool that may be used to determine necessary software failsafe techniques is a fault coverage matrix. The focus of this analysis is on determining the best set of controls (e.g., software failsafe techniques) to cover an identified set of failure classes (e.g., ECU hardware failure classes such as ALU miscalculations, and memory errors) such that adequate coverage is provided for each failure class.

The analysis can be performed using a spreadsheet similar to the one shown in Table 1.

Table 1: Fault Coverage Matrix

	Selected.	**Potential Failure 1**	**Potential Failure 2**	**...**	**Potential Failure N**
Potential Risk		*Critical*	*Moderate*		*Low*
SW FS Tech. 1	Yes	H	H		H
SW FS Tech. 2	No	M	H		N
...					
SW FS Tech. n	Yes	L	N		L
Coverage Metric		M	H		H

Known potential failures and associated risk levels are captured across the top of the spreadsheet. A list of known controls (e.g., failsafe methods) relevant to the potential failures is captured in the first column of the spreadsheet. The spreadsheet is filled out such that the coverage (e.g., High, Medium, Low, None) provided by each control for each of the potential failure classes is specified in the cells of the matrix. The controls that are currently selected for implementation are identified in the second column of the spreadsheet. The spreadsheet sums up the coverage level for each potential failure based on the coverage provided by each of the controls selected for implementation. The coverage metric depends on the potential risk associated with a potential hazard, such that high risk implies that higher coverage is required.

A significant advantage of a fault coverage matrix is that all failsafe techniques are considered at the same time, instead of individually, as is typically the case with FTA or FMEA. This global view helps verify that the best set of overall techniques is selected. Taken together, the PHA identifies the initial list of techniques, FTA and FMEA provide complementary detailed analysis to help verify that identified failsafe techniques cover all faults, and fault coverage matrix helps identify a final optimized set of failsafe techniques.

SOFTWARE FAILSAFE TECHNIQUES

This section provides information about different software failsafe techniques. For each technique a description, discussion of the major failures a given test will detect, and design limitations are given. In general, each of the techniques described may detect multiple types of failures. In some cases, the root cause of the failure may not be determined. However, detection of a failure is sufficient to trigger the appropriate failsafe action for the system. Eleven techniques are described in total. Table A1, found in the appendix, provides a comprehensive summary of these methods.

COMPLEMENT DATA READ/WRITE

Complement data read/write is useful for assuring the integrity of data being stored to memory (RAM). The data that is to be retained is stored as the actual value in one part of memory. The one's complement of the data is calculated and then stored in a separate part of memory. For example, if the data to be retained is 0xB136, then 0xB136 is stored in one part of memory, and 0x4EC9, the one's complement, is stored in a different part of memory. When the data is to be used, the two stored values are summed. If the summation is not zero, then a degradation in the memory has occurred.

Specific data storage errors that can be detected using this method include individual bits that are hard stuck to a value, and values that decay over time. Since this test is run prior to a value being used, even the long-term decay of values can be detected.

The major limitation on implementing the complement data method is memory size. If every data value is stored with its complement, the amount of RAM needed would double. To address the size requirements, data values can be partitioned into safety-critical data and non-safety-critical data. Only those variables identified as safety critical are stored as complements. In addition to the size limitation, if the complement values are stored in close physical proximity in memory, then a failure to a section of memory could cause both values to fail. A solution to this problem can be to store the complements in different physical locations, either on different pages of memory, if available, or in physically separated areas of the memory structure.

CHECKSUM COMPARES

The basic idea behind the checksum technique is to verify the integrity of program or calibration memory (ROM / Flash). The checksum method sums all of the data in memory, then truncates the sum to give the checksum value. The one's complement of the sum may be taken for easier comparison, however, two's complement or other formats are also common. When the checksum is verified, the data is again summed, and the original checksum value is added to the new sum of the data. A successful test results in zero. The length of a checksum can vary. In some applications words are used, and in other applications bytes are used.

A checksum test can be done at various times during program execution. One common time is at initialization. An initialization checksum test may be implemented in two ways. One of these is done mainly in ROM where the code will not change from cycle to cycle. In the original coding, the checksum values are calculated, stored, and then compared with the value calculated during initialization. For memory that will change from cycle to cycle, like EEPROM, a checksum can be calculated and stored at shutdown after all of the new values are written, then compared with the value calculated at initialization.

A ROM checksum may also be verified during runtime. This test may be done as a background task that takes many loop-times to test the entire code. Since verifying the entire ROM may take many loop-times, an error may persist for many control cycles before it is detected. To reduce the likelihood that an error in a safety-critical section of the code persists beyond a certain time, a separate checksum can be performed at a faster rate for the safety-critical code portions. This is called a fast compare.

The fast compare method detects failures in the ROM and EEPROM. Checksums are able to detect permanent errors in memory, such as flipped bits, and other changes in values. Since the calculation of the checksum requires the use of the ALU, this method also provides some fault detection coverage for the ALU.

The largest limitation related to checksum tests is time. During runtime, the background test may be too slow to detect all errors in time to prevent a failure from leading to a potential hazard. Therefore, the code may be partitioned into safety-critical and non-safety-critical code, and the fast compare method may be used for the safety-critical code sections. This method helps confirm that a fault occurring in the safety-critical code is detected fast enough to prevent a failure from leading to a potential hazard. Since the tests performed during runtime are executed in a background task, there is typically not a large burden on the CPU resources.

REDUNDANT CODING

Redundant coding, or dual path software, is a methodology to store critical code in the program memory identically in 2 different memory areas. During runtime, both sets of code are run using the same inputs and the results are compared. The two results should be the same (or within some specified tolerance), so that a difference indicates an error. One method to improve redundant coding is to store the different pieces of code on separate pages of program memory. This way, if there is a failure on a particular page of memory, the failure will not manifest itself in the second copy of the code.

This technique can detect changes in memory (either ROM, RAM or EEPROM), and intermittent faults in the ALU, such as faults caused by EMI.

The largest limitation for redundant coding is it doubles the amount of code and processor time needed to implement a function. Another limitation is only transient or intermittent faults in the ALU can be detected.

REDUNDANT “ORTHOGONAL” CODING

Orthogonal coding is a process where safety-critical code is implemented two times using different processes or processor resources for each implementation. Orthogonal coding may be done using a different algorithm for the calculation, using the same hardware resources, or using a different algorithm and different hardware resources. Since the orthogonal coding method relies on the use of different methods of calculations, the two results may not be exactly equal to each other. Therefore, when a comparison is done, a tolerance may be required to determine if the results match.

The major failures that can be detected by orthogonal coding are failures in memory or the ALU. Orthogonal coding may be effective at detecting a number of ALU failures depending on how it is implemented; for example, using different instructions of the ALU or using different hardware.

This orthogonal coding method may be memory intensive as it doubles the amount of memory required to implement a function. It may also double the amount of CPU time required. In addition, this method requires more development time since two different algorithms have to be created and maintained throughout development. Finally, the tolerances must be validated to help confirm that they are not too constrained, thereby leading to false positives, and that they are not too unconstrained, resulting in false negatives (i.e., no failures are identified, when a failure actually exists.)

Redundant “Orthogonal” Coding Example: MAC vs ALU

An Arithmetic Logic Unit (ALU) in parallel with a Digital Signal Processor (DSP) peripheral is one example of the redundant “orthogonal” coding technique appropriate for providing coverage of arithmetic intensive control algorithms. The ST Microelectronics ST10 processor features a Multiply and Accumulate (MAC) DSP peripheral in combination with the ALU within the CPU core. The configuration of the CPU core and MAC peripheral within the ST10 microcontroller is shown in the block diagram given in Figure 1.

The MAC and ALU have different instruction sets for mathematical operations. Several operations are possible within the MAC, however the unit is designed to optimize multiply, accumulate, and digital filtering operations.

A strategy has been developed for use within brake control applications to perform fixed point multiply instructions in parallel both in the ALU and in the MAC for each usage of the multiply operation. The products from the MAC and ALU are compared and should always be equivalent. A detected error indicates an issue in one of the peripherals. The basic data flow of this strategy is shown in Figure 2.

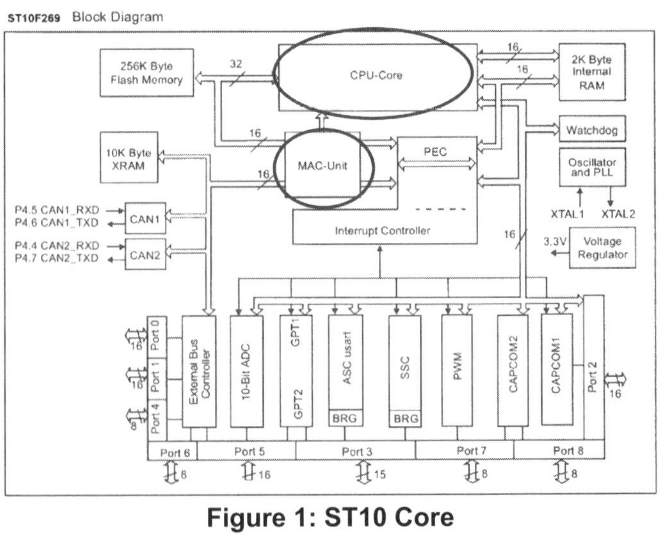

Figure 1: ST10 Core

The coverage of this strategy may be evaluated by identifying the number of multiplication operations used within an algorithm per execution loop. The MAC vs ALU compare will occur for each multiplication operation or macro that is executed. For a typical embedded controls fixed-point implementation, several types of multiplication macros may be used. A coverage matrix may be developed to identify which functions make use of certain multiply operations and how many multiplies are required per execution loop. Failsafe coverage is provided for the ALU during each usage of the MAC vs ALU compare. The redundant coding technique may be combined with other techniques to maximize the overall system failsafe coverage.

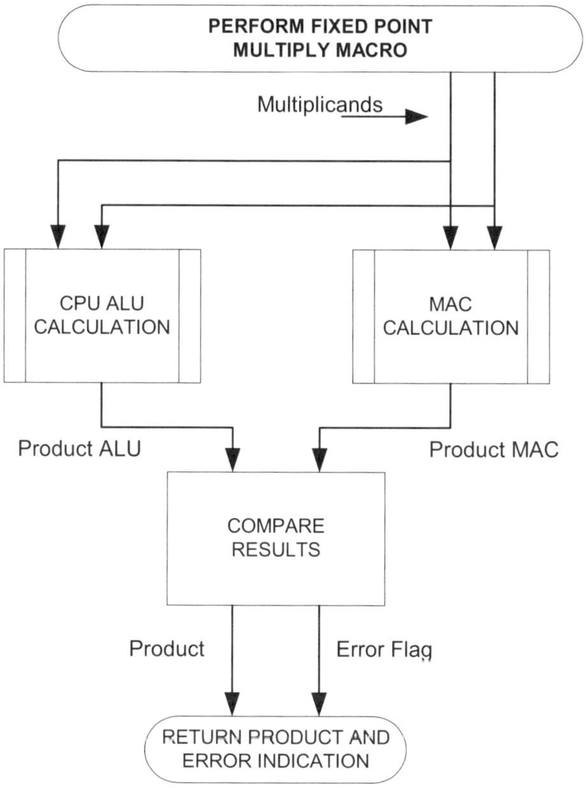

Figure 2: MAC vs ALU Compare

PROGRAM FLOW MONITORING

Program flow monitoring (PFM) or process sequencing, is a technique to include a specific seed (initial key value) and key (final value/result) process within the program function to assure that the program execution has completed the major parts of the program, and that it has completed them in the correct order. Typically, the program being monitored will contain specific update points throughout the program flow. The update points are specific functions that operate on a parameter being supplied to them. This parameter may be referred to as the key value. At regular update points, or at the end of the program execution, the resultant key value is compared to the pre-calculated acceptable value. If a mismatch is discovered, then a program execution error has occurred. PFM may be implemented in two ways: application independent or application dependent.

The application independent method works by having a PFM update point between each function call. A consequence or side effect of this approach is that the point can be updated without the function having actually been called. However, this approach also provides greater flexibility and opportunity for re-use. For example, assume there are common functions A, B, C, and D across applications, and that for a particular application only functions A and D are needed. Using the application independent implementation allows the program flow monitoring code to be used without modification across both applications.

The application dependent implementation is more tightly integrated into the program execution. The actual PFM update points are coded within the functions themselves. This approach helps assure that all of the functions are called and that they are called in the correct order.

If specific functions need to be called within a certain window of time in relation to other functions, the application independent or application dependent methods of PFM may be enhanced to help verify the correct timing requirements. This enhanced method is known as time dependent PFM. This method helps confirm not only that the functions are called and that they are called in the correct order, but also that they are called within the required window of time. This task is accomplished by requiring the PFM update to occur at a specific time during the program execution. A flow chart showing the differences in the implementation is shown in Figure A1 of the appendix.

At each update point in the program execution, a function is executed to update the PFM variable. Various algorithms can be utilized for updating the key value. A simple version of an update function is:

PFM_key=PFM_key+PFM_ID
PFM_key=PFM_key*PFM_ID

- PFM_key is the value carried throughout the loop that becomes the key
- PFM_ID is the ID of the update point. If there were four update points, then they would be numbered 1 to 4.

Therefore, as long as all of these updates or entry points, are run in the right order, the key will be correct. It is also beneficial to have multiple seed and key pairs so that the test cannot be passed merely because the key value is stuck at the correct value, or just never rewritten.

There are multiple ways to design PFM. One of these ways is with a single microprocessor design. The microprocessor can check the value of the PFM key at the end of a loop. This is equivalent to having the microprocessor check itself, and thus, not all failures related to PFM will be detected. Another design strategy uses an asymmetric microprocessor. An illustration of PFM data flow for an asymmetric design is shown in Appendix Figure A2. The monitoring microprocessor can query the main microprocessor every other loop for the PFM key. Since the monitoring microprocessor is an independent piece of hardware, it will be able to pick up most failures related to PFM. Another design strategy can be used if the controller is part of a distributed system. One of the other controllers in the system can take the place of the monitoring microprocessor in querying the main controller.

PFM can detect process errors such as the program skipping an important part of the program calculation. The extent to which program flow monitoring can detect errors is dependent on how many update points there are in the program and where the updates occur within the program (i.e., within the functions, or between function calls).

The biggest limitation for program flow monitoring is the amount of processor time consumed by the technique. If there are many PFM updates within a program performing a number of calculations, the amount of processor time PFM requires can be significant. Consequently, there is a trade-off inherent with PFM; the deeper the updates or thread depth, the better the detection ability of the method, but the more processor time is required.

Another design decision is which type of PFM to use. The benefit of an application independent approach is increased flexibility; the PFM code may be used over multiple applications. However, the coverage is limited and provides less confidence that a skipped function will be detected. Using an application dependent approach allows for better coverage and more confidence that a skipped function will be detected, but requires more maintenance as different applications may require a different set of functions to be used requiring all of the PFM routines to be reworked for each application. The time dependent approach used in conjunction with the application independent or application dependent methods helps assure that the program is flowing within the desired time frame, however, this method may not be feasible for applications with interrupts, since the interrupts may disrupt the timing.

RAM TESTS

RAM tests may be performed at initialization or during system runtime. A RAM initialization test is typically a set of tests to determine if the RAM of the microprocessor is functioning correctly before any application program tasks are started. On initialization, the RAM is tested to make sure that it can be written to and that it can hold a value for a short period of time. This is accomplished by writing a specific value or pattern to all RAM locations and then reading it back and comparing the read values to the written values. This operation is done twice using different values each time. Typically, the hex numbers 0xAA and 0x55 are used. These numbers are chosen so that all bits will have a '1' and then a '0' written to them. Other methods, such as "walking ones" method, where a single bit is systematically written and cleared are also commonly used.

There are two major failures of RAM that can be detected with this test: bits stuck as either a '1' or a '0', and decaying RAM cells. Some decaying faults may still pass depending on how long it takes the value to decay.

RAM tests may also be performed during system runtime. This test method is similar to the test at initialization, where a specific pattern is written and read to RAM values. The runtime test must be designed as to not interfere with normal operation of the system since test values written to RAM, if read and used by the application during the test, could cause improper operation. This can be accomplished by performing the test during a background task or disabling other system resources while performing the test. The runtime test will take longer to check all RAM than the test at initialization. During runtime, RAM must be checked in small segments incrementally per application loop in order to minimize impact on system resources.

POWER UP/DOWN MEMORY WRITE TESTS

Power up/down memory write tests are used to determine if a controller has shut down properly. Information critical to the proper operation of the system may need to be stored in nonvolatile (NVM) or "keep-alive" (KAM) memory between ignition cycles. Typically, this information is stored during the shutdown sequence of the controller.

During controller initialization, a specific data pattern is written to a NVM location (e.g. 0x55). During the shutdown sequence of a controller a different pattern is written to the same NVM location (e.g. 0xAA). A compare of the memory location is made at the next initialization sequence. If the data matches the data pattern written at the previous shutdown (e.g. 0xAA) then the test indicates that the controller had shutdown properly. A data read of the initialization pattern (e.g. 0x55) indicates that the controller had not gone through shutdown properly.

The power up/down sequence is effective in identifying when the controller has been abnormally reset or when system power is lost prior to completion of a shutdown sequence. Safety-critical processes or data may need to be reinitialized upon detecting an abnormal shutdown sequence. The design of a power up/down memory sequence must be coordinated with the overall power moding and software task execution of the controller design. In addition, NVM or KAM hardware resources must be present in the hardware design.

TEST CASES

Test cases or test vectors are used to exercise the instructions of the ALU to detect ALU faults. Independent hardware is required to perform the test cases. Either an asymmetric processor or a secondary processor in a distributed system can be used to perform test cases.

The ALU operations are tested using an algorithm written to access all of the ALU instructions used in the main program. This algorithm is called by the independent hardware using a seed and the output is compared to an output key. The seed is the initial starting value to be input into the test case calculation. There are multiple seed values. After all of the test calculations are completed, the output should be equal to the key that is appropriate for the given seed.

The algorithm can be split into multiple parts. Each part can be called at different times during a loop execution or the different parts may be called over multiple loops. Ideally the algorithm will cover all of the instructions of a microprocessor, but since the instruction set may be large (over 200 instructions for a Motorola HC12), including only those instructions used in the program is generally acceptable.

There are two ways to implement test cases. One is to have a sequenced query, such that the order of the seeds is the same every time the program is run. Another method is to have a random query. In the random query, the monitoring unit has the ability to vary the order of the test cases.

The major types of ALU failures that can be detected using test cases include register failures and individual instruction failures.

The test case method requires independent hardware to perform the test cases, so it can only be used in a design that will have either a monitoring unit or multiple processors as in a distributed system. Since the majority of safety-critical automotive software is written in higher level languages such as C, C++, Modula, etc., it is useful to know which low-level instructions are used to implement the high-level instructions, so these instructions can be adequately tested. If the program changes and new instructions are utilized, then the test cases will need to be modified to include the new instructions.

COMPUTER OPERATING PROPERLY (COP) WATCHDOG TIMER

A watchdog timer is a device that helps assure that the microcontroller is operating properly. A watchdog timer may be internal or external to the system. It is a mechanism that begins to count down once it has been initiated. The device needs to be toggled / refreshed by software within a certain period of time to prevent a microcontroller reset. For an internal watchdog timer implementation, the counter and refresh circuitry are built into the microprocessor chip. For an external implementation, the counter and refresh circuitry are external to the microprocessor chip. An external watchdog timer is typically built using an external RC circuit to perform the timing function. The external timer is toggled or refreshed via an output line from the microprocessor, and a reset is triggered via a reset input to the microprocessor in the event the timer function reaches the pre-set watchdog time.

Watchdog timers are useful for detecting failures such as timing delays, infinite loops, and hung interrupts. Depending on the implementation (i.e., the toggle values or refresh mechanism), watchdog timers may also trigger a reset if the program skips certain steps; i.e., if the toggle values are sent out of order.

If a watchdog is to be used, a key decision is whether an external or internal watchdog should be selected. External watchdog timers are more robust than internal watchdog timers in that they can detect more failures. For example, an internal watchdog timer will not continue to function, and thus will not reset the microprocessor, in the event that the system clock malfunctions. This could happen if the power is reduced to a level that does not cause the micro to reset, but that causes it to cease to function properly. In this situation, an external watchdog would still trigger a reset of the micro. However, external watchdogs require additional hardware which must be designed to interface with the micro. Application and customer safety requirements, as well as other failsafe design methods must be considered in determining which type of watchdog timer is feasible.

COMPONENT/PERIPHERAL TESTS

Software techniques may be used to determine if a specific hardware peripheral or driver is operating properly. For example, a controller output may be driven during a specific initialization sequence and monitored for correct operation. Another example is the comparison of data from two redundant peripherals, where an invalid comparison within a magnitude and/or time tolerance will indicate a failure.

Component/peripheral tests are specific to a hardware design. Often, redundant components are needed for a sufficient failsafe strategy. The design strategy may use additional tests beyond a compare of two inputs to isolate the exact component that is faulty. Synchronization and detection tolerance issues must be taken into account to help assure that the test is accurately identifying failed components.

REASONABLENESS TESTS

Reasonableness tests are methods in which a simplified model is developed for a control variable. The simplified model receives system inputs and determines an estimate of the expected output value. The actual value is compared to the expected value. If the two values differ by some pre-specified tolerance, then it is assumed that there is an error somewhere in the process.

These tests are high-level process checks. They do not detect a specific fault, but rather detect a problem in a calculated output value. They detect that the actual value is out of range with respect to the expected or estimated value. In general, this method provides a sanity check of the overall process.

This method is application dependent; therefore the limitations of this method depend on the specific application.

EXAMPLE REFERENCE: DELPHI ELECTRIC BRAKE SYSTEM 3.0 DESIGN

This section illustrates the application of certain hazard analysis and software failsafe techniques as applied to the Delphi Electric Brake System 3.0 design.

DELPHI ELECTRIC BRAKE 3.0 ARCHICTECTURE

Appendix Figure A3 shows a system mechanization of the Delphi Electric Brake (DEB) 3.0 system. The DEB 3.0 system is a hybrid braking system that contains 2 electric calipers, one on each rear wheel of the vehicle, while the front brakes maintain a conventional hydraulic apply system. The electric calipers receive commands from the brake system controller via a CAN link. The system controller receives all the inputs to the system, and provides the controls for the front hydraulic modulator as well as the processing for all the higher order functions (Anti-Lock Braking, Traction Control, Electronic Stability Control, etc.).

Figure A4 in the appendix shows a mechanization for the controller that is attached to the electric caliper. This controller receives commands from the system controller and provides the positioning of a brushless motor to actuate the rear brake. In addition to the control of the motor/actuator, a park brake mechanism is included in the brake that is controlled by the electric caliper controller. For design space and cost imperatives, each rear controller contains a single microcontroller. It was important during the design of this system that the safety implications of independent electronic control of each rear brake be managed appropriately.

DEB SYSTEM AND SOFTWARE ANALYSIS

Several of the system analysis methods discussed throughout this paper have been applied to the development phase of the DEB controller. Specifically Preliminary Hazard Analysis (PHA) and Fault Tree Analysis (FTA) were used to identify potential hazards and causes of these potential hazards for the DEB system. A coverage matrix was developed to consider which software failsafe techniques would be appropriate to detect potential controller failures that have the possibility of leading to hazards.

Table A2 in the appendix provides an example portion of the PHA. Failure to provide acceleration consistent with driver intent has been identified as a high level potential hazard within the DEB system. Several possible mishap scenarios are described which could result from the occurrence of this potential hazard. One item listed as a cause for such a potential hazard is that of controller failure.

To investigate effectiveness of strategies to detect possible controller failures a coverage matrix was developed. Potential severity and likelihood to occur were assessed for various types of potential controller failures such as memory failures, CPU failures, software processing errors, interface failures and communication failures. Proposed software failsafe techniques were considered for each controller failure category to determine if the coverage is strong (probable) or weak (less effective). Items identified as strong coverage would be considered as part of the failsafe software design. Table A3 shown in the appendix illustrates an abbreviated example of a portion of the coverage matrix.

FTA was used to identify causes of potential hazards of the rear electric brake system. A false apply of the DEB was analyzed to determine its possible causes. A DEB false apply was defined as too much caliper apply. The goal of this analysis is to work the graphical fault tree down to sufficient levels of detail that would identify undesirable causes for failures within the software design. Once these areas were identified, the appropriate software failsafe techniques were applied in order to diagnose these conditions and take the appropriate failsafe action.

A simplified example of a FTA diagram for the DEB 3.0 brake system is shown in Figure 3. It should be noted that this could be expanded to several more levels of detail, however, a general example is shown here. Several causes are identified as factors that given a potential failure could lead to a DEB false apply. Items represented by a transfer symbol (triangle) represent areas that may be further detailed on a separate page of the Fault Tree. Two areas identified as functional elements that could cause a false brake apply are improper behavior of the CAN transceiver and associated software, and improper behavior of the DEB controller software in its entirety.

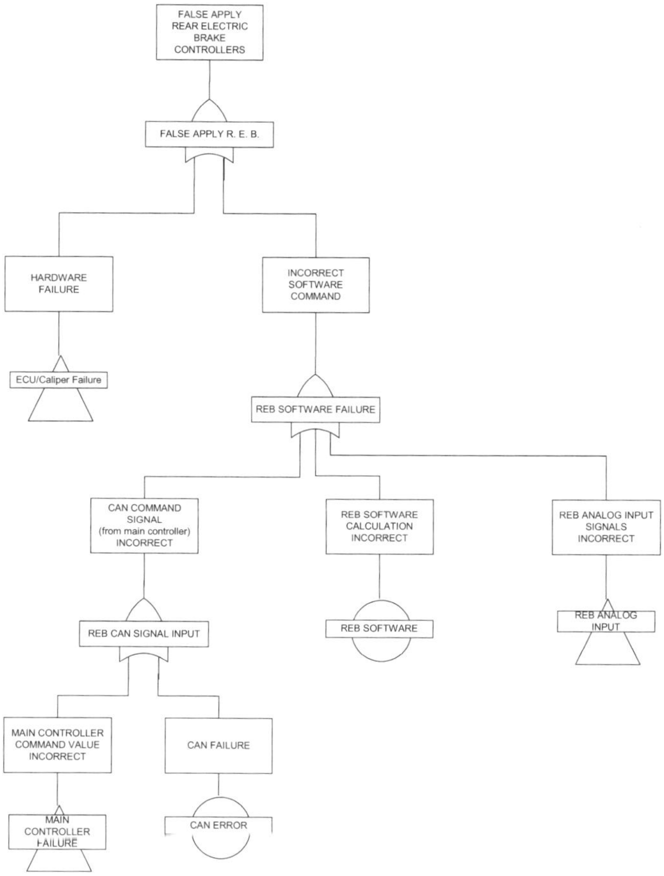

Figure 3: DEB False Apply Fault Tree Analysis

To mitigate the risk of these elements causing a false brake apply, Program Flow Monitoring and Reference Model Reasonable Tests are applied to the design. The following sections describe the tests that were applied to the DEB system.

PROGRAM FLOW MONITORING EXAMPLE

Given that DEB 3.0 is a distributed system, the PFM strategy for this application was to use the multiple controllers to crosscheck program flow. As the two rear controllers run the same software, the primary check is between these two controllers. Every other loop, a rear controller will query the other rear controller to request the key. A rolling seed is used, such that if the key received by the second controller is correct, the controller then sends the next seed. If the second controller thinks there is a problem, instead of shutting down both of the rear controllers, and thus shutting down the rear brakes, the controller will send back a message indicating that the key is wrong. At this time the system controller, which monitors all PFM communications, compares the key of the controller with what it believes the key should be. If the system controller does not agree with the key value, then the controller being tested will fail PFM and appropriate action will be taken. If the controller finds that the key is correct, the controller that initiated the query will fail PFM. The flow of events is summarized in Figure 4.

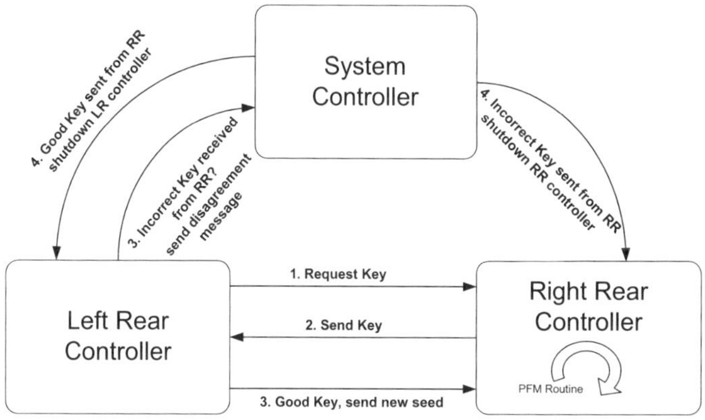

Figure 4: PFM Communications for DEB 3.0

The algorithm for PFM implements test cases to integrate the two techniques. Prior to this application the only experience with program flow monitoring known within Delphi had been using an asymmetric design. Therefore, to work out the exact procedure of the program flow monitoring, a computer simulation was created. The simulation consisted of three computers connected over a CAN link with the CAN traffic being monitored. Each computer simulated a different controller in the system. The goal of the simulation was to develop the messages that were needed to implement PFM and make sure that the idea would work over a CAN bus. To make the program easier to work with, the algorithm implemented for this test was a simple addition and multiplication routine instead of a comprehensive test algorithm.

The simulation demonstrated that the process could detect bit errors as long as they occurred in the correct loop. Since the key is only checked every other loop, it is possible for bit errors, such as a stuck bit, to go undetected by this test. Permanent bit errors were detected during the testing. The simulation program was also able to demonstrate the capability of PFM to detect program execution out of its intended sequence.

FORCE TO POSITION REFERENCE MODEL

For DEB 3.0 system, the output position of the motor is the physical variable that is controlled. The desired position of the motor is based on the force command given by the system controller. The entire process entails the performance of numerous calculations, thus, there are many places for errors to occur. To provide broad coverage of the entire process, a reasonableness test was developed for the position output.

The reasonableness test is set up so the system controller takes its force command and uses a non-linear lookup table to find the desired position. Next it uses a set of second order transfer functions to estimate the actual output of the motor. The transfer functions are used to model the dynamics of the motor. The output of these transfer functions is then compared to the actual motor position sent by the rear controller.

The system controller is only able to get an estimate for the motor position, so the comparison needs to have a tolerance. This tolerance needs to be based on the worst part of the model, which is a step-input for the force. Since the slope of the position curve is so high, a small error in time creates a large error in position. The output of the simulation and the error is presented in Figure 5.

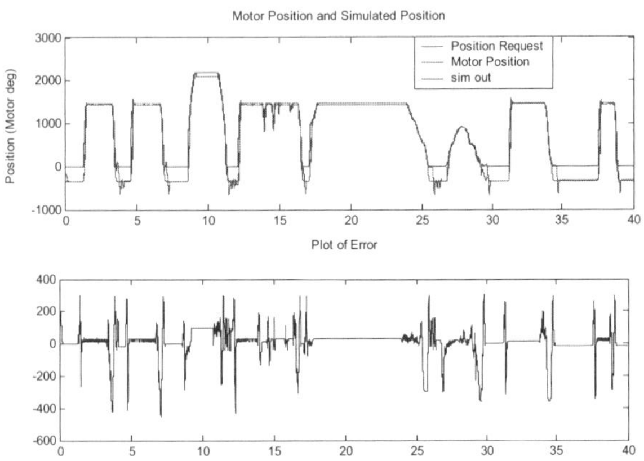

Figure 5: Plot of actual and simulated position with a plot of the error

From the simulation it was concluded that significant errors in position would be caught prior to these errors leading to a potential hazard.

CONCLUSION

The development of advanced safety-critical automotive systems is driving the development of new tools and processes to help verify that these systems operate safely and that they are reliable and predictable. For these systems, product safety needs to be considered up front and addressed as part of the overall design process. This paper summarizes many of the available techniques to help analyze and implement a safe embedded system design. Based on our application experience, the analysis and failsafe techniques described here may be considered sound and beneficial. These techniques will continue to evolve as new technological challenges are recognized and addressed.

REFERENCES

1. Delphi Secured Microcontroller Architecture SAE# 2000-01-1052
2. A Safety System Process For By Wire Automotive Systems SAE# 2000-01-1056
3. A Comprehensive Hazard Analysis Technique for Safety-Critical Automotive Systems SAE#2001-01-0674
4. Diagnostic Development for an Electric Power Steering System SAE# 2000-01-0819
5. The BRAKE Project – Centralized Versus Distributed Redundancy for Brake-by-Wire Systems SAE# 2002-01-0266
6. Delphi ETC Systems for Model Year 2000; Driver Features, System Security, and OEM Benefits . . . SAE# 2000-01-0556
7. Standardized EGAS Monitoring Concept Ver 1.0
8. SW FMEA Methodology Presentation
9. B. J. Czerny, J. G. D'Ambrosio, Paravila O. Jacob, et. al. *A Software Safety Process for Safety-Critical Advanced Automotive Systems,* Proceedings of The International System Safety Conference, August 2003.

CONTACT

Eldon G. Leaphart, Engineering Manager – Diagnostics, Communications & System Software / Controlled Brakes, Delphi Corp., 12501 E. Grand River, MC 483-3DB-210, Brighton, MI, 48116-8326 Phone: (810)-494-4767, Fax:(810)494-4458 email: eldon.g.leaphart@delphi.com

APPENDIX

Table A1: Summary of Software Failsafe Techniques - Criteria Selection Matrix (Part 1)

	Memory Failures	CPU Failure	Software Processing Errors	Interface (I/O) Failures	Communication Failure	System Failsafe
Complement Data R/W Duplicate storage of variables as data and complement value. Complement values are checked for correctness prior to data usage	✓ Memory intensive. Will require duplicate memory allocation for each parameter. Also increases CPU time load for complement check routine.	✓ Generally targeted toward memory failures, however, miscompare could indicate CPU failure to access data correctly	X n/a	X n/a	X n/a	Possible input to fail action decision
Checksum Compares Add sections of memory together to get the checksum value. When checked memory readded and sums compared.	✓ Could be slow to catch a fault depending on method chosen: Continuous background (slower) vs Fast compare. Fast compare requires specific placement of data.	X n/a	X n/a	X n/a	✓ Checksum methods used to verify serial data integrity between processors / controllers	Possible input to fail action decision
Redundant Coding Run a duplicate copy of a section of code and compare the answers prior to using.	✓ Memory intensive. Requires twice as much memory to implement a function.	✓ Doubles the amount of processing time to implement a function	✓ Incorrect result could indicate software processing error within a single path	X n/a	X n/a	Possible input to fail action decision
Redundant Orthogonal Implement a section of code using a different method or processor resources. Run both sections of code and compare answers prior to using.	✓ Memory intensive. Requires twice as much memory to implement a function.	✓ May doubles the amount of processing time to implement a function. Could be hardware or micro architecture dependent.	Incorrect result could indicate software processing error within a single path	X n/a	X n/a	Possible input to fail action decision
Program Flow Monitoring Uses a thread imbedded in important functions to assure all of the functions were called and in the right order. Implies asymetrical or symetrical hardware architecture.	X n/a	✓ Thread algorithm should be designed to have minimal effect on CPU load. Assumption that CPU failure may impact normal sequence of code execution.	✓ Effective method for identifying some synchronization code issues. Coverage of execution sequence is a function of thread "depth".	X n/a	✓ Depending on architecture employed, could indicate issues with interprocessor synchronization	Possible input to fail action decision
Initialization Test RAM or ROM test at initialization	✓ Fast check of memory resources during initialization. Application must take into consideration system startup timing requirements	X n/a	X n/a	X n/a	X n/a	Possible input to fail action decision
Power Up/Down R/W Write a pattern to memory for proper shutdown, and then write a different pattern at start-up	✓ Keap-Alive (KAM) or Non Volatile (NVM) memory required as part of design.	✓ n/a	✓ Effective method for showing that orderly shutdown was obtained. Should be coordinated with overall system moding strategy.	X n/a	X n/a	Possible input to fail action decision
Test Cases Give the controller a set of calculations to test the ALU. Implies asymetrical or symetrical hardware architecture.	✓ Provides some coverage of memory locations assuming that test case memory access failure would impact computed result.	✓ Test cases must be designed to consume a minimal amount of time relative to application. Test cases should be representative of methods / machine instructions used throughout application. Difficult to guarantee 100% coverage.	X n/a	X n/a	✓ Depending on architecture employed, could indicate issues with interprocessor synchronization	Possible input to fail action decision
COP Watchdog Timer that will cause a reset if it is allowed to zero	X n/a	X n/a	✓ Effectice in identifying software / task execution errors. Analysis required to choose watchdog frequency relative to system failure requirements.	X n/a	X n/a	Possible input to fail action decision

Table A1: Summary of Software Failsafe Techniques - Criteria Selection Matrix (Part 2)

	Memory Failures	CPU Failure	Software Processing Errors	Interface (I/O) Failures	Communication Failure	System Failsafe
Peripheral Test Software routine designed to monitor output	X n/a	X n/a	X n/a	✓ Dependent on hardware architecture of system. Implies checking redundant inputs or monitoring output feedback. Synchronization of comparison or tolerances must be considered	X n/a	✓ Typically includes mechanism to provide actuator failsafe for system
Reasonableness Test Uses a simplified model of the controlled variable to assure that the variable is in a reasonable area.	X n/a	X n/a	X n/a	✓ May need to determine regions of operation where model is valid prior to usage. Apply to variables driving controlled output	X n/a	Possible input to fail action decision

Table A2: Example Section - Delphi Electric Brakes 3.0 Preliminary Hazard Analysis

Concept									Projected System					
Num.	Hazard	Major	Minor	Accident Scenario	Causes	Sev.	Lik.	Haz. Risk	Hazard Controls	Causes	Sev. w/ Cntl.	Lik. w/ Cntls	Hazard Risk of System	Recommendations and Comments
HAZ-01.0	**Failure to Provide Desired Acceleration**	Vehicle does not provide acceleration consistent with driver intent						High					Moderate	
HAZ-01.1		*Total Loss*	Park brake fails to release	Driver attempts to drive vehicle with locked park brake, pulls out into traffic resulting in a minor collision	Bad PB Switch (D), wiring, connectors, or failed controller (D); failed PB motor (D)	III	E	Low	Fault Tolerant PB Switch; Actuator Diagnostics; Driver warning		III	E	Low	
HAZ-01.2		*Total Loss*	Failed interlock signal prevents driver from shifting into gear when desired	Driver unable to move vehicle after emergency stop at intersection or railroad crossing, vehicle hit by on coming vehicle or train	Failed brake determination (D)	I	E	Moderate	Redundant & diverse sensors w/ diagnostics; Driver Warning	Failed Pedal Travel sensor (E)	I	E	Moderate	
HAZ-01.3		*Degraded*	Reduced accleration capability due to undesired apply of braking system	Brake system inadvertantly applied while vehicle stopped, driver attempts to pull out into traffic, resulting in severe collision	Bad PB switch (D), common mode controller fault (D)	I	D	High	Redundant & diverse sensors w/ diagnostics; Fault Tolerant PB Switch; Driver warning;	Bad PB switch (mechanically faulted) (D), common mode controller fault (D)	I	E	Moderate	
HAZ-01.4		*Degraded*	Reduced accleration due to undesired traction control request	Vehicle does not accelerate as expected during a passing manuever; Vehicle unable to accelerate through an intersection	Improper Wheel Speed signals (D), specific controller failure (E)	II	D	High	Command voting; Redundant & diverse sensors w/ diagnostics; Watchdog; Fail silent components; Driver Warning	Improper Wheel Speed signals (D), controller failure (E)	II	E	Moderate	
HAZ-01.5		*Unwanted*	Undesired acceleration (e.g., negative vehicle acceleration (roll back) on incline due to loss of hill hold capability)	Vehicle is stopped on a hill, driver releases brakes to depresses the gas pedal, vehicle rolls back into another vehicle	Loss of higher level functions (controller failure (D))	III	E	Low	Command voting; Redundant & diverse sensors w/ diagnostics; Watchdog; Fail silent components; Driver Warning	Loss of higher level functions (controller failure (E))	III	E	Low	

Table A3: Coverage Matrix for DEB Controller

	Used?	Memory Failures	CPU Failure		Software Processing Errors
Potential Severity					
Likelihood to Occur					
Safety metric Code					
Program Flow Monitoring	yes	weak	strong	. . .	strong
Internal watchdog	yes		weak	. . .	strong
External watchdog	no		strong	. . .	strong
Flash checksummed during runtime and startup	yes	strong	weak	. . .	
Safety critical code fast checksum	no	strong	weak	. . .	
Software well written and verified	yes			. . .	strong
Key ROM locations tested at start-up	no	strong	weak		
.					
.					
.					
EEPROM checksummed at startup	yes	strong	weak		
Algorithm using complement values for safety critical values	yes	strong	weak		

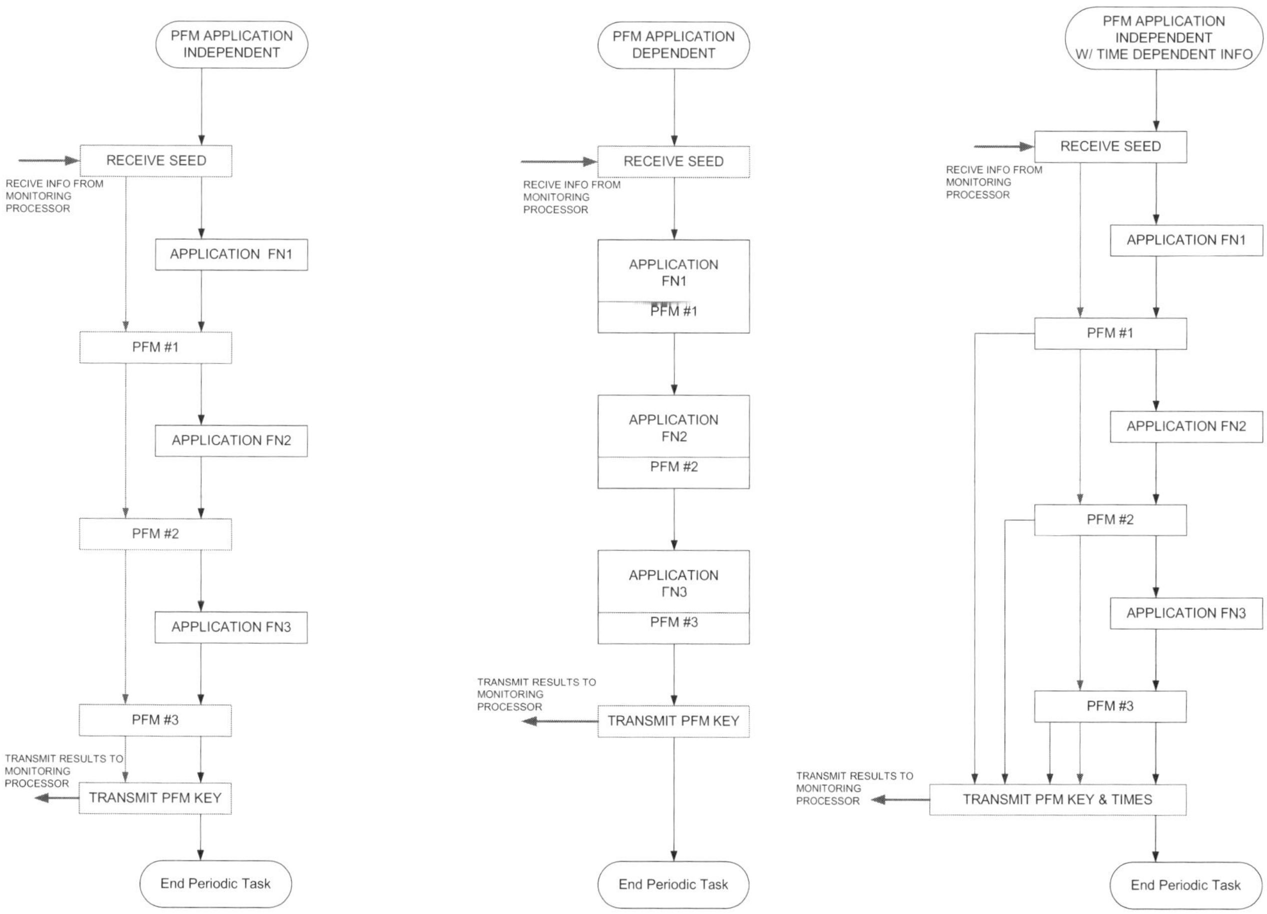

Figure A1: Example of Program Flow Monitoring Data Flow

ASYMMETRIC OR SYMMETRIC DESIGN DATA FLOW : PROGRAM FLOW MONITORING & TEST CASES

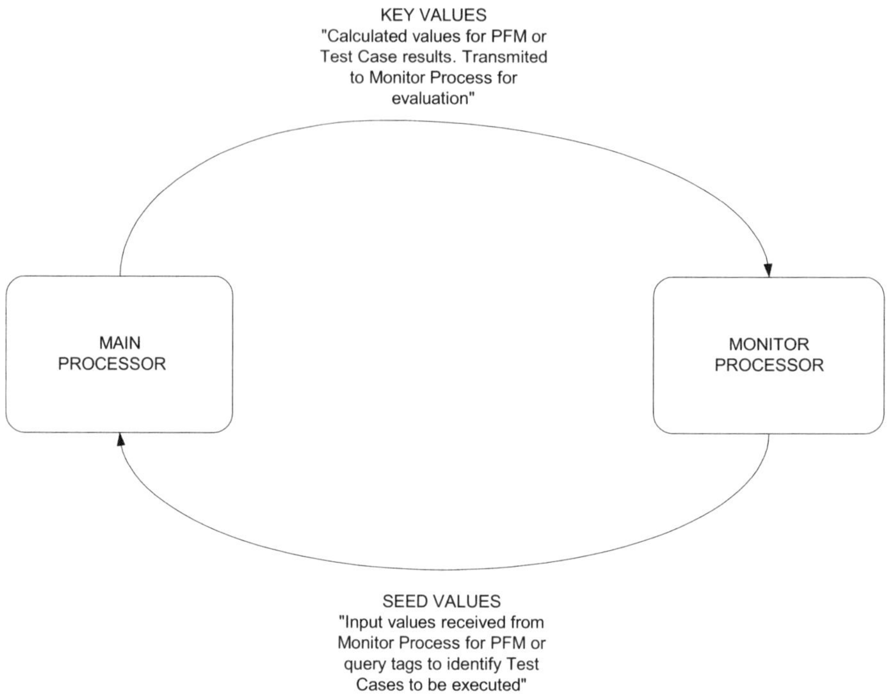

Figure A2: Example of Program Flow Monitoring Data Flow

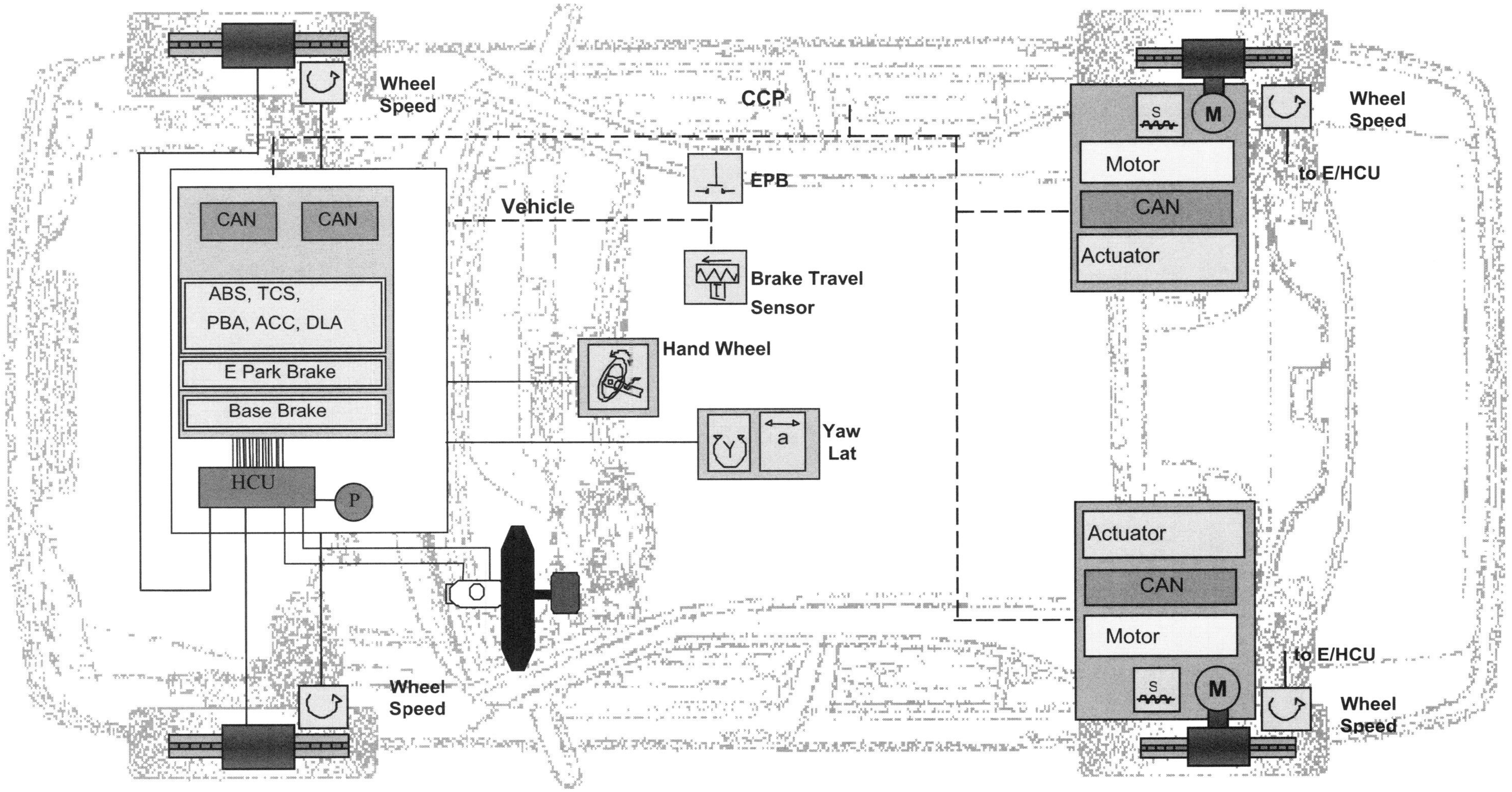

Figure A3: Delphi Electric Brakes 3.0 System Mechanization

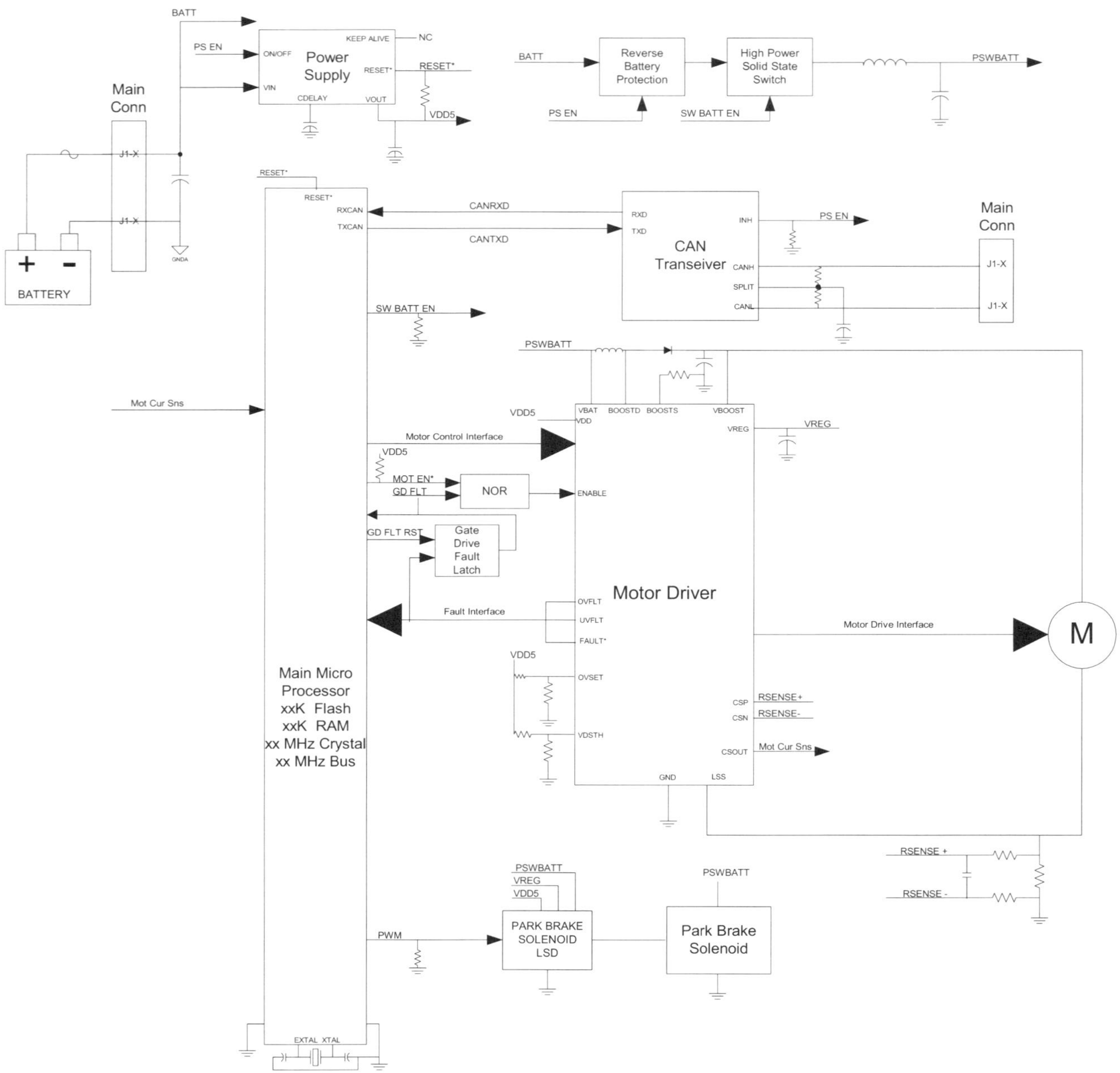

Figure A4: Delphi Electric Brakes 3.0 Controller Mechanization

2004-01-1666

An Adaptable Software Safety Process for Automotive Safety-Critical Systems

Barbara J. Czerny, Joseph G. D'Ambrosio, Paravila O. Jacob, Brian T. Murray and Padma Sundaram
Delphi Corporation

ABSTRACT

In this paper, we review existing software safety standards, guidelines, and other software safety documents. Common software safety elements from these documents are identified. We then describe an adaptable software safety process for automotive safety-critical systems based on these common elements. The process specifies high-level requirements and recommended methods for satisfying the requirements. In addition, we describe how the proposed process may be integrated into a proposed system safety process, and how it may be integrated with an existing software development process.

INTRODUCTION

As new, often complex, advanced automotive systems are implemented to enhance vehicle safety, performance, and comfort, system safety programs are being utilized to help eliminate potential hazards. Software is a key component in these systems. Software is increasingly controlling essential vehicle functions such as steering and braking, and on some levels, independently of the driver. These systems help provide potential improvements in vehicle safety, however, unexpected interactions between software and the system and its environment may lead to potentially hazardous situations.

The potential software hazards that may lead to these situations must be identified and mitigated by the system safety program. Although potential software hazards must be considered during system safety analyses, the unique aspects of software warrant that software-specific safety engineering techniques be applied. As such, following a software safety process integrated within a system safety process helps confirm that best-practice software safety engineering techniques are performed, thus, providing increased confidence that the software does not create or contribute to potentially hazardous situations at the system level.

There are many software safety process standards, guidelines, and methods that exist today, including the Motor Industry Software Reliability Association (MISRA) guidelines [1]. The existing standards and guidelines, however, do not fully address software safety in the automotive domain, nor do they address the unique aspects of the automotive domain that make an automotive-specific software safety process desirable. These unique aspects include the following:

1. Automotive suppliers work with customers from different parts of the world. Each of these may require different system and software safety standards to follow, and they may specify the specific types and levels of analyses to perform.
2. The development environment is based almost exclusively on the C programming language, unlike the military, aerospace, and nuclear power industries. This means the levels and types of analyses that can be performed are not as broad.
3. Certification according to quality standards such as QS 9000 and ISO 16949 is required in order to do business. Periodic audits are required to confirm strict adherence to internally defined procedures.

These unique issues can be addressed by a software safety process based on a set of required high-level tasks, with a corresponding set of recommended methods to implement the tasks. It is not possible to directly incorporate a rigid process standard (e.g., IEC-61508) as a required procedure without industry-wide agreement, since any divergence to meet unique project needs could result in a quality audit non-compliance. Thus, an automotive-specific software safety process must be flexible enough to accommodate different customer desires and requirements and to accommodate the varying needs during the different stages of product development, while at the same time enforcing a structured process that helps lead to software safety. The goal is to integrate best-practice elements from existing documents into a structured process that satisfies the flexibility requirements. This is achieved by a process that includes a set of high-level required tasks and recommended methods to implement the high-level tasks. The process assumes that a good underlying software development process is in place and that software safety cannot be considered apart from system safety; the software safety process should be integrated into and compatible with established system safety and software development processes.

In the next section, we provide an overview of existing software safety documents that we reviewed in developing our proposed software safety process for the automotive domain. Next we identify and describe the common best-practice elements of a software safety process. Finally, we describe our proposed software safety process for safety-critical advanced automotive systems, discuss our experience applying the proposed process, and present our conclusions.

EXISTING SOFTWARE SAFETY DOCUMENTS

A number of software safety standards and guidelines documents and methods from various organizations exist today. This section provides a brief overview of several of these standards, guidelines, and methods.

National Aeronautics and Space Administration (NASA): NASA-STD-8719.13A provides the requirements to implement a systematic approach to software safety as an integral part of the overall system safety program [2]. The standard is intended to be applied to software that could cause or contribute to the system reaching a specific hazardous state, software that is intended to detect or take corrective action if the system reaches a specific hazardous state, and software that is intended to help mitigate damage if a mishap occurs. Safety-critical software is identified during the system and subsystem safety analyses. The level of required software safety effort for a system is determined by its system category and the hazard severity level. The NASA Guidebook, NASA-GB-1740.13-96, provides more detailed information to assist in applying the standard [3].

U.S. Department of Defense: MIL-STD-882C is primarily geared toward system safety, so a detailed software safety process is not addressed [4]. It does, however, provide a software hazard risk assessment process that considers the potential hazard severity and the degree of control that the software exercises over the hardware. The software control categories are based on the level of control the software has over the hazardous function being assessed. It does not provide guidance or recommendations on the tasks or levels of analysis to perform for the determined software criticality.

Radio Technical Commission for Aeronautics (RTCA): RTCA/DO-178B provides guidelines for the production of software for airborne systems and equipment that performs its intended function with a level of confidence in safety that complies with airworthiness requirements [5]. This standard provides a means of categorizing software, provides a good description of software development tasks, and links the system safety assessment process with the software development process. No specific safety tasks are detailed and the software categories are not directly translatable to the automotive domain.

Joint Software System Safety Committee (JSSSC): The JSSSC Software System Safety Handbook, a Technical & Managerial Team Approach, provides management and engineering guidelines to achieve a reasonable level of assurance that software will execute within the system context with an acceptable level of safety risk [6]. The document was a joint effort between U.S. armed forces branches, with cooperation from several governmental agencies, academia, and defense industry contractors. It was intended to capture the "best practices" pertaining to software system safety. The software safety process is integrated with the system safety and software development processes. The process includes identifying generic and system safety-critical software requirements, performing appropriate software safety analyses during each stage of the software lifecycle, verifying that the software was developed in agreement with applicable standards and criteria, and developing a software safety assessment. No specific guidance is provided on determining the level of software safety effort required.

U.K. Ministry of Defense (MOD): MOD DEF STAN 00-55 emphasizes the procedures necessary for specification, design, coding, production, and in-service maintenance and modification of safety-critical software [7]. The standard identifies two categories of software: safety-related software and safety-critical software. Safety-related software is software that relates to a safety function or system and encompasses all Safety Integrity Levels (SILs). Safety-critical software is software that relates to a safety-critical function or system; this is software of the highest SIL (S4), the failure of which could cause the highest risk to human life. It provides guidance and recommendations on the requirements for developing software at the various SILs.

International Electrotechnical Commission (IEC): IEC 61508 part 3 describes the software requirements of the IEC 61508 standard and is intended to be used only after a thorough understanding of parts 1 and 2 of the standard [8]. Part 3 applies to any software forming part of or used to develop a safety-related system as described within the scope of 61508 parts 1 and 2. The level of analysis detail required is dependent on the determined software SIL. The standard includes requirements for safety lifecycle phases and activities to be applied during the design and development of the safety-related software, and requirements for software safety validation. It includes guidance and recommendations for the selection of techniques and measures to use to satisfy the determined SIL. IEC 61508 was intended to be a generic standard from which application specific standards would be developed.

Motor Industry Software Reliability Association (MISRA): MISRA compiled eight detailed reports containing information on specific issues relating to automotive software. The reports are summarized in a single document: *Development Guidelines for Vehicle Based Software* [1]. The summary report contains information on the software lifecycle and describes three approaches for determining the integrity associated with an ECU (detailed in report 2 on Integrity [9]). Although the report does not provide an explicit process for software safety that could be directly implemented, it does provide a good overview of issues that need to be addressed when developing vehicle based software, and it contains good recommendations.

APT Research, Inc.: APT's 15 Step Process for Definition and Verification of Critical Safety Functions in Software was presented at the 2001 International System Safety Conference [10]. The 15 steps include identifying system hazards, identifying software safety functional requirements, and tailoring the safety effort to criticality. The method shows the integration of the 15 step process for software system safety into the system safety process and the software lifecycle.

Given the safety-critical nature of some advanced automotive systems, application of techniques above and beyond existing software development techniques should be considered. Suppliers of automotive systems, and automobile manufacturers currently apply various safety analyses in varying degrees to systems being developed. At this point, no single software safety standard has been adopted by the automotive industry. However, there are elements common to most of the identified processes and methods, and this set of common elements can form the basis of a software safety process that provides adequate flexibility. These elements include: software safety planning, requirements analysis, architecture design analysis, detailed design analysis, code analysis, test planning, testing, test analysis, and assessments.

COMMON ELEMENTS OF A SOFTWARE SAFETY PROCESS

This section provides an overview of some typical elements of a software safety process. The software safety process proposed in this paper is an integration and adaptation of these elements.

Software Safety Planning: Software safety planning begins in the conceptual design phase of development. Inputs into this task may include the conceptual design, the System Safety Program Plan (SSPP), the Preliminary Hazard List (PHL), and the Preliminary Hazard Analysis (PHA). During this task, a plan is developed for carrying out the software safety program for the project. The Software Safety Program Plan (SWSPP) identifies the software safety activities deemed necessary for the project, and is developed in conjunction with, and may be part of the System Safety Program Plan. The plan may evolve during the development process.

Software Safety Requirements Analysis: Software safety requirements analysis begins in the system and software requirements analysis phases of development. Inputs into this task may include the system safety requirements, the system safety concept, the PHA, and the software requirements. The purpose of this task is to identify safety-critical software requirements, to help validate that the decomposition of the system level safety requirements to the software safety requirements is complete and consistent, and to provide safety-related recommendations to the design and testing phases. Safety-critical software requirements are identified during the system PHA to eliminate, mitigate, or control hazards related to software. Software safety requirements may also stem from government regulations, customer requirements, or internal requirements. Information for eliminating, mitigating, or controlling hazards related to software, and safety-related design and testing recommendations are passed on to the architecture design phase. A matrix identifying software safety requirements may be initiated to track the requirements throughout the development process.

Software Safety Architecture Design Analysis: The software safety architecture design analysis task begins in the system and software architecture design phases. Inputs into this task may include the system architecture design, the system hazard analyses outputs (e.g., the PHA and safety concept), the safety-related design and testing recommendations from the software safety requirements analysis task, the software architecture design, the software safety requirements, and software criticality and tailoring guidelines. The PHA and software safety requirements are reviewed to determine if additional hazards related to software can be eliminated, mitigated, or controlled, and additional information on eliminating, mitigating, or controlling the hazards is passed on to the detailed design phase. Software components and functions are identified in the software architecture design phase. The software components and functions that implement the software safety requirements or that affect the output of the software safety requirements are identified as safety-critical. A software criticality level may be determined for the safety-critical software components and functions. Software criticality levels indicate the potential level of risk associated with different software components. The correctness and completeness of the software architecture design as it is related to the software safety requirements and the safety-related design recommendations is analyzed to help ensure that the design satisfies the software safety requirements. Safety-related recommendations for the detailed design and test procedures are provided, and test coverage of the software safety requirements is verified.

Software Safety Effort Tailoring: Tailoring the software safety effort is an umbrella task that begins when the safety-critical software components and functions are identified and assigned criticality levels. This task is relevant to all software safety tasks. Inputs into this task include the system safety requirements, system safety hazard analysis outputs, software safety requirements, and the software safety architecture and detailed design analysis outputs. Appropriate levels and types of analyses are identified based on the determined criticality level from the software safety architecture design analysis. Information on suggested levels and types of analyses, testing, and verification and validation for the identified criticality levels may be obtained from tailoring guidelines tables if they exist. The output from this task is the tailoring recommendations. Criticality levels and tailoring recommendations may be tracked in the software safety requirements matrix.

Software Safety Detailed Design Analysis: Software safety detailed design analysis begins in the software detailed design phase. Inputs into this task may include the system hazard analyses (e.g., the detailed hazard analysis and hazard control specifications), the system and software detailed designs, the software safety requirements, software safety architecture design analysis output, safety-related detailed design recommendations, and tailoring recommendations. The identified safety-critical software components and functions that implement the software safety requirements are refined to the unit level software components and functions. The system and software detailed designs are analyzed to help ensure that the software detailed design satisfies the software safety requirements. Subsystem interfaces may be analyzed to identify potential hazards related to interfacing subsystems, such as hazardous interface failure modes and data errors. Test coverage of software safety requirements is verified, and safety-related recommendations for the software implementation are provided. Software safety detailed design analysis continues during a portion of the software implementation and unit testing phases. Outputs from this task may include the identified safety-critical unit level software components and functions, the identified subsystem interfacing hazards, and safety-related software implementation and test coverage recommendations. Any identified subsystem interface hazards are input back to the relevant system hazard analyses.

Software Safety Code Analysis: This task begins in the software implementation and unit-testing phase. Inputs into this task may include the system hazard analyses outputs (i.e., detailed hazard analyses outputs), software safety requirements, software detailed design, software safety detailed design analysis output, safety-related software implementation recommendations, software implementation, and the tailoring recommendations. The completeness and correctness of the code as related to the software safety requirements, software detailed design, and safety-related software implementation recommendations is analyzed to help ensure that the software safety requirements are satisfied in the implementation. The analysis may check for potentially unsafe states that may be caused by I/O timing, out-of-sequence events, adverse environments, hardware failure sensitivities, failure of events, etc. Test coverage of the software safety requirements is analyzed. This information may be tracked in the software safety requirements matrix. A software implementation report describing the types of analyses performed and the results of the analyses, and software implementation test coverage recommendations may be written.

Software Safety Test Planning: Software safety test planning begins in the software architecture design phase and continues through the software integration and acceptance testing phase. Inputs into this task may include system hazard analyses outputs (i.e., PHA, safety concept, detailed hazard analyses, and hazard control specifications), system safety test plans and test procedures, software safety requirements, software test plans and test procedures, and tailoring recommendations. During this task, appropriate software safety tests that address all identified potential hazards related to or affected by the software are incorporated into the software safety test plan. The software safety test plan is developed to help ensure that all identified safety requirements related to or affected by software will be adequately tested. Test procedures should include both nominal and off-nominal conditions. System safety test plans and test procedures related to software are examined and additional system safety test plans and procedures are developed as required. The software safety test plan may be part of the software test plan. Testing and verification requirements may be included in the software safety requirements matrix. This facilitates the tracking and verification process to help ensure that the software safety requirements are satisfied and appropriately tested and verified.

Software Safety Testing and Test Analysis: These tasks begin in the software implementation and unit testing phase. Inputs into the software safety testing task include the system and software safety test plans and procedures. Inputs into the software safety test analysis task may include the software safety requirements, system safety program plan, software safety program plan, system and software safety test plans and procedures, and the system and software safety test results. Planned software safety tests are performed, and test results are reviewed to help ensure that safety requirements have been satisfied. This helps ensure that the identified potential hazards have been eliminated or controlled to an acceptable level of risk according to the system and software safety program plans. Any software problems identified are corrected, and follow-up tests are defined to verify that the identified software problems were corrected and that no additional problems were introduced into the system. The appropriate systems people are notified if a system corrective action is required.

Verify Software Developed in Agreement With Applicable Standards / Guidelines: This task begins in the software architecture design phase and continues through to the latter stages of development. Inputs into this task may include the system safety program plan, software safety program plan, and applicable standards / guidelines. The standards / guidelines may come from the customer, government regulations, or they may be internal standards / guidelines. The goal of this task is to help ensure that all applicable safety-related standards and guidelines identified in the program plans have been adhered to in the development of the safety-critical software components and functions. Any discrepancies in adhering to the standards / guidelines are identified and recommendations are made for alleviating the discrepancies.

Software Safety Assessments and the Software Safety Case: Software safety assessments begin in the early stages of development, and a software safety assessment is completed at each major gate review during the project development process. This assessment describes the current state of safety in the software being developed. The assessment indicates any known issues and what will be done to resolve them. Issues from previous safety assessments that have been closed are identified and marked as closed. The software safety case provides supporting documentation and justification as to why the developers believe the software as developed is safe. This documentation is developed from the final software safety assessment. All open issues from the final software safety assessment should be closed. If issues remain open, justification must be provided as to why the open issues are acceptable. Any residual risk associated with software that remains in the system and that has been determined acceptable, is justified in the software safety case. Inputs into these tasks include outputs from all system and software safety tasks, and previous system and software safety assessments. The software safety assessments and the software safety case may be part of the system safety assessments / case respectively.

In the next section we describe the proposed software safety process that combines the elements described in this section.

A SOFTWARE SAFETY PROCESS FOR SAFETY-CRITICAL ADVANCED AUTOMOTIVE SYSTEMS

In addition to the previously discussed criteria that need to be satisfied for the automotive industry, Delphi has certain internal requirements that need to be satisfied. The internal requirements are that the software safety process:

- must be integrated into and compatible with the proposed system safety process for safety-critical advanced automotive systems,
- must be integrated into and compatible with established software development processes and compatible with established coding standards, and
- must be adaptable to different projects in different stages of the development process.

To support different stages of development, the levels and types of analyses chosen are typically more rigorous for a production design than for a prototype design.

The software safety process described here satisfies both the internal and external criteria by building upon common best-practice elements from several existing standards, guidelines, and methods. These elements have been tailored and combined in an effort to make them applicable to the automotive domain; specifically to safety-critical, advanced automotive systems. The main sources used are the JSSSC Software System Safety Handbook, NASA-STD-8719.13A, and APT Research, Inc's. 15 step process (Figure 1). The other documents contain much useful information as well and are used to provide additional information.

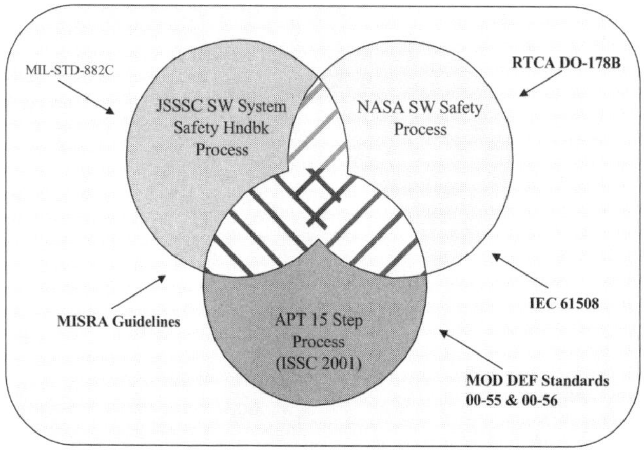

Figure 1: Integration of software safety documents.

SOFTWARE SAFETY PROCESS ARCHITECTURE

Figure 2 shows the software safety process architecture being developed within Delphi for safety-critical, advanced automotive systems. The process consists of a software safety procedure that contains a set of high-level required tasks, and a set of recommended methods to implement the high-level tasks. The software safety procedure high-level tasks are generic enough to be applicable to all safety-critical systems. The set of recommended methods to implement the high-level tasks may be tailored for specific projects in different stages of development and for different customer requirements.

Figure 3 shows the common best-practice elements that make up the foundation of our proposed software safety process. The diagram shows the software safety tasks, general process flow of the tasks, the relationships between tasks, and the duration of the tasks. Task boxes that overlap vertically, may be carried out simultaneously. For example, software safety test planning begins during the software safety architecture design analysis and continues into software safety testing and test analysis. Software safety assessments begin during the software requirements analysis task and continue throughout the process, ending in the final software safety case. Not all of the process flow options are shown since the figure would become too cluttered. Only the main flows are shown; the same will be true for other process description figures that follow.

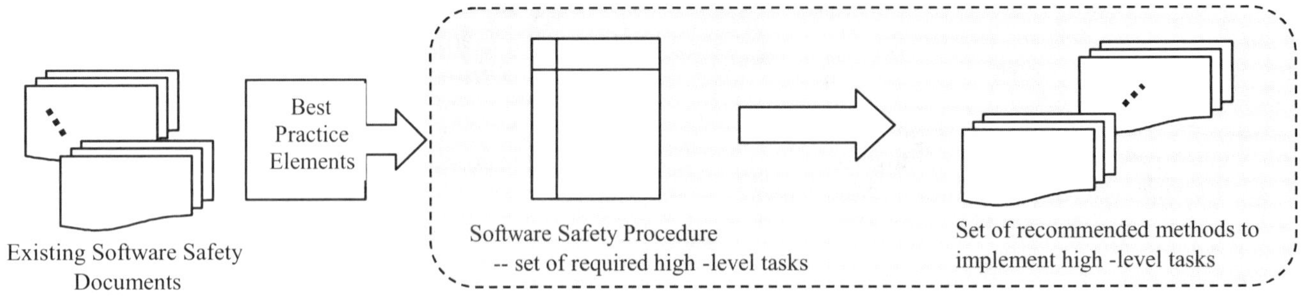

Figure 2: Software safety process architecture.

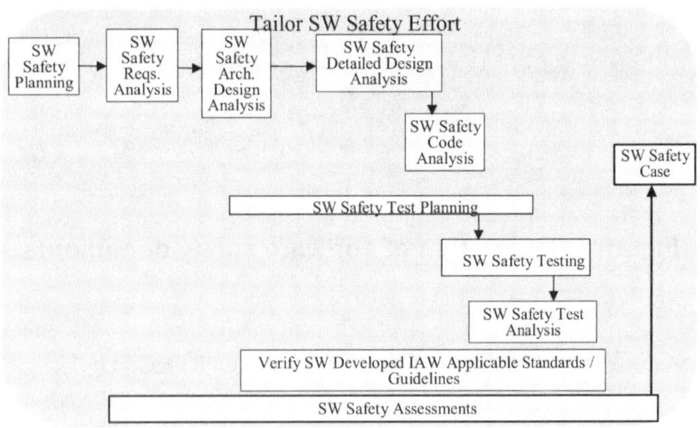

Figure 3: Software safety tasks and relationships.

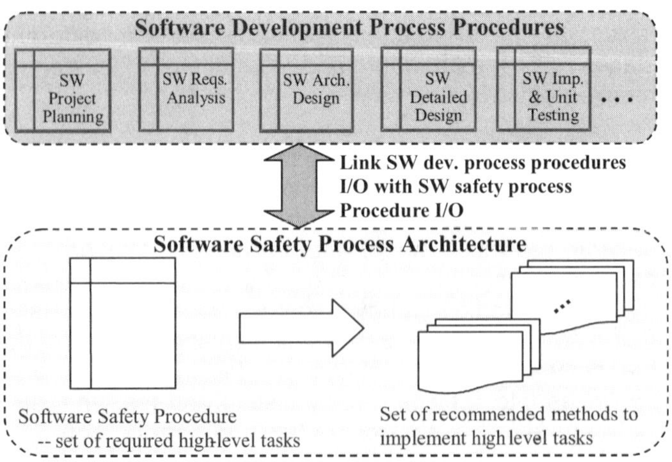

Figure 4: Phase one integration with software development process.

INTEGRATION WITH SOFTWARE DEVELOPMENT PROCESS

The software safety process will be used in conjunction with an existing software development process within Delphi. The software safety process will be integrated into the software development process in two phases. The initial phase one integration (Figure 4) will be a loose integration of the two processes and consists of identifying the procedures required to satisfy the software development process and linking their inputs and outputs with the software safety process inputs and outputs. Phase one integration will be an evaluation and modification stage. The software safety process will be applied on various safety-critical advanced automotive systems and adapted based on feedback from the application of the process. Once we are fairly confident that the process works well for various applications, we will move to the phase two integration with the existing software development process.

Since any software safety process developed is closely integrated with the software development process, the goal is to ultimately incorporate the software safety process aspects directly into the software development procedures. The phase two integration will be a tight integration of the two processes, where the actual high-level required tasks from the software safety process will be directly incorporated into the software development process procedures (Figure 5). For example, the software safety planning task will be directly integrated into the software planning procedure. Within the software planning procedure will be a requirement that if the system being developed is safety-critical, then software safety planning must be performed. The same requirement will be present in the other software development procedures for their corresponding software safety tasks.

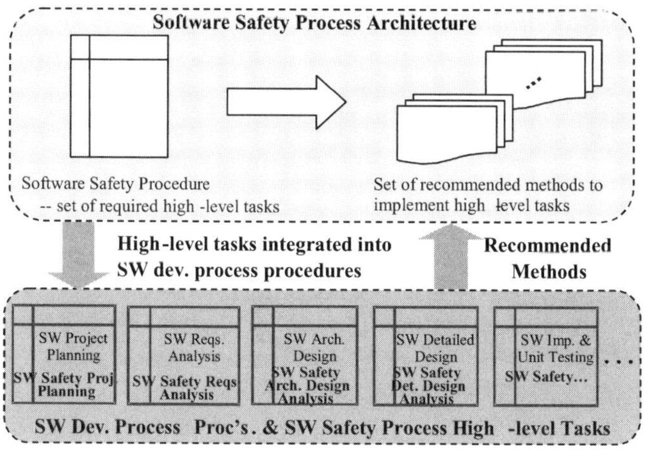

Figure 5: Phase two integration with software development process.

Figure 6 shows a phase one, loose integration view of the tasks of the software development process and the tasks of the software safety process and the relationships between the tasks. The tasks in the upper shaded area are tasks in the software development process. Tasks shown outside of boxes in the shaded areas are umbrella tasks and are performed throughout the corresponding process (i.e., SW Requirements Traceability, SW Configuration Management, etc. are applicable throughout the software development process, and the Tailor SW Safety Effort task is applicable throughout the software safety process). As previously stated, the lengths of the boxes in the figure show the duration of the tasks and the vertical relation of boxes shows the various tasks that may overlap. The software development process tasks are outside the scope of this paper and will not be discussed in detail. The tasks that occur during each of the software safety process steps are the same as those described under the Elements of a Software Safety Process section and will not be further described here.

INTEGRATION WITH SYSTEM SAFETY PROCESS

Since software safety cannot be considered apart from system safety, the software safety process should be part of an established system safety process as well. Figure 7 shows the relationships between all three processes, the proposed system safety process for safety-critical advanced automotive systems, the established software development process, and the proposed software safety process. The figure shows the phase one loose integration of the three processes. Software safety requirements are initially obtained from the system safety PHA when potential hazards are identified that may be related to software. Additional safety requirements may be identified during later system and software process steps.

Results affecting the software safety process obtained from system safety analyses are communicated to the software developers during the appropriate stage of development. For example, the appropriate levels and types of analyses for the high-level software safety process tasks are determined based on identified software criticality levels that follow from the results of the system safety analyses, the project's development phase, and customer requirements. Likewise, information obtained during the software safety process that affects the system safety analyses results is communicated to the appropriate system developers.

Software and system safety assessments start during the early stages of both the system and software safety processes, and continue through the development of the final system and software safety case. A system and software safety assessment is presented at each major gate review during the development process, with the final safety assessment forming the basis for the safety case.

In addition to defining the software safety tasks and the relationships between the tasks of the software development, software safety, and system safety processes, we have also identified the inputs and outputs that may be acted on, generated by, and used between the processes. Figure 8 shows possible inputs and outputs that may be acted on and generated by the software safety process. For example, outputs generated by the SW Safety Requirements Analysis task are the software safety requirements, and design and testing recommendations. The software safety requirements become inputs into and may be acted on by the SW Safety Architecture Design Analysis, Detailed Design Analysis, Code Analysis, and so on. Similar diagrams exist for the inputs and outputs that may be exchanged between the system safety and software safety processes, and between the software development and software safety processes.

DISCUSSION

The proposed software safety process has been applied to projects in different stages of development (e.g., prototype and production) [11]. Application of the proposed process to these projects confirmed the need for flexibility and adaptability of the process. Prototype projects are generally attempting to prove a concept and are characterized by rapidly changing requirements. A prototype vehicle is typically built, and operating restrictions can be placed on the use of the prototype vehicle. In these cases, it is not necessary that a safety case and associated validation testing be fully completed. However, the development team may decide that a basic understanding of software safety issues is important if these issues will have a significant impact on the later phases of product development. For a production project, a product is developed for a specific customer application and put into production. This level of development requires a more rigorous application of established safety processes.

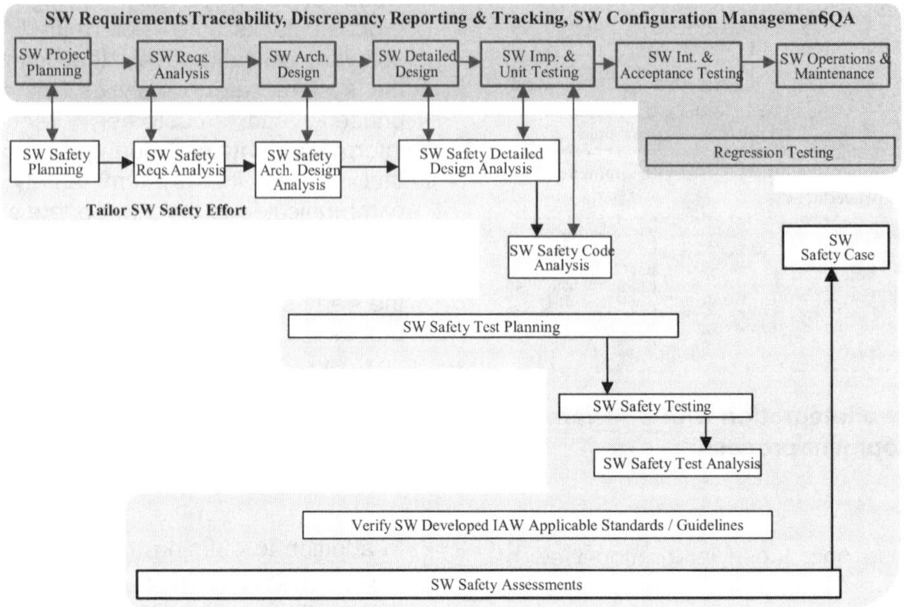

Figure 6: SW development and SW safety process task relationships.

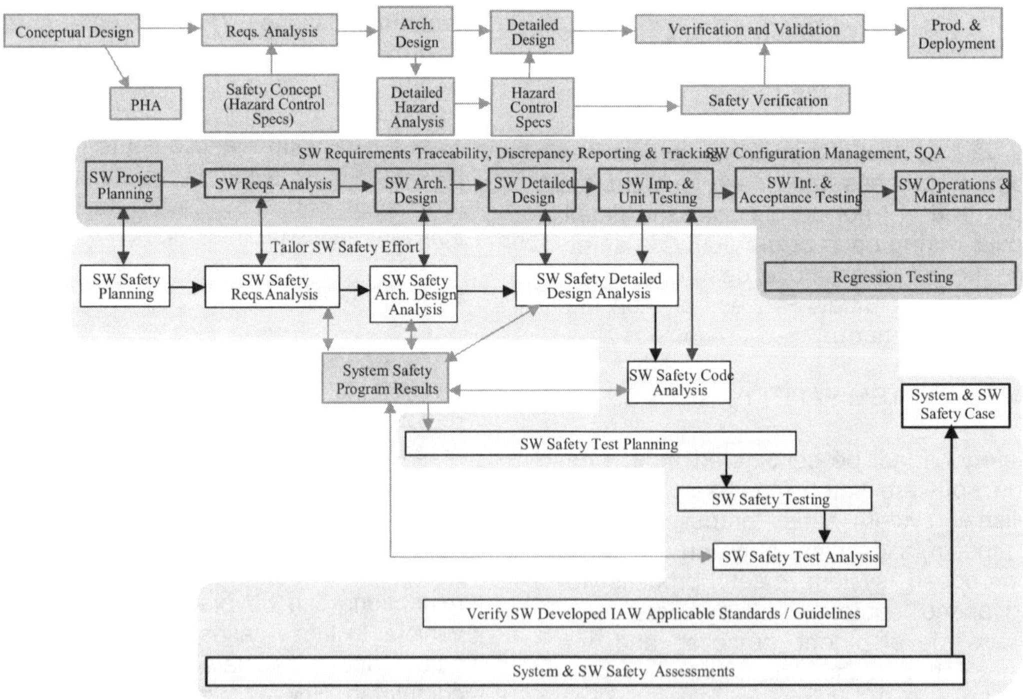

Figure 7: System safety, SW development, and SW safety process task relationships.

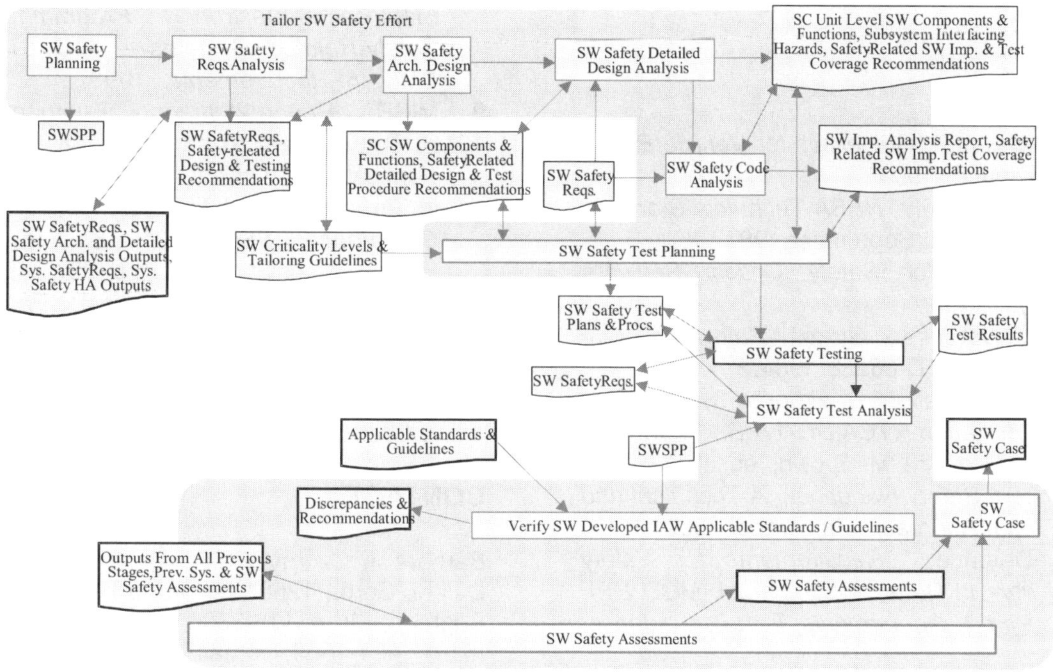

Figure 8: Software safety process inputs and outputs.

For the projects the proposed process was applied to, the levels and types of analyses selected for the various software safety process tasks varied depending on the project's stage of development, however, there was some overlap in the selected levels and types of analyses chosen. For the selected methods that overlapped, some proved beneficial across both types of projects, while others were beneficial only on production projects. For example, the more detailed level analysis applied to a prototype project during the software safety detailed design analysis proved to be too time consuming for this application and was never fully completed. In contrast, the software safety detailed design analysis applied on a production project proved effective in analyzing the integrity of the software design and its potential impact on system hazards. In all cases, the need for a software safety activity was confirmed, however, detailed analysis was deemed most suitable for production projects. These and other lessons learned have been incorporated into the process definition.

Overall, the process appears to be efficient and to meet Delphi's needs. It provides a structured software safety process that fosters consistent application of best-practice software safety engineering techniques to safety-critical advanced automotive systems. It is adaptable so that Delphi can satisfy different customer desires and requirements, and it is flexible enough to work in different stages of product development.

CONCLUSION

Software safety is important for safety-critical advanced automotive systems, especially since software is increasingly controlling essential vehicle functions such as steering and braking, and on some levels, independently of the driver. A software safety process used in conjunction with a system safety process can help confirm that appropriate software-specific safety engineering techniques are applied. As discussed in the paper, it is not possible for Delphi to adopt an existing rigid software safety process. A software safety process feasible for Delphi must be adaptable to different customers and to projects in different stages of development.

In this paper we reviewed existing standards, guidelines, and methods for software safety. In addition, we described some common elements of a software safety process derived from the reviewed documents, and then presented how these elements are integrated to form the proposed automotive, software safety process. The process is based on a set of required high-level tasks with a corresponding set of recommended methods for implementing the tasks based on the determined software criticality level. The recommended methods are adaptable to the specific needs of individual projects. We demonstrated how the software safety process may be integrated with the existing software development process, and how it may be integrated with our proposed system safety process.

Initial applications of the process have been positive. As we move forward, the set of recommended methods will be revised appropriately based on the results of future applications. When we have gained experience with the overall process and its generic applicability, we will be in a position to decide if the process should become a required procedure.

REFERENCES

1. MISRA. *Development Guidelines for Vehicle Based Software*. November 1994.
2. NASA. *Software Safety: NASA Technical Standard NASA-STD-8719.13A.* September 1987.
3. NASA. *Guidebook for Safety Critical Software NASA-GB-1740.13-96.*
4. Department of Defense. *System Safety Program Requirements, MIL-STD-882C.* 1984.
5. RTCA. *SW Considerations in Airborne Systems and Equipment Certification RTCA/DO-178B.* 1994.
6. D. Alberico, J. Bozarth, M. Brown, et. al. *JSSSC Software System Safety Handbook; A Technical and Managerial Team Approach.* December 1999.
7. Ministry of Defence. *Requirements for Safety Related Software in Defence Equipment; MOD DEF STD 00-55.*; *Part 1: Requirments; Part 2: Guidance.* August 1997.
8. IEC. *International Standard; Functional Safety of Electrical / Electronic / Programmable Electronic Safety-Related Systems – IEC 61508-3; Part 3: Software Requirements*. 1998.
9. MISRA. *Report 2; Integrity*. February 1995.
10. H. D. Kuettner, Jr. and P. R. Owen, "Definition and Verification of Critical Safety Functions in Software", in *Proceedings of the International System Safety Conference (ISSC) 2001.* System Safety Society, Unionville, Virginia 2001. pp. 337-346.
11. B. J. Czerny, J. G. D'Ambrosio, Paravila O. Jacob, et. al. *A Software Safety Process for Safety-Critical Advanced Automotive Systems,* Proceedings of The International System Safety Conference, August 2003.

CONTACT

Barbara J. Czerny, Ph.D., Sr. System Safety Engineer, Delphi, Corp., 12501 E. Grand River, MC 483-3DB-210, Brighton, MI, 48116-8326; Phone: (810-494-5894), Fax: (810) 494-4689, email: barbara.j.czerny@delphi.com

2004-01-1665

A Design Methodology for Safety-Relevant Automotive Electronic Systems

Stefan Benz, Elmar Dilger and Werner Dieterle
Robert Bosch GmbH

Klaus D. Müller-Glaser
University of Karlsruhe

ABSTRACT

For the development of future safety-relevant automotive electronic systems a thorough adaptation of the existing design process is necessary to consider safety and reliability in a more systematic way.

In this paper an approach for a new design methodology is presented. It is based on the V-Model which is the established process model for the development of electronic and software systems in the automotive domain. For an advanced consideration of safety and reliability the existing process is extended by a second V (with process elements that have a special focus on safety and reliability) to a "Double V". The new elements are interconnected with the existing ones at several points of time during the development process. By a defined information exchange between the two Vs continuity in the methodology is guaranteed. Basis for the extension are experiences of the aerospace domain that were adopted to automotive conditions.

INTRODUCTION

In the automotive industry there is a clear trend to an increase in the number of electronic systems in a vehicle. More and more often mechanical implementations of vehicle functions are replaced by electronics or software systems.

One can find several reasons for this trend. Many functions in today's automobiles cannot be implemented without the extensive use of electronics. Systems that provide more comfort or additional functions such as navigation or "infotainment" rely heavily on electronics in the car. The recent achievements concerning lower fuel consumption, lower emissions, and at the same time greater engine power and performance due to better engine control are also to some extent the result of more and better powertrain control electronics. Other consequences of the extensive use of electronics are lower costs for the manufacturer and the supplier and also lower running costs for the customer. Finally the most important achievement in the context of this paper is definitely the increase of passive and especially active vehicle safety [1].

But it is clear that passive safety systems such as airbags or special body concepts tap their full potential. Further major improvements of vehicle safety through passive safety systems do not seem very likely. Future safety enhancements will have to rely mostly on active safety systems that actively avoid accidents before they can occur. One of the first active safety systems was the antilock braking system (ABS), state of the art today is the electronic stability program (ESP). Additional active safety functions that we will see in the future will culminate eventually in autonomous driving.

Active safety systems of the next generation will incorporate advanced direct interventions in brakes and also in steering. These functions will be implemented by so called x-by-wire systems, vehicle systems in which the transmission of energy and information is done in an electrical way only, i.e. mechanical or hydraulic system parts are replaced by electronics [2].

Until recently electronics was used mostly for non safety-relevant systems[1] (with the exception of antilock braking systems), therefore the design process had mainly the function of the system in focus. Safety was one of the many non-functional properties.

As the system functions in a vehicle are getting more and more complex, a more methodical approach for the development of these systems is necessary. New design approaches such as TITUS (see [3]) that focus on the functional design concentrate on the control of complexity.

[1] Safety-relevant systems are defined here as systems where failure conditions can lead directly to serious injury or death of one or several persons.

It is obvious that safety-relevant electronic systems impose strong requirements for safety and reliability. They therefore demand also new methods for the system development and design. In the past the function of a system was in the center of focus and other, non-functional conditions were considered as an addition. But for safety-relevant systems the safety of the vehicle system will have to be considered on the same level as its function, thus a thorough adaptation of the existing design process and its methodology is necessary.

The standard process for the development of automotive electronic systems is based on the V-Model '97, a life cycle process model that is the development standard for IT systems in the Federal Republic of Germany (see [4]). Originally intended for IT systems it was adapted for other domains, including the automotive domain. Most design processes in the automotive industry of today's are based on the V-Model (see for example [5]). However the methodical procedure is neither standardized nor formalized throughout the automotive domain, almost every manufacturer and supplier uses a slightly different design method.

In the automotive world some efforts are made to gain additional experience in the development of safety-relevant electronic systems. In the aerospace industry (a domain with very standardized and formal development processes) there is a lot of experience with safety-relevant systems. Aerospace development processes have the function and on an equal footing the safety of an aerospace system as center of focus (see [6] or [7]). However these processes cannot be used directly in the automotive domain, as the boundary conditions and the requirements for safety and reliability in these two domains are quite different.

In this paper an approach for a design methodology for the development of safety-relevant automotive electronic systems is presented. The methodology considers safety equally with function and can be described as a "Double V-Model". It is based on the basic principle of the V-Model. A second V with elements that have a special focus on safety and reliability is added to the original V and connected to it at several appropriate points of time during the development process. The concepts underlying the additional elements were taken from the aerospace domain and adapted to automotive conditions. [8]

This paper is structured as follows: In the next chapter safety and reliability are defined as they are used in this paper. Then some short highlights on the state of the art of development methods in different domains of industry are outlined. That chapter is followed by a description of the requirements for a new design methodology. Finally the new design methodology is described in detail.

SAFETY AND RELIABILITY

Safety is defined as a circumstance where the risk is not higher than the limiting risk [9]. An exact estimation and quantization of the limiting risk is elaborate and difficult to obtain in most cases. Normally the limiting risk is defined by a social consensus.

The risk that is connected to a technical activity or state is described by a probability statement consisting of the expected frequency of occurrence of an event that leads to harm on the one hand and the possible degree of harm in the case of occurrence on the other hand.

The mathematical connection between these two properties is not defined, a quantitative comparison between two different risks is therefore difficult. Thus safety cannot always be described quantitatively. To overcome this problem a risk estimation is usually used, i.e. the system is categorized into risk levels (for example into "safety integrity levels" in IEC 61508 [10]).

Reliability is defined as the ability of a system to work correctly for a given period of time under specified conditions [11]. Mathematically reliability is usually described by a survival probability. A very common probability distribution function is the Weibull distribution. For electronic components a special case of the Weibull distribution is usually applicable, the exponential distribution:

In this case the reliability R for a given time t is

$$R(t) = e^{-\lambda t}; t \geq 0$$

where λ is the failure rate.

A very similar but still different term in this context is *availability*. The availability at a given time is defined as the probability to find a unit at that given point of time in a state of functionality [12].

The mean availability A can be defined as

$$A = \frac{MTBF}{MTBF + MDT}$$

where $MTBF$ denotes the mean time between failures and MDT the mean downtime.

Despite these precise definitions the common usage of reliability and safety sometimes seems unclear, the terms are occasionally confused. But one must be aware that they define very contrary system properties.

In the safety world as well as in the reliability world similar concepts are used (for example redundancy), but the goals differ strongly. When designing a safe system the main focus is on minimizing risk and possibly dangerous effects of potential systems faults. Under no circumstances the system shall harm passengers or the environment. This is accomplished by measures for fault avoidance, error detection and control. In case of doubt the system is shut down when inconsistencies are discovered.

This is different for highly reliable systems. For these priority is on the function of the system; problems such as an incorrect signal or timing problems can possibly be tolerated for a short time.

So safety and reliability are system properties with quite different boundary conditions. But in some special cases a clear differentiation is not easily possible.

This distinction is impossible or not useful for some safety-relevant automotive electronic systems (and also not for many safety-relevant systems in other domains). System functions like steering or braking have very high requirements for safety. Simultaneously that safety property is also the system's requirement for reliability, as a failure of a steering or braking system is a safety critical condition for the vehicle as a whole. A principle for the design of such a system is that it may never fail; and if parts of it still do, they have to fail in a way that the environment of the system is not endangered[2].

The main reason for requesting that system property is the fact that for these systems a single safe system condition cannot be defined for all driving situations. Therefore the system under consideration simultaneously has the same requirements for safety as for reliability; i.e. the vehicle is only safe if e.g. its steering system is reliable.

The consequence for systems with this particular property is that in the same context requirements for safety can be replaced by requirements for reliability. The safety assessment process in the aerospace industry is based on that fact. For systems without a safe system state the safety requirements can be transferred into reliability requirements. And though safety cannot be assessed quantitatively reliability can.

STATE OF THE ART

In the following sections a short overview of the state of the art concerning development processes and design methodologies in several different industrial domains is given.

AUTOMOTIVE DOMAIN

Concerning the development of electronic systems in the automotive context there are several boundary conditions one has to take into consideration.

In comparison to other industrial domains - particularly to those with similar requirements concerning safety and reliability - automotive electronic systems have very high production volumes. This means that adaptations to special characteristics of a system pay off. This results in many variations of one product, also due to customer adaption or country specific solutions. Additionally the related mechanical components are subject to wear as maintenance levels are difficult to assure. And passenger car drivers receive little or no training compared with other users of computer-based products.

An additional aspect is the demand for composability of electronic system which will play an ever increasing role especially in the collaboration between manufacturer and several component or system suppliers.

The main focus in the development of vehicles and vehicle systems is on system test. Over several months the systems and in the end the whole vehicle is tested and tried intensively.

As an abstract model for the description of the development process of automotive electronic systems the so called "Automotive V-Model" is widely used. It is based on the V-Model '97 [4]. The V-Model '97 consists of four submodels *project management*, *system development*, *quality assurance* and *configuration management*. In the development standard the procedure, methods and tool requirements for all four submodels are defined. The graphical representation of the main submodel *system development* is shaped like a V, this is the source for the name "V-Model".

In the V-Model '97 the chronological order of the process elements is not defined. The standard uses the term "activities" for the smallest process elements. They interact with each other, a chronological order is defined implicitly.

Originally the V-Model '97 was designed for IT systems, in the meantime it was adapted for other domains including the automotive domain. Here usually several consecutive Vs are used (e.g. one each for the A, B and C sample). Many stages of the development process are supported by tools, also "rapid prototyping" is occasionally used to shorten the development process.

Detailed descriptions of automotive design processes can be found in [5], [13] or [14]. In figure 1 on the following page one can see a typical automotive system design engineering process.

AEROSPACE DOMAIN

Except for several military based standards the standard SAE ARP 4754[3] [15] is most commonly used for the development of civil and also military electronic systems in the aerospace industry (a survey of design processes in the aerospace domain can be found in [6]).

In the standard SAE ARP 4754 the system development process is described on a very abstract level, for more

[2]Systems with this property are called "fail operational systems", i.e. the system function is still provided despite of a failure in the system.

[3]ARP: Aerospace Recommended Practice

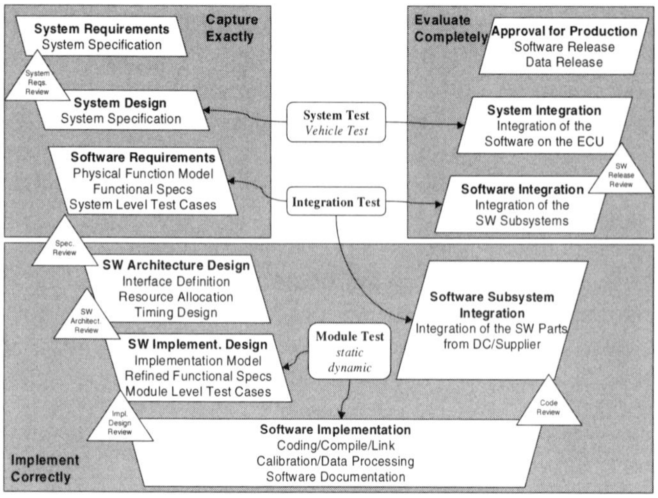

Figure 1: The "Automotive V-Model" according to [5]

details other documents are referenced. So for the software development life cycle DO-178B [16] is used. The system development process and the corresponding standards are shown in figure 2.

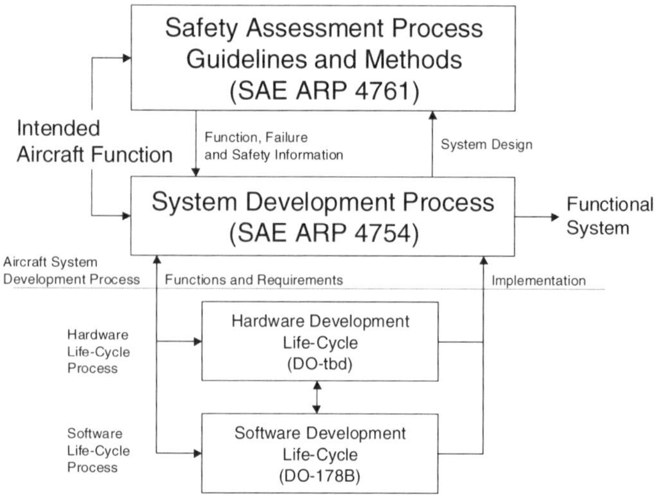

Figure 2: Development standards in the aerospace industry as described in SAE ARP 4754

For the consideration of safety during the system development process SAE ARP 4754 references SAE ARP 4761 [17]. These two standards are deeply interconnected, the consideration of system safety in SAE ARP 4761 takes place in a process parallel to the development of the system function in SAE ARP 4754. The safety assessment process of SAE ARP 4761 can be seen in figure 3.

Both standards describe a system safety life cycle process that - apart from a functional development of the system - puts a special focus on the treatment of safety and reliability of the system. Essential elements of the safety assessment process are steps called *functional hazard assessment*, *preliminary system safety assessment* and *system safety assessment*. They guarantee the fulfillment of the safety requirements. Quantitative analysis and assessment methods are used intensively.

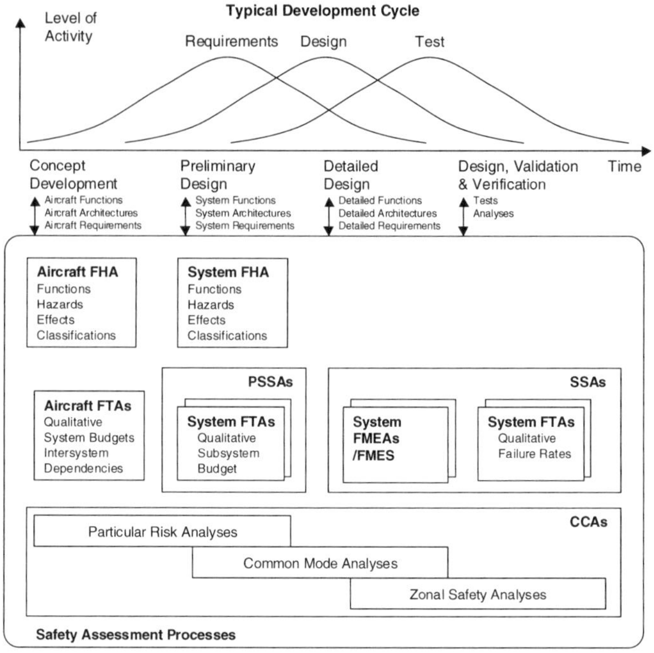

Figure 3: Overview of the safety assessment process as described in SAE ARP 4761

A basic property of system development in the aerospace domain is the certification of the process and of the designed system. Usually a third party such as the FAA[4] certifies the developed product at the end of the system development. The manufacturer can be sure that his product is safe according to the state of the art, but this procedure makes the development life cycles very long and expensive, and only a small number of system varieties can be used.

As the boundary conditions and the system requirements in the aerospace domain are quite different from those in the automotive domain, the process models of the aerospace domain cannot be transferred directly. Core contents can be reused, but they have to be adapted to the conditions in the automotive domain.

A very interesting property for the development of safety-relevant electronic systems that could be reused in the automotive domain is the clear distinction between function and safety in the process and the usage of different methods and tools on a quantitative level. Thus the correct development of a safety-relevant system can be guaranteed and verified.

AUTOMATION ENGINEERING DOMAIN

A third industrial domain where safety-relevant electronic systems are widely used is automation engineering. Usually so called programmable logic controllers (PLCs) are used for the control of assembly plants or power stations. Typically these PLCs are certified or are based on certified concepts to construct a certified total system [18].

[4]FAA: Federal Aviation Administration

In the automation engineering domain there are several established standards for the development of safety-relevant electronic systems, such as EN 954 [19] for safety related parts of control systems or IEC 61508 [10] for general electrical or electronic safety related systems.

IEC 61508 is a generic process standard that has its roots in the automation of chemical plants. For other industrial domains this standard was adapted to the appropriate boundary conditions (so IEC 62061 [20] is the adaption of IEC 61508 for the mechanical engineering domain). A characteristic feature of IEC 61508 is the categorization of the particular functional units of the system in four so called "safety integrity levels" (SILs). The process is tailored depending on that SIL classification.

Whether IEC 61508 will also play a major role for automotive electronic systems in the future cannot be foreseen today. Currently work is in progress to investigate on this issue and to possibly develop an automotive development standard based on IEC 61508.

IEC 61508 or EN 954 cannot be used in the automotive domain without major adaptations. Also the concept of certified components is mainly not usable in the automotive domain.

The overall safety life cycle of IEC 61508 can be seen in figure 4.

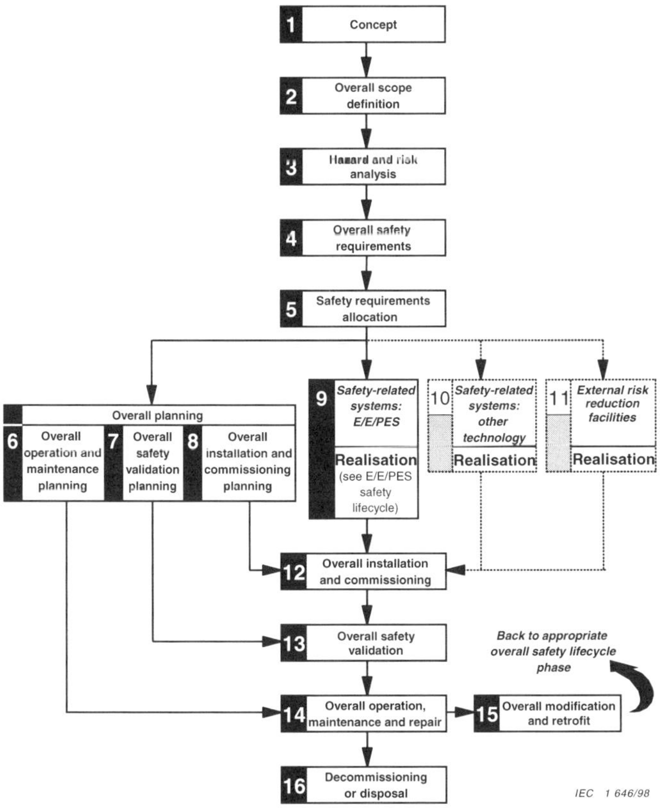

Figure 4: Overall safety life cycle of IEC 61508

REQUIREMENTS OF A NEW DESIGN METHODOLOGY

A new design methodology to support the development of safety-relevant automotive electronic systems has to fulfill additional requirements.

It should support the design engineer in collecting, conceiving and understanding the requirements for safety and reliability of the system. So as a very first step in the development process there should be a phase in which the design engineer decides on the relevant potential hazards of the system and based on that decision defines the system requirements concerning safety.

For the design of the system, the system architecture, and the implementation of the system an a priori knowledge whether the chosen design is able to fulfill the system safety requirements would also be helpful. This would allow the design of a system architecture that perfectly fits to its requirements; development time could be saved and costs cut.

BASIC IDEA OF THE PROPOSED DESIGN METHODOLOGY

The basic idea of the new approach for a design engineering methodology presented in this paper is the extension of the V-Model which is proven and well introduced in the automotive domain. As already stated above the V-Model is extended by process elements with a special focus on safety and reliability to a "Double V-Model" (see figure 5).

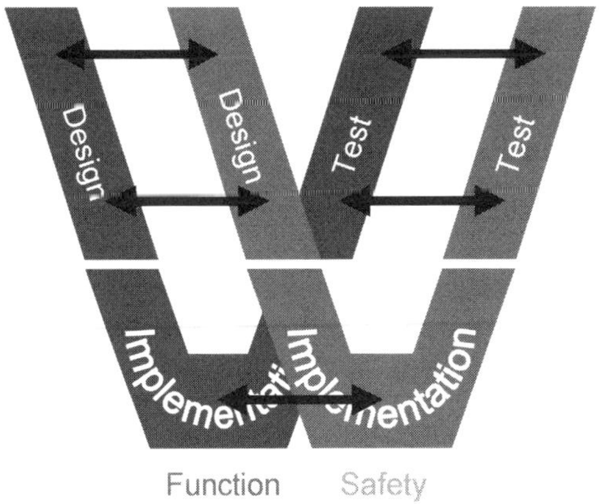

Figure 5: Basic idea for the "Double V-Model"

Two equally important Vs shall indicate that the additional elements concerning safety and reliability have the same significance as the old ones concerning function. Until now the focus was primarily on the function of the system, now function and safety have to be considered on an equal footing.

The basic idea of the methodology is to take the advantages of the development processes in the aerospace industry, align them to automotive demands and include them in the existing process. So roughly speaking the additional elements here are taken from the aerospace standard SAE ARP 4761 and adapted to the special requirements of the automotive domain. In figure 6 on the following page the proposed design methodology can be seen in an overview.

The rectangular elements in figure 6 are process steps that focus on the functional aspect of the system development. Essentially they are identical to those of the "automotive V-Model". The rounded elements are the newly added ones and primarily have the safety-relevant aspects of the system development in focus.

Arrows in the figure represent an information flow between the process steps, double arrows an information exchange or an iteration. Not all arrows and therefore not all connections between the process elements are shown as they would make the figure too complex.

The development of a system starts with a *system requirements analysis*. Based on the requirements in the *functional design* the system is specified from a functional point of view, parallel this functional design is analyzed in the *functional hazard analysis* and safety requirements are mapped to functional units.

Based on the functional description the *system design* takes place, here the functional description is mapped on a system architecture. In parallel the proposed architecture is assessed on the fulfillment of the safety requirements. This takes place in the *design accompanying system safety assessment*. If the system design is complete, the system components are implemented.

After the *implementation* of the system the system parts are integrated and tested (*integration and test*). Again, in a parallel step (*system safety assessment*) safety issues are assessed. At the end of the system development procedure there is the final step *approval and commissioning*.

THE DESIGN METHODOLOGY IN DETAIL

In the following the process elements of the "Double V-Model" are described in detail. Special emphasis is put on the newly added process steps.

SYSTEM REQUIREMENTS ANALYSIS

The core input for the development of a new system consists of ideas about a specific function, sometimes also already detailed thoughts about the structure of the system. So the concept of a new system function may be already expressed or there are prototypes of a new system that can be the basis for the development of a new system. In other cases a new system is a further development of an existing system, so functions are added or removed, or the boundary conditions of the system change.

The system requirements usually come from the customer, additionally they can be based on internal assessments and also on requirements of a third category. These are for example technical conditions (size, interfaces, platform requirements, etc.) or statutory demands (mostly standards or regulations of approving institutions).

The system requirements can be divided into so called functional requirements (everything concerning the system function directly) and non-functional requirements (all other requirements). Non-functional requirements are economic requirements, safety, reliability, maintainability or availability, system platforms to be used, the periphery and interfaces of the system or mechanical requirements.

In the *system requirements analysis* these system requirements are collected and processed systematically. This step is to be carried out independent of ideas on the implementation of the system.

As a first step in the *system requirements analysis* the actual status is analyzed. The state of the art in the context of the proposed system is examined, also the usual boundary conditions and whether one has to take special conditions into account. Additionally already existing systems in the same context are regarded.

The next step is the description of the system in the form of a system specification. Here the system functions, boundary conditions, non-functional requirements, etc. are defined. This specification is usually submitted by the customer or created in cooperation. After that specification the overall boundary conditions of the system should be determined and defined.

An important next step is the system requirements controlling. Here an assessment concerning the technical and economical feasibility of the system takes place. If the system cannot be constructed considering all requirements the project has to be stopped at that point.

As a last step in the *system requirements analysis* the system validation procedure is specified.

FUNCTIONAL DESIGN

In the design methodology presented in this paper the paradigm that function drives design is used (often also called "form follows function"). So the system function is the main driver for the system design. Equally important but depending on the system function is safety. Other system properties are junior to those.

The design of the system starts with a *functional design*. Here the functional system requirements based on the results of the *system requirements analysis* are systematically specified in further detail and the system function is determined.

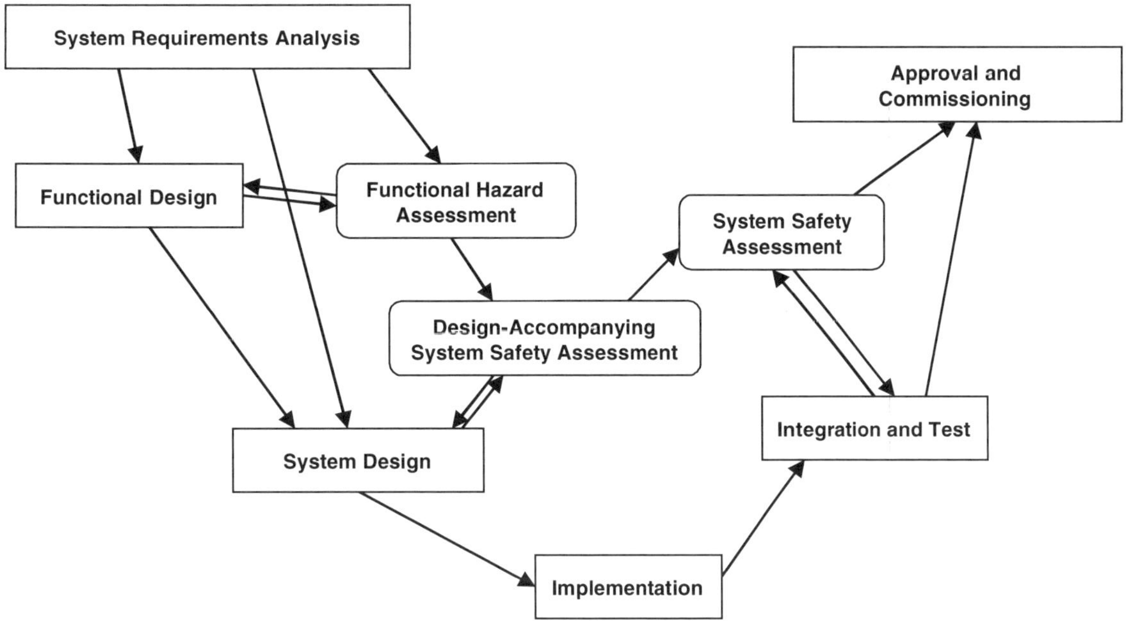

Figure 6: The proposed design methodology in an overview

Building up on a first sketch of the system function in the *system requirements analysis* the system functions are further detailed, divided into sub functions and based on that procedure the conceptual design is iteratively improved. The functional architecture is determined, a complete functional design and the structure of the system with function definition, partitioning and also test cases is the result of *functional design*.

This detailing of the conceptual outline of the system takes place in close iterative interconnection with the *functional hazard assessment*. After every detailing step in *functional design* a new *functional hazard assessment* for that particular level of detail takes place.

During this phase the function of the system is defined independently of any realization consideration, i.e. the physical architecture and implementation aspects are not yet considered here. The functional design contains the consideration of the functional flow, from the reception of a stimulating input signal (for example the signal of a vehicle system or the action of the driver) over the processing to the creation of a new action (for example the control of an actuator or the display of messages).

The procedure here is supported by established methods for the functional description. Apart from the classic SA/RT[5] or the rather new UML[6], CARTRONIC (see [21]) as a special method for the functional design of automotive systems is an alternative. The functional description in this step is independent from the description method used, the result is a detailed function list for the usage in the parallel *functional hazard analysis* and in the consecutive *system design*.

[5]SA/RT: Structured Analysis / Real Time Analysis

[6]UML: Unified Modeling Language

Functional design is repeated for the functional description on lower levels, the results are exchanged iteratively with the *functional hazard analysis*. The output of these steps is the complete functional description of the system paired with safety and reliability requirements of its functional units.

FUNCTIONAL HAZARD ASSESSMENT

In the *functional hazard assessment* the functional safety requirements for the system are assessed and defined in further detail. For this purpose the potential hazards of system functions are determined and classified and, if possible, reliability requirements are derived. Non-functional safety requirements are not considered in the *functional hazard assessment*. By a very systematic procedure in the early steps of the system development process continuity to subsequent process steps is assured.

Based on the *system requirements analysis* in the *functional hazard assessment* the system is categorized into risk classes. For this purpose the system is brought in context with its environment, then impacts on the environment and of the environment on the system are analyzed.

Like in the parallel step *functional design* the assessment here focuses on the system's functions only. Based on the function list that was generated in the *functional design*, the corresponding failure conditions are identified and described. For every function there may be several possible failure conditions depending on the system's environmental conditions. The effects of these failure conditions result in possible risks and hazards. These hazards are classified into risk classes based on their severity.

This classification is a very critical point in the functional hazard assessment. It should be carried out using a methodology that is adapted to the boundary conditions of the automotive industry (for example "CARTRONIC safety analysis" (CSA) according to [22]). Usually a risk graph or decision tree is used; the risk graph of CSA with 5 safety levels ("SL") can be found in figure 7.

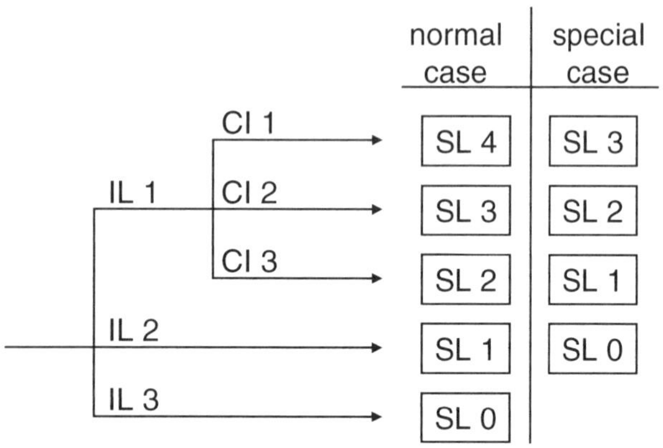

Figure 7: Risk graph of CSA

"IL" is an abbreviation for the question whether there is an immediate danger for body and life, "CI" stands for the question if there is an ability to control or influence the hazard. Additional explanations can be found in table 1.

Table 1: Abbreviations used in figure 7

Abbrev.	Explanation
IL 1	Immediate danger to body and life
IL 2	Slight injuries at most
IL 3	No immediate danger to body and life
CI 1	Neither controllable nor influencable
CI 2	Difficult to control or to influence
CI 3	Controllable or influencable

For systems without a safe system state it is possible to allocate reliability requirements to the different risk classes. Results of the risk classification are therefore concrete quantitative safety and reliability requirements for the implementation of that particular functional unit.

Parallel to that assessment the functional representation of the system is detailed incrementally. Through iterative process steps between the *functional design* and the *functional hazard assessment* also sub functions are analyzed concerning hazards. The risk classes are bequeathed to the sub functions and newly assessed. So step by step an extensive representation of the hazard classification with detailed quantitative requirements for reliability of all function and sub functions evolves.

Thereafter methods for the verification of these safety requirements later in the process are specified. The results of the hazard analysis are compared with previous experiences to see whether there is consistence in different system designs.

SYSTEM DESIGN

Based on the *functional design* in the step *system design* the so far purely functional description of the system is mapped on a system architecture. The result is a description of the implementation of the functional system representation considering the non-functional requirements. This description comprises for example interface descriptions or specifications of the system integration.

The phase *system design* alternates iteratively with *design-accompanying system safety assessment* where an assessment of the proposed architecture alternatives takes place. If the requirements for reliability are not fulfilled a redesign must take place.

The first step of *system design* is a rough system design, containing the basic architecture and interfaces. Based on that description the feasibility of the underlying system concept is investigated. The assessment of safety issues takes place in the parallel *design-accompanying system safety assessment*. If the concept does not fulfill its requirements the system architecture has to be redesigned.

Next the system requirements are analyzed for completeness and on whether new requirements or boundary conditions are generated by the system architecture itself. Then the system's internal and external interfaces are described. Also the integration and verification of the system is specified here.

Further on single system components are selected and fixed tasks are assigned to them. This is repeated on the lower levels of abstraction. So, after the physical structuring of the system, the concrete design of the system components takes place, the system is split in hardware and software, HW and SW requirements are derived. Then the software and hardware structure are designed. The design of the structural redundancies and the determination of the redundant communication channels takes place here. This description extends to the level of task scheduling and to the planning of communication between distributed system elements, concerning hardware to the selection of processor components or details of construction.

All aspects concerning safety are analyzed in parallel during the *design-accompanying system safety assessment*.

Usually there are various different architecture design alternatives that fulfill the functional requirements. The selection of an alternative results from a comparison and assessment of the system design concerning the requirements for reliability and the common boundary conditions. This selection of a proper architecture can be supported using sets of standard components or design patterns (an example for the usage of software design patterns can be found in [23]).

DESIGN-ACCOMPANYING SYSTEM SAFETY ASSESSMENT

The *design-accompanying system safety assessment* is a systematic examination of the system architectures that were proposed during *system design*. The goal of this step is the assessment of these recommended architectures concerning their applicability especially with respect to safety. The main idea of the *design-accompanying system safety assessment* is a thorough examination whether the architecture and the proposed system design fulfill the quantitative and qualitative requirements of the *functional hazard analysis* or if faults or failures can possibly lead to system hazards.

The basic idea of the quantitative assessment is the replacement of safety requirements by corresponding reliability requirements in the *functional hazard assessment*. But also non-quantitative safety requirements are assessed in this step.

As stated above the *design-accompanying system safety assessment* alternates iteratively with *system design*. Analysis and assessment of the design concepts is an action that takes place parallel to the definition of the system architecture.

The first step of *design-accompanying system safety assessment* is the evaluation whether the system safety requirements are complete. Then a thorough examination of the system architecture concerning safety takes place. For the identification of faults and combinations of faults that could lead to critical failure conditions in the system or to hazards resulting from them the architecture concepts and the implementation of the system are analyzed by corresponding quantitative assessment methods. If no quantitative assessment is possible the examination has to take place qualitatively and quantitative results have to be estimated.

It is evaluated whether the system architecture and the planned concept design can reasonably be expected to meet the safety requirements and objectives. The methods used are *fault tree analysis* (FTA) and *markov analysis* (MA) for reliability issues and *failure mode and effects analysis* (FMEA) for safety conditions. Common cause errors are examined by *common cause analysis* (CCA).

In addition also other potential hazards than functional hazards are examined. So it is determined whether the system concept design contains additional hazards, especially hazards that are generated by the system design itself. The method used for this purpose is FMEA.

Based on the results of these analyses safety requirements for the design on lower levels (also for hardware and software) and for other systems involved are derived. That way it is ensured that the implementation of the systems fully fulfills its requirements for safety and reliability. Then the *design-accompanying systems assessment* is repeated for lower system levels.

This procedure is supported by several data sources. The most famous database is the MIL-Handbook 217 [24]; though officially discontinued for several years it is still widely used in the civil and military aviation industry. Other databases include the MIL-Handbook 978 "NASA Parts Application Handbook", RAC's[7] "Nonelectronic Parts Reliability Data" and "Failure Mode/Mechanism Distribution" and Rome Laboratory's "Reliability Engineer's Toolkit", which all provide failure rates for many component types.

A major database which is used in the automotive domain is the "Reliability Data Handbook" of the "Union Technique de l'Electricité" (UTE) [25]. It is to be expected that it will be the successor of the MIL-Handbook 217. An example for the application of the UTE handbook can be seen in [26].

IMPLEMENTATION

The results of *system design* and *design-accompanying system safety assessment* are definite directions for the system's *implementation*. The system is realized according to these directions, i.e. software code is created and hardware (electronics as well as mechanics) is realized.

In practice the *implementation* of a system is a crucial step. The realization of an electronic system has to be done with great care. Usually this step is supported by several specialized tool chains.

A characteristic feature of today's electronic systems is the possibility to integrate a complex hardware/software system on a single chip as a so called "system on chip" (SoC). In this case it is basically not relevant whether a function is implemented in electronic hardware or software.

INTEGRATION AND TEST

After the *implementation* of the system parts the next step is the *integration* of the single system components and parts and then the functional *test*. The system's software and hardware are integrated and tested against their functional specification (it is also checked whether the interfaces work correctly etc.). Then the system parts are integrated to systems and tested iteratively. By repeating these integration steps the whole and complete system comes to life.

Methods used here are verification ("are we building the system right", i.e. check against the results of *system design*) and validation ("are we building the right system", i.e. check against the results of *system requirements analysis*).

[7]RAC: Reliability Analysis Center

This procedure is accompanied by the *system safety assessment* where the fulfillment of the safety requirements is verified and validated.

SYSTEM SAFETY ASSESSMENT

In parallel to the process step *integration and test* in *system safety assessment* the system's conformance to its safety requirements is checked. This is accomplished by the verification of the system's design requirements established in the *functional hazard assessment*, by a review of the hazard classification of the *functional hazard assessment* and by verification and validation of other safety requirements.

The process of verification and validation as it is known from the V-Model takes place in three steps. Here the functional part as well as the non-functional but non-safety part is assessed during *integration and test*, the issues concerning safety are tested in the *system safety assessment*.

Generally speaking the procedure in the *system safety assessment* is very similar to the one during the *design-accompanying system safety assessment*. The same tools are used, but the main difference is now that the system in focus is a physically implemented, real and running system. The same fault tree as in the *design-accompanying system safety assessment* can be reused, but the failure rates of the elements in the fault tree no longer come from a database but from the experiences with the real system components.

As with *system design* and *design-accompanying system safety assessment* the process switches here iteratively back and forth between *integration and test* and *system safety assessment*. The system is integrated and tested on its correct function, parallel the system's safety properties are proven. For this purpose the system parts have to fulfill the requirements of the *system requirements analysis* and especially the quantitative reliability requirements of the *functional hazard assessment*.

The review is done by a systematic examination of the system, its architecture and its realization by various methods of verification and validation (for example testing, fault injection or formal methods). So it is possible to prove that the system fulfills its safety requirements.

APPROVAL AND COMMISSIONING

At the end of the automotive development process there is the type *approval* and the *commissioning* of the vehicle system. Relevant standards for approval for road traffic in Europe are the appropriate European approval regulations, for example for a braking system the regulation UN ECE-R 13 (see [27]) or for a steering system the regulation UN ECE-R 79 (see [28]). Especially the appendices concerning electronics in both guidelines apply to safety-relevant automotive electronic systems.

Fundamental contents of these appendices are that during a type approval the approving institution has to receive a thorough documentation of the system safety concept. In [27]: *"the manufacturer shall provide a documentation package ... [in which] the safety concept ... shall be explained"*. For this purpose the documentation of the *system safety assessment* can be used.

Furthermore the approving institution may check the correct system safety behavior (as it is described in the system safety concept) by methods such as fault injection. *"... the reaction of "The System" shall, at the discretion of the Type Approval Authority, be checked under the influence of a failure in any individual Unit by applying corresponding output signals to electrical Units or mechanical elements in order to simulate the effects of internal faults within the unit."*

If the system was designed using the methodology presented here all the relevant documents were produced during the system development process. Therefore the vehicle system should pass this test phase easily.

CONCLUSION

In this paper a conceptual scheme for a new design methodology for the development of safety-relevant automotive applications was presented. A special focus was put on the methodical consideration of safety and reliability. The concepts for the methodical steps were taken from experiences of the aerospace industry and were adopted to the special conditions in the automotive domain.

The separation of function and safety during the design process is a key feature for the development of safe electronic systems.

The level of detail in the single steps of the process is limited due the restricted number of pages of this paper. As this approach is based primarily on theory it needs to be evaluated in practice. At the present point of time this methodology is evaluated on the example of a steer-by-wire system.

ACKNOWLEDGEMENTS

We wish to thank Dr. Bernd Müller, Dr. Dieter Lienert and Prof. Dr.-Ing. Winfried Görke for numerous ideas in discussion that led to the contents of this paper.

REFERENCES

[1] S. DAIS. Electronics and Sensors: Basics of Safety. In *Technical Congress*. VDA - German Association of the Automotive Industry. 2002.

[2] E. DILGER and W. DIETERLE. Fault-Tolerant Electronic Architectures for Safety Relevant Automotive Systems. *at - Automatisierungstechnik*, pages 375–381. August 2002.

[3] U. FREUND, T. RIEGRAF, M. HEMPRICH and K. WERTHER. Interface Based Design of Distributed Embedded Automotive Software - The TITUS Approach. In *VDI-Berichte 1547: Electronic Systems for Vehicles in Baden-Baden*, pages 105–123. VDI. October 2000.

[4] V-MODEL '97. Development Standard for IT-Systems of the Federal Republic of Germany - Lifecycle Process Model. www.v-modell-iabg.de, IABG. 1997.

[5] J. BORTOLAZZI. *Scriptum for the lecture "Systems Engineering for Automotive Electronics"*. Institut für Technik der Informationsverarbeitung (Institute for Information Processing Technology) at the University of Karlsruhe, Germany. 2002.

[6] R. KNEPPER. The Safety and Reliability Process in the Civil Aircraft Industry. DaimlerChrysler Aerospace Airbus GmbH, Hamburg.

[7] J. BEAUFAYS. Air Navigation System Safety Assessment Methodology. Eurocontrol. 2000.

[8] S. BENZ. Eine Entwicklungsmethodik für sicherheitsrelevante Elektronikysteme im Automobil. In *15th Workshop on Testmethods and Reliability of Circuits and Systems*. Kooperationsgemeinschaft Rechnergestützter Schaltungs- und Systementwurf (GI FA 3.5.5 / ITG FA 8.2 / GMM FB 8). March 2003.

[9] DIN VDE 31000-2. Allgemeine Leitsätze für das sicherheitsgerechte Gestalten technischer Erzeugnisse. Begriffe der Sicherheitstechnik, Grundbegriffe. 1987.

[10] IEC 61508. Functional Safety of electrical / electronic / programmable electronic safety-related systems. 1998.

[11] VDI/VDE 3542. Zuverlässigkeit und Sicherheit komplexer Systeme (Begriffe). VDI Handbuch Technische Zuverlässigkeit. VDI-Gesellschaft Systementwicklung und Projektgestaltung. 1995.

[12] DIN 40041. Zuverlässigkeit - Begriffe. 1990.

[13] Bosch Research Info: Rapid System Development. Robert Bosch GmbH. 3rd issue 1999.

[14] B. HEDENETZ. *Entwurf von verteilten, fehlertoleranten Elektronikarchitekturen in Kraftfahrzeugen*. Ph.D. thesis, University of Tübingen, Germany. 2001.

[15] SAE AEROSPACE RECOMMENDED PRACTICE 4754. Certification Considerations for Highly-Integrated or Complex Aircraft Systems. 1996.

[16] DO 178 B. Software Considerations in Airborne Systems and Equipment Certification. RTCA. 1992.

[17] SAE AEROSPACE RECOMMENDED PRACTICE 4761. Guidelines and Methods for Conducting the Safety Assessment Process on Civil Airborne Systems and Equipment. 1996.

[18] N. STOREY. *Safety-Critical Computer Systems*. Addison-Wesley. 1996.

[19] DIN EN 954. Safety of machinery - Safety-related parts of control systems. 1997.

[20] IEC 62061. Safety of machinery - Functional safety of electrical, electronic and programmable control systems for machinery. 2002.

[21] T. BERTRAM, R. BITZER, R. MAYER and A. VOLKART. CARTRONIC - An Open Architecture for Networking the Control Systems of an Automobile. In *SAE World Congress*. 1998.

[22] T. BERTRAM, P. DOMINKE and B. MÜLLER. The Safety-Related Aspect of CARTRONIC. In *SAE World Congress*. 1999.

[23] M. AUERSWALD, M. HERRMANN, S. KOWALEWSKI and V. SCHULTE-COERNE. Design Patterns for Fault-Tolerant Software-Intensive Systems. *at - Automatisierungstechnik*, pages 389–398. August 2002.

[24] MIL-HANDBOOK 217. Reliability Prediction of Electronic Equipment. DoD. 1995.

[25] RELIABILITY DATA HANDBOOK. Union Technique de l'Electricité. 2000.

[26] H. SCHWAB, A. KLÖNNE, S. RECK, I. RAMESOHL, G. STURTZER and B. KEITH. Reliability evaluation of a permanent magnet synchronous motor drive for an automotive application. In *10th European Conference on Power Electronics and Applications*. September 2003.

[27] UN ECE-R 13. Uniform provisions concerning the approval of vehicles of categories M, N and O with regard to braking. 2001.

[28] UN ECE-R 79. Uniform provisions concerning the approval of vehicles with regard to steering equipment. Draft. 2001.

CONTACT

Stefan Benz
Robert Bosch GmbH, FV/SLI
P. O. Box 10 60 50
70049 Stuttgart
Germany

Phone: +49 711 811 38329
Fax: +49 711 811 7136
E-Mail: Stefan.Benz@de.bosch.com

2004-01-1663

Preserving System Safety Across the Boundary Between System Integrator and Software Contractor

Jeffrey Howard
Safeware Engineering Corporation

ABSTRACT

Complex automotive systems are not developed entirely by one organization. OEMs purchase subsystems from integrators who, in turn, purchase hardware components from suppliers and contract for the development of software components. Safety is an emergent property of the system as a whole, making it difficult to preserve safety-related information across the organizational boundaries between OEMs, integrators, and contractors. We propose the *intent specification*, an improved specification format, and *SpecTRM-RL* (SpecTRM Requirements Language), a readable component requirements modeling language, to communicate requirements, design, and safety information across organizational boundaries in a form that promotes its effective use.

INTRODUCTION

Although engineering processes vary between organizations, every sensible engineering effort shares some steps. A business goal drives the development of a new system or the evolution of an existing system. System engineers develop system-level requirements to achieve those business goals and make design decisions allocating responsibilities to components. Specialists, such as mechanical, electrical, and software engineers design components according to the requirements given to them by system engineers. These pieces are tested individually, integrated into a finished system, and tested together.

System safety engineering begins with identifying hazards in the system and developing constraints on the system design that mitigate these hazards. For example, cruise control systems have the constraint that the system must disengage when the brake pedal is pressed. System safety continues by ensuring that components are designed in such a way as to enforce the safety constraints. Accidents occur when a safety constraint is not adequately enforced by the system's components.

Ensuring that components enforce system safety constraints is a difficult task. Safety engineers often perform safety analyses separated from the system engineers, who are meanwhile making critical design decisions. The results of safety analyses are often presented later, as a design critique. This information is frequently ignored or rationalized away because it is too costly and too time-consuming to change the design that late in the process.

The problem is compounded during component development. Safety is an emergent property of the whole system. Component developers do not have sufficient perspective on the system as a whole to fully evaluate the safety of their component. They must rely on information communicated by the system and safety engineers. If this safety information is incomplete or ambiguous, risk increases. Some companies have tried to solve this problem with integrated product teams. While a step in the right direction, there are additional communication barriers that must be overcome. Different engineering disciplines receive very different training. Miscommunication is common because of differing backgrounds, perspectives, and assumptions.

This problem of communicating safety information to component engineers is exacerbated when the engineering effort is spread across multiple organizations. Today's complex automotive systems are built by original equipment manufacturers (OEMs). OEMs purchase subsystems from integrators who in turn purchase parts and contract software development from suppliers. The sophistication of these subsystems and components requires specialized engineering knowledge. Working with external suppliers is a cost effective strategy. However, working with suppliers greatly increases the difficulty of building a system with the right system-level properties, such as safety.

This paper describes software's susceptibility to difficulties in enforcing safety constraints across organizational boundaries. The vehicle for communicating technical information about software to component developers is the software requirements specification. Common problems with specifications are used to develop a list of properties a requirements document should possess for successfully communicating safety information to suppliers. Lastly, a form of specification, called an *intent specification*, and a component requirements modeling language, called *SpecTRM-RL*, are introduced. Intent specifications and SpecTRM-RL models help ensure complete and unambiguous communication about system and component properties, including safety.

Throughout the introduction of intent specifications and SpecTRM-RL, we use an example of a simple cruise control system. The software controller takes input from the driver as well as from the vehicle. During operation, information about the vehicle's speed is used to generate commands for the throttle control.

SAFETY AND SOFTWARE

Complex, safety-critical systems are becoming increasingly software intensive. Software delivers substantial benefits because of it its flexibility. System designs can be changed without retooling or remanufacturing. Software changes do not adversely impact subsystem weight or space constraints. Once developed, the per-copy production cost of software is negligible. Other than memory and performance requirements on the underlying hardware, software imposes no physical constraints on the designs that can be implemented.

The same flexibility that makes software so useful gives rise to its greatest flaws. Software lacks the physical constraints that enforce discipline on the design process and control complexity. A mechanical part of a given volume can only support so many interconnections with other parts. Software has no limiting sense of distance between components; only the designer's decisions limit coupling and complexity.

The flexibility of software also induces a sense of complacency and convenience. Problems discovered in hardware are often solved by adding compensating features, and thus complexity, to software. Because software is so easy to adapt and change, frequently software development efforts begin before there's a sufficient understanding of what the software needs to do. Engineers presume they can tweak software later if their initial guesses don't work out.

These problems are exacerbated by dividing system and software engineering between separate organizations. Software vendors are frequently instructed to begin work before the system's needs are well defined. Software requirements are changed as the project progresses. In some cases, there may not be a good record of the changes requested, meaning that the software requirements documents fall behind the behavior actually required of the software. If the requirements are incomplete or ambiguous, there is a far greater risk that the supplier will produce software that makes the system unsafe.

Typically, software returned by suppliers comes with little accompanying documentation beyond what was initially provided by the integrator. System integrators who have not kept their requirements up to date or who produced incomplete requirements initially may not have enough information to predict the behavior of the software. Should the software be reused in a future system, out of date specifications make it difficult and costly to predict what changes in the surrounding system will require modifications to the software. In some cases, companies are forced to throw away software developed at great expense because they do not sufficiently understand how it works to make any claim about the safety of a system employing that software. Worse, companies may keep using software components even though their engineers are afraid to make any changes.

A successful system safety effort identifies constraints that will maintain system safety and ensures those constraints are enforced by system and component designs. Software is particularly difficult because there are no intrinsic physical properties to limit the complexity of a software system. If complexity renders the software intellectually unmanageable, ensuring that appropriate safety constraints are present becomes impossible. The challenge is to find a way to make software requirements and constraints intellectually manageable for everyone involved in the engineering effort.

COMMON SPECIFICATION PROBLEMS - Software requirements specifications are used to communicate information about software requirements and constraints between organizations. Given that software-related accidents occur when the behavior of software is not understood well enough to ensure it enforces system safety constraints, it should not be surprising that as much as 90% of the decision-making related to safety occurs in the requirements portion of the life cycle. In fact, almost all software-related incidents and accidents are traced back to flaws in the requirements.

Ambiguous Requirements - There are many notations for precisely describing software requirements, but most of them require a strong background in discrete mathematics. This makes them unsuitable as a medium of exchange between different engineering disciplines and organizations. Most specifications continue to be written in narrative text. Unfortunately, narrative textual documents often leave readers in disagreement on the meaning of the requirements. An engineering team that must debate the correct reading of a requirement rather than the correctness of the requirement is working with an ambiguous specification.

Incorrect Requirements - Informal, textual descriptions of component behavior cannot be executed or analyzed. Extensive human review catches many problems, but the process is time consuming and expensive. Schedule pressure often causes specification review to be cut. Too often, the requirements for software are validated by running the finished code on a test bench or in a simulated environment and observing whether anything goes wrong. By this point, fixing any requirements errors found will incur serious cost and schedule problems.

Contracting an outside organization to supply software magnifies this problem. Undoubtedly the requirements for the software will change as the system is designed and developed. Many system integrators lob changes over the fence to the contractor; it's treated as the contractor's responsibility to make sure the changes will work. When software is returned to the customer, the requirements document may have fallen so far behind that it does not reflect the behavior of the software. An outdated specification can cripple attempts to analyze the safety of later changes or reuse the software in another product. On a project suffering from this problem, engineers will go to great lengths to avoid modifying existing software because they no longer understand how the software works.

Incomplete Requirements - Incompleteness is one of the most common problems in software specifications. Omissions from specifications come in a variety of forms. The requirements may lack a provision for when expected inputs fail to arrive. The system may be missing a feedback path to detect the effects of an output. There may be no description of how to handle bad values. More subtly, different parts of the specification may prescribe conflicting behaviors. In the best of circumstances, software developers will recognize the incompleteness and contact their customer to resolve the problem. More often, the omission goes unnoticed, creating the possibility of an accident.

Over-specified Requirements - This error is often caused by a reaction to previous three problems. System integrators, plagued by ambiguous, incorrect, and incomplete requirements, decide to solve the problem by specifying requirements with an extraordinary degree of precision. Instead of focusing on the behavior the component should exhibit, the specification includes algorithms and data structures, possibly to the extent of including bit lengths for numeric data. All of this extra information belongs in the design and implementation of the component. While it may provide the system integrator a short-term sense of security, this solution is ultimately self-defeating and dangerous.

Software developers are quite good at writing software that satisfies its requirements. Requirements with design information mixed in remove the flexibility that software developers use to accomplish their goals. Providing a supplier over-specified requirements is similar to handing a construction bill of materials to an architect and demanding that the architect design a house that will use exactly those materials: no more, no less. Software developed for a specification with excess design information will generally cost more and take longer, as the software designers must meet both the intent of the requirements and the literal written requirement statements.

Beyond the cost and schedule impact of demanding unnecessary design features from the software supplier, there is a safety risk as well. Additional design information adds complexity. Complexity makes it difficult to analyze, understand, and predict the behavior of the software in the context of the system as a whole. Intellectually unmanageable requirements are much less likely to correctly enforce the safety constraints required to maintain safe system operation.

CRITERIA FOR A GOOD SPECIFICATION - The four problem categories above can be used to develop criteria for specifications that will enhance communication of important design information between suppliers and integrators.

1. **Clarity** – Requirements must be unambiguous. System engineers, software engineers, domain experts, and managers should all be capable of reading the specification and deriving the same understanding of what the software component is to do.
2. **Correctness** – A requirements specification should accurately describe the software to be built. Initially, this means that the specification should lend itself to validation techniques such as simulation, analysis, and review. Later, this means that changes to the requirements should be easy to record in the requirements specification, not simply applied to the software and forgotten.
3. **Completeness** – Incomplete requirements are a repeated cause of incidents and accidents. In this context, a software specification is complete when a reader can distinguish between acceptable and unacceptable software implementations.
4. **Conciseness** – Software requirements specifications should contain only as much information as necessary to describe the relationship between inputs to the software and outputs the software produces. Systems engineers may think of this as describing the transfer function for the software. A completely black box view of the behavior of the software allows software developers the freedom to meet project goals. Additional information about the design of the software hampers safety analysis efforts.

A specification that meets the above criteria will enhance communication between system integrators and the suppliers of their software. Safety constraints on the software design are far easier for software suppliers to use in a specification with the above properties. Such a specification also makes it easier to verify that the software developed enforces the safety constraints.

These same characteristics are helpful in other situations as well. If system integration and software development are within the same organization, clarity, correctness, completeness, and conciseness in the software specification will still benefit the project. Specifications that can be read and understood across engineering disciplines facilitate communication between software developers, system engineers, and safety engineers. A high quality specification is also an asset when a system is passed from research to production, leading to reduced costs and faster time to market.

INTENT SPECIFICATIONS

Software developers cannot evaluate the safety implications of software design choices without safety-related design constraints from system safety engineering. To develop good safety constraints, system engineers and safety engineers must have a good understanding of the system as a whole. Intent specifications are an improved format for system specifications, covering the whole life cycle of system development [1].

Intent specifications do not require additional effort compared to traditional specification formats. Most of the information to be included in an intent specification is already generated in any reasonable system specification effort. Intent specifications simply organize this information in a manner that supports the use of that information by engineers.

Most systems have substantial amounts of documentation. Often, system documentation contains a great deal of redundancy, missing information, and inconsistency. As the system evolves during development and once deployed, system documentation is often allowed to fall out of date. If more than one organization is involved, it is difficult to ensure all have consistent and up to date views of the system. Evaluating the safety impact of changes becomes difficult and expensive, when possible at all. Organizing information in an intent specification helps with these problems.

	Environment	Operator	System and components	V&V
Level 0	Project management plans, status information, safety plan, etc.			
Level 1 System Purpose	Assumptions Constraints	Responsibilities Requirements I/F requirements	System goals, high–level requirements, design constraints, limitations	Preliminary Hazard Analysis Reviews
Level 2 System Principles	External interfaces	Task analyses Task allocation Controls, displays	Logic principles, control laws, functional decomposition and allocation	Validation plan and results, System Hazard Analysis
Level 3 Blackbox Models	Environment models	Operator Task models HCI models	Blackbox functional models Interface specifications	Analysis plans and results, Subsystem Hazard Analysis
Level 4 Design Rep.		HCI design	Software and hardware design specs	Test plans and results
Level 5 Physical Rep.		GUI design, physical controls design	Software code, hardware assembly instructions	Test plans and results
Level 6 Operations	Audit procedures	Operator manuals Maintenance Training materials	Error reports, change requests, etc.	Performance monitoring and audits

Figure 1 - Structure of an intent specification. Particular contents will vary based on the scope and nature of a project.

The structure of an intent specification, shown in figure 1, differs from traditional specification formats. Current specification methodologies manage complexity by employing two forms of abstraction. The first is part-whole decomposition: the whole is broken down into parts that are simpler when considered individually. The horizontal headings at the top of figure 1 show this kind of breakdown. The intent specification records information about the environment in which the system will operate, the operator's interactions with the system, the system itself and its components, and verification and validation procedures.

The second form of abstraction already used in system specifications is refinement. Refinement is also a form of part-whole decomposition: a complicated function is broken down into simpler steps. Most specifications are stratified into levels of refinement. Levels above describe what function is to be accomplished; levels below describe how the function will be accomplished by presenting a refinement of the function.

In an intent specification, the major level divisions in the specification do not represent refinement. Each level describes the whole system from a different view, supporting different tasks. A significant advantage of this structure is that it answers not just questions of what is to be done and how, but also *why*. Each level provides motivation and intent for the decisions at the level below were made the way they were. This information about what was intended is often the most difficult to reconstruct or deduce from traditional specification formats. Understanding why decisions were made as they were is critical for ascertaining the safety impact of proposed changes. Many accidents have been caused because engineers or workers making changes – different from those that built the system – did not realize they were altering a design choice originally made for safety reasons. Part-whole decomposition and refinement are still present in an intent specification, but they're found within each level.

Level 0 is the top level of the intent specification. It is the program management view of the system. Program management plans, including a system safety plan, are kept at this level of the specification. Traceability links from this level point to the documented results of activities described in the plans.

Level 1 is the system requirements level of the specification. This level includes any introductory material or historical information about previous similar systems. The goals of the system are recorded here, as well as the high level functional requirements. The following is an example requirement from the cruise control software:

HLR0 The cruise control system shall maintain a speed set by the driver (← G0) (↓ 2.2).

The arrows at the end of the requirement indicate links to other areas of the specification. Left and right arrows indicate information on the same level of the specification. Up and down arrows indicate information at different levels of the intent specification. In this case, the reference to G0 is to a goal the system is intended to accomplish. The other reference, to section 2.2, is a reference to the design decisions motivated by this requirement. With tool support, the labels G0 and 2.2 would be hyperlinks to navigate to those entries. This system-level requirement will become the basis for design decisions made at level 2.

The early results of the system safety effort are recorded at level one as well. Even as the functional requirements are being decided, a hazard log is started, noting hazards discovered during preliminary hazard analysis.

H1 Cruise control does not maintain correct speed (→ SC1).

These hazards are used to develop safety constraints. Requirements describe what the system must do in order to achieve functional goals. Safety constraints describe what the system must not do in order to maintain safe operations. Just as requirements must be satisfied, these safety constraints must be enforced by the design decisions at lower levels.

SC1 Cruise control must maintain the correct speed (← H1 (↓ 2.2).

By keeping both safety constraints and system requirements in the same level of the same document, the safety information is not "out of sight, out of mind" when system engineers are making important decisions about the system. This integration of system and safety engineering makes it easier to convey safety-related information to component developers in other organizations.

Level 2 is the system engineering design view of the system. This level assists system engineers in reasoning about the physical principles and system-level design decisions that will shape the system. External interfaces are described, tasks are partitioned between the operator and the automation, and functions are decomposed and allocated to components. Design decisions at this level are given traceability links back up to the requirements and constraints at level one that motivate them. Design decisions are also linked down to the component requirements specifications at the next level.

2.2 When the driver issues a set command, the cruise control will set its speed to maintain at the current vehicle speed and engage control of the throttle

(↑ HLR0, HLR1, SC1)

(↓ 3.Throttle Delta, 3.Set Next Speed Set Point, 3.Cruise Control, 3.Compute Throttle Delta, 3.Speed Set Point, 3.Vehicle Speed, 3.Speed Error Threshold).

2.2.1 Vehicle speed will be maintained by calculating the difference between current speed and desired speed, then issuing a command to the throttle control to increase or decrease engine torque by a small step size (↑ SC3) (↓ 3.Throttle Delta, 3.Compute Throttle Delta).

Rationale: *Changing the throttle in small increments will lead to smooth acceleration and deceleration that will not jar the driver and passengers.*

The refinement of the design decision in 2.2.1 occurs at the same level of the intent specification as the larger concept being refined. Note also that rationale and assumptions are recorded next to the design decisions they support. Recording rationale and assumptions is invaluable for evolution of any system. A system that is reused in another environment may not be able to rely on the assumptions that were originally made. Without a record of where assumptions were made, it is difficult to reconstruct why some design decisions were made the way they were.

At level 3, the behaviors allocated to components are specified rigorously. Component requirements are written as models of black box behavior. Black box models support reasoning about individual component behavior and the interaction between components. This level provides an unambiguous interface between system engineers and component engineers, even across organizational boundaries. The language used at this level is SpecTRM-RL, a requirements modeling language. The modeling language is based on finite automata theory, meaning that it can be executed and formally analyzed. However, the language was designed for readability, and reviewers can be taught the language with about 15 minutes of training.

Level 4 is the design representation of components for the system. The notation used here varies across projects and component types. If the volume of design information is too great to include in one document, or if the supplier is not obligated to deliver complete design information, pointers to the location and owners of the design information are included at this level instead.

Level 5 is the physical representation of the components of the system. For hardware components, these would be hardware assembly instructions. For software, this would be the source code. For very small safety-critical software components, such as the shutdown code for a nuclear reactor, it might be possible to include a listing of the code in this section. For larger projects, such as the controller in an automotive system, the specification would contain a pointer to the location of the configuration management system housing the code.

The final level of an intent specification, level 6, describes the operations of the system. This will include any audits performed on the system in the field. Should there be any incidents or accidents involving the system, information about them would be recorded at this level of the specification. Furthermore, any changes made to the system after it is fielded should have some change impact analysis recorded here.

Taken as a whole, the intent specification provides a comprehensive view of the entire life cycle of a system development effort. Although the specification is presented in a logical ordering from initial planning through operations, a lockstep progression from one level to the next is neither necessary nor practical. In any system engineering process, there will be iteration, skipping around, and updating of old material as new information is discovered.

Large projects may find it impractical to keep the entire intent specification in one document. Using the intent specification format still provides an excellent interface to project information. In large project, the intent specification is simply spread across multiple documents. References, preferably hyperlinks, point between the documents. Similarly, system integrators may give a supplier only a portion of the intent specification, such as the SpecTRM-RL requirements model for the component to be built by that supplier.

Traceability hyperlinks are used extensively within and between levels of the specification. Links point both directions, from higher levels to lower levels and back up again. Following these links provides a clear of view of the rationale behind design decisions. Often, this rationale is most difficult to reconstruct from traditional specifications. The accessibility of this rationale is a key factor in successfully writing a component specification that preserves system safety information as component development is passed from the system integrator to a software supplier.

SPECTRM-RL

SpecTRM-RL is a formal modeling language for component requirements [2]. These models are used at on level 3 of an intent specification. SpecTRM-RL models clearly describe the component behavior that will effect the design decisions at level 2 of the intent specification. The language describes only externally visible behavior, which can be thought of as the transfer function across the component. Details about the design and implementation of the component greatly complicate review and needlessly hinder the efforts of software suppliers. Because externally visible behavior alone is described, components may be implemented in hardware, software, or as tasks for a human operator. Software may be designed and implemented with any combination of methodology, notation, and language desirable.

SpecTRM-RL is based on a theory of finite automata, state machines, so models are executable and analyzable. Component behavior can be evaluated in a high fidelity system simulation or in conjunction with the behavior of other components, before turning over the specification to a supplier. If behavior is discovered that has a negative safety impact, it can be corrected before the cost and schedule impact of the change would have to be negotiated with the development organization. Specifications can also be automatically analyzed to detect some undesirable properties such as nondeterminism.

Despite the formal model underneath the language, SpecTRM-RL was designed to be readable and reviewable. Many formal specification languages are difficult to read and review. In these cases, the requirements must be maintained separately from the formal model. When changes are made, as is inevitable, the formal model must also be updated. Most projects do not have the resources for separate specification and modeling efforts. SpecTRM-RL, requiring only fifteen minutes training to read, solves this problem by being both the readable specification and formal model.

In addition to being executable, analyzable, and readable, the language also strongly supports development of complete specifications. Over sixty criteria for completeness of specifications have been identified [3] and validated at JPL [4]. These criteria cover all areas of component behavior including startup and shutdown, mode changes, acceptable values, timing, load and capacity, human-machine interface, data age, feedback, reversibility, and path robustness.

Originally, these criteria were expressed in a checklist format. The criteria were found to be difficult to apply as a checklist. As a result, the criteria were built into the syntax of the SpecTRM-RL modeling language [5].

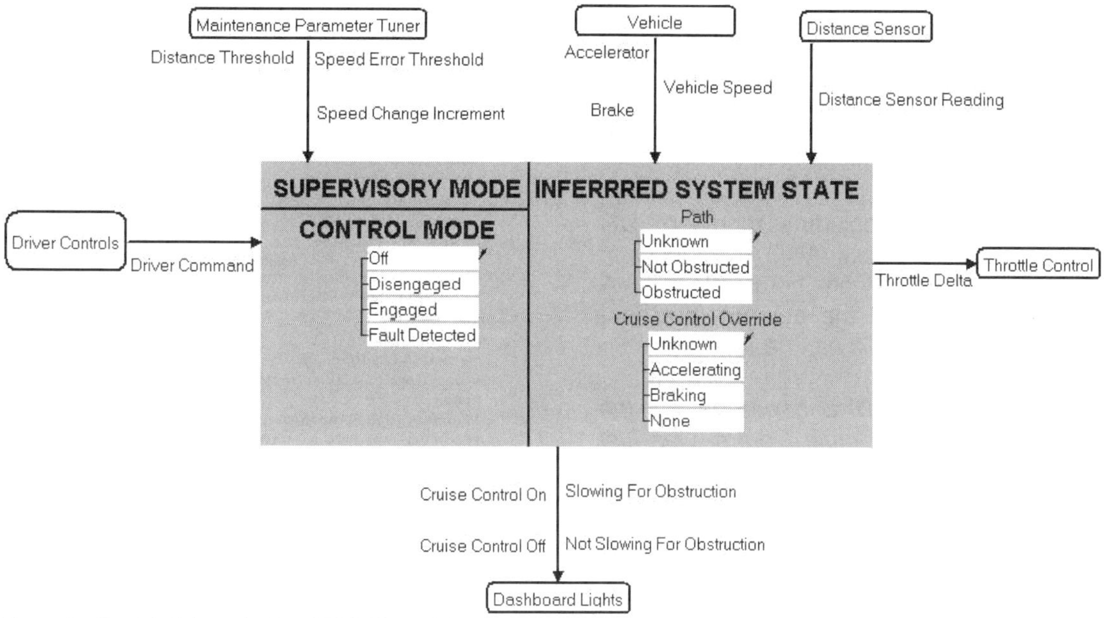

Figure 2 - Graphical view of a SpecTRM-RL model.

Building the criteria into the modeling language keeps the criteria in mind as the component requirements are developed. Almost all of the completeness criteria are included in the syntax of the SpecTRM-RL modeling language. Most of those that could not be directly included can be checked using automated tools [6]. The few remaining criteria must be checked by human review.

Figure 2, below, shows the graphical notation used in SpecTRM-RL. The notation is similar to the engineering depiction of a control loop. Inputs to the component are shown to the left and above the box. Outputs from the component are drawn to the right and down. The box in the center depicts the component for which requirements are being modeled. The box is divided into sections for modes and for inferred state.

Modes are used for coarse-grained partitioning of system behaviors. Engineers frequently use modes in describing real-time embedded control software because they're a natural expression of system behavior. The behavior of control software in a maintenance or standby mode is often entirely different from that of the operational mode, even for identical sequences of input.

The other portion of the component box shows inferred state. Every controller must have a model of the state of the system it controls, means to affect the system, and rules to predict the affects of changes it makes to the system. A thermostat is a simple example. A thermostat has a model of the current temperature and desired set point. The thermostat is connected to a furnace or air condition system, giving it the means to affect the temperature. Lastly, the thermostat controller is given rules to predict the how turning on or off the furnace or air conditioning will affect the temperature. Using these things, the thermostat maintains temperature. The inferred state variables in the thermostat controller are the set point and the current temperature. Note that inferred state may differ from the actual state of the system. The thermostat may have an inaccurate representation of the true temperature. Accidents frequently occur when the controller incorrectly infers the state of the system.

AND/OR TABLES - The behavior of each of the inputs, outputs, modes, and states depicted in the graphical view is defined by AND/OR tables. Figure 3 shows a portion of the cruise control's logic for determining the mode of the software. The figure shows the logic used to decide that the cruise control should be in the disengaged mode.

= Disengaged

System Start	F	F	F
Off	T	*	*
Disengaged	*	T	*
Engaged	*	*	T
Driver Command is Off	*	F	*
Driver Command is On	T	*	*
Driver Command is Set Speed	*	F	*
Cruise Control Override in state None	*	*	F

Figure 3 - AND/OR table

Each table evaluates, as a whole, to true or false. If the table is true, then the state machine takes the transition to *Disengaged*. If the table is false, the *Disengaged* transition is not taken. To evaluate whether the table is true or false, begin by looking at the expressions in the first column. Each of these expressions will be true or false. For example, the first expression, "System Start," will be true when the system is first activated, and never thereafter. The expression in the second row, “Off,” is true if the cruise control was in the off mode just before now. The truth values of these expressions are matched against values in the columns to the right. Columns on the right may contain ‘F’ for false, ‘T’ for true, or ‘*’ for a don’t care which matches anything.

When matching truth values, columns represent OR relationships, while rows represent AND relationships. The table as a whole is true if any one column is true. A column is true if every row in the column matches. Thus, the AND/OR table in figure 3 can be read as:

Cruise Control transitions to Disengaged if (1) the system is not just starting up, Cruise Control was just previously in Off mode, and the Driver Command is On, or (2) if the system is not just starting up, Cruise Control was just previously in Disengaged mode, Driver Command is not Off, and Driver Command is not Set Speed, or (3) if the system is not just starting up, Cruise Control was in the Engaged mode, and the Cruise Control Override state value is not in a state of None.

To fully understand the behavior of the mode change described above, one would want a more information about the meaning of Cruise Control Override, the various Driver Command values, and the intended effects of the driver commands. All of this information is readily available from the full model, and the table makes very clear and readable how these concepts relate to determine whether the mode change to Disengaged takes place.

The AND/OR table format is easy enough to read that reviewers quickly can be taught to understand the language syntax. Readers often find AND/OR tables clearer and easier to understand than English textual descriptions. The tables lack the ambiguity of textual descriptions. Arguments over the meaning of requirements are replaced by disagreements over what appropriate behavior for a component really is. The input, output, mode, and state definitions demonstrated below all make use of AND/OR tables to describe the behavior of the component.

OUTPUTS - An output specification from the cruise control example is shown in figure 4. This output is a command to the throttle control. Throttle control commands are sent to keep the car moving at its set speed despite variations in the slope of the road. The first section of the output is a series of attribute value pairs. These attributes provide information about the output beyond what can be expressed in an AND/OR table. Much of this information enforces the completeness of the requirements specification.

Output Command

Throttle Delta

Destination: Throttle Control
Fields:
Name: Throttle Delta
Type: Real
Acceptable Values: Any
Units: Unknown
Granularity: Unknown
Hazardous Values: Unknown
Exception-Handling: None
Description: This output commands the throttle control to keep vehicle speed at the cruise control's set point.
Comments:
Timing Behavior:
Initiation Delay: .2 seconds
Completion Deadline: .25 seconds
Output Capacity Assumptions:
Load:
Minimum Time Between Outputs: .5 seconds
Maximum Time Between Outputs: None
Hazardous Timing Behavior:
Exception-Handling:
Feedback Information:
Variables: Vehicle Speed
Values: Real (km/h)
Relationship: As the throttle control takes action based on these commands, vehicle speed will change. Measuring vehicle speed provides feedback on the effects of this command.
Minimum Latency:
Relationship: As the throttle control takes action based on these commands, vehicle speed will change. Measuring vehicle speed provides feedback on the effects of this command.
Minimum Latency:
Maximum Latency:
Exception-Handling:
Reversed By:
Description: This output sends changes to the throttle control for the vehicle. These changes are used to keep the vehicle speed at the set point desired by the driver.
Comments:
References: (↑ 2.2, 2.2.1)(→ 3.Cruise Control, 3.Speed Set Point, 3.Vehicle Speed, 3.Speed Error Threshold, 3.Compute Throttle Delta)

TRIGGERING CONDITION

Cruise Control in mode Engaged	T
|(Speed Set Point – Vehicle Speed)| > Speed Error Threshold	T

MESSAGE CONTENTS

Field:	Value:
Throttle Delta	Compute Throttle Delta()

Figure 4 - Output command definition.

Each of the fields in the output must be described. The type is noted, as are acceptable values within that type. Granularity, units, and any possible hazardous values for the output are recorded. Simple mistakes such as writing software to use the wrong system of units do occur. Providing a clear, complete specification to the organization supplying the software will help avoid these problems.

Timing attributes record how long an actuator will take to act on a command and how long the action is expected to take. Load information is also important. In the Three Mile Island nuclear reactor accident, a line printer was used to report errors as alarms tripped [7]. At one point during the accident, so many alarm messages were queued up to print that the printer output was running three hours behind the state of the reactor. At times the printer jammed, losing some information altogether. With a specification in SpecTRM-RL, development organizations can be made aware of limitations in the software’s environment without requiring full knowledge of the rest of the system.

The attributes for feedback information ensure that the controller has some way of detecting the effect of the outputs. The last few attributes show how the output can be reversed and provide a place for a general description of and comments on the output. The references attribute is a convenience for navigating through the model. It lists the states, modes, and inputs used to define this output. Reviewers use this information to find related portions of the model.

As shown in figure 4, it may not be possible to fill in all the attributes associated with an output. When reverse engineering an existing incomplete specification, the information may not be available. Researchers who are working with a proof of concept system do not need the same level of completeness necessary for productizing a system. In these cases, marking the attribute as unknown is sufficient. Doing so has the advantage of making it obvious where information is missing.

The second portion of the output specification is labeled, "Definition." An AND/OR table is used to describe when the output will be issued by the component. If the AND/OR table for the triggering condition evaluates to true, then the output is sent. If the triggering condition is false, then the output is not sent. Since the AND/OR table in figure 4 has only a single column, it is a simple AND of all the conditions in the table. To paraphrase the table, a throttle delta command is sent whenever the cruise control is in the "Engaged" mode and the difference between the speed set point and vehicle speed exceeds an error threshold.

The final portion of the output definition is a table showing the contents of the output. For this output, the only field is the delta in the throttle setting to be sent to the throttle control. Other outputs could be more complex, such as telemetry data or the results of internal diagnostic checks on the component.

MODES - The behavior of most controllers divides into large groupings. Engineers are used to thinking of these divisions in terms of modes. For example, as shown in figure 4 above, the cruise control behaves very differently depending on whether the system is engaged. When not engaged, the triggering condition can never be true, so the throttle delta output cannot be sent. SpecTRM-RL directly represents mode logic. An example of a mode definition is shown in figure 5.

Control Mode

Cruise Control

Description: Cruise Control is the main control mode for the cruise control system. The most complicated logic in the system is determining when the cruise control should engage and disengage in order to maintain the safety of the system.

Comment:

References: (→ 3.Driver Command, 3.Cruise Control Override, 3.Vehicle Speed)

Appears In: (↑2.1, 2.2, 2.5, 2.6)(← 3.Cruise Control On, 3.Cruise Control Off, 3.Slowing For Obstruction, 3.Not Slowing For Obstruction, 3.Throttle Delta, 3.Set Next Speed Set Point, 3.Decrease Next Speed Set Point, 3.Increase Next Speed Set Point)

DEFINITION

= Off

System Start	T	*
Driver Command is Off	*	T

= Disengaged

System Start	F	F	F
Off	T	*	*
Disengaged	*	T	*
Engaged	*	*	T
Driver Command is Off	*	F	*
Driver Command is On	T	*	*
Driver Command is Set Speed	*	F	*
Cruise Control Override in state None	*	*	F

= Engaged

System Start	F	F
Disengaged	T	*
Engaged	*	T
Driver Command is Off	*	F
Driver Command is Set Speed	T	*
Cruise Control Override in state None	*	T

= Fault Detected

System Start	F	F	F	F
Fault Detected	T	*	*	*
Time Since Cruise Control Last Entered Off > 5 seconds	*	T	*	T
Driver Command is Off	F	F	F	F
Vehicle Speed was Never Received	*	T	F	*
Vehicle Speed is Obsolete	*	*	T	*
Cruise Control Override in state Unknown	*	*	*	T

Figure 5 - Example of a SpecTRM-RL mode definition.

The mode definition begins with a set of attributes, including a description of the mode and any comments on it. The "References," and "Appears In," attributes are conveniences for navigation, listing all of the other definitions in the model that this mode uses and all of the definitions that use this mode, respectively. When the model is developed using a software tool, these listings can be hyperlinked.

The definition section of the mode shows how the component transitions between different modes. When one of the AND/OR tables is true, the system transitions to that mode. In figure 5, the cruise control transitions to the startup mode whenever the system first starts. The system transitions to the Engaged mode when (1) driver sets a speed to maintain or (2) the system was just in the Engaged mode and the driver has not accelerated or braked, overriding the cruise control system. Lastly, the system transitions to an internal fault detected mode if the system is not receiving the inputs it needs.

STATES - An inferred state is something the component observes about the thing it controls: the plant, in control theory terminology. Figure 6 shows the attributes for an example from the cruise control, the state that determines whether an obstacle is in the car's path. The cruise control system uses a sensor to detect other vehicles in the driving path. If an obstruction is found within a threshold distance, the cruise control system automatically reduces the set speed. The attributes on the state value are similar to the mode attributes.

State Value

Path

Obsolescence: If the distance sensor input becomes obsolete, the Path state transitions to Unknown.

Exception-Handling:

Related Inputs: (→ 3.Distance Sensor, 3.Distance Threshold)

Description: The cruise control system uses a distance sensor to detect objects in front of the car. When an object is close enough (closer than the distance threshold), the car's path is inferred to be obstructed.

Comments:

References: (→ 3.Distance Sensor Reading, 3.Distance Threshold)

Appears In: (↑ 2.7)(← 3.Slowing for Obstruction, 3.Not Slowing for Obstruction, 3.Decrease Next Speed Set Point)

DEFINITION

= Unknown

System Start	T	*	*
Distance Threshold is Obsolete	*	T	*
Distance Sensor Reading is Obsolete	*	*	T

= Not Obstructed

Distance Sensor Reading > Distance Threshold	T

= Obstructed

Distance Sensor Reading <= Distance Threshold	T

Figure 6 - Definition of a state value.

Figure 6 shows the definition of the *Path* state from the cruise control example. Every state value definition in a SpecTRM-RL model must have an unknown state. Inferred system state is not guaranteed to correspond to the true state of the system. Many accidents have been caused by a software controller's model becoming inconsistent with the true state of the system. One of the first systems modeled with a precursor language to SpecTRM-RL was an air traffic collision avoidance system. In an early version of that system, when reset, the software assumed the plane was on the ground, where warnings are not issued.

Accidents have also been caused when systems, taken offline for maintenance, were put back online and resumed operations exactly where they left off. State values should include logic to transition to the unknown state when the system is first started or returns from a mode that suspends input processing. Beginning in the unknown state forces the software to rebuild its model with the true state of the system.

INPUTS - Figure 7 shows the attribute portion of the input definition for a reading of the vehicle's speed. The attributes for input definitions enforce a number of completeness criteria. The type of the input is recorded, as are its possible values within that type, units, and granularity. Timing behavior is also covered, including the minimum and maximum time between consecutive inputs. The obsolescence attribute handles stale data. No input is good forever, and the obsolescence attribute describes how long data remains valid. Accidents have resulted from systems working with outdated data. Obsolescence also affects the definition of the input. There are three possible ways for the input to take on a value. First, when new data arrives for the input, the input takes on the value of the new data. Second, when new data has not arrived, but the existing data is still valid, the input keeps its value. Third, after enough time has passed, the input becomes obsolete.

Input Value

Vehicle Speed

Source: Vehicle
Type: Real
Possible Values (Expected Range): Any
Units: km/h (kilometers per hour)
Granularity: Unknown
Exception-Handling: None
Timing Behavior:
Load: Unknown
Minimum Time Between Inputs: Unknown
Maximum Time Between Inputs: Unknown
Maximum Time Before First Input: Unknown
Related Outputs: Throttle Delta
Latency: Unknown
Time After Output: Unknown
Exception-Handling: Unknown
Obsolescence: .1 seconds
Exception-Handling:
Description: This input is the speed at which the vehicle is moving.
Comments:
References:
Appears In: (↑ 2.2)(← 3.Throttle Delta, 3.Set Next Speed Set Point, 3.Compute Throttle Delta)

DEFINITION

= New Data for Vehicle Speed

Vehicle Speed was Received	T

= Previous Value of Vehicle Speed

Vehicle Speed was Received	F
Time Since Vehicle Speed was Last Received <= 100 milliseconds	T

= Obsolete

System Start	T	*	*
Vehicle Speed was Never Received	T	T	*
Time Since Vehicle Speed was Last Received > 100 milliseconds	*	*	T

Figure 7 - Definition of an input

EXECUTION AND ANALYSIS - With outputs, modes, states, and inputs for a component all defined in SpecTRM-RL, the component has a clear, complete, and concise definition. SpecTRM-RL also assists in validating that the requirements are correct. Without correct requirements, a good software supplier will provide exactly the software asked for: the wrong software.

SpecTRM-RL is readily reviewable by domain experts. Pilots, air traffic controllers, doctors, project leaders, and office managers have all been trained to read SpecTRM-RL models in fifteen minutes. Human review remains one of the most effective techniques for achieving correctness of requirements.

In addition to human review, the formal model underlying SpecTRM-RL requirements specifications allows the model to be parsed, executed, and analyzed. The behavior of the software component can be validated before the specification is delivered to a software development contractor. If the software is found not to enforce required safety constraints, it can be modified with little cost or schedule impact.

Although simulation is a powerful tool for exploring the behavior of software, it is heavily dependent on the data used to drive the simulation. Automated analyses of the model are possible as well. For example, one of the completeness criteria states that for every possible combination of system states and inputs, at least one table in a state, mode, or input should be true. This is a kind of robustness: no matter what happens to the system, some response is defined. If the robustness criterion is not met, there is some combination of system state and input that the requirements don't handle. Cases where a specification is not robust can be identified by an automated analysis.

Another completeness criterion states that specifications should be deterministic. Determinism means that for any combination of states and inputs, each state, mode, or input should have no more than one table true. If more than one table were true, it would indicate that more than one behavior was defined as valid for a particular combination of states and inputs. Predicting the behavior of nondeterministic systems is difficult, making it hard to assure that the system is safe. Nondeterminism in a SpecTRM-RL specification can be identified by automated analysis.

CONCLUSION

It is all too easy for important safety information to be lost at the interface between system engineering and software development. System engineers and software engineers come from different backgrounds, have different training, and view the system from different perspectives. Additionally, software developers do not have a holistic system perspective to work from; they are dependent on clear communication of safety constraints from system engineering. These problems are magnified when software development is outsourced to a separate supplier organization.

Safety-related decision-making occurs very early in the development life cycle. Almost all software-related incidents and accidents have been traced back to flaws in the requirements for the software. An improved format for writing specifications can help engineers and domain experts reason about a system's behavior and build desirable properties, such as safety, into the system from the beginning.

A good specification will be clear, correct, complete, and concise. The specification must be clear so that readers with a variety of backgrounds, including system engineering, system safety engineering, software safety engineering, software engineering, and domain expertise, can all read the specification and form similar mental models of how the system will operate. With a clear specification, contractors outside the integrator's organization will have a similar understanding of the software to be built as the system engineers who developed the requirements for the software. A clear understanding of the safety-critical design constraints will assist the software developer in ensuring that the software delivered is safe. Intent specifications increase the clarity of a system specification by organizing information so that engineers can find what they're looking for quickly and see the rationale behind engineering decisions. SpecTRM-RL models make component behavior clear with the use of easily readable AND/OR tables, which offer an unambiguous description of system behavior.

Correctness is a necessary property of a specification that will act as a medium for preserving safety information between suppliers and customers. Software developed to an incorrect specification will not correctly enforce constraints required for system safety. A very common source of incorrectness in requirements specifications comes from asymmetric evolution of the system and its documentation. If the specification is not up to date with the system, it cannot be relied upon to help evaluate the safety of proposed changes. Intent specifications assist in evaluating correctness of the specification by including information on the verification and validation at every level of the system development effort. The SpecTRM-RL modeling language assists in developing correct specifications in several ways. The models are conducive to expert review, allowing many errors to be caught early. Due to the formal nature of the models, they can also be simulated and analyzed, allowing violations of safety constraints to be detected before the generation of a component design and implementation.

Completeness is necessary in a specification that preserves safety information. Software suppliers that are provided incomplete specifications may not realize what information is missing. The software delivered meet the requirements, such as they are, but if an unhandled system state is encountered, the software may not behave in a safe manner. Many accidents have been caused by incompleteness in specifications. The strong traceability in intent specification assists in developing complete by ensuring that important design and safety information is not overlooked. The syntax of the SpecTRM-RL requirements modeling language was carefully developed to incorporate over sixty completeness criteria. Where insufficient information is available, the attribute templates call attention to what is missing.

A specification that preserves safety information between organizations needs to be concise. Large volumes of irrelevant information distract from what is important. By focusing only on true requirements and design constraints for a software development effort, the specification assures that important constraints will not be lost in excessive documentation. Additionally, by avoiding over-specification, the software supplier is afforded as much flexibility as possible to meet the goals of the development project as quickly and economically as possible. The intent specification is a concise representation of a system, offering an organizational structure that reduces duplication and conflicting information. SpecTRM-RL models are a concise representation of component requirements. No internal design information is included in a SpecTRM-RL model. The model describes purely black box behavior, the relationship between the inputs and outputs.

Intent specifications and SpecTRM-RL have been developed through decades of academic research and industrial project experience. The language has been used on projects in the medical device, aerospace, and automotive industries. Intent specifications and SpecTRM-RL models enhance reasoning about system and component behavior, act as bridges between engineering disciplines, and help preserve safety information across organizational boundaries.

ACKNOWLEDGMENTS

This paper would not have been possible without the valuable comments of Mr. Grady Lee of Safeware Engineering Corporation, Professor Nancy Leveson of MIT, and the SAE World Congress' reviewers.

REFERENCES

1. Nancy G. Leveson. Intent Specifications: An Approach to Building Human-Centered Specifications, IEEE Transactions on Software Engineering, January 2000.
2. Nancy G. Leveson, Mats Heimdahl, and Jon Damon Reese. Designing Specification Languages for Process Control Systems: Lessons Learned and Steps to the Future, Presented at SIGSOFT FOSE '99 (Foundations of Software Engineering), Toulouse, September 1999.
3. Nancy G. Leveson. Safeware: System Safety and Computers, Addison-Wesley Publishing Company, Reading Massachusetts, 1995.
4. Robyn R. Lutz. Targeting Safety-Related Errors during the Software Requirements Analysis, Proceedings of SIGSOFT '93: Foundations of Software Engineering, ACM Press, New York, 1993.
5. Nancy G. Leveson. Completeness in Formal Specification Language Design for Process Control Systems, Proceedings of Formal Methods in Software Practice Conference, August 2000.
6. Mats Heimdahl and Nancy G. Leveson. Completeness and Consistency Analysis of State-Based Requirements, IEEE Transactions on Software Engineering, May 1996.
7. John G. Kemeny. Saving American Democracy: The Lessons of Three Mile Island, Technology Review, June 1980.

CONTACT

Jeffrey Howard is a systems engineer for Safeware Engineering Corporation. He graduated a master of engineering from MIT in computer science. Currently, he works on the development of tools to support working with intent specifications and SpecTRM-RL models. He can be reached at howard@safeware-eng.com.

2004-01-0708

Development of Safety-Critical Software Using Automatic Code Generation

Michael Beine and Rainer Otterbach
dSPACE GmbH

Michael Jungmann
MTU Aero Engines GmbH

ABSTRACT

In future cars, mechanical and hydraulic components will be replaced by new electronic systems (x-by-wire). A failure of such a system constitutes a safety hazard for the passengers as well as for the environment of the car. Thus electronics and in particular software are taking over more responsibility and safety-critical tasks. To minimize the risk of failure in such systems safety standards are applied for their development. The safety standard IEC 61508 has been established for automotive electronic systems.

At the same time, automatic code generation is increasingly being used for automotive software development. This is to cope with today s increasing requirements concerning cost reduction and time needed for ECU development combined with growing complexity.

However, automatic code generation is hardly ever used today for the development of safety-critical systems. Reasons for this are the specific requirements on the code as well as inadequate experience in the development of safety-critical software itself.

This paper deals with the application of automatic code generation for the development of safety-critical systems. It describes the role and benefits of automatic code generation in a safety-critical software development process. The requirements imposed on an automatic code generator by a safety standard such as the IEC 61508 are examined. The pros and cons of using a certified code generator and possible alternatives are discussed. The benefits and know-how gained from many years of experience in developing software according to safety standards such as RTCA DO-178B in the aerospace industry is taken into consideration.

The paper uses dSPACE s production code generator TargetLink as an example. The use of TargetLink at ATENA Engineering for the development of IEC 61508 SIL 3 software is described. The experiences and accomplishments made at ATENA are shown.

INTRODUCTION

SAFETY-CRITICAL SYSTEMS

The number of safety-critical systems in vehicles is rapidly increasing. A few years ago, the failure of a computer system in a vehicle would in the worst case only mean the loss of a function, but in the systems of the future, a wrong reaction to a fault could pose a safety hazard for the vehicle s occupants and other road users.

It is natural to rely on experience from the aviation industry when developing safety-critical systems for automotive applications. In the aviation industry programmable systems have been used for several decades for flight control, aircraft engine control, landing gear control, etc. The safety and reliability requirements of these systems are comparable with steer-by-wire or brake-by-wire systems which are currently under development in the automotive industry.

There are standards and methods for the development and manufacturing of avionics systems that help to meet an acceptable level of safety. In the area of software development the safety standard RTCA DO-178B has been established. It is also a suitable detailed standard to fulfill the requirements of IEC 61508 with regard to software.

AUTOMOTIVE SOFTWARE DEVELOPMENT

To cope with rising demands, such as the growing number of electronic systems in a vehicle, increasing complexity and shorter time-to-market, the automotive industry is increasingly adopting model-based design methods and using automatic code generators for software development.

In contrast, automatic code generators are hardly ever used for the development of safety-critical systems. Firstly, very special requirements are imposed on the code for safety-critical systems. Secondly, many software suppliers are only just beginning to apply appropriate development standards, so they cannot tackle the introduction of automatic code generation at the same time.

However, especially the high complexity and functional requirements of safety-critical systems demand the use of modern tools for developing, designing, implementing, verifying and validating such systems. Automatic code generation is a key player, and helps to cope with these growing demands.

SAFETY-CRITICAL SYSTEMS AND THEIR SOFTWARE

To minimize the dangers of safety-critical systems, special development standards and processes have been designed for use in such applications.

STANDARDS

All safety standards define a set of activities that have to be carried out in order to achieve a desired safety level. These activities can generally be grouped into three categories: selecting development methods and tools, implementing the system, and verifying and validating the system.

The standards differ with regard to their perspective on the system to be developed. Some standards cover the development of the overall system. Among them are IEC 61508 and ARINC 653. Other standards like RTCA DO-178B and DoD-2167A only deal with the development of the software, but are more detailed.

The established standard in automotive electronics is IEC 61508 [1]. This is a generic safety standard that requires the definition of more detailed standards for specific industries and projects. Software engineering studies have shown that the RTCA DO-178B [2] software development standard, originally defined for the aviation industry, is also a suitable detailed standard that corresponds to the IEC 61508 safety standard [3 et al].

SOFTWARE DEVELOPMENT PROCESS FOR SAFETY-CRITICAL SYSTEMS

Though not specifically requested and not mandatory, most of these standards follow the V-cycle, since the number of listed requirements barely leaves room for choice.

Figure 1 shows the software development process for safety-critical systems arranged according to the well-known V-cycle.

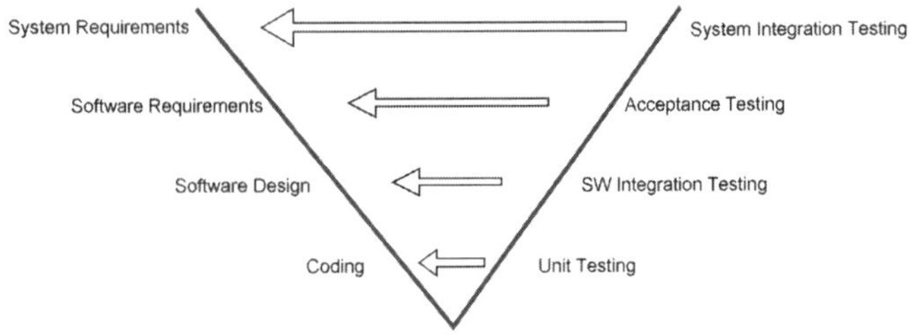

Figure 1: The V-Model for software development

The left side of the V-cycle describes the implementation path, starting out from high-level requirements and becoming more detailed at every step up to the creation of actual production code, while the right side represents the verification path in which each verification phase is shown opposite its corresponding implementation phase.

Using model-based design methods and tools for implementation has tangible benefits. System requirements analysis is the first step on the implementation path. All functional and non-functional requirements for the system are expressed. Using model-based design tools already in this phase results in an executable specification. Such a model-based executable specification has many advantages over a purely textual specification. First of all, a model-based specification meets the requirement of IEC 61508 to use a semi-formal method for specification. This form of specification is required because it leaves less room for interpretation than a textual one, thus reducing the possibility of misinterpretation and error. It also allows a seamless continuation of the implementation path; the software requirements, software design and coding phase using automatic code generation are naturally connected. With model-based design the system behavior can be assessed and tested in every phase of the implementation, which allows undesired behavior and errors to be found and fixed early on. Thus the result of every development phase can be secured by means of verification and validation.

The first phase of the verification path is the unit or module test phase to verify the production code and the smallest functional blocks in the executable software. The activities in this phase include static analysis and dynamic testing of functionality. In conjunction with model-based design and automatic code generation even the unit test phase can become considerably more efficient. This is possible because the test frame software can also be automatically generated. Software integration testing, acceptance testing and system integration testing complete the verification path. However, testing itself is not the main object of this paper.

The activities to be performed in each phase differ only slightly in the individual standards. However, all of these standards have one thing in common: The objective of safety can be achieved only by systematically performing all the activities. Taken on its own, no individual step within this process allows the quality of the developed software to be assessed.

BENEFITS OF AUTOMATIC CODE GENERATION

In any study of automatic code generation the focus is on the coding and the unit testing phase. Coding marks a cornerstone in the software development process; automatic code generation is the key technology in this development phase. There are several reasons for using an automatic code generator in the development of safety-critical systems.

Using automatic code generation is natural once the software design is already available as an executable specification. A code generator can convert this executable specification with a lower implementation error rate than a human programmer. Each manual translation step is prone to errors and is time consuming. The potential for introducing errors is high, especially when going through iterations and making successive changes, and consistency between the software design specification and the code is often lost. This is not the case when using automatic code generation. A code generator translates the model and the changes made to the model into code reliably, reproducibly, and constantly day after day. It ensures that the generated code is always consistent with the software design. Furthermore, since the documentation can typically also be generated along with the code, it is easily kept up to date as well. Thus with automatic code generation the software design, the implementation and the documentation are automatically kept in sync.

Using a code generator is no guarantee for getting error-free software. There is no formal proof that there will never be a coding error, since there is an infinite number of combinations of modeling constructs and parameters. On the other hand, a well designed and tested code generator produces significantly fewer programming errors than human software programmers do. While software programmers might have good and bad days, the automatic code generator performs its task with the same level of quality every day. The code generator learns with every improvement made and from every bug that has been fixed an d it never forgets.

REQUIREMENTS IMPOSED ON THE CODE GENERATOR

The quality and reliability of the code generator and the generated code are of paramount importance when using automatic code generation for the development of safety-critical systems.

TOOL CERTIFICATION

IEC 61508 requires that any tools that are used must be either certified or, proven in use, or that a justification for their use must be given. This requirement must be addressed before using a code generator. Certification involves undergoing a procedure under a national or international standard, which is not specified in greater detail. One applicable standard is the above-mentioned RTCA DO-178B.

This gives users two alternatives. One is to have the code generator itself certified according to RTCA DO-178B, which might enable developers to cut down on the volume of verification activities. However, such cuts would apply only to individual activities in the unit or module testing phase and are project-specific. General omission of a verification phase is normally not possible. Certifying the code generator presupposes that a certifiable code generator exists, i.e., the development documentation for the code generator must have been created and supplied by the tool vendor according to RTCA DO-178B at the same level that is applicable for the application software. This means that using the code generator for a project according to IEC 61508 SIL (Safety Integrity Level) 3 would require a RTCA DO-178B Level A development process for the code generator. Moreover, the code generator requires project-specific certification in which the interaction between the code generator, compilation system and target processor is tested. Consequently this form of certification is configuration-specific.

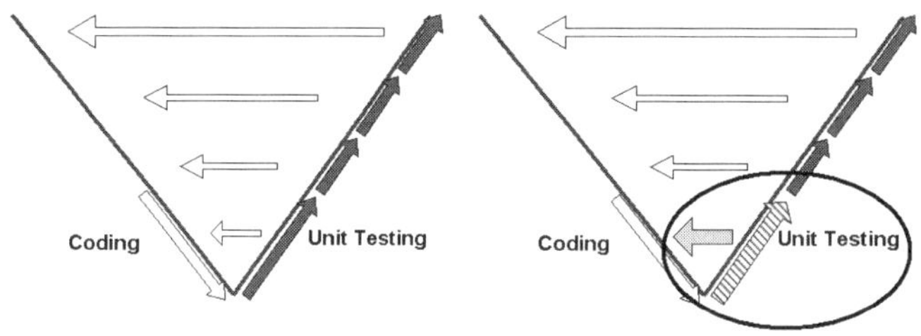

Figure 2: Work reduction potential when using a certified code generator

This all involves an enormous workload, which is reflected in the cost of a certifiable code generator and of project-specific certification. The price of a certifiable code generator is between five and ten times higher than that of an uncertified one. Project-specific certification involves costs between 50,000 and 10 0,000. Moreover, the certification procedure for a code generator is also very time-consuming, and the maintenance cycle for such a generator is currently around two years, while approximately three months are sufficient to produce a corrected version for an uncertified code generator.

The alternative method is to subject the application software to all the prescribed verification activities as if no code generator had been used in its creation. The verification process used for this can also find errors that might be introduced by an automatic code generator. This approach does not involve the disadvantages described above, and is also more cost effective for most projects.

SOFTWARE QUALITY STANDARDS

Since the code generation tool is a complex piece of software, systematic development and system testing is of paramount importance. Process-oriented development standards such as CMM, ISO 9001 and ISO/IEC 15504 provide a framework for managing the increasing complexity in software development.

A suitable standard to be applied during the development of a code generator is ISO/IEC 15504, also known as SPICE ("Software Process Improvement Capability Determination") [4]. It has been defined by a working group of ISO/IEC on the basis of existing quality standards and combines ideas from ISO 9001 and CMM. SPICE requires that formal process assessments be performed by an independent organization on a regular basis. The assessment results are then used to derive process improvement activities. When the standard is followed and a high level of compliance is achieved there is a good chance that the software produced will be of high quality.

Therefore the "Hersteller Initiative Software" (HIS), a German automotive OEM initiative consisting of Audi, BMW, DaimlerChrysler, Porsche, and Volkswagen, demands that ECU development and the development of major software tools comply with SPICE [5].

CODE REQUIREMENTS

Multiple, very special requirements are imposed on the source code for safety-critical systems. Among them are, for example:

- the restriction to a subset of the programming language used which is deemed safe
- restricting control and dataflow structures to precisely specified variants
- following accurately specified rules regarding the scope of functions and data
- following predefined complexity measures, and
- the readability, maintainability, and testability of the source code.

Such demands on the code quality are defined in the MISRA C guidelines, for instance [6]. This is a generally accessible standard for the use of C language in ECU projects up to SIL 3 produced by the "British Motor Industry Software Reliability Association". The standard s title is "Guidelines for the Use of the C Language in Vehicle-Based Software" but it is commonly referred to as "MISRA C". Depending on the certification process chosen, these requirements also have to be met by the automatically generated code.

Furthermore, there is the demand that the generated code must not use significantly more memory and execution time than hand code. Inadequate efficiency has for a long time been the reason why automatic code generation has not been applied in production.

PROCESS INTEGRATION

To fully enjoy the benefits of automatic code generation, the code generator has to be optimally integrated into the development process and tool chain in use. Seamless integration and interaction with the model-based design tool is self-explanatory. Furthermore, open interfaces and the possibility for tool automation are useful to connect with other tools, automate work steps, and to prepare and support later development phases, e.g., the succeeding unit and module test.

TARGETLINK AS A CODE GENERATION TOOL FOR SOFTWARE DEVELOPMENT IN SAFETY-CRITICAL APPLICATIONS

The design and simulation tools most widely used in the automotive industry are MATLAB®, Simulink® and Stateflow®. Data flow models are described in Simulink, while Stateflow is used for the control flow parts. Simulink and Stateflow are integrated, and allow a mix of both types in one model. TargetLink, the production code generator from dSPACE, is seamlessly integrated into MATLAB and allows reliable conversion into C code of software designs that are available as Simulink/Stateflow models.

SOFTWARE VERIFICATION

One major advantage of using model-based design methods and tools is the capability of early verification by means of simulation. TargetLink supports different simulation modes which allow the correctness of the implementation, i.e., the code generated by TargetLink, to be tested directly. This is done by comparing simulation results with results from reference simulations, frequently referred to as model-in-the-loop simulations (MIL). This verification can be performed stepwise:

- First, offline simulation is performed. The generated code is compiled with a host compiler and executed on the host PC; this is also known as software-in-the-loop simulation (SIL).
- Then simulation is performed on an evaluation board equipped with the target processor so that the generated code is compiled with the actual target compiler; this simulation mode is called processor-in-the-loop simulation (PIL).

In all supported simulation modes, TargetLink allows signal traces of block outputs to be logged. These signal traces can be saved and plotted on top of each other, thus providing direct visual feedback and allowing further analysis.

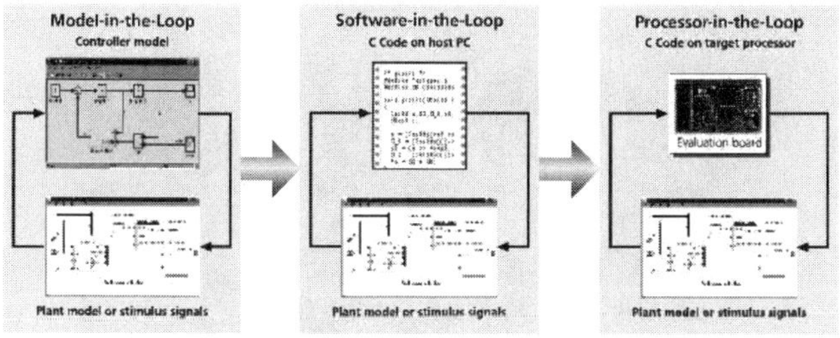

Figure 3: Model-, Software- and Processor-in-the-Loop simulation support the unit testing phase

The user has full control over file and function partitioning. This allows code for specific model parts to be generated in separate C functions and files. These model parts can also be implemented and tested incrementally. This has the advantage that when changes are made, it is not necessary to regenerate and test the code for model parts and software modules that have not changed.

SIL and PIL simulation are complemented by code coverage analysis. The coverage types currently supported are statement and decision coverage. All this means that extensive support is given to the unit or module testing phase, including model parts that are generated incrementally.

CODE REQUIREMENTS

Code Performance

TargetLink has been specifically designed for production coding and can well match human programmers' efficiency in terms of memory consumption and execution speed. Numerous benchmarks and user experience reports show that in most cases TargetLink is within a narrow range of what the human programmers produce.

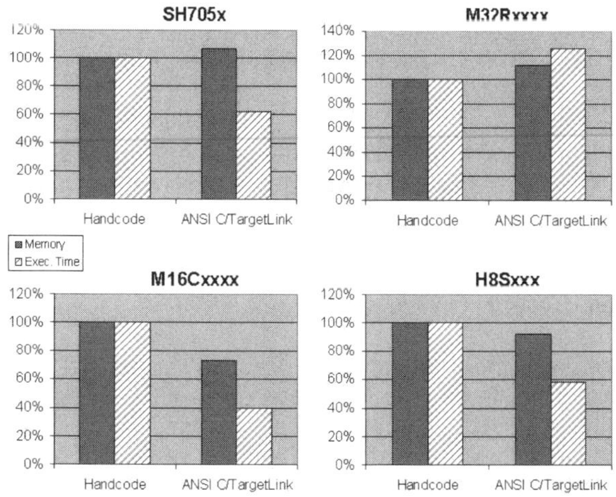

Figure 4: Results of a customer benchmark comparing TargetLink code with handwritten code on different processors

Readability

TargetLink code can be subjected to code inspections and reviews. The generated code is easily readable and has many comments, enabling the user to go back from the code to the model. Unnecessary macros, function calls, cryptic naming, etc. are avoided. Comprehensive configuration options give the user full control over variable, function and file naming, as well as the flexibility of partitioning the code into functions and files to keep the structure logical and manageable.

MISRA

TargetLink-generated code complies with most of the 127 MISRA rules. MISRA expressly permits deviations from the standard if they are technically justified and documented. Such a compliance document that describes deviations from the MISRA standard is published by dSPACE [8].

PROCESS INTEGRATION

TargetLink comes with a comprehensive and fully documented API. It grants access to all TargetLink properties and settings and allows all processes to be automated, while at the same time providing options for intervention in the individual process phases. For example, "hook functions" allow user tasks to be performed at all stages of the build process, directly before and after code generation, before and after compilation for the host and for the target, before and after download to the evaluation board for PIL simulation, and before and after ASAP2 or calibration file generation.

SOFTWARE QUALITY ASSURANCE

Special attention is paid to quality assurance during TargetLink product development.

Release Testing

For maximum reliability and quality, comprehensive tests are performed before each release of a TargetLink version.

- Automatic test engine run: Several hundred thousand code patterns are tested and several million test points are run through. These tests are repeated for every processor that is supported separately.
- Automatic test suite run: This test is run on several thousand models. Different input values are used and parameters are varied to produce more than a hundred thousand test cases, whose results are compared with expected values.
- Semiautomatic system test: This is to check the installation and correct functioning in different PC configurations and in interaction with different MATLAB and compiler versions.

- Manual testing with customer models: Tests are performed using large, real ECU models and evaluated manually.

In addition, a beta test phase is carried out with selected customers before official product release. Thus, when TargetLink is delivered to customers, it is already to a certain extent "proven in use".

SPICE

Particular emphasis is placed on a mature development process for developing TargetLink. To secure, improve and certify systematic development TargetLink is developed according to SPICE, as required by HIS, the German manufacturers initiative for software. Compliance is monitored by an independent auditor. The scope of auditing is also specified by HIS.

Additionally, both the development process and the release tests are supervised by an independent in-house software quality management team.

APPLICATION EXAMPLE

Since its introduction to the market in 1999 TargetLink has been used for many control applications worldwide. TargetLink code is in production in various applications, some of which are safety-related. For example, TargetLink has been used to develop a cabin pressure control system. TargetLink generated code was certified according to RTCA DO-178B Level A, the highest safety level, meaning that a failure would have catastrophic consequences for the aircraft [7].

In the following, the use of TargetLink at ATENA Engineering for the development of safety-critical software according to IEC 61508 SIL 3 is described.

EXPERIENCES IN SAFETY-CRITICAL SYSTEMS DEVELOPMENT AT ATENA

Through close cooperation with its parent company, MTU Aero Engines GmbH, ATENA Engineering has decades of experience in developing safety-critical systems in the aviation sector to fall back on. For example, MTU Aero Engines GmbH has developed and produced the engine controllers for a number of European aviation projects, and is still actively involved in such work. The aircraft engine controllers are multichannel electronic control units (ECUs) with between 4 and 10 processors. The response times needed for control are in a range of 2 ms.

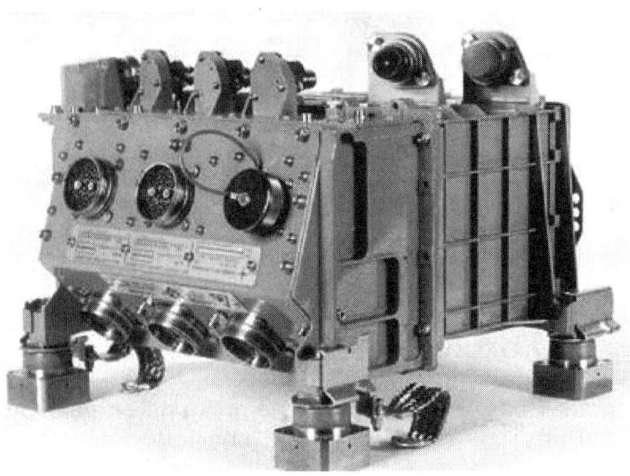

Figure 5: Multi-channel electronic control unit developed by MTU Aero Engines

RTCA DO-178A [9], along with its predecessors and project-specific derivations, has been the standard for developing such systems for many years. ATENA, as a subsidiary of MTU Aero Engines, offers engineering services both to the aviation industry and to automobile manufacturers and their suppliers. This constellation means that ATENA is in a position to apply software development standards comprehensively to safety-critical systems in automobiles.

EXPERIENCES WITH AUTOMATIC CODE GENERATION

MTU Aero Engines as well as ATENA have made comprehensive experience with automatic code generation. Both have performed detailed evaluations of different code generators in the past. Criteria for these evaluations were mostly independent of the question of whether the code generator was to be used in avionics or automotive projects. In both cases the focus was on the applicability for the development of safety-critical systems according to RTCA DO-178B Level A and IEC 61508 SIL3, respectively. Central points of the evaluations were:

- Integration into the overall development process
- Quality and efficiency of the generated code
- Flexible and comprehensive configuration options of the code generator

APPLICATION OF TARGETLINK AT ATENA

Because of the disadvantages described earlier, ATENA decided not to use certified code generators for safety-critical systems. Instead, complete verification of the application software is performed, and the verification process itself is automated as much as possible. Nevertheless, great importance is placed on the quality and reliability of the code generator used. Quality characteristics that do not involve the disadvantages described above, such as a development process that complies with ISO/IEC15504 (SPICE), are regarded as important.

At ATENA, model-based design using the MathWorks tools MATLAB, Simulink, Stateflow was already established. Therefore a search was made for a code generator that integrates well with this modeling tool suite.

A thorough evaluation of the available code generation tools led to the decision to use TargetLink for the software implementation of a safety-critical system according to IEC 61508 SIL3. The system involved is an automotive application related to alternative fuel concepts. The volume of the related Simulink model is approximately 120,000 subsystems. The software components are up to 25,000 code lines in size. The main reasons for selecting TargetLink were its technical features as well as its high product quality.

In order to integrate TargetLink into the software development process, ATENA made several adaptations to the code generation process. These were aimed at further automating the generation process and achieving the necessary software quality for safety-critical applications. The adaptations included the greatly reduced use of pointers and interrupts, compliance with various complexity criteria, and the replacement of function-like macros and functions from the standard libraries with user-defined functions. Work was also performed on enabling such complex systems to be broken down into subsystems and translated separately, and on allowing the generation of standardized description files for calibration tools for the subsystems. The test frame software for testing the individual modules was also automatically generated for preparing the unit testing phase. The adaptations were supported by the TargetLink API. It allows all the processes to be automated; while at the same time provides options for intervention in the individual process phases and access to all TargetLink properties and settings.

The software development process, whose implementation phase is decisively supported by dSPACE s TargetLink, has now been in use at ATENA since November 2002. Automatic code generation plays a major role, as the company has succeeded in generating up to 80% of the entire production code from Simulink models. The code generator is embedded in a project-specific tool chain. This guarantees compliance with the quality criteria for safety-critical applications.

Recently, ATENA delivered the final product for the first development phase for integration tests at the OEM s facilities. From this first development phase the following experience was gained:

- The model-based requirement analysis and design phase led to an executable specification which allowed early system verification. This resulted in a mature specification. Later software tests revealed only a few errors that originated from the specification.
- The automatic code generation allowed the implementation of the software design with a low error rate and with high efficiency, especially for minor modifications.
- The source code generated by ATENA s software development process with dSPACE s TargetLink as a key item is of acceptable quality for safety-critical applications.

Based on the evaluation results also made at MTU Aero Engines and the project experiences made at ATENA, it is planned to use TargetLink also for future avionics projects at MTU Aero Engines.

CONCLUSION

Automatic code generation complementing model-based design provides many benefits for the user and can be applied for the development of safety-critical systems. It is suitable for software that has to comply with safety standards such as IEC 61508 SIL3 when embedded into an adequate development process. There are several requirements imposed on the code generator relating to quality and technical features that have to be met.

TargetLink, the production code generator by dSPACE, meets these requirements and is therefore also a suitable code generator for the development of safety-critical systems.

ATENA Engineering opted for TargetLink due to its technical features and quality level. At the same time ATENA decided to use this code generator with fewer qualification options, but to perform complete qualification on application software. This alternative is considered to be more cost effective than using a code generator which is fully qualified according to RTCA DO-178B Level A.

ATENA and dSPACE are cooperating closely to extend integration into the tool chain in future versions of TargetLink and to reinforce the support given to safety-critical aspects of code generation.

REFERENCES

1. IEC 61508 Functional Safety of Electrical/Electronic/Programmable Electronic Safety Related Systems, IEC 1998
2. RTCA/DO-178B Software Considerations in Airborne Systems and Equipment Certification, RTCA Inc., 1st Dec 1992
3. Bauer,C.; Plawecki,D.: IEC 61508, Part 3 vs. RTCA/DO-178B A Comparative Study. Konferenz Anwendung des internationalen Standards IEC 61508 in der Praxis, Januar 2003
4. ISO/IEC TR 15504:1998, Information Technology - Software Process Assessment
5. Wagner, Merkle, Bortolazzi, Marx, Lange: Hersteller Initiative Software, Automotive Electronics 1/2003
6. MISRA Guidelines for the Use of the C Language in Vehicle Based Systems, April 1998
7. Aloui, Andreas: C Code Reaches New Heights at Nord-Micro, dSPACE Newsletter, 1/2002
8. Thomsen, T.: Integration of International Standards for Production Code Generation, SAE Technical Paper 03AE-32, 2003
9. RTCA/DO-178A Software Considerations in Airborne Systems and Equipment Certification, RTCA Inc., 22nd Mar 1985

CONTACT

Michael Beine
dSPACE GmbH Pro duct Manager TargetLink
Technologiepark 25
33100 Paderborn
Germany
mailto:mbeine@dspace.de

Dr. Rainer Otterbach
dSPACE GmbH Leader Product Management
Technologiepark 25
33100 Paderborn
Germany
mailto:rotterbach@dspace.de

Michael Jungmann
MTU Aero Engines GmbH
Engine Systems Design/Software Development
Dachauer Straße 665
80995 Munich
Germany
mailto:Michael.Jungmann@muc.mtu.de

SOFTWARE FOR MODELING

2005-01-1884

A Dynamic Model of Automotive Air Conditioning Systems

Zheng David Lou
Visteon Corporation

ABSTRACT

A dynamic computer model of automotive air conditioning systems was developed. The model uses simulation software for the coding of 1-D heat transfer, thermodynamics, fluid flow, and control valves. The same software is used to model 3-D solid dynamics associated with mechanical mechanisms of the compressor. The dynamics of the entire AC system is thus simulated within the same software environment. The results will show the models potential applications in component and system design, calibration and control.

INTRODUCTION

Automotive A/C systems have been used for generations, and their complete analysis has eluded us for as long because of their complexity, such as 2-phase nature of the refrigerant and its coupling with a complicated mechanical system. With the global drive for better fuel economy, many new A/C systems are equipped with variable displacement compressors, the control of which demands clearer understanding of the system dynamics to avoid instabilities [1].

This study is an effort to develop an A/C system model that has enough detail in the compressor mechanism and associated control devices that it will serve as an effective tool in the system and control strategy development. The compressor dynamics are also intricately tied with the rest of the system. A flow rate variation out of the compressor for example will immediately introduce charge migration and pressure change throughout the system. A fair amount of attention is therefore needed to model the condenser and evaporator to account for phase changes between liquid and vapor phases.

A/C SYSTEM MODEL

A/C SYSTEM

The A/C system studied in this paper, as illustrated in Figure 1, includes a variable displacement compressor, a discharge hose, a condenser, a receiver, a thermal expansion valve (TXV), an evaporator, and a suction hose.

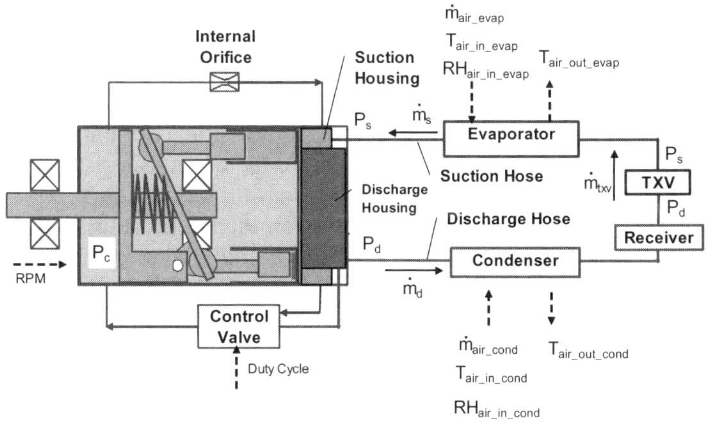

Figure 1 -- A/C System

The variable displacement compressor further includes an internal orifice and a control valve, which function as half a hydraulic bridge to regulate the compressor crankcase pressure *Pc* to a desired value between those of the discharge pressure *Pd* and suction pressure *Ps*. The compressor swashplate angle and thus its displacement are largely a function of the crankcase pressure *Pc* along with the discharge pressure *Pd*, the suction pressure *Ps*, and the rotational speed *RPM*. The internal orifice is a fixed flow resistance, and the control valve offers variable metering effect.

The compressor delivers the refrigerant, at a mass flow rate of $\dot{m}_d$, to the rest of the A/C system through the discharge hose and draws it back, at a mass flow rate of $\dot{m}_s$ through the suction hose. The values of $\dot{m}_d$ and $\dot{m}_s$ are generally not equal during dynamic periods because of the charge migration throughout the system, which this model is able to monitor. Under steady state, the superheated refrigerant enters the condenser to release heat to the vehicle front air and condenses to sub-cooled liquid, expands through the TXV into a two-phase state at a lower pressure *Ps*, and passes the evaporator to extract the heat from the cabin air and vaporize into a superheated vapor.

COMPRESSOR MODEL

The performance of a variable displacement compressor is dominated by its driving and displacement mechanism, pumping mechanism, and control valve.

Driving and Displacement Mechanism

The compressor driving and displacement mechanism, illustrated in Figure 2, includes a shaft, a spring, a sleeve, a swashplate, seven pairs of pistons and shoes. The shaft and swashplate are connected through a guiding pin (fixed on the swashplate) and slot (cut into the shaft), which allow the former to slide almost linearly inside the latter in a direction that makes an angle β with the Z-axis. At the same time, the swashplate revolves with the shaft along the Z-axis and pivots in the X-Z plane at point A, where it is pinned on the sleeve, resulting in a swashplate angle α, the magnitude of which is roughly proportional to the compressor displacement. The sleeve slides over the shaft in the Z direction. There are seven (7) pairs of axially arranged pistons and cylinders, with the former being connected with the swashplate through the seven pairs of shoes, which slide along the swashplate surface. The spring is to resist the leftward motion of the sleeve, thus helping reduce the swashplate angle.

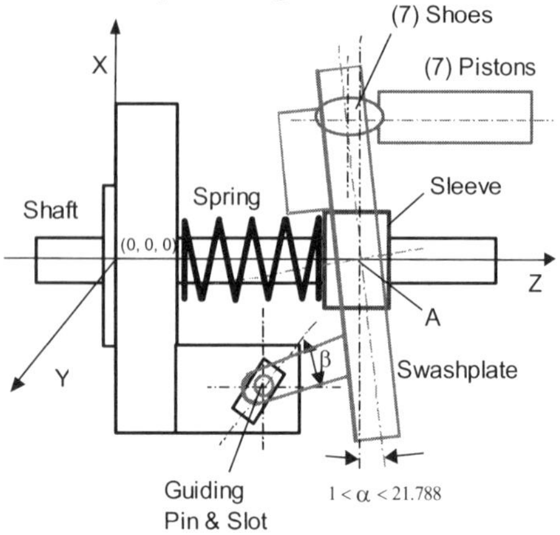

Figure 2. Compressor Driving and Displacement Mechanism

The dynamics of the driving and displacement mechanism are simulated using SimMechanics, a part of MatLab/Simulink software family, which has various elements for rigid bodies, constraints & drivers, joints, and sensors & actuators as illustrated in Figure 3. For each rigid body, for example, the program inputs include mass, inertia tensor, coordinate origins and axes for center of gravity, other user-specified body coordinate systems, and body initial position and orientation.

Pumping Mechanism

The model of the pumping mechanism as illustrated in Figure 4 includes the calculation of specific volume, mass, internal energy, enthalpy, pressure and temperature inside each of the seven cylinders through the application of the mass conservation, the first law, and thermal dynamic relations and properties. The entire pumping mechanism is modeled with the Simulink module. The piston displacement is directly from the driving and displacement mechanism model, which in return obtain the pressure force on the piston top surface from the pumping mechanism model.

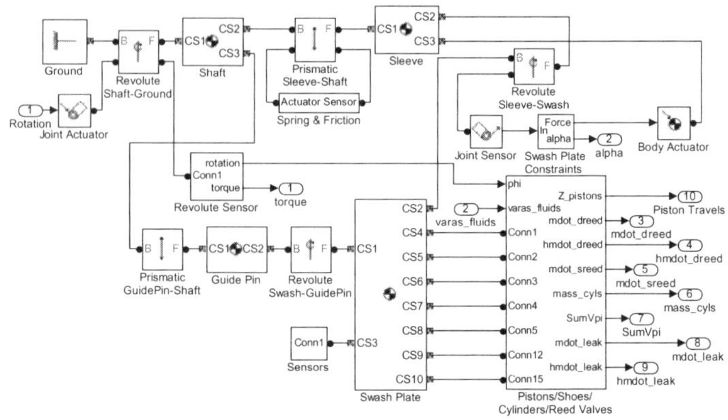

Figure 3. SimMechanics Model of the Compressor Driving and Displacement Mechanism

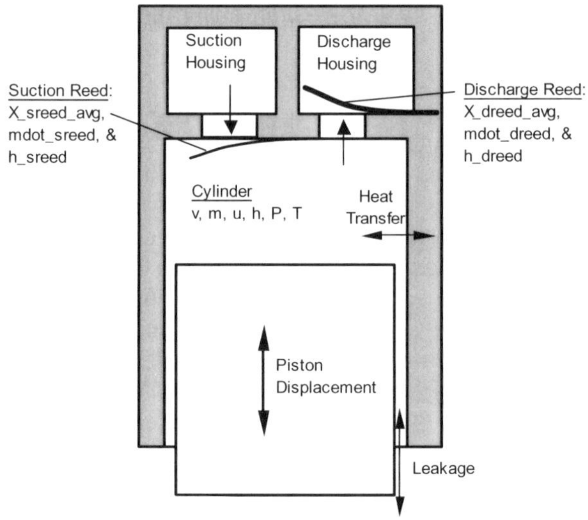

Figure 4. Compressor Pumping Mechanism Schematic

Both discharge and suction reed valves are dynamically simulated, which includes inertia, stiffness, stiction, and differential pressures. The resulting time-varying average discharge & suction valve displacements (*X_dreed_avg & Xsreed_avg*) are used to calculate the valve openings and mass flows (*mdot_dreed and mdot_sreed*).

The model also accounts for the leakage through the clearance between the piston and cylinder wall and heat transfer between the refrigerant and the cylinder wall. The primary impact of the leakage is a reduction in the compressor volumetric efficiency. If the leakage rate gets too high, the leak flow interferes with that from the control valve and causes severe control problem.

Control Valve

The control valve in Figure 1 is a proportional solenoid valve, also called ECV (Externally Controlled Valve), and is regulated by the duty cycle of a PWM (Pulse Width Modulation) signal. The electrical current at the solenoid is close to a DC signal because of system inductance, and its amplitude is substantially proportional to the duty cycle. The ECV has a much higher natural frequency than the rest of the system and is therefore modeled using a steady state solution. If the flow forces are neglected, one has:

*Ps = (Fo - Ks*X – Pc*Ac - Fem)/As*

Fem = f(I, X)

Where *Ps* is the suction pressure, *Fo* the spring preload, *Ks* the spring stiffness, *X* the valve displacement, *Pc* the crankcase pressure, *Ac* the *Pc* acting surface, *As* the *Ps* acting surface, and Fem the electromagnetic force, which is a function of the valve displacement *X* and the current *I*. Because *As* >> *Ac* and *Pc* is close to *Ps*, the ECV is able to observe the control relationship as illustrated in Figure 5,

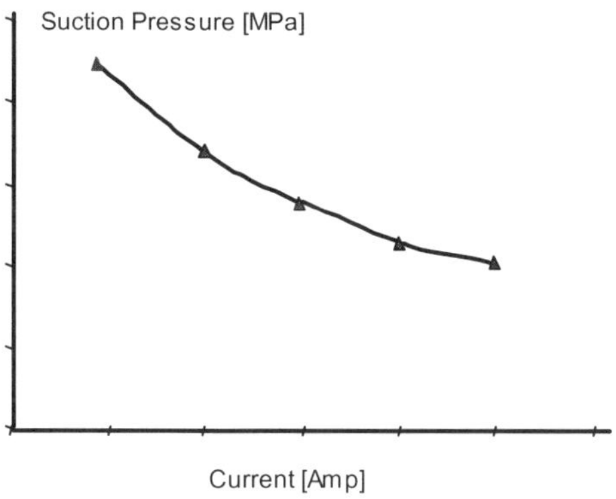

Figure 5. ECV Operation Characteristics

HEAT EXCHANGERS

In this study, two-phase flows are simulated using the equilibrium slip flow approach, which allows different phases at a tube location to be at the same temperature but different axial velocities. Flow tubes are divided into control volumes. Within each control volume *CV* as illustrated in Figure 6,

- the mass is calculated from the mass conservation, the mass change in the control volume *CV* being equal to the inlet mass flow rate minus the exit mass flow rate:
$$\frac{\partial m_{CV}}{\partial t} = \dot{m}_i - \dot{m}_o$$
- the internal energy is calculated from the first law, the internal energy change in the control volume *CV* being equal to the inlet enthalpy flow, minus the exit enthalpy flow, and plus the heat transfer:
$$\frac{\partial (mu)_{CV}}{\partial t} = \dot{m}_i h_i - \dot{m}_o h_o + \dot{Q}$$
- the specific volume and density are calculated from the mass and volume,
- the enthalpy *h*, pressure *P*, temperature *T*, and quality *X* are derived from the thermodynamic relations and properties,
- the inlet flow and associated properties are from the upstream control volume, and
- the exit flow and associated properties including velocity slip ratio are based on the properties of the current control volume.

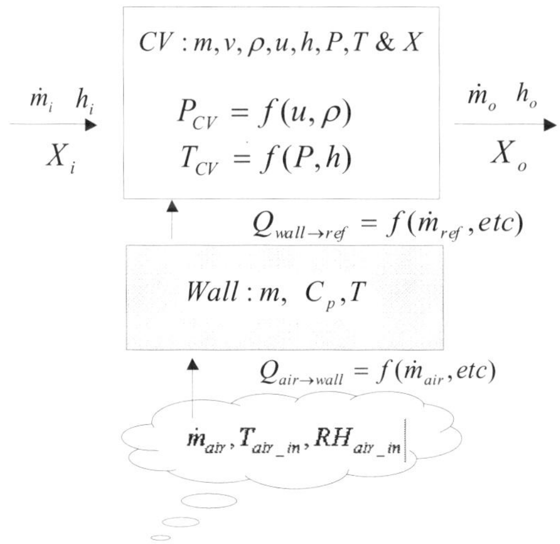

Figure 6. Physics at a Control Volume

The convection heat transfer from the tube wall to the refrigerant is treated differently depending on whether the refrigerant is in superheated, subcooled, or 2-phase state. The wall temperature is assumed uniform within the same control volume, i.e., no conduction heat transfer is considered.

In a single-phase region (either superheated or subcooled), the model uses Dittus-Boelter correlation [6]:

$$Nu = 0.023 \operatorname{Re}^{0.8} \Pr^{n}$$

where Nu is the Nusselt number, Re the Reynolds number, Pr the Prandtl number, and n an exponent with a value equal to 0.3 and 0.4 for cooling and heating, respectively.

In a two-phase condensation region, it adopts a correlation by Shah [4]

$$Nu = Nu_{LO}[(1-X)^{0.8} + \frac{3.8X^{0.76}(1-X)^{0.04}}{(P/P_{cr})^{0.38}}]$$

where Nu_{LO} is the liquid only Nusselt number per above Dittus-Boelter correlation, X the vapor quality, P the pressure, and Pcr the Critical pressure.

In a two-phase evaporation region, the model adopts a correlation by Klimenko [5]

$$h = 0.087 \, \mathrm{Re}_m^{\ 0.6} \, \mathrm{Pr}_l^{\ 1/6} \left(\frac{\rho_v}{\rho_l}\right)^{0.2} \left(\frac{k_v}{k_l}\right)^{0.09} \left(\frac{k_l}{D_L}\right)$$

$$V_m = \frac{G}{\rho_l}[1 + X(\frac{\rho_l}{\rho_v} - 1)]$$

$$D_L = \sqrt{\frac{\sigma}{g(\rho_l - \rho_v)}}$$

where V_m and Re_m are the mixture velocity and Reynolds numbers, respectively, D_L Laplace constant, σ the surface tension, ρ_l liquid density, ρ_v vapor density, g gravitational acceleration, and G mass flux.

The convection heat transfer from the air to the wall, including the fins, is based either on experimental data or an in-house CAE code. Relative humidity and associated condensation are considered.

In an A/C loop, unsteady nature of the refrigerant flow is more dominated by the charge migration caused by heat transfer and phase changes than by transient pressure forces. No fluid inertia is therefore considered in the model. Still the steady state pressure losses are estimated at each time point. In the two-phase region, the vapor fraction is as important as the overall flow rate or velocity and flow, and the following equation by Yan & Lin [2] is used to calculate the pressure drop

$$\Delta P = -\frac{0.11 \mathrm{Re}_{eq}^{-0.1}(2G^2 v_m L)}{D}$$

$$\mathrm{Re}_{eq} = \frac{GD}{\mu_l}[(1 - X) + X(\frac{\rho_l}{\rho_g})^{0.5}]$$

Where Re_{eq} is an equivalent Reynolds number [3], D hydraulic diameter, L length of the tube, v_m average specific volume of two-phase mixture, μ_l liquid viscosity.

A sample condenser as shown in Figure 7 has 16, 14, 7, 4 tubes in its four passes, respectively. It is divided into 10 control volumes along the flow direction. Each control volume physically covers multiple parallel tubes, 16 of them in the cases of CV1 and CV2. In the model however, it is a single tube with the same hydraulic diameter and flow rate but its length equal to the sum of all the tubes in the pass.

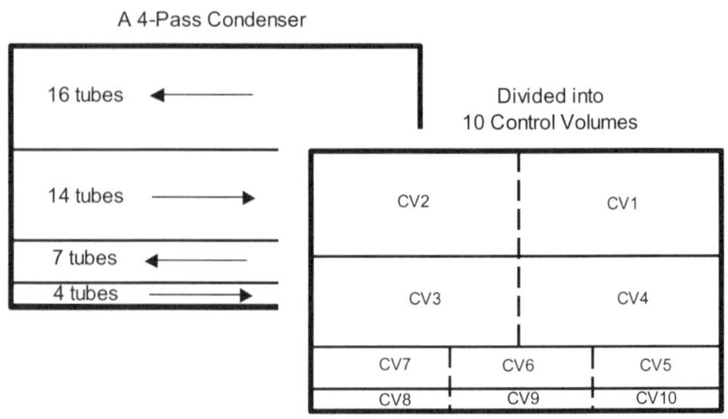

Figure 7. A 4-Pass Condenser and its Control Volumes

The sizes or number of control volumes should be determined to balance the needs for the simulation accuracy and efficiency. If one is interested in the system dynamics, ten elements may be sufficient. If one is interested in the component design of the condenser itself, more elements definitely will help. The primary impact of the control volume size is to the refrigerant-side heat transfer, i.e. that between the refrigerant and wall. In this model, selection of the heat transfer correlation is based on the average quality within a control volume. If the refrigerant enters a control volume, e.g. CV7 in Figure 7, as a two-phase fluid with a low quality value, e.g. 0.05, and exits as a subcooled liquid. The average state may be in two-phase or subcool depending on whether the exit state is slightly or substantially subcooled, and thus either a two-phase or subcool heat transfer correlation is applied. Heat transfer is generally much more intense in a two-phase region than in a single-phase region. This difference is not substantial or abrupt if the two-phase quality is either close to 0 or 1. This model uses the average state in a control volume and does not track the exact location of the transition point between the two-phase and single-phase regions. With this approach, the resulting inaccuracy in average refrigerant-side heat transfer coefficient may be of the first order for a large control volume like the control volume CV7. But because of the air-side limiting nature of heat transfer in a condenser or evaporator, the error in the overall heat transfer coefficient and thus the heat transfer rate may be of the second order.

THERMAL EXPANSION VALVE

An R134a gas charge TXV (Thermal Expansion Valve) is modeled in this study. The TXV continuously balances the refrigerant flow rate into the evaporator and the system load by monitoring the superheat level exiting the evaporator. The valve stem is balanced primarily by three pressures in the following manner

$$P_b = P_e + P_s$$

where *Pe* is the evaporator exit pressure, *Ps* the equivalent spring pressure, and *Pb* the bulb or charge pressure, which is equal to the saturation pressure corresponding to the temperature *Tb* of the refrigerant in the bulb or power assembly. The power assembly is exposed to the evaporator exit flow, and the bulb temperature *Tb* is therefore substantially equal to the evaporator exit temperature *Te* at steady state. These two temperatures are not equal dynamically because of the thermal resistance of the power assembly. In this study, the following equations are used,

$$\frac{dT_b}{dt} = (T_e - T_b)/\tau \quad \text{and}$$

$$\tau = \frac{mC_p}{\alpha A}$$

where τ is a time constant that is a function of the thermal mass m, specific heat C_p, surface heat transfer coefficient α, and surface area A of the power assembly. In the study, the time constant τ is estimated based on the supplier data.

RECEIVER

The receiver is modeled with the following principles and assumptions:

- the mass conservation,
- the energy conservation without heat transfer and mechanical losses,
- uniform temperature within the control volume,
- only the liquid phase when available exiting the control volume with the liquid phase enthalpy, and
- various thermodynamic relationships and properties.

SUCTION AND DISCHARGE HOSES

The suction and discharge hoses are modeled with the following principles and assumptions:

- the mass conservation,
- the energy conservation without heat transfer and pressure losses,
- uniform temperature within the control volume, and
- various thermodynamic relationships and properties.

SIMULATION RESULTS

Condenser Performance

The condenser model is calibrated with the calorimeter data of a four-pass micro-port condenser with a face area of 624 x 350 mm^2. The condenser is divided into 10 control volumes as illustrated in Figure 7 although its tube distribution among four passes is somewhat different from that illustrated in the figure. In both experiment and simulation, the refrigerant inlet conditions are fixed at 1690 kPa and 93.3 deg. C, and the outlet condition is fixed at a subcooling of 5.6 deg. C. The inlet air conditions are 44.1 deg. C and 40% relative humidity. As shown in Table 1, the capacity prediction error is within 3% while the refrigerant-side pressure drop has larger errors. Capacity values are normalized against the experimental value at 4.0 m/s air speed.

Table 1. Condenser Calibration: Experimental vs. Simulation

Air Speed	Capacity [Normalized]			Refrigerant Pressure Drop [kPa]		
[m/s]	Experiment	Simulation	Error [%]	Experiment	Simulation	Error [%]
2.0	0.64	0.63	-2.6	68	66	-1.8
3.0	0.84	0.86	2.7	113	120	6.1
4.0	1.00	0.98	-2.0	152	146	-4.5

Reed Valve Dynamics

One of the most important aspects of compressor design is the reed valve dynamics. As shown in Figure 8, this model is able to provide details on many compressor variables such as discharge reed valve displacement and mass flow rate, suction reed valve displacement and mass flow rate, and cylinder volume and pressure. One can for example use this tool to select a proper stiffness for a reed valve based on its dynamic profile.

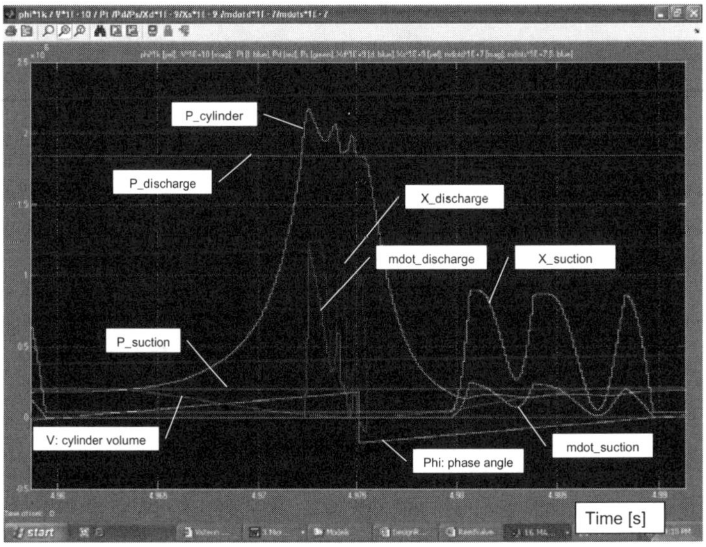

Figure 8. Reed Valve Dynamics

Compressor and System Responses at Duty Cycle Changes

One important A/C system performance measure is how fast the compressor torque decreases after the compressor is switched off. The simulation confirms that the torque is reduced to less than 20% of the peak value within one second.

In addition, the model is generally able to predict the dynamic pattern of the collapsing of the system pressures as shown in Figure 9. In this case, the experimental data are from a 190-cc compressor while a numerical model is based on a 160-cc compressor. But during the simulation, the heat transfer coefficients and the mass airflows are adjusted so that the numerical pre-switch system pressures are in agreement with those of the experimental data. Within 10 seconds after the switch, both numerical and experimental pressures collapse substantially close to their steady state values.

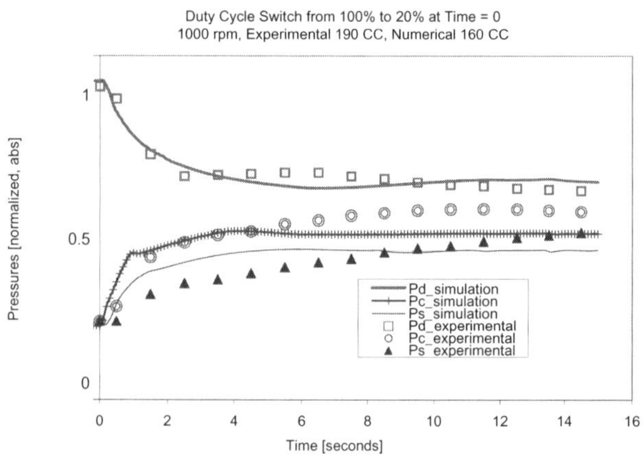

Figure 9. System Responses to Duty Cycle Switch from On to Off

Compressor and System Responses at Engine Key-On

With a clutchless compressor, OEMs are concerned with the key-on torque from a variable displacement compressor. With the cold start engine RPM shown in Figure 10, this model is able to predict a swashplate swing and corresponding torque overshoot as shown in Figure 11. The initial and surrounding temperatures are set at 24 deg C. The pulley ratio is 1.4. The engine is started with no intention for A/C mode thus with 0% duty cycle for the control valve and free convections for the condenser and evaporator. The torque overshoot is at 1.5 seconds after the start, consistent with some physical experience in a test vehicle. This overshoot is caused by a dynamic swing of the swashplate from 1 degree up to 6 degrees. Also, the discharge and suction pressures start spreading out from its initial value of 5.5 barG.

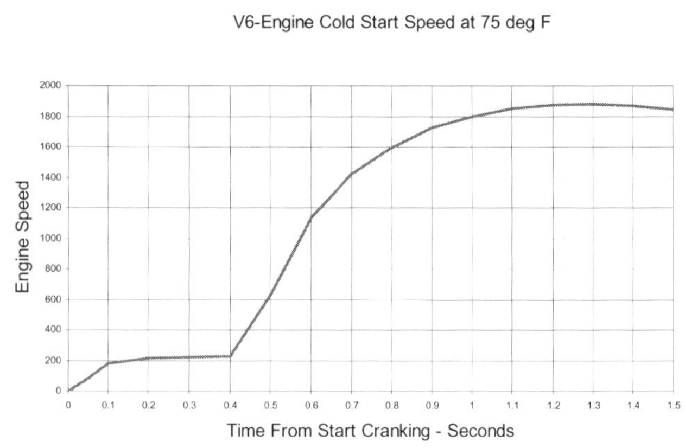

Figure 10. Cold Start Engine RPM

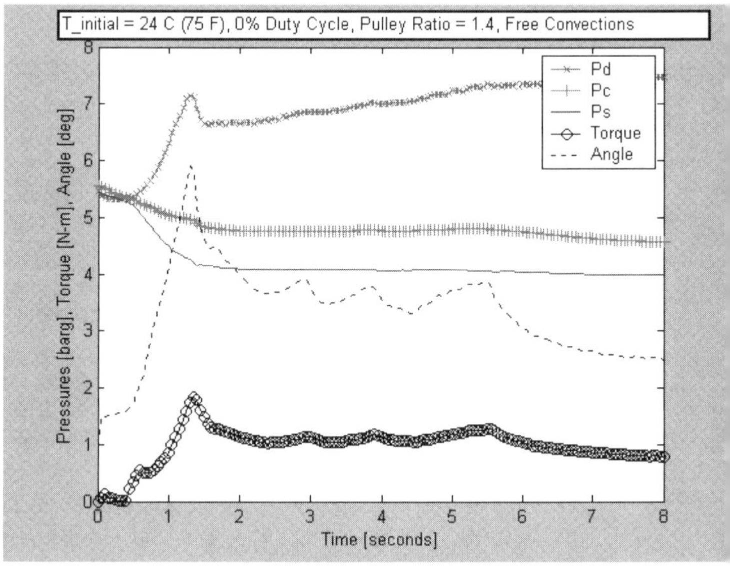

Figure 11. Clutchless Compressor Responses during Cold Start

Application in Model-in-the-Loop Simulations

The A/C system model from this study has been successfully incorporated in a model-in-the-loop simulation, where it is combined with HVAC, cabin and EATC (electronic automatic temperature control) models to help design and pre-calibrate EATC systems and to investigate static and dynamic properties as well as simulate the influence of the system on the whole vehicle already in the early states of the development process [7].

CONCLUSION

This study has shown that it is possible to develop an integrated computer model for the entire automotive air conditioning system, including 1-D models of 2-phase heat transfer, thermodynamics, fluid flow, and control valves and a 3-D model of solid dynamics associated with mechanical mechanisms of the compressor. The dynamics of the entire AC system are thus simulated within the same software environment. The model can be used in component design and system design, calibration and control.

ACKNOWLEDGMENTS

The author wishes to thank Yong Huang, Thomas Finn, Michael Theodore Jr., Kanwal Bhatia, Erik Lundberg, John Meyer, George Yang, Jerry Kuo, Arif Khan, Chao Zhang, and many others for their technical inputs. The author also wants to thank the Air Conditioning and Refrigeration Center (ACRC) at the University of Illinois at Urbana-Champaign that provided R134a thermodynamic properties in the Simulink format.

REFERENCES

1. Hamery B., Liu J.M., Riviere C. "Instabilities occurring in an automotive A/C loop equipped with an externally controlled compressor and a thermal expansion valve," SAE Technical Paper Series 200-01-1717.
2. Yan Y.Y. & Lin T.F. "Evaporation heat transfer and pressure drop of refrigerant R-134a in a small pipe," Int. J. of Heat and Mass Transfer, 41 (1998) 4183-4194.
3. Akers W.W., Deans H.A., Crosser O.K. "Condensation heat transfer within horizontal tubes," Chem. Eng. Prog. Symposium Ser. 55(1959), 171-176.
4. Shah M.M. "A new correlation for heat transfer during boiling flow through pipes." ASHRAE Transactions 88(1):185-96, 1976.
5. Klimenko VV "A generalized correlation for two-phase forced flow heat transfer," Int. J. Heat Mass Transfer 1988, 31(3):541-52
6. Dittus SJ, Boelter LMK. University of California Publications on Engineering, 2:443, 1930
7. Domschke R, & Matthes M "Model-in-the-loop simulations for climate control optimisation" VDI-Berichte Nr. 1846 'Berechnung und Simulation im Fahrzeugbau' (Numerical Analysis and Simulation in Vehicle Engineering) Conference, Wuerzburg 29.-30. September 2004, pp. 495-504.

CONTACT

Zheng David Lou, Visteon Corporation, 45000 Helm Street, Plymouth, MI 48170. zlou@visteon.com

2005-01-1350

Advances in Rapid Control Prototyping - Results of a Pilot Project for Engine Control -

Frank Schuette, Dirk Berneck and Martin Eckmann
dSPACE GmbH

Shigeaki Kakizaki
NISSAN MOTOR CO., LTD

ABSTRACT

The technological development in the field of automotive electronics is proceeding at almost break-neck speed. The functions being developed and integrated into cars are growing in complexity and volume. With the increasing number and variety of sensors and actuators, electronics have to handle a greater amount of data, and the acquisition and generation of I/O signals is also growing in complexity, for example, in engine management applications. Moreover, intelligent and complex algorithms need to be processed in a minimum of time. This all intensifies the need for Rapid Control Prototyping (RCP), a proven method of decisively speeding up the model-based software development process of automotive electronic control units (ECUs) [1],[2]. All these demanding tasks, including connecting sensors and actuators to the RCP system, need to be performed within a standard prototyping environment.

The first part of the paper presents a new modular hardware platform for signal conditioning and power stages. The second part describes the different phases of a field trial of this new hardware platform. This was a pilot project in which the NISSAN MOTOR CO., LTD, Japan [1] used the new signal conditioning and power stage hardware in a fullpass application to control its well established VQ engine [3],[4].

INTRODUCTION

In fullpass applications, the RCP system completely replaces the production ECU and has full authority to control the plant. The fullpass approach is typically used for proof-of-concept decisions, where the production ECU is not yet available. Besides the necessary real-time computation power, easy access to I/O and the corresponding software tool-chain, all sensors and actuators need to be connected to the real world.

However, the huge variety of different sensors and actuators used in the automotive field, especially during the rapid prototyping phase, means that each sensor and actuator often requires its own interface circuit. The design and implementation of such circuits is often seen as a necessary minor chore. In reality, it is often a major source of costs and development time.

In order to meet the different requirements, dSPACE has developed a new modular and flexible hardware platform called RapidPro, which is introduced in the section "Fullpass Control Applications on RCP Systems". The RapidPro hardware is an extension to dSPACE Prototyper, such as MicroAutoBox and AutoBox, and consists of three different unit types:

- The RapidPro SC Unit, a modular, intelligent signal conditioning unit
- The RapidPro Power Unit, a modular, intelligent power stage unit
- The RapidPro Control Unit, a scalable, intelligent I/O subsystem

The aim of the pilot project with NISSAN from dSPACE's point of view was to validate the new architecture and features of the RapidPro SC and Power Units as early as possible in an engine control application. NISSAN wanted to develop new control functions with the help of the fullpass approach completely based on the dSPACE hardware and software tool chain.

During kick-off of the pilot project, the two companies prepared a rough concept design of the NISSAN-specific RapidPro System, which was to be used to control NISSAN's well established VQ engine.

The resulting system overview, including a brief description of the VQ engine, is described in the section "Nissan-specific RapidPro System".

[1] Called NISSAN for short below.

The pilot project involved the following steps:

To detail the concept design, NISSAN provided the specifications for sensor inputs and actuator outputs, which were investigated and mapped against the RapidPro modules. Afterwards the RapidPro modules were configured and the wiring harness was specified. ("Phase 1: Detailed System Specification and Configuration").

After assembly of the system and the wiring harness, initial tests with real loads connected to RapidPro were performed in the laboratory. Synchronization to crankshaft and camshaft, and correct ignition and injection, were tested by hardware-in-the-loop simulation (HIL). All these tests were performed with a pure stimulus model running on the MicroAutoBox ("Phase 2: Commissioning and Real Load Tests").

After these steps the RapidPro System was shipped to NISSAN, where it was connected to an existing dSPACE hardware-in-the-loop simulator which matches the actuators and sensors of the real engine, and completely set into operation based on open-loop stimulation. NISSAN's engine controller was integrated in the stimulus model, after which the final HIL tests were completed ("Phase 3: HIL Tests on a Test Bench").

Finally, the system ran in closed loop at a real engine in the vehicle ("Phase 4: In-Vehicle-Tests").

FULLPASS CONTROL APPLICATIONS ON RCP SYSTEMS

In a model-based development process, the controller software of an ECU is designed graphically with block diagrams and state charts [2]. The resulting model is a formal specification that unambiguously and completely defines the functionality of the system throughout the whole development process. The models can be converted into executable specifications that can be tested in real-world scenarios. This requires an application with appropriate I/O interfaces to be generated from the model and executed on real-time hardware physically connected to the controlled system. The software tool chain of an RCP system can be characterized as follows:

- Modeling environment like Simulink®/Stateflow® with extensive block and state diagram support.
- A code generator which automatically translates block diagrams and state charts from the model into C code.
- Use of a non-production real-time operating system or scheduler.
- Comprehensive, generic I/O driver functionality that allows users to easily connect or generate I/O software and to set all parameters.
- Comprehensive virtual instrumentation software for capturing real-time data and changing any desired control parameter during real-time experiments.

To avoid hardware restrictions during the design phase, RCP hardware is usually more powerful than production ECUs. For example, dSPACE's AutoBox can be equipped with high-performance, double-precision floating-point processor boards (PowerPC at 1 GHz) that can run complex algorithms in microseconds. It also leaves room for instrumentation tasks, capturing model variables, or changing control parameters during run time to analyze the control system behavior. In addition, ample memory and integrated flight recorder functionality are provided for long-term data acquisition.

In cases where such systems are intended to replace an actual ECU, they should be designed as in-vehicle devices, operable without user intervention, just like an ECU. dSPACE's MicroAutoBox is an example of such a system, where high computing performance (PowerPC at 800 MHz) is combined with full in-vehicle capability. The application program is stored in nonvolatile memory, allowing the system to start up autonomously after power-up. A PC or notebook can be connected temporarily for program download and data analysis (hot plugging).

In addition, the I/O electronics must cover most needs with A/D and D/A converters, digital signals like bit I/O, pulse width modulated (PWM) signals for power electronics, communications like CAN, LIN, FlexRay, and interfaces for sensors such as speed or position, incremental encoders, and so on.

SIGNAL CONDITIONING AND POWER STAGES

Most real-world sensors and transducers generate signals that have to be conditioned before an RCP system can acquire them reliably and accurately. This front-end processing, referred to as signal conditioning, includes functions such as signal protection, amplification, offset, filtering, electrical isolation, and so on. Most RCP systems therefore require some form of external signal conditioning, as shown in Figure 1.

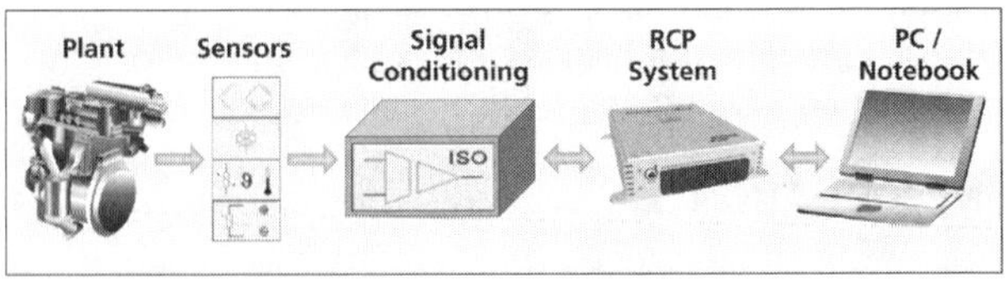

Figure 1: Front-end signal conditioning as an external component of the RCP system.

Unlike sensors, actuators such as electrical drives, solenoid valves, injectors, relays and so on require high-current and/or high-voltage output drivers to be driven. Moreover, detailed diagnostic functions of the status of the output driver circuit are essential for on-board diagnosis (OBD) and in safety-critical applications.

Most RCP systems therefore also require some form of external power stages, as shown in Figure 2.

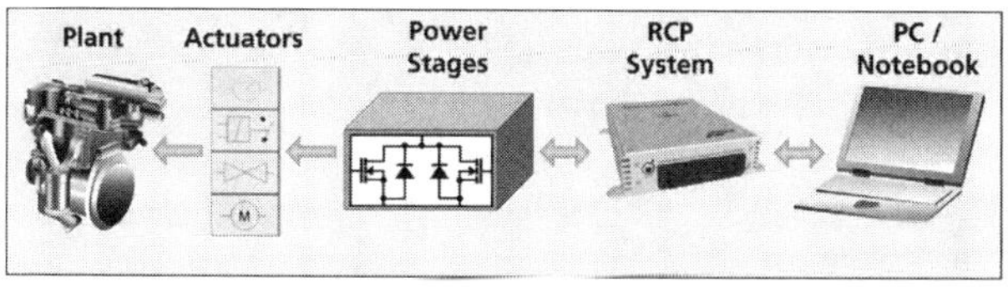

Figure 2: Front-end power stages as an external component of the RCP system.

Most choices of front-end signal conditioning and power stage units available on the market are nonintelligent configurations that offer the bare minimum of functionality. Many discussions with customers have revealed that some basic requirements regarding signal conditioning and power stage front ends are similar, but others are very application-specific. Thus, items such as modularity and expandability are obviously at the top of the list of the basic requirements. Small, compact, and robust housing for use in vehicles, on test benches, and in laboratories is also essential. Automotive protection and extensive diagnostic capabilities are especially required for power stages.

The output driver circuit itself should support detailed load failure functions such as: overcurrent, short circuit (to ground and battery), open load, overtemperature and undervoltage detection. Besides all this, power supply and temperature monitoring are required for the whole unit, as well as remote control inputs. Other aspects were tight integration with software for easy handling of configuration settings, and finally, cost-effectiveness, reusability and adaptability to customer-specific requirements.

EXPANDING RCP SYSTEMS WITH SIGNAL CONDITIONING AND POWER STAGES

The dSPACE RapidPro SC Unit and Power Unit are two add-on front ends to expand RCP systems with signal conditioning and power stages (see Figure 3).

The units are connected to the RCP system via analog and digital signals. For diagnostic capabilities, the SC and Power Units can be additionally connected to the MicroAutoBox via a serial peripheral interface (SPI) link. A diagnostic blockset available for the MicroAutoBox supports diagnostics of the status of the SC and Power Units themselves and of the mounted modules, including the output driver circuits. For hardware configuration of all RapidPro Units, there is a new Windows-based tool called ConfigurationDesk available, briefly described later in the section "Hardware Configuration". Each of the units can be used separately, or they can form a stack for use as one physical unit. The mechanical design allows any combination (see Figure 4).

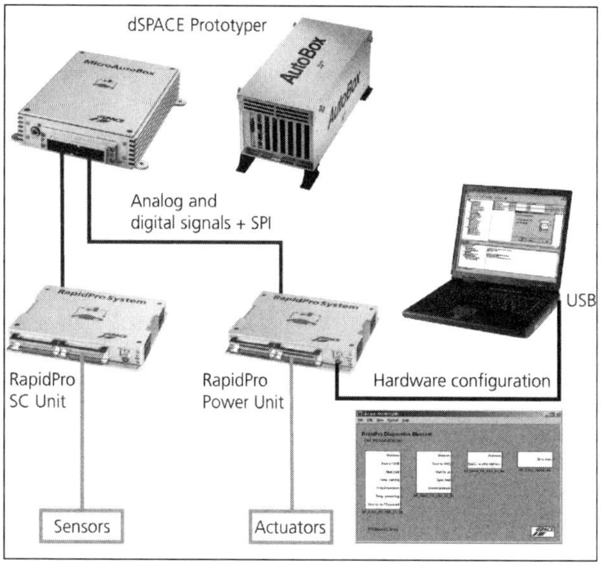

Figure 3: Expanding RCP systems with signal conditioning and power stages.

Figure 4: RapidPro Units form a stack.

The RapidPro SC Unit is a highly modular, configurable signal conditioning unit, especially designed for rapid prototyping in automotive applications. All inputs and outputs are therefore automotive-protected. The compact and robust unit consists of single- and/or multi-channel signal conditioning modules (SCM) mounted on a carrier board with space for up to eight SCMs. Each SCM has up to eight channels, independently of whether they are analog or digital. Mixed channel configurations on one module are also possible. Up to 64 channels overall can be realized per SC unit.

Taking all the different electrical vehicle systems into account, such as 12 V in passenger cars, 24 V in trucks and perhaps 42-V vehicle electrical systems in the future, an isolated on-board power supply supports a wide input voltage range of between 6 V and 60 V. The isolated power supply and internal separation of ground lines makes it possible to use the units in measurement applications. Fully 3-way isolated channels can be supported.

To fulfill the above requirements regarding diagnostic functions and configuration tasks, each unit contains an integrated on-board microcontroller. This also handles communication with the host PC via a USB interface and provides an SPI interface to the RCP system for diagnostics. The carrier board also includes integrated temperature and power supply monitoring. The RCP system can request the status via SPI bus and in the event of an error, it can switch off the SC Unit via a remote control input. Real-time diagnostic functionality is supported by a blockset on the MicroAutoBox. Because it is becoming even more difficult to put complex systems into operation, and to simplify testing, the carrier board contains a reference voltage and an on-board A/D converter. This makes it possible to display all the measurement values of each channel of an SCM on the host PC, and allows support of SCMs with self-calibration. Each SCM contains a simple microcontroller, including a nonvolatile on-chip memory for storing the module-specific configuration settings and/or calibration values. The intelligent configuration concept allows the use of both software-programmable SCMs with adjustable input settings and simple nonintelligent, hardware-configurable modules. The modular design makes it possible to adapt almost any type of sensor, such as temperature, pressure, accelerometers, speed, position and so on, to the RCP system. There is a wide range of off-the-shelf modules available to cover applications such as chassis, drives and engine control, including specific modules for lambda probes and knock sensors based on custom ASICs. Some SCMs can be equipped with electronic devices according to customer-specific requirements during assembly. Fully customized SCMs are also possible.

The RapidPro Power Unit is a front-end power stage unit based on the same architecture as the SC Unit. It generates high-current signals to drive actuators and loads. The carrier board offers space for up to six power stage modules (PSM). This allows up to 48 power driver channels per Power Unit. A wide range of hardware- and software-configurable, off-the-shelf PSMs is available. Fully customized PSMs are also possible. Special focus has been put on extensive support of the diagnostic functionality of the output driver circuit.

ACQUISITION AND GENERATION OF COMPLEX I/O SIGNALS

Some applications, such as engine or vehicle dynamics control, require the acquisition and generation of complex I/O signals (e.g., crankshaft, camshaft, ignition, injection) independently of the main CPU and the simulation step size of the model.

The RapidPro Control Unit is based on an MPC565 microcontroller, which is also utilized in production ECUs in powertrain applications. In this case, the micro-controller is used as a slave to expand RCP systems with additional I/O functionality according to developers' requirements. The Control Unit is based on the same architecture as the other RapidPro Units and offers space for up to six SCMs and two communication modules (COM). The Control Unit uses the same SCMs as the SC Unit. To build an intelligent I/O subsystem tailored to a particular application, the Control Unit can be used separately or in combination with the other units. An integrated unit connection bus (UCB) allows one or more Power and/or SC Units to be connected directly to the Control Unit without external wiring (see Figure 5). This increases reliability and improves handling.

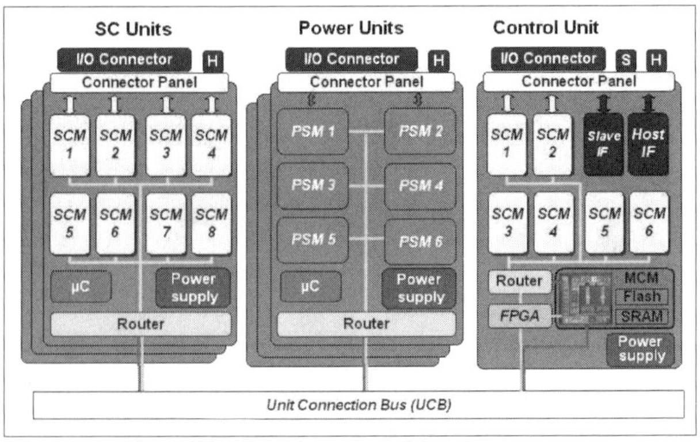

Figure 5: RapidPro hardware architecture.

An integrated routing concept on all carrier boards allows flexible assignment of both analog and digital signals between the SCMs/PSMs and the I/O ports of the MPC565. The digital I/O signals of modules have to pass through an on-board FPGA on the Control Unit before they are routed to the MPC565. This allows additional functionality to be added, for example, crankshaft/camshaft signal analysis for engine control, depending on the particular application.

Communication between the RapidPro Control Unit and the RCP system is realized via a dual-port memory interface and a high-speed LVDS link with a transfer rate of up to 250 Mbit/s (10 MB/s effective in both directions). In addition, there is a generic RapidPro Control Unit blockset available for MicroAutoBox and modular systems based on the DS1005 Processor Board and the DS4121 ECU Interface Board. This offers extensive functions for both, standard I/O such as PWM, bit I/O and A/D, and application-specific functions such as engine, chassis and drives control (Figure 6).

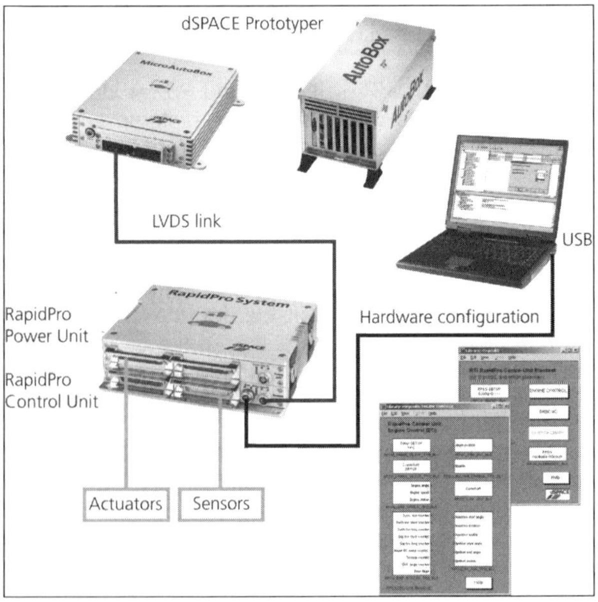

Figure 6: Adding additional I/O functionality to dSPACE RCP systems with the RapidPro Control Unit.

HARDWARE CONFIGURATION

To configure the RapidPro Units, a notebook is connected via a USB interface. After a hardware scan, ConfigurationDesk shows all the available units and modules in a tree view for easy navigation. The available hardware channels are listed in a channel list and can be easily configured on the channels property page (see Figure 7). All settings are saved on the hardware and in a project folder on the hard disk. Additionally, it is possible to export the hardware topology including configured signal labels. Importing the file in the RapidPro Control Unit blockset ensures intuitive and easy handling of the available hardware resources in the Simulink model.

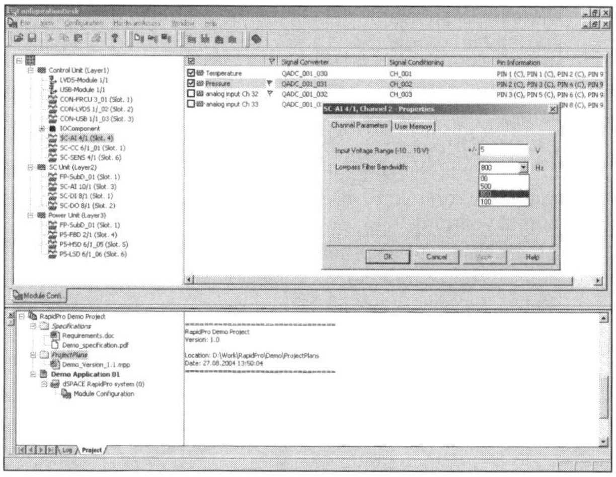

Figure 7: ConfigurationDesk used for configuration and monitoring of RapidPro hardware.

When the RapidPro is running, ConfigurationDesk monitors the module temperatures of the units and can also display diagnostic messages like open load, undervoltage and short circuits. To support the wiring of the hardware system to actuators and sensors, a pinout can be generated and viewed or exported to Excel.

NISSAN-SPECIFIC RAPIDPRO SYSTEM

The pilot project with NISSAN started at the end of 2003. To evaluate the RapidPro System under real conditions, NISSAN selected the MAXIMA as test vehicle. This was powered by the latest 3.5L VQ engine, Figure 8.

Figure 8: NISSAN VQ engine.

NISSAN VQ ENGINE

The VQ engine series is NISSAN's mainstream V6 engine lineup that has been listed in the World's "10 Best Engines" for the 10th straight year because of its high performance [3]. NISSAN has kept improving the VQ engine's performances continuously. It is now in its 3rd generation and has a new Engine Management System (EMS).

The EMS supports many variable devices such as continuously variable valve timing and a variable air induction system, in order to achieve high-output [4]. System function redundancies are reduced by using a CAN network and by reading multiple information from one sensor (e.g. cam position sensor).

Another special feature of the EMS is that it provides efficient emission reduction control by using an advanced air/fuel mixture control strategy and lambda probes. NISSAN implemented a "Sliding Mode Control" strategy – a modern control theory - in the emission control.

SYSTEM OVERVIEW

The NISSAN-specific RapidPro System (*Figure 9*) for controlling the VQ engine in a fullpass approach consists of the dSPACE MicroAutoBox, connected via analog and digital signals to two RapidPro SC Units and one Power Unit. The units are equipped with the modules listed in *Table 1*, which are necessary for the first closed-loop tests. A detailed description of the units and modules is available in [6].

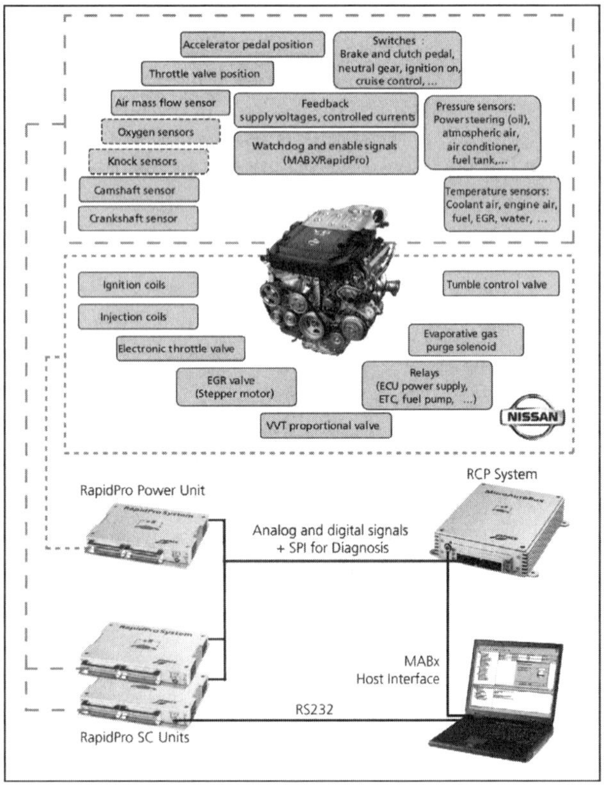

Figure 9: System overview.

An engine controller model running on the MicroAutoBox was available from previous projects at NISSAN. The proven software tool chain consisting of MATLAB/Simulink, a Simulink I/O blockset for the MicroAutoBox (dSPACE Real-Time Interface (RTI) for basic I/O and Extended Engine Control) and ControlDesk for calibration and measurement purposes (a sophisticated graphical user interface tool, seamlessly integrated in the tool chain), was used.

A terminal application connected to the RapidPro System via RS232 was used for hardware configuration during the pilot project, because ConfigurationDesk was not available at that time.

STEPS DURING THE PILOT PROJECT

After the rough concept design, the pilot project was divided into four major steps:

1. Detailed system specification and configuration
2. Commissioning and real load test
3. HIL tests on a test bench
4. In-vehicle tests

PHASE 1: DETAILED SYSTEM SPECIFICATION AND CONFIGURATION

After project kick-off, NISSAN provided a detailed actuator and sensor specification and a connector pinout of the original ECU connector.

No.	Module Description	Example Application
1	**SC-AI4/1** (4-channel analog input module, software-configurable input and output range, software-selectable filter frequencies, hardware-configurable pull-up and pull-down circuit)	Throttle position and pressure sensor signals which must be amplified
3	**SC-AI10/1** (10-channel analog input module, on-board sensor supply 1 W/5 V, hardware-configurable pull-up/pull-down and 1st- or 2nd- order lowpass filter)	Accelerator pedal position, pressure sensors, temperature sensors, air mass flow sensor, sensor supply and battery voltage measurements
2	**SC-DI8/1** (8-channel digital input module, comparator thresholds software adjustable, hardware-configurable pull-up/pull-down, input voltage range: +/- 60 V)	Crankshaft/camshaft sensors, switches (e.g., brake, neutral gear, etc.)
1	**SC-DO8/1** (8-channel digital output module, 1-A total output current, up to 40-V output voltage)	Relays, ignition coils
1	**SC-SENS4/1** (4-channel sensor supply module, software-configurable, 2.5-20 V, 1.2 W per channel, thermal overcurrent protection)	Supply for sensors and ignition coils
1	**PS-FBD2/1** (2-channel full-bridge driver module, up to 5 A per channel, output voltage up to 40 V, software-configurable current limitation and switch-off time, current measurement, load failure diagnosis	Throttle valve, tumble control valve
5	**PS-LSD6/1** (6-channel low side driver module, 4 channels up to 5 A, 2 channels up to 1 A, output voltage up to 37 V, clamping voltage 45 V or 75 V, load failure diagnosis, current measurement on two 5-A channels)	Evaporative gas purge solenoid, VVT valve solenoids, EGR stepper motor, fuel injectors, heater, O_2 sensor

Table 1: Overview of project related modules.

These were the basis for dSPACE to specify and configure the RapidPro System in more detail. The result of the analysis was a "RapidPro signal list" (Figure 10), which contains the mapping between actuators/sensors, RapidPro and MicroAutoBox. The final version derived after a few iterations with NISSAN contains the whole signal path from actuator/sensor to ECU connector pin to RapidPro front connector (module slot and pin) to RapidPro rear connector to MicroAutoBox connector (I/O channel) to I/O model (Simulink / Real-Time Interface).

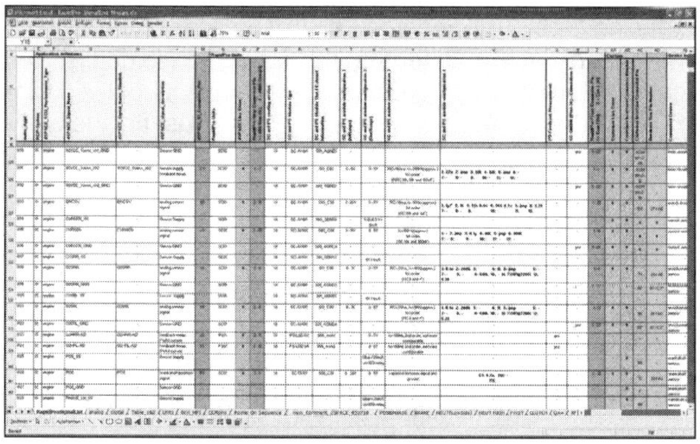

Figure 10: RapidPro signal list.

Further information such as signal names, ground lines, and the detailed hardware configuration of the modules is included in addition to the signal path.

The two available analog input modules are briefly described to give an impression of the possible module configurations and the kind of information which has to be detailed before assembling and commissioning the final system.

For example, the SC-AI10/1 module is an analog input module with ten single-ended input channels which requires two slots on the carrier board. All input channels are referenced to one common, "isolated" ground. A downstream high common-mode voltage difference amplifier is used to reference the "isolated" input signal to analog ground and to translate a high input impedance into a lower output impedance. The amplifier does not affect the characteristics of the user-defined network because its input impedance is higher than 1 GOhm. The module can be adapted to different sensor types by soldering resistors or capacitors on a user-definable network for each input channel (Figure 11). For example, the input range of the module can be changed by adding a voltage divider. Also it is possible to add a 1st- or 2nd-order lowpass filter and/or a pull-up or pull-down resistors to the input. In total up to 12 components per channel can be soldered to the module, which results in a customer-specific configuration like that shown in Figure 12.

The SC-AI4/1 module is an analog input module with four differential input channels. The input and output voltage ranges and the cut-off frequencies of a 2nd-order lowpass filter are software-configurable. In addition pull-up or pull-down resistors can be soldered on the user-definable network, see Figure 13.

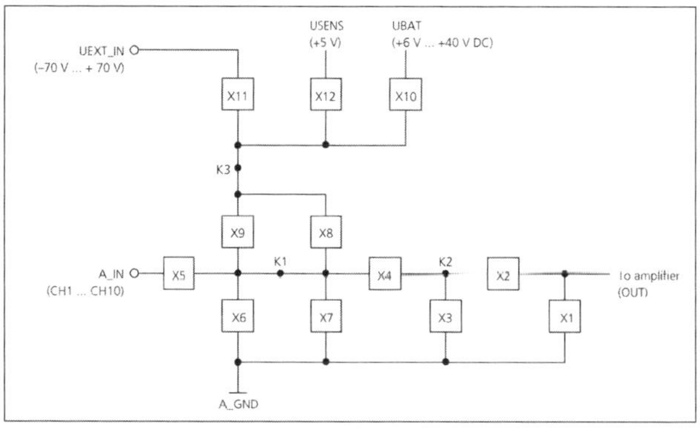

Figure 11: User-definable network on SC-AI10/1 module (one network per channel).

Figure 12: Customer-specific configured SC-AI10/1 module.

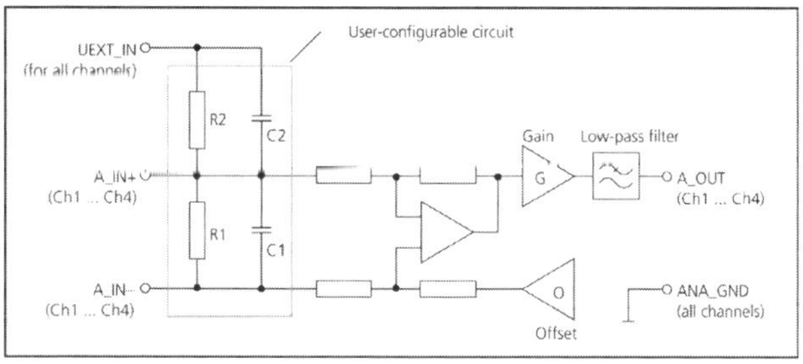

Figure 13: User-definable network on SC-AI4/1 module (one network per channel).

The application-specific GND concept has to be taken into account in designing and specifying the wiring harness. The GND concept depicted in Figure 17 was realized for this project.

When all the configuration settings and specifications had been done, dSPACE configured the modules and assembled the two wiring harnesses between the ECU and RapidPro's front connectors and between RapidPro's rear connectors and the MicroAutoBox connectors. The resulting system, shown in Figure 14, includes some spare signals, which can be used to connect additional sensors and actuators for future use during further development of NISSAN's VQ engine. In the maximum configuration, the I/O channels of the MicroAutoBox are fully used.

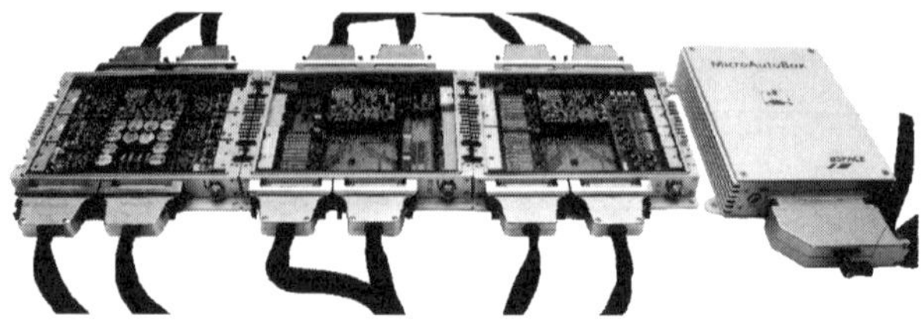

Figure 14: NISSAN specific RapidPro System connected to dSPACE MicroAutoBox.

PHASE 2: COMMISSIONING AND REAL LOAD TESTS

Initial system commissioning was done at dSPACE. A Simulink I/O model, shown in Figure 15, for the MicroAutoBox was designed using RTI for basic I/O and the Extended Engine Control blockset. Channel by channel, the RapidPro System was connected to real loads which were made available by NISSAN. Each signal was tested against NISSAN's actuator and sensor specification. Some problems occurred during this phase, mainly because several components of the hardware and the RapidPro firmware had to interact correctly in this early phase compared to the release of RapidPro. When the problems were solved, some basic tests were run on a hardware-in-the-loop (HIL) test bench at dSPACE, focusing mainly on synchronization to crankshaft and camshaft signals and on correct ignition and injection.

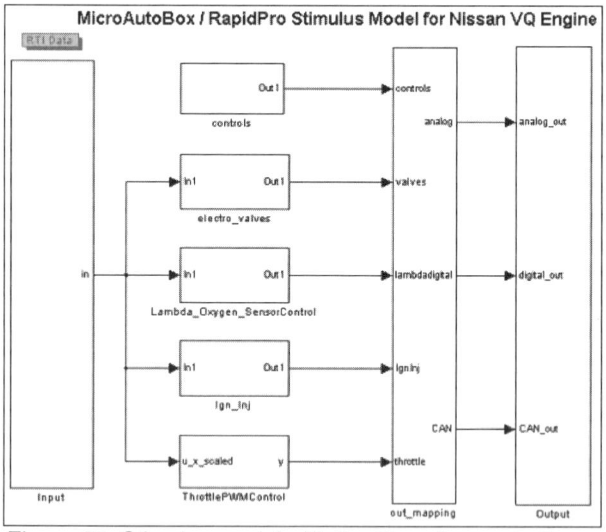

Figure 15: Stimulus model with throttle valve controller.

During the commissioning of the system, especially the wiring harness, the need for a compact and easy-to-handle RapidPro-specific break-out box (BOB) was obvious. In additon to typical features like closing / opening all the relevant signal connections, inserting stimulus signals, and signal measurement, this RapidPro specific BOB will have the advantage that RapidPro specific Sub-D connectors are used, see *Figure 16*. If wiring errors are detected during commissioning, they can be temporarily corrected by rewiring directly on the BOB. When the commissioning phase is finished, the RapidPro BOB can be removed and direct connection to the RapidPro System can be made with the corrected wiring harness.

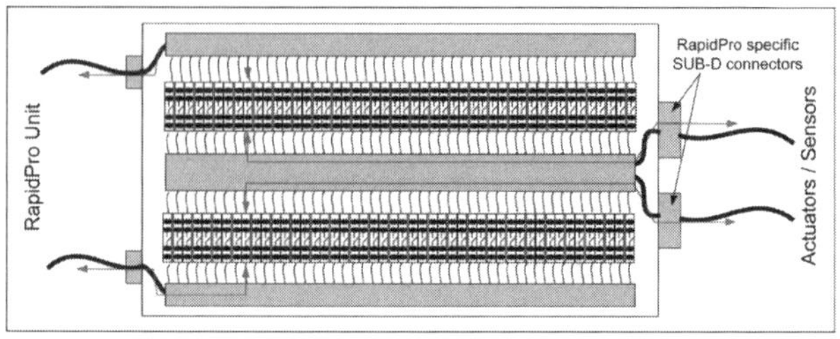

Figure 16: RapidPro-specific break-out box.

PHASE 3: HIL TESTS ON A TEST BENCH

The next step, done at NISSAN together with dSPACE, was to integrate the NISSAN fullpass controller model in the I/O model, followed by detailed HIL tests. The HIL tests were available from tests performed for the original ECU in a previous phase. The final configuration of the RapidPro System was done via the terminal application during these HIL tests. For example, the upper and lower threshold values of some digital input channels on the SC-DI8/1 modules and the cut-off frequencies of some channels on the SC-AI4/1 module were adjusted.

The overall system passed the test without the hardware having to be changed. After this success NISSAN decided to go on directly with in-vehicle tests with the NISSAN MAXIMA on a roller dynamometer.

PHASE 4: IN-VEHICLE TESTS

After the integration of the RapidPro System into the vehicle (see Figure 18), the most important signals were checked without the engine running. In the first trial the engine started and ran without trouble. During the tests on the real engine, some problems occurred because the implementation on the real engine was different to the specified GND concept (Figure 17), which was designed for a new prototyping version of the engine agreed on during the specification phase and which worked fine during all prior tests. The sensors got their GND contact via the sensor cases, which were mounted in the engine block directly and not, as was specified, via separate GND lines to the main star point. The engine block was directly connected to the car battery. All analog measurements were still referenced to the star point, so that an additional, pulsating voltage drop at the star point, caused by high current flow of real components, lead to incorrect measurement results.

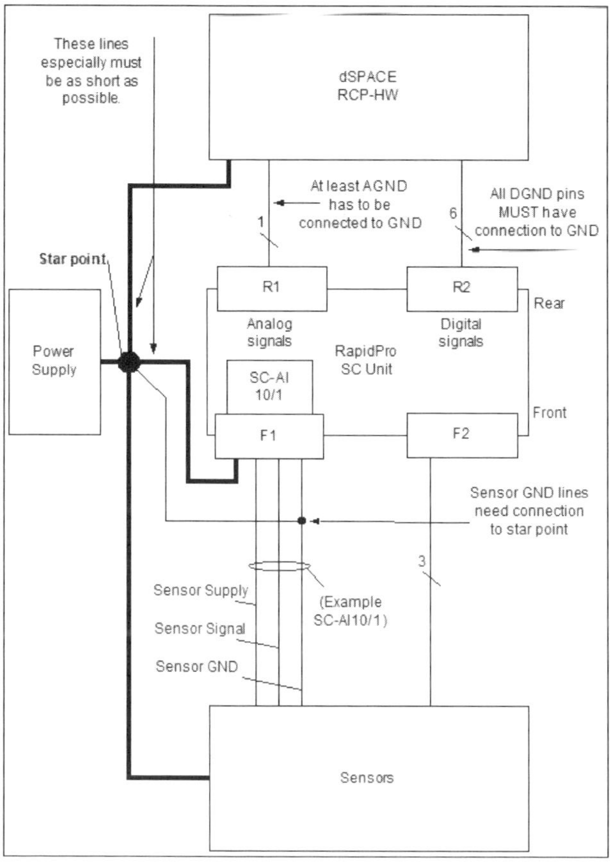

Figure 17: Application-specific GND concept.

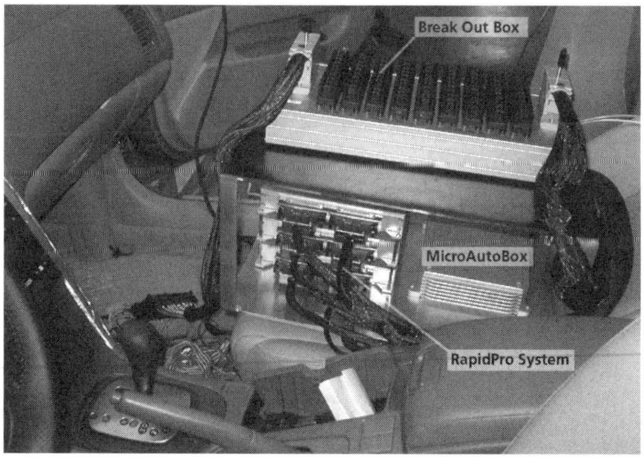

Figure 18: Use of RapidPro inside NISSAN's MAXIMA prototyping car.

These signal disturbances were finally eliminated by modifications to the wiring harness in which the original star point was divided up into supply GND lines and sensor GND lines. The sensor GND lines were bypassed and connected directly to the battery.

With the engine running, all the sensors, actuators and signals were successfully tested for functionality, plausibility, and noise. The input signals were validated by measurements done with an oscilloscope and compared to the input signals of the model measured with ControlDesk.

During phase 4, the engineers from NISSAN were trained in using and configuring the RapidPro System.

NISSAN'S EXPERIENCE OF USING THE RAPIDPRO SYSTEM

NISSAN has been able to use the prototype of the RapidPro System since the summer of 2004, to evaluate it and to develop new functions for engine control in Simulink based on the fullpass concept. The configurable spare channels on the RapidPro Units also make it possible to evaluate new sensor and actuator types in parallel to the current ones.

The RapidPro units are still working under real conditions without any problems. The RapidPro prototype will be replaced with units of the release version when they become available.

The fullpass approach to full control of the VQ engine has to be expanded by lambda control and knock detection. The necessary modules will become available in Q2/2005 and will be used in a successor project. Because MicroAutoBox's I/O resources are already exceeded, a Control Unit will be used additionally to expand the I/O functionallity. An overview of the resulting system is shown in Figure 6. An additional advantage of this system architecture is that no customer-specific wiring harness is needed between MicroAutoBox and RapidPro. Only the high-speed serial link is used for communication.

CONCLUSION

dSPACE RCP systems are in widespread use in fullpass and bypass applications. Besides the necessary real-time computation power, easy access to I/O and the corresponding software tool-chain, all sensors and actuators need to be connected to the RCP system. Up to now dSPACE customers had to close this gap themselves. They either used some kind of signal conditioning and power stage modules available on the market, built their own custom-made solutions, or tried to reuse existing ECU interfaces. This involved varying amounts of adaptation work. The new RapidPro System, using modules that are hardware- and software-configurable, closes the gap. All the components can be reused, reconfigured, and extended, for example in later projects, with a minimum of effort. After some pilot projects done with NISSAN, DaimlerChrysler [5] et al. the "tried and tested" version of the RapidPro System is now available. It has not only been tested in the laboratory to verify the concept, but some prototypes of the system have also been used in real applications by customers. Customer feedback and early experience gained from real projects with customers have been used to stabilize the system and to provide a ready-to-use product.

Customers like NISSAN have been able to find out whether the system fulfills their requirements. The decision by NISSAN to go on with RapidPro for fullpass applications and to gain more experience with the new hardware is shown by an agreement to do a successor project to investigate modules for knock detection and lambda control.

TYPICAL SCENARIOS FOR RAPIDPRO CUSTOMERS

These will be the typical scenarios for RapidPro customers:

RapidPro off the shelf:

a) Customers can order single components such as units and modules and configure them by themselves.

b) Alternatively, customers can get a complete, configured, customer-specific system which is ready to use.

RapidPro as a turn-key system (or individual items from the list below): dSPACE offers engineering services to provide turn-key solutions, including

a) Concept design

b) Detailed system specification and configuration

c) Specification and assembly of a customer-specific wiring harness

d) On-site engineering for system commissioning

e) Training

f) Design of customer-specific modules if the available and planned modules cannot fulfill the customer's requirements.

REFERENCES

1. H. Hanselmann: Development Speed-Up for Electronic Control Systems, Convergence International Congress on Transportation Electronics, Dearborn, October 19-21, 1998.
2. H. Hanselmann, F. Schütte: Control System Prototyping And Testing With Modern Tools, PCIM, Nürnberg, June 19-21, 2001.
3. Tsuyoshi Michishita, Eiichi Matsumoto: Introduction of VQ engine – Straight 10 years awards of Word's 10 Best Engines, NISSAN TECHNICAL REVIEW No. 54.
4. Tsuyoshi Michishita, Naoki Nakada, Takeshi Yamagiwa, Keiichi Murata: Third Generation of High-Response and High-Output 3.5L V-6 Engine, SAE World Congress, Detroit, March, 2002.
5. DaimlerChrysler: ABC with RapidPro, dSPACE NEWS 03/2004.
6. dSPACE: RapidPro System – Installation and Configuration Guide.

CONTACT

For more information, please contact:

Dr. Frank Schuette
Lead Engineer Applications

dSPACE GmbH
Technologiepark 25
D-33100 Paderborn
Germany

Phone: +49 5251 1638-644
Fax: +49 5251 16198-644
Email: fschuette@dspace.de

2005-01-1281

AutoMoDe – Notations, Methods, and Tools for Model-Based Development of Automotive Software

Andreas Bauer, Manfred Broy, Jan Romberg and Bernhard Schätz
Institut für Informatik, Technische Universität München

Peter Braun
Validas AG

Ulrich Freund and Nuria Mata
ETAS Engineering Tools GmbH

Robert Sandner
BMW AG

Dirk Ziegenbein
Robert Bosch GmbH

ABSTRACT

This paper describes the first results from the AutoMoDe project (Automotive Model-based Development), where an integrated methodology for model-based development of automotive control software is being developed. The results presented include a number of problem-oriented graphical notations, based on a formally defined operational model, which are associated with system views for various degrees of abstraction. It is shown how the approach can be used for partitioning comprehensive system designs for subsequent implementation-related tasks. Recent experiences from a case study of an engine management system, specific issues related to reengineering, and the current status of CASE-tool support are also presented.

1. INTRODUCTION

AutoMoDe is a joint research project consisting of members of the Software & Systems Engineering group at the Technische Universität München, Validas AG, ETAS GmbH, Robert Bosch GmbH, and BMW AG. The overall goal of the project is to develop an integrated methodology for model-based development of automotive control software based on custom, problem-specific design notations with an explicit formal foundation. A series of prototypical tools is being developed which builds on the existing AutoFOCUS [HSE97] framework in order to illustrate the key elements of the methodology presented.

1.1 BACKGROUND

Current challenges in automotive control systems design include quickly rising system complexity across all domains, tight time-to-market constraints (necessitating better predictability for design efforts such as integration), the transition from realization of control logic in mechanical/electrical systems to software implementations, and heterogeneous design chains crossing several technical disciplines and organizations or companies.

Traditionally, the focus of embedded software engineering has been on the later and thus more detailed abstraction levels, which deal mostly with implementation-related issues. More abstract system descriptions typically take a back seat in the design process because they lack suitable notations, methodologies, and integration between abstraction layers. However, working at higher levels of abstraction will be a key factor in tackling the prevalent complexity issues in automotive software engineering, and in catering to different stakeholders in the design chain. For these reasons, the organization of design artefacts along abstraction layers tailored for different stakeholders and different phases has been identified as a possible remedy in the past [BBRS03][Thu03]. A related method should then provide support for easy transitions between layers, e.g. for restructuring designs. Notations and underlying models, such as notations for architectural and behavioural design, should be well-integrated.

Consequently, AutoMoDe aims to address the obvious need to organize such artefacts along various abstract levels, tailored for different stakeholders and phases in the overall systems development process. Additionally, transitions between abstraction levels, like restructuring operations, are supported by the accompanying tools.

Exchanging design information in a heterogeneous setting is aided by well-established, intuitive, and unambiguous notations for design artefacts. The AutoMoDe tools aim to provide for such support with formally founded notations, with powerful consistency checks suited to the different abstraction levels, and by providing the ability to validate a design's behavior with built-in simulators.

1.2 OVERVIEW

Section 2 of the paper introduces the operational model of AutoFOCUS designs, which are based on a formally defined system model using explicit data-flow and discrete-time semantics. The accompanying tool support facilitates early system validation through simulation and verification capabilities, as well as powerful consistency checks. In Section 3, we detail the graphical representation of AutoMoDe designs, which are specified using a number of different views for differently abstract system levels – each of which targets specific aspects of the design of automotive control systems. Section 4 explains the results and experiences gained during a reengineering case study using the AutoMoDe approach and tool prototypes. The conclusions of this paper are summarized in Section 5, which also gives an outlook and discusses future activities.

1.3 RELATED WORK

In the related research project Automotive [BBRS03] BMW, Bosch, ETAS, Telelogic and Technische Universität München have developed a model-based language and method for developing automotive control software. To achieve practicable usability, the method was based upon relevant and commercially available tools like ASCET, DOORS and the UML Suite. The Automotive Modeling Language (AML) is based on a definition of distinct abstraction levels similar to the AutoMoDe abstraction levels presented in this paper, a concrete syntax based on the UML 1.x and the ASCET notations for structural system descriptions, and an easily usable concept for variants and configurations.

AutoMoDe addresses two inadequacies in Automotive's results: firstly, instead of UML 1.x, AutoMoDe uses the AutoFOCUS [HSE97] notation with an explicit concept of components and their composition for the description of the structures of embedded systems. AutoFOCUS is also very closely related to selected UML 2.0 concepts, so possible standard conformance in the future is not regarded as critical. Secondly, beyond the purely structural designs considered in Automotive, AutoMoDe is also concerned with behavioral aspects.

EAST-EEA: The European automotive industry started this project running under the ITEA banner in 2001. The project was finished in July 2004. An Architecture Description language, the so-called EAST-ADL, has been defined, along with a definition of a domain-specific middleware [Thu03]. The EAST-ADL [Fre04] is structured into several abstraction levels by defining the following architectures:

- Vehicle Project (VP)
- Functional Analysis Architecture (FAA)
- Functional Design Architecture (FDA)
- Logical Architecture (LA)
- Operational Architecture (OA)
- Hardware Architecture (HA)
- Technical Architecture (TA)

The vehicle project describes the electronic features of a vehicle and all resulting variants from a customer's point of view. The structure of electronic features themselves is described by using inputs and outputs in the functional analysis architecture. If necessary, the structure can be reinforced by behavioral models. The functional design architecture is the most abstract description of software structures to be found later in the ECU software implementation. This architecture consists of a functional hierarchy with a focus on typed structures and signal exchange. In addition, it is possible to associate behavior with functions of the hierarchy. However, only elementary functions, i.e. the leafs of a functional hierarchy, can carry behavior to be flexible for the final placement of functions on ECUs. Behavior can be described using finite state machines, difference equations, or plain source code. The logical architecture describes the system on a pure instance level. Instances can be mapped to ECUs and OS tasks while still respecting all timing and memory requirements. ECU properties with their sensors and actuators as well as the bus systems are described in the hardware architecture. Configurable basic software components for the ECUs like OS, HAL and the EAST middleware form the technical architecture, while the operational architecture describes the running system. The latter is conceived by mapping the instances of the logical architecture to ECUs, OS tasks and bus messages and then by applying code generation of the application software and by configuring the basic software.

The use of explicit operational modes for decomposition has also been brought forward by other authors, e.g. [MR98]. In addition to the idea of using explicit notations for operational modes, our approach employs such mode representations across several levels of abstraction, especially for coarse-grained structuring of systems, and in particular investigates transformations between different mode representations suited for different abstraction levels.

The concept of expressing frequencies and event patterns as Boolean expressions (clocks), and the idea of providing a type system for such clocks, originates from the field of synchronous programming languages [IEE91].

2. OPERATIONAL MODEL

AutoFocus uses a message-based, discrete-time communication scheme as its core semantic model. AutoFocus designs are built from networks of components or blocks (drawn graphically as a rectangle) exchanging messages with each other and with the environment via explicit interfaces (drawn as small circles) and connectors between interfaces. Messages are time stamped with respect to a global, discrete time base. This computational model supports a high degree of modularity by making component interfaces complete and explicit. It also provides a reduced degree of complexity: Because the discrete time base abstracts from implementation details such as detailed timing or communication mechanisms, the use of timing information below the chosen granularity of observable discrete clock ticks is avoided. Examples for such detailed assumptions include the ordering of message arrivals within one time slot, or duration and delays of transfer. Real-time intervals of the implementation are therefore abstracted by logical-time intervals.

Note that this message-based time-synchronous communication model does cater to both periodic and sporadic communication as required for a mixed modeling of time-triggered and event-triggered behavior. As shown in Figure 1, each channel in the abstract model either holds a message represented by an explicit value or the "√" ("tick") value indicating the absence of a message. Thus modeling of event-triggered behavior is naturally covered by the AutoFocus notation by reacting explicitly depending on the presence (or absence) of a message.

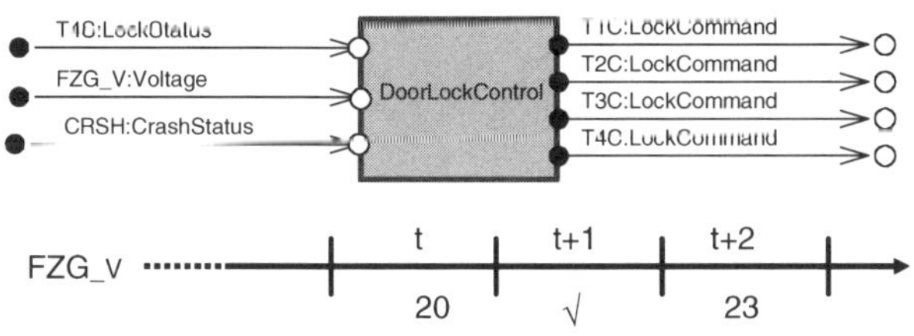

Figure 1: Message-based, time-synchronous communication.

3. ABSTRACTION LEVELS AND VIEWS

The different system abstractions and their supported views on the system (see Fig. 2) are central to the model-based approach of AutoMoDe. The system abstractions chosen are similar to those defined in [Thu03] (see also Sec. 1.3), but are adapted to match the model-based AutoMoDe development process. The abstraction levels and the corresponding use of the AutoFocus notations are introduced in the following.

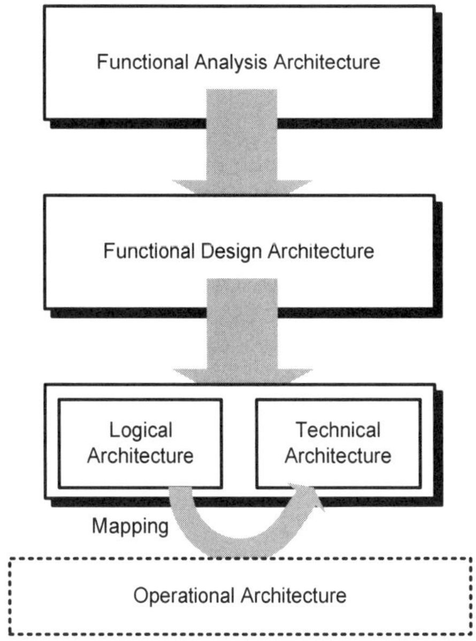

Figure 2: AutoMoDe abstraction levels.

3.1 FUNCTIONAL ANALYSIS ARCHITECTURE

The Functional Analysis Architecture (FAA) is the most abstract level considered in AutoMoDe. The FAA provides a system-level abstraction representing the vehicle functionalities to be implemented in either hardware or software.

Use Cases, Feature Tree and Hierarchy Diagrams

Today a vehicle may have more than 2,500 software based functions. Typically, it is up to the requirements engineer to decide which functions are required and how these should be structured and realized in terms of user interactions.

Feature trees and *feature hierarchies* provide a structural view of these functions. The structuring of feature hierarchies is strictly use case oriented: at each level of the hierarchy, "function families" are structured into sub-functions until atomic functions are reached.

We introduce special relationships between functions that indicate dependencies. Functions without dependencies are distinguished explicitly and are required to remain independent. The behavior of atomic functions can be described by using scenarios in terms of interaction diagrams (message sequence charts), or in a more thorough way via state machines.

An FAA-level description is typically complete as to the functionalities being considered and the functional dependencies between them. It enables the identification of functional dependencies and potential conflicts between vehicle functions to be identified, and the validation of functional concepts based on prototypical behavioral descriptions.

Means to achieve these goals include rules as well as model simulation. Based on the functional structure and dependencies, rules identify possible conflicts and suggest suitable countermeasures to resolve them. An exemplary rule is the introduction of specific arbitration functionality wherever two vehicle functions access the same actuator. These rules are E/E-architecture-driven and have been developed in the context of the CARTRONIC framework [LTS01]. The simulation also considers the prototypical behavioral descriptions. These descriptions are not optimized for efficient implementation and abstract from details such as concrete data types.

System Structure Diagrams

The dominating notation used on the FAA level is called System Structure Diagram (SSD). SSDs are used for describing a high-level architectural decomposition of a system, similar to UML 2.0 component diagrams [UML]. SSDs consist of a network of *components,* shown as rectangles, with statically typed message-passing interfaces (*ports*), shown as black and white circles. Explicit directed connectors (*channels*) connect ports and indicate the direction of message flow between components. Components can be either recursively defined by other SSDs, or by a number of specifically suited notations for behavioral description (see Sec. 3.2). On the FAA level, it may be perfectly adequate to leave the detailed behavior unspecified. For an example SSD, see Fig. 3.

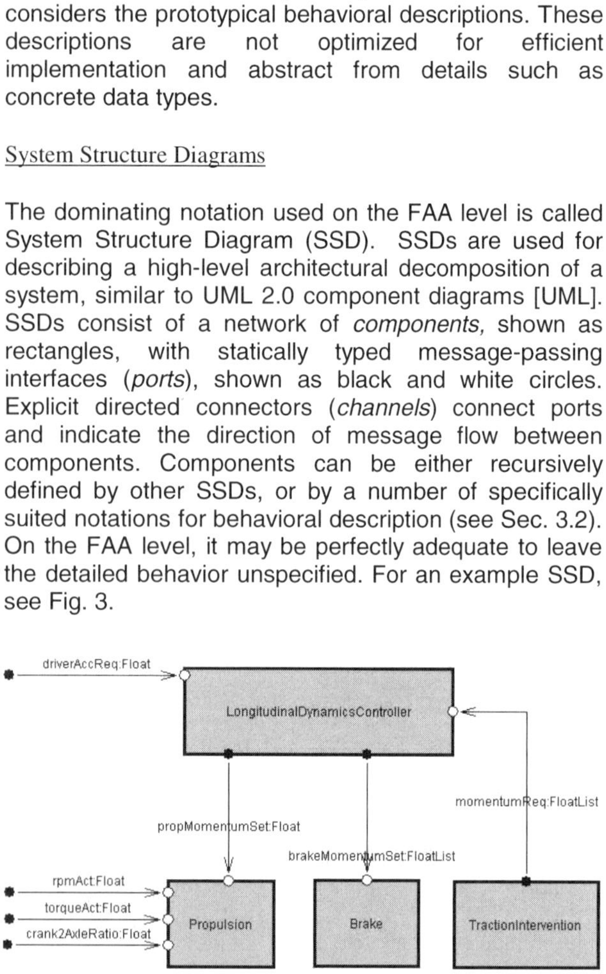

Figure 3: Example SSD component network on the FAA level.

The component boundaries introduced with SSDs have semantic implications as well – each SSD-level channel introduces a delay in the communication between components. Because of AutoMoDe's global discrete-time semantics, such implicit introduction of delays is a prerequisite for later partitioning with reduced revalidation effort (see Sec. 3.4).

Note that SSDs are not unique to the FAA, but will be used throughout this text on other abstract system levels as well (see Sec. 3.2 – 3.3).

3.2 FUNCTIONAL DESIGN ARCHITECTURE

The AutoMoDe system abstraction Functional Design Architecture (FDA) is a structurally as well as behaviorally complete description of the software part of the system or a subsystem. The description is in terms of actual software components that can be instantiated in later phases of the development process. In its current version, AutoFocus supports classification and instantiation of components through the *shared components* mechanism, which groups structurally and behaviorally equivalent components without using a concept of an explicit component type or class, such as in UML. Structural or behavioral modifications to one component in the group are automatically propagated to all other components in the group. The general question of whether explicit component types are beneficial for automotive control systems development is the subject of ongoing research.

In contrast to FAA-level functionalities, atomic SSD components in the FDA are required to have a well-defined behavior. Behavior specifications of atomic components are allowed in terms of *Data Flow Diagrams*, which specify algorithms in terms of blocks communicating through data flows, *Mode Transition Diagrams*, which decompose the component's behavior into distinct operational modes, or *State Transition Diagrams*, which specify reactive, event-driven behavior in an automaton-like style.

Data Flow Diagrams

Data Flow Diagrams (DFD) define an algorithmic computation of a component. Graphically, DFDs are similar to SSDs (see Fig. 4): DFDs are built from individual *blocks* with ports connected by channels. Typing of ports is dynamic, using type inference properties of operators. A block may be recursively defined by another DFD. The behavior of atomic DFD blocks is given either through a Mode Transition Diagram (MTD), through a State Transition Diagram (STD), or directly through an expression (*function*) in AutoFocus's base language [HSE97]. For example, block "Minus" in Fig. 4 is defined by the function "a - b", where "a" and "b" are port identifiers (not shown). It is possible to define adequate block libraries for discrete-time computations with this mechanism.

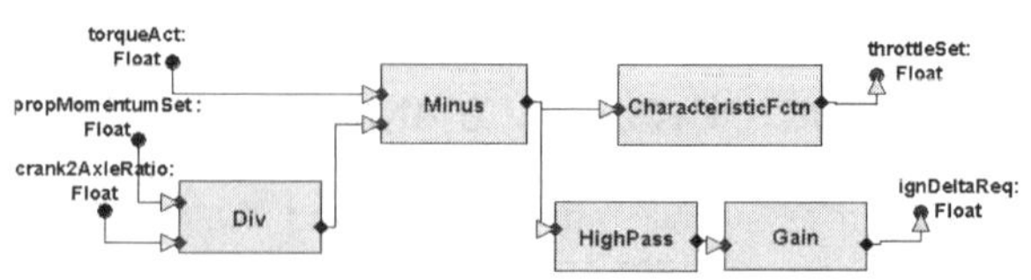

Figure 4: Example DFD for a longitudinal momentum controller component.

In contrast to the delayed composition primitives in SSDs, the semantics of DFD composition is "instantaneous", in the spirit of synchronous languages [IEE91]. In the AutoFocus tool, instantaneous communication primitives are accompanied by a causality check for detecting instantaneous loops. Note that computations "happening at the same time" in FAA-, FDA- or LA-level models are perfectly valid abstractions of sequential, time-consuming computations on the level of the Operational Architecture (OA) if the abstract model's computations are observed with a delay, such as the delays introduced by SSD composition. The duration of the delay then defines the deadline for the sequential computation on the OA level.

Mode Transition Diagrams

Mode Transition Diagrams (MTDs) are used to represent explicit system modes and alternate behaviors within modes (see Fig. 5). MTDs consist of *modes* and *transitions* between modes. Transitions are triggered by certain combinations of messages arriving at the MTD's component. The behavior of the component within a mode is then defined by a subordinate DFD or SSD associated with the mode. As illustrated by the detailed example in Sec. 4, MTDs provide a valuable means of architectural decomposition specifically suited for embedded control systems.

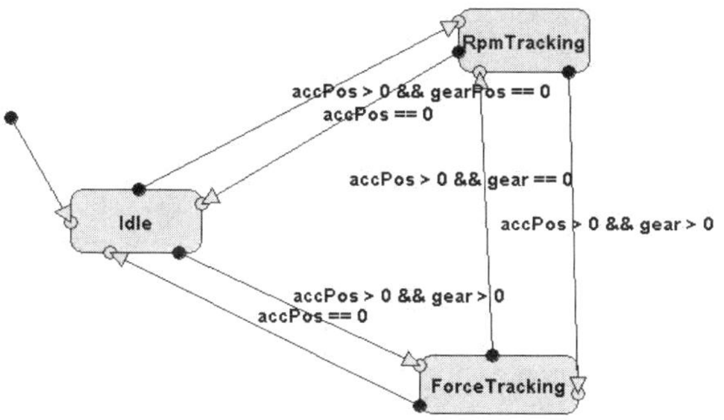

Figure 5: Example MTD for operational modes of an engine.

State Transition Diagrams

STDs are extended finite state machines with *states* and *transitions* between states (see Fig. 6). STDs are similar to the popular Statecharts notation, but with some syntactic restrictions. Through the restrictions chosen - no AND_states, no inter-level transitions, restricted preemption primitives - semantic ambiguities allowed by some standard Statecharts dialects [vdBeeck94] are avoided.

3.3 LOGICAL AND TECHNICAL ARCHITECTURE

The Logical and Technical Architecture (LA, TA) is the most implementation-oriented abstraction level supported by the AutoMoDe method. FDA-level components are instantiated and grouped into *clusters* at the LA level. The TA represents hardware and platform components (ECUs, buses, message frames) used to implement the system.

A cluster can be thought of as a "smallest deployable unit" in a software system. Consequently, several clusters may be mapped to a given operating system task on the OA level, but a given cluster will not be split across several tasks.

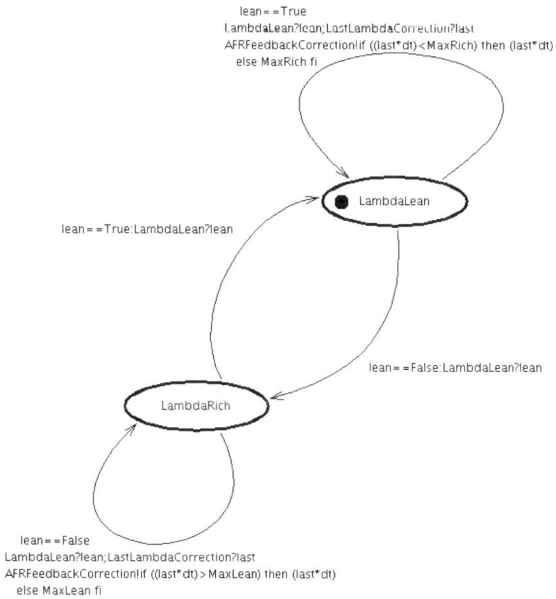

Figure 6: Example STD for exhaust control of an engine.

Cluster Communication Diagrams

The notation used for top-level definition of the LA structure is called Cluster Communication Diagrams (CCD). Like SSD components, clusters have statically typed interfaces. In contrast to the recursive definition of SSDs and DFDs, CCD clusters must not be defined by other CCDs. On the other hand, hierarchical DFD descriptions are perfectly adequate for clusters. The type system at the LA level is extended by *implementation types* which capture the more or less platform-related constraints associated with implementation. Signal frequencies and event patterns are required to be explicit on the LA level, but not at the FAA and FDA levels.

Signal frequencies and event patterns are represented in the AutoFOCUS notation as *clocks*: Each message flow in AutoFOCUS is associated with such a clock. The clock for any given flow indicates either the frequency of message exchange (periodic case), or a condition describing the event pattern (aperiodic case). Syntactically, a clock is simply a Boolean expression evaluating to logical "true" whenever a message is present on the clock's flow. Graphically, clocks may be represented as an element of a channel's or port's label, which has the form *<id>:<type-expr>:<clock-expr>*. Clocks are supported within the AutoFOCUS2 tool through by explicit sampling operators (indicated as "w" blocks in the examples) and an inference system, similar to type inference in programming languages.

The graphical representation of CCDs is identical to DFDs (see Fig. 7). In particular, the half-shaded diamonds indicate explicit delay operators.

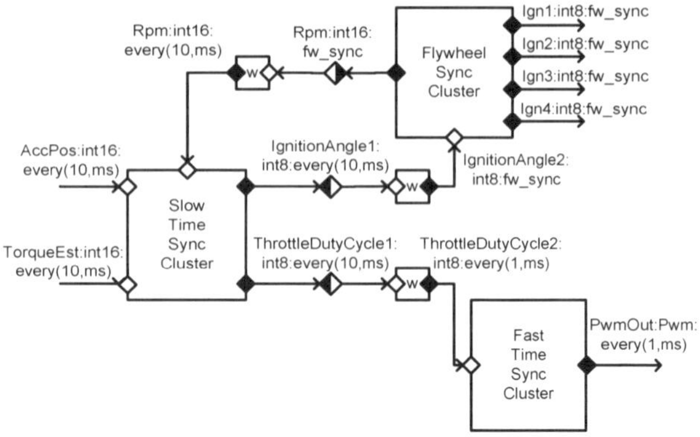

Figure 7: Example CCD for engine controller (simplified).

3.4 TRANSITIONS BETWEEN ABSTRACTION LEVELS

Transition FDA->LA

To make the transition from an SSD representation on the FDA level to an LA-level CCD, some of the topmost SSD hierarchy may be dissolved in favor of a flat CCD representation. Clusters can then be defined in terms of (hierarchical) DFDs, MTDs, and STDs. In addition, abstract data types such as "int" are typically mapped to an implementation, e.g. "int8" or "int16". Similarly, a floating-point message on the FDA level may be mapped to a fixed-point or integer message on the LA level.

In order to represent high-level MTDs as a network of clusters on the LA level, the AutoFOCUS2 tool prototype provides a built-in utility to transform an MTD into a semantically equivalent, partitionable dataflow model.

Partitioning in AutoMoDe involves grouping an FDA-level description into clusters on the LA level. Because of the many design tradeoffs involved, we expect that this task can be automated chiefly on the level of elementary transformations, while essential design decisions are left to the engineer. As an example, the following two heuristics are supported by elementary AutoMoDe refactoring steps:

(1) Partitioning along the SSD structure decomposition of FDA-level components. This strategy provides a clear one-to-one correspondence between FDA and LA descriptions.

(2) Partitioning according to common signal and communication frequencies. This strategy may be preferable for technical reasons, e.g. reduced number of control flow (if-then-else) statements, reduced execution time jitter, better utilization of resources.

With regard to required delays and dedicated conditions concerning the syntactic and semantic validity of CCDs, clusters may depend on the characteristics of a given Technical Architecture. As an example, consider an OSEK-conforming operating system as a target platform, with inter-task communication between tasks using data integrity mechanisms [PMS+95] and fixed-priority, preemptive scheduling. In this framework, communication from "slower" clusters to a "faster" cluster necessitates the introduction of at least one delay operator in the direction of data flow. On the other hand, communication in the opposite direction ("fast" to "slow" cluster communication) does not require introduction of delays in the CCD.

Note that implementation-driven introduction of delays may significantly alter a system's behavior, depending on the nature of an application. Therefore, early introduction of delays from the top down between SSD components on the FAA and FDA levels is made with the intention of avoiding a costly revalidation of models after the transition to the (more implementation-driven) LA/TA level.

4. REENGINEERING

The aforementioned concepts and descriptions (see Sec. 2 – 4) have been applied to an extensive automotive case study where the engine controller for a four-stroke gasoline engine was modeled. Originally, this case study [Bea99] was provided in terms of a detailed ASCET [ETAS] design and has been reengineered in important parts using an early AutoMoDe tool prototype, AutoFocus (see Sec. 5), along with the related notations and underlying semantics.

4.1 OPERATING MODES

Compared to ASCET, AutoFOCUS provides a richer set of control flow primitives. As it turns out, the AutoMoDe concept of modes and MTDs can capture and encapsulate implicit operation modes of the original ASCET design especially well. What is more, implicit modes of ASCET processes can be made explicit to the developer by using MTDs rather than control flow operators such as if-then-else (see Fig. 8).

In other words, MTDs support a comprehensible design not only because they hide parts of a complex computation in hierarchical components, but also because they make clear which mode a certain part of the system is in that is actually being modeled by the user at any time. The latter is an especially strong argument for the use of modes from a methodological point of view, since they prevent potentially conflicting control flow statements based on a wrong evaluation of Boolean variables, or flags.

For example, the purpose of the component named "ThrottleRateOfChange" is to determine the rate at which the throttle valve position changes, not only depending on its current and the desired position, but also depending on very specific states of the entire engine (control management). Other approaches such as MATLAB Simulink or ASCET would typically use a separate block for handling the control logic of the engine modes. This control logic block would then communicate a large number of flags and control values to the remaining blocks of the subsystem to influence their behavior based on the current operating mode. The control values, however, are evaluated inside the control logic block as well as in the respective subsystems, which may yield (deadlock-like) conflicts due to subtle inconsistencies in the control flow.

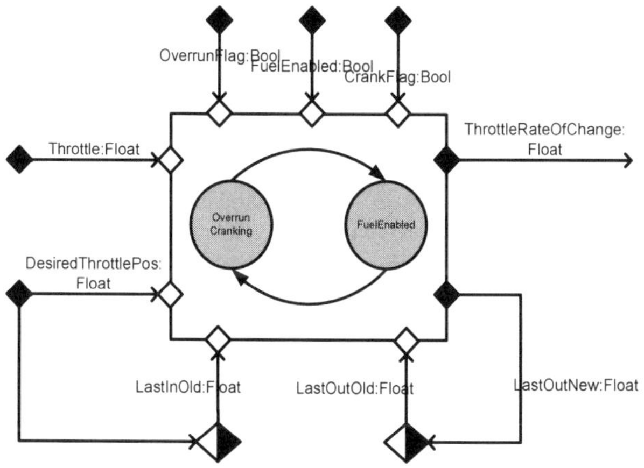

Figure 8: AutoFocus component, ThrottleRateOfChange, with an embedded MTD which consists of two states: "FuelEnabled" and "CrankingOverrun".

Modeling a subsystem such as "ThrottleRateOfChange" with MTDs and modes, on the other hand, separates the component into distinctive modes – even using distinctive name tags for those – which are modeled and viewed separately in the tool depending on the respective engine state (see Fig. 8). That is, an MTD design treats modes as first-class citizens while purely control-flow oriented modelling usually lacks the concept of mode representation.

In the case of ThrottleRateOfChange, the top-level MTD is attached to two separate DFD descriptions which, when executed, each consume incoming signals, *if* the mode being dealt with is currently active. Clearly, the calculation of the throttle position change rate is different in cranking mode compared to "normal" engine operation while driving. Consequently, the incoming Boolean flags for mode control can also be thought of as DFD trigger signals.

4.2 DATA- VS. CONTROL-FLOW-ORIENTED DESIGN

On the other hand, MTDs also provide a great amount of flexibility, since they do not impose any restrictions on the level of abstraction where modes can be used. Hence, MTDs may appear on the structural level inside SSD components, or in the FDA and LA/TA inside DFD blocks. Additionally, MTDs may be hierarchic, meaning one single mode may consist of further MTDs with additional modes.

It is then up to the systems engineer to decide whether MTDs should capture a more *data*-oriented or *controlflow* oriented view of the system.

A purely dataflow-oriented approach to MTDs would result in MTDs at very low levels of abstraction inside isolated DFD blocks. This essentially results in a high number of single MTDs and yields a greater effort in constructing a consistent global state model.

On the other hand, the purely control flow oriented way of using MTDs will impose only a small number of MTDs in top-level components, which then more or less represent the global state model of the system. The downside of this approach, however, is there is not much modularity and greater redundancy in the model, since a lot of the underlying computations would be made in more than one mode.

5. TOOL SUPPORT

The goal is to validate all concepts developed in AutoMoDe by at least some prototypical tools. An enhanced tool for specifications used in AutoMoDe is currently being developed called AutoFOCUS2, which is based on the existing AutoFOCUS framework. The tool includes a generic logic programming language interpreter for checking and manipulating specifications. Fig. 9 shows AutoFOCUS2's logic language interpreter with the model browser and two editor windows. The consistency check mechanism also includes rules to identify possible violations of consistency conditions.

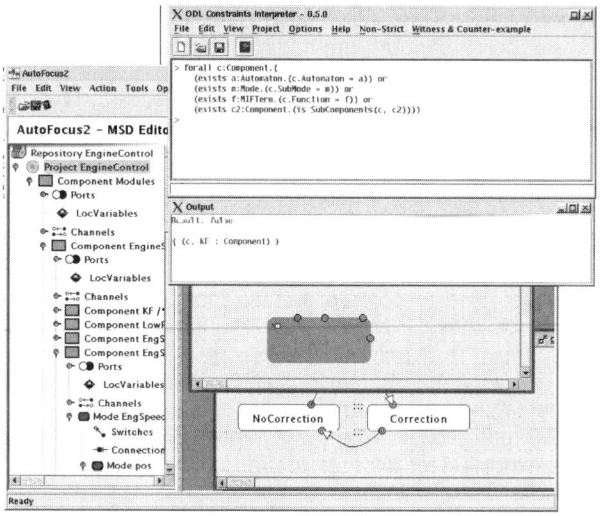

Figure 9: Screenshot AutoFocus2 tool.

The existing AutoFocus tool [HSE97] and the AutoFOCUS2 tool include further extensions, in addition to basic functionalities like simulation or code generation. A concept to describe the implementation of data types based on the experiences gained from ASCET was developed and included into the AutoFOCUS framework. This means that an abstract number data type like integer or real could be used in early specifications.

These data types will be refined in later phases, for example into a 16-bit integer or a standard fixed or floating point representation. AutoFocus supports simulation and type consistency mechanisms for abstract data types as well as implementation data types, and even for specifications using a mix of both.

The model often has to be re-factored when automotive control software is designed. For example, a component within one component hierarchy has to be moved into another hierarchy. Typical re-factorings are defined and implemented in AutoMoDe. For this reason the AutoFocus framework was extended by a generic language to specify model transformations.

A central part of the AutoMoDe tool prototype is the connection with ASCET. We show the practical applicability by integrating a currently used tool for automotive control software design. On the one hand, AutoMoDe models should be automatically transformed into ASCET models so that it is possible to further refine these models to generate code for different embedded platforms. On the other hand, ASCET models are converted into AutoMoDe models to support reengineering of existing models. Naturally, reengineering is the only tool supported, but it generally does not automatically produce reasonable results without user interaction. The integration with ASCET described here is currently ongoing research.

6. CONCLUSION

This paper has presented some of the early results of the AutoMoDe project, including a definition of abstraction layers, some support, methodically motivated transitions between layers, the graphical notations supported by AutoFocus tools, a case study from the engine management domain, and the status of tool support for AutoMoDe.

The complex relationships between single design artefacts of a typical automotive software design call for a rich set of possible structuring relations between artefacts. The AutoMoDe domain model currently lacks some of these necessary relationships, such as a mechanism for component typing, or for describing product variants. Extensions of the AutoMoDe domain model with simple typing and variant mechanisms, similar to the results of the Automotive project [BBRS03], are envisioned for the near future.

Obviously, the combination of a globally clocked operational model with typical event-triggered communication media such as CAN, which is not tightly synchronized, raises some interesting questions for research. We present in [RB04] a proposal on how to use event-triggered media for firm real-time deployment of globally clocked models with comparatively small implementation overhead. This topic will also be a subject of further investigation.

REFERENCES

1. [vdB94] M. v. d. Beeck: Comparison of Statecharts Variants. Third Int'l Symposium on Formal Techniques in Real-Time and Fault-Tolerant Systems (FTRTFT). pp. 128 – 148, 1994
2. [Bea99] A. J. Beaumont et al.: *Automation of ECU Software Development : From Concept to Production Level Code*, SAE-Paper 1999-01-1174
3. [BBRS03] M. v. d. Beeck, P. Braun, M. Rappl and C. Schröder. UML for Real: Design of Embedded Real-Time Systems, Automotive UML. In: Bran Selic, Grant Martin and Luciano Lavagno (eds), Kluwer Academic Publishers, ISBN 1-4020-7501-4, May 2003
4. [Thu03] Thurner, T., et al., The EAST-EEA project – a middleware based software architecture for networked electronic control units in vehicles. In: Electronic Systems for Vehicles (VDI Berichte 1789), p 545 ff. VDI-Verlag, Düsseldorf, 2003.
5. [Fre04] U. Freund et al.: The EAST-ADL: A Joint Effort of the European Automotive Industry to Structure Distributed Automotive Embedded Control Software. 2nd Workshop on Embedded Real-Time Systems, Toulouse 2004.
6. [ETAS] ETAS Engineering Tools GmbH. ASCET User Manual Version 5.0. 2004. ETAS GmbH.
7. [HSE97] F. Huber, B. Schätz, G. Einert. Consistent Graphical Specification of Distributed Systems. FME '97, LNCS 1313, pp. 122 - 141, Springer.
8. [IEE91] Another look at real-time programming. Special Section of the Proceedings of the IEEE, 79(9), September 1991
9. [Kop93] H. Kopetz. Should Responsive Systems be Event-Triggered or Time-Triggered? IEICE Trans. Inf. & Syst., Vol. E76-D (11), 1993
10. [LTS01] A. Lapp, P. Torre Flores, J. Schirmer, D. Kraft, W. Hermsen, T. Bertram, J. Petersen. Softwareentwicklung für Steuergeräte im Systemverbund – Von der CARTRONIC-Domänenstruktur zum Steuergerätecode. VDI-Berichte 1646 "Elektronik im Kraftfahrzeug". pp. 249-276. 2001
11. [MR98] F. Maraninchi and Y. Rémond. Mode-Automata: About Modes and States for Reactive Systems. Proc. European Symposium on Programming, Lisbon, Portugal, 1998
12. [PMS+95] S. Poledna, T. Mocken, J. Scheimann, T. Beck. ERCOS: An Operating System for Automotive Applications. SAE International Congress, 1995
13. [RB04] J. Romberg and A. Bauer. Loose Synchronization of Event-Triggered Networks for Distribution of Synchronous Programs. EMSOFT, Pisa, Italy, 2004

2005-01-0781

Formal Verification for Model-Based Development

Amar Bouali and Bernard Dion
Esterel Technologies

ABSTRACT

Formal verification is increasingly used for checking and proving the correctness of digital systems. In this paper, we present formal verification as a cost-effective technique for the verification and validation of model-based safety-critical embedded systems.

We start by explaining how formal verification can be easily integrated in a model-based development methodology for critical embedded software. In the methodology examined, the development methods are based upon a formal and deterministic language representation and a correct-by-construction automatic code generation. In this methodology, formal verification proves that what you execute conforms to safety requirements, and what you execute is exactly what you embed. We show the impacts and benefits of using formal verification in software development that must be compliant with the IEC 61508 standards, especially for SIL 3 and SIL 4 software development. We conclude by detailing specific formal verification techniques and tools available today for use in a state-of-the-art model-based development environment. We focus on the verification of safety requirements, involving control-logic aspects as well as data computation aspects of embedded software. We explain how to relate this model-checking activity with the objectives of the software life cycle process described in the IEC 61508 standards.

Figures on the use of the SCADE product and its Design Verifier module on several realistic safety-related automotive applications illustrate the presentation.

INTRODUCTION

Several industrial domains develop safety and mission critical systems (SMC), systems in which a failure during their functioning may induce high damages in terms of human lives or economics. This is the case for the industries of transportation, nuclear plants, medical devices, etc. Different international organizations and commissions work at defining the guidelines and certification criteria to ensure the maximum reliability and robustness for such SMC systems. The guidelines and certification criteria are defined on top of processes and objectives for the development of an entire industrial system. These systems increasingly embed electronic digital systems to control several of their parts, some being highly critical such as the control of the engine of a car or its braking system, the flight control commands of an airplane, etc. An important part of the guidelines is dedicated to the development process for the hardware and software components of the embedded electronic systems and its relation to safety. In this paper, we focus on the standards developed for the automotive industry. The International Electrotechnical Commission publishes the IEC 61508 international standard for the functional safety of electrical / electronic / programmable electronic safety-related systems (E/E/PES), which is relevant also for the development of automotive electric / electronic systems today, and which may in an adapted form, become a standard for the automotive industry soon.

THE IEC 61508 STANDARD

In this document, we assume that the reader is familiar with IEC 61508 end its terminology ([1]). The IEC 61508 standard is separated into 7 parts identified from Part 1 to Part 7. In this document we focus on Part 3, which is concerned with the software requirements of the standard, and whose scope:

- *"Applies to any software forming part of a safety related system or used to develop a safety related system [...]"*
- *"Establishes requirements for safety lifecycle phases and activities which shall be applied during the design and development of the safety-related software (the software safety lifecycle model)."*
- *"Provides requirements for information relating to the software safety validation to be passed to the organization carrying out the E/E/PES integration."*
- *"Provides requirements for the preparation of information and procedures concerning software needed by the user for the operation and maintenance of the E/E/PES safety related system."*
- *"Provides requirements to be met by the organization carrying out modifications to safety related software."*

- *"Provides requirements for support tools such as development and design tools, language translators, testing and debugging tools, configuration management tools."*

The software process model described in the IEC 61508-3 is illustrated in Figure 1.

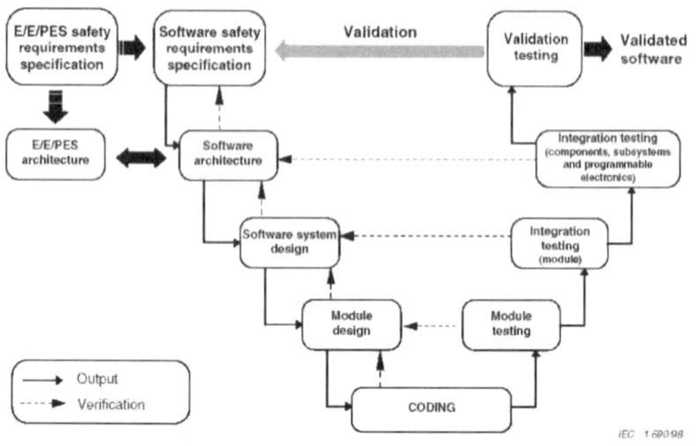

Figure 1: The IEC 61508 software process model

Model-based development is an emerging methodology for the development of safety-critical embedded software. Model-based development relies on the use of unambiguous formalisms for specifying embedded systems and brings techniques and tools to fulfill the objectives of the software process model at least at the specification phase. SCADE (Safety Critical Application Development Environment) is both a graphical notation for modeling critical embedded software and a suite of tools for the aerospace and automotive industries. SCADE is a developed and commercialized by Esterel technologies [7]. SCADE is known as a cost effective model-based development method relying on formal methods for the development of safety critical embedded software ([2]). SCADE methodology is fully compliant with the IEC 61508 standard. In this paper, we focus on the formal verification module of SCADE as a powerful technique for the verification of safety requirements that can significantly reduce the cost of verification.

The paper's presentation is organized as follows. The first section presents the SCADE development method and its formal verification module called Design Verifier; we explain the aim of formal verification and show the benefits of using Design Verifier in an IEC 61508 development context. The next section details how to perform formal verification with Design Verifier in SCADE and gives figures on the cost of usage and on its performances. The next section gives information about the formal verification technology used by Design Verifier.

FORMAL MODELING AND VERIFICATION WITH SCADE

THE SCADE DEVELOMENT PROCESS

SCADE is a model-based development method recognized as an efficient and cost-effective way to develop critical embedded software. SCADE relies on classical graphical notations for modeling block diagrams and state machines. SCADE is used for airborne software with DO-178B Level A objectives (see [2], [4], [5]) and in transport/automotive with IEC 61508 SIL 3/ SIL 4 objectives, for which SCADE is particularly well-suited as it is a formal method; its graphical modeling formalism benefits from deterministic formal semantics, allowing the derivation of a clean mathematical model from a SCADE design. The same deterministic model is then used for correct-by-construction automatic code generation and formal verification. There is no gap between the model Design Verifier uses for its analysis and the model that is executed at simulation time or on target. Thanks to this fundamental characteristic, the SCADE methodology offers a correct-by-construction development flow, which saves a high amount of effort by simplifying the verification and validation activities, easing the achievement of the objectives of these activities. In particular, the testing activities after code generation can be eliminated as all the verification and validation activities can be shifted to the model specification level.

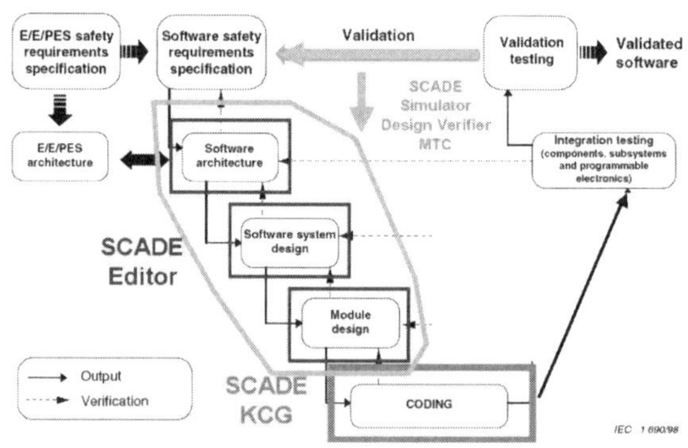

Figure 2: SCADE in the IEC 61508 software process

SCADE provides an original and powerful verification technique based on Formal Verification technologies. In this paper we show how this technique can further enhance the efficiency of verification and validation activities to fulfill the objectives of the IEC 61508 process model.

AIM OF FORMAL VERIFICATION IN SCADE

Formal verification of computer systems is a set of activities consisting of using a mathematical framework to reason about system behaviors and properties in a rigorous way. The recipe for formal verification is:

1. Define a formal model of the system; that is, a mathematical model representing the states of a system and its behaviors.

2. Define, over the formal model, a set of mathematical methods and tools to reason about the systems behaviors and properties.

The mathematical model for a SCADE description is systems of data-flow equations. Formal verification is meant to help during the phase of verification and validation of the requirements related to safety. Being rigorous, formal verification can significantly reduce the efforts spent during the software lifecycle.

A Simple Example

To introduce formal verification, let's consider a simple example, which counts the number of times some event has been seen as true. Figure 3 shows how these requirements are captured in the SCADE graphical language as a SCADE node called `CountEvent`.

The event is represented as an input flow called `Event`. The counted value is stored in the output flow `Count`, which is set to 0 at the initial cycle.

SCADE simulation lets you test and verify the correctness of the design. However, with testing, we are never 100% sure that your design is correct as we usually never test all the possible scenarios that may happened.

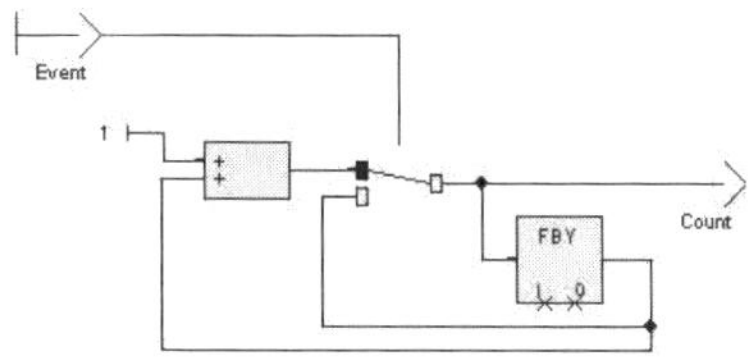

Figure 3: the CountEvent SCADE model

The set of equations that follows is the mathematical model of our SCADE design, where the parameter t stands for the discrete time clock variable:

$$\begin{cases} Count(0) = 0 \\ \forall t > 0,\ Count(t) = \begin{cases} Count(t-1)+1, & \text{if } Event(t) = true \\ Count(t-1), & otherwise \end{cases} \end{cases}$$

Given this mathematical representation, we can prove, in the mathematical sense, the following requirement property:

*(**Prop_P1**) For all cycle t, Count(t) is always greater or equal than Count(t-1)*

With such a proof, we are 100% sure that the design fulfils this property for all possible scenarios. As a result, we can say that we formally verify the correctness of the program.

Easy Access to Formal Verification

Expressing a property and finding a proof for a real system containing complex algorithms and control logic may require a high amount of time and expertise in mathematics. Hence, the big challenge of formal verification is to provide system engineers and software developers with an efficient, easy-to-use and friendly framework, which not only does not require a lot of time to use, but also enables increased confidence in the system.

For this challenge, SCADE brings a solution for easy access to formal verification for a wide range of users, with the following characteristics:

Property Expression: The SCADE language itself expresses properties. There is no need to learn a mathematical dialect to express the property requirements you want your design to fulfill.

Verification of a Property: This is a push-button feature of the SCADE tool, which basically provides and Yes/No answer. Moreover, in the case of a "No" answer, the tool lets you discover, in an automatic and user-friendly way, why a "No" answer was reached.

DESIGN VERIFIER IN THE SCADE PROCESS

The software module by which SCADE enables formal verification is called Design Verifier. Design Verifier helps the SCADE designer to check the correctness of SCADE specifications at different stages of the SCADE development process. Design Verifier helps detect specification and coding errors at the early phase of the software flow, minimizing the risk of discovering these errors during the final validation at the integration phase. The input to Design Verifier is a set of *safety properties* that have to be checked for correctness on the design. This set of safety properties is extracted and specified from the safety requirements.

Design Verifier Workflow

Figure 4 sketches the Design Verifier's workflow. It consists of successive tasks that may be iterated. There are three kinds of tasks:

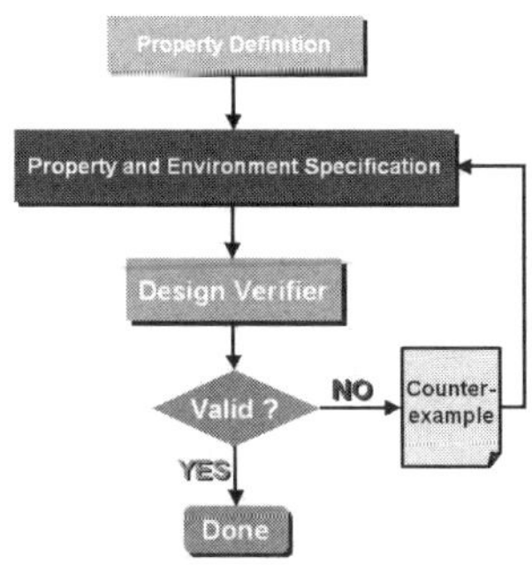

Figure 4: Design Verifier workflow

Property Definition: This task consists of extracting from the safety requirements to check with Design Verifier.

Property and Environment Specification: This task consists of formally describing, as SCADE observer properties, the safety requirement extracted as properties in SCADE. Necessary information from the environment of the design must also be specified formally in SCADE as well.

Design Verifier Execution: This task corresponds to the usage of Design Verifier.

Safety Property and Environment Specification

Safety property and environment specification means selecting the software safety and safety integrity requirements that have to be verified and validated and which Design Verifier can handle. For a given requirement, a property will state that something bad will never happen with respect to that requirement. We call such a property a safety property. Here are some examples of such safety properties:

- *The lift cannot move while the doors are open.*
- *In landing mode, if altitude is lower than 100, then the landing gear must be opened.*
- *If the brake pedal is pressed, then the cruise control regulation mode must be off within 100 ms.*

DESIGN VERIFIER BENEFITS

To achieve the verification and validation objectives of the software safety life cycle, IEC 61508 recommends performing code inspection and review, code analysis, and testing. Design Verifier adds cost effectiveness and efficiency in the verification and validation activities of the software safety life cycle. Also, formal verification can add efficiency in the communication between the safety life cycle and the software safety life cycle. Design Verifier detects specification and modeling errors very early in the development process. By proving the design correct with respect to the safety requirements, Design Verifier allows for higher confidence in the developed software. We detail the benefits of Design verifier in the different phases of the software process:

Benefits in the Safety Requirements Specification Phase

Table A-1 of IEC 61508-3 recommends the use of a formal notation for specifying the software safety requirement, especially for SIL 4. In this table, R and HR means recommended and highly recommended respectively.

Technique/Measure	SIL 1	SIL 2	SIL 3	SIL 4
Computer-aided specification tools	R	R	HR	HR
Semi-formal methods	R	R	HR	HR
Formal methods	-	R	R	HR

Design Verifier uses the SCADE notation to formally specify the safety requirements, hence fully complying with the recommended techniques. When testing is a recommended technique for some objective, IEC 61508 stipulates: "*Where the development uses formal methods (see C.2.4 of IEC 61508-7), formal proofs (see C.5.13 of IEC 61508-7) or assertions (see C.3.3 of IEC 61508-7), such tests may be reduced in scope.*"

Benefits in the Software Module Testing Phase

The objectives of the software module testing are "*to verify that the requirements for software safety (in terms of the required software safety functions and the software safety integrity) have been achieved -- to show that each software module performs its intended function and does not perform unintended functions*".

We propose Design Verifier as a formal verification technique that significantly reduces the scope of the tests needing to be written to verify that the software modules comply with the safety requirements. Design Verifier can be particularly useful for verifying safety properties related to boundary value analysis and control flow analysis.

Benefits in the Software Integration Testing Phase

The objectives of the software integration testing are "*to verify that the requirements for software safety (in terms of the required software safety functions and the software safety integrity) have been achieved – to show that all software modules, components and subsystems interact correctly to perform their intended function and do not perform unintended functions*".

We propose Design Verifier as a formal verification technique that significantly reduces the scope of the tests needing to be written to verify that the integrated modules interact correctly, preserving the safety requirements.

Benefits in the Relationships with other Activities

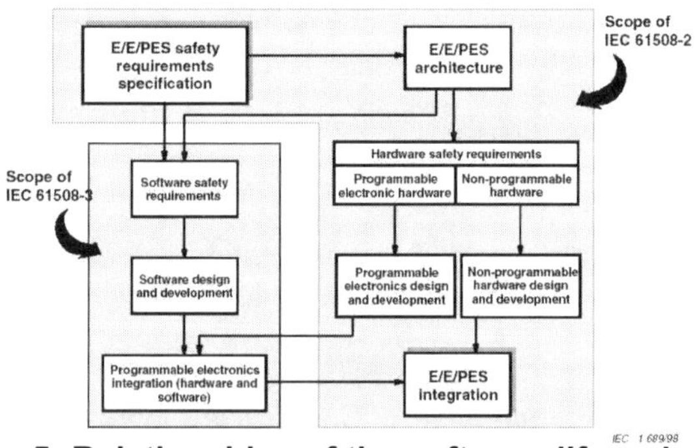

Figure 5: Relationships of the software lifecycle with other E/E/PES activites

Figure 5 illustrates the relationships between the software lifecycle and the other E/E/PES development activities. Design Verifier helps in reducing the validation efforts when considering the verification of the integration of the software with the hardware by contributing in delivering more reliable and trusted software.

Benefits in the Costs of Development

Figure 6 shows the estimated gain of using Design Verifier in the SCADE software process. This estimation has been made from numbers extracted from several evaluation and projects at customers.

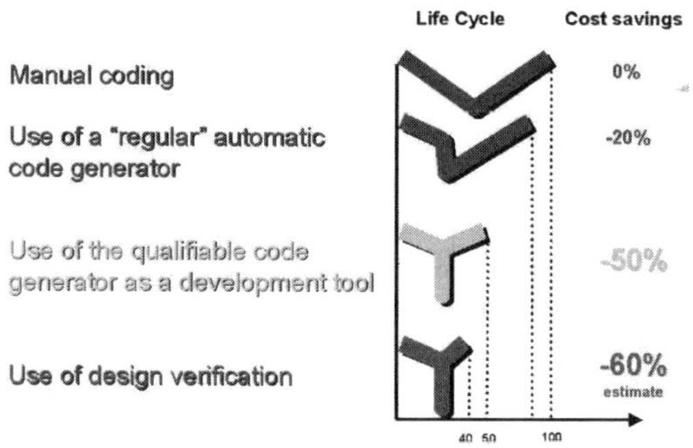

Figure 6: Costs savings with SCADE and Design Verifier

WORKING WITH DESIGN VERIFIER

VERIFICATION BY MEANS OF OBSERVERS

The Principle of SCADE Observers

An observer is a SCADE node that has as inputs, input and output flows from a design, and has as output, a single Boolean flow. An observer node is meant to model a property. Figure 7 illustrates the principle of verification of a property modeled as an observer node, which observes the input and output of a SCADE design and implements the condition expressed by the property. The observer produces a Boolean output flag that should always be true.

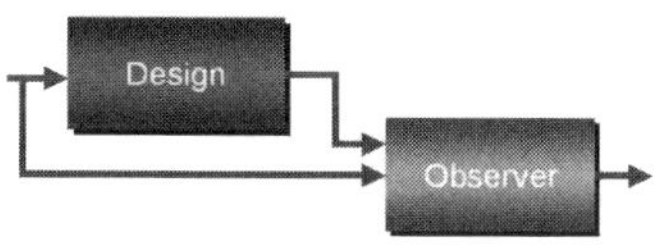

Figure 7: SCADE observer principle

Example of Properties

The properties that can be modeled as a SCADE observer are called safety properties. Safety properties aim at expressing that something bad will never happen. The notion of "bad" is related to the safety requirements of the system.

Here are some examples of Properties:

Prop1 = (Lift_Motor_On = true) implies (Doors_Are_Closed = true)

This property expresses that if *Lift_Motor_On* is true then *Doors_Are_Closed* must be true. This property is modeled as SCADE observer as shown below.

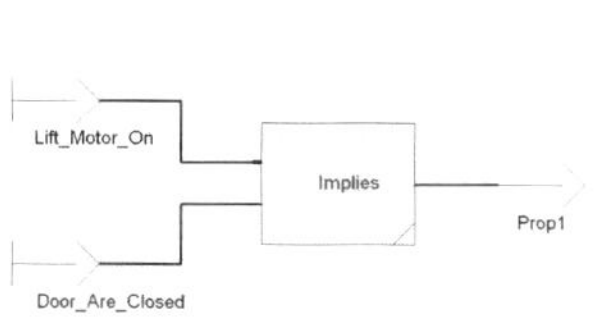

This SCADE node uses a node called "Implies" implementing the logical implies Boolean operator. This node is available in a standard SCADE library "libverification".

Prop2 = (Water_Level > 100 during 5 cycles) implies (Alarm_On = true)

This property expresses that if the water level is greater than 100 during 5 cycles then the alarm must be on. This property is modeled as a SCADE observer as shown below.

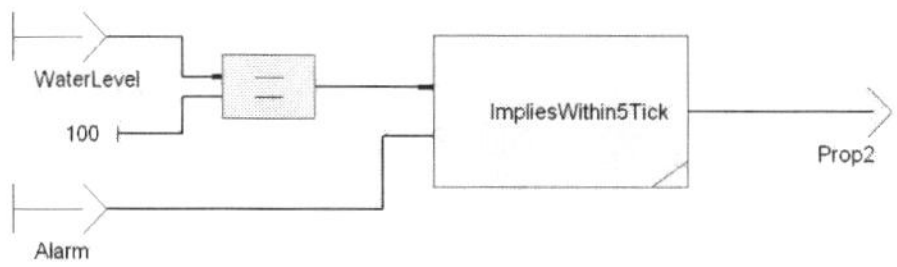

This SCADE node uses a node called "ImpliesWithin5Tick", which is available in a standard SCADE library libverification.

Prop3 = (Sensor – pre(Sensor)) > Threshold) implies (SensorValid = false) after 2 cycles

This example considers a sensor "*Sensor*". It compares the delta difference between the current value of *Sensor* and the previous value of *Sensor* with a threshold value *SensorThreshold*. If the difference is greater than this threshold, the sensor must be considered as not valid. This property is modeled as a SCADE observer as shown below.

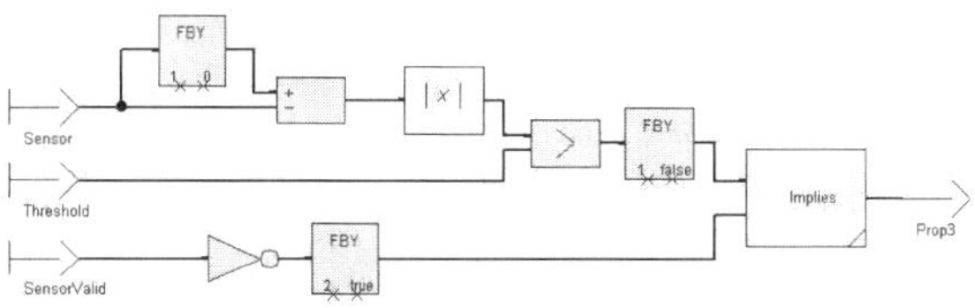

As illustrated in the previous section, a property is modeled as SCADE node. Such a SCADE node is called an "*observer*", as it observes the inputs and outputs of a design model producing a false flag whenever some condition is not respected.

For example, property *(Prop_P1)* can be expressed as the following SCADE node, called `Prop_P1`:

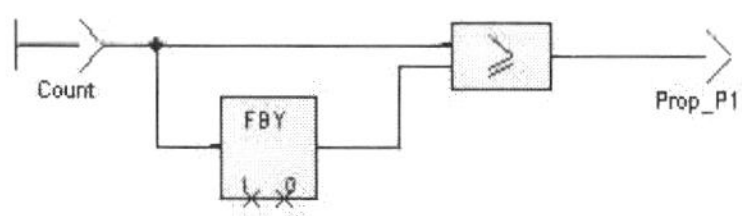

This property observer test the value of a flow called `Count` received as input. It produces a Boolean flow called `Prop_P1` that is true if the value of `Count` is greater or equal than the previous value of `Count`, and is false otherwise.

To be able to verify the property, we have to connect the observer node to the design node as illustrated below:

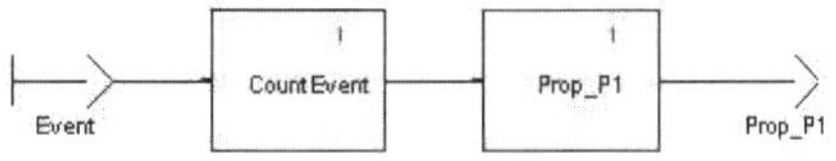

Verifying the property involves merely asking Design Verifier whether the Boolean output flow `Prop_P1` is always true or not by a simple button click.

Scope of Design Verifier Properties

An embedded system usually has two kinds of computational parts:

Continuous Control Computation: This corresponds to the algorithms implementing the control laws of the system. This part is mostly data-flow computing mathematical functions.

Discrete Control Logic Computation: This part corresponds to the logical aspects of the design such as mode management, alarm handling, and communication protocols. This part essentially involves Boolean computation and a few arithmetic operations.

1. The first kind of computation is usually verified through intensive testing and validated by simulation. These parts form the core knowledge of the company that develops it and are re-used in different projects. Very few errors are located in the algorithms themselves.

2. The second computation is more delicate for designers because it is first hard to model and implement, and then even harder to verify. The main reason is that this part involves a high combination of cases and situations and there is no efficient way to test it, as it requires a lot of time and effort to cover the high number of cases. Our experience shows that, even though this part tends to represent 20% of the overall design, it is likely to contain 80% of the defects and errors introduced either during the specification phase or during the software implementation phase.

SCADE helps simplify the design of these logical parts with appropriate language constructs to capture the design specification efficiently and accurately. The goal of Design Verifier is to help efficiently verify and cover the verification and validation of the parts of the design of the second kind of computation.

THE ENVIRONMENT SPECIFICATION

Most of the time, performing a verification of a property requires some additional information about the environment of the design under analysis. The situation is as follows: If a property is found valid, even with no information about the environment, then it is valid whatever the environment is. If a property is found falsifiable, then the generated counter-example may correspond to unrealistic scenario; this can happen when Design Verifier lacks information about some values from the environment, which causes Design Verifier to choose non-realistic values. For instance, suppose you want to verify a property involving a temperature value in a car or in a plane cockpit. If no information is given about the physical range of this value, Design Verifier can set this value to –1000000°C, which is of course a non-realistic value. Hence, it may be necessary to inform Design Verifier about the fact that the temperature is always between 15°C and 35°C.

This kind of environment specification is fundamental for Design Verifier. First, it prevents Design Verifier from considering unrealistic values in the possible scenarios of the design. Second, it helps reduce the state space search making the overall verification process more efficient.

The accuracy of the environment specification depends on the property. It corresponds to the amount of information needed to verify the property. A priori, there is no indication of the amount of information needed to verify the property. The procedure is an iterative procedure consisting of refining the environment characterization as long as Design Verifier generates unrealistic counter-examples. The procedure stops as soon as the property is found valid, or a counter-example corresponds to a possible scenario. See [3] for a practical example capturing the environment on a real-life industrial case.

CONFIGURING AND LAUNCHING DESIGN VERIFIER

Given a property and its environment described in SCADE, Design Verifier can decide whether or not this property is valid. The property is valid if it is true for every possible sequence of inputs. In case the property is not valid, Design Verifier can produce a counter-example that demonstrates how the property can be falsified. The problem of deciding the validity of a property is a complex one; in some cases the analysis can be very lengthy, or the validity of the property is simply not possible to decide. In most cases, Design Verifier will detect such impossibility and raise an error for the property. To give a precise definition of what properties are to difficult to handle is not trivial. On a large scale system, normally the best way to find out is to try the property and let Design Verifier decide if it is too complex or not. Once properties and their environment are specified, the user may attach a specific strategy to a property, either a Debug strategy or a Proof strategy. If none is attached, Design Verifier uses a default Proof strategy.

The user launches Design Verifier by a simple button-click. The user is kept informed about the progress of Design Verifier during the analysis of the properties to verify. At the end of the analysis, a report gathering all the results is generated.

Outcome of the Analysis

The possible outcomes of Design Verifier's analysis of a property are:

- **Valid**: The property is true for every possible sequence of values assigned to the input of the design.

- **Falsifiable**: There is a sequence of values that falsifies the property.

- **Indeterminate**: The strategy has detected that the given property cannot be decided by the strategy and terminates.

- **Non-convergence**: The property is not decidable by the strategy, but the strategy does not detect this; it will continue forever.

Falsifiable Result
There exists a counter-example scenario that falsifies the property. Design Verifier generates such a scenario for use in the Simulator for debugging. By playing the scenario, one of the following situations may be revealed:

- There is a bug in the model. The counter-example helps in understanding the root cause of such a bug and in fixing it.

- There is a problem in the specification. The counter-example reveals an ambiguity in the specification leading to a contradiction in the design. The counter-example helps in recovering the error in the specification.

- The counter-example scenario falsifies the property with unrealistic values for the inputs of the design. This helps in adding constraints in the environment specification to restrict the possible values of the inputs in order to avoid Design Verifier generating unrealistic scenarios and eventually proving the property.

- The property is badly specified. Some errors might be introduced in the specification of the property itself.

Indeterminate Result
There are several reasons why Design Verifier may return an indeterminate result. We briefly list the different cases:

- Design Verifier has reached some user-defined timeout or bound in the depth of the analysis.

- The user stopped the analysis.

- An error occurred during the analysis, due for instance to the presence of a *non-linear* expression in the design that cannot be handled by Design Verifier. Section "Design Verifier Technology" (p. 9) details the technical background of such a case.

Design Verifier provides mapping groups as a way to prove properties for which we obtain an indeterminate result.

Non-Convergence
In practice, of course, in the non-convergence case the strategy will eventually be stopped by the user, or after a given timeout. Note that a lengthy analysis is not necessarily due to non-convergence; it may simply be a problem and convergence towards valid or falsified is very slow.

In case of non-convergence or an indeterminate result, there are ways to work around the incompleteness of the strategy by modifying parameters, changing strategy, or by changing the problem formulation.

Tuning the Strategies

At the early stage of design, we expect that the design contains errors. It is recommended then to use Debug strategies to chase errors and bugs in the design very efficiently. Once no more counter-examples are found with the Debug strategies up to some bound and the confidence in the design gets higher, the Proof strategy can be used to prove the property.

There are some parameters that can be tuned for the Debug and Proof strategies, summarized in the following table.

Option	**Debug**	**Proof**	**Usage**
Integer size (bits)	✓	✓	Declare a finite range for integers. Reduce the space search and making the analysis more efficient.
Depth (cycle)	✓		Limit the depth in number of execution cycle of the analysis.
Timeout (sec)	✓	✓	Limit the time for the analysis.

Using Mapping Groups

A mapping group is a substitution mechanism consisting of replacing a constant or a node of your design by another constant or node one for the purpose of property verification by Design Verifier, with the following constraint:

- Replacing constants should have the same type as the replaced constant.

- Replacing nodes should have exactly the same interface as the replaced node.

Typically, imported constants and function nodes may be replaced with a mapping group. Also, the complex SCADE nodes containing non-linear expressions may be replaced through a mapping group. For example, we can replace a trigonometric function like *sin* by a node that computes a linear form of *sin* using a split-case structure.

FACTS AND FIGURES

Costs of Usage

We give a series of numbers about the costs implied by the usage of Design Verifier for newly trained users and expert users. These numbers give the amount of effort spent to achieve the verification of different kinds of properties, from simple ones related to a simple SCADE module to more complex ones at the system integration level. The numbers have been extracted from several evaluations of Design Verifier in real projects and are reported in the following table:

	Simple		**Medium**		**Complex**	
	# Iter	time	# Iter	time	# Iter	time
Trainee	3	0,5h to 1h	4 to 5	1 to 2 days	7+	-
Expert	1	0,25h to 0,5h	1 to 2	2h	3	6h to 5 days

This table contains two lines. The Trainee line corresponds to numbers extracted from the workload of a beginner user, newly trained to Design Verifier and knowledgeable in SCADE. The Expert line corresponds to an expert user in Design Verifier, which means a person who has some experience in Design Verifier usage in real projects. The table contains three columns, Simple, Medium, and Complex:

- The **Simple** column is related to the effort spent for simple properties regarding a single SCADE node. The property has no, or very few, timing dependencies.
- The **Medium** column reflects the effort for properties involving complex timing constructs and several modules connected together.
- The **Complex** column is related to a complete system made up of several sub-systems communicating through communication networks. The effort includes the need for adding modeling parts in the design to model the communication media and the degree of asynchronous behavior resulting from the communication operated on the used communication networks.

Each main column contains two sub-columns. The first gives the number of iterative steps of the workflow iteration described in Figure 4. The second gives the corresponding time spent to complete the work. This table shows the relatively low work effort compared to the testing activity for achieving successful results with Design Verifier, even for beginners. Moreover, the learning curve if relatively short and users quickly become experts.

Performances

The table below gives some figures obtained by applying Design Verifier on several properties of different designs from customers

Name	**Size**	**#Prop**	**Time**	**Comment**
flap	1905	3	5	Properties containing arithmetic expressions and time operators. All found valid.
Sensor voter	3454	13	1335	Properties verifying the correctness of a sensor voter algorithm, including a model of an environment and fault-injection. See [3] a paper describing this case-study.
Blink	1610	1	< 1	Falsifiable property of length 82 cycles, confirmed by the customer.
Fcs	4267	1	< 1	Valid
fcc	N/A	19	3600	16 Boolean logic properties, 3 having numerical aspects. 2 specification errors found.

Here is the meaning of each column:

- **Name**: The name of the design
- **Size**: The size, when available, is given in number of lines of textual SCADE code.
- **#Prop**: The number of properties to verify in the design.
- **Time**: The overall time in seconds spent by Design Verifier to analyze all the properties of the design.
- **Comment**: additional information about the design case

DESIGN VERIFIER TECHNOLOGY

THE CONE OF INFLUENCE OF A PROPERTY

Roughly, the code of influence (COI) of a property is the structural part of the design, which the property depends on. As an illustration, consider thc following schematic design example.

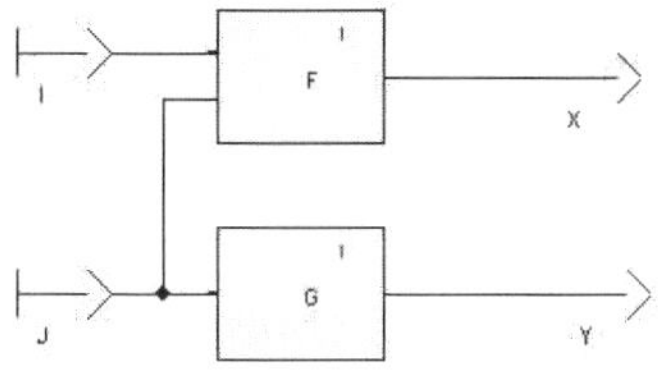

The design has two input I and J, and two outputs, X and Y, as the result of two sub-nodes F and G respectively. Figure shows an instance of the Design node connected to an observer node Prop, which takes as input the input I and the output X from the design and produces an output called Prop. The COI of the observer property is then the structural part of the design concerned with the elements involved in the property, that's, the output X, the sub-node F that computes X, and the inputs I and J. The rest of the structure of the design, that is the output Y and the sub-node G, does not belong to the COI of the observer property.

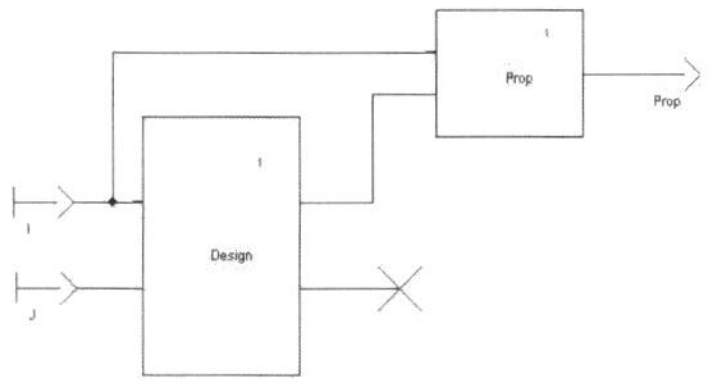

The computation of the structural COI of a property is a straightforward and easy operation. Before starting an analysis, Design Verifier computes the COI to get remove the part of the design that has no influence in the property under analysis

THE ARITHMETIC SUBSET HANDLED BY DESIGN VERIFIER

Design Verifier fully supports the SCADE language. However, there are properties that cannot be decided by the Design Verifier. For these properties, the Design Verifier returns an *indeterminate* result. The following section describes under what circumstances such properties can arise, and presents workarounds that can be used to avoid them. It is established that solving arithmetic problems on integers and real numbers in general is not decidable. That means that there is no way to define an algorithm that is able to decide whether a formula containing conditions expressions over real number and integer number variables is true or not. Nevertheless, there are decidable subsets of mathematical problems over real numbers and integer numbers, which Design Verifier supports. These subsets are known as linear arithmetic for integer number and linear arithmetic for real number problems. Surprisingly, problems on integer numbers are harder than problems on real numbers, so that the linear arithmetic for integer numbers is more restrictive than linear arithmetic for real numbers.

Design Verifier will produce an *indeterminate* result for a property that has in its COI some arithmetic expression not belonging to the supported decidable subset. We detail hereafter this subset.

Linear Arithmetic for Integer Numbers

In the linear arithmetic for integer numbers, we are allowed to use only addition and subtraction of integer variables. Multiplication is allowed only if one operand is a constant such as $3 \times X$, which is in fact a short-hand for $X + X + X$. Non-linearity appears as soon as division or modulo division is used, or when using a non-linear mathematical function such as exponential, square root, trigonometric functions, etc.

Linear arithmetic		**Design Verifier**
Addition (+)	Yes	Fully supported
Subtraction (-)	Yes	Fully supported
Multiplication (x)	Partially	One operand must be evaluated as a constant
Division / Modulo	No	

Linear Arithmetic for Real Numbers

The linear arithmetic for real numbers is the set of expressions of the form $A \times X + B$, where A and B are rational numbers. Non-linearity appears as soon as an expression contains a product or a division of two or more variables. Moreover, like for in the integer case, an expression is non-linear whenever it contains a non-linear mathematical.

Linear arithmetic		**Design Verifier**
Addition (+)	Yes	Fully supported
Subtraction (-)	Yes	Fully supported
Multiplication (x)	Partially	One operand must be evaluated as a constant
Division / Modulo	Partially	One operand must be evaluated as a constant

DESIGN VERIFIER'S PROOF ENGINES

Design Verifier is built around Proof engines developed by our technology partner Prover Technology [8]. The proof engine implements algorithms and strategies for checking if some logical formulation in SCADE is always true. The engine contains:

- A satisfiability solver (SAT) for solving Boolean formulae.

- A sequential induction scheme based on SAT to analyze a property on all the execution cycles of a design.

- Decision procedure and constraint solving algorithms to deal with the arithmetic on integer and real numbers.

Details on the SAT-based technology can be found in [6]. Design Verifier offers two families of strategies for the verification of properties, called Debug and Proof strategies.

THE STRATEGIES AND ALGORITHMS

The Debug Strategies

A debug strategy is a partial strategy, which will only try to falsify a property; hence, it cannot be used to prove a property as valid. A debug strategy has a parameter *depth*, which is an integer number. The strategy will search for counter-example of a length less than or equal to the depth parameter. If there is no such a counter-example that falsifies the property, the debug strategy will terminate and give the result *indeterminate*. In this situation there are two possible ways to proceed:

- Increase the depth parameter of the debug strategy to look for a longer input sequence, which may enable the strategy to falsify the property.

- Switch to a proof strategy, since the property may in fact be valid, and a debug strategy can never detect this, but a proof strategy can.

The Proof Strategies

The proof strategies are used to prove that a property is valid. Proof strategies apply different algorithms depending on the data types involved in the COI of a property to prove, essentially Boolean values, integer numbers, or real numbers. Design Verifier adopts divide-and-conquer strategies for the analysis of properties. For each type of data, that is Boolean, integer, or real, there are specific algorithms to deal with them. These algorithms do not have the same cost, and some are fairly expensive. Design Verifier tries to use the most effective techniques first on a subset of the data in the COI of a property. If using this subset and these techniques is enough, then we are done obtaining a fast response. Otherwise, Design Verifier uses the more complex techniques taking into account more data from the more complex part.

Arithmetic

When dealing with the arithmetic part of a design, Design Verifier performs the following steps:

1. Try proving the property only considering the constructs of the design that can be decided and ignoring the rest of the design.

2. If the property is valid we are done, or else we have a partial counter-example.

3. Try to extend the counter-example for the non-linear arithmetic ignored part.

4. If the extended counter-example is a good one, then terminate with this counter-example and a falsifiable result, or else go to step 5.

5. Terminate with an *indeterminate* result.

The reason that we terminate with an *indeterminate* result if the extension fails is that there are potentially infinitely many counter-examples, so, in many cases, such a strategy would loop forever.

Constraint Propagation

While analyzing a property, Design Verifier propagates constraints on the data values. These constraints correspond to the environment constraints specified by the user. Design Verifier can deduce new constraints from these constraints for other data. For instance, if the user specifies that a height `H` coded as an integer number is always in the range [0:10], and if we have the following equalities in a design:

```
G = 10
V = 2*G*H
if (V > 300) then Flag = true else Flag = false
```

By propagating the knowledge on `H`, we obtain that `V` is in the range [0:200]. This makes Design Verifier deduce that `(V > 300)` is always false, hence Design Verifier deduces that `Flag` is always false.

If the data involved in a property are integers defined within some range or within some bit size (e.g. 8, 16, or 32 bits), then the proof strategy always gives a Valid or Falsifiable result.

CONCLUSION

In this paper we show the motivation of adding formal verification activities in model-based development. We highlight the clear benefits when using an environment such as SCADE, which is equipped with a specification language using formal semantics, and with a qualifiable code generator, and a formal verification tool called Design Verifier.

Design Verifier helps in gaining efficiency when performing the verification and validation activities as described in the IEC 61508 standards for the software life cycle of E/E/PES. For the safety requirements that can be expressed as formal safety properties, Design Verifier has a clear advantage compared to testing as there is no need to write tests to verify a particular property. Moreover, Design Verifier performs an exhaustive search of the design execution space to check a given property, offering a 100% confidence when a property is proved valid.

REFERENCES

1. IEC 61508:1998 Functional safety of electrical / electronic / programmable electronic safety-related systems (Parts 1-7)
2. Bernard Dion, *Correct-By-Construction Methods for the Development of Safety-Critical Applications*, SAE World Congress 2003, Paper Number 04AE-129.
3. Amar Bouali, Darren Coffer, and Samar Dajani-Brown: *Formal Verification of an Avionics Sensor Voter*, to appear in Proc. of the international conference on Formal Techniques for Real-Time and Fault-Tolerant Systems (FTRTFT), Grenoble, France, Sep. 2004.
4. Jean-Louis Camus and Bernard Dion: *Efficient Development of Airborne Software with SCADE Suite*, Esterel Technologies, 2003
5. François Pilarski: Cost effectiveness of formal methods in the development of avionics systems at Aerospatiale, 17th Digital Avionics Conference, Nov. 1st-5th 1998, Seattle, WA.
6. Mary Sheeran, Satnam Singh, and Gunnar Stålmarck: *Checking safety properties using induction and a SAT-solver*, in Proc. Formal Methods in Computer Aided Design (FMCAD 2000), Springer LNCS, Nov. 2000.
7. Esterel Technologies, http://www.esterel-technologies.com
8. Prover Technology, http://www.prover.com

CONTACT

Amar Bouali (Amar.Bouali@esterel-technologies.com). Amar Bouali is Deputy CTO at Esterel Technologies (http://www.esterel-technologies.com), heading the Embedded Software Expert Group.

Amar Bouali received a PhD in computer sciences from University of Paris 7. Before joining Esterel Technologies, Amar Bouali spent 10 years in academic research activities, working in formal methods for embedded systems at INRIA and Ecole des Mines.

2005-01-0700

Model Reduction for Automotive Engine to Enhance Thermal Management of European Modern Cars

C. Garnier, J. Bellettre and M. Tazerout
Ecole des Mines de Nantes

R. Haller and G. Guyonvarch
Valeo Climate Control Inc.

ABSTRACT

This paper focuses on the prediction of thermal losses and indicated performance in modern automotive engines. In a previous study, a complete simulation software was developed in order to both predict the car cabin blown air temperature and simulate the fluid circuits temperature. The two-zone, 0-dimensionnal combustion model presented in this paper aims to enhance this software. Theoretical overview reveals that thermal losses can be deduced from a predictive correlation of indicated performance. This correlation is established with a statistical tool and empirical coefficients are proposed. As a result of this study, the simulation software becomes a real-time computing tool that considers variable parameters previously neglected.

INTRODUCTION

Efficiency of combustion in automotive engines has significantly improved in the last decade, decreasing both fuel consumption and pollutant emissions. High-pressure injection systems optimize the engine performance, and reduce the thermal losses in the transfer to the coolant loop. In the case of a cold start, the drop is particularly pronounced. The equipment suppliers now use Additional Heating Systems (*AHS*) in order to make up for this thermal deficit and to achieve an acceptable comfort level for passengers. These systems can be electrical, mechanical or chemical.

Equipment suppliers design software that are able to accurately simulate the engine behaviour during the warm-up phase. In this way, Pirotais [1,2,3] developed a simulation software able to predict the heater core output air temperature. This software also considers atmospheric conditions and transient operations. The software simulates a typical driving cycle to predict the fluid circuit overtime considering both electrical and mechanical suppliers. The nodal method is used to describe the structure and the heat transfer between the car components. The architecture of this tool is presented in Fig. 1.

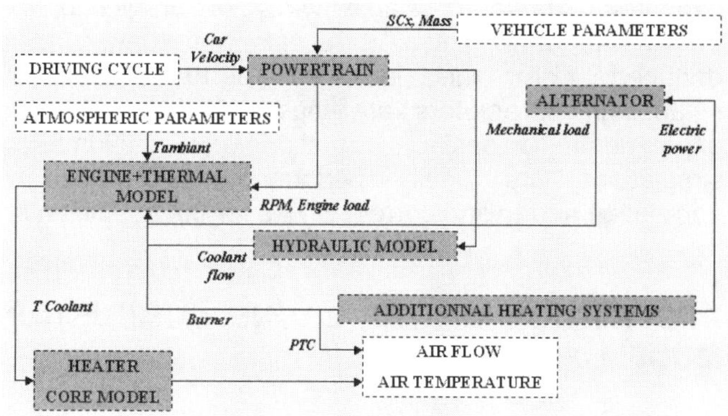

Fig.1 - Simulation software structure [2]

The aim of this tool is to predict the efficiency and consumption of the *AHS* simulated by additional power injected in the appropriate fluid circuit. Technical designers thus have an evaluation of the effects on the passengers' comfort. The first step of the simulation process uses a "1-zone" combustion model to plot the engine heat losses cartographies, used as data for the software (Fig. 2).

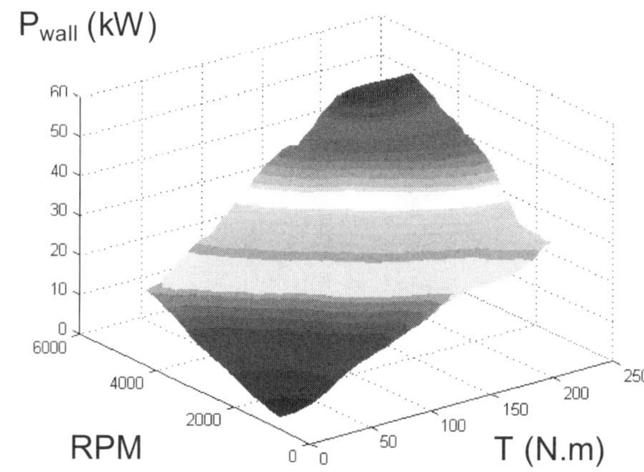

Fig.2 - Heat losses cartography from the single zone combustion model [2]

This pretreatment obviously requires time for a sufficient number of simulations representing the entire range of running conditions. It thus considers a uniform wall temperature and assumes that all other parameters are constant. If the injection timing, injection diagram, EGR rate or other parameters were included, the number of simulations would quickly become impossible to realize.

The purpose of this study is primarily to enhance the software with a “2-zone” combustion model. This model is used to simulate the cylinder pressure diagrams and indicated performance. Secondary an empirical correlation is developed to directly obtain the heat losses from the adapted engine parameters. The calculation of cartographies are thus unnecessary, the software computes all engine combustion cycles.

The first part of this paper describes the combustion model and hypothesis used to represent the heat transfer towards the coolant loop. The model is then validated from experimental data performed on a laboratory direct injection diesel engine. A theoretical overview exposes possible ways of predicting the engines thermal losses. It reveals that they can be deduced from the indicated performance. The influencing parameters are thus determined by the experimental design methodology. Numerical simulations are finally performed to propose a correlation and to find correct values for the coefficients.

DESCRIPTION OF THE 2-ZONE COMBUSTION MODEL

GENERAL OUTLINE

Contrary to the single zone combustion model, the 2-zone model considers chemical reactions and consequently pollutant generation. The chamber is crossed by a front flame, which divides the mass of gases into two parts, the burned and unburned zones. Considering a homogeneous cylinder pressure, each zone is characterised by a temperature, a volume and chemical composition (Fig. 3).

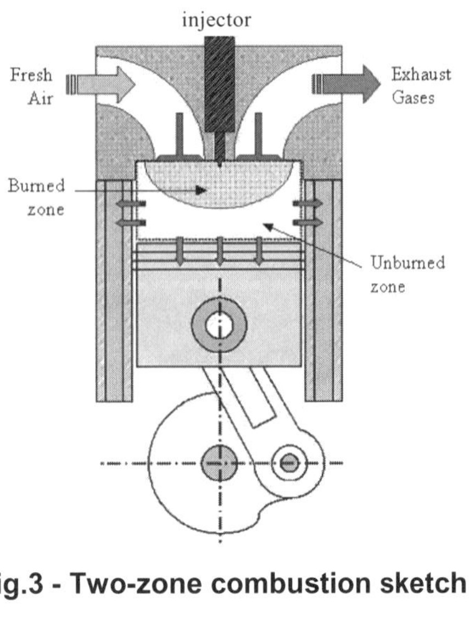

Fig.3 - Two-zone combustion sketch

During admission, compression and exhaust phases, a homogeneous gas is at work. Its composition should be constant after combustion. The chemical reactions, which may appear in the expansion phase and after, are neglected as explained in [5].

GOVERNING EQUATIONS

The energy conservation applied to each zone gives the differential equations of burned and unburned zones temperature.

$$\frac{dT_u}{d\theta} = \frac{\left(-P\frac{dV_u}{d\theta} - \frac{dQ_{w,u}}{d\theta} - u_u\frac{dm_u}{d\theta} - h_u.m_{cyl}\frac{dx_b}{d\theta}\right)}{m_u.c_{v,u}} \quad (1)$$

$$\frac{dT_b}{d\theta} = \frac{\left(-P\frac{dV_b}{d\theta} - \frac{dQ_{w,b}}{d\theta} - u_b\frac{dm_b}{d\theta} + h_b.m_{cyl}\frac{dx_b}{d\theta}\right)}{m_b.c_{v,b}} \quad (2)$$

where T represent the temperature, V the volume and P the cylinder pressure of burned (b) and unburned (u) gases. Q_w is the heat loss through the cylinder walls. m, h and u are the mass, enthalpy and internal energy per mass respectively. The burned fraction x_b is described below.

Thermodynamic properties of the gases as specific heat and molar enthalpy expressions are temperature dependent according to results of Kee *et al.* [4]. Other correlations between variables are determined by the ideal gas law and mass and volume conservation (Eq. 3 to 5).

$$P = \frac{m_u.R_u.T_u + m_b.R_b.T_b}{V_{cyl}} \quad (3)$$

$$V_u + V_b = V_{cyl} \quad (4)$$

$$m_u + m_b = m_{cyl} \quad (5)$$

The burned fraction x_b describes the combustion process. Several laws are available in the literature. The most appropriate was developed by Wiebe and used by Tazerout [5], Rousseau [6] and Bilcan [7] on different kind of engines. The ratio between the burned and global masses is seen in Eq. 6.

$$x_b = 1 - \exp\left(-a_w.\left(\frac{\theta-\theta_0}{\Delta\theta}\right)^{M_w+1}\right) \quad (6)$$

This equation needs 4 parameters for evaluation. a_w is fixed at 6.908 for a final burned fraction of 0.999 [7]. θ_0 represents the start of combustion and $\Delta\theta$ the combustion duration, expressed in crank angle degrees. The form factor M_w describes the energy distribution during combustion: Gaussian one if equal to unity. In the present case M_w = 0.9.

Ignition Delay

Ignition Delay (*ID*) is defined as the time between start of injection and start of combustion. Several predictive correlations are described in the literature. They are also validated on different kind of engines and running conditions. Assanis *et al.* [8] developed a correlation based both on steady state and transient operations for direct injection diesel engines. That is applicable to the present study because of the interest in cold-starting conditions.

τ_{ID} depends on the equivalence ratio, the mean temperature *T (K)* of the homogeneous zone and the pressure *P (bar)* over the ignition interval. So, to calculate this correlation (Eq. 7) a computed cycle or an accurate estimation of the cylinder pressure at the beginning of the combustion is required.

$$\tau_{ID} = 2{,}4.10^{-3}.\phi^{-0{,}2}.\overline{P}^{-1{,}02}.\exp\left(\frac{E_a}{R_u.\overline{T}}\right) \quad (7)$$

where E_a/R_u is held constant at a value given by Watson (1980) and used by Assanis *et al.* [8]. The combustion starts when the condition described in Eq. 8 is reached.

$$\int_{t_{inj}}^{t_{inj}+\tau_{ID}} \frac{dt}{\tau_{ID}(t)} \approx 1 \quad (8)$$

Wall Temperature

The combustion occurs near the Top Dead Center and the wall temperature is consequently non-uniform in the chamber [9]. However, Pirotais *et al.* [3] showed that an average temperature of the 3 main regions of the chamber is acceptable. The spatial average considers the piston, the cylinder liner and the cylinder head (with valves). The time average is obtained with integration of this mean spatial temperature throughout the entire cycle.

According to this method, a new mean wall temperature is considered for each simulated engine cycle in the software. This temperature thus takes into account previous heat transfer from the combustion chamber to the liner, the piston and the head.

Heat Transfer

The heat transfer through the cylinder liner, the piston and the cylinder head are assumed to be of convective type. Indeed, the Hohenberg correlation for the convective coefficient is well-adapted to direct injection diesel engines since high-pressure injection systems authorize reduction in soot particle formation. The expression of the power transfered to the coolant fluid and the convective coefficient are presented in Eq. 9 and 10.

$$\frac{dQ_{w,i}}{dt} = h_{g,i}.S_{p,i}.(T_{g,i} - <\overline{T}_w>) \quad (9)$$

$$h_g = \alpha_h.100.(10^{-5}.P)^{0{,}8}.(u_{pis}+1{,}4)^{0{,}8}.T_g^{-0{,}4}.V^{-0{,}06} \quad (10)$$

were the index *i* corresponds to the burned or unburned zone. $h_{g,i}$ is the convective coefficient and $S_{w,i}$ the wall area. The first coefficient α_h depends on experimental calibration. Initial value proposed by Hohenberg is 1.3.

Wall areas of burned and unburned zones are calculated using the burned fraction correlation [5] (Eq. 6).

$$S_{w,u} = \left(\frac{\pi.d_s^2}{2} + \frac{4.V_{cyl}}{d_s}\right).\left(1-\sqrt{x_b}\right) \quad (11)$$

$$S_{w,b} = \left(\frac{\pi.d_s^2}{2} + \frac{4.V_{cyl}}{d_s}\right).\sqrt{x_b} \quad (12)$$

Chemistry of Combustion

The combustion reaction was written for a $C_{10,8}H_{18,7}$ fuel with properties found in [11]. The combustion equation is described in Eq. 13.

$$\begin{aligned}&\phi\varepsilon.C_{10,8}H_{18,7} + 0{,}21.O_2 + 0{,}79.N_2 \rightarrow \nu_1.CO_2 + \\ &\nu_2.H_2O + \nu_3.N_2 + \nu_4.O_2 + \nu_5.CO + \nu_6.H_2 + \\ &\nu_7.H + \nu_8.O + \nu_9.OH + \nu_{10}.NO\end{aligned} \quad (13)$$

where Φ is the equivalence ratio. CO_2, H_2O, N_2 and O_2 result from the complete combustion, CO, H_2, H, O, OH, and NO from dissociation and recombination reactions. υ_i represent the mole fraction of the component i, and ε is the stoechiometric fuel quantity for one mole of products. The fundamental chemical reactions considered in this study are described in the Eq. 14 to 19. Conservation laws give additional equations to solve the non-linear system. Kinetic parameters are taken from [5] and [9].

$$\frac{1}{2}H_2 \Leftrightarrow H \quad (14)$$

$$\frac{1}{2}O_2 \Leftrightarrow O \quad (15)$$

$$\frac{1}{2}H_2 + \frac{1}{2}O_2 \Leftrightarrow OH \quad (16)$$

$$\frac{1}{2}O_2 + \frac{1}{2}N_2 \Leftrightarrow NO \quad (17)$$

$$H_2 + \frac{1}{2}O_2 \Leftrightarrow H_2O \quad (18)$$

$$CO + \frac{1}{2}O_2 \Leftrightarrow CO_2 \quad (19)$$

Numerical Method

MATLAB is used to solve equations. Conservation of the chemical elements relations combined with equilibrium constant expressions give a non-linear system solvable with different methods. Olikara and Borman [12] described a complete method using a Newton–Raphson algorithm. This method is also used in this study. A 0.1 crank angle resolution gives a good compromise between computing time and accuracy of the results.

EXPERIMENTAL VALIDATION

Experimental tests on a naturally aspired single-cylinder direct-injection diesel engine are performed in order to validate the numerical model. The main design parameters are presented in Tab. 1. The complete description of tests can be found in [7].

Constructor	Listter-Petter
Bore	95,2 mm
Stroke	88,9 mm
Displacement	633 cm^3
Compression ratio	19.2 : 1
Cooling system	air circulation

Tab. 1 Engine technical features

Experimental data were taken from two studies [7,13]. The model was calibrated with 30% to 80% loads. Furthermore, diesel engines rarely run above 80% because of increased pollutant and soot emissions.

The α_h corrective coefficient is calibrated on the first and last parts of the pressure curves, corresponding to admission/compression and expansion/exhaust phases of the engine cycle. Indeed, the heat transfer occurs during the entire cycle but is mainly during the expansion phase. According to previous studies [3,7] and the calibration tests of the 2-zone combustion model, the corrective coefficient is fixed to 1.

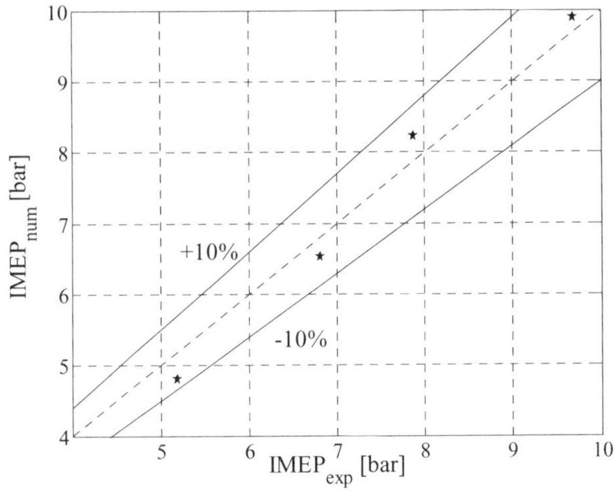

Fig. 4 - Comparison of experimental and numerical *IMEP* for 1500 rev/min

Numerical and experimental IMEP are compared and presented in Fig. 4. The relative error does not exceed 7 % and validation of the model is confirmed. Fig. 5 shows an example of a comparison of pressure curves. Computer modeling gives good correlation with experimental diagrams in the whole range of tests.

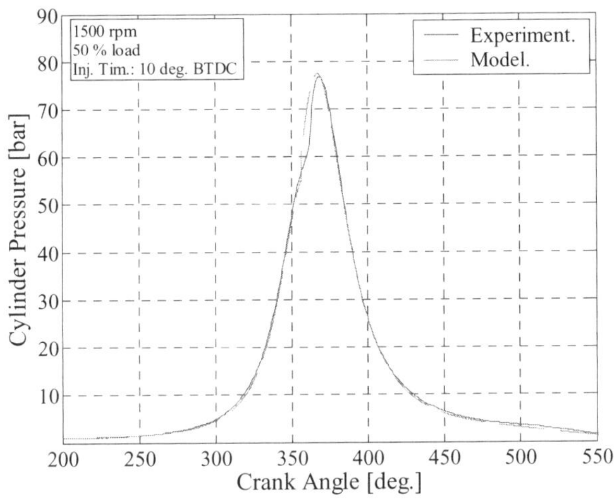

Fig. 5 - Comparison between experimental and numerical pressure diagrams

MODEL REDUCTION

OBJECTIVE

As previously stated, the final objective of this study is to avoid the use of the cartographies and thus directly obtain the heat transfer from the cylinder. Furthermore, the cartographies assume almost all the parameters as constants, whereas the new method has to consider the variation of all required parameters at each engine cycle simulation if possible.

THEORETICAL ANALYSIS

The *IMEP* can be expressed from the volumetric efficiency η_v, the equivalence ratio Φ and the indicated efficiency η_i with the following well known relations (Eq. 20 to 22).

$$IMEP = \eta_i . \eta_v . \phi . \frac{LHV}{\sigma} . \frac{P_{atm}}{R_{atm} . T_{atm}} \quad (20)$$

with

$$IMEP = \frac{W_{ind}}{V_c} \quad (21)$$

$$\eta_i = \frac{W_{ind}}{m_{diesel,\ injected} . LHV} \quad (22)$$

where *LHV* and σ are fuel properties. The atmospheric conditions are easily measurable, whereas the equivalence ratio is known thanks to the predefined injection diagrams. Accurate measures of the volumetric efficiency are also possible. Thus only η_i needs to be modeled.

In addition, the global energy balance of an entire engine cycle can be expressed as in Eq. 23.

$$m_f . LHV = W_i + Q_w + Q_{ex} \quad (23)$$

where m_f is the mass of fuel introduced in the chamber (assuming a complete combustion), Q_w the thermal losses and Q_{ex} the heat rejected in exhaust gases. Moreover, Q_{ex} can be deduced on-line from the measured mass flow rate and temperature in the exhaust pipe.

Finaly, the knowledge of η_i leads to Q_w (Q_{ex} and $m_f.LHV$ being known) via the Eq. 22 and 23. Q_w can thus be deduced from the indicated work via the *IMEP* or the indicated efficiency (Eq. (16) and (17)). The method used to predict the thermal losses is summed up in Fig. 6.

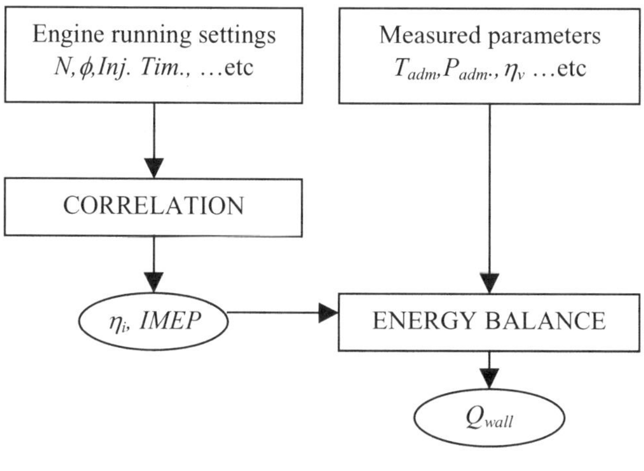

Fig. 6 - Structure of the method

STATISTICAL ANALYSIS

Numerous of studies enumerate parameters influencing indicated performance [3,5,9]. A statistical method is here used to determine these parameters and calculate their effect. The experimental design methodology is a statistical tool often used to evaluate influencing factors on measured variables. The parameters and their extreme values are presented in the Tab. 1.

Parameter	Lower value	Higher value
Rotation Speed [rpm]	1000	3000
Equivalence ratio [-]	0.3	0.8
Inj. Tim. [deg. BTDC]	5	20
$< T_{wall} >$ [K]	273	473
Inj. Duration [deg.]	10	30

Tab. 1 - Parameters used in the statistical test

These parameters were chosen for their association with global heat transfer towards the coolant loop. Rotation speed and load are first taken because of their natural influence on the indicated performance as shown in Fig. 7.

Advanced injection timing usually improves the indicated efficiency because of the increased pressure peak value [9,14] (Fig. 8). This necessarily leads to the modification of the thermal losses. For angle higher than 20 deg. BTDC, the model predicts a decrease in the indicated efficiency (Fig. 9), usually described in the literature [9].

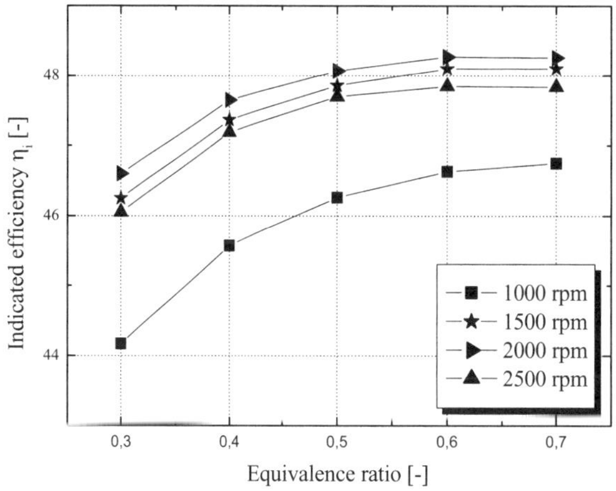

Fig. 7 - Evolution of the indicated efficiency with the rotation speed and load variations (Inj. Tim. 10 deg. BTDC)

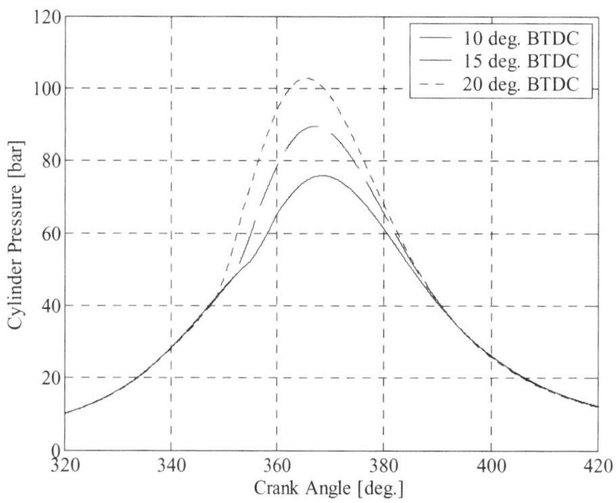

Fig. 8 - Pressure diagrams for different injection timings (1500 rpm, 50% load)

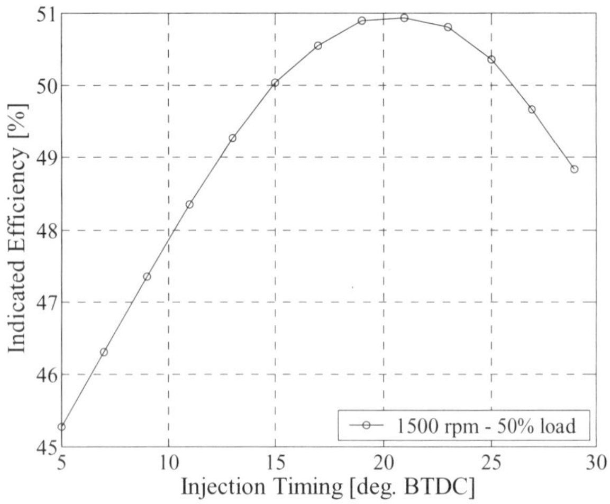

Fig. 9 - Evolution of the indicated efficiency with the injection timing variation

The mean wall temperature is assumed to vary between 0 and 200°C over the range of tested loads. Indeed, the study concerns the evolution of the car cabin blown air temperature in typical driving cycles (MVEG, EUDC...). These cycles include several cold starts, the first with a low ambiant temperature. The mean wall temperature variation thus must start from 0°C or below. Finally, injection duration is also taken into account to confirm the injection system can influence the pressure diagram.

The statistical design methodology is thus used to examine the effect of these 5 parameters, on the *IMEP* and the indicated efficiency (Tab. 2). According to the method, the values have no physical validity but the comparison between them leads to conclusions.

Parameter	Effect on IMEP	Effect on η_i
Rotation Speed [rpm]	-0.78	-0.27
Equivalence ratio [-]	3.83	1.38
Inj. Tim. [deg. BTDC]	0.45	2.66
$< T_{wall} >$ [K]	-0.09	0.66
Inj. Duration [deg.]	-0.01	-0.06

Tab. 2 - Effect of single parameters

The first three parameters influence the variables more. Interesting results appear concerning the Equivalence ratio and the injection timing effects on both *IMEP* and efficiency values. The difference may be linked to the fact that the indicated efficiency considers the fuel consumption whereas the *IMEP* only resumed the indicated work. These parameters are then chosen to establish the predictive correlation. The mean wall temperature also has a major effect on η_i. This parameter must be taken into account in the future studies.

Modelling of the effect of Speed and Load Variations

Rotation speed and load effects are firstly studied considering constant injection timing (Fig. 7). Fitting studies are performed to determine the mathematical trend of the curves. Fig. 10 presents a comparison between the numerical and fitted curves. A first order exponential decay behaviour is proposed (Eq. 24).

$$\eta_i = \eta_0 - \alpha \,.\exp\left(-\frac{\phi}{\phi_0}\right) \quad (\phi < 0.7) \tag{24}$$

This method is applied for each rotation speed and the same trend is confirmed.

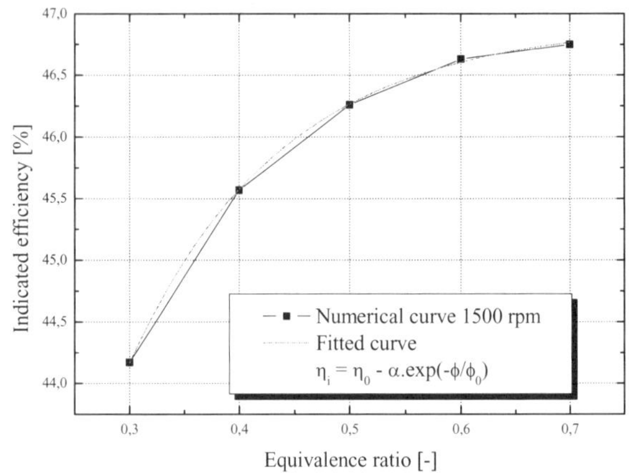

Fig. 10 - Fitting result from a numerical pressure curve

A more extensive observation reveals that the coefficients η_0, α and ϕ_0 follow a linear behaviour with the engine speed. Each coefficient can thus be deduced from Eq. 25.

$$p_i = A_i + B_i.N \tag{25}$$

where p_i represents these three coefficients. The A_i and B_i values are exposed in Tab. 3.

Parameter	A_i	B_i
η_0	46.7	$6.2\ 10^{-4}$
α	14.4	$8.5\ 10^{-3}$
Φ_0	0.16	$-2.6\ 10^{-5}$

Tab. 3 - Values of linear coefficients

From Eq. 24 and 25, and the empirical values of Tab. 3, the relative error between the numerical and correlated results can be calculated. Results are shown in Fig. 11. The maximum absolute value remains under 2%, proving the accuracy of the correlation.

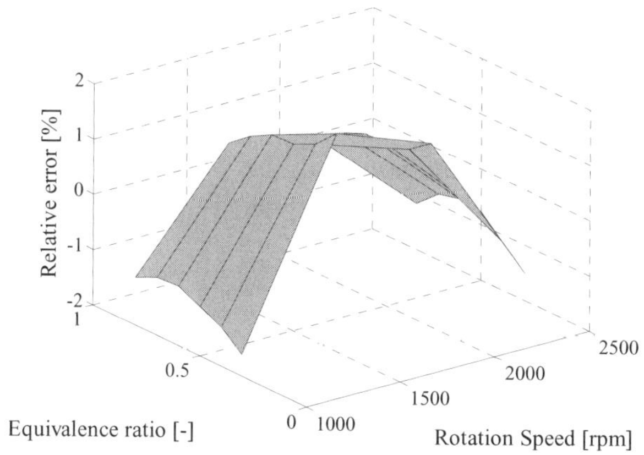

Fig. 11 - Relative error between the numerical and predicted indicated efficiencies

Influence of Injection Timing

The effect of the injection timing on the indicated efficiency was previously shown in Fig. 9. The efficiency increases until reaching a maximum at about 20 deg. BTDC, and then decreases. In the next figures, η_0, α and ϕ_0 are calculated for different injection timings, comprised between 10 and 20 deg. BTDC, considering two different rotation speeds (1000 rpm and 3000 rpm) (Fig. 12, 13 and 14).

Different conclusions can be deduced from the present curves. The coefficient η_0 increases with injection timing (Fig 12). Moreover, η_0 is not very influenced by the variation of the rotation. Reaction of α with the two variables is different (Fig. 13). The values vary in a larger range if the speed is high. Fig. 14 presents the ϕ_0 variation. In this case the decrease seems to be unchanged by *N*, and the curve is simply displaced to inferior values if the speed is increased.

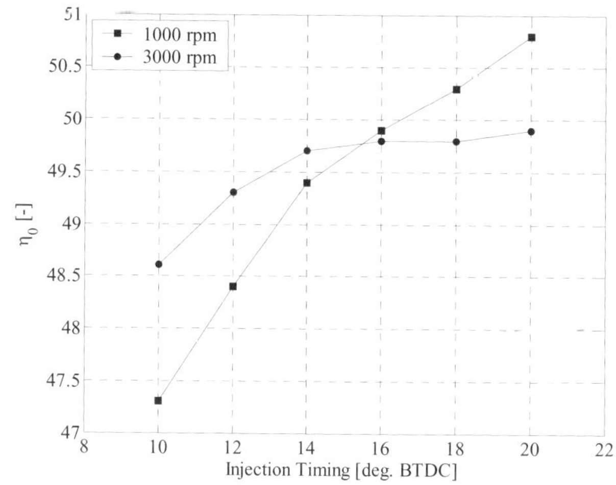

Fig. 12 - Evolution of η_0 with the injection timing and the rotation speed

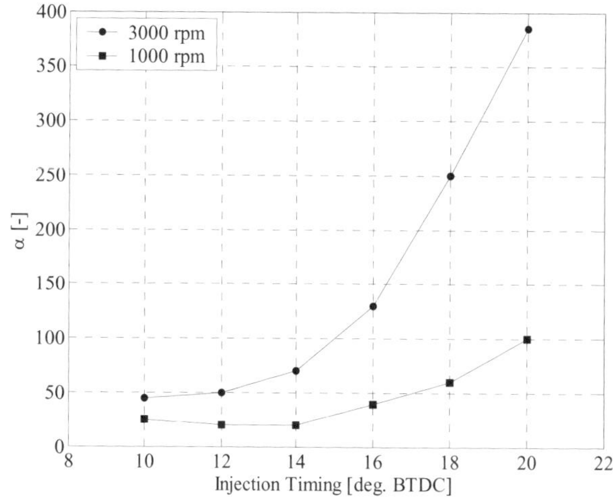

Fig. 13 - Evolution of α with the injection timing and the rotation speed

The study of these three coefficients gives characterisation of the influence of the injection timing. The behaviours are various and the speed does not act in the same ways. The relations obtained would give a very complex global correlation, if injection timing effect must be included.

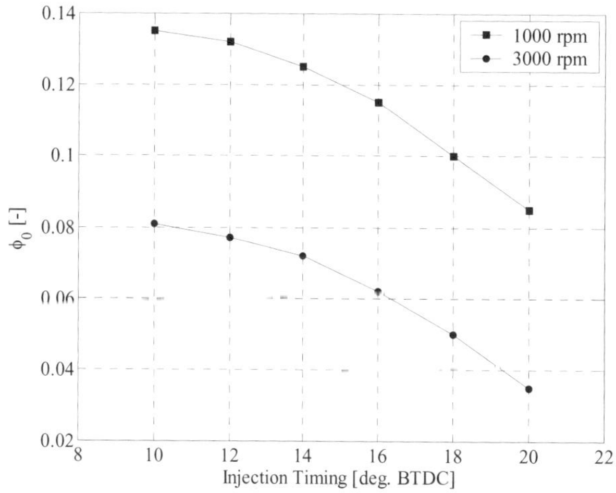

Fig. 14 - Evolution of Φ_0 with the injection timing and the rotation speed

To conclude, the first correlation with coefficients of Tab. 3 gives good results for the prediction of the indicated efficiency at fixed injection timing. Fig. 15 confirms that the trend of η_i is unvarying with the injection timing. Only η_0 variation should be taken into account in first approximation.

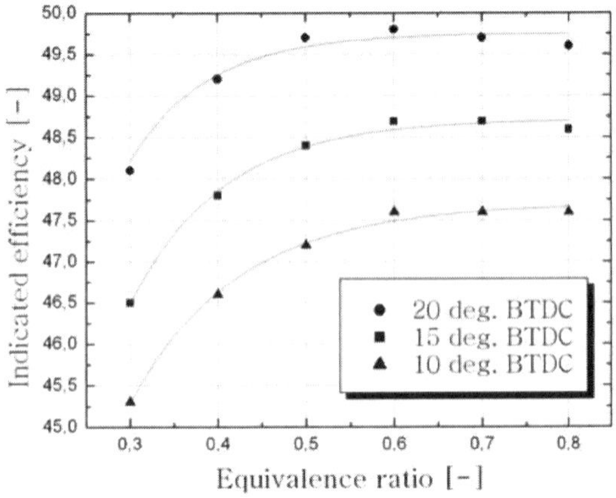

Fig. 15 - Evolution of the correlation trend for η_i with the injection timing (N = 1500 rpm)

CONCLUSION AND PERSPECTIVES

This study aims to develop a predictive correlation of the indicated efficiency of automotive engines. A 2-zone combustion model is detailed in this paper. This model describes the combustion processes and allows pollutant generation modelization.

Several parameter effects are studied thanks to the experimental design methodology. The model primarily describes the efficiency with the load and the rotation speed. The predictive correlation can be composed of three coefficients, linearly linked with the speed, which provide accurate results.

The complementary studies on the injection timing give other information on the behaviour of the efficiency. However they are complex compared with the previous forms and may provide a minimal gain of accuracy. The coefficients of the correlation have to be adapted to the variation of the injection timing if needed in the future.

Future studies aim to validate the 2-zone combustion model on real automotive engines. In this way, a model for the turbocharger will be developed and coupled with the engine model. The experiments on engines will allow to study the thermal losses and their relation with the settings. The final correlation should offer also emission simulation for the global simulation tool.

REFERENCES

[1] F. Pirotais, J. Bellettre, O. Le Corre, M. Tazerout, G. de Pelsemaeker and G. Guyonvarch. *A model of energetic interactions between a car engine, the cabin heating system and the electrical system*. SAE Paper N° 2002-01-2224, 2002.

[2] F. Pirotais, J. Bellettre, O. Le Corre, M. Tazerout, G. de Pelsemaeker and G. Guyonvarch. *A diesel engine thermal transient simulation. Coupling between a combustion model and a thermal model*. SAE Paper N° 2003-01-0224, 2003.

[3] F. Pirotais. *Contribution à la modélisation du flux thermique disponible pour le chauffage d'un habitacle d'automobile après un démarrage à froid.* Université de Nantes, PhD Thesis (in french), 2004.

[4] R.J. Kee, F.M. Rupley, J.A. Miller. *The chemkin thermodynamic database.* Sandia Report, SAND87-8215B – UC-4, 1992.

[5] M.L. Tazerout. *Etude des possibilités d'amélioration du rendement à charge partielle des moteurs à allumage commandé.* Université Catholique de Louvain, PhD Thesis (in french), 1991.

[6] S. Rousseau, B. Lemoult, M. Tazerout. *Combustion characterization of natural gas in a lean burn spark ignition engine*. ImechE, Proc. Instn. Mech. Engrs., volume 213 Paert D, 481-489, 1999.

[7] A. Bilcan, O. Le Corre, M. Tazerout, A. Ramesh and S. Ganesan. *Characterization of the LPG-Diesel dual fuel combustion.* Proceedings of the 2nd International SAEINDIA Mobility Conference, Technology Directions for clean, safe and efficient vehicles, SAE Paper 2001-28-0036, 249-258, Chenmai, India, 2002.

[8] D.N. Assanis, Z.S. Filipi, S.B. Fiveland, M. Syrimis. *A predictive ignition delay correlation under steady state and transient operation of a direct injection diesel engine.* Paper n°00-ICE-231, ICE-Vol. 33-2, 1999 Fall Technical Conference ASME, 1999.

[9] J.B. Heywood. *Internal Combustion Engine Fundamentals.* Number ISBN 0-07-100499-8, McGraw-Hill Editions, 1988.

[10] G.F. Hohenberg. *Advanced approaches for heat transfer calculations.* SAE Paper N° 790825, 1979.

[11] S.R. Turns. *An introduction to combustion. Concepts and applications.*McGraw-Hill International Editions, 1996.

[12] C. Olikara, G.L. Borman. *A computer program for calculating properties of equilibrium combustion products with some applications to I.C. engines.* SAE Paper N° 750468, 1975.

[13] A. Kerihuel, M. Senthil Kumar, J. Bellettre, M. Tazerout. *Investigations on a CI engine using animal fat and its emulsions with water and methanol as fuel.* SAE Paper 05P-95 (submitted).

[14] C.D. Rakopoulos, D.C. Rakopoulos, E.G. Giakoumis, D.C. Kyritsis. *Validation and sensitivity analysis of a two-zone diesel engine model for combustion and emissions prediction.*

Energy Conversion and Management 45 (2004) 1471-1495, 2004.

NOMENCLATURE

Roman letters

c_p	Specific heat	$(J.kg^{-1}.K^{-1})$
c_v	Specific heat	$(J.kg^{-1}.K^{-1})$
N	Rotation speed	$(rev.min^{-1})$
P	Pressure	(Pa)
Q	Heat energy	(J)
S	Surface	(m^2)
T	Temperature	(K)
V	Volume	(m^3)
h_g	Conv. heat trans. coef.	$(W.m^{-2}.K^{-1})$

Greek letters

θ	Crankshaft angle	(degree)
τ	Time constant	(s)
η	Efficiency	(%)
σ	Quantity of air at stoechiometry	(kg/kg)

Abbreviation

AHS	Additional Heating System	
BTDC	Before Top Dead Center	
ID	Ignition delay	
LHV	Low Heating Value	(J/kg)

2005-01-0056

Running Real-Time Engine Model Simulation with Hardware-in-the-Loop for Diesel Engine Development

P. J. Shayler and A. J. Allen
The University of Nottingham

A. L. Roberts
Jaguar Cars Ltd.

ABSTRACT

The paper reports the design of a model and HIL system produced to support the development and testing of Electronic Control Unit/Engine Management System (ECU/EMS) software for a V6 turbo-charged automotive diesel engine. The engine model, developed in Simulink, is compiled to execute on a dSpace platform and interacts with the ECU/EMS module in real time. The main features of the engine model are outlined. The configuration of the model and HIL components are described, and the performance of the system is illustrated and discussed. Practical decisions on the inclusion of real or virtual sensors and actuators, and other implementation issues, are explained. Recent and potential future applications of the system are described.

INTRODUCTION

In the automotive industry, the introduction and increasing role of engine electronic management systems has transformed many aspects of engine control, on-board monitoring and diagnostics over the last thirty years. Electronic components were first used on spark ignition engines to improve spark timing and air/fuel mixture control but have since expanded to include many other functions. Recent years have witnessed a similar growth in the use of microprocessor modules and electronic sensors and actuators on light duty diesel engines, particularly with the introduction of electronically controlled common rail fuel injection systems. In many areas of engine control and systems monitoring it is now impracticable to use anything other than electronic solutions, and the development and testing of software for Electronic Control Modules/Engine Management System (ECU/EMS or simply ECU from hereafter) is a substantial and key part of engine development programmes. There is considerable interest in reducing the time and resources expended on this and reducing the need for engine prototypes to work with. Linking the real-time simulation of engine (and vehicle) behaviour to a live ECU in a hardware-in-the-loop (HIL) arrangement is one approach to achieving these goals. HIL technology is maturing rapidly and examples of other recent applications in the automotive sector are reported in [1], [2] and [3]. In the current study, handling the interactions between a physical ECU and a virtual engine simulation with a HIL set-up has proven to be very effective and the design and implementation of this facility will be described.

The model used in the real-time simulation is a version of NuSim [4]. This is an integrated systems model (ISM), which is built from a set of systems level sub-models to describe the behaviour of the whole engine or the performance attributes of a vehicle. Examples of other ISM s reported in the literature include Simcar [5], Advisor [6] and GT Power [7]. As simulation tools for internal combustion engine investigations, ISM s typically combine elements of thermodynamic codes with those of real time simulators and plant models. Thermodynamic codes such as Wave, Boost, Merlin and Merrimee have, as noted in [8], physics-based submodels which meet requirements for conservation of mass and energy applied to simply connected zones and assuming one dimensional flow conditions. Real time simulators and plant models used most commonly in control applications are generally zero dimensional with very simplified physics. In some cases, the use of simple mathematical emulations allow real or near real time simulations to be achieved using very limited computing power. From the outset, NuSim was intended to be run in near real-time on high specification personal computers without this extreme level of simplification. The model was produced in Matlab Simulink and designed to incorporate engine management system software supplied as, or converted into, Simulink code.

An early application of NuSim described in [4], was to investigate the robustness of performance to uncontrolled sources of build variation. The study outlined in this case was carried out towards the end of the product development process. ECU software was well advanced but still undergoing evaluation and improvements to strategy features and calibration settings. Experience showed that it was practicable to translate the ECU code into Simulink for embedding into NuSim but that this process was non-trivial and time consuming. Later applications, to I4 and V8 gasoline engines and I4 turbocharged diesel engines, have revealed other disadvantages. The calibration of the controller can be modified in the target environment but often the strategy cannot, which makes it difficult to maintain consistency of strategy versions if there is parallel and ongoing development using test bed or vehicle hardware. If, as is often the case, the ECU software is developed by a Tier 1 supplier, confidentiality agreements can impede the re-implementation and distribution of the software. Controller diagnostic and monitoring tools (development aids or dev-aids) are generally designed to connect with physical ECU controllers and are not easily interfaced to a non real-time Simulink controller. Without these aids, diagnostic information on the internal state of the controller is unavailable. In principle many of these disadvantages or problems can be overcome by running NuSim in a HIL environment. Although there is a growing body of literature on hardware-in-the-loop applications, most are relatively simple compared to running an engine model in real time with a physical ECU linked via hardware-in-the-loop. This presents various challenges and the aim of the following is to describe solutions which have been identified. The work has been carried out for a particular application, namely for the development of a turbocharged V6 direct injection diesel engine.

NuSIM-DV6

In common with other versions of NuSim, the engine model is the main component. NuSim-DV6 represents a six cylinder V-type direct injection diesel engine with twin variable geometry turbochargers and twin external exhaust gas recirculation (EGR) systems. The engine model is comprised of a set of submodels describing the behaviour of various subsystems and processes. Simulation output provides cycle-averaged values of pressures, temperatures and flow rates for intake and exhaust gas flows, values of indicated and brake mean effective pressures describing work output characteristics, specific fuel consumption and flow rates of engine-out pollutant emission. The thermal behaviour and friction characteristics of the engine from cold start-up through to fully-warm operating conditions is described using an embedded version of PROMETS [9] and aftertreatment characteristics can be described using a development of PROMEX [10]. The engine model is coupled to models of the transmission and the vehicle tractive load characteristics, together with a driver model and a description of the vehicle target driving pattern. The driver model is essentially a PID controller which adjusts pedal position to eliminate the error between target and actual vehicle speed. Thus, the model can be represented at the simplest level as shown in Figure 1. Manual and automatic transmission models have been developed and the most common pattern of vehicle operation of interest is that for the New European Drive Cycle (NEDC) because of its association with benchmark figures for fuel economy assessments and the evaluation of tailpipe pollutant emissions.

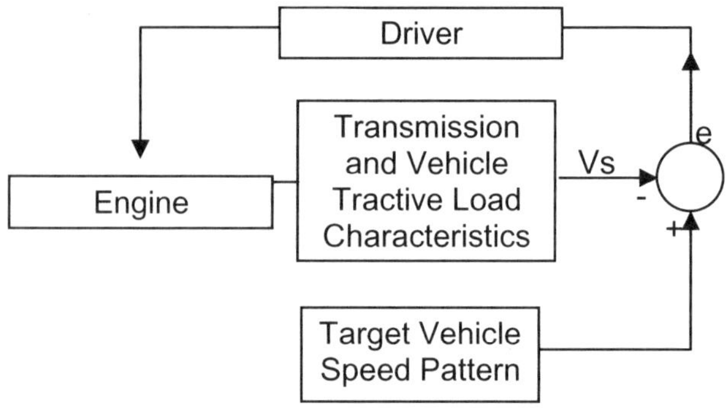

Figure 1: The virtual vehicle is driven by a tuned PID controller which adjusts pedal input to follow target pattern of vehicle speed, Vs. Additional features describe energy transfers during braking and over-run conditions.

The various submodels used in this version of NuSim are a mix of new or pre-existing, but modified, submodels drawn from other versions of NuSim. Component specifications, including turbine and compressor maps for the turbochargers, were provided by suppliers. Evaluation of the model was based on comparisons with experimental data recorded under steady state operating conditions from test bed investigations. An indication of the quality of predictions is given in Figure 2, which shows the agreement between simulation results and test bed data for representative parameters of interest. The range of speed and load covered by these test results included those associated with NEDC operating conditions. Results for the simulation of a NEDC test are shown in Figure 3. The results illustrate some of the thermal and friction behaviour characteristics which are of interest when making predictions of fuel economy for example, and also show how EGR rate varies dramatically throughout the drive cycle. The HC, NO_x and CO characteristics illustrated are cumulative engine-out values not tailpipe values, but as noted there is the capability to predict tailpipe emissions by using PROMEX.

Simulations are generated by a time marching solution of the governing set of equations using Euler 1, which is a standard solver option within Simulink. The execution time step is set to be 2ms of simulated time. The step size avoids excessively high computational work and is less than the minimum required for simulation results to be independent of step size. It does limit the bandwidth of information which can be extracted from NuSim, however, as indicated diagrammatically in Figure 4. The model does not give crank angle resolved information on the development of cylinder pressure for example, but is able to resolve inter-cycle phenomenon down to timescales of approximately one piston stroke. The model is comfortably able to resolve features of behaviour on longer timescales, such as turbocharger lag, which is typically of the order of one second, and the time taken for engine warm up to be completed, which is typically several hundred seconds.

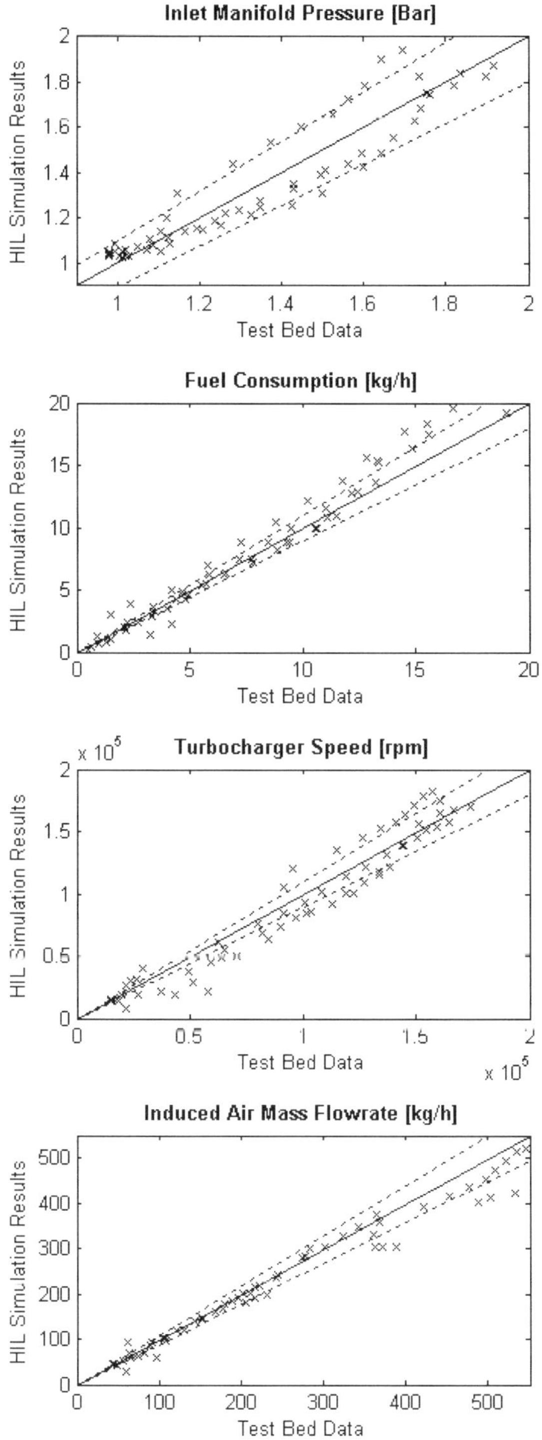

Figure 2: Examples of the correlation between simulated and test bed data for NuSim-DV6 in steady state operating conditions. Graphs include a 1-1 correspondence line, and 10% error margins. The data set includes fully warm engine operation from idle up to 75% of full load.

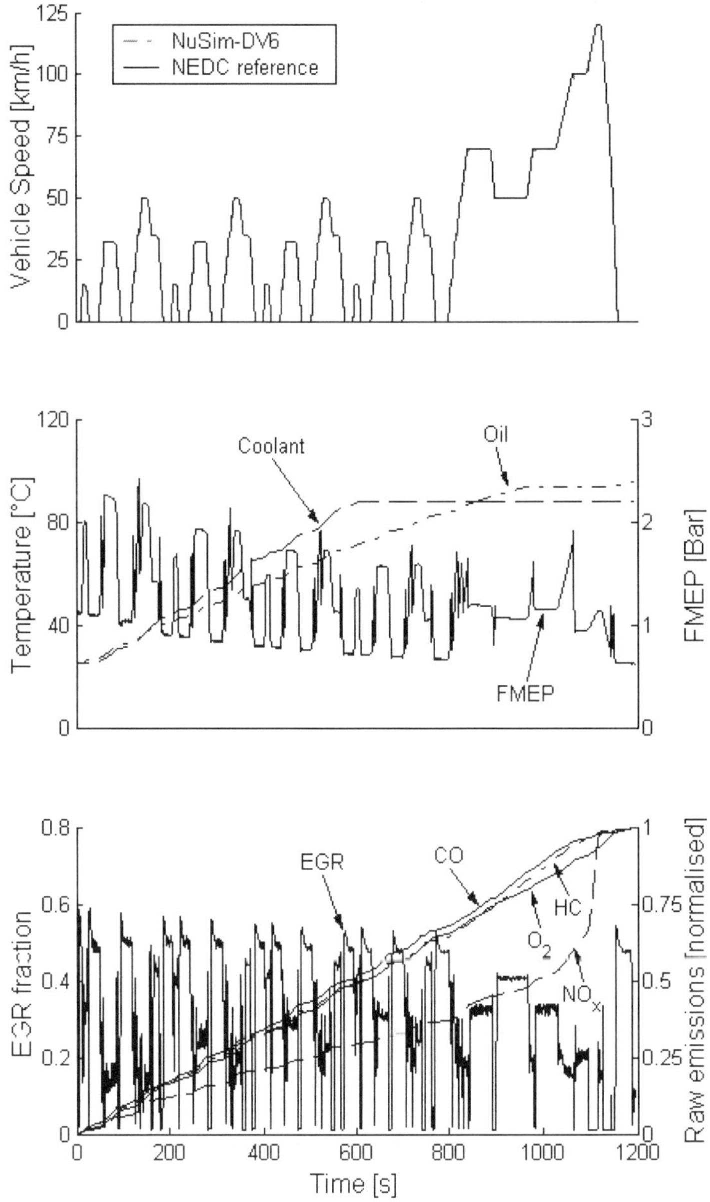

Figure 3: Typical results from an ambient start NEDC simulation. The plots show a sample of the available data related to vehicle, thermal, and emissions related features.

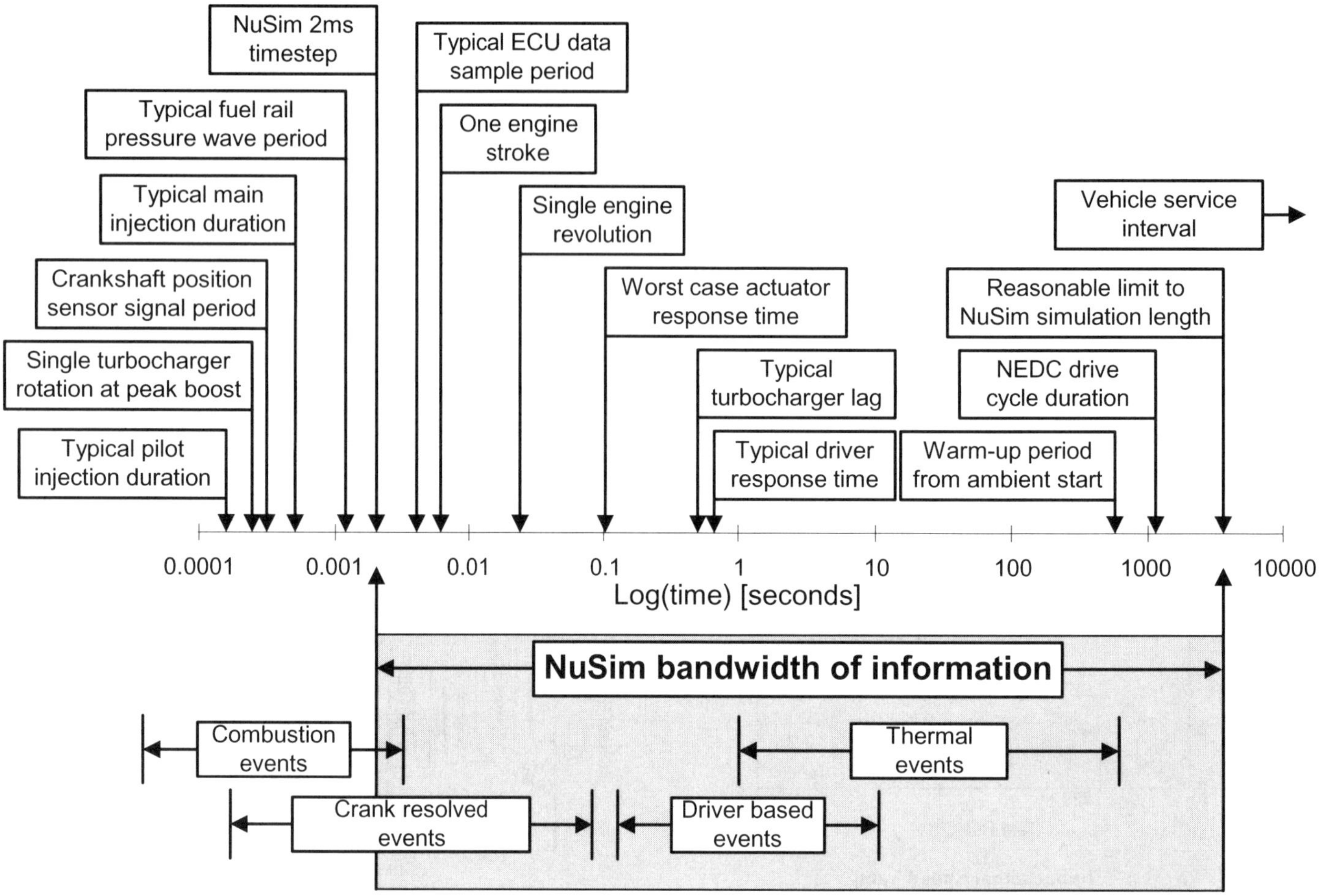

Figure 4: The bandwidth of information available to NuSim as dictated by the 2ms simulation step size.

INTEGRATION OF NUSIM AND A HIL RIG

Previously NuSim has been run as a closed set of models, which include both sensors and actuators. In the HIL set-up with a real ECU, there exists the possibility of using real actuators. Here, a mix of real and virtual actuators has been used. A schematic of the divide between virtual and physical hardware is shown in Figure 5. This layout is similar to that utilized in [8]. Above the software/hardware dividing line is the NuSim simulation software. The virtual sensors block contains models of sensors providing inputs to the ECU, including manifold absolute pressure (MAP), mass air flow-rate (MAF), temperature sensors and driver/vehicle sensors such as accelerator pedal position (APP) and vehicle speed. The set of virtual actuators includes the fuel rail pressure control valve (PCV). The set of physical actuators includes the six fuel injectors, the actuators for guide vane settings on the turbines of each turbocharger, two EGR valves and an intake air throttle. The functionality of these actuators could have been simulated in NuSim at some cost to computational time, but the use of physical actuators provided the most simple and cost effective way to meet closed-loop control and diagnostic monitoring requirements described in the next paragraph. The interface between the virtual and physical parts of the system is provided by dSpace hardware. The Genix unit provides voltage and current scaling appropriate for the ECU and actuator hardware. Information passes between the real actuators and ECU directly and signal feedback is also passed to dSpace from the positional sensors built into each actuator. The feedback can be modified in the virtual sensors block, located in NuSim, above the software/hardware divide.

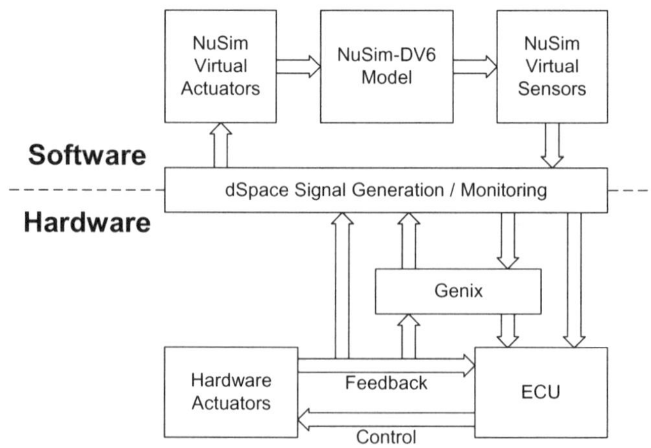

Figure 5: Schematic of the divide between hardware and software in the HIL/NuSim environment.

The physical set-up is shown in Figure 6 and Figure 7. The physical actuators are wired in to the ECU loom, as would be the case for the real engine. In the HIL environment the ECU performs various diagnostics checks on its peripherals and expects them to behave according to specific and time limited criteria. Most of the sensors provide voltage, resistive or pulse width modulated (PWM) inputs to the ECU. Several of the actuators have integral sensors which provide position feedback signals to the ECU, allowing closed loop control of their individual settings with high speed response. Actuators of this sort include the variable geometry turbocharger (VGT) actuators, the EGR valves and air throttles. Other actuators which include the fuel injectors are expected to draw a specific amount of current from the controller unit and diagnostic faults will be triggered if the current is outside of a specific behaviour pattern.

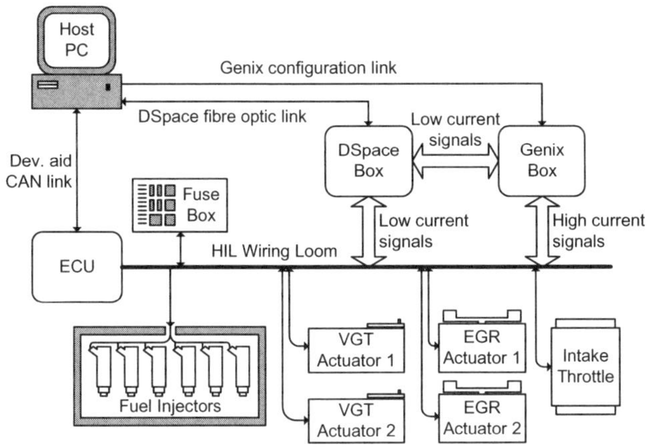

Figure 6: Physical components of the NuSim-DV6 HIL rig. Set-up and data acquisition is controlled on the host PC, with real-time simulation carried out on the dSpace processor box. The Genix unit provides voltage and current scaling appropriate for the ECU and actuator hardware. The fuel injectors are fitted into an enclosure to reduce the noise generated during simulation.

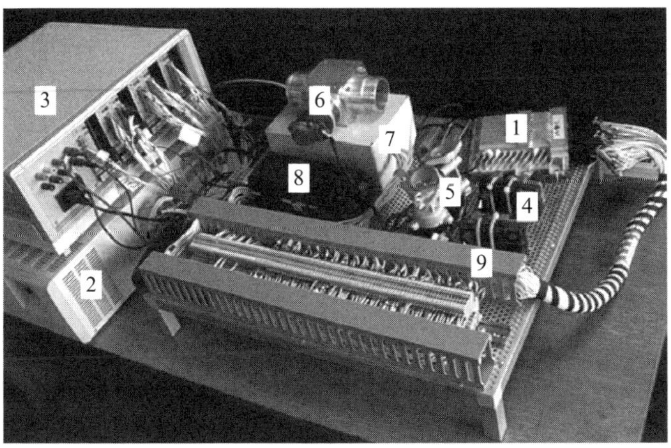

Figure 7: Photograph of the HIL rig showing 1) ECU, 2) dSpace processor box, 3) Genix signal conditioning unit, 4) VGT actuators, 5) EGR valve actuators, 6) Intake throttle, 7) Fuel injectors (Enclosed), 8) Fuse / relay box, 9) HIL wiring loom.

MODEL AWARENESS OF INTER-CYCLE EVENTS

In most cases the hardware is concerned with all events occurring on a high speed (less than 2ms) time base. Generally, the model provides time or cycle averaged information on events which occur over longer timescales (2ms or greater). A dSpace DS2210 signal processing board is used to handle the event resolution problems this simple division gives rise to. A time-averaged engine speed is an input to the board from NuSim. Using this, the board generates two signals which are inputs to the ECU. The first simulates the output of a toothwheel encoder rotating at crankshaft speed. This has a resolution of 6°CA and has a missing tooth to identify crank top-dead-centre position. The second simulates the output of the camshaft position sensor. The ECU defines pilot and main injection settings for each cylinder and sends actuating signals to the real injectors. These signals are monitored by the DS2210 board which relays the information back into NuSim. This is used in calculations of gross work produced by each cylinder as it completes compression and power strokes of its cycle. The fidelity of the simulation depends upon the injection settings being assigned to the correct cylinder and the correct firing order being maintained within NuSim. If these conditions are not met, the accuracy of responses to idle speed control and other dynamic control features of the ECU software deteriorate. In principle, the 2ms time step for NuSim allows each cylinder to be identified uniquely at engine speeds up to 3600 rpm, corresponding to 120° CA rotation during the time step. At speeds higher than 3600 rpm, a shorter time step can be used to retain the resolution, but the majority of tests are at lower speeds. Identification problems can arise when engine speed is changing, however, because the crank position determined from the inputs to the ECU can diverge from the value implied by the mean engine speed used in NuSim. This problem is solved by resetting crank position in NuSim to the common reference given by the last known injection event. The estimated crank rotation from this point is then always sufficiently accurate to identify in NuSim which cylinder the ECU has been assigned to fire next.

REAL TIME SIMULATION

The compilation of Simulink models for real-time execution on dSpace hardware is carried out using the Matlab Real-Time Workshop tools chain, and is a relatively straightforward process taking around 10 minutes to complete for NuSim-DV6 on a modern desktop PC. On a DS1005 processor board, the execution of a 2ms time step for NuSim takes around 0.4ms of processor time, and updates of PROMETS at 100ms time intervals take around 7ms. In the non real-time environment model execution can slow down without consequences at 100ms intervals as PROMETS is updated, but this is unsuitable for the real-time environment. The NuSim simulation step must be completed every 2ms if the deadlines of real-time simulation are to be met.

The problem was solved by the introduction of task-based model execution. Each task is allocated a priority and can be suspended to allow a higher-priority task to meet its real-time execution deadline. This is managed by a real-time operating system, with the execution of NuSim having a higher priority than that of PROMETS. The separate tasks are uniquely identified within the Simulink modelling environment, and the allocation of a higher priority to tasks carried out at shorter time intervals is straightforward. The real-time operating system and scheduling algorithm consume a relatively small amount of processor time, as do housekeeping operations such as task suspension, when only a small number of tasks are to be managed. The effect on the scheduling and execution of NuSim and PROMETS is shown diagrammatically in Figure 8. In the real-time environment execution of NuSim alone leaves the processor idle for around 80% of the available time, but this 80% is split into segments of around 1.6ms each in size. PROMETS, running at a lower priority than NuSim, is scheduled to run in these previously idle time slots.

There is a slight difference in data transfer between tasks at run time, as shown in Figure 8. In the non real-time environment NuSim is immediately updated with the results of a PROMETS calculation, and outputs received from PROMETS are always consistent with the inputs taken from the previous simulation step of NuSim. In the real-time configuration, results are only transferred to NuSim once PROMETS execution has completed, which is several milliseconds after the inputs to PROMETS were gathered. Given that PROMETS simulates thermal behaviour, which is slowly varying, the effect of the delay is insignificant.

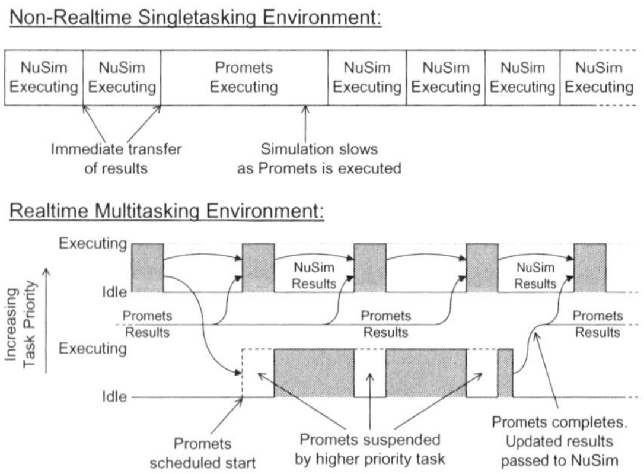

Figure 8: Execution of NuSim and Promets in the non real-time and real-time environments.

INITIATION OF SIMULATION

For most applications of NuSim outside of the real-time HIL environment, cold start behaviour is unimportant because the ECU code is implemented in Simulink and the various validation and consistency checks this performs can be suppressed. The simulation can be started from, say, a stable idle running condition. In the HIL environment with real ECU hardware present the start condition has to be described in a more realistic way, from key-on with the engine initialised to zero rotation speed. This required a number of modifications to NuSim to ensure stability and internal consistency. The turbochargers begin at zero rpm, fuel quantity is zero, as are the initial air and EGR flow rates. The equivalence ratio is taken to be zero, and hence the overall air / fuel ratio is infinitely large, and work output is zero. A basic starter motor model was added as a subsystem to NuSim and at key-on this delivers work input to the crankshaft and transmission submodels. Engine inertia and frictional losses are taken into account when computing the gain in engine speed which this work transfer produces. As engine speed increases, air flow through the engine occurs, due to engine pumping work. A pressure change across the turbochargers develops causing these to rotate once their frictional resistance is overcome. After one or two engine cycles fuel injection commences and the combustion system model delivers an indicated work output to the crankshaft giving rise to a run up of engine speed. Eventually an idling condition is reached, at which point the starter motor is disabled and the ECU begins to control idle speed.

This physics-based description of the start process provides a reasonably good representation of pressure, turbocharger speed and engine speed variations during the start-up process. A comparison between simulated and recorded start data is given in Figure 9.

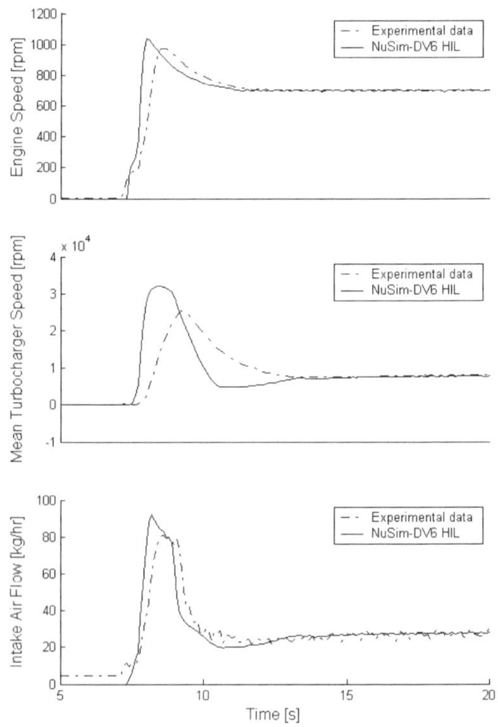

Figure 9: Comparison of model behaviour during a key-on event with data taken from an engine test bed.

Given the basic nature of the start-up model the results are in good agreement. Most importantly, this provides a realistic key-on event for the ECU to monitor and allows the normal state of engine running to be reached without triggering OBD flags or registering internal fault codes within the ECU. All potential divide by zero errors have been avoided and infinities such as those produced by initial values of zero air flow rate have been identified and dealt with appropriately.

APPLICATION TO OBD DEVELOPMENTS

An early application of the NuSim/HIL facility has been to support developments in the area of On Board Diagnostics (OBD). North America and European regulations specify minimum requirements for OBD to ensure vehicles in service continue to meet restrictions on emissions. The OBD system must detect faults that might affect these emissions, whilst avoiding false registers. The OBD software in the ECU has fault-detected triggers which must be set to meet both requirements. The influence of tolerance stack-ups on this can be critical. The potential for variations in sensor and actuator behaviour to trigger false detections an important concern. The potential was investigated by modifying sensor and actuator characteristics within the virtual environment of NuSim, and examining the consequences to simulated behaviour.

A standard test sequence entailed placing the ECU and model into a desired state, introducing a specific fault or modified behaviour pattern across the set of sensors and/or actuators, and recording both model and ECU performance over a specified pattern of operating conditions. This procedure is common to that employed by others, including [2]. Here, the NEDC was used to define the operating conditions, and a test array requiring over 200 NEDC drive cycle simulations, each taking approximately 20 minutes to complete, was completed in around 70 hours of continuous HIL rig running. This was made possible by the use of test automation scripts which execute at run-time on both the dSpace processor board, and on the host PC. The scripts were written to manage the recording of simulation data and to co-ordinate the automatic resetting of both the model and the ECU.

DISCUSSION AND CONCLUSIONS

The combination of HIL environment and NuSim real-time simulation overcomes the restrictions associated with using simulation models whilst retaining the benefits of not needing test bed or vehicle hardware to run with. Use of a physical ECU has numerous advantages over any conversion or re-implementation of the ECU s software in the modelling environment. Extensive testing of the ECU module, both hardware and software, covering strategy and calibration upgrades, compliance with OBD and emissions regulations, performance robustness to build variations and so on, can be carried out. Potential future applications include the simulation of component failure conditions and performance consequences, and the investigation of factors which influence tail-pipe emissions.

The notion of task separation and prioritisation within the model could be extended to cover more intra-cycle events and features of actuator behaviour within the plant model. Medium and longer duration events, such as regeneration for a diesel particulate filter in the aftertreatment system, can also be modelled in this way. The only limiting factor to expansion of the model is the availability of previous dead-time on the DS1005 processor board, though this limit itself can be overcome by employing multi-processor dSpace hardware systems.

ACKNOWLEDGEMENTS

The authors are pleased to acknowledge the support of Jaguar Cars for this study and thank the company for permission to report the work in this paper.

REFERENCES

1. Ramaswamy D et al, A Case Study in Hardware-In-the-Loop Testing: Development of an ECU for a Hybrid Electric Vehicle , SAE Paper 2004-01-0303, SAE Congress Detroit, 8th-11th March 2004.
2. Nabi S, Balike M, Allen J, Rzemien K, An Overview of Hardware-In-the-Loop Testing Systems at Visteon , SAE Paper 2004-01-1240, SAE Congress Detroit, 8th-11th March 2004.
3. Ploger M, Saucer J, Budenbender M, Held J, Costanzo F, De Manes, M, Di Mare G, Ferrara F, Montieri A, Testing Networked ECUs in a Virtual Car Environment , SAE Paper 2004-01-1724, SAE Congress Detroit, 8th-11th March 2004.
4. Shayler P J, Dow P I, Davies M T, A model and methodology used to assess the robustness of vehicle emissions and fuel economy characteristics , IMechE Paper C606/013/2002, 2002.
5. Moskwa J J, Weeks R W, Automotive engine modelling for real-time control using Matlab Simulink , SAE Paper No 950417, 1995.
6. Anon, ADVISOR , www.ctts.nrel.gov/analysis/advisor.html
7. Ciesla C, Keriban R, Morel T, Engine/powertrain/vehicle modelling tool applicable to all stages of the design process , SAE Paper No 2000-01-0934, 2000.
8. Tauzia X, Chesse P, Hetet J-F, Bonin A, MERIMEE: A simulation software to study diesel engines used for military propulsion , ASME Paper ICEF 2002-494, ASME Conference Proceedings ICE-Vol 39, 2002.
9. Shayler P J, Chick J P, Hayden D, Yuen R, Ma T, Progress on modelling engine thermal behaviour for VTMS applications , SAE Transactions, Journal of Engines, Section 3, Vol 106, pp 2008-2019, 1997.
10. Shayler P J, Hayden D, Ma T, Exhaust system heat transfer and catalytic convertor performance , SAE Paper 1999-01-0453, SAE Congress Detroit, 1st-4th March 1999. Also SAE Transactions, Journal of Fuels and Lubricants, Vol. 108, 1999.

2004-01-1618

Feasibility of Reusable Vehicle Modeling: Application to Hybrid Vehicles

A. Rousseau, P. Sharer and F. Besnier
Argonne National Laboratory

ABSTRACT

Many of today s vehicle modeling tools are good for simulation, but they provide rather limited support for model building and management. Setting up a simulation model requires more than writing down state equations and running them on a computer. The role of a model library is to manage the physics of the system and allow users to share and reuse component models. In this paper, we describe how modern software techniques can be used to support modeling and design activities; the objective is to provide better system models in less time by assembling these system models in a plug and play architecture. With the introduction of hybrid electric vehicles, the number of components that can populate a model has increased considerably, and more components translates into more drivetrain configurations. To address these needs, we explain how users can simulate a large number of drivetrain configurations. The proposed approach could be used to establish standards within the automotive modeling community.

INTRODUCTION

In a world of growing competitiveness, the role of simulation in vehicle development is constantly increasing. Because of the number of possible advanced powertrain architectures such as hybrid or fuel cell that can be employed, the development of the next generation of vehicles will require accurate, flexible simulation tools. Such tools are necessary to quickly narrow the technology focus to those configurations and components that are best able to reduce fuel consumption and emissions. The simulation tools must be flexible enough to encompass a wide variety of components and drivetrain configurations.

With improvements in computer performance, many researchers started developing their own vehicle models. But often, computers in simulation are used only to crunch numbers. Moreover, model complexity is not the same as model quality. Using wrong assumptions can lead to erroneous conclusions; errors can come from modeling assumptions or from data. To answer the right questions, users need to have the right modeling tools modeling. For instance, one common mistake is to study engine emissions by using a steady-state model or to study component transient behavior by using a backward model. Indeed, specific component models and modeling philosophies should be used for specific applications.

In this article, we describe how a graphical user interface (GUI), combined with an innovative software architecture, can be used to support powertrain modeling. It is important to separate modeling from simulation: We will focus on component model management and powertrain building management. that the paper will address ways in which component model management involves much more than assigning specific folders for each component and discuss how powertrain building management is more complicated than just manually connecting components together. The Powertrain System Analysis Toolkit (PSAT) developed at Argonne National Laboratory will be used to explain the methodology.

PSAT INTRODUCTION

PSAT [1, 2] is a powerful modeling tool that allows users to realistically evaluate not only fuel consumption but also vehicle performance. One of the most important characteristics of PSAT is that it is a forward-looking model meaning that PSAT allows users to model real-world conditions by using real commands. For this reason, PSAT is called a command-based model. A driver model estimates the wheel torque necessary to achieve the desired vehicle speed. The powertrain controller then sends real commands to the different components: throttle for engine, displacement for clutch, gear number for transmission, or mechanical braking for wheels to achieve the desired wheel torque. Because the components react to the commands as they would under real-world conditions, researchers can implement advanced component models (based on physics rather than lookup tables), take into account transient effects (e.g., engine starting, clutch engagement/disengagement or shifting), or develop realistic control strategies (which can be used later to control hardware).

PSAT, developed under Matlab/Simulink [3], allows the simulation of more than 150 predefined configurations, including conventional, electric, parallel hybrid, series hybrid, fuel cell, fuel cell hybrid, and power-split hybrid vehicles. Users can also choose two wheel drive (2wd), four wheel drive (4wd), or two-times-two-wheel drive (2t2wd). Such a capability is only possible by building all these drivetrain configurations according to a user s inputs and component models from libraries. PSAT takes additional advantage of the Matlab/Simulink environment by allowing both control strategy and component models to be directly coupled in the same environment (which is not the case for C or FORTRAN codes), as well as providing the option to integrate any code using S-functions.

PSAT flexibility and reusability are based upon several characteristics, which are discussed in the following section.

SOFTWARE ARCHITECTURE

NAMING NOMENCLATURE A well-defined nomenclature is fundamental to allowing users to easily understand the tool and quickly access the results. Once users are familiar with the nomenclature, they can access parameters just by deducing their names. The rules governing PSAT variable names are defined as follows:

- Begin with the type of component.
- Next provide type of data, which can have up to two elements.
- Up to 63 total characters are allowed by MATLAB.
- Output variables end in hist.

No uppercase is used in the code. Examples of parameter names are provided in Table 1.

USE OF STRUCTURE Structures are MATLAB arrays with named "data containers" called fields. The fields of a structure can contain any kind of data. For example, one field might contain a text string representing a name, another might contain a scalar representing a fuel economy result, a third might hold an efficiency matrix, and so on. These structures allow the software to be better organized and, consequently, provide quicker access to information for users.

PSAT uses several structures that not only store predefined powertrain configurations that the users can access, but also store the user choices and the simulation results. Table 2 describes the fields used to define the drivetrain configurations.

POWERTRAIN BUILDING A significant number of advanced vehicle configurations are available; in fact, a count of only the most popular options yields more than one thousand. Because of time and money constraints, it is impossible to build and test every one of these configurations. In addition, for each configuration, users need to be able to choose among different component models. To be able to make the right decisions, users need a flexible simulation tool that allows easy drivetrain options and component model comparison.

Table 1. PSAT Naming Nomenclature

Parameter	Type of component	Type of data #1	Type of data #2
eng_spd_hist	"eng" for engine	"spd" for speed	
mc_volt_hist	"mc" for motor controller	"volt" for voltage	
ptc_eng_trq_max_hist	Engine information used in the controller ("ptc")	"trq" for torque	"max" for maximum

Table 2. PSAT Structure for Powertrain Configurations

Structure	Field name	Description
config	name	Name of the powertrain (example: "par_2wd_p2_ct")
	pwt	Hybrid Family (example: "Parallel Hybrid")
	axle	Number of axles (example: "2 wheel drive")
	trans	Transmission technology (example: "ct" for continuous variable transmission)
	name_compo	List of the component used in the powertrain (example: {'drv', 'eng', 'mc', 'wh')
	ver_compo	List of component versions the user can select for this powertrain
	pos_compo	Location of each component in the powertrain and component it is connected to
	prop_strat	List of control strategies available for the powertrain. Users will choose one.
	trs	List of transient needed for the powertrain

Two options are commonly used within the modeling community: a rigid, predefined, saved-model option and a tedious, user-defined, component-by-component-assembled-model option. The first option has the advantage of speed but lacks drivetrain diversity because of the large number of drivetrain models that need to be independently saved. A change of a single component model results in a new drivetrain model. The second option has the advantage of conserving library space and allows flexibility in drivetrain type, but it requires inordinate amounts of the user s time to assemble the drivetrain models from the component libraries. Both options quickly lead to versioning and space issues. Having a couple hundred powertrain models saved or building them by hand are obviously not optimal solutions.

Within PSAT, the powertrain configurations are not saved, but rather, they are automatically built. On the basis of the user's choices, the information from the structure is used to select the proper model for each component, put it in the proper location, and connect all the components together. Adding a configuration is then as simple as adding a new field in the structure config (Table 2).

POWERTRAIN MODEL

As an example, it is interesting to look at a PSAT parallel configuration model to understand the interest in using a standard format (Figure 1). The driver output is a torque demand at the wheels, which is proportional to an accelerator or brake pedal command. This demand is sent to the powertrain controller (PTC), which decides how each component of the drivetrain should operate. Indeed, we make choices about the blending among the different energy sources and when and how we start the engine or shift a gear. The PTC sends specific commands to the component control unit so that they can be understood by the models. For instance, the PTC

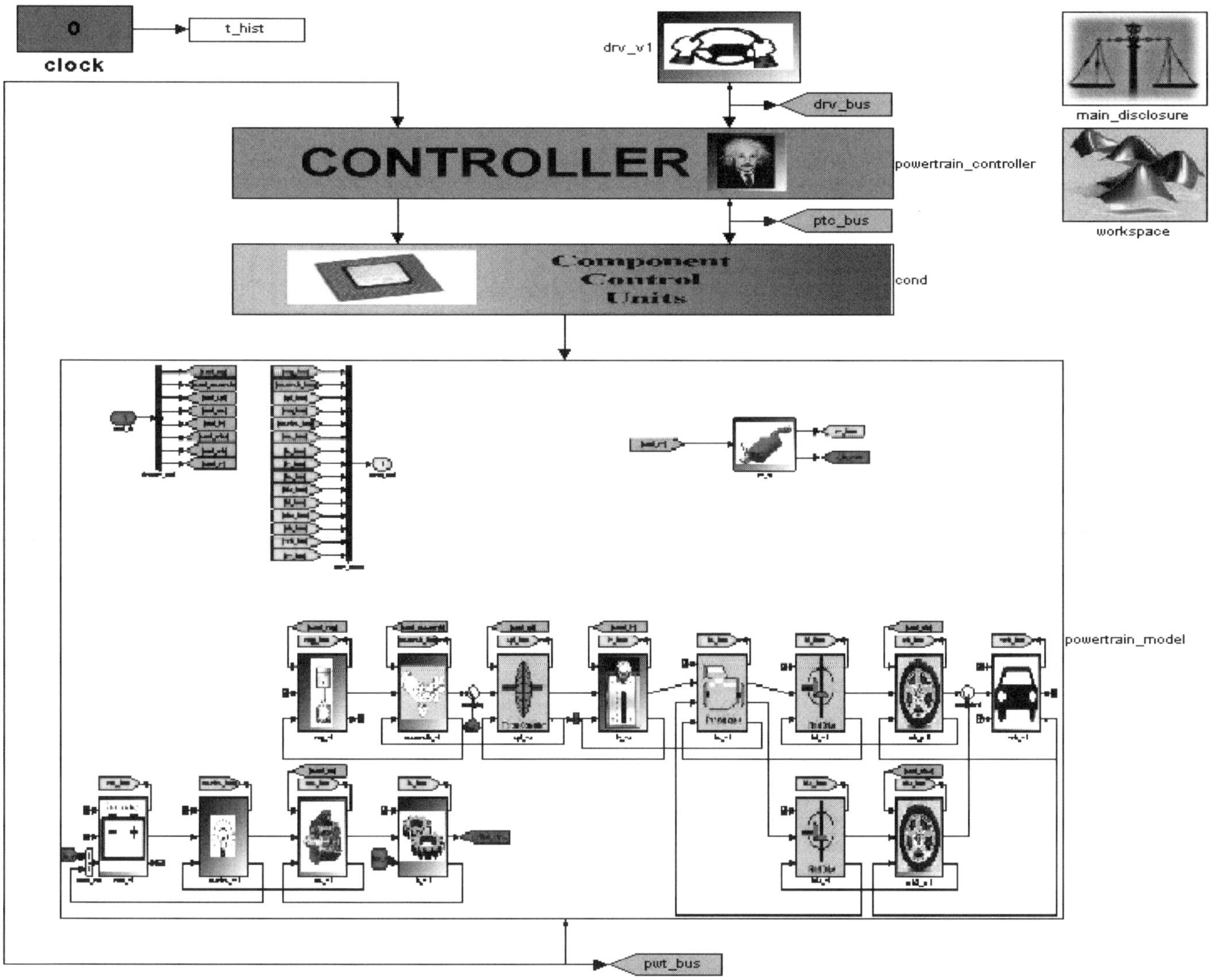

Figure 1. Example of PSAT Powertrain Model

asks for a specific torque to the engine, and the engine control unit (ECU) block within the component control unit transforms the torque into a throttle demand that the engine model can process. Then, the mechanical power from the engine and the electrical power from the motor (via the battery) are summed. In fact, both mechanical and electrical power are used to propel the vehicle. The component s information is collected (via sensors), and a bus is created (pwt_bus) to enable the system to use the information back in the controller to make the next decision.

COMPONENT MODEL

As shown in Figure 2, each component model is saved in one of three specific libraries:

- The component model: models the physics of the system.
- The constraints block: used to define the limits of the component (for instance, the maximum engine torque at the current speed).
- The signal conditioning block: used to send the proper command to the component in the component control unit (Figure 1).

The name of the library, as well as each block, also follows naming convention rules based upon the component name ("eng") as well as the model version (1 in our example).

ORGANIZATION FORMAT To easily exchange the models and implement new ones, a common format, based on Bond Graph [4], is used between the

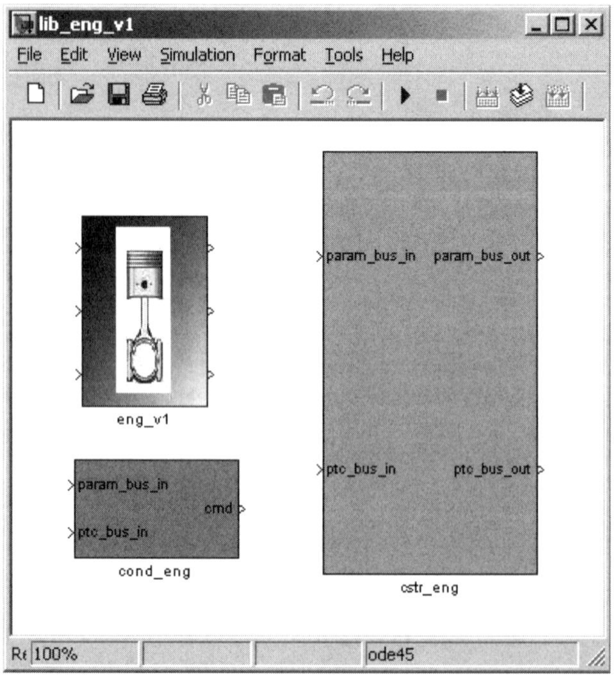

Figure 2. Example of Component Library — Engine

input/output of the power ports, as shown in Figure 1. The first ports are used for the information:

- Input: components commands (on/off engine, gear number, etc.)
- Output (sensors): simulated measures (torque, rotational speed, current, voltage, etc.)

The second ports carry the effort (e.g., voltage, torque); the last ones carry the flow (e.g., current, speed).

This format allows users to select different levels of component models depending upon the goal of the simulation (i.e., if the user is interested in the fuel cell component, he/she can use a very detailed fuel cell model while the rest of the models are based upon look-up tables). It is very important to notice that the first input and output are vectors and can have any desired size: a simple engine model can have only two inputs (such as engine on/off and engine command), while a detailed engine model can have five or more inputs.

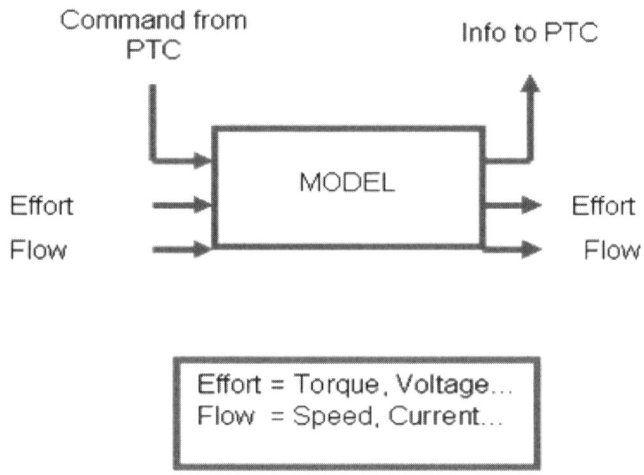

Figure 3. Global Formalism for the I/O of the Models Using Bond Graph

USE OF GOTO-FROM FORMAT As shown in Figure 4, to simplify the component models, we decided to use the GOTO-FROM format. As far as the models are concerned, all of the GOTO-FROM blocks are local and are located at the upper level of the model (no blocks are located in the subsystems). To facilitate the work for Hardware in the Loop (Control Desk access to the parameters and variables by using the Tags), the names of the Tags are defined in accordance with certain rules.

Other rules apply when developing a new component model:

- Colors are used to simplify model understanding: inports are in red, outports in cyan, GOTO-FROM in green, and constants in yellow.

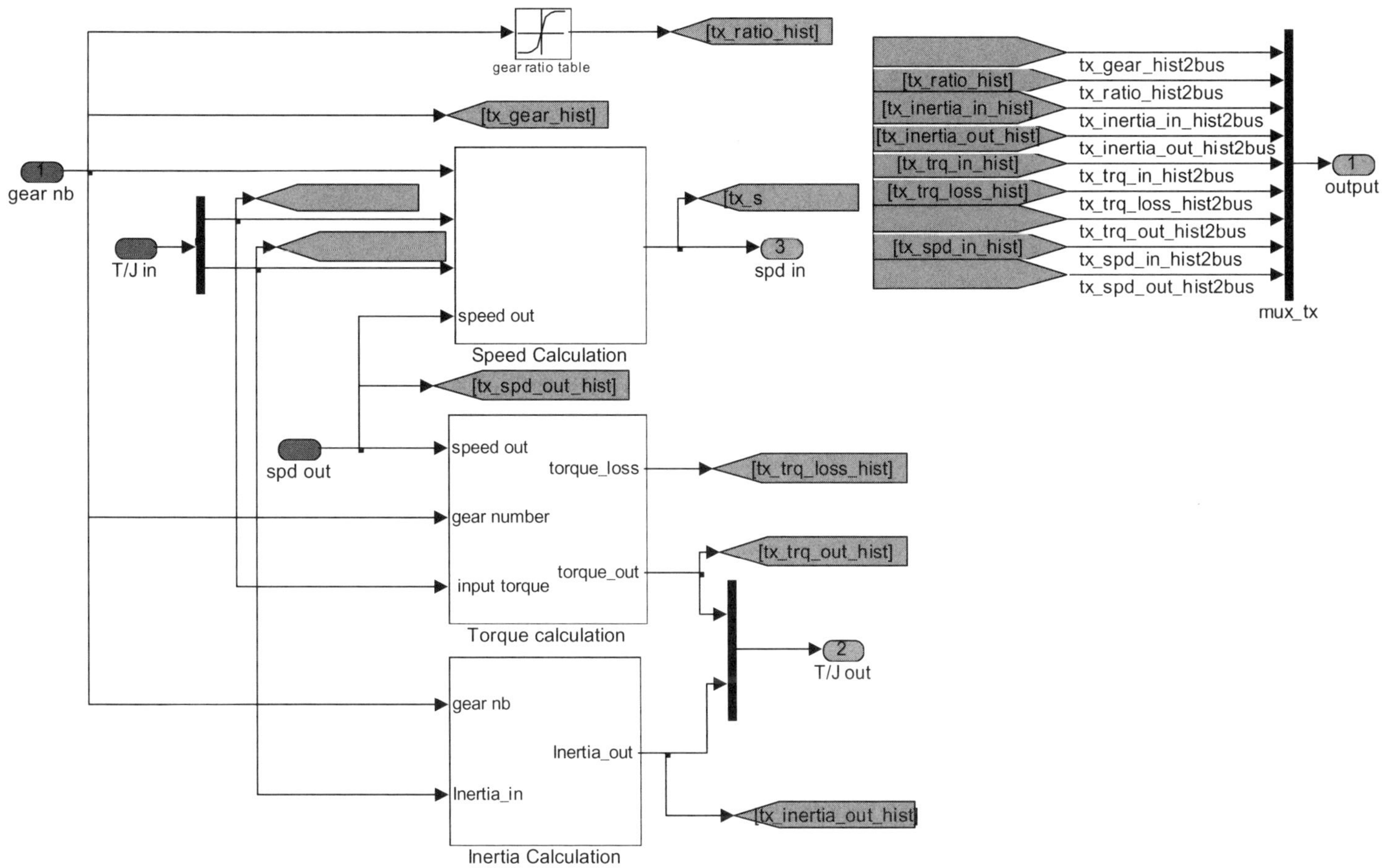

Figure 4. Example of Transmission Component Model

- Three blocks are used within each model to calculate speed, torque, and inertia.
- Lines to connect the information to the bus are named (parameter'2bus). These names are used so that users automatically know where each parameter is located in the buses.

Once the buses are created, users can access the parameters simply by using their names, as shown in Figure 5. For example, if the user wants to access the engine speed (parameter "eng_spd_hist"), he/she will use the parameter "nb_" followed by the name of the parameter. Accessing the wrong information is a major cause for mistakes and, as most simulation model users know, one of the most difficult to find. Another advantage of using this parameterized bus structure is that no major revision of the drivetrain model s structure is necessary when swapping between engine models with different numbers of output parameters, because the size of the bus is automatically updated.

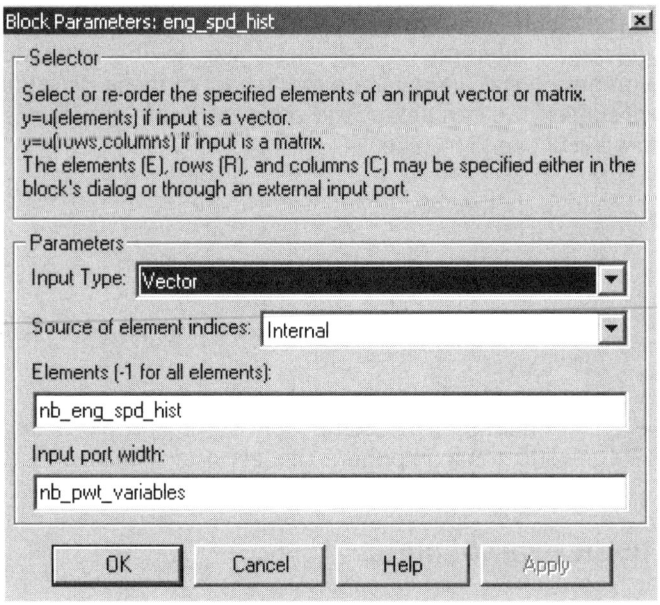

Figure 5. Parameter Access in the Buses

CONTROL STRATEGY

PSAT powertrain controllers, which are in charge of commanding the different components, have a generic structure common to all configurations, as shown in Figure 6. By using the accelerator/brake pedals and the information coming (via sensors) from the component models, we evaluate the constraints of the system, such as the maximum available torque of the engine. We then take those limits into account to define the optimized control strategy, which allows us to use the powertrain

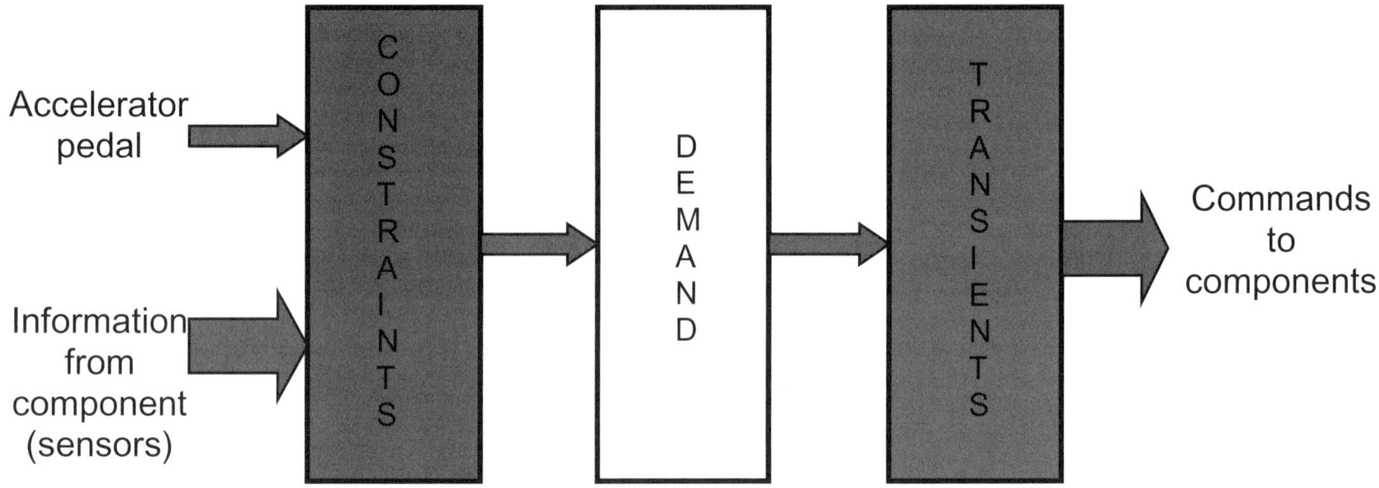

Figure 6. Powertrain Control Strategy Organization

components to minimize fuel consumption and emissions. Finally, we take the transients into account by defining the actions required to satisfy the control strategy demands. For instance, if the control strategy decides to shift gears with a manual transmission, we have to cut off the engine ignition, declutch, engage the neutral gear, engage the new gear, clutch, and inject once again. These steps have to happen successively and lead to a modification of the demands previously sent by the demand block.

Within the PSAT powertrain controller, different strategies can be selected within a particular powertrain model. Indeed, because the strategy has an important impact on the fuel consumption, it is interesting to switch between different control strategies to be able to compare them. To evaluate the impact of these different strategies, we can select and compare them through the graphical user interface.

Parameter nomenclature:

- The outputs of the constraints block end in cstr_hist.
- The outputs of the demand block end in dmd_hist (strategy).
- The outputs of the transient block end by trs_hist.
- The outputs of the component command block that goes to the component models end by cmd_hist (command).

USER FRIENDLINESS

GRAPHICAL USER INTERFACE Development of a graphical user interface (GUI) is very important to facilitate user choices in terms of drivetrain configuration, initialization files, and cycles.

Initialization Window Figure 7 shows an example of the initialization window.

1) Because of its flexibility, PSAT allows users to choose more than 150 pre-selected configurations. When looking at hybrids, it is difficult to talk about a parallel, because there are probably several hundred of them. So we decided to provide a picture of the exact drivetrain configuration (as shown on the upper left). Because the user also has the option of changing the location of the electric motor(s), a popup menu has been added (position 1 to 4).

2) Because of the number of components available in PSAT, it was impossible to keep the list of possible plots in a single popup menu. We decided to have a separate list for each component, as shown on the bottom left.

3) Several other choices were made available to facilitate user's decisions:

- Checkboxes allow users to choose their particular configurations.
- Because several levels of modeling can be available for a component (e.g., look-up table, neural network, or physical-based for engine), a new column is used to allow users to choose the version.
- Question marks allow users to directly open the right part of the documentation to provide information on the different levels of modeling available.
- A last popup menu has been added to provide information on the technology of the component (e.g., spark ignition [SI] or compressed ignition [CI] for engine).
- An option to choose between 2WD, 2t2WD, and 4WD configurations has been added.
- If we can also change any look-up table by scaling the different components, specific parameters can be changed by using the variable list. The parameters are listed by component.

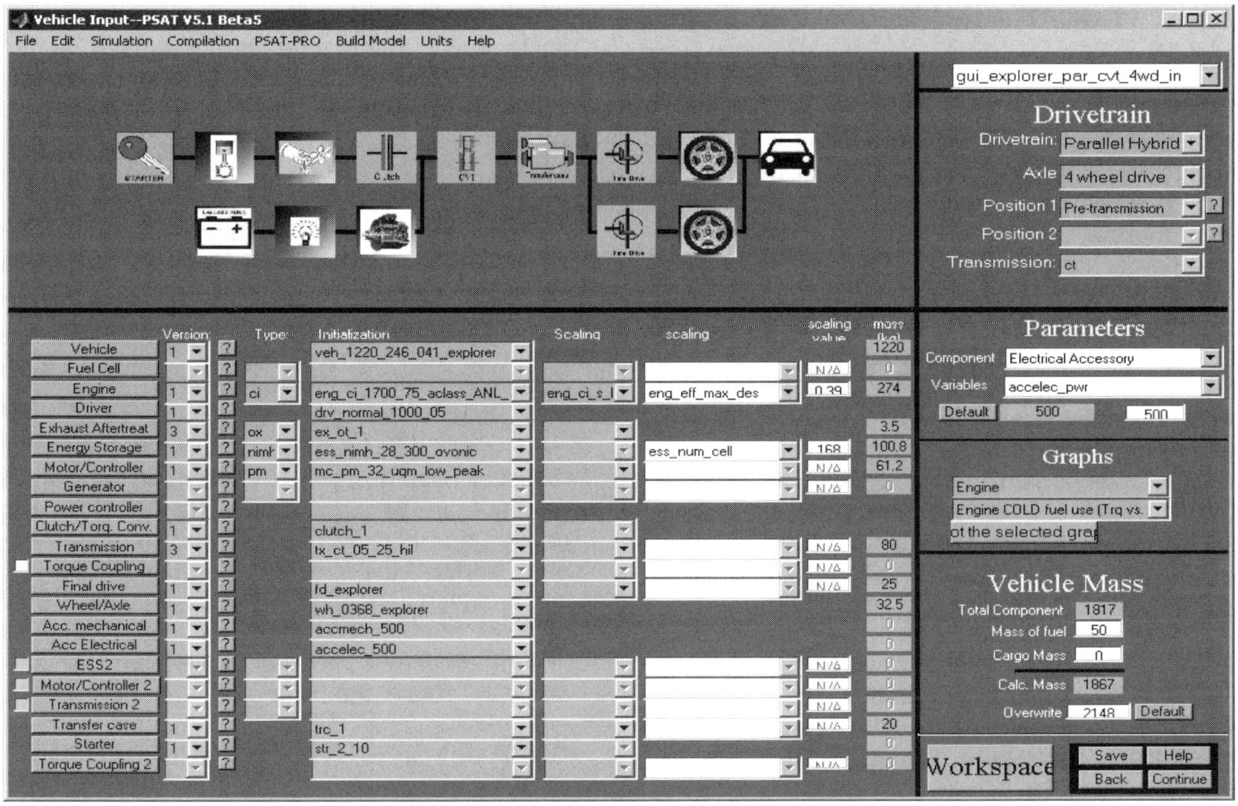

Figure 7. Input for Graphical User Interface

4) The main menu also allows users to:

- Change the simulation algorithm (variable or fixed step size); and
- Choose the units they want on the GUI (e.g., Standard International [SI] or U.S. units).

5) Several other specific windows have been developed to easily and automatically integrate new component models (version) or types without opening any MATLAB m-files. It is our intention that the user can do everything through the GUI without opening even one file.

Cycle Choice Window The second window of the GUI allows users to choose the type of simulation to be performed; in addition to choosing among a large number of driving cycles, simulating the performance of the drivetrains, and conducting a parametric study, users can employ specific tools to run several simulations in a row. These tools are critical because they allow engineers to spend time analyzing results, instead of waiting in front of their computers until the end of a simulation. Users can run dozens of simulations during the night and analyze the results in the morning.

When developing a control strategy, engineers always use standard cycles. However, standard cycles have limited benefits because they do not usually allow users to check the system behavior close to its limits (e.g., battery state-of-charge). It is then necessary to validate the strategies by using real trips (about one to two hours long in real time) rather than cycles (10 to 20 minutes long in real time). We have developed an innovative GUI that allows users to build their own cycles, which could be several hours long.

Post-Processing Windows Because of the complexity of hybrid electric vehicles, the post-processing information obtained after each simulation has been completed is crucial. PSAT naturally provides the final results of each simulation and the capability to plot each parameter. Users then have the option of easily comparing, in a couple of clicks, the same parameters from different simulations in order to, for example, study the influence of a powertrain configuration on fuel consumption. But more than the plots, detailed post-processing data including energy, power, efficiency, torque, speed, current, and voltage are very useful to users.

Moreover, to better understand and improve the drivetrain control strategy, PSAT provides all of this information for the four different conditions of operation (acceleration/deceleration and charging/discharging).

In order to run several simulations in a row and access them later, each simulation is saved by using four different files:

- A document, including the initial conditions and the final results;
- A MAT file with all the variables from the simulation;
- A file with all the post-processing calculations (e.g., energies, efficiencies); and

- An m-file to be able to rerun the exact same simulation.

IMPLEMENTATION OF NEW DATA OR MODELS

Because most simulation tools are developed for outside use, one of the most important characteristics of a tool is the option to easily implement proprietary data sets, component models, or control strategies. Using the structured approach previously described, we developed a specific GUI that enables the user to implement anything without modifying a single line of code, as shown in Figure 8.

By using this window, users can add, view, or delete data files, scaling algorithms, calculation files (for pre-processing), or component models, as well as change the picture. Component compatibilities are also taken into account, which is an important but uncommon capability for this type of software. In the PSAT model, both the compatibility with the drivetrain configuration and with other component models are taken into account. For instance, one torque converter can only be used with a specific automatic transmission of an engine technology (or type) with a specific after-treatment. Because software developers cannot expect users to know or remember all the different compatibilities, PSAT makes sure that only compatible choices are available for selection in the input window (Figure 7).

COMPARISON OF SIMULATIONS As previously mentioned, the number of advanced powertrain configurations is almost endless. To be able to make the right decision, users need to be able to easily compare different options. Several features have been implemented in the code that allow users to run several simulations in a row and later access the results. For the same configuration, users can run several driving schedules in a row, as well as performing parametric studies. To allow comparison among different powertrain or control strategies, we incorporated the ability to automatically create batch runs. This is only possible by saving the simulation parameters as well as the initial conditions and final results.

Figure 9 shows an example of a comparison between powertrain options. In that example, we ran a Toyota Prius and a Honda Insight on the Japan 1015 cycle. Each simulation can be accessed through a popup menu, and parameters, such as engine torque (bottom graph), can be compared. The first plot shows the desired and obtained vehicle speeds (m/s), the second one shows the engine torques (Nm) for each configuration. The figure shows that the Toyota Prius requires more torque from the engine than does the Insight because of the lower weight and better aerodynamics of the Insight.

COMPARISON BETWEEN SIMULATION AND TEST

In order to be sure we select the right configuration or control strategy, both the component and the drivetrain models need to be validated. Validation is a very important aspect of software development because it demonstrates to users the degree of accuracy of the software. Modeling tools can be validated by using different data sources, including vehicle, component, or drivetrain tests.

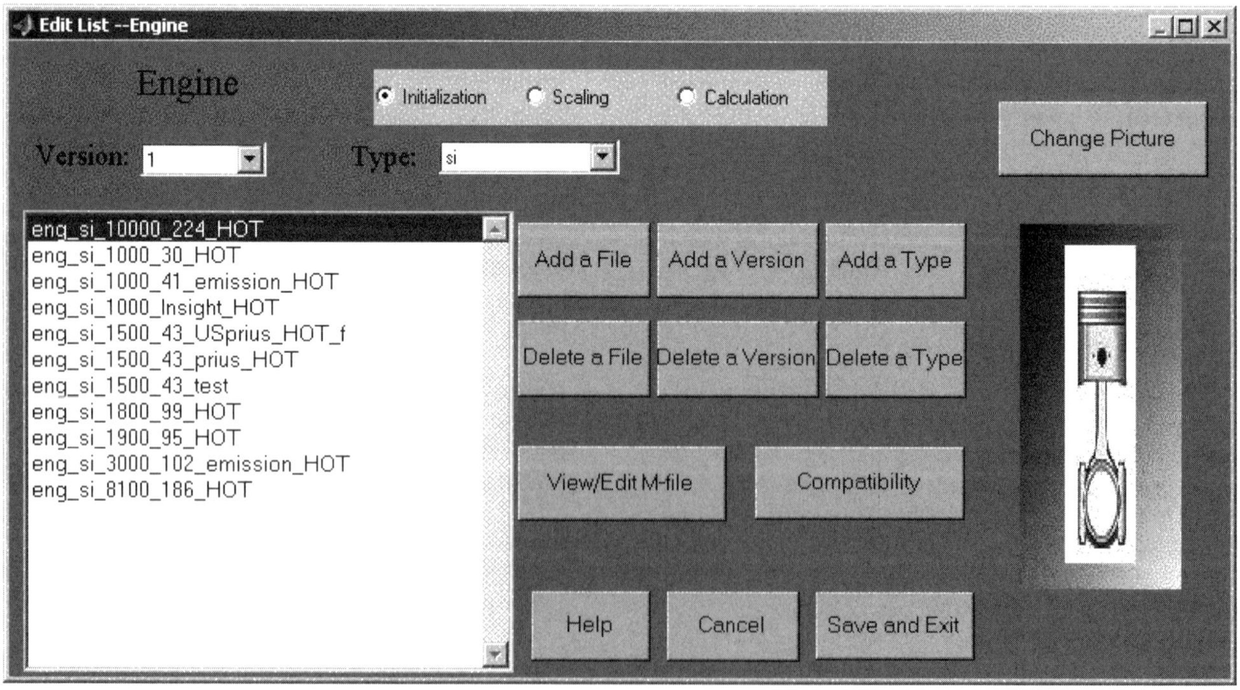

Figure 8: Integration of Data and Model

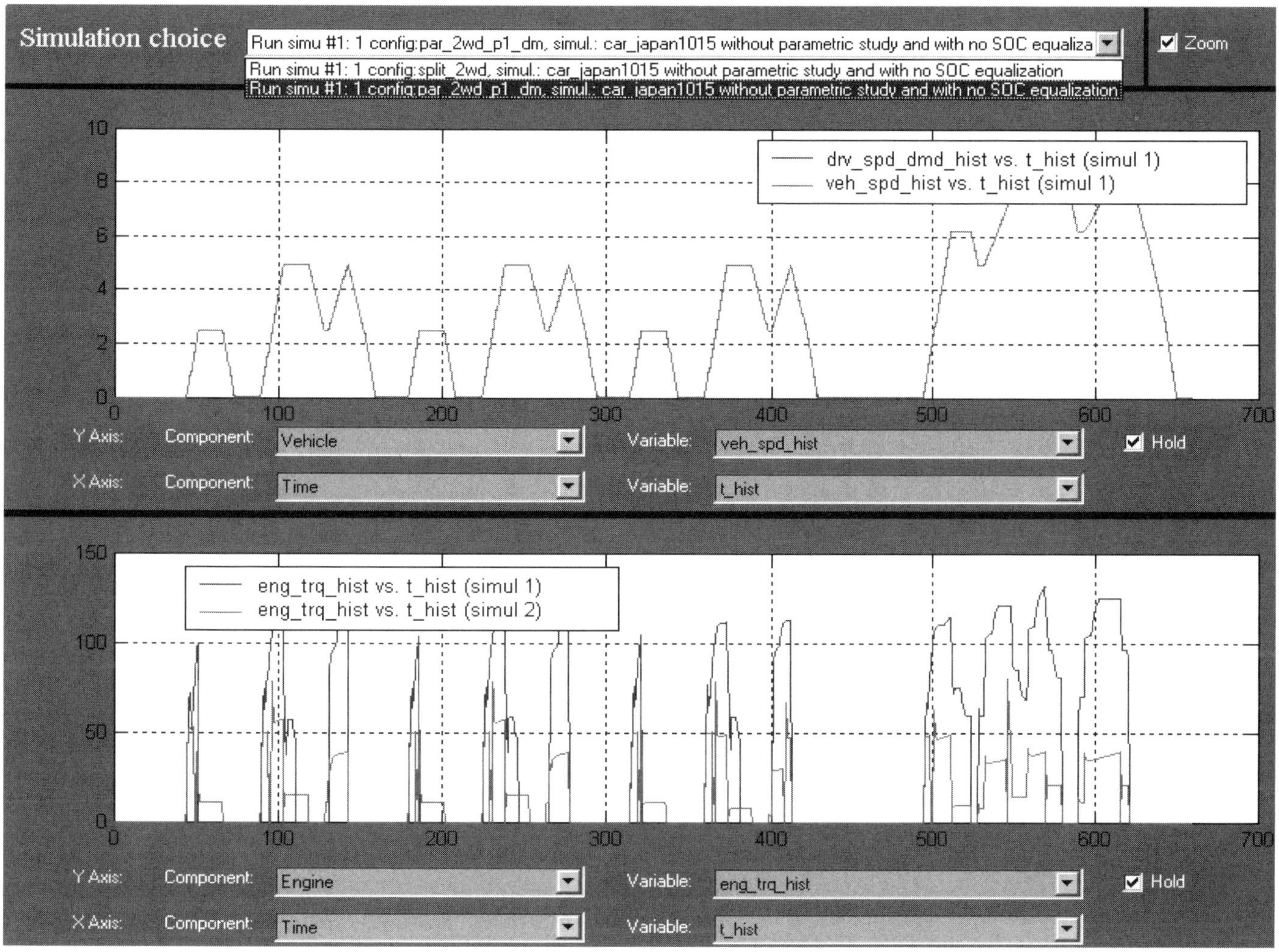

Figure 9. Prius and Insight Simulation Comparison on Japan 1015 Cycle

Argonne used all of these methodologies to validate PSAT. However, although Argonne s Advanced Powertrain Research Facility (APRF) is sufficient in the two first cases, the development of a specific tool dedicated to prototyping was necessary for drivetrain testing. To answer U.S. Department of Energy (DOE) and FreedomCAR Partnership needs, Argonne developed PSAT-PRO, the extension of PSAT for prototyping.

In order to easily compare test and simulation data (from APRF or PSAT-PRO), a specific window has been developed, as shown in Figure 10, to be able to dynamically replay tests, as well as simulation. Simultaneously examining both data sets allows users to process much more information than with static plots. Users can compare at every sample time the different powertrain parameters. In this example, the vehicle speed is shown, as well as the engine, motor, and generator maps. The tool allows users to quickly understand where the engine operates and, most importantly, why (i.e., deceleration, acceleration). In addition to being useful for understanding the control strategy of a particular vehicle from test data, this GUI can also be used to improve a control strategy from simulation.

CONCLUSION

Because of the number of possible hybrid architectures, the development of the next generation of vehicles will require advanced and innovative simulation tools. Model complexity does not mean model quality: flexibility, reusability, and user friendliness are key characteristics to model quality. By using a well-defined nomenclature, a structured approach, and an innovative algorithm, we are able to allow users to choose among more predefined drivetrain configurations than any other tool. Easy implementation of component data and models (including handing compatibility issues), as well as control strategies, is possible because we used a unified component model approach and a graphical user interface. Finally, comparison between simulations or between test data and simulation is facilitated by innovative dynamic interfaces. The structured, yet flexible, approach used in PSAT could be used as a base to establish industry standards within the automotive modeling community, where each institution implements its own data or model in a common generic software architecture.

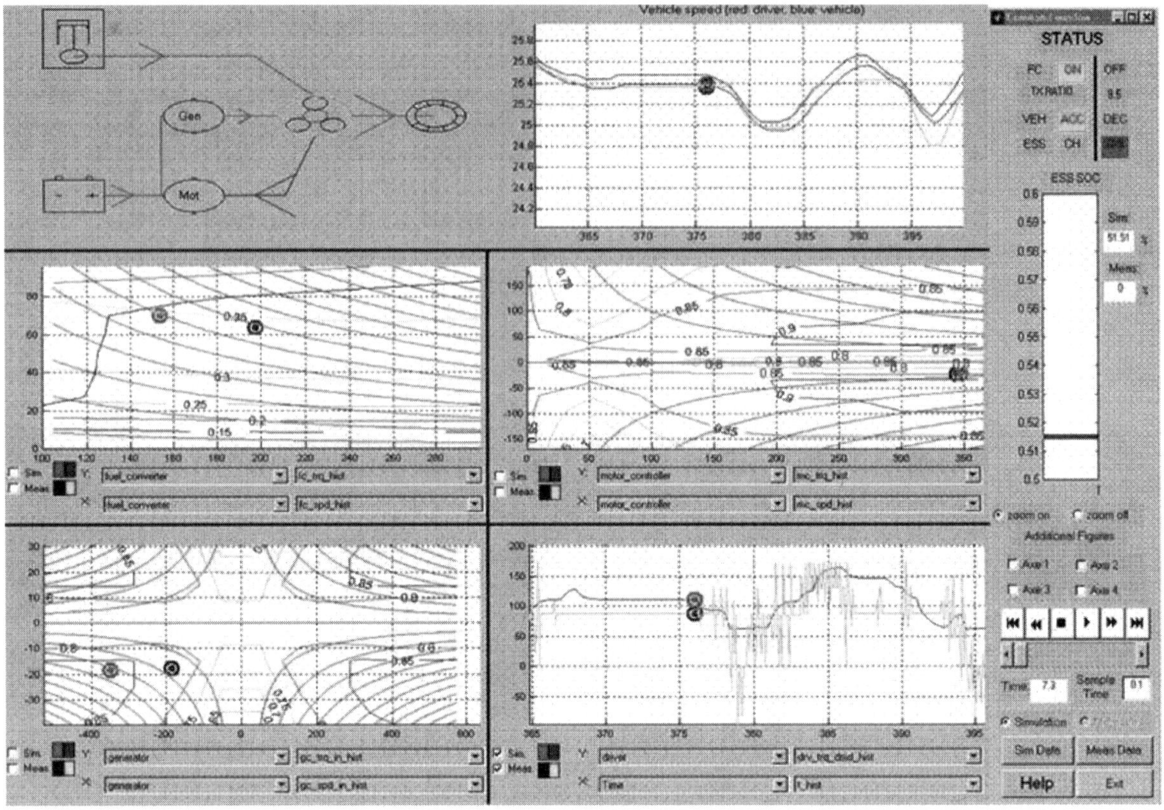

Figure 10. Dynamic Comparison of Test and Simulation — Prius Example

ACKNOWLEDGMENTS

This work was supported by the U.S. Department of Energy, under contract W-31-109-Eng-38. The authors would like to thank Bob Kost and Lee Slezak of DOE, who sponsored this activity.

REFERENCES

1. Argonne National Laboratory, PSAT (Powertrain Systems Analysis Toolkit), www.psat.anl.gov, last updated October 15, 2003.
2. Rousseau, A., S. Pagerit, and G. Monnet, The New PNGV System Analysis Toolkit PSAT V4.1 Evolution and Improvements, SAE paper 01-2536, Future Transportation Technology Conference, Costa Mesa, Calif., August 2001.
3. The Mathworks, Inc., Matlab Release 13, User's Guide, 2003.
4. Karnopp, D., D. Margolis, and R. Rosenberg, System Dynamics: A Unified Approach, 2nd edition, John Wiley & Sons, Inc., New York, 1990.

CONTACT

Aymeric Rousseau
(630) 252-7261
E-mail: arousseau@anl.gov

2004-01-1593

Model-based Testing of Embedded Automotive Software using MTest

Klaus Lamberg, Michael Beine, Mario Eschmann and Rainer Otterbach
dSPACE GmbH

Mirko Conrad and Ines Fey
DaimlerChrysler AG

ABSTRACT

Permanently increasing software complexity of today's electronic control units (ECUs) makes testing a central and significant task within embedded software development. While new software functions are still being developed or optimized, other functions already undergo certain tests, mostly on module level but also on system and integration level.

Testing must be done as early as possible within the automotive development process. Typically ECU software developers test new function modules by stimulating the code with test data and capturing the modules' output behavior to compare it with reference data.

This paper presents a new and systematic way of testing embedded software for automotive electronics, called MTest. MTest combines the classical module test with model-based development. The central element of MTest is the classification-tree method, which has originally been developed by the DaimlerChrysler research department. The classification-tree method exists for several years now and is mostly used for C-code testing. Now, it has been adopted to the needs of a model-based development process for embedded systems.

MTest opens a new way of assuring quality for embedded software, that is especially designated for automotive software developers.

This paper demonstrates, how MTest is used to test automotive software from model-in-the-loop over software-in-the-loop down to processor-in-the-loop testing. Additionally, test scenarios once developed using MTest can be reused in a hardware-in-the-loop environment. Thus, MTest provides a means to automatically test automotive software within the whole development process.

MODEL-BASED SOFTWARE DEVELOPMENT

Within automotive electronics development, a model-based development process has been established over the last years. Using modelling, simulation and code generation tools is a common way to develop and implement new vehicle functions.

Therefore the control function to be developed is described by the means of simulation tools like MATLAB/Simulink/Stateflow (function design). Such tools provide a graphical way of describing functions and systems. This includes block diagram notations as well as state charts. Using Rapid Control Prototyping (RCP) systems, the new functions can be proven in the real vehicle or on a dynamometer. Therefore, automatic code generation is used to generate C code from the function model. This code is run on powerful real-time systems. Such systems are connected to the real plant by special I/O. Changes can be made directly to the function model and tried out immediately by generating code once again.

Implementation of the function on a production ECU is done by automatic production code generation. However, the requirements on a production code generator are much higher then for RCP. The generated code must be highly efficient, error-free, reproducible, and well documented. An example for a production code generator is TargetLink [1].

MODEL-BASED TESTING

Today's way of automotive function and software developing using RCP is characterized by an experimentational way of working. Systematic and automated testing doesn't play an important role so far. Additionally, testing tools are missing today, which provide special methods for the testing tasks in the specific process stages. This is true especially for the early stages of function and software development. The model-based testing process as described in the following lays a major focus on systematic and automated testing in the early stages. It also includes ECU testing activities which are typical for the later development stages.

THE MODEL-BASED TESTING PROCESS

The model-based testing process (Figure 1) describes the different activities within the whole automotive electronics development process from a testing point of view. This includes testing in the early function development as well as ECU testing later in the process.

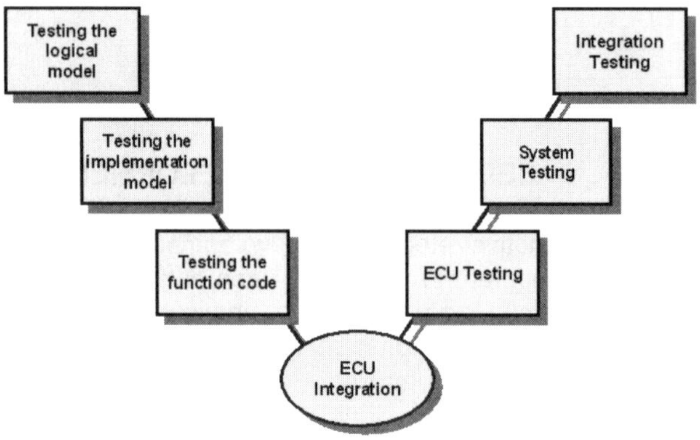

Figure 1: Model-based testing process

Testing the Logical Model

Testing the logical model means systematic and automatic testing of an executable model of the function or controller to be developed. This model is the test object or unit under test (UUT). The test can be done open loop or closed loop using a model of the plant (model-in-the-loop, MIL).

Testing the Implementation Model

The functional model has to be prepared for implementation. Software design information has to be added. Functional models are usually floating-point models whereas the implementation in C is often realized using fixed-point arithmetic. Thus scaling information, implementation data type, LSB and offset, is to be specified for each block output signal and parameter. The behaviour of the fixed-point implementation has to be compared with the behaviour of the functional model. It has to be checked if the quantization effects introduced are acceptable. This verification is done by simulation since the equivalence between the two representations can not be formally proved. The implementation model can also be tested in a closed loop environment (MIL).

Testing the Function Code

The next step is testing the actual function code. This can be done on a host PC (software-in-the-loop, SIL) or on the target processor (processor-in-the-loop, PIL).

ECU Integration

The generated function code has to be integrated with the overall ECU software. ECU integration means integration with other function modules, the operating system and I/O drivers. Although, this step isn't a test step in the sense of the model-based test process, the integration is already to be taken into consideration and prepared when creating the implementation model. Special focus is laid on functions and global variables that are defined or to be called and re-used outside of the implementation model. Their definitions and declarations must match the ones in the external code.

The operating system integration depends on the scope of the model. If the model only describes a single feature that will be part of one task in the ECU then the call of the generated function is usually manually implemented in the OS frame. If the model has a wider scope and consists of multiple functions and tasks then operating system objects are already specified in the model. For example, dSPACE offers a special module for its production code generator TargetLink to support and automate the integration with OSEK operating systems. Task distribution, intertask communication and other OS properties can be specified directly in the model. The generated code then already contains the corresponding OSEK keywords and can be integrated with the OSEK OS without any further manual integration work [2].

ECU Testing

ECU testing typically is done using hardware-in-the-loop simulation (HIL). Therefore the ECU prototype is connected to a real-time simulation system simulating the plant. Corresponding ECUs are also simulated rest bus simulation). Almost always, ECU testing is black box testing where the inputs are stimulated and the outputs are monitored.

System Testing

System testing means to test the ECU in its direct technical environment using HIL simulation. Therefore the ECU is at least partially integrated with other ECUs and its behavior is tested in conjunction with other ECUs.

Integration Testing

Finally, all ECU of a single vehicle are integrated and the whole network system is tested. This is called integration testing. HIL simulation is used more and more also for integration testing.

The testing tasks described above are different from each other. In the following, a testing methodology is described, which supports especially the early phases, i.e. testing the logical model, testing the implementation model and testing the function code. Additionally, test data once developed using this methodology can also be used in hardware-in-the-loop simulation.

THE MTEST APPROACH

The MTest (MTest = "Model-based Testing") methodology complements model-based development with a method for systematic test definition. The starting point of the MTest testing process is a model of the function or controller to be developed, implemented in Simulink or TargetLink. Based on the interface of the logical model, and by using the classification-tree method, the function developer can derive test scenarios systematically and describe them graphically. With the graphical representation the user gets visual information about the test coverage. Test coverage indicates, how well the test cases cover the range of possible test input combinations and is therefore the most important test metrics.

The MTest process consists of different testing activities building the base for systematic testing and therefore support for a systematic testing procedure. The testing activities are shown in Figure 2 schematically. They are described more detailed in the following paragraphs.

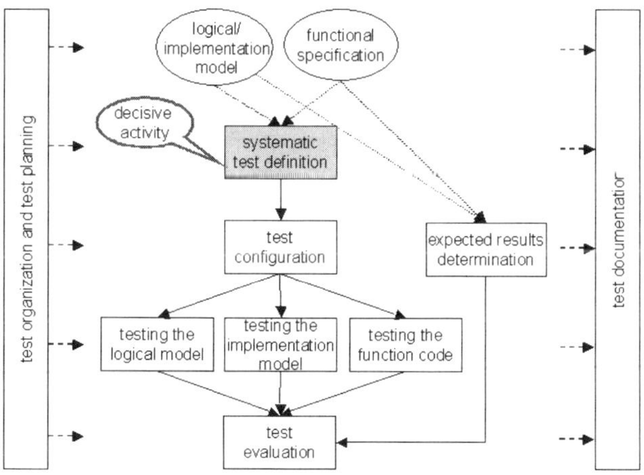

Figure 2: Model-based testing activities

As an alternative to using the classification tree method, it is possible to use existing data as test data. This is called "direct testing".

To illustrate the MTest approach to model-based testing we use the example of a vehicle dynamics control (VDC) system that controls the vehicle motion in physical limit conditions (cf. [3]).

SYSTEMATIC TEST DEFINITION

The MTest approach to model-based testing utilizes a specific instance of the classification-tree method with extensions for embedded systems (CTM/ES) for the systematic test definition, [4], [5]. The classification-tree method is a black-box partition test design technique, where the input domain of the test object is split up under different aspects usually corresponding to different input data sources. The different partitions, called classifications, are subdivided into (input data) equivalence classes. Finally different combinations of input data classes are selected and arranged into test sequences.

SELECTING THE TEST INTERFACE

A subsystem within an .mdl file is to be selected first in order to be able to subsequently relate the test scenarios to the respective units under test (UUT), i.e. Simulink or TargetLink model subsystems.

The interface of the subsystem to be tested will then be analyzed automatically, and the information relevant to testing will be extracted (model extraction).

The input of the subsystem to be tested forms the potential input variables for the test. It is consequently called a 'potential test interface'. There is no necessity, however, of using the potential test interface for the test object stimulation on a one-to-one basis: Fed back values, for example, do not need to be predetermined as they are generated by the system environment. On the other hand, it is often easier to describe a complex input signal by means of (additive or multiplicative) superposition of two sub-signals. In this case, instead of the potential interface signal the two sub-signals would be described. The values actually used for the simulation are referred to as 'effective input interface'. If there are differences between the potential and the effective interface for a certain test object they have to be mapped onto each other.

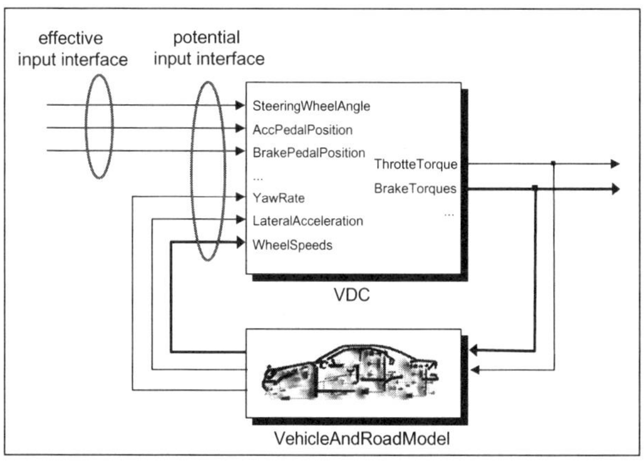

Figure 3: Potential and effective test interface

The VDC-software's behavior is determined, among other things, by the steering-wheel angle, the accelerator and brake pedal positions, the yaw rate, and the four-wheel speeds. These signals form the example's input interface. If a closed loop test with typical driving maneuver is to be performed, however, only the driver inputs SteeringWheelAngle, AcceleratorPedalPosition and BrakePedalPosition have to be stimulated. These values form the effective input interface. The remaining input values are implicitly determined by the vehicle/road model. (Figure 3).

CREATING THE CLASSIFICATION-TREE

Based on the effective test interface MTest automatically outputs a first, incomplete instance of a classification-tree called basic tree (Figure 4): The name of the unit under test itself forms its root node (here: "VDC"), the signals of the effective input interface (e.g Steering-WheelAngle) are denoted as classifications below the root node. In a second step, the generated classifications must be disjointedly and completely partitioned into (equivalence) classes which are suitable abstractions of individual input values for testing purposes.

The partitioning aims to achieve a selection of the individual classes in such a way that they behave homogeneously with respect to the detection of potential errors. That is, the unit under test behaves either correctly or erroneously for all the values of one class (uniformity hypothesis).

An heuristic procedure has proved successful in approaching this ideal partitioning as much as possible in practice. The inputs' data type and value range provide first valuable clues to partitioning: Where real-valued data types with minimum and maximum values established are concerned, it is possible, for example, to create a standard class each for the boundary values, for the value of zero and for those intervals in between. Alternatively, real-valued data types could be partitioned into same size sub-intervals. Similar data-type specific standard classifications can also be utilized for different data types.

As soon as information on the data types or value ranges of the input variables is available to MTest, the data-type specific standard classifications for different data types can be generated automatically (cf. [6]).

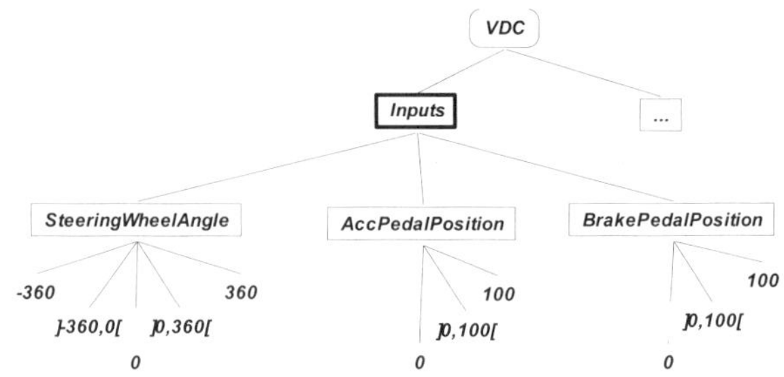

Figure 4: Basic classification-tree

Figure 4 depicts the automatically generated basic classification tree for the VDC example: It contains the automatically generated standard partitioning for real-valued signals for the 3 effective interface input signals. The following five classes arise for the signal Steering-WheelAngle, which can take on values from the range of –360° to 360°: –360,]-360, 0[, 0,]0, 360[and 360. In this case,]x, y[denotes an interval, open on both sides, with the boundary values x and y.

As a rule, the data-type specific standard classifications are not detailed enough for a systematic test. They have to be refined or modified manually in order to approach a partitioning according to the uniformity hypothesis. The quality of the specification and the tester's experience are crucial in this respect.

The evaluation of the pedal positions (described as percentages in the VDC software) recognizes a pedal as depressed only if it is activated above a certain threshold value PedMin. The pedal values above and below the threshold should therefore be considered separately because behavior is expected to differ. As the acceleration force also influences the system behavior, there has to be an additional distinction between light (pedal position< 50%) and strong (pedal position ≥ 50%) pedal operation. The result is a final partitioning of the pedal positions into the classes 0,]0, PedMin[, PedMin,]PedMin, 50[, [50, 100[and 100. In this case, [x, y[denotes an interval which is closed on the left side and open on the right side. The partitioning of SteeringWheelAngle has been refined as to subsume all 90° sections in one class respectively (Figure 6).

DEFINING TEST SEQUENCES

Based on the input partitions, test sequences can be determined. These sequences specify how the behavior of the regarded unit under test should be tested. The domain for the description of test scenarios is provided by the completed classification-tree. The tree is used as the head of the combination table. Each sequence captures a data abstraction of the unit under test's inputs. Hence, it describes – largely independent from detailed or precise data – what is to be tested. In order to represent test sequences in an abstract way, they are decomposed into individual test steps. According to their temporal order, the steps build the rows of the combination table. Such a sequence of test steps is called a test sequence. Each test step defines the inputs of the UUT over a certain time span. The time spans are listed in a separate column on the right-hand side of the combination table. The beginning and end points of these time intervals are called synchronization points, since they synchronize the stimuli signals at the beginning and end of every test step.

The description of the values of the single stimuli signals for each test step takes place by marking a class defined for this signal in the classification tree. This is indicated in the middle part of the combination table. The stimulus signal in the respective test step is thus restricted to the part-interval or single value of the marked class. The combination of the marked input classes of a test step determines the input of the UUT at the respective synchronization point.

The values of the stimuli signals between the synchronization points are described by basic signal shapes. Different signal shapes (e.g. ramp, step function, sine) are represented by different arrow types which connect two successive markings in the combination table. In this way stimuli signals can be described in an abstract manner by the means of parameterized, stepwise defined functions [4].

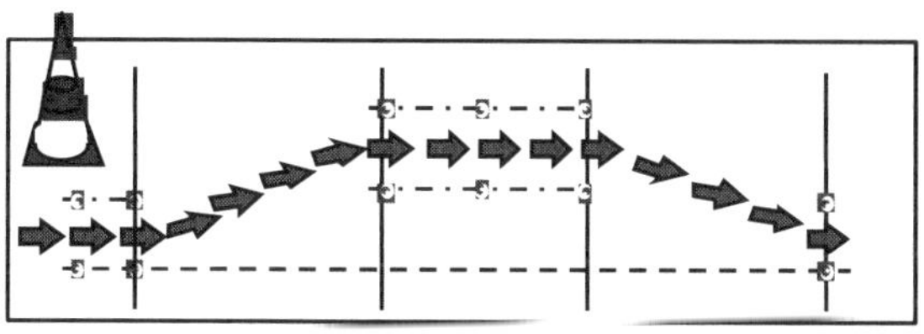

Figure 5: Driving maneuver "lane change"

Figure 6 shows the lane changing maneuver of Figure 5 as a test sequence in the combination table. After an acceleration phase the steering-wheel is first turned in a 90-degree angle to the left, then, the steering wheel is turned to 90-degree in the opposite direction and back into the original position. After a hold phase the wheel is turned by 90-degrees to the right, back to the left and back into the neutral position. Here, a solid line as arrow type means a ramp shaped change of the signal value, no (visible) transition means a skip of the signal value at the end of the interval.

The accelerator pedal ramps up during the acceleration phase. The adjusted pedal position is held and at the end of the test sequence the pedal is released again. At the same time, the brakes areactivated. A dashed line as an arrow type denotes a change of the signal value in the form of a sine half-wave.

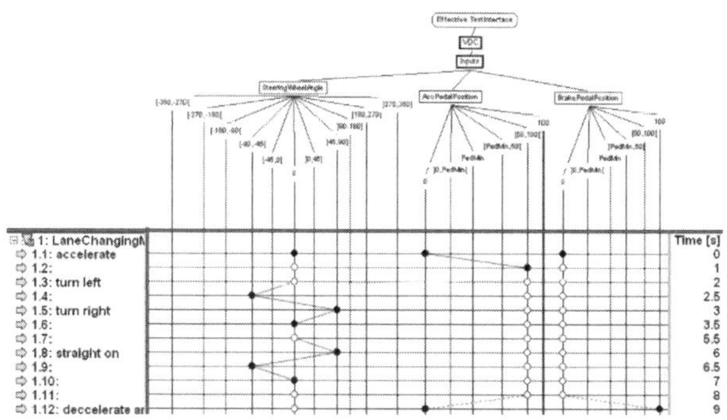

Figure 6: Classification-tree with test sequence

Further test sequences can be described underneath the classification tree by using the procedure above mentioned.

After the determination of test sequences has been completed, it is necessary to check if they ensure a sufficient test coverage. At this early stage of the testing process, the CTM/ES already allows the determination of different abstract coverage criteria based on the classification tree and the test sequences:

A requirements coverage analysis can verify whether all requirements of the requirements specification are covered sufficiently by the test sequences. In general, a n:m-relationship exists between requirements and test scenarios. In the course of the analysis it is necessary to prove that every requirement is being checked by at least one test scenario and that the existing test scenarios are adequate to test the respective requirements.

Furthermore, the CTM/ES supports a range coverage analysis. This analysis checks the sufficient consideration of all equivalence classes defined in the classification tree in the test sequences. This check can be executed, according to the respective application case, by using different, so-called classification-tree coverage criteria (CTC) (cf. [5]):

- The minimum criterion (CTC_{MIN}) requires every class of the tree to be selected in at least one test step. The minimum criterion is usually accomplishable with a few test sequences, the error detection rate, however, is rather low.
- The maximum criterion (CTC_{MAX}) requires every possible class combination to be selected in at least one test step. The fullfillment of the maximum criterion should bring a high error detection rate. Because of the combinatorical "explosion", this criterion is only practicable when having a small number of classes.
- The n-wise combination criteria (CTC_n) present a compromise. Here, it is necessary to ensure that every possible combination of n classes is selected in at least one test step. For example, a pair-wise combination of classes (CTC_2) is practicable.

The selection of appropriate criteria has to take place in a problem-specific way within the frame of the test planning. If the criteria defined beforehand are not sufficiently fulfilled, additional test steps / test sequences need to be added until the required criteria are fulfilled.

TEST CONFIGURATION

TEST DATA REFINEMENT

The test scenarios gained by using the classification method only contain abstracted stimulus information, because only equivalent classes have been used, but no specific data. Thus, in a second step, the test data is instantiated by the means of definite numbers. The borders of the classes of the classification tree are used as signal constraints, wherein the actual signal traces can vary.

Instantiating the test data is done using the signal editor shown in Figure 7. The borders of the equivalent classes form the constraints of the value ranges at the respective sample points. By default, MTest uses the mean values of the intervals in the classification tree. In this example at the sample point *3 sec*, the value has been edited within the constraints (cf. the marked cell at the bottom).

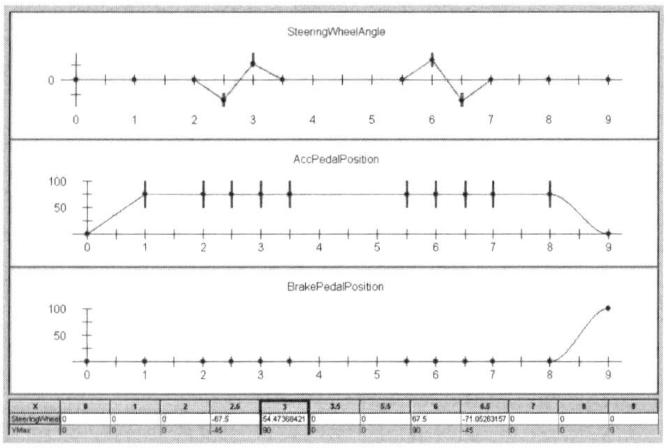

Figure 7: Signal editor with interval borders

As an alternative to the classification tree method, MTest allows to use real data as test data. This is called “direct testing”. Using direct testing, it is possible to import real measurement data, gained e.g. from a driving or a dynamometer experiment. This data can be used to stimulate the UUT (Figure 8).

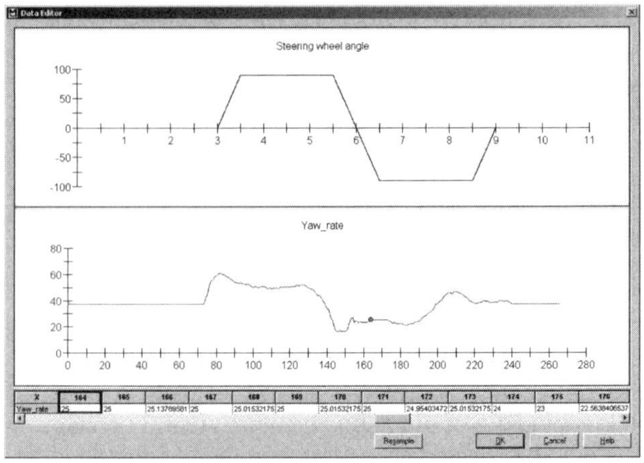

Figure 8: Editing imported test data

ASSIGNING TEST DATA TO THE MODEL INTERFACE

When using the classification tree method, the assignment of the generated test data to the inputs of the UUT is done automatically. Using direct testing, the data must be explicitly assigned to the inputs of the UUT (Figure 9).

Once the assignment has been defined by the user, MTest checks for the consistency in terms of signal type, signal complexity and signal dimension.

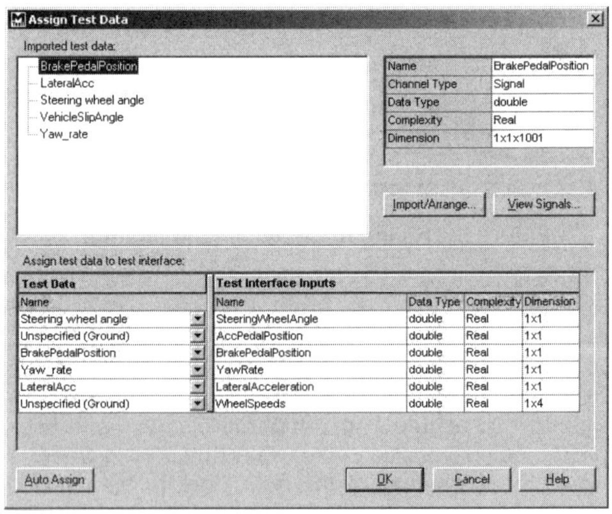

Figure 9: Assignment of test data

DEFINE REFERENCE DATA AND EVALUATION CRITERIA

Test and simulation results can also be compared to reference data. Reference data can be a result of former test runs, or even be any kind of measurement data which can be imported.

In a further configuration step, the user can define the evaluation rule, i.e. how the results shall be compared with reference data. Therefore the user can select out of a set of evaluation criteria, including absolute and relative difference. This set can also be extended by the user.

TEST EXECUTION AND EVALUATION

GENERATE TEST HARNESS

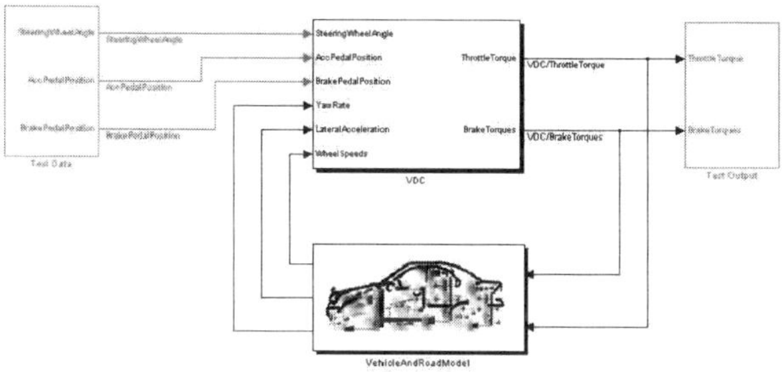

Figure 10: Generated test harness in Simulink

The test scenarios which have been defined using the classification tree method or by importing data can be applied to all of the three representation forms of the UUT: to the logical model, to the implementation model and to the function code. Therefore, MTest supports executing the tests in Simulink and in the different TargetLink modes. For the execution, a test harness in generated automatically in Simulink (Figure 10) and a copy of the Simulink or TargetLink model of the UUT is inserted into it.

MTest can activate the required TargetLink simulation mode – floating-point simulation on the host PC („TargetLink MIL"), production code simulation on the host PC („TargetLink SIL") and production code simulation on the target processor („TargetLink PIL") – and if necessary start the TargetLink production code generation.

Figure 11 shows, how the desired modes in which a test sequence is to be executed, can be selected by the user. Figure 12 shows how the different modes are represented in an MTest project tree.

Figure 11: Dialog to define the simulation modes

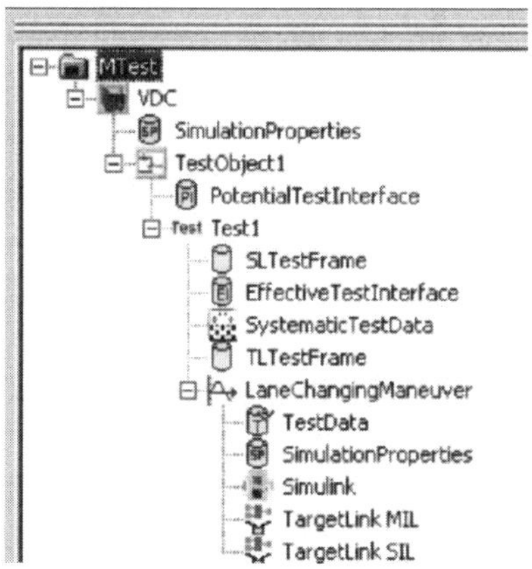

Figure 12: MTest project tree with different simulation modes

RESULT MANAGEMENT AND REPORT GENERATION

Once the test has been executed, the test results are collected automatically and displayed instantly. The amount and the depth of result information can be adjusted by the user.

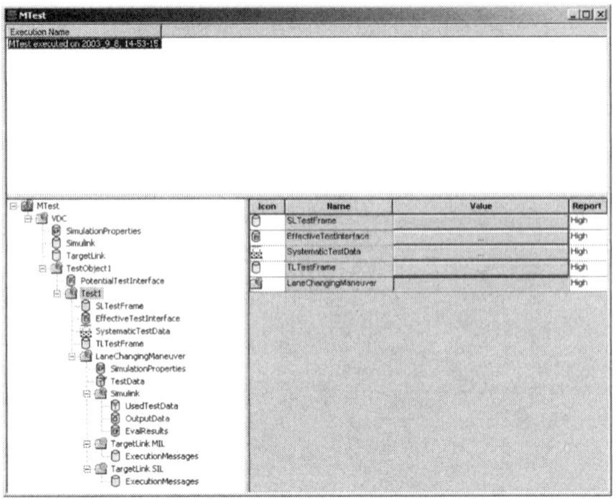

Figure 13: Result browser

The results are structured hierarchically and displayed as a tree (cf. Figure 13). The result tree can include any data item and any test which has been done in the different simulation modes. The user can navigate through the tree and view all details. For signal traces, the user can also generate plots immediately.

Finally, it is possible to generate test reports based on the result information. Test reports can be produced in different formats, e.g. HTML or PDF.

MTEST IN AUTOMATIONDESK

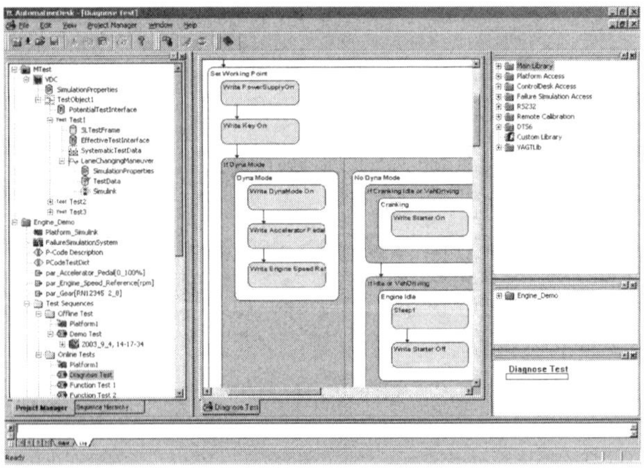

Figure 14: AutomationDesk

Although the testing tasks and activities within the different process stages vary from each other very much, a testing environment must combine and integrate all these approaches under one common roof. this can be achieved, if all the necessary elements of the process are provided by one testing tool. An example is the tool AutomationDesk, [7], [8] (Figure 14).

The enormous number of tests for a single function or ECU is not only to be developed and executed. All the tests must be stored and administrated consistently, so that they can be performed repeatedly ("regression testing") and reproduced at any time. The large amount of test results – each test run produces a new result instance – must be stored persistently. Storing, maintaining and administrating this large amount of tests together with the test data and the test results require powerful means to manage test projects.

Figure 15 shows an example of a test project structure in AutomationDesk. The test data and the test results together with the test sequences are displayed in project trees. The upper part of the tree contains an MTest project, the lower part contains a typical HIL project. It is obvious that test data which have been gained e.g. by using MTest can be reused in an HIL simulation.

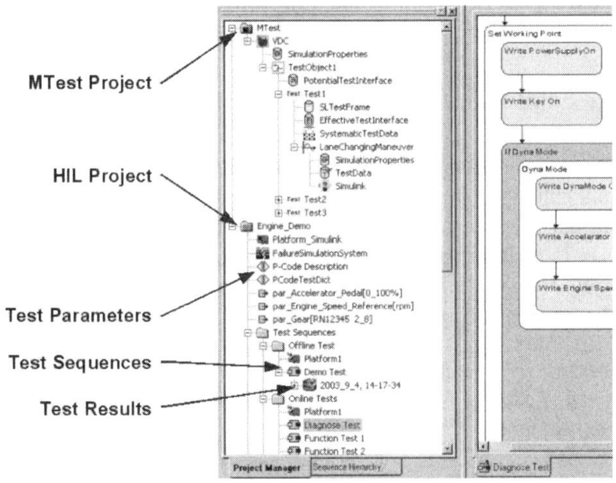

Figure 15: Test project examples in AutomationDesk

CONCLUSION

This paper describes a method and a tool for systematic and automated testing, called MTest. Based on a model-based testing process, MTest especially allows for model-based testing in early function and software development. The core of MTest is the classification-tree method, providing a systematic way of developing test scenarios graphically. Since MTest is an integral part of the test automation environment AutomationDesk, test scenarios once developed using MTest can be reused in later development stages, e.g. when testing real ECUs or ECU prototypes by the means of hardware-in-the-loop simulation. AutomationDesk together with MTest therefore build a testing environment supporting the whole model-based development process.

REFERENCES

1. dSPACE TargetLink poduct information: http://www.dspaceinc.com.
2. Köster, L.; Thomsen, t.; Stracke, R.: Connecting Simulink to Osek: Automatic Code Generation for Real-Time Operating Systems with TargetLink. SAE 2001,March 5-8,2001,Detroit, Michigan, USA Technical Paper 2001-01-0024
3. v. Zanten, A.; Erhardt, R.; Landesfeind, K.; Pfaff, G.: Stability Control. In: R. K. Jurgen (Ed.): Automotive Electronics Handbook. 2nd edition, Mc Graw-Hill, 1999
4. Broekman, E.; Notenboom, E.: Testing Embedded Software. Addison-Wesley, 2003
5. Grochtmann, M.; Grimm, K.: Classification Trees for Partition Testing. Software Testing, Verification and Reliability, 3, 63-82, 1993
6. Conrad, M.;Dörr, H.; Stuermer, I.; Schuerr, A.: Graph Transformations for Model-based Testing. Proc. of Modellierung 2002, Tutzing (D), März 2002
7. Lamberg, K., Richert, J.; Rasche, R.: A New Environment for Integrated Development and Management of ECU Tests. SAE 2003-01-1024, 2003.
8. dSPACE AutomationDesk product information: http://www.dspaceinc.com .

CONTACT

Dr. Klaus Lamberg is responsible for the product strategy, product planning, and product launches of test and experiment software at dSPACE GmbH, Paderborn, Germany.

E-mail: klamberg@dspace.de
Web: http://www.dspaceinc.com

2004-01-0900

Integration of a Common Rail Diesel Engine Model into an Industrial Engine Software Development Process

J. Baumann, D. D. Torkzadeh and U. Kiencke
Institute of Industrial Information Technology, University of Karlsruhe (TH), Germany

T. Schlegl and W. Oestreicher
Siemens VDO Automotive AG, Germany

ABSTRACT

In this paper we show the benefits of integrating a sophisticated engine model into an engine software development process. The core goal is the simulation based tuning of engine control parameters. The work reported here is resulting out of a prolonged co-operation between Siemens VDO Automotive AG and the Institute of Industrial Information Technology, University of Karlsruhe (TH), Germany. The approach is based on a model of the variable energy conversion process within a Diesel engine. The model features phenomenological fuel spray and vaporization models as well as cylinder individual mechanical aspects and fully copes with multiple injection systems. To be useful for an industrial function development process it provides a flexible and modular structure and features computational efficiency - considering real-time capability. The model is matched with the behavior of an engine of interest and connected with a control function under development. The control performance is evaluated by stimulation of the control function with test signals while being calculated in closed-loop with the engine model. Tuneable parameters are adapted to optimize engine performance until meeting given requirements. Simulation results of the achieved improvement in control performance will be presented.

INTRODUCTION

Engine developers focus on maximizing the efficiency of the energy conversion process in their engines and minimizing raw emissions. Additionally, engines shall reach performance goals like e.g. simultaneously featuring good dynamics and smooth running in idle conditions. Diesel engines suppliers support this process by providing sophisticated injection systems. The latest generation of them can realize multiple injection patterns. To fully utilize these systems high performance engine control functions are provided by the suppliers as software in addition to the system hardware. Both get more and more complex. To reduce time and costs for control design the performance of control functions has to be quantitatively evaluated at an early stage of development. As the number of tuneable control parameters increases tremendously it is crucial to evaluate control functions and optimize their parameters in an iterative process during development at the developers desk. Judging control performance functions needs simulations of the function itself together with an appropriate dynamical model of the engine.

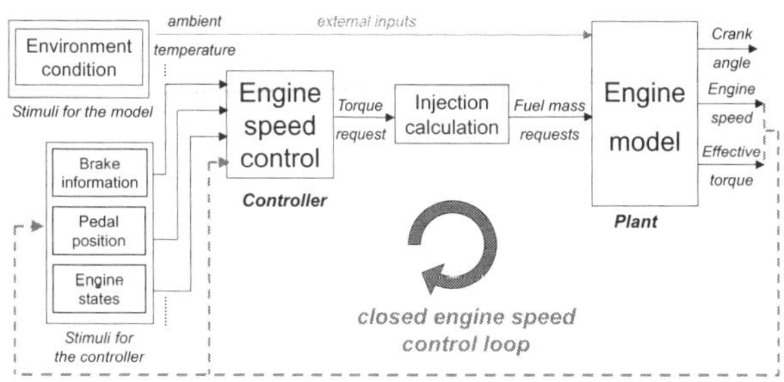

Figure 1: Control Loop Structure

A typical control architecture is shown in Figure 1. To be useful for both development and parameterization of the engine speed control part, all elements of the depicted structure are implemented in the Matlab/Simulink[1] based computer aided control system design tool SDA[2] used by Siemens VDO. The architecture comprises the engine model, the engine speed controller and stimuli generators. The latter provide physical values consumed by controller and model, e.g. the ambient temperature or driver actions on brake or accelerator pedal. Additionally, crankshaft synchronous and time synchronous trigger signals are generated to schedule the calculation of the different modules within the engine speed controller, as if implemented in an ECU[3]. Output signals of the engine model are fed into the engine speed controller as measured values to close the loop. Furthermore they are put into the stimuli generator to calculate the crankshaft synchronous trigger and the current state of the engine which triggers switching events in the controller.

With the use of pre-, main-, and post-injection or even multiple injections common rail systems allow to shape injection curves as necessary. As every type of injection curve can be applied, the energy conversion process has to be calculated in a flexible way. The Vibe- or Double-Vibe[4] functions formerly used as an approximation for the energy conversion process cannot be employed any longer.

In this paper, a phenomenological approach for evaporation is used as input to the thermodynamic equations involved. Thus, influence of the injection curve on pressure and temperature inside the combustion chamber can be derived. The information on in-cylinder variables can be taken as input for several sophisticated controllers applied in ECU's today. For usage in Hardware-in-the-Loop applications, the combustion process has to be calculated in realtime. The modeling language is Matlab/Simulink and C/C++.

In this project, different engine models have been designed for various fields of applications. There is a model of a single cylinder for realtime calculation as well as a four-cylinder model of the complete engine including frictional forces. The four-cylinder model is used in functionality development to examine controllers in an early stage of the development process. Another idea for the engine model is to fit stricter emission laws by changing the shape of the injection curve. For this purpose, an emission model may be used [7].

At first, the structure of the model will be explained. Reflecting the importance, next an evaporation model is presented. Finally, exemplary results of the use of the model in an industrial development process are presented.

[1]Matlab and Simulink are registered trademarks of the Mathworks

[2]SDA = System Design Automation. SDA is a registered trademark by Siemens VDO.

[3]ECU = engine control unit

[4]For more information see [4].

MODEL STRUCTURE

In this section the structure of the engine model will be explained. Figure 2 gives a conceptional overview of the different models, designed as part of this project. The injection-rate and the idle-speed serve as input to the model. Further models may be connected to the engine model via interfaces, e.g. a hydraulic model of the common-rail itself or a model for the intake and outlet air-ways.

Included in the model is a central unit for process-control. Another unit is responsible for the calculation of engine friction and, with help of crankshaft geometry, as output signal the engine torque. In addition, an emission-model for NOx , OCH, and $Soot$ is available. Due to partly extremely complex chemical reactions, these emissions cannot be calculated in real-time.

Engine-speed, engine-torque, pressure and temperature are the output signals of the model. With an engine-speed feedback, the model may run autonomously except for single input injection-rate. A drivetrain model can be connected to the output to simulate the complete powertrain section.

To adapt the model to real engine conditions, all important engine parameters are adaptable. Example are drill, upstroke, crankshaft geometry, or number of drills of the injector. Furthermore, some parameters can be changed dynamically, during simulation, e.g. EGR[5]-rate, fuel pressure inside the common-rail or turbo-charge pressure. The degree of freedom of the model is 34.

EVAPORATION AND COMBUSTION MODEL

In this section a short introduction of the evaporation model and the thermodynamics will be presented.

EVAPORATION Modern diesel engines with a common rail injection system can use many different forms of injection curves. For realistic simulation results and a flexible simulation, the fuel spray model of Constien [1] is used. The fuel spray model is a basic approach that calculates a phenomenological burning function according to a given injection curve.

[5]EGR = exhaust gas recirculation, used to avoid emissions

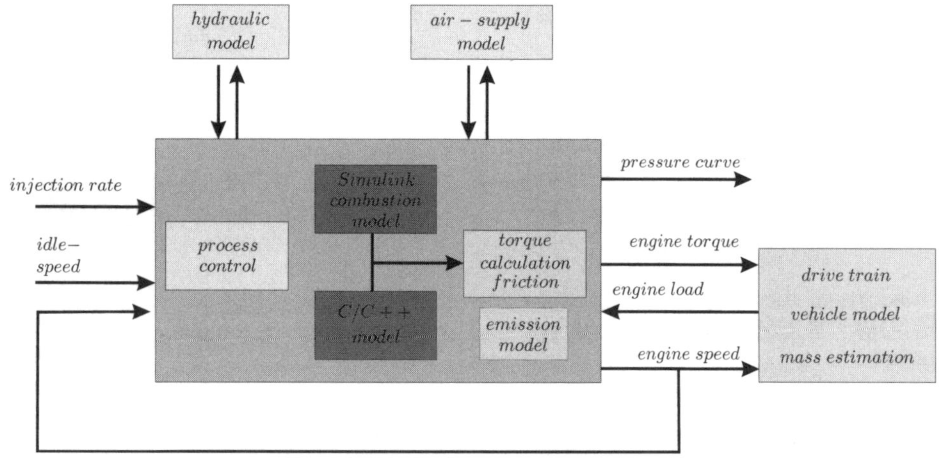

Figure 2: Structure of the engine-model-toolbox.

The injection is controlled by an injection-curve. Along this curve, fuel is injected into the virtual combustion chamber over the crankshaft angle. For every injected fuel portion the amount of liquid, gaseous and burned components will be determined as function of the crankshaft angle. Figure 3 shows the different states of the fuel portions injected at different crankshaft angles.

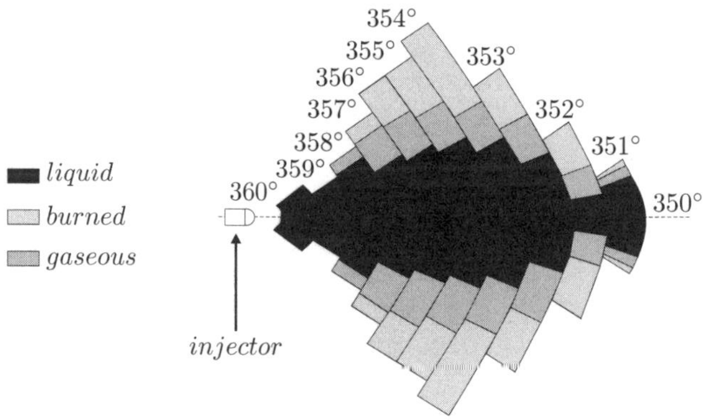

Figure 3: Fuel Spray Model

As it is impossible to analytically describe the multitude of drop diameters, an average drop diameter dependent from injection pressure and from mass of the fuel portion is adopted according the Sauter [1] approach:

$$d_{32} = \frac{\sum_i d_{Ti}^3 \cdot H_i}{\sum_i d_{Ti}^2 \cdot H_i} \quad . \tag{1}$$

H is the number of drops with identical diameter, d_T is the drop diameter and i represents the injection time. To calculate the Sauter diameter d_{32} the Varde/Popa/Varde [2] approach is used:

$$d_{32} = 16.58\,(Re \cdot We)^{-0.28} \quad , \tag{2}$$

where Re is the Reynolds number and We the Weber number of the flow. The Sauter diameter finally is calculated in dependence of air density ρ_L, fuel density ρ_K, fuel surface strain σ_K, kinetic fuel viscosity ϑ_K and diameter of the injector nozzle d_T:

$$d_{32} = 12.392 \cdot \frac{d_T^{0.44} \cdot \rho_K^{0.42} \cdot (\sigma_K \cdot \vartheta_K)^{0.28}}{\Delta p^{0.42} \cdot \rho_L^{0.28}} \quad . \tag{3}$$

Furthermore, the number of drops $N_{T,i}$ as function of injected fuel portion $m_{K,i}$ is required to compute the evaporation:

$$N_{T,i} = \frac{m_{K,i}}{\frac{\pi}{6} \cdot d_{32}^3 \cdot \rho_K} \quad . \tag{4}$$

By means of equations 2 and 4, the surface $A_{K,i}$ of all drops accrued in time step i can be calculated:

$$A_{K,i} = N_{T,i} \cdot \pi \cdot d_{32}^3 \quad . \tag{5}$$

Finally, the amount of fuel $m_{K,V,i}$ evaporating in every time step can be calculated in dependence of drop diameter d_{32}, surface $A_{K,i}$, in cylinder pressure p, engine speed n and a diffusion constant C_{Diff}:

$$m_{K,V,i} = C_{Diff} \cdot A_{K,i} \cdot p^{m_p} \cdot \frac{1}{60 \cdot n \cdot d_{32}} \quad . \tag{6}$$

The amount of liquid fuel at the next step $i+1$ is calculated by subtracting the evaporating fuel of equation 6 from the liquid proportion of the last step i

$$m_{K,i+1} = m_{K,i} - m_{K,V,i} \quad . \tag{7}$$

The drop diameter d_T needs to be determined in every time step:

$$d_{T,i} = \sqrt[3]{\frac{m_{K,i}}{\frac{\pi}{6} \cdot N_{T,i}}} \quad . \tag{8}$$

The inflammation delay time τ_{ZV} is calculated with an empirical formula used by Constien [1]:

$$\tau_{ZV} = 2.1 \cdot \overline{p}^{-1.02} \cdot e^{\frac{2100\,K}{\overline{T}}} \quad , \tag{9}$$

where $\overline{p}$ represents the mean pressure and $\overline{T}$ the mean temperature for a given time interval.

COMBUSTION The basis for the combustion model is a two-zone model, with one zone filled with fresh air and the other zone filled with the combustion products. For the simulation the same pressure is assumed throughout the whole combustion chamber. Furthermore, the mixture is meant to be homogeneous inside each zone. There is no simulation of swirl or blow-by effects.

For the calculation of the combustion process itself, the thermodynamic equation for the combustion chamber has to be solved:

$$dW_t + dQ_a + \sum_i dm_i \cdot (h_i + e_{ai}) = dU + dE_a \qquad (10)$$

where W_t is technical work, Q_a external heat, m_i mass crossing the system border with the enthalpy h_i and specific external energy e_{ai}. U is the internal energy of the system, E_a the external energy e.g. kinetic or potential energy. In addition, the flame process and the air-fuel ratio in the flame front have to be derived.

After charge exchange, fuel injection and evaporation process start, being calculated as described above. As soon as the cylinder temperature is high enough and inflammation delay time is expired, the combustion process will begin in the flame front. The combustion process may already have started while fuel is still being injected. The mass air flow from the unburned area through the flame front into the burned area has to be calculated. The volume of the unburned area decreases and the one of the burned area increases until the entire combustion chamber is filled with burned gas and the combustion ends.

The instantaneous air-fuel ratio λ_V in the combusted zone results from following equation:

$$\lambda_V = \frac{m_V - m_B - m_{Br}}{L_{st} \cdot (m_B + m_{Br})} \ , \qquad (11)$$

where m_V is mass of the burned mixture, m_B total amount of burned fuel, m_{Br} residual gas of the burned area and L_{st} stoichiometric air-fuel ratio.

The air-fuel ratio λ_i in the flame front represents how much fuel and air merge out of the unburned zone into the flame front to be burned. The evaporation model yields the input values for the flame front. In diesel engines, the inflammation starts first with a rich mixture getting leaner during the combustion. A definition of the air-fuel ration λ_i in the flame front can be found in [4]:

$$\lambda_i = \frac{\frac{dm_V}{d\varphi} - \frac{dm_B}{d\varphi} - \frac{dm_{Br}}{d\varphi}}{L_{st} \cdot \left(\frac{dm_B}{d\varphi} + \frac{dm_{Br}}{d\varphi}\right)} \ . \qquad (12)$$

According to the mass- and energy-balance of the two-dimensional model, the pressure change in the combustion chamber may now be derived (cp. equation 10). The following simplifications were established for simulation:

$$\frac{\partial u_V}{\partial p} = \frac{\partial R_V}{\partial T_V} = \frac{\partial R_V}{\partial p} = \frac{\partial R_V}{\partial \lambda_V} = 0 \ . \qquad (13)$$

This simplification is applied because it has not been possible to solve the dependence from the internal energy u on the pressure and from the ideal gas constant R_V on pressure, temperature and air-fuel ratio.

In order to solve equation 10, the masses of the burned and unburned areas are necessary. The mass change dm_u of the unburned area is

$$\frac{dm_u}{d\varphi} = m \cdot \left(\frac{\frac{dm_{inj}}{d\varphi} - (1 + \lambda_i \cdot L_{st}) \cdot \frac{dm_B}{d\varphi}}{m - m_{R0} \cdot (1 + \lambda_i \cdot L_{st})} \right) \ , \qquad (14)$$

the mass change of the burned area is

$$\frac{dm_V}{d\varphi} = (1 + \lambda_i \cdot L_{st}) \cdot \frac{m \cdot \frac{dm_B}{d\varphi} - m_{R0} \cdot \frac{dm_{inj}}{d\varphi}}{m - m_{R0} \cdot (1 + \lambda_i \cdot L_{st})} \qquad (15)$$

in dependence from injected fuel mass m_{inj} and from mass of residual gas m_{R0} within the combustion chamber. The volume of the unburned area V_U can be calculated, considering specific heat capacity c_{vu} and heat loss Q_{Wu} throughout the covered combustion chamber wall by the unburned area:

$$\frac{dV_U}{d\varphi} = \frac{-\lambda_i \cdot L_{st}}{p \cdot (1 + \frac{c_{vu}}{R})} \cdot T_V \cdot (c_{vu} + R) \cdot \frac{m}{m - m_{Br} \cdot (1 + \lambda_i \cdot L_{st})} \cdot \frac{dm_B}{d\varphi} - \frac{c_{vu} \cdot V_u}{R} \cdot \frac{dp}{d\varphi} - \frac{dQ_{Wu}}{d\varphi} \ . \qquad (16)$$

With equation 16, the volume of the burned area V_{Br} is now determinable:

$$V_{CC}(\varphi) = V_U + V_{Br} \ . \qquad (17)$$

The volume of the flame front V_x mab be derived as follows:

$$V_x = \frac{\left(m_V^{(n)} - m_V^{(n+1)}\right) \cdot R_m \cdot T_V}{10^5 \frac{N}{m^2 \cdot bar} \cdot p_Z} \ , \qquad (18)$$

where $m_V^{(n)}$ is the mass of the burned area in step n and p_Z the combustion chamber pressure.

REALTIME CALCULATION

For Hardware-in-the-Loop applications the model has to be calculated in realtime. The realtime hardware used in this project is a so-called $alpha - combo$ from dSPACE[6]. The alpha-combo includes two controllers, a Texas Instrument signal processor and a DEC-alpha processor. The alpha-combo is connected to the ECU via CAN-bus. Figure 4 shows the implementation of the model into a real car environment for a 4-cylinder engine. Each time a crankshaft angle of 180° is reached, the model calculates all important variables in parallel to the real combustion process.

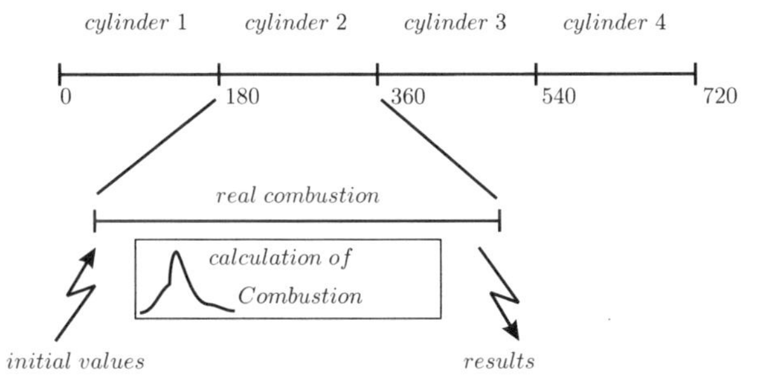

Figure 4: Model in realtime environment.

In order to get satisfying simulation results, it is necessary to provide a resolution, in terms of crankshaft angle, of at least 1°. This means for an engine speed of $3000\ rpm$ a step size of at least $56\mu s$. Due to a very computational model, some precautions like avoidance of divisions, inclusion of optimized C/C++ code, and Taylor approximations of power and root operations had to be made [6]. The result is a decrease of step size vom $100\mu s$ to nearly $50\mu s$ which allows a realtime simulation of the combustion process up to ~ 3500 crankshaft revolutions per minute.

APPLICATION IN ENGINE DEVELOPMENT PROCESS

In this section simulation results are presented to demonstrate the accuracy and usefulness of the proposed engine model for development and pre-tuning of engine control functions. At first, simulations are compared with data measured at an engine test bench which show how exact the model matches the behavior of a real engine. Further results are related to the application of the model for simulation based pre-tuning of an engine speed controller. The designed engine has a displacement volume of $490\ ccm$ per cylinder ($10\ cm$ upstroke, $7.9\ cm$ drill). The injection pressure is $800\ bar$ and includes a pre-injection.

[6]dSpace GmbH, Paderborn, Germany.

The engine simulation is executed for idle-speed condition with approximately 960 crankshaft revolutions per minute.

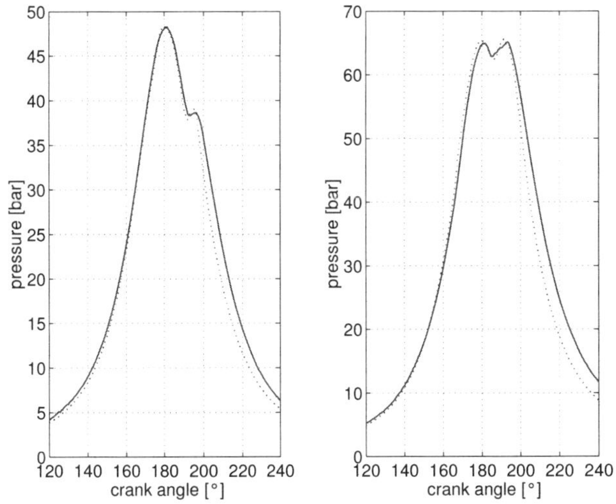

Figure 5: Comparison of measured pressure data (solid) and simulation results (dashed).

Figure 5 shows the comparison between measured pressure data of an engine test bench and simulation results. It is obvious that the calculated pressure curves nearly fit to the measured data when using identical boundary conditions. The model was adjusted by allying real engine parameters.

Figure 6 and Figure 7 show exemplary results when using the model to find a feasible pre-tuning of an engine speed controller. Now, the model additionally features a unit to calculate frictional forces and the geometry of the crankshaft. A generic Siemens VDO engine speed controller used in several serial projects is used to guide the engine speed of the model towards different setpoints. It comprises a PI control law along with additional modules for activation/deactivation of the controller, for filtering the engine speed setpoint and for control limitation and anti wind-up. The controller delivers a desired mean engine torque per combustion cycle which is converted into desired fuel masses for pilot and main injections. Therefore, a simplified efficiency model for the conversion from mean engine torque to fuel mass is used.

Both engine model and a discrete-time version of the engine speed controller are embedded into SDA which is the tool for development of engine control functions used by Siemens VDO as mentioned before. SDA allows for calculating control functions at given execution rates and at given positions of the crankshaft. The engine speed controller under consideration is calculated once before every firing event. The trigger for this calculation is derived from the engine speed and position signal which are delivered by the engine model. The engine speed signal fed into the controller is calculated from the time between the top-dead center positions of two cylinders firing consecutively. These border conditions for triggering and calculation of the controller closely resemble the situation in a real Siemens VDO ECU.

The behavior of the engine speed controller is determined by about 70 logical and continuous tuneable parameters. In the following, we focus on finding appropriate proportional gain and integral action time for the PI control core. For this application the Simulink model comprising controller and engine model was solved using the Simulink `ode15s` solver with a minimum admissible step-size of 1 μs and a maximum one of 50 μs.

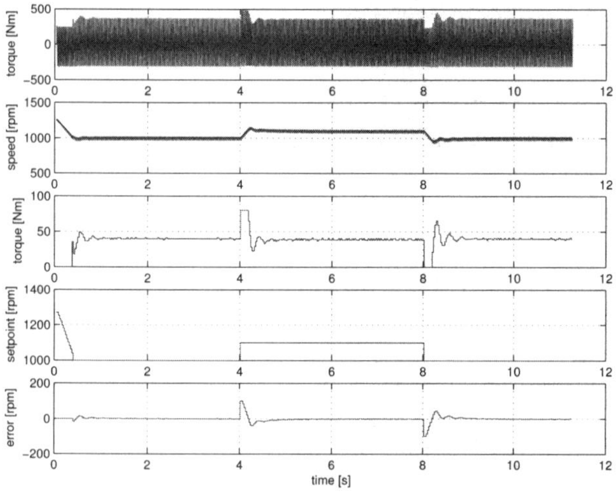

Figure 6: Results of 4 cylinder engine model showing bad control performance.

Figure 6 displays results of a first simulation run. The uppermost plot shows the engine torque. The control parameters focused at were set to the following values:

- proportional gain $K_R = 1Nm/rpm$
- integral action time $T_N = 0.5s$

At the beginning of the simulation there is no fueling which causes the engine speed to decrease, see the second plot of Figure 6. At an engine speed value of 1050 rpm the controller is initialized with an output torque of $35Nm$. After some oscillations around the setpoint, which can be seen in the speed control error trace in the lowest plot, engine speed settles at the setpoint of $1000rpm$. The engine mean torque commanded by the engine speed controller is depicted in the third plot. At $t = 4s$ the setpoint steps up to $1100rpm$, cf. fourth plot, causing the controller output torque to jump to its upper limit which is set to $80Nm$, see third plot. The engine speed reaches the setpoint after some oscillations around the setpoint. The setpoint step down towards $1000rpm$ shows a similar behavior of the controller. Because of the tendency to cause oscillations the control behavior is rated insufficient.

Figure 7 displays results of a second simulation run with the same border conditions as the first one. However, the proportional gain was set to $K_R = 0.25Nm/rpm$ and the integral action time to $T_N = 1s$. In contrast to the first simulation run the controller does not induce engine speed oscillations and behaves smoother. Consequently, the behavior is judged to be better than in the first case. Indeed, the control parameters which were used in the second simulation run show a good behavior when being applied in a vehicle, too. The results prove how advantageous the engine model can be used for simulation based pre-tuning of an engine speed controller.

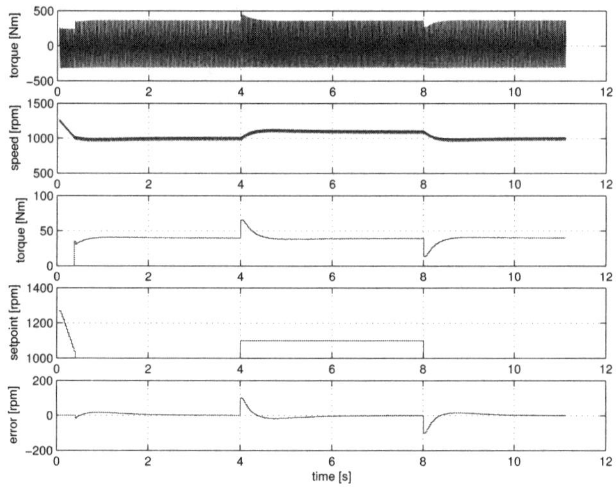

Figure 7: Results of 4 cylinder engine model showing good control performance.

SUMMARY AND OUTLOOK

In this project engine models for the use in ECU development and realtime calculation have been designed. The phenomenological fuel-spray model is the central element of all models. It provides a flexible energy conversion function which fits to all kinds of injection curves.

The realtime model of the combustion process can be employed as a state observer during operation of the engine. The Matlab/Simulink models are suitable to simulate the complete engine, including frictional forces and crankshaft geometry. Interfaces allow extensions of the model, e.g. with a drivetrain model or a hydraulic model of the common rail itself.

With help of the model it is possible to calculate in-cylinder variables and use them as input for sophisticated controllers. Therefore, a pre-tuning of controller parameters based on this highly precise engine model can be done on the engine developer's desktop. The model was used in closed-loop with an engine speed controller to find appropriate parameter values of proportional gain and integral action time. The achieved results are highly satisfying. The model-based pre-tuning of controller parameters saves expensive engine test bench time and efficiently accelerates the development process of engine control functions.

Future work will concentrate on the automation of the pre-tuning process by use of cost-functions. The realtime capability of the model may be improved by further precautions to simplify the computational effort while not decreasing the quality of the results. Furthermore, with help of a drivetrain model, the simulation of the complete powertrain may be achievable in realtime.

ANNOTATION

This project is a co-operation between Siemens VDO Automotive AG, Regensburg, Germany, and the Institute of Industrial Information Technology, University of Karlsruhe (TH), Germany. The presented work is trademarked by international patents.

CONTACT

Julian Baumann, Dara D. Torkzadeh and Uwe Kiencke can be reached through the Institute of Industrial Information Technology, University of Karlsruhe (TH), Germany at +49(721)608-4518, or {baumann, torkzadeh, kiencke}@iiit.uni-karlsruhe.de. Thomas Schlegl and Wolfgang Oestreicher can be reached through Siemens VDO Automotive AG, Germany at +49(941)790-61612, or Th.Schlegl@ieee.org and Wolfgang.Oestreicher@siemens.com.

REFERENCES

[1] Martin Constien. *Bestimmung von Eispritz- und Brennverlauf eines direkt einspritzenden Dieselmotors.* PhD thesis, Technische Universität München, 1991.

[2] Peter Herzog. *Möglichkeiten, Grenzen und Vorausberechnung der einspritzspezifischen Gemischbildung bei schnelllaufenden Dieselmotoren mit direkter luftverteilender Kraftstoffeinspritzung*, volume 12. VDI Verlag, Düsseldorf, 1989.

[3] Uwe Kiencke and Lars Nielsen. *Automotive Control Systems.* Springer Verlag, Berlin, 2000.

[4] Rudolf Pischinger, Manfred Klell, and Theodor Sams. *Thermodynamik der Verbrennungskraftmaschine.* Springer Verlag, Wien New York, 2. edition, 2002.

[5] Dara D. Torkzadeh. *Echtzeitsimulation der Verbrennung und modellbasierte Reglersynthese am Common-Rail Dieselmotor.* Logos Verlag, Berlin, 2003.

[6] Dara D. Torkzadeh and Uwe Kiencke. Introduction of a realtime diesel-engine model for controller design. *15th Triennial World Congress of IFAC, Barcelona*, 2002.

[7] Dara D. Torkzadeh, Wolfgang Längst, and Uwe Kiencke. Combustion and exhaust gas modeling of a common rail diesel engine - an approach. *SAE Technical Paper 2001-01-1243*, 2001.

2004-01-0704

A Model For Electronic Control Units Software Requirements Specification

Massimo Annunziata, Ferdinando De Cristofaro, Carlo Di Giuseppe, Agostino Natale and Stefano Scala
Elasis S.C.p.A - Control Systems Department

ABSTRACT

In the automotive world the “electronics and software” are continuously increasing. In this scenario the correct definition and, more in general, correct management of requirements in the development process of software is a key factor for continuous success. The real-time characteristic of control functions software makes the development process more complex and articulated.

This paper describes a more formal approach in the software requirements specification which is based on a requirements model.

The purpose of the model is to define the syntax and semantic elements in an unambiguous way. These elements are based on three classes of requirements: “simple requirement”, “composite requirement” and “composite with a finite state machine requirement”.

The paper gives a description of every class of requirement, and the capability of model elements composition to describe more complex requirements is shown.

A specialization of the model for the Engine Management System requirements specification is described, and an application of the specialized model for a Fiat EMS real case is presented.

INTRODUCTION

Modeling and simulating vehicle electronic system/software functionality has been used increasingly over the last few years. The complexity of vehicle software is growing remarkably, and functionality is becoming more distributed. Since more and more software development is out-sourced, stakeholders involved in product development increase and a considerable number of product variants arises. So the need to define correct, precise and unambiguous specifications becomes critical.

A more formal requirement specification reduces misunderstanding between stakeholders and improves acceptance testing from the customer. Therefore the customer can detect failures much earlier to fulfill the requirements, thus reducing development costs. Change requests can also be more efficiently assessed and responded to, thus providing more timely feedback.

A requirement specification can vary according to the development process level, so the decomposition of the requirement into its sub-requirements can be subjective. Using a textual description for requirement specifications can lead to poor results. First, it is very difficult to make the text precise. Secondly, it is labor-intensive and error-prone to ensure accurate and internal consistency. Lastly, as designs evolve and changes are made to the system, there is no reliable way to ensure that these changes are properly reflected in the textual documentation.

In order to solve the problem of using a simple textual requirement specification a more formal syntax and semantic element model has been adopted.

A combination of textual documentation to capture explanations and motivations a state machine model and a data flow diagram to capture system content have been explored...:

A Requirement Management Tools (RMT) that allows a categorization (for example by means of an assignment of particular attributes to a requirement) is very useful to keep track of requirements.
Existing commercial RMTs often support basic database management for individual requirements, their attributes, and their history, but differ widely with respect to the user interface, tracing and view management, distributed work, extensibility mechanisms, and import-export interfaces. Some of these tools are: Doors, RequisitePro, RDT, RTM, SLATE,....

RequisitePro as an RMT, Rational Rose RealTime for model designing and Soda as a report generator have been tested.

SOME DEFINITIONS

Before introducing our model, or better meta-model, for requirement specifications, some definitions are necessary.

A *requirement* is every externally visible aspect of a system that will be necessary/desirable for the customer or a condition/capability to which the system must be compliant.

Generally we can identify different kinds of requirements:

- Functional requirements: the functionality desired from the product; services the system should provide, how the system should react to particular inputs and how the system should behave in particular situations.
- Non Functional requirements: quality, reliability that the product must have; constraints on the services or functions offered by the system, such as timing constraints, constraints on the development process, standards, etc.
- Domain requirements: requirements of the application domain or that reflect characteristics of it.

The real meaning of requirements described depends on the context in which they are used.

Any model used for requirement specifications should therefore be:

- readable to all people involved in product realization: the customer, the project team, the development team, testers, the final user, etc.
- able to produce different levels of abstraction avoiding the detailed overload that often occurs when data is passed between different domains.
- Maintainable, allowing traceability and fitting all different types of requirement.

The model that is presented here accomplishes this objective.

MODEL DESCRIPTION

Automotive electronically controlled systems must often provide a set of functionalities by managing a set of components, both hardware and software, such as sensors, actuators, controlled system, components of the electronic control unit, On Board Diagnosis software (OBD), etc.

In order to reduce the set of possible syntax and semantic elements to be used to specify requirements, a requirement model of three classes has been defined.

The three classes of requirements are: **Simple, Composite** and **Composite with FSM**.

SIMPLE REQUIREMENT

A simple requirement will be used when it is atomic and cannot be decomposed further.

It has the following attributes:

- Unique Identification
- Type: functional, non functional and domain constraints
- Class: Simple
- Description: Textual, Graphical, External document references
- Ranking: Mandatory, Conditional, Optional

It can have additional information:

- Input: Input data list
- Output: Output data list

COMPOSITE REQUIREMENT

A composite requirement will be used when it is further decomposable into sub requirements.

It has all the attributes of a simple requirement, plus:

- Local dictionary: List of all data defined and used inside itself
- Sub-requirements: List of all sub-requirements

COMPOSITE WITH FSM REQUIREMENT

A composite with FSM requirement will be used when it is further decomposable into sub requirements and presents a behavior which is dependent on a state.
It is a composite requirement, plus:

- States: list of all states
- Transitions: list of all transitions between states

where:

- every state has at least one sub-requirement
- every transition has:
  One Event Requirement
  Zero or more Action Requirements

Definition

ID	
TYPE	
CLASS	Simple, Composite, Composite with FSM
DESCRIPTION	
RANKING	Mandatory, Conditional, Optional

Input

NAME	SOURCE

Local Dictionary

NAME	SOURCE

Output

NAME	DESCRIPTION	TYPE AND VALUE

SubRequirements

NAME

States

NAME	REQUIREMENTS

Transictions

NAME	EVENT REQUIREMENT	ACTION REQUIREMENTS

Figure 1: The diagram representation of three requirements' basic class

Each of these can be decomposed further (as needed) to provide specification of the software processes. See Fig. 2 for an example of requirements decomposition.

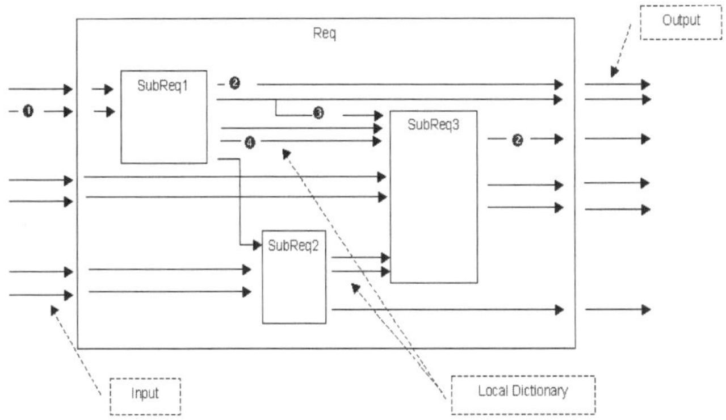

Figure 2: An example of requirement decomposition: input data ❶, output data ❷ and ❸, local data ❹;

Further very useful decomposition of requirements can be level-oriented.

The simplest level oriented decomposition is based on only two levels:

Level 1 requirements

Level >1 requirements.

In a Level 1 requirements specification, the product is specified describing the required behavior of the overall controlled system.

In Level >1 requirements the product is specified in more details than in Level 1 requirements, i.e. at the sub-system level.

Levels of test definitions are defined in accordance with the requirements level definition.

In Level 1 requirements the product is tested looking at the controlled system. For examples for testing the EMS it is enough to measure a set of physical variables on the Engine and/or Vehicle.

In Level >1 requirements testing we test the system looking at its internal variables. For examples for testing the EMS, the EMS output and internal variables must be measured.

The above vertical decomposition can be easily obtained using the described model.

The model allows the user

1. to follow the Information flow (by means of Input/Output/Local dictionary/Sub-requirements details).
2. to focus on single sub-function. A different description of behavioral can be used: textual, pseudo-code, link to other requirements, link to legal requirements, etc..

The model is flexible and customizable to a particular application. The attributes presented here are a minimal set, but their number and meaning can be modified. For examples "owner" and "working state" attributes can be added to indicate the specific authorities, people, organization or the domain expert that is the requirement's owner and progress toward finishing the requirement. The minimal set has been conceived to cover basic information needs that can be easily agreed on.

MODEL APPLICATION

An application of the described model to an Engine Management system (EMS) is described in the following.

An EMS is a system that produces the engine actuation variables (Throttle Angle, Ignition, Injection Timing,...), in order to produce a given torque to satisfy requests by the driver or other auxiliary systems (A/C, CC, VDC,...) with minimum fuel consumption, ensuring safety, driveability, emissions and legal constraints.

The main characteristics of the considered EMS:

- torque-based with DBW throttle system
- speed density air measurement system
- single “pencil coils” per cylinder

The development process starts with the requirements specification. The system model and its environment are specified by block diagram, state machine and DFD.

Three tables have been used (see Fig. 1 for most general one) to represent the three requirement classes. Thus the requirements' document is a sequence of tables organized in a tree structure.

Because the EMS will interact with the engine + vehicle, it must be classified as a hard real-time embedded system.

The high level requirements have been organized into three general types:

Functional requirements
- Performance: Torque, Idle, Fuel Consumption
- Comfort

Non Functional requirements
- Protection
- Emission
- Interconnection
- Diagnosis
- Safety
- Reliability
- Legal issues (e.g. EMC)

Domain requirements
- Plant definition

The level of decomposition applied to EMS consists of:

- Level 1: The EMS is specified describing the required behavior of the controlled system (EMS plus Engine plus Vehicle);
- Level >1: The EMS is specified in more details than level 1 requirements, i.e. at the sub-system level.

From the test point of view the difference between Level 1 and Level >1 requirements is:

- Level 1 requirements testing: To test the EMS is enough to measure the output of engine and/or vehicle “turn key System testing”
- Level > 1 requirements testing: To test the EMS is necessary to measure the EMS output and internal variables.

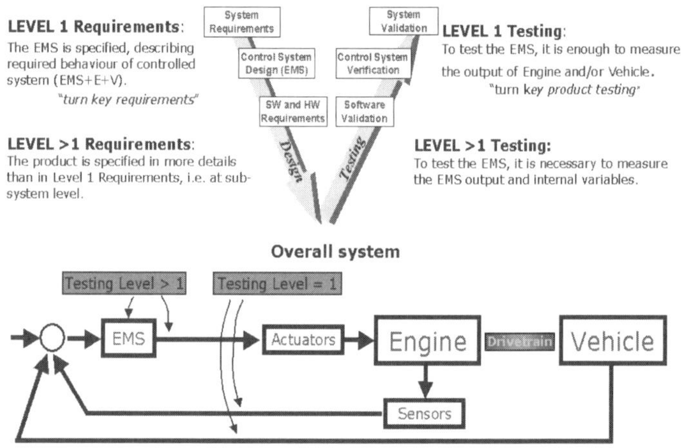

Figure 3: V model with requirements' level decomposition

Typically, the sensor/device reads a physical variable and converts it into electric signals. These signals are then treated by I/O drivers and an A/D converter. Other kinds of interfaces are implemented by means of K line, W line and CAN.

Thus the top level requirements architecture and signals' elaboration flow are shown in Fig. 4 and Fig. 5 respectively.

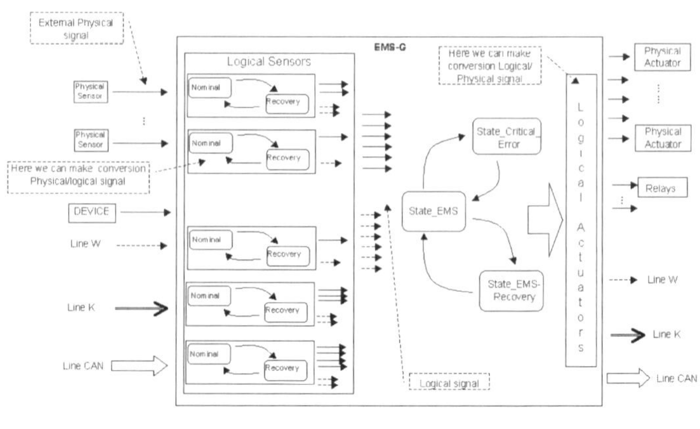

Figure 4. Overall top level requirement architecture

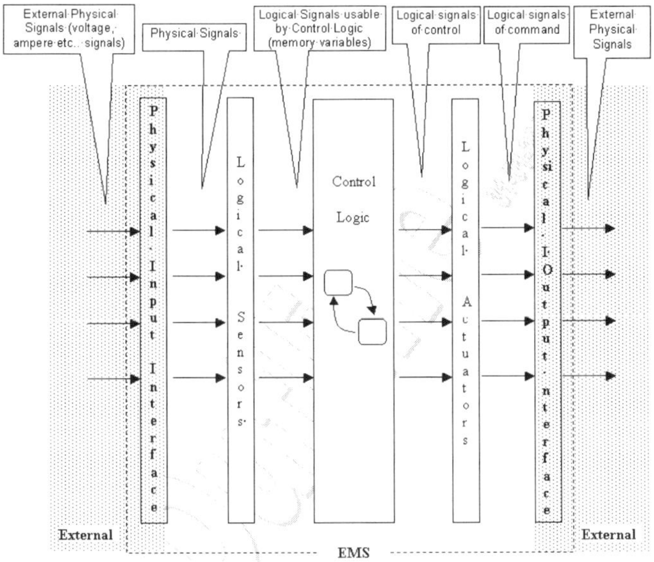

Figure 5. Signals' Elaboration Flow

The physical interface with external systems (sensors, actuators, devices, other ECUs and diagnostic instruments) manages the signals coming from sensors by adapting and converting them.

Logical Sensors and Logical Actuators requirements include the logical treatment of signals that come from the Input Physical Interface and go to the Output Physical Interface.

The requirements "Logical Sensors" and "Logical Actuators" manage devices, sensors, actuators signals of the vehicle in order to produce both the signal value and additional information on the validity of this value, such as a valid data bit or a value of the level of confidence for the generated signal value. They have sub-requirements to deal with both nominal and recovery behavior, to be activated for example in fault conditions.

Control Logic is the main part of EMS that manages the control strategies.

Every requirement at every abstraction level has input/output/local dictionary signals, so it is isolated from others. If we correctly specify its interface completely, then its implementation can change with a low cost impact. This encourages the reuse and maintainability of requirements.

CONCLUSION

The Requirements Specification Model, as defined in this document:

- captures all requirements using text, graphic, state chart, data flow diagram, and pseudo-code description
- captures interface information by specifying input and output information
- harmonizes the internal sub-division of requirements, if necessary
- includes a local dictionary of data used and consumed in requirements
- enables verification of requirements and their mutual interaction

REFERENCES

1. S. Robertson and J. Robertson, *Mastering the Requirements Process*, Addison-Wesley, Boston, 1999.
2. I. Sommerville and P. Sawyer, *Requirements Engineering: A Good Practice Guide*, John Wiley & Sons, New York, 1997.
3. M. Weber and J. Weisbrod, *Requirements Engineering in Automotive Development: Experiences and Challenges*, IEEE.
4. *Requirements Categorization* from INCOSE, Editor: Andrew Gabb, 04 February 2001
5. M. Mutz, M. Huhn, U. Goltz, C. Kr mke, *Model Based System Development in Automotive,* SAE2002, SAE press 03B-128

DEFINITIONS, ACRONYMS, ABBREVIATIONS

INCOSE: International Council on Systems Engineering: Is an international authoritative body promoting the application of an interdisciplinary approach and means to enable the realization of successful systems.

EMS: Engine Management System.

DFD: Data Flow Diagram

2004-01-0300

Model-Based System Development – Is it the Solution to Control the Expanding System Complexity In The Vehicle?

Roland Jeutter and Bernd Heppner
ETAS GmbH

ABSTRACT

Already today the car is a complex embedded system with a multitude of linked subsystems and components. In future these distributed systems have to be developed faster and with high quality via integrated, optimized design process. Scalable systems with an increased maintainability can be generated, if an agreement on a standardized technical architecture (hard- and software) is made at the beginning of the development. The challenges in the design of such distributed systems can be met through advanced automotive systems and software engineering in conjunction with suitable processes, methods and tools. Because the designers that must collaborate are distributed in different divisions or companies, it is essential that an overarching model based design methodology is used.

INTRODUCTION

The automotive industry is undergoing a technological change, and technical innovations. Especially electrical and electronic systems and software will continue to characterize the next 15 years. The trend from hardware to software solutions will continue. 90% of innovations depend on software. Software design allows a greater degree of freedom than any other technology. Software also provides enormous benefits and potential regarding cost reduction, reduced weight, reduced space requirements, improved reliability, etc..

The following constraints govern electrical/electronic systems design [1]:

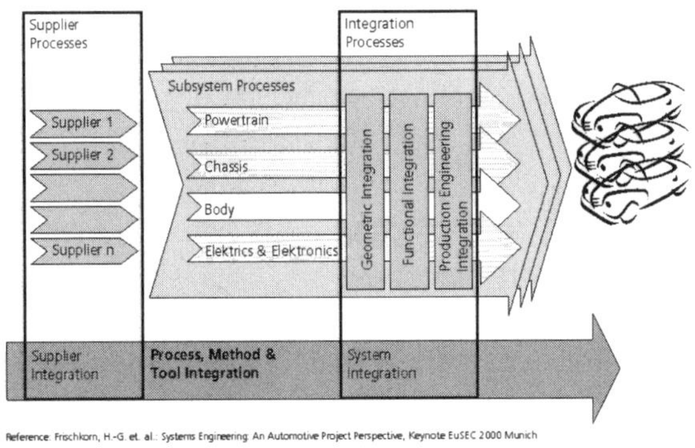

Figure 1.

- The design of complex functionality with tight requirements on safety and correctness.
- The design of distributed architectures consisting of several subsystems
- The mapping of the functionality onto the components of a distributed architecture with tight real-time and communication constraints.
- High-cost pressure restricts the usage of control unit resources in the components of a distributed architecture; SW implementations must be very efficient (performance, memory).
- Since system and component hardware and software have different product life cycles than the vehicle, it is necessary to standardize the technical system architectures (HW and SW) in the distributed system to enable maintainability, service and re-use.

- The high-time pressure and growing complexity of electronic systems conceal high risks in quality thus adequate system and software engineering process need to be implemented.

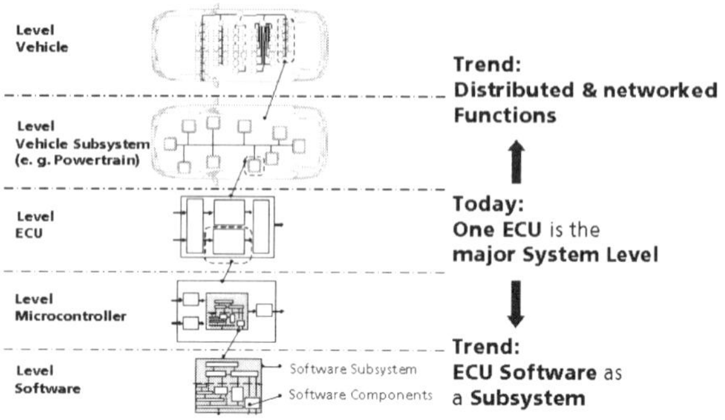

Figure 2.

AUTOMOTIVE SYSTEM AND SOFTWARE ENGINEERING

Vehicle functions can be defined as something that the customer perceives as and considers as valuable. Today, vehicle functions are typically implemented through systems that consist of electronics, software, sensors, and actuators [2].

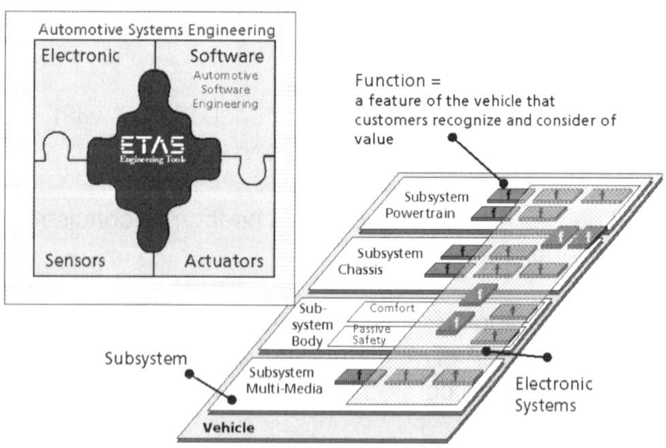

Figure 3.

In practice, a separate development process exists for each of these areas, and these different processes must be integrated into the systems engineering process.

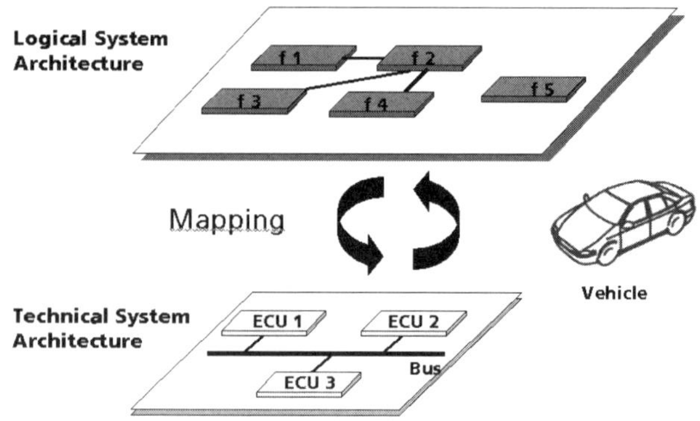

Figure 4.

The set of functions and the interoperability of the functions are called the Logical System Architecture.

The Technical System Architecture must fulfill all requirements derived from technical, economical, organizational as well as production constraints.

The development steps required to define the Technical System Architecture and to implement the Logical Architecture is called Automotive System Engineering.

The implementation of software on the Technical System Architecture is called Automotive Software Engineering [2].

MODEL BASED SYSTEM DESIGN

The Model Based System Engineering Process is the technical oriented core process embedded in the overall design process of a vehicle. It needs to be extended with the understanding of the organizational structures in the industry. Today the intellectual property required for the design , development and production of electrical/electronic systems is distributed between Tier2, Tier1 and OEM. The competencies, know-how and expertise in the automotive industry in distributed system design methodologies, system engineering and embedded realtime software engineering need to be improved in a very short time and the engineers have to systematically become trained and supported in new methodologies. In addition infrastructures and IT systems for collaboration, information sharing and distributed project management are required to achieve a consistent flow of information and expertise. Best in class management processes (e.g. CMMI, SPICE,..) have to be put in place to achieve high quality results and to control the risk in technical and financial areas.

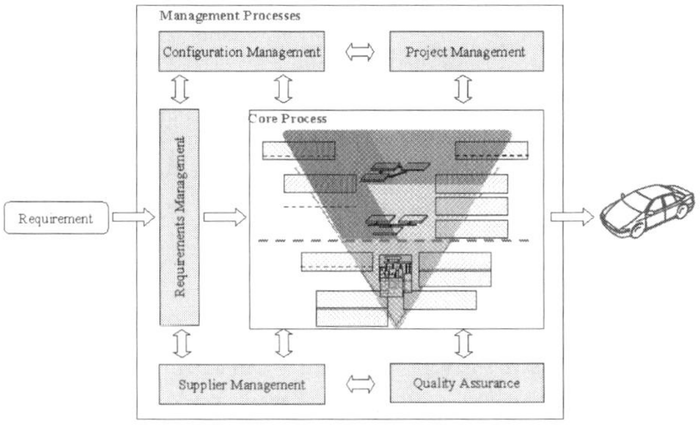

Figure 5.

The need to re-use functions throughout all levels of the design hierarchy to meet time-to market requirements, to assure the high requirements in safety and reliability and to cope with the increasing variety of car-platforms and models will directly lead to Standardized Technical System Architectures.

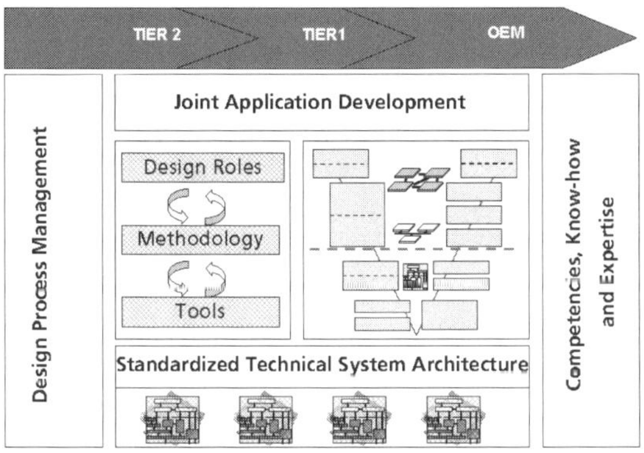

Figure 6.

Model Based System Engineering - In Model Based Engineering, the model is the central artifact and is used and systematically refined throughout the entire development process. The model contains continuous and/or discrete elements that capture the behavior of the system under development. Its structure reflects the functional decomposition of the system as well as the technical architecture in hardware and software. Furthermore the model must be executable and can be simulated. Tools exist to also animate the model to explore the behavior of the system to support recursive and iterative development style through the different levels of abstraction.

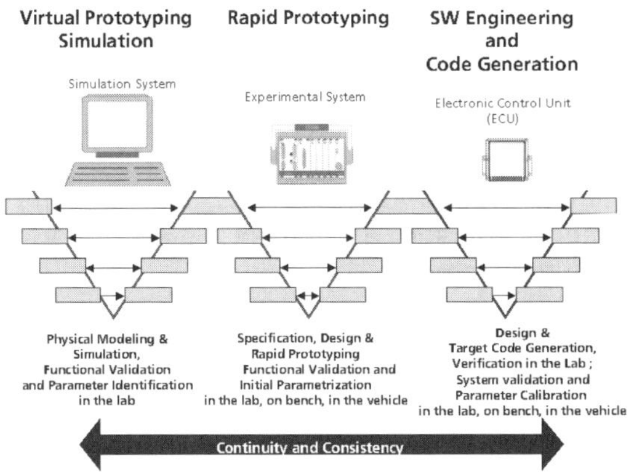

Figure 7.

The V-Model - The V-model visualizes the core process in the development of vehicle functions. It contains the following major development steps (excerpt from [1]):

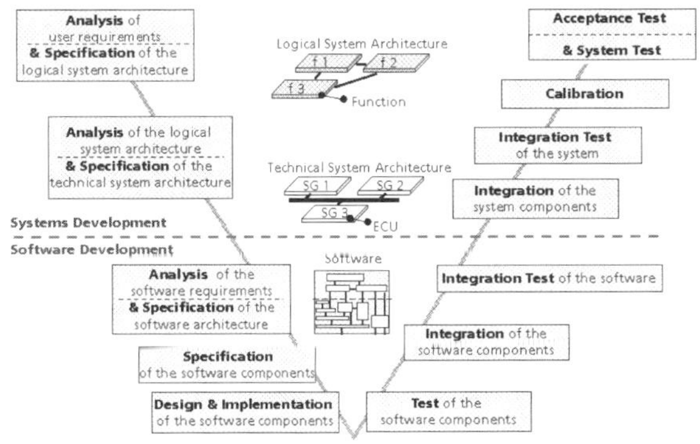

Figure 8.

- Analyzing user requirements and specifying the logical system architecture:Based on the user requirements to be considered, the goal of this process step is to specify the logical system architecture, i.e., the function network, the function interfaces, and the communication between the functions for the entire vehicle or for a subsystem. No decisions for the technical implementation are made yet during this step.

- Analyzing the logical system architecture and specifying the technical system architecture:

The logical system architecture forms the basis for specifying the actual technical system architecture. Various methodologies from the technical disciplines involved are used to analyze the alternative implementations that are based on a uniform logical system architecture. The technical system architecture also determines what functions or sub functions will be implemented by software. These are also referred to as software requirements

- Analyzing the software requirements and specifying the software architecture:

The next step is to analyze the software requirements and specify the software architecture, i.e., determine the boundaries and interfaces for the software system, software components, software layers and operating modes.

- Specifying the software components:

It is now possible to specify the software components. This is initially based on the assumption of an “ideal world.” This means that implementation details, such as implementation in integer arithmetic, are neglected at this stage.

- Designing, implementing and testing the software components:

“Real world” aspects are now taken into consideration in the design. It is necessary to determine all details pertaining to implementation. The software components are then implemented and tested based on these design decisions.

- Integrating software components and software integration testing:

Developing and testing the software components, which is frequently done by separate employee teams, is followed by the integration of the software components into the software system, and subsequent integration testing of the software.

- Integrating the system components and system integration testing:

This step consists of combining the software with the control unit hardware to obtain a functional electronic control unit. The electronic control units must then be integrated with the other components in the electronic system, i.e., set-point adjusters, sensors and actuators, so that a system integration test can be performed.

- Calibration:

The calibration of the software functions of the control units consists of setting the parameters of those software functions that often have to be determined individually for each type and variant of a vehicle. The parameters may be implemented in the software in the form of characteristic values, curves and maps.

- System test and acceptance test:

At this point, it is finally possible to conduct a system test against the logical system architecture and an acceptance test against the user requirements.

A number of tool vendors provide Computer Aided Design Tools for the different development stages in the V-model. E.g. ETAS offers tools and services that assist systems and software development in a consistent way. The model-based function development process is supported by the different product families in ETAS' portfolio as follows [2]:

- Analysis and specification of functions: the ASCET-SD product family
- System design, implementation and integration of functions (software engineering): ASCET-SD target code generators, DDS data management, OSEK operating system, realtime Architect, the ERCOSEK real-time operating system and Automotive Services
- Function testing: the LabCar product family
- Calibration, application (data feed and testing) of functions: the INCA product family

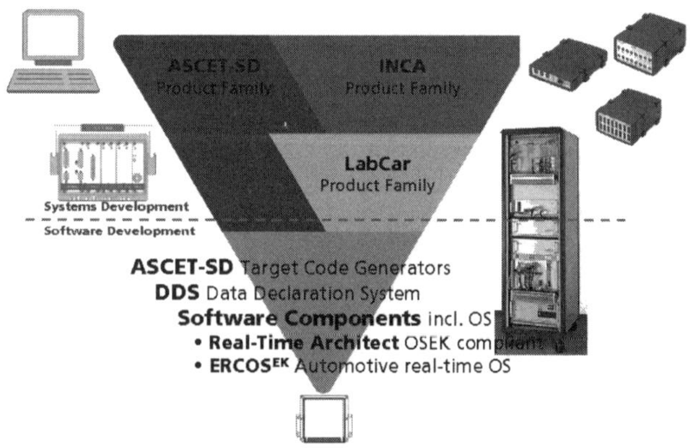

Figure 9.

Standardized Technical Architectures for Reusable Software Functions - The Logical System Architecture describes the complete functional and logical behavior of a system.

newly formed consortium of automotive OEMs and Tier 1 suppliers [3].

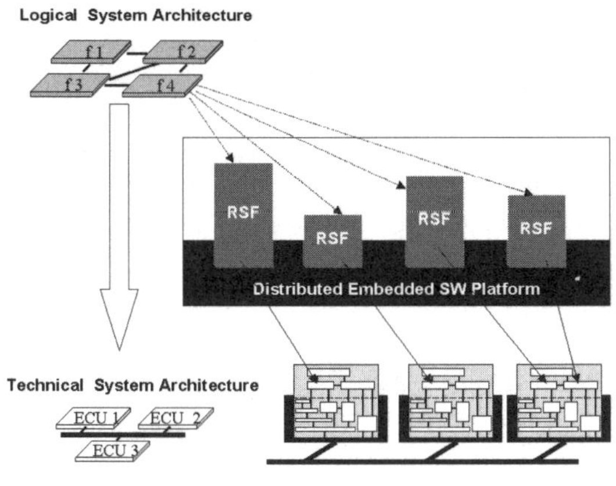

Figure 10.

It consists of functions with inputs and outputs and communication relationships between each other independent from the technical realization. The functions of the Logical System Architecture get realized in various SW modules (Reusable Software Functions RSF) that interact in a given order.

The Technical System Architecture is the real physical network of ECU's, that is supposed to provide the execution power for the Logical System Architecture. It is based on standardized hardware components and provides the physical interface between the ECU's and between the ECU's and the vehicle. Components of the Technical System Architecture are e.g. micro controller, input and output devices, memory, network components (CAN, LIN, MOST, FlexRay, ..) , sensors and actuators.

The Distributed Embedded Platform Software separates the RSF from the hardware and enables a standardized interoperability between the RSFs. The RSF must be developed in isolation from the final hardware and the RSF must interface, in the end, to real inputs and outputs. The RSF must therefore be written against abstract hardware, the inputs and outputs must be hidden behind an Standardized Application Programming Interface. The Distributed Embedded Platform Software is a layered SW architecture and is based on the standardized operating system OSEK, software components that abstract the network and hardware architecture and service routines, that allow the administration and maintenance of the distributed system. The top level of the platform software often called middleware enables the hardware independent communication between the RSFs and the hardware with real inputs and outputs. Enabling re-use and interoperability are driving the agenda of AUTOSAR, a newly formed consortium of automotive OEMs and Tier 1 suppliers [3].

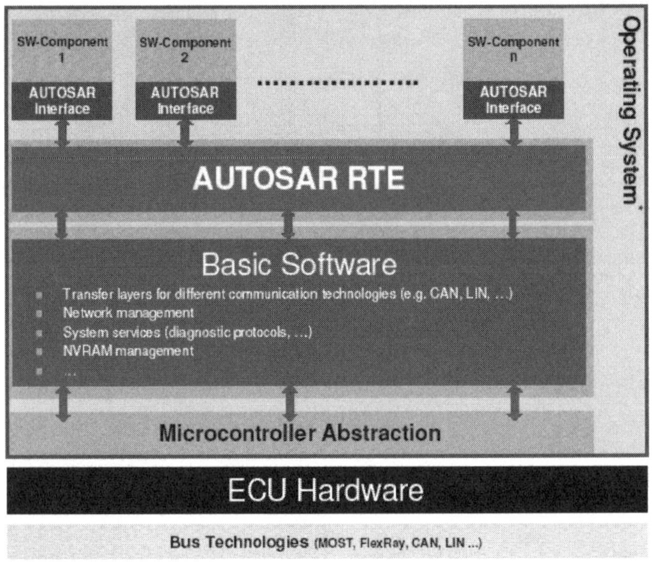

Figure 11.

<u>RSF Component Based Design Process</u> - To meet-time-to market requirements, to increase reliability and to cope with the increasing variety of car-platforms and models the re-use of the Logical System Architecture refined by RSFs is a proven approach. With reusable, interoperable RSF components, the entire industry can then share commodity functions while concentrating on novel, high-value added competitive hardware and software parts.

The Big Goal is to support Reusable Software Functions (RSFs) that allow functions to be developed once, put in a library and used again and again without changes.

The Standardized Distributed Embedded Software Platform must be defined and specified at the beginning of the design process and provides the stable foundation for the RSF Component Based Design Process.

Required New Roles

- RSFSuppliers - People who write RSFs against a requirement using a Specification Contract.
- ECU Integrators - People who take a set of RSFs and put them into an ECU, mapping abstract I/O to physical I/O.
- System Integrator - The person responsible for managing the ECU Integrators to connect all the ECU's in the car.

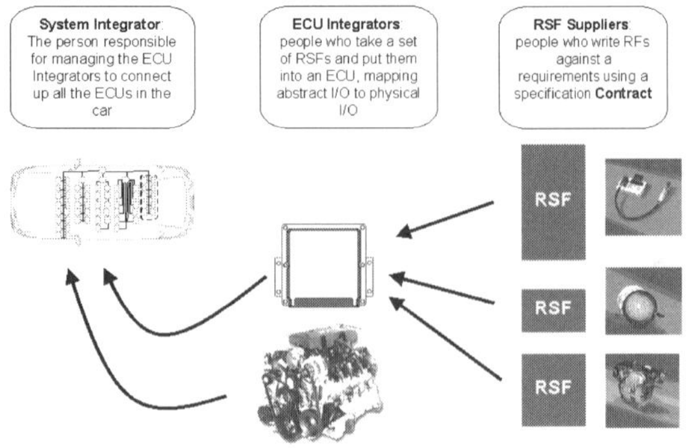

Figure 12.

New Processes - There needs to be a new way of managing the specification, test and integration of RSFs.

- RSF Development - The developer of an RSF doesn't know what other RSFs are running on the ECU. He doesn't do the integration work, the ECU may not exist yet. The developer must therefore make assumptions about the amount of CPU time and resources available to his RSF.

This leads to the idea of a Specification Contract for RSF that states what the specified behavior is and what permitted resource usage is of the RSF.

- Specification Contract for RSFs - The Contract is a promise that the RSF will do certain things, e.g. it promises to produce outputs with a certain timeliness, to have a certain functional behavior, to take only a certain CPU time, to take only a certain amount of memory.The Contract also contains obligations on the person using the RSF, e.g. they must give the tasks in the RSF certain CPU times, they must make sure the inputs have a certain timeliness.

The person integrating RSFs into a single ECU (i.e. the ECU Integrator) must respect these assumptions. The ECU Integrator can use this assumptions in selecting RSFs and designing the system by contract and building the system based on RSF components.

MANAGEMENT OF THE SYSTEM DESIGN PROCESS

The efficient use of distributed architectures implies that functions do not necessarily correspond one-to-one with technical architecture components. This implies that a much tighter relationship among design teams of different companies of the design supply chain must be established. Because designers that must collaborate are distributed in different companies, it is essential that an overarching design methodology be devised with the adequate support of tools and modeling techniques . The underlying premise of Design Process Management is that the quality of the final product is largely determined by the quality of the process used to develop and maintain it. An effective process ties together people, tools and methods into an integrated whole [4].

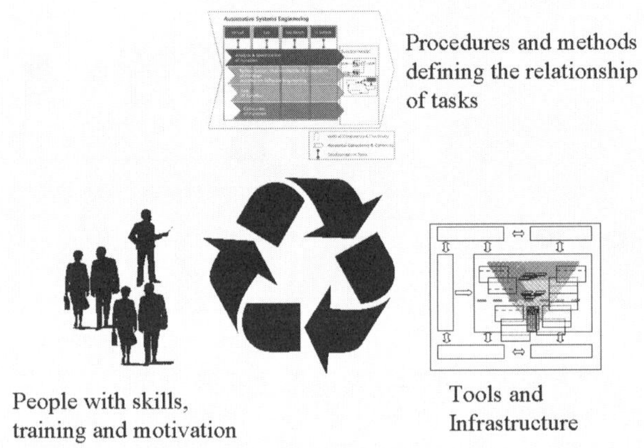

Figure 13.

<u>Key Management Processes and Ability to Perform</u> - The purpose of Requirements Management is to establish a common understanding between the customer and the system design project of the customer's requirements that will be addressed [4]. The requirements management incorporates the task [1]

- Collection of requirements
- Tracking of change requests

It is important that the applied requirements management process addresses the specific needs in the Automotive System Design:

- Cooperation between companies
- Co-working between teams
- Refinement of the requirements because not all requirements can be fixed at the start of the project
- Long Lifecycle of the final product with requirements for service, maintenance, recycling ,...

Configuration Management involves identifying the configuration of the system (e.g. selected components, their descriptions, specifications, implementations, ..) at given points in time, systematically controlling changes to the configuration and maintaining the traceability of the configuration throughout the life cycle. It addresses the different life cycles of the system components as well as the generation of the components, artifacts/results by different teams, e.g. specification, behavior models, RSFs with their Contracts, the integration and validation plans and experiment results.

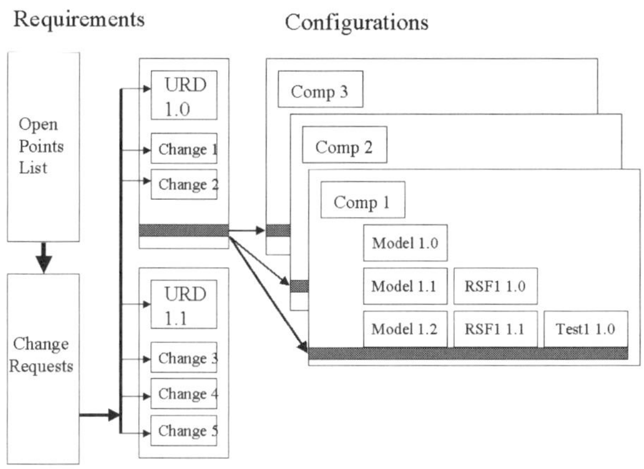

Figure 14.

The Project Management consists of 2 important tasks: the planning and the controlling of the individual projects to achieve the objectives in terms of Quality, Cost and Time. Planning is the most important task, because it incorporates the constraints in organizational structures and in available resources (people, money and technologies) by setting individual work packages, that contribute to the overall success of the project. Controlling on the other side consists of tracking the progress of the work packages and analyzing the differences. Immediate definition of countermeasure to minimize the risk for the quality, cost and time objectives is the most difficult job and leads to successful and not successful project results.

The work packages in the design of distributed systems are executed in different teams and across company boundaries. Thus the above described challenges in the Design Process Management can only be controlled by an integrated approach of Requirements, Configuration and Project Management based on standardized communication interfaces and a common understanding of the project management model.

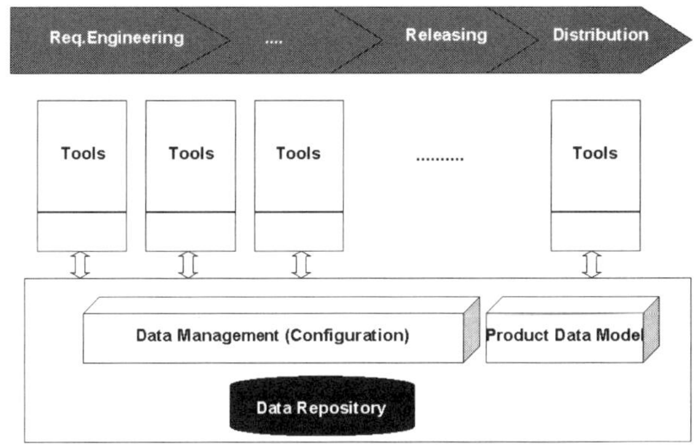

Figure 15.

At any time of the project the actual valid state of the requirements needs to be accessible for all members of the design project team. Changes need to be collected, analyzed for the impact and released in a systematic way. Beside the tracking of changes a consistent link from requirement to the realization/implementation in functions and components needs to be established to allow efficient and consistent co-working and collaboration.

In addition specific training programs be put in place to develop the skills and knowledge of individuals so they can perform their roles effectively. For each role the future skills needs are identified and how these skills are obtained. Some skills might be developed through informal vehicles (e.g. on-the-job training and mentoring), whereas other skills need more formal training vehicles (e.g. classroom training) to become effectively established .

CONCLUSION

We are convinced that the challenges to design distributed systems for vehicles can be met through advanced Automotive Systems and Software Engineering in conjunction with suitable processes, methods and tools. Standardized Technical System Architecture and on standards based infrastructures for collaboration and co-working become as important as the mature management of the System Design Process. Design in the past was treated as an art but needs to be managed as a consistent design process in future. The solutions to those challenges will have a great impact on the way how the vehicle's electrical and electronic architecture are designed.

REFERENCES

(1) AlbertoSangiovanni-Vincentelli Integrated Electronics in the Car and the Design Chain: Revolution or Evolution? - Electronic Systems for Vehicles, 25 and 26 th September 2003Type any references over these paragraphs.
(2) Zurawka T., Schäuffele J., ETAS GmbH - Automotive Software Engineering, ATZ/MTZ technical book
(3) AUTOSAR, www.autosar.org
(4) Carnegie Mellon University Software Engineering InstituteThe Capability Maturity Model, Addison-Wesley

CONTACT

Roland Jeutter
Vice-President
German Operations
ETAS GmbH, Postfach 30 02 40, 70442 Stuttgart, roland.jeutter@etas.de

Bernd Heppner
Director Business Development
German Operations
ETAS GmbH, Postfach 30 02 40, 70442 Stuttgart, bernd.heppner@etas.de

2003-01-3131

Modeling of Steady and Quasi-Steady Flows within a Flat Disc Type Armature Fuel Injector

M. H. Shojaeefard and M. Shariati
Iran University of Science and Technology, Automotive Engineering Department,
Supported by Iran Khodro Company

ABSTRACT

The internal flow in the fully open stage, opening and closing stage within an automotive fuel injector with flat disc type armature are studied in this work. The physical domain includes the region from top of the armature to the exit of the injector orifice, where most of the pressure drop occurs. Because of symmetry, only one quarter of the physical domain is considered. The FLUENT software was applied to obtain the numerical solutions.

In this model, the flow was assumed to be isothermal and incompressible.

Predicted static flow rates provide good agreements with the experimental measurements. For the transient condition, 20, 40, 60 and 80 micrometer of armature lift was modeled. The main results included comparisons between pressure and velocity contours in the various armature lifts and also variation of outlet flow velocity and flow rate, Discharge coefficient vs. armature lifts were studied.

INTRODUCTION

The ability to control the quantity and timing of fuel delivered by an electronic fuel injection system has provided the automotive engineers with a means to improve the performance and economy of today's automobile. However, with the increasing demand for improvements in fuel economy, emission reductions, and the consumers expectations in vehicle drivability and engine starting, significant emphasis must still be placed on the continued development of electronic fuel injection [1].

The essential component in the fuel delivery system is the fuel injector. An electro-magnetic device, the fuel injector is the primary mechanism for the delivery of the proper quantity of fuel and the overall preparation of the fuel for mixing with the inlet air. Controlled by the engine management system, the fuel delivery is mapped for varying engine conditions to provide an optimum fuel delivery over the entire operating regime. As the injector performance varies with changes in fuel properties, temperature, operating voltage and over life, it is essential to understand these interactions and provide a means to compensate or to minimize the effects [1].

The major components of an injector consist of valve body, spring, solenoid coil, armature, seat and orifice. A cross-sectional view of the SAGEM F type fuel injector is shown in figure 1, listing major components.

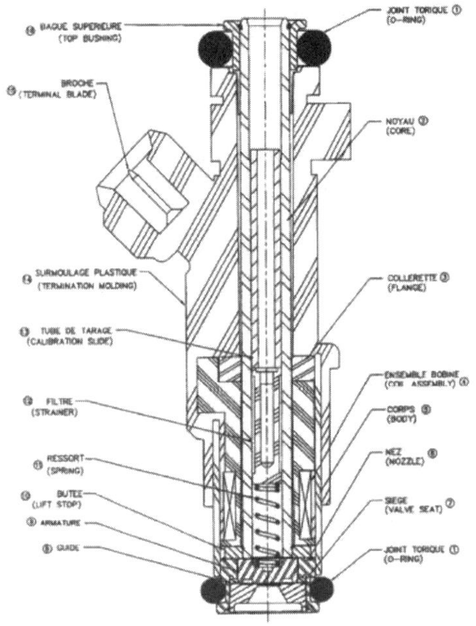

Fig. 1- SAGEM F Type Injector [2]

This injector was designed primarily as a high pressure (200-400 kpa) top feed injector. The key internal design features are a low-mass flat armature and an open orifice valve seat [2]. This injector with its open metering orifice is resistant to deposit buildup.

When the injector is in the unenergized (valve closed) position, the armature is held against the valve seat by the hydraulic force. As an electrical pulse from the control unit is passed through the injector coil assembly, a magnetic field is created. The armature will remain against the valve seat until the magnetic forces become sufficient to overcome the forces generated by the spring and hydraulic force. Once this is achieved, the armature starts to lift away from the sealing ring on the valve seat towards the core allowing pressurized fuel to enter the metering orifice located within the sealing ring of the valve seat. In reverse, once the electrical pulse from the electronic control unit is stopped, the magnetic forces starts to decay. However, the armature will remain in the energized (valve open) position until the injector spring forces overcome the decaying magnetic forces. When the spring force becomes the dominant force, the armature will move away from the stopper and return back to the valve seat interrupting fuel flow [3].

When the injector is in operation, the flow is driven out of the injector by the pressure drop between the intake manifold and the fuel rail. The flow is mainly metered by the smallest flow area inside the injector. Orifice and valve are two possible locations of the smallest flow area during the fuel injection process. In the fully open stage, the orifice area is usually smaller than the valve area, and thus the orifice area becomes the major metering area. The area ratio of valve area over orifice area is a parameter affecting the static flow in the fully open stage. A higher ratio $\left(A_{valve}/A_{orifice}\right)$ generally decreases this effect [4].

In the initial opening and final closing stage, the armature is only slightly lifted and most pressure drop occurs across the valve and scat. It is desirable that the static flow of an injector is only determined by fuel pressure and orifice size. However, lift can not always be increased due to the opening time and closing time concerns and minimum operating voltage. All of these factors must be balanced in an injector design. Therefore, for a large orifice injector, small changes in armature lift may cause non-negligible effects in the static flow due to the valve/seat area is in the same order of the orifice area [1].

In the past, interesting improvements in injector design have been achieved. However, previous works are mainly based on trial and error which has heavily relied upon the experimental database. The state of the art with respect to theoretically understanding of flow phenomena inside an injector is relatively poor due to the tiny flow passage and extremely fast transients during the opening and closing periods. Only a few investigators have made the efforts to study the internal flow. Recent works in the fuel spray research indicate that flow structure at the exit of the orifice has a profound influence on the jet atomization and the character of the spray. There is a need to understand the flow field inside a fuel injector, and the effect of internal geometry [1].

In this study, two different operating conditions, one geometric model for the fully opened position and four other models for studying of opening and closing period have been considered.

NUMERICAL MODELS

The SAGEM F type injector was studied in this work. The physical domain includes the region from the upstream of the armature to the downstream of the orifice where most of the pressure drop occurs.

Because of symmetry, only one quarter of the physical domain is considered for the computational domain.

Figure 2a and 2b shows the armature model and Fig. 3 shows the armature seat and its sealing rings.

Figure 4 shows the one quarter of physical domain. The completed mesh is shown in Fig. 4. Non-uniform grid is used, with a total of 44094 computational cells in fully open stage model. Such grid finesse, especially in the region of the orifice, was found to be necessary to obtain the accurate results.

In the main model for determining of static flow rate, armature is in the maximum stroke ($100\,\mu m$) and four other models were considered for the analysis of opening and closing period when the armature lift was in 20, 40, 60, 80 micrometer of its seat where quasi-steady boundary conditions were applied in each model. The analysis of opening and closing stage two times performed, once in the stationary condition with ignorance of armature velocity and again with consideration of armature velocity.

The solver of the FLUENT software was applied to obtain the numerical solutions.

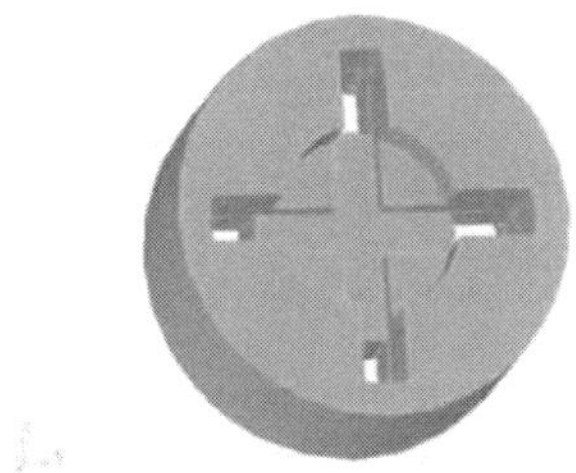

Fig. 2a –Top View of Armature

Fig. 2b –Bottom View of Armature

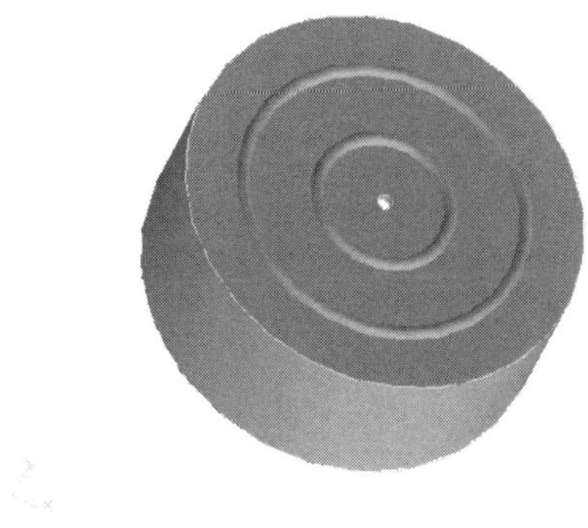

Fig. 3 – Armature Seat with its Sealing Rings

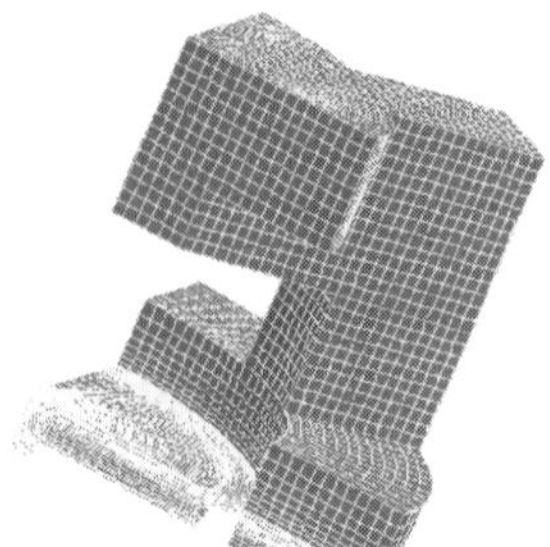

Fig. 4 - Mesh and Physical Domain

Assumptions: The Flow is assumed to be turbulent and isothermal. The $\kappa-\varepsilon$ turbulence model was used to account for turbulence effect, and the flow temperature is assumed to be at $20^\circ C$.

Working Fluid Properties: The gasoline fluid with the density of $730\,kg/m^3$ and the viscosity of $4.38\times10^{-4}\ pa-s$ was used for the working fluid.

Boundary Conditions: Ambient pressure was applied at the injector exit and a pressure drop 350kpa was applied at the inlet of the armature. Periodic boundary conditions were applied to the symmetry plane.

The Mass and Momentum Conservation Equations: The differential equations which represent the conservation laws of mass and momentum in Cartesian tensor notation are as follows [5]:

$$\frac{\partial u}{\partial x}+\frac{\partial v}{\partial y}+\frac{\partial w}{\partial z}=0 \quad (1)$$

$$\rho div(uU)=\frac{\partial(-p+\tau_{xx})}{\partial x}+\frac{\partial \tau_{yx}}{\partial y}+\frac{\partial \tau_{zx}}{\partial z}+S_{Mx} \quad (2)$$

$$\rho div(vU)=\frac{\partial \tau_{xy}}{\partial x}+\frac{\partial(-p+\tau_{yy})}{\partial y}+\frac{\partial \tau_{zy}}{\partial z}+S_{My} \quad (3)$$

$$\rho div(wU)=\frac{\partial \tau_{zx}}{\partial x}+\frac{\partial \tau_{yz}}{\partial y}+\frac{\partial(-p+\tau_{zz})}{\partial z}+S_{Mz} \quad (4)$$

In these equations, p is pressure, ρ is density and τ_{ij} is the shear stress tensor.

Turbulence Model: The turbulent kinetic energy κ and dissipation ε are predicted via solution of transport equations [5]:

$$div(\rho\kappa U)=div\left[\frac{\mu_t}{\sigma_\kappa}grad\kappa\right]+2\mu_t E_{ij}\cdot E_{ij}-\rho\varepsilon \quad (5)$$

$$div(\rho\varepsilon U)=div\left[\frac{\mu_t}{\sigma_\varepsilon}grad\varepsilon\right]+C_{1\varepsilon}\frac{\varepsilon}{\kappa}2\mu_t E_{ij}\cdot E_{ij}-C_{2\varepsilon}\rho\frac{\varepsilon^2}{\kappa} \quad (6)$$

Where μ_t is the turbulent viscosity, given by the expression

$$\mu_t=\rho C_\mu\frac{\kappa^2}{\varepsilon} \quad (7)$$

The coefficients C_μ, $C_{1\varepsilon}$, $C_{2\varepsilon}$, σ_κ and σ_ε are empirical constants whose values can be found in table 1.

C_μ	σ_κ	σ_ε	$C_{1\varepsilon}$	$C_{2\varepsilon}$
0.09	1.00	1.30	1.44	1.92

Table1- $\kappa-\varepsilon$ model parameters

RESULTS AND DISCUSSIONS

The static pressure distribution and velocity magnitude at static condition when the armature is in the maximum stroke was first analyzed. Figure 5 and 6 depict pressure distribution and velocity magnitude, respectively, at armature lift of $100\,\mu m$ and orifice diameter $0.34\,mm$, in the middle surface of physical domain that is shown in Fig.4. A pressure drop 350kpa is applied across the injector, with the inlet at the top of the armature and the outlet (orifice exit) at the bottom. Predicted pressure distributions show that most of the pressure drop does occur at the orifice. The maximum value of velocity is $31\,m/s$ and it is in the orifice. At this condition, the calculated mass flow rate has 2.1% discrepancy relative to experimental mass flow rate.

The experimental data of SAGEM Co. has been used in this work. Table 2 shows SAGEM Co. test condition and results.

Test Fuel Pressure	3.5 bar
Test Temperature	$20^{\circ}C$
Coil Resistance	12.25 Ohm
Static Flow Rate	$85.65\ gr/\min$

Table 2–Test Results of SAGEM Co.[2]

In order to determine the pressure distribution and velocity magnitude at opening and closing stage, figure 7 and 8 compare these items at armature lift of 20, 40, 60 and $80\,\mu m$ with stationary condition. The pressure gradient at the valve is bigger at the smaller armature lift indicating the effect of valve area. At a higher armature lift, the valve area is big enough to produce insignificant influence on flow field. The pressure drop percent in valve area at different armature lifts is shown in Table 3.

The armature lift has a similar effect on the flow velocity. Flow velocity at the valve is higher at a smaller armature lift; it is easier for the flow to feel the existence and influence of the valve. A higher armature lift produces a smaller flow velocity because of a smaller flow resistance and larger discharge coefficient.

Flow separations are observed due to shape-edged seat. Finally, it is worth noting that the predicted static pressure close to the corner of the orifice entry is

Armature Lift (μm)	Pressure Drop Percent in Valve Area
20	31.5%
40	7.31%
60	2.96%
80	1.94%
100	1%

Table 3 - Pressure Drop Percent in Valve Area at Different Armature Lift

lower than the vapor pressure of gasoline which suggests that cavitations may occur locally. The cavitations model is not included in the current work due to the lack of a general theory for cavitating process and numerical difficulties to simulate a free surface. The local negative pressures are believed to be attributed to the lack of two-phase flow model.

Two lines including orifice center line and valve center line have been considered in the middle surface of physical domain to clarify of pressure and velocity variation at valve and orifice area which have been shown in Fig. 9.

Fig. 10 and 11 show static pressure and velocity variations for various armature lift on the orifice center line, respectively. When fuel flow comes to the exit surface, pressure drop and fuel velocity increase and static pressure reaches zero level at the exit.

In Fig. 12 and 13 also as later explained figures, static pressure and velocity variations have been shown on the valve center line.

Of course, it should be mentioned that three-dimensional pressure variation and velocity investigation as one-dimensional can not explain variation property.

The flow rate increases rapidly as the armature begins to open the valve, showing part of the metering process is still controlled by the valve area at this stage and the curve gradually flatten off at large armature lift. Similar trends can also be seen in the mean velocity values at orifice exit are plotted against the armature lift.

In fuel injectors, fluid flows through a restriction or reduction in flow area, i.e., at the orifice or valve. For this kind of flow, the discharge coefficient is an important parameter in determining flow characteristics. Physically, the discharge coefficient is defined as the ratio of the actual mass flow over the ideal mass flow. Mathematically, the discharge coefficient for a flow of liquid going through an orifice can be expressed as [7]:

$$C_D = \frac{\dot{m}_{real}}{\dot{m}_{ideal}} = \frac{\dot{m}_{real}}{A_2\left[\frac{2\rho(p_1 - p_2)}{1-\left(A_2/A_1\right)^2}\right]^{1/2}} \quad (8)$$

Where the subscripts 1 and 2 denote the upstream and downstream locations of the orifice. The ideal flow in equation (8) is derived from the Bernoulli's equation and the continuity equation for a frictionless, incompressible flow. In injectors with valve fully opened, A_2 can be approximated to the orifice area and A_1 the injector inlet area. Since A_2 is usually much smaller than A_1 , equation (9) can be further approximated to the following equation without losing much accuracy.

$$C_D = \frac{\dot{m}_{real}}{A_o\left[2\rho\left(p_{rail} - p_{\text{int}\,ake}\right)\right]^{1/2}} \quad (9)$$

where A_o is the orifice area of the injector, p_{rail} and $p_{\text{int}\,ake}$ and are the fuel rail pressure and the intake manifold pressure, respectively.

Variations of discharge coefficient with armature lifts are shown in Fig. 14. The discharge coefficient is smaller at a lower armature lift showing the effect of valve area in restricting the flow. Since for a given injector operated at the same pressure drop, the ideal mass flow will remain the same, the decrease in discharge coefficient is directly from the reduction of actual mass flow. Computational result of discharge coefficient at static condition show good agreement with the experimental discharge coefficient at this condition, with the discrepancy about 2.8%. Table 4 shows the comparison of experimental and computational discharge coefficient.

Experimental C_D	0.69
Computational C_D	0.71
Discrepancy	2.8%

Table 4 – Comparison of Experimental and Computational Discharge Coefficient

The analysis in the quasi-steady conditions performed again in four models with consideration $0.1\,m/s$ average velocity of armature during of opening and closing stage (With assumption of 1 ms as an opening period time of injector and consideration of $100\,\mu m$ armature stroke). With these boundary conditions the discrepancy of computational data was about 0.01% relative to stationary condition. Therefore, with ignorance to this little error, results of stationary condition can be used for analysis of fuel flow in the opening and closing period of injector.

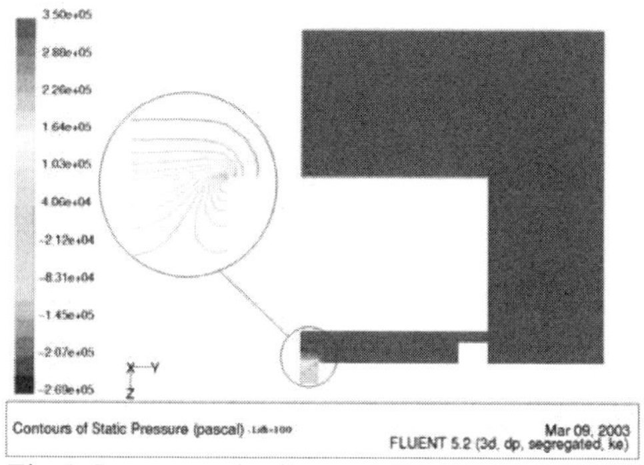

Fig.5- Pressure Distribution at Fully Open Stage

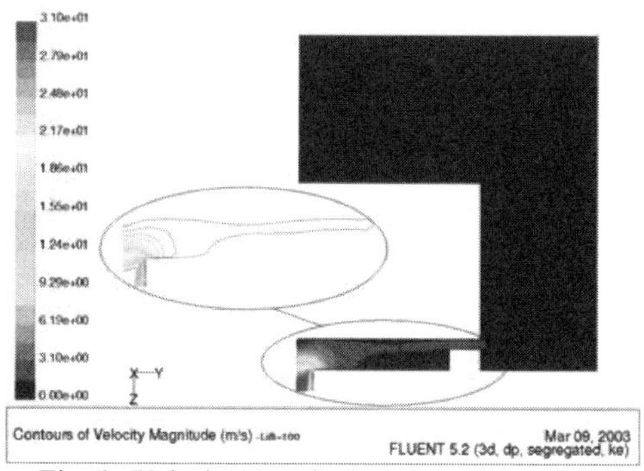

Fig. 6 - Velocity Magnitude at Fully Open Stage

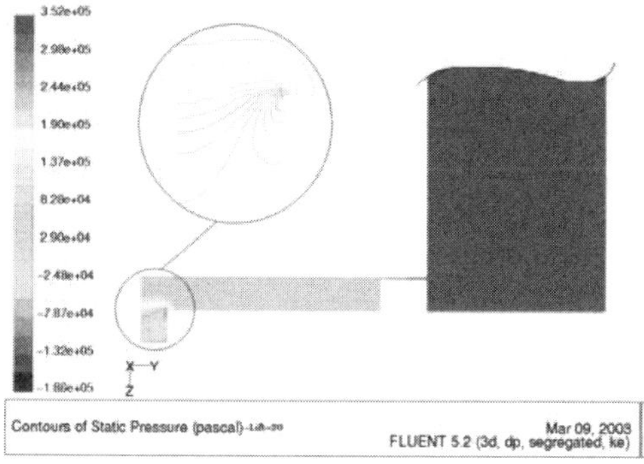

Lift=20 μm

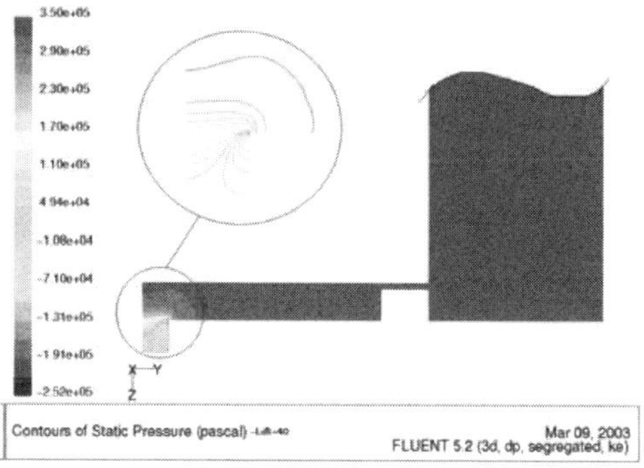

Lift=40 μm

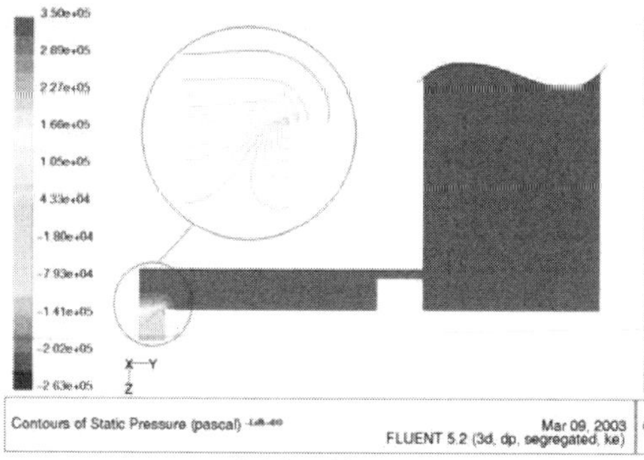

Lift=60 μm

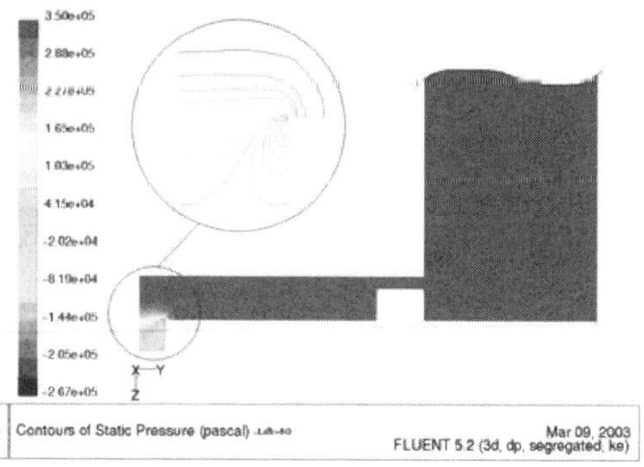

Lift=80 μm

Fig. 7 – Comparison of Pressure Distributions at Various Armature Lift

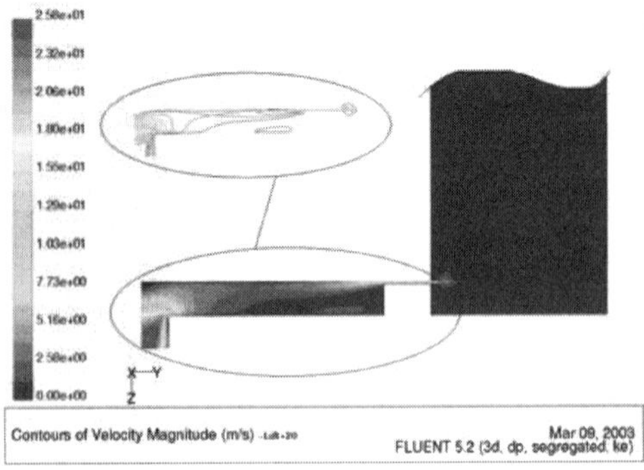

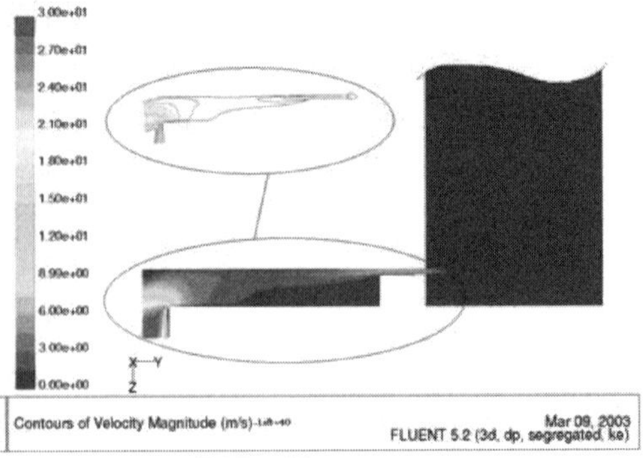

Lift=20 μm Lift=40 μm

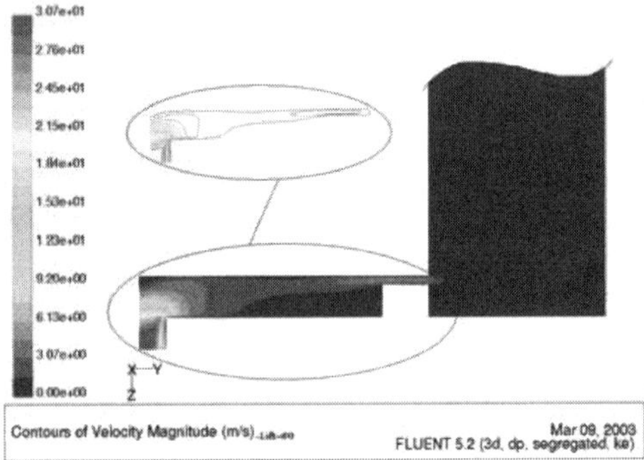

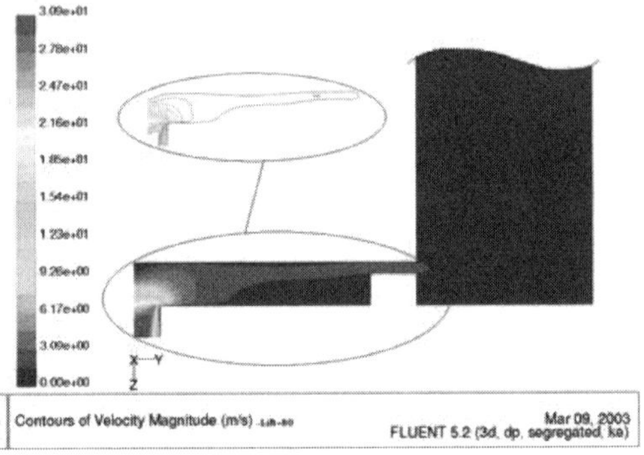

Lift=60 μm Lift=80 μm

Fig. 8 – Comparison of Velocity Magnitudes at Various Armature Lift

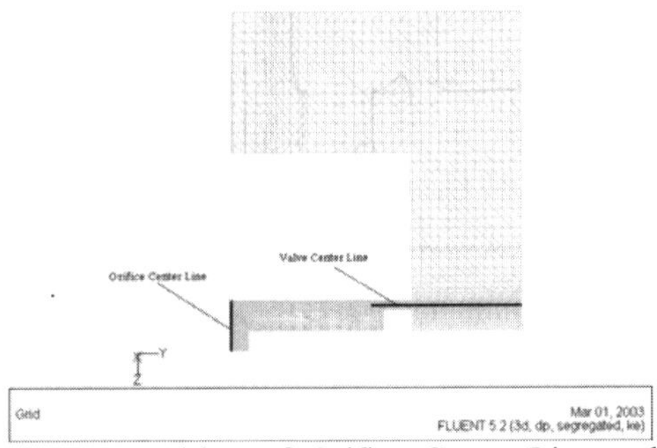

Fig. 9 –Position of Orifice Center Line and Valve center Line at Middle Surface of Physical Domain

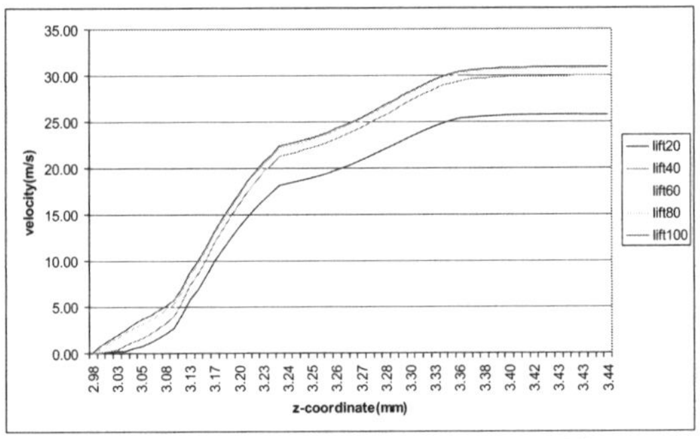

Fig. 11 - Variation of Velocity in the Orifice Center Line

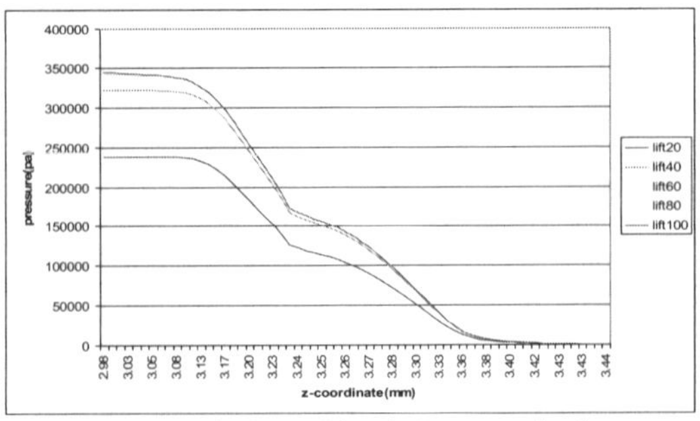

Fig. 10 – Variation of Static Pressure in the Orifice Center Line

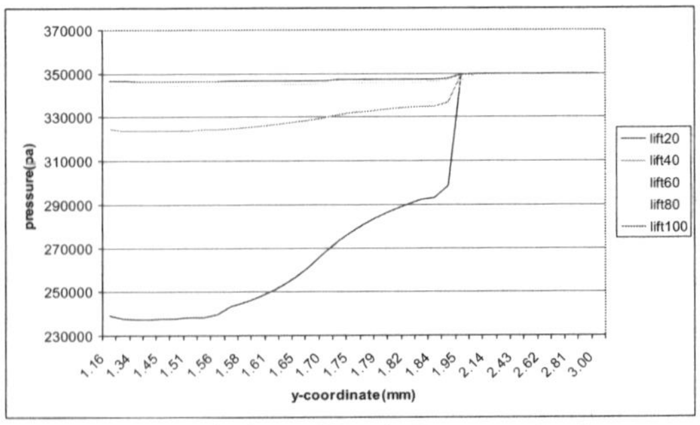

Fig. 12 - Variation of Static Pressure in the Valve Center Line

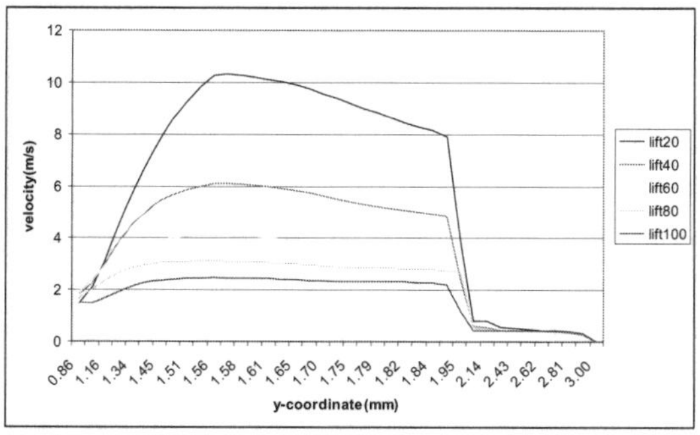

Fig. 13 - Variation of Velocity in the Valve Center Line

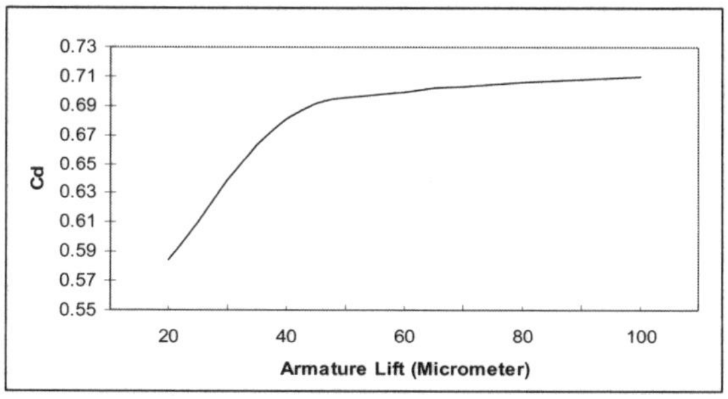

Fig.14 –Discharge Coefficients at Various Armature Lifts

CONCLUSIONS

Steady and quasi-steady flows inside a flat disc type armature fuel injector have been investigated. The results are summarized as follows:

- In the static condition, most of the pressure drop does occur at the orifice.
- Static flow rate and discharge coefficient predicted by computational model compared well with the test data.
- Flow rate and velocity of flow in orifice exit are found sensitive to the changes of armature lift. In the opening and closing stage, larger armature lift leads to a higher flow rate and velocity of flow in orifice exit.
- Higher armature lift leads to larger discharge coefficient.
- Because of the little error, results of boundary condition with ignorance of armature velocity can be used for analysis of fuel flow in opening and closing stage.

ACKNOWLEDGMENTS

The test data that provided by SAGEM company has been used in this paper.

REFRENCES

1. Ren, W.M., Nally Jr., J.F., "Computer Modeling of Steady and Transient Flows Within a Gasoline Fuel Injector", Proceeding of ASME Fluids Engineering Division, 1996.
2. SAGEM SA Automotive Division Company Data Sheet, 01F002A Injector, 2000.
3. Andrighetti, J.P. and Gallup, D. R., "Design-Development of the Lucas CAV Multipoint Gasoline Injector", SAE Paper 870127, 1987.
4. Chen J.L., DeVriese, D., Chen, G., and Creehan, J.L., "Influence of Needle Lift on Gasoline Injector Static Flow", SAE Paper 961121, 1996.
5. Versteeg, H.K., Malalasekera, W., "An Introduction to Computational Fluid Dynamics: The Finite Volume Method", Addison-Wesley Pub Co, 1996.
6. Stone, R., "Introduction to Internal Combustion Engines", 2nd Edition, MacMillan, Hong Kong, 1992.
7. Heywood, J.B., "Internal Combustion Engine Fundamentals", 1st Edition, McGraw-Hill, 1988.

2003-01-2711

Three Dimensional Finite Element Analysis of Crankshaft Torsional Vibrations using Parametric Modeling Techniques

Ravi Kumar Burla, P. Seshu and H. Hirani
Dept. of Mechanical Eng., Indian Institute of Technology, Bombay

P. R. Sajanpawar and H. S. Suresh
Mahindra and Mahindra Ltd, Tractor Division, Bombay

ABSTRACT

Automotive crankshafts are subjected to fluctuating torques due to periodic explosions in the cylinders. Accurate three dimensional finite element modeling is often time consuming. The present research aims to reduce the pre-processing efforts by developing parametric software. A three-dimensional parametric finite element model of crankshaft is developed using brick and wedge elements. Crankshaft main journal bearings are modeled as linear springs and dashpots. The piston and reciprocating masses are lumped at the ends of the crank pins. Viscous damper as well as shaft material damping has been modeled. Results from the three-dimensional analysis have been compared with those obtained using beam element models to assess the capabilities and limitations of such simplified models. It has been demonstrated that the simplified beam element models result in significant errors and 3-dimensional finite element analysis is essential for accurate predictions.

INTRODUCTION

Torsional vibrations result from the twisting reaction created in rotating shafts due to a fluctuating torque. Torsional response of automobile driveline components should be accurately estimated for optimal design. Many researchers modeled the crankshaft and driveline components as a set of lumped masses and springs. Kamath et. al. [1] and Chen and Chang [2] modeled each throw of crankshaft as a set of one mass and two springs. Drouin et. al. [3] modeled all the components of an entire tractor driveline using a set of masses and springs. Each throw of crankshaft was modeled as a set of three masses and four springs. In their work, inertia was calculated using the formulae given by Ker Wilson [4]. Petkus and Clark [5] presented an algorithm based on a generalization of classical Holzer method for torsional vibration analysis to calculate the natural frequencies and forced response of crankshafts and drivelines. They too modeled the given system as a set of masses and springs. Tecco and Grohnke [6] and Szadkowski and Naganathan [7] developed torsional vibration analysis software using matrix methods and lumped parameter modeling. Szadkowski and Naganathan [7] accounted for clutch hysteresis and universal joint disturbances. Birkett et. al. [8] extended the work of Tecco and Grohnke [6] to study the effect of torsional fluctuations due to universal joints. Observing that, in practical crankshafts, longitudinal, bending and torsional vibrations are coupled, Shimoyamada et. al. [9] presented a numerical method for calculating the waveforms for stresses in crankshafts. They used the Transfer Matrix Method with lumped mass and stiffness modeling of the crankshafts.

The models described above are restricted to analyze the torsional vibrations only. Generally, torsion modes are coupled with the bending modes in practical multi-cylinder engine crankshafts due to their complex geometry. In an inline crankshaft, the torsion modes are coupled with the out-of-plane-bending modes, and the in-plane-bending modes are coupled with the axial modes. Modeling of crankshaft for 3D-vibration analysis was done by Okamura et. al. [10] and Smaili et. al. [11]. Okamura et. al. [10] modeled the crankshaft as a set of rigidly jointed structures consisting of round bars and blocks of rectangular cross-section. Journal bearings were modeled as a set of linear springs and dashpots. The pulley and flywheel were modeled as lumped masses. A dynamic stiffness matrix was formed for this structure. On a four-cylinder engine crankshaft example, their simple model predicted the natural frequencies to be 120Hz, 320Hz and 419Hz as against 3D finite element predictions of 156Hz, 382Hz and 551Hz respectively. Morita and Okamura [12] modified the model developed in Okamura et. al. [10] by modeling the flywheel with plate finite elements, taking into account bending only in the direction of the crankshaft axis. Dynamic analysis was carried out to study the engine noise levels and pulley was re-designed to minimize the noise.

Smaili et. al. [11] developed a four node, 6 d.o.f. per node, line finite element and used it to analyze a crankshaft for vibration behavior. Timoshenko beam theory was used to account for the shear deformation. Crank webs were modeled as a set of three rectangular blocks. The journal bearings were modeled as an elastic foundation. The equivalent mass for piston and connecting rod were lumped at the crank-ends. Dynamic response was not analyzed in their study. Although these models predict the 3-dimensional vibration response, they are not based on three-dimensional finite elements and hence cannot be reliably used to calculate the dynamic stresses at critical locations such as the fillets.

Henry et. al. [13] developed a new tool for crankshaft durability assessment, based on a completely 3-dimensional numerical analysis. The stress calculations involved an initial 3-dimensional finite element analysis followed by a local Boundary element based fillet zoom technique. They performed dynamic analysis of a crankshaft to calculate fillet stresses and used these results in the modification of web design. Prakash et. al. [14] studied the effect of bearing stiffness on the natural frequencies of crankshafts using 8-node brick elements. Prakash et. al. [15] used a combination of classical methods and finite element method to calculate the dynamic response and life estimate of the crankshaft. ANSYS software was used for finite element modeling of the crankshaft using 3-dimensional solid element. Although the above 3D models could be used for analysis and design of real-life crankshafts, a lot of time would be consumed in the preprocessing stage. This difficulty can be overcome by developing a parametric model of the crankshaft. Athavale and Sajanpawar [16] developed a finite element model generator. This was used to analyze crankshafts for static stresses in the fillets, but dynamic analysis of crankshafts was not carried out in their study.

The present work extends the work of Athavale and Sajanpawar [16] and aims at developing a complete parametric model of typical automotive crankshafts using 3D finite elements. The finite element model has been developed using APDL [17] (ANSYS Parametric Design Language). In addition to the "geometry parameters" for describing the complex geometry of the crankshafts, "finite element parameters" have been used to enable generation of a user controlled finite element mesh yielding accurate results. The complete 3D model has been used to predict the dynamic stresses in an existing crankshaft of a commercial vehicle under operating loads. The parametric model developed here, while closely modeling the real life crankshafts, saves significant design cycle time as a range of crankshafts can be readily modeled using suitable numerical values for the parameters.

Finite element model of crankshaft has been developed using 3-dimensional elements. One dimensional beam element models have also been used for comparative study. Crankshaft main bearings have been modeled as linear springs and viscous dampers for the dynamic analysis. The details of the models are described in the following section.

2. FINITE ELEMENT MODEL OF CRANKSHAFT ASSEMBLY

2.1 Parameterization Of Crankshaft Assembly

2.1.1 Three Dimensional Finite Element Model

A typical crankshaft used in automobiles is shown in Fig. 1. The parameterization of crank web, crankpins and journals for 3-dimensional analysis is shown in Figures 2 and 3. The flywheel and pulley are assumed to be cylindrical in shape, so the parameters chosen for them are radius and thickness. A total of approximately 20 – 25 parameters have been used to model the complete geometry of the multi-cylinder crankshaft along with the pulley and the flywheel. These also include the meshing parameters (size of the element) at critical and non-critical regions in crankshaft assembly. The crankshaft system is modeled using 8-noded brick and 6-noded wedge finite elements. Details of a representative 3-dimensional finite element model are shown in Figures 4 - 6.

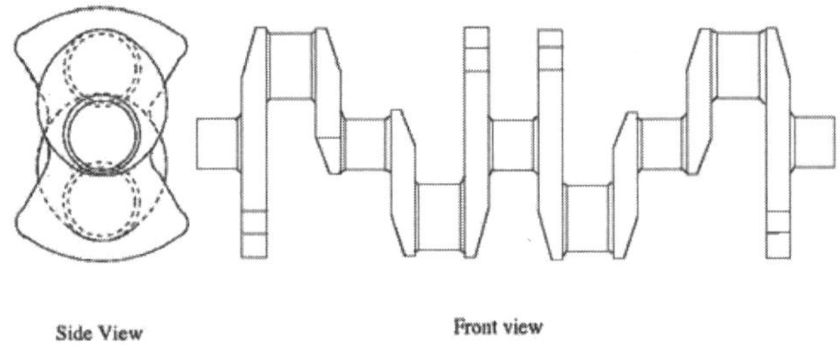

Fig. 1: Typical automotive crankshaft

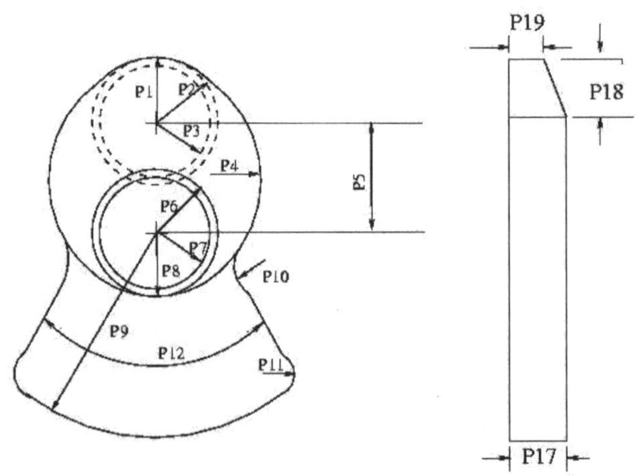

Fig. 2: Parameterization of Crank-web

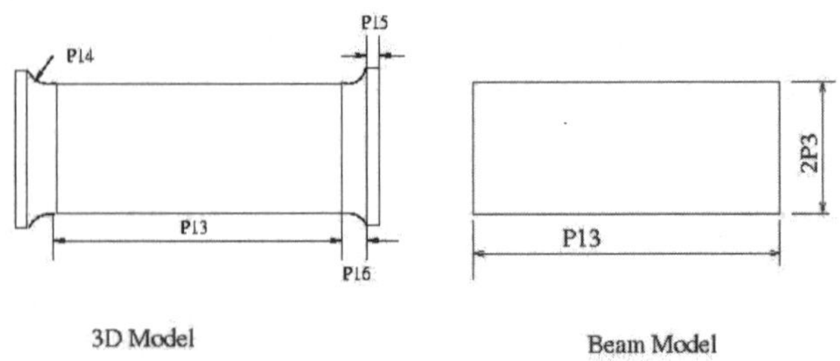

Fig. 3: Parameterization of Crank pin

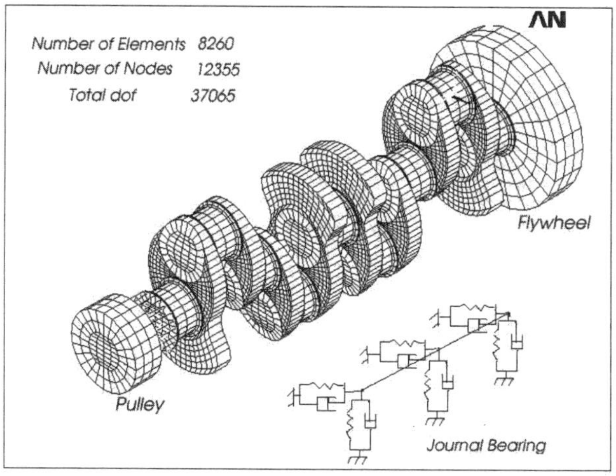

Fig. 4: Representative 3D Finite Element Model

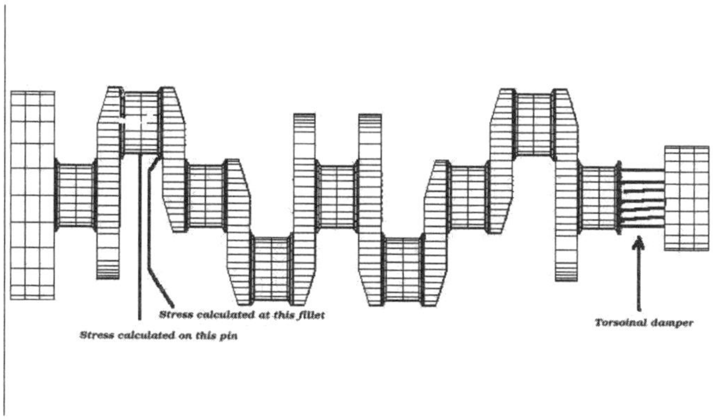

Fig. 5: A View of the Finite Element Model Indicating the Fillet and Crank-Pin Locations

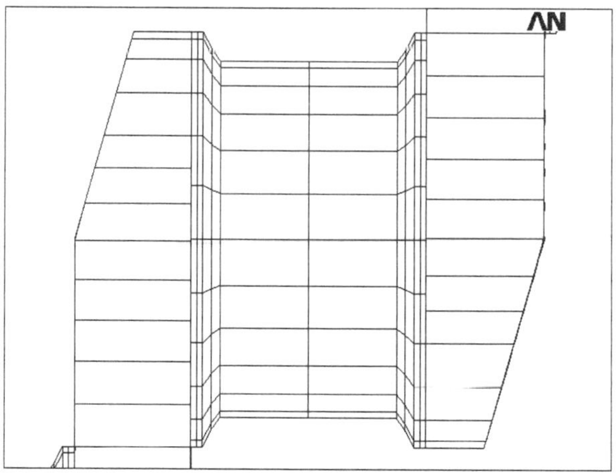

Fig. 6: Typical Model of Fillet (Element Size approx. 1mm)

Software has been developed which uses these parameters for building the complete finite element model of the crankshaft assembly. This enables the designer to readily build a finite element model of any crankshaft assembly by providing numerical values for necessary parameters, thus saving a significant amount of time in preprocessing. Repetitive modeling (e.g., for parametric studies) becomes simpler, as the designer needs only to provide a new set of parameters and the software builds the finite element model. The stresses at critical regions can be studied more accurately by changing the parameter for meshing at critical regions.

2.1.2 Beam Element Model

The beam element model developed in the present work is based on the model presented in Okamura et al. [10] and Smaili [11]. The web without counter weight has been modeled by one rectangular block and the web with counter weight has been modeled by three rectangular blocks as shown in Fig. 7. The dimensions of the rectangular blocks are chosen such that total area of these blocks equals the area of the web profile. This simplification enables us to use beam elements with uniform cross sections. The crankpins (journals) are modeled using beam elements whose cross-sectional area and moment of inertia are taken to be equal to the original crankpins (journals). The pulley and flywheel have been modeled as lumped masses. Journal bearings have been modeled as linear springs. A representative beam finite element model is shown in Fig. 8.

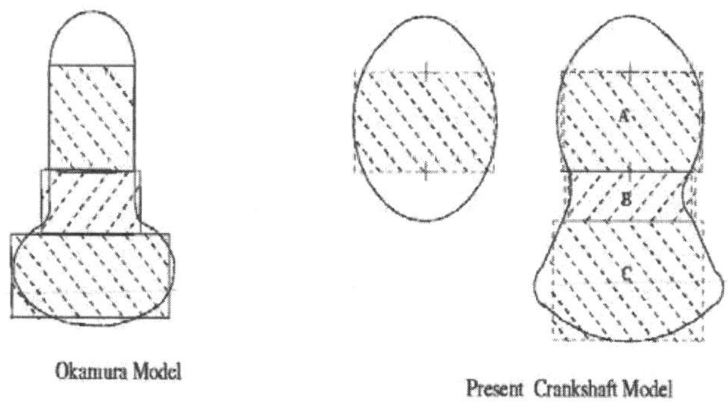

Fig. 7: Modeling of Crankweb (for Beam Elements)

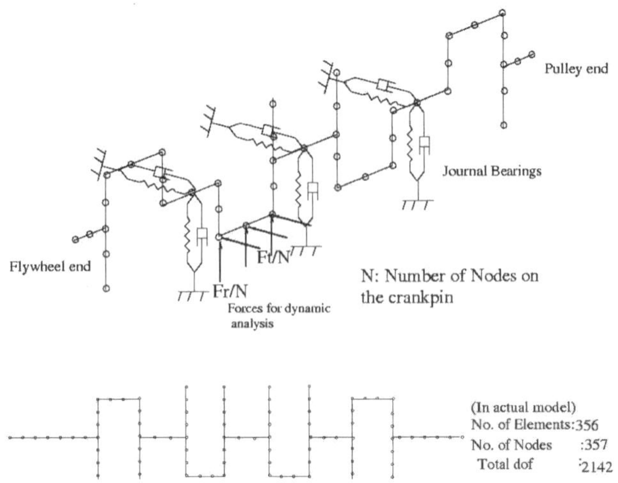

Fig. 8: Representative Beam Element Model of Crankshaft

2.2 Modeling Of Bearings And Dampers

The Crankshaft main bearings are modeled as three sets of linear springs and dash-pots in the directions normal to the crankshaft axis. They are attached at the middle of the respective crank-journals. It is assumed that the central set of spring-dashpot system supports 50% of the bearing load, the remaining two support 25% each, because of the parabolic distribution of the oil film pressure along the crankshaft axis (Morita and Okamura [12]). The gas force and inertia force for a given cylinder are assumed to be equally supported by the adjacent crank journals. The spring stiffness and damping coefficients are calculated using the formulae given in Rao [18].

3. RESULTS AND DISCUSSION

3.1 Free Vibration Analysis

The beam element model (Fig. 8) is used to analyze the crankshaft of Okamura et al. [10]. The present finite element model predicts the fundamental frequency as 1072.9 Hz as against the result of 1109 Hz (Transfer Matrix Method) and 1075 Hz (Experimental) quoted in Okamura et al. [10]. Thus our model's prediction is seen to be closer to the experimentally observed frequency than their prediction. The 3-dimensional finite element model is used to analyze a simple crankshaft without flywheel, pulley and bearings. Our results are compared with those obtained using the equivalent shaft model of Ker Wilson [4]. The equivalent mass of the crankweb was calculated numerically and that for crankpins and journals were calculated from standard formulae given in Ker Wilson [4]. These masses are lumped at the ends of the equivalent shafts. The equivalent shaft model for a crankthrow is shown in Fig. 9. The diameter of the equivalent shaft is taken to be the diameter of the journal. The results for the first five torsional natural frequencies obtained by the equivalent shaft model and 3-dimensional finite element model are given in Table 1. It is observed that the classical equivalent shaft model under/over-estimates some of the frequencies by as much as 16%.

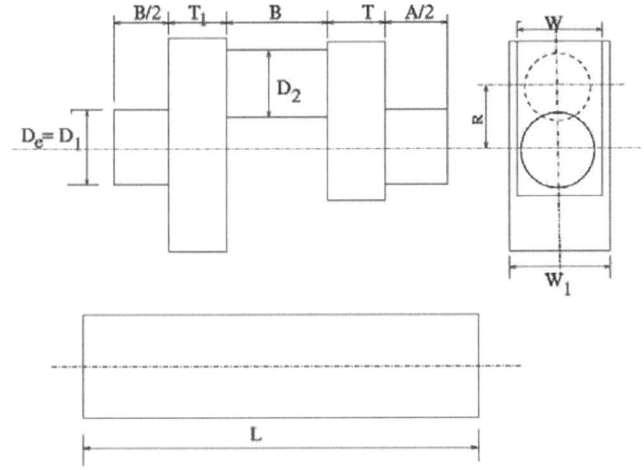

$$L = D_e^4 \left[\frac{A+0.8T}{D_1^4} + \frac{1.5B}{D_2^4} + 1.5R \left\{ \frac{1}{T^3 W^3} + \frac{1}{T_1^3 W_1^3} \right\} \right]$$

Fig. 9: Equivalent Shaft for Crankthrow

Table 1 Comparison of torsional frequencies (Hz) obtained from 3D FE Model and Equivalent Shaft Model

Mode No	Wilson Model (1956)	Present 3D FE Model	Deviation (%)
1	687	764	11.20
2	1438	1206	-16.13
3	1947	1941	-.3091
4	2423	2129	-12.1
5	2841	3023	6.42

The beam and 3-dimensional element models are further used to analyze the following five systems for their free vibration behavior:

- System 1: A simple crankshaft (free-free) without flywheel, pulley and bearings.
- System 2: System 1 is modified by including the journal bearings
- System 3: System 1 is modified by incorporating the flywheel and pulley
- System 4: System 3 is modified by modeling the journal bearings as linear springs

- System 5: System 4 is modified by modeling the reciprocating masses of the pistons. In both the 3D model and the beam model, the mass is lumped on the crankpins. This system is further used in dynamic response analysis.

The above systems are analyzed in order to study the effect of journal bearings, pulley and flywheel on the natural frequencies and mode shapes of the crankshaft system. The results of free vibration analysis for these cases are compared in Tables 2, 3 and 4. Conventional analysis of crankshaft vibrations considers pure torsional modes. The mode shape plots (Figures 10-12) show both the torsional and bending deflections in the model. It has also been corroborated by observation of the animation of mode shapes that the torsion modes are coupled with bending modes and lateral vibrations in the crankshaft webs are also observed. Hence, present finite element models yield an insight into the complexities of real-life crankshaft vibrations and lead to accurate prediction of coupled mode shapes and corresponding natural frequencies.

Table 2 Comparison of torsional frequencies (Hz) for beam and 3D models

	System 1			System 2		
S No	3D Model	Beam Model	Deviation (%)	3D Model	Beam Model	Deviation (%)
1	764	693	-9.3	790	710	10.3
2	1206	1291	7.0	1280	1370	7.03
3	1941	1612	-17.0	1968	1646	-16.4
4	2129	1673	-21.4	2132	1673	-21.5
5	3023	2335	-22.8	3027	2338	-22.8

Table3 Comparison of torsional frequencies (Hz) for beam and 3D models

	System 3			System 4		
S. No.	3D Model	Beam Model	Deviation (%)	3D Model	Beam Model	Deviation (%)
1	487	562	15.4	516	606	17.5
2	1093	1172	7.2	1132	1213	7.15
3	1487	1505	1.2	1491	1513	1.5
4	1892	2207	16.7	1922	2216	15.3
5	2924	3469	18.6	2966	3472	17.1

Table 4 Comparison of torsional frequencies (Hz) for beam and 3D models for System 5

S. No.	3D Model	Beam Model	Deviation (%)
1	484	346	28.5
2	1097	1286	17.2
3	1423	1487	4.5
4	1828	1951	6.7
5	2896	3228	11.5

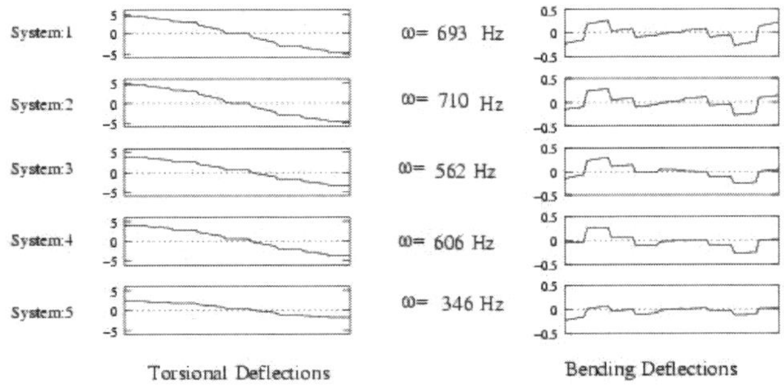

Fig. 10: First Mode Shape for Beam Element Model

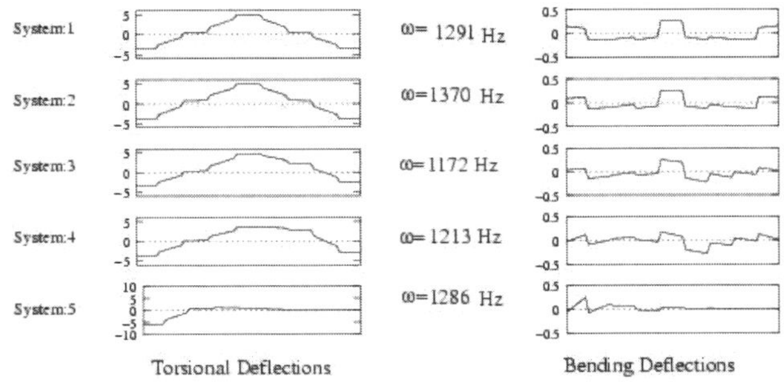

Fig. 11: Second Mode Shape for Beam Element Model

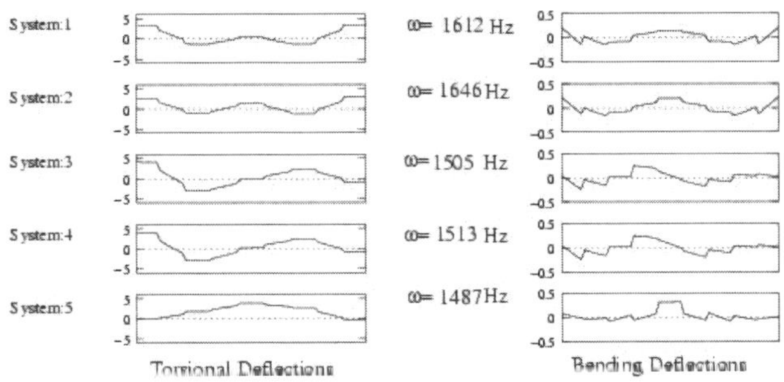

Fig. 12: Third Mode Shape for Beam Element Model

Addition of the flywheel and pulley to the crankshaft significantly reduce the natural frequencies of the system as can be observed from the results of System 1 and System 3. The reduction in the natural frequency could be as much as 18%. It can be observed from the mode shape plots that with the introduction of pulley and flywheel, the nodes get shifted towards the pulley, but the mode shapes do not change significantly. Incorporation of the reciprocating mass of the pistons significantly reduces the natural frequencies of System 4. The reduction in the natural frequencies is around 16% for beam model and 6% for the 3D model. In view of the coupling between the torsion and the bending modes, it is expected that the bearings will also significantly alter the natural frequencies. From a comparison of results obtained for systems 1&2 (or 3&4), we observe that the natural frequencies could increase by as much as 6-7%. It can be observed from the mode shape plots that for a given mode shape, the bending mode shape alters due to the introduction of bearings while the torsion part remains unaffected. Hence, for accurate study of crankshaft vibrations, bearings should also be taken in consideration.

Campbell diagrams as obtained based on beam element and 3-dimensional model results have been plotted in Figures 13-14. It can be observed that the beam element model significantly overestimates (or) underestimates the resonance by as much as (16%-20%). Thus the 3-dimensional finite element model, though requiring greater computational resources, provides accurate estimate of the resonance frequencies and also provides greater insight into the coupled vibration behavior.

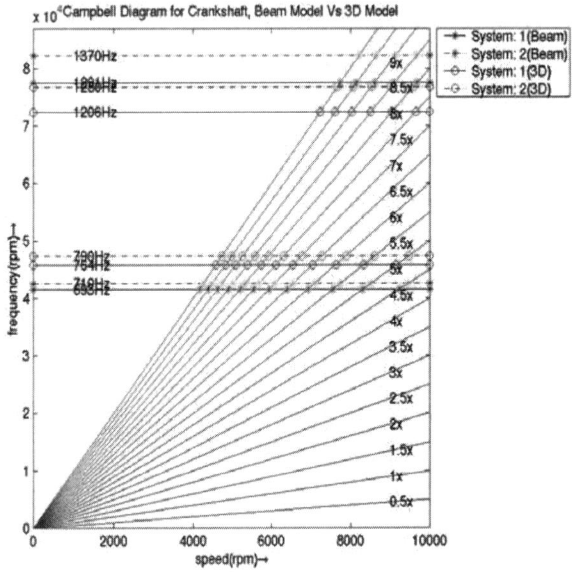

Fig. 13: Campbell Diagrams for Beam and 3D Models for Systems 1 and 2

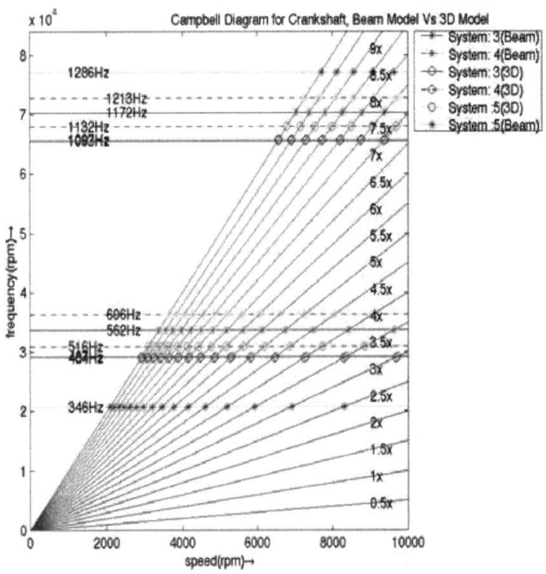

Fig. 14: Campbell Diagram for Beam and 3D Models for Systems 3, 4and 5

3.2 Dynamic Response Analysis

The gas forces and the inertial forces due to the reciprocating masses will contribute to the excitation forces on the crankshaft system. Fourier analysis is performed to calculate the first 20 harmonics of the excitation forces (Fig. 15). The radial and tangential components of the forces are applied on the crankpins as shown in Fig. 8. These components are then used to perform harmonic analysis of crankshaft to calculate the steady state vibration behavior. Dynamic response to individual harmonics is first obtained and then superposed to estimate the total response. Viscous damping in the torsional damper (2000 Ns/m) as well as structural damping in the material (damping ratio = 0.01) has been taken into account while calculating the steady state response.

The results for a given excitation order are presented in the form of stress as a function of the speed of rotation of the crankshaft. In order to compare the predictions from beam element model and the 3-dimensional finite element model, stresses at two locations have been used viz., at the crank pin and at a fillet. The former is a non-critical stress region and the latter is a representative critical section. The stresses in the fillet are highly localized and the stress gradients are very high. A very fine finite element mesh, as shown in fig. 6, has been used to model the fillets. The von mises stress at these two locations is depicted in figures 16 - 18.

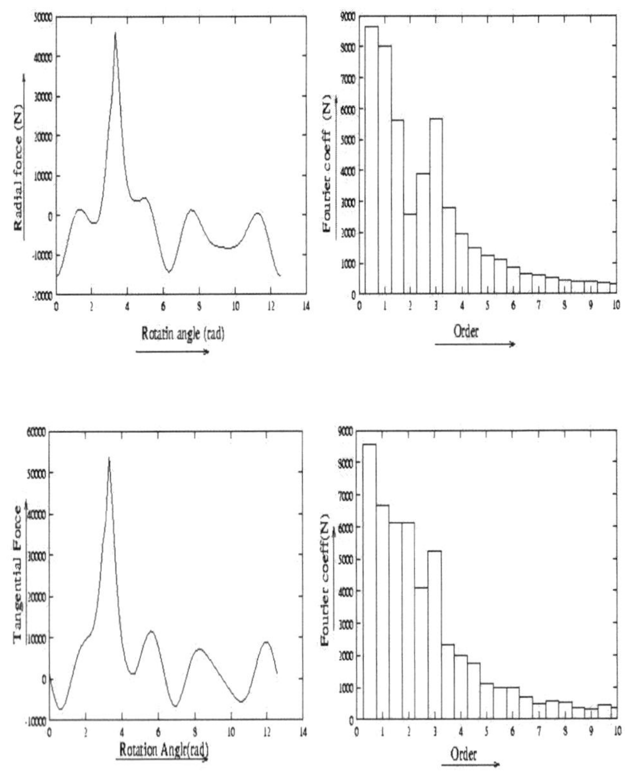

Fig. 15: Dynamic forces on the crankshaft and their Fourier decomposition

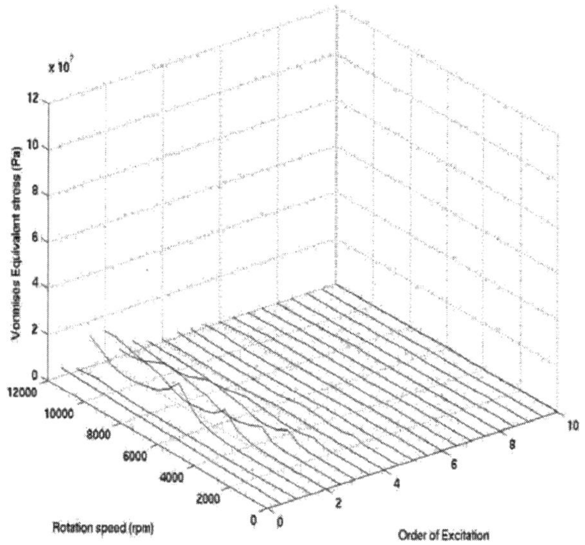

Fig. 16: Dynamic Stresses on Crank-Pin (Beam Element Model)

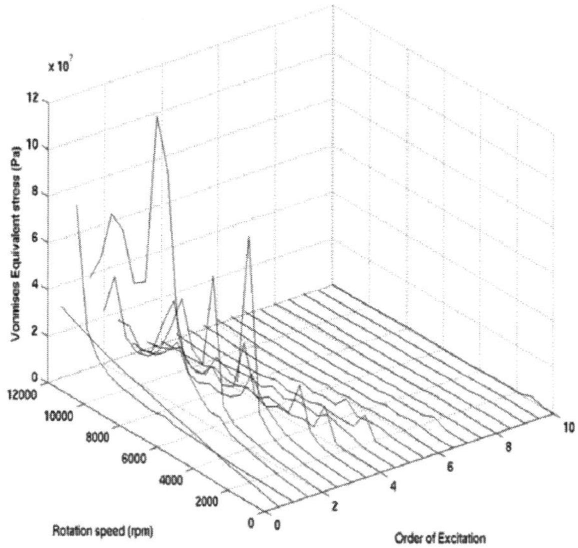

FIG. **18.** Dynamic stresses at fillet (3d model)

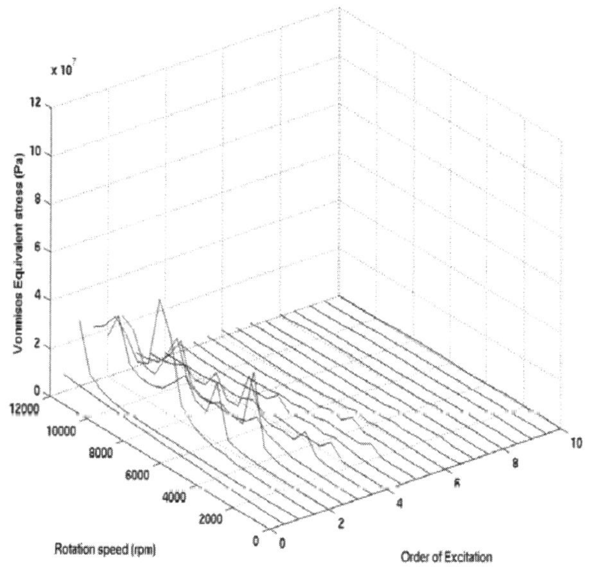

FIG. **17**. Dynamic stresses on crank-pin (3d model)

The critical frequencies of excitation can be estimated based on peak response and it can be observed that there is a shift in the critical frequencies as compared to the Campbell diagrams. It can be observed that maximum stresses correspond to the 3.5 order of excitation, which agrees with the results presented by Morita and Okamura [12].

The CPU time consumed for complete dynamic analysis of 3d model on sun ultrasparc machine was approximately 12 hours, while that for beam model was just about 12 minutes. However it is observed that the beam element model overestimates the stresses even in non-critical regions such as crankpin. It is also incapable of estimating stresses at a critical location such as a fillet. Thus a 3-dimensional finite element model yields a much better insight into the critical stress regions of practical automotive crankshafts. With the parametric modeling developed in the present work, the detailed 3-dimensional modeling of real life complex crankshafts has been considerably simplified. The tools developed here can be used further for design optimization

4. CONCLUSION

A parametric model has been developed for detailed three dimensional finite element analyses of typical real life automobile crankshafts. Results from the 3-dimensional analysis have been compared with those obtained using beam element models to assess the capabilities and limitations of such simplified models. Based on the dynamic analysis of an existing commercial vehicle crankshaft system, it has been demonstrated that the simplified beam element models incur an error of the order of 15% in natural frequency estimation and are incapable of predicting limiting stresses in critical regions. For practical multi-cylinder engine crankshafts, torsion modes are coupled with the bending modes due to their complex geometry. Three dimensional finite element model is able to capture the complex mode shape accurately and hence gives a better estimate of system frequencies.

It has been observed that the addition of flywheel and pulley significantly reduces the natural frequencies of the system and the nodes are shifted towards the pulley. While, the addition of journal bearings in the model increases the natural frequencies of the system, the flexural components in the mode shapes also change. Hence, flywheel, pulley and journal bearings should be taken into account for an accurate study of crankshaft vibration behavior.

Three-dimensional finite element modeling, however, consumes significant computational effort. Present parametric model reduces the model preparation effort significantly and can be readily extended for parametric study and design optimization.

REFERENCES

1. Kamath, M., Narasimhan, R., Nataraj, C. and Ramamurti, V., "Dynamic response of a Multicylinder Engine with a Viscous or hysteric crankshaft damper", Journal of Sound and Vibration, Vol. 81, no 3, pp 448-452, 1982.
2. Chen, S. K. and Chang, T., "Crankshaft Torsional and Damping Simulation - An Update and Correlation with Test Results", SAE Paper No: 861226, SAE Transactions Section 4, 4.964-4.985, 1986.
3. Drouin, B., Goupillon, J. F., Brassart, F. and Gublin, F., "Dynamic Modeling of the Transmission Line of an Agricultural Tractor", SAE Paper No: 911779, SAE Transactions, Section 2, Journal of Commercial Vehicles, 100:189-199, 1991.
4. Wilson, W. K., Practical Solutions to Torsional Vibration Problems, Vol. 1, Chapman and Hall Ltd, London, 1956.
5. Petkus, E. P. and Clark, S. F., "A Simple Algorithm for Torsional Vibration Analysis", SAE Paper No: 870996, SAE Transactions, Section 3, 96:3.156-3.163, 1987.
6. Tecco, T. C. and Grohnke, D. A., "Computer Simulation of Drive train Torsional Vibration in Heavy and Medium Duty Trucks", SAE 861960, SAE Transactions, Section 5, 5.967-5.974, 1986.
7. Szadkowski, A. and Naganathan, N. G., "TORAN: A Comprehensive tool for Driveline Torsionals", SAE Paper No: 942322, SAE Transactions, Section 2, Journal of Commercial Vehicles 103:587-596, 1994.
8. Birkett, C.A., Tecco, T. C. and Grohnke, D. A., "Computer Simulation of Driveline Vibration due to Universal Joints in Heavy and Medium Duty Trucks", SAE 912700, SAE Transactions, 642-647, 1991.
9. Shimoyamada, K., Iwamoto, S., Kodama, T., Honda, Y. and Wakabayashi K., "A Numerical Simulation Method for Vibration Stress Waveforms of High-Speed Diesel Engine Crankshaft System", SAE 910631, SAE Transactions, 933-953, 1991.
10. Okamura, H., Shinno, A., Yamanaka, T., Suzuki, A. and Sogabe, K., "Simple Modeling and Analysis for Crankshaft Three-Dimensional Vibrations, Part1: Background and Application to free Vibrations", Transactions of ASME, Journal of Vibration and Acoustics, 117:70-79, 1995.
11. Smaili, A. A., and Khetawat, M. P., "Dynamic Modeling of Automotive Engine Crankshafts", Mechanisms and Machine Theory, Vol. 29, No.7, pp 995-1006, 1993.
12. Morita, T. and Okamura, H., "Simple Modeling and Analysis for Crankshaft Three- Dimensional Vibrations, Part2: Application to an Operating Engine Crankshaft", Transactions of ASME, Journal of Vibration and Acoustics, 117:80-86, 1995.
13. Henry, J. P., Toplosky, J. and Abramczuk, M., "Crankshaft Durability Prediction- A New 3-Dimensional Approach", SAE 920087, SAE Transactions, Section 5, Journal of Materials and Manufacturing, 16-26, 1992.
14. Prakash, V., Venkatesh, D.N. and Shrinivasa, U., 1994, "The Effect of Bearing Stiffnesses on the Crankshaft Natural Frequencies", SAE Paper 940697, SAE Transactions, 1994.
15. Prakash, V., Aprameyan, K. and Shrinivasa, U., 1998, "An FEM Based Approach to Crankshaft Dynamics and Life Estimation", SAE 980565, SAE Transactions, Section 3, Journal of Engines, 107:825-837, 1998.
16. Athavale, S. and Sajanpawar, P. R., "Finite Element Model Generator for Assessment and Optimization of Crankshaft Design", SAE paper no: 912494, Presented in 6th International Pacific Conference, SAE, Seoul, Korea, Oct 28-Nov 1, 1991
17. ANSYS, "ANSYS Parametric Design Language", ANSYS Manual, ANSYS Inc., 1997
18. Rao, J. S., Rotor Dynamics, New Age International Publication, New Delhi-India, 1996

CONTACT

Prof. P. SESHU, Associate Professor
Mechanical Engineering Department
Indian Institute of Technology
Powai, Mumbai – 400 076, India.
Email: seshu@me.iitb.ac.in

2003-01-1031

Defect Identification With Model-Based Test Automation

Mark Blackburn, Aaron Nauman and Bob Busser
Software Productivity Consortium

Bryan Stensvad
T-VEC Technologies, Inc.

ABSTRACT

Software is an integral part of automotive products, but organizations face many problems that impede rapid development of software systems critical to their operations and growth. Manual processes to generate tests for software will become increasingly insufficient as automotive software becomes more complex, and more safety-critical. A method exists to develop tests automatically from formal, precise requirement and design models. A model-based approach allows teams to build software systems with measurably higher quality, in less time than with non model-based approaches. This paper discusses a Test Automation Framework (TAF) combining tools and methods to automate comprehensive test generation based on models. Automatic generation of software tests leads to dramatic performance and quality gains relative to manual test generation.

INTRODUCTION

Improvements in software testing are needed throughout industry. The National Institute of Science and Technology (NIST) estimated the cost of insufficient software testing processes in the United States during 2000 at $59 billion. The National Highway Traffic Safety Administration has conducted several recalls in recent years due to software defects. [1, 2] As software functionality drives a larger portion of the value of the automobile, the software development and verification processes must be improved to address the added complexities created by software-intensive automotive systems. One of the practices finding adoption is the use of formal, precise models of software requirements and design to support the development and verification processes.

This paper discusses model-based test automation methods and tools, referred to collectively as the Test Automation Framework (TAF) that reduce the time and resources necessary to develop high quality and high assurance systems. TAF has been effective in locating and correcting requirement defects early in the development process, reducing manual test development effort, and reducing rework. TAF integrates various model development and test generation tools to support defect prevention and automated testing of systems and software. TAF has been used for modeling and testing system, software integration, software unit, and some hardware/software integration functionality.

BACKGROUND

The core test generation tools underlying TAF were developed and used in verification of safety-critical avionics software in the late 1980s. Teams led by the Software Productivity Consortium (Consortium) have applied the method in a wide variety of software projects since 1996. It has been applied to critical software applications in various domains including spacecraft, automotive applications, medical devices, autopilots, display systems, engine controls, and airborne traffic and collision avoidance systems. TAF has also been applied to non-critical applications such as databases, client-server, web-based, automotive, and telecommunication applications. The method is compatible with many languages, such as C, C++, Java, Ada, Perl, PL/I, SQL, and proprietary languages, with various commercial test injection products, such as DynaComm and WinRunner, and several test environments. Most users of the approach have reduced their verification and test effort by approximately 50 percent [3, 4, 5].

RELATED WORK

There are papers that describe requirement-modeling [6, 7, 8], and others with examples that support automated test generation [9-13]. Ranville provides a summary of the uses of model-based automation in the automotive industry [14]. Asisi provides a historical perspective on test vector generation and describes some of the leading commercial tools [15]. Pretschner and Lotzbeyer briefly discuss Extreme Modeling that includes model-based test generation [16]. There are various approaches to model-based testing and Robinson hosts a website that provides useful links to authors, tools and papers [17].

CURRENT SOFTWARE TESTING PRACTICES

While it is common for software tests to be executed automatically, the tests themselves are often created manually. This process relies on the judgment and experience of testing professionals. It is impossible in practice for such teams to anticipate and develop a test for every way that a program might fail. When done manually, the tasks related to test design are typically slow, error prone, and highly variable. They can account for 60 percent of testing effort. Organizations have reported spending nearly 50 percent of their test effort just developing and debugging test scripts. The difficulties of such test development are compounded by the fact that the testing process is often done under extreme time pressures, since it must wait until coding is complete and usually starts near release deadlines. Significantly, testers find that about half of the defects found in the final code are due to defects in the underlying requirements.

MODELING CONCEPTS AND TOOLS

Models are constructed to represent and emphasize some aspects of a given system while ignoring or hiding others. They are often used to manage complexity or focus on elements of a system particularly important for analysis. They help engineers perform reviews and analyses of specific system properties by removing details irrelevant to the review or analysis being conducted. Engineers have used different types of models and modeling approaches for designing systems throughout history.

There are various approaches to modeling software, such as functional tabular requirement models, models based on Unified Modeling Language (UML), control system modeling, and hybrids. Software tools are available to support each of these approaches.

TYPES OF MODELING APPROACHES

Software developers can follow two broad modeling approaches: the modeling of software *requirements*, which describes *what the system is supposed to do*, and the modeling of software *design*, which describes *how the system is supposed to function*. Requirements models can be built using tools such as the Naval Research Laboratory s Software Cost Reduction (SCR) method, while design models can be built using tools such as Simulink, MatrixX, or UML-based tools. Modeling provides a means of formalizing the requirements. The discipline and structure of the modeling process helps eliminate incompleteness, and the resulting models provide a basis for tools to assist in detecting incompleteness and contradictions early in the development process. TAF supports both modeling approaches with tools that translate either requirement-based or design-based models into test specifications.

MODELING PERSPECTIVES

Models are described using specification languages, usually supported through graphical modeling environments. Specification languages provide abstract descriptions of system and software requirement and design information. Cooke et al. developed a scheme that classified specification language characteristics [18]. Independent of any specification language, Figure 1 illustrates three categories of specifications based on the purpose of the specification. Cooke et al. indicates that most specification languages usually are based on a hybrid approach that integrates different classes of specifications.

Requirement specifications define the boundaries between the environment and the system and, as a result, impose constraints on the system. Functional specifications define behavior in terms of the interfaces between components, and design specifies the component itself. A specification may include behavioral, structural, and qualitative properties. Behavioral properties define the relationships between inputs and outputs of the system. Structural properties provide the basis for the composition of the system components. Qualitative requirements [19] define nonfunctional requirements. Often, languages support certain elements of requirement and functional specifications and are termed functional requirements, as opposed to nonfunctional requirements [20].

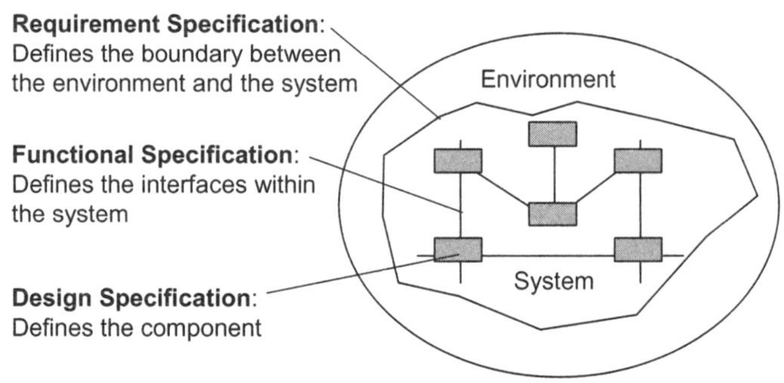

D. Cooke et. al; 1996

Figure 1. Specification Purposes

BENEFITS OF MODEL-BASED AUTOMATED TESTING

The TAF approach leverages models to support requirement defect analysis and to automate test design. It checks the model for various properties, like consistency and guides their refinement. Once the models are refined, TAF uses them to generate tests to verify the code. Eliminating model defects before coding begins, automating the design of tests, and generating the test drivers or scripts result in a more efficient process, significant cost savings, and higher quality code.

IMPROVED REQUIREMENTS

In order to be testable, a requirement must be complete, consistent and unambiguous. While any potential misinterpretation of the requirement due to incompleteness is a defect, TAF focuses on another form of requirement defect, referred to as a contradiction or feature interaction defect. These types of defects arise from inconsistencies or contradictions within requirements or between them. Such defects can be introduced when more than one individual develops or maintains the requirements. Often the information necessary to diagnose requirement contradictions spans many pages of one or more documents. Such defects are difficult to identify manually when requirements are documented in informal or semi-formal manners, such as textual documents. Although rigorous manual inspection techniques have been developed to minimize incompleteness and contradictions, there are practical limits to their effectiveness. These limits relate to human cognition and depend on the number and experience of people involved. TAF supports more thorough requirement testability analysis, allowing developers to iteratively refine and clarify models until they are free of defects.

REQUIREMENT VALIDATION

Requirement validation ensures captured requirements reflect the functionality desired by the customer and other stakeholders. Although requirement validation does not focus specifically on requirement testability analysis, it does support it. Requirement validation involves an engineer, user or customer judging the validity of each requirement. Models provide a means for stakeholders to precisely understand the requirements and assist in recognizing omissions. Tests automatically derived from the model support requirement validation through manual inspection or execution within simulation or host environments.

COMPREHENSIVE TESTS

Theoretical and empirical studies have shown that programming errors occur at boundaries, logical points in software at which decisions are made or data is passed from one subroutine to another. TAF uses the model to traverse the logical paths through the program, determining the locations of boundaries and identifying reachability problems, where a particular thread through a model may not be achievable in the program itself. TAF uses test selection criteria based on domain testing theory. TAF supports test vector generation, test driver generation, test coverage analysis, and the checking and reporting of test results. Test vectors include inputs, expected outputs, and identification information that traces to the associated requirement.

DEFECT DISCOVERY

Defect discovery using model-based test automation is both more effective and more efficient than using only manual methods. One pilot study, conducted by a Consortium member company, comparing formal Fagan inspections with TAF requirement verification, revealed that Fagan inspections uncovered 33 defects. In comparison, TAF uncovered all 33 of the Fagan inspection defects plus 56 more. Attempting to repeat the Fagan inspection did not improve its results. The improved defect detection of TAF prevented nearly two-thirds more defects from entering the rest of the development lifecycle.

Following manual test generation practices, defects are not identified until late in the process, sometimes after release, when they are most expensive to fix. Automating test generation based on models, defects are found earlier in the process and faster. The rate of defect discovery increases early in the process, but quickly curtails. Many defects are found in the requirements phase, before they propagate to later development phases. Defect prevention is most effective during the requirements phase when it costs two orders of magnitude less than after the coding process.

Figure 2 represents the conceptual differences between manual and automatic test generation. The existing process of discovering and eliminating software defects is represented by the curve labeled "Old" while the effects of early defect discovery aided by automation is illustrated by the trend curve labeled "New." Industrial applications have demonstrated that TAF directly supports early defect identification and defect prevention through the use of requirement testability analysis [4].

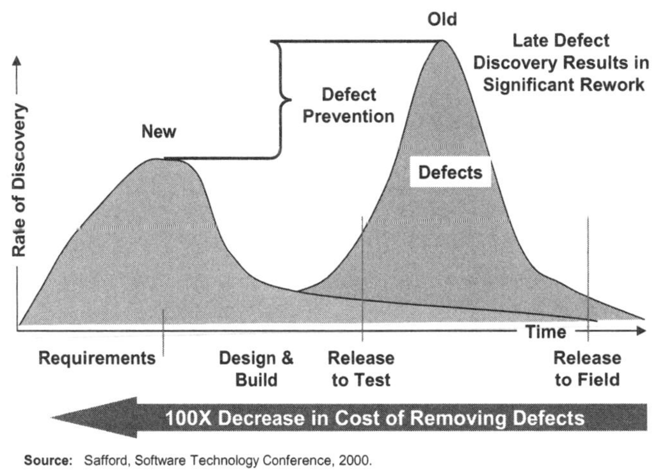

Figure 2. Early Defect Identification and Prevention

Rockwell Collins used a requirement modeling method to develop the mode control logic of a Flight Guidance System (FGS) for a General Aviation class aircraft [21, 22]. Rockwell Collins later used an early version of the TAF approach for model-based analysis and test automation to analyze the requirement model and generate tests for a new implementation of the FGS system [23]. As reflected in Figure 3, the FGS was first specified by hand using the Consortium Requirement Engineering Method (CoRE). It was then inspected, and about a year later it entered into a tool supporting the

SCR method provided by the Naval Research Laboratory (NRL). Despite careful review and correction of 33 errors in the CoRE model, the SCRtool s analysis capabilities revealed an additional 27 errors [22]. Statezni later used an early TAF translator and the T-VEC toolset to analyze the SCR model, generate test vectors and test drivers. The test drivers were executed against a java implementation of the FGS requirements [23] and revealed six errors. Offutt applied his tool to the FGS model and found two errors [24]. The latest TAF toolset, described in this paper, identified 25 errors more than the original 27 errors.

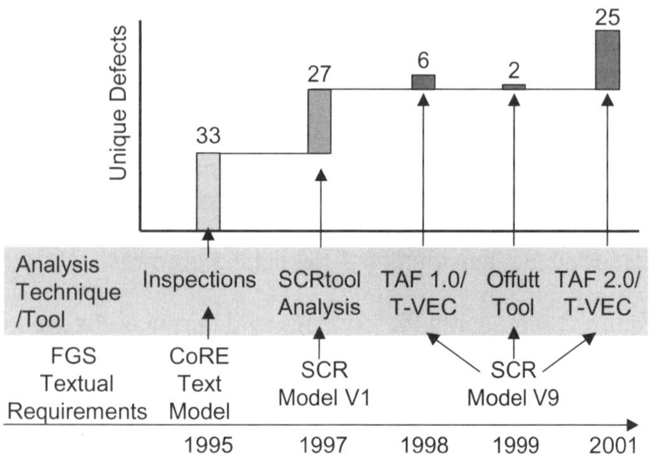

Figure 3. Model Evolution and Analysis

DEFECT ANALYSIS CONCEPTS

Requirement clarification during model development can uncover requirement problems such as ambiguities and inconsistencies. However, subtle errors or errors resulting from inherent system complexity can hide defects in a model or implementation. This section briefly describes defect types and how automated model analysis identifies them.

DEFECT TYPES

There are two types of errors: computation errors and domain errors. As defined by Howden a **computation error** occurs when the correct path through the program is taken, but the output is incorrect due to faults in the computation along the path. A **domain error** occurs when an incorrect output is generated due to executing the wrong path through a program [25]. Such errors can be introduced in a model as a result of errors in the requirements or during the requirement clarification process.

DOMAIN ERROR EXAMPLE

The concept of a program path and its related output computation is analogous to a requirement or design thread of a model. A domain error for a model thread means that there is no input set that satisfies the model constraints. Consider the following trivial example (and graphic insert):

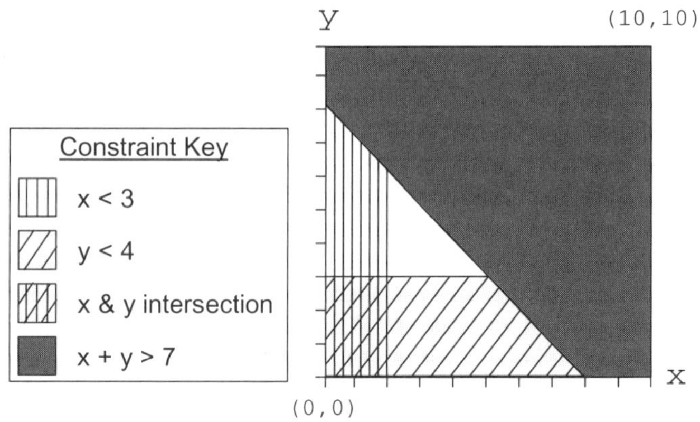


```
x: Integer with domain from 0 to 10
y: Integer with domain from 0 to 10
z: Integer with domain from 0 to 10
```

If there is a requirement that specifies

```
z = 0 when
      x < 3 AND
      y < 4 AND
      x + y > 7

then
   maximum value for x is 2
   maximum value for y is 3
   minimum value for x + y is 8
```

The region represented by the intersection of x & y does not overlap the constraint region defined by x + y > 7. The constraint expression is contradictory and cannot be satisfied. The contradiction results in a domain error, because the variable z will never be assigned a value of 0 through this requirement. Thus, the requirement is untestable. Real-world problems typically include complex constraints that span many modules or components of an application. In these situations it can be difficult to isolate these types of errors through manual processes. Automated model analysis provides a tool for locating these errors.

COMPUTATIONAL ERROR

Computational errors can result from various root causes such as an expression with incorrect variables, incorrect operators (+ instead of -), missing or incorrect parenthesis, or incorrect constants. Erroneous expressions can result in range errors, either underflows or overflows for the data type of the object. During test generation, low-bound and high-bound values are selected for the variables used in the computation in an attempt to stimulate range errors that can be traced to an expression with a defect. Blackburn provides examples of several computational errors that result from common errors in developing expressions for scaled arithmetic [26].

TEST AUTOMATION FROM DESIGN MODELS

TAF supports model analysis and test generation for design-based modeling, simulation and code generation tools, like MATRIXx and the MathWorks Simulink tools. Statezni, from Rockwell Collins, describes how TAF with the MATRIXx-based modeling tool was used to support the development and verification of safety-critical avionics software that must meet the rigorous guidelines of the Federal Aviation Administration [27]. After applying TAF on a pilot project, the Rockwell Collins lead engineer estimated that a Level A Flight critical project could save up to 52 percent of its total software development costs using a full model-based development environment, including auto-code and auto-test. [28]

Similar TAF support is available for analysis and test generation of Simulink design models. Figure 4 depicts the process and dataflow for TAF with Simulink. Project engineers use Simulink to develop a model of the software system. Source code can be generated automatically from the model or developed manually. A component of TAF produces test specifications directly from the Simulink model, then analyzes the test specifications to generate an optimal set of test vectors. It is during this process that the T-VEC component of TAF identifies model errors, such as contradictions or inconsistencies, which can result in unachievable conditions or other undesirable properties. Test driver generation transforms the test vectors from a generic format and produces a test driver and associated test harness that is compatible with the code generated by the Mathworks system. These test drivers are compiled in the same environment as the source code to generate a test program. This test program is executed in conjunction with the source code in the target environment. When the test harness executes, the results of each test are stored for comparison with the expected results. Finally, TAF tools compare test results to expected outputs to verify that the code implementation satisfies the model.

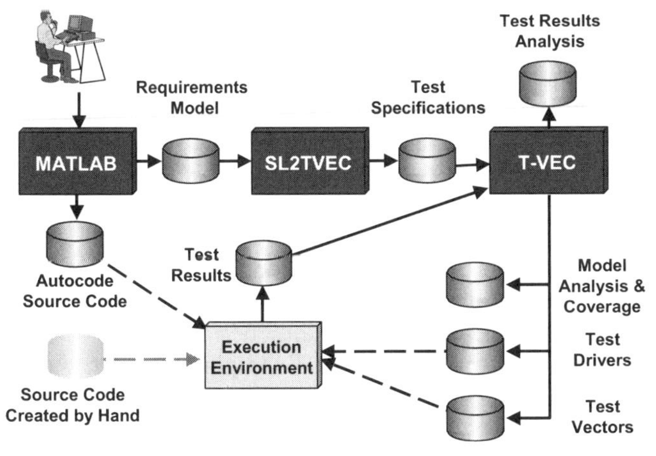

Figure 4. TAF Process and Dataflow Using MATLAB

TEST GENERATION EFFECTIVENESS

Test generation capabilities can be more effective than manual testing as was demonstrated when TAF was used to identify the fault that is believed to have caused the Mars Polar Lander (MPL) crash. NASA started the MPL project on February 7, 1994. Six years later, on December 3, 1999, after the MPL had traveled 11 months and over 35 million miles, all contact with the craft was lost just 12 minutes from its scheduled landing on Mars. The MPL cost $165 million to develop and deploy.

In fewer than 24 hours, TAF tools were used to identify an error in the software controlling the MPL s landing procedures [5]. This software monitored the touchdown legs of the craft and controlled the engine. It is believed this software falsely indicated the MPL had landed while it was still 40 meters above Mar s surface. Consequently, the engine was shut down prematurely, and the craft crashed. This defect could have been identified, and fixed, with TAF tools, if the requirements for the landing features of the MPL had been modeled and tested early in the process.

ORGANIZATIONAL BEST PRACTICES

Design modeling using tools like MATRIXx and Mathwork s Simulink is performed continuously, but when design models are not the basis for verification, requirement models can be developed. Requirement modeling can be applied after development is complete, however, significant benefits have been realized when it is applied during development. Ideally, test engineers work in parallel with developers to stabilize interfaces, refine requirements, and build models to support iterative test and development. Test engineers write the requirements for the products (which in some cases are very poorly documented) in the form of models, as opposed to hundreds or thousands of lines of test scripts. They generate the test vectors and test drivers automatically During iterative development, if the component behavior, the interface, or the requirements change, the models are modified, and test cases and test drivers are regenerated, and re-executed. The key advantage is that testing proceeds in parallel with development. Users like Lockheed Martin state that test is being reduced by about fifty percent or more, while describing how early requirement analysis significantly reduces rework through elimination of requirement defects (i.e., contradiction, inconsistencies, feature interaction problems) [2, 3]. This typical and pragmatic use of TAF parallels eXtreme Programming (XP) [29] where tests are created before the program. However, others refer to this model-based method as Extreme Modeling (XM) [16, 30], which applies the principles to write tests prior to coding. With XP, test code is developed manually, but with XM the requirements are modeled and tests are generated.

CONCLUSION

This paper describes a model-based Test Automation Framework (TAF) for software verification that supports defect prevention and automatic generation of tests. Such an approach reduces requirement defects, manual test development effort, and rework involved in developing and testing both software and systems. Analysis of software requirement or design models helps reduce the cost of developing high quality and reliable software. Static and dynamic analysis validates requirements models and comprehensive tests for code can be developed from models.

Development teams have reported significant savings using this approach. These teams have found that requirement modeling takes no longer than traditional test planning, while reducing redundancy and building a reusable model library capturing the organization s key intellectual assets. Because testing activities occur in parallel to development efforts, testing teams get involved from the beginning and stay involved throughout the process, reducing the risk of schedule overruns. Defect prevention is a key benefit of the approach. It is achieved using model analysis to detect and correct requirements defects early in the development process. The verification models enable automated test generation. This eliminates the typically manual and error-prone test design activities while providing measurable requirement-based test coverage. Organizations have demonstrated that the approach can be integrated into existing processes to achieve significant cost and schedule savings.

REFERENCES

1. National Institute of Standards and Technology. The Economic Impact of Inadequate Infrastructure for Software Testing. May 2002.
2. National Highway Traffic Safety Administration recall database. ftp://ftp.nhtsa.dot.gov/rev_recalls.
3. Kelly, V. E.L. Safford, M. Siok, M. Blackburn, Requirements Testability and Test Automation, Lockheed Martin Joint Symposium, June 2001.
4. Safford, Ed L. Test Automation Framework, State-based and Signal Flow Examples, Twelfth Annual Software Technology Conference, April 30 - May 5, 2000.
5. Blackburn, M.R., R.D. Busser, A.M. Nauman, R. Knickerbocker, R. Kasuda, Mars Polar Lander Fault Identification Using Model-based Testing, Proceeding in IEEE/NASA 26th Software Engineering Workshop, November 2001.
6. Heitmeyer, C., R. Jeffords, B. Labaw, Automated Consistency Checking of Requirements Specifications. ACM TOSEM, 5(3):231-261, 1996.
7. Parnas, D., J. Madley, Functional Decomposition for Computer Systems Engineering (Version 2), TR CRL 237, Telecommunication Research Inst. of Ontario, McMaster University, 1991.
8. van Schouwen, A.J., The A-7 Requirements Model: Re-Examination for Real-Time System and an Application for Monitoring Systems. TR 90-276, Queen's University, Kinston, Ontario, 1990.
9. Blackburn, M.R., R.D. Busser, A.M. Nauman, Removing Requirement Defects and Automating Test, STAREAST, May 2001.
10. Blackburn, M. R., R.D. Busser, A.M. Nauman, How To Develop Models For Requirement Analysis And Test Automation, Software Technology Conference, May 2001.
11. Blackburn, M. R., R.D. Busser, A.M. Nauman, Eliminating Requirement Defects and Automating Test, Test Computer Software Conference, June 2001.
12. Blackburn, M.R., R.D. Busser, A.M. Nauman, R. Chandramouli, Model-based Approach to Security Test Automation, In Proceeding of Quality Week 2001, June 2001.
13. Busser, R. D., M. R. Blackburn, A. M. Nauman, Automated Model Analysis and Test Generation for Flight Guidance Mode Logic, Digital Avionics System Conference, 2001.
14. Ranville, S. Practical Application of Model-Based Software Design for Automotive. SAE Paper 2002-01-0876.
15. Aissi, S.,Test Vector Generation: Current Status and Future Trends, Software Quality Professional, Volume 4, Issue 2, March 2002.
16. Pretschner, A., H. Lotzbeyer, Model Based Testing with Constraint Logic Programming: First Results and Challenges, Proc. 2nd ICSE Intl. Workshop on Automated Program Analysis, Testing and Verification (WAPATV'01), Toronto, May 2001.
17. Robinson, H., http://www.model-based-testing.org/.
18. Cooke, D., A. Gates, E. Demirors, O.Demirors, M. Tankik, B. Kramer, Languages for the Specification of Software, Journal of Systems Software, 32:269-308, 1996.
19. Yeh, R.T., P. Zave, A.P. Conn, G.E. Cole, Software Requirements: New Directions and Perspectives, in Handbook of Software Engineering, Editors C. R. Vick and C. V. Ramamoorthy, Van Nostrand Reinhold, 1984.
20. Roman, G.C., A Taxonomy of Current Issues in Requirements Engineering, IEEE Computer, 18(4):14-23, 1985.
21. Miller, S. P. K. F. Hoech, Specifying the Mode Logic of a Flight Guidance System in CoRE.
22. Miller, S. P., Specifying the Mode Logic of a Flight Guidance System in CoRE and SCR. Second Workshop on Formal Methods in Software Practice (FMSP'98), Clearwater Beach, Florida, March, 1998.
23. Statezni, David. Test Automation Framework, State-based and Signal Flow Examples, Twelfth Annual Software Technology Conference, April 30 - May 5, 2000.
24. Offutt, A.J., Generating Test Data From Requirements/Specifications: Phase III Final Report, George Mason University, November 24, 1999.
25. Howden, W.E., Reliability of the Path Analysis Testing Strategy. IEEE Transactions on Software Engineering, 2(9):208-215, 1976.

26. Blackburn, M. R., Using Models For Test Generation And Analysis, Digital Avionics System Conference, October, 1998.
27. Statezni, David. T-VEC's Test Vector Generation System, Software Testing & Quality Engineering, May/June 2001.
28. Software Productivity Consortium, Rockwell Pilot Project Technical Note, SPC-2000045-MC. 2000.
29. Beck, K., Extreme Programming Explained: Embrace Change. Addison Wesley, 1999.
30. Boger, M., T. Baier, F. Wienberg, and W. Lamersdorf. Extreme modeling. In Proc. Extreme Programming and Flexible Processes in SW Engineering (XP'00), 2000.

CONTACT

Mark Blackburn is a fellow of the Software Productivity Consortium, a not-for-profit organization established to improve the productivity of software and system development.

DEFINITIONS, ACRONYMS, ABBREVIATIONS

COTS: Commercial off-the-shelf

GUI: Graphical user interface

Java: High-level programming language

MPL: Mars Polar Lander

NRL: Naval Research Laboratory

Perl: High-level programming language

SCR: Software Cost Reduction

SQL: Structured Query Language

TAF: Test Automation Framework

TTM: T-VEC Tabular Modeler

UML: Unified Modeling Language

2003-01-1017

Model Based System Development in Automotive

Martin Mutz, Michaela Huhn and Ursula Goltz
Institute for Software, Program Development Department

Carsten Krömke
Volkswagen Inc., Electric/Electronics

ABSTRACT

The paper presents a major part of the STEP-X project (Structured Development Process by the example of X-By-Wire-Application in the automotive), namely a seamless, model based software development process in automotive engineering.

Our process is model based and supported by a tool chain. The tool DOORS is used for requirements management and engineering whereas the CASE tool Artisan RtS based on the Unified Modeling Language (UML) and the CASE tool Ascet SD are used for specification and design purposes. Each of these tools has its particular strength in a certain design phase.

We propose designing rules and modeling guidelines for the development of state based behavior which conforms to seamless model transformation in our tool chain. The rules are checked by an embedded rule-checker.

Additionally we illustrate our approach in a case study on a subsystem of the Volkswagen car electronics. The case study is characterized by state-oriented and concurrent behavior as well as time and value-discrete information processing.

INTRODUCTION

In the last years the role of software in vehicles has grown dramatically. Body electronic features and driver support functions turn out to be crucial for the success of modern vehicles. Automotive software will become even more complex and expensive in future. It is planned to extend established functions and to implement new functionality which will use the formerly isolated ECUs (electronic control units) as an integrated network. These functions will be realized as cooperating distributed real-time tasks. Thus, on the one hand system complexity will grow, on the other hand short time-to-market cycles and a considerable number of variants needs a well-structured development process supporting the reuse of components at all phases. We aim for a seamless development process for automotive functions with structured methods supported by case tools.

We start with the informal requirements documents as they are presently in use. In a first step, the textual requirements are analyzed and improved w.r.t. aspects like structure, completeness, and consistency. This is done using the requirements management and engineering tool DOORS. Next a model based function specification and the system architecture are elaborated using the UML tool Artisan RtS. When it comes to implementation, the design is transferred to Ascet SD. Ascet SD allows automated code generation optimized for controllers specific in the automotive area. We consider not only all phases from requirements engineering to code generation, but also the integration of tools and methods proven in practice. This raises applicability and acceptance of our approach, but for the prize of a more heterogeneous tool chain.

Although many tools for embedded software design offer object-oriented concepts for structuring and state-based behavioral models (e.g. statecharts), the tools differ in the modeling elements and semantical details. Thus the exchange of models between different tools is not a question of the file format, but a difficult problem of semantic transformations. We have implemented an automated transformation from UML statecharts modeling the design on an abstract level to more detailed „implementation statecharts“ in Ascet SD notation. We present design rules and modeling guidelines for UML-statecharts, which are a prerequisite for the transformation and which can be automatically checked by a rule checker module integrated in a Java program. We demonstrate our model based design process and the tool chain with a case study on a window lift system.

STRUCTURED DEVELOPMENT PROCESS

In the STEP-X project we follow a model based approach. Different graphical notations are used in our development process for embedded automotive systems. A system description contains information about functions, architecture, distribution of ECUs, actors, sensors, and the environment. The information is introduced in different phases of the design process. For a seamless development process, those information has to be integrated tightly.

In the next subsections we present some phases of the structured development process of the STEP-X project (see fig. 1) in more detail.

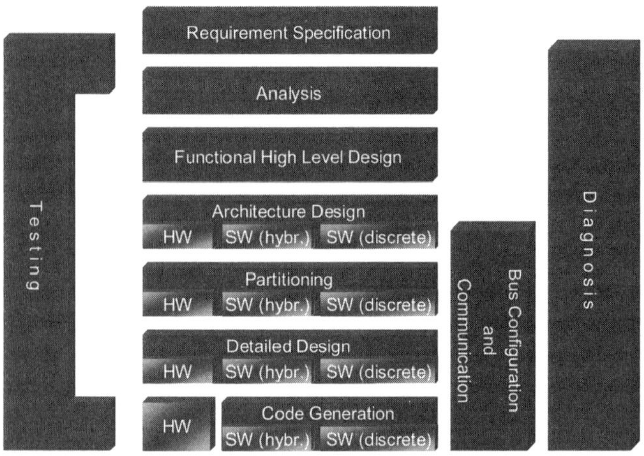

Figure 1: Structured Development Process of STEP-X

CASE STUDY – WINDOW LIFT SYSTEM

In order to illustrate our approach, we consider a window lift system. The system model and its environment are specified by UML diagrams, block diagrams and state machines as supported by Artisan RtS and Ascet SD.

The case study is based on a specification of a Volkswagen window lift system. It is characterized by state-oriented and concurrent behavior as well as time- and value-discrete information processing.

The following functionality has to be considered:

- Manual / automatic control of all 4 windows
- Comfort closing / opening[1] through key-operated switch and radio remote control
- Child protection
- Primitive shut protection
- Restriction / extension of the functionality depending on the status of other components (ignition, door position, etc.).
- Timing constraints, e.g. delayed closing of all windows during comfort closing
- Multiple functionality of a switch

The software control of the four windows is distributed over five different ECUs. An ECU processing the local functions is assigned to each window. A central module executes functionality common to all windows. Variants specific for countries or vehicle types are not considered in the model.

REQUIREMENT SPECIFICATION

The development process starts with the requirements specification which is divided into two parts:

1. Functional requirements
2. Non-functional requirements

Functional requirements describe what the system shall do, in particular the tasks of the electronic control. Non-functional requirements concern quality criteria, all kinds of restrictions w.r.t. the design, performance and maintainability, reuse of components, and the conformance to standards [2]. Non-functional requirements are given as text or references to other documents. Functional requirements are additionally described by suitable UML diagrams. The textual requirements and the corresponding UML models are managed by DOORS.

DOORS

The requirements management and engineering tool DOORS by telelogic [6] is one of the leading tools in the field and wide spread in automotive. It supports extensive features for analysis and tracing of requirements. DOORS offers a multi-user platform for the development of structured requirements documents. DOORS offers several structuring concepts like modules and headings. In DOORS, elementary requirements are considered as objects. They can be classified by freely definable attributes. Related requirements are connected by links, no matter whether they belong to the same or another document. Via links, one can trace the dependencies in requirements documents. DOORS offers comprehensive support for managing these links. The evolution of requirements during the development process can be traced by various history functions. The embedded language DXL is used to implement interfaces to other tools. The interfaces can be modified or extended by application programmers. Thereby UML-models or block diagrams from continuous modeling can be linked to DOORS objects.

[1] simultaneous closing / opening of all windows

Requirements Analysis

Based on existing documents, a new specific structure for automotive requirements was worked out. Different kinds of information like functional requirements, parameter lists or the glossary are split in separate documents (DOORS modules). Each document has its specific internal hierarchical structure (see fig. 2). If necessary, attributes are assigned to the requirements to classify aspects like quality of service. Related requirements and graphical representations are referenced via links. Also UML models of particular aspects or components are linked to the corresponding requirements. Thereby the designer can navigate through the textual and model based system descriptions and gets a better understanding of the system under development. The links are also used for completeness analysis.

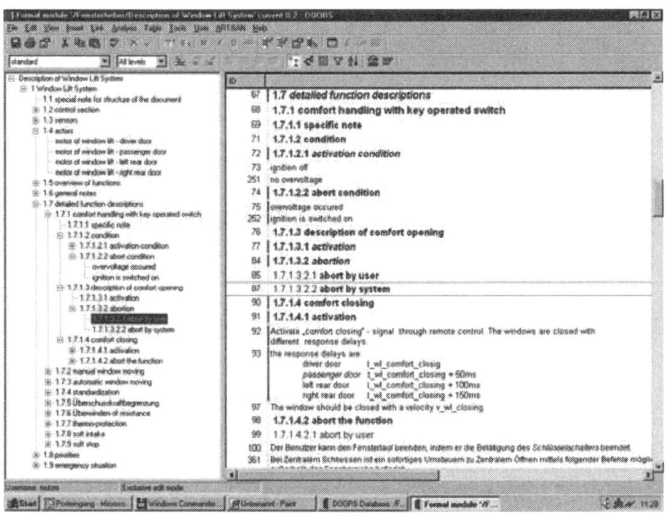

Figure 2: Functional requirements with DOORS

In the STEP-X development process- as in other projects on automotive software design like FORSOFT II [3] - links are inserted for the following purposes:

1. Improving the understanding
2. Representing mutual dependencies on requirements
3. Navigation to models
4. Tracing of requirements

ANALYSIS

Next the system functions are defined. In the object-oriented analysis based on the DOORS requirements documents, the objects and their methods are identified and communication with the environment is specified. In this phase, Artisan RtS is used for modeling.

Artisan RtS

Artisan Realtime Studio by ARTiSAN is a CASE tool based on the UML. The tool supports most of the UML diagrams. Additionally it offers diagram types for the modeling of the system architecture, the distribution of tasks and their communication (concurrency diagrams) and non-functional aspects. The discrete system behavior is modeled by UML statecharts which can be simulated. Artisan RtS provides automated code generation and reverse engineering. The repository handles the access to the models and the configuration management in a multi-user project.

Analysis models

Scenarios show typical interactions between the user and the system, exceptionary behavior or test cases. They are represented by use case diagrams. Use cases are still a mainly informal notation. (see fig. 3).

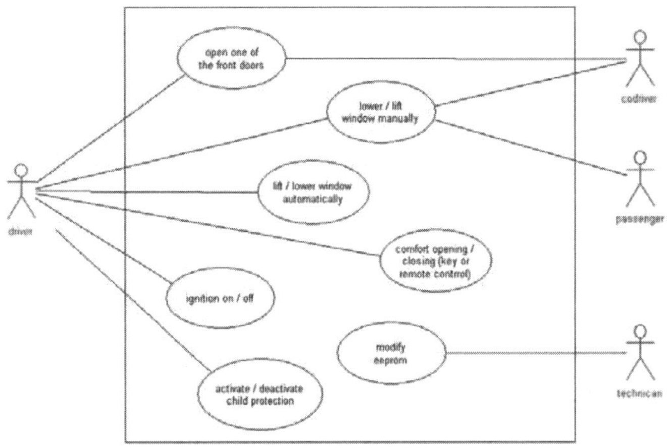

Figure 3. Scenario of the window lift system in a Use Case Diagram

Complex scenarios are refined into more detailed use case diagrams. In case timing constraints or the ordering of events is of importance, we introduce sequence diagrams.

A major goal of this step is to improve the understanding of the system under development. We consider the system boundary at the interface to the user, the interactions are determined by the sensors, buttons and other hardware.

Functional Decomposition

Afterwards the functions of the system are specified. The functions are modeled as modular logical entities on an abstract level. Complex functions are decomposed to increase the reuse of components. Then the data flow is analyzed and specified using interfaces to ensure modularity.

The communication of a complex function with other system components or the environment is realized by a coordinator class unique for each complex function. This pattern was applied uniformly for the decomposition of all components and all hierarchical layers.

Functions, interfaces, and signals are represented in class diagrams. To improve readability, functions are structured hierarchically. Signals sent and received by functions on the same hierarchical layer are modeled by dependency arrows.

Signals sent between functions from different hierarchical layers are modeled by associations which are annotated with the signal type (see fig. 4). Interfaces are modeled using the UML lollipop notation.

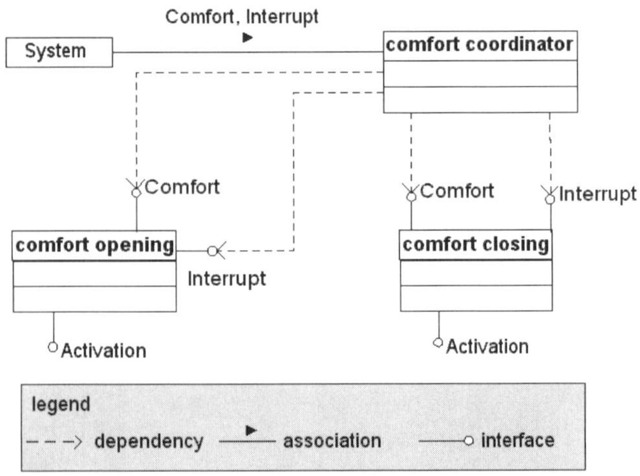

Figure 4: Decomposition of comfort window movement

Coordinator classes control the communication of the sub-functions among each other and with the rest of the system. At this early development phase, all information about dependencies between functions is encapsulated in the coordinator class. Later it may be spread in other elements of the design.

To summarize the steps of the analysis:

1. Use case analysis: Identification of actors and scenarios and use cases
2. Refinement of scenarios by textual descriptions and sequence diagrams
3. Definition of system functions
4. Functional decomposition
5. Analysis of data flow and definition of interfaces
6. Modeling of hierarchical functions, interfaces and data flows in class diagrams

FUNCTIONAL HIGH LEVEL DESIGN

In the high level design the variants are instantiated, signals are refined and assigned to the instances, and the behavior is specified using UML state diagrams. In the following these steps are described in detail.

Variants

Now we build the variants of the functions and instantiate them. Variants are introduced if we need several functions with similar structure and behavior. For instance, in the window lift system we build three variants of a control logic: one for the driver door, one for the passenger and one for the rear doors. The control at the driver door differs from the others as it allows to run all windows. The control at the rear doors contains an additional sub-function for child protection. Variants are modeled by class diagrams. Fig. 5 shows the variants of the window lift system.

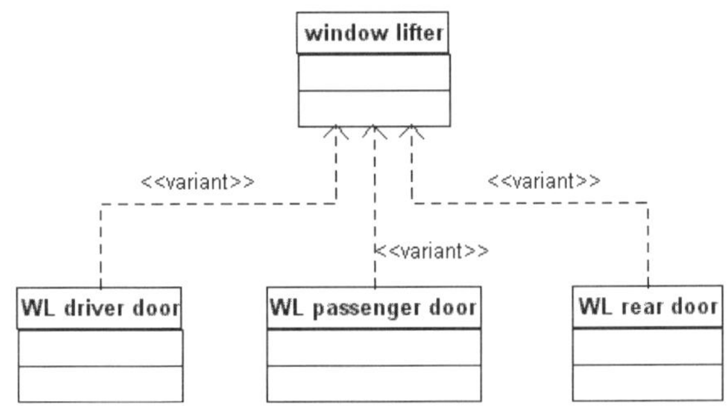

Figure 5: Variants of the window lift class

Variants are linked to the original class through dependencies annotated by <<variant>>. In the window lift system the variants only differ with respect to their coordinator class.

It is possible that building variants generates new mutual dependencies on the instances which were not known in the analysis phase. These are added to the existing class diagrams. Moreover, the interfaces are refined appropriately. Interfaces are modeled by parameterized classes which are also called *template classes*. They handle variable aspects of classes. A template class may not be a super class or the target of an association. First the parameters have to be bound to values. The parameters are noted in a dashed box at the upper right corner of a class symbol (see Fig. 6).

To keep communication more flexible, not only the signals itself but also their source are first specified in a parameterized way. By instantiating the templates, information on the source of signals may be added to the interfaces.

E.g. the top level coordinator class of the window lift system evaluates all sensor signals from the car environment like sensors, buttons, remote control etc. The signals are sent to the control units of the corresponding window. At the local control units the coordinator class determines how to react on a signal.

Analysis of signals from the environment

Next the signals from other parts of the vehicle are analyzed in detail to refine actors and their interfaces. A structured representation of signals from sensors, buttons and switches is given in a class diagram (see fig. 7). Signals are usually named X_Y_Z or X_Z where X is the source of the signal, Y the addressed window and Z the type of the signal. In particular, D stands for driver, P for passenger, RL/RR for rear left/right, and CP for child protection. UEKB-stands for the overpressure exception handling. O means Open, C Close, and UE-overpressure. The buttons have different states to trigger manual opening and closing or the automatic drive of the corresponding window.

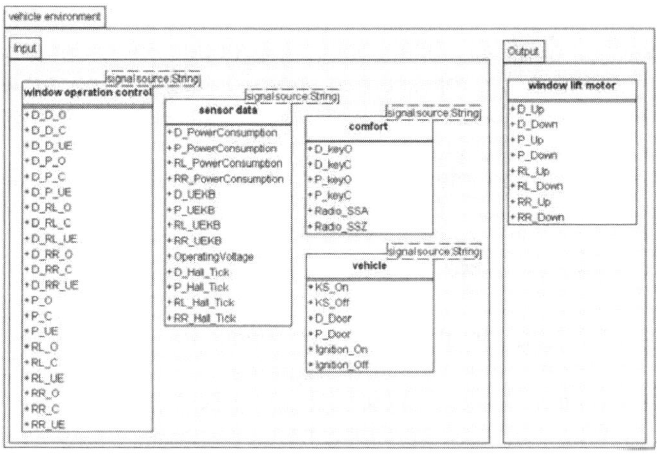

Figure 6: In- and output signals from the environment

System behavior modeled by statecharts

The behavior of the system is modeled already in the high level design by UML statecharts [4] to allow for early simulation and testing of the design. Some remarks on state charts are given in the section on design vs. implementation automata. The last step of the high level design is to assign state diagrams to the functions and coordinator classes. All signals received from the environment are handled by the top level coordinator class which sends them to the addressed subordinate function or coordinator. Figure 7 shows the state diagram of a coordinator.

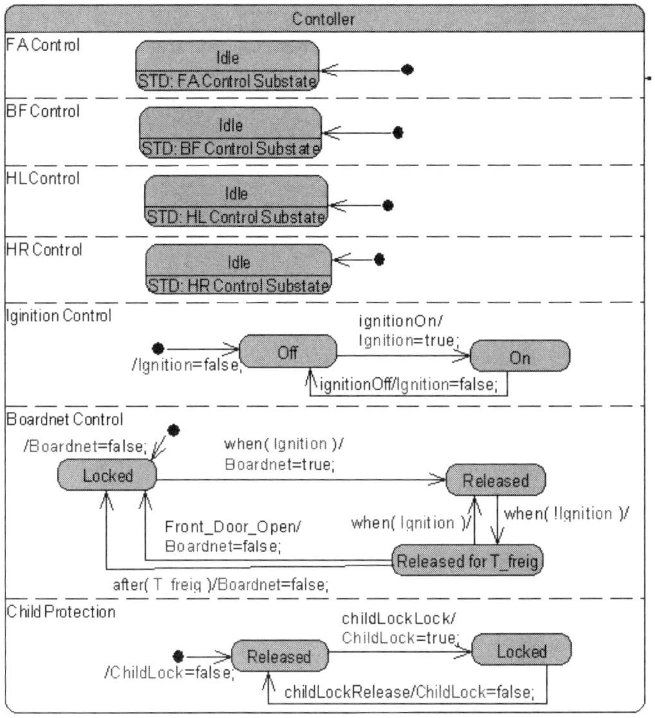

Figure 7: Coordinator Statechart of window lift system

The coordinator consists of seven sub-states executed in parallel. The upper four realize the window control for the driver and the passenger windows and the two rear windows. Each sub-state itself contains a statechart describing the function in more detail. Figure 8 shows the control of the driver window. By the buttons the driver window can be opened or closed manually (the window moves as long as the button is pressed) or in an automatic drive until an upper or lower stop position is reached.

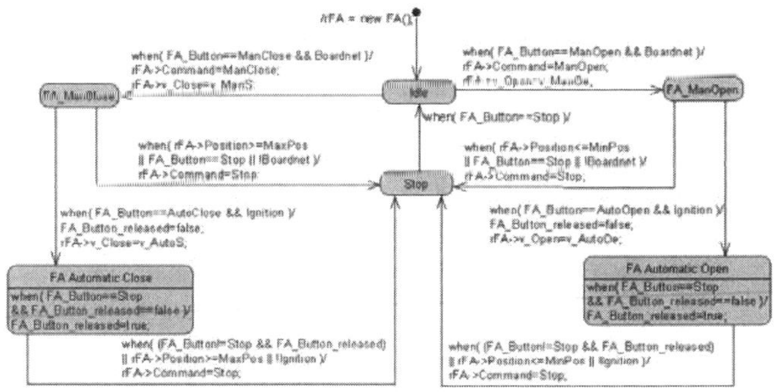

Figure 8: Behavior of the control in the driver door

Depending on the status of the ignition, a release signal for the window lift system is set, which activates or deactivates certain functionality. For instance an automatic window drive is only possible if the board net is activated. The comfort closing is only activated if the board net is down. The lowest parallel (orthogonal) sub-state in figure 8 models the child protection. If the child protection is active the buttons at the rear doors are deactivated.

As described earlier the control is located at the coordinator class. The movement of the windows is controlled by local classes which sent and receive signals from the motor of the corresponding window. A statechart for the local control of a window lift motor is shown in figure 9.

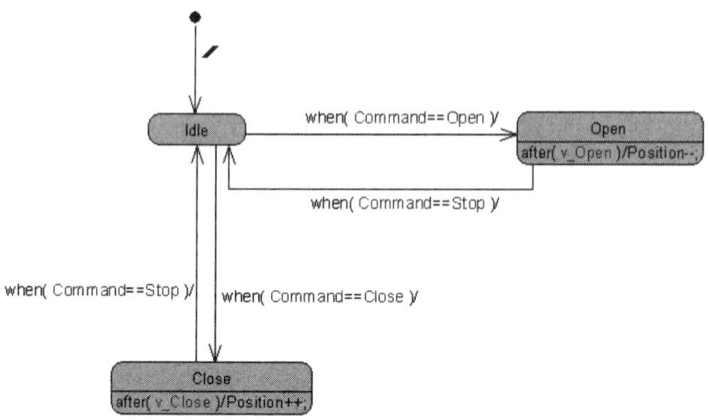

Figure 9: Behavior of any window

Here the advantages of object-orientation show up, since the motor control is only modeled once and then instantiated for each window. Thus not only the design (and the derived code) will be smaller, but also it becomes more readable, quality assurance is more efficient and maintainability is improved.

To summarize the steps of the high level design:

1. Building of variants and analysis of new data flows
2. Classification of signal sources
3. Instantiation of variants
4. Refinement of data flows and binding of interface template classes
5. Refinement of signals and the environment
6. Modeling of the refined class diagrams
7. Modeling of the system behavior by statecharts

ARCHITECTURE DESIGN

In this phase instances are defined and distributed over the logic electronic control units. The system architecture (in particular some restrictions from the physical distribution) has to be considered in the instance creation. Depending on, whether a function is implemented in a distributed manner or centrally, it must be instantiated once or several times.

The instantiated functions and variants are assigned to the ECUs. Deployment diagrams are used for visualization. These diagrams show ECUs as nodes (arithmetic unit and storage capacity) and the software components available on them (see figure 10).

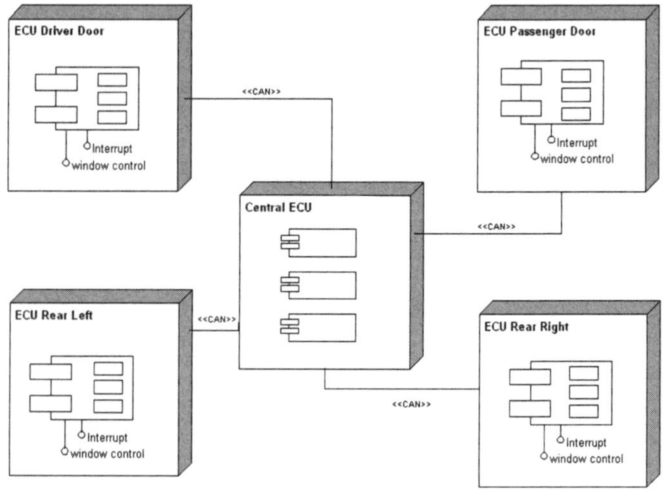

Figure 10: Distribution of instances on five ECUs

Within the components, the instantiated functions are represented as objects. In order to improve the readability of the architecture we abstain from the representation of all instances in the nodes. If necessary nodes can be represented more detailed in separate diagrams. Increasing the degree of detail requires an additional structuring. Software components communicate through interfaces which are represented as lollipop notation (see figure 11).

To summarize the steps of the architecture design:

1. Instantiation of variants and functions; representation in class diagrams
2. Distribution of instances on logic ECUs; representation by deployment diagrams

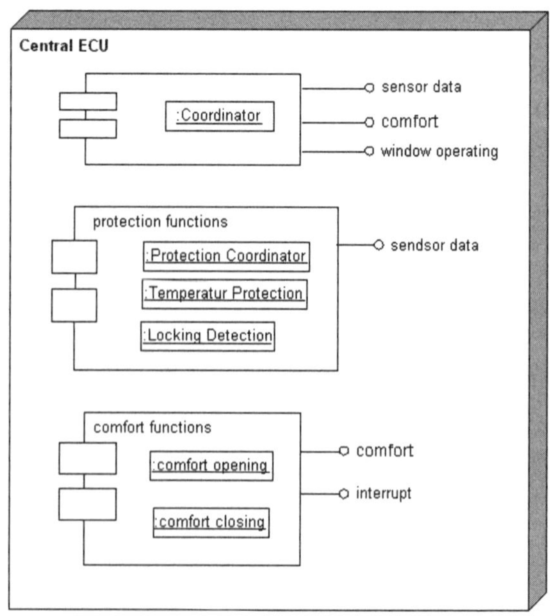

Figure 11: Details of the central ECU

DETAILED DESIGN

In the early development phases, the Unified Modeling Language was applied for modeling, because on the one hand it supports the object-oriented approach and on the other hand it is a quasi industrial standard. Similar approaches are taken in other automotive projects [3,5,6]. A further benefit of the UML is that numerous tools are available for modeling. These tools do not only allow to draw models for documentation purposes but also to simulate and partly verify the design. Code generation from class diagrams and state charts is supported by several tools. Unfortunately, the generated code is presently not optimized for micro controllers. Thus the tools are not suitable for the implementation phase. Another drawback is that design models for continuous components are often difficult to integrate.

Modeling with Ascet SD

In our STEP-X development process, Ascet SD by ETAS [8] is used in the implementation phase. It offers a developing environment for ECUs. In particular Ascet SD supports the phases from detailed design up to code generation. Block diagrams are mainly used to structure the system. In order to describe the behavior, state machines can be used as well as C-code and the specific language ESDL. In addition Ascet SD offers elements for the specification of real-time behavior. Particular advantages of the tool Ascet SD are the simulation facilities for the state machines and the code generator which generates optimized code for different micro controller targets that are commonly used in automotive applications.

Modeling architecture and behavior

The architectural model from the high level design and the attached statecharts are the basis for modeling the system structure and behavior. The architecture is represented by Ascet SD block diagrams. Single functional variants are represented by classes to which state machines are assigned. The state machines are discussed in detail in the following chapter where we consider the transformation of UML state diagrams to Ascet SD state machines.

The communication between functions is achieved by method calls, similar as in the UML. An other and more common way to realize communication is to embed the classes and/or their instances into blocks. Thereby, information can be exchanged by messages (global variables). Contrasting methods which may only have a single return value, in blocks several signals may be sent to different components. Blocks may contain several blocks and instances so that a hierarchical structure is possible. However, blocks cannot be inherited. This is a difference between Ascet SD classes and object-oriented approach.

The deployment diagram from the high level is transferred into Ascet SD. Blocks represent the UML components and have fixed inputs and outputs. A vehicle environment is modeled for simulation. For this purpose, two different blocks are introduced: First a block *environment* that contains all keyboard switches and lights of the comfort system and secondly, the block *sensor-coordinator*, which receives signals from the sensors and thereby influences the behavior of the window (e.g. shut protection). The functions and their variants are modeled as classes and automata like in the high level design. A part of the architecture of the window lift system is shown in figure 12.

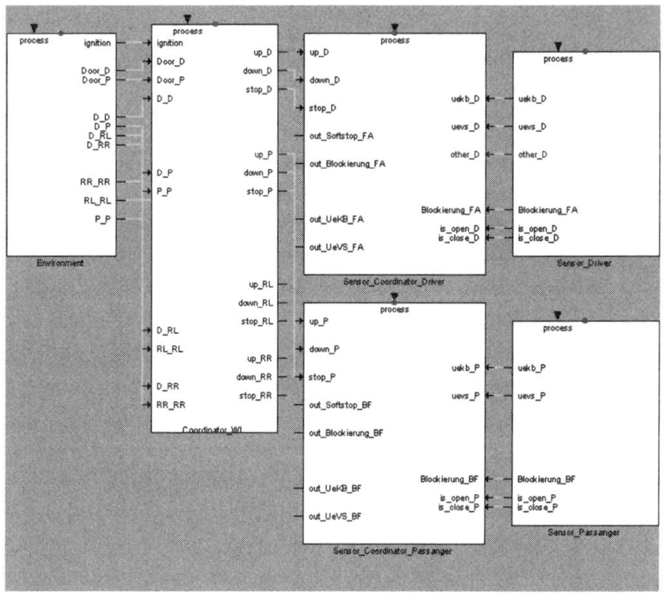

Figure 12: Part of architecture of the window lift system in Ascet SD

The behavior of the functions is modeled by Ascet SD state machines. The signals received by a block are transmitted automatically to the embedded state machines where possibly a transition is triggered. However, the semantics of UML and Ascet SD statecharts differ in many points. In particular UML statecharts offer powerful modeling elements that are not available in Ascet SD. How to adapt the two statechart semantics is discussed in the next chapter.

To summarize the steps of the detailed design:

1. Designing the architecture with block diagrams
2. Definition of the communication channels
3. Modeling behavior by state machines

AUTOMATIC MODEL TRANSFORMATION

One of the major problems to be solved for the seamless tool support of our development process is the transition from the UML tool Artisan to Ascet SD, which generates small and efficient code for ECUs from simple statecharts. On the one hand, we use comfortable modeling features from UML statecharts which are not available in Ascet SD, on the other hand the semantics implemented in these tools differ even for common features. We aim at developing tool support for a model transformation to solve this problem. The following subproblems are being solved:

- A rule checker has been implemented which checks statecharts with respect to user defined rules.
- A tool for the (semi-)automatic transformation of Artisan statecharts in Ascet SD state machines is being developed. Basis of the transformation is an analysis of the semantic differences.
- A model optimizer is supposed to improve the model which is generated by this transformation with the aim of obtaining more compact code.
- Finally, we will develop a tool for checking the equivalence between Artisan and Ascet SD models with respect to certain properties. One application of this tool could be to check models used for implementation against the models given in the digital requirements specification.

Fig. 13 shows the structure of the model transformation tool being developed. In the following, we explain the parts in more detail.

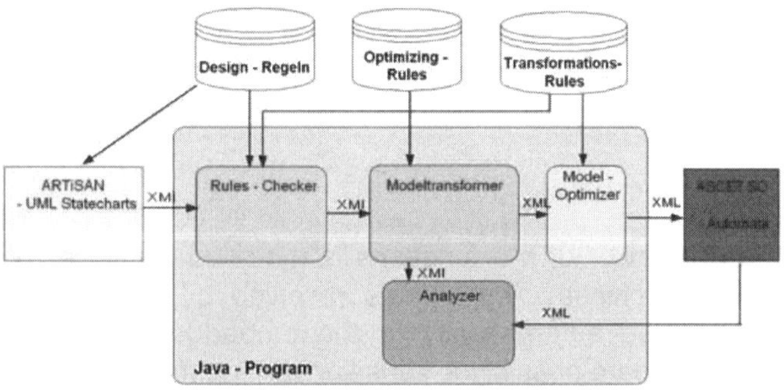

Figure 13: Concept of the (semi-)automatic model transformation

DESIGN AUTOMATA VERSUS IMPLEMENTATION AUTOMATA

One of the main difficulties for obtaining seamless tool support is dealing with differences in the semantics of models in the tools being used. It is important for our methodology to be able to use the interplay between different UML models and the comfortable modeling features in statecharts as offered by Artisan RtS.

On the other hand, it is indispensable to use a tool specialized for generating code for ECUs as Ascet SD to be able to obtain compact code. To be able to integrate both tools in our process, we thoroughly analyzed the semantic features for statecharts in Artisan, respectively state machines in Ascet SD.

Statecharts concept of Artisan RtS

UML state diagrams are an extension of transition diagrams. The statecharts support hierarchy, structure and orthogonality. UML statecharts consist of states and transitions. A state can be *basic*, *concurrent* or *composite*. If a state is *basic* it has no sub-states. A *composite* state has sub-states and exactly one of them is active at a certain point of time. A *concurrent* state has composite sub-states. All of them are active if the parent state is active. Composite, concurrent and basic states form a tree structure and this hierarchy allows for stepwise refinement of the behavior of complex systems. States are connected by transitions.

In the simple case transitions are connected directly with a source and a target state. UML allows the connection of transition segments by different kinds of connectors (join, fork, condition, selection and junction connectors). They can be used e.g. to split an incoming transition into multiple outgoing transitions with different guard conditions. For further details we refer to [4]. Transitions may be labeled by *events, guards* and *actions*. *Events* trigger transitions: if the corresponding event occurs and the transition *guard* evaluates to *true*, the transition has to be taken immediately if there is no conflict with another transition. A conflict happens e.g. if two transitions leave the same state and both are enabled. Then only one of them can fire. Transitions also may be in conflict if their source states are related hierarchically. In this case the UML priority scheme gives priority to transitions having source states lower in the state hierarchy. *Actions* are performed when a transition is taken; Examples for actions are an assignment or a sending of events.

The UML assumes a data structure like a queue to store events which shall be received in the next step by a statechart. A dispatching mechanism selects one event of the queue if processing of the previous event is fully completed (run-to-completion). Temporal behavior can be modeled through time events (events are sent delayed by a certain time) and scheduled actions (actions are executed delayed). State transitions of the system occur in a step-wise manner. A step contains a maximal non-conflicting set of transitions that fire simultaneously.

State Machines concept of Ascet SD

Ascet SD state machines consist of simple and of hierarchical states. Substates are supported in the same way as in Artisan RtS, however there is no orthogonality, i.e. parallelism, in state machines (this will be one of the main tasks in the model transformation).

Transitions are labeled with events, conditions and actions, whereby events are required. Each transitions must have an explicit priority such that the model will be deterministic. For hierarchical states the priority concept is opposite to UML: transitions higher in the hierarchy have higher priority. Fork, join and junction connectors are unified as one type of connector. Time aspects cannot be specified by transition labels because all Ascet SD transitions are time triggered.

RULE CHECKER

Our rule checker is part of the model transformation entity we are implementing for a seamless tool chain. It verifies whether user defined design rules on state charts have been observed in a given model. For our application, these rules are currently being designed. On the one hand they aim for a compact, consistent and uniform modeling on the UML layer, on the other hand appropriate models for the transformation to Ascet SD state machine shall be achieved.

The rule checker is parametric with respect to the rules being checked, so it may be used also in other contexts.

The rule checker allows various input formats, based on XMI (XML Metadata Interchange). The model is checked according to user defined rules. A set of predefined rules (49 rules ordered in 4 categories) is offered. A selection is given below:

Design check

- Checks if the statechart contains two or more states with equal names
- Detects illegal denotation of statechart elements, including state names, transition and trigger names
- Detects transitions without any trigger, guard and action
- Detects isolated states (states without incoming and outgoing transitions)

Consistency check

- Ensures that the diagram does not contain empty orthogonal components.
- Detects defined events that are not used as a trigger in any transition.
- Checks if the root state contains an initial state.
- Detects miracle states (states without incoming transitions but with outgoing transition).

The rules can be changed or redefined by the user. Due to a special format, they are independent of the implementation in the tool. Rules can be activated or deactivated by the user of the rule checker.

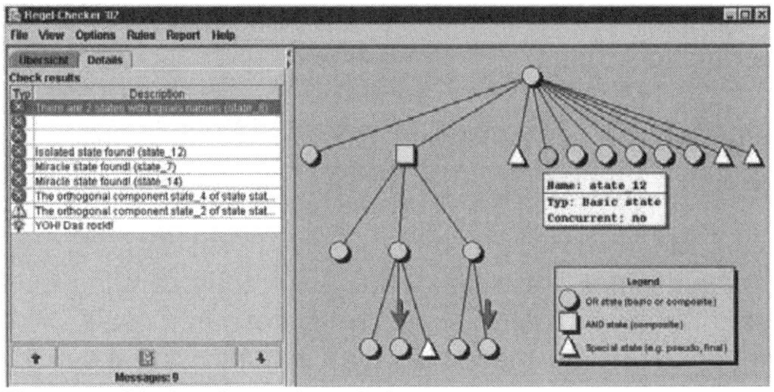

Figure 14: Display of error through the tree

Fig. 14 shows the user interface of the rule checker and the result of an analysis for an example statechart. In the right window, the statechart is shown as a tree structure. Different types of states (AND- , OR- and pseudostates) are represented by different shapes of nodes. Using the shift button one may obtain detailed information for each node in a yellow box. In the left window, all messages (warnings, errors and suggestions) are listed. Selecting a message results in a red arrow pointing to the relevant part of the tree.

If the model contains no errors, it may be stored and delivered to the next module (e.g. the kernel of the model transformation).

MODELTRANSFORMATION AND VERIFICATION

The remaining parts of our framework (model transformator, optimizer and verificator) are currently being developed. For the transformator and the verificator, we have obtained first results from preliminary studies which are shortly outlined below.

For the transformation of statecharts into hierarchical state machines, the semantic studies exhibited that a number of features may not be transformed. These are for example

- timed events and actions,
- transitions with negated events,
- create- and destroy-events,
- nondeterminism,
- C-functions.

These features may be forbidden by the rule checker and will result in error messages.

The model transformator then has to solve the following tasks:

- Transform parallelism into sequential execution.
- Modifying priorities according to the respective semantics.
- Adapt transformable constructs.

To prepare for verification, we will also allow to flatten hierarchical states.

The resulting state machine will be optimized to obtain compact target code (this part of our approach is still future work).

The verificator, for which we are currently carrying out prototypical studies, will compare statechart models with hierarchical state machines. One way of using it would be to compare state machine on the implementation level with statechart specifications. In particular when dealing with variables, we will face problems with complexity and state explosion. One way of overcoming these problems will be to use abstraction methods.

CONCLUSION

This paper presented the seamless structured development process with the corresponding tool chain and the design methodology being developed in the STEP-X project. We used the window lifter as a case study. The models were generated and simulated with the UML tool Artisan RtS. For the system architecture, we used UML deployment diagrams.

We outlined a tool supported transformation to obtain Ascet SD state machines from Artisan RtS Statecharts. The code generation for the electronic control units, the partitioning of the system and the communication between components is then obtained with Ascet SD. Further research will concentrate on the integration of testing and information for diagnosis in this process.

ACKNOWLEDGMENTS

This work has partially been funded by the Volkswagen Inc. within the STEP-X project.

REFERENCES

1. C. Schröder, U.Pansa: UML@Automotive - Ein durchgängiges und adaptives Vorgehensmodell für den Softwareentwicklungsprozess in der Automobilindustrie, Praxis Profiline, IN-CAR-COMPUTING, 2000
2. B. Gebhard and M. Rappl: Requirements Management for Automotive Systems Development, SAE2000, SAE Press 00P-107, 2000
3. M. von der Beeck, P. Braun, M. Rappl, C. Schröder: Modellbasierte Softwareentwicklung für automobilspezifische Steuergerätenetzwerke, VDI, 2000
4. OMG Unified Modeling Language Specification, www.omg.org
5. M. Götze, W. Kattanek: Experiences with the UML in the Design of Automotive ECUs, University Ilmenau
6. AUTOMOTIVE-UML: DaimlerChrysler AG http://www.automotive-uml.com/.
7. http://www.telelogic.com/products/doorsers/doors/
8. www.etas.de/

CONTACT

The authors' e-mail addresses:

carsten.kroemke@volkswagen.de
goltz@ips.cs.tu-bs.de
huhn@ips.cs.tu-bs.de
mutz@ips.cs.tu-bs.de

Web Address of the STEP-X project:

www.step-x.de

2003-01-0355

Implementation-Conscious Rapid Control Prototyping Platform for Advanced Model-Based Engine Control

Minsuk Shin, Wootaik Lee and Myoungho Sunwoo
Hanyang University

ABSTRACT

In the extremely cost-sensitive and competitive automotive industry, manufacturers and suppliers are constantly searching for means to reduce both time-to-market and development costs. As a kind of solution to satisfy these requirements, Rapid-control prototyping (RCP) has established itself as viable technology, albeit not everywhere yet. One big gap in the development process, however, is the transfer form RCP to a target implementation with all its limitations. This paper presents a new RCP platform, which aims to provide a consistent environment both at the RCP step and at the target code implementation step. To achieve this goal, the proposed prototyping system is designed very similar to the real production ECU as much as possible, and it supports all the features, which are needed in the RCP. This prototyping system strictly adheres to the layered architecture of the final production ECU, and separates the automatically generated part of software, or the application area, from the hand coded area, which generally carefully designed and tested.

INTRODUCTION

In automotive industry, rapid control prototyping (RCP) has established to satisfy the requirement to reduce both time-to-market and development costs. Hanselmann [1], Dorey [2], and many others have described a possible role for RCP especially in powertrain control system development and addressed the issues of RCP method.

In RCP platform, automatic code generation is absolutely essential. Automatic code generation permits the abstracted model, which is represented by computer-aided control system design (CACSD) tools, to be transformed into a software language implementation and consequently eliminates tedious and error-prone hand coding procedure. It has potential for tremendous productivity improvements for automotive control application software implementation.

Up to recent days, a simultaneous real-time capability of the closed loop engine control has only been possible when using expensive special hard- and software system of the industrial companies, such as dSPACE®, ETAS®. Furthermore these tools separate RCP phase and target implementation phase strictly. Narrowing the technical gap between these phases is thought as one of the most challenging works [3,4,5]. This paper presents a new RCP platform to bridge the gap between RCP step and target implementation step. This RCP environment is designed very similar to final production ECU with the help of layered architecture of software and powerful microcontrollers capable of floating-point calculation. The MATLAB® tool-chain has been selected as a base CACSD environment. A newly developed customized target of REAL-TIME WORKSHOP® converts a graphically represented control algorithm into optimized code of target processor. This platform also enables the function developers to integrate some legacy code easily. To measure the running data of target processor and to calibrate some control parameters in real-time, CAN calibration protocol (CCP) is adopted, and this feature make this RCP more similar to a production ECU.

TARGET-IDENTICAL RCP PLATFORM

RCP is originally designed to test and validate control algorithm easily in real-time experiment environment, and target code implementation issues are allotted to the software engineers later. This approach deepens the gap between these two phases and makes it more difficult to cooperate concurrently. The different computing power between the prototyping and production processors, and the different abstraction level of the development stages cause difficulties in seamless transitions from RCP to target.

If the RCP hardware is designed on the basis of the final ECU with the full support of RCP tools, the RCP and target implementation stages are more easily integrated. Such a target-identical RCP solves a lot of problems to bridge the technical gap, which is mainly caused by the different requirements of each phase, such as level of abstraction, computing power of processor, fixed or floating-point arithmetic capability and others.

A new implementation-conscious RCP platform for the engine control system is developed, based on MATHWORKS® tools and MPC555®. This platform is designed for target implementation as well as RCP. To achieve this goal, layered architecture is strictly maintained and RCP platform is designed similar to the production controller as much as possible.

SYSTEM ARCHITECTURE

Figure 1 shows hardware block diagram of this platform. This RCP platform hardware mainly consists of a PC host and a microcontroller target. At the PC host, control algorithm is developed, tested, and converted to target code under SIMULINK®, and at the target system, this code is executed in real-time. MPC555® microcontroller of MOTOROLA® is selected as a target processor. Each subsystem is specially designed to minimize the CPU intervention and should be specially programmed as device drivers.

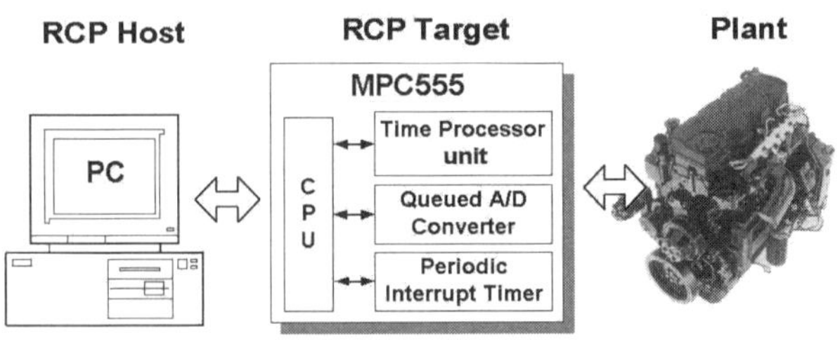

Figure 1 Hardware block diagram RCP platform

Figure 2 shows software architecture of RCP platform. This platform adheres to the layered architecture similar to the target implementation.

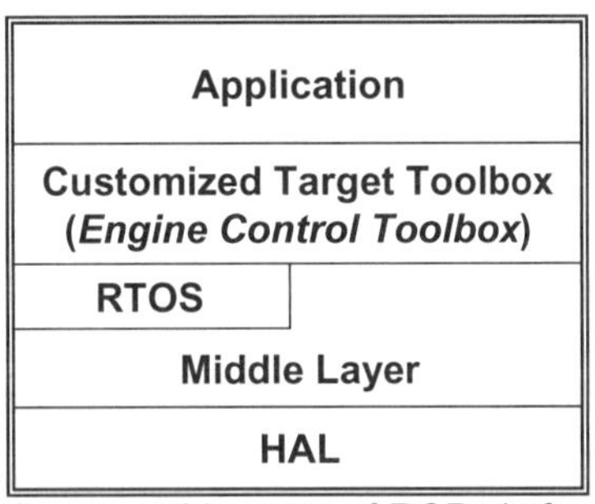

Figure 2 Software architecture of RCP platform

To utilize all the features of subsystem fully, hardware abstraction layer(HAL) is specially designed just like implementation of target code. To help the programmer to manipulate the lowest level I/O easily, a middle layer is inserted on top of HAL. The highest layer, which is normally expressed as an application, is composed of the control algorithms, some diagnosis functions and so on. In this layer, a set of abstract principles is helpful to define the complicated algorithms. For the graphical definition of algorithms, SIMULINK®/STATEFLOW® are used. REAL-TIME WORKSHOP® works to customize the code from SIMULINK® model and to generate optimal code for SIMULINK® blocks. And Engine Control Toolbox is designed and it inserted between the application layer and the middle layer to provide some abstracted SIMULINK® blocks and to incorporate with existing legacy codes. The CCP is selected as measurement and calibration tool in this study.

HARDWARE ABSTRACTION LAYER (HAL)

A HAL is the lowest component of software that functions something like an API. In strict technical architecture, HAL resides at the device level, a layer below the standard API level. The HAL is a set of low-level software that harnesses the power of the hardware into an easy-to-use API library. It allows programmers to write applications with all the device-independent advantages of writing to an API, but without the large processing overhead that APIs normally demand. In other words, it isolates the other parts of software from the hardware variations, and exports an API to upper layers to handle hardware dependent issues such as i) processor initialization, ii) device driver support, iii) timing and interrupt functions, iv) firmware interface functions, v) low level error handling, and others.

The device driver in the HAL is designed to simplify use of the sub-modules of MPC555® such as TPU, QADC, PIT and TouCAN.

MIDDLE LAYER

The middle layer provides more tailored programming environment for the programmer. In general, the functions and their parameters in HAL have not only less abstracted meaning for the programmer, but they are also mutually dependent on others. For example, the fuel injection driver needs the injection duration parameters, which is represented in the multiples of the microcontroller's system clock, but an application program wants to interface the fuel amount in time domain representation. The middle layer converts the fuel injection time to the appropriate value according to the system clock. Therefore, this layer encapsulates the HAL to alleviate these difficulties and to enhance the convenience of the programmer. In this layer, device drivers and interrupt service routines, which are independently designed and tested, are integrated and modified based on the need of application program. In other words, main objectives of HAL are generally to maximize the efficiency of CPU and to provide primitive functions and variables, but the middle layer aims to provide an unified way of accessing or handling lower level I/O and more comfortable programming environment including RTOS. Traditionally this layer is used as a base layer for hand coding, so the pre-described objectives are the most important factors. But in this study, all the features are carefully organized and designed for the TARGET LANGUAGE COMPILER® realizations. Also primitive scheduler is implemented in this layer. By use of this scheduler, any functions of application program can be easily activated in accordance with the pre-defined condition, such as event or time period.

CAN CALIBRATION PROTOCOL

ASAP working group tried to reach mutual agreement and standardization in automation, modularization and compatibility of all equipment to do measurement, calibration and diagnosis, and defined the CCP as a part of its standard.

The CCP calibration protocol provides the following features for the development and testing phase as well as for end-of line programming: i) read and write access to a very wide variety of memory locations, ii) simultaneous calibration and measurement, iii) flash programming, and iv) simultaneous handling of different control units. CCP is currently widely used in automotive industry and is becoming a standard tool more and more. Therefore, the CCP is selected as measurement and calibration tool in this study.

CCP provides the ability to access data at such a fast rate that it is possible to run an application at the same time.

ENGINE CONTROL TOOLBOX

To generate seamlessly the execution code from the graphically represented control algorithm, it is necessary to design some new graphical blocks, which correspond to the functions in the lower layer, and to link the generated code with the legacy code of the lower layer. For this purpose, a new customized target toolbox of REAL-TIME WORKSHOP®, called Engine Control Toolbox, is designed. This toolbox works as a channel from the graphical block diagram to the lower level hardware dependent functions. With the help of this toolbox, the developers can easily modify the off-line simulation model of the control algorithm to the appropriate form in the same way of off-line simulation, and generate the execution code and download it to the target processor just a few mouse clicks.

Each block of the toolbox is designed with the S-function of SIMULINK®, and has its own mask window, which enables the developers to easily configure the parameters of the block. Figure 3 shows all the blocks of the customized target toolbox. All the blocks are categorized as initialization, input functions, output functions and scheduling functions. *System Initialize* block contains the parameters, which are needed at the board initialization step, such as time trigger period, angle trigger period, and so on. I/O signals, which are generally interacted with the engine sensors and actuators at the off-line simulation phase, are replaced with the appropriate sink and source blocks. *RPM*, *Lambda*, *TPS*, *MAP*, and *MAF* block correspond to the input signals which are generally used in engine control application, and *Event-Trig ADC* and *Time-Trig ADC* block can be used to interface other analog signals. Update blocks, such as *Fuel Duration*, *Spark Timing*, and so on, are used to interface the output signals of the controller model to the lower layer functions. The control algorithm can be executed in multi-rate according to the pre-defined conditions by the use of *Time trigger* or *Event trigger* blocks and *Task* block. *Time trigger* or *Event trigger* blocks can be configured in the multiple of the minimum time interval or angle interval, which are configured in the *System initialization* block.

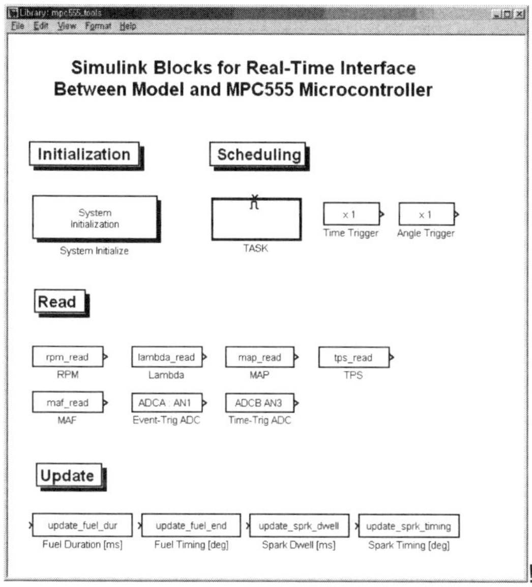

Figure 3 Engine Control Toolbox of RCP platform

AUTOMATIC CODE GENERATION PROCEDURE

The automatic code generation is the most important characteristic of the RCP. Figure 4 shows the automatic code generation procedure and the related utilities and files when using the REAL-TIME WORKSHOP® and the developed toolbox. This toolbox consists of the *template makefile*, *system target file*, *block target file*, *main program code*, and *lower layer code*. In the first step of the code generation procedure, the control algorithm, which is composed of SIMULINK® block libraries, is converted to the *intermediate model description file* by REAL-TIME WORKSHOP®. This intermediate file contains model information, which can be easily handled by TARGET LANGUAGE COMPILER®. And the next step, TARGET LANGUAGE COMPILER® generates the C-code files by analyzing the intermediate file and integrating the *function library*, *system target file*, and *block target file*. The *function library* contains the code information of the SIMULINK® block components, and the *system target file* has the system configuration information, such as code size, execution speed optimization, and so on. As the each SIMULINK® block has its own code generation part in the function library, so the block of the customized target toolbox has its own code information in the *block target file*. In the meanwhile, a customized makefile is created by the REAL-TIME WORKSHOP®, based on the *template makefile*. With the help of this customized makefile and the make utility, the generated model codes, *main program code*, and the *lower layer code*, such as HAL and middle layer, can be compiled and linked together, and the executable program file is created at last.

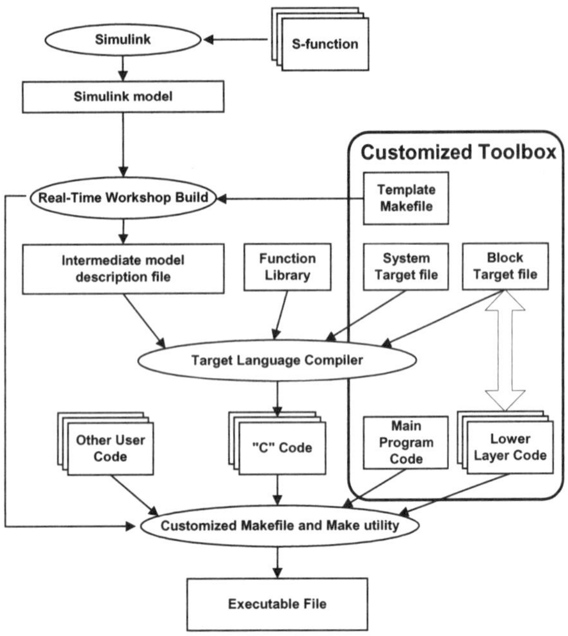

Figure 4 Automatic code generation procedure

In this code generation procedure, there exist two methods to import the legacy code. If the code is corresponding to the control algorithm, it can be represented in the form of S-function. S-function is easily integrated in the automatic code generation procedure as shown in Figure 4. On the other hand, if the code is related to the lower layer functions, it can be integrated using the customized makefile, as in the case of the importing the *lower layer code* of the toolbox.

ENGINE CONTROL EXPERIMENTS

To prove the feasibility of the proposed RCP platform, some air-to-fuel ratio (AFR) control applications are performed using the developed platform as shown in Figure 5 and the simulation results are presented in Figure 6 and Figure 7. The AFR control algorithm employs PI control law and sliding mode control law with a Smith predictor based on the measurement of a wide-band oxygen sensor [6]. These control algorithms are implemented by using the introduced RCP platform, and the control experiments are conducted in transient engine operation.

For the performance evaluation of the proposed controller, the throttle angle is changed as shown in Figure 6 (a) and Figure 7 (a), to simulate a fast tip-in and tip-out situation that allows the engine to be operated between 2000 and 4000 rpm. Also the engine is assumed to operate under constant load condition. UEGO sensor is assumed to have measurement delay of two engine cycle because of the event-based nature of the engine, and it is assumed to have band limited white noise.

Figure 6 (c) and Figure 7 (c) represent the off-line simulation results of the models. The controller implemented using RCP platform is also tested under a virtual experimental environment and the result is shown in Figure 6 (d) and Figure 7 (d). Compared with the off-line simulation, the performance of the experiment using the RCP platform degrades slightly. Non-zero execution time of the plant model and the control algorithm, measurement noise, quantization and others in the experiment also cause this degradation. And the different execution period of the control task also degrades the control performance. These problems, which may occur in the target implementation stage and the validation stage, can be efficiently handled with the help of this virtual development environment.

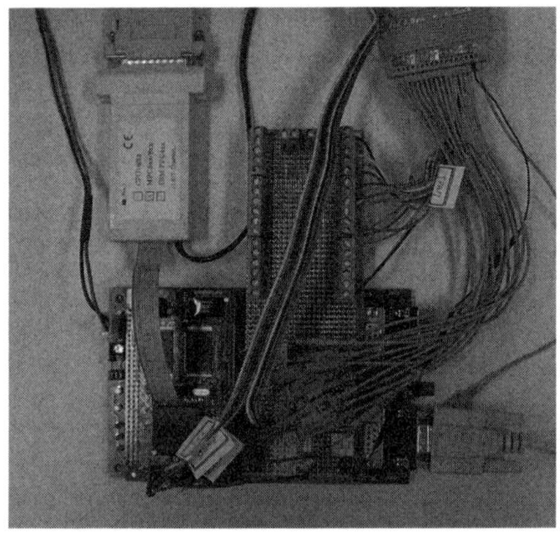

Figure 5 MPC555 board and debugging tools in RCP platform

The simulation results show that the proposed RCP platform with the virtual experiment environment can handle efficiently various problems caused by transitions among the development steps.

CONCLUSION

A new target-identical RCP platform is developed for the automotive engine control application. This RCP aims to provide a consistent environment both at the RCP step and at the target code implementation step. To achieve this goal, the proposed prototyping system is designed very similar to the real production ECU as much as possible, and it supports all the features, which are needed in the RCP.

This prototyping system strictly adheres to the layered architecture of the final production ECU, and separates the automatically generated part of software, or the application area, from the hand coded area, which generally carefully designed and tested because of the hardware dependency and the efficiency of microcontroller. The MATLAB® tool-chain has been selected as a base CACSD environment in this study. A newly developed Engine Control Toolbox of REAL-TIME WORKSHOP® converts a graphically represented control algorithm into optimized application codes and

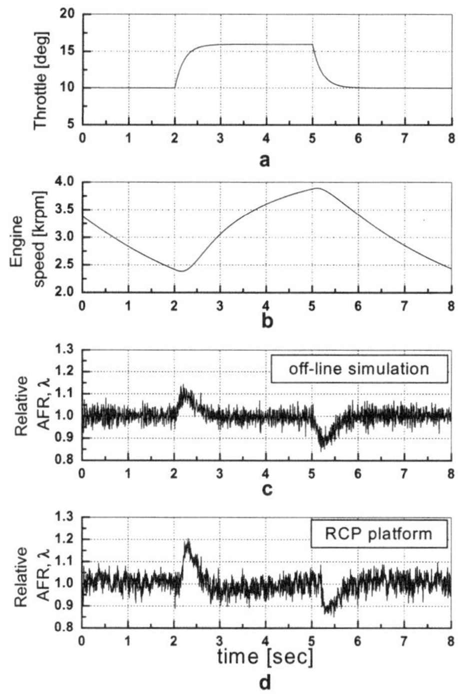

Figure 6 Comparison of off-line simulation and experiment using RCP platform (PI control system, 60-2 type crank signal, time-based execution)

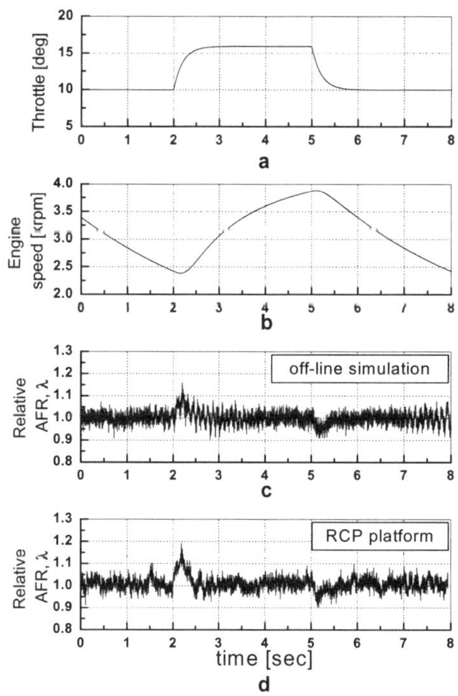

Figure 7 Comparison of off-line simulation and experiment using RCP platform (sliding mode control system, 36-1 type crank signal, event-based execution)

links them with other parts of the software to generate executable code for the target processor. To measure the running data of target processor and to calibrate some control parameters in real-time, CCP is adopted, and this feature make this rapid prototyping controller more identical to production ECU.

The developed prototyping system proposed here is advantageous because the controller design, the automatic generation and the implementation of the control program on the target processor can be carried out under an integrated environment.

REFERENCES

1. Hanselmann, H., 1996, "Automotive control: from concept to experiment to product", *Proc. of the 1996 IEEE Int'l Sym. on CACSD*, Dearborn, MI, USA, pp. 129 –134
2. Dorey, R.E.; Maclay, D., "Rapid prototyping for the development of powertrain control systems", *Proc. of the 1996 IEEE Int'l Sym. on CACSD*, Dearborn, MI, USA, pp. 135 –140
3. Stylo, A.; Diana, G., 1999, "An advanced real-time research and teaching tool for the design and analysis of control", *Africon, IEEE* , vol. 1 , pp. 511 - 516
4. Rebeschiess, S., 1999, "MIRCOS - microcontroller-based real time control system toolbox for use with Matlab/Simulink", *Proc. of the 1999 IEEE Int'l Sym. on CACSD*, Hawaii, USA, pp. 267 –272
5. Slomka, F., Dorfel, M., Munzenberger, R., Hofmann, R., 2000, "Hardware/software codesign and rapid prototyping of embedded systems", *IEEE Design & Test of Computers*, Vol. 17, pp. 28 –38
6. Yoon, P.J., 2000, "Nonlinear Dynamic Modeling and Control of Spark Ignition Engines", Ph.D Thesis, Department of Mechanical Engineering, Hanyang University
7. Dorey, R.E.; Scarisbrick, A.D., 1997, "Rapid prototyping methodology applied to the powertrain control system", *IEE Colloquium on System Control Integration and Rapid Prototyping in the Automotive Industry*, pp. 4/1 -4/4
8. Furry, S., Kainz J., 1998, "Rapid Algorithm Development Tools Applied to Engine Management Systems", *SAE paper* 980799
9. Yacob, Y., Chevalier A., 2001, "Rapid Prototyping with the Controller Area Network (CAN)", *SAE Paper* 2001-01-1224
10. Koster, L, Thomsen, T., Stracke, R., 2001, "Connectiong Simulink to OSEK: Automatic Code Generation for Real-Time Operating System with TargetLink", *SAE Paper* 2001-01-0024

SOFTWARE FOR TESTING

2005-01-1044

Integrated Test Platforms: Taking Advantage of Advances in Computer Hardware and Software

Mark D. Robison
Uson L.P.

ABSTRACT

Ongoing hardware, software, and networking advances in low-cost, general-purpose computing platforms have opened the door for powerful, highly usable, integrated test platforms for demanding industrial applications. With a focus on the automotive industry, this paper reviews the pros and cons of integrated test platforms versus single-purpose and stand-alone testers. Potential improvements in in-process testing are discussed along with techniques for effectively using such testing to improve daily production quality, to maintain high production rates, to avoid unplanned downtime, and to facilitate process and product improvements and refinements through the use of monitoring, data collection, and analysis tools.

INTRODUCTION

Manufacturing test platforms come in many shapes and sizes, as does the concept of integration. Integration can be in the form of combining different test technologies onto a single platform in order to improve production rates, or it can be in the form of sharing or multi-tasking advanced test controllers in order to minimize capital investment. Alternatively, effective integration can exist only at the conceptual level by integrating only the test data, resulting in a greater understanding of the total picture. This paper touches on each of these in turn.

IN PROCESS TESTING

In-process testing is key to product quality and consistency. The standard focus of in-process testing is accuracy and repeatability, which is key to maximizing product quality while minimizing costs. Often overlooked though, is the opportunity for an organization to extract the maximum value from its test data for the benefit of both the manufacturing process and the product design. Cost effective improvements that result in more accurate test data, more efficient collection of test data, or better use of test data can be leveraged to create a competitive advantage and to improve profits. Several enabling technologies now exist to do just that.

ENABLING TECHNOLOGIES FOR IMPROVED TESTING

Technological advances in personal and office computing continue to create new opportunities for extending the capability and value of in-process testing. These advances include:

- Low cost, high speed computing platforms
- Low cost, high speed data networks
- Low cost, ultra high capacity storage devices
- Advanced, easy to use operating systems
- Software advances that facilitate the rapid creation of solid products that simplify even the most complex of tests.

The resulting improvement opportunities exist at four levels:

1. Individual tests
2. Individual test stations
3. The complete production line
4. Multiple production lines; remote or local

TEST LEVEL IMPROVEMENT OPPORTUNITIES

Individual tests can be improved by improving the test method, the test implementation, or by some combination of these. Improved test methods sometimes result from revolutionary scientific advances, but, more often than not, improvements are evolutionary in nature. Advanced leak testing[1] is one example of the later. By using high-resolution sensors, increased data collection, and incrementally smarter algorithms, test time can be reduced while simultaneously increasing test accuracy. For some applications, reduced test time can mean the elimination of an entire test station, for example, two stations to perform a given test instead of three stations. Table 1 summarizes test accuracy and test duration improvements achieved for several applications when advanced leak testing technology has been applied.

Table 1: Classical Leak Detection versus Advanced Leak Detection							
Part	Component	Leak Test Method	Trials	% Time Reduction	% Tests Within		
					± 2.5%	± 5%	± 10%
Engine Model 1	Cylinder Head	Classical	400	Baseline	N.A.	N.A.	89
		ALD	400	3%	N.A.	N.A.	99
Engine Model 2	Oil Cavity	Classical	100	Baseline	50	52	99
		ALD	100	40%	90	99	100
Non-engine	N.A.	Classical	60	Baseline	N.A.	N.A.	N.A.
		ALD	60	27%	60	90	100

Implementation improvements facilitated by the aforementioned enabling technologies include:

- Clear, intuitive user interfaces to minimize training costs and reduce operator error
- Test-to-test consistency using presentation standards, again to minimize training costs and reduce operator error
- Built-in diagnostics and troubleshooting tools for rapid detection and resolution of problems
- Advanced logging and data export options that enable process and product improvements through both real-time and after-the-fact data analysis.

STATION LEVEL IMPROVEMENT OPPORTUNITIES

Very significant, on-going savings can be achieved by doing more at a particular test station. Such station-level integration, where multiple tests are performed either concurrently or nearly concurrently by overlapping portions of different tests, have the potential for higher production rates through reduced overall test time and reduced inter-station transfer time. Direct savings also result from fewer test stations and reduced real estate requirements.

Disadvantages of station-level integration include increased test station complexity, which is primarily a development concern, and increased dependency upon a single station, which can be addressed by an adequate spares policy.

Example: Station-level integration

One example of station-level integration is to combine leak testing of the water jacket and oil cavity with green-engine compression testing.

During the machining of the engine components, cylinder block, and cylinder heads, they are leak tested to assure there are no leaks in the water cavities or in the oil cavities.

As the engine is assembled, the engine must be tested to verify the assembly of the seals, gaskets, and plugs.

The main areas to be tested are the water cavity, the oil cavity, and the compression of the power stroke of the engine. These tests can all be performed in one test station using state-of-the-art multi-channel testers, such as the Uson Vector.

The test process first rotates the engine to establish the initial conditions for the compression test. Transducers at each spark plug opening measure the pressure rise due to the compression of the cylinder while torque and position sensors monitor the crankshaft. Pre-programmed limits for each cylinder as well as a detailed master "signature" of all sensor signals are compared against results for the part under test as each cylinder experiences its compression cycle, yielding quick test results. A failed compression test can abort the test sequence or continue with the leak tests in order to gather more information about the nature of the defect.

The compression test is exited with the crankshaft properly positioned for the subsequent leak tests of the water cavity and the oil cavity. Portions of these tests can be overlapped to minimize test time.

Because the oil cavity is the larger of the two cavities, the tester fills the oil cavity first. During this step the tester monitors the water cavity for a pressure increase in order to detect cross-wall leakage between the two cavities. Following this step, normal leak tests are performed concurrently on each cavity. Figure 1 illustrates the timeline for the test steps.

LINE LEVEL IMPROVEMENT OPPORTUNITIES

Test controllers, running gigahertz-class processors, are capable of supporting multiple test stations concurrently. This approach diverges significantly from the current business model used by large manufacturers, but deserves consideration, as the potential exists for reducing initial deployment costs by tens of thousands of dollars. For example, a ten-channel test platform could support ten distinct tests. These tests could be distributed between one to ten different test stations.

Figure 1. Integrated Test Station: Engine Compression Test, Water Cavity and Oil Cavity Leak Tests

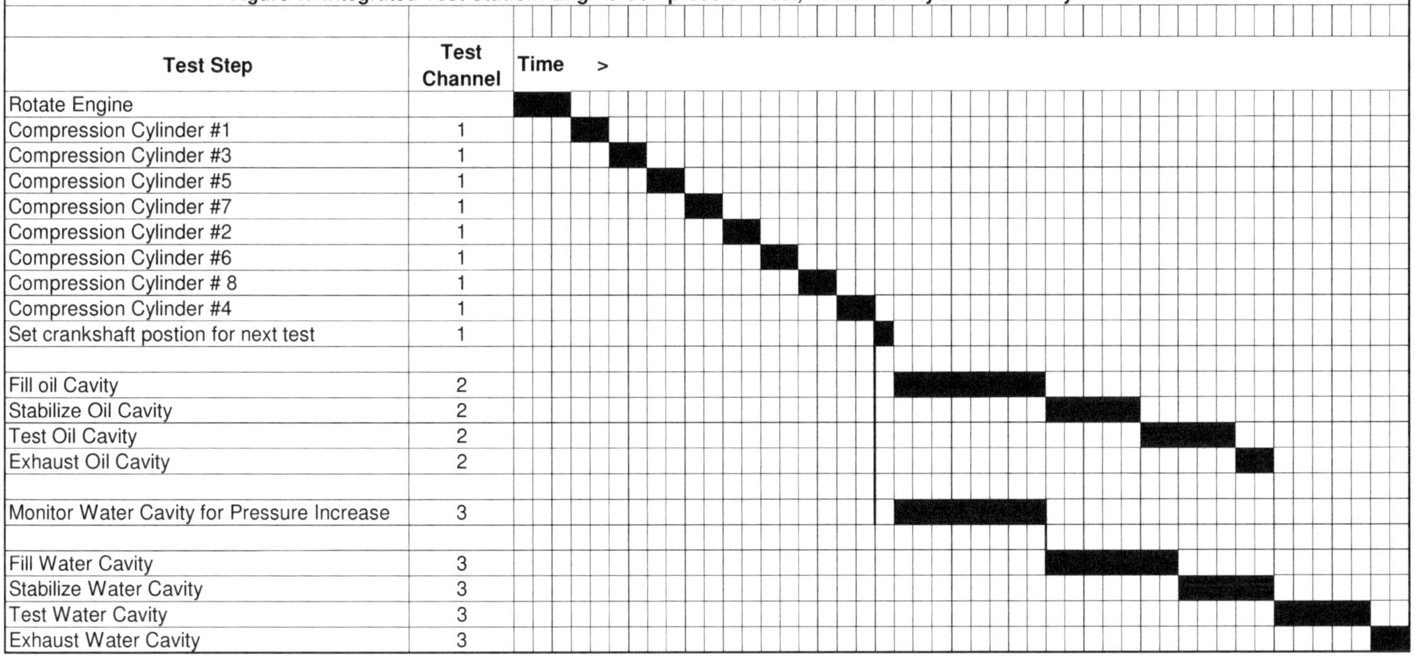

The economic benefit of replacing ten $20,000 testers with a single $150,000 tester is obvious.

The introduction of a single point of failure for the affected test stations merits concern, but an adequate spares policy can mitigate this risk. Because of the fewer number of components, the actual mean time between failures for the system as a whole is reduced.

A second form of integration at the production line level is data integration: the creation of an integrated, global view of the test results and associated test data. From such a perspective, subtle trends can be detected that would otherwise go unnoticed, enabling emerging problems to be preemptively identified and corrected.

Data integration has only modest costs associated with it, yet it can result in tremendous savings both short term and long term by minimizing down time and facilitating product changes for improved manufacturability.

RESULTS MONITORING AND ANALYSIS

Regardless of the level of integration, monitoring and analyzing test results can yield significant gains. Benefits arise from monitoring test results real time as well as from after-the-fact analysis of data collected over weeks, months, or even years.

REAL-TIME MONITORING

A well-implemented and utilized information system can help improve production quality by enabling the timely detection of process level problems, and having a clear indication of the source of process problems can minimize the time needed to correct a problem. By proactively monitoring quality-impacting trends as they happen, higher production rates can be achieved by correcting emerging problems before they result in product rejects.

Example: Early Detection Of Trends

Assume that for a particular test, a part is completely acceptable if its test results are below 100. However, by design, and confirmed by historically collected data, good parts, on average, pass the test with a value of 80. Furthermore, the tests are repeatable to within ±10% of the average value. By continuously monitoring this average over a period of hours, days or weeks, a monitoring system can detect emerging problems, such as excessive machine wear. When the running average reaches a configured threshold, say 90 for this example, the monitoring system automatically notifies a supervisor or maintenance personnel via email or other mechanism, supplying sufficient details to enable investigation of the situation. With such early detection, corrective action can be taken before *any* parts are rejected. All of the parts may still be passing the test but by a much smaller margin. Because of the normal variability in the results, casual observation of the data by an operator would likely not detect the worsening condition.

HISTORICAL DATA ANALYSIS

An important trait of a test results monitoring and analysis system is the ability to archive detailed test data and interim test results. By collecting and analyzing interim test results, as well as final test results, valuable insight can be gained into the types of problems that occur and the relative frequency at which they occur. Armed with this information, product improvements aimed specifically at reducing or eliminating those problems can be implemented resulting in increased manufacturability and fewer product rejects.

The accumulation of historical data also creates a detailed audit trail that can be beneficial for in-depth studies of process or product defects. The data archive can also be used for management reports and for tracking various improvement initiatives. Zoom-like capability in the analysis tool can facilitate rapid identification of problems, such as a misaligned pallet, by highlighting deviations at a high level, and then allowing an increasingly granular view of the test results. For example, if a weekly report summarizing the percent rejects by day shows an anomaly on Friday, then a "drill-down" feature would allow Friday's test results to be readily viewed by operator, by batch, or by pallet, allowing the underlying cause to surface.

Perhaps one of the greatest yet underappreciated values to accumulating a large body of test results is that the statistical distribution of those results can be more fully understood. Understanding the true distribution of test results is key to optimally setting pass/fail criteria, which in turn affect the number of good parts falsely rejected and the number of defective parts erroneously accepted.

Example: Needless Part Reworking

To minimize falsely accepted parts, reject limits are typically set lower than the true design requirements dictate. If this limit is not set optimally, an excessive number of false rejects can occur causing needless rework and retesting. By reviewing historical data to fully understand the distribution of test results, an optimal reject limit can be determined to minimize wasted effort.

CONCLUSION

The right combination of test equipment and tools can provide insight into process problems and potential product improvements that would otherwise be unavailable. Having insight into such subtle trends gives you the ability to preemptively solve problems for greater profitability. While low cost and stand-alone testers have their place, combining powerful and flexible test equipment with integrated information management will be the distinguishing hallmark of companies that survive today's competitive climate.

ACKNOWLEDGMENTS

Special thanks go to Carl Aquilino, President of Uson, and Dan McCauley, Director of Automotive Programs at Uson, and Alan Campbell, Vector Product Manger at Uson, all of whom have contributed to the concepts and content presented here.

REFERENCES

1. System and Method for Leak Rate Testing During Adiabatic Cooling, U.S. Patent 6,741,955 B2 dated May 25, 2004.

CONTACT

Mark D. Robison, V.P. of Engineering
Uson L. P.
8640 North Eldridge Parkway
Houston, TX 77041
USA
Phone: 281-671-2000
Email: mrobison@uson.com

2005-01-1041

Next Generation Instrumentation and Testing Software built from the .NET Framework

Steven E Kuznicki
Accurate Technologies Inc.

ABSTRACT

This paper discusses various aspects of Microsoft's newest programming framework and how the new technology that this framework provides is being used in next generation in-vehicle and lab automotive instrumentation. Included in this paper are challenges that in-vehicle and lab instrumentation software faces today and how the .NET technologies solve these challenges. Some of the technologies discussed in this paper include the C# language, .NET reflection, and application architecture.

The paper specifically focuses on a new CAN analysis tool that addresses, through the .NET technology, many of the challenges faced by bus analysis tools today. Some of these topics include:

- Conformance to instrumentation and test system standards (e.g. ASAM).
- Legacy system integration.
- Customer specific needs that may be proprietary.
- The many communication protocols that exist in the automotive area. Some of the protocols include CCP, XCP, KWP2000, J1939 and others.

This satisfies a need for localized custom feature development in a global marketplace.

INTRODUCTION

There are numerous tool solutions available for CAN network development. Many of these tools support raw CAN network traffic analysis as well as options for standard higher-level-protocol (HLP) support. Different functional features are available as well, from simple monitoring of network traffic to injecting (replaying) messages back onto the bus.

In-house tools provide the most specific functionality but take development resources and may not be general enough to work across multiple groups. Off the shelf tools provide adequate features for most testing scenarios, but may be too complex for non-programmers to adjust to. There is a need for a mix of these that can satisfy all levels of users. It needs to be general enough for day to day testing, and also able to handle demanding testing and verification scenarios.

1.0 SOFTWARE TECHNOLOGY

As computing power increases, there are more ways to exploit software development. Programming languages have evolved to take advantage of the ever changing hardware resources made available. This in turn provides enriched development environments with better tools and compilers. This results in faster software development cycles which translates into getting the right tools to the engineers that need them quicker and more reliably.

1.1 NEW FRAMEWORK MODEL

The latest software technology from Microsoft is the .NET Framework (pronounced dot-net). This is an important new component of the Microsoft family of operating systems which form the foundation of next generation, Windows-based applications. The .NET Framework helps software tool developers more easily build, maintain, and deploy new software products.

One of the first of many languages supported, C# (C-Sharp) became also one of the most popular ones. The first widely distributed implementation of C# was released by Microsoft in July 2000, as part of its .NET Framework initiative. Some other popular programming languages that support the .NET Framework include: Java, Visual Basic, and C++.

There are two (2) main parts of the .NET Framework: the common language runtime (CLR) and the .NET Framework class library.

- **Common Language Runtime.** The CLR provides common services for programs written in a variety of languages (e.g. C++, C#, Visual Basic, Java). It takes on the responsibility for memory management, error handling, and security management.

- **.NET Framework Class Library.** This prepackaged library provides developers a rich set of components that help rapidly extend capabilities of their software.

Taking advantage of the .NET Framework can benefit developers that provide similar tools to more than one client.

1.2 .NET REFLECTION

In the .NET Framework, all functionality is exposed via assemblies. Assemblies are the building blocks of every application built for the .NET Framework. Assemblies can take the form of Dynamic Link Library files (DLL). They contain modules which in turn contain various (user defined) types (structures, enumerations and classes, etc.)

Reflection is the ability to find information about these data types contained in an assembly at time of execution. This powerful mechanism not only provides a way for introspecting assemblies and objects, but also to create and invoke methods and properties on those types at runtime.

The powerful .NET Reflection mechanism provides many creative possibilities to software tool developers. One possibility is providing a Base Application Shell which can query for available common functionality that is contained in local assemblies. Next is a simple example of how such a tool could be developed.

2.0 APPLICATION EXAMPLE

Presented here is an example that uses Reflection to determine what components are available at application runtime. It uses this information to create and invoke methods on found types.

Suppose that there exists, an application that acquires data from an external source. The software tool application is to display signal information being acquired, but with the requirement that it can be displayed in different ways. All the application needs to know is a common type that has defined methods and properties. A *base* class is defined for this purpose:

```
public class BaseGauge {
public virtual void
UpdateValue (double value) {
    return; // base does nothing
  }
}
```

The method UpdateValue is defined in the base class takes a double type value and is responsible for 'updating' the visual control representation of the incoming signal. Now, a derived class will need to be developed which will implement the specific visual representation for displaying the signal value.

```
public class RoundGauge : BaseGauge {
public override void
UpdateValue (double value) {
    AdjustNeedleGraphic(value);
  }
}
```

Figure 1 : Round Gauge Control

The private function `AdjustNeedleGraphic` provides the specific graphic rendering necessary for this control's implementation.

A second derived class can be developed that gives the user an alternative visual representation of the signal value:

```
public class LinearGauge : BaseGauge {
public override void
UpdateValue (double value) {
    MoveBarGraphic(value);
  }
}
```

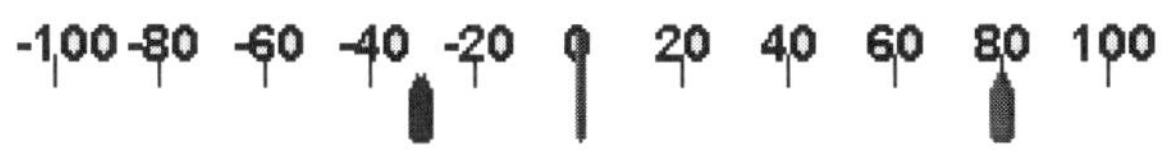

Figure 2 : Linear Gauge Control

The application shell needs to provide basic services in order to accept and support these controls. These include:

- Ability to detect and enumerate types that are derived from `BaseGauge` (using Reflection).
- Container control to house the gauge control(s).
- Mechanism to call into the constructed object's UpdateValue() method with a valid (double) value.

A simple application is shown here:

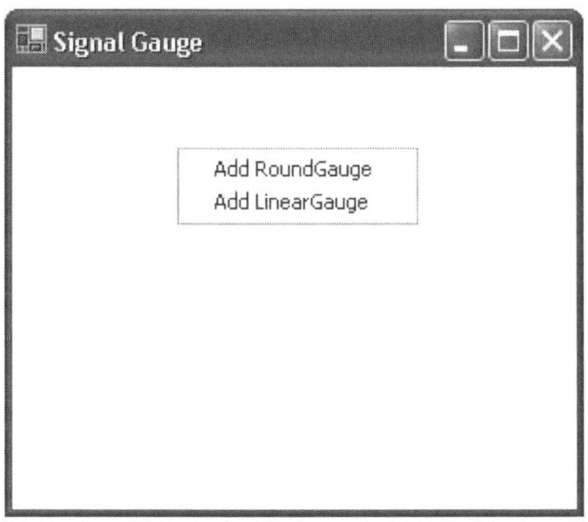

Figure 3 : Simple Gauge Application Shell

The context menu (rectangle) that pops up (user right-clicks in the window area) is created by introspecting available local assemblies. The type names of the derived `BaseGauge` classes are used as menu options.

Deriving from base classes is nothing new to object oriented development. The ability to investigate runtime objects has been widely done using Microsoft's RTTI technology (Runtime Type Information). Usually, the object you are investigating is already instantiated. The major drawback of RTTI is that it requires full knowledge of the needed types at compile time. Reflection can inspect all methods and properties at runtime.

In the same way the base class is used to define common functionality, specific '`Interface`' constructs can be used to do the same. The difference in using an interface and a base class is that interfaces can not provide a default implementation of methods or default values for properties. If a .NET class type implements an interface, then that class needs to supply complete implementation for every method and property defined by the interface.

The use of `Interfaces` can be used in the same way in regards to .NET Reflection. Assemblies can be queried for certain class types that support a certain published interface. In this way, the .NET Framework is providing similar functionality to the COM/ATL and ActiveX/OLE controls that are very popular Windows programming components.

3.0 EXTENDING CONCEPTS

The reflection technique that is illustrated in the previous section can be used in many other areas of tool software development. Discovering types at runtime is beneficial for

1) deployment control (excluding/including certain modules for select customers)
2) customer developed modules (providing proprietary custom developed modules)
3) intellectual property protection (since developed code stays in-house)
4) code reusability (providing a generic application shell)
5) discovering unique capabilities (creating runtime application menus and options)
6) global distributed development (developers do not need entire application solution to deliver derived types)

In the development of software for instrumentation, specific examples are now given.

3.1 COMMUNICATIONS SUPPORT

By defining a base class (or interface) that includes common functionality for communications, an application can be developed that determines which ones to make available to the user at runtime.

A base (common) device definition may look like this:

```
public class BaseCommDevice {
  public virtual void PrepareDevice();
  public virtual int InitCommunications();
  public virtual bool Connect();
}
```

In this sense, this architecture is very similar to the ASAM-GDI (Generic Device Interface) specification. An application can be constructed independent of the working principle of physical devices or to the way communication objects access the devices. There is no need for the application to use device specific commands or controller sequences for acceptance of measurement data. This responsibility is left up the specific implementation of the `BaseCommDevice` derived object and/or contained components.

Derived implementations can be developed that interface with specific hardware like CAN, LIN, and other popular communication networks. Likewise can be done for defining the protocol objects used to decode/encode message/data packets. For example:

```
public class BaseCommandSet {
  public virtual void
    DecodeBuffer(byte[] buffer);
  public virtual int MessageID();
}
```

A derived class (e.g. `J1939CommandSet`) would provide a specific way to decode the incoming buffer and provide protocol specific information to the application.

3.2 CUSTOM PLUG-IN / TOOL-KIT

Designing an architecture that can take advantage of the .NET Framework technology can lead to a powerful tool option. This creates an environment that can be enhanced for future requirements with the use of add-on or plug-in software components. This type of architecture demands a good base product design.

The user can then pick and choose which options are needed for their purposes only. This cuts down on the cost of unneeded modules. Far more important, the development cost of special functionality is reduced since the support for plug-ins is built into the application.

The next figure shows the relationship of the Application Shell to the Base Objects and the specific object implementations (or plug-ins).

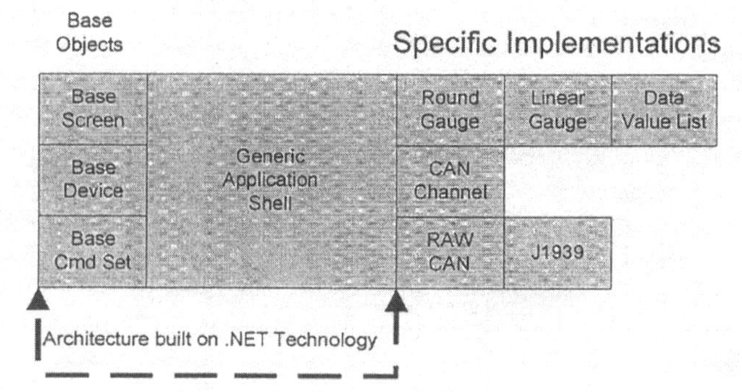

Figure 4 : Custom Architecture

4.0 CANLAB

These concepts have been put into practical use with Accurate Technologies' CANLab product. CANLab provides all the necessary functionality for basic investigation for CAN Network analysis. This includes network traffic monitoring and transmission of users own messages out on the network. Ways that CANLab uses the .NET Framework is outlined in this section.

4.1 BASE FUNCTIONALITY

CANLab provides base functionality by defining the behavior and roles of several base components. Here we give a few examples of these components and what service they provide to the main application.

Component relationship and function in Figure 5:

1) The Generic Application Shell – CANLab: the application knows the behavior and purpose behind Base defined components. It also has industry level knowledge of how things should work.
2) Base Devices Collection – application maintains collection of Base Device components that it has found via Reflection upon startup. These instantiated objects support common behaviors such as 'Connect' and 'Disconnect'.

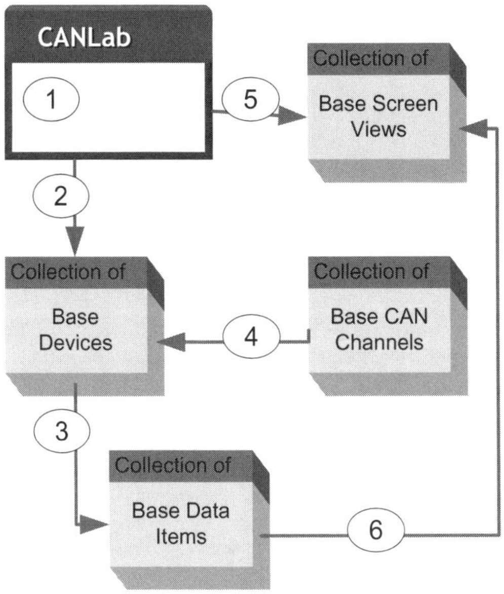

Figure 5 : CANLab and Base Component Relationships

3) Base Data Items Collection – each device stores a collection of Data Items which represent value based objects. For CAN Networks this means CAN Messages and Signals (Network Nodes are supported as well for organization purposes). The application can show these items in a tree list control.
4) CAN Channel Object – used by the device implementation if the device is CAN based. There can be several vendors that support interfacing to CAN Networks. A Base CAN Channel abstracts the specific details needed to make this hardware connection work.
5) Base Screen View Collection – application maintains a collection of possible view objects which are needed to visualize Data Item values. All screen objects are discovered at start-up time using Reflection.
6) Base Screen and Base Data Items – Screen objects display values represented by Data Items. Screens can selectively accept certain types of Data Items depending on the display functionality of that view e.g. Signal values in a 'Scope' style view versus Message values in a 'Data List' style view.

Creating and integrating another Screen View, Device, Data Item or CAN Channel can be accomplished without ever re-building the main application. This way, the base product software project stays untouched. Not only does this save time in testing subsequent release versions but lowers the risk that a new introduced component will adversely affect the behavior of current ones.

In the same manner above, CANLab supports (base) components for importing and exporting different file formats and printing capabilities.

In order to derive from a base component and implement specific behavior, the developer will need access to the assemblies (dll files) which are delivered with CANLab.

4.2 .NET PROPERTY GRID COMPONENT

One of the .NET Components that uses Reflection extensively is the Property Grid Dialog. CANLab uses this component to let the user manipulate several different components including devices and data items.

The property grid control is one of the main window views of CANLab. By using reflection, the property grid extracts the properties of the class and displays its values. It can determine the name and type of the properties and display them in a grid style view. In this way, the application does not need to develop a custom dialog for every component that it uses. This cuts down on development costs and provides a common and often used interface which the user becomes familiar. A sample of the property grid is shown below.

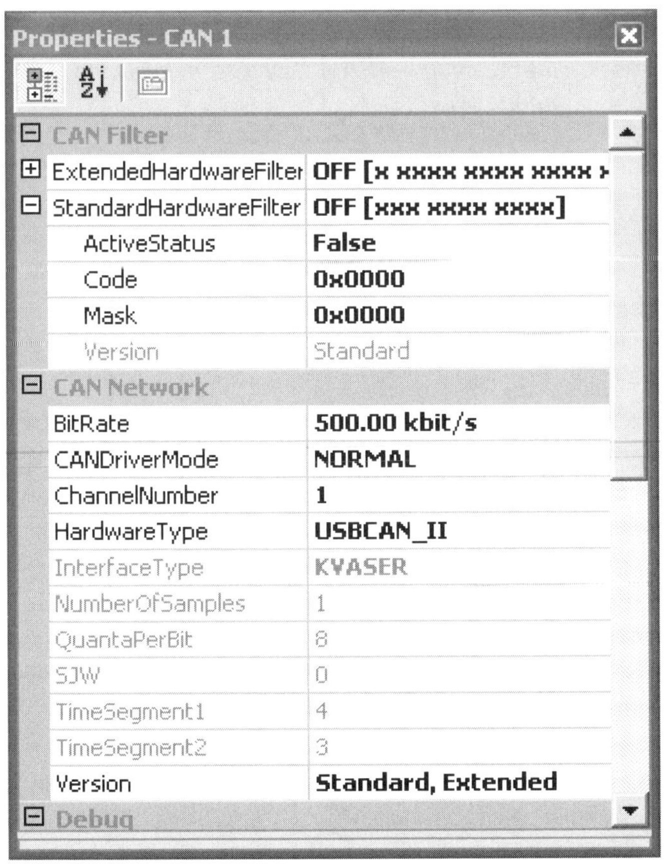

Figure 6 : Property Grid Example

5.0 CONCLUSIONS

This paper has shown how .NET reflection can be used to enable localization of feature development and even provide a framework where customers can completely customize instrumentation software including adding customer specific communications protocols, file formats, and any other aspects of instrumentation software. In addition, the new C# language is also discussed. Specific ways that the C# language can be used to improve efficiency and product quality are given.

REFERENCES

1. *Compiling for the .NET Common Language Runtime (CLR)*. John Gough. Prentice Hall 2002
2. *C# Language Specification Final Draft October 2002* – ECMA (Technical Committee 39 (TC39) Task Group 2 (TG2)
3. *Common Language Infrastructure (CLI) Partitions I to V* - Standard ECMA-335 2nd Edition 2002
4. *Inside Microsoft .NET IL Assembler* – Serge Lidin, Microsoft Press 2002
5. *Introduction – ASAM-GDI (Generic Device Interface)* Version 1.0; Release (GDI_OVW_ENG.DOC) from ASAM - Association for Standardization of Automation and Measuring Systems.

2004-01-1762

The Bus Crusher and The Armageddon Device

Part I

Ronald P. Brombach
Ford Motor Company

ABSTRACT

Most testing is done in a clean or static environment. The electrical power is connected to a constant voltage power supply. This is not very representative of the automotive environment. Electrically the automobile is very noisy, has dynamic behavior, and can be difficult to model and emulate.

Microprocessors can make mistakes at a rate of 1 million mistakes per second. If we can't test the software 100%, we need methods that will give us some level of confidence that the product is of high quality and will satisfy the customer. The test methods detailed in this paper can help gauge the robustness of the software.

These extraordinary tests are not part of standard test methods. Standard test methods use traceability of the functional requirements to develop the test. Each function is tested once per test cycle. Most of these tests are "Happy Path". [3] Occasionally, failure modes and the effects are included in the test plan.

Most of the tests defined in this document are either based on lessons learned or are trying to find ways to expose non-robust software designs. These tests may also exacerbate software and hardware signal race conditions.

The background of the author is in the realm of body module controller behavior; therefore examples detailed in this document are related to body module controllers. There is no reason the methods can't be applied to other ECUs (Electronic Control Unit) on the vehicle.

This paper discusses the evolution and development of a method and process to stress automotive embedded software. The intended process will look for abnormal or undesired operation in the ECU.

INTRODUCTION

Lessons learned, design reviews and testing inspired development of these test methods. During design reviews weaknesses are identified. Tests were run to make sure no issues were observable to the customer.

Microprocessors controlled systems behave as signal amplifiers. They have the same design sensitivities as other hardware-only electronic devices in the vehicle. Noise may cause the electronic device to become non-operational or break the hardware. The tests try to expose weaknesses in the basic software architecture.

This paper describes the four tests of the Armageddon device

- **Bus Crusher** – This device is basically a "chattering" relay that is directly connected to the bus communication wires. The "chattering" relay was a technique used by the EMC (Electro-magnetic Compliance) department to make sure an ECU is robust to the frequency spectrum. The original EMC test was nonconductive. The test was modified to actually short the bus to ground and introduce the noise directly on the bus wires.
- **RF Blaster** – This is a noise generator that sends a 315 Megahertz modulated signal. It is intended to interfere with the Remote Key Entry (RKE) and tire pressure signals (TPM).
- **Reset cycle test** – When a microcontroller runs through reset, it should take the same time every time to go to a specific point in the software. If this is true, a repetitive test can be developed to measure the reset time variation. A microprocessor digital port pin can be "toggled" as a way to measure this. Too much variation can be a sign of instability in the software.

- **Software Key-life test** – Base assumption: what does a computer do all the time, every time? Execute the same function over and over, in the same amount of time, each time.

  Most test plans test the function only once. i.e., "Open the door, interior lights come on, PASS". How about a test that opens and closes the door as many times as the driver might in the life of the vehicle?

TESTING MINDSET

One of the main goals of a test engineer is to develop and conduct tests. "I'm not happy until I can break the software", should be the goal of the test engineer. The person developing these tests needs to review the design, identify possible weaknesses, and develop tests to push the limits of the design. Interrupt service routines (ISR) are a great place to look for these weaknesses. Questions the test engineer may ask:

1. How can I break the design?
2. What are the lessons learned from previous vehicle programs?
3. What are the standard and customary tests that are currently conducted?
4. Are there any unusual conditions the vehicle may be subjected to?
5. What is the ECU supposed to do? Review the functional requirements.

A WAY TO APPROACH TESTING

In a previous paper [1], [3], it was stated that, "Interrupts are inherently evil". One of the reasons for stating this is that interrupts can execute at an inopportune time. Interrupts can be a weakness in the software architecture. Areas to focus on are:

- **Communication Network noise** target messages to the ECU at a much faster rate than specified. When the bus crusher is connected to the network wires, the bus faults cause the network to engage its re-try strategy.

  An ABS module went off-line when the bus was intermittently faulted. This did not affect the safe operation of the vehicle since the base brakes and power assist still operated correctly. The driver was notified of this condition with an ABS warning lamp.

  The software was designed to handle a fixed amount of processing requirements any more would indicate an exception. The software did not account for additional CPU chronometrics in its control loop. It was not found in pre-production testing because an intermittent network fault was not in the test plan.

  The supplier did a great job of trying to understand the CPU chronometrics, but the software did not account for network noise. An error on the network controller looked like a network message. The exception handling software in the device driver was not written correctly.

- **Weaknesses in the ISR structure:** identify weaknesses in the software interrupt structure and note how the microcontroller manages them. The RF (Radio Frequency) Blaster may expose the RF receiver to hardware/software interface issues. The ISR may run more often than it was designed to run. The next step is to review the hardware to see if the right amount of filtering is correct for the application.

- **Determinism Testing for repeatability must include:** Built-in test measurement capabilities. Design-in testability. Testing to the worst-case CPU chronometrics. There is a memory read operation (PID $C950) that is used to retrieve the CPU chronometrics. A Parameter Identifier (PID) can be defined to report if the software is unstable and not able to meet its real-time requirements [2].

- **Poor functional specifications:** functional specifications may be incomplete or ambiguous. The test engineer should test the limits of the design even though it is not required to meet that limit. Test to failure must be their objective. The failure can be reviewed later to determine if it is a real failure that would yield any customer impact.

- **Boundary testing example**: if the requirement for a digital input debounce is defined to be 5 consecutive samples at 10 milliseconds each, the boundary test will send a pulse of 38 milliseconds and make sure the device does not activate. Wait for 300 milliseconds, then keep incrementing the pulse width by 1 millisecond until the device activates. There will be a curve that shows a 40ms input pulse will sometimes cause the output device to activate and with a 51 milliseconds pulse, the output device activates 100% of the time.

Figure 1 shows typical elements of a body module ECU.

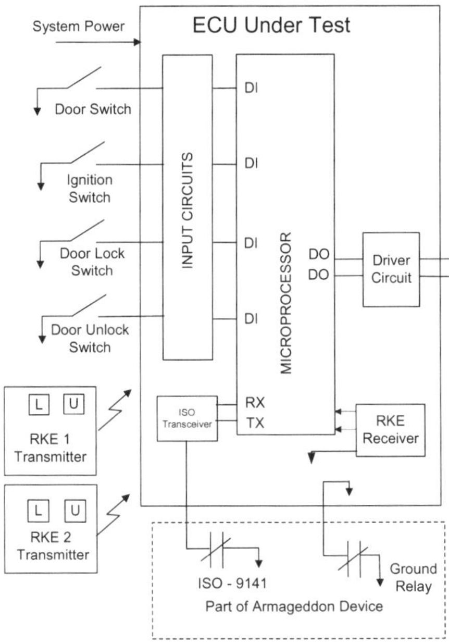

Figure 1: Block schematic of the ECU under test

SYSTEM STABILITY IN A NOISEY ENVIRONMENT

System stability is a good metric. In real-time embedded control systems, a measure of the system stability is based on the repeatability of the software or the stability of the software overtime. Determinism must be designed into the software and tested. Stability may be difficult to measure; however, the delayed responses to stimuli can be measured. More attention should be given to this area of deterministic behavior.

The response time could be any transfer function **(Figure 2)**. For example: If a door ajar signal requests the ECU to turn on an interior lamp, the response time can be easily measured. The system response should not vary by any more than the supplier has indicated as "Dither".

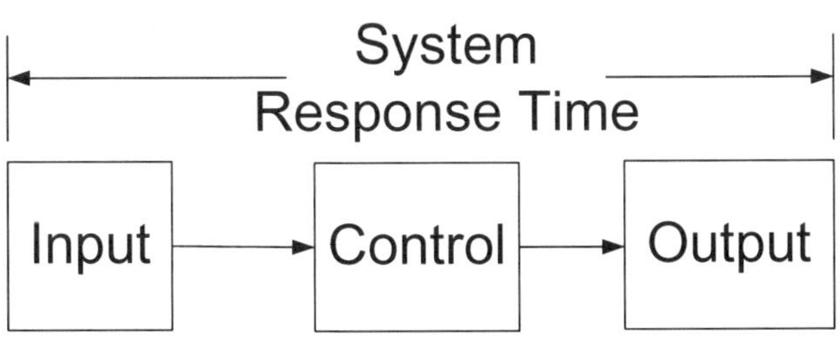

Figure 2: Input - Control - Output function

CALCULATING THE WORST-CASE AND BEST-CASE SYSTEM RESPONSE

System Response Time measurements must be taken without the noise condition and then measured with the noise present. Compare the time response difference, and make sure it is less than the worst-case dither the supplier specifies. Figure 3 shows the timing analysis for the best-case system response and **Figure 4** shows the worst-case system response. A dual trace oscilloscope is used to measure these times as shown in **Figure 5**. The pass-fail criteria are based on the worst-case and best case throughput calculation.

When noise is introduced into the system, the worst-case system response is not 83 milliseconds but 90 milliseconds. The ISRs may take up 6 milliseconds. One millisecond or 10% is needed for the design margin. The pass-fail criteria of the system response is 43 to 89 milliseconds. Anything measured outside of this limit is a sign the software is unstable.

The analysis is based on the following assumptions:

1. The Input algorithm is evaluated every 10 milliseconds with an execution time of 1 millisecond.
2. Five consecutive samples are required before a new debounce value is determined. This gives 40 to 50 ms debounce tolerance.
3. The Control algorithm is evaluated every 40 milliseconds with an execution time of 1 millisecond.
4. The Output algorithm is evaluated every 10 milliseconds with an execution time of 1 millisecond. The output algorithm is executed after the Control algorithm. (When it is scheduled to execute.)

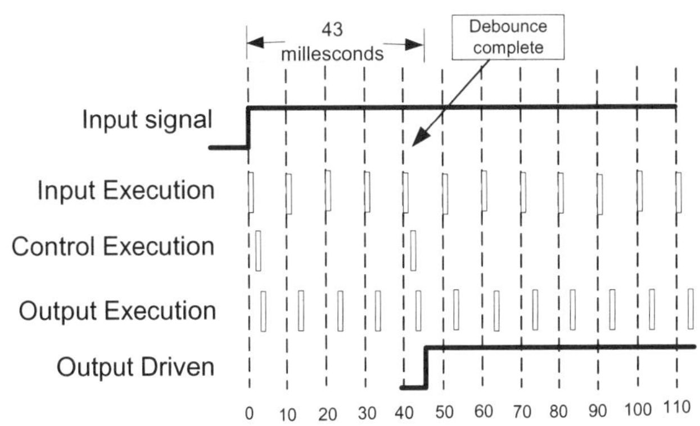

Figure 3: Best-case system responses without noise

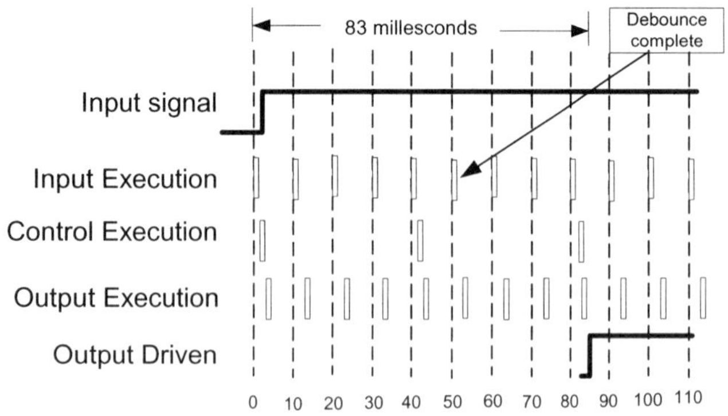

Figure 4: Worst-case system responses without noise

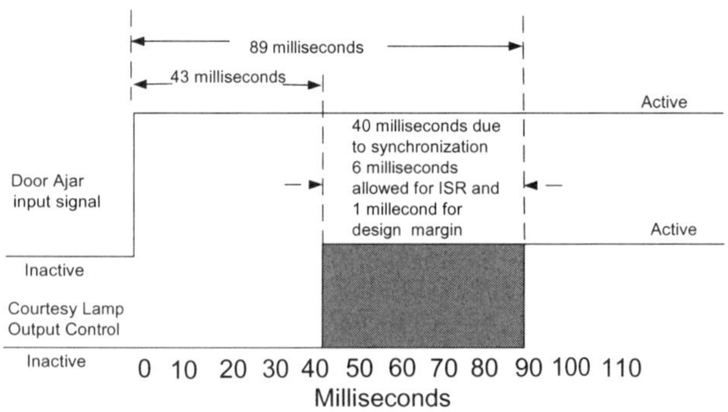

Figure 5: Measuring System response and delay

The test set up to measure the system response is found in **Figure 6**. The test person will activate the Door Ajar Switch. The response time is measured by reading the difference in between the rising edges of the door ajar signal and the interior lamp control output on the dual-trace scope.

To measure the system response for the best-case response time, use the following sequence:

1) Turn off the Bus crusher
2) Open the door
3) Read the scope and record the results
4) Repeat Steps 1 and 2 ten times
5) Use the smallest delay time as the minimum value (Time must be $43 \leq t \leq 89$)

To stress the software and attempt to get the worst-case timing, use the following sequence:

1) Turn on the Bus Crusher
2) Open the door
3) Read the scope and record the results
4) Repeat Steps 1 and 2 ten times
5) Use the largest delay time as the maximum value (Time must be $43 \leq t \leq 89$)

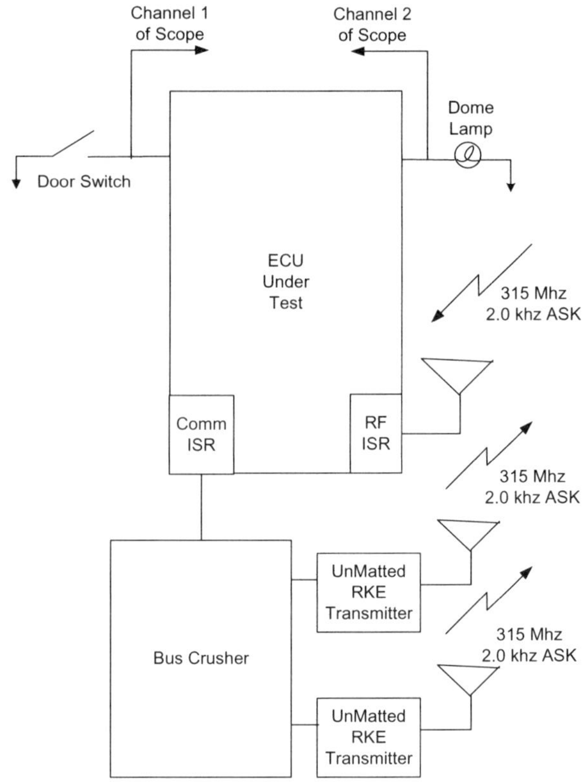

Figure 6: Test setup to measure the System Response time with a noisy RF environment.

DEVELOPING THE BUS CRUSHER

The bus crusher was first developed as a way to find defects in the J1850-PWM protocol. The basic idea was to see if the protocol could handle a noisy environment. A vehicle example of this shorted condition is a bus wire chaffing to ground. (A "Copper seeking screw" has found the network wires). Since the J1850–PWM protocol is a two-wire fault-tolerant protocol, it should be tolerant of a single wire fault. The test was to see if the network could operate with faults present and no undetected data errors on the network.

The Bus Crusher test needed a device that could cause faults on each wire and the fault tolerance of the network could be verified successfully.

Figure 9 shows the bus connection to the test vehicle's network. The picture of the Bus Crusher is shown in **Figure 7. Figure 8** shows the schematic of the Bus Crusher.

The test vehicle had ten specially designed ECUs, dedicated to test the J1850-PWM protocol. This vehicle was one of a fleet of 60 that was dedicated to prove-out the protocol. The Bus Crusher caused a ground short on each bus wire, but not both at the same time. The Bus Crusher did not find any issues with the J1850-PWM protocol after it ran for several weeks. No undetected errors were generated in the protocol.

In 1993, a single wire protocol called UBP (UART Based Protocol) was developed. It was time to blow off the dust and put the Bus Crusher to work again. This time a defect was found in the software. The Bus Crusher caused message aliasing. A truncated message could "look" like a valid message. A weakness was found in the protocol timing calculation. The protocol timers were adjusted and the aliasing issue was resolved. Until the root cause of the problem was discovered, engineers would say, "This is not a real world test. There will never be that much noise on the vehicle". The response was "Does it really matter if the test method uncovered a design defect?" This means the root cause must be identified. If the issue does not cause a degradation of network performance, then maybe it does not need to be repaired. In any case, the reason for the misbehavior must be discussed.

There were some performance issues with the Bus Crusher components. The fast mode would operate until the relays heated up. The Bus Crusher would stop operating after about 5 minutes of operation. The Slow mode operated continuously without a problem.

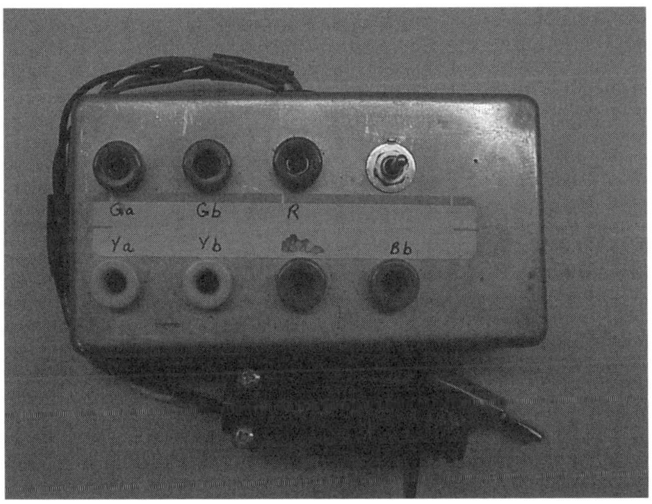

Figure 7: Bus Crusher Version 1.0

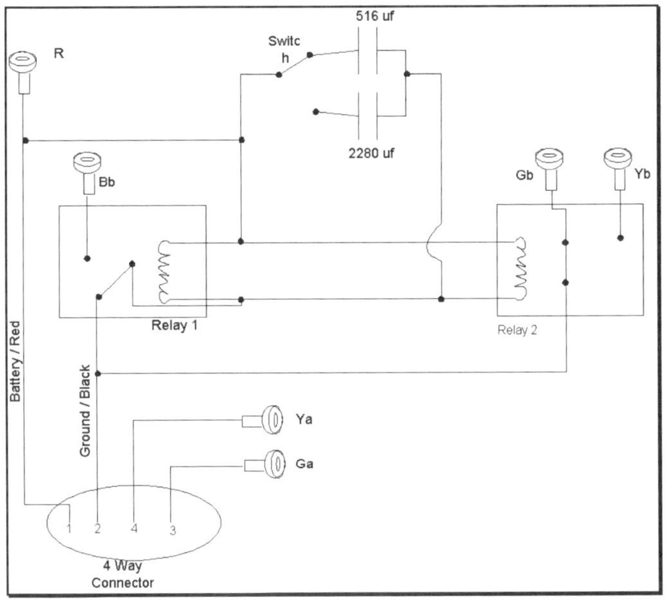

Figure 8 : Schematic of the bus crusher version 1

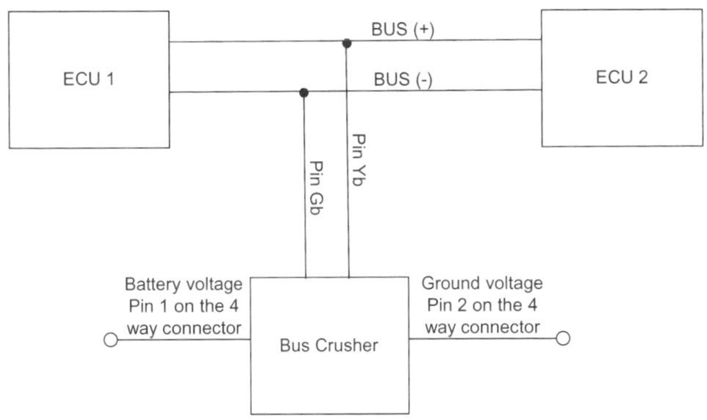

Figure 9: Connection to the Vehicle J1850 PWM communication bus

WAVE FORM INJECTED BY THE BUS CRUSHER

A digital scope was connected to the ISO-9141 circuit with the Bus Crusher connected. Channel 1 of the oscilloscope is connected to Gb and Channel 2 is connected to Yb.

Figure 10 shows the Bus Crusher in slow mode. The frequency of the noise event is approximately 400 milliseconds apart.

Figure 11 shows a zoomed-in view of the noise pulse in the slow mode. The pulse width is approximately 3 milliseconds. As you can see, there is ringing and other harmonic distortions. This is the noise that the test engineer needs to introduce.

Figure 12 shows the Bus Crusher in fast mode. The pulse width is about 1.5 milliseconds. The frequency of the noise event is approximately 110 milliseconds apart.

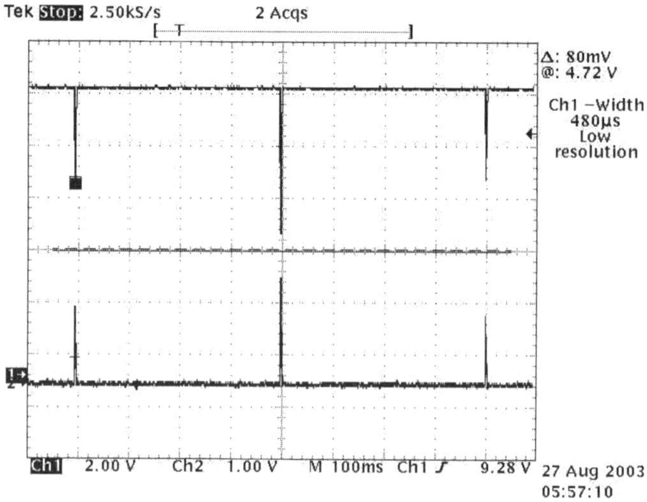

Figure 10: Bus Crusher in slow mode.

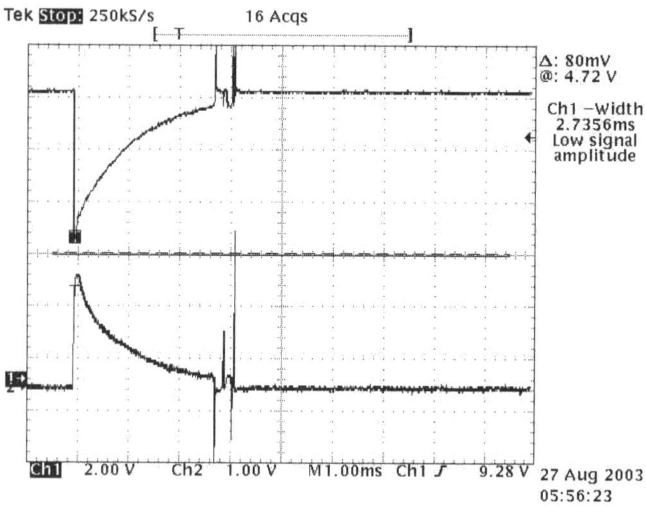

Figure 11 Zoomed-in view of the Bus Crusher in slow mode noise pulse.

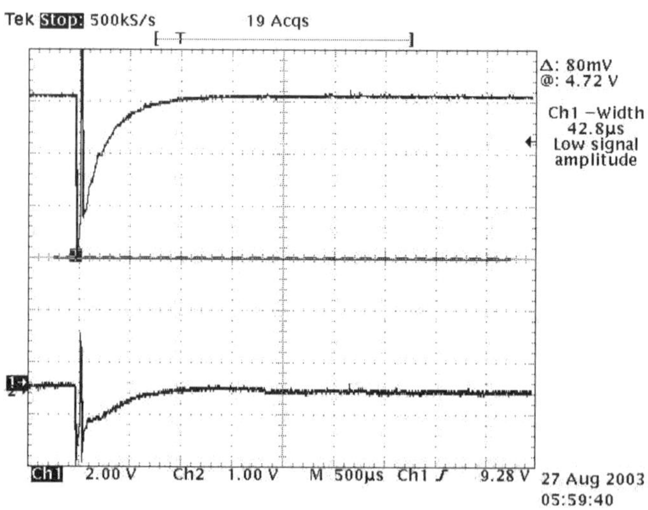

Figure 12 Zoomed-in view of the Bus Crusher in fast mode noise pulse.

EVOLVING THE BUS CRUSHER TO THE ARMAGEDDON DEVICE VERSION 1

The Bus Crusher has evolved as a way to introduce more noise on other circuits in the vehicle. It was named the Armageddon Device because people would comment, "These conditions will never happen. Why do we have to design for the end of the world?" The response was, "The design is not required to survive the end of the world, but we should know the behavior of the ECU if these odd events were to concurrently happen."

Figure 13 is a diagram that shows the connections of a mated RKE transmitter (the ECU will respond to a key press), an unmated RKE transmitter, and the Armageddon Device. The relay contacts are connected to the unlock button. It is safer to use unlock than lock to reduce the risk of burning out the lock motors and not being able to get in the vehicle. **Figure 14** shows the schematic of the Armageddon device.

A design weakness was found on a prototype vehicle that uses software as the overheat algorithm for the power lock motors. The sequence was:

1. Press the matted RKE transmitter unlock several time,
2. Cause the ECU to reset,
3. Repeat Steps 1 and 2.

The overheat algorithm restarts after a microprocessor reset, losing the data on the number of unlocking cycles that have been completed.

If this cycle is run for three minutes, all the lock motors heated up and burned out. Here is the "end-of-the-world" scenario. In this case the software was not changed because it was extremely unlikely that all these events would occur at the same time and the complexity of the software change was not worth the risk.

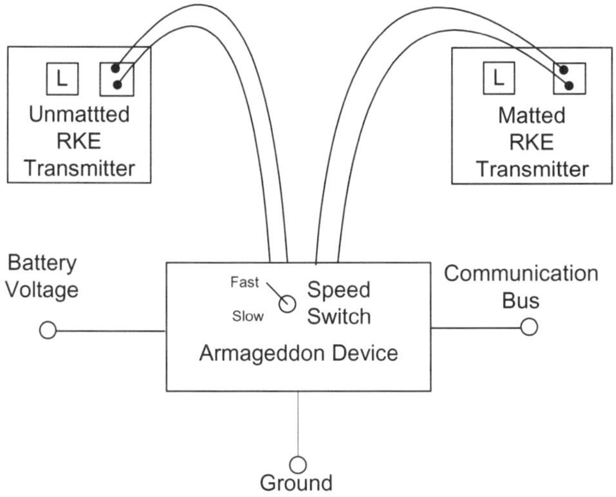

Figure 13 Armageddon Device version 1 with mated and unmated RKE transmitters.

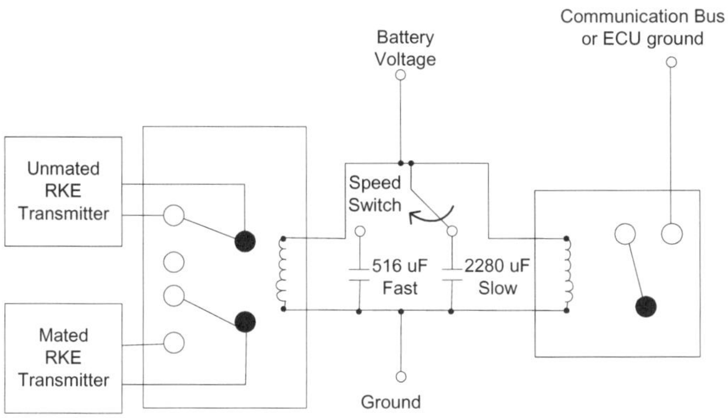

Figure 14 Block Diagram of the Armageddon Device Version 1

FUTURE TEST DEVELOPMENT PART II

This section establishes the framework for automating and integrating the previously discussed tests. The plan is to develop an automated test tool that will cycle the feature and introduce noise to the ECU. These tests stress the major operating modes of the software.

Most of the tests described in the following sections have been done on a manual basis. In the upcoming year, the plan is to integrate these individual tests into a PC-based controller. This controller will orchestrate the tests and generate a report. **Figure 18** shows the proposed setup with the ECU. The Armageddon Device Version 2 will integrate the following independent tests:

1. Reset Cycle Test
2. MSCAN Bus Crusher
3. RF Blaster
4. Software Key-Life test

THE RESET CYCLE TEST

A repeatable input–to-output transfer must be identified for the reset test. Under a reset condition, the headlamps turn on first before the input switch is debounced. This fail-safe method reduces the amount of time the headlamps may be off during a reset condition. Measuring the time between the reset line being unasserted and headlamps coming on should be the same time every time. The test sequence is as follows:

1. Be sure Headlamp switch is off
2. Cycle the power causing an ECU reset by opening the ground circuit.
3. Measure time between the ground circuit being connected and the time the headlamps turn on.
4. Delay 300 ms to 500ms. (Randomly)
5. Repeat Steps 1, 2, 3 and 4 for 10,000 times.

A pass-fail criterion is a variation of a few microseconds between samples. If this test is automated, it will take about 6 hours based on a 2 second cycle time.

THE MSCAN CRUSHER, A MODIFIED BUS CRUSHER TEST

The first version of the Bus Crusher was designed to cause faults on the communication channel. It was designed to stress the protocol under test. The modified Bus Crusher is intended to overload the ECU with data that it normally receives and prove the MSCAN interface will not get overloaded or show signs of ECU performance degradation.

The test engineer needs to have a message list for the ECU. These messages are transmitted to the ECU at the fastest possible rate. (If one PC is not capable of transmitting the data, an additional PC is needed to make sure the MSCAN network is 100 % loaded.)

The MSCAN Crusher sends messages at the lowest priority. Normal network messages must not be blocked. With the MSCAN Crusher messages at the lowest priority will be arbitrated off the bus.

The state of health message is also included to the MSCAN Crusher message set. These messages are looking for faults in the software.

SAE paper [2] describes a PID $C950 and $C951 that captures these events:

1. Number of CPU resets
2. Number of Illegal OP code
3. Number Watchdog timer
4. Number Stack Overflows

5. Number of scheduler time overflows

Use the diagnostic mode to retrieve continuous DTCs

In addition to sending out the "Normal" network messages, a fault is introduced periodically on the network. This will make sure the fault management strategy in the ECU behaves correctly and it can recover from the network faults. The Bus Fault Control Relay is energized, causing a 20-millisecond fault every 200 milliseconds. Both bus wires are shorted to ground for this test.

THE RF BLASTER TEST

The RKE RF receiver can be a tricky device to control in software. Hardware and software must play together nicely. There are a lot of interactions with ISR and base signal processing. The intent of the RF Blaster test is to flood the 315 MHz RF with noise and verify that the receiver can handle the noise.

Three lessons learned from RF noise:

1. RF noise causes the microcontroller to reset. A 9.6 KHz signal ASK (Amplitude Shift Key) signal with a 315 Mhz carrier caused the microcontroller to spend too much time servicing the ISR. The ISR was capable of handling an 8 KHz signal. The worst-case modulated signal was assumed to be 2.2 KHz. No one asked if the hardware was going to filter out the high frequency noise.

2. A body module ECU had a problem with the ISR overload condition. A combination of RF data input band pass filter being set too low and a lack of an RF ISR protection strategy caused the RF ISR to be overloaded and spending too much time servicing the ISR. The watchdog was not being serviced often enough, resulting in a microprocessor reset. The problem was compounded when the microprocessor was accessing EEPROM and the watchdog caused the reset. The EERPOM was corrupted, losing the matted RKE transmitter data. The associated RKE transmitter was erased and had to be re-associated. The problem first appeared as a random RKE Transmitter "getting out of sync" until the root cause was found.

3. An ECU had a problem where the RF receiver would not come out of low power mode and recognize that an RKE transmitter button was pressed. Errors in the low power mode strategy caused the ECU to disable interrupts when it entered low power mode. This ISR was needed to wake the ECU from the low power mode. The problem was that if a new RKE message was received between two instructions (2 microseconds), the module entered sleep mode with interrupts disabled. Swapping two instructions in the low power mode routine solved the problem.

<u>**"Highway to Hell" test**</u> - There was a problem with a high power audio amplifier that would go into thermal overload. The overload condition was not identified using classical music. If the volume was turned up to the maximum limit when "Highway to Hell" by A/C D/C was played, the amplifier went into thermal overload. When the CD is played through the RF transmitter, it will generate a random noise spectrum that is between 20 Hz to 17 KHz The audio output of the CD player will be connected to the input of the RF amplifier. **Figure 15** shows the connections from the CD player to the RF amplifier.

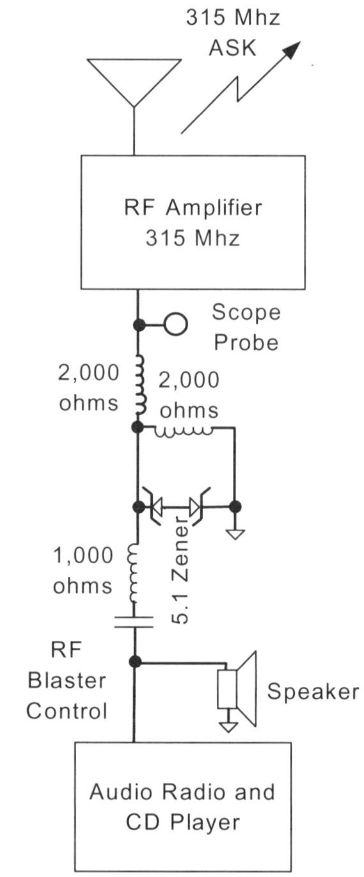

Figure 15 315 Mhz RF Blaster with Radio/CD controlled

THE TPMS TRANSMITTER

The TPMS transmitter, shown in **Figure 16 ,** sends a valid TPMS message on command. The carrier is 315 MHz FSK (Frequency Shift Key). The 9.6 KHz digital signal contains the TMPS data. There shall be more than 100 millisecond delay in the TPMS transmission from when it is requested.

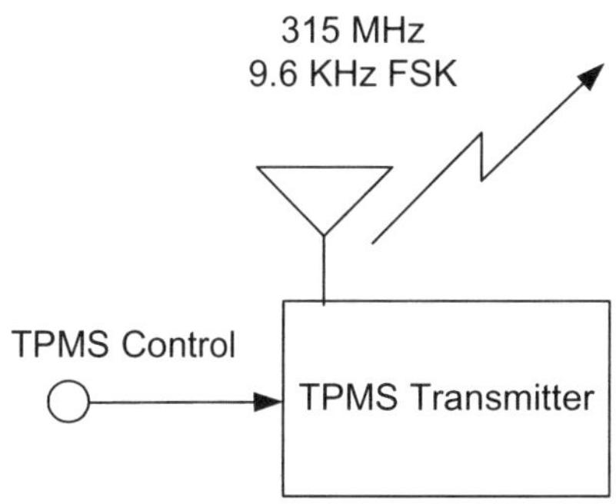

Figure 16 TPMS Transmitter

THE RKE TRANSIMTTER

Figure 17 shows the RKE Transmitter set up. This transmitter is the same device that is used for production and sold to the customers. This unit will generate valid lock and unlock RF messages. The carrier is 315 MHz ASK (Amplitude Shift Key). The 2.0 KHz digital signals contain the RKE data. This RKE transmitter is matted to the ECU. The lock motors will activate when a lock or unlock is requested by the PC controller.

The RKE transmitter the customer receives is battery operated. This unit must be powered by a 3-volt power supply in order to guarantee fidelity of the RF transmission. This should minimize a false test failure.

The connections to the RKE transmitter are made by soldering leads from the lock and unlock switches in the RKE transmitter to the relay contacts of the I/O controller.

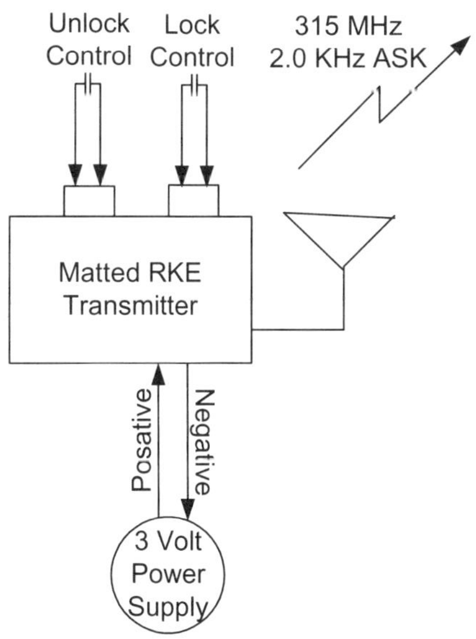

Figure 17 RKE Transmitter

DEVELOPING THE SOFTWARE KEY-LIFE TEST

The key-life test is based on a usage profile that the product will experience during the life of the product. For 10 years 150,000 miles the lock and unlock may be cycled as much as 10,000 times.

This test should be conducted in an EMC screen room to eliminate the possibility of other RF noise interfering with the ECU receiver. The ECU is expected to respond to every lock and unlock event for 10,000 times. It takes about 25 seconds to run one cycle. For 10,000 cycles it would take approximately 70 hours.

Figure 18 shows the ECU under test and the components that make up the Armageddon Device 2. The Happy Path test sequence is as follows:

1) RF Blaster disabled, TPMS transmitter disabled, RKE transmitter disabled.
2) Request the RKE Unlock to be transmitted
    a. Result: Lock motor cycle to the unlock position
    b. Result: Interior lights turn on
3) Request the TPMS signal to be transmitted.
4) Wait 20 seconds
5) Request RKE Lock
    a. Result: lock motors cycle to the lock position
    b. Result: Interior lights turn off
6) Request TPMS count PID
    a. Result: Verify that a new TPMS message was received

Repeat Steps 1 and 6 for 10,000 times.

It s time to introduce noise after the ECU passes the preceding sequence of tests. This is where the Armageddon Device 2 concepts come into play with numerous noise components. It takes about 30 seconds to run one cycle. For 10,000 cycles it would take approximately 83 hours. The noisy environment is introduced with the following steps:

1) Disable the TPMS transmitter, disable the RKE transmitter and enable the RF Blaster
2) Enable the MSCAN Crusher
3) Wait 5 seconds
4) Disable the RF Blaster
5) Wait 100 milliseconds
6) Request RKE Unlock transmission
    a. Result: Lock motors cycle to the unlock position.
    b. Result: Interior lights turns on
7) Wait 500 milliseconds (RKE will transmit for 400 milliseconds)
8) Request the TPMS signal to be transmitted.
9) Wait 200 Milliseconds
10) Enable the RF Blaster

11) Wait 18 seconds

12) Disable the RF Blaster

13) Wait 100 Milliseconds

14) Request the RKE Lock transmission

    a. Result: lock motors cycle to the lock position

    b. Result: Interior lights turn off.

15) Request the TPMS count PID

    a. Verify that a new TPMS message was received.

Repeat Steps 1 and 15 for 10,000 times.

There is enough time between the test steps to have the RF overload be unloaded to make sure the ECU can receive the TPMS and RKE data. The pass-fail criterion is that the TPMS and RKE should not lose any information.

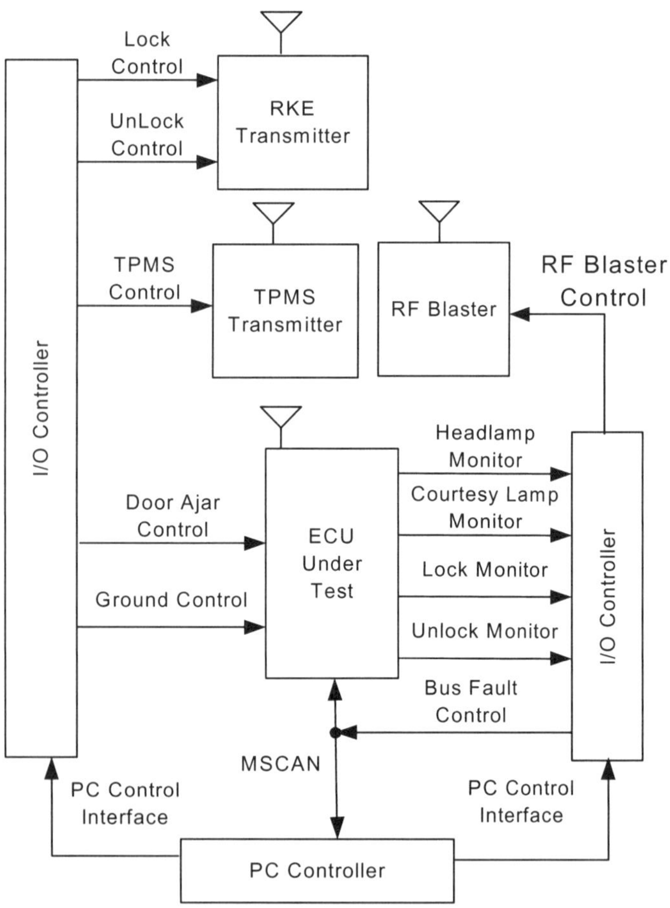

Figure 18 The Armageddon Device version 2

CALIBRATING THE SET-UP

An RF spectrum analyzer may be used to ensure that the RF transmitters are working. A spectrum analyzer is used to make sure the RF signals are being transmitted and the signal strength is correct.

Connect a scope to the input of the RF generator labeled SCOPE PROBE in **FIGURE 15**. Adjust the CD audio amplifier signal so the signal is not being clipped. The RF generator is set to 315 MHz Frequency Modulated carrier.

If a spectrum analyzer is not available, then the matted RKE transmitter can be used. The ECU does not respond to an unlock command when pressing the RKE transmitter unlock button while turning up the signal strength. Once this threshold is established, turn the signal up 3db more to make sure the RF Blaster is the dominant signal.

The next step is to disable the RF Blaster, press "unlock" on the RKE transmitter and make sure the lock motors activate. **FIGURE 19** shows the block diagram of the spectrum analyzer that is tuned for 315 MHz. If a RF Spectrum analyzer is used, record the field strength for the test report.

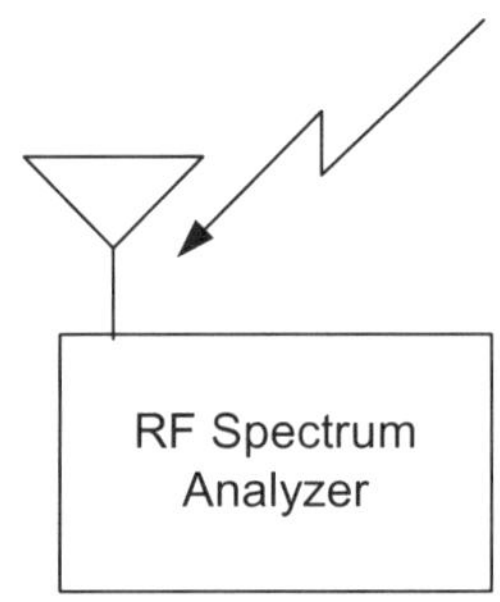

Figure 19 RF calibration tool

CONCLUSION

Don't assume an operating condition will never happen. Test all possible conditions. Determine the probability of such an event after the testing is completed. If the failure could be safety-related or has damaged the hardware, the software may be modified to mitigate this condition.

Product quality is directly related to the quality of the testing. Happy Path testing does not assure that the product will work as intended when it is under stress or in the presence of a faulted condition.

The Armageddon Device is a tool that introduces noise, which exacerbates software and hardware race conditions. If the Armageddon device does not find issues in the design, it does not mean that there aren't any. Identifying issues in software takes extraordinary efforts due to highly complex logic.

The test engineer must go beyond Happy Path testing. Reading the functional specification and testing every path is not enough. Make sure possible failure modes are thoroughly tested as well. Identify weaknesses in the design by reviewing the requirements, design and past testing methods.

Design tests that go beyond the defined boundaries by reviewing fault management strategies. Don't be limited to what is written in the specifications.

Calculate the system response time to see if noise affects the response time. The real world automotive vehicle contains lots of noise. Make sure the designs can tolerate and operate in the real world.

The ECU functionality is generally tested once during the test cycle. Repetitive cycling of a function may find defects in the software algorithms. These Key-life tests can be done manually, but a computer-controlled sequencer is more appropriate and provides repeatability.

ABOUT THE AUTHOR

Ron Brombach has 21 years of experience in the computer field. He has worked in the steel industry, industrial automation design, and for the past 17 years, at Ford Motor Company. Ron has been an active member of SAE since 1992. Ron has earned a B.S. in electrical engineering from Oakland University in Rochester, Michigan and a M.S. in Computer Control Systems from Wayne State University in Detroit, Mi. Ron currently is a supervisor in the body module software area. Ron's Email address is AR15@peoplepc.com

REFERENCES

[1] SAE paper 01PC-345 ***Lessons Learned the Hard Way*** March 2001 by James Weinfuther, Allen Fenderson, Daniel King and Ronald P. Brombach P.E. Ford Motor Company Dearborn, MI

[2] SAE paper March 2002 02PC-01-873 ***Robust Embedded Software Begins With High-Quality Requirements*** Ronald P. Brombach P.E., James M. Weinfurther, Allen E. Fenderson, Daniel M. King Ford Motor Company, Dearborn MI.

[3] SAE paper March 2000 00PC-317 ***Automotive Software Development Evaluation*** Ronald P. Brombach P.E. Ford Motor Company, Dearborn MI.

2003-01-1024

A New Environment for Integrated Development and Management of ECU Tests

Klaus Lamberg, Jobst Richert and Rainer Rasche
dSPACE GmbH

ABSTRACT

Due to the rapidly increasing number of electronic control units (ECUs) in modern vehicles, software and ECU testing plays a major role within the development of automotive electronics. To ensure effective as well as efficient testing within the whole development process, a seamless transition in terms of the reusability of tests and test data as well as powerful and efficient means for developing and describing tests are required. This paper therefore presents a new integration approach for modern test development and test management. Besides a very easy-to-use way of describing tests graphically, the main focus of the new approach is on the management of a large number of tests, test data, and test results, allowing close integration into the automotive development processes.

INTRODUCTION

The constantly increasing software complexity of today's electronic control units (ECUs) requires new ways of developing automotive electronics. While new software functions are still being developed or optimized, other functions are already undergoing certain tests, mostly on module level but also on system and integration level. Test and quality gates have to be verified repeatedly by regression testing. The immense effort that is invested nowadays in the development and execution of tests necessitates new methods and tools to cope with the increased time-to-market pressure.

Because of the continuously increasing number of electronic control units (ECUs) in modern vehicles, their connection via bus systems and the resulting growing complexity, new development methodologies are becoming more and more important. A significant characteristic of modern development processes is the increasing need for testing. As a result, there is a rapidly growing demand for developing and managing large amounts of tests, test data and test results.

In automotive development, and in particular in the development of electrics/electronics (E/E), development processes are increasingly following the V-cycle (cf. Figure 1).

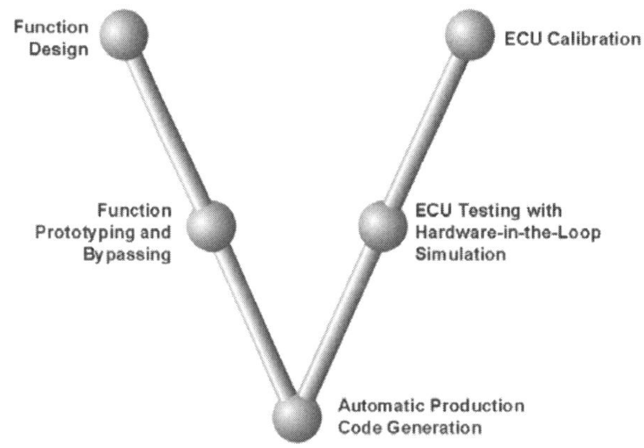

Figure 1: ECU development process

The V-cycle mainly consists of the following steps:

- Function design: Developing control functions and algorithms
- Function prototyping and bypassing: Trying out the developed functions in an actual vehicle or on a test bench
- Automatic production code generation: Implementing the functions via automatic code generation for microcontrollers in the ECU
- ECU testing with hardware-in-the-loop simulation: Testing the ECU and its functions in a virtual/simulated environment
- ECU calibration: Fine-tuning of functions and algorithms by adjusting ECU parameters during test drives on the test bench or in the vehicle.

It must be pointed out that the V-cycle is applied repeatedly throughout the development phase and is not straightforward. As a rule, the entire V-cycle is run through at least once from one ECU development procedure to the next.

Today, the testing of ECUs as well as ECU networks is mainly performed by using hardware-in-the-loop simulation [1]. But with testing becoming an increasingly important step in automotive electronics development, new challenges arise. These challenges mainly deal with the extension of testing tasks from HIL simulation to other development stages. But also there is an increasing demand for improvements of test development and management.

ECU TESTING IN AUTOMOTIVE ELECTRONICS DEVELOPMENT

HARDWARE-IN-THE-LOOP-SIMULATION AND TESTING

In hardware-in-the-loop simulation (HIL simulation), the behavior of the vehicle is simulated by software and hardware models. Real vehicle components (real parts) are then connected, via their electrical interfaces, to a simulator, which reproduces the behavior of the real environment [2].

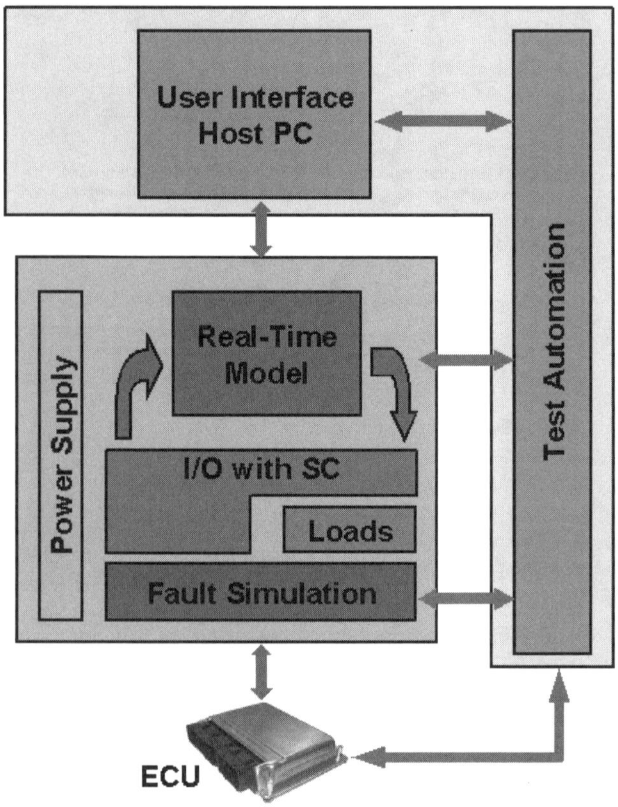

Figure 2: Typical hardware-in-the-loop system architecture

Figure 2 shows the fundamental design of HIL systems. Instead of being connected to an actual vehicle, the ECU to be tested is connected to a simulation system. This runs a model of the vehicle process and associated sensors and actuators that will usually have been developed and implemented with suitable modeling tools such as MATLAB/Simulink. C code is generated automatically from the representation within the modeling tools and then downloaded to special real-time hardware for execution. The real-time hardware is in turn connected to the ECU's electrical interface via special I/O boards and suitable signal conditioning for level adjustment, using either simulated or real loads.

Using hardware-in-the-loop simulation has several advantages. The main ones are as follows:

- Controller functions can be tested in the early stages of development, even before a test carrier (prototype vehicle) has been produced. The electronics system therefore reaches a high degree of maturity very early on, providing an improved starting-point for later development phases.
- The objective of using simultaneous engineering, or simultaneous development of ECU and vehicle, to cut the time needed for development, can only be achieved by methods such as HIL simulation.
- Expensive field trials or experiments in borderline zones and hazardous situations can partly be replaced by laboratory or desktop experiments.
- Extreme or unusual ambient conditions can be adjusted at will by varying the parameters in the model. Thus typical winter test drives under low-μ conditions (snow and ice) can be carried out in summer, and cold-start tests can be performed repeatedly.
- Failures and errors that could have devastating effects in a real vehicle (sensor failures, line breaks, ground errors, error frames on the CAN bus etc.) can be simulated and tested systematically.
- The experiments performed on the HIL system can be reproduced precisely, and automatically repeated as often as required.

Thus testing by using HIL technology is increasingly important for the efficient integration of production type engine ECUs. With more and more automotive manufacturers relying on independent ECU suppliers, the need for automatically testing black box systems, including closed-loop and open-loop functions, self-diagnosis, failure memory management and the triggering of failure lamps, becomes paramount. Since malfunctions in these areas cause a high recall potential for vehicle manufacturers, especially when they are discovered by the customer first, they must be avoided by all means.

TESTING ECU FUNCTIONS AND CONTROLLERS

Today, HIL simulation is used for nearly every type of ECU within modern vehicles. Typical examples include testing single ECU functions, for example, engine or vehicle dynamics controllers, as well as whole ECUs, for example, as part of the powertrain [3]. Usually, this is done by stimulating the respective ECU function (directly at the function's interface, if possible, or indirectly by driving the whole system into the specific operating scenario) and monitoring its behavior. Stimulation and monitoring can require access to model variables as well as ECU internal variables. However, due to the fact that usually only the outer interface (inputs and outputs, no internal values) of the function to be tested is accessed, this kind of test is also called black box testing.

Accessing ECU internal variables usually requires the HIL system to make use of a calibration tool. The calibration tool provides the interface to the internal values of the ECU and enables static parameter labels (such as threshold values in the form of scalars, curves, or characteristics maps) to be accessed. Furthermore, it is possible to measure dynamically changeable values with time stamps. The host PC software of the HIL system remotely controls the calibration tool, which reads and writes ECU internal variables. A widely used interface for remote-controlling calibration tools within such an automation scenario is the ASAM MCD 3 MC standard, defined by the Association for Standardization of Automation and Measuring Systems (ASAM, [4]).

ONBOARD DIAGNOSIS TESTS

During the past few years, the software for engine ECUs in particular has shown an almost exponential growth in range and complexity, since they must fulfill the customers' demands for comfort as well as the government's requirements on emissions reduction and onboard diagnosis (OBD-I and OBD-II). In addition, the number of control loops within engine control systems is increasing, which means that proper testing of ECU operation requires all control loops to be in place and closed. Onboard diagnosis tests are therefore a typical application area for HIL simulation [5]. A typical diagnostic test consists of different steps performed sequentially, as follows:

1. Drive into specified operating point
2. Activate electrical fault by remote-control relay
3. Read out ECU diagnostic memory
4. Evaluate test by comparing the detected fault with the expected fault
5. Generate report automatically

While steps 1 to 4 are typically performed repeatedly for each possible fault, step 5 can be performed subsequently after having tested all fault combinations at least once.

Since this kind of test has a very uniform structure and can easily be reused for different kinds of electrical faults, it has very great potential for savings in time and therefore cost.

TESTING NETWORKED ECUS

Today, automobiles can contain up to 70 electronic control units for powertrain control, chassis control and comfort systems. Since these ECUs are interconnected (for example, via CAN), and the functions are distributed throughout the ECUs, it is necessary to test the entire ECU network before in-vehicle validation and calibration can be performed. For example, when a driver wants to start his car on a slippery road, the traction control becomes active, i.e. the vehicle dynamics control unit (ESP) and the engine management system cooperate in order to keep the car on the road. Since there are several important functions like this, particularly in powertrain control systems, it is obvious that these functions have to be tested before going into a real vehicle. Figure 3 shows an HIL system consisting of four stand-alone HIL simulators connected to each other [6]. This system is used for testing all the powertrain ECUs of one vehicle operating together in a network. But HIL simulation for testing networked ECUs is also used for body electronics [7].

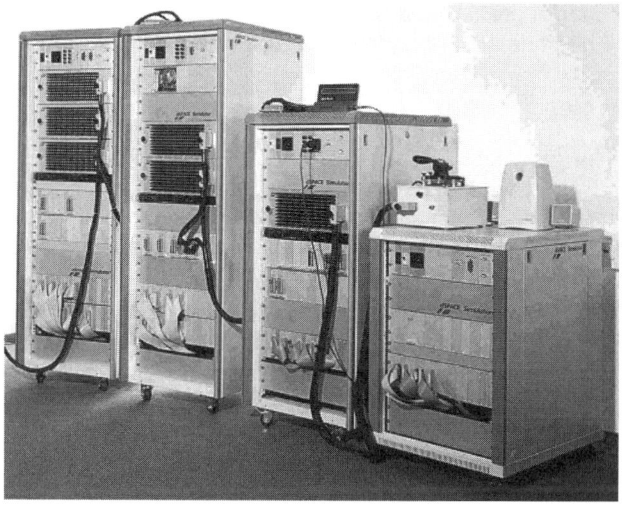

Figure 3: HIL simulator for testing networked powertrain ECUs

The large and ever increasing number of ECUs within a vehicle not only requires functional testing on network level. Another important item is power consumption. Therefore in body electronics especially, ECUs are required to provide features like sleep mode and wake-up, as part of overall network management. Sleep mode, for example, means that all ECUs can switch into an energy-saving operating mode during parking and can be woken up as soon as the driver unlocks the vehicle to use it once again. If only one single ECU fails to switch into sleep mode correctly, none of the others do, and the battery will be flat within a few hours.

SEAMLESS TESTING OVER THE WHOLE PROCESS

As already mentioned earlier, new challenges are coming up today with respect to testing automotive electronics. Accepting them will change the way of testing in the near future. The most significant trends to be observed are described below.

Automated Testing

Today, HIL simulation and testing are mostly performed manually. This means that a user interacts with the entire/total system consisting of real ECUs and HIL simulator to perform virtual test drives in the laboratory. But more and more, interaction is being replaced by automation, and users are developing test programs. Such test programs replace the user in the task of interacting with the simulation process.

Automatic testing increases the benefit of HIL simulation significantly. For example, it allows testing overnight and on weekends ("lights-out-tests") and reduces the overall time needed for testing. For example, there are reports that automatically performing diagnostic tests for transmission ECUs using HIL technology has decreased the workload by a factor of ten [8]. Additionally, automated testing leads to broader test coverage and a greater test depth, meaning that more tests can be done and more details can be tested.

Early Testing

In the future, testing will not be limited to HIL-simulation. While HIL simulation and testing always mean that at least one ECU prototype must be available, testing can also be applied in earlier development stages. This is the case, for example, if the ECU function is represented by a simulation model, for example, implemented in Simulink and/or Stateflow. The function model runs against a plant model of the vehicle part to be controlled. While the function model is the so-called unit under test (UUT), the plant model can be understood as a virtual environment for the UUT. Because the function model is integrated into the control loop, this is often called model-in-the-loop (MIL). Even at this early stage, a large amount of functional testing can be performed. Additionally, structural tests like model coverage can be done.

In contrast to MIL, in a software-in-the-loop (SIL) environment the UUT is not represented by a simulation model, but by the C code which later in the process will be implemented on the ECU. The C code can also be run against the plant model. This also allows functional tests to be run to compare the C code behavior to that of the function model. Additionally, even in an SIL stage, structural testing is possible, for example, code coverage tests.

The last step within this process is hardware-in-the-loop simulation and testing as described above. It is of great importance for an efficient development process that the functional tests developed and performed in earlier stages can be done once again using the HIL system. In addition, further test can be done, for example, diagnostic tests, ECU network tests etc..

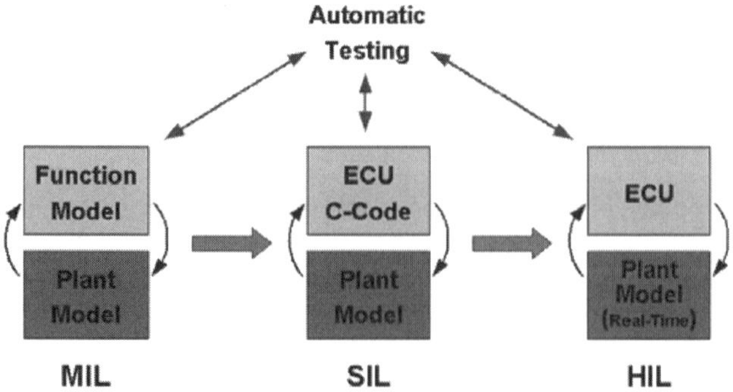

Figure 4: Automatic, model-based testing in different process stages

Figure 4 illustrates the approach of automatic testing in different development stages.

EFFICIENT TEST DEVELOPMENT

USER TYPES

Working within a test project consists of two different tasks: developing tests on the one hand and performing tests on the other hand. Thus different use cases and therefore two different types of users can be identified (see Figure 5).

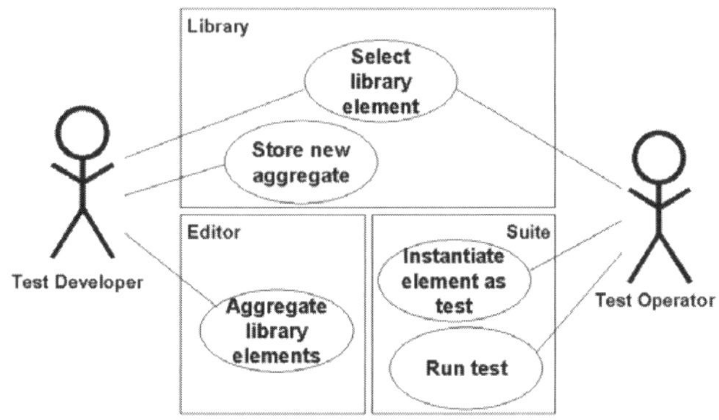

Figure 5: Use case diagram

Test Developer

Test developers are typically experts on developing and implementing tests. They are familiar with the appropriate testing methods, with the test language to use and the tools for implementing and managing tests. Given a library with basic test steps (basic building blocks, for example, reading and writing model variables, activating electrical short circuits) and also higher-level test routines (for example, driving cycles), test developer typically extend the library of reusable tests. They select the necessary elements from the library and combine them to produce new and higher-level automation sequences using an appropriate editor. Afterwards they save the new sequences back to the library for future reuse or for usage by others.

Test Operator

Given a certain function or real ECU to be tested (UUT) and a respective test job (testing task), operators know what tests to perform. They select the necessary tests from a test library provided by the test developer and create executable instances of them. After having selected and parameterized the tests to be done, the operator runs them. Automatically, test results are generated which the operator then archives and – if required – reports to those responsible for the development and release of the UUT.

GRAPHICAL TEST DEVELOPMENT

Today, tests are mostly written using script languages. Examples of widely used script languages for such purposes are Python or Visual Basic. Sometimes even C or C++ are used, but because of the necessity to compile such programs instead of directly interpreting them, they are not widely used for writing test programs.

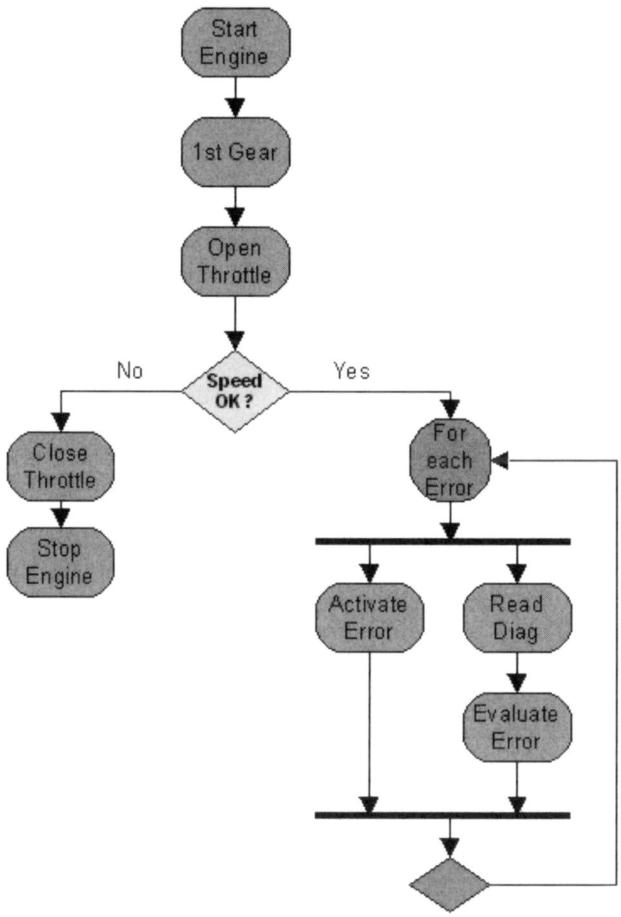

Figure 6: Graphical test chart

Writing test programs by using script languages is very flexible. For example, advanced users can implement their own libraries and therefore make functionality reusable. But on the other hand, manual programming means that the script language has to be learnt first. For occasional users script programming is inconvenient and error prone. It would therefore be better if a test developer could develop tests on a higher level, providing a more convenient and efficient way of describing tests. This can be achieved by graphical test development. A graphical and easy-to-use test sequence editor ensures a short and steep learning curve. Developing tests becomes a fast and efficient task. In addition, a graphical representation allows easy navigation even in very complex test sequences with several hierarchy levels.

Figure 6 shows an example of what a graphical test chart can look like. The semantics used are based on UML activity charts. UML (Unified Modeling Language, [9]) is becoming more and more important in automotive electronics development, particularly for software architecture design, but also for test specification.

AUTOMATION INTERFACES

The effectiveness of each automation concept is fundamentally driven by the automation capabilities that the concept offers. This means that a user not only needs to automate a certain procedure, but that there must be a wide range of functionality that the procedure is able to perform. This functionality is mainly determined by the interfaces that can be accessed from within a test. Besides many other interfaces that are typically used for testing ECUs, the following are the most important ones.

Real-time Model

A fundamental element of model-based testing is the interface between the test program and the simulation model. Typically, this interface is specified by model variables. This means that the test can access the models of the simulation model directly. This includes reading from and writing to model variables. The access mechanism must be independent of the model running in real time or non-real time. This can be achieved by a transparent platform concept, which implements an appropriate software layer abstracting from different simulation environments. The benefit is that the test can address the same variable on different platforms in the same way, so no migration effort is required when changing the platform from one development stage to the next. This enables a seamless transition from the offline phase (i.e. MIL and SIL) to the online phase (HIL), without the necessity of modifying the tests.

Electrical Fault-Simulation

Nearly every HIL simulation system is equipped with relay boards for electrical fault simulation. The relays are used to generate real electrical short circuits on the ECU input and output pins. The ECU diagnostic software is required to detect such faults and to correctly handle the fault, for example, by limp home functions, and to store the information about the fault in the ECU internal diagnostic memory. Typical electrical faults to be simulated using relays are short circuits to battery voltage and to ground or broken wires. Newer HIL systems are even able to simulate transition resistances or bleeding resistances, or to enable multiple faults at the same time.

Due to its uniform procedure, testing diagnostic functions has very great potential for saving time and cost. Therefore automatic fault stimulation is provided by most HIL systems by the ability to remotely and automatically control the fault simulation relays. This requires an automation interface for sending switching commands, for example via the serial interface (RS232) or via the CAN bus.

Diagnostic Interface

As already explained earlier, testing diagnostic functions of course requires the ability to access and read the diagnostic memory of the ECU.

There are different ways of accessing the diagnostic memory of ECUs from within an HIL environment. One is to implement the diagnostic protocol as a software component within the HIL system itself. The test can then use the protocol layer, which provides an appropriate API (application programming interface). But on the other hand, in many cases it might be more efficient to make use of commercial diagnostic tools, like Diagnostic Tool Set (DTS, [10]). Such tools can also be controlled from within a test and serve as an interface to the ECU diagnostic functionality.

LIBRARIES

The aim of providing the different automation interfaces described above and their range of functionality is that a test developer can use them to efficiently create new tests based on them. The new approach therefore includes a library concept supporting a high degree of reusability on different levels of complexity. In general, three types of libraries can be distinguished:

- Global built-in libraries provide basic functionality to be used by a test developer to implement new automation sequences. Built-in libraries contain control structures, calculation functions, algebraic and logical operators, etc. They may also provide the automation interfaces described above.
- Global custom libraries are custom-made extensions of the available automation functionality. For example, automation sequences used very often (for example, driving into a specific operating point) can be described once by a test developer and reused in different test sequences. Like the global built-in libraries, the global custom libraries are project-independent.
- Project-specific libraries exist only within a project. They also provide test sequences developed by a test developer. But in contrast to global libraries, the test sequences in project-specific (local) libraries contain test sequences which only make sense when performed within a specific project context. The sequences are so closely related to the ECUs and their functions that performing them outside the project makes no sense at all. Typically, the test sequences in project-specific libraries are used, instantiated, parameterized, and executed by the test operator by dragging them from the library to so-called suites.

Such a library concept provides an important advantage: not only can test developers use predefined automation steps and aggregate them to higher-level sequences. They can also put the sequences back into the library so that they can be reused in other tests or other projects, or by other users. By doing so, test developer make their own know-how reusable, because their knowledge on how to test and how to write tests has been built into the sequence. Over time, the library contains more and more standardized and reusable test sequences and test templates. As a result, test development becomes increasingly efficient.

TEST PROJECT MANAGEMENT

Often, testing ECUs requires hundreds or even thousands of tests to be developed, maintained and performed. All the tests must be stored and administrated consistently, so that they can be performed repeatedly ("regression testing") and reproduced at any time. The large amount of test results – each test run produces a new result instance – must be stored persistently. Based on these results, reports can be generated automatically. The storage, maintenance, and administration of this large amount of tests together with the test data and the test results require powerful means to manage test projects.

STRUCTURING PROJECTS

In general, the larger a test project is, the more important it is to structure it. Structuring in this context means grouping a number of tests according to a specific criterion. Such criteria include the different functions to be tested, the ECUs in a network, different development stages (MIL, SIL, HIL, see above), different users involved in the project, etc. Tests which belong to the same group according to a specific criterion are then simply grouped under it.

In practice, such a project structure can be represented by a tree. While the project itself is represented by the tree root, the tests are the leaves of the tree. Figure 7 shows an example of a test project structure.

Beyond the project node on the highest level and the test programs on the lowest level there are further items within the project structure. Together with the tests, the test data and the test results are depicted separately. Additionally, it can be seen from the figure that a distinction is made between libraries and suites. This correlates with the different use cases identified and explained before. The libraries are intended to pick up the test sequences provided by the test developer. In software terms, the libraries hold templates of the test sequences. The suites contain the test sequence instances which the test operator is intended to perform when testing a specific function or ECU. This can be done by simply dragging a sequence from a library and dropping it into a suite.

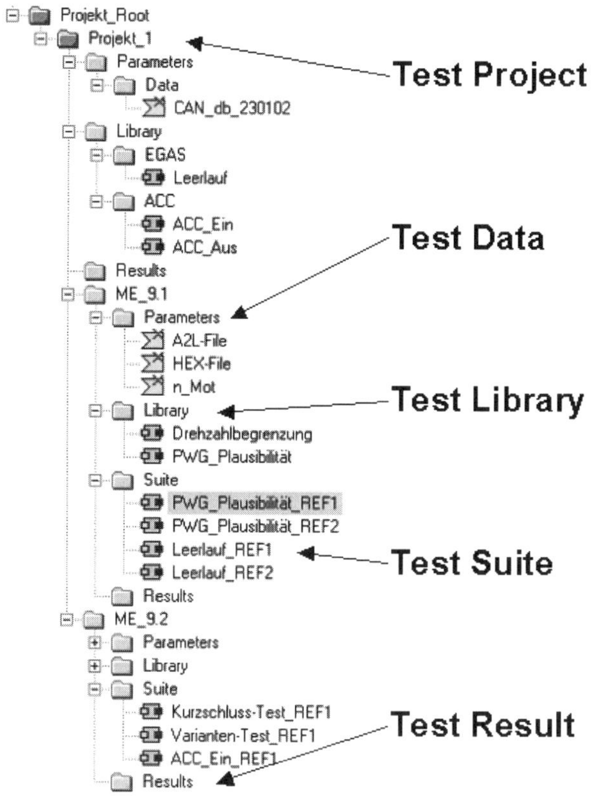

Figure 7:Example of a test project structure

PROCESS INTEGRATION

User Management

A suitable user management allowing to define user profiles is mandatory for a modern development and testing tool. These profiles can be used either to define access rights or user-specific tool settings. If many users are working in co-projects, it is essential to be able to prohibit access to certain development results while others are already released for team use.

A modern user management system also provides features to define a so-called operator version by blocking the test editor, for example, only allowing to aggregate predefined library elements or to parameterize predefined test sequences. Tailor-made tool features are mandatory to meet the different user types, requirements and expectations.

To keep any administration effort as minimal as possible, it is extremely important to interlock the user management of the test environment very tightly with the used version control or configuration management system.

Version Control

It is not only important for liability reasons that tests and test results can be reproduced at any time. This may even happen years after the tests have been done for the first time and the respective vehicle and its electronics system have been released for series production and launched to market. This requires, that each test version has to be stored separately. As soon as a test implementation is modified, this must lead to the creation of a new version of the test. This requires a version management system as the backbone of the project management system and an appropriate interface between the project management software and the version management system.

Applicability Criteria

Since HIL simulation and testing enables simultaneous engineering very well, it might occur that tests are already being developed and performed, while the respective ECU functions, for example, diagnostic functions, are not yet implemented. Therefore, it must be possible to assign so-called applicability criteria to each test. Examples of such test applicability criteria are the ECU's hardware and software versions. If the ECU with a specific software version is not yet defined to implement the diagnostic functions, testing these does not make sense and produces valueless results. But attaching a minimum software version number to the diagnostic tests can prevent them from being executed automatically, as long as the ECU version number is less than the number attached to the test. This is particularly important for large test projects, where tests often are executed all at once and at the same time using batch execution. In such cases, a test operator cannot check the applicability of each test manually.

Requirements Management

To an increasing degree today so-called requirement management tools, like DOORS are applied to collect the ECU requirements. These requirements are normally structured by using hierarchies which reflect the ECU's function or aggregate groups. It is self-evident that structures and names already declared in the requirement management tool must be transferred to the test environment even if the tools or their documents are not directly connected.

NEW ENVIRONMENT FOR THE DEVELOPMENT AND MANAGEMENT OF TESTS

The above described concept has been implemented in a new dSPACE tool named AutomationDesk. To achieve a wide acceptance as well as a broad range of applicability, it is necessary to provide an open software product for customization. But also application-specific functions which need to be integrated lead to the requirement of open interfaces. This allows even the user to implement the newly requested features and thus to extend the functional range of a tool.

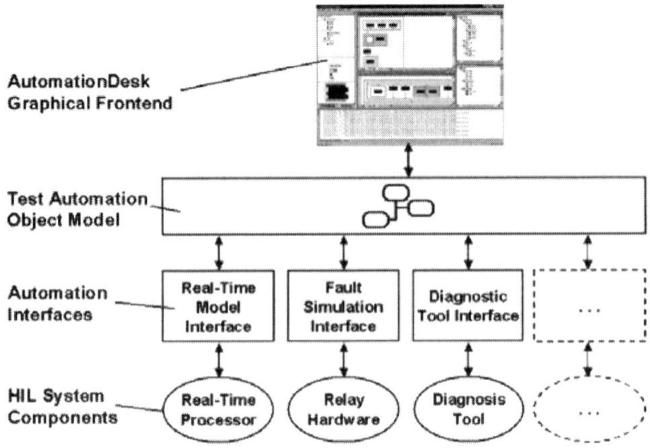

Figure 8:Overall system architecture

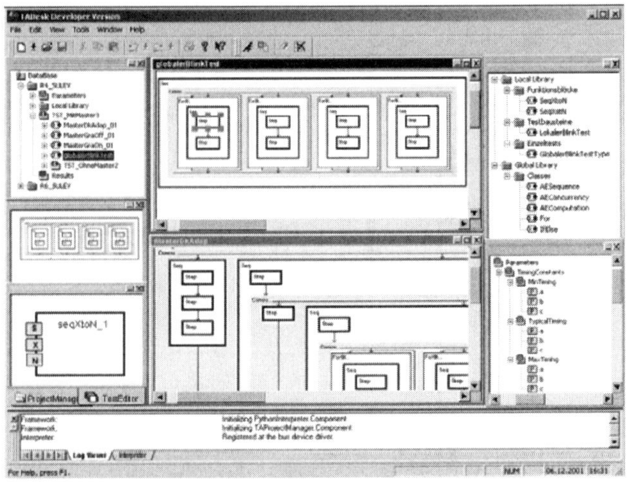

Figure 9:AutomationDesk

Figure 8 sketches the open architecture of dSPACE AutomationDesk. The frontend, providing features for graphical test development and project administration, interacts with the underlying Test Automation Object Model. This object-oriented framework provides various classes representing the graphical objects the user is working with by using the frontend, for example, by dragging and dropping objects. The classes of the Test Automation Object Model build a unifying and normalizing layer and therefore cover the underlying functionality like control structures (for example, sequence, for, while, repeat), data objects and the automation interfaces such as Real-Time Model Access, Electrical Fault Simulation Interfaces or Diagnostic Tool Interfaces.

The user can add new classes to the open framework by extending the Test Automation Object Model on his own. Examples of the need to customize the object model can be the necessity to access other hardware or software tools. Therefore the user can define his own classes, which in the object-oriented sense are derived from predefined abstract classes. After having done so, he can use these custom-specific classes within the graphical frontend of dSPACE AutomationDesk.

Figure 9 shows a screenshot of the graphical frontend. On the upper left-hand side, the project management component and its project tree can be seen. In the middle, two tests are shown being represented by a graphical chart. And on the right-hand side, the library browser is shown, which contains the global built-in library and the global custom libraries.

Figure 10 shows the interrelationship between the sequence editor and the library. Library elements are taken from the library (1) and inserted into any test description using drag & drop operations. After building up a new sequence from existing library elements, this new element can be put back into the library using the same mechanism (2). Afterwards, the new element can be reused in other tests or projects.

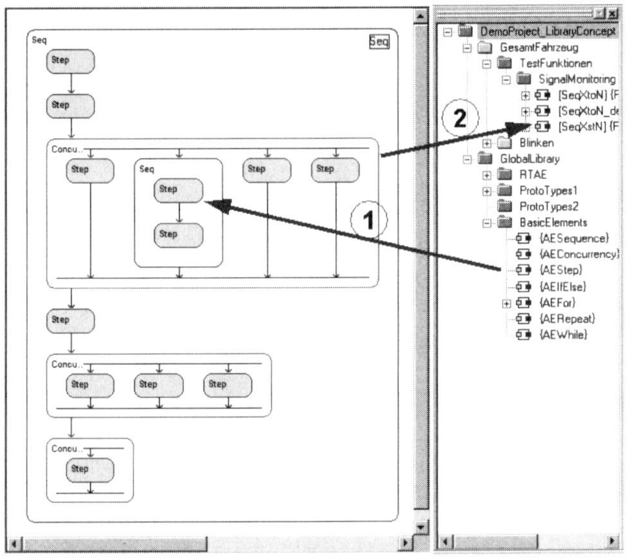

Figure 10:Library mechanism

CONCLUSION

The ever increasing complexity of electronic systems in modern vehicles requires new ways of developing and testing electronic control units and their functionality. It has been shown that automatic testing is becoming increasingly important. This not only holds in conjunction with hardware-in-the-loop simulation, but more and more for earlier development phases. The key terms in this context are model-in-the-loop and software-in-the-loop. However, the growing demand for automated testing in all stages of the development process requires a new approach for developing and managing large test projects over the whole development process.

As a solution, a new integrated environment for test development and management – dSPACE AutomationDesk – has been introduced. The key features of dSPACE AutomationDesk are

- A graphical sequence editor, enabling a fast and efficient way of developing tests.
- An extendable automation library providing the basic functionality.
- A project management component for storing and managing the large amount of tests, test data, and test results in a structured project representation.

Finally, it has been discussed how the new environment can be integrated into existing development processes.

AutomationDesk is being developed in close cooperation with major German automotive OEMs. Since first pilot applications have already been done successfully, it is planned to launch AutomationDesk 1.0 to be used in productive testing projects starting mid 2003.

REFERENCES

1. Schütte, H.; Plöger, M.; Diekstall, K.; Wältermann, P.; Michalsky, Th.: Testsysteme im Steuergeräte-Entwicklungsprozess. Automotive Electronics, pp. 16-21, March, 2001
2. Lamberg, K.; Wältermann, P.: Using HIL-Simulation to test Mechatronic Components in Automotive Engineering. 22. Tagung Mechatronik im Automobil, Munich, Germany, 2000
3. Amorim, J.: Renault – Validation of Powertrain ECUs. DSPACE News 1/2002, Paderborn
4. ASAM: http://www.asam.net/docs/MCD-18-3MC-SP-R-020101-E.pdf
5. Boot, R.; Richert, J.; Schütte, H.: Automated Test of ECUs in a Hardware-in-the-Loop Simulation Environment. CACSD, Kona, Hawaii, August 22-27, 1999
6. Gehring, J.; Schütte, H.: A Hardware-in-the-Loop Test Bench for the Validation of Complex ECU Networks. SAE 2002, Detroit, USA
7. Lemp, D.: Opel Vectra Heading for its World Premiere. DSPACE News 1/2002, Paderborn
8. Gühmann, C.; Riese, J.: Testautomatisierung in der Hardware-in-the-Loop Simulation. VDI-Berichte Nr. 1672, Germany 2002
9. OMG: http://www.uml.org
10. Softing AG: Diagnostic Tool Set Product Information.
    http://www.softing.com/en/ae/products/dts.htm, Munich, Germany, 2002

CONTACT

Dr. Klaus Lamberg is responsible for the product strategy, product planning, and product launches in the area of hardware-in-the-loop simulators and test-automation at dSPACE GmbH, Paderborn, Germany.

E-mail: klamberg@dspace.de
Web: http://www.dspaceinc.com

SOFTWARE SOURCE CODES

2005-01-1665

Verifying Code Automatically Generated From an Executable Model

Cheryl A. Williams, Michael A. Kropinski, Onassis Matthews and Michael A. Steele
General Motors Powertrain

ABSTRACT

Currently in the automotive industry, most software source code is manually generated (*i.e.*, hand written). This manually generated code is written to satisfy requirements that are generally specified or captured in an algorithm document. However, this process can be very error prone since errors can be introduced during the manual translation of the algorithm document to code. A better method would be to automatically generate code directly from the algorithm document. Therefore, the automotive industry is striving to model new and existing algorithms in an executable-modeling paradigm where code can be automatically generated. The advent of executable models together with automatic code generation should allow the translation of model to code to be error free, and this error-free status can be confirmed through testing. A three-stage process is presented to functionally verify the model, functionally verify the automatically-generated code, and structurally verify the code.

INTRODUCTION

Software source code that is created to satisfy an algorithm specification is intended to be error free. After an algorithm specified as an executable model is determined to be functionally correct, source code can be automatically generated from the model. Ideally, the automatic translation process would be error free, but this should be confirmed through testing the source code against the original model. Automatically-created test cases can facilitate this confirmation.

A three-stage process can be used for verifying automatically generated code against an executable model. In Stage 1, test cases can be used to determine the correctness of an executable model. In Stage 2, the test cases established in Stage 1 can be executed against the code that is automatically generated from the model. Finally, in Stage 3, new test cases can be automatically generated for the executable model and then used to test automatically generated code. Figure 1 depicts the typical three-stage process tool chain. This paper will describe each stage generally, followed by specific usage employed by General Motors Powertrain.

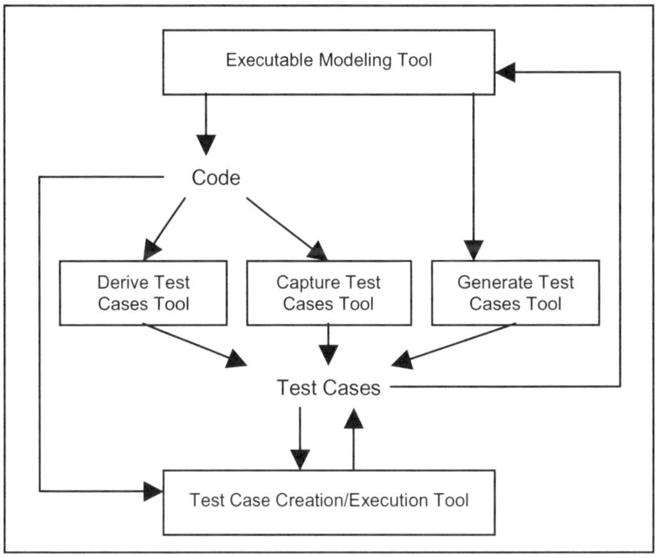

Figure 1: Three-Stage Process Tool Chain

STAGE 1: VERIFY EXECUTABLE MODEL

As mentioned previously, the automotive industry is currently creating executable models from existing algorithm documents. One of the reasons for creating these executable models is so that they can be eventually used to automatically generate source code. Therefore, it is imperative that newly created executable models accurately reflect the original algorithm document.

In general, the functional correctness of the executable model can be determined by executing test cases against the model. These test cases can be manually generated, automatically captured, or automatically derived.

MANUALLY GENERATED TEST CASES

When existing hand-written code for an algorithm exists, one way in which the correctness of the executable model of the algorithm can be established is to take test cases that have been manually generated for testing existing hand-written code and use them to verify the correctness of the executable model. In most executable modeling environments, it is relatively easy to take these manually generated test cases and execute them against the model.

General Motors Powertrain currently tests much of its software using a tool known as RiBeTT. RiBeTT is an acronym for Ring Behavioral Test Tool. The term "ring" refers to a software subsystem, where each ring is responsible for providing one or more of the behaviors required in the target embedded controller. General Motors Powertrain software architecture is made up of these rings. RiBeTT, in general, "enables users to define software behavioral tests; it executes the defined tests on user-selected software; it captures, displays, and reports the resulting software behavior, as directed by the user; and it allows the user to configuration manage the test definition, the software under test, and the test results" [1].

The test cases executed in RiBeTT are typically manually written and behavioral in nature. Thus, they can be used to ensure that the model, behaviorally, is an accurate representation of the existing code.

AUTOMATICALLY CAPTURED TEST CASES

Another way in which the correctness of the executable model can be established (when there is existing code) is to use test cases captured from the execution of the code in its target.

General Motors Powertrain currently uses data logging tools, such as INCA® by ETAS GmbH, to capture the input-output relationship for a ring that is running in a target embedded controller. These test cases are then imported into MATLAB® to be ex ecuted against the model.

AUTOMATICALLY DERIVED TEST CASES

Yet another way in which the correctness of the model can be established is to automatically generate test cases from the manually generated source code and execute these against the model.

Although General Motors Powertrain has not investigated this method, a tool to automatically create such test cases from code would be a useful addition to the industry s tool suite.

STAGE 2: VERIFY AUTO-CODE

One of the main reasons for creating executable models is that they, in turn, can be used to automatically generate source code. The test cases used in Stage 1 can then be directly used to test the automatically generated source code.

General Motors Powertrain is automatically generating source code from MATLAB® (Simulink®/Stateflow®) models using Real-Time Workshop® Embedded Coder. This automatically generated code is then tested with RiBeTT using test cases from Stage 1. In addition, General Motors Powertrain is currently investigating the idea of using the MATLAB® environment to execute the test cases, in parallel, against both the model and the automatically generated code.

STAGE 3: AUTO-VERIFY AUTO-CODE

Ideally, the process of verifying the code generated from the executable model would be automated. If the test cases comprising the basis for comparison can be automatically created from the model, the correctness of the generated code can be assured with a minimum of effort during the iterative model revision process.

Structural tests created from the model can be used to perform black-box functional testing of the generated code. If the coverage of the model is complete, and the execution of these tests on the code yields no errors, then the correlation of the generated code to the model is good, and the confidence level in the error-free status of the code is high. The use of structural tests generated from the model presumes that the model has already been verified to be functionally correct. Thus, any errors detected using this process may be attributed to inaccuracy in the code generation process, or inconsistencies between test environments.

One way to do this is through the use of an external tool specifically designed for test case generation. General Motors Powertrain has completed a case study detailing the iterative process of creating test cases from a model, and using these test cases to verify that the generated code corresponds to the executable model. A summary of the case study and the investigated model is presented here. In the example, the executable model is again specified in the MATLAB® environment. The tool used to create the test cases is Reactisfi, a s oftware program by Reactive Systems, Inc.

At the time the following example was developed, the Reactis® tool suite lacked certain features that have since been added or improved. Therefore, quite a bit of model preparation was required. Using more recent versions of the tool, the necessity for the user to perform many of these tasks has been eliminated, or the tasks themselves simplified. The full tasks are presented here for the sake of completeness.

MODEL PREPARATION

Several steps were required to prepare a model to be used to derive test cases. These steps were performed in the executable modeling environment (MATLAB® in this case).

<u>Model Preparation Procedure</u>

The following procedure was developed to prepare a model for the automated test case creation process.

The procedure includes several steps designed to address model issues described in greater detail below.

1. Open the executable model for editing.
2. Attach any data files to the executable model in such a way that when the model file is referenced, all requisite data files are loaded as well.
3. Isolate the algorithm (*e.g.*, delete the plant).
4. Place constant parameter blocks as the source for any input signals that should remain constant during each test case. Ensure that initial values for these parameters are stored in the data files referenced in Step 2.
5. Zero the values of any input signals that are not consumed in the model.
6. Place an input port (data type "auto") as the source for each remaining input signal.
7. Pass any input signals requiring rescale through gain blocks with the appropriate scaling factors.
8. Pass the input signals into the algorithm portion of the model.
9. Pass all output signals through any required type conversion blocks (*e.g.*, to align scaling if desired).
10. Simplify model execution order and select a fixed-step solver. For example, if there are two processes in the model, one periodic at 100 ms and one asynchronous (variable period, anticipated to be faster than 100ms and of higher priority), the asynchronous process might be rescheduled. The task might be changed to execute periodically at a rate that is a factor of 100 ms (*e.g.*, 50 ms) and give this task higher priority.

<u>General System-Level Input and Output Signals</u>

One requirement for an executable model to be processed by Reactis® was for all system-level input and output signals to be represented at the highest level of the model using input and output ports. The original model used in this example was comprised of a plant driven by a control algorithm in a feedback loop. Therefore, the original model needed to be modified by deleting the plant, and connecting the control algorithm input signals to named input ports.

The following three categories of specialized inputs needing particular consideration were identified.

1. Constant Input Signals

The values of these input signals must change only once per test, as opposed to once per time step in each test. For example, the input representing the number of cylinders in an engine may be an input variable that was initially set equal to a calibrated constant value. Changing the value of this input on sequential time steps should not be allowed. Instead, this value should only change from test to test in order to obtain better coverage of the model. As a result, input signals falling into this category had to be connected to constant parameter blocks instead of top-level input ports. The values of these parameters could then be set once at the start of each test.

2. Unconsumed Input Signals

Signals of this type might have been input to the model for a notational purpose, but were not consumed anywhere in the model. No benefit would be derived from varying such input values. Therefore, they were simply connected to zeroed constant value blocks.

3. Rescaled Input Signals

These input values were provided at the interface with a different scaling than that expected by the model. Since a rescale was expected, this was accomplished by inserting an appropriately scaled gain block between the input value and control algorithm.

TEST CASE CREATION

At this point, a prepared model would be ready to import into the test case creation tool. While most of the work in this step could be performed in Reactis®, several tool deficiencies at the time required workarounds involving further revision of the prepared model in MATLAB®.

First, the range and resolution of the input values were determined. Inspection of the existing model documentation, data dictionary entries, and legacy code was used to establish these values. Ideally, this process step would also be automated.

<u>Model Interfaces</u>

The interfaces to the example model were provided in the legacy code using fixed-point data types exclusively. However, the executable model used floating-point types. If the full range of all floating-point values were specified as allowable input values, this would not correctly indicate the resolution of these values. The version of Reactis® used did not yet support the specification of fixed-point resolution for input values, but it did allow the individual enumeration of all allowed input values for a given signal.

For example, if the model were to consume a variable with range [0, 8192) and resolution 0.125, one option was to specify to the test case creation tool that the allowed values, explicitly, were {0, 0.125, 0.25, 0.375, ..., 8191.75, 8191.875}. Explicitly specifying all 65536 allowed values did correctly work, but increased the overhead of the test case creation process such that it became prohibitively lengthy.

Alternately, the range could be set to the continuous range 0-8192, with all double-precision values in that range allowed. This might be suitable for general model execution and testing, but would be too imprecise for using the resulting test cases to verify the automatically generated code.

Instead, the technique employed was to specify the full floating-point range as allowable input values, and then place unit conversion blocks on the input signals in the executable model itself so the input signals would be converted to the closest fixed-point value (of appropriate scaling) before reaching the control algorithm. The outputs of these quantizer blocks, in addition to feeding the control algorithm, were connected to output ports so the converted values of the input signals could be recorded (albeit as an output of the model at each time step). These values could then be post-processed in order for them to be used when executing the test cases against code.

A sample unit conversion block for range: [0, 256) and resolution: 0.003906 ap pears in Figure 2.

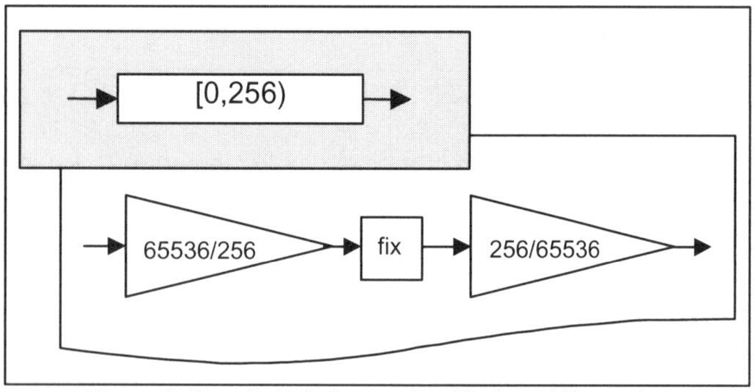

Figure 2: Sample Unit Conversion Block and Expansion

Data Type Consistency

An important requirement in the test case creation phase concerned data types in the model. Not only was it necessary that the data types of the input signals be accurately represented, but also that data type propagation through the model match the originally specified operation of the executable model. However, at the time this case study was completed, the test-case creation tool employed was incapable of supporting single-precision floating-point math. The version of Reactis® employed did not yet allow the specification of single-type input values, nor did it support simulation of single-precision math operations. The example model used single typing. It could still be executed in Reactis® and test cases generated, but math would be performed using double precision instead.

Therefore, two new tasks were required in order to create test cases with single-type inputs, and output values reflecting data type use consistent with that specified in the original model.

The first involved additional modification of the executable model. Explicit typecasts to single type were placed on any input signals whose input values would be specified using double type instead of single type.

The second introduced a post-processing step after test case creation. This step is detailed in the section describing execution of the automatically created test cases.

Model Coverage

At this point, test cases could be automatically created for the example model. Test cases were inspected and iteratively improved until the desired level of model coverage was achieved. Various coverage metrics could be monitored to determine the completeness of a set of test cases. These included subsystem, branch, state, condition, and transition coverage, as well as MC/DC coverage. For the model under test, a test suite with 5 tests and 111 total test steps (20, 5, 39, 42, and 5 steps for each test) was eventually created. Coverage statistics for this test suite appear in Table 1.

	Total	Number Covered	Number Unreachable	Number Uncovered	% Coverage
Subsystems	15	15	0	0	100%
Branches	289	228	4	57	80%
States	0	0	0	0	100%
Condition Actions	0	0	0	0	100%
Transition Actions	0	0	0	0	100%

Table 1: Reported Model Coverage

Note that several measures of coverage can be used to determine how completely a given set of test cases covers the executable model. These test cases can be viewed as a set of structural tests that should exercise all branches, loops, etc. in the model. Only then can the test cases be declared to be complete.

In the coverage statistics presented in Table 1, the branch coverage appears to be incomplete at 80%. However, this was still considered to be sufficient completion. The only parts of the model remaining uncovered corresponded to the condition of exceeding the upper and lower limits of the axes of lookup tables. For example, if a lookup table had an axis with values ranging from 0-6400, but the input signal corresponded to an unsigned model input value that was only allowed to range from 0-8192, the lower limit of the axis would never be saturated. This was reported as an uncovered part of the model in the coverage statistics, although the result was acceptable.

TEST CASE EXECUTION

When the automatically created test cases were ready to be executed against the automatically generated code, they could be exported from Reactis® in a v ariety of formats. The format chosen by General Motors Powertrain for this case study was a text-based native MATLAB® format that specified input and output data on a vector-per-signal over time basis, arranged by test step. In addition to being easily readable by the modeling tool, it was a format easily parseable into the XML format required for import into RiBeTT, the code test tool. This conversion between non-standard formats was necessary in the absence of an industry standard for test case file format.

Test Case Post-Processing

As discussed in the section on type propagation, post-processing of the test cases was required to update the expected output values to correctly reflect the propagation of any single-precision model typing. This was done in the original modeling environment, MATLAB®. A script was written to execute the exported test cases against the executable model using all type requirements originally specified in the model. The script then collected the exercised model output values, and updated the expected output values originally reported in each test case. The new values could be recorded using hexadecimal representation or engineering units.

Better type support in the test case creation tool would have eliminated the necessity for this step. However, it might still be desirable to re-execute the test cases in the original model environment to eliminate variations in output value caused by model interpretation differences.

Using the newly collected output values, the script then updated the "expected output" values originally reported in each of the test cases, with the option of recording these values in hexadecimal representation instead of engineering units. While better type support in the test case creation tool would have eliminated this step, it may still be desirable to execute the test cases in the original model environment to increase confidence in the precision of the reported expected output values.

Test Case Format Translation

For execution of the test cases on the automatically generated code, RiBeTT requires that the test input values be specified using the same data types provided by the legacy code interfacing with the algorithm. The inputs to the example model were provided using fixed-point data types in the legacy code, so the test input values were expected to be provided using fixed-point types. However, the automatically-generated test cases specified floating-point model input values. These could not be copied directly into the RiBeTT XML test case file. Instead, the fixed-point approximations of the model input values determined by Reactisfi ha d to be specified. This could be accomplished either by quantizing each input value as it was being copied into the XML file, or by simply copying the converted values reported as outputs of the unit conversion blocks. Recall that in the test case creation phase, these values were also recorded as outputs of the model, specifically for this purpose.

Test Case Execution

The automatically created test cases were then ready to be executed in RiBeTT against the automatically generated code. What started as structural tests of the model designed to achieve 100% model coverage, now become the input and expected outputs for performing functional testing of the code. The goal was 100% conformance of the reported code results to the expected results; if this was not the case, corrections could be made and test cases re-executed.

TEST CASE RESULTS

Problems were found at two phases in this case study. First, problems were discovered during the test case generation phase and the efforts to achieve good model coverage. Second, differences were encountered during the test case execution phase between expected results and code execution results. The issues discovered fell into one or more of the following categories: problems in the executable model, problems in the automatically generated code, or problems in the tools used. Issues falling into the former two categories could be resolved through an iterative process of correction, test case update, and code execution. Issues in the last category, however, could merely be noted and taken into account when evaluating the results.

The following examples illustrate sample issues uncovered in the case study, and their categorization.

Model Problem: Constant Table Definitions

The original model contained several two-dimensional arrays serving as lookup tables. However, the input signals used to index into these arrays had been connected in reverse order, so the arrays were transposed relative to the input signals. This was only discovered after the test case creation tool reported that the coverage of these tables was incomplete when exercising the input values within the allowed ranges.

Code Problem: Constant Table Syntax

The automatically generated code contained a function call to a custom legacy code lookup routine for these tables, and the arguments were being inserted into the function call in the incorrect order.

Model/Code Problem: Divide By Zero Behavior

The default behavior in Simulink®, which the generated code emulated, is such that X/0 yields infinity or 'inf,' and 0/0 yields 'Not a Number' or 'NaN.' In contrast, the exception handler on the target microprocessor for the generated code specified that X/0 yield 'max of range' (appropriately signed) and 0/0 yield 0. A protected division operation had to be modeled in order for the behavior of the generated code to meet the specification.

Tool Problem: Precision

Because of underlying differences in the implementations of floating point employed by the three tools used in this case study (MATLAB®, Reactis®, and RiBeTT), the calculation of the same math operation in the different environments occasionally yielded results differing by one bit. For example, the addition of four signals together in one expression might yield different results in MATLAB® and RiBeTT. The precision of intermediate values carried during extended math operations varied by tool, and even by math operation. Some of the differences between MATLAB®(and hence Reactis®) and RiBeTT could be eliminated by replacing compound math expressions with cascaded math functions. Thus, fewer implicit intermediate values were left to create differences. For example, an addition operation "++++" could be replaced with three cascaded "++" operations.

Generally, errors reported due to precision differences between the model and code were quite small; however, some errors propagated throughout the model and could result in differences of several thousand engineering units at the output of the model. Efforts were made to minimize the occurrences of such differences; however, some could not be eliminated. This made it more difficult to evaluate test results in RiBeTT, as tests failing for these reasons had to be justified.

Results

After much iteration through the test case creation, execution, and revision process, the final set of automatically created test cases was used to verify the automatically generated code. Expected output values were specified using tolerances of 0.01 engineering units on all outputs expect those for which this was not possible due to the confirmed propagation of precision differences. Therefore, it was possible to report a high confidence level in the code-to-model correlation and the error-free status of the code.

CONCLUSION

An accurate executable algorithm model is critical for successful development of new and existing algorithms. If a model is determined to be behaviorally correct using Stage 1 of the verification process, it can then become the basis for further code development. If code is automatically generated from this model, it can be verified to be behaviorally correct (Stage 2). However, once a model is known to be behaviorally correct, a more thorough comparison of the generated code to the specifying model can be completed by creating a new, exhaustive set of test cases that exercise the entire model structurally. These test cases are then executed against the generated code. When the set of test cases completely cover the specifying model, and the results of testing the code are good, and the application of these test cases to the generated code yields accurate results, then the code can be found to be not only structurally correct, but also behaviorally correct.

The advantage of automating the test case creation process as detailed in Stage 3 is that this may be done quickly, repeatably, and with known coverage statistics. The test cases can then be used during the development phase of a model instead of only at the completion of such development. This also instills confidence in the accuracy of the automatically generated code. These two advantages allow the opportunity to more confidently use models as the basis for future algorithm development.

Recent improvements have greatly extended both the functionality and usability of tools used to generate code from executable models, as well as tools used to automatically create test cases from these models. Taking advantage of these improvements, the processes described in this paper could be greatly simplified and increasingly automated.

ACKNOWLEDGMENTS

The authors wish to thank Reactive Systems, Inc. for the technical and developmental support provided during completion of the Stage 3 case study and in support of the resulting process.

The authors also wish to thank The MathWorks, Inc. for the continued support in pursuit of error-free code generated automatically from models.

REFERENCES

1. Constance, D., et al., "Software Subsystem Behavioral Testing for Real-Time Embedded Controller Applications," SAE Paper 2004-01-0264.

CONTACT

Cheryl A. Williams
General Motors Powertrain
cheryl.williams@gm.com

2005-01-1360

Automatic Code Generation and Platform Based Design Methodology: An Engine Management System Design Case Study

Alberto Ferrari
PARADES EEIG

Giovanni Gaviani, Giacomo Gentile, Monti Stefano and Luigi Romagnoli
Magneti Marelli Powertrain

Michael Beine
dSPACE

ABSTRACT

The design of a complex real-time embedded system requires the specification of its functionality, the design of the hardware and software architectures, the implementation of hardware and software components and finally the system validation. The designer, starting from the specification, refines the solution trying to minimize the system cost while satisfying functional and non functional requirements. The automatic code generation from models and the introduction of the platform-based design methodology can drastically improve the design efficiency of the software partition, while maintaining acceptable the cost overhead of the final system. In this approach, both top-down and bottom-up aspects are considered and solutions are found by a meet-in-the-middle approach that couples model refinement and platform modeling. In more details, given a model of the implementation platform, which describes the available services and data types, the algorithms captured by models are refined and then automatically translated to software components. These components are integrated with handwritten (e.g. legacy) software modules together with the software platform. A final validation phase on the real target is performed to finally validate the functionality and to guarantee that the performance constraints are met.

The methodology described in this paper has proven in the years of deployment its validity and maturity level. The effective results are the improvement of the time-to-market and the capability to cope with the complexity of modern embedded controllers for power-train. The selected automatic code generation environment (the model compiler) has been instrumental in implementing our model based design methodology.

In the future, the platform based design methodology will allow an easy accommodation of the new automotive software architecture standard promoted by the AUTOSAR consortium.

INTRODUCTION

The design of an engine management system is a very challenging problem in automotive electronics because of the complexity of the functions to be implemented and the real-time and cost constraints. In brief, an engine management system controls a combustion engine, or more recently a hybrid engine (i.e. combustion engine coupled with an electrical motor), to offer appropriate driving performance (e.g. drivability, comfort, and safety) while minimizing fuel consumption and pollutant emissions. The behavior of the controlled system is achieved by actuating several control inputs, such as throttle position, fuel injection and spark ignition that are tightly synchronized with the rotation of mechanical components (e.g. crankshaft and camshaft). Hence, several control algorithms must be designed to correctly control those inputs. Recently, the implementation of these algorithms has been migrated from hardware to software due to the always shrinking time-to-market, the continuously changing specifications and the high hardware implementation cost. Hence, several control algorithms must be executed by one or more computing platforms with hard real-time constraints.

The complexity of the nowadays control algorithms and the tight design dependency between the plant (engine), provided by the car maker, the electrical control unit (ECU), provided by the sub-system maker, and the hardware and software components (CPU, DSP and

RTOS) are such that there is the need of an integrated design chain and of a common standardized platform to fully explore the design space and better utilize the available technologies. The integrated design chain should be supported by a common design methodology and design flow, with a shared understanding of specifications and implementation constraints between all the actors of the design.

In the automotive domain, there are initiatives mainly focused on common standardized computing platforms. In Europe the AUTOSAR (15) initiative is aimed to define a common software platform to enable a sharing of software components and an easy software integration. Similar initiative has been taken in Japan by the automotive industry with the JASPAR initiative. In our understanding these initiatives will provide an important breakthrough in the design of automotive controllers and coupled with an integrated design chain will provide the backbone for the design and implementation of the next generation engine management systems.

In this paper we present the model-based design methodology that has been introduced in Magneti Marelli Powertrain and a gasoline direct injection case study (Ferrari, et al., 2004). The design methodology is based on a meet-in-the-middle (Vincentelli and Ferrari, 1999) approach with a defined set of abstraction layers and successive refinements (Balluchi, et al., 2002)

The requirements of the highest level of abstraction, the system, are expressed in terms of functionality and performance indexes. These requirements are captured by executable models and are shared between car makers and sub system makers, drastically reducing ambiguity, i.e. possible interpretation errors. These models are then refined down to implementation.

The design of control algorithms is a fundamental part of the design flow. It starts from a functional specification and ends up with a detailed description of the algorithms. In the model-based design methodology, the part of the control algorithm that is mapped to the software partition is automatically translated from a model representation to a set of software components. The software architecture of the application will accommodate and compose together those software components such that the real-time requirements are met. In the proposed design flow, the control algorithms are captured using the MATLAB/Simulink (The Mathworks, 2002) design environment and the automatic translation of the model to C-language code is performed with the production code generator TargetLink (dSPACE, 2002), in the sequel called model compiler and described in one of the following sections.

To correctly accommodate automatically and manually generated software components, a software architecture composed of different layers has been totally specified and implemented. The layer closest to the hardware is the basic input output system (BIOS). The upper layer is containing the device drivers that encapsulate electrical drivers of sensors and actuators. The RTOS and communication services are common to all layers. All these layers implement the software platform that supports the application software. The latter has structural and semantic aspects that allow the correct integration of the software components implementing the entire set of control algorithms.

To obtain the final implementation of the power-train controller, several players belonging to different organizations within a company and/or different companies have to cooperate during the design of each component and of the entire system: car makers must provide and share controller specifications, plant models and calibration sets, silicon suppliers must provide performance models of the micro-controllers. A common design methodology and tool chain is the key of success in coping with the design complexity and constraints.

The paper is organized as follows: the first part describes the methodology framework defined by Magneti Marelli Powertrain. The second part describes the TargetLink environment by dSPACE. The third part is dedicated to the description of a real design case. The fourth one indicates future work and conclusions.

DESIGN METHODOLOGY

The basic tenets of the Platform-based Design Methodology as exposed in (Vincentelli and Ferrari, 1999) are:

- Regarding design as a ``meeting-in-the-middle process'' where successive refinements of specifications meet with abstractions of potential implementations;
- The identification of precisely defined layers where the refinement and abstraction process take place.

The layers then support designs built upon them isolating from lower-level details but letting enough information transpire about lower levels of abstraction to allow design space exploration with a fairly accurate prediction of the properties of the final implementation. The information should be incorporated in appropriate parameters that annotate design choices at the present layer of abstraction. These layers of abstraction are called *Platforms*. In this paper, a platform is defined to be *an abstraction layer in the design flow that facilitates a number of possible refinements into a subsequent abstraction layer (platform) in the design flow*. The abstraction layer contains several possible design solutions but limits the design exploration space. During the design process, at every step we choose a *platform instance* in the platform space. Every pair of platforms, the tools and methods that are used to map the upper layer of abstraction into the lower level one is a *platform stack*. Key to the application of the design principle is the careful definition of the platform layers. Platforms can be defined at several points of the design process. Some levels of abstractions are more important than others in the overall design trade-off space. In particular, the articulation point between system definition and implementation is a critical one for design quality and time.

In the proposed approach, five main levels of abstraction are identified: system level, function level, operation level, architecture level, and component level (Balluchi, et al., 2002). Figure 1 shows the design methodology and the five level of abstraction. In each single design step (Figure 2), the platform is abstracted and captured by a platform description (platform model). The function requirements at each level of abstraction are captured (functional model) and mapped to the selected platform to analyse performances and verify the design step. A synthesis step is then performed to generate the particular instance of the components of the platforms.

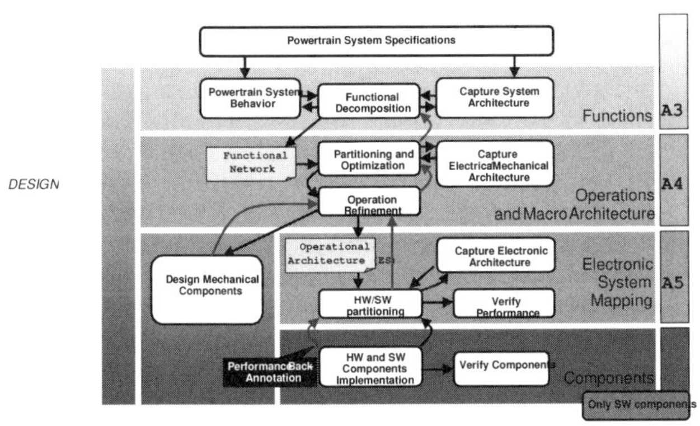

Figure 1 Platform-based design methodology for automotive

The five levels of the platform stack are described in the sequel.

System: car manufactures define the specifications of power-train control systems in terms of desired performances of the vehicle in response to driver's commands. Additional requested specifications, defined by governments or car manufacturers associations, are concerned with fuel consumption, noise and tail pipe emissions. At the system level, the given specifications are analyzed and expressed in an analytical formalism. Specifications have to be clearly stated and negotiated between customer and supplier to make sure that they are realizable within the budget and time allowed for completing the design.

Functions: the design of the functionality to be realized by the control system to meet the system specifications described above is very complex. A good quality of the design is obtained by decomposing the system into interacting sub-systems, referred to as functions. The decomposition allows designers to address the complexity if it leads to a design process that can be carried out as independently as possible for each component. The structure of the functions is the model of the platform at this level of abstraction. System specifications are spread out among the functional components so that the composition of the behaviors of the components is guaranteed to meet the requested objectives and constraints. The output of the functional level design is a desired behavior for each function.

Operation: at the operation level, the desired behaviors have to be obtained, satisfying also some local objectives and constraints. Solutions are expressed in terms of basic building blocks, called operations. In a first design attempt, for each function, control strategies achieving the given specifications are devised and captured with executive models. The control strategies operate on variables that are measured on the physical domain and produce values of variables that act on the physical domain. Then, each control strategy is refined by introducing chains of elementary operations, so that the set of all solutions can be integrated in a unique operations network.

Architecture and Electronic System Mapping: the design step at the architectural level produces a mapping between the behavior that the system must realize (operations) and the platform representing the chosen system architecture, i.e. an interconnection of mechanical and electrical components (e.g., sensors, actuators, microprocessors and ASICs). The set of components either are available in a library of existing parts or must be designed ex novo. This architecture and component-selection task is the subject of intense research by the system design community.

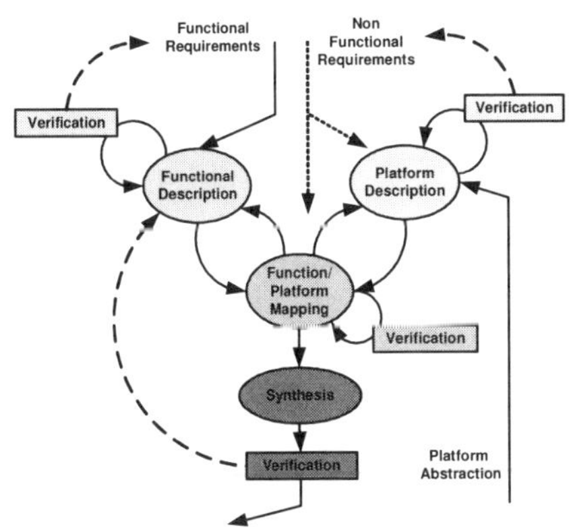

Figure 2 Platform based design methodology

The integration of the components and its validation is performed as done in the classical V cycle design methodology (with usage of models for virtual plants).

At the end of the integration process, the control algorithms are tuned adjusting the control parameters (i.e. model parameters) and final validation of the plant assumptions is carried out. This task is performed in collaboration with car makers and provides back to the control designers fundamental information of the actual behavior of the system and assumptions made during the design.

In the past, the correctness of the algorithms and of the final implementation was validated by prototypes of the target ECU employed during the entire design. The algorithms were frequently described only on documents and C language and the physical validation was detecting, algorithms, coding and architectural errors. This is fundamentally different from the current methodology. In this case the correctness of the final implementation is strictly connected to the correctness of the control algorithms (operations), captured by models from which the implementation is automatically derived. The use of prototype ECUs is limited to the algorithm exploration phase to validate the assumption on the plant and the correctness of the control algorithms. Prototypes are not used anymore to explore the control algorithm and to validate the algorithm implementation. Moreover, since in the past the algorithm exploration and coding validation were carried out mainly at the software level, the final specification of the control algorithm was known only at C level, stability and other analysises were not carried out and the final solution was very difficult to be reused. Instead, in the proposed model-based design methodology, validation starts as soon as the designer conceives the controllers with the use of complex models of the plant, describing the engine, driveline and the driver, or in a simpler way set of recorded input/output traces. If a control algorithm is subject to changes, the model is first modified and then the new software is generated, resulting in a natural synchronization between the model representation of the control algorithm and its implementation. This methodology shift drastically reduces the use of prototypes of target ECUs, resulting in a strong reduction of design time and cost which we can account for around 30-40% in our case study.

MODEL REFINEMENT AND SOFTWARE PLATFORM

To support the methodology, the tool chain must handle the refinement of components from one level of abstraction to another one. Ideally, it should be possible to support, in the same design framework, the refinement of system requirements to control algorithms, and then to software or hardware components. If we consider data refinements, system specification and control algorithms might be provided in floating point notation of quantities, e.g. the quantity of fuel injected in the cylinder, while at the software level this data might be refined to 16 bits fixed point representation.

In the design process, different actors (working even in different companies) will interpret the data in different ways: as floating point or fixed point notation. For example, car maker will provide requirements in floating point notation and will perform calibrations and measurements in a coherent manner, but at the implementation level the power-train controller executes operation only in fixed point data, hence software engineers have to manage software in fixed point notation.

Not a single tool today in the market easily supports refinement in a general meaning. The more advance research project on this topic is the Metropolis project at the University of Berkeley, see (Burch, et al., 2002) and (The Metropolis Project).

In our approach, the algorithm specifications are captured in Simulink and are refined with different models, expressing different level of details, consistently linked by a configuration management system. The verification of the refinement is obtained only via functional simulation.

TARGETLINK: FROM MODEL TO C CODE

The translation of Simulink/Stateflow models to C code is performed by the production code generator TargetLink from dSPACE (Hanselmann, et al, 1999). This model compiler creates the software modules (component level) from refined algorithms as described in the previous sections. Code efficiency is one of the primary requirements for a model compiler. Another important necessity is process efficiency and flexibility. This term refers to the capability of the tool to support and adjust to an existing development process and how safe it can be used within that process. The fulfillment of these and several other requirements make TargetLink a tool that is highly integrateable within the environment of model-based design methodology.

USER-GUIDED MODEL REFINEMENT

As outlined in the previous section, the center focus of working with a model compiler is model refinement. The Simulink and Stateflow models represent the complete functional specification of an algorithm. This specification has to be prepared for code generation. This basically means that data for implementation have to be added to each block and each subsystem of the model specification.

If micro-controllers with fixed-point arithmetic are being used, then variable scaling is the inevitable first step of the refinement process. TargetLink supports the user by providing comprehensive scaling options including automatic scaling support. Automatic scaling helps the user to quickly move from floating point to a fixed point implementation which can then be further refined throughout the process. The user can select one of two different approaches: Automatic scaling calculation based on signal ranges recorded during simulation or scaling derived from value ranges computed based on worst-case assumptions, also referred to as worst-case auto scaling (TargetLink Production Code Generation Guide).

Further data for variable definitions and declarations have to be specified, such as names, storage classes, scope and other attributes. Function and task partitioning takes place, and some model optimizations can be carried out to support efficient code generation. The designer has different options for entering the code generation related data. One of these is based on a data dictionary: a common storage location for data objects that are being referenced in models and used for code generation.

Data dictionaries typically store data objects such as the definition of global variables, OS messages, function interfaces and macros. In traditional manual programming, such data is directly coded in C. This is why data dictionaries are seldom used in manual programming. However, together with a model-based design methodology, data dictionaries are very beneficial. Their major advantages are:

- a common project data source for large ECU projects;
- support for multi-model projects, allowing projects to be spread among different models, with their shared data objects stored in one location;
- protection of intellectual property by a systematic separation of the model-based algorithm specification from the data dictionary-based implementation specification;
- variant handling by supporting multiple values for a single property or by switching complete data dictionaries or branches of one data dictionary in the background.

A dialog-guided model refinement process, utilizing the user interfaces as described above, relieves the implementation specialist from a lot of tedious detailed work. He still specifies the implementation on a bit-accurate level, but does not write C code any longer. This reduces implementation errors and significantly increases software quality.

CONFIGURABILITY OF THE CODE GENERATION PROCESS

Model refinements are not just limited to data typing and fixed-point scaling. Properties that directly impact programming language aspects of the generated code are equally important when code is to be implemented on a production ECU:

- naming conventions have to be followed,
- variables have to be properly declared and put into the right memory sections,
- there need to be efficient ways to link to external code,
- the code output format should comply to company-specific standards.

Generated code for real production projects will always have to interface external code, specifically to components of the lower software layers or to the proven legacy code of the application layer. TargetLink has a wide variety of specification means on the block diagram level to easily interface with non-generated code. In particular, these are:

- inclusion of existing header files,
- use of external global variables,
- use of externally defined macros,
- call to imported functions,
- call to access functions or macros,
- definition of Custom Code blocks which contain hand-written C code.

Related to this is TargetLink's full support of the OSEK/VDX operating system, see (OSEK/VDX, 2001) and (Thomsen, 2002). TargetLink provides an extended library of special OSEK blocks which make the operation of system objects, such as tasks, alarms or critical sections, available at the block diagram level.

The code can be generated in a format that exactly matches company-specific C code templates. Code output formatting is possible through XML and a XSLT style sheet. This allows the user to define the format of file and function headers, the format of code comments and the inclusion of specific header files. Furthermore, TargetLink-generated code complies with the MISRA C standard, see (MISRA, 1998) and (Thomsen, 2002).

The link between the code and the calibration tools is based on parameter description files which are standardized by the ASAM-MCD 2MC standard (formerly called ASAP2), see (ASAM-MCD 2MC, 2000). Modern code generators can all generate this format. Should there be a calibration system in use which applies a proprietary standard, then the user can write his own export filter. This can be done based on detailed information on the generated code which can be accessed via the TargetLink API. Alternatively the TargetLink-generated ASAP2 file could be post-processed within the MATLAB environment to any other format.

SIMULATION-BASED TESTING

Code generators and simulation environments complement each other in an almost symbiotic relationship. An integrated environment, such as Simulink together with TargetLink, allows a variety of significant development tasks to be completed. Simulation results are used for

- automatic fixed-point scaling,
- code testing and verification,
- benchmarking.

Although code generators work virtually flawlessly in comparison to manual programming, the generated code still needs to be tested. The strength of an integrated environment is that code tests are performed in the same environment that was used to specify the underlying simulation model. Functional identity is achieved when simulation results match. The validity of tests is documented by code coverage measurement. TargetLink provides the environment for a three-step verification process, which shows that the model and the generated software components have identical behaviour.

The first step of this verification process is called model-in-the-loop simulation (MIL). It captures the specified behavior of the model by recording block output and block state data to an internal data server. The minimum and maximum values can be used for the automatic scaling of fixed-point data types mentioned before. The traces from MIL simulation are the basis for the subsequent steps.

Software-in-the-loop simulation (SIL) is the next step. Code is generated and compiled with a host compiler and executed in the same simulation environment.

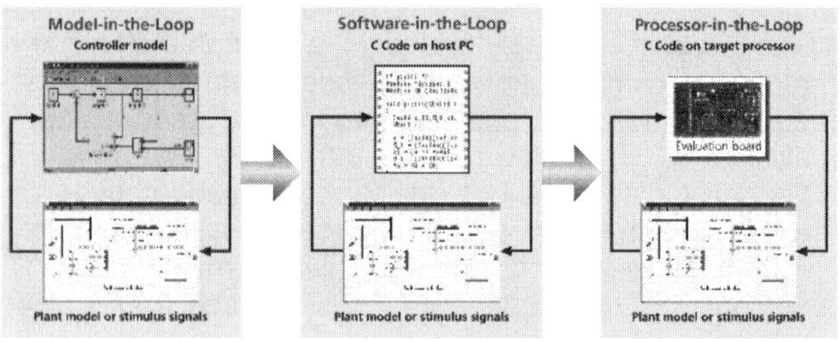

Figure 2: MIL, SIL and PIL simulation modes: a three-step process to verify generated code

Code that runs correctly on the PC can still cause trouble on the target processor. Therefore, the final checks need to be done with processor-in-the-loop simulation (PIL). An off-the-shelf evaluation board equipped with the target processor is connected to the host PC; the generated code is compiled with the target compiler and downloaded to the evaluation board. TargetLink manages communication between the host PC and the processor board.

Performing MIL, SIL, PIL simulation and switching between the different simulation modes is completely automated and does not require any user interaction. In all supported simulation modes TargetLink allows signal traces of block outputs to be logged. These signal traces can be saved and plotted on top of each other, thus providing direct visual feedback and allowing further analysis. This is especially helpful for inspecting quantization effects and verifying the float-to-fixed point transformation.

If plots from the PIL simulation deviate from those in the SIL simulation, then the most likely cause is a problem with the target compiler, or in rare cases a problem with the processor. Both, SIL and PIL simulation, directly support code coverage analysis. The user can select between statement or decision coverage. If the plots match each other and a sufficient coverage of the generated code has been achieved, then the behaviour of the generated software component is with a high level of certainty equivalent with the behaviour of the specification model. This three-step simulation approach is easy, intuitive and quick – and, as a result, it is a safe testing method.

PIL simulation can also be used to profile the generated code and to further refine the implementation. During simulation, TargetLink automatically measures the execution time and stack consumption of the generated C functions directly on the target processor. Furthermore, code summaries list the RAM and ROM usage detailed for each function. These features allow the user to quickly try out implementation options, immediately measure the impact of the change on the generated code, and make logical implementation decisions for the most efficient implementation of a software component.

GASOLINE DIRECT INJECTION CASE STUDY

The model-based methodology has been applied to the gasoline direct injection (GDI) engine control. The most innovative concept of a GDI engine that requires new control algorithm is the ability to inject the gasoline directly in the combustion chamber trough an injector. This capability removes the restriction of introducing fuel into the combustion chamber only when induction valves are open, and as a result a GDI engine has better performance and fuel economy and less pollution than traditional gasoline one. The complexity of a GDI engine resides to the need of a more precise control on the fuel-air mixture and combustion. In particular, the system differs from traditional one for the presence of a high-pressure fuel pump (to inject fuel directly into the cylinder), injectors that support a high pressure flux of gasoline and generate adapted spray pattern (Pontoppidan and Gaviani, 1997) an intake port that generates the desired vortex in the combustion chamber, a more complex treatment of exhaust gas. The engine runs with two different independent combustion modes: homogeneous and stratified. The former being the traditional combustion mode, the latter presenting a non homogeneous air to fuel ratio (A/R) in the combustion chamber.

The high complexity and the presence of innovative control algorithms make this system a perfect case study. Moreover, the strong dependency between the design of the combustion chamber and the design of the combustion control algorithm requires a deep analysis (prior implementation) and a strong interaction, with exchange of models, between car makers and sub system suppliers. The most important component of a GDI engine management system are: the air control by electronic throttle (DBW), variable valve timing or exhaust gas recirculation (EGR), self diagnosis for sensors and actuators and for emission regulations (EOBD/OBDII), safety control, exhaust emission control with Lambda sensor and linear lambda sensor to control the A/R, NoX sensor for NoX trap control, high pressure injector control.

Starting from the car maker requirements, the system has been decomposed and refined into 125 operations decomposed as shown in Table 1.

Table 1 Platform and application breakdown

	Components Models	SLOC	%Model Compiler
PLATFORM	26	26500	0%
APPLICATION	86 with MC 13 HC	93600	90%

A set of 86 operations has been completely modeled and automatically translated to C code via TargetLink, resulting in 94% of the total application code. The application component accounts for more than 2/3 of the total lines of code, while the remaining part is related to the software platform. As expected the system design cycle has been reduced compared to the traditional approach by 20-40%. However, the time of the first design cycle was comparable (or even longer) of the traditional one. This was mainly due to the complexity to harnesses the design process and build the modeling library. The subsequent design cycles have been drastically faster and the final number of design cycles has been reduced. If we consider the number of software engineers involved in the production of the code, the productivity increased by a factor of 4, in terms of line of source code per hour (SLOC/hour). This improvement has been confirmed also in design cycles with strong modification of functional requirements. As a result of this increased software productivity, the ratio between the number of software engineers and control engineers in the design team was lower than in previous product developments.

The model compiler has been applied only to models that are mapped to the application partition. This partition contains mainly new control strategies and some legacy software components implemented by hand. The former have high level of reworks due to the strong interaction with car makers to finalize the requirements and are subject to several design cycles during the system development. The reuse of these components across different products might not be very high.

The software platform defined in the design methodology is instead shared among several products and the variation of the constituting components follows the variation of the hardware components. The platform code has been written by hand since it is highly efficient code and its functionality it is not typically subject to variation during the development time. The strong separation between application and platform and the enforcement of a software architecture also for the platform have been instrumental to manage a variety of ECU configurations and different hardware platforms. In particular, the number of cylinders has been parameterized and can vary from 2 to 6. At the same time, the platform handles the different lists of sensors and actuators needed to support the different engine configurations.

The resulting encapsulation of the hardware platform has been proven by the variations of the set of custom ICs, during the design time that did not require any modification of the control algorithms.

In details, the platform has allowed to manage up to 7 different types of engines and ECUs belonging to 2 different car makers without modifying the control models.

This flexibility is the result of the adopted methodology that encapsulates these variants with the minimum amount of software differentiation. In particular, all the hardware and engine configuration variants have been captured in the lower level of the layered software architecture, respectively BIOS and device drivers, while the software application has been composed with the automatically generated or hand written software components. This flexibility introduced by the software layering has also encapsulated the evolution of the ECU from the first hardware prototype (A) to the start of production.

The model based software components have been also delivered to multi-point injection engine managements, today in productions.

CONCLUSION AND FUTURE WORKS

The methodology described in this paper has shown in the years of use in the GDI product development its validity and the maturity level of the tools. The application to a real product has shown the improvement of the time-to-market and the capability to cope with the complexity of modern power-train controllers. The TargetLink model compiler has been instrumental in implementing our model-based design methodology.

To further improve the cost reduction, a tremendous effort in modeling power-train physical processes and other electro-mechanical components (sensors and actuators) will be required. We feel that in the future, the creation and use of plant models will play a strategic role in the automotive domain.

Other improvements must be done to better cover some important design aspects, such as requirements tracking at the model level, unified framework for refinement and model protection to support the exchange of intellectual properties between car makers and subsystem makers.

An important problem of future investigations is the capability to support the development of ECUs based on models captured with different semantics and tool environments (ASCET-SD).

In the future we expect:

- to have more data related to the process to quantify the advantages of the approach in large scale development;
- to start a formalization of architectural aspects, such as the description of the software platform with architectural description language and UML, based on the AUTOSAR standard;
- to improve the integration in the design chain.

In the near future, we plan to extend the use of the model-based design methodology to other power-train applications and to exploit the new coming features of the new releases of TargetLink.

Finally, the application of the model-based design methodology is expected to drastically decrease the time-to-market of new power-train controllers. In conclusion, the definition of a common design methodology and tool chain is the key of success in coping with the complexity and constraints

ACKNOWLEDGMENT

We would like to thank Alberto Sangiovanni-Vincentelli for the foundation of the platform-based design methodology and A. Balluchi from PARADES for the main conception and contribution to the methodology. Cesare Pancotti, Giovanni Stara, Giovanni Reggiani, Walter Nesci and Paolo Marceca from Magneti Marelli Powertrain for their contribution to the GDI project and Software Architecture.

REFERENCES

1. ASAM-MCD 2MC (2000), Version 1.
2. Balluchi A, et al. (1999). Functional and Architectural Specification for Power--train Control System Design.
3. Balarin F. (1997) The POLIS Approach In Hardware-Software Co-Design of Embedded Systems. Kluwer Academic Publishers
4. Burch J. R, et al (2002) Modeling Techniques in Design-by-Refinement Methodologies In Integrated Design and Process Technology
5. dSPACE (2002) TargetLink http://www.dspace.de
6. Hanselmann, H.Kiffmeier, U.Köster, L.Meyer (1999) Automatic Generation of Production Quality Code for ECUs, In SAE Technical Paper 99P-12
7. TheMathworks(2002) MATLAB/Simulink,http://www.mathworks.com
8. TheMetropolisProject http://www.gigascale.org/metropolis
9. MISRA (1998) Guidelines for the Use of the C Language
10. OSEK/VDX Operating System (2001), Version 2.2
11. Pontoppidan, M., Gaviani, G. (1997) Direct Fuel Injection, a Study of Injector Requirements for Different Mixture Preparation Concepts, In SAE Paper, No: 970628
12. Thomsen T, Stracke R, Köster L.(2001) Connecting Simulink to OSEK: Automatic Code Generation for Real-Time Operating Systems with TargetLink SAE Technical Paper 01PC-117
13. Thomsen T (2002) Integration of International Standards for Production Code Generation In SAE Technical Paper 2003-01-0855
14. Vincentelli A. Sangiovanni, A. Ferrari (1999) System Design: Traditional concepts and new paradigms. In Proceedings of ICCD
15. AUTOSAR, www.autosar.org
16. A. Ferrari, G. Gaviani, G. Gentile, G. Stara, L. Romagnoli, and T. Thomsen. From Conception to Implementation: a Model Based Design Approach. In *IFAC Symposium on Advances in Automotive Control (IFAC-AAC'04)*, April 2004.
17. ASCET-SD, www.etas.de

CONTACT

Alberto Ferrari
PARADES EEIG
Via San Pantaleo, 66
00185 Roma, Italy
mailto:aferrari@parades.rm.cnr.it

Michael Beine
dSPACE GmbH – Product Manager TargetLink
Technologiepark 25
33100 Paderborn, Germany
mailto:mbeine@dspace.de

Giovanni Gaviani
Magneti Marelli Powertrain – R&D Director
Via Timavo 33
40100 Bologna
Italy
mailto: Giovanni.Gaviani@bologna.marelli.it

Giacomo Gentile
Magneti Marelli Powertrain – Design Methodologies
Via Timavo 33
40100 Bologna
Italy
mailto: Giacomo.Gentile@bologna.marelli.it

Stefano Monti
Magneti Marelli Powertrain – Sw Architectures
Via Timavo 33
40100 Bologna
Italy
mailto: Stefano.Monti@bologna.marelli.it

Luigi Romagnoli
Magneti Marelli Powertrain – Sw Architectures
Via Timavo 33
40100 Bologna
Italy
mailto: Luigi.Romagnoli@bologna.marelli.it

2004-01-0677

A Source Code Generator Approach to Implementing Diagnostics in Vehicle Control Units

Christoph Rätz
Vector Infomatik GmbH

ABSTRACT

Implementing diagnostic functionality in automotive ECUs is usually an expensive, time-consuming and inefficient process. Computer-generated source code based on ECU-specific diagnostic data can dramatically decrease costs and development time and increase quality and efficiency.

By considering the weaknesses in the typical ECU diagnostic development process, a clear set of objectives emerges. Defining a source code generation system that addresses these weaknesses lays the groundwork for implementing a successful solution. A case study using the Vector CANdesc (CAN Diagnostic Embedded Software Component) is presented as a proof of concept.

INTRODUCTION

It is common knowledge that the electronic content of automobiles is increasing with each year. ECUs are becoming more complex and their software is too. Some of the most complex software found in an ECU goes to support diagnostic functions and communications. Diagnostics can make up as much as half of an ECU's software. Given the amount of software that goes into implementing diagnostics in an ECU and a vehicle, an efficient implementation process is essential. Reuse of standard software components is a well-known approach to reducing development time while improving quality. Diagnostics software has been slow to move in this direction and still offers opportunities for significant gains.

In order to take advantage of reusable diagnostic software components, it is necessary to examine the diagnostic software structure and identify the parts that are ECU-independent and the parts that are ECU-specific. The ECU-independent parts are reusable across different ECUs and can be written once up front and deployed over multiple ECUs in multiple vehicle programs that share common high-level diagnostic requirements. These ECU-independent parts will implement diagnostic requirements common to all ECUs – like protocol handling, session management and communication control. The ECU-specific parts are not reusable across different ECUs in a single vehicle. These ECU-specific parts will implement diagnostic requirements unique to the ECU – like data access and control algorithms.

This ECU-specific software will not fit the write-once-use-many model of the ECU-independent software, but it can still be done in a more efficient way through a source code generation process that automates as much of the ECU-specific software development as possible. This generated software implements the interface between the ECU-independent component and the ECU-specific data and logic. This ECU-specific data and logic tend to be independent of a vehicle program and can be reused across multiple applications of the same ECU in different vehicle programs – even across vehicle programs for different OEMs using different diagnostic protocols.

TYPICAL DIAGNOSTIC SOFTWARE DEVELOPMENT PROCESS

Most OEMs and suppliers work together in a common diagnostic development process. This development process starts with the OEM documenting requirements and passing these documents off to the supplier for implementation. Some of these requirements documents will specify the OEM diagnostic strategy and implementation requirements common to all ECUs. The remainder of the documents will specify the ECU-specific requirements. From these documents, the supplier will manually develop the diagnostic software for their ECU. The OEM tests the completed ECU in order to verify proper implementation of all requirements – protocol handling, message processing, proper sequencing, data access and ECU control.

TYPICAL DEVELOPMENT PROCESS ISSUES

The typical diagnostic development process is laden with inefficiencies that drive costs up and require large efforts over several months to complete. Each step along the way offers different opportunities for process improvement at both the OEM and the supplier.

PROCESS ISSUES FROM THE OEM PERSPECTIVE

The process begins with the authoring of requirements documents. The diagnostic requirements are captured in a word processor document. Even if a document template is used to help guide the structure and description of the requirements, this authoring process requires a great deal of time and effort and produces a document which is difficult to verify as either complete or correct. Even an excellent document provides limited value as only humans can process the information in this format. Automated processing of such word processor documents is generally not possible.

Once these documents are completed, they are delivered to the supplier. The supplier engineers must read and interpret the requirements documents. A supplier that is familiar with the OEM's diagnostic strategy may comprehend the documents quite well. Less experienced suppliers will not fully understand the requirements on their own and will require significant support from the OEM in getting up the learning curve. The biggest drawback in this situation is that no matter how much attention is given to detail by both the OEM and the supplier, the entire diagnostic implementation must be based on human interpretation. Where documents are even slightly ambiguous or wording is not perfectly clear, opportunities for misinterpretation are everywhere and almost always lead to non-compliant implementations. The worst-case scenario is when ECU-independent requirements leave room for interpretation. In this situation, not only are non-compliant implementations likely, but also come in the form of multiple suppliers having the same misunderstandings and producing the same issues across multiple ECUs.

Once the ECU software is implemented and delivered to the OEM, the OEM must test each ECU for compliance to requirements. Given the process, the OEM has no choice other than to test each and every requirement on each and every ECU. This means redundant and parallel testing of all ECUs. As described above, these tests are likely to identify failures that are shared by multiple ECUs. In the end, many engineers are all working to find and resolve similar issues across multiple ECUs.

Confounding the situation is the possibility that what appears to be a non-compliant ECU implementation may really be a non-compliant diagnostic test system creating an invalid test case. As the diagnostic test system developers must create their software from the same documents as the ECU suppliers, this situation of non-compliant testers can be almost as common as non-compliant ECUs. This leaves the ECU release engineer with the responsibility of validating not only the ECU implementation, but also the implementation of multiple test systems used in engineering, manufacturing and dealership service activities.

Non-compliances in ECUs and diagnostic test systems not only mean failed tests, they mean reworks and retesting. These cycles of reworks and retesting are extremely time-consuming, place a large burden on resources and extend development cycle times.

A correctly implemented ECU is the goal. Despite the issues discussed here, many ECUs do eventually reach a fully validated state. Unfortunately, these ECUs tend to remain islands of quality. There is no means for enforcing the reuse of correct software from one supplier in the ECU of another supplier. Some suppliers are so big that there is no means for enforcing the reuse of software across ECUs from their different divisions. Without reuse, there is no alternative to the endless cycle of repetitive testing and repetitive resolution of the same issues found over and over again.

Unrelated to the issues of human interpretation of diagnostic requirements is the traditional practice of scheduling diagnostic implementation late in the ECU development process. Frequently, diagnostic functions are the last implemented in an ECU. There are two problems with this situation. The obvious problem is that the first time the diagnostics can be tested is near the end of the development cycle when there is little or no room for error and reworks and retesting are difficult to accommodate. The other, less obvious, problem is that diagnostics are an important means of testing all functional requirements of an ECU. Without diagnostics, many basic ECU functional requirements are difficult or impossible to test. As a result, the late availability of diagnostics forces many tests to wait until the last minute, even for functional requirements implemented early in the development cycle. Postponing important testing dramatically increases the risk of identifying significant issues very late in the development cycle and putting program timing at risk.

PROCESS ISSUES FROM THE SUPPLIER PERSPECTIVE

The process issues the supplier must manage are similar in nature to those of the OEM. Development engineers are left to learn and apply requirements spanning numerous documents, from different organizations and different authors; each with their own points of view, sets of terminology and levels of completeness. Incomplete definitions, ambiguous wording, inconsistent terminology and requirements that are just plain assumed and not specified make the job nearly impossible to execute right the first time.

Suppliers familiar with the OEM's diagnostic strategy and requirements have an advantage and may manage their development without significant issues. New or less experienced suppliers face the task with little hope for a smooth experience. Without the opportunity to take advantage of the lessons learned by others, all suppliers must reproduce the efforts of all others. These parallel and redundant efforts are likely to have issues in the same places and require reworks that again must be done without the lessons learned by others.

Suppliers must also deal with the possibility of changes in diagnostic requirements from the OEM. Changes in requirements come in the form of new documentation. These new documents not only mean reworks, but also further increase the opportunity for misinterpretation of requirements that can lead to reworks of the reworks. Such a circle of dependencies encourages the scheduling of diagnostic implementation to occur late in the development cycle. Resource estimation and deployment are especially difficult under these circumstances. Unexpected issues identified late in development magnify the need for effective program management and raises the issue from the engineering level to the supervisory and managerial levels.

DIAGNOSTIC SOURCE CODE GENERATOR

From office productivity software to web applications, most modern software packages profit from including pre-built components. This enables the engineer to focus on the core tasks of the specific application and to reduce the overall complexity. This also appeals to the project manager as the development proceeds faster and risks are reduced.

You can differentiate two different areas of diagnostics in embedded software applications. One area is detecting and handling malfunction situations. The other area is making diagnostic information accessible to the outside world. While the first task is unique to each control unit, the second – the transport of diagnostic information – is a perfect candidate to be implemented in terms of a pre-built component.

Again, the objects to be transported are specific to a control unit. This means that ECU-specific information must be taken into account when creating such a tailored diagnostic component. A C-source code generator that uses an ECU-specific diagnostic database as input lends itself best to this situation.

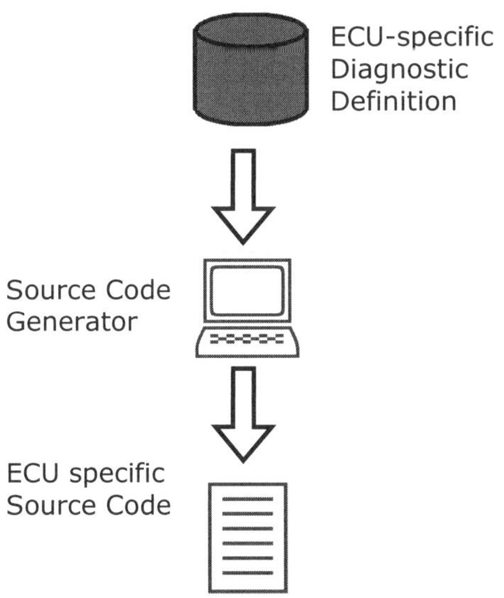

Figure 1: ECU Source Code Generation

The diagnostic component approach has another significant advantage. Different diagnostic specifications from different OEMs can share a common application-programming interface. This interface not only hides the complexity of the extensive ECU-independent diagnostic specifications from the programmer, but also keeps ECU application software design OEM-independent.

A component developer creates a framework that covers the diagnostic specification in detail, but is independent of any ECU-specific requirement. The design and implementation of a diagnostic component is still labor intensive – even more so than the one-time implementation of diagnostics for a single ECU. In addition, the component must be tested thoroughly, in each configuration it can be used. But this is a one-time effort for the reusable diagnostic component and not a job that must be repeated for each ECU.

The OEM defines ECU-specific requirements as input data to the source code generator. These formal requirements replace the current word-processing based documents used today as a basis for software development. However, word-processing documents can still be automatically generated from the input data to the source code generator. The source code generator creates an ECU-specific diagnostic software component according to the OEM's diagnostic requirements.

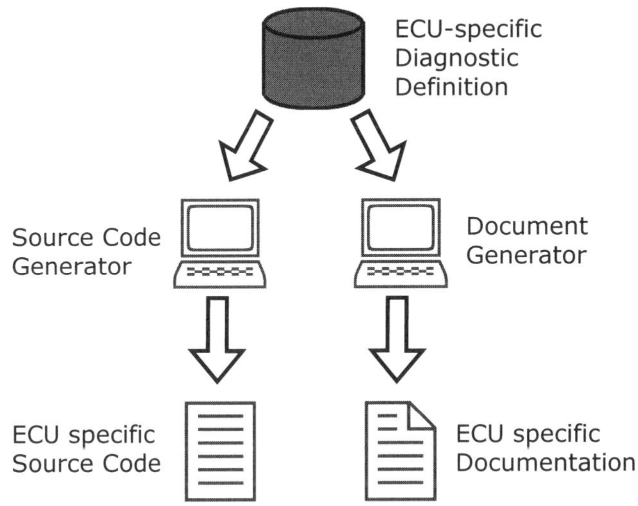

Figure 2: Word-Processing Documents Generation

Next, the supplier develops the ECU-specific parts of the diagnostic software. The complexity of the diagnostic protocol is hidden by the application-programming interface that connects the communications to the diagnostic functions in the ECU. There are three levels of application support:

- Some diagnostic services are completely implemented in generated code. The application is not involved at all. This is possible for all ECU-independent tasks related to communication and session management. Even the tasks of data access can be completely automated, if all application data is directly accessible as global variables.
- Some services are partially implemented in generated code and call the application for help. The component splits down any service request into atomic requests. The application is notified on event when actions must be performed.

- Some services are only handled on the communication level and rely on the application for processing. This level is ideal for low-level services, as it offers maximum flexibility to the application.

The ECU-specific source code fills in the ECU-independent framework to complete the diagnostic implementation. It fills in blanks that the code generator could not know how to fill and manages the interactions with complex I/O ports and device drivers (like off-board EEPROM). Finally, it provides access to any data that is transported to the outside world via the vehicle network. Consequently, the ECU developer can concentrate on the malfunction detection, analysis and reaction.

ADVANTAGES FROM THE OEM PERSPECTIVE

The component developer manually develops a framework that covers all ECU-independent requirements. This is the same process done today by any supplier and the OEM. The main difference is that parallel and redundant efforts for implementation and testing are eliminated. Redundant implementation errors are simply avoided. This also eliminates common and parallel misunderstandings of the diagnostic specifications. When the work is done and the component is ready for use, the same framework is shared by all suppliers.

The OEM must define ECU-specific requirements as input to the source code generator. This formal specification of requirements can be used for much more than just tailoring embedded software components. It can be reused to generate a word-processor specification document and to generate input data to test systems. If an appropriate tool supports the data entry process, the entire diagnostic process will be more efficient than it is today.

Another advantage of software components is their availability. As the software component is typically ready for use at the beginning of product development, the very first implementation contains diagnostics. This improves the quality of the complete diagnostic implementation as late changes are reduced or avoided completely.

In the end, the diagnostic protocol is considered to be just an information transport vehicle – and that's just what it is.

ADVANTAGES FROM THE SUPPLIER PERSPECTIVE

The supplier generates source code for all ECU-specific requirements. He does not need to interpret the OEM's diagnostic protocol specification at all. Inconsistent interpretations and implementations of the protocol are left out of the process, which leads to a significant reduction of development iterations.

The diagnostic component is typically available at the beginning of the project. This helps to predict the processor and memory resource requirements of the ECU. Reliable numbers for the diagnostic functional block are available, the resources requiring estimation are reduced, and project risks are reduced.

The ECU-specific source code is filled in the blanks of the ECU-independent framework to complete the diagnostic functional block. Because the component's interface mirrors the functional requirements, any ECU-specific requirements are insulated from OEM specifications. This means that ECU-specific source code can be reused for other vehicle programs with other OEMs without significant modifications. This structure produces ECU application code that is highly portable and not tied to a specific vehicle program.

REQUIREMENTS FOR EFFECTIVE SOURCE CODE GENERATION

For each manufacturer, the requirements concerning protocol implementation are already accounted for in the ECU-independent framework. But there is a need for ECU-specific information when implementing an effective diagnostic component.

The framework needs to know which services are supported and which sub-functions or identifiers are valid for each service. All other services and sub-functions are rejected by the ECU.

To enable the component to process a request autonomously, detailed information about each service is required. What length is valid for this service request? What information is exchanged? What are the encoding, byte order and length of this information? In terms of ECU-features: which PIDs, DTCs and test routines are supported?

But there is still another aspect to be considered - ECU diagnostic states. For some services, the response to a request depends on the current diagnostic state (like active session and security access). Special services manage and control these diagnostic states (like StartDiagnosticSession and SecurityAccess). In addition, some diagnostic requests may only be executed in certain operational states (like key-on engine-off).

All of this information must be provided to the source code generator in order to produce a diagnostic component that is as reusable as possible.

CASE STUDY – CANDESC

There is already a working implementation of this approach to implementing diagnostics in an ECU. CANdesc, a part of the Vector embedded software family, implements exactly this approach. In its current version, CANdesc works with FNOS, GMLAN and FIATLAN. CANdesc is already used in production ECUs appearing in vehicles from Opel.

CANdelaStudio defines the ECU-specific data. There is no need to memorize any message formats with

CANdelaStudio. An ECU-feature-oriented user interface provides and intuitive view of diagnostic information and allows easy and comfortable editing. The CANdelaStudio data model supports all of the data requirements necessary for effective source code generation as described in this document.

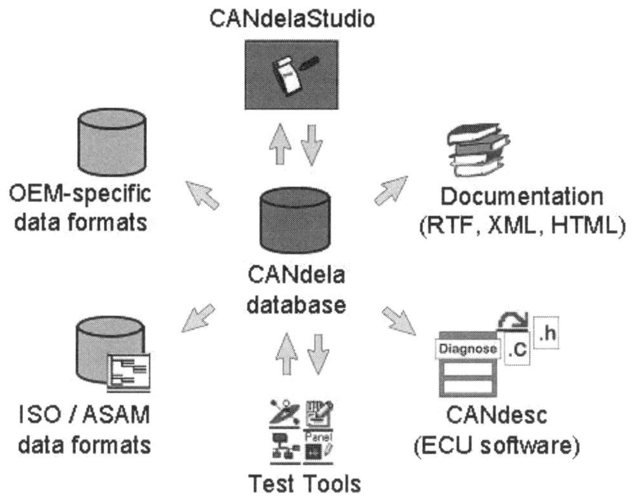

Figure 3: Vector CANdela Unified Database

CANgen is the source code generation tool that outputs the component source code. CANgen does not concentrate on diagnostics alone and generates many embedded software components used in vehicle control units. The CANgen configuration tool is in wide use and generates embedded ECU software frameworks for many different OEMs. The diagnostic source code generator in CANgen exploits the comprehensive data created in CANdelaStudio to produce as complete a diagnostic framework as possible.

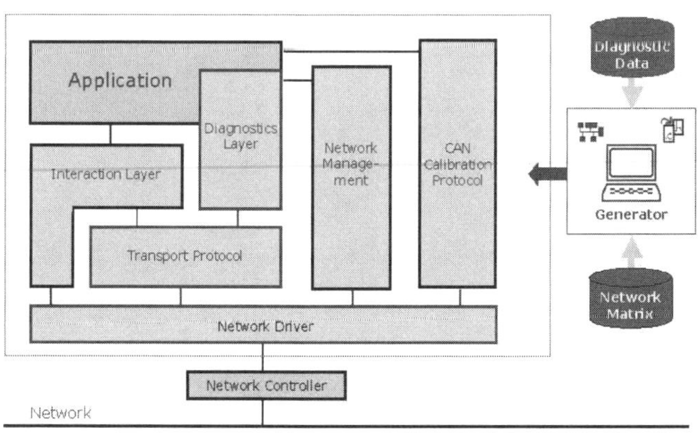

Figure 4: Vector CANbedded Component Architecture

Currently, CANdesc implements three levels of generation: full, assisted and basic. These generation levels mirror the different levels of application support described earlier. There are meaningful default settings defined for each service, but any configuration can be adapted to any ECU-specific needs.

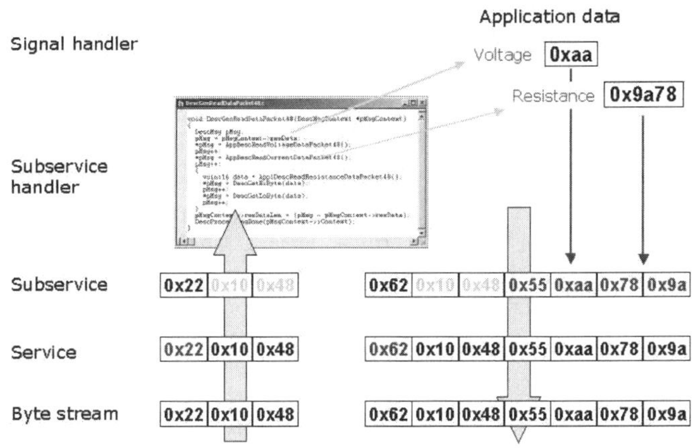

Figure 5: Service processing in CANdesc

Resource requirements are a very sensitive point in automotive embedded software. CANdesc was designed with these points in mind. Exact resource consumption depends on microcontroller and compiler selection and so does the optimal generation configuration, but typical values according to real world experiences are shown here. Please note that these numbers represent a replacement for the resources consumed in a typical diagnostic implementation, not an addition to them.

- ROM: about 8k for an ECU of moderate complexity (like a climate control or transmission unit)
- ROM: about 5k for even the smallest of ECUs
- RAM: about 128 bytes

The experiences with CANdesc are impressive. From the start of the component generation and integration, all diagnostic services are working in a few hours. Most of the features work right on the first try and many features are running by the end of the first day. This rapid development allows for a complete diagnostic implementation before the first bench delivery. The total diagnostic development time is reduced by up to 80%, which corresponds to a savings in effort of about 4 man months.

However, experiences show that there is room to further advance this approach. The next steps will be to provide optional add-ons to the base CANdesc component for management of fault memory and dynamically definable data packets. These add-ons not only extend the scope of the reusable diagnostic component, but also produce significant additional savings by dramatically reducing the amount of ECU-specific source code that must be manually developed in the current implementation.

CONCLUSION

Diagnostic implementation is a significant part of the ECU development process. The diagnostic world has been slow to move in the direction of standard software components for implementing the ECU embedded software. There are opportunities for significant gains in both efficiency and quality.

The malfunction detection strategies and operational diagnostics strongly depend on an individual ECU – these tasks continue to be part of the ECU application. But the ECU-independent part is a perfect candidate to be implemented as a diagnostic framework.

To optimize the implementation, this framework must consider ECU-specific data. The framework source code is to be generated by a computer based generation tool, processing a formal description of all diagnostic requirements. Moreover, this formal description of diagnostic data has significant potential to simplify the complete diagnostic development process. Generated specifications, documentation and tester data replace their hand-made counterparts of the traditional diagnostic development process.

The Vector CANdela product family shows that the concept of computer generated diagnostic software components not only works in real world applications, but also delivers on the promises made by such an approach.

CONTACT

Christoph Rätz, Vector Informatik GmbH

Ingersheimer Strasse 24
70499 Stuttgart
Germany

christoph.raetz@vector-informatik.de

DEFINITIONS, ACRONYMS, ABBREVIATIONS

API – Application Programming Interface

ECU – Electronic Control Unit

OEM – Original Equipment Manufacturer; carmaker.

2003-01-0863

Auto-Generated Production Code Development for Ford/Think Fuel Cell Vehicle Programme

C. E. Wartnaby, S. M. Bennett and M. Ellims
Pi Technology

R. R. Raju, M. S. Mohammed, B. Patel and S. C. Jones
Ford Motor Company

ABSTRACT

Pi Technology and the Ford Motor Company are using MATLAB Simulink/Stateflow model based design and automatic code generation in C, for the main software development for three electronic control units targeted at the Ford Focus fuel cell vehicle.

The automatic generation of code for embedded automotive applications offers a number of potential advantages over traditional methods. These include faster development, the avoidance of coding errors and avoiding inconsistencies with the design specification. However, the use of automatically generated code in production-intent safety-related systems requires at least the same standard of validation and verification. If code generation were perfect, one could validate only the design. However, it is impractical to require that the code generator must be validated for all possible input designs. Furthermore it must be assumed that the compiler and the hardware can also introduce faults. Therefore we adopt the approach of testing output code for the particular designs we wish to implement, in the same manner as we would test hand-written code for production systems [1]. This retains the additional benefits of exposing the design to further detailed scrutiny in test preparation, and encouraging designs that are straightforward to test.

This paper discusses the development lifecycle employed on this project, highlighting the particular benefits, issues, and challenges surrounding the use of automatically generated code for these production-quality safety-related automotive controllers.

INTRODUCTION

The Ford Focus fuel cell vehicle (FCV) is targeted for limited production in 2004. It is a hybrid vehicle with a hydrogen fuel cell providing primary power for an electric drivetrain. Around a dozen electronic control units (ECUs) must interact via CAN to manage the vehicle systems correctly. Of these, three are being developed by Pi Technology to meet Ford's requirements.

They are: the vehicle system controller (VSC), energy management module (EMM) and thermal system controller (TSC). Building three brand new applications and knowing that system requirements were likely to evolve during the lifetime of the project, Pi has attempted a new and powerful development philosophy for these controllers, based on automatic code generation.

Modified versions of an existing Motorola MPC555-based ECU with identical processor cores are used to support the three applications. Pi has taken advantage of this common hardware by writing a custom set of Simulink library blocks, tailored for automotive use, to provide input/output, communications and diagnostic services. The ECU and libraries together make a reusable *platform* for application development[1].

The platform software is a mix of Simulink and traditional C code, and is entirely generic. Any of the three applications can be built on it, as all application-specific functionality resides in the high-level Simulink model. This allows rapid application changes to be made, or entirely new applications to be built and run: one mouse-click turns the whole application model into an executable ready for download.

THE VEHICLE CONTROL APPLICATIONS

The functions of the three ECUs under development for this project are briefly described here.

The VSC is the key supervisory controller in the vehicle. It coordinates the complex startup and shutdown sequencing of the vehicle systems necessary to bring the fuel cell up to a working state and return it to a state suitable for being entirely switched off. It is also

[1] The platform software has been reused and extended in Pi Technology's OpenECU™ project.

responsible for balancing electrical supplies and demands at all times. This central rôle requires that it often acts as a communications hub, with very high CAN message throughput. The VSC also reads key driver inputs such as the three-track accelerator pedal sensor, ignition key inputs, brake pedal inputs and cruise control switches. It computes the driver-requested torque and coordinates the electric motor and regenerative braking control modules to achieve this. Finally, it contains extensive fault mitigation logic to maintain the vehicle in the optimum safe state should any of a variety of faults occur.

The EMM has responsibility for managing the vehicle energy stores. It maintains the high voltage battery state of charge during driving and manages the "sinking" of surplus electrical energy during regenerative braking. It also oversees the hydrogen refuelling and optional electrical recharging operations, and monitors hydrogen levels and leakage sensors. It has a significant number of digital inputs and outputs to support those functions as well as some miscellaneous vehicle device control. It supports a significant amount of CAN messaging on two independent buses, and acts as a gateway between them.

The TSC has responsibility for thermal management. Cooling a low-temperature automotive fuel cell is more challenging than cooling a comparable combustion engine as there is little heat rejection to atmosphere by the exhaust, and a relatively small temperature difference between the heat source and ambient. Heating can also be required for fuel cell start-up. To achieve all this the TSC actively controls a number of fans and pumps, requiring digital and pulse-width modulated outputs, based on numerous analogue temperature sensor measurements. It communicates with the VSC via the main CAN bus.

The VSC is the most demanding of the three applications in terms of computational resource requirements, and so provides the key challenge for the autocode process we have adopted.

I. AUTOMATIC CODE GENERATION: BENEFITS, CHALLENGES AND SOLUTIONS

Our goal was to not only use automatic code, but to do so in an effective way consistent with safety-related automotive systems. This requires both good technical solutions, and good process. In this section we discusses the technical solutions.

WHAT IS AUTOCODE?

Automatically generated code or "autocode" here consists of C language source files that are derived from a higher-level design by a software program. These source files are compiled and linked using a normal compiler to produce an executable software application which can run on target hardware.

The action of the code generator on the models to produce C code is analogous to the action of a C compiler on C code to produce assembly language files. Just as we hope that a C compiler will produce technically perfect assembly files that require no editing by hand, given correct C source, so we hope the code generator to produce technically perfect C files from the design models, if those models are error-free. In each case the human designer is liberated from working in low-level code in favour of using higher-level constructs closer to the application engineering level (though the choice of high-level constructs may be influenced by the lower-level code that results). And in each case we test the output code, to verify that it actually works as intended.

ADVANTAGES OF AUTOMATIC CODE GENERATION

In the formal software development of safety-related systems, Pi produces a detailed design model using some methodology (such as Simulink modelling) even if the code is to be written by hand. So the automatic code generation step saves some labour (and hence time and cost) in translating models to code. Coding is a small but very significant component of a traditional hand-coded development.

The real gain comes when many iterations of design and coding are made, especially in prototyping work, when the proportion of time that would have been spent hand coding increases and so the cost saving of automatic code generation is increased. Many design iterations and extensions have indeed been necessary for the ECUs described here, due to the novelty of the applications and the vehicle system in general.

Furthermore the effects of design changes can be much more quickly assessed and fed back to further design changes using automatic code generation; this is rapid prototyping. The size and complexity of these applications precluded thrashing out a complete design using rapid prototyping techniques alone at the model level, but has allowed quick experimental work on some features at this level by building on the stable base of the formally specified and systematically tested main development work.

There are benefits beyond time saving and cost reduction, however. The automatic generation of code from the design removes the error-prone process of maintaining design and code in synchronicity; we can be sure that the code always corresponds exactly to what is in the design models, with nothing left out and no obsolete code left in.

A further great benefit on this particular project is portability: the application models can be run on different target hardware, so long as a compatible Simulink block library is provided for that hardware. On this project the first prototype applications were run not on the MPC555-based final target hardware, but on PC-based systems using MathWorks' xPC product and our own I/O drivers.

Once the MPC555 target hardware and software platform was available, porting the applications was straightforward. Most of the work involved (e.g. data type changes and multitasking support) was not specific to the hardware, but was required for production code efficiency.

DISADVANTAGES OF AUTOMATIC CODE GENERATION

Using an automatic code generator removes our ability to write the C code exactly as we wish, just as using a C compiler takes away our control of the assembly output (and hence machine code) it produces. Disadvantages include:

- the code may be inefficient in terms of ROM, RAM or CPU consumption;
- the code may not be structured to facilitate unit testing, static analysis or inspection, which are recommended [2][3] for safety-related work;
- the constants and variables may be named or declared in a way that makes the use of a calibration or diagnostic tools to access them difficult, yet such access is vital for automotive development and service;
- the code may make calls to unvalidated library functions which we do not wish to include in the build;
- the difficulty of performing low-level I/O and communications operations;
- significant work may still be involved in combining automatically generated code sections with each other and hand-written code;
- the danger of being "locked in" to particular tools.

These are all problems we have had to address on this project, as detailed in the following sections.

AUTOCODE APPROACH

For this project we have not chosen to autocode sections of functionality and combine these through manually-written "glue" code to interface them to each other and the hardware. To do so would require the storage and maintenance of the resulting autocode, removing much of the advantage in automatically generating it.

Instead we have adopted a fully-blown autocode structure, similar in convenience and power to commercial rapid prototyping systems, yet running on the actual production target hardware with all of the advantages of its low unit cost, production hardware dimensions, connections and signal conditioning, low power consumption etc. Our approach is to carry this "rapid prototype on production target" code through to production by the application of suitable process and controls, so that no porting or conversion exercise is required for production.

The MPC555 microcontroller, running at 40 MHz, is a relatively powerful embedded platform with native floating-point support. This removes the need for processor-specific arithmetic optimizations which have been performed by others [4].

CHOICE OF TOOLS

Pi chooses tools for particular projects on a case-by-case basis. For this project we chose The MathWorks' MATLAB version R12 with its Simulink and Stateflow toolboxes.

By designing application software in Simulink and Stateflow, we can (with some care) achieve several things simultaneously:

- considering it as a CASE tool, we have a formal detailed design from which software can be written by hand if necessary;
- as a code generator specification, target code can be derived from it automatically;
- as an executable design, offline simulation can be performed as part of the design process, increasing the chance that it will work first time;
- as documentation for the software and calibration work, given suitable structure and commenting.

Other than the ability to run simulations, Simulink has the particular advantage over a traditional Yourdon CASE tool of allowing reusable library blocks, which is important for minimising complexity and maximising maintainability. It is also sufficiently mainstream that we are unlikely to be left using an obsolete format, and if necessary could port from it to tools from other vendors.

Several code generators are available for Simulink or similar modelling tools, including TargetLink [4][5], MatrixX SystemBuild [6], ADI Beacon and AscetSD [7].

For this particular project, Real-Time Workshop was selected for the following reasons:

- Pi already had in-house expertise with it, including customisation of the code generation process and adding new library blocks using Target Language Compiler (TLC) code;
- it allowed reasonably straightforward early prototyping work using MathWorks xPC target, before the production target hardware was available;

- after some experiments it was found to be adequately efficient with RAM, ROM and CPU, if not necessarily the best in class;
- it allowed the use of quite "natural" Simulink models for code generation, without extra configuration information or custom block libraries (other than to perform I/O).

We have chosen to keep using the MATLAB R12 release versions of MathWorks tools, as available at the start of the project, even though more recent versions would provide significant benefits for autocoding (using the now separate Embedded Coder product). This is because the R12 version of RTW is adequate for these applications and the production hardware they must run on. However, to port our platform code to a newer release would involve a significant amount of work due to changes in the core scheduling scheme and some changes in TLC code interfaces, which would consume unnecessary project effort. But for future projects we might well adopt a newer release of MATLAB tools.

HARDWARE INTERFACE: THE PLATFORM CONCEPT

The three different ECU applications (the VSC, EMM and TSC) have been implemented on essentially the same hardware, with only minor I/O differences. This hardware is little changed from a previous Ford/Visteon powertrain control module used for a different production vehicle programme.

Pi's *platform* is the combination of this hardware, and software libraries, that allow any of the supported applications to operate. This is similar to a rapid prototyping system. The software libraries take the form of Simulink library blocks which provide access to hardware inputs, outputs and configuration options, together with support code to run a given autocode application. An example library block is shown in Figure 1.

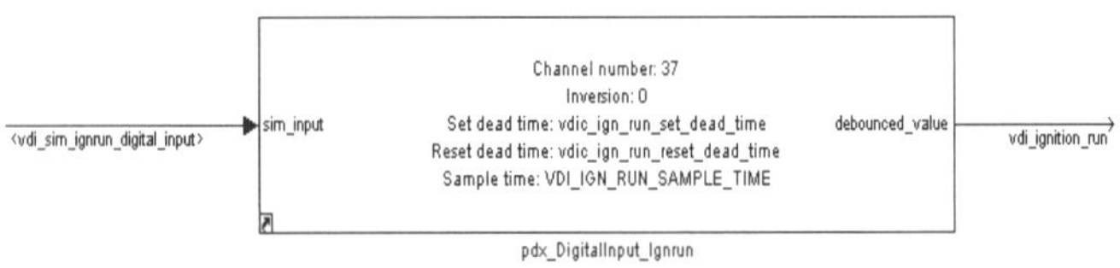

Figure 1: an example platform Simulink library block; this one is used to read a real digital input value into a Simulink-based application.

To take full advantage of the common hardware, and to allow as yet unknown future applications to be supported in the same way with little or no additional investment, a strict partitioning has been adopted between each application and the platform: the platform is entirely generic, with no special knowledge of any of the applications, while all application-specific information is contained only in the Simulink models that constitute that application. Necessary application-specific information such as channel numbers and message ID values is supplied to the platform only via Simulink block parameters as seen in Figure 1.

Currently each application consists only of Simulink/Stateflow models and textual data dictionaries; there is no application-specific hand-written code (though this could be used to optimize portions if necessary). Real-Time Workshop is used to convert each application into C code, and as part of this process calls to hand-written C functions are automatically inserted where platform library blocks were used. Those calls depend on Target Language Compiler (TLC) files which define the automatic code details for our custom blocks. The hand-written platform functions called perform actual hardware input and output activity by accessing microcontroller special function registers. This structure is shown in Figure 2.

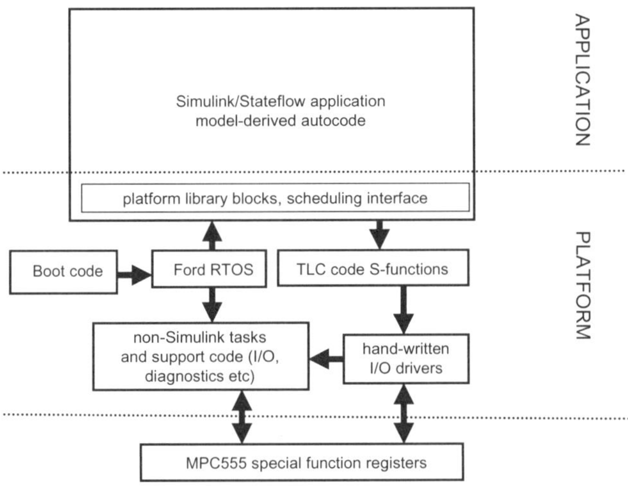

Figure 2: application/platform architecture

Pi decided to use automatic code generation only for the application part of each ECU; that is, validation, control algorithms, high-level decision logic etc. We consider that many lower-level operations, though feasible in model-derived autocode, can be handled far more effectively by hand-written C code. C code is notably stronger than Simulink in any work driven by tables of data or involving bitwise operations, hardware registers or requiring operating system primitives. Note that other authors [8] have attempted to combine application and device driver autocoding, but not using Simulink for the device driver part.

Deciding the boundary between what is done in Simulink and what is done in C code was a delicate balancing act considering code efficiency, design details, and how often each facility was expected to be reused amongst the supported applications.

CAN messaging is particularly important for these applications, so special care was taken in this interface. The platform CAN library blocks are implemented entirely in C code. They perform all transmission, reception, bitfield packaging and unpackaging, and expose convenient message-dependent inputs and outputs at the Simulink level which can be simply written to or read by other blocks. These inputs and outputs are given Boolean, integer or floating-point types as appropriate, to optimise CPU and RAM utilization.

Other operations which we felt should not be attempted at the Simulink model level include responding to diagnostic and calibration tool communications protocols, parsing a serial data stream from a global positioning service (GPS) receiver, and coordinating access to non-volatile memory.

A notable weakness of the Real-Time Workshop R12 release[2] is that it presently generates multiple copies of code corresponding to multiple library block instances, instead of generating several calls to a single function as would be natural and obvious in hand code. This has been tolerated generally but has influenced our implementation in some cases. For example, we have defined a library block which takes an integer value from a CAN message, checks it against special 'error' values, range-checks it, and scales it to floating-point form in engineering units. This was initially implemented as a pure Simulink library subsystem, but had to be re-implemented as C code because the large number of code instances generated by the Simulink version led to excessive ROM consumption. In general any library block that is used tens or hundreds of times has required implementation in C to reduce code size.

On the other hand, some work within platform blocks is actually implemented in Simulink; for example, the validation and filtering of analogue input values is well-suited to Simulink, and the number of block instances in any one application is reasonably small. So while hand code is called to actually read the analogue to digital converter result, autocode is used to process that value, within the platform library block.

CODE STRUCTURE AND SIMULINK/STATEFLOW STYLE GUIDE

A style guide is used to control modelling with the aim of improving code efficiency, maintenance, clarity, consistency and testability.

The style guide directs engineers to adopt certain modelling practices and ban constructs which are deemed unsuitable for some reason. This encourages the bulk of processing, including decision logic, to be performed in Simulink, but using Stateflow charts where appropriate for state-based or complex combinatorial logic. Overall, the modelling style is chosen to coincide as closely as possible with the natural strengths of Simulink and Stateflow, keeping the models as simple as possible overall. This tends to lead to the generation of efficient code, and models that are as maintainable and straightforward as possible.

We have avoided adopting a suggested modelling style which mandates the separation of "control" and "processing", by performing all decisions in Stateflow charts, as this dramatically increases model complexity without any apparent advantage. Nor have we adopted a style which encourages the implementation of most equations and logic in Stateflow chart code. While providing an efficient translation to C code, the latter obviates much of the advantage of using model-based design because the design is almost C code already.

A style is also chosen such that our models could be used, as a fallback position, as designs for hand code should that ever become necessary. The style guide is updated as new issues are found; it provides a distillation of our experience in generating good models.

Particular Simulink blocks that are known to generate unwanted library function calls, which depend on absolute time or which generate particularly inefficient code are banned. For example, the Simulink "initial condition" block is useful for performing special action at the first iteration of the model. However, it generates automatic code with unnecessary function calls and makes use of a problematic 'minus infinity' constant. Using a discrete unit delay (1/z) block fed by a constant (e.g. 0), but with some different initial value (e.g. 1), can usually be used instead, with efficient and straightforward code generated.

Simulink, Stateflow and Real-Time Workshop options are set to improve code efficiency and testability; sometimes there is a trade-off between these aims. For example, the 'atomic subsystem' option is applied to Simulink subsystems to break the generated code into separate functions of a size suitable for unit testing. However, this results in a deeper call tree and limits local variable reuse, increasing stack consumption.

The 'Boolean logic signals' option is selected to represent flags as small integers instead of floating-point numbers, dramatically improving RAM and CPU use. Where appropriate, integers are used for other quantities such as enumerations and "raw" CAN values. Floating-point signals are used for "engineering units" quantities, but single-precision (32-bit) format is used instead of the default double precision.

Other options to control the storage of signals are carefully chosen to minimize RAM consumption. The RTW code is structured such that the same routines are

[2] Later releases of RTW allow "code reuse", but we have stayed with R12 for this project to avoid significant porting work to other core code.

called from different rate tasks, with only part running in each task. With a multi-tasking system this effectively means that any stack space required for local variables is multiplied by the number of rate tasks, each of which has its own stack. It is therefore preferable to encourage RTW to statically allocate storage for signals, rather than using stack (temporary) variables as would be intuitive. Statically allocated signal variables are also essential for unit testing and convenient for access using the calibration tool or debugger. The simplistic and repetitive nature of the generated code makes the effect of compiler options very predictable and their careful selection important.

TASK SCHEDULING

In a typical powertrain control application, a number of tasks must be performed at various periodic rates, while other processing must be done in response to events. Internal combustion engine controllers typically have to perform calculations at particular engine angles, usually by means of interrupt-triggered processing based on a crankshaft position sensor signal. The three applications considered here are simpler than combustion engine controllers in this respect, because all of the required application processing is periodic, linked to the requirement to send outputs as periodic CAN messages. This lack of event-based processing is a very useful simplification for applying automatic code generation, because Simulink models naturally run on a periodic tick or multiples of that period. (Note however that event-based processing is quite possible in Simulink-derived code.)

To allow fast periodic tasks to run on time even in the presence of lengthy slower tasks, a pre-emptive multitasking model was adopted. The existing and well-proven Ford proprietary "PTEC" real-time operating system was used as the basis. Custom code was written to automatically generate a different operating system task for each Simulink-derived model task rate, with priorities descending with increasing period length (rate monotonic ordering) [9].

As it happens, present chronometric data indicate that a single-tasking model would suffice, as the work of all 'slow' tasks is completed sufficiently fast to be inserted between invocations of the fastest (10 ms) code. Adopting a non-preemptive approach would reduce model complexity and RAM consumption. However, a multi-tasking model is maintained to guard against slow tasks growing to the point at which they must straddle invocations of the fastest tasks as development proceeds. See also "Code Performance" below.

The standard Simulink mechanism of specifying sample rates for data source blocks (typically hardware or CAN inputs), with other downstream blocks generally inheriting those sample rates, is used to control which processes are performed at what rates. No custom configuration is used to specify tasks. This ensures that the application, with its various periodic tasks, executes on the target hardware in just the same way as it does in Simulink simulation.

Simulink conventions were also used to ensure data consistency across rate transitions, and indeed are enforced by Simulink.

Non-autogenerated tasks and interrupt-driven processes within the platform are also set up to handle necessary background work and serial communications, as for any comparable hand-coded project.

CALIBRATION SUPPORT

To support development work, the ATI/Vision CAN calibration tool is used. While later versions of MATLAB provide some support for ASAP2, this was not available early in the project. Pi therefore wrote a custom tool to generate an ASAP2 description file. RAM variables (signals) and adjustable calibration constants to which the calibration tool must have access are explicitly named in the model and identified in the data dictionary. At build time the Pi tool parses the data dictionary to produce a template ASAP2 file, the storage type of signals or constants identified as displayable or calibratible in the data-dictionary are set to "exported global", and finally after the model is compiled the tool extracts the addresses from the linker output and completes the ASAP2 file. Adopting this route enables Pi to include other data into the ASAP2 file such as a description, min-max range and resolution.

DIAGNOSTIC SUPPORT

Particular platform library blocks are provided to represent elements that should be accessible to Ford's standard Worldwide Diagnostic System service tool. This interrogates ECUs via CAN to discover which diagnostic trouble codes (DTCs) might be set and to inspect variable values accessible as Parameter Identifiers (PIDs), using a Keyword Protocol 2000 based protocol. This is a specific automotive facility that no ordinary rapid-prototyping system provides.

The platform has no prior knowledge of what DTCs and PIDs are to be included in each application. Instead the application designer includes DTC block instances in the Simulink model, specifying the identifier number as a parameter and feeding in a suitable Boolean flag representing the fault. Similarly PIDs are handled as library blocks. An additional library block allows the application designer to specify the CAN message identities which should be used for diagnostic communication.

CODE PERFORMANCE

It is difficult to compare the resources used by these autocoded applications with hand-coded versions, as they are completely novel, and no hand-coded equivalent versions exist. However, as an indication of resource utilization, some data are presented here for the most demanding of the three applications: the VSC. As explained above, this is the most demanding of the three applications; as an indication, it processes the values of some 1300 CAN messages per second.

The available hardware resources are as follows on the MPC555-based ECU:

Internal flash memory (ROM) for code	448 KB
Off-chip flash memory for calibration data, and future expansion of the code	1024 KB
TOTAL ROM AVAILABLE	**1472 KB**
Internal RAM	26 KB
Off-chip RAM	32 KB
TOTAL RAM AVAILABLE	**58 KB**
PowerPC CPU core with native floating-point support operating at	**40 MHz**

The actual size and timing statistics for two particular releases of the VSC application are as follows; the first is a comparatively early build, while the latter is closer to full production functionality:

	Software Version	
	V1.4	V1.9
Application periodic tasks	7	7
Application features (some rather large)	16	16
Simulink blocks	4715	8000
Non-trivial subsystems	700	1200
Code size	232 KB	330 KB
Calibrations and Constants	23 KB	26.5KB
TOTAL ROM REQUIREMENT	**250 KB**	**356.5KB**
Application/platform RAM	11.5 KB	18KB
Stack/RTOS RAM	15 KB	20KB
TOTAL RAM REQUIREMENT	**25.9 KB**	**38KB**
Approximate average CPU load	**30%**	**45%**
Approximate peak CPU load in worst 10 ms 'tick'	50%	60%

The hardware modules here are specified with plentiful RAM and ROM to allow some contingency, but these data show that quite demanding applications based on this platform could fit in only the on-board MPC555 resources. For our applications and modules, we have headroom to allow for future expansion due to new functionality or changes in build procedures or options to aid testing.

From informal code inspection, we acknowledge that at least some automatically coded portions of these applications would consume somewhat less RAM and fewer CPU cycles if coded by hand; however, the advantages in automatic code generation for this project easily outweigh any potential savings in terms of reducing available hardware resources in production, and we cannot say that RAM would be saved if *all* the code were written by hand. RTW's active re-use of storage may actually give it an edge over handwritten code.

Some preliminary static analysis of the RTW code (using PCLint) did not reveal any significant issues with the code produced, but formal static code analysis remains to be done on this project.

II. DEVELOPMENT PROCESS FRAMEWORK

This section describes how the Simulink modelling and autogeneration of code described above fits into the complete development process for this project. We require not just autocode that runs in production, but autocode which we are confident is suitable for use in a safety-related production system. The development process, and in particular the testing, is vital in achieving this.

Pi Technology projects for safety-related software-based systems generally follow a V-model formal development lifecycle. However, the detailed processes and tools employed at each development stage are tailored to suit the circumstances, safety integrity level [2] and customer preferences of particular projects, as documented by a project-specific quality plan [13]. The FCV project is no exception in this regard. The major difference from traditional Pi projects is the automatic generation of C code from application design models; the usual steps of requirements capture, specification, design and various levels of testing, together with formal change and configuration control, are all present, as appropriate for a safety-related software product. The platform consists to a large extent of hand-written code, for which the development process is comparable with other projects within Pi Technology.

The work for all stages is maintained under configuration control, changes are coordinated through a formal change control system, and all work is peer reviewed.

See also [14] for a general discussion of process and tools as applied to model-based design and automatic code generation.

REQUIREMENTS CAPTURE FOR EACH ECU

Vehicle-system-wide requirements documentation and communications protocol specifications are maintained by Ford in a dedicated DOORS database. For the particular control modules under development by Pi, Ford provided requirements documents early in the project. Subsequently, requirements have been received as written or verbal requests, subsidiary requirements documents addressing particular functional areas, and as Ford change requests. All Ford-requested changes are captured in the Pi change control database and then scheduled for incorporation into phased software releases. Ultimately, requirements are incorporated into the subsystem design specification (SDS) for each ECU.

SUBSYSTEM DESIGN SPECIFICATION (FUNCTIONAL SPECIFICATION)

In the SDS for each ECU, the customer requirements are decomposed (where necessary) into small statements of required functionality which are:

- clear and unambiguous;
- assigned a unique alphanumeric tag to allow subsequent traceability;
- related to their source (person, change request, document reference);
- testable from the external interface to the ECU (including communications messages).

The SDS is largely a textual document, but with tables and diagrams where appropriate. Its main purpose is to provide detailed requirements for design work and functional testing. However, it is also a key component of end-user documentation in that it specifies the externally observable functionality of the ECU.

Pi's Pixref tool [12] is used to automatically generate traceability matrices linking each functional requirement in the SDS to the Simulink model location where it is implemented, and the test script location where it is tested (see 'system testing' below). It works simply by searching for and correlating the incidence of requirement tags in different files to generate each matrix, highlighting any requirements that have not been referenced or obsolete ones that are still present in error.

A separate SDS is written for the platform in isolation, defining what services it provides and the detailed functionality of the library blocks it makes available. This is used to design and test against, and as a user manual for engineers working on the application models.

ARCHITECTURAL DESIGN

A *feature* on this project is a software module relating to some functional area, whose boundary is chosen to keep the content logically consistent, reasonably self-contained and of a size manageable for a single engineer to work on at one time.

For the platform, which is responsible for all hardware interaction and is largely hand-coded, the architectural design document contains detailed explanations and policies spanning a number of areas relating to hardware interfacing, naming, build process, the memory map, etc., as well as partitioning the software into features.

For each ECU application however, the architectural design consists of two components, each of which is brief. Firstly, a textual document describes any overall design policies or the way in which key application-specific problems are solved, and the functional requirements of the SDS are allocated to features. Secondly, an application model is constructed in Simulink which links together the (initially empty) subsidiary models corresponding to each feature. The "wiring" of this model defines the data interfaces between feature models, which is subsequently refined through data dictionary entries for the data flows. See Figure 3.

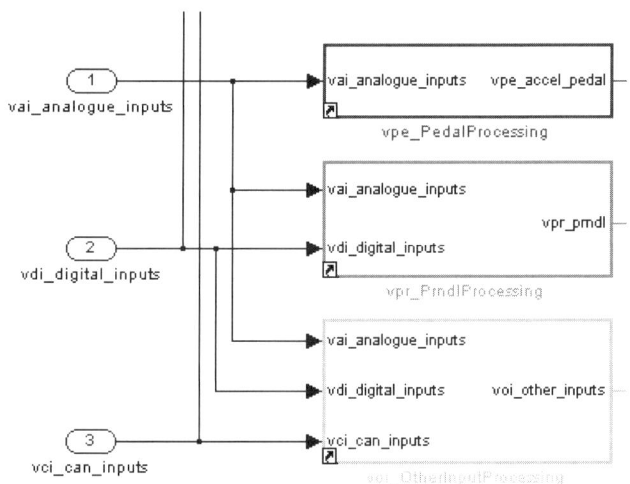

Figure 3: a fragment of the application model showing three feature blocks

A generic application architecture document explains how and why to structure an application in general, to avoid repetition in the architecture documents for the three applications considered here (and any additional applications in the future).

DETAILED DESIGN (FEATURE MODELLING)

Each application feature is constructed as a separate Simulink library model. This is to facilitate configuration management and parallel working within the team, simply by making each model a separate file in the version control system. These are all linked to from the application model, which allows the entire application to be navigated as a whole though the component parts are constructed as separate entities. As explained previously, the style guide is used to give a consistent and effective modelling style. Software requirement tags are included in comments to provide traceability to the SDS, which can be checked using Pi's Pixref tool. See Figure 4.

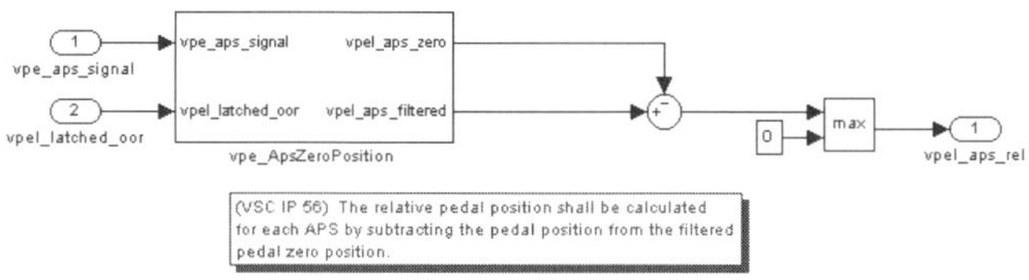

Figure 4: an example model fragment using standard Simulink blocks, subsystems and comments for traceability and understanding

For the platform software a detailed design document is written for each feature, containing a structure chart, an explanation of the design considerations, and pseudocode of (just) sufficient detail to allow subsequent coding and unit testing.

CODING

The process of generating code from the application model is entirely automatic. No modification or integration of the Real-Time Workshop code is performed by hand. The entire application model is simply loaded in Simulink and the 'Build' menu item selected, resulting in a downloadable file in a few minutes. This is of paramount importance in prototyping, development and testing work, where changes can be made at the model level and run on the production-intent target hardware almost immediately. This ensures that all implementation changes are made at the model level, with which the code is rigorously consistent; there is no temptation to "tweak", and then have to maintain separately, the derived code.

Simulink parts of the platform are built at the same time as the application that uses them. Hand-coded C files are written and maintained by hand in the traditional way.

TESTING

As with other Pi software development projects, testing is divided into different activities with different complementary aims and strengths. These are explained in turn below.

UNIT TESTING

To unit test in Pi's terms is to exercise thoroughly small *units* (typically C functions) of implementation, meeting certain structural and data coverage goals, to ensure that those units operate correctly over a wide permutation of input data [1]. The purpose is twofold: to ensure that the code operates in all cases in accordance with the design, but secondly (and perhaps less obviously) to ensure that the design behaviour is reasonable for all cases. Unit testing finds both design problems and coding errors. Where automatic code generation is used, we expect few coding errors, but still expect to find detailed design errors.

Unit testing uncovers errors of detail that are unlikely to be identified in higher-level testing; however, it will not uncover problems that depend on the successful integration of different areas of software, and the actual hardware.

The code generator and compiler are not *trusted*. At present, validated code generation and compilation tools do not exist for Simulink and the MPC555 processor and in any event are not even theoretically possible[3]. Indeed during development some examples of erroneous auto-generated C code have been found. Some issues occur which prevent models building or generated code compiling. However, we have experienced issues in the code including the details of the precise behaviour of enabled subsystems in multirate code and the silent non-support for some bitwise operations.

Our approach therefore is to test the generated code, rather than relying on model-level testing and assuming 'perfect' translation by the tools. This code testing differs little from software unit testing on Pi's hand coded projects, using Pi's test harness generator (THG) to exercise code units in a test harness automatically derived from an input spreadsheet which lists test vectors and provides expected output values. The application models provide an executable detailed design for the code, defining its proper operation. In preparing test vectors and expected results, which are compared with the actual results generated by the implementation, unexpected outcomes from that design are encountered and fixed.

Only a small trial amount of unit testing has been performed so far on this development project, but full unit testing is required for the production releases.

[3] We can never be sure that the specification is correct. No verification system can verify every correct program. We can never be certain the verification system is correct [10].

FEATURE (MODULE) TESTING

The bulk of functional testing is performed at a hardware-in-the-loop level (see "System Testing" below). Many of the system tests are targeted at specific areas of functionality rather than overall system validation, as they must be to achieve high functional coverage. However, simulation-based testing in MATLAB has also employed informally as part of the model-based design process, and sometimes formally to validate small areas of functionality that are inaccessible at the hardware level in a production-intent software build. This fills the potential gap between unit testing and system testing, by detecting defects that depend on the interaction between smaller units but which would still be difficult to detect from the external hardware interface.

Formal simulation tests take the form of spreadsheets which list the feature inputs (including calibration "constant" values) and expected outputs as a function of time. A custom Pi tool, the Simulink Harness Generator (SHG) automatically runs some arbitrary model subsystem through the specified input regime and checks whether the observed outputs are within the defined expected tolerances.

This form of testing allows inputs to be set to combinations that would be difficult or impossible to achieve when running the feature embedded in the whole application, including different calibration value permutations, while allowing time-dependent behaviour to be exercised. The bulk of code and data coverage is achieved through code unit testing, however, as code testing is required anyway to meet safety-related process and coverage goals.

For platform code, feature tests are used to exercise otherwise inaccessible behaviour (e.g. CAN error handling), sometimes requiring intrusion on the code under test using the debugger.

INTEGRATION TESTING

Integration testing here is limited to performing checks that CPU load and memory usage is within acceptable limits such that the application executes reliably on the target hardware. The other testing stages are relied upon to show the proper execution of algorithms and translation of inputs to outputs.

Additional facilities have been included in the platform software to allow the run-time accumulation of statistics relating to task execution times and maximum achieved stack depths. These are vital for application/platform integration testing, as the platform has no prior knowledge of the computational demands of the application, and an application might be specified that exceeds the available CPU resources.

SYSTEM TESTING

The SDS for each application, which provides the definition for system-level testing, does not intrude into the internal software architecture of each ECU. It is written specifically to define only behaviour that is observable at a hardware level. As a result, system testing for these autocoded applications does not differ from that of hand-coded systems, and so this level of testing is entirely conventional for Pi.

System testing detects defects that stem from any software or hardware issues resulting in an ECU failing to meet its functional requirements. Its great strength is that it builds confidence that the ECU actually meets its requirements in real-world conditions, when all of the software and hardware is integrated as a whole. However, system testing can never achieve full code or data coverage as unit testing can. Qualitatively, system test coverage is actually greater for these applications than for a typical combustion engine controller, because of the large number of CAN data interfaces which expose much of the detailed internal processing to relatively direct test scrutiny.

A Pi AutoSim hardware-in-the-loop (HIL) simulator [11] is configured to mimic the rest of the vehicle from the perspective of the ECU under test. It provides 12V digital signals and simulated loads, voltage and resistance inputs, and most importantly for these applications, a full simulation of the vehicle CAN traffic. The system of CAN inputs and outputs is derived automatically from the vehicle protocol specification using a custom software tool written for this project.

In the same way that modelling effort is divided into features to allow parallel working, so system tests are divided into a number of automatically executed *scripts*, each written to test some collection of functional requirements detailed in the SDS. The test scripts set both hardware and CAN inputs, and test hardware and CAN outputs, and also control the ECU power supplies. Traceability to the SDS is ensured by commenting the requirements tags in the test scripts. Pixref is again used to generate a traceability matrix showing where each requirement is tested in some script, and that no obsolete requirements are tested for.

A brief overall test plan explains general testing strategy, documents the procedure for formal test execution and the recording of results, and explains the hardware set-up and organisation of files. However, the detailed thinking behind each test is contained only in the test script itself (and the resulting output files), to minimise maintenance problems and allow parallel working by different engineers. All test scripts are run automatically in sequence before any formal software release as regression tests.

For the platform, special test applications are devised which map inputs to outputs (both hardware and communications signals) in a predefined way. These are then built and tested in the same way as the "real" applications.

For prototype development work, only system tests and model review are employed. Unit testing is generally omitted due to the high rate of requirements change on such a novel project. This achieves acceptable quality for the limited scope of development work by the customer.

CONCLUSION

Together Pi Technology and Ford have successfully developed three novel and challenging production-intent control applications using model-based design and automatic code generation from MATLAB/Simulink. The resulting code currently consumes only a modest proportion of the available RAM, ROM and CPU cycles available on the target Motorola MPC555-based production hardware.

Key elements in the success of the project so far are:

- formal controlled development of requirements, models and tests;
- the use of a modelling *style guide* which encourages consistent, structured and yet straightforward use of Simulink and Stateflow and constructs which lead to efficient generated code;
- investing in the development of a generic *platform* to support these and possible future applications;
- implementing low-level driver functionality using hand-written code, not autocode;
- completely embracing fully automatic code generation for the applications, with no hand-coded adjustments or fixes to the application code;
- automated hardware-in-the-loop system testing.

The development of these applications is presently still in progress, but successive versions have been running successfully on target hardware in a number of test vehicles for many months.

REFERENCES

1. Michael Ellims and Richard P. Parkins, "Unit Testing Techniques and Tool Support", SAE 1999-01-2842.
2. "Development Guidelines for Vehicle-Based Software", Motor Industry Software Reliability Association (MISRA), available from http://www.misra.org.uk.
3. IEC 61508, Functional Safety: Safety Related Systems . Parts 1 to 7.
4. Giuseppe Amato and Lutz K ster, "High Performance Code Generation for Audo, an Automotive micro-Controller from Infineon Technologies", SAE 2000-01-0393.
5. Lutz K ster, Thomas Thomsen and Ralf Stracke, "Connecting Simulink to OSEK: Automatic Code Generation for Real-Time Operating Systems with TargetLink", SAE 2001-01-0024.
6. Walton Fehr, Todd Martin, Robert Lapkass and Danilo Viazzo, "Graphical Modeling and Code Generation for Distributed Automotive Control Systems", SAE 2000-01-3061.
7. Andreas Greff and Torsten G nther, "A New Approach for a Multi-Fuel, Torque Based ECU Concept using Automatic Code Generation", SAE 2001-01-0267.
8. George Saikalis, Shigeru Oho and Steffen Zunft, "Zero Hand Coding Approach for Controller Development", SAE 2002-01-0142.
9. C.L. Liu, James, W. Layland, Scheduling algorithms for multirogramming in a hard real-time environment , Journal of the ACM, 20(1) January 1973, pp 46-61.
10. Beizer, B, oftware Testing Techniques international Thomson Computer Press 1990.
11. Ellims, M. Hardware in the Loop Testing of Embedded Control Software , ImechE Symposium on Engine Control Systems, IEE Control 2000, University of Cambridge 6 September 2000. Available at http://www.pitechnology.com.
12. The Pixref tool will be available for download from http://www.pitechnology.com.
13. Mike Ellims and Keith Jackson, "IS0 9001: Making the Right Mistakes", SAE 2000-01-0714.
14. Scott Ranville, "Practical Application of Model-Based Software Design for Automotive", SAE 2002-01-0876.

CONTACT

The authors can be emailed at charlie.wartnaby@pitechnology.com. See also http://www.pitechnology.com.

DEFINITIONS, ACRONYMS, ABBREVIATIONS

ASAP: Arbeitskreis zur Standardisierung von Applikationssystemen (workgroup for the standardization of application systems, see http://www.asam.net/) (calibration tool description file format)

CAN: Controller Area Network

DTC: Diagnostic Trouble Code

ECU: Electronic Control Unit

EMM: Energy Management Module

FCV: Fuel Cell Vehicle

GPS: Global Positioning Service

HIL: Hardware-In-the-Loop

IEC : International Electrotechnical Committee.

PID: Parameter IDentifier

SDS: Subsystem Design Specification

SHG: Simulink Harness Generator

THG: Test Harness Generator

TLC: Target Language Compiler

TSC: Thermal Systems Controller

VSC: Vehicle System Controller

MISCELLANEOUS SOFTWARE APPLICATIONS

2005-01-2368

Noise Cancellation Technique for Automotive Intake Noise Using a Manifold Bridging Technique

Colin Novak, Helen Ule and Robert Gaspar
University of Windsor

ABSTRACT

Due to considerable efforts of automobile manufacturers to attenuate various noise sources within the passenger compartment, other sources, including induction noise have become more noticeable. The present study investigates the feasibility of using a non-conventional noise cancellation technique to improve the acoustic performance of an automotive induction system by using acoustic energy derived from the exhaust manifold as the dynamic noise source to cancel intake noise.

The validity of this technique was first investigated analytically using a computational engine simulation software program. Using these results, a physical model of the bridge was installed and tested on a motored engine. The realized attenuation of the intake noise was evaluated using conventional FFT analysis techniques as well as psychoacoustic metrics including loudness, sharpness, roughness and fluctuation strength.

While good correlation was found between the numerical and experimental results, additional work is recommended before implementation of a manifold bridge can be considered commercially viable.

INTRODUCTION

Due to the competitive nature of the automotive industry, a greater focus has been given to the increased need for better crash, emissions and acoustic performance of automobiles in the past 10 to 15 years. This has been influenced by the end consumer's demand for improved performance in terms of efficiency, safety, acceleration and comfort. It has been accepted that both the amount of noise generated by a car and the perceived quality of that noise is important. These are both paramount to the satisfaction of the end user. Thus, many challenges exist in refining the acoustic comfort of today's automobiles.

Due to the influence of the many moving parts associated with the operation of today's vehicles, one should not be surprised by the amount of noise that can be heard within the passenger compartment of the vehicle. Sources of this noise include exterior wind noise, tire noise as well as the combustion process in the engine. Unfortunately, given the efforts that have been done to attenuate some of these noise sources, other potential sources such as induction noise have become more noticeable. Studies have shown that approximately 11% of the overall sound level produced by the average automobile is caused by the air intake system. [1]

The traditional method of controlling intake noise is through the implementation of Helmholtz resonators which target specific problem frequencies. Also used are adaptive passive systems, which allow for the resonator volume to vary according to the RPM of the engine. The disadvantage of such systems is that it is becoming increasingly difficult to fit these systems under an already crowded underhood environment.

Given the above, automotive engineers are pursuing new innovative methods for controlling and attenuating induction noise. Active noise cancellation (ANC), which has its own disadvantages, is one such method that has shown promising results in the effort to attenuate intake noise.

It was the objective of this work to investigate the feasibility of attenuating induction noise through a cancellation technique through implementation of tuned exhaust noise feedback through the intake system. Noise cancellation techniques usually involve a computer controlled loudspeaker as a negating noise source to cancel the unwanted noise. In this study, the reduction of the acoustical energy in the intake system is realized by using the engine's exhaust noise, instead of a speaker, as the cancelling dynamic noise source. To facilitate this, a physical bridge between the exhaust and intake manifold of the engine is introduced to allow the transfer of the acoustic energy. This bridge took the form of four individual ducts of diameter greater than 1 inch attached between the exhaust manifold runners and the intake manifold runners. The specific configuration of which corresponding manifold runners were connected is detailed in a future section.

AUTOMOTIVE AIR INDUCTION NOISE

Automotive inlet noise is the result of several parameters including the movement of the intake valves, the physical dimensions and orientation of the manifold ducting and any attached accessories. It is the combination of two processes that result in the noise emission from the induction system.

The first process is the propagation of pressure pulses caused when the intake valve opens to the cylinder. This cylinder has a pressure greater than atmospheric. A second pulse occurs when the intake valve closes. Given the repetition of these pulses, oscillation of intake air at the natural frequency of the inlet passage column results. This noise is further reduced to approximately 80 to 150 Hz due to the influence of the intake system ducting, silencers and air cleaner package. These oscillations with respect to the timing of the inlet valve are illustrated in Figure 1.

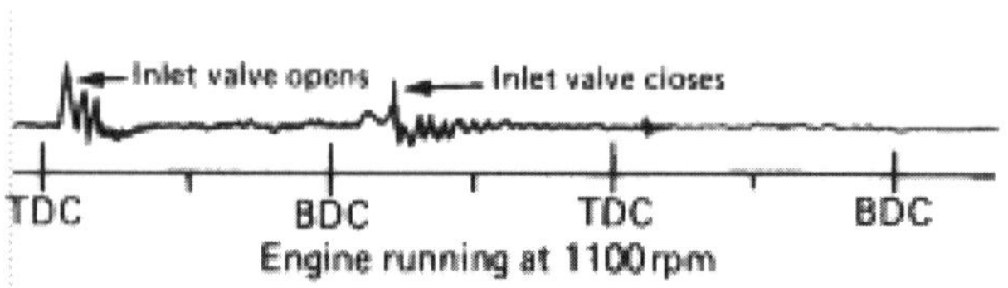

Figure 1: Inlet Noise Oscillogram [2]

The second process is the result of flow, or gas generated noise, which is the result of turbulence being created as the mean flow of air travels across the valve seat. A high frequency broad spectrum noise above 1000 Hz is generated by this high velocity flow but consequently becomes attenuated by the inlet ducting, air cleaner and transmission path between the engine and passenger compartment.

This study investigates the attenuation of induction noise through a noise cancelling technique both through numerical modelling and experimentation. For the purpose of numerical modelling, the propagation of the noise through the intake system is assumed to be a one-dimensional wave. Previous studies have shown that consideration of this noise as plane acoustic propagation has been able to provide reliable prediction for low frequency noise such as that for automotive intake noise measured at the inlet orifice [3].

ATTENUATION TECHNIQUES FOR AUTOMOTIVE INDUCTION NOISE

In order to understand what merit can be realized by this investigation, a discussion of the more conventional method of controlling intake noise is warranted. The most popular of these include passive techniques such as mufflers and resonators but alternative solutions using active noise cancellation are becoming more realistic.

PASSIVE ATTENUATION TECHNIQUES

Automotive intake noise is most often attenuated through the application of passive control techniques. These techniques are usually the simplest and least expensive form of attenuation but do not always yield the best results. The primary method of passive noise control for induction noise works by reducing the acoustic energy flow. This is accomplished by changing the acoustic impedance of the power output often through the use of a sudden cross section change. Much has been published of these methods so only a brief description is given here.

For automotive induction noise, the most common passive noise control technique is the Helmholtz resonator. This control technique is most effective for targeting specific frequencies of unwanted noise. When acoustic energy travels down a tube or pipe, a specifically chosen attached volume can be used to attenuate the traveling noise. This is accomplished by providing an alternative path with negligible impedance for the energy at the target frequency. The volume, or resonator, is appropriately sized to this specific unwanted target frequency. A reflected acoustic wave results, which bounces back toward the source and effectively cancels the unwanted noise. A schematic of a resonator for an automotive induction system is given in Figure 2.

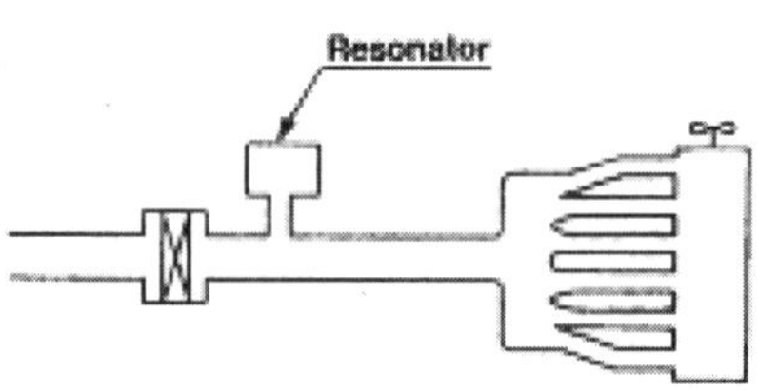

Figure 2: Schematic of Automotive Induction Resonator [4]

Mufflers that use an expansion chamber are another example of a passive noise control technique. These may or may not include absorption material as an integral part of the attenuation mechanism. If absorbing material is not used, the muffler is a reactive muffler otherwise the muffler is called a dissipative muffler. Expansion chamber mufflers rely on the geometric shape of the expansion chamber to provide an impedance change for the travelling acoustic wave. This impedance change results in some of the acoustic energy to reflect back and cancel the incoming energy, thus providing attenuation. A cutaway of a typical multi-chambered muffler is given in Figure 3.

ACTIVE ATTENUATION TECHNIQUES

In recent years, active noise control (ANC) for the attenuation of automotive induction noise has been a topic of interest. This is a technique that attenuates an unwanted noise wave by cancelling it through the introduction of a second noise wave equal in amplitude but opposite in phase to the unwanted noise. This method most often uses a computer controlled loudspeaker to generate the cancelling sound field.

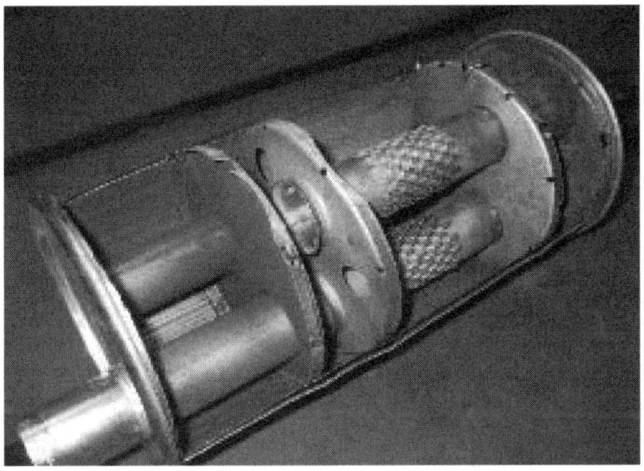

Figure 3: Cutaway of a Typical Multi-Chambered Muffler

Active noise control systems are most often adaptive feedforward, adaptive feedback or the wave synthesis system. The most common, the feedforward system, is shown as a schematic in Figure 4.

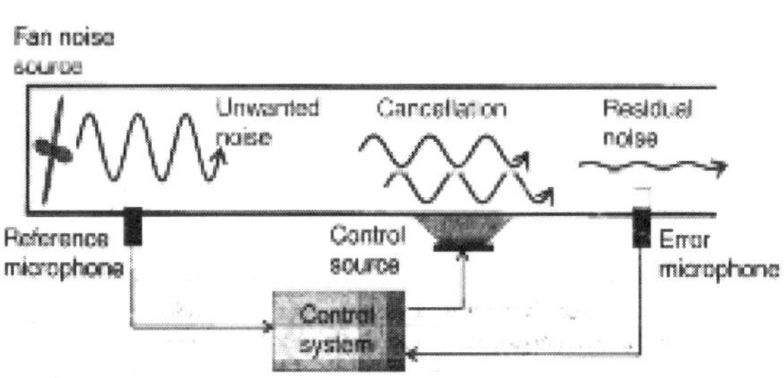

Figure 4: Schematic of Active Noise Control Feedforward System [5]

A successful application of ANC in an automotive induction system was demonstrated by McLean.[7] Here, a loudspeaker was place both co-axially and co-planar with the termination of the air intake duct. It was found that by placing the speaker co-axially, this arrangement provided greater attenuation than the case where the speaker is aligned orthogonal to the intake ducting.

METHOD

This study began with modelling the problem using a computer software package to optimize a design. This was followed by experimental simulations to verify the results. As stated, the objective of this work was to investigate the feasibility of attenuating induction noise through a noise cancellation technique by feeding tuned exhaust noise into the intake system. This physical link between the exhaust and intake manifold of the engine allows for the transfer of the acoustic energy. This is referred to as a manifold bridge.

THEORETICAL MODELLING

To model the proposed manifold bridge, an engine modelling software program called Ricardo WAVE was used. "WAVE is a computer-aided engineering code developed by Ricardo to analyse the dynamics of pressure waves, mass flows and energy losses in ducts plenums and the intake and exhaust manifolds of various systems" [6]. WAVE uses a one-dimensional finite difference approach of the theoretical thermo-fluid equations of the working fluids of the defined system.

Before the manifold bridge could be designed, a proven model of the engine to be used in the experimental verification had to be created. This included modelling the geometric data of the engine and it's components. Also inputted was engine data such as timing and valve lift profile as well as the operating conditions of the engine. These include the inlet and exhaust wall temperature, operating speed, head, piston and cylinder temperatures as well as any applicable ambient conditions.

Once the parameters of the engine were determined, the model was created using WAVEBUILD. This postprocessor provides the ability to create and synthesize all the building blocks representing the various ducts, volumes and other engine components. Figure 5 is an illustration of the original unbridged model of the engine used in this investigation. Once the model was created, experimental acoustic measurements were conducted on the actual engine motored on a dynamometer in a semi-anechoic room to verify the accuracy of the computer model. The details of the experimental set-up are discussed in a later section.

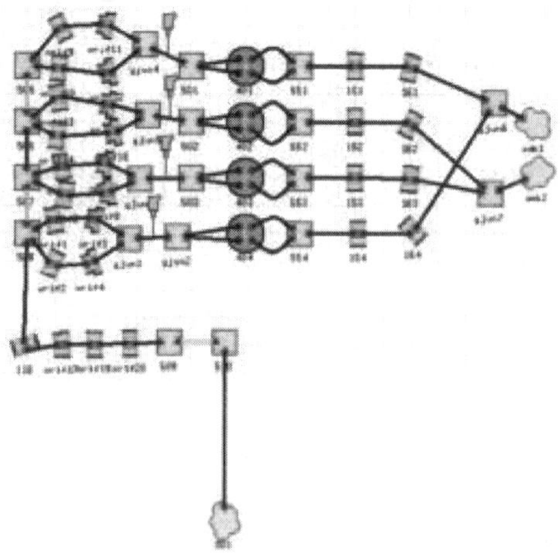

Figure 5: WAVE Model of Unbridged Engine

Once a verified engine model was created, optimization of the bridge design was pursued. This involved the determination of the optimum physical parameters of the manifold bridge that would produce the greatest acoustical attenuation at the intake orifice as well as the best improvement in sound quality. This meant that both the configuration of the bridge runners as well as their lengths and diameters had to be determined.

Different configurations of the manifold bridge runners were investigated and evaluated using the numerical analysis software to determine the best overall noise attenuation. The first configuration used a single bridge from the exhaust manifold output to the intake manifold plenum. The second case used four bridging ducts running from each exhaust manifold runner to their corresponding intake manifold runners. In other words, a bridging duct attached to the exhaust runner associated with cylinder number one was attached to the intake runner, also for cylinder number one. The third configuration had each bridging runner connected to each of the four exhaust manifold runners. These were routed to each of the corresponding intake manifold runners that were associated with cylinders that were 180 degrees out of phase with respect to the firing order of the engine.

Given the analysis of the three bridge configurations, it was determined that the second approach provided the best attenuation measured at the air induction inlet. That is, the configuration where the bridging ducts were linked from the exhaust manifold runners to their corresponding intake manifold runners achieved the greatest noise reduction. From there, the length and diameters of the bridge runners where then also optimized until the greatest inlet noise attenuation was realized.

EXPERIMENTAL SETUP

Using the bridge configuration design described above, the bridge was constructed and installed on the engine. The material used for the bridging duct was polyvinyl chloride chosen for its availability and ease of manipulation into the various required shapes.

First, the intake and exhaust manifolds were modified with ports to accept the bridging ducts at the locations specified by the analytical model. Next, the ducting material was cut to length and manipulated between the two manifolds.

Once the engine was modified with the bridge, it was installed in the semi-anechoic room and attached to the dynamometer. For testing purposes, the engine was motored on the dynamometer since the facilities were not equipped at this time to accommodate a fired engine. Consequently, all modelled results were also determined with the case of the engine being motored. For both the model and experiments, the microphone was placed at a distance of 100 mm from the opening of the intake system ducting.

DISCUSSION OF ANALYSIS PARAMETERS

The ability of the manifold bridge to improve the acoustical performance of an automotive intake system was determined by measuring both the realized attenuation with the bridge as well as any improvement in the measured sound quality. The later of these two was accomplished through application of several psychoacoustic metrics.

The standard measured parameters included A-weighted sound pressure level and frequency spectra measured for various steady rpm's of the engine's operating range. The psychoacoustic metrics used included loudness, sharpness, fluctuation strength and roughness. In order to put the declared values of these metrics into perspective, a definition of psychoacoustics and a description of the sound quality metrics used is included.

In the evaluation of the acoustic comfort of a sound, fundamental metrics such as sound pressure level are not adequate to truly represent the actual hearing sensations. The science of psychoacoustics involves the quantitative evaluation of these subjective sensations using sound quality metrics. The application of sound quality metrics allow for the repeatable visualization of the relationship that exists between the physical and perceptual acoustic quantities.

Zwicker Loudness is a standardized metric that describes the human perception of loudness as opposed to a simply reported sound pressure level. This loudness value takes into account the temporal processing of sounds as well as any audiological masking effects[7]. The unit of loudness is sones and is given across Bark, or critical, bands, as opposed to conventional frequency bands.

Sharpness describes the high frequency annoyance of noise by applying a weighting factor on sounds above 2 kHz. This overall measurement, which has units of acum, is useful for such sounds as broadband sources, wind or rushing air noise and gear meshing or grinding sounds. When the intake air travels across the valve seat at a high velocity, a high frequency component of intake noise is created. Because of this, sharpness was thought to be an appropriate metric for the evaluation of the merits of the manifold bridge.

Fluctuation strength and roughness are two metrics used to describe the annoyance of modulating sounds. The fluctuation strength metric focuses on sounds that modulate at frequencies between 0.5 Hz and 20 Hz, with 4 Hz being the most annoying fluctuation. The unit of amplitude for fluctuation strength is the vacil. Roughness focuses on noise that is modulating at frequencies between 20 Hz and 300 Hz, with the most annoying modulation here being 70 Hz. The unit of amplitude for roughness is the asper. It has been found that when sounds modulate faster than 300 Hz, the human ear is not able to distinguish this from a normal pure tone. Examples of modulating sources include beating sounds, sirens and fan blades.

DISCUSSION OF RESULTS

As part of the analysis of the modelled results to optimize the bridge design, a transient simulation was performed on the unmodified and bridged engines. Like the steady state simulations to be discussed later, the transient runs ranged from 1000 to 6500 rpm. No transient analysis was performed on the experimental engine since the dynamometer used in the study was not capable of such operation.

Figures 6 and 7 are colour map representations of the induction noise during these transient simulations modelled 100 millimetres from the intake orifice. Figure 6 shows the frequency of the intake noise for the rpm range of the unmodified engine while Figure 7 illustrates the same for the engine modified with the manifold bridge. The acoustic shortcomings of the unmodified engine are incontrovertibly obvious. The yellow and orange streaks representing the fundamental and subsequent harmonic frequencies are more apparent with more red showing on the map of the unmodified engine. This illustrates the presence of higher amplitudes of noise at the fundamental frequencies that is associated with the speed of the engine. Also, the bridged engine simulation has less of the higher sound pressure level represented by the green colour. Similarly, it has more of the lower sound pressure represented by the mid and dark blue shades.

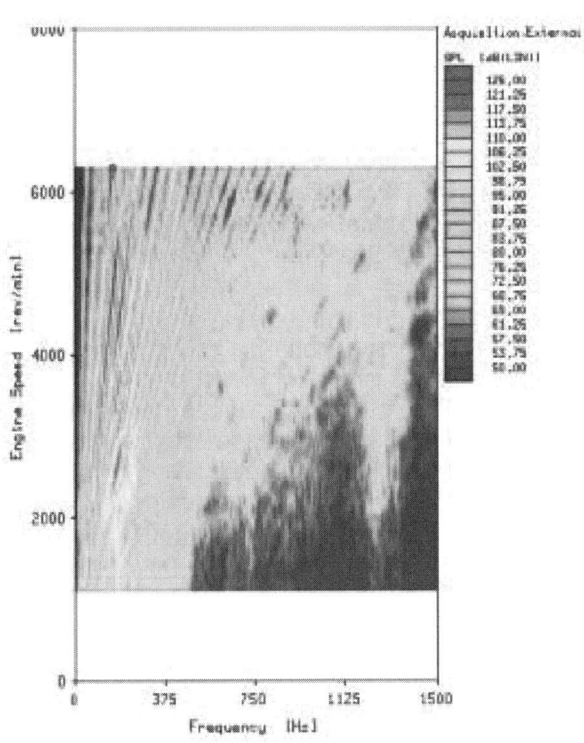

Figure 6: Colour Map of Intake Noise of Modelled Unmodified Engine

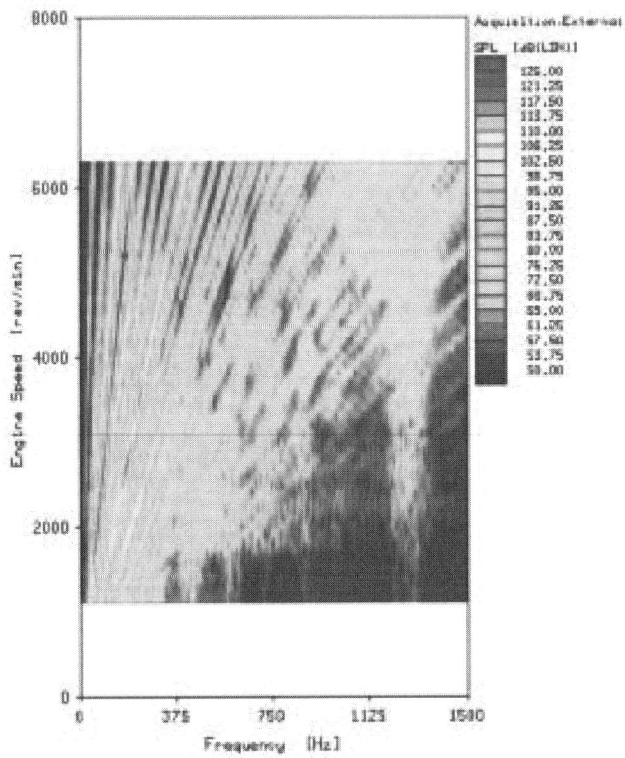

Figure 7: Colour Map of Intake Noise of Modelled Bridged Engine

Once the bridge parameters were established and the numerical results were obtained for the chosen design, the physical bridge was built and mounted on the test engine for acoustical evaluation. Specifically, the realized attenuation between the original and bridged engines was determined for steady state conditions for engine speeds from 1000 to 6500 rpm for both the modelled and actual engines. Sound quality analyses were also carried out for the modelled and motored engines.

Figure 8 is an illustration of the sound pressure levels for the steady state engine speeds ranging from 1000 to 6500 rpm. The four shown curves represent the sound pressure levels for the case of the modelled and experimental unmodified engine along with the case of engine modified with the manifold bridge. It can be seen that the addition of the bridging device resulted in greater attenuation for both the numerical and experimental measurements. Also noted is that while both the modelled and experimental cases exhibited overall noise attenuation with the implementation of the manifold bridge, the experimental results showed a greater difference with the most attenuation occurring in the engine operating range from approximately 2800 to 3400 rpm.

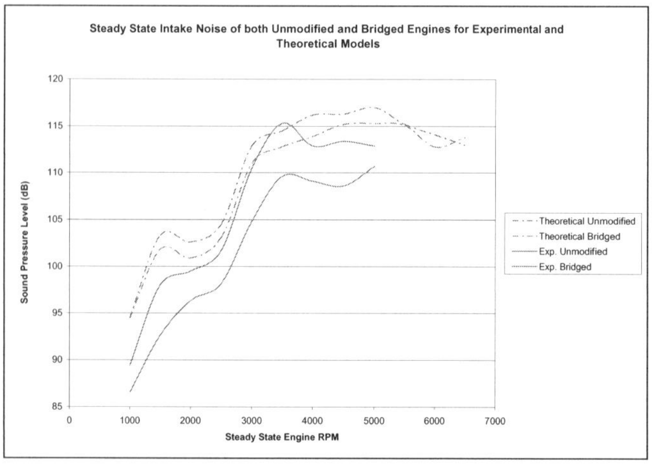

Figure 8: Predicted Intake Noise of Modelled and Experimental Unmodified and Bridged Engines.

Figure 9 shows the predicted A-weighted intake noise of the four engine cases. Good attenuation with the manifold bridge is demonstrated by the experimental results. The theoretical results, however, do not fair as well. It can be seen that these modified and unmodified engine results cross each other several times throughout the operating speeds tested. This is similarly noted in the loudness diagram shown in Figure 10. This is due to a higher frequency content in the modelled results that did not materialize in the experimental measurements. It is suspected that the analytical model was not as capable to accurately predict the results at the higher engine speeds.

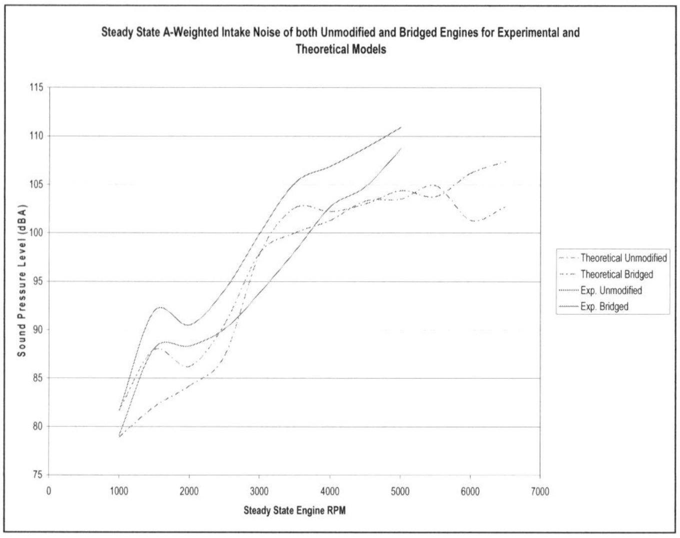

Figure 9: Predicted A-Weighted Intake Noise of Modelled and Experimental Unmodified and Bridged Engines

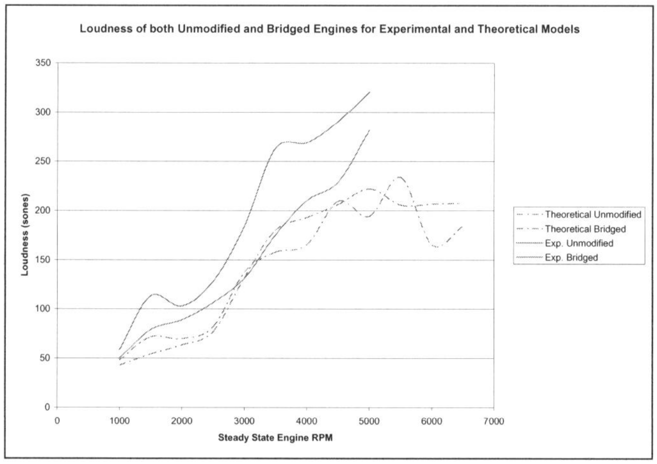

Figure 10: Predicted Loudness of Modelled and Experimental Unmodified and Bridged Engines

Figure 11 shows the sharpness results of the four engine models which is a measure of the high frequency content of the noise. Again, the modelled and experimental results differ. Like in the case of the A-weighted results, it is felt that the numerical model is not accurately predicting the high frequency portion of the results. This aside, it was found that despite what was originally postulated, very little sharpness was apparent in the intake noise. This is due to the attenuation of the higher frequencies by the intake ducting and air filter element.

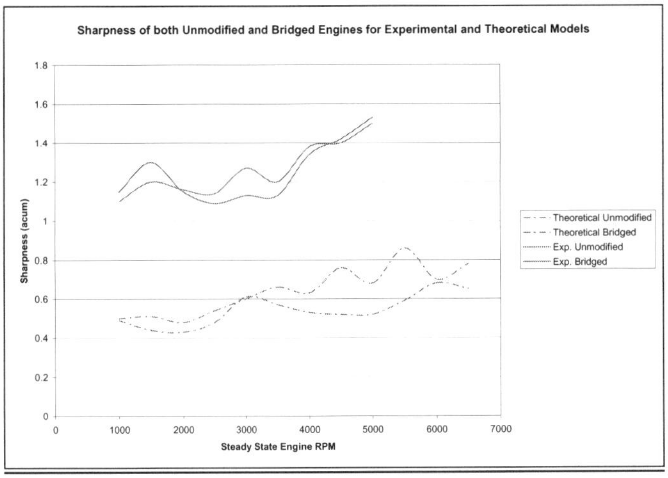

Figure 11: Predicted Sharpness of the Modelled and Experimental Unmodified and Bridged Engines

Figure 12 illustrates the roughness results of the four engines. The resulting patterns for the experimental and theoretical results are for the most part similar, only with a greater dynamic range given to the experimental results.

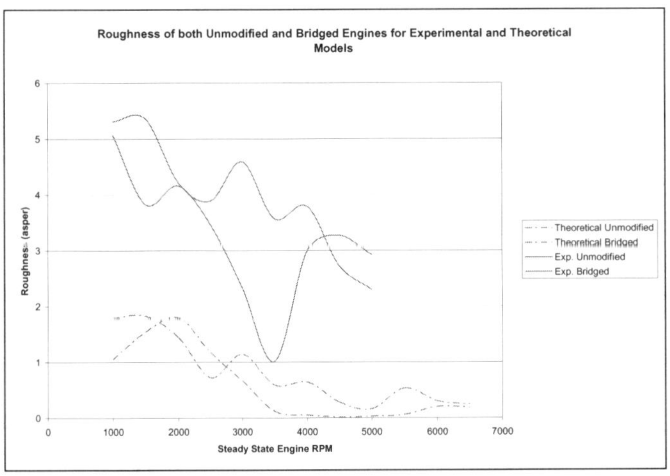

Figure 12: Predicted Roughness of Modelled and Experimental Unmodified and Bridged Engines

Similar to above, Figure 13 shows the fluctuation strength results of the four engines. Again the resulting pattern of the experimental and theoretical results are for the most part similar, however, this time a slight phase shift is present between the experimental and theoretical results, with the experimental curve features occurring at lower rpm than the theoretical. In this circumstance, the modulation in the 20 to 300 Hz range causing the fluctuation signal is occurring first at lower engine speeds with the actual engine when compared to the analytical model.

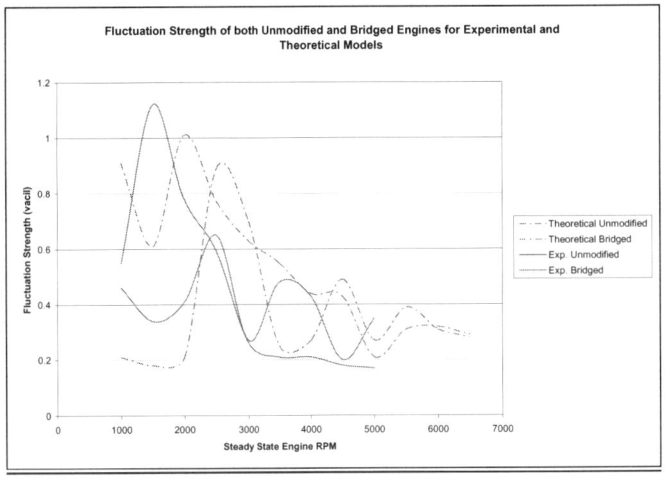

Figure 13: Predicted Fluctuation Strength of Modelled and Experimental Unmodified and Bridged Engines

CONCLUSION

For the conditions investigated, specifically for this four cylinder motored engine, it has been shown that the implementation of the manifold bridge has a positive influence on both the amplitude and the sound quality of induction noise. While this investigation showed the merits of the bridge for both the theoretical modelling and experimental measurements, it is felt that greater credibility should be given to the later of the two. The focus of the material given here was the realized acoustical results of the bridge implementation and did not report on the other engine performance criteria. It should be realized that the addition of the manifold bridge will affect some of these other criteria. Of particular importance would be the influence of the additional exhaust gas recirculation. Given further investigation, if this was found to be detrimental to the engine performance, it is proposed that the exhaust and intake systems could be isolated from each other through the addition of a membrane or a dual walled bladder system. This investigation, however, does demonstrate the merits in pursuing further refinements of this unique noise control approach.

REFERENCES

1. Pricken, Franc. "Active Noise Cancellation in Future Air Intake Systems." *Powertrain Systems NVH*. SAE 2000 World Congress. Detroit, Michigan: Society of Automotive Engineers, 06/03, 2000.

2. Nelson, P., 1987, *Transportation Noise Reference Book*, Cambridge, Great Britain : Butterworth & Co.
3. Chiatti, G. and Chiavola, O., "Engine Intake Noise Modelling by Using a Time/Frequency Approach," *SAE;* 2001-01-1440.
4. Nishio, Y., Kohama, T. And Kuroda, O., "New Approach to Low-Noise Air Intake System Development," *SAE*; 911042.
5. Snyder, S., 2000, *Active Noise Control Primer*, Adelaide, Australia : Springer.
6. . Ricardo Software, *WAVE Basic User Manual*, 2001.
7. Zwicker, E., and Fastl, H., 1999, *Psycho-acoustics Facts and Models, second edition*, Berlin, Germany : Springer.

CONTACT

Prof. Colin Novak, P.Eng.
Lecturer, Faculty of Engineering
Department of Mechanical, Automotive and Materials Engineering
University of Windsor
401 Sunset Ave.
Windsor, Ontario N9B 3P4
Phone: (519) 253-3000 ext. 2634
Fax: (800) 241-9149

2005-01-0083

A Benchmark Test for Springback: Experimental Procedures and Results of a Slit-Ring Test

Z. Cedric Xia, Craig E. Miller and Maurice Lou
Ford Motor Company

Ming F. Shi, A. Konieczny and X. M. Chen
United States Steel Corporation

Thomas Gnaeupel-Herold
National Institute of Standards and Technology

ABSTRACT

Experimental procedures and results of a benchmark test for springback are reported and a complete suite of obtained data is provided for the validation of forming and springback simulation software. The test is usually referred as the Slit-Ring test where a cylindrical cup is first formed by deep drawing and then a ring is cut from the mid-section of the cup. The opening of the ring upon slitting releases the residual stresses in the formed cup and provides a valuable set of easy-to-measure, easy-to-characterize springback data. The test represents a realistic deep draw stamping operation with stretching and bending deformation, and is highly repeatable in a laboratory environment. In this study, six different automotive materials are evaluated. They included one aluminum alloy (AA6022-T4), one deep drawing quality and special killed (DQSK) mild steel, one bake hardenable (BH) medium strength steel, a conventional high-strength low-alloy (HSLA) steel, and two advanced high-strength steels (AHSS) represented by one dual-phase (DP) steel, and one TRansformation Induced Plasticity (TRIP) steel. A particularly interesting aspect of this experiment is the direct measurement of residual stresses by diffractive stress analysis in collaboration with NIST Center for Neutron Research, and is believed to be the first application of this technique to sheet metal forming. Complete material data and experimental results are documented, including punch force trajectories, amount of draw-in, ring opening displacement, axial and hoop stresses before and after the rings were slit. The data is ideal for the evaluation and improvement of current forming and springback simulation capabilities. Efforts for the correlation of simulation with the obtained experimental data are underway and will be reported in follow-up studies.

INTRODUCTION

Springback has been a serious problem for automotive sheet metal stamping, especially with increasing usage of lightweight materials such as aluminum and advanced high-strength steels (AHSS) for vehicle body structures and closures. Applications of those materials pose a particular challenge because of their severe and sometimes peculiar springback behavior. Tremendous efforts have been devoted to correct springback related problems during die tryout, which are both costly and time consuming. More recently, attempts have been made to predict springback behavior with numerical simulations thanks to the rapid advancements of the finite element technology coupled with ever expanding computing power through the Massively Parallel Processors (MPP) and the Symmetric Multi-Processors (SMP). The progress has been exciting and has offered a real opportunity for applying the technology to daily production work. However, the accuracy of springback prediction remains mixed: it performs remarkably well for some parts but proves to be unsatisfactory for others, due to the fact that springback is such a complex behavior. It is influenced not only by the die design but also by the physical stamping process. In order to characterize springback and correlate it with numerical simulation, one has to design careful tests in a controlled environment to eliminate process variables commonly existing in a production setting.

The draw/bend test proposed in [1-2] is an excellent test for springback characterization and great insights have been gained in finding deficiencies of current predictive techniques such as material modeling and integration algorithms [3]. Flanging tests with either straight flanging or curved flanging are two other examples for the characterization of springback [4]. However, while those tests are preferred in understanding certain aspects of springback, they are not truly deep drawing operations usually seen in industrial stamping operations.

This study documents an experimental procedure and results for a benchmark springback test first proposed by Demeri *et. al* [5, 6], and is often referred as the Slit-Ring Test. The test consists of four steps: (a). Deep draw a cylindrical cup from a circular blank with a constant blankholder force; (b). Cut a circular ring from the mid-section of the drawn cup; (c). Slit the ring along certain direction to release residual stresses introduced from drawing operation, and (d). Measure the opening of the ring (springback). The test is highly effective and objective for the following reasons:

• The experimental setup and procedure for the test are relatively simple and highly repeatable.

• The forming deformation involves both bending and stretching, which mimics actual stamping operation.

• The springback amount is relatively large and easy to measure, thus avoiding experimental errors found in some other tests.

• The simulation effort required is manageable, and there are no difficulties associated with springback characterization.

Experiments were conducted in this study for six grades of automotive materials, including one aluminum alloy (6022-T4), one deep drawing quality and special killed mild steel (DQSK), one bake hardenable medium strength steel (BH210, also sometimes referred as BH33), a conventional high-strength low-alloy steel (HSLA350, also sometimes referred as HSLA50), one Dual-Phase steel (DP600) and one TRansformation Induced Plasticity steel (TRIP600). Circular metal blanks were prepared at United States Steel Corporation's Automotive Center, and the cylinder cups were formed at Ford Research Laboratory. Ring cutouts and measurements were performed at the Metallurgy Division of National Institute of Standards and Technology (NIST).

The paper is organized as follows. The experimental procedure for the cup drawing is first detailed, along with mechanical properties of the six different sheet metals used. Experimental results for punch force trajectory and the rim periphery of the drawn cup are plotted for each material. The procedure for the ring slitting is documented in the following section, and the ring opening displacements due to springback are measured. A particular interesting aspect of this experiment is the direct measurement of residual stresses by diffractive stress analysis in collaboration with NIST Center for Neutron Research, and is believed to be the first application of this technique to sheet metal forming. Axial and hoop stresses before and after the ring was slit are then presented in this section. Finally, the paper concludes with discussions on how the obtained data can be effectively used to improve existing capabilities of springback simulation tools.

CUP DRAWING

EXPERIMENTAL SETUP

The experimental setup for the cup drawing is illustrated in Figure 1 with all dimensions shown. The circular blank was first held by the binder ring with a specified blankholder force F, and formed into a cup as the punch traveled upward. The total punch travel is defined as the distance between positions when the punch first contacts the blank to its final stop. In this study, the blankholder force used was 88.9kN and the maximum punch travel was set at 56mm using a steady punch travel speed of 5mm/s.

An oil lubricant is applied manually to both sides of blanks with a density of approximately 1.25g/m^2. In addition a sheet of solid lubricant is also applied to the die side of blank to further reduce friction. It is recommended that a smaller friction coefficient such as 0.07 ~ 0.08 be used for the Coulomb friction model between punch and sheet metal in this case.

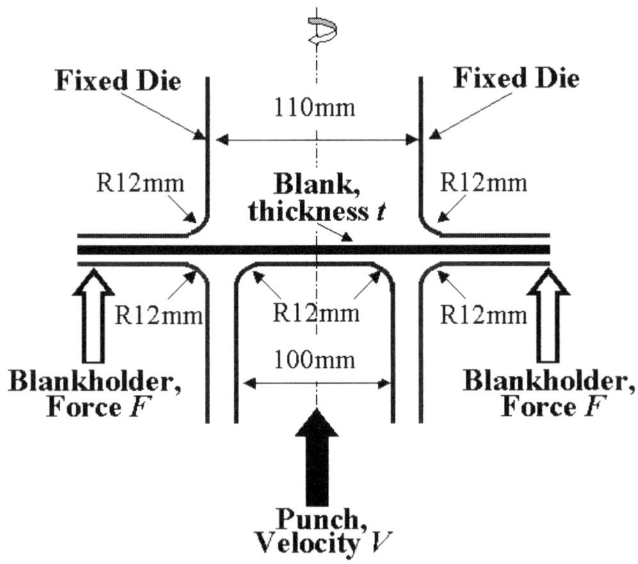

Figure 1. Experimental Setup for Cup Drawing

MATERIAL PROPERTIES

Six different automotive sheet metals were selected for the study, which were outlined earlier. Their respective coatings and gauges are listed in Table 1 (see Appendix for all table listings). All circular blanks have a diameter of 195mm. The aluminum blanks were water-jetted and all steel blanks were milled.

The tensile properties of those materials were tested along three orientations, namely the rolling direction (0^0), the diagonal direction (45^0), and the transverse direction (90^0). Their stress – effective plastic strain curves are plotted below in Figures 2a-2f.

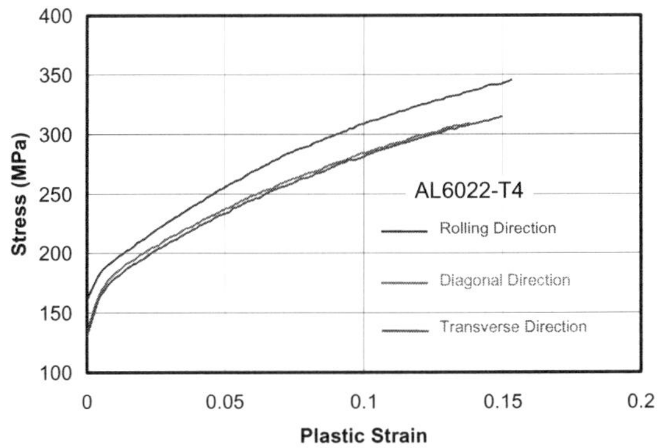

Figure 2a. Tensile Curves for AA6022-T4

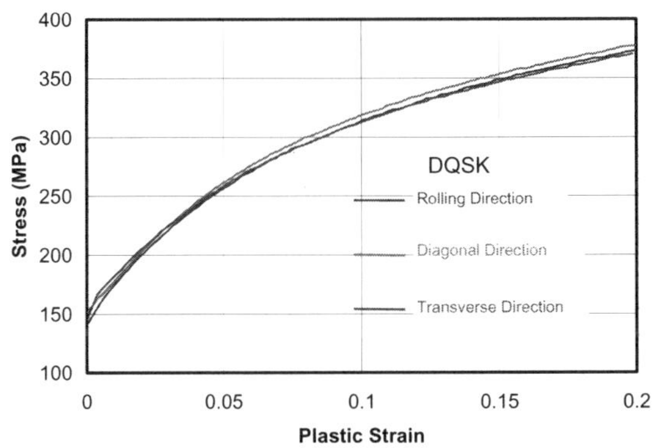

Figure 2b. Tensile Curves for DQSK

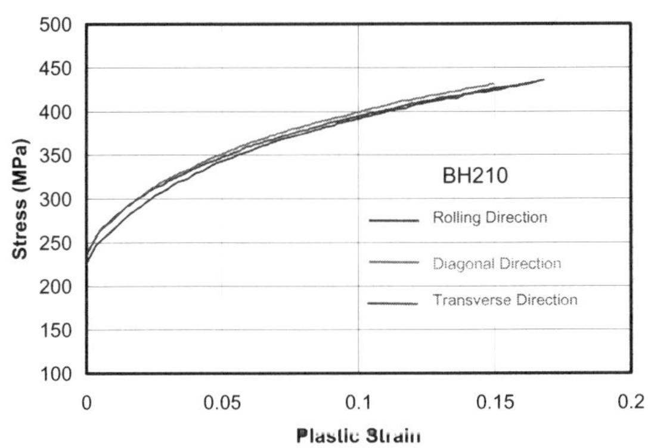

Figure 2c. Tensile Curves for BH210 (BH33)

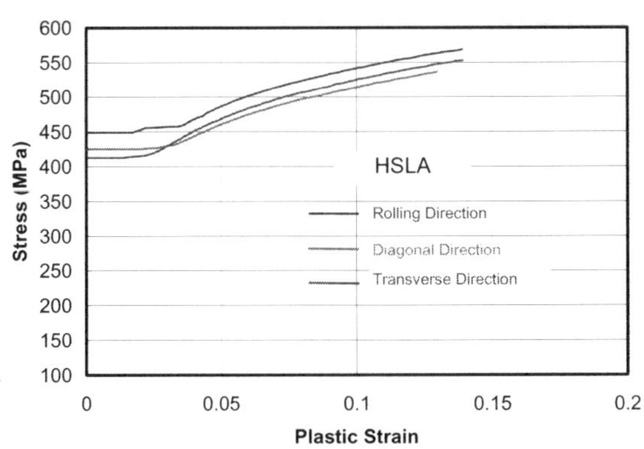

Figure 2d. Tensile Curves for HSLA350

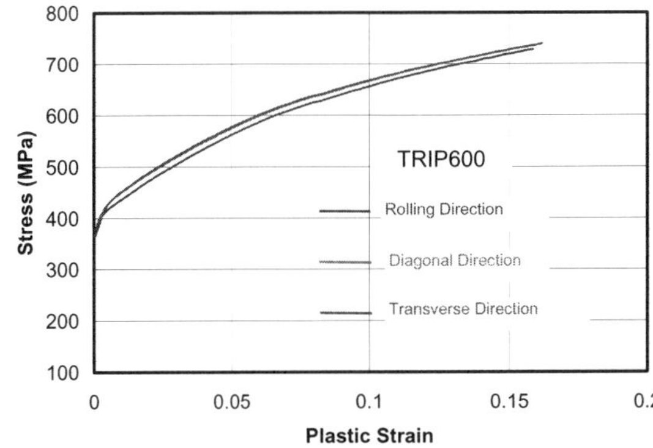

Figure 2e. Tensile Curves for TRIP600

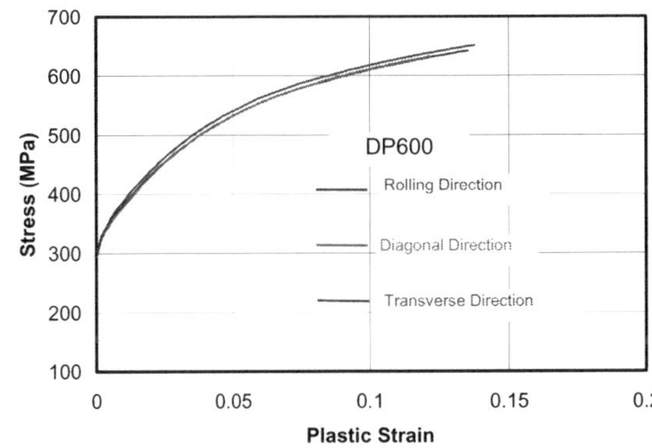

Figure 2f. Tensile Curves for DP600

Figure 2. True Stress-Strain Curves for Six Tested Alloys

The anisotropic properties are characterized by the Hill's R-values and were measured in all three orientations. Their values are listed in Table 2 along with the strains at which they are measured.

TEST MATRIX

Although the test equipment is axisymmetric, cups drawn are not necessarily in perfect cylindrical shape due to the anisotropic properties of sheet metals. The unique springback results obtained as a consequence of slitting the ring might have orientation dependence. To investigate this effect, DQSK was selected as a case study because of its relatively large anisotropic property. Permanent marks were applied to all blanks prior to forming to identify the rolling direction (0^0) and transverse direction (90^0). The number of cups drawn from each material is listed in Table 3. Notice that 3 extra DQSK cups were drawn for the orientation study. The DQSK sheet does not require special care during forming operations because of the axisymmetric nature of the test setup.

CUP DRAWING RESULTS

Pictures of formed cups are shown in Figure 3 for all six different sheet metals. The punch force trajectory as a function of punch displacement for each tested material is plotted in Figure 4. The force required to form a cup depends not only on the strength of the material but also on its thickness. This is clearly reflected in the figure where the thicker and stronger HSLA340, DP600 and TRIP600 exhibit significant higher punch forces than those for AA6022, DQSK and BH210.

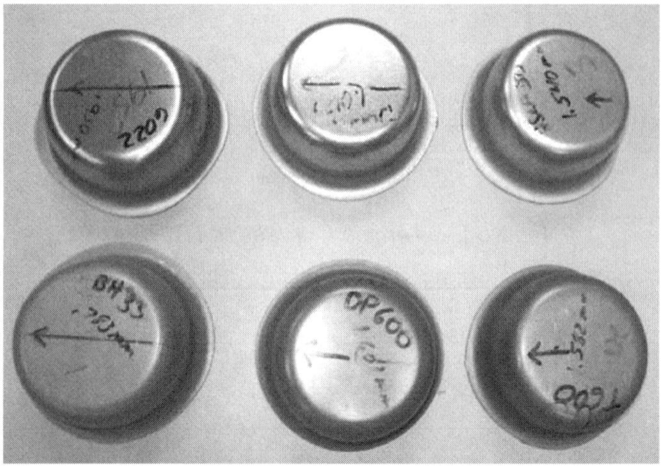

Figure 3. Six Formed Cups

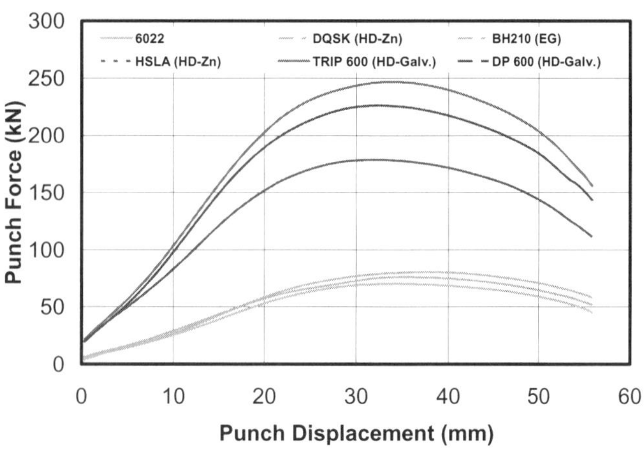

Figure 4. Punch Force vs. Punch Displacement for the Cup Drawing of Six Alloys. (With Blankholder Clamping Force 88.9kN)

Because of the anisotropic nature of those materials, the height of the flange above the base was not uniform, as would be expected in a symmetrical operation on an isotropic blank. Instead "ears" were developed in positions symmetrically situated with respect to the rolling direction in the original sheet [7]. One formed cup is selected for each material and the periphery of the cup flange is measured and presented in Figure 5. Such measurements are of particular importance since it contains information about the amount of draw-in during forming as well as characteristics of anisotropic deformation, which are valuable for correlating simulation results with physical testing.

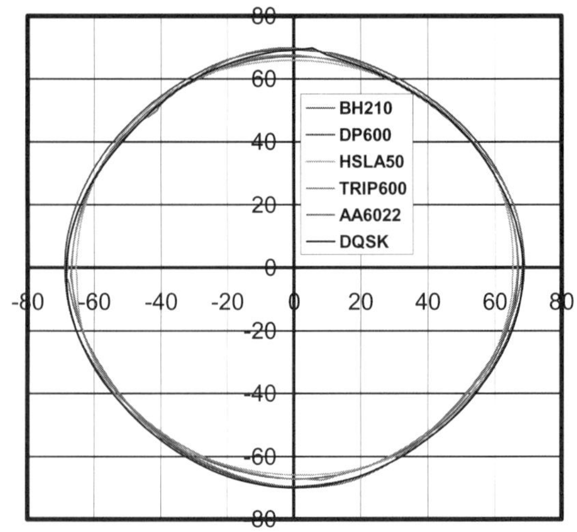

Figure 5. The Rim Periphery of drawn Cups

SLIT RING

RING OPENING

The experiment provides an easy-to-measure, easy-to-characterize springback test with a realistic deep drawing deformation history. In order to obtain consistent data, a ring was cut from the middle section of the cup where the residual stresses were expected to have minimal variations. The procedure is as follows:

(1). Cup Cutting: A circular ring is cut from the formed cup by EDM (or alternatively by laser cutting), measured 25mm and 40mm respectively from the outer surface of the cup bottom, with the width of the ring to be 15mm, as illustrated in Figure 6a.

(2). Ring Slitting: Each ring is slit along either the rolling direction (0^0) or the transverse direction (90^0) as marked according to specifications in Table 3.

(3). Springback Measurement: The ring will be opening up upon slitting due to the release of residual stresses introduced during forming operations. If the metal blank is perfectly isotropic, the ring will have the same opening across the cut section, and its deformation will remain planar. The linear opening distance D will be sufficient to characterize the springback (Figure 6b). If the alloy is strongly anisotropic, the opened ring might also exhibit twisting and/or warping behavior.

The measured data for the opening of the slit ring are entered in Table 4, and again plotted in Figure 7.

It should be noted that the six sheet metals used in this study are all different in thickness. The springback amount is highly dependent on material's strength as well as thickness. Therefore readers should take into account both effects when interpreting results in Figure 7.

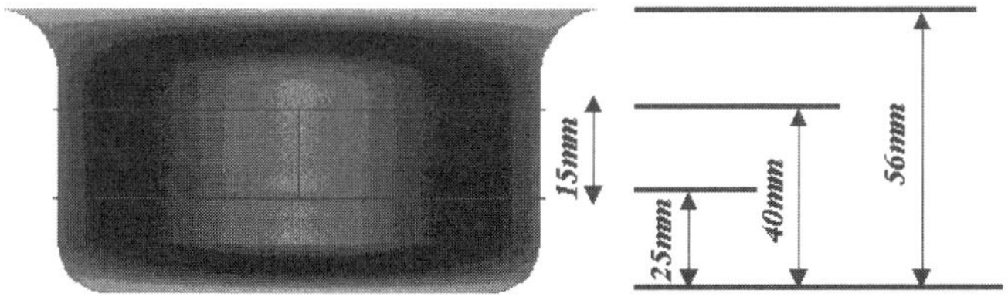

Figure 6a. Ring Cut Position

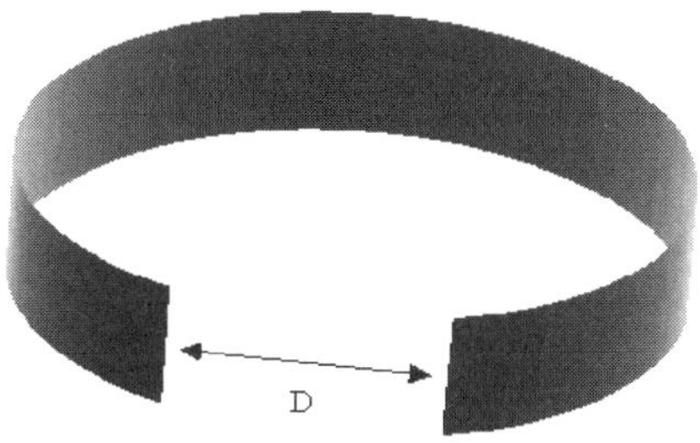

Figure 6b. Ring Opening

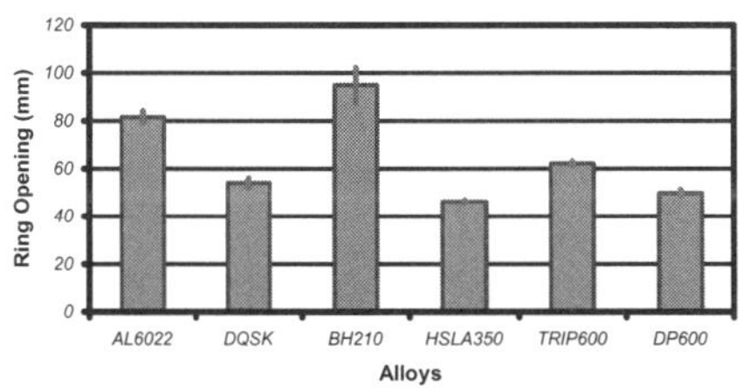

Figure 7. Opening Distance of the Slit Rings

RESIDUAL STRESSES

The springback is due to the release of residual stresses from forming. It is essential to have accurate prediction of residual stresses in order to obtain satisfactory springback results. Recent advancements in diffractive stress analysis make it possible to measure stresses with non-destructive methods. It relies on the measure of changes in inter-atomic spacings with neutrons or X-rays, and then relates the lattice spacing to elastic deformation. The described benchmark test has the advantage that the residual stresses are still present in the ring in its freestanding state before it is cut open, making it ideal for diffractive stress analysis.

The diffractive stress analysis is conducted with synchrotron measurement at the Advanced Photon Source of NIST for the aluminum alloy AA6022-T4. The techniques employed are detailed in [8] and [9], and a brief summary of measured results are presented here. Readers are encouraged to study the article in [9] for an in-depth analysis. The precision of stress measurement is believed to within ±10MPa.

A circular ring is first cut from the mid-section of the formed cup with a width of 15mm, as illustrated in Figure 8. The orientation of the original blank was carefully marked before forming and is indicated in the Figure.

A variety of stresses are measured for the ring, including axial and hoop stresses in the intact ring (Figures 9 and 10), in the slit ring (Figure 11), and stress comparison at rolling and transverse directions.

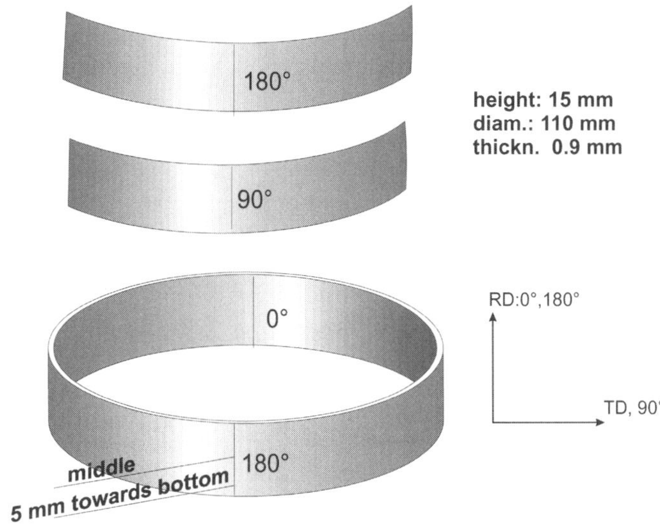

Figure 8. Illustration of the Ring for Stress Measurement

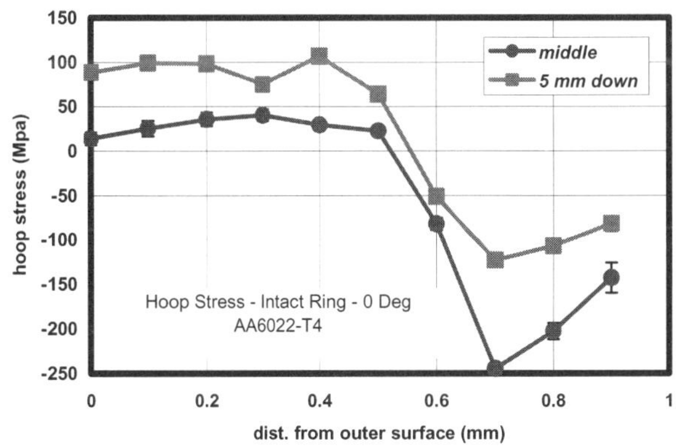

(a). 0° Mark

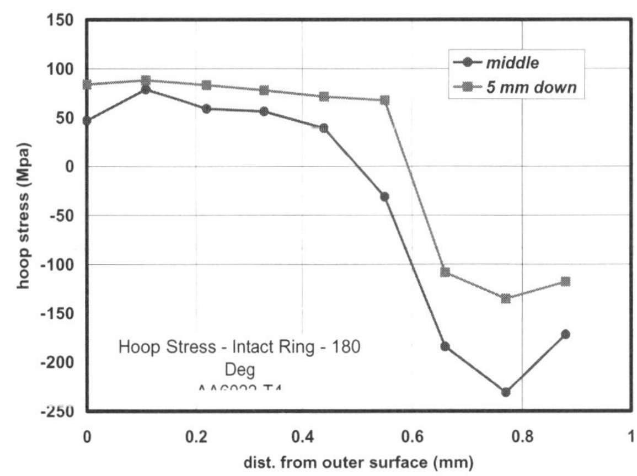

(b). 180° Mark

Figure 9. Hoop Stress on the Intact Ring

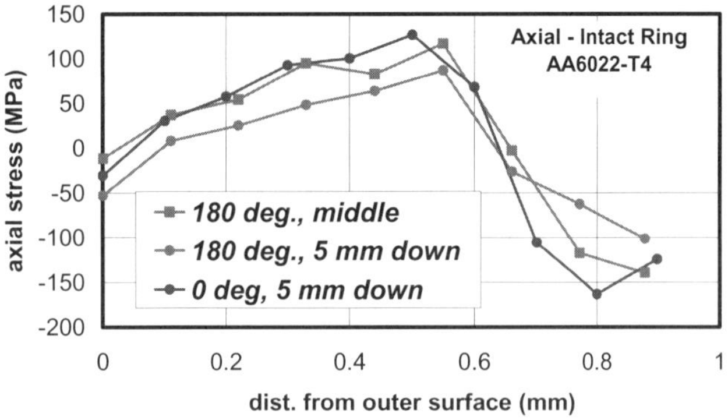

Figure 10. Axial Stress on the Intact Ring

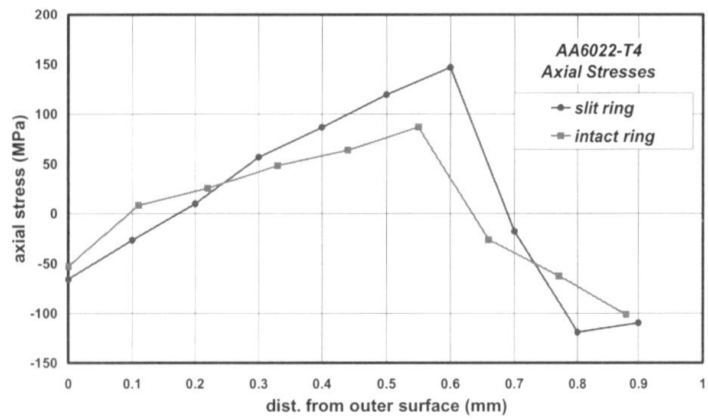

(a). Axial Stress

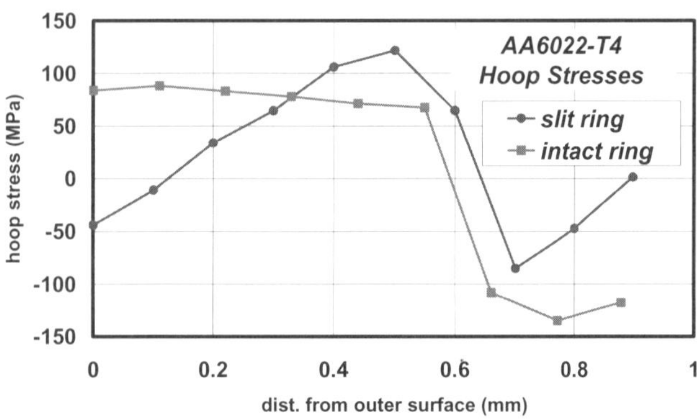

(b). Hoop Stress

Figure 11. Comparison of Stresses before and After Ring Slit

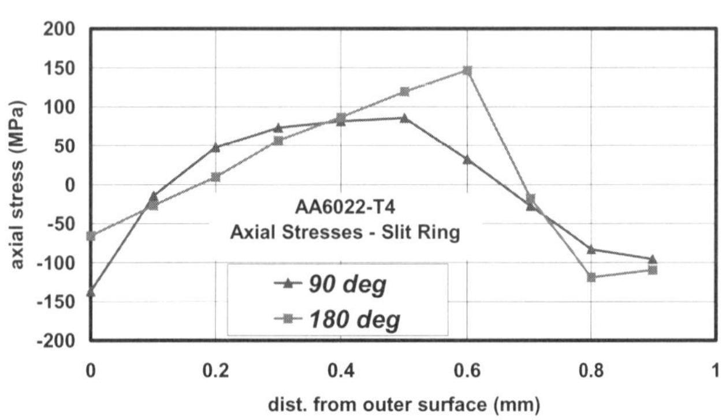

(a). Axial Stress

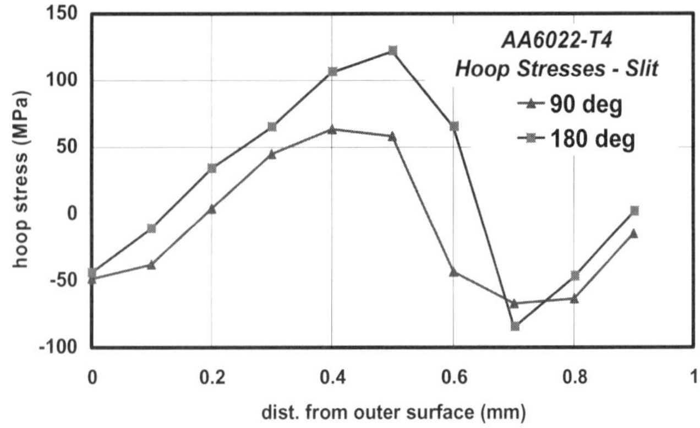

(b). Hoop Stress

Figure 12. Comparison of Stresses at 90° Mark and 180^0 Mark

DISCUSSION

The experimental procedure and results summarized in this study provided a complete suite of benchmark data for the simulation of springback in a meaningful way. Efforts are underway to conduct springback simulation for the conducted test with the aim to understand springback characteristics during different forming stages, to identify deficiencies in existing predictive methods, and to develop better simulation techniques suitable for production applications. Special attention will be devoted to simulation procedures and parameters to examine their effects on the accuracy of predictions. Among other things, the current treatment of intermediate steps during the process and the role of material anisotropy and hardening behavior will be scrutinized. The availability of residual stress measurements will enable better validation of simulation results and shed light into how stress relaxation occurs during springback. The simulation results and correlation with the data obtained in this study will be reported later.

CONCLUSIONS

Some conclusions can be made based on the above study:

- A springback benchmark test has been developed using the slit ring test and springback benchmark data are developed for various automotive sheet metals including advanced high strength steels and an aluminum alloy.

- The slit ring test is not only simple and easy to carry out but is also a repeatable and reproducible test.

- The data developed in this study from various automotive sheet metals can be used to assess the accuracy of computer simulation software in springback predictions.

- The diffractive stress analysis technique was used to measure the residual stress before and after springback. Residual stress results are very valuable to assess the effectiveness of computer simulation codes in stress prediction.

ACKNOWLEDGMENTS

This study was initiated as part of an industrial consortium effort on springback following successful completion of NIST-ATP "Springback Predictability Project". The authors would like to thank all team members from participating companies, universities and national labs for their helpful discussions during the course of the study. We are grateful to United States Steel Corporation for providing steel sheets and for milling steel blanks; the Scientific Research Laboratories of Ford Motor Company for water-jetting aluminum sheets and for forming the cups, The Center for Neutron Research at NIST for measuring residual stresses in the cups, and ALCOA for providing aluminum alloys used in this work.

REFERENCES

1. D.W. Vallance and D.K. Matlock (1992) "Application of the bending-under-tension friction test to coated sheet metals", Journal of Material Engineering and Performance, 1 (5), 685-694.
2. R.H. Wagoner, W.D. Caden, W.P. Carden, and D.K. Matlock (1997) "Springback after drawing and bending of metal sheets", THERMEC'97, Australia.
3. K. Li and R.H. Wagoner (1998) "Simulation of springback", NUMIFORM'98, Simulation of Materials Processing: Theory, Methods and Applications, Edited by J. Huetink and F.P.T. Baaijens, Netherlands.
4. N. Song (2000) "Springback Prediction of Straight Flanging Operation", Master Thesis, Department of Mechanical Engineering, Northwestern University.
5. M.Y. Demeri, M. Lou and M.J. Saran (2000) "A Benchmark Test for Springback Simulation in Sheet Metal Forming", SAE 2000-01-2657, International Body Engineering Conference, Detroit, Michigan.
6. M.Y. Demeri (2002) "Residual Stresses in Cup Drawing of Automotive Alloys", International Body Engineering Conference & Exhibition and Automotive & Transportation Technology Conference, Paris, France.
7. R. Hill (1950) "The Mathematical Theory of Plasticity", Oxford University Press.
8. T. Gnaeupel-Herold, T.J. Foecke, H. J. Prask and R.J. Fields (2004) "An investigation of springback stresses in an AISI-1010 deep drawn cup", *Materials Science and Engineering, in press.*
9. T. Gnaeupel-Herold, H. J. Prask, R.J. Fields, T.J. Foecke, Z.C. Xia and U. Lienert (2004) "A synchrotron study of residual stresses in a Al6022 deep drawn cup", *Materials Science and Engineering* **A366**, 104-113.

CONTACT

For further information or questions and comments regarding this study, please contact Cedric Xia of Ford Motor Company at zxia@ford.com or 313-845-2322.

APPENDIX: Tables Listings

Table 1. Tested Alloys

Material Grade	**AA6022**	**DQSK**	**BH210**	**HSLA340**	**TRIP600**	**DP600**
Coating	-	HDGI	EG	HDGI	HDGA	HDGA
Gauge (mm)	0.93	1.01	0.78	1.54	1.58	1.60

Table 2. List of R Values

Alloy / *Orientation*	**AA6022**	**DQSK**	**BH210**	**HSLA340**	**TRIP600**	**DP600**
Rolling (0^0)	0.758	1.734	1.48	0.909	0.916	0.843
Diagonal (45^0)	0.426	1.515	1.349	1.064	0.833	0.912

Transverse (90^0)	0.475	2.085	2.173	0.68	0.967	1.018
Average	0.521	1.712	1.588	0.929	0.887	0.921
Strains at which R values are measured	17%	17%	17%	15%	17%	15%

Table 3. Number of Cups Drawn for Each Material

TEST ID	**AA6022**		**DQSK**		**BH210**		**HSLA340**		**TRIP600**		**DP600**	
	0^0	90^0	0^0	90^0	0^0	90^0	0^0	90^0	0^0	90^0	0^0	90^0
C7-9	4											
C10-12		0										
C13-17			4									
C18-21				3								
C22-25					4							
C26-29							4					
C30-33									4			
C34-37											4	

Table 4. Opening Distance D for Slit Rings (in mm)

Material	**AA6022**	**DQSK**	**BH210**	**HSLA340**	**TRIP600**	**DP600**
test #1	80.2	56	96.1	46.7	61.5	49.2
test #2	82.5	54.3	94.6	46.5	62.8	49.6
test #3	79.25	54.1	87.6	46.8	61.8	48.79
test #4	84.33	52.1	102.2	45.6	63.1	51.2
test #5	NA	53.5	NA	NA	NA	NA
test #6	NA	55.1	NA	NA	NA	NA
test #7	NA	53.8	NA	NA	NA	NA
Mean Value	81.57	54.2	95.1	46.4	62.3	49.7
Upper deviation	2.76	1.87	7.08	0.4	0.8	1.50
Lower Deviation	2.32	2.03	7.53	0.8	0.8	0.91

2004-01-1656

Intelligent Fault Diagnosis System of Automotive Power Assembly Based on Sound Intensity Identification

Chen Xiaohua and Wei Shaohua
School of Mechanical Engineering, Nanjing University of Science & Technology, China

Zhang Bingjun and Zhu Xuehua
Nanjing Automotive Technological Center, China

ABSTRACT

According to the situation analysis of automotive dynamic assembly in using, and the fault diagnosis and inspection in production, a set of intelligent fault diagnosis and inspection system was developed based on sound intensity identification. By analyzing the system requirements and defining the design rules, the whole architecture was given, and its work principle was studied systematically. The software that was used in data process, control and fault diagnosis was developed, several key problems were discussed during software was developed, including of building knowledge base, designing the intelligent fault diagnosis model, and describing input/output mode etc. All of these work would provide guarantee for putting this system in practice.

1 INTRODUCTION

Traditionally, noise-based fault diagnosis was based on testing surrounding sound pressure as the research object, by which the results were the average values of sound pressure in those tested point, but this method required a good environment condition, and the frequency distinguishability is lower. However, sound intensity based on cross spectrum technique is a vector, which describes energy current per area in sound fields. So sound power can be measured without reverberation chamber and anechoic chamber, and sound power of machinery can be measured accurately in general circumstance or plant, so as to queue and orient the sound sources expediently and so on. Thus, many research institutes go in for such application research, and some products have been in use. Those research of fault diagnosis based on sound intensity are paid more and more attention. So, using the method of sound intensity to measure sound power of sound source under nonideal conditions, and continuing to carry out theoretic and applied research would have a great influence on the industry and service of automobile.

The automotive dynamic assembly is the biggest contributor to automotive noise, and its exceptional noise often induces unexpected consequence. Therefore, it will be important that to identify the main noise sources and exceptional noise of the automotive dynamic assembly.

2 THE SCHEME OF INTELLIGENT FAULT DIAGNOSIS SYSTEM

2.1 System Requirement Analysis

The intelligent fault diagnosis system of automotive power assembly based on sound intensity identification is a set of equipment which gathers sound intensity measurement technology, computer analysis and control technology etc. It would be able to realize these main functions as below:

(1) The nephogram of sound intensity could be drawn by scanning a side plane near the tested object.

(2) The main noise source of the tested object would be identified automatically, and it could provide a condition for reducing noise.

(3) The noise frequency structure could be identified by analyzing the spectrum.

(4) It could be used as a supervise equipment, to find faults and table a proposal to improve product quality.

The system makes the sound intensity probe move fast in a two dimension plane, and ensure position accurately through stepping motor controlled by a computer, many kinds of analysis can be done automatically by computer software to sound intensity signals measured by probes. Besides, it should have the functions of storing and maintaining data, displaying graphics, printing the testing result tables, and so on.

2.2 System Design Rule

To establish the fault diagnosis system of automotive power assembly, the rules as listed below must be complied:

(1) The system must be safe and reliable when running. It must have the measures of safety and protection; the bump between slide block and track carrying sensors carrier should be reduced when the sensor got to the end of track, and this system also has the ability to obey the instructions from the software.

(2) The system should have strong adaptability to environment.

(3) This system should have enough measurement accuracy, including both the accuracy of magnitude and phase, orientating accuracy and repetition accuracy too.

(4) Easy maneuverability, convenient usage, high automation, high maintainability, and so on.

(5) The software should have complete functions.

(6) Rational economic effect.

2.3 System Whole Scheme

According to above requirement analysis and design rules, the intelligent fault diagnosis system of automotive power assembly was established. the work principle of hardware is presented in Figure 1. According to ISO9614-1 established in 1992, this system adopts continuous scanning method to test sound intensity, and utilizes relative equipments to control the executive mechanism automatically.

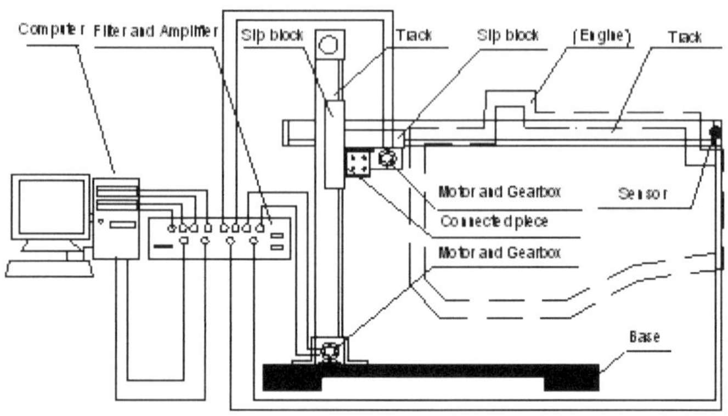

Fig.1. testing system principle

2.4 System Work Principle

It is clearly, sound intensity sensor is fixed on the end of horizontal track and moves with the track in a vertical plane. The analog signals collected by sensors are sent to dynamic signal acquisition and then transferred to digital signals by A/D card, through filter and amplifier, then the data of sound intensity can be processed by computer software and fault diagnosis can be realized at the same time.

User can operate the functional software to send the instructions out to port cards, and digital signals then are translated into analog signals, and then analog signals are enlarged by amplifier, by which the stepping motors can be moved or stopped, the bracket then real-time moves on horizontal and vertical directions, the sensors follows the bracket to move on the same 2D plane. By this software, user can control the sound intensity sensors the scanning step length and scanning velocity.

2.5 Main Components and Performance

Main components of this fault diagnosis system include:

(1) Sound intensity sensor: It is the type 3599, made by B&K, Denmark.

(2) Preamplifier and filter: adopting the type AF-2, made by COINV, China.

(3) Dynamic signal acquisition and generation: adopting the type NI-4451, with 2 AI/2 AO, 16-bit, made in USA.

(4) Track and its components: MAX line system, type MZK060, Germany.

(5) BERGER LAHR 3-phase stepping motor: type VRDM-364/LHA, rated torque 0.45N.m.

(6) Gearbox: type LP050, made by MAX, Germany, ratio 5:1.

3 DESIGNING SOFTWARE FUNCTIONS

Processing signals gathered by sensor, controlling the executive mechanism and diagnosing faults to automotive power assembly, all would be realized by applied software. The structure of fault diagnosis software based on sound intensity identification is shown as Figure 2.

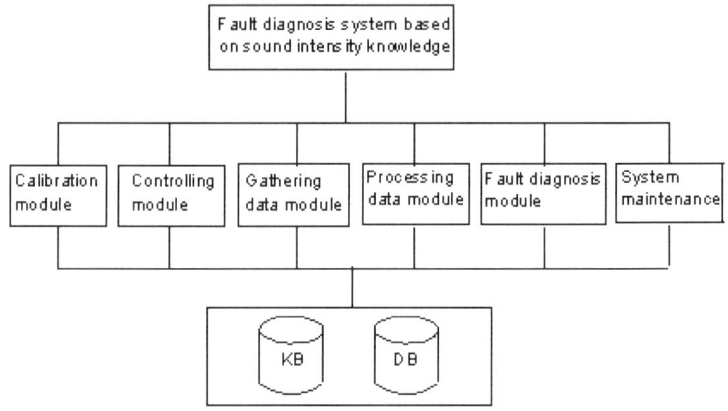

Fig.2. software function structure

It can be see from Figure 2, system software consists of 6 functional modules:

(1) Calibration module: initializing hardware and software of the system.

(2) Controlling module: controlling hardware running and testing modes.

(3) Gathering data module: acquiring sound intensity and sound pressure signals by different methods.

(4) Processing data module: processing sound intensity and sound pressure signals.

(5) Fault diagnosis module: processing and analyzing those acquired data, finding exceptional phenomena, and diagnosing the faults of automotive dynamic assembly by neural network and Knowledge Base.

(6) System maintenance: maintaining KB and DB, including of adding, deleting and modifying to information, it also can provide essential explanation of function, performance, operation.

The system kernel is KB and DB. DB is used for storing and managing initial data information, result information, middle process information and experienced information in correlative reference and so on. The relation data structure was adopted to establish DB.

The tool, developing the system software, is VB language, and the interface between hardware and software will realize information alternation by using the interface function.

4 DESIGNING KNOWLEDGE BASE

KB stores and manages present fault modes of automotive power assembly, which mainly deals with 3 aspects: knowledge acquirement, knowledge denotation and knowledge management.

4.1 Knowledge Acquirement

Knowledge stored in KB includes common fault phenomenon, fault state, fault sources of automotive power assembly and those measures taken to eliminate faults. These knowledge mainly derives from practical experience accumulation, consequence to analyze pre-setting faults, relative references, and knowledge obtained from machine self-study. Practical working experience accumulation means to consult technicians and operators, and then record fault phenomenon, sources and measures to eliminate faults during practically producing automotive power assembly, and then to amend them is necessary. To analyze pre-setting faults means to setup potential faults to automotive power assemblies, make assemblies running, and observe the characteristics showed by them, and then store them in KB as samples. Fault modes from references and measures to eliminate faults are important sources for KB, which are the accumulation from scholar and expert experiments.

4.2 Knowledge Denotation

Knowledge denotation is the process of signifying and formalizing the knowledge. At present, the productive rule denotation is the main methods for knowledge denotation, and knowledge description can be done by relation datasheet, those increasing conception and rules can also be denoted by using its expansible items easily. This design method of knowledge system is mature and convenient, so it is used in our research.

The productive rule is denoted as following.

IF a and b or c THEN d and e CF=k

In this expression, a,b and c are conditions, d and e is conclusion, CF denotes the credibility of this rule, k ranges from 0 to 100. In this research, a,b and c are the fault phenomena or keywords, d is the fault sources, and e is the measures to eliminate faults.

The relation datasheet structure is shown as Table 1.

Table 1 Fault diagnosis rule table

Expression	Meaning	Denotation	Meaning
RuleID	Rule mark	SourceOfTrouble	Fault source
FailureModeID	Exceptional mode mark	WeightingFactor	Fault source weight
FailureMode	Exceptional mode	DiagnosticTool	Diagnostic tool
FailureDegree	Exceptional degree	MaintainWay	Maintain description
SourceTroubleID	Fault source mark	JudgeMark	Judge mark

5 FAULT DIAGNOSIS PRINCIPLE FOR AUTOMOTIVE POWER ASSEMBLY

Developing fault diagnosis system was final goal, that is to say, the final diagnosis results can be given by theoretical modeling and numerical analysis. Neural network had been used to establish the diagnostic model of automobile dynamic assembly in this research.

5.1 Model of Intelligent Fault Diagnosis

In this research, the knowledge neural network model is defined as an aggregation composed of knowledge process and artificial neural network. First, the mechanism of knowledge acquirement and knowledge transaction are set up, knowledge processed are stored in KB, and a set of training samples is provided to neural network. Input vectors of this neural network are acquired from KB directly, the new mapping relation between network input and output can be written into KB again by self-study module. The diagnostic efficiency could be evaluated by the module of efficiency evaluation, then the network weight value and threshold are fixed automatically to get to the best efficiency of the knowledge neural network. The model of knowledge network is shown as Figure 3.

5.2 System Input/Output Mode Description

The prototype developed in this research have 3 input modes (shown as Figure 4): □ inputting directly the exceptional mode codes (exceptional phenomenon codes), but user must know the corresponding relation between codes and faults when using this input mode; □ inputting the captions or description of fault phenomenon. In this mode, user can input several keywords or integral description of fault phenomenon, the system will recognize

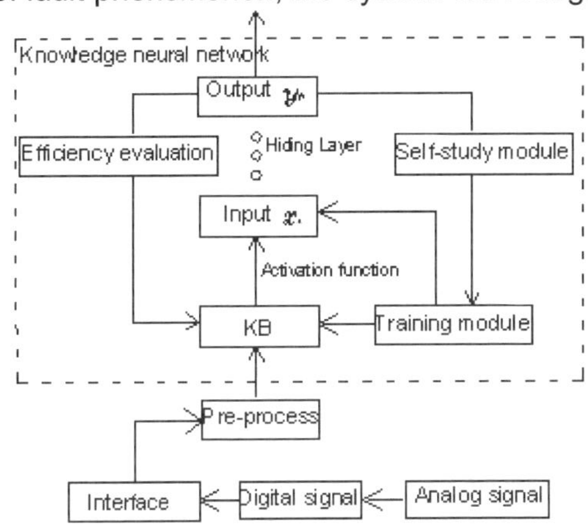

Fig.3. Knowledge network model

automatically the keywords, and then find out relevant input vectors in the KB for the network to analyze; □ getting information directly from tested objects through connecting to signal acquirement equipments, But the knowledge is *rough*, and it must be preprocessed before being written into KB, it is possible for the intelligent fault diagnosis to realize absolute automation.

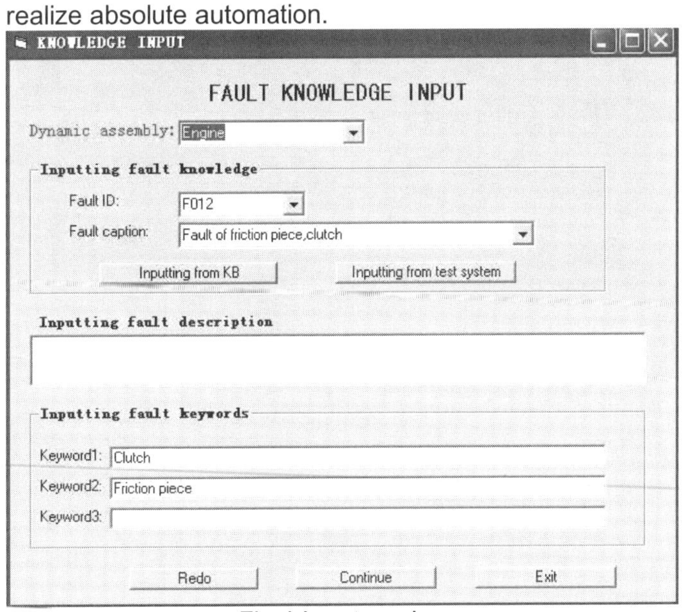

Fig.4.Input modes

In order to meet practicality, the system output modes are designed visually, vividly and directly, shown as Figure 5. According to the sorts of automotive assembly, its sketch can be was shown on the output interface, and according to the reasoning results, the fault positions can be displayed in the sketch, and the fault causes and the eliminating fault methods are marked near the sketch. For example, a fault source of a clutch and the eliminating its method are described in Figure 5.

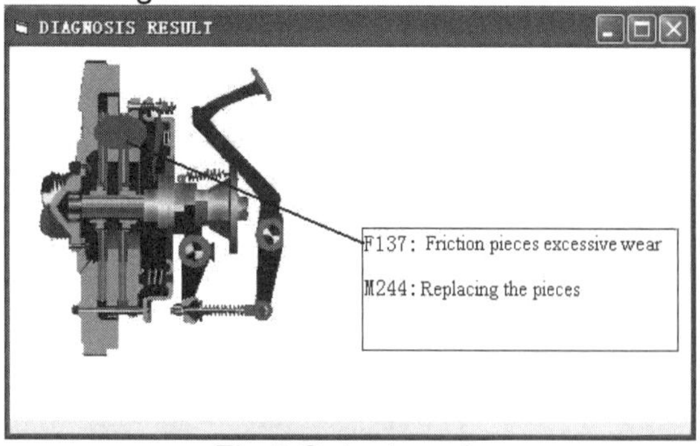

Fig.5.Output mode

6 CONCLUSION

Automotive fault diagnosis based on sound intensity identification is one of the focuses in the field of automotive research. By describing sound intensity test method, the hardware and software architectures of this automotive fault diagnosis system were analyzed systematically, and several key techniques during establishing the system were discussed in this paper. This discussion is the conceptual design for intelligent fault diagnosis system of automotive power assembly, it is the description of system designed thought, which will also provide technical guidance to the detailed design and putting the intelligent fault diagnosis of automobile in practice.

REFERENCE

1 M. Shakeri et al., Sequential Testing Algorithms for Multiple Fault Diagnosis. IEEE Transactions on SMC: Part A-Systems and Applications, Vol. 30, No. 1, pp. 1-14, January 2000.

2 P.M. Frank. Fault diagnosis in dynamic systems using analytical and knowledge based redundancy A su rvey and new results. Automatic, vol. 26, pp. 459-474, 1990.

CONTACT

I received my B.A.Sc. of Mechanical Engineering in East-China Institute of Technology in 1990, and Ph.D. in School of Mechanical Engineering in Nanjing University of Science & Technology in 1995.

My research interests include fault diagnosis, FEA, intelligent design for vehicle, analytical and experimental modeling on vehicle etc.

E-Mail: cxhua@mail.njust.edu.cn

or: hongzh68@sohu.com

2004-01-0859

Highly Responsive Mechatronic Differential for Maximizing Safety and Driver Benefits of Adaptive Control Strategies

Stuart Hamilton and Mircea Gradu
Timken Automotive

ABSTRACT

Sophisticated AWD technologies start to significantly penetrate the passenger car and crossover SUV market segments, aiming to enhance safety and fun-to-drive, rather than only focusing on the traction characteristics of the vehicle. This paper introduces a new active differential concept, which takes advantage simultaneously of a very comprehensive vehicle dynamics control software strategy and of advanced hardware capable to provide outstanding torque modulation characteristics. The differential assembly developed by Timken Automotive consists of magnetic particle clutches, mounted in a quasi-static torque split arrangement with planetary gear sets. This new mechatronic module offers simultaneously good power and torque density, which makes it suitable in particular for front wheel drive based vehicle platforms converted to all wheel drive configurations. A major advantage is the fact that the torque transfer characteristic does not depend on the input-output relative speed. Also, the differential actuation is purely electrical, allowing a high flexibility for the control algorithms. The paper presents a summary of the mechatronic module development process, its integration in the vehicle demonstrator and the test results validating its superior performance.

INTRODUCTION

There is a consensus between the automotive analysts regarding the fact that, although introduced as a niche product, the AWD systems will evolve similarly to ABS, becoming mainstream in the near future, and the emphasis will be on their interaction with the rest of the vehicle. Driveline configurations will be based on existing, proven technology, combined and utilized in an innovative manner.

The shift of priorities among different drivers and enablers for the 4WD technologies is illustrated in Fig.1. While traction improvements were the focus in the 1995-2000 time frame, safety, handling and fun-to-drive are becoming more important in the future. From the technical standpoint, providing a cost effective solution with very good torque modulation is essential (Fig.2). The control of the vehicle dynamics by using active differentials creates exciting technical and commercial opportunities. The share of the on demand active technologies increases continuously, and will reach an estimate of 40% of the total 4WD market by 2006 (Fig.3).

MECHATRONIC MODULE DESCRIPTION AND CONCEPT VALIDATION

The majority of the current design solutions for torque bias couplings used in 4WD applications are based on wet-plate clutches (Fig.4). The amount of torque transferred by the coupling can be varied by modifying the number of friction plates engaged or by modulating the applied pressure. Several mechanisms are used for generating and controlling the pressure, including ball-ramp arrangements, gear-pumps, axial pistons / cam plates, hydraulic valves etc. The characteristics of the wet-plate clutch depend on the friction coefficient, relative speed, pressure and temperature, with the friction coefficient at its turn being a function of pressure, relative speed and temperature. The pumping mechanism operates also strongly depending on relative speed, fluid viscosity and temperature. These influence factors lead to complicated control algorithms involving multi-dimensional maps that have to be defined and then utilized by the electronic control units (Fig.5a).

The Timken mechatronic solution addresses the stringent requirement for controllability by employing a magnetic particle clutch, coupled, in a quasi-static torque split arrangement with a planetary gear system (Fig.6). Magnetic particle clutches are well known for very good torque modulation capabilities and for torque characteristics that are independent of the differential speed in the clutch. The main control parameter of the clutch is the applied current (Fig.5b), which energizes a coil and aligns the steel particles similarly to bridges in the gap between the clutch rotors. The torque is transferred between the rotors due to the friction with the micron-sized steel particles. The torque split arrangement increases significantly the torque capacity of the coupling, directing only a fraction of the torque through the magnetic particle clutch. This principle is confirmed by the measurement results presented in Fig.7, which shows also the torque balance equations.

Magnetic particle clutches were used in several production or concept automotive driveline applications, without employing torque split concepts (Fig.8), which led to relatively large and heavy designs. In case of applying the torque split principle, based on the power conservation, beside the torque reduction, the magnetic particle clutch can benefit from an increase of the relative speed between its rotors, compared to the coupling input / output differential speed. Accordingly to the clutch manufacturers, differential speeds over 10 RPM contribute significantly to the elimination of any torque fluctuations.

Beside the coupling configuration, a transfer case equivalent design (Fig.9) was prototyped and used for concept validation testing. In this arrangement, the low torque component accommodating the magnetic particle clutch is the carrier of a double planet epicyclical gearbox, with two outputs, corresponding to the front and rear axles. The prototype was tested on a rig allowing for the simulation of AWD specific conditions, including continuous slip (Fig.10). The main characteristics determined on the test rig were hysteresis (torque vs. current), torque vs. slip speed and torque ramp-up / ramp-down response and modulation capabilities (Fig.11). Based on the data published by the magnetic particle clutch manufacturer regarding the heat dissipation characteristics of the clutch, similar charts can be derived for the Timken torque bias coupling. In Fig.12, the heat dissipation curve under intermittent slip represents the upper limit for the area defining the torque differential speed allowable operating conditions.

INTEGRATION IN THE DRIVELINE OF THE VEHICLE DEMONSTRATOR

Controlled torque distribution within the vehicle driveline can be realized by employing the Timken mechatronic module at the transfer case location or at the axle center, in a side-to-side (twin) arrangement or in a front-to-rear (in-line) arrangement. In order to maximize the benefits resulting from the control strategy and obtain both front-to-rear and side-to-side torque bias, the twin arrangement at the rear axle was selected (Fig.13).

A torque vs. vehicle speed chart can be used to determine the coupling torque capacity requirements. At low speed / high torque, the emphasis is on traction enhancements, and one possibility is to use a locking device in parallel with the coupling, for maximizing the torque transfer to the secondary drive axle. At higher speed, lower torque, vehicle dynamic control prevails, and this is the operating range where the good torque modulation capability of the Timken coupling can bring a significant advantage. The chart in Fig.14 illustrates the fact that the coupling torque capacity requirement can be based on the maximum transmissible wheel torque for different driving surface conditions and/or on the maximum available torque at the wheel at a certain vehicle speed.

The vehicle demonstrator (Fig.15) based on a Ford Maverick/Escape 3.0L (V6 with automatic transmission) was build with the main goal of proving the mechatronic module enhanced functionality and the corresponding benefits in terms of vehicle dynamics and traction. Accordingly, due to an extremely tight development schedule, some of the components of the active differential prototype were adapted from other automotive applications. The magnetic particle clutches are from the shelf available devices, used primarily in industrial environments, with harsher duty cycles, especially with respect to load magnitude and duration. This is an interesting aspect, because it justifies a significant downsizing potential of the clutch, when designed specifically for an automotive application. The axle center (Fig.16) can be split in 4 units: the left and right clutch assemblies, the epicyclical gear assembly, which replaces a conventional bevel gear differential cluster and the pinion cartridge assembly (Fig.17), which in a more integrated format is part of the current product offering [2]. This elegant design solution permits the complete separation of the gear components requiring specific lubrication from the magnetic particle clutch components, requiring a dry sealed environment. It also ensures the ease of servicing the clutches and provides better temperature monitoring opportunities in the experimental phases.

Fig.18 indicates the location of the different mechanical components and sensors belonging to the active differential system within the vehicle. The electronic control unit Proteus, developed by Prodrive, centralizes all the sensor inputs (Fig.19) and makes the decision regarding the current applied on the clutches, which univocally determines the torque at the corresponding wheel.

A 4 positions switch was installed on the dashboard (Fig.20), allowing the driver to choose between 4 predefined torque control strategies. For demonstration purposes, the following modes of operation were considered:

1. Front Wheel Drive only,
2. AWD with traction enhancements only,
3. AWD with traction and yaw authority enhancements,
4. AWD with traction and yaw authority enhancements more sporty oriented.

VEHICLE DEMONSTRATOR TEST RESULTS

The different aspects of the torque control strategy employed in the vehicle demonstrator are presented in Fig.21. Some of the key tests performed to date at the

Prodrive proving facilities and MIRA in UK are listed below:

- Traction Lo w friction B asalt Drive full throttle
- Traction Sp lit mu Basal t / concrete 1 st gear full throttle
- Traction Sp lit mu front / rear Ba salt / concrete 1 st gear full throttle
- Yaw Authority / traction - Traction from a junction Bri dport pebble full throttle
    - Start on full lock
    - Start straight then turn
- Yaw Authority - Power turn Basalt
- Traction, Yaw Damping and Yaw Authority Low friction Bas alt / concrete Drive
    - Slalom through cones at 20 mph.
- Yaw Damping and Yaw Authority Hig h friction Basalt / concrete Drive
    - Lane change or obstacle avoidance manoeuvre at 65 mph.

The same manoeuvres were performed with the standard vehicle, before the integration of the Timken active differential system, in order to provide a base line, against which the Timken vehicle demonstrator results can be evaluated.
Typical measurement results are shown in Fig.22.

CONCLUSIONS AND FUTURE DEVELOPMENT ACTIVITIES

The improved torque control offered by the Timken mechatronic module in an AWD application was fully confirmed by vehicle test results. The system demonstrated minimal hysteresis, which allowed a fine degree of control to be exercised. Although only in a prototype stage, the Timken mechatronic module performed at the NVH level of a refined production design in conjunction with the Prodrive developed Active Torque Dynamics strategy, proving the fact that the basic concept is correct for the application. The system is able to demonstrate strong yaw authority and yaw damping, maintaining simultaneously a very good level of traction enhancement. The steering control over the standard vehicle is improved in variable traction events.
Future development work will also concentrate on mechanically re-designing the entire rear axle unit as a more cohesive design as opposed to a set of individual components matched together. This exercise will greatly increase the torque and power density of the unit.
In conclusion, the Timken mechatronic module combined with an advanced control strategy represents a very promising technology for enhanced safety and fun-to-drive in AWD automotive applications.

REFERENCES

1. Wesley M. Dick : All-Whell and Four-Wheel-Drive Vehicle Systems. SAE SP-1063.
2. Mircea Gradu : Investigation of Package Bearings to Improve Driveline Performance. SAE Paper 2000-01-1785.
3. SAE TopTec: Innovations in Four Wheel Drive / All Wheel Drive Systems. South Bend, Indiana. April 12-14, 1999.
4. Advances in 4 Wheel Drive Vehicle Systems Toptec. Ypsilanti, Michigan, June 4-5, 2002.
5. Randolph C. Williams : 4WD Market Trends in Vehicles & the Adaptations of New Technology in the International Automotive Community. SAE Paper 1999-01-1264.
6. Michael Hoeck, Christian Gasch : The Influence of Various 4WD Driveline Configurations on Handling and Traction on Low Friction Surfaces. SAE Paper 1999-01-0743.
7. US Patent 6,098,770. Clutch Assembly Having Reaction Force Circuit. Aug.8, 2000.
8. US Patent 5,979,631. Device for Transmitting Torque Between Two Rotatable Shafts. Nov.9, 1999.
9. U.S. Patent 4,656,889. Transmission Sytem with Lamella Coupling for All Wheel Drive Vehicle. Apr.14, 1987.
10. U.S. Patent 4,417,641. System for Controlling Two-Wheel and Four-Wheel Drives. Nov.29, 1983.
11. U.S. Patent 5,888,163. Hydraulic Coupling for Vehicle Drivetrain. Mar.30, 1999.
12. U.S. Patent 4,871,049. Coupling Device of Magnetic Particle Type. Oct.3, 1989.
13. What s the Diff ? Automotive Industries, June 2000.
14. Torque transfer design targets cost and weight. European Automotive Design. May, 2002

ACKNOWLEDGEMENTS

The authors would like to thank Mr. Rich Adams (Vice President Timken Automotive Chassis Group) and Mr. Russ Folger (Vice President Timken Automotive Chassis Engineering) for their continuous support and encouragement on this ambitious project.

CONTACT

Stuart Hamilton
Global Market Unit Manager New Products and Processes, Timken Automotive
Tel.: (330) 471-7280
E-mail: hamiltst@timken.com

Dr. Mircea Gradu - Manager Advanced Technologies
Timken Automotive
Tel.: (330) 471-2279
E-mail: gradum@timken.com

Fig.1 - New niches are being created by changing customer preferences:

↓ Priority level ↓

Drivers / Enablers for the 4WD Technologies	1995-2000	2000-2006
Safety (ESP, VDC)	3	1
Traction Improvements	1	5
Handling / Fun-to-drive	6	2
Fuel Economy / Efficiency	4	4
Price Differential by Year	2	3
Weight / Compactness	5	4
Adaptive Controllability & Responsiveness	7	4

Fig.2 - While system cost remains important, system responsiveness and controllability is one of the highest technical priorities

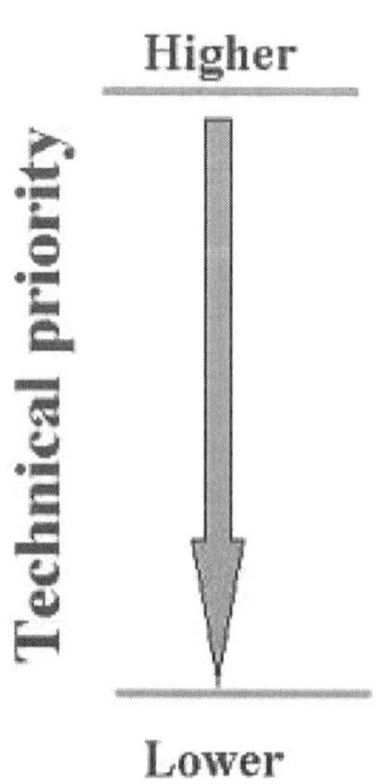

- System Cost
- Adaptive Controllability Based on Very Good Torque Modulation Capability
- Short Engagement / Disengagement Time (ABS/ESP Compatibility)
- High Efficiency
- Compact Envelope for Vehicle Packaging
- Reduced Weight
- High Torque Capacity

Fig.3 - Dynamic Vehicle Control through Active Differentials technology will create exciting market opportunity

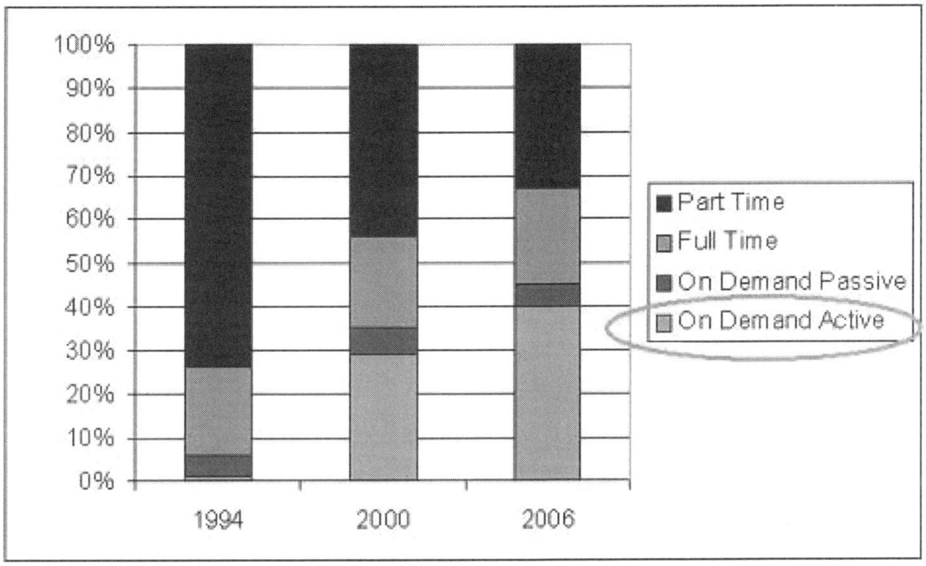

Fig.4 - The majority of the current Torque Bias Coupling designs are based on Wet-plate Clutches

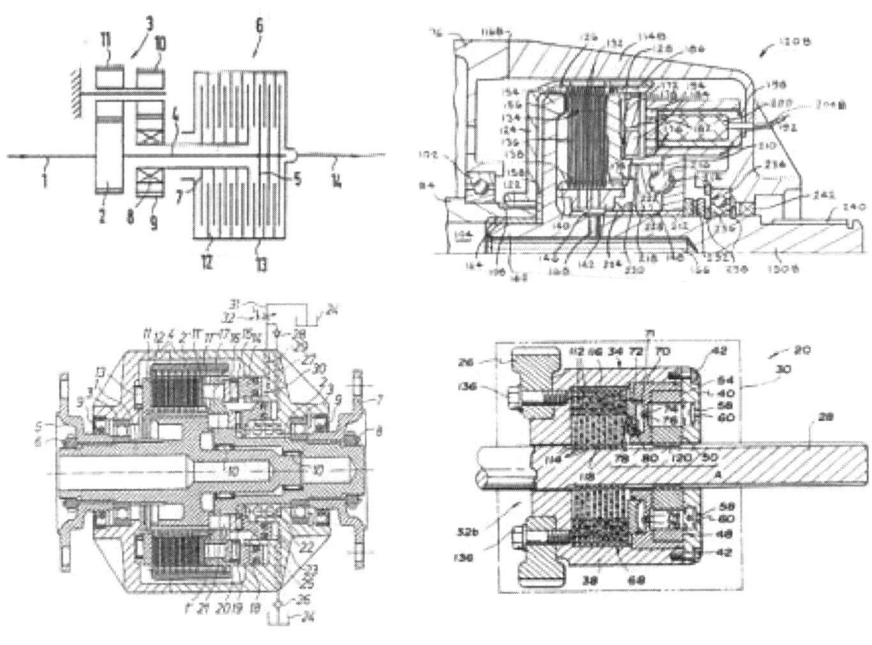

Fig.5 - Torque Bias Coupling Control Parameters

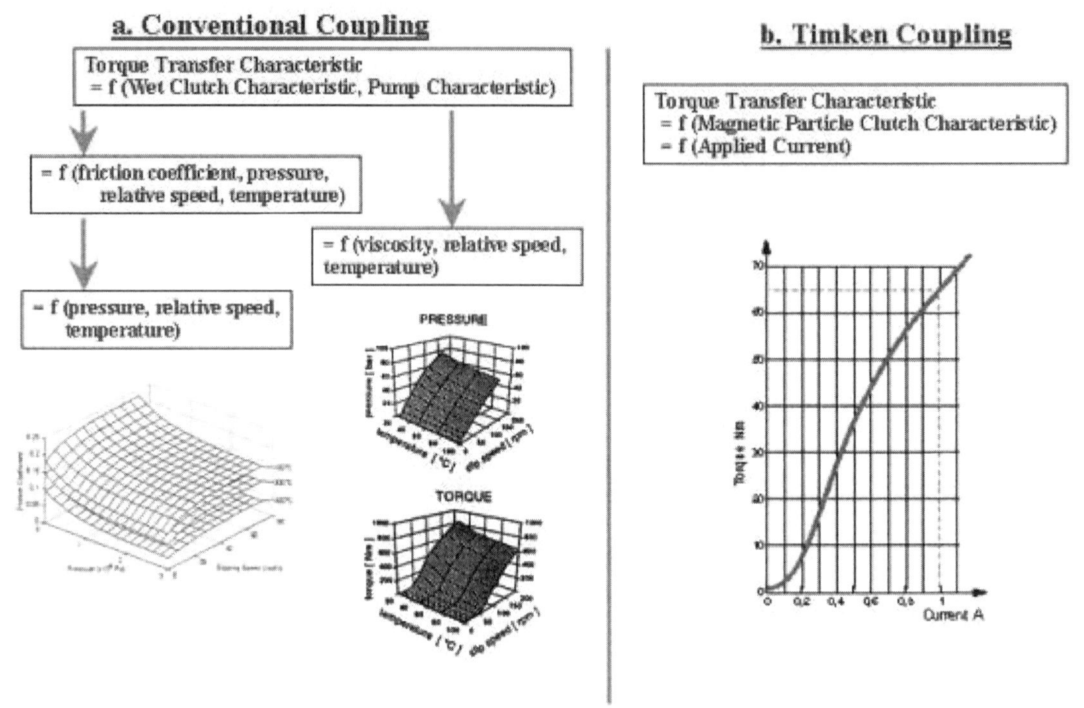

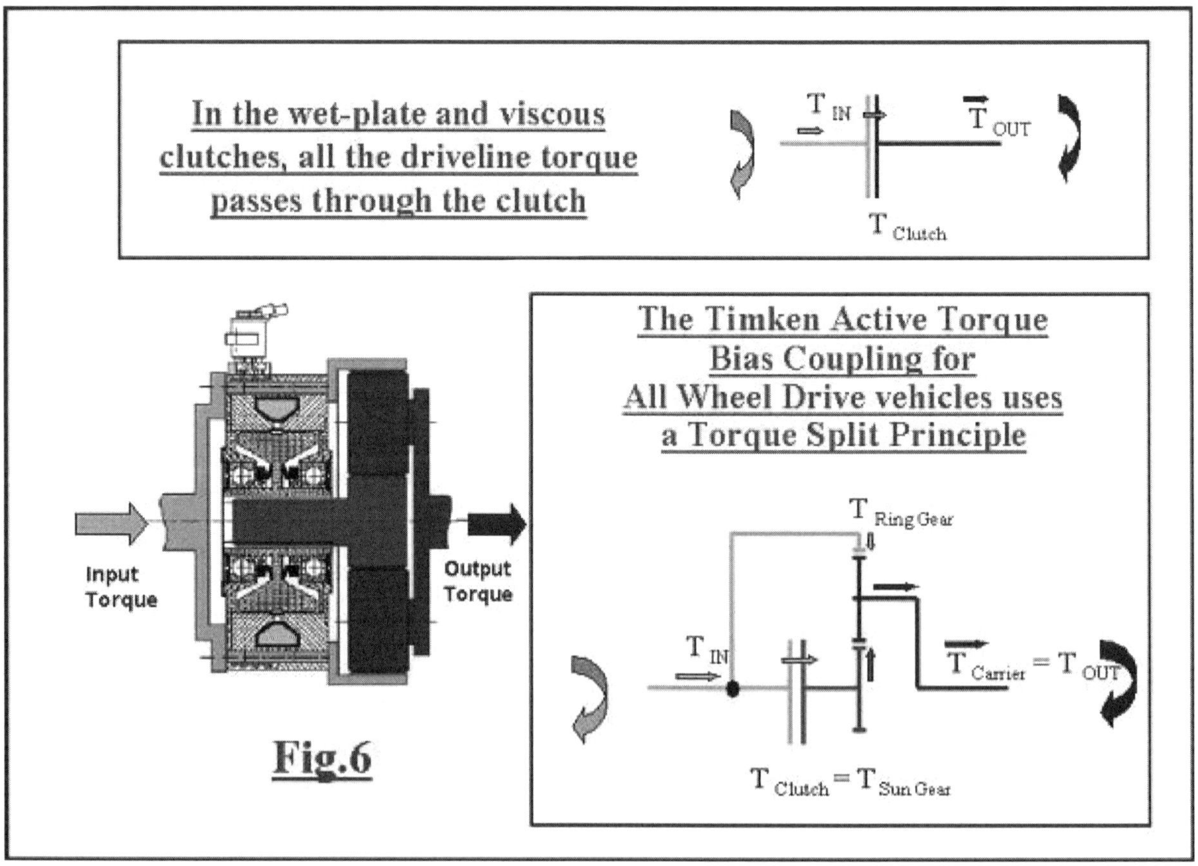

Fig.6

Fig.7

Measured Low Range Torque Characteristics of the Magnetic Particle Clutch (MPC) and of the Torque Bias Coupling (TBC) consisting of MPC + Planetary Gear with K=3.34

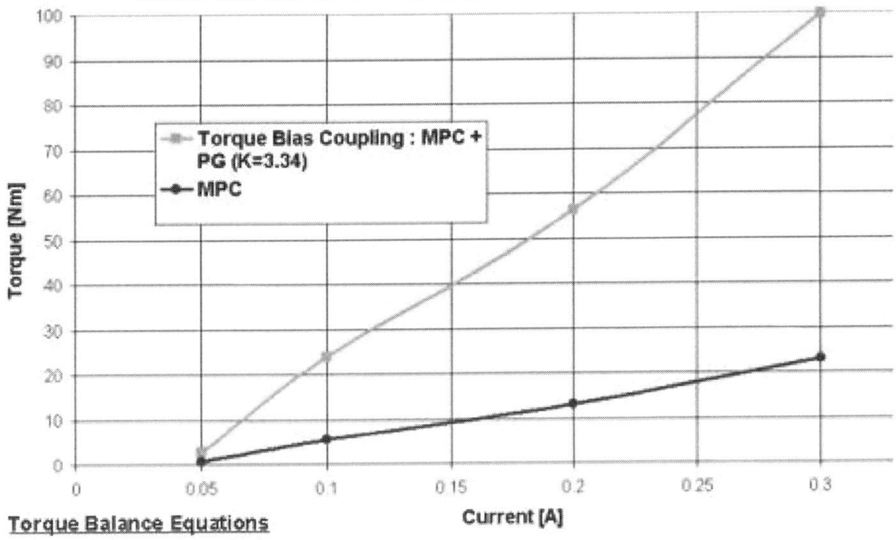

Torque Balance Equations

[1] $T_{Carrier} = T_{Ring\ Gear} + T_{Sun\ Gear}$

[2] $T_{Clutch} = T_{Sun\ Gear}$

[3] $T_{OUT} = T_{Carrier}$

[4] $(T_{Ring\ Gear} / T_{Sun\ Gear}) = (\text{Teeth}\ \#_{Ring\ Gear} / \text{Teeth}\ \#_{Sun\ Gear}) = u$

Fig. 8 - Magnetic Particle Clutches were used in several automotive driveline applications, without employing torque split concepts

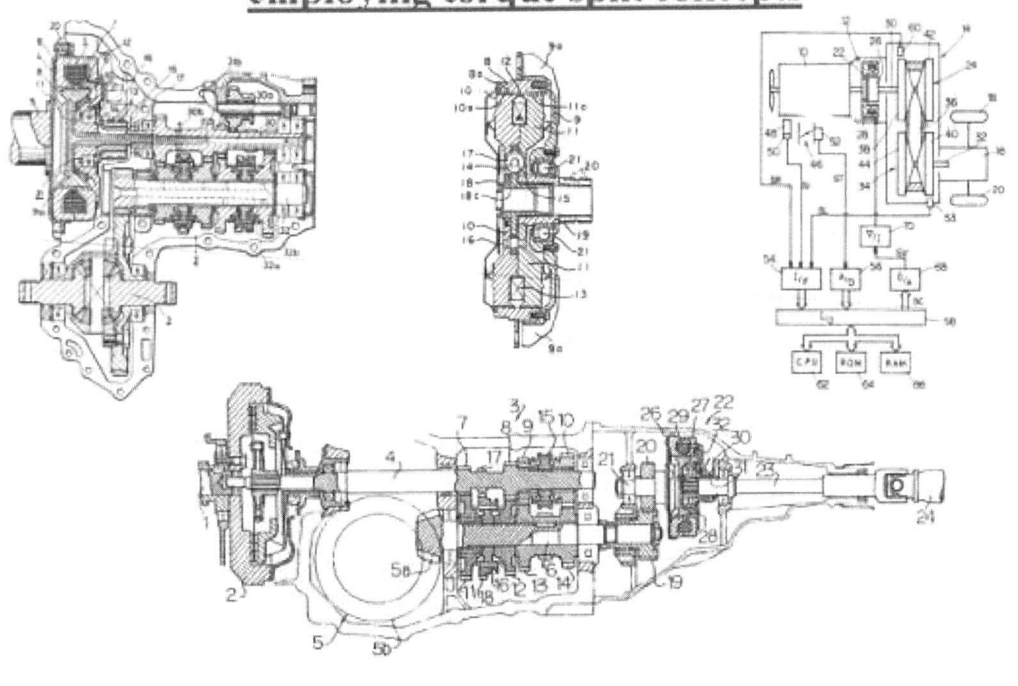

Fig.9 - The Timken Phase I Prototype of the Torque Bias Coupling in a Transfer Case Configuration

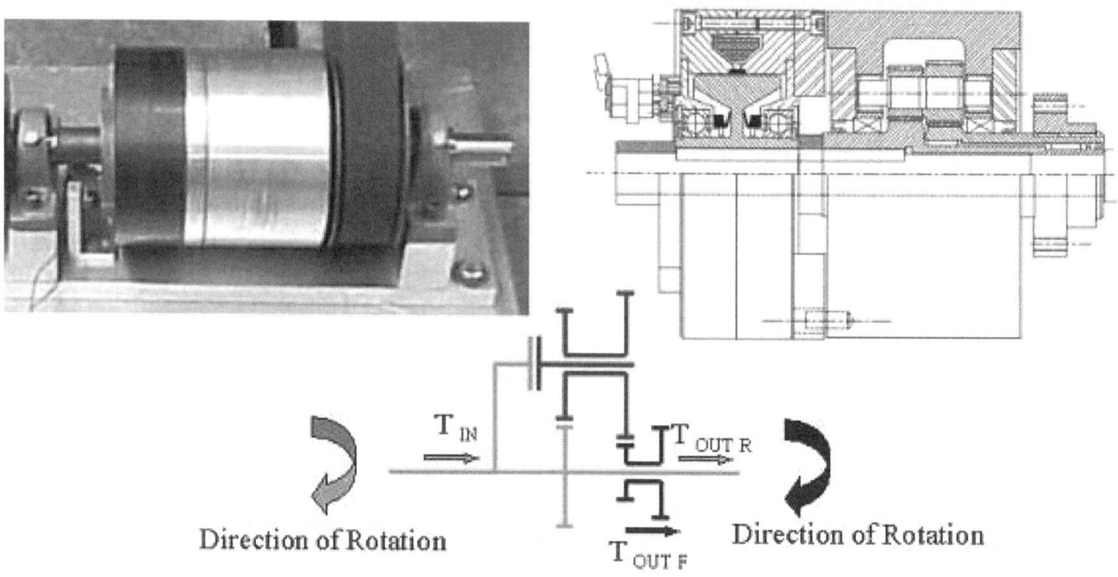

Fig.10 - The Timken Test Rig allows for the simulation of AWD specific operating conditions

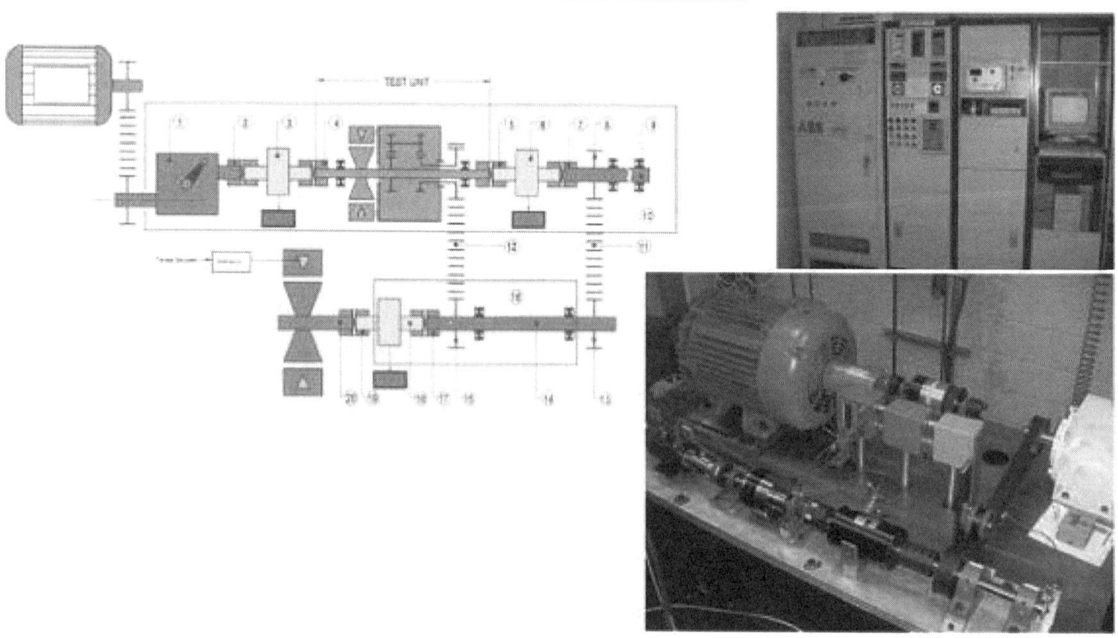

Fig.11 - The Hysteresis (Torque vs. Current), Torque vs. Slip Speed, and Acceleration – Deceleration Characteristics were determined on the test rig

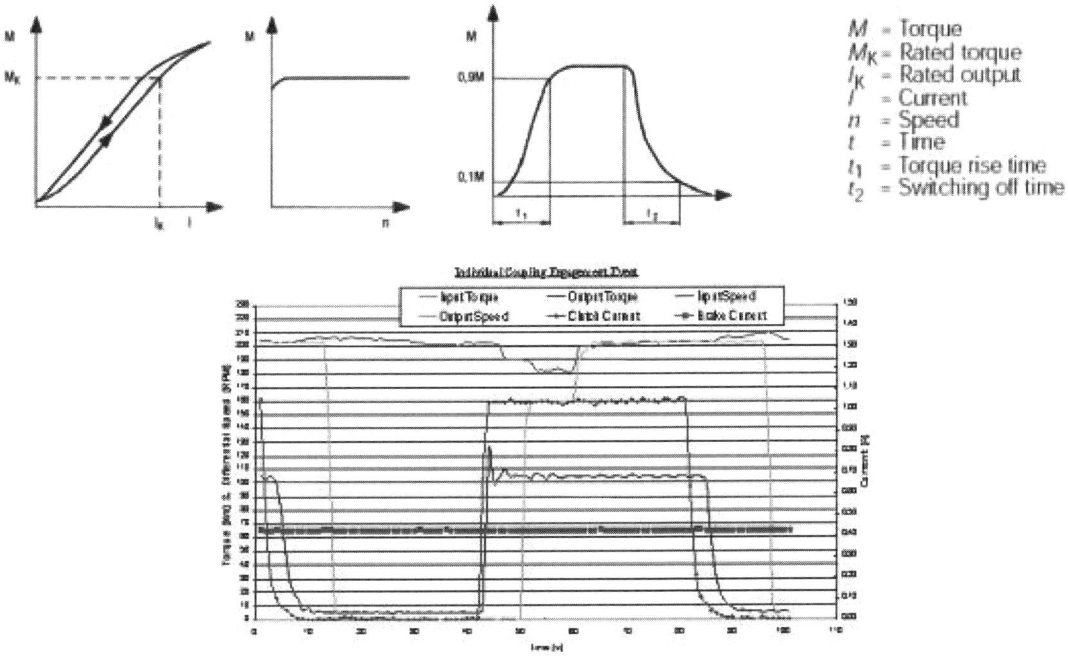

Fig.12 - A Torque - Slip Speed chart is used to establish the TBC operating range

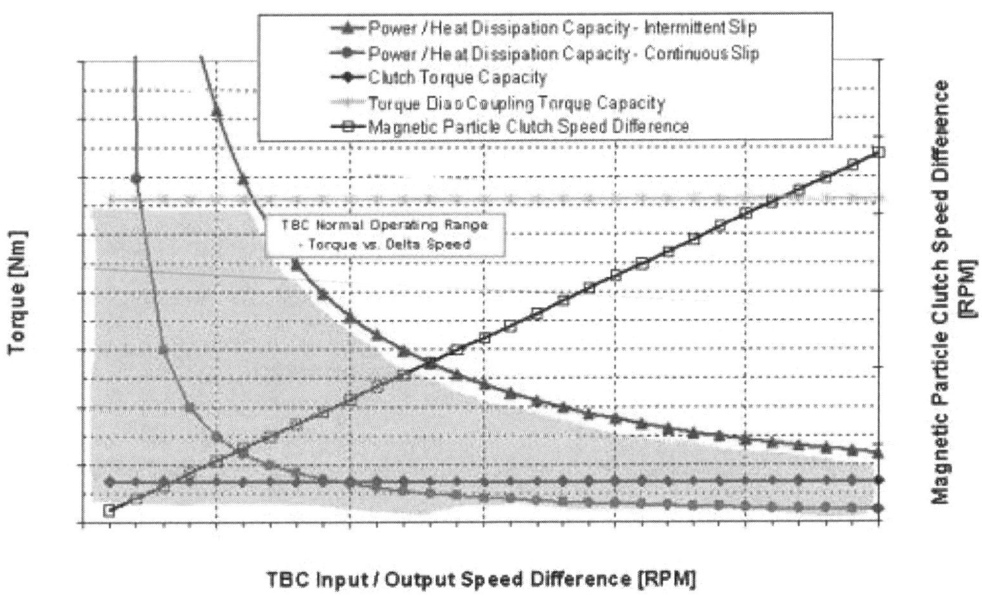

Fig. 13 - Controlled Torque Bias can be realized at the Transfer Case location or at each Axle Center

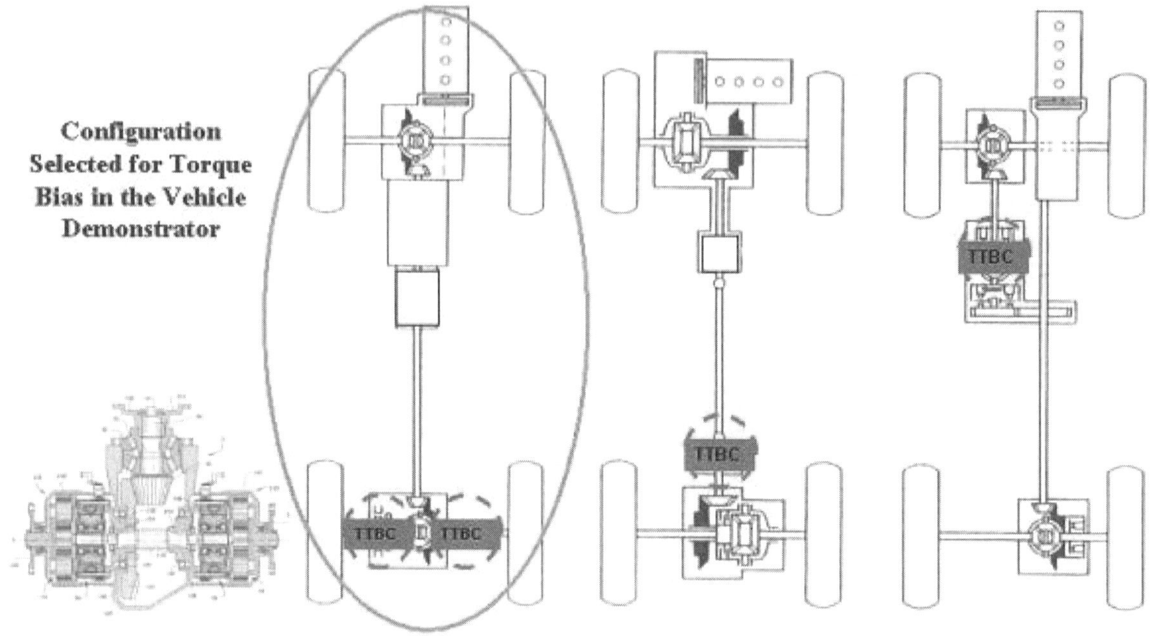

Fig.14 - A Torque - Vehicle Speed chart can be used to establish the Torque Bias Coupling (TBC) operating range

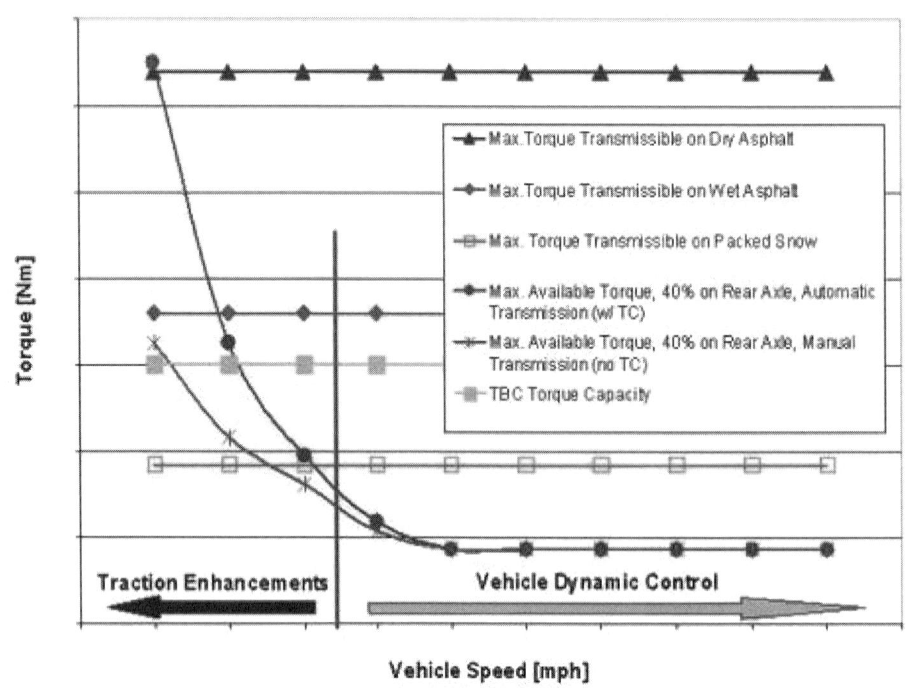

Fig.15 – The Timken-Prodrive Vehicle Demonstrator uses a Ford Maverick/Escape 3.0L V6 platform

Fig.16 – The Rear Axle Center of the Vehicle Demonstrator can be split in 4 units:

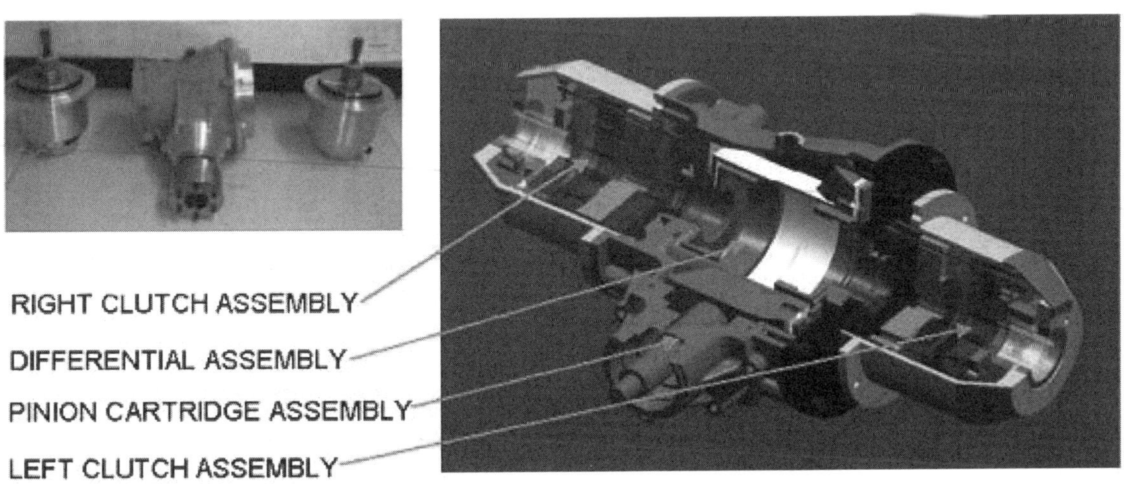

Fig.17 – The Pinion Shaft is Supported by a Bolt-on Cartridge Assembly of Pinion-Pac Type

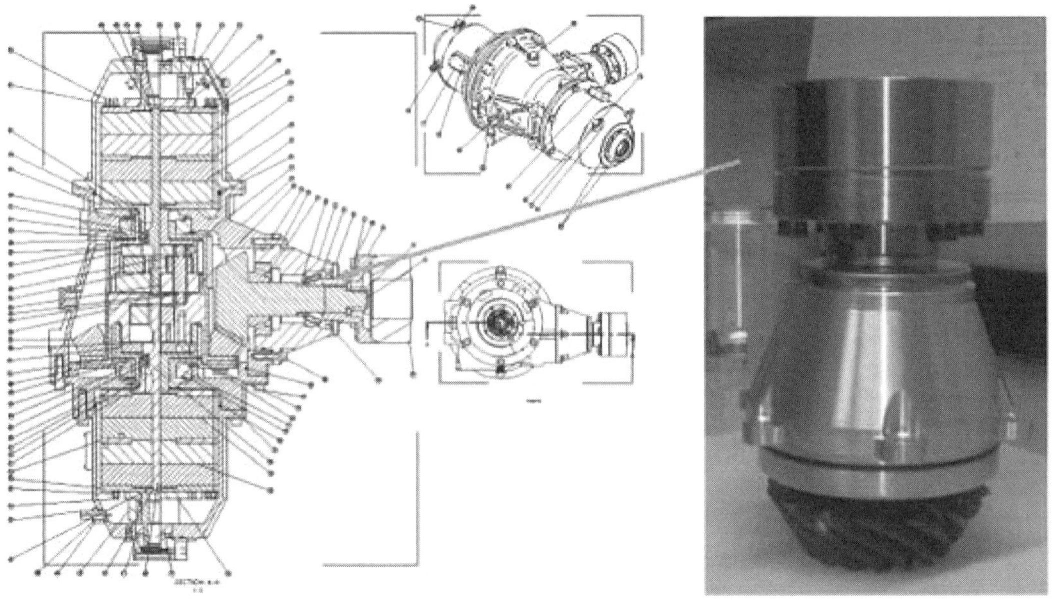

Fig.18 – Location of Sensors and Different System Components within the Vehicle Demonstrator

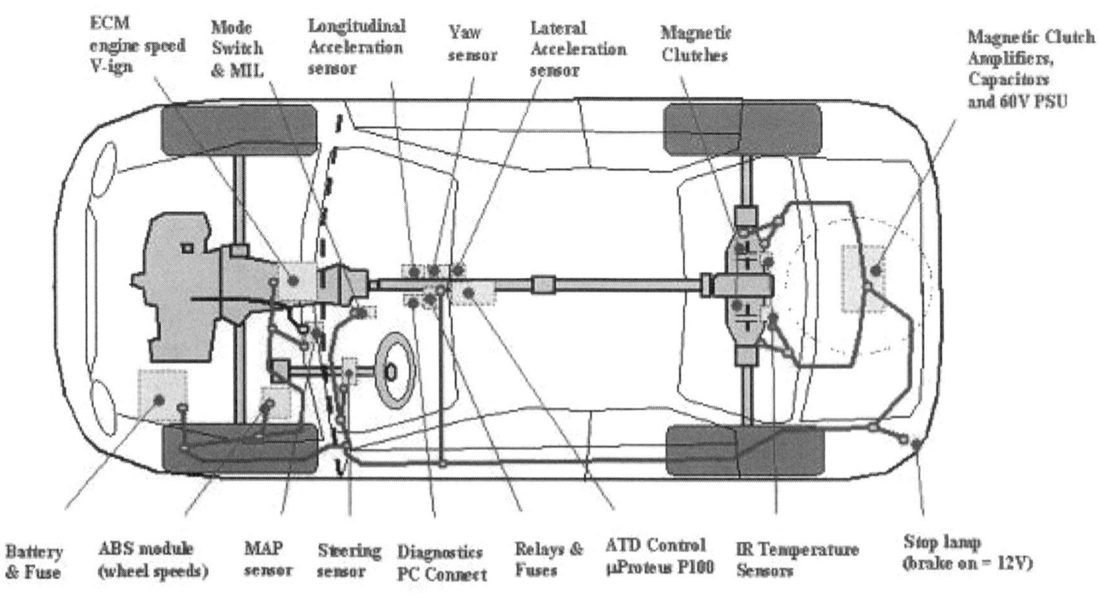

Fig.19 – The Proteus ECU centralizes all the Sensor Inputs and determines the Current applied on the Magnetic Particle Clutches

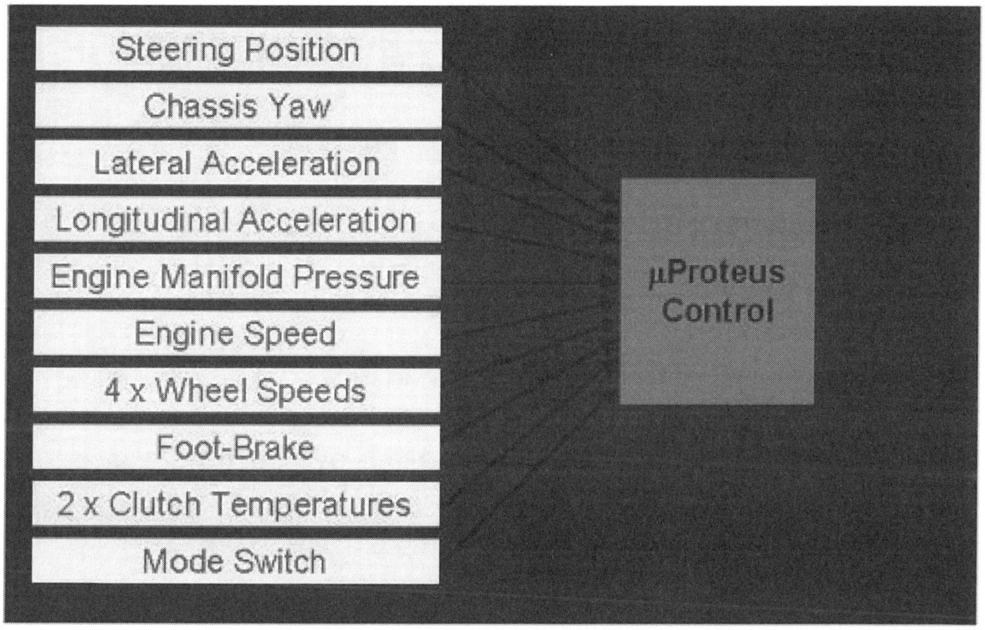

Fig.20 – The Proteus ECU and the Operating Mode Switch allow easy access for further testing and development work using the Vehicle Demonstrator

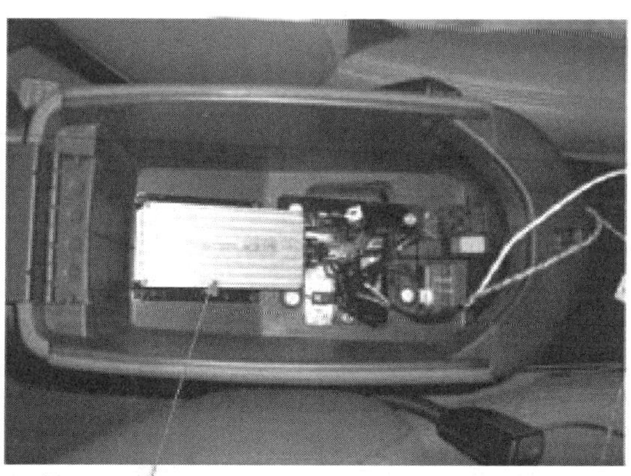

μ Proteus control unit in centre console

Active Torque Dynamic control switch

Fig.21 – Highlights of the Active Differential Control Strategy

- Three calibrations: Traction (only), Normal and Sport
- Monitoring of chassis movement and steering demand
- Detection of over and under-steer conditions
- Detection of wheel-slip
- Engine Load approximation using a M.A.P. sensor
- Independent control of clutches to improve vehicle handling
- Independent control of clutches to improve traction
- Failure Mode Detection and Strategy

Fig.22 – Example of Vehicle Demonstrator Test Results

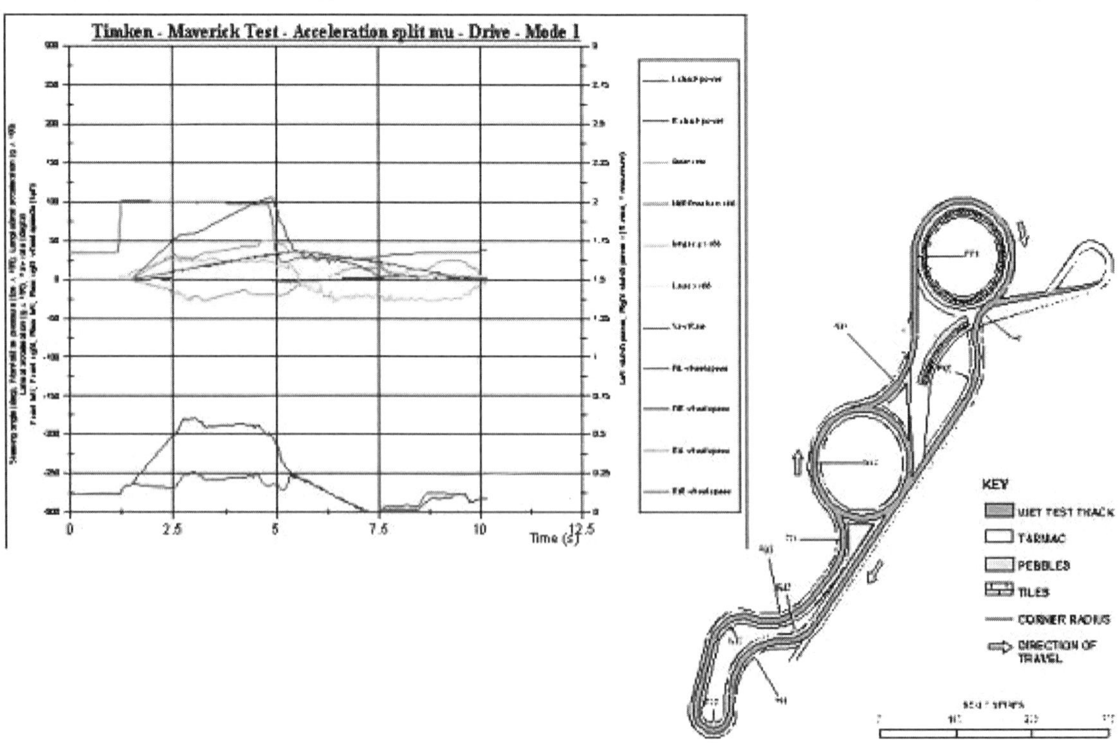

2004-01-0685

Optimization and Robust Design of Heat Sinks for Automotive Electronics Applications

Fatma Kocer, Sid Medina, Balaji Bharadwaj, Rodolfo Palma and Roger Keranen
Visteon Corporation

ABSTRACT

The increasing power requirement for automotive electronics (radios, etc.), combined with ever-shrinking size and weight allowances, is creating a greater need for optimization and robust design of heat sinks. Not only does a heat sink directly affect the overall performance and reliability of a specific electronics application, but a well-designed, optimized heat sink can have other benefits such as eliminating the requirement for special fans, reducing weight of the application, eliminating additional heat sink support structures, etc.

Optimizing heat sink efficiency and thermal performance offers a challenge, due to the many competing design requirements. These requirements include effecting greater temperature reductions, accommodating vehicle packaging requirements and size limitations, generating a uniform heat distribution, etc., and all the while reducing the heat sink cost. Furthermore, a good design would also consider the possible effects on performance of dimensional variations, which result from the manufacturing processes in use - such as die casting, extrusion and stamping.

This paper examines these design issues, and outlines an optimization and robust design framework for heat sink design, using the capabilities of iSight, and additional in-house heat sink design software. The case of an optimization, and robust design studies, on a radio heat sink will be examined in this paper.

INTRODUCTION

The increasing power requirements for automotive electronics (radios, etc.), combined with ever-shrinking size and weight allowances, is creating a greater need for optimization and robust design of heat sinks designs which include minimal cost trade-offs in material and manufacturing.

One reason the heat sink plays an important role in electronics application design is the requirement for semiconductor device junction temperature control. Junction temperature of a semiconductor device is a key factor in the performance reliability and life expectancy of the electronics application. In fact, studies of the relationship between reliability and the operating temperature of a typical silicon semi-conductor device show that a reduction in the junction temperature corresponds to an exponential increase in the reliability and life expectancy of the device. Therefore, controlling the device operating temperature within the limits set by the device manufacturer is essential for long life and reliable performance of a component and this control is directly dependent on the heat sink.

Heat sinks are devices that assist in dissipating heat from a semiconductor device to a cooler ambient; usually air (Figure 1). In most situations, heat transfer across the interface between the device and the cooler ambient is the least efficient within the system, and the solid-air interface represents the greatest barrier for heat dissipation. A heat sink lowers this barrier by increasing the surface area that is in direct contact with the cooler ambient. This allows more heat to be dissipated and lowers the device operating temperature. The primary purpose of a heat sink is to maintain the device temperature below the maximum allowable temperature specified by the device manufacturer.

Figure 1: Heat Sink of a Radio

The heat sink design impacts the performance of other components in the following manner:

(i) eliminating the requirement for fans and other cooling methods,

(ii) weight reduction of the entire application in order to pass shock and vibration tests,

(iii) elimination of extra heat sink support (additional holes and real estate in the PCB), etc.

The optimization of the heat sink design for best thermal performance should take into the account the following considerations:

(i) variations in power levels and operating conditions of the semiconductor device have a direct impact on the amount of heat generated.

(ii) increasing the surface area of the heat sink would improve heat dissipation while increasing cost and may not satisfy the requirements of the vehicle package,

(iii) increasing the base thickness of the heat sink would result in uniform distribution of heat while increasing the overall weight of the application.

(iv) dimensional variations and limitations such as fin height-to-gap aspect ratio, minimum fin thickness-to-height, and minimum base-to-fin thickness influenced by manufacturing processes such as extrusion, stamping and die-casting influence the thermal performance of the heat sink. Tight tolerances on the heat sink dimensions would result in achieving desired junction temperature, but would increase the cost of manufacturing.

In this paper, the above issues are examined and an optimization and robust design framework for heat sink design using the capabilities of iSight and internally developed heat sink design software is outlined.

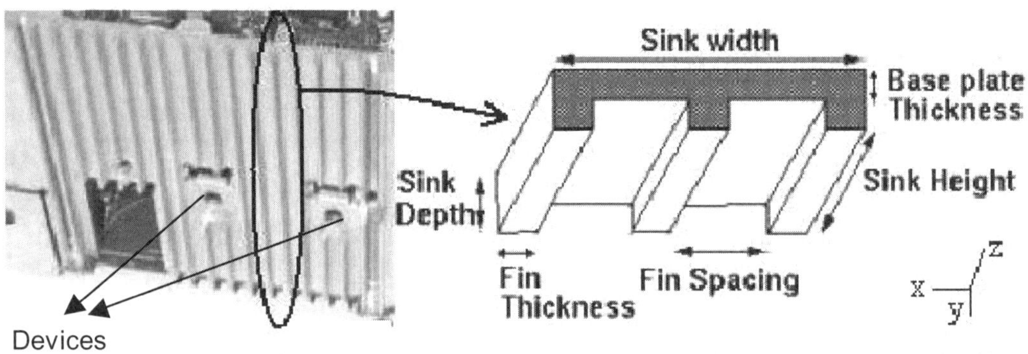

Figure 2: Heat Sink Modeling

PROBLEM IMPLEMENTATION

Heat sinks are modeled as in Figure 2. As seen in this Figure, heat sink geometry is determined by sink width, sink depth, sink height, base plate thickness, number of fins and fin thickness. Heat sink performance is a function of these variables as well as device locations.

Device locations are functions of sink width and sink height. However, for a given sink width and sink height, they can be placed in a number of locations. Heat sink surface are divided into a number of grids using the sink width, sink height and device dimensions. Each grid is numbered. The devices can be located at any of these grids but most likely it will be required to locate them in the row closest to the board such that wiring is kept to a minimum. For a schematic explanation of this see Figure 3. As seen in this Figure, the devices can be located in grids 0 to 27, however since the heat sink is attached to the board in the bottom of the sink; it may be required that the devices can only be located at grids 0 to 6.

The objective of heat sink design is to minimize the heat sink weight, while still assuring that the device temperatures are less than the allowable temperatures, and all the manufacturing requirements are met.

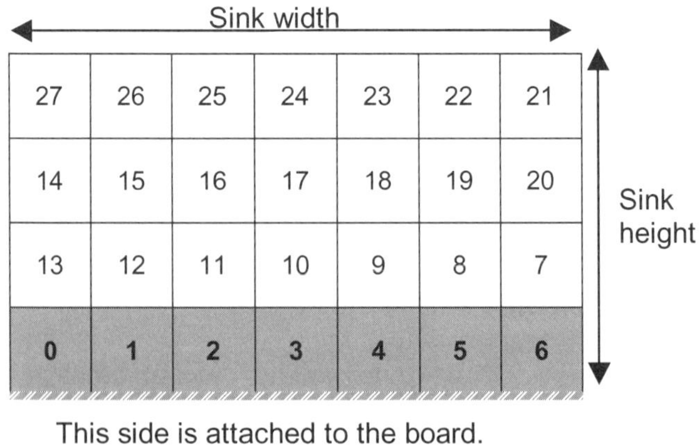

Figure 3: Possible Device Locations

The implementation of this problem to iSight Version 7.0 can be seen in Figures 4 and 5. In Figure 4, design parameters that do not vary - such as device power, ambient temperature, etc. are specified within the *HSTOOptimization* task. The geometrical variables sink width, sink depth, fin thickness and base thickness are specified within the *MainLoop* task. Number of fins are determined within the heat sink design software. Based on the *MainLoop task, DeviceLocationLoop* task iteratively finds the grids at which devices can be located. Figure 5 illustrates the process integration used for heat sink design optimization.

In the next section, the implemented process is applied to an example heat sink. First, the deterministic optimal design formulation and results are given. Afterwards, the robust design problem formulation and results are given.

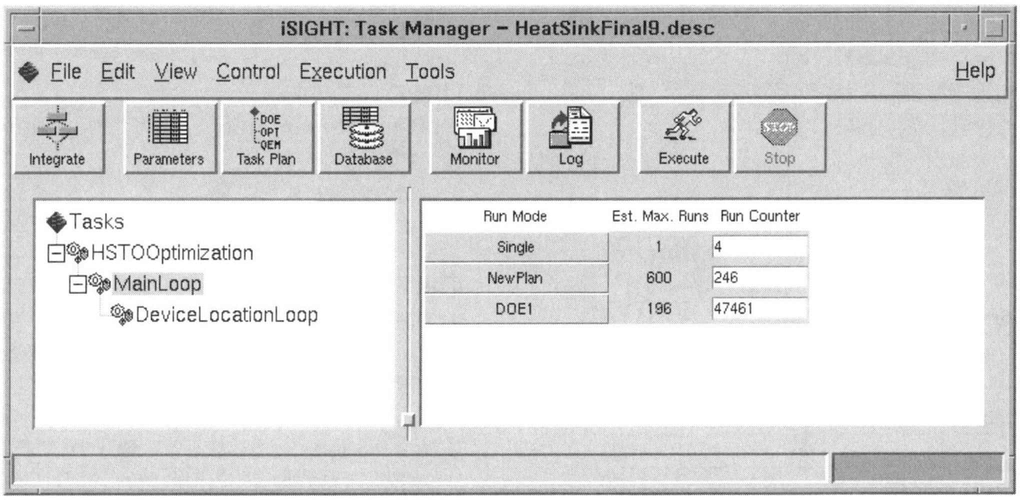

Figure 4: iSight Task Manager for Heat Sink Optimization

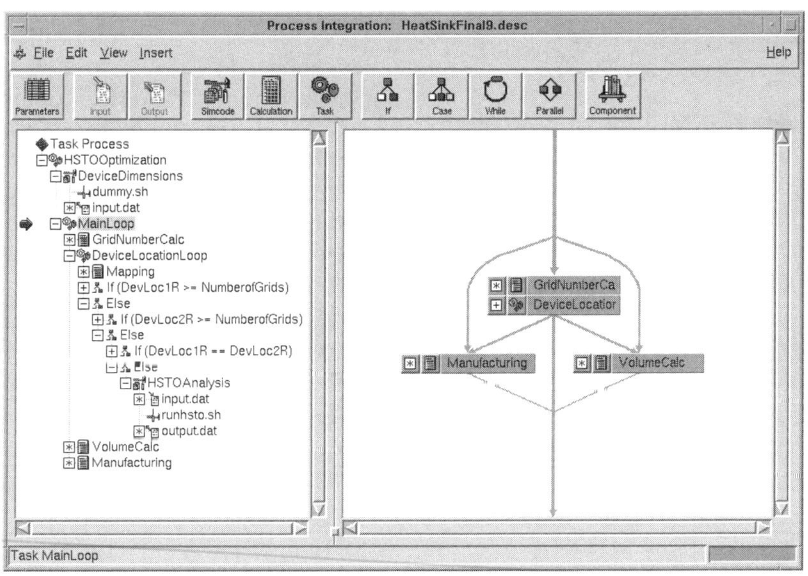

Figure 5: iSight Process Integration for Heat Sink Optimization

EXAMPLE PROBLEM

The heat sink that is considered here is for an automotive radio application. The sink height is fixed to the radio height as 115 mm. There are two T0220 devices with 20W power in each. The devices are 10mm 15mm. They can only be placed in the first row in order to minimize the wiring. The fins are vertical and analysis is free convection. The ambient temperature is 25°C.

OPTIMIZATION PROBLEM FORMULATION:

Design Variables:

$80.0 \leq \text{SinkWidth} \leq 140.0$ mm

$6.0 \leq \text{SinkDepth} \leq 20.0$ mm

$1.0 \leq \text{FinThickness} \leq 2.0$ mm

$1.0 \leq \text{BaseThickness} \leq 4.0$ mm

Device Location 1 $\in$ GridSet

Device Location 2 $\in$ GridSet

where GridSet={0,1,2,3,4,5,6,7,8,9,10,11,12,13}

Design Objectives:

Minimize sink volume (mm^3)

Design Constraints:

Temperature ≤ 150°C

Manufacturing. Constraint:

$$\frac{\text{SinkDepth} - \text{BaseThickness}}{\text{OptFinSpacing}} \leq 6.0$$

Optimum design results are given in Table 1. As seen here, the initial design did not satisfy the temperature constraint. Optimum design found with Genetic Algorithm and Sequential Quadratic Programming satisfies the temperature constraint however is 76% heavier than the initial design.

Table 1: Optimization Results

Design Variables	Initial Design	Optimum Design
SinkWidth (mm)	102.50	134.00
Sink Depth (mm)	12.58	17.00
Fin Thickness (mm)	1.00	1.27
Base Thick. (mm)	3.97	1.86
Dev. Loc. 1	62	81
Dev. Loc. 2	67	87
Temperature (C)	183.	149.
Manufacturing Const.	1.5	2.2
Sink Volume (mm^3)	147,536	260,268
Method Specifics	-	GA, SQP

ROBUST DESIGN PROBLEM FORMULATION:

The objective in robust design is to design the heat sink such that for given variation in power and tolerances in dimensions the design satisfies the constraints with 3σ reliability.

Random Variables:

130 ≤ SinkWidth ≤ 140.0 mm	normal, σ = 0.86 mm
SinkHeight = 115.0 mm	normal, σ = 0.86 mm
12.0 ≤ SinkDepth ≤ 20.0 mm	normal, σ = 0.23 mm
1.0 ≤ FinThickness ≤ 2.0 mm	normal, σ = 0.15 mm
1.0 ≤ BaseThickness ≤ 4.0 mm	normal, σ = 0.15 mm
Power = 20 W	log-norm, σ = 3.0 W

The information for dimensional variations is from reference [1].

Quality Objectives:

min sink volume (mm^3) (both the μ and σ)

Quality Constraints:

Temperature ≤ 150 °C	3σ, 1350 ppm
Manufacturing Const. ≤ 6.	3σ, 1350 ppm

The robust design results are given in Table 2. As seen in this table, the design found using optimization is only 0.7σ. Robust design process improves the design by increasing the reliability of the temperature constraint to 1.2σ however this is still below the required reliability of 3σ. This could result from any of a number of factors - design variable ranges, random variable variations, starting point, etc. Further cases would need to be run to satisfy the quality constraints. And, it is possible that no satisfactory design exists, for the given quality constraints. Manufacturing constraint is 10σ in both the designs. Heat sink from the robust design process is 12% heavier than the optimum design however the standard deviation is 15% lower in robust design compared to optimum design. Probability distributions for power and temperature are given in Figure 6. Pareto plot for temperature is given in Figure 7. As can be observed in this Figure, power is the dominant factor in temperature followed by base thickness and sink depth.

Table 2: Robust Design Results

Design Variables	Deterministic Optimum	Robust Analysis	Robust Optimum
Sink Width (mm)	134.00		132.40
SinkHeigth (mm)	115.00		115.00
Sink Depth (mm)	17.00		19.36
Fin Thickness (mm)	1.27		1.74
Base Thick. (mm)	1.86		2.59
Power	20.00		20.00
Temperature	149	σ = 149; 0.7σ	σ = 143; 1.2σ
Manufacturing Const.	2.2	10σ, $\sigma = 0.1$	10σ, $\sigma = 0.1$
Sink Volume (mm^3)	260268	260268∓4515	2900000∓3856
Method Specifics		Monte Carlo 150 analysis	Monte Carlo SQP

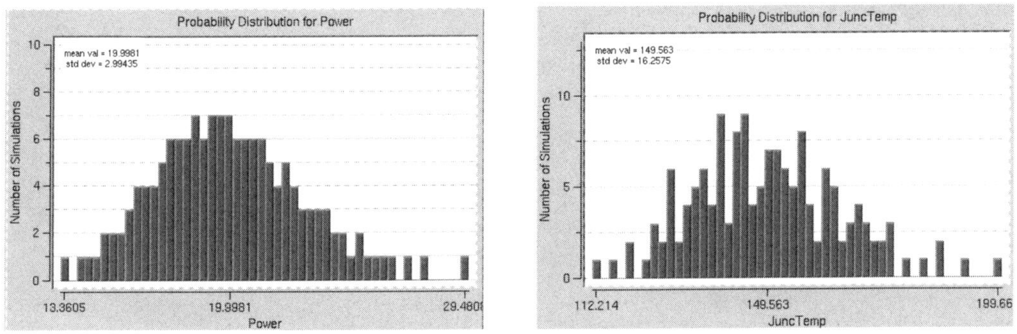

Figure 6: Probability Distributions for Power and Temperature

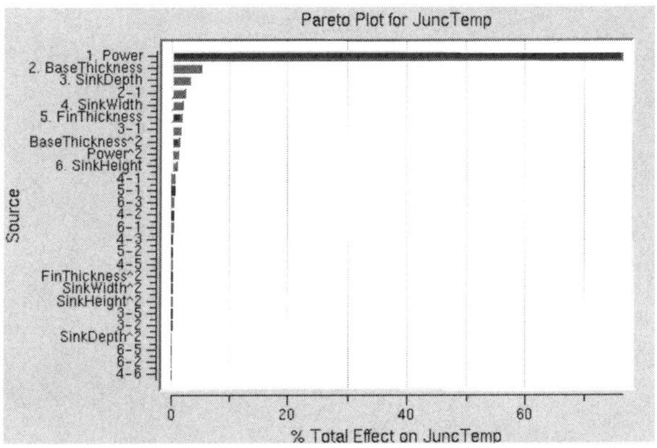

Figure 7: Pareto Plot for Temperature

CONCLUSIONS

This paper has illustrated an optimization and robust design method for heat sink design, using the capabilities of the iSight V7.0, along with internally developed heat sink design software. The method allows us to find designs that meet the specifications. We applied this framework to an automotive radio application, and observed from the results, that optimization improved the performance of the heat sink. The method also allows to find robust designs. For the example case, when manufacturing variations are considered, the design is not as robust. Robust design methods did improve the design quality in this case, but not as much as required. The problem formulation might be modified to improve results, but it is also possible that no design satisfying the quality constraints exists.

ACKNOWLEDGMENTS

The authors would like to thank to Joe Tillo, Visteon Interior/Exterior Department, and Simon Yuan from Engineous Software for their help.

REFERENCES

[1] AAVID Thermalloy, "http://www.aavidthermalloy.com/products/extrusion/dimtol.shtml", 2002

2004-01-0388

Automotive Body Structure Enhancement for Buzz, Squeak and Rattle

Raj Sohmshetty and Ramana Kappagantu
Ford Motor Company

Basavapatna P. Naganarayana and S. Shankar
Lohitsa Inc.

ABSTRACT

Today, the interior noise perceived by the occupants is becoming an important factor driving the design standards for the design of most of the interior assemblies in an automotive vehicle. Buzz, Squeak and Rattle (BSR) is a major contributor towards the perceived noise of annoyance to the vehicle occupants. An automotive vehicle consists of many assemblies such as instrumentation panel, doors, sun/moon-roof, deck lids, hood, etc. which are the potential sources of BSR noise. The potential locations of critical BSR noise could be contained within such assemblies as well as across their boundaries. An extensive study is made regarding the overall structural behavior as well as their interaction under typical road loads to come up with enhanced design for improved quality from the BSR noise perspective.

The alternative designs were comparatively evaluated for their relative noise level from buzz, squeak and rattle perspective using an analytical tool - N-hance.BSR. Critical noise sources both at the system as well as the assembly levels were identified and the relative noise levels were compared critically to determine the influence of the design changes on the BSR quality of the system and its assemblies. In this paper, a brief introduction is provided regarding the typical product background and the noise quality requirements, typical design changes that influence the BSR characteristics followed by a brief introduction to the software N-hance.BSR. The results of such a comparative design evaluation from N-hance.BSR are presented at the end. The observations of the critical squeak and rattle locations were confirmed by physical tests on the baseline and enhanced models.

INTRODUCTION

Squeak & Rattle is an important warranty sensitive attribute for OEMs. It also affects perceived quality of the vehicle. Structure-borne noise has been considered as one of the top FIVE issues from customer satisfaction point of view in J D Powers Review [1]. Therefore, it is imperative that the vehicle development process should incorporate squeak & rattle prediction and prevention up-front.

Traditionally, product assessment from BSR perspective heavily depends on physical prototype tests and feedback from subjective driving tests. The quality of such an assessment, therefore, is subjected to the wide variance that is natural in human perception and experience. The physical tests are accepted as conclusive in the industry. However, up-front robust design minimizes non-redundant physical tests by providing positive and conclusive assessment of the BSR characteristics of the product. In effect, requests for design changes after the testing are also reduced.

Figure 1 presents a generic squeak & rattle prevention strategy and analytical metrics that may be used for enhancing the product design from squeak and rattle perspective.

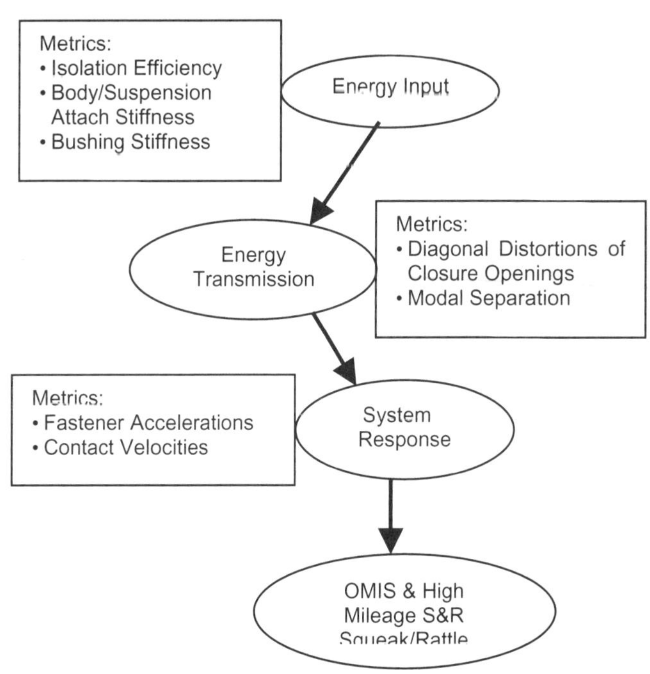

Figure 1: Squeak & Rattle Metrics

Currently, one has to depend on explicit analyses and extensive engineering judgment and intuition to predict squeak and rattle critical locations for designing product enhancements. An analytical tool to conduct quick what-if studies to validate and alter the interim design changes is required to make such a design cycle more efficient, productive and reliable.

Finite element method has been the accepted analytical tool for most of the analytical experiments in the industry today. The modeling of nonlinear contact dynamics and nonlinearities in material and other properties makes squeak and rattle analysis complex. This complexity makes, the job of doing a full-fledged analysis upfront in the product cycle difficult, more so, in a real turn-around time. In addition, the large amount of data generated from the analysis of even a typical automotive assembly makes the problem formidable. Numerous possibilities of squeak and rattle issues within assemblies as well as across assemblies makes the analysis even more complex as well as confusing. In essence, the industry needs a robust tool that can be employed in real-time product cycles for reliable prediction of BSR issues upfront to help designers to enhance product design. The fundamental requirements for an analytical tool for real-time usage are:

- The solution process should incorporate robust mathematical tools as well as reliable engineering perspective and experience in the field to provide consistent results of practical relevance and acceptance;
- The process should be based upon simplistic solutions that the industry depends upon in a regular basis;
- The analysis paradigm should be concise with minimal solution space requirements for a fast churning of results;
- The process should encompass the major internal as well as external factors influencing the phenomena;
- The process should be automated involving minimal user input and interactions;
- The analytical tool should be flexible for continuous as well as intermittent usage covering global as well as local solution perspective; and
- The analytical tool should incorporate reporting and presentation mechanisms catering to the industrial needs and requiring least user interaction without loosing clarity.

In this paper, it is demonstrated how an analytical tool N-hance.BSR can be efficiently employed for real-time verification of design modifications from squeak and rattle perspective in system and sub-system levels in an integrated approach.

The critical squeak and rattle locations identified by the analytical method presented in this paper were confirmed by real life data.

VEHICLE TRIM BODY MODEL

The study was primarily aimed at understanding the effect of design changes for improved isolation efficiency on squeak and rattle. Two trimmed body finite element models of a luxury car from an advanced research project were considered in this effort (Figure 2). The first (baseline) version corresponds to a production version. The second version was derived from the baseline with design actions to improve body isolation efficiency for different load paths improvement in attachment stiffness by about 60%. Both models have approximately 3 million degrees of freedom with enough detail in the structural components. The models were well correlated with hardware builds static stiffness (within 3%) and dynamic modes at body-in-prime and trimmed body levels.

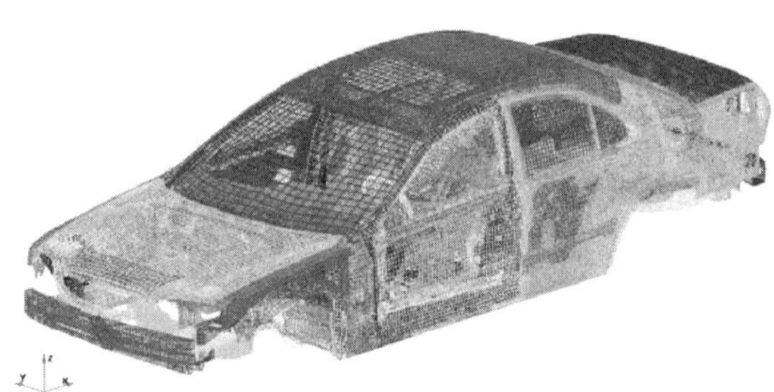

Figure 2. Baseline Model

The definition of plastic trim components such as IP, glove box, door trim, etc, would help in a refined study of the corresponding assemblies. However for want of resources and to stick to the emphasis on effects of body structure improvement, the models were kept devoid of these plastic trim enhancements, while considering the design modifications from the baseline design to the enhanced design. However, one of the objectives of this paper is to examine the influence of the body structure design changes on the squeak and rattle characteristics of the trim components.

The trimmed body models were integrated with other modal components to build a full vehicle model in VSIGN (a Ford proprietary software) and the acoustic and tactile responses to road loads were evaluated and correlated with those from test. The assemblies included in the model for study are listed in Table-1.

For the loading, a fixed amplitude sinusoidal sweeping excitation on 4 tires with diagonal out-of-phase, i.e. out-of-phase between left & right tires and front & rear tires was used. This load case is known to excite the twisting mode and is considered the most severe excitation for squeak & rattle evaluation. The loads were applied at the tire patches (in the full vehicle model) and the corresponding attachment forces at the chassis to trimmed body attachment degrees of freedom were evaluated. These cascaded loads were used in the BSR evaluation. Typical loads applied at the attachment points are shown in Figure 3.

Table-1: Assemblies	
7	Front Door (Left)
8	Rear Door (Left)
9	Instrument Panel (IP)
10	Body-in-White
11	Hood
12	Deck Lid
15	Seats
16	Front Door (Right)
17	Rear Door (Right)
18	Fuel Tank
19	Moon Roof
20	Fenders

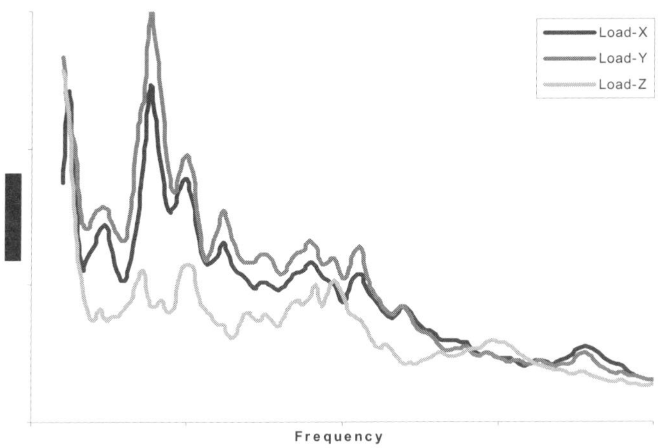

Figure 3. Typical Attachment Loads

THE ANALYTICAL METHOD AND ITS SCOPE

Recently, a methodology based on analytical solutions was developed for catering to the BSR needs in the industry intended for real-time applications in typical product design-validation cycle [3]. The method is built on the simple linear solutions (static, eigen and dynamic response) that any design validation team uses in the industry. Input from the user is minimized to a finite element model and its linear static, eigen and dynamic responses at a predetermined set of potential critical BSR spots. The solution space is minimized based on a combination of mathematical and heuristic rules encompassing the geometrical and modal characteristics of the model. This enables the tool to provide a realistic turn-around time without compromising reliability of the results. The potential BSR spots are then evaluated by projecting the linear solutions to determine the relative loss of energy and momentum to which the noise is directly related. The analysis is carried out in complex domain as well as real time domain to include influence of frequency content as well as history of deformation. Possible mathematical redundancies and impossibilities are identified and eliminated from the solution by a set of rules based on engineering experience and judgment. The technology is found to be quite reliable as against physical tests for many automotive assemblies such as instrument panel and doors [4]. The method accommodates various loads random, simple harmonic (sine sweep), time history, complex fields, etc. that a re commonly encountered in automotive engineering. It also provides optional inclusion of the influence of environmental fields, material properties (impact and frictional), and tolerance and stack-up factors on the BSR characteristics of the assembly in an *a priori* as well as in an *a posteriori* sense.

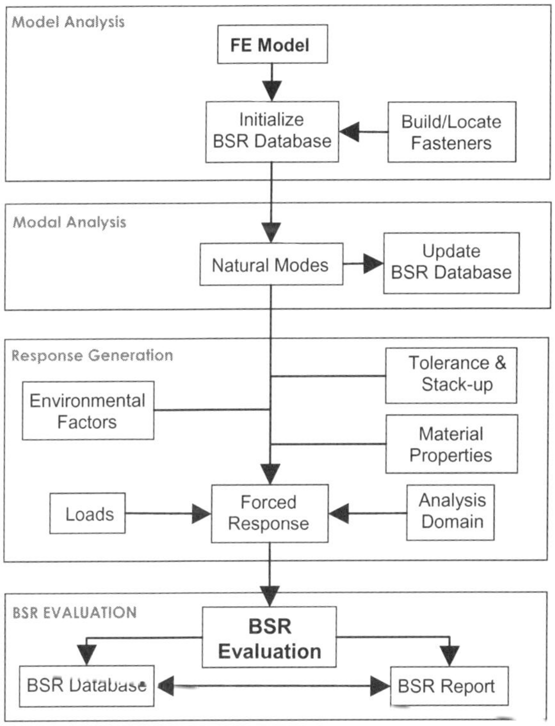

Figure 4: The critical BSR location prediction system A n Over View [1]

The above-described process has been incorporated into an object-oriented software N-ha nce.BSR [5] to provide quick turn-around times for BSR characteristics evaluation. The software is geared for comparative evaluation of various design alternatives in the virtual domain before changing the product design and the subsequent tooling. It can analyze the vehicle body system results and evaluate the system as well as assemblies/sub-assemblies individually as well as collectively. N-hance.BSR fully automates managing the model and data requests for various solutions, handling the gigabytes of finite element solutions, analyzing the finite element solutions for BSR evaluation, and finally generating comprehensive reports in various formats. N-hance.BSR consists of the following main modules (Figure 4):

1) Model analysis read a Finite Element Model, analyze its geometric content and initialize the BSR database.
2) Modal Analysis g enerate eigen solution request, read the eigen value solution, analyze the modal solution and update the BSR database.
3) Influencing Parameters: Read the influencing parameters (environmental, material and geometric) OPTIONAL .
4) Domain of Interest: Read domain of interest (system level and the specific assemblies of interest) OPTIONAL.
5) Generate the potential BSR location set.
6) Dynamic Response: Read the dynamic loads, generate dynamic response request and read the forced response.
7) BSR Evaluation: Evaluate all the predetermined BSR locations from buzz, squeak and rattle perspectives and update the BSR database with buzz, squeak and rattle results.
8) BSR Report: Extract BSR data from the database as per users preferences, generate BSR report for presentation and/or documentation.

A typical design cycle would involve evaluation of the baseline model, design enhancement based on N-hance results and other sources, evaluation of the enhanced model in comparison with the baseline model as shown in Figure 5.

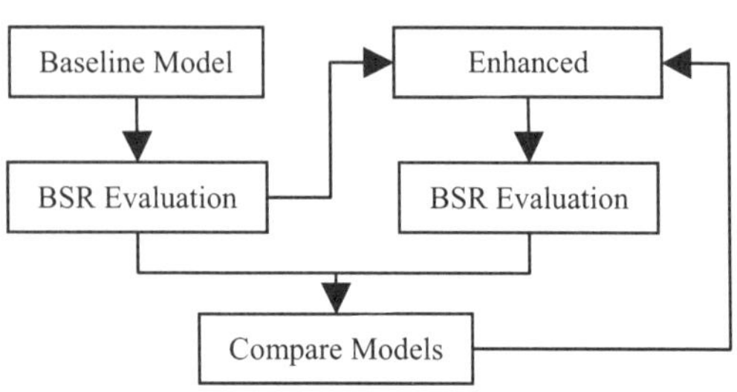

Figure 5: Comparison of Alternative Designs for BSR

BSR EVALUATION OF BASELINE MODEL

The baseline model of the trim body under consideration is evaluated for buzz, squeak and rattle using the linear vibration solutions from commercial analysis software, MSC/NASTRAN.

The baseline model is evaluated for squeak and rattle at system level for predicting critical locations between the assemblies as well as between parts in certain pre-specified assemblies (IP and Front Door). Potential squeak and rattle spots between assemblies as well as between the internal parts of the pre-selected assemblies are generated based on the analysis of the modal characteristics. Finally, the dynamic response of the model to the prescribed loads is evaluated for squeak and rattle. The resource requirements for such an analysis is provided in Table-2.

Table-2: Resource Requirements	
Model Size	4-500,000 Nodes
	100 MB
Eigen Results (OP2)	2,000 MB
Dynamic Response (OP2)	30,000 Nodes
	300 MB
N-hance: BSR Database	350 MB
Total Memory / Evaluation	2,800 MB
N-hance: Model Processing	5 min
N-hance: BSR Database Generation	20 min
N-hance: BSR Evaluation	25 min
N-hance: Report Generation	< 1 min
NASTRAN: Eigen Solution	7 hr 30 min
NASTRAN: Dynamic Response (RESTART)	30 min
Total Time / Evaluation #	< 9 hrs

Reduces with faster machine for NASTRAN solutions

The map of the rattle points with normalized propensity index is shown in Figure 6.

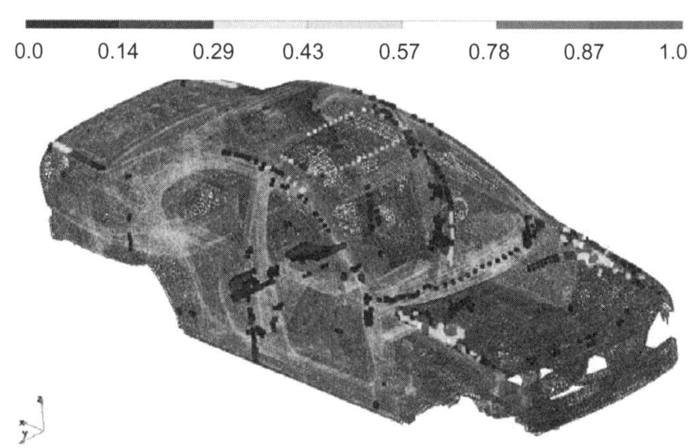

Figure 6: Rattle Point Map Baseli ne Model

The rattle point distribution and the rattle fields (at most critical point) are presented for the most critical assembly pair (hood and body-in-white) in Figure 7.

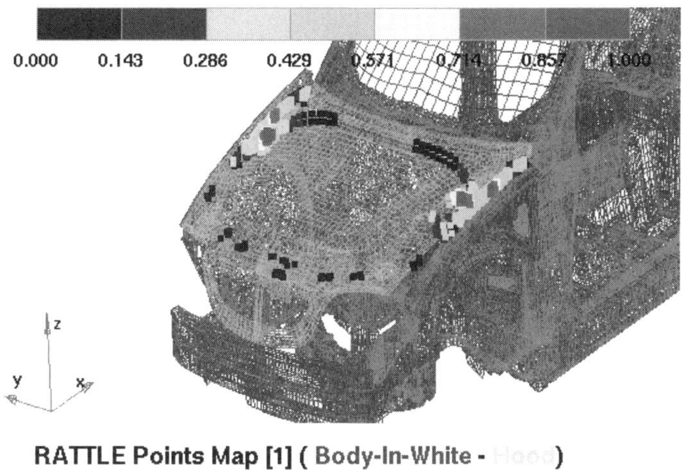

(a) Rattle Point Map Hood and Body

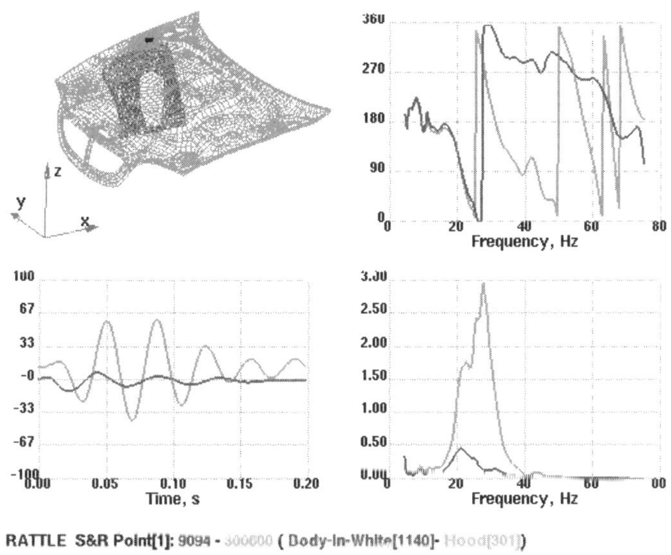

(b) Critical Rattle Field Hood and Body.

Figure 7: Critical Rattling Assemblies Baseline Model

The rattle propensity index for the assemblies are presented along with the population index (number of rattle points between each assembly pair) is presented in Figure 8. The assembly pairs Body-Hood, Body-Moon Roof and Hood-IP are found most critical for rattle intensity while assembly pairs Body-Hood and Body-Fuel Tank are found to have highest propensity (number of critical rattle locations).

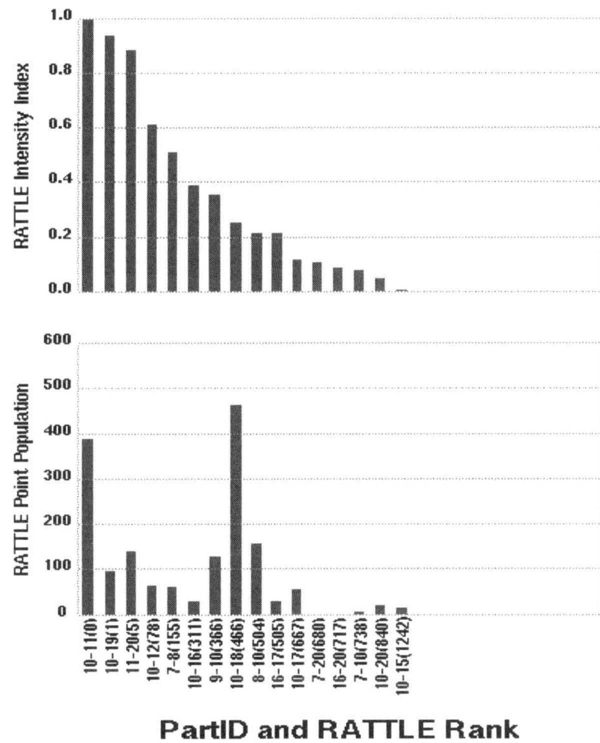

Figure 8: Rattle Index Baseline Model

As an example of integrated system and subsystem analysis, results are presented for the IP Assembly rattle point map (Figure 9), the rattle fields at the most critical point (Figure 10) and Rattle Index (Figure 11).

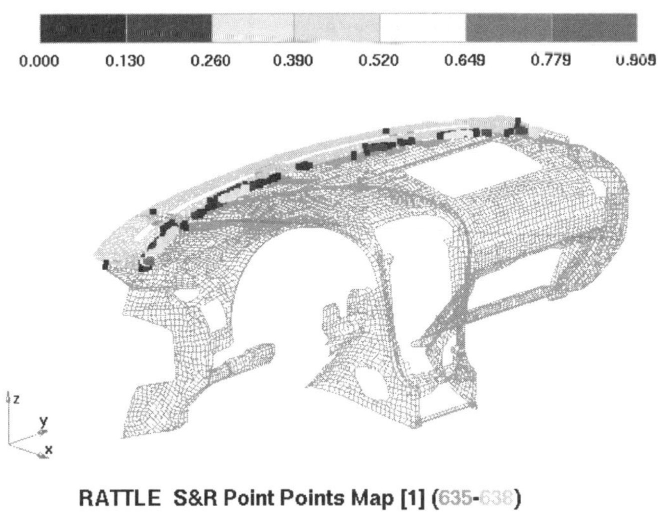

Figure 9: Rattle Point Map IP, Baseline Model

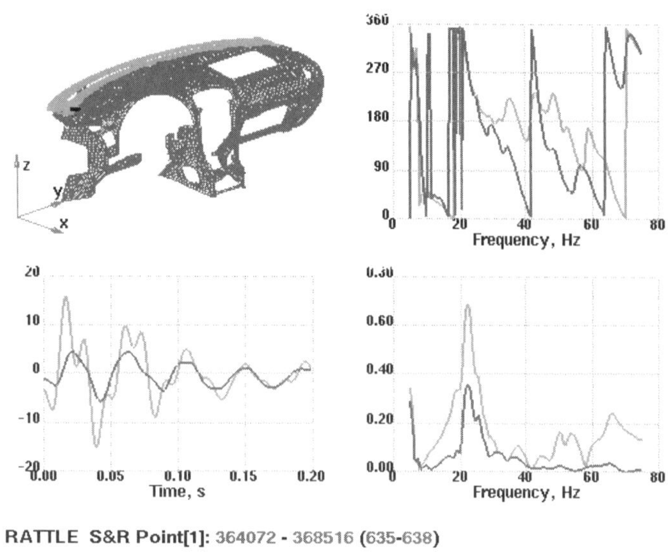

Figure 10: Critical Rattling Part IP, Ba seline Model

The critical squeak and rattle zones, and the associated assemblies and parts are examined closely to come up with suitable design enhancements for bringing down the squeak and rattle propensity in the model.

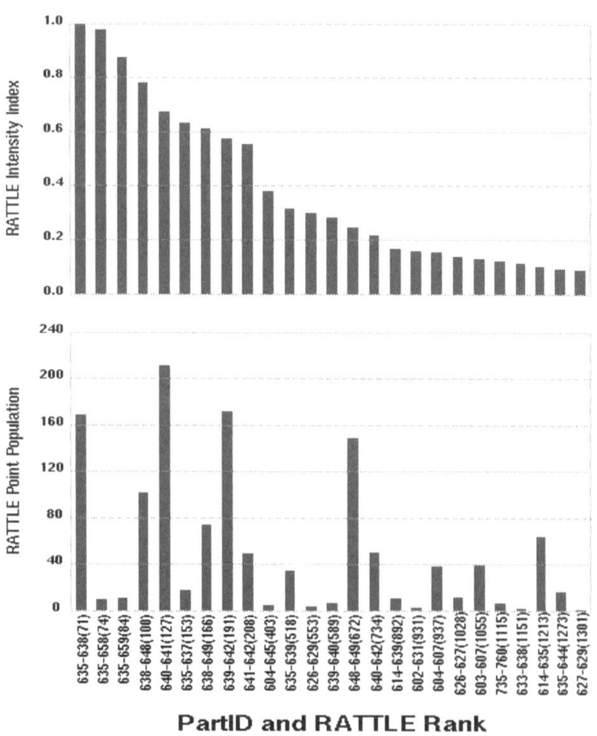

Figure 11: Rattle Index IP, Baseline Mo del

DESIGN ENHANCEMENTS

Squeak and rattle problems can be corrected broadly at three levels 1) modifying the design to mitigate loads at major load paths, 2) balancing the body structure stiffness towards improving load distribution and 3) treating the responders. The first two involves an upfront good body structure design. The last is more of a fix mechanism and is expensive.

The design strategy adopted in this project was to improve isolation efficiency and alleviate the loads in the major noise path and at the same time improve the stiffness. This improvement called for identifying the major noise paths and identifying changes in body structure design to reflect on what we call *attachment stiffness* at all the attachment DOF. Though the name has a local connotation, this stiffness metric is highly influenced by global modes. Thus working on the attachment stiffness automatically implies working on global modes. In addition the improved attachment stiffness allows for lowering the bushing/mount rates and still maintain the overall stiffness at the attachment specified by vehicle dynamics community. In order to improve the squeak and rattle responses at the critical locations, the dominant squeak and rattle frequency ranges are first identified from the squeak and rattle response. Then, the modes participating in these response frequencies are identified for improvement.

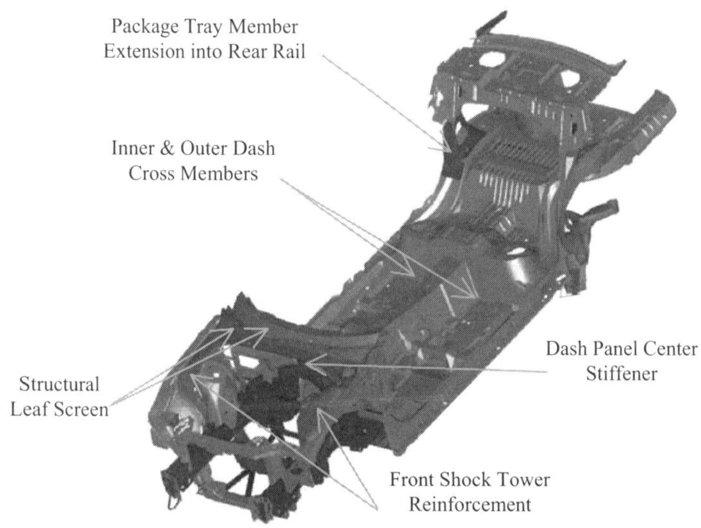

Figure 12 Typical Design Enhancements for Squeak and Rattle

Major changes between the two models are depicted in Figure 12. Though the changes seem to be global, these are purely driven by the diagnostics from attachment stiffness at major load paths.

BSR EVALUATION OF ENHANCED MODEL

The enhanced model of the trim body under consideration is evaluated for buzz, squeak and rattle using the linear vibration solutions from commercial analysis software, MSC/NASTRAN. The results are presented in this section in comparison with the baseline results. The map of rattle points for the enhanced model with index normalized with reference to the baseline maximum is presented in Figure 13. In comparison with Figure 6, it can be observed that most of the critical rattle locations are alleviated in the enhanced design except between the moon-roof and body-in-white assemblies.

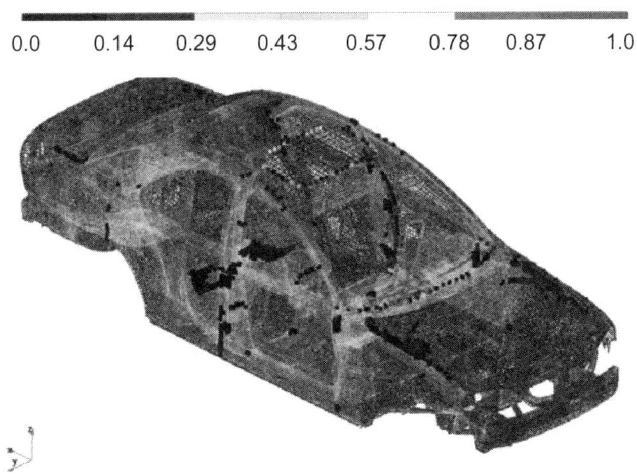

Figure 13: Rattle Point Map Enhanced Model

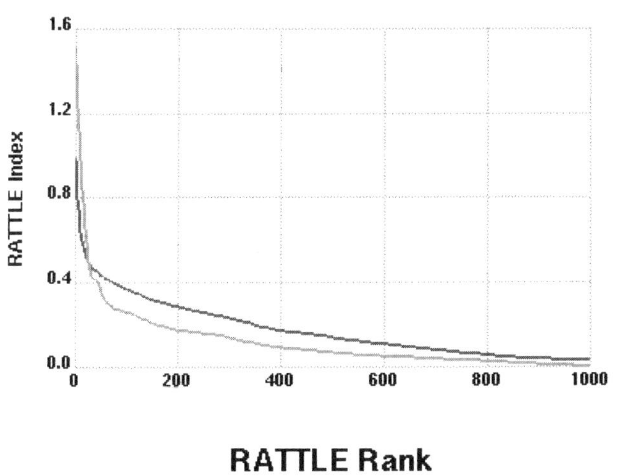

Figure 14: Rattle Index Enhan ced vs. Baseline Models

The rattle and squeak indices for the enhanced model are then shown in comparison with the baseline model, all indices being normalized with reference to the corresponding baseline maximum (Figure 14 and Figure 15). The deterioration of rattle performance between the two assemblies is confirmed in Figure 14. However, the enhanced design is found to be more robust in squeak at all the locations. Enhanced model shows overall improvement of 10-15% in rattle and 10-20% in squeak, in-spite of a deterioration of about 50% in rattle between body-in-white and moon-roof.

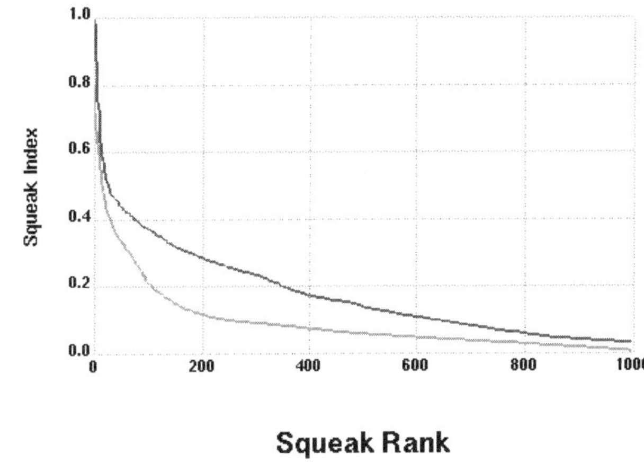

Figure 15: Squeak Index Enhan ced vs. Baseline Models

To demonstrate the sub-system level design evaluation, similar results are presented for the IP (from the system-level analysis) in Figures 16-18. The results show remarkable improvement in rattle and squeak characteristics within the IP assembly demonstrating how the system-level design changes influence the sub-system level squeak and rattle performance positively. An enhancement of about 20% in rattle and 15% in squeak can be observed at the critical zones in the IP assembly.

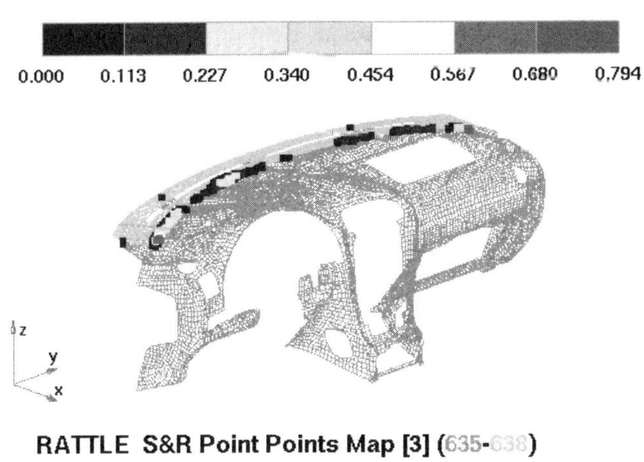

Figure 16: Rattle Point Map IP, Enhanced M odel

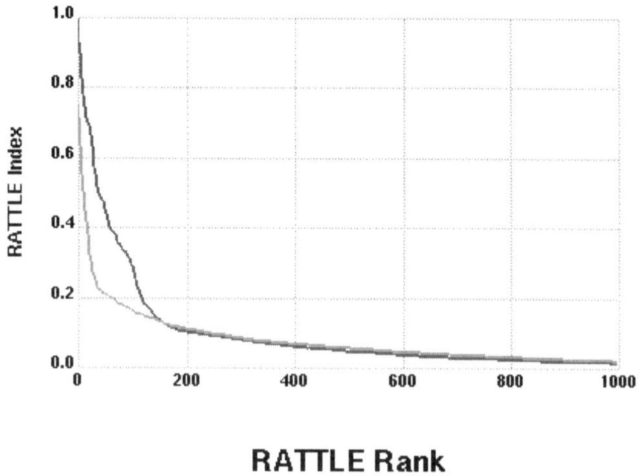

Figure 17: Rattle Index IP, Enhanced vs. Baseli ne Models

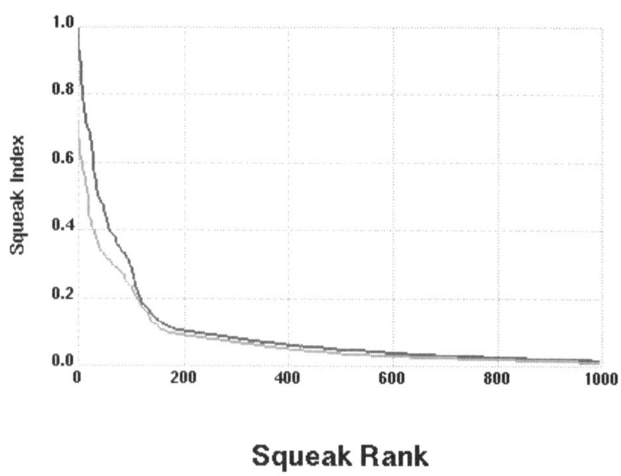

Figure 18: Squeak Index IP, Enhanced vs. Baseline Models

CONCLUSIONS

This paper demonstrated how squeak and rattle issues in an automotive trim body can be reduced by certain design modifications. The study indicated that improving suspension to body attachment stiffness along with bushing rate reduction (thereby, improving isolation efficiency) reduces a vehicle's propensity for squeaks and rattles.

The study also illustrated the usefulness of an analytical tool and software in the assessment of the design changes for identification of critical locations and zones for further deign enhancements from S&R perspective. The study demonstrated that system level structure improvement to improve structure-borne noise also improves sub-system level S&R performance. It also illustrated how an integrated examination of the S&R characteristics in system and sub-system levels can be efficiently used for assessing a product design.

The analytical tool used in this study N-hance.BSR is found to be efficient and user-friendly to achieve quick turn-around time for validating design modifications providing reliable indicators for potential critical squeak and rattle spots.

Critical squeak and rattle locations identified in the model were confirmed based on real life data.

ACKNOWLEDGMENTS

At Ford Motor Company: Everett Kuo and Shih-Emn Chen for expert consultation; Ron Quaglia, Bijan Shahidi, Matt Zaluzec, Carl Johnson, and Charles Wu for review and approval of the paper.

At Lohitsa: Ruslan V Dashko for generating results.

REFERENCES

1. Farokh Kavarana and Benny Rediers, "Squeak and Rattle State of the Art and Beyond", SAE paper 1999-01-1728, Moise and Vibration Conference, Traverse City, MI, May 17-20, 1999.
2. VSIGN, Rel. 02.10.00, 2001. "Ford internal NVH software for full vehicle analysis"
3. B P Naganarayana, S Shankar and V S Bhattachar, "**Structural Noise Predictor**", Patent (US and International) Pending.
4. Naganarayana B. P., Shankar S., Bhattachar V S, Brines R S and Rao S., **N-hance: Software for identification of critical BSR locations in automotive assemblies using finite element models",** #03NVC-283, Noise & Vibration Conference, SAE-2003, May 5-8, Grand Traverse Resort, Traverse City, MI, USA, 2003.
5. B P Naganarayana, "**N-hance.BSR – An analytical tool for predicting buzz, squeak and rattle in structural assemblies: Users' Manual**", Lohitsa Inc., Bloomfield Hills, MI, USA, Nov. 2002.

CONTACT

Raj Sohmshetty at rsohmshe@ford.com.
Basavapatna P Naganarayana at naga@lohitsa.com.

2004-01-0383

Software Tools for Programming High-Quality Haptic Interfaces

Christophe Ramstein
Immersion Corporation

Henry da Costa and Danny Grant
Immersion Canada Inc.

ABSTRACT

Haptics refers to the sense of touch. The challenge of designing and integrating high-quality programmable haptic interfaces requires technical knowledge, usability experience and software tools. This paper provides design guidelines for software tools intended to facilitate the design and integration of programmable haptic controls and describes a suite of fundamental tools to which the design guidelines have been applied. *Immersion Studio® for Automotive* (ISA) is a user-friendly software application for interactively designing and programming haptic sensations. ISA supports a large variety of devices including 1D and 2D force feedback devices. Together with the *Immersion API for Automotive* and firmware, ISA constitutes the basis for creating high-quality programmable haptic systems.

INTRODUCTION

Haptics refers to the sense of touch. Programmable haptic controls are input/output devices capable of producing physical sensations. Programmable haptic systems are emerging systems suitable for many applications including gaming, medical, 3D, consumer and automotive applications. In 2000, BMW created a paradigm shift by introducing the first fully programmable haptic rotary control [2].

Compared with traditional mechanical controls, programmable haptic devices can vary their feel under the proper digital control, thus offering increased interface functionality and run-time customization [4,5,6].

As programmable haptic devices are making their way into car interiors, there is a need to develop software tools APIs and authoring tools to suppo rt the control and use of these hardware devices [8,9,10]. This paper describes requirements and design guidelines for building high-quality haptic interfaces for car interiors. To support the argument, two software products are described: *Immersion Studio® for Automotive* (ISA) and the *Immersion API*. These are the first authoring tool and API available to enable car designers and car manufacturers to design the feel of their new generation of programmable haptic interfaces and integrate them with their existing systems.

STATE OF THE ART

Historically and physiologically, people have been using their sense of touch to precisely operate dashboard controls, steering wheels, gear shifters and gas and brake pedals. Provided by mechanical controls or calculated by software simulation, haptic feedback is a natural and efficient sensory channel to provide feedback while manipulating a physical system. The haptic experience is also an integral part of the enjoyment a user experiences while driving.

FEEL THE MENU!

Let us consider two simple examples to understand what haptic controls, in this case rotary controls, can do and how mechanical controls differ from programmable haptic controls. Figure 1 shows a hypothetical menu displayed by the car s on-board computer. Two haptic effects are discussed: endless detents and detents with surrounding barriers.

<u>A. Endless detents</u>

Here the rotary control features endless detents. While rotating the knob, the user feels each detent. There are no hard stops; therefore, the user can continue turning clockwise or counter-clockwise indefinitely. Endless detents are mechanical detents on a real physical knob, or simulated detents on a programmable rotary control.

Typically, each detent corresponds with a menu item. While turning the knob, the GUI highlights the previous or the next menu item.

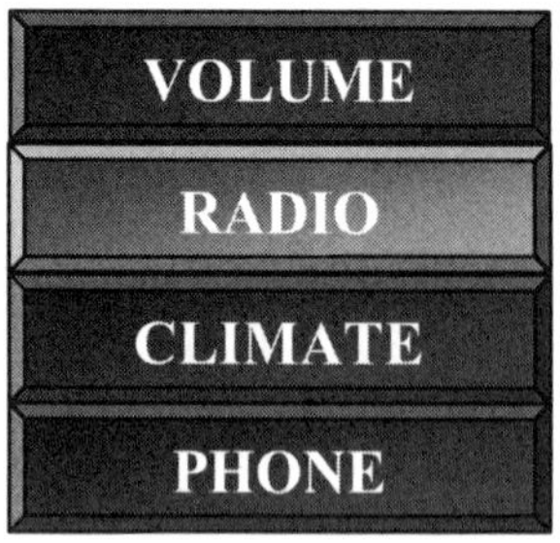

Figure 1. Hypothetical menu with four items

A limitation with a mechanical implementation of the endless detents is that the number of detents per revolution is fixed; for example, the control may have 8 detents per revolution. During the design stage of the HMI or for run-time customization, one may want to change the number of detents per revolution and have, for example, 16 detents per revolution. With physical controls, this would require hardware changes. With programmable controls, this modification would be done by simple software changes.

Another limitation with the mechanical control is that the rendering of detents, including the magnitude and friction, are fixed. Again, for design purposes or for run-time customization, one may prefer stronger or weaker detents or perhaps detents with a completely different type of feel. Programmable haptics allows modifications of the type, magnitude and friction of the detents by simple software changes.

B. Detents with surrounding barriers

In this second case, the rotary control features two hard stops that limit travel in either direction with four detents placed in between. As before, the user is able to navigate the menu by feeling the transition from one menu item to the next. In addition, due to the barriers, the user knows without looking at the screen that they have reached the top or bottom item. Barriers are useful for indicating the beginning and end of a list, thus providing information by tactile cues alone on where the cursor focus is currently located.

A limitation with a mechanical implementation of a barrier bounded detent having four detents is that it works only for lists having four items. If the menu has more or fewer than four items, the user will be confused or unable to select all the items. Programmable haptics, on the other hand, allows the designer to modify the number of detents placed in between the barriers thereby allowing the computer to handle menus with different numbers of items.

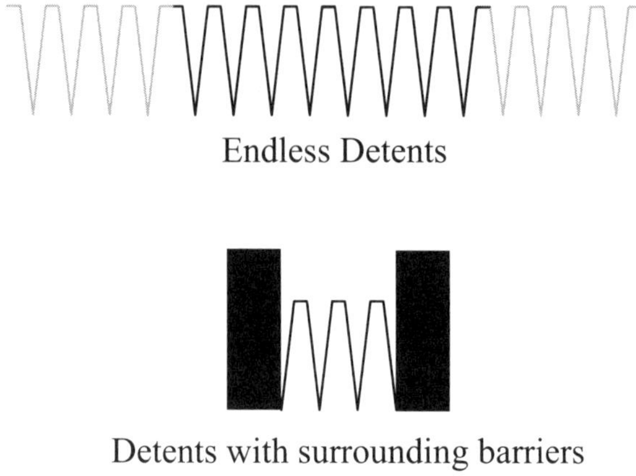

Figure 2. Two haptic effects: endless detents (Case A) and barrier bounded detents (Case B)

HAPTIC ARCHITECTURE

Haptic systems are complex systems, and building such systems involves multi-disciplinary expertise [1]. A typical programmable haptic system is comprised of hardware and software components. Hardware components include a mechanical interface such as a knob, actuators to create output forces, sensors to get device data, and electronic components. Software components include a haptic kernel, haptic API, drivers and applications.

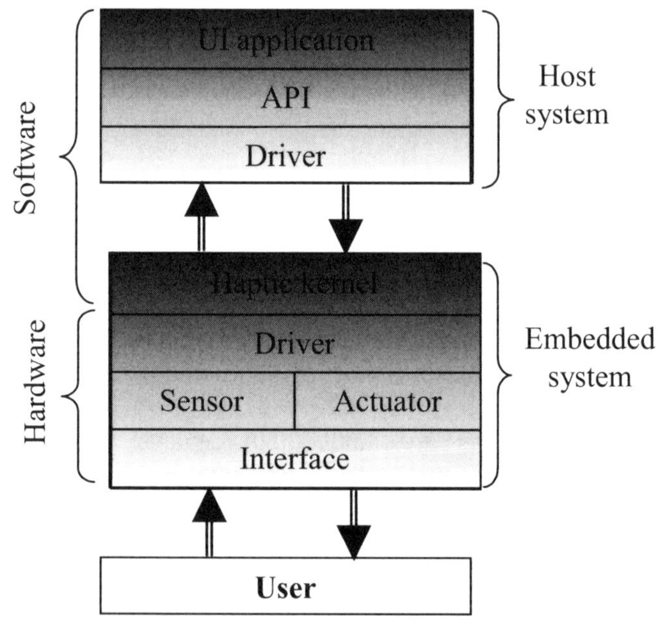

Figure 3. Haptic system architecture

The UI application implements the user interface with menus, lists and others widgets. To enable this graphical interface with haptics, the application calls haptic API functions.

The API provides functions to create effects, modify their parameters, and play and stop the effects. The API also provides input information on device positions and switch states. Device drivers may be required when the

API needs to communicate with lower level layers located on a second local processor.

The haptic kernel performs all of the real-time force calculations for rendering the different haptic effects. The kernel also handles the communication with the API and with the input and output electronics to read the sensors and to write to the motor outputs.

BMW IDRIVE

The iDrive control [2] is a rotary haptic device intended to provide enhanced user interaction with the man-machine interface. The iDrive interface consists of a rotating knob with push-to-select and lateral-select functionality. The uniqueness and novelty of the iDrive device is that in addition to accepting user input, it can create programmable touch feedback; that is, it can present different tactile sensations to the user.

Figure 4. iDrive rotary control in the BMW 7 series

With an iDrive-type control, users can navigate menus, scroll through lists, control sliders, etc. while feeling unique and programmable haptic sensations per widget and action. A device micro-controller calculates and renders forces in real-time to give the user the feeling of detents, barriers, hills, vibrations, pops, damping and even springs. These effects can be played at run-time depending on the current displayed menu or device function; for example, a specific menu may require a different number of detents or different magnitude detents than another menu. The iDrive is available in the BMW 7 series, 5 series and Rolls-Royce Phantom.

PROGRAMMABLE HAPTICS IN DEMAND

Programmable haptics is experiencing increased demand in the automotive market with already more than one manufacturer featuring haptics. In addition to BMW, Volkswagen has developed its own version of a rotary haptic control. The VW device consists of a knob approximately 30mm in diameter and a push-to-select feature with no lateral motion. The user interface is simple and uses detents and barriers to reinforce the graphical user interface widgets. Nissan has also demonstrated a haptic scroll wheel integrated in the steering wheel. Other automotive manufacturers are currently considering integrating this technology. [3]

DESIGN GUIDELINES FOR API AND AUTHORING TOOLS

THE EARLY DAYS OF HAPTICS

Programmable haptic devices require real-time algorithms to calculate the force output according to the device position or speed. For true haptic devices such as the iDrive controller, this calculation is done at a 1000Hz frequency. This high frequency rate is required to render high-quality force sensations such as good detents, friction or spring effects but also to minimize instabilities.

Once the real-time algorithms are done and running in the haptic kernel (see Figure 3), haptic designers have to come up with the UI application and associated haptic effects. For a given user interface screen, the designer has to define how the menu or the button should feel, create the haptic effects, and set the haptic parameters.

In the early days of programmable haptics, the task of designing haptics was rudimentary. The real-time algorithms were hard-coded in the haptic kernel with very limited flexibility to change the effects, and no high-level tools were available to make the design process user-friendlier. Haptic designers were engineers figuring out to the best of their ability how the menu should feel. Those same engineers then programmed the sensations in the target application.

GUIDELINES AND BENEFITS FOR A HAPTIC API

A useful generic API targeted at programmable haptic devices for car applications should have the following features:

- There are many computer operating systems for cars; therefore, the API would benefit form a system-independent design. The API should be easily portable with minimal and well-identified changes.
- There can be many implementations of haptic systems involving different devices such as scroll wheels, rotary knobs and 2-D joysticks. A haptic API should support different types of devices to allow flexibility and reusability. The device handling code to get/set data from/to haptic devices should be isolated into device-specific driver modules. The application should not have to deal with device handling details. In addition, spatial information should be processed in user coordinates and percentages independent of device coordinates.
- The API should benefit from a C or C++ design and architecture. Additional API wrappers, including Java wrappers, should be provided with the basic API.
- The API should take advantage of the composite nature of haptic effects. It should support basic types of effects such as detents and barriers, as well as composite effects being the superposition of basic effects.
- The API should allow applications to create and locate haptic effects in a scene. A scene is like a document and may contain several haptic effects

that can be created, modified, saved and opened later.

- The application and the API should exchange certain events. The API should report events to the application such as device positions, switches and index changes for detents, while the application should notify the API of relevant changes such as setting the index of a detent. The API should be optimized to limit bandwidth usage in the car.
- The API should provide standard, comprehensive documentation.

One of the main benefits of the generic API would be to encapsulate the complexity of programming haptics and let the designer/programmer focus on enabling their target applications with haptic effects instead of struggling with the mechanics of programming the real-time algorithms and handling the communication between computers and devices. This includes:

- Hiding the low-level communication protocol. There is no need for the application developer to implement, test and maintain the communication protocol if it is handled by the API.
- Abstracting the communication layer handling specific input/output ports. Lower-level communication modules should be available for CANbus, MOST Bus, serial bus, USB bus, etc.
- Encapsulating the complexity of scrolling lists or scrolling grids, event handling such as detent index changes, time-based effects, etc.
- Transparently handling the position calculations and effect localization.
- Supporting multiple devices. A target application may need to use multiple devices simultaneously; for example, two rotary devices together with one 2-D joystick.

GUIDELINES AND BENEFITS FOR A HAPTIC AUTHORING TOOL

HMI haptic designers should be concerned with the feel of the haptic effects and how the effects correlate with the user interface. Designers should not have to deal with the implementation details of haptics, how it works at the low-level or how it is programmed. Haptics should be considered like other media, such as graphical and audio media, and have user-friendly authoring tools available.

The authoring tool is the software application allowing designers to create, modify, experience and save/restore haptic sensations for a given haptic device. This is the equivalent of a CAD application but for haptics.

An ideal authoring tool for designing haptics should be a user-friendly software application allowing designers to create haptic sensations. Various effect parameters could be defined and modified, and the result immediately experienced. The authoring tool should run on a mainstream operating system such as Microsoft Windows and support a large variety of haptic controls including rotary controls and 2-D devices such as trackballs and joysticks. The authoring tool should support archiving features and possibly include an automatic code generator for providing an easy way to port the created effects into the target automotive application.

The main benefits of the authoring tool would be to allow non-technical designers to precisely design effects with a desired feel. The authoring tool should include the following features.

- Natural, easy-to-use user interface. Users should be able to drag effect graphics to modify parameters including width and amplitude. Keyboard entry is supported for setting parameters.
- What you see is what you feel users should feel the effect played on the device and see a cursor moving on the screen according to device motions. This applies to 1-D and 2-D devices.
- An archiving mechanism should be available making cross-design and cross-team tasks efficient.
- A code generator provides the source code for what has been designed. This includes the exact parameters given in the GUI. The code needs to be copied into the target application and will compile as is or can be modified to refine some of the behaviors.

IMMERSION TOOLS FOR AUTOMOTIVE

To address the lack of support for designing and integrating high-quality haptic effects for car applications, Immersion Corporation developed software tools, two of which are presented here are: *Immersion Studio® for Automotive* (ISA) and the *Immersion API for Automotive*. ISA is the first commercial authoring tool available for automotive haptic content authoring. ISA has been designed and developed to address the needs identified in the previous section for a high-level user-friendly authoring tool.

EFFECTS SUPPORTED BY ISA/API

There are several groups of programmable haptic devices in automotive applications, including those that have one degree of freedom, such as rotary knobs or scroll wheels; and those that have two degrees of freedom, such as joysticks or trackballs. ISA supports both 1-D and 2-D devices and each type of device has a defined set of effects with corresponding parameters

ISA and the underlying API include a collection of pre-defined haptic effects divided into two groups: 1-D haptic effects and 2-D haptic effects. Within each of these groups, there are basic effects and composite effects, the latter being combinations of two or more basic effects.

The following section describes some of these effects and provides examples of how these effects could be used.

1-D HAPTIC EFFECTS

The haptic control implements an alphabet of basic effect types. For a 1-D haptic device such as a knob or a slider, this includes detents, barriers, time-based effects and a variety of other basic building blocks not described in this paper. (See [7] for a detailed description of the 1-D basic effects.)

Regular detents

Detents allow a designer to implement virtual preferred locations of a knob or slider. A physical example would be in the control of a fan, where the high, medium and low settings each have a preferred position. This effect can be used to clearly correlate device positions with on-screen graphics. Parameters in the detent effect allow the location, shape, magnitude, width, dead band and number of detents to be adjusted. The detent force profile, or shape, dictates the tactile feel as the user moves from one position to the other. The wide variety of design parameters offered by ISA allows the software designer to create a rich array of differing haptic experiences that can be presented to the end user.

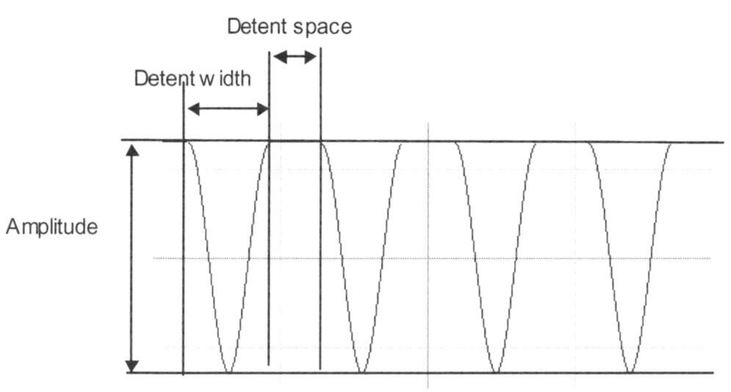

Figure 5. Four equally spaced detents

An example profile is given in Figure 5, where four equally spaced detents are displayed with some of their adjustable parameters.

Barriers

The Barrier effect is used to indicate a hard stop by generating a force designed to limit the travel of the device within a requested range. The Barrier effect is used in many applications such as at the ends of a radio tuner control or for balance and treble control limits. Barrier effects can be either one-sided (left or right) or two-sided (left and right).

The achievable stiffness of the barrier depends on the hardware limitations of the device. With an iDrive-type system, one can create the impression of a stiff wall with 50mN.m peak torque.

Time based effects

The periodic effect creates a vibration with a given period and amplitude. This effect could be used to outline list items, menu items or buttons. Time-based effects are defined by several parameters such as the duration, the frequency of vibration, the shape of the waveform, and the rising and falling magnitude envelope.

A time-based effect can be triggered by the application in several ways including: 1) when the cursor enters a detent, 2) when a button is pressed or 3) to signal a user interface event.

1-D COMPOSITE EFFECTS

ISA and the API also support composite effects. Composite effects are built from the basic effects to make likely combinations required for the interface design.

Figure 6 shows an example of a composite effect made up of four basic elements: 1) barrier, left and right; 2) large detent on the left, 3) large detent on the right and 4) three small detents.

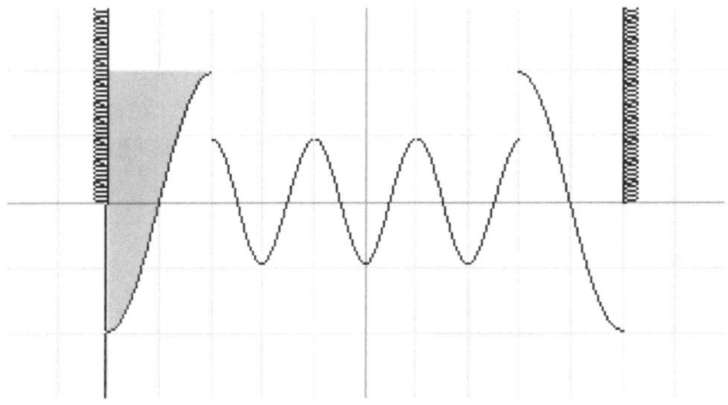

Figure 6. Example of a composite effect

A user exploring this force profile on a knob device would feel three light force detents in the center with two stronger detents on the ends. The two barriers would limit the user motion to the detents.

A number of composite effects that have likely use in a standard GUI application have been predefined in ISA. They are not described in this paper.

2-D EFFECTS

Haptic devices, such as trackballs and joysticks, have two active degrees of freedom. ISA and the API support a number of 2-D effects. This includes matrix, enclosures, periodics, springs and many more. (See [7] for a detailed description of the 2-D basic effects.)

Matrix

The Matrix is an M x N grid of basic effects with or without a surrounding wall. As a user navigates within the grid, a force pulls them to the center of the grid cells. If the matrix has a surrounding wall, forces will be exerted against the user to keep the user from venturing beyond the boundaries of the matrix.

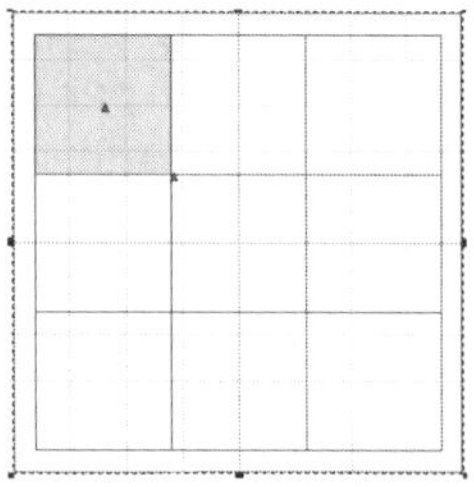

Figure 7. A 2D matrix

The matrix can be given a direction, expressed in degrees, which specifies its angle of rotation. The detents within each cell have variable parameters as described in the 1-D effect. An optional center dead band, which can mask selected central cells, can also be defined in order to create a menu that can be navigated in a rotational fashion.

IMMERSION STUDIOfi FOR AUT OMOTIVE

Figure 8 is a snapshot of ISA and shows a haptic scene with two haptic effects: a periodic effect at the bottom of the right hand window and four equally spaced detents in the right-hand top window. The interface is divided into several parts: main toolbar, scene window, spatial effects window, temporal effects window. The scene property window allows users to change all haptic parameters including the amplitude, origin of the effect, width or duration of the effect, type of the force profile, etc.

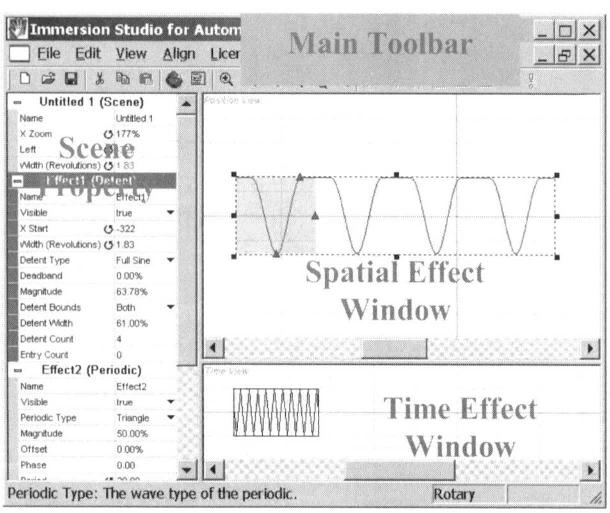

Figure 8. Snapshot of *Immersion Studio® for Automotive* (ISA)

SOURCE CODE GENERATION

ISA comes with a source code generator feature that produces the C++ code to play the effects designed with ISA. The generator creates a header file (.h) and a source file (.cpp). By clicking on a simple menu button, ISA will generate the code for the current haptic scene. ISA users can define the names of these files.

```
//{{ISA_HOBB_DEFINITION
CImaKnobBbDetent g_Detent1;

IMA_KNOB_DETENT_DATA g_Detent1Data
(
  (IMA_KNOB_DETENT_TYPE)3             // DetentType
, (int)(0.00)                         // Deadband
, (int)(49.61)                        // Magnitude
, (IMA_KNOB_DETENT_BOUNDS)0x3         // DetentBounds
, 4                                   // DetentCount
, 0                                   // EntryCount
, (int)(0.00)                         // DetentSpacing
, (int)(100.00)                       // DetentWidth
);
IMA_TPTR_TO(IMA_KNOB_DETENT_DATA)
g_pDetent1Data1(g_Detent1Data);
//}}ISA_HOBB_DEFINITION
```

Figure 9. Part of the source code generated by ISA's code generator feature for four detents

The files contain all the parameters and effects designed in ISA and the class and methods to be called by the target application to initialize, terminate, start, stop, get and set the haptic parameters.

Figure 9 gives a portion of the code generated by ISA to create four detents. A detent class is instantiated named g_Detent1, and its parameters are in a structure called Detent1Data. The parameters of detents include a type such as Full Sine, detent magnitude, number of detents, as well as the width and spacing between detents.

API DETAILS

The API is the software that manages haptic effects on a force-feedback device. As shown in Figure 3, the API runs on the host system typically o n the same CPU that runs the user interface and communicates with the embedded system via a driver and transport mechanism such as CANBus, MOST bus, etc. Applications use the API to access the force-feedback device, define haptic effects and scenes, start and stop a scene, receive event notification from the device and control the device. The API provides a collection of haptic effect classes that can be used by applications, and hides the low-level implementation details.

The API consists of two parts: a kernel and an effect library. The kernel contains functions to initialize and terminate the API, open one or more force-feedback devices, create and locate haptic effects and scenes, get information about the created effects and scenes, start and stop a scene on a device, and send control notifications to the device. The kernel is also responsible for notifying the application of the events that occur on the device through an event handling mechanism. The application is responsible for handling the events in an application-specific manner, such as updating the user interface, current scene or device state.

The effect library component of the API contains C++ classes that can be instantiated by the application to implement the desired haptic effects; for example, if the application requires a barrier bounded detent effect such as that shown in Figure 2b, the application can create an instance of the ClmaKnobBbBDB class. In all, the API provides 20 1-D effect classes and 13 2-D effect classes representing a collection of useful effects. Some of the effects are intended as building blocks that should be combined with other effect building blocks into more sophisticated haptic scenes. Because applications can combine several effects into a scene, there is much flexibility and room for creativity and experimentation. In addition, each effect has several parameters that can be adjusted to achieve a precise feel.

TOWARDS INTEGRATING HAPTICS INTO A HMI

We will conclude the description of the design guidelines and haptic tools by describing how these tools are used by automotive HMI designers to integrate haptics in their applications. Let us consider the menu in Figure 1 and describe how the HMI designer would enable this menu with haptic effects.

DESIGN CONSIDERATIONS

Assume that the designer intends to create detents for navigating the menu. A good starting point would be to consider 12 detents per revolution. This would take 120 degrees to traverse the four items of the menu. The designer will have to consider force amplitude, profile shape as well as damping add-ons:

- The shape will give the style of the detent; for example, triangular detents feel crisper than full sine detents.
- Finding the right amplitude is challenging and will typically require experience and usability testing. If the detent is too strong, users may dislike the feel or overshoot the target. If the amplitude is too weak, users may not feel the detent.
- The designer may want to add some damping or friction to get more accurate control.

Finally, the designer may consider barriers to surround the detents instead of having endless detents. By doing this, the target application will need to update the number of detents between the surrounding barriers according to the number of items in the menu.

CONSTRUCTING THE HAPTIC GUI

In order to obtain an appreciation for how the authoring tool can help the designer to quickly add haptics to existing GUIs, let us describe the steps required to implement the four-item menu described above. The necessary steps to create tactile sensations are the following:

1. Create a new scene for the targeted device; in this case a rotary device.
2. Insert a pre-defined haptic detent. The detent will have initial default parameters including 5 detents per revolution, bounded, 50% magnitude and full sine force profile.
3. Modify the parameters to obtain the desired feel; for example, set the detent count to four, set the bounding to endless, change the width of the effect, change the force profile to triangle and change the amplitude.
4. Operate the device and experience in real-time the haptic detent. Repeat steps 3 and 4 until satisfied with the user experience.
5. Save the effect to a file for future use.
6. Generate source code related to the detent and save the code in header and source files.

The steps necessary to integrate this effect into a target application displaying the four-item menu are:

1. Include the header files of the Immersion API in the target application.
2. Optionally make changes to the generated code.
3. Modify the target application to start and stop the haptic scene at the desired times in response to user interface state changes.
4. Add event handling to respond to detent index change events that are sent by the Immersion API to the target application as the user rotates the haptic device.
5. Compile and link with the application and API libraries.

With these simple steps the designer has created a desirable haptic experience and added it to the target application. The ease with which haptic sensations can be added to applications will greatly reduce the amount of time it takes to implement haptics in automotive interfaces.

CONCLUSION

The challenge of programming for force feedback is not the act of coding; it is the act of *designing* touch sensations that appropriately match interface menus and events. Designing touch sensations requires a creative and interactive process where parameters are defined, experienced and modified until the desired tactile experience is obtained. More importantly, designers do not have to be concerned with the details of the haptics as there are predefined effects that only need to be adjusted and correlated to the graphics.

With tools such as *Immersion Studio™ for Automotive* and the *Immersion API*, designers can quickly try out the tactile feel of designed effects, visually see the results of parameter changes on tactile sensation and easily change parameters with the GUI.

The challenge of introducing haptic technologies in the automotive field is no longer the act of an innovative leader. It has become the mark of any manufacturer interested in making substantially better interfaces where touch is a must-have to gain comfort, safety and efficiency.

REFERENCES

[1] Ramstein, C., *Toward Architecture Models for Multimodal Interfaces with Force Feedback*, in proc. of international conference HCI' 95 on Human Factors in Computing Systems, Tokyo, Japan, July 1995.

[2] BMW iDrive Control: http://www.bmwworld.com/models/e65.htm

[3] Levin, M. et al., *Control Knob with Multiple Degrees of Freedom and Force Feedback*, US006,154,201, November 2000.

[4] Badescu, Wampler and Mavroidis, *Rotary Haptic Knob for Vehicular Instrument Controls*, Proceedings of the 10th Symp. On Haptic Interfaces For Virtual Envir. & Teleoperator Systs. 2002.

[5] Mauter, Katki, *The Application of Operation Haptics in Automotive Engineering*, General Automotive Manufacturing and Technology 2003, April 2003.

[6] Burdea, G., *Force and Touch Feedback for Virtual Reality,* New York: John Wiley and Sons, 1996.

[7] Online help file for Immersion Studio and Immersion API. Available upon request. Immersion Corporation.

[8] MacLean, K. E. (2000). *Designing with Haptic Feedback*, in Proceedings of IEEE Robotics and Automation (ICRA'2000), San Francisco, CA, April 22-28

[9] Münch, S., Dillmann, R., *Haptic output in multimodal user interfaces,* Proceedings of the 1997 International Conference on Intelligent User Interfaces, pp. 105-112, 1997.

[10] Enriquez, M. J., MacLean, K. E. (2003). *The Hapticon Editor: A Tool in Support of Haptic Communication Research*, in Proc of the 11th Annual Symposium on Haptic Interfaces for Virtual Environments and Teleoperator Systems, IEEE-VR2003, Los Angeles, CA, 2003.

CONTACT

Christophe Ramstein has been promoting haptics since 1989. He completed a Ph.D. in computer-haptics in 1991, designed the first mouse-based haptic system for blind users in 1993, designed and validated haptic systems for applications in 3D, aerospace, medical, remote education, automotive and communication. He is the VP, Engineering of Immersion since 2002.

Site address: http://www.immersion.com

Email: cramstein@immersion.com

2003-01-1288

Software Development Process and Software-Components for X-by-Wire Systems

Andreas Krüger, Dietmar Kant and Markus Buhlmann
Audi AG

ABSTRACT

The term *X-by-Wire* is commonly used in the automotive industry to describe the notion of replacing current mechanical or hydraulic chassis and powertrain systems with pure electro-mechanical systems.

The paper describes the current trends and the architecture of future chassis electronics systems. The first part of the paper covers the systems architecture of x-by-wire electronics systems. We describe the network and the software architecture in more detail. The paper also explains some of the software components, in particular the operating system and the communication layer.

The second part of the paper gives a description of the current state of the development process for software intended for safety-relevant systems. A possible tool chain for this development process, current possibilities as well as limitations and challenges are described.

INTRODUCTION

The intensive competition among car manufacturers world-wide forces these companies to deploy ever new functions in the car to keep up with their competitors, and satisfy customer demand. Examples for this can be observed in the area of body electronics (think about smart central locking, or seat memories) or infotainment systems (e.g., navigation aids).

Another big differentiator is the "safety argument". The passive safety (airbags and seat belt systems, ...) is not unlimited. Particularly the construction methods, but also the number of airbags will soon reach a limit. Therefore, the manufacturers are turning towards active safety systems, i.e., vehicle dynamics control systems or electronic stability systems, to push the driving behavior of the vehicle even closer to the physical limits.

The next step in extending the active safety features of a car is the direct control of vehicle functions like steering or braking without driver interaction [1,2]. Distributed digital control system that replace the conventional mechanics and hydraulics (commonly called by-wire systems) are becoming an accepted solution in the car industry, after they have already been established for some time in the aircraft industry. Besides the direct control over vehicle functions, these systems offer a number of additional advantages.

Obviously, by-wire systems place stringent demands on system design: high reliability and the distribution of control that requires timely communication between the connected modules are the most important.

The rest of this paper is structured as follows: The next Section describes the roadmap from current individual, networked systems to truly distributed x-by-wire applications. The following Section gives an overview of the systems architecture as currently planned by Audi. A final Section discusses some of the standard software components required for by-wire systems.

MULTIPLEXING IN CHASSIS ELECTRONICS – THE WAY TOWARDS X-BY-WIRE SYSTEMS

Today's vehicle electronics is characterized by a number of relatively autonomous electronic control units (ECUs). These ECUs communicate by means of a bus system (CAN [3] has established itself as today's de-facto standard), and are thus loosely coupled to each other.

In our view, the next step towards x-by-wire systems will most likely be higher-level, global control systems that coordinate all chassis systems (braking, steering, damping, powertrain). In the first step, this will be another separate ECU implementing some central control algorithms. Later, these control algorithms will move into the various chassis ECUs, and thus be implemented as a distributed control system (Fig. 1).

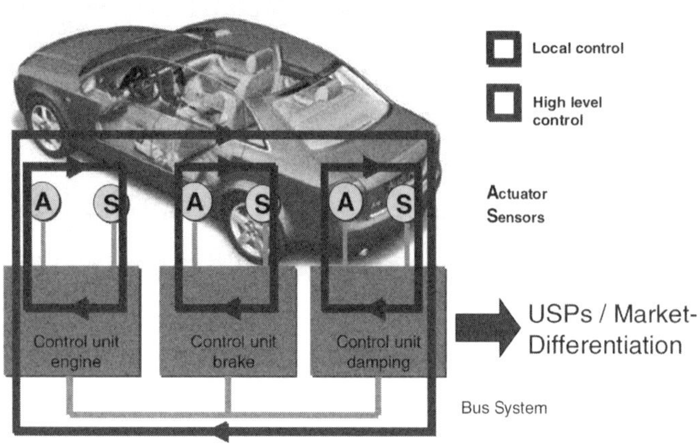

Figure 1: Global, distributed control algorithms.

These *Global Chassis Control* Systems will couple systems that today operate almost autonomously much closer to each other and coordinate the algorithms they implement. These system include

1. Electronic Stability Program (ESP),
2. steering support, and
3. active chassis / damping.

An example could be the coordination of ESP control with dedicated steering actions to further improve vehicle stability.

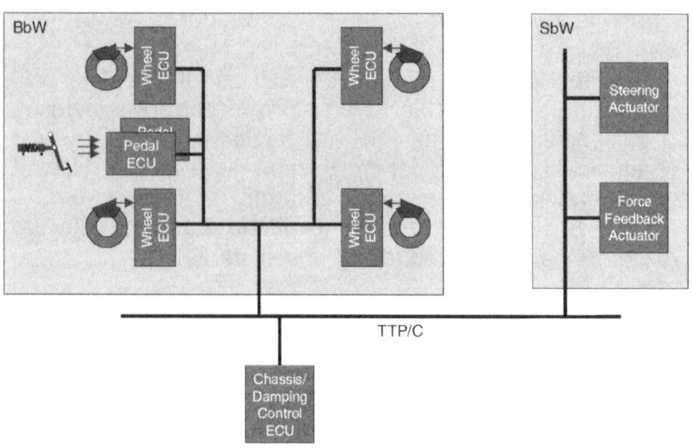

Figure 2: Schematic diagram of a networked by-wire system with distributed braking and steering system.

As a further step, x-by-wire systems will not only couple single functions into one distributed control loop, but will distribute the functions themselves into a number of ECUs. An example for this is the often mentioned brake-by-wire, where at least four wheel nodes and pedal sensors communicate via a real-time bus system. Figure 2 shows an example of a distributed brake- and steer-by-wire system.

SYSTEM ARCHITECTURE

This Section deals with electronics architectures for safety-relevant electronics systems. In particular, we cover the network and software architectures. Since the development process plays a significant role in highly dependable systems, we also cover this topic.

DEVELOPMENT PROCESS

There already exists a number of standards and regulations for the development of safety-relevant systems, all of them in domains other than the automotive industry. The standard IEC 61508 [4], which is a generic, cross-domain process directive, or DO-178B [5], the US aerospace regulation for software systems, are both well-known standards.

The development and safety processes in the aerospace and railway industries are both very rigid. They require a well defined acceptance and certification procedure for each system or system variant. The automotive industry cannot simply adopt such a process, because there are a lot more variants to handle in a single car line than there are in railway or airplane systems. A certification procedure for all systems that result from the combination of all variants (considering all extra equipment combinations) would be very time consuming and incur prohibitive costs. Therefore, no car manufacturer would risk such a paralysis of his development efforts. The solution instead is most probably to develop subsystems with very narrow and well-defined interfaces to the other car sub-systems.

When developing one of these sub-systems, a generally accepted safety development process is nevertheless required, in analogy to the above-mentioned regulations. These existing standards are a very good guideline, and much of their contents must be adopted, thereby considering and extending already existing processes.

NETWORK ARCHITECTURE

Current automotive control systems for safety applications (e.g., anti-lock braking system – ABS) consist of a single central ECU performing the entire control algorithm for a single mechanical or hydraulic subsystem. They are furthermore characterized by the following properties:

1. There is a star-topology wiring to the sensors and actuators connecting the ECU to the controlled system.
2. There is a safe state that can be reached if the ECU fails.

The micro-controller of such an ECU is usually equipped with self-checking mechanisms (watchdog or dual CPU,

software checks). If one of these mechanisms detects an error, the controller resets and remains passive. This is perfectly feasible, since the (hydraulic) basic system can operate without the electronics system's functionality. Systems that exhibit such a behavior are called *fail-safe* systems.

An x-by-wire system replaces this conventional architecture by a distributed computer system controlling an electro-mechanic subsystem: a network of autonomous node computers communicating via a broadcast bus system replaces the previous star topology. Furthermore, such a system must be *fail-operational:* the functionality must be maintained even in the presence of a fault, since there will be no mechanical or hydraulic backup any more.

To achieve this property, fault-tolerance is a mandatory requirement. The type and number of faults that the system shall tolerate must be defined in a fault hypothesis. Fault tolerance in turn is usually implemented by adding a certain degree of redundancy to the system. This and the fact that any of the envisioned systems will control the mechanics at different locations within the car is the main reason for the implementation as a distributed solution.

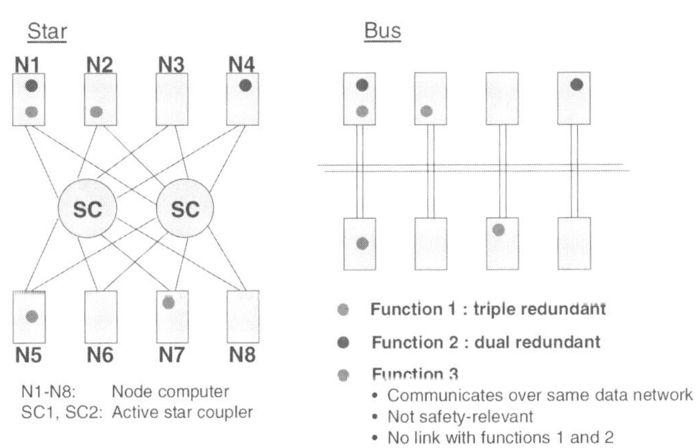

Figure 3: Star and bus topology data networks with redundant software functions.

A recently performed comparison between bus and star topologies (Figure 3) shows that the star topology has a much better fault resilience [6]. The reason is that in a star topology a single component failure may only affect one connection between a node computer and the central star coupler. Furthermore, a failure of the central star coupler – an argument that has long been used against star topologies – can be tolerated by appropriate redundancy strategies (e.g., replicating the communication channels and the star coupler and adding dedicated communication lines between the star couplers).

Figure 3 also shows an example of how single safety-relevant functions may be replicated. The degree of redundancy, i.e., how many instantiations of a function are actually implemented in the system, is scalable. If a node computer fails, the safety relevant functions are still available on other nodes. This kind of *hot redundancy* is best suited for the required short control loops and the nature of the statically configured ECUs.

SOFTWARE ARCHITECTURE

Apart from the previously discussed network architecture, the architecture of the software that is executed within the ECUs is of central importance.

The main objective of the discussed software architecture is the re-use of individual functions, which are stored as software modules in a *function repository.* The function repository contains modules that are generated only once as executable models – either as a modeled control system (e.g., using Matlab/Simulink or ASCET/SD), or as a state machine (e.g., using Stateflow or Rhapsody), depending on the nature of the application. This will result in software carry-over parts (COP) that can be used multiple times across car lines and vehicle variants. Figure 4 visualizes the idea of a function repository.

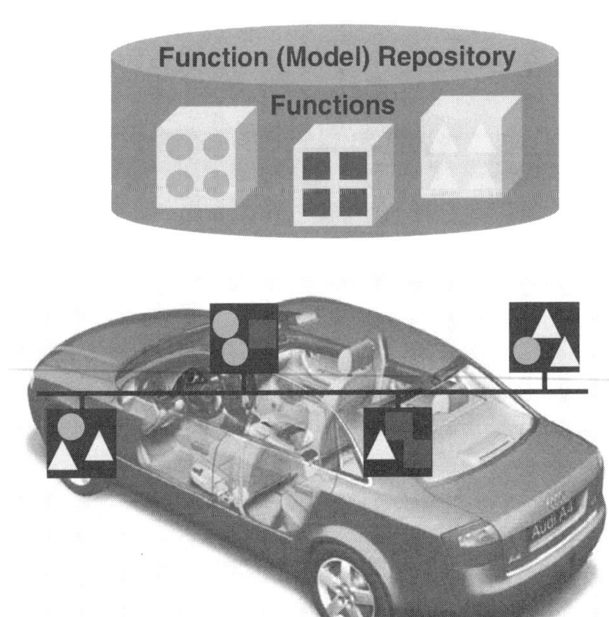

Figure 4: A *function repository* contains reusable software components.

Admittedly, this vision is a challenge for the automotive industry, and there are a number of obstacles to overcome. Nevertheless, it is of high importance for automotive OEMs to adapt and to apply some concepts from the area of software engineering for the automotive industry.

A necessary prerequisite for the objective we just described is an open software architecture, which enables portability and re-use of individual software modules, ensures correct diagnosis of faults, and which supports safety and fault tolerance functions. These properties can be fulfilled if at least the following software components are part of the software architecture:

- An operating system, which controls the task execution and administrates the system resources.
- A communication layer (middleware layer), which abstracts from the communication over the underlying data bus, makes the distribution of functions transparent to the application, and also implements some fault tolerance functions.
- A hardware abstraction layer (HAL), which hides the access to the peripheral hardware modules (pulse width modulation, analog-to-digital conversion, serial interfaces, etc.) behind a standardized interface.

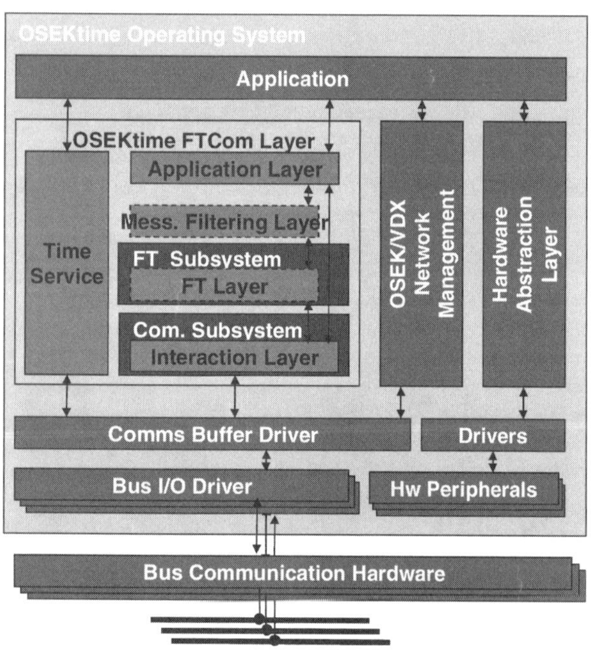

Figure 5: Software architecture for x-by-wire systems.

Figure 5 shows the interaction of these and other components. The operating system, which conforms to the OSEK specification *OSEKtime OS* [7,9], coordinates the execution of the application tasks and the tasks of the other standard components. The application accesses the data network via a *fault-tolerant communication layer (FTCom)* [8,9]. Similarly, the application accesses to peripheral modules of the hardware via a standardized *hardware abstraction layer.* Network management [10] should be mentioned at this place for the sake of completeness, but its detailed discussion is beyond the scope of this paper.

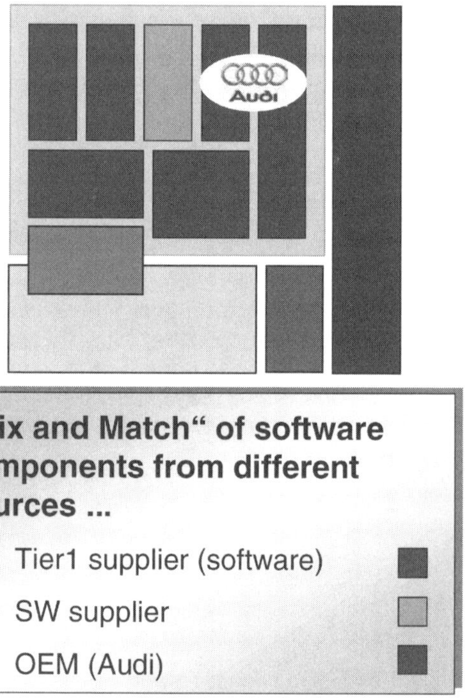

Figure 6: Combination of software components from different suppliers.

Once portability and re-usability of software components is ensured by an open software architecture, it is also technically feasible to combine software components from different suppliers within a single ECU. An ECU supplier, for instance, may provide the basic application functions (e.g., local control algorithms). Other functions that require know-how about the integration of the ECU in the overall vehicle system and about its interaction with other ECUs, like global control algorithms, may be implemented by the car manufacturer. Finally, it is also possible that independent software suppliers provide basic functions, which are not specific to a certain ECU (e.g., diagnosis functions). Figure 6 illustrates this vision.

STANDARD SOFTWARE COMPONENTS FOR X-BY-WIRE SYSTEMS

This Section provides a more detailed description of the standard software components for a x-by-wire run-time system.

OPERATING SYSTEM

The OSEKtime operating system specification follows the time-triggered paradigm of real-time system design. In a time-triggered system all safety-related communication and computation actions are triggered by the progression of time. The operating system offers all basic services for real-time applications, i.e., dispatching of tasks, system time and clock synchronization, local message handling, error detection mechanisms, and - if necessary - interrupt handling.

Task Management

A task defines an autonomous single threaded piece of application software that is designed to potentially run in parallel with other tasks. Tasks are executed sequentially starting at the entry point and running to the exit point. In a time-triggered application activation events originate from entries in a timetable only. The timetable, which contains the activation times and deadlines for each task, is generated by an external scheduling tool before the run-time of the system. Based on these entries, the OS can also detect a deadline overrun of a task. Further, there are no blocking mechanisms through events or resource management, since potential access conflicts can be solved off-line by the scheduling tool.

Processing Levels

Each task and interrupt service routine (ISR) is statically assigned to one *processing level* that defines its priority. There are three ranges of processing levels in an OSEKtime system:

1. Non-maskable interrupt service routines and the OSEKtime dispatcher (highest processing range),
2. maskable interrupt service routines and time-triggered tasks (time-triggered processing range),
3. background task or OSEK/VDX subsystem (lowest processing range).

For the lowest processing range two implementations are possible, an OSEKtime idle task or a full OSEK/VDX OS subsystem (see Figure 7). The use of an OSEK/VDX OS subsystem yields the advantages of compatibility to existing applications and support of event-triggered application parts (e.g. diagnosis, gateways, ...). In a safety-related application, however, the OSEK/VDX subsystem should only be available, if the micro-controller provides a sufficient number of interrupt levels for the implementation of the above model and if there is hardware support for memory protection (e.g., a memory management unit, MMU). Only if these two requirements are fulfilled it can be guaranteed that an event-driven task of the OSEK/VDX subsystem cannot interfere with a safety-related time-driven task located in the OSEKtime subsystem. In case no OSEK/VDX subsystem is required or the above requirements cannot be met, an OSEKtime operating system can also be implemented without an OSEK/VDX subsystem. The functionality of the OSEKtime subsystem is not limited by this.

Further mechanisms of the OSEKtime operating system, such as interrupt handling, provision of a system time, local message handling, and task deadline enforcement are not covered in this paper. Details can be found in the respective literature [7,9].

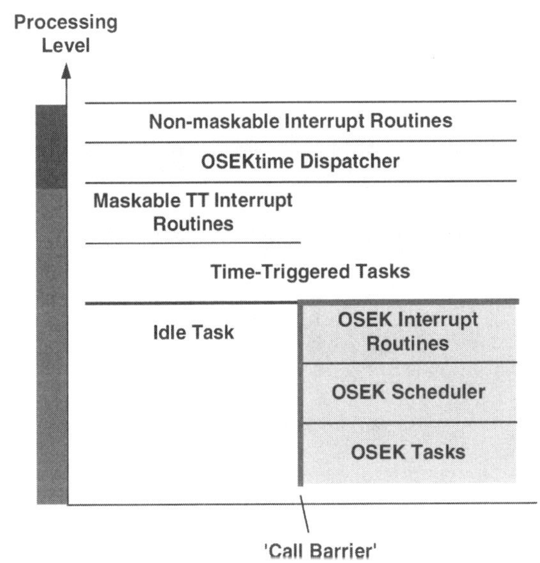

Figure 7: OSEKtime OS processing levels.

MIDDLEWARE LAYER

A middleware layer is responsible for the interaction between the communication network and the application software, and should provide an abstraction from the network access to the application. For the application, communication across a data network must be equivalent to the inter-task communication that is done locally within a single ECU. For these purposes, the OSEKtime working group has specified a fault-tolerant communication layer (FTCom) [8,9]. It provides the necessary services to support fault-tolerant hard real-time distributed applications. In the following these services are described.

Message Transmission

Transmission of messages and its related services is the most important task of the fault-tolerant communication layer. At this point it is important to distinguish between *signals*, which are application level data elements such as engine speed or brake force, and *frames*, which are entities that are exchanged via the communication bus.

FTCom provides a signal-oriented interface, i.e., it provides relevant, ready-to-use application data to the application tasks. It is the task of the FTCom layer to pack the application-level signals into the corresponding frames, in which they are transmitted on the communication network, and to unpack them again at the receiver. In this context it is also the task of the FTCom layer to consider possible different byte orders (e.g., little Endian, big Endian) between sender and receiver.

Transmission of messages is thus comparable to using global state variables, which makes the communication system transparent to the application.

Redundancy Management

In safety-related applications, certain messages are transmitted multiple times, in order to prevent message loss and to tolerate transient faults. In this case, the FTCom layer replicates the signals at the sending ECU. Accordingly, it forwards only one copy of the signal to the application at the receiving ECUs.

Agreement and Message Filtering

The individual signals of a single data item that is transmitted redundantly may contain different values. This may, e.g., be the case when raw sensor values are transmitted. In this case it is necessary to perform an agreement algorithm with these signals. For this purpose, the FTCom layer provides predefined agreement algorithms and a framework for user defined agreement algorithms.

Signals may also be filtered according to certain pre-configured criteria. In this case, only those signals that fulfill the filter criterion that is individually pre-configured for this signal are forwarded to the application. Usage of the agreement algorithm and of message filtering is optional.

HARDWARE ABSTRACTION LAYER

As a third standard software component the hardware abstraction layer should be mentioned at this point. The primary objective for using such a standardized interface to the hardware peripherals is to ensure portability of application functions between different hardware platforms. This is achieved by placing a standardized API (Application Program Interface) between application tasks and hardware drivers.

The German *OEM Initiative Software* (Hersteller-Initiative Software – HIS) [11] is currently establishing such a standardized interface, with activities in most of the participating companies. First prototype implementations do exist, and experiences from using these will find its way into revised versions of the software.

CONCLUSION

This paper has described the evolution that is currently happening in the automotive industry in the field of chassis systems, in particular braking or steering systems. The trend towards by-wire systems, i.e., replacing mechanical and hydraulic components with distributed electronic control systems, has been shown. The electronics architecture that is necessary to implement such safety-relevant systems was discussed from the viewpoint of multiplexing (network topologies), and – in some detail –from the viewpoint of the software architecture.

Especially the area of software engineering is a key technology for the implementation of all future electronics systems in the car. In particular, this is true for x-by-wire systems. Therefore, we covered the software architecture for ECUs in such systems, and briefly touched on some ideas to make software a re-usable vehicle component.

The last section described some standard software components in more detail. In particular, we covered the operating system (OSEKtime OS) and the communication layer (OSEKtime FTCom). We also pointed out the requirement for a hardware abstraction layer (HAL).

The automotive industry is still only beginning to get a grip on mastering the software development process. Software engineering methods, re-use of software, and software quality management – to name but a few – find their way into the automotive development process only very slowly. Therefore, the concepts we just discussed are only snapshots seen from today's point of view. One can expect that the described technologies will undergo further massive advancements in the future.

ACKNOWLEDGMENTS

This work has been supported by the European IST project "Next TTA" under project no. IST-2001-32111.

REFERENCES

1. Dilger, E., Johansson, L.Å., Kopetz, H., Krug, M., Lidén, P., McCall, G., Mortara, P., Müller, B., Panizza, U., Poledna, S., Schedl, A.V., Söderberg, J., Strömberg, M., Thurner, T. (1997). *Towards an Architecture for Safety-Related Fault-Tolerant Systems in Vehicles.* European Conference on Safety and Reliability, Lissabon, Portugal, 1997.

2. Führer, T. and Schedl, A. (1999). *The Steer-by-Wire Prototype Implementation: Realizing Time-Triggered System Design, Fail Silence Behavior and Active Replication with Fault-Tolerance Support.* SAE Congress & Exhibition, Detroit, MI, USA, 1999. SAE paper no. 99PC216.
3. International Standards Organization (1993). *Road Vehicles – Interchange of Digital Information – Controller Area Network (CAN) for High Speed Communication.* ISO11898, 1993.
4. International Electrotechnical Commission (1998). *Functional safety of electrical / electronic / programmable electronic safety-related systems.* IEC 61508, 1998.
5. *RTCA*, Inc. (1992). *Software considerations in airborne systems and equipment certification.* RTCA/DO-178B, 1992.
6. Kopetz, H., Bauer, G., Poledna, S. (2001). *Tolerating Arbitrary Node Failures in the Time-Triggered Architecture.* SAE Congress & Exhibition, Detroit, MI, USA, 2001. SAE paper no. 2001-01-0677
7. OSEK/VDX (2001). *OSEKtime Operating System*, Version 1.0, OSEK/VDX Homepage, URL: http://www.osek-vdx.org/, 2001.
8. OSEK/VDX (2001). *OSEKtime Fault-Tolerant Communication (FTCom)*, Version 1.0, OSEK/VDX Homepage, URL: http://www.osek-vdx.org/, 2001.
9. Krüger, A., Domaratsky, Y., Holzmann, B., Schedl, A., Ebner, C., Belschner, R., Hedenetz, B., Fuchs, E., Zahir, A., Boutin, S., Dilger, E., Führer, T., Nossal, R., Pfaffeneder, B., Poledna, S., Glück, M., Tanzer, C., Ringler, T. (2000). *OSEKtime: Highly Dependable Applications - Objectives, Basics and Concepts*, OSEK/VDX 3rd International Workshop, Bad Homburg, 2000.
10. OSEK/VDX (2000). *OSEK Network Management*, Version 2.5.1, OSEK/VDX Homepage, URL: http://www.osek-vdx.org/, 2000.
11. Lange, K., Bortolazzi, J., Brangs, P., Marx, D., Wagner, G. (2001). *Hersteller-Initiative Software.* 10. Internationale Tagung Elektronik im Kraftfahrzeug, Baden-Baden, VDI-Berichte Nr. 1646, 2001.

CONTACT

Dr. Andreas Krüger, MBA
AUDI AG
I/EE-93
D-85045 Ingolstadt
Germany
Email: Andreas.Krueger@audi.de

2003-01-1197

The Software for a New Electric Clutch Actuator Concept

Reinhard Ludes
AFT Atlas Fahrzeugtechnik GmbH

Thomas Pfund
LuK GmbH & Co.

ABSTRACT

Software plays a very significant role in automotive technology. The number and importance of mechatronic systems has increased greatly. The system functions and the vehicle's characteristics are being carried out more and more by the software. This trend is further encouraged by high-performance processor systems, which are becoming more affordable and offer ever more functions and increased storage space. However, more effort must be invested in software specification, implementation, testing and the validation of the entire mechatronic system. The various methods of generating software have not kept pace with the demands made on software systems, nor with the complexities created or caused daily by large development teams. The present project takes an alternative route in developing the software for an electrical central release bearing. The required control software is generated automatically by using TDC (Total Development Chain) by AFT.

INTRODUCTION

Software plays a very significant role in automotive technology these days (Figure 1).

software in automotive technology	
control & automation of functions, tasks and procedures	
in the vehicle:	
• automatic control functions	injection, ignition, engine management, ABS, ASR, ESP, etc.
• automation	clutch (TCI), transmission (ASG)
in development & production:	
• methods & tools	CAD, CAE, CAM
• automation	robotics

Figure 1: Software in Automotive Technology

The number and importance of mechatronic systems whose mechanical and electronic software-based functions are highly integrated has greatly increased, particularly in the vehicle itself. The software here performs automatic control functions or automates behavior previously performed by the driver.

The systems' functionality is growing increasingly more complex, and this complexity of system behavior, as a result, is found more and more in the software. While ever more platforms are being used in the area of mechanical and electronic hardware, even across brands, the tasks of individualizing the system and the vehicle s characteristics are increasingly being given to the software.

This trend is furthered by the apparent ease with which the software can be modified. It is also encouraged by the emergence of high-performance processor systems, which are not only becoming more affordable, but are offering ever more extensive functions and increased storage space. This is an irresistible temptation for ambitious development teams.

But what is often overlooked is the fact that ever greater effort must be invested in software specification, implementation and testing and the validation of the entire mechatronic system, both on the test-bed and in the vehicle. This is because the different methods to generate software have not kept pace with the demands made on software systems or with the complexities created or caused daily by large development teams.

Of course there has been progress in programming languages: several testing methods have been established to optimize the generation process and language extensions like object-oriented programming (OOP), which are also slowly becoming established in the area of real-time programming. The fact remains, however, that while it is making significant contributions to control and automation, software itself is generally still written by hand. Software development is thus still a classical manufacturing process.

The present project takes an alternative route in developing the software for an electrical central release bearing. The required control software is generated automatically by using TDC, the abbreviation for Total Development Chain, by AFT (AFT Atlas Fahrzeugtechnik). A full description follows.

THE STANDARD DEVELOPMENT PROCESS FOR SOFTWARE

The standard software development process is outlined in the next section. Naturally, there are many variations, but we will mention only the major and generally typical steps here (Figure 2).

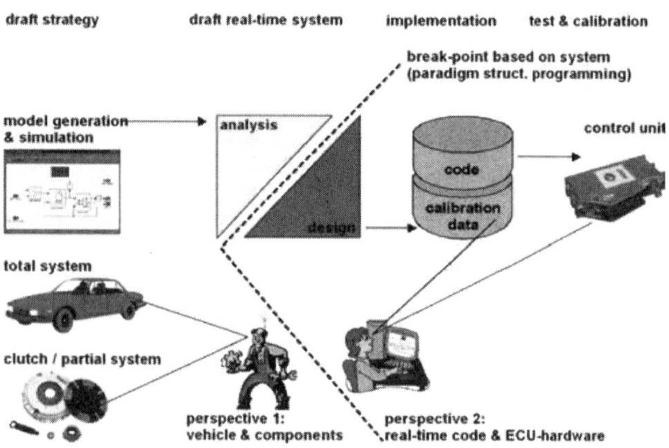

Fig. 2: Standard Development Process for Software

First, the control strategy is formulated based on an idea. This can be accomplished by writing a performance specification for the control strategy. An advanced procedure includes the specification and all types of detailing of the controller design using a simulation program whose structure and configuration is continuously improved based on models. The advantage of this method concerning written specifications and drafts is that it is easy for professionals specializing in different areas, such as systems analysis, control and simulation to discuss the control unit model.

In the classical method, once the control or automation function at leas t in the area of simulation has reached a sufficient level, it is now the software expert whose qualifications come from the area of real-time software and micro controller design w ho is ultimately responsible for the implementation.

The basic control design concept is the beginning of a software expert s work. They must also first analyze the task given to them and create an appropriate design concept for the real-time system before they can even begin to think about implementation, i.e. translating the functions into concrete code. The reason for this break, or gap in the process, also seen in the diagram, is that there are different ways to approach the model.

The main consideration in controller design is the object to be controlled. In our case, the control motor is the actuator and the clutch is the controlled system.

All models are oriented in this way. In real-time software design, as in any other software design, the architecture of the computer or microcontroller system is the main issue. The models used as the basis are thus primarily oriented toward the standard architecture of computer cores or processors (Von-Neumann machines), and all standard procedural programming languages are designed on this basis.

The size of this gap varies and depends on the languages and methods used. The gap is, however, part of the basic nature of the task. Only object-oriented programming shows a way to eliminate this.

There are several approaches for bridging this gap. Above all, the outcome depends on the development process; how well the communications between the two basic developer types work.

At LuK, for example, we have been successful in providing both qualifications to a team or to individual developers. In addition, code segments are transferred from the offline simulation to the source code level in the control software.

CODE GENERATION IN THE DEVELOPMENT PROCESS

In this project we have taken the road of automatic code generation. The code generator used here takes the controller s model, or simulation-based specification, and generates close-to-production real-time code for the microcontroller in the automotive control unit. This creates a continuous path from the specification of the general, to the detailed strategy, to the real-time code in the control unit (Figure 3).

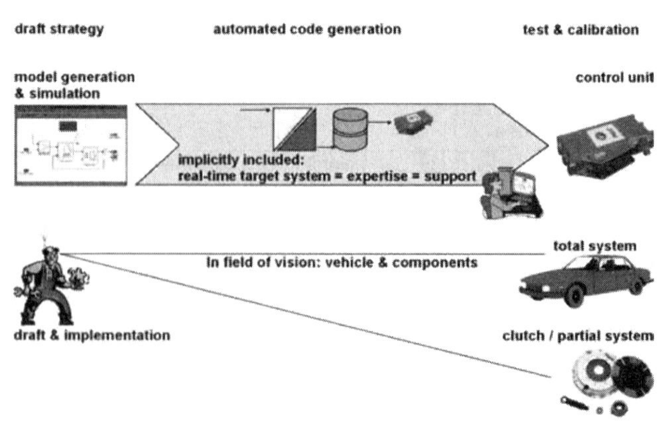

Fig. 3: AFT s TDC in the Development Process

This approach is in line with LuK's philosophy of using only one expert or group of experts to design and implement the entire system. In contrast to the classical method, the system engineer also designs the control unit. Since the knowledge of how to generate the optimum code is implicitly present in the code generator, the functions relating to the vehicle or one of its components are thus kept in mind from the beginning on. The processor, the programming language and the operating system that are used in the concrete application remain in the background.

The gap in the specification or the entire development chain has thus been bridged.

USING TDC IN ELECTRIC CENTRAL RELEASE BEARING ENGINEERING: THE CLOSED TOOL CHAIN

The TDC by AFT is more than just a code generator. TDC is a heterogeneous but closed chain of tools which supports the entire development process. Heterogeneous, because different tools are used together in fact the exact tools that are best suited for the specific project. Closed, because the results of the test are fed back in to gradually improve the original model (Figure 4).

In this project, we used MATLAB®/Simulink® for simulation, TargetLink for code generation, AFT PROST as a suitable prototype control unit, and MARC I for recalibration in the test phase. It is also worth noting that this data, which is used for calibration, i.e. fine-tuning, is already specified in the simulation software. Description files according to ASAP2 are also generated and automatically read into the calibration system, so no additional configuration by the test engineer is needed.

The data obtained during calibration can also be worked into the simulation model to fine-tune it. Furthermore, the measurement data obtained in the test can be fed back into the simulation from the calibration system or from additional measuring systems. The data loggers RAMboX and TORnadO are available at LuK for this purpose and are currently in use.

To form the development chain as described, only the code generator and a series of expansions to the calibration system had to be installed. All additional tools were on hand and known to the developers.

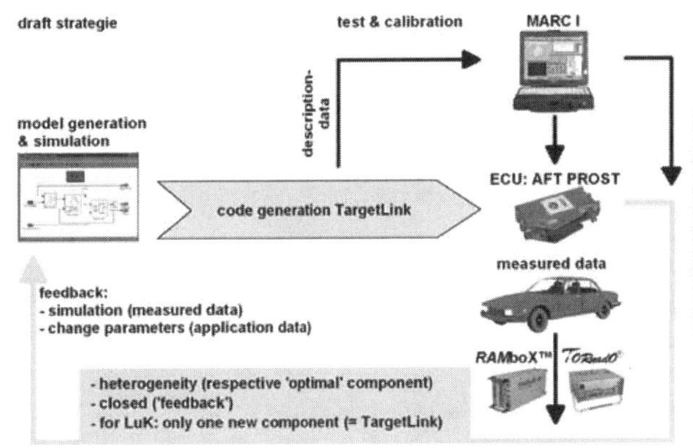

Fig. 4: Characteristics of AFT s TDC: Example LuK Electric Central Release Bearing

THE ELECTRIC CENTRAL RELEASE BEARING PROJECT AS REFLECTED IN THE V MODEL

As indicated in the previous illustration, TDC reflects a development method that goes beyond pure code generation. It is a simulation-based development method with executable specifications. The entire chain is based on the V model. Figure 5 shows the implementation of the V model with all components that are used at AFT to develop software for control devices.

The following are characteristics of the V model as the basis of the development strategy:

- Each design step on the left has a corresponding test and validation step shown opposite it on the right.
- Transfers and tests are required between the steps on the left side to ensure that the specifications given in the previous step are also fulfilled.

This also means that the later an error is detected, the more steps must be rerun during the re-design, i.e. the more difficult and expensive correction becomes. Rectifying basic specifications that are inadequate or not met which can only be discovered in the final step is thus especially critical.

The left side shows the two significant design steps:

1. Specifications phase the s ystem specifications are made to provide clear definitions, in some cases based on the simulation models and automatic condition generators.
2. Model generation bas ed on the system specification and any existing partial models, a structure that can be simulated is generated, embedded in a more or less exact plant model.

Both steps use MATLAB®/Simulink®, in the first step for the rough system structures and specifications, and in the second step to define the entire model of the control unit and controlled system in detail.

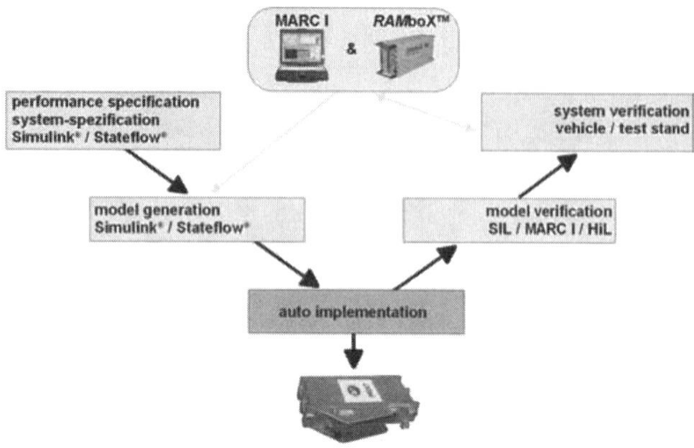

Fig. 5: Electric Central Release Bearing Project Reflected in the V Model

The process ends in the peak of the V model with auto-code generation on the specific target system. Thus, the original control unit specification is not only executable in a simulated environment, but also on a control unit as the target system. The entire chain is continuous.

On the right side, across from the two design steps are the corresponding test and validation steps:

1. Model-verification, i.e. validation of the basic control function, in the SiL simulation (SiL = software-in-the-loop) or if applicable already in a HiL system (HiL = hardware-in-the-loop).
2. Total system verification on the test stand or test section.

The controller is first tested in an SiL simulation. In the SiL simulation, all interfaces to the specified software are depicted as models, so that the simulation can run without any additional hardware expense. The system supports this process in that 1) the I/O interfaces to the controller-operating system are available as TargetLink-blocks and 2) a simulation is possible with both floating decimal and fixed decimal arithmetic.

An HiL system is used for model verification when individual components of the controlled system cannot be represented in the simulation environment, or only can with great effort. At the same time, unfortunately, a test in the entire system, is not yet economically feasible, e.g. due to high application costs. It must also be decided on a case-by-case basis which parts in the controlled system must be physically present and which can be represented as a software model.

It is not necessary to decide whether the control unit will be 'operated' for real or in the real-time simulated controlled system. It is worth noting that in this simulation-based development process, the system model required for the HiL simulator is largely already available from the system specification step in the V model. In the project described here, the main partial systems were installed on a test stand (central release bearing against non-rotating clutch, PROST control unit and higher-level TCI control unit).

System verification is based on the overall system design. If the use of the clutch is determined in the vehicle, the validation must also be completed in the vehicle or in a vehicle-adequate system environment. In this case the final verification was also performed in the vehicle.

Thus, all stages of the V model were completed in this project.

The data measured in the control design stage and in the final test was also fed back.

RESULTS, PART I: THE CONTROLLER STRUCTURE

The controller specification at first consisted of rough blocks, which were then filled in by refinements and parameter configurations. The blocks can be output graphically directly with their links (signal paths) and thus can support an optimal specification and documentation: These blocks are:

1. the target and actual value calculation block for the controller
2. the state controller as the core
3. the pre-control block to raise the adjustment dynamics
4. the logic block, which differentiates different basic conditions (e.g. normal, limp-home and test operation)
5. the follow-up functions, which encompass different engagement and disengagement procedures
6. the set outputs block for checking limit values and adjusting to the hardware
7. the measured values on CAN block as a control function during the test phase

A graphic representation will be created automatically. The integration at this point, however, is far too extensive and detailed. Each block can be further broken down to show more detail. The controller as an entire system is once more a block in the offline simulation model, in which the remaining components of the controlled system are also represented: end stage, electric motor, the belt gears and the clutch itself.

Since the controller s specifications finally result in a controller model that can run both 'against' the model of the controlled system in the simulation and 'against' the real controlled system after code has been generated in the control unit, it is an executable specification.

When the controller is in a real control unit after code generation, the generated software accesses the same input and output interfaces as already shown in the model. The interface to the system software then makes up the AFT controller interface (ACI).

RESULTS, PART II: SAVINGS POTENTIAL

Three man-months were originally planned to generate the controller for the electric central release bearing; the three phases were to design the simulation, implement the code and for initial testing. The project was changed to the new method and tools two weeks after the project s start. The new method also required a total of three man-months.

No control group was defined to solve the same problem using conventional methods. There is, however, sufficient experience with the definition and execution of such projects that we can assume that approximately the same amount of work was required.

Some of the effort certainly went into breaking in the new tools. The first decision was made after two weeks to change the method. This cost a few extra hours, since the simulation models had to be readjusted to the real-time requirements.

It is estimated that the development time for trained employees can be reduced about 50 % to 1.5 man-months. This results first from an analysis of the time records and secondly from the calculation check: According to experience, 1.5 man-months are required to write the code for the controller, test and configure the calibration system. This expense is eliminated with the new method.

The possible effect of this ratio is thus a good 50 % and results from eliminating the need to generate the controller code and the time required to configure the calibration system by fine-tuning it in the test phase.

It should be mentioned that validation steps required for production release (long-time test operation, code inspection, etc.) were not included in the time estimate. If these were taken into account, the ratio effect would be reduced.

It is very important to note here that this is not a purely functional representation on a high-performance automotive test system.

The AFT PROST prototype control unit includes a standard microcontroller with fixed-point arithmetic, suitable for production. The generated code is transferable to a conventional production unit without any great effort.

This design, because it is closer to production than other systems that are based on high-performance floating-point processors and which only allow a functional depiction in a prototype vehicle, is thus far superior.

ANSWERS TO FREQUENTLY ASKED QUESTIONS

WHAT IS THE SOURCE OF THE EFFECTIVENESS OF THE CODE GENERATOR? WHAT IS IT BASED ON?

The effectiveness is based on the execution time of the generated code. It is between 0.9 and 1.2 times that of hand-written code, according to the manufacturer s specifications. This is an excellent amount, since previous code-generators generated code with 2.5 to 3 times the run time. Even in the worst case scenario, 20 % longer run time is certainly acceptable. It requires approximately the same memory space as hand-written code.

This information was not tested during this project; the same tasks would have had to be performed by experienced programmers at the same time. However, the run times and file sizes are within the expected ranges.

There are essentially three reasons for this run-time efficiency:

First the target processor uses special codes that are not available to C programmers. Second, the code generator implicitly has special knowledge of the efficient programming of the target processor and the operating system environment used, which a code-writer would first have to painstakingly establish. Furthermore, with proper handling of the tools, the variables are scaled to the value range used. This allows the use of favorable multiplication and division operations, which would otherwise require a great deal of time. The scaling can be done automatically or by indicating the value range.

IS THE GENERATED CODE READABLE?

The code is readable. Variable names are also taken over in a changed, but readable form. A prefix or suffix is simply added to connect the data structure with the higher-level system structure.

Certainly this question is related to the fact that developers have a certain level of distrust of generated code. Even code generated by high-level language compilers is not generally inspected by the developers.

HOW CAN I WORK WITH PROJECTS THAT GO BACK TO EXISTING CODE?

AFT's TDC has particular advantages over other development tools in this area: There are generally three sources (Figure 6):

1. Code generation from the Simulink® models as previously described.
2. Adding hand-written code.
3. The option to integrate code from other projects.

In all cases, if the conventions are followed, even the parameters can be generated for the calibration system, i.e. the description data of the control unit .

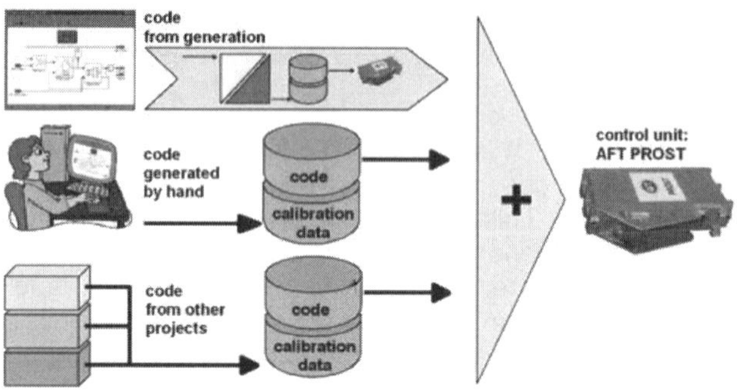

Fig. 6: An Advantage of AFT s TDC: Code Mixing

TO WHAT EXTENT IS THE GENERATED CODE SUITABLE FOR PRODUCTION?

As mentioned previously, the code is nearly production-ready, since it is generated for conventional micro-processors as the target system. The function unit used the automotive prototype control unit AFT PROST is likewise close to being production-ready. The results are thus transferable to subsequent 1:1 series development.

SUMMARY AND OUTLOOK

Figure 7 summarizes the general advantages of a closed development chain and the specific advantages of AFT's TDC. These advantages result from the features shown on the left.

The development of the central release bearing software was a pre-development pilot project covered by the performance capabilities of the AFT's TDC for the design and generation of the control unit production code. At LuK Buehl the current thinking is to also run a pilot project during production development.

Other AFT projects have been completed successfully in the past. These include projects for DaimlerChrysler research, who af ter appropriate training of their employees have since used the same tools and methods for their own development projects.

AFT carries out other engineering service projects, using AFT's TDC as the method and tool to complete projects quickly and cost-efficiently. These also include the development of control unit software for LuK-FH in Bad Homburg.

Summary	
general features of TDC:	resulting advantages of TDC:
• General simulation-based development process from initial design to implementation, including all test phases (SiL, HiL, application and testing)	• Shorter development times
• High-level design (model-based graphics)	• Reduction of possible error sources
• Executable specification	• Cost-reductions
• Implicit documentation	• Shifts the required expertise to the system engineer
• Expertise real-time system, partially in code generator	• Intermediate test results in each development step
basic characteristics of AFT s TDC:	specific advantages of AFT s TDC:
• Heterogeneous system	• Adaptable to changing standards
• Combination of the best components that have proved to be standard: MATLABfi/ SimuLinkfi, code generator by dSPACE, MARC I, RAMboX, TORnadO	• Can be integrated into existing projects: combination or mixture of hand-coded and automatically generated code • Configurable to customer requirements • Production-ready code

Fig. 7: Summary

CONTACT

Reinhard Ludes
AFT Atlas Fahrzeugtechnik GmbH
Gewerbestrasse 14
58791 WERDOHL
GERMANY
Phone: +49 23 92 8 09-2 00
Fax: +49 23 92 8 09-1 00
Email: r.ludes@aft-werdohl.de

Thomas Pfund
LuK GmbH & Co.
Industriestrasse 3
77815 BUEHL
GERMANY
Tel: +49 72 23 9 41-90 50
Fax: +49 72 23 9 41-90 53
Email: thomas.pfund@luk.de

FUTURE SOFTWARE TRENDS

Future Software Trends

Software's Ever Increasing Role

Ronald K. Jurgen, Editor

In *SAE 100 Future Look,* several contributors expressed their views on software trends. Christopher Cook of Infineon Technologies had this to say:

"Increased intelligence throughout the system presents exciting opportunities, as well as challenges, in the development of software. The principal opportunity lies in the ability to differentiate a standard platform through software. Accomplishing this while achieving quality and reliability goals calls for constant improvement in the available development tools. Fortunately, the industry's move to integrated development environment and autocode generation supports a higher level of collaboration between chip suppliers, integrators, and vehicle OEMs, with each business partner contributing his critical expertise at different levels of system design."

Marty Thall of the Microsoft Automotive Business Unit said:

"Today, software's role is expanding beyond just vehicle design and manufacturing to serve as an important gateway for enabling the integration of consumer electronics into the car. However, the growing complexity of software challenges the industry to determine how to harness powerful technology most efficiently and cost effectively in the future. . . At the heart of our approach, we envision a future in which every person drives a 'connected car.' This vision provides automakers with the ability to reduce cost and complexity through the adoption of a standard software platform."

In the papers herein, these thoughts on future trends in software were expressed:

- "Currently developing a microcontroller adapter is a time-consuming task. However, additional software tools may make the adapter development more efficient and feasible to users. This may be considered as future work." – (2005-01-1430)

- "Currently in the automotive industry, most software source code is manually generated (i.e., handwritten). This manually generated code is written to satisfy requirements that are generally specified or captured in an algorithm document. However, this process can be very error prone since errors can be introduced during the manual translation of the algorithm document to code. A better method would be to automatically generate code directly from the algorithm document. Therefore, the automotive industry is striving to model new and existing algorithms in an executable-modeling paradigm where code can be automatically generated." – (2005-01-1665)

- “Especially the area of software engineering is a key technology for the implementation of all future electronic systems in the car. In particular, this is true for x-by-wire systems. . . . The automotive industry is still only beginning to get a grip on mastering the software development process. Software engineering methods, reuse of software, and software quality management—to name but a few—find their way into the automotive development process only very slowly.” – (2003-01-1288)

- “We are convinced that the challenges to design distributed systems for vehicles can be met through advanced Automotive Systems and Software Engineering in conjunction with suitable processes, methods and tools. Standardized Technical System Architecture and standards-based infrastructures for collaboration and co-working become as important as the mature management of the System Design Process. Design in the past was treated as an art but needs to be managed as a consistent design process in the future. The solutions to these challenges will have a great impact on the way the vehicle’s electrical and electronic architecture are designed.” – (2004-01-0300)

- “Because of the number of possible hybrid architectures, the development of the next generation of vehicles will require advanced and innovative simulation tools. Model complexity does not mean model quality; flexibility, reusability, and user friendliness are key characteristics to model quality. . . The structured, yet flexible, approach used in PSAT could be used as a base to establish industry standards within the automotive modeling community, where each institution implements its own data or model in a common generic software architecture.” – (2004-01-1618)